# ENGINEERING CHEMISTRY

# ENGINEERING CHEMISTRY

*By*

**Dr. B.S. CHAUHAN**

*M.Sc., Ph.D.*

*Professor and Head*

*Noida Institute of Engg. & Tech.*

*Greater Noida (U.P.)*

UNIVERSITY SCIENCE PRESS

(An Imprint of Laxmi Publications Pvt. Ltd.)

An ISO 9001:2015 Company

**BENGALURU • CHENNAI • GUWAHATI • HYDERABAD • JALANDHAR**

**KOCHI • KOLKATA • LUCKNOW • MUMBAI • RANCHI**

**NEW DELHI**

**ENGINEERING CHEMISTRY**



Printed and bound in India
*Typeset at* : Goswami Associates, Delhi.
*First Edition* : 2006, *Second Edition* : 2008, *Third Edition* : 2009, *Reprint* : 2014
*ISBN* : 978-81-318-0579-4



**Branches**

| | |
|---|---|
| Bengaluru | 080-26 75 69 30 |
| Chennai | 044-24 34 47 26 |
| Guwahati | 0361-254 36 69 |
| Hyderabad | 040-27 55 53 83 |
| Jalandhar | 0181-222 12 72 |
| Kochi | 0484-405 13 03 |
| Kolkata | 033-40 04 77 79 |
| Lucknow | 0522-430 36 13 |
| Ranchi | 0651-224 24 64 |

PUBLISHED IN INDIA BY

UNIVERSITY SCIENCE PRESS
(An Imprint of Laxmi Publications Pvt. Ltd.)

An ISO 9001:2015 Company
113, GOLDEN HOUSE, GURUDWARA ROAD, DARYAGANJ,
NEW DELHI - 110002, INDIA
Telephone : 91-11-4353 2500, 4353 2501
www.laxmipublications.com
info@laxmipublications.com

C—R/021/06
Printed at: Mehra Offset Press, Delhi.

*To My Parents With*
*Immense Love and Regard*

# CONTENTS

# PREFACE TO THE THIRD EDITION

This book has been divided into seventeen chapters covering all the disciplines of engineering chemistry such as inorganic, organic, synthetic, physical, applied, industrial, spectroscopic, and environmental chemistry. The book has been written with the requirements of engineering students in mind; *i.e.*, every aspect of a topic is dealt with, keeping the focus on engineering science. Key features of the book include a simple and holistic approach with a large number of illustrations and tables.

All the chapters embodied in the book are of basic importance in the curriculum of engineering students and provides a core course of engineering chemistry for all branches of engineering. The book has indepth and complete discussion on chapters like *Instrument Methods of Analysis, Polymer Chemistry, Lubricants, Products of Silica Industry, Explosives and Abrasives, New Engineering Materials, Chemical Kinetics, Phase Equilibrium—the Phase Rule, Electrochemistry and Corrosion, Water and Water Treatment, Fuels, Environentaal Pollution, Advance Theory of Chemical Bonding, Co-ordinations Chemistry, Solid State Chemistry, General Organic Chemistry, Chemical Thermodynamics.*

Each Chapter consists of methadical introduction, historical background, indepth discussion, practical applications and significance. A large number of numerical problems, drawn from various university examinations, have been let at the end of theoretical part of each chapter.

I hope the book will prove a useful tool for students of undergraduate engineering classes. Critical suggestions from users on material presented are welcomed for the improvement of the book.

**—Author**

# PREFACE TO THE FIRST EDITION

This book has been divided into sixteen chapters covering all the disciplines of engineering chemistry such as inorganic, organic, synthetic, physical, applied, industrial, spectroscopic, and environmental chemistry. The book is written with the requirements of engineering students in mind; *i.e.*, every aspect of a topic is dealt with, keeping the focus on engineering science. Key features of the book include a simple and holistic approach with a large number of illustrations and tables.

Chapter 1, titled *Spectroscopic Methods of Analysis*, describes basic principles of spectroscopic analysis in brief and UV, IR, NMR Mass, and AAS spectroscopy in depth with a number of solved examples. Chapter 2, *Polymer Chemistry,* discusses classification, characteristics and properties, molecular weight, crystallinity, structure and properties of polymers, commercially important polymers and biopolymers. Preparation, properties, and applications of *conducting polymers* have also been described. *Chapter 3* describes general principles involved in organic chemistry, reaction intermediates and types of organic reactions with energy profile diagrams. *Chapter 4* provides complete mechanism and applications of organic name reactions provided in syllabus. *Chapter 5* describes various aspects of stereochemistry. Optical, geometrical and conformational isomerism in organic molecules is explained in a simple manner with energy profile diagrams. *Chapter 6* elaborates on electrochemical principles, galvanic cells, electrode potential, electrochemical series, chemical cells, concentration cells, relation of thermodynamic functions with cell potential and the Nernst equation with a large number of numerical problems and examples. *Chapter 7* deals with concepts of chemical kinetics such as rate of reaction, rate law, order and molecularity of reactions, first and second order rate equation, activation energy, theories of reaction rate, and complex reactions. *Chapter 8* explains Gibbs Phase rule and its applications to water and sulphur systems. *Chapter 9* deals with the theories of corrosion, factors affecting corrosion, types of corrosion, and corrosion control. *Chapter 10* describes water chemistry, explaining water hardness, water softening by zeolite, lime-soda method and ion exchange resins, boiler feed water, waste water treatment, and reverse osmosis. *Chapter 11* discusses classification and calorific value of fuels, coal analysis, liquid and gaseous fuels, and nonconventional energy sources such as solar energy, biomass and biogas. *Chapter 12* describes environmental pollution, different pollutants, air pollution, acid rain, smog, and ozone chemistry. *Chapter 13* explains valence bond theory, molecular orbital theory, metallic bonding, semiconductors, Born-Haber cycle and hydrogen bonding. *Chapter 14* describes the concept of coordination compounds. *Chapter 15* discusses solid state chemistry covering the topics crystal structure, unit cell, space lattice, cubic systems, laws of crystallography, Bragg's Law, graphite and fullerenes.

*Chapter 16,* titled *Experimental Chemistry,* explains basic principles of volumetric analysis, types of titrations, preparation of solutions, and experiments, including organic analysis. Exhaustive discussion is given at the end of each experiment as well as general discussion questions and answers on volumetry and organic analysis. Complete information regarding laboratory accidents and first aid is provided in the same chapter.

At the end, useful appendices and sample exams are provided.

We hope the book will prove a useful tool for students of undergraduate engineering classes. Critical suggestions from users on material presented are welcomed for the improvement of the book.

**—Authors**

# Chapter 1 INSTRUMENTAL METHODS OF ANALYSIS

## 1.1 INTRODUCTION

Instrumentation is an exciting and fascinating part of chemical analysis that interacts with all the areas of chemistry, biochemistry, engineering, geology, health related sciences, polymer science, material science and environmental science. Analytical instrumentation plays an important role in the production and evaluation of new products and in the protection of consumers and the environment. This instrumentation provides the lower detection limits required to assure safe foods, drugs, water and air. These techniques include molecular spectroscopy, chromatography, microwave spectroscopy, thermal analysis etc.

Molecular spectroscopy is defined as *the study of interaction of electromagnetic radiations with matter*. It provides valuable information regarding molecular structure such as molecular symmetry, bond distances, bond angles; chemical structure such as chemical properties, electronic distribution, bond strength, molecular reactions; and thermodynamic properties.

## 1.2 ELECTROMAGNETIC SPECTRUM

*An electromagnetic radiation is a form of energy that is transmitted through space at an enormous velocity.* The quantum mechanics suggests that electromagnetic radiation has a dual character, *i.e.*, exhibiting

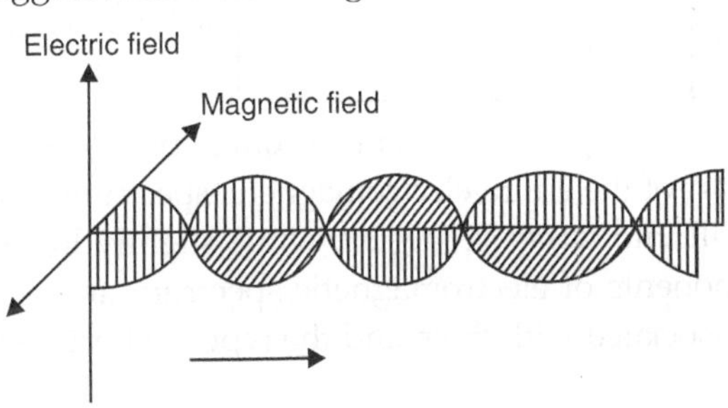


*FIGURE 1.1* **Beam of electromagnetic radiation according to wave nature concept.**

wave as well as particle behavior. According to wave nature of radiations, the electromagnetic wave consists of oscillating electric and magnetic field. The vectors of electric and magnetic field are perpendicular to the direction of propagation of radiations (Figure 1.1).

The electromagnetic spectrum extends from the radio wave to the microwave, infrared, visible, and ultraviolet regions. Beyond these are X-ray, gamma ray and cosmic ray region shown in Figure. 1.2.

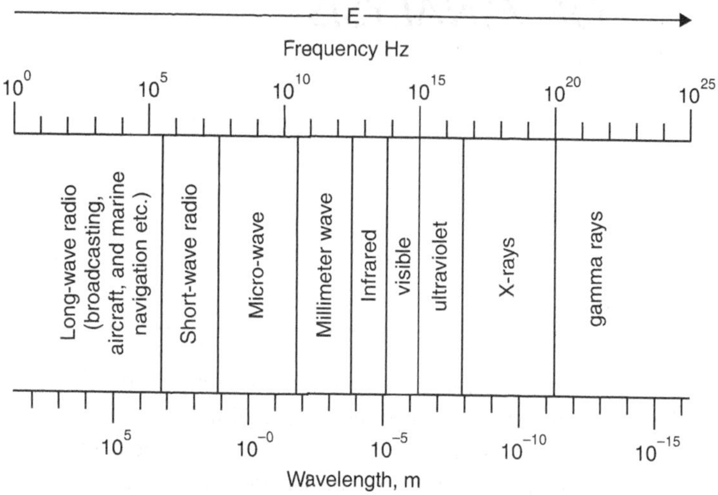


***FIGURE 1.2*** **Various regions of electromagnetic spectrum.**

**TABLE 1.1 Some Components of Electromagnetic Spectrum**

| *Type of Spectra* | *Wavelength* | *Energy in kcal/mol* | *Effect on the Molecule* |
|---|---|---|---|
| 1. Ultraviolet | 200–400 nm | 143–82 | Changes in electronic energy levels within the molecule. |
| 2. Visible | 400–800 nm | 82–36 | Changes in electronic energy levels within the molecule. |
| 3. Infrared | 2–16 μm | 14.3–1.8 | Changes in vibrational and rotational movements of the molecule |
| 4. Nuclear magnetic resonance | | | Induces changes in the magnetic energy levels of certain nuclei. |

The wavelengths of visible light range from 400 nm (violet) to 800 nm (red). The *visible region* however, is a very small part of the entire electromagnetic spectrum. Wavelengths slightly shorter than those of the visible fall into *ultraviolet region,* while slightly longer wavelengths fall intwo, *infrared region.* Some components of electromagnetic spectrum are listed in Table 1.1 alongwith their wavelength, energy associated with them and the type of effect they are capable of producing in a molecule.

Common symbols used in spectroscopy are given in Table 1.2.

**Table 1.2 Symbols Used in Spectroscopy**

| *Symbol* | *Definition* |
|---|---|
| ν | Frequency in Hz cycles per second |
| λ | Wavelength. |
| μm | Micrometre, same as micron (μ), $10^{-6}$ m. |
| Å | Angstrom, $10^{-10}$ m or $10^{-12}$ cm. |
| $cm^{-1}$ | Wave number, frequency in reciprocal or 1/λ. |

## 1.3 LAMBERT-BEER OR BEER'S LAW

Lambert's law states that, *when a monochromatic light is passed through a solution, the decrease in the intensity of light with the thickness of the solution is directly proportional to the intensity of incident light.*

$$I_t = I_o e^{-kx} \tag{1.1}$$

where, $I_t$ = intensity of the transmitted light

$I_o$ = intensity of the incident light

$x$ = thickness of the medium or the length of the solution

$k$ = absorption coefficient

Beer's law an extension of Lambert's law states that, *when a monochromatic light is passed through a solution, the decrease in the intensity of light with the thickness of the solution is directly proportional to intensity of incident light as well as the concentration of the solution.*

$$I = I_o e^{-\varepsilon c x} \tag{1.2}$$

where, ε = Molar absorption coefficient

$c$ = concentration of solution

The equation (1.2) may be written

$$\frac{I}{I_o} = e^{-\varepsilon c x}$$

$$\log \frac{I}{I_o} = -\varepsilon c x$$

where, $\frac{I}{I_o}$ is called *transmittance,* T and, $-\log \frac{I}{I_o} = -\log T$ and is called *absorbance.*

$$-\log \frac{I}{I_o} = -\log T = A$$

$$\therefore \qquad A = \varepsilon c x \qquad (1.3)$$

$$\varepsilon = \frac{A}{cx}$$

where, $\varepsilon$ = molar absorbing coefficient

$c$ = molar concentration

$x$ = path length in centimetres

The magnitude of the molar absorption coefficient is directly related to the probability of a particular transition.

$$\varepsilon_{max} = 0.87 \times 10^{20} \text{ P.a.}$$

P = Transition probability with values ranging from 0 to 1.

a = Target area of the absorbing system.

Generally transitions with $\varepsilon_{max}$ value more than $10^4$ are called *allowed transition* and $\varepsilon_{max}$ value less than $10^4$ are called *forbidden transitions.*

The instrument used for absorption measurements are made with of various instruments. For example, colorimeters or spectrometers are constructed according to the following general scheme (Figure 1.3). It can be studied under as:

1. **Light source.** In the infrared region, the source of heat radiations is Nernst glower or globar. In the Nernst glower, the filament consist of a mixture of oxides of the cerium and thorium, which are kept at high temperature (~ 1500°C) electrically. For more rarely studied regions (25 to 1000 μ) sources like the Welsbach lamp (a quartz-jacketed mercury arc), strongly heated glass etc. have been used. However, the radiations are not monochromatic and for use in spectrographs, it must be followed by some dispersing element.

   In the visible-UV region, usual sources are incandescent lamps or many types of discharge tubes such as hydrogen discharge lamps, deuterium lamps, and mercury arc. In Raman spectrometer, the source is mercury arc.

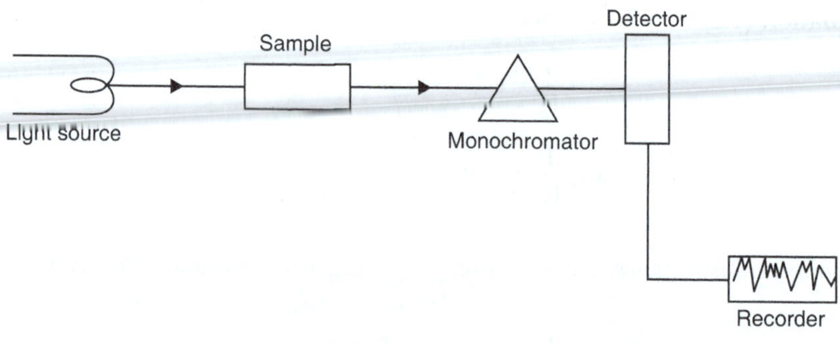


FIGURE 1.3 A simple sketch of absorption measurement instruments.

2. **The sample container.** The sample is —placed in the sample container. The windows of the sample must be chemically resistant to the sample. In the visible-ultraviolet regions,

glass or quartz windows are used. In the infrared region, NaCl and KBr windows are used. For the study of aqueous solutions, AgCl windows are necessary. The length of sample container is 0.91 mm 01 mm for solid and liquid samples. Gas samples are usually contained in cells either 5 cm to 10 cm long.

3. **The monochromator.** Except for microwave spectroscopy, some element is necessary by means of which the radiations can be separated in space according to wavelength, after it has passed the sample. This part of the spectrometer is known as monochromator. Its principal part is a dispersing element, *viz*, a prism or grating. Sometimes filters transparent over limited wavelength ranges can be used in combination with gratings.

4. **The detector.** In all spectrometers, the emitted radiations from the sample must be analyzed with regard to radiation intensity as a function. The radiation energy is measured by transforming it to some other forms of energy, usually electrical energy. The "transformer" is called the detector.

   In the microwave region, the detector is a crystal rectifier. At the higher frequencies in the infrared region, thermocouples are used. In the visible and the ultraviolet region the photocells are used.

5. **Spectrum recording.** When there is no sample present, the detector output will be constant over the range of frequencies covered by the instrument. However, when the radiations are passed through a sample, having just two possible energy levels $E_1$ and $E_2$, the detector output will show a sudden fall at frequency given by $\nu = [(E_2 - E_1)/h]$. It means some energy at this frequency will be absorbed by the sample and it no longer reaches the detector. The resulting trace on chart paper records the detector output on the percentage of the energy absorbed by the sample on the light (I). The spectrum has been scanned between the beginning and ending frequencies. Such a spectrum is referred as 'frequency domain', which indicates record of detector output against frequency.

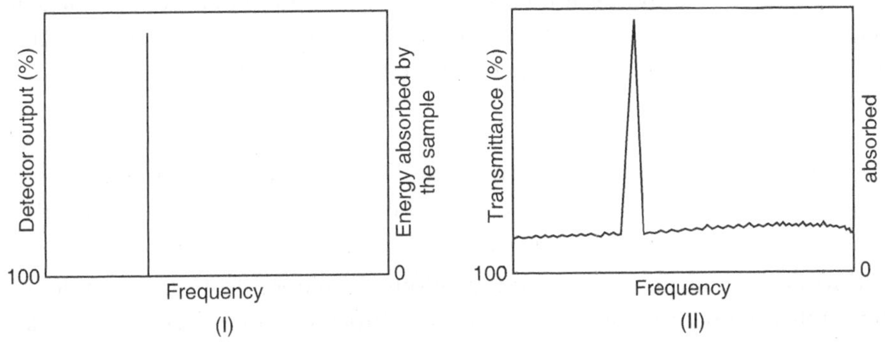


*FIGURE 1.4*

It is evident from the record (II) the ideal situation represented in (I) is seldom attained. Thus, the base line is seldom horizontal (as represented in ideal situation). The reasons for this are:

- It is impossible to make the slits infinitely narrow. Therefore, a range of frequencies rather than single frequency reaches the detector.
- Energy transitions in atoms or molecules are not absolutely sharp but always occur over a range of frequencies.

- The sensitivity of the detector is also frequency dependent. These factors cause a broadening of the spectral absorption. Besides this, there occur small peaks and troughs arising from randomly generated electric signals in the detector, the amplifier, or the pen recorder, collectively referred as "noise" on the spectrum.

## 1.3a COLORIMETRY

Colorimetry is based on the variation of the colour of a system with change in concentrations. In this method, the concentration of a substance is determined by measurement of the relative absorption of light with respect to a known concentration of the substance. In visual colorimetry natural or artificial white light, is generally used as a source of light, and determinations are made with the help of a simple instrument called a *colorimeter*. If the eye is replaced by a photoelectric cell, the instrument is called **photoelectric colorimeter**. This instrument makes use of relatively narrow range of wavelengths of light obtained by passing white light through filters (for example, plates of coloured glass, gelation etc.) and hence such an instrument is called a *filter photometer*. If a source of visible radiation extending upto the UV region of the spectrum and if radiations upto a bandwidth of less than 1 nm can be obtained with a suitable device (*e.g.*, spectrometer) and if it contains a device (*e.g.*, photometer) with which measurement of intensity of transmitted radiation can be made, then such an instrument is called a *spectro-photometer*.

**Principle:** The fundamental principle of colorimetry is to compare, under well designed conditions, the colour produced by the substance in unknown amount with the same colour produced by a known amount of material being determined. The following methods are available for the quantitative comparison of colours of known and unknown solutions :

1. **Standard Series method.** In this method, Nessler tubes are employed. There are nothing but colourless glass tubes of uniform cross section and flat bottoms. The solution of the substance being determined is made up to a known volume, and the colour is compared with that of a series of standards in the same way starting from known amounts of the substance being determined. 100 mL of the solutions of the unknown and each of the standard solutions are placed in Nessler tubes, and the solutions are viewed vertically through the length of the liquid columns. The concentration of the unknown solution is equal to that of the standard solution having the same colour. LOVIBOND—2000 comparator works on this principle.
2. **Dilution method.** In this method, a colour developed from the sample and standard solution are taken in two identical glass tubes of same diameter and are observed horizontally through the tubes. The more concentrated solution is diluted until the colours in the tubes are of identical intensity when observed horizontally through the same thickness of solutions. The relative concentrations of the two original solutions (known and unknown) are then proportional to the heights of the matched solutions in the tubes.
3. **Balancing method.** In this method, comparison is made in two tubes, and the height of the liquid in one tube is so adjusted that when both the tubes are observed vertically, the colour intensities in the tubes are equal. If the concentration of solution in one of the tubes is known then the concentration in the other can be calculated from the relation

$$c_1 I_1 = c_2 I_2$$

where $c_1$ and $c_2$ are the known and unknown concentrations and $I_1$ and $I_2$ are their respective lengths of the two liquid columns.

$$c_2 = c_1 I_1/I_2$$

Duboscq colorimeter works on this principle.

4. **Photoelectric photometer method.** In this method, the human eye is replaced by a photoelectric cell. Instruments which incorporate photoelectric cells measure the light absorption and not the colour of the substance, hence the term *'photoelectric colorimeters'* is a misnomer. They may be better called as *photoelectric photometers* or *photoelectric comparators* or *absorptiometers*.

These instruments essentially consist of a light source, a suitable light filter to provide nearly monochromatic light, a glass cell for the sample/standard solution, and a measuring device to determine the response of the photoelectric cell. The photometer is first calibrated by measuring the absorbance of a series of solutions of known concentrations (at the wavelength at which the coloured species has maximum absorbance). A calibration curve is plotted connecting the concentration with the readings of the measuring device employed (optical density). The concentration of the unknown solution is then determined by noting the cell response and referring it to the calibration curve.

The photoelectric photometers are available in different type incorporating one or two photocells. In the one-cell type of instruments, the absorption of light by the solution is usually measured directly by determining the current output of the photoelectric cell in relation to the value obtained with the pure solvent. In instruments of the one-cell type, it is essential to ensure that the light source is of constant intensity and the photocell used doesn't exhibit fatigue effect.

The two-cell type of photoelectric photometer is considered to be more trustworthy because any fluctuation of the intensity of light source will affect both cells alike as they are matched for their spectral response. In this type of instrument, the two photocells, which were illuminated by the same source of light, are balanced against each other through a galvanometer. The test solution is placed before one cell and the pure solvent before the other, and the difference in current output is measured.

## 1.4 UV SPECTROSCOPY

### 1.4.1 Introduction

UV absorption spectra arises from the transition of electrons within a molecule or ion from a lower to higher energy level of the molecule or ion. When a molecule absorbs ultraviolet radiation of frequency ($\nu$), the electrons in that molecule may undergo transition from a lower to higher energy level, the energy difference between the two states is given by,

$$\Delta E = h\nu$$

where, $h$ is Planck's constant. Also,

$$\Delta E = E_2 - E_1$$
$E_1$ = Energy of the ground state
$E_2$ = Energy of the excited state

We know that the total energy of the molecule is equal to the sum of the electronic, vibrational, and rotational energies. Therefore,

$$E_{total} = E_{electronic} + E_{vibrational} + E_{rotational}$$

The absorption in the UV region involves single electronic transition, from lower energy state to higher energy state, so the spectrum must contain a single discrete line. But in actual practice, a broad absorption band appears. This is because the electronic transitions are also accompanied by vibrational and rotational transitions. As the energy required for these transitions is much less, so the broad band appears.

In UV range, the absorbed energy produces changes in the valence electrons of the molecules. The electrons involved are:

(*i*) *π-electrons.* These are involved in unsaturated compounds such as dienes, trienes, aromatic compounds, etc.

(*ii*) *σ-electrons.* These involves the electrons of the saturated bond between carbon and hydrogen. Energy required to excite these electrons is very high.

(*iii*) *n-electrons.* These electrons are not involved in bonding between atoms in the molecules, for example, lone pair of electrons in nitrogen, oxygen, sulphur, or halogens.

### 1.4.2 Electronic Transitions

When a molecule absorbs energy in the UV or visible region, its electrons are promoted from a bonding molecular orbital to an antibonding molecular orbital. The electronic transitions involved in the UV and visible region are of the following types (Figure 1.5):

(*i*) $\sigma \rightarrow \sigma*$ transitions (*ii*) $n \rightarrow \sigma*$ transitions
(*iii*) $n \rightarrow \pi*$ transitions (*iv*) $\pi \rightarrow \pi*$ transitions

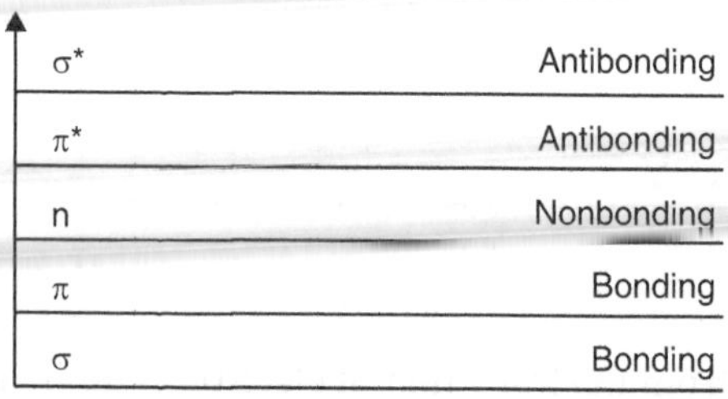


*FIGURE 1.5*

(*i*) $\sigma \rightarrow \sigma*$ *transitions.* These transitions occur by the promotion of sigma electrons from the bonding to antibonding (σ*) orbitals. These transitions require high energy and thus absorb radiation in far region (126 nm–136 nm). These transitions are seen in saturated hydrocarbons, for example, propane shows absorption around 135 nm, and methane shows $\lambda_{max}$ at 121.9 nm.

(*ii*) $n \rightarrow \sigma^*$ *transitions.* These transitions are seen in saturated compounds, containing heteroatom in addition to $\sigma \rightarrow \sigma^*$ transitions. The energy required for these transitions is less than that required for $\sigma \rightarrow \sigma^*$ transitions. For example, $CH_3OH$ absorbs at ordinary UV region and $\lambda_{max}$ is 183 nm, for methyl chloride $\lambda_{max}$ is 173 nm and water absorbs at 167 nm.

(*iii*) $n \rightarrow \pi^*$ *transitions.* Unsaturated compounds containing heteroatoms such as O, N and S show this type of transition. The electron from $n$-orbitals are transferred to $\pi^*$ orbitals and require less amount of energy and occur at longer wavelength.

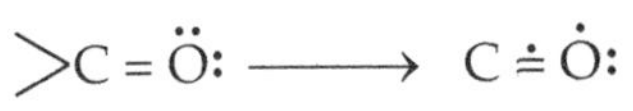

$n \rightarrow \pi^*$ transition

(*iv*) $\pi \rightarrow \pi^*$ *transitions.* The electron transfer occurs from the bonding $\pi$-orbital to the antibonding $\pi^*$ orbital. These transitions are only associated with compounds containing double or triple bond and aromatic compounds. These transitions require less energy and occur at higher wavelength.

$\pi \rightarrow \pi^*$

For example, unconjugated alkenes show an absorption band around 170 nm to 190 nm and saturated aldehydes and lactones absorb at about 180 nm.

### 1.4.3 Chromophores and Auxochromes

A group of atoms in a molecule that is repsonsible for the absorption of radiation is called a *chromophore.* A compound containing a chromophore is called *chromogen*. Chromophores are of two types:

1. Chromophores which contains $\pi$-electrons only and undergo $\pi \rightarrow \pi^*$ transition. For example, ethylenes, acetylenes, etc.
2. Chromophore which contain both $\pi$-electrons and $n$-electrons. Such chromophores undergo $\pi \rightarrow \pi^*$ and $n \rightarrow \pi^*$ transitions. For example, carbonyl, nitriles, azo compounds, nitro compounds, etc. The carbonyl group in acetone shows absorption at 190 nm ($\pi \rightarrow \pi^*$) and 270 nm ($n \rightarrow \pi^*$). Examples of some of the molecular types in which electronic transitions occur are given in Table 1.3.

**TABLE 1.3 Examples of Electronic Transitions**

| *Compound* | $\lambda_{max}$ *(nm)* |
|---|---|
| $\sigma \rightarrow \sigma^*$ | |
| $CH_4$ | 122 |
| $CH_3-CH_3$ | 135 |
| $\pi \rightarrow \pi^*$ | |
| $CH_2=CH_2$ | 162 |
| $CH \equiv CH$ | 173 |

| | |
|---|---|
| $(CH_3)_2C = O$ | 190 |
| 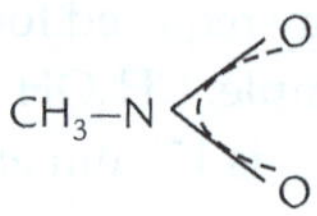 | 201 |
| $n \rightarrow \sigma^*$ | |
| $H_2\ddot{O}:$ | 167 |
| $CH_3\ddot{\underset{\cdot\cdot}{Cl}}:$ | 172 |
| $(CH_3)N:$ | 227 |
| $n \rightarrow \pi^*$ | |
| $(CH_3)_2\ddot{C}:$ | 280 |

Identical functional groups in different molecules will not necessarily absorb at exactly the same wavelength. The energy change for particular transition determines the position of absorption for a given group. The energy change depend upon the structural environments of the molecule. For the qualitative interpretation of the spectra, only the region above 200 nm is important. The position and intensity of an absorption band of a simple chromophore may be modified by the attachment of certain groups in the basic chromophore system known as *auxochromes.* The auxochromes produce the following shifts:

(*a*) **Bathochromic Shift.** This effect causes the absorption maximum to shift toward longer wavelength ($\lambda$) or lower frequencies. This is called red shift and is due to substitution or solvent effect.

(*b*) **Hypsochromic Shift.** This is the shift to lower wavelength. This is caused by the removal of conjugation or change in the solvent polarity. For example, aniline shows $\lambda_{max}$ 280 nm, whereas anilinium ion (acidic solution of aniline) shows $\lambda_{max}$ 254 nm. This hypsochromic shift is due to the removal of $n \rightarrow \pi$ conjugation of the lone pair of electrons of the nitrogen atom of aniline with the $\pi$-bonded system of the benzene ring on protonation because the protonated aniline (anilinium ion) has no lone pair of electrons for conjugation. Similarly, there is a hypsochromic shift of 10-20 nm in the $\lambda_{max}$ of $\pi \rightarrow \pi^*$ bands of carbonyl compounds on going from ethanal as solvent to hexane, *i.e.*, an decreasing solvent polarity.

The bathochromic shift can be brought about in two ways:

(*i*) *By conjugation of chromophoric groups.* When two chromophoric groups present in the molecule are separated by two or more single bonds, their effect on the spectrum is usually additive and there is little interaction between the isolated chromophoric groups.

When the two chromophoric groups are conjugated, separated by only single bond, the electron system is spread over at least four atomic centres. This results in shifting the absorption bands to longer wavelength region. The shift is of 15 nm–45 nm to longer wavelength, for example,

| | | $\lambda_{max}$ | Transition |
|---|---|---|---|
| (a) | $-\overset{\vert}{C}=\overset{\vert}{C}-\overset{\vert}{C}=\overset{\vert}{C}-$ <br> 1, 3-Butadiene | 217 | $\pi \rightarrow \pi^*$ |
| (b) | $-\overset{\vert}{\underset{\vert}{C}}-\overset{\vert}{C}=\overset{\vert}{C}-\overset{\vert}{C}=O$ <br> But-2-enal | 218 <br> 320 | $\pi \rightarrow \pi^*$ <br> $n \rightarrow \pi^*$ |
| (c) | $-\overset{\vert}{C}=\overset{\vert}{C}-COOH$ | 260 <br> 242 | $\pi \rightarrow \pi^*$ <br> $n \rightarrow \pi^*$ |

(*ii*) *By attaching auxochromes to the system.* Auxochromes generally deepen the colour of a chromophore but cannot by themselves impart colour to a compound. These lead to shifts in the absorption to a longer wavelength and result in increase in the intensity of the absorption peak. Some typical auxochromes are:

$-NH_2$, $-NHCH_3$, $-N(CH_3)_2$, $-SO_3H$, $-OH$, $-COOH$, $-Cl$, $-CH_3$, $-OCH_3$, $-CN$, $-COCH_3$ and $-CONH_2$. These groups contain non-bonding electrons and the transitions involving non-bonding electrons are responsible for this effect. For example, benzene shows an absorption band $\lambda_{max}$ at 255 nm and aniline shows $\lambda_{max}$ at 280 nm.

$\lambda_{max} = 255$ nm

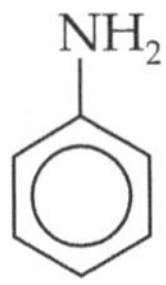

$\lambda_{max} = 280$ nm

**Solvent used in UV Spectroscopy**

The solvent must be transparent within the wavelength range being examined. Generally, the solvent used is 95% ethanol. It is a cheap and a good solvent and is transparent down to 210 nm. The following table (Table 1.4) shows a list of some commonly used solvents, and the minimum wavelength from which they may be used in 1 cm cells.

**TABLE 1.4 Common Solvents used in UV Spectroscopy**

| *Solvent* | *Wavelength (nm)* | *Solvent* | *Wavelength (nm)* |
|---|---|---|---|
| Water | 191 | Ethanol | 204 |
| Cyclohexane | 196 | Ether | 215 |
| Hexane | 200 | Chloroform | 240 |
| Methanol | 203 | Carbon tetrachloride | 258 |

### 1.4.4 Woodward-Fieser Rules for Calculating $\lambda_{max}$ in Conjugated Dienes and Trienes

Woodward (1941) formulated a set of empirical rules for calculating or predicting $\lambda_{max}$, in conjugated acyclic and six-membered ring dienes. These rules, modified by Fieser and Scott on the basis of wide experience with dienes and trienes, are called Woodward-Fieser rules and are summarized in Table 1.5. First, we discuss the following terms used in Woodward-Fieser rules.

**TABLE 1.5. Woodward-Fieser rules for calculating $\lambda_{max}$ in conjugated dienes and trienes**

| | |
|---|---|
| Base value for acyclic or heteroannular diene | 214 nm |
| Base value for homoannular diene | 253 nm |
| Increment for each: | |
| Alkyl substituent or ring residue | 5 nm |
| Exocyclic conjugated double bond | 5 nm |
| Double bond extending conjugation | 30 nm |
| —OR (alkoxy) | 6 nm |
| —Cl, — Br | 6 nm |
| —OCOR (acyloxy) | 0 nm |
| —SR (alkylthio) | 30 nm |
| —$NR_2$ (dialkylamino) | 60 nm |
| In the same double bond is exocyclic to two rings simultaneously | 10 nm |
| Solvent correction | 0 nm |
| Calculated* $\lambda_{max}$ of the compound | Total nm |

*For $\pi \rightarrow \pi^*$ transition (K-band).

(*i*) **Homoannular Dienes.** In homoannular dienes, conjugated double bonds are present in the same ring and having *s-cis* (*cisoid*) configuration (*s* = single bond joining the two doubly bonded carbon atoms):

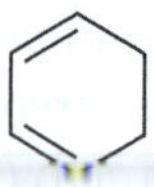

(1)

The *s-cis* configuration causes strain which raises the ground state energy level of the molecule leaving the high energy excited state relatively unchanged. Thus, the transition energy is lowered resulting in the shift of absorption position to a longer wavelength. Acyclic dienes exist mostly in the strainless *s-trans* (*transoid*) conformation with relatively lower ground state energy level. Thus their absorptions appear at shorter wavelengths. For example, 1, 3-cyclohexadiene (I) shows $\lambda_{max}$ 256 nm, whereas 1, 3-butadiene shows $\lambda_{max}$ 217 nm. Also, due to the shorter distance between the two ends of the chromophore, *s-cis* dienes give lower $\varepsilon_{max}$ (~ 10,000) than that of the *s-trans* dienes (~ 20, 000).

*s-trans* 1, 3-butadiene ~ 97.5% $\lambda_{max}$ 217 nm

*x-cis* 1, 3-butadiene ~ 2.5%

(*ii*) **Heteroannular Dienes.** In heteroannular dienes, conjugated double bonds are not present in the same ring and these have *s-trans* (*transoid*) configuration.

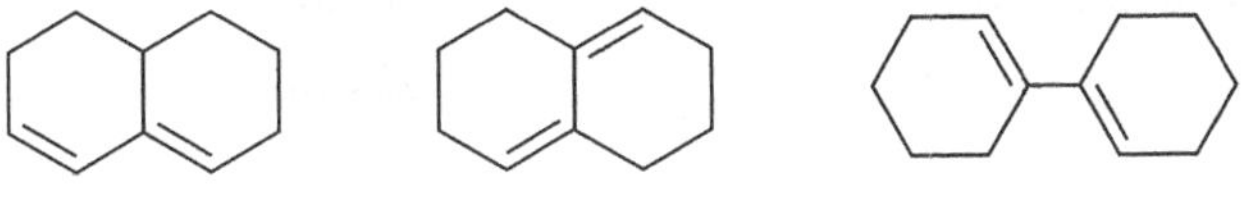

(*iii*) **Exocyclic Conjugated Double Bonds.** The carbon carbon double bonds projecting outside a ring are called *exocyclic double bonds*. For example

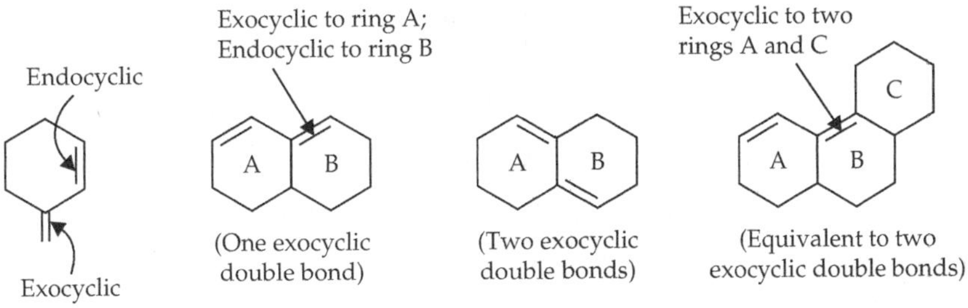


Note that the same double bond may be exocyclic to one ring, while endocyclic to the other and sometimes the same double bond may be exocyclic to two rings simultaneously.

(*iv*) **Alkyl Substituents and Ring Residues.** Only the alkyl substituents and ring residues attached to the carbon atoms constituting the conjugated system of the compound are taken into account. Following examples indicate such carbon atoms by numbers and the alkyl substituents and ring residues by dotted lines.

$$CH_3—CH_2 \vdots \overset{4}{C}H=\overset{3}{C}H—\overset{2}{C}H=\overset{1}{C}H \vdots CH_3$$

Two alkyl substituents

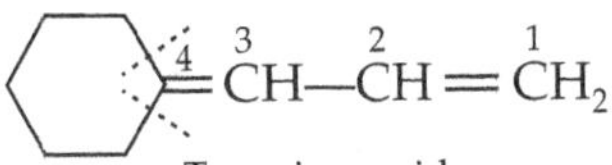


Two ring residues

In compounds containing both homoanular and heteroannular diene systems, the calculations are based on the longer wavelength (253 nm), *i.e.* the homoannular diene system.

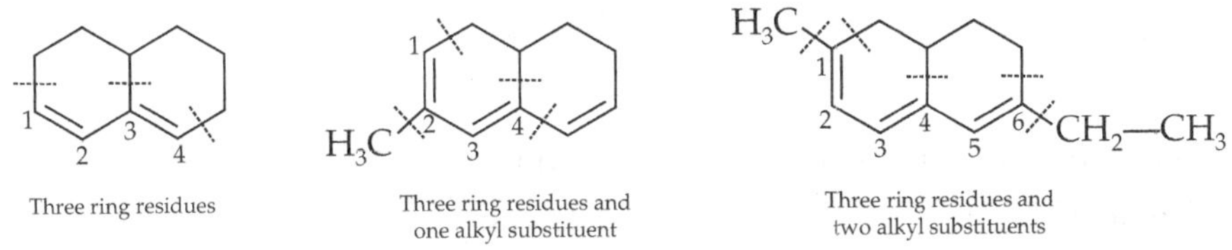


The calculated and observed values of $\lambda_{max}$ usually match within ±5 nm as shown in the following examples illustrating the applications of Woodward-Fieser rules (Table 1.5).

**Example 1.** *Using Woodward-Fieser rules, calculate* $\lambda_{max}$ *for 2, 3-dimethyl-1, 3-butadiene.*

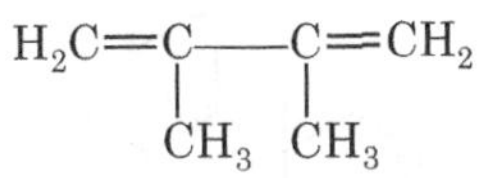


It is an acyclic diene with two alkyl substituents.

Thus, $\lambda_{max}$ of this compound is

| | |
|---|---|
| Base value | 214 nm |
| Two alkyl substituents (2 × 5) | 10 nm |
| Calculated $\lambda_{max}$ | 224 nm |
| Observed $\lambda_{max}$ | 226 nm |

**Example 2.** *Calculate the wavelength of the maximum UV absorption for*

Myrcene

Since, it is an acyclic diene with one alkyl substituent, thus

| | |
|---|---|
| Base value | 214 nm |
| One alkyl substituent | 5 nm |
| Calculated $\lambda_{max}$ | 219 nm |
| Observed $\lambda_{max}$ | 224 nm |

**Example 3.** *Predict the value of* $\lambda_{max}$ *for*

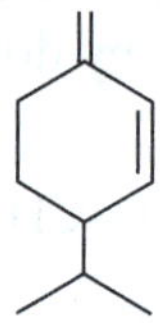

β-phellandrene

This is a heteroannular diene (conjugated double bonds are not in the same ring) with two ring residues and one exocyclic double bond, hence

| | |
|---|---|
| Base value | 214 nm |
| Two alkyl substituents ( 2 × 5) | 10 nm |
| One exocyclic double bond | 5 nm |
| Predicted $\lambda_{max}$ | 229 nm |
| Observed $\lambda_{max}$ | 232 nm |

**Example 4.** *Applying Woodward-Fieser rules, calculate the value of absorption maximum for the ethanolic solution of*

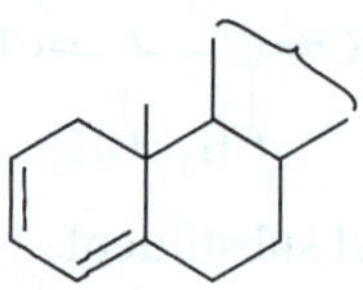

This is a homoannular diene with three ring residues and one exocyclic double bond, thus

| | |
|---|---|
| Base value | 253 nm |
| Three ring residues (3 × 5) | 15 nm |
| One exocyclic double bond | 5 nm |
| Calculated $\lambda_{max}$ | 273 nm |
| Observed $\lambda_{max}$ | 275 nm ($\varepsilon_{max}$ 10,000) |

**Example 5.** *Calculate $\lambda_{max}$ for the ethanolic solution of*

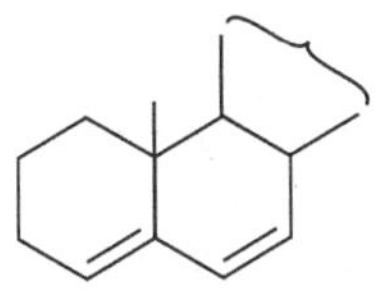

This is a heteroannular diene with three ring residues and one exocyclic double bond, thus

| | |
|---|---|
| Base value | 214 nm |
| Three ring residues (3 × 5) | 15 nm |
| One exocyclic double bond | 5 nm |
| Calculated $\lambda_{max}$ | 234 nm |
| Observed $\lambda_{max}$ | 235 nm ($\varepsilon_{max}$ 19,000) |

**Example 6.** *Hydrogenation of one mole of the triene A with one mole of $H_2$ gives isomeric dienes having molecular formula $C_{10}H_{14}$. Show how UV spectroscopy and the expected $\lambda_{max}$ values could distinguish these isomers:*

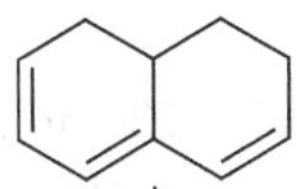

A

The hydrogenation of one mole of the triene A with one mole of $H_2$ may give three isomeric dienes B, C and D with molecular formula $C_{10}H_{14}$:

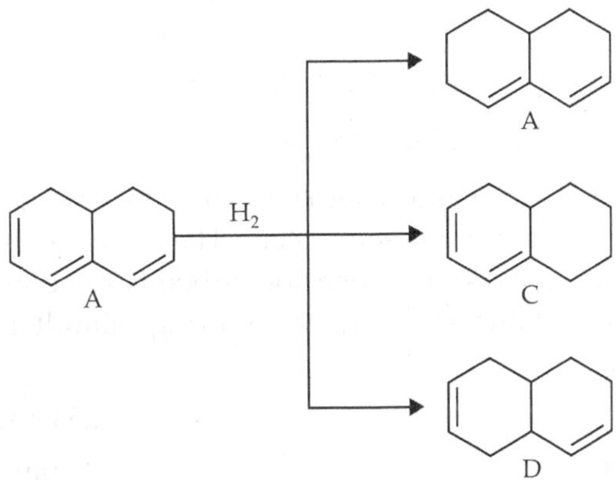

Diene B is a heteroannular conjugated diene with three ring residues and one exocyclic double bond, thus its expected $\lambda_{max} = 214 + 3 \times 5 + 5 = 234$ nm.

Diene C being a homoannular conjugated diene with three ring residues and one exocyclic double bond, thus its expected $\lambda_{max} = 253 + 3 \times 5 + 5 = 273$ nm.

Diene D is an unconjugated diene, hence $\lambda_{max} < 200$ nm.

Thus, by comparing the expected $\lambda_{max}$ values of the isomeric dienes B, C and D with their observed values of $\lambda_{max}$, we can distinguish these isomers.

**Example 7.** *An organic compound can have one of the following structures:*

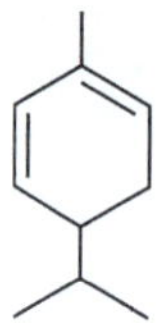

(*a*)

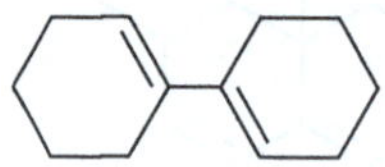

(*b*)

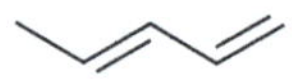

(*c*)

The $\lambda_{max}$ of the compound is 236 nm. Which is the most likely structure of the compound ? Explain your choice.

Let us calculate the $\lambda_{max}$ for each of these structures:

Structure (*a*) is a homoannular diene with two ring residues and one alkyl substituent, hence its calculated $\lambda_{max} = 253 + 3 \times 5 = 268$ nm.

Structure (*b*) represents a heteroannular diene with four ring residues, hence its calculated $\lambda_{max} = 214 + 4 \times 5 = 234$ nm.

Structure (*c*) shows an acyclic conjugated diene with one alkyl substituent, hence its calculated $\lambda_{max} = 214 + 5 = 219$ nm.

Since the given $\lambda_{max}$ of the compound is 236 nm, its most likely structure is (*b*) because the calculated $\lambda_{max}$ (234 nm) for this structure is very close to the given value (236 nm).

**Example 8.** *Applying Woodward-Fieser rules, calculate the value of absorption maximum for*

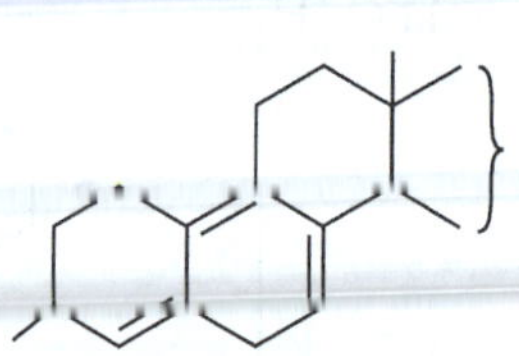

It contains both homoannular and heteroannular diene systems but the calculation of its $\lambda_{max}$ will be based on the homoannular diene system. There are six ring residues attached to the carbon atoms of the entire conjugated system, one double bond extending conjugation, two exocyclic double bonds and one double bond exocyclic to two rings simultaneously. Thus, $\lambda_{max}$ of this compound is calculated as:

| | |
|---|---|
| Base value | 253 nm |
| Six ring residues (6 × 5) | 30 nm |
| One double bond extending conjugation | 30 nm |

| | |
|---|---|
| Two exocyclic double bonds (2 × 5) | 10 nm |
| One double bond exocyclic to two rings simultaneously | 10 nm |
| Calculated $\lambda_{max}$ | 333 nm |

**Example 9.** *The following dienes have the experimental $\lambda_{max}$ 243 and 265 nm in ethanol. Giving reasons, correlate the $\lambda_{max}$ values of the structures (a) and (b):*

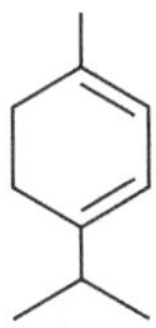
(*a*)

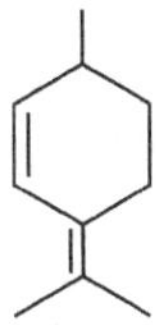
(*b*)

The calculated $\lambda_{max}$ for (*a*) (a homoannular diene with two ring residues and two alkyl substituents) is 253 + 2 × 5 + 2 × 5 = 273 nm. Thus, structure (*a*) is the diene having experimental $\lambda_{max}$ 265 nm as this value is in fair agreement with the calculated $\lambda_{max}$ for (*a*).

The calculated $\lambda_{max}$ for (*b*) (a heteroannular diene with two ring residues, two alkyl substituents and one exocyclic double bond) is 214 + 2 × 5 + 2 × 5 + 5 = 239 nm. Thus, the structure (*b*) is the diene having the experimental $\lambda_{max}$ 243 nm because this value is in agreement with the calculated $\lambda_{max}$ for (*b*).

**Example 10.** *Which of the following compounds is expected to have higher value of $\lambda_{max}$ and why ?*

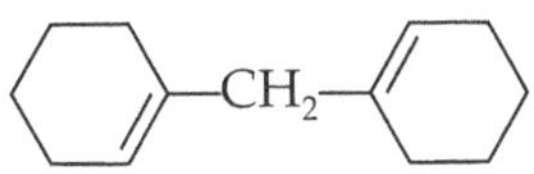

(*a*)

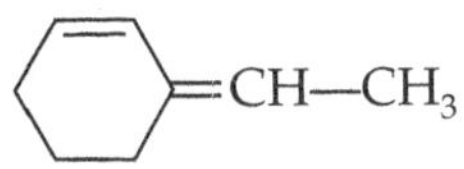

(*b*)

Compound (*a*) is an unconjugated diene. Hence, it will have $\lambda_{max}$ < 200 nm. The compound (*b*) is a heteroannular conjugated diene with two ring residues, one alkyl substituent and one exocyclic double bond. Thus, it is expected to have $\lambda_{max}$ (214 + 2 × 5 + 5 + 5) = 234 nm, *i.e.* higher than that of the compound (*a*).

### 1.4.5 Woodward-Fieser Rules for Calculating $\lambda_{max}$ in α,β-Unsaturated Carbonyl Compounds

Compounds containing a carbonyl group (C=O) in conjugation with an ethylenic groups (C=C) are called enones. UV spectra of enones are characterized by an intense absorption band (*K*-band) due to π → π* transition in the range 215-250 nm ($\varepsilon_{max}$ usually 10,000–20,000) and a weak *R*-band due to π → π* transition in 310-330 nm region ($\varepsilon_{max}$ usually 10-100).

Similar to dienes and trienes, there are set rules called Woodward-Fieser rules for calculating or predicting $\lambda_{max}$ in α, β-unsaturated carbonyl compounds (enones). These rules first framed by Woodward and modified by Fieser and by Scott are given in Table 1.6.

**TABLE 1.6. Woodward-Fieser rules for calculating $\lambda_{max}$ in $\alpha$, $\beta$-unsaturated carbonyl compounds**

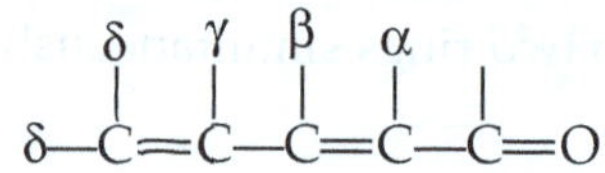

| | | |
|---|---|---|
| Base values for | | |
| Acyclic $\alpha$, $\beta$-unsaturated ketones | | 215 nm |
| Six-membered cyclic $\alpha$, $\beta$-unsaturated ketones | | 215 nm |
| Five-membered cyclic $\alpha$, $\beta$-unsaturated ketones | | 202 nm |
| $\alpha$, $\beta$-unsaturated aldehydes | | 207 nm |
| Increment for each | | |
| Double bond extending conjugation | | 30 nm |
| Alkyl group or ring residue $\alpha$ | | 10 nm |
| | $\beta$ | 12 nm |
| | $\gamma$ and higher | 18 nm |
| –OH (hydroxy) | $\alpha$ | 35 nm |
| | $\beta$ | 30 nm |
| | $\delta$ | 50 nm |
| –OAc (acyloxy) | $\alpha$, $\beta$, $\delta$ | 6 nm |
| –OMe (methoxy) | $\alpha$ | 35 nm |
| | $\beta$ | 30 nm |
| | $\gamma$ | 17 nm |
| | $\delta$ | 31 nm |
| –SR (alkylthio) | $\beta$ | 85 nm |
| –Cl | $\alpha$ | 15 nm |
| | $\beta$ | 12 nm |
| –Br | $\alpha$ | 25 nm |
| | $\beta$ | 30 nm |
| $-NR_2$ | $\beta$ | 95 nm |
| Exocyclic carbon-carbon double bond | | 5 nm |
| Homoannular diene component | | 39 nm |
| Calculated† $\lambda_{max}$ (in EtOH) | | Total nm |

†For $\pi \rightarrow \pi^*$ transition (*K*-band).

For calculated $\lambda_{max}$ in other solvents, a solvent correction given in Table 1.7 must be carried out.

**TABLE 1.7 Solvent corrections**

| *Solvent* | *Correction (nm)* |
|---|---|
| Ethanol | 0 |
| Methanol | 0 |
| Dioxane | +5 |
| Chloroform | +1 |
| Ether | +7 |
| Hexane | +11 |
| Cyclohexane | +11 |
| Water | -8 |

Since carbonyl compounds are polar, the positions of the *K*- and *R*-bands of enones are dependent on the solvent. Hence, solvent corrections are required (Table 1.7) to obtain the calculated values of $\lambda_{max}$ in a particular solvent. The $\varepsilon_{max}$ for *cisoid* enones are usually <10,000, while that of transoid are > 10,000. The calculated values of $\lambda_{max}$ are usually within ±5 nm of the observed values as shown in the following examples illustrating the applications of Woodward-Fieser rules (Table 1.7).

**Example 11.** *Applying Woodward-Fieser rules, calculate the $\lambda_{max}$ for the ethanolic solution of mesityl oxide*

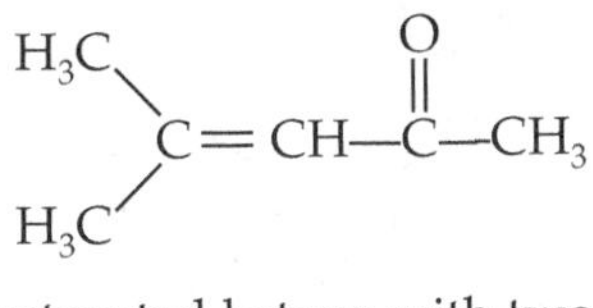

This being an acyclic α, β-unsaturated ketone with two β-alkyl substituents, the $\lambda_{max}$ of this compound is calculated as

| | |
|---|---|
| Base value | 215 nm |
| Two β-alkyl substituents (2 × 12) | 24 nm |
| Calculated $\lambda_{max}$ | 239 nm |
| Observed $\lambda_{max}$ | 237 nm |

**Example 12.** *Predict the value of $\lambda_{max}$ (hexane) for*

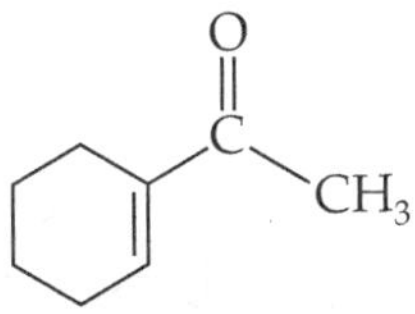

This is a six-membered cyclic α, β-unsaturated ketone with one α- and one β- alkyl substituents. Hence,

| | |
|---|---|
| Base value | 215 nm |
| One α-alkyl substituent | 10 nm |
| One β-alkyl substituent | 12 nm |
| Predicted $\lambda_{max}$ (EtOH) | 237 nm |
| Observed $\lambda_{max}$ (EtOH) | 249 nm |
| Calculated $\lambda_{max}$ (hexane) | 237 (EtOH) |
| Solvent correction | + 11 |
| | 248 nm |

**Example 13.** *Calculate $\lambda_{max}$ (EtOH) for*

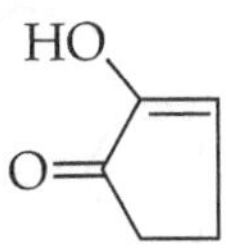

It is a five-membered cyclic α, β-unsaturated ketone with one α-hydroxy and one β-ring residue. Thus,

| | |
|---|---|
| Base value | 202 nm |
| One α-hydroxy groups | 35 nm |
| One β-ring residue | 12 nm |
| Calculated $\lambda_{max}$ (EtOH) | 249 nm |
| Observed $\lambda_{max}$ (EtOH) | 247 nm |

**Example 14.** *What is the expected $\lambda_{max}$ for*

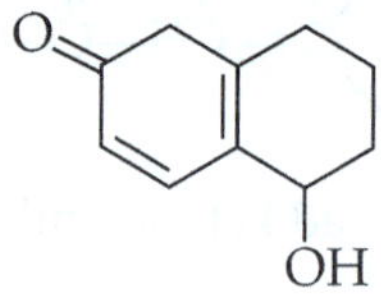


This being an α, β-unsaturated cyclohexenone system, hence

| | |
|---|---|
| Base value | 215 nm |
| Ring residues γ(1) | 18 nm |
| δ(2) | 36 nm |
| One homoannular diene component | 39 nm |
| One double bond extending conjugation | 30 nm |
| Expected $\lambda_{max}$ (EtOH) | 338 nm |

**Example 15.** *Calculate $\lambda_{max}$ for*

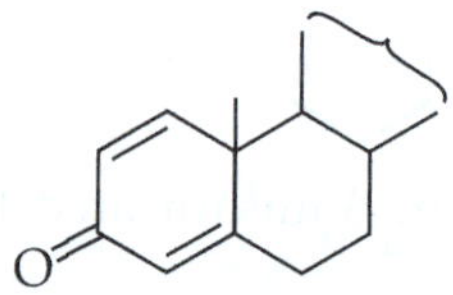


It is a six-membered cyclic α, β-unsaturated ketone in which the carbonyl group has α, β-unsaturation on either side, *i.e.* there is cross (branched) conjugation in this compound. In such cases, $\lambda_{max}$ is calculated by considering most highly substituted conjugated system which gives higher value of $\lambda_{max}$. Thus,

| | |
|---|---|
| Base value | 215 nm |
| Two β-ring residues (2 × 12) | 24 nm |
| One exocyclic carbon-carbon double bond | 5 nm |
| Calculated $\lambda_{max}$ (EtOH) | 244 nm |
| Observed $\lambda_{max}$ (EtOH) | 245 nm |

**Example 16.** *Calculate the wavelength of the maximum UV absorption for the ethanolic solution of*

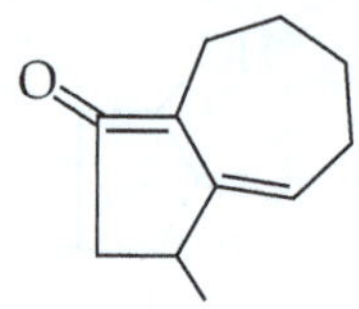

It is an α, β-unsaturated cyclopentenone system. Hence,

| | |
|---|---|
| Base value | 202 nm |
| One β-alkyl substituent | 12 nm |
| One double bond extending conjugation | 30 nm |
| One exocyclic carbon-carbon double bond | 5 nm |
| One γ-ring residue | 18 nm |
| One δ-ring residue | 18 nm |
| Calculated $\lambda_{max}$ (EtOH) | 285 nm |
| Observed value (EtOH) | 287 nm |

**Example 17.** *An organic compound can have one of the following structures:*

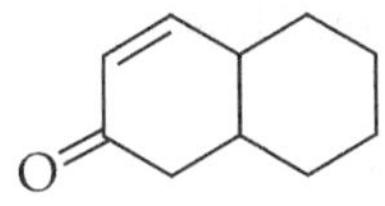
(a)

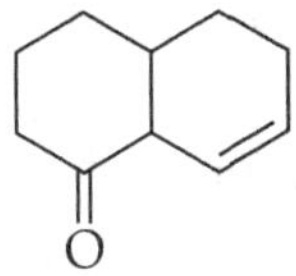
(b)

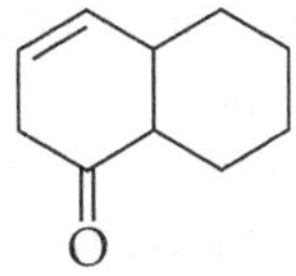
(c)

The $\lambda_{max}$ of ethanolic solution of the compound is 242 nm. Which is the most likely structure of the compound ? Explain your choice.

Let us calculate $\lambda_{max}$ for each of the above structures.

The calculated $\lambda_{max}$ (EtOH) for:

(*a*) 215 (base) + 12(1 β-ring residue) = 227 nm.

(*b*) 215 (base) + 10(1 α-ring residue) + 12 (1 β-ring residue) + 5(1 exocyclic carbon-carbon double bond) = 242 nm.

(*c*) It has an unconjugated system. Hence its C═C will show $\lambda_{max}$ < 200 nm due to $\pi \rightarrow \pi^*$ transition, and C═O will show two bands at ~ 180 and 285 nm due to $\pi \rightarrow \pi^*$ and $n \rightarrow \pi^*$ transitions, respectively.

Since the given $\lambda_{max}$ of the compound is 242 nm, its most likely structure is (*b*) because the calculated $\lambda_{max}$ for this structure is equal to the given value.

**Example 18.** *The following α, β-unsaturated ketones have $\lambda_{max}$ 236, 256 and 270 nm in ethanol. Explain which is which ?*

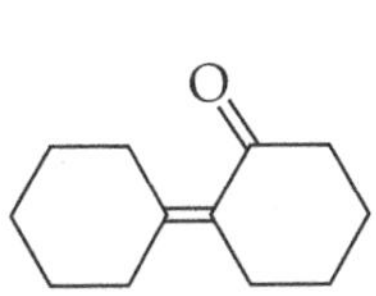
(a)

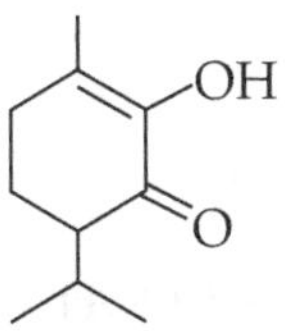
(b)

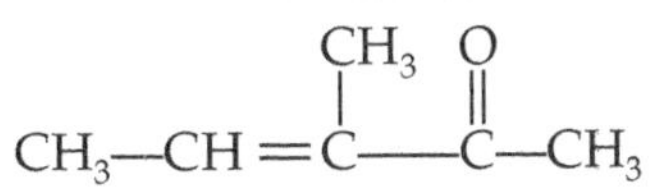
(c)

The calculated $\lambda_{max}$ for (*a*) is 215 (base) + 10 (1 α-ring residue) + 2 × 12 (2β-ring residues) + 10 (the same C═C exocyclic to two rings) = 259 nm (EtOH). Thus, (*a*) is the α, β-unsaturated ketone having $\lambda_{max}$ 256 nm because this value agrees with the calculated $\lambda_{max}$ for (*a*).

The calculated $\lambda_{max}$ for (*b*) is 215 (base) + 35(1 α-OH) + 2 × 12 (2 β-ring residues) = 274 nm (EtOH). Thus, (*b*) is the α, β-unsaturated ketone having $\lambda_{max}$ 270 nm because this value agrees with the calculated $\lambda_{max}$ for (*b*).

The calculated $\lambda_{max}$ for (*c*) is 215 (base) + 10 (1 α-alkyl substituent) + 12 (1β-alkyl substituent) = 237 nm (EtOH). Thus, (*c*) is the α, β-unsaturated ketone having $\lambda_{max}$ 236 nm because this value agrees with the calculated $\lambda_{max}$ for (*c*).

**Example 19.** *Predict the $\lambda_{max}$ for the ethanolic solution of*

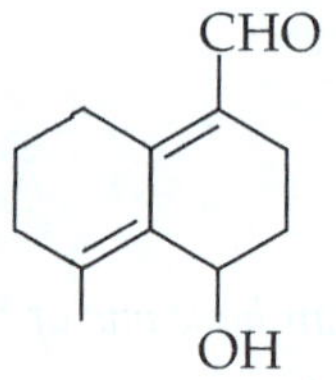


This compound has an α, β-unsaturated aldehyde system. Thus,

| | |
|---|---|
| Base value | 207 nm |
| One double bond extending conjugation | 30 nm |
| One α-ring residue | 10 nm |
| One β-ring residue | 12 nm |
| One γ-ring residue | 18 nm |
| One δ-ring residue | 18 nm |
| One δ-alkyl substituent | 18 nm |
| Two carbon-carbon exocyclic double bonds (2 × 5) | 10 nm |
| Predicted $\lambda_{max}$ (EtOH) | 323 nm |

### 1.4.6 α, β-Unsaturated Carboxylic Acids and Esters

The attachment of groups containing a lone pair of electrons to the carbonyl group, as in carboxylic acids, esters and amides, shifts the $n \rightarrow \pi^*$ transition to shorter wavelengths (200–220 nm) region with little effect on intensity. This hypsochromic shift is due to combined inductive and resonance effects. For example, compared to the *R*-band of acetaldehyde ($\lambda_{max}$ 290 nm) that of acetic acid, ethyl acetate and acetamide appear at 204, 211 and 220 nm, respectively.

α, β-unsaturated acids, esters and amides display strong *K*-bands characteristic of the conjugated system. α, β-unsaturated acids and esters follow a trend similar to that of enones. Thus, we can calculate $\lambda_{max}$ of their *K*-bands taking the base value 195 nm and using the data given in Table 1.6. As shown by the following examples, the calculated and observed values of $\lambda_{max}$ match within ±5 nm.

**Example 20.** *Calculate $\lambda_{max}$ for*

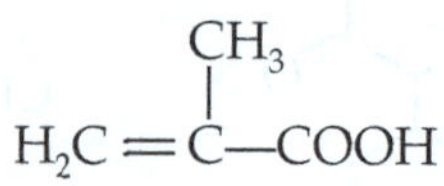


| | |
|---|---|
| Base value | 195 nm |
| One α-alkyl substituent | 10 nm |
| Calculated $\lambda_{max}$ (EtOH) | 205 nm |
| Observed $\lambda_{max}$ (EtOH) | 210 nm |

**Example 21.** *Predict the $\lambda_{max}$ for*

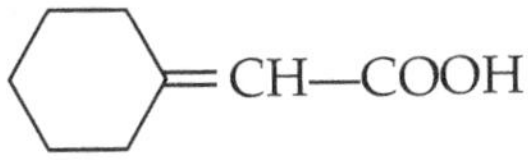

| | |
|---|---|
| Base value | 195 nm |
| Two β-ring residues (2 × 12) | 24 nm |
| One exocyclic C = C | 5 nm |
| Predicted $\lambda_{max}$ (EtOH) | 224 nm |
| Observed $\lambda_{max}$ (EtOH) | 220 nm |

### 1.4.7 UV Spectra of Aromatic Compounds

The UV spectrum of benzene shows three major maxima at 184 nm, 204 nm and 254 nm. Evidently the first maxima is in the far UV region and other two maxima are in weak region. Due to substitution in the benzene ring, there is a bathochromic shift of the benzene bands as shown below:

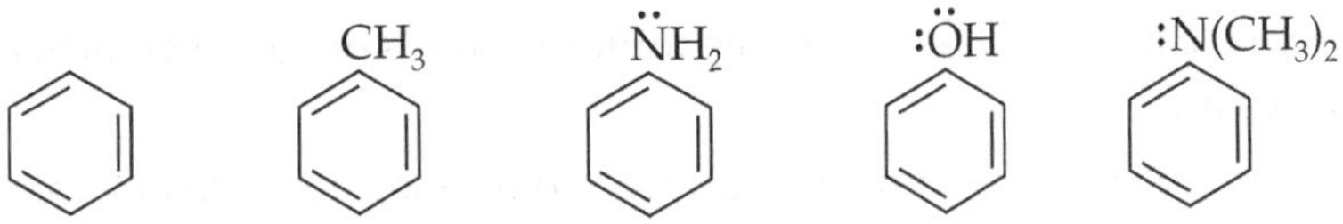

| | | $CH_3$ | $\ddot{N}H_2$ | $:\ddot{O}H$ | $:N(CH_3)_2$ |
|---|---|---|---|---|---|
| $\lambda_{max}$ | 184 nm | 208 nm | 230 nm | 212 nm | 250 nm |
| $\lambda_{max}$ | 204 nm | 260 nm | 280 nm | 260 nm | 300 nm |
| $\lambda_{max}$ | 254 nm | | | | |

The value $\lambda_{max}$ depends upon the nature of substituents and resonance interaction, for example,

*(i)*

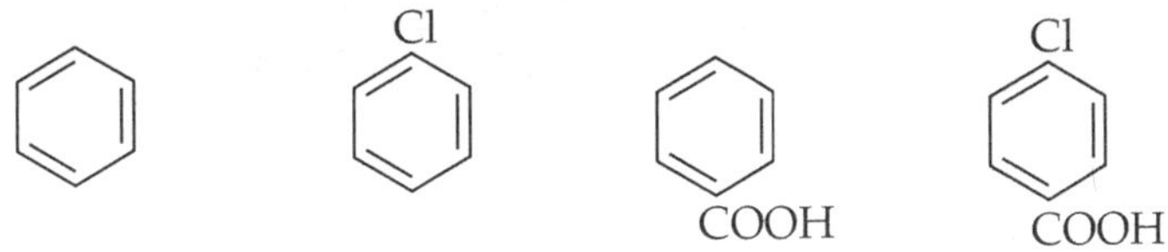

| | | | | |
|---|---|---|---|---|
| $\lambda_{max}$ (nm) | 204 | 210 | 230 | 242 |

*(ii)*

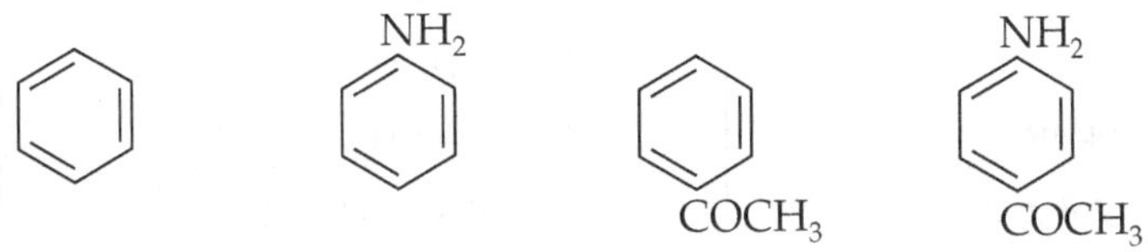

| | | | | |
|---|---|---|---|---|
| $\lambda_{max}$ (nm) | 204 | 230 | 246 | 320 |

In the first case, there is no resonance interaction and both Cl and COOH groups show additive effect in $\lambda_{max}$. Therefore, $\lambda_{max}$ for *p*-chlorobenzoic acid should be 210 + 230 – 204 = 236 nm, which is very close to the observed value. On the other hand, in the second case, the —$NH_2$ and —$COCH_3$ groups have opposing electrical effect. Therefore $\lambda_{max}$ for *p*-amino acetophenone should be 230 + 246 – 272 nm. But its observed value is quite higher. The UV spectrum of diphenyl suggests considerable resonance interaction between rings as shown below:

 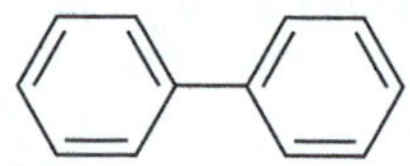

$\lambda_{max}$ 240 nm 250 nm

If two rings carry substituents in a suitable position, the molar extinction coefficient diminishes due to hinderance of the coplanarity of the two rings.

But hydrocarbon, like diphenyl methane, has a typical benzene-like spectrum.

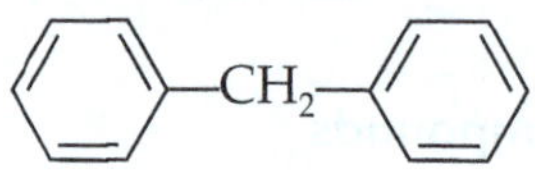

$\lambda_{max}$ 260 nm

On the basis of the above discussion we may conclude that UV spectroscopy is useful only in compounds which have extended conjugation between carbon-carbon double bonds and between carbon-carbon or carbon-oxygen/nitrogen multiple bonds or between carbon-carbon double bonds and a benzene ring.

The values of $\lambda_{max}$ for some unsaturated componds are given in Table 1.8.

**TABLE 1.8 $\lambda_{max}$ of Some Unsaturated Compounds**

| *Compound* | *Structure* | *$\lambda_{max}$(nm)* |
|---|---|---|
| Ethane | $CH_2 = CH_2$ | 171 |
| *Trans*-3-hexene | 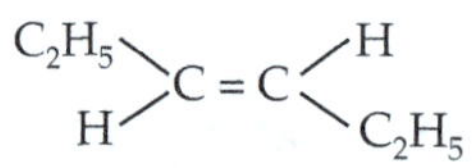 | 184 |
| Cyclohexene |  | 182 |
| 1-Octene | $CH_3(CH_2)_5 - CH = CH_2$ | 177 |
| 1-Octyne | $CH_3(CH_2)_5 - C \equiv CH$ | 185 |
| Propyne | $CH_3 - C \equiv CH$ | 187 |
| 1, 3 Butadiene | $CH_2 = CH - CH = CH_2$ | 217 |
| 1, 4-pentadiene | $CH_2 = CH - CH_2 - CH = CH_2$ | 178 |
| 1-Butene-3-yne | $CH_2 = CH - C \equiv CH$ | 228 |
| *Cis*-1, 3-pentadiene | 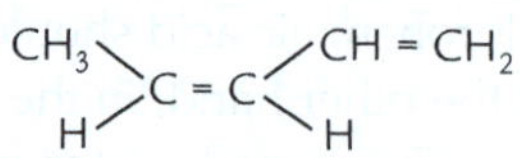 | 223 |
| *Trans*-1, 3-pentadiene | 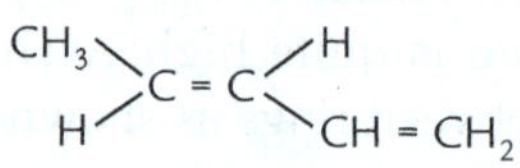 | 223.5 |

### 1.4.8 Applications of UV Spectroscopy

In the application of the electronic spectroscopy for structural analysis, only the region above 200 nm is really useful and the region below 200 nm is hardly useful for this purpose. Some important applications of UV and visible spectroscopy to organic chemistry are summarized as follows.

(*i*) **Detection of a Functional Group (Chromophore).** The presence or absence of a particular chromophore may be indicated by the presence or absence of an absorption band in the expected wavelength region. For example, the presence of a low intensity band in the region 270-300 nm indicates the presence of an aldehydic or ketonic carbonyl group. If the spectrum is transparent above 200 nm, it shows the absence of

(*a*) an aldehydic or ketonic carbonyl group.

(*b*) a conjugated-system.

(*c*) an aromatic ring.

(*d*) a bromine or iodine atom in the molecule.

However, an unconjugated C=C bond or some other atoms or groups may be present in the molecule if it does not absorb above 200 nm. Thus, no definite conclusion can be drawn regarding the structure of the molecule if it absorbs below 200 nm.

(*ii*) **Detection of Conjugation and Elucidation of Its Nature.** Compounds containing a conjugated system including aromatics are characterized by their absorptions above 200 nm. We have noted that the longer the conjugated system, the longer is the wavelength of absorption, usually accompanied by the increased intensity. Also, substitutions on a conjugated system generally cause bathochromic and hyperchromic effects. Thus, we can elucidate the nature of conjugation by comparing the values of $\lambda_{max}$ and $\varepsilon_{max}$ for the compound under study with that of a probable analogous compound.

UV spectral data for a variety of compounds are available for comparison, *e.g.* acyclic conjugated dienes show intense *K*-bands in region 215-230 nm ($\varepsilon_{max}$ ~ 21,000). Similarly, α, β-unsaturated aldehydes and ketones show an intense *K*-band in the 215-250 nm region with $\varepsilon_{max}$ usually 10,000–20,000, and a weak *R*-band in the region 310-330 nm ($\varepsilon_{max}$ 10–100). Aromatic compounds are characterized by an intense band ($E_2$–band) near 200 nm ($\varepsilon_{max}$ ~8000), and a weak *B*-band with fine structure near 260 nm ($\varepsilon_{max}$ ~200). In all the above cases, additional conjugation as well as substitution generally causes bath chromic and hypsochromic effects.

(*iii*) **Study of Extent of Conjugation.** The values of $\lambda_{max}$ and $\varepsilon_{max}$ increase as the number of conjugated multiple bonds increases, thus the extent of conjugation can be estimated. It has been found that the absorption occurs in the visible region if a polyene has eight or more conjugated double bonds. A polyene with such a sufficient conjugation becomes coloured, *e.g.* β-carotene (orange) and lycopene (red) having eleven conjugated double bonds are coloured and absorb in the visible region at 450 nm, $\varepsilon_{max}$ $14 \times 10^4$ and 474 nm, $\varepsilon_{max}$ $18.6 \times 10^4$, respectively.

(*iv*) **Distinction Between Conjugated and Unconjugated Compounds.** In general, electronic spectroscopy can distinguish isomeric conjugated and unconjugated compounds. For example, isomeric dienes 2, 4-hexadiene and 2, 5-hexadiene can be readily differentiated because the former being a conjugated diene will absorb above 200 nm (227 nm), whereas the latter being an unconjugated diene will absorb below 200 nm (~ 170 nm). Similarly, an α, β-unsaturated ketone can be readily distinguished from its β, γ-isomer because the former having conjugated system will show both the $\pi \rightarrow \pi^*$ and $n \rightarrow \pi^*$ transition bands at longer wavelengths compared to that of the latter which is an unconjugated compound.

(*v*) **Study of Strain.** In molecules like 2-substituted biphenyls, there is steric strain which forces the rings out of coplanarity resulting in the loss of conjugation. This causes hypsochromic and hypochromic effects which are measures of steric strain in such molecules, *i.e.* the larger are these effects, the greater will be the steric strain.

(*vi*) **Determination of Configurations of Geometrical Isomers.** It is possible when there is loss of coplanarity of one isomer due to steric hindrance resulting in the loss of conjugation accompanied by hypsochromic and hypochromic effects. Obviously, this isomer is the *cis*-isomer in which groups are closer to each other to cause steric and to force the groups out of coplanarity. Thus, *cis*-isomers absorb at shorter wavelengths and have lower intensity than the *trans*-isomers, *e.g. cis*-stilbene shows $\lambda_{max}$ 283 nm, $\varepsilon_{max}$ 12,300, whereas *trans*-stilbene shows $\lambda_{max}$ 295 nm, $\varepsilon_{max}$ 25,000 (Section 2.18).

(*vii*) **Study of Tautomerism.** UV spectroscopy can be used for identifying the predominant (stable) tautomer. For example, 2-hydroxypyridine (IX, *R*=H) and 2-pyridone (*X*, *R*= H) equilibrium has been shown to lie far to the right, *i.e.* 2-pyridone predominates, because the UV spectrum of the solution resembles with that of a solution of N-methyl-2-pyridone (*X*, *R* = Me) but not with that of 2-methoxypyridine (IX, *R* = Me).

(*viii*) **Confirmation of Suspected Phenols and Aromatic Amines.** The spectral characteristic of phenols and aromatic amines change with the change of pH of the solution. Thus, suspected phenols and aniline derivatives may be confirmed by comparison of UV spectra recorded in neutral and alkaline or acid solutions (for details, see section 2.18).

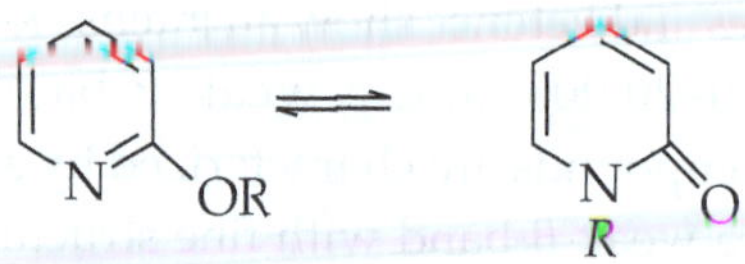

(IX)
R = Me
$\lambda_{max}$ < 205 nm ($\varepsilon_{max}$ > 5300)
369 nm ($\varepsilon_{max}$ = 3230)

(X)
R = Me
$\lambda_{max}$ = 226 nm ($\varepsilon_{max}$ = 6100)
297 nm ($\varepsilon_{max}$ = 5700)
R = H
$\lambda_{max}$ = 224 nm ($\varepsilon_{max}$ = 7230)
293 nm ($\varepsilon_{max}$ = 5900)

### 1.4.9 Solved Problems on UV Spectroscopy

**Problem 1.** *Calculate the value of $\varepsilon_{max}$ for compound X. $2 \times 10^{-5}$ M solution of an organic compound X placed in a 1 cm tube has an absorbance (A) of 0.63 at $\lambda_{max}$ = 237 nm.*

**Solution.** We know that $\varepsilon_{max} = \dfrac{A}{l.c} = \dfrac{0.63}{1 \times 2 \times 10^{-5}} = 31{,}500.$

**Problem 2.** *The UV spectrum of acetone shows two peaks at $\lambda_{max}$ 189 nm, $\varepsilon_{max}$ = 900 and $\lambda_{max}$ ~ 273 nm ($\varepsilon_{max}$= 15). Identify the electronic transitions for each peak.*

**Solution.** The following two peaks are possible in acetone:

($\pi \to \pi^*$)

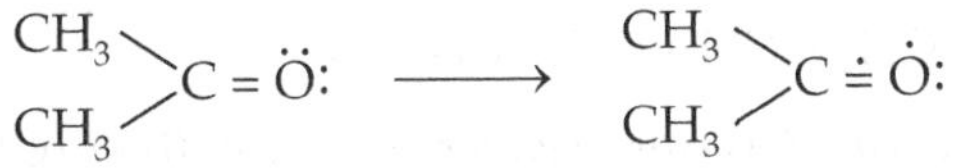

($n \to \pi^*$)

Since $n \to \pi^*$ transition requires less energy than the $\pi \to \pi^*$ transition, hence $n \to \pi^*$ occurs relatively at higher wavelength than $\pi \to \pi^*$. In other words, the peak at $\lambda_{max}$ – 273 nm is due to $n \to \pi^*$ transition and the peak $\lambda_{max}$ ~ 189 nm is due to $\pi \to \pi^*$ transition.

**Problem 3.** *Select the compounds that will absorb UV radiations.*

*Acetone, benzene, benzoic acid, butadiene, chloro hexane, ethonol, ethylene, heptane, heptene, nitrobenzene and water.*

**Solution.** UV radiations are absorbed by unsaturated compounds, preferably those with extended conjugation. Therefore, the following compounds are expected to absorb UV radiation:

Acetone, benzene, benzoic acid, butadiene, ethylene, and nitrobenzene.

However, ethylene and acetone containing only an isolated double bond (C = C or C = O) would absorb appreciably only in the far UV region, *i.e.*, below 200 nm.

**Problem 4.** *Calculated $\lambda_{max}$ for the following compounds:*

(*a*) $CH_3—CH = CH — CH = CH — CH_3$

(*b*) 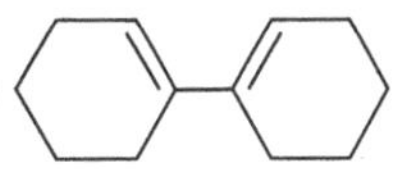

(*c*) 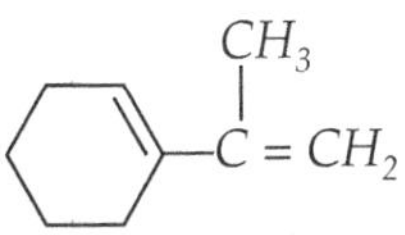

**Solution.** (*a*)

Basic value = 217 nm
For two methyls = 10 nm
Calculated $\lambda_{max}$ = 227 nm

(*b*)

Basic value = 217 nm
For four rings residues = 20 nm
Calculated $\lambda_{max}$ = 237 nm

(*c*)

Basic value = 217 nm
For two rings residues = 10 nm
For one exocyclic double bond = 5 nm
Calculated $\lambda_{max}$ = 232 nm.

**Problem 5.** *A diene has $\lambda_{max}$ at 175 nm. What do you know from this data?*

**Solution.** Since the value of $\lambda_{max}$ is in the far UV region hence the two double bonds are isolated double bonds in the given diene. It may be:

$$CH_2 = CH - CH_2 - CH = CH_2.$$

**Problem 6.** *A diene of molecular formula $C_4H_6$ has an intense peak at 217 nm in its UV spectrum. Deduce the structure of $C_4H_6$.*

**Solution.** Since the value of $\lambda_{max}$ is 217 nm, it suggests that the diene is a conjugated diene.

Hence, the structural formula of $C_4H_6$ is:

$$CH_2 = CH - CH = CH_2$$

**Problem 7.** *The enol acetate of structure (A) can be either structure (B) or (C). The observed $\lambda_{max}$ is 238 nm. Deduce its structure, assuming that the acetate group has no significant effect.*

**Solution.** Since, the $\lambda_{max}$ of the cyclic diene structure is less than 253 nm, the homoannular diene structure (C) is ruled out. The calculated value of $\lambda_{max}$ for structure (B) is 214 + 3 × 5 + 5M = 234 nm, which is close to the observed value (238 nm).

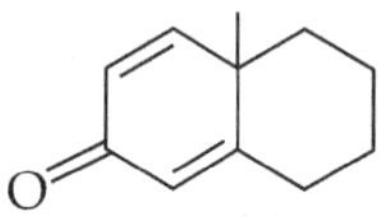


(a)

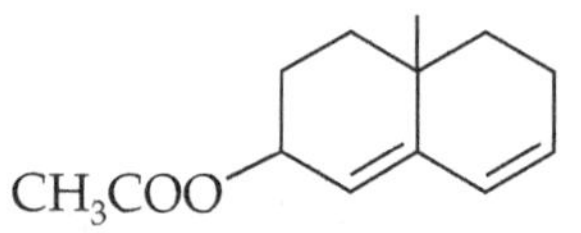


(b)

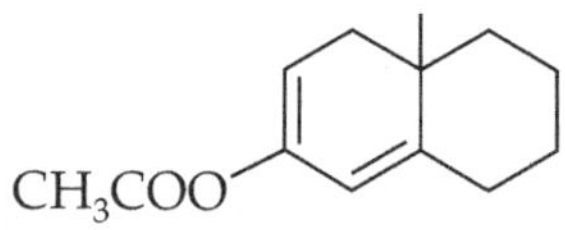


(c)

**Problem 8.** *Calculate $\lambda_{max}$ for*

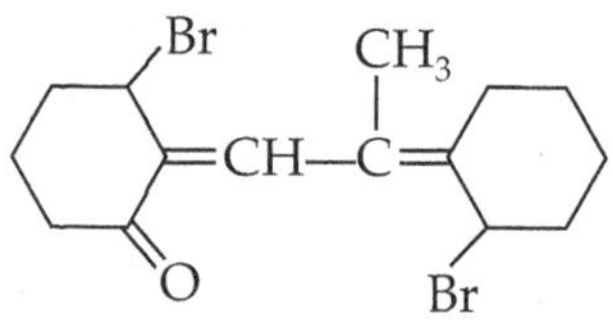


**Solution.** The $\lambda_{max}$ for the above compound can be calculated as:

| | |
|---|---|
| Base value | 215 nm |
| One double bond extending conjugation | 30 nm |
| One α-ring residue | 10 nm |
| One γ-alkyl substituent | 18 nm |
| Two δ-ring residues (2 × 18) | 36 nm |
| Two exocyclic C = C (2 × 5) | 10 nm |
| Calculated $\lambda_{max}$ (EtOH) | 319 nm |

**Problem 9.** *Calculate $\lambda_{max}$ for the ethanolic solution of p-chloroacetophenone.*

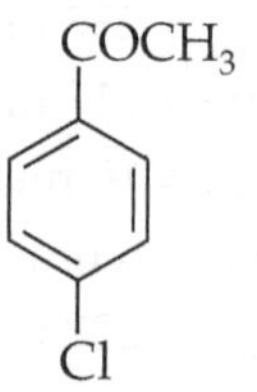

**Solution.** In this compound G is an alkyl group. Thus,

| | |
|---|---|
| Base value | 246 nm |
| Para Cl | 10 nm |
| Calculated $\lambda_{max}$ (EtOH) | 256 nm |
| Observed $\lambda_{max}$ (EtOH) | 254 nm |

**Problem 10.** *Predict the absorption maximum for the ethanolic solution of 3,4-dihydroxyacetophenone.*

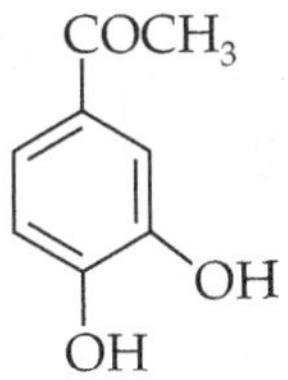


**Solution.** In this compound G is an alkyl group. Thus,

| | |
|---|---|
| Base value | 246 nm |
| *m*-OH | 7 nm |
| *p*-OH | 25 nm |
| Calcualted $\lambda_{max}$ (EtOH) | 278 nm |
| Observed $\lambda_{max}$ (EtOH) | 281 nm |

**Problem 11.** *Calculate the absorption maximum for p-bromobenzoic acid.*

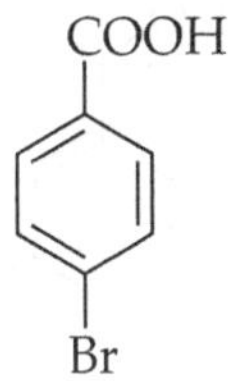


In this case G is OH. Thus,

| | |
|---|---|
| Base value | 230 nm |
| *p*-Br | 15 nm |
| Calculated $\lambda_{max}$ (EtOH) | 245 nm |
| Observed $\lambda_{max}$ (EtOH) | 245 nm |

**Problem 12.** *Calculate $\lambda_{max}$ for 6-methoxytetralone.*

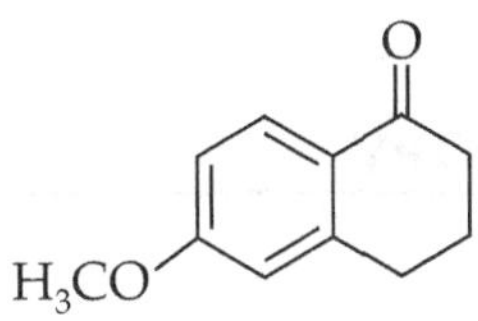


**Solution.** In this case G is a ring residue. Thus,

| | |
|---|---|
| Base value | 246 nm |
| *o*-ring residue | 3 nm |
| *p*-methoxyl | 25 nm |
| Calculated $\lambda_{max}$ (EtOH) | 274 nm |
| Observed $\lambda_{max}$ (EtOH) | 276 nm |

**Problem 13.** *Calculate the absorption maximum for 3-carbethoxy-5-chloro-8-hydroxyl-4-methyltetralone.*

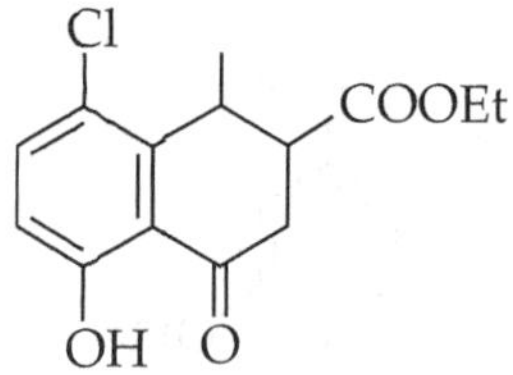


In this case *G* is a ring residue. Thus,

| | |
|---|---|
| Base value | 246 nm |
| *o*-ring residue | 3 nm |
| *o*-hydroxyl | 7 nm |
| Calculated $\lambda_{max}$ (EtOH) | 256 nm |
| Observed $\lambda_{max}$ (EtOH) | 257 nm |

**Example 14.** *Predict the $\lambda_{max}$ for the ethanolic solution of 3, 4-dimethoxy-10-oxooctahydrophenanthrene.*

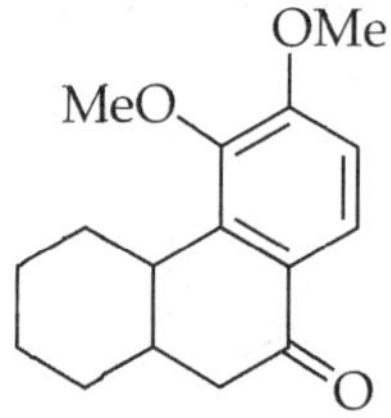


**Solution.** In this compound *G* is a ring residue. Thus,

| | |
|---|---|
| Base value | 246 nm |
| *m*-methoxyl | 7 nm |
| *p*-methoxyl | 25 nm |
| *o*-ring residue | 3 nm |
| Predicted $\lambda_{max}$ (EtOH) | 281 nm |
| Observed $\lambda_{max}$ (EtOH) | 278 nm |

## 1.5 INFRARED SPECTROSCOPY

### 1.5.1 Introduction

IR spectroscopy is the most powerful analytical technique for the identification of chemical constitution. IR radiations range from 2 μm to 16 μm and are not enough energetic to produce electronic excitations in most organic molecules. These radiations cause stretching and bending of organic bonds. After absorption of IR radiations, the molecule thus give close-packed absorption bands known as *infrared absorption spectrum.* A number of bands may be present in the IR spectrum corresponding to the characteristic functional group and bonds present in the substance.

### 1.5.2 Theory (Origin) of Infrared Spectroscopy

IR absorption spectra originate from transitions in vibrational and rotational energy levels within a molecule. On absorption of IR radiation, vibrational and rotational energies of the molecule are increased. When a molecule absorbs IR radiation below 100 $cm^{-1}$, the absorbed radiation causes transitions in its rotational energy levels. Since these energy levels are quantized, a molecular rotational spectrum consists of discrete lines.

When a molecule absorbs IR radiation in the range 100–10,000 $cm^{-1}$, the absorbed radiation causes transitions in its vibrational energy levels. These energy levels are also quantized, but vibratonal spectra appear as bands rather than discrete lines. The energy differences between various rotational energy levels of a molecule are far less than that between its vibrational energy levels. Thus, a single transition in vibrational energy levels is accompanied by a large number of transitions in rotational energy levels and so the vibrational spectra appear as vibrational-rotational bands instead of discrete lines. Organic chemists are mainly concerned with these vibrational-rotational bands, especially with those occurring in the regin 4000–667 $cm^{-1}$ (2.5–15 μ).

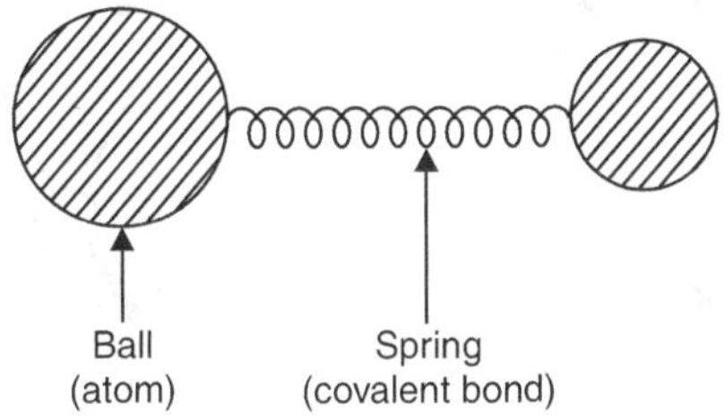


Various atoms in a molecule may be regarded as balls of different masses and the covalent bonds between them as weightless tiny springs holding such balls together. Atoms in a molecule are not still but they vibrate. The two types (modes) of fundamental molecular vibrations known are:

(*i*) **Stretching Vibrations.** In stretching vibrations, the distance between two atoms increases or decreases, but the atoms remain in the same bond axis. Stretching vibrations are of two types:

(*a*) ***Symmetrical stretching.*** In this mode of vibration, the movement of atoms with respect to the common (or central) atom is simultaneously in the same direction along the same bond axis (Fig. 1.6(*a*)).

(*b*) ***Asymmetrical stretching.*** In this vibration, one atom approaches the common atom while the other departs from it (Fig. 1.6(*b*)).

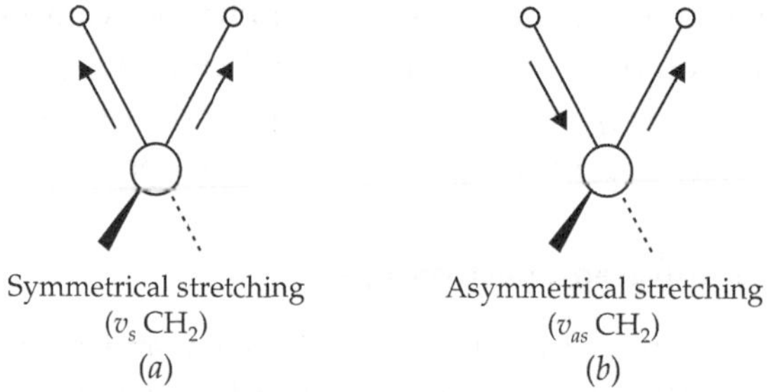


***FIGURE 1.6*** **Stretching vibrations of a $CH_2$ group (v $CH_2$)**

(*ii*) **Bending Vibrations (Deformations).** In such vibrations, the positions of the atoms change with respect to their original bond axes. Bending vibrations are of four types:

(*a*) ***Scissoring.*** In this mode of vibration, the movement of atoms is in the opposite direction with change in their bond axes as well as in the bond angle they form with the central atom (Fig. 1.7 (*a*)).

(*b*) ***Rocking.*** In this vibration, the movement of atoms takes place in the same direction with change in their bond axes (Fig. 1.7(*b*)).

(*c*) ***Wagging.*** In this vibration, two atoms simulaneously move above and below the plane with respect to the common atom (Fig. 1.7(*c*)).

(*d*) ***Twisting.*** In this mode of vibration, one of the atom moves up and the other moves down the plane with respect to the common atom (Fig. 1.7(*d*)).

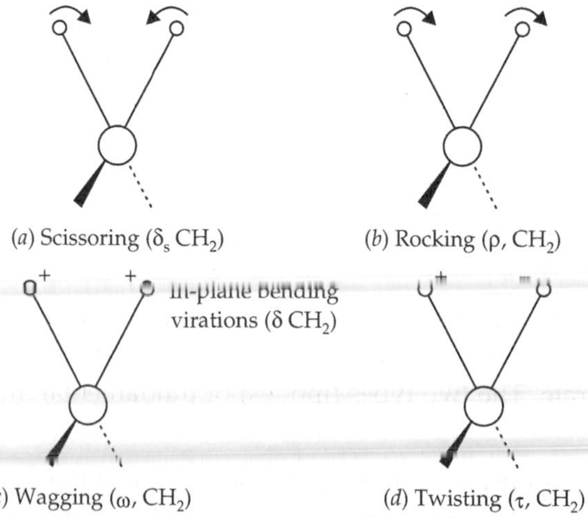


***FIGURE 1.7*** **Bending vibrations (deformations) of a $CH_2$ group (+ and – signs indicate movements perpendicular to the plane of the paper)**

Infrared radiation is absorbed when the oscillating dipole moment, due to a molecular vibration, interacts with the oscillating electric field of the infrared beam. This interaction occurs and hence, an absorption band appears only when a molecular vibration produces a change in the dipole moment of the molecule. Otherwise, the vibration is said to be *infrared inactive* and it will show no absorption band in the infrared spectrum. Usually, larger the change in dipole moment, the higher is intensity of absorption. It is not necessary for a molecule to have a permanent dipole moment for IR absorption.

It may be noted that various stretching and bending vibartions of a bond have a definite and quantized frequency of their own. When infrared radiation of the same frequency is incident on the molecule, there is absorption of energy resulting in the increase of the amplitude of that particular absorption; now the molecule is in the excited state. Such an absorption gives rise to a peak in the IR spectrum. From the excited state, the molecule returns to the ground state by release of extra energy through rotational, collision and translational processes resulting in the increase of temperature of the sample under investigation.

We know that lesser energy is required to bend a spring than that required for stretching it. Analogously, bending vibrations of a bond require lesser energy than its stretching vibrations. Thus, the absorption bands resulting from bending vibrations appear at lower wavenumbers those that resulting from stretching vibrations of the same bond. Stretching vibrations usually cause peaks of high intensity.

### 1.5.3 Number of Fundamental Vibrations

The IR spectra of polyatomic molecules may exhibit more than one vibrational absorption bands. The number of these bands corresponds to the number of fundamental vibrations in the molecule which can be calculated from the degrees of freedom of the molecule. The degrees of freedom of a molecule are equal to the total degrees of freedom of its individual atoms. Each atom has three degrees of freedom corresponding to the three Cartesian coordinates ($x$, $y$ and $z$) necessary to describe its position relative to other atoms in the molecule. Therefore, a molecule having $n$ atoms will have $3n$ degrees of freedom. In case of a non-linear molecule, three of the degrees of freedom describe rotation and three describe translation. Thus, the remaining $(3n - 3 - 3) = 3n - 6$ degrees of freedom are its vibrational degrees of freedom or fundamental vibrations, because

Total degress of freedom ($3n$) = Translational + Rotational
+ Vibrational degrees of freedom

In case of a linear molecule, only two degrees of freedom describe rotation (because rotation about its axis of linearity does not change the positions of the atom) and three describe translation. Thus, the remaining $(3n - 2 - 3) = 3n - 5$ degrees of freedom are vibrational degrees of freedom or fundamental vibrations.

The number of vibrational degrees of freedom for the linear carbon dioxide molecule can be calculated as follows:

Number of atoms ($n$) = 3

Total degrees of freedom ($3n$) = $3 \times 3 = 9$

Rotational degrees of freedom = 2

Translational degrees of freedom = 3

Therefore, vibrational degrees of freedom = $9 - 2 - 3 = 4$

Since each vibrational degree of freedom corresponds to a fundamental vibration and each fundamental vibration corresponds to an absorption band, for carbon dioxide molecule there should be four theoretical fundamental bands.

Similarly, for a non-linear molecule ethane ($C_2H_6$), the vibrational degrees of freedom can be calculated as:

Number of atoms ($n$) = 8

Total degrees of freedom ($3n$) = $3 \times 8 = 24$

Rotational degrees of freedom = 3

Translational degrees of freedom = 3

Hence, vibrational degrees of freedom = 24 – 3 – 3 = 18

Thus, theoretically there should be 18 absorption bands in the IR spectrum of ethane.

In case of benzene ($C_6H_6$), the number of vibrational degrees of freedom can be calculated as follows:

Number of atoms ($n$) = 12

Total degrees of freedom = ($3n$) = 3 × 12 × = 36

Rotational degrees of freedom = 3

Translational degrees of freedom = 3

Therefore, vibrational degrees of freedom = 36 – 3 – 3 = 30

Thus, theoretically, there should be 30 fundamental vibrational bands in the IR spectrum of benzene.

It has been observed that in actual IR spectra, the theoretical number of fundamental bands are seldom obtained because there are certain factors which decrease, whereas certain factors increase the number of bands. The following factors decrease the theoretical number of fundamental bands:

(*i*) The frequencies of fundamental vibrations which fall outside of the region 4000–667 $cm^{-1}$.

(*ii*) Fundamental bands which are so weak that they are not observed.

(*iii*) Fundamental bands which are so close that they coalesce.

(*iv*) The occurrence of a degenerate band from several absorptions of the same frequency in highly symmetrical molecules such as carbon dioxide.

(*v*) Certain fundamental vibrational bands which do not appear in the infrared spectrum due to lack of the required change in dipole-moment of the molecule, *e.g.* in carbon dioxide molecule.

The carbon dioxide molecule is linear and has four fundamental vibrations (3 × 3) – 5 = 4 (Fig. 1.8). Thus, four theoretical fundamental bands are expected but actually it shows only two. The symmetrical stretching vibration in carbon dioxide is IR inactive because it produces no change in the dipole moment of the molecule. The two bending vibrations are equivalent and absorb at the same wavenumber (667.3 $cm^{-1}$). Thus, the IR spectrum of carbon dioxide shows only two fundamental absorption bands, one at 2350 $cm^{-1}$ due to asymmetrical stretching vibration, and the other at 667.3 $cm^{-1}$ due to the two bending vibrations.

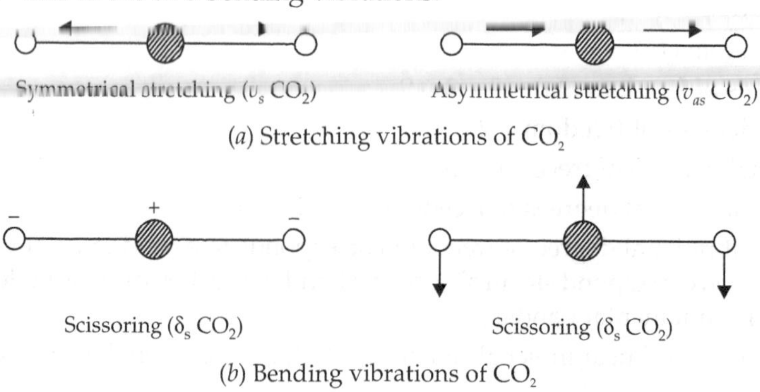


*FIGURE 1.8* **Fundamental vibrations of the linear carbon dioxide molecule (+ and – signs indicate movements perpendicular to the plane of paper)**

In a molecule having a centre of symmetry, the vibrations which are symmetrical about the centre of symmetry are IR inactive. For example, the symmetrical stretching vibrations of carbon dioxide and the symmetrical C = C stretching vibrations of ethylene do not result in an absorption band in the IR spectra. It should be noted that ethylene shows bands for C—H stretching vibrations. Similarly, symmetrical diatomic molecules such as $H_2$, $N_2$, $O_2$ and $Cl_2$ do not absorb IR radiation, but unsymmetrical diatomic molecules like carbon monoxide and iodine chloride (I—Cl) absorb IR radiation.

### 1.5.4 Calculation of Vibrational Frequencies

The stretching vibrations of two bonded atoms may be regarded as the vibration of two balls connected by a spring, a situation for which Hooke's law applies. Thus, an approximate value for the stretching vibrational frequency, *i.e.*, the stretching absorption frequency of a bond, can be calculated by Hooke's law

$$\bar{\upsilon} = \frac{1}{2\pi c}\sqrt{\frac{f}{\mu}}$$

where $\mu = \dfrac{m_1 m_2}{m_1 + m_2}$ is the reduced mass, $m_1$ and $m_2$ are the masses (*g*) of the atoms linked to the particular bond, *f* the force constant of the bond in dynes/cm, *c* the velocity of light ($2.998 \times 10^{10}$ cm/sec) and $\bar{\upsilon}$ the stretching vibrational frequency of the bond in $cm^{-1}$ (wavenumber).

The force constant of a bond is related to its bond strength. The value of force constant is approximately $5 \times 10^5$ dynes/cm for single bonds and approximately twice and thrice of this value for double and triple bonds, respectively.

The force constant is a characteristic property of a bond and like other physical constants, it is another physical constant. The frequencies of IR absorptions are commonly used to calculate the force constants of various bonds.

Applying Hooke's law, the frequency of the C—H stretching vibration can be calculated as follows:

The atomic mass of carbon is 12 and that of hydrogen is 1. Thus,

$$\bar{\upsilon} = \frac{1}{2\pi \times 2.998 \times 10^{10}\ \text{cm sec}^{-1}} \times \sqrt{\frac{5 \times 10^5\ \text{dynes cm}^{-1}\left(\dfrac{12}{6.023} + \dfrac{1}{6.023}\right) \times 10^{-23}\ \text{g}}{\left(\dfrac{12}{6.023} \times 10^{-23}\ \text{g}\right)\left(\dfrac{1}{6.023} \times 10^{-23}\ \text{g}\right)}}$$

Therefore, $\bar{\upsilon} = 3032\ cm^{-1}$.

Actually, C—H stretching vibrations of $CH_3$ and $CH_2$ groups are generally observed in the range 2850–3000 $cm^{-1}$.

The value of stretching vibrational frequency depends on the bond strength and the reduced mass. Obviously, the stretching vibrational frequency of a bond increases with an increase in the bond strength or decrease in the reduced mass of the bonded atoms. Thus, due to the greater bond strength (value of *f*) of C═C and C═O bonds, they are expected to absorb at higher stretching frequencies than that of C—C and C—O, respectively. Similarly, O—H stretching absorption appears at a higher frequency than that of C—C because of the lower value of reduced mass for O—H compared to that for C—C. For the same reason, we would predict O—H stretching vibration to be of higher frequency than O—D stretching vibration.

### 1.5.5 Effect of Hydrogen Bonding Vibrational Frequencies

Hydrogen bonding remarkably lowers the stretching frequencies of both the groups involved in it, and also changes the shape and intensity of the absorption bands. Usually, absorption bands become more intense and broad on hydrogen bonding. The stronger the hydrogen bond, lower is the O—H stretching frequency. Thus, the value of O—H stretching frequency is a test for hydrogen bonding as well as a measure of the strength of hydrogen bonds.

Non-hydrogen-bonded (free) O—H group of alcohols and phenols show sharp and strong absorption bands in the region 3590–3650 $cm^{-1}$. Sharp, non-hydrogen-bonded O—H bands are observed only in the vapour phase, in very dilute solution, in non-polar solvents or when hydrogen bonding is prevented by steric hindrance. Pure samples and concentrated solutions of alcohols and phenols show broad O—H stretching bands in the region about 3200–3600 $cm^{-1}$ due to intermolecular hydrogen bonding. The N—H stretching frequencies of amines are also affected by hydrogen bonding in the same way as that of the hydroxyl group but frequency shifts for amines are lesser than that for hydroxyl compounds. This is because nitrogen is less electronegative than oxygen and so the hydrogen bonding in amines is weaker than that in hydroxy compounds. For example, non-hydrogen-bonded primary amines exhibit two bands, one near 3400 $cm^{-1}$ and the other near 3500 $cm^{-1}$ due to symmetrical and asymmetrical N—H stretching modes, respectively. In pure amines, these bands respectively appear in the range 3250–3330 $cm^{-1}$ and 3330–3400 $cm^{-1}$ due to intermolecular hydrogen bonding.

In liquid or solid state, and in concentrated solutions, carboxylic acids exist as dimers due to strong intermolecular hydrogen bonding.

*FIGURE 1.9*

The strong hydrogen bonding between C═O and OH groups lowers their stretching frequency. Because of more polar O—H bond, carboxylic acids form stronger hydrogen bond than alcohols. Carboxylic acid dimers show very broad, intense O—H stretching absorption in the region 2500–3000 $cm^{-1}$, whereas non-hydrogen-bonded carboxylic acids (monomers) show the O—H absorption near 3550 $cm^{-1}$. Hydrogen bonding weakens the C═O bond, *i.e.* its force constant is reduced, resulting in absorption at a lower frequency than the monomer. Thus, C═O stretching bands of monomers of saturated carboxylic acids appear at 1760 $cm^{-1}$, whereas that of dimers appear at about 1710 $cm^{-1}$.

In many cases, some O—H groups are hydrogen bonded and some are free. Hence, bands due to both may be observed.

***Explanation for lowering the absorption frequency due to hydrogen bonding.*** Let us take the example of hydroxy compounds. On hydrogen bonding, the original O—H bond is lengthend (weakened) due to electrostaic attraction between the hydrogen atoms of one molecule and oxygen atom of the other, and thus the force constant of the O—H bond is reduced resulting in a dcrease in its stretching frequency.

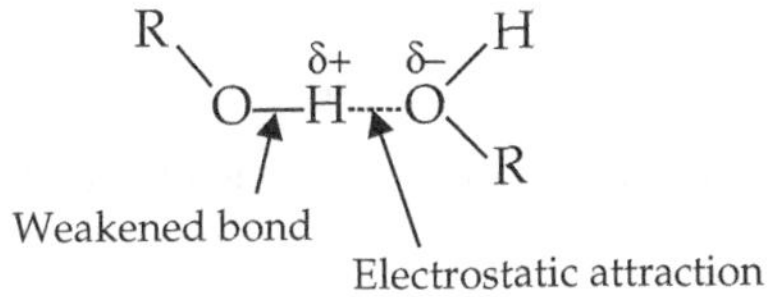


Further, the hydrogen bond may be regarded as a resonance hybrid of the following resonating structures, consequently the O—H bond is weakened and its stretching frequency is lowered.

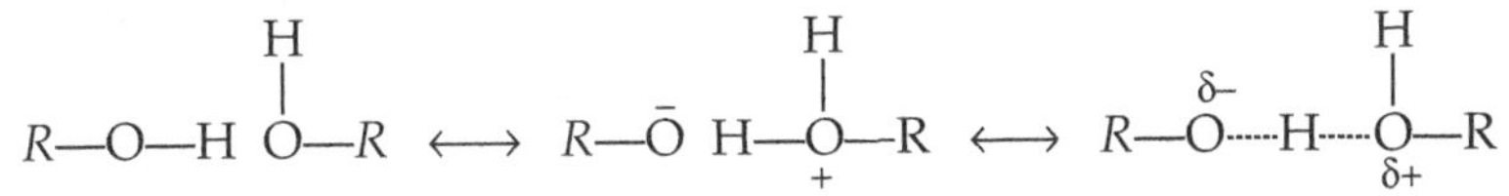

***Distinction between inter- and intramolecular hydrogen bondings.*** In very dilute solutions (in non-polar solvents), intermolecular distances are too large to form intermolecualr hydrogen bonds. On the other hand, in pure liquids, solids and concentrated solutions, the molecules are closer to form intermolecualr hydrogen bond. Thus, intermolecular hydrogen bonding is concentration dependent. On dilution with non-polar solvents, intermolecular hydrogen bonds are broken. Hence, there is decrease in intensity or disappearance of the hydrogen-bonded O—H stretching band and increase in intensity or appearance of free O—H stretching absorption. Thus, in very dulute solutions (in non-polar solvents, concentration < ~ 0.01 M), the O—H stretching frequency is shifted to a higher frequency in case of intermolecular hydrogen bonding.

Intramolecular hydrogen bonding is within the same molecule, hence it is not affected by change in intermolecular distances. Thus, intramolecular hydrogen bonds are unaffected by dilution, and so the absorption band is also unaffected. Intramolecular hydrogen bonding (chelation) is very strong in enols and compounds like methyl salicylate etc., due to resonance stabilization of the chelate ring. For example, in enols, the O—H group involved in chelation shows broad absorption band in the range 2500–3200 $cm^{-1}$, whereas the C=O stretching band in enolic form occurs at 1630 $cm^{-1}$ and that in the *keto* form at 1725 $cm^{-1}$. From the relative intensities of the two bands, it is possible to determine ratio of the *keto* and *enol* forms.

*enol* of $RCOCH_2COR$

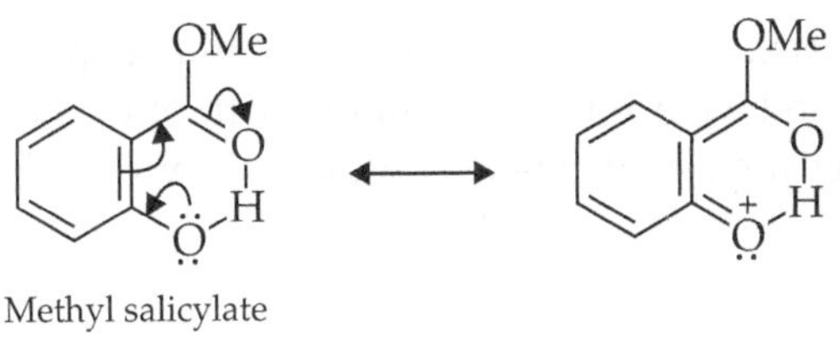

Methyl salicylate

Due to interactions of π electrons of Lewis bases (such as alkenes and benzene) with acidic hydrogen, lengthening and hence weakening of O—H bonds has also been observed. For example, the O—H stretching frequency of phenols is lowered by 40–100 $cm^{-1}$ when the spectrum is recorded in benzene solution as compared to that recorded in $CCl_4$ solution.

### 1.5.6 Conditions for IR Absorption

The conditions for the absorption of IR radiation are:

1. The natural frequency of vibration of a molecule must be equal to the frequency of incident radiation.
2. Changes in vibration must stimulate changes in the dipole moment of the molecule.
3. The frequency of the vibration must satisfy the equation, $E = h\nu$.
4. The intensity of absorption must be proportional to the square of the rate of change of the dipole.

### 1.5.7 Instrumentation

(*a*) **The Instrument.** The spectrophotometer consists of a source of infrared light emitting radiation throughout the whole frequency range of the instrument. This light is passed through two beams of equal intensity and one beam is arranged to pass through the sample to be examined. If the frequency of a vibration of the sample molecule falls within the range of the instrument, the molecule may absorb energy of this frequency from the light. The spectrum is thus scanned by comparing the intensity of the two beams after one has passed through the sample to be examined. A simplified working of the instrument is shown in Figure 1.10.

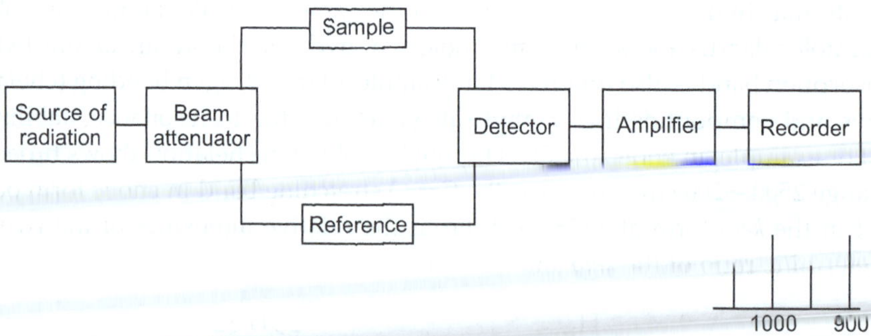


**FIGURE 1.10 Schematic diagram of a recording spectrophotometer.**

(*b*) **Sampling Methods.** The compounds may be examined in the vapour phase, as pure liquids, in solution, and in the solid state.

(*i*) *In the vapor phase.* The vapor is introduced into a special cell, usually about 10 cm long, which can be placed directly in the path of the cone of the IR beams. The end walls of the cell are usually made of sodium chloride, which is transparent to infrared light.

(*ii*) *As a liquid.* A drop of the liquid is squeezed between the flat plates of sodium chloride.

(*iii*) *In solution.* The compound is dissolved to give 1 to 5 per cent solution in carbon tetrachloride. This solution is introduced into a special cell, 0.1 mm to 1 mm thick, made up of sodium chloride. A second cell of equal thickness but containing pure solvent is placed in the path of other beam of spectrophotometer in order that solvent absorptions should be balanced. Spectra taken in such dilute solutions in non-polar solvents are most desirable, because they are better resolved than spectra taken on solids.

(*iv*) *In the solid state.* Solids are examined in crystalline form by dilution in mull. The most widely used method is that of *nujol mull.* About 1 mg of a solid is finely ground in a small agate mortar with a drop of liquid hydrocarbon (Nujol, kaydal). The mull is then pressed between flat plates of sodium chloride.

*Nujol* is a high-boiling fraction from petroleum and has absorption bands at 2520 nm to 2860 nm (C—H Stretch) and around 1460 and 1380 $cm^{-1}$. So, when nujol is used as a mulling agent, no information can be obtained in the regions near these bands. The spectra obtained of the solid sample have different band positions than those of the corresponding solution spectra. The spectra of the solid sample is not better resolved to strong intermolecular forces.

### 1.5.8 Group Frequencies and Analysis

A particular part of the infrared spectrum is referred to either by wavelength or, more preferably, by its frequency. Frequency is not expressed in hertz but in wave numbers, $cm^{-1}$. The wave number is simply the number of waves per centimeter, and is equal to the reciprocal of the wavelength in centimeters. The IR spectrum may be divided into two parts:

(*a*) **Characteristic frequencies.** The region 4000 $cm^{-1}$ to 1400 $cm^{-1}$ (approx.) contains the peak characteristic of individual groups. The position of these peaks is only slightly altered by the rest of the molecule. Hence, the frequencies corresponding to such groups are known as *characteristic frequencies.*

(*b*) **Fingerprint region.** The region 1400 $cm^{-1}$ to 650 $cm^{-1}$ is referred as the fingerprint region. The absorption pattern in this region is complex. This portion is extremely valuable when examined with reference to other regions, for example, if alcoholic or phenolic (O—H) stretching frequency appears in the high frequency region and absorption also occurs at 1260 $cm^{-1}$ to 1000 $cm^{-1}$ (fingerprint region). The latter is used to confirm the assignment of the former, *i.e.*, alcoholic or phenolic nature with highly specific structures. This is the region usually checked for identification, since it is associated with vibrational (and rotational) energy changes of the molecular skeleton, and so is characteristic of the compound.

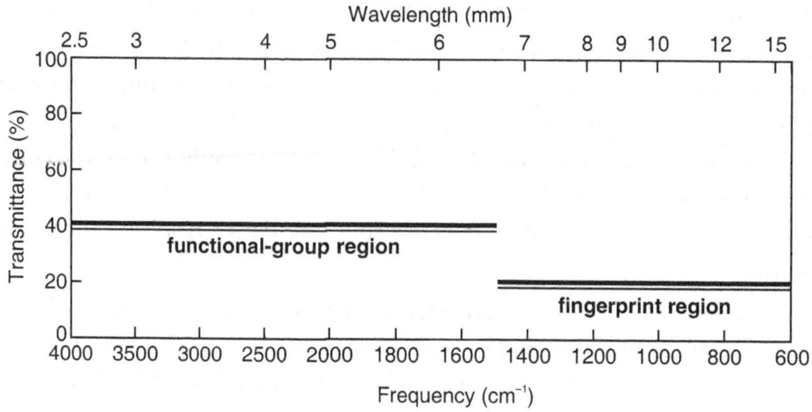


*FIGURE 1.10 (a)*

**TABLE 1.9 Common Infrared Absorption Frequencies**

| *Functional group* | *Found in* | *Range ($cm^{-1}$)* | *No. of bands* |
|---|---|---|---|
| C = O | Acyclic ketones and cyclohexanones | 1725–1705 | |
| C=O<br>\|<br>H | Aldehydes | 1740–1720 | |
| C = O | Cyclopentanones | 1750–1720 | |
| C = O | Cyclobutanones | ~ 1775 | |
| C = C – C = O | αβ-Unsaturated ketones | 1685–1665 | |
| ArC = O | Aromatic aldehydes and ketones | 1700–1680 | |
| C=O<br>\|<br>OH | Carboxylic acids | 1710–1700 | |
| C=O<br>\|<br>OR | Esters and δ-lactones,<br>γ-electrons | 1750–1735<br>1780–1760 | |
| C = C – C=O<br>\|<br>OR | Aromatic and conjugated esters | 1725–1710 | |
| C=O<br>\|<br>$NH_2$ | Primary amide | 1700–1650 | |
| O=C–O–C=O | Anhydrides | 1850–1800<br>1790–1740 | Two bands |

| | | | |
|---|---|---|---|
| C ≡ N | Nitriles | 2280–2200 | |
| C ≡ C | Alkynes | 2300–2100 | |
| OH | Alcohols (free) | 3650–3590 | |
| OH | Alcohols (hydrogen-bonded) | 3400–3200 | |
| OH | Carboxylic acids (hydrogen-bonded) | 2700–2500 | Several bands |
| NH | Secondary amines | 3500–3300 | One band |
| $NO_2$ | Nitro compounds (aromatic) | 1570–1550 | Two |
| | | 1370–1300 | bands |
| $NO_2$ | Nitro compounds (aliphatic) | 1570–1550<br>1380–1370 | Two bands |

**Characteristic Infrared Bands**

The absorption ranges of the major functional groups in organic compounds are summarized in Table 1.9. All the bands in the IR spectrum need not be attempted, but the presence or absence of functional groups such as carbonyl, hydroxyl, amino, etc. is ascertained by this technique and these finding should be used in conjugation with the information obtained from UV, NMR, and mass spectra.

1. *Alkane and alkene residue.* Precise position of the various CH, $CH_2$ and $CH_3$, symmetrical and unsymmetrical vibration frequencies is well known. Since most organic molecules possess alkane residues, the groups of saturated C—H absorption bands are of little diagnostic value, whereas the absence of saturated C—H absorption in a spectrum is of diagnostic value. Unsaturated and aromatic C—H stretching frequencies can be distinguished from the saturated C—H absorption since the latter occurs below 3000 $cm^{-1}$ while the former gives rise to much less intense absorption above 3000 $cm^{-1}$.
2. *Alcohol and phenols.* Compounds containing O—H bands (alcohols, etc.) exhibit a peak between 3700 and 3200 $cm^{-1}$. The position of this band is dependent on the existence and strength of a hydrogen bond. The IR spectrum of ethanol in vapor state (no hydrogen band) shows a sharp band near 3700 $cm^{-1}$ which is described as a free hydroxyl group (Figure 1.11).

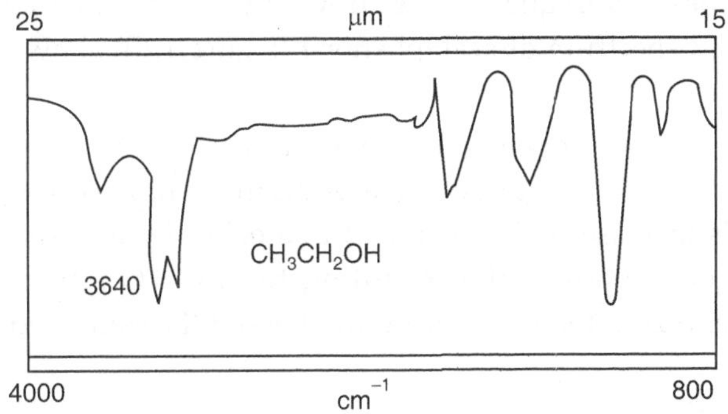


*FIGURE 1.11* **The IR spectrum of ethanol (in vapour state).**

However, when the spectrum is run in solution, this sharp band is almost replaced by a broad band at 3350 cm$^{-1}$ (Figure 1.12). On further increasing the concentration, the band, at 3700 cm$^{-1}$, completely disappears. The broad band at 3350 cm$^{-1}$ is characteristic of the hydrogen bonded hydroxyl group and since the intensity of this band changes with concentration, it is a case of intermolecular hydrogen bonding.

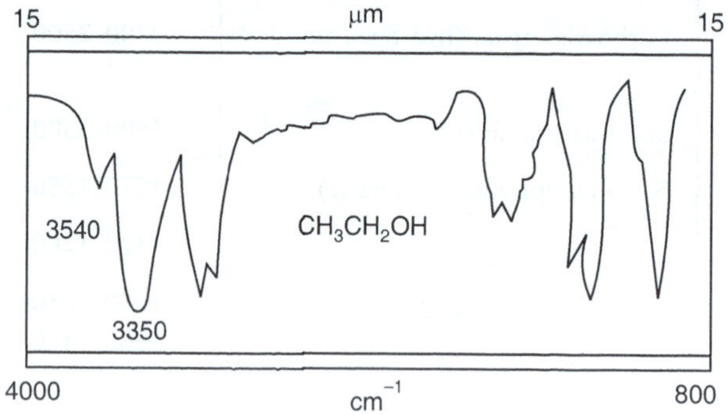


***FIGURE 1.12*** **The IR spectrum of ethanol (concentrated solution).**

3. *Absorption frequencies of carbonyl bond.* The absorption caused by a carbonyl group is the strongest band in the IR spectra of carbonyl compounds. The general trends of structural variation on the position of C = O stretching frequencies may be summarized as:
   (*i*) The more electronegative the group X in the system R—CO—X, the higher the frequency.
   (*ii*) αβ-unsaturation causes a lowering of frequency of 15 cm$^{-1}$ to 40 cm$^{-1}$, except in amides.
   (*iii*) Further conjugation has relatively little effect.
   (*iv*) Hydrogen bonding to a carbonyl group causes a shift to lower frequency of 40 cm$^{-1}$ to 60 cm$^{-1}$. Acids, amides, enolized β-keto carbonyl system, and *o*-hydroxy- and *o*-aminophenyl carbonyl compounds show this effect.
   (*v*) When more than one of the structural influences on a particular carbonyl group is operating, the net effect is usually close to additive.
   (*vi*) Decrease in the ring size increases the frequency of the carbonyl stretching vibration.

   Acyclic ketones show absorption bands at around 1725 cm$^{-1}$ to 1705 cm$^{-1}$ whereas in aromatic ketones this frequency goes down by 30 cm$^{-1}$ to a lower wave number. For example, the IR spectrum of acetophenone (Figure 1.13) shows carbonyl absorption at 1690 cm$^{-1}$.

   The IR spectrum of cyclopentanone shows absorption band at 1750 cm$^{-1}$ (Figure 1.14) showing that decrease in ring size increases the frequency of carbonyl stretching vibration. However, this increase in frequency can be offset by introducing conjugation in the molecule; thus, the absorption resulting from C = O stretching of cyclohexanone (Figure 1.15) and 2-cyclopentanone occur at about the same frequency.

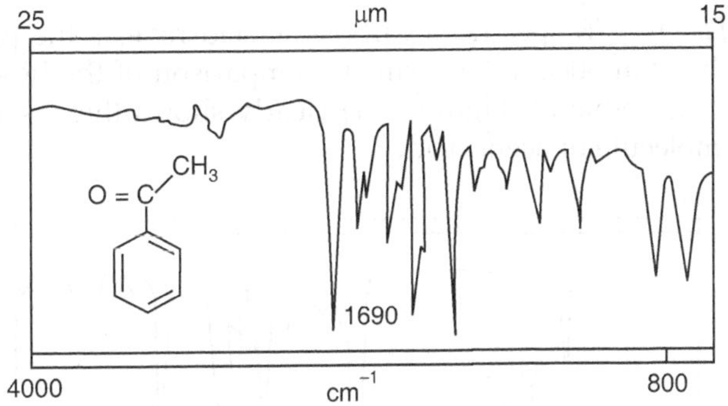


*FIGURE 1.13* **The IR spectrum of acetophenone.**

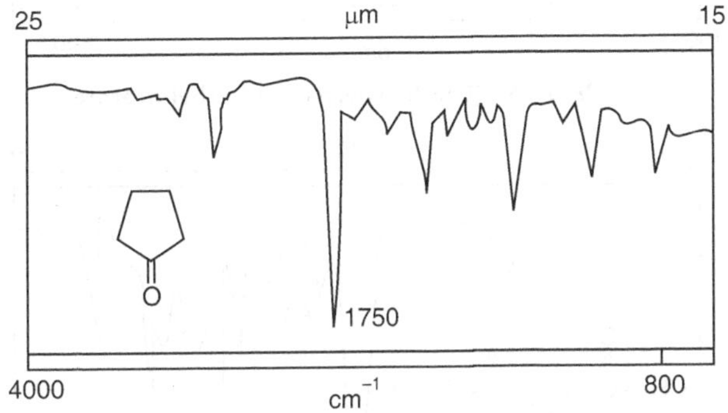


*FIGURE 1.14* **The IR spectrum of cyclopentanone.**

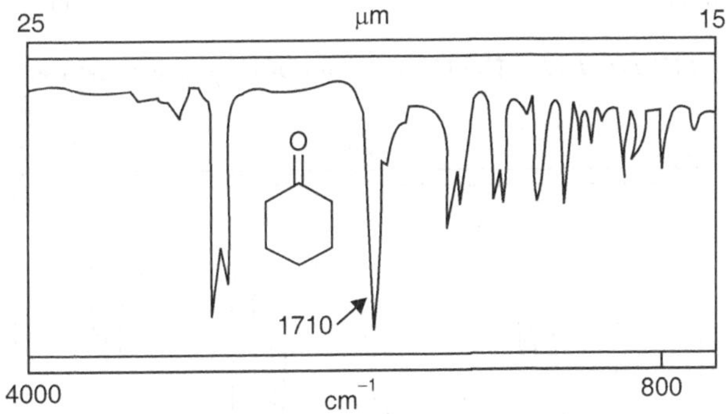


*FIGURE 1.15* **The IR spectrum of cyclohexanone.**

The position of absorption of the carbonyl stretching of esters (Figure 1.15) is somewhat higher than ketones and is dependent, as with ketones, on conjugation, unsaturation, and ring size. Thus, aromatic and conjugated esters show carbonyl absorption at about 1725 $cm^{-1}$ to 1710 $cm^{-1}$. An interesting example is that of ethyl acetoacetate, which is an equilibrium mixture of keto form (92%) and enol form (8%). The IR spectrum (Figure 1.16), in addition to showing keto and ester carbonyl frequencies, also shows a band at 1650 $cm^{-1}$ which has been ascribed to β-hydroxy-α, β-unsaturated ester carbonyl.

4. *Carboxylic acids.* The IR spectra of carboxylic acid reveals the presence of a hydrogen bond (broad band at 3000 to 2500 cm$^{-1}$). Comparison of the IR spectrum of acetic acid (Figure 1.18) with ethanol (Figure 1.11) clearly shows the absence of a free hydroxyl group in the molecule of acetic acid.

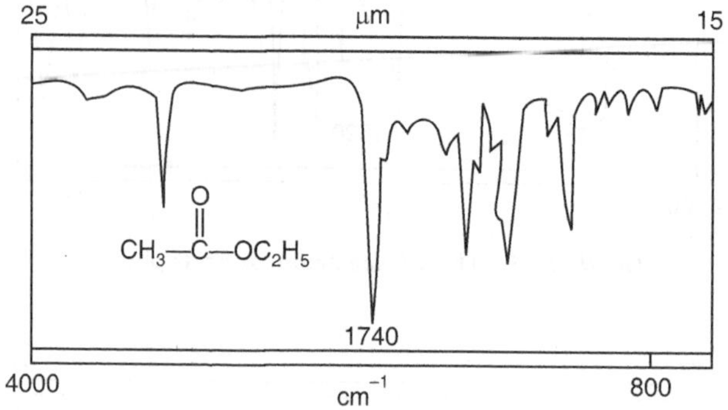


*FIGURE 1.16* **The IR spectrum of ethyl acetate.**

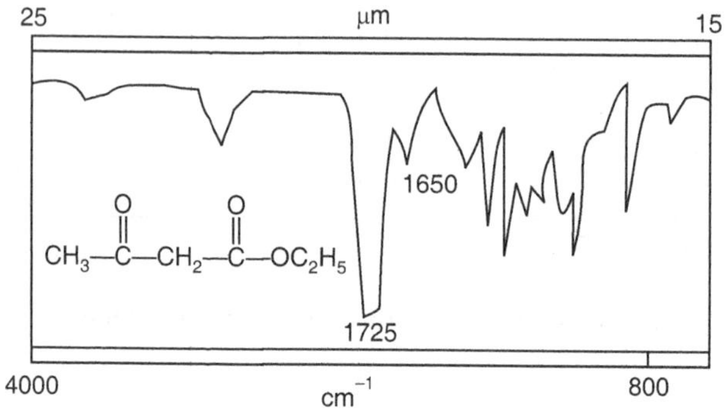


*FIGURE 1.17* **The IR spectrum of ethyl acetoacetate.**

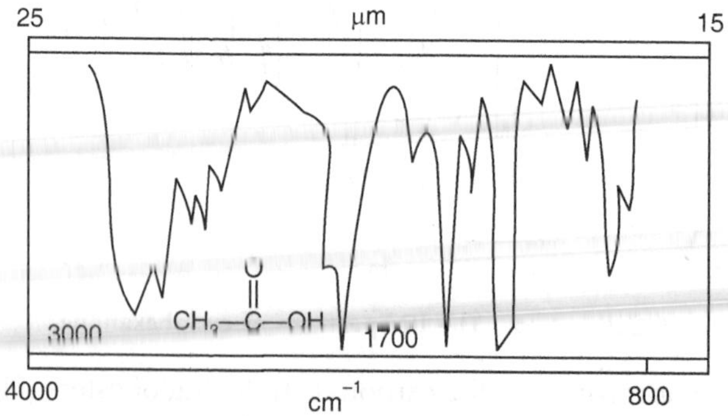


*FIGURE 1.18* **The IR spectrum of acetic acid.**

### 1.5.9 Applications of IR Spectroscopy

1. **Qualitative analysis.** The qualitative analysis can easily be done by comparing the spectrum of the compound under examination and the standard one.

2. **Quantitative analysis.** The quantitative analysis is based on the determination of the concentration of one of the functional group of compound being estimated, *e.g.*, the concentration of hexanol may be determined in the mixture of hexane and hexanol by measuring absorption of OH bond. The following formula is used to calculate the concentration:

$$A = -\log \frac{I}{I_o} = \varepsilon c x$$

where A = absorbance, I = intensity of radiation after leaving the sample under consideration, $I_o$= intensity of radiation initially, ε = absorptivity to the cell, $x$ = path length to the sample cell, and $c$ = concentration of solution.

Since ε and $x$ are constant for a particular cell hence.

$$A \propto c$$

With the help of a calibration curve, the concentration of unknown sample may be determined easily.

3. **Determination of purity of the sample.** Since impurity reduces the sharpness of the individual bands as well as extra band appear in the IR spectra hence, purity of the sample may be estimated easily.
4. **Presence of water in the sample.** If sample is crystallized by water, it may be possible some water remains in the sample. In that case, some bands at 3600 $cm^{-1}$–3200 $cm^{-1}$, ~ 1650 $cm^{-1}$ and 600 $cm^{-1}$ to 300 $cm^{-1}$ region appear. But if water molecules are present as ligand and coordinated with metal ion then a band appears at 880 $cm^{-1}$ to 650 $cm^{-1}$ region.
5. **Study of reaction kinetics.** Since IR spectra are very helpful in deciding the structure of the molecule easily and rapidly, this technique may be helpful in the study of reaction mechanism (or kinetics). In the study of reaction kinetics, IR spectra is used for the detection and estimation of intermediate compounds, *e.g.*, the progress of oxidation of secondary alcohol to ketone is studied by getting the IR spectra of the test portion at different times. In this case, as expected, there is disappearance of O—H stretching (~ 3570 $cm^{-1}$) and appearance of O—H stretching (~ 1725 $cm^{-1}$) takes place.
6. **Shape of molecule.** This technique has been proved very successful in deducing the shape of the molecule, *e.g.*, $N_2O$ and $H_2O$ both are triatomic molecules. Since both these molecules give three bands, they are bent molecules. If these molecules are linear, then only two bands will be obtained in IR spectrum.
7. Determination of force constant from vibrational spectrum.
8. Distinguishing intra and intermolecular hydrogen bonding.
9. Analysis of a mixture of aromatic hydrocarbons.
10. Distinguishing positional isomers of a compound.

### 1.5.10 Solved Problems in IR Spectroscopy

**Problem 15.** *Write down the approximate position of the characteristic IR bands in the following compounds:*

*(i)* $CH_3CH_2OH$ *(ii)* $CH_3COCH_3$ *(iii)* $CH_2 = CHCOCH_3$

*(iv)* $C_6H_5COCH_3$ *(v)* $CF_3OCF_3$ *(vi)* $CH_3OCH_3$

**Solution.**

| *Compound* | *Characteristic IR bands* |
|---|---|
| (*i*) $CH_3CH_2OH$ | A strong and broad hydrogen bonded O—H stretching band at 3400 $cm^{-1}$–3200 $cm^{-1}$ s. |
| (*ii*) $CH_3COCH_3$ | An intense C = O stretching band at ~ 1710 $cm^{-1}$. |
| (*iii*) $CH_2 = CHCOCH_3$ | An intense C = O stretching band of an $\alpha$, $\beta$–unsaturated carbonyl group at ~ 1680 $cm^{-1}$. |
| (*iv*) $C_6H_5COCH_3$ | An intense C = O stretching band at ~ 1680 $cm^{-1}$ due to benzene ring. |
| (*v*) $CF_3COCF_3$ | An intense C = O stretching band at ~ 1790 $cm^{-1}$ due to the presence of electron withdrawing groups. |
| (*vi*) $CH_3OCH_3$ | A characteristic C—O stretching band for methoxy group. |

**Problem 16.** *How many bands due to fundamental vibrations do you expect to observe in the IR spectrum of water ?*

**Solution.** Water molecule is a nonlinear molecule, hence its vibrational degree of freedom, *i.e.* fundamental vibrations will be equal to $3n - 6$ and can be calculated as follows:

Number of atoms $(n) = 3$

Total degrees of freedom $(3n) = 3 \times 3 = 9$

Rotational degrees of freedom = 3

Translational degrees of feedom = 3

Hence, vibration degrees of freedom $(3n - 6) = 9 - 3 - 3 = 3$

Thus, there will be three fundamental modes of vibration. These are symmetrical stretching, asymmetrical stretching and bending vibrations. All these modes of vibrations are IR active and are non-degenerate. Thus, three absorption bands corresponding to these three fundamental modes of vibration are expected in the IR spectrum of water (the observed absorption bands occur at 3652, 3756 and 1595 $cm^{-1}$).

**Problem 17.** *How would you identify the following pairs of compounds with the help of IR spectra:*

*(i) Ethanol and propanone*

*(ii) Methanol and ethanoic acid*

*(iii) Ethyl methyl ketone and methyl vinyl ketone*

*(iv) Cyclobutanone and cyclohexanone.*

**Solution.** (*i*) Ethanol ($CH_3CH_2$—OH) would show a broad hydrogen bonded OH-stretching band at 3400 $cm^{-1}$ to 3200 $cm^{-1}$ while propanone ($CH_3COCH_3$) would show a strong $>C=O$ stretching at ~ 1710 $cm^{-1}$.

(*ii*) Both methanol ($CH_3OH$) and ethanoic acid ($CH_3COOH$) would show a OH stretching band at 3300 $cm^{-1}$ and 3000 $cm^{-1}$ respectively. But ethanoic acid will also show an intense $>C=O$ stretching band at about 1725 $cm^{-1}$.

(*iii*) Both ethyl methyl ketone ($CH_3CH_2COCH_3$) and methyl vinyl ketone ($CH_3 - CO - CH = CH_2$) show strong $>C=O$ stretching band. This band appears at 1710 $cm^{-1}$ in the former compound while at 1680 $cm^{-1}$ in the latter compound due to $\alpha$, $\beta$ unsaturation.

(*iv*) Cyclohexanone will show a strong $>C=O$ stretching band at ~ 1780 $cm^{-1}$.

**Problem 18.** *How would you differentiate between the following pairs of compounds using IR spectra:*

*(i)* $CH_3CH_2CHO$ *and* $CH_2 = CH—CH_2OH$,

*(ii)* 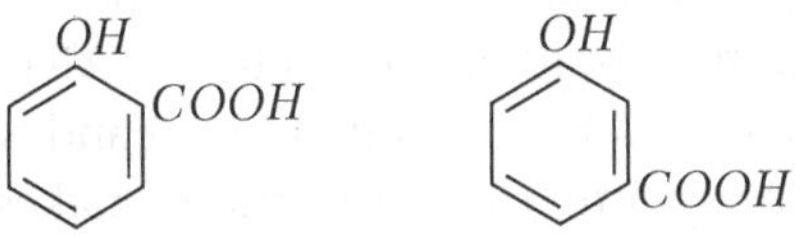

**Solution.** (*i*) The compound $CH_3CH_2CHO$ will show strong characteristic band at ~ 1740 $cm^{-1}$ to 1720 $cm^{-1}$ for $>C=O$ group. The compound $CH_2 = CH - CH_2 - OH$ will show a strong and broad hydrogen bonded OH stretching band at 3400 $cm^{-1}$—3200 $cm^{-1}$, ~ 1050 $cm^{-1}$ due to C—O stretching and bending, 3100 $cm^{-1}$—3000 $cm^{-1}$ for C = C bond and ~ 905 $cm^{-1}$ for vinyl group $CH_2 = CH$.

(*ii*) In case of 2-hydroxy benzoic acid —OH and —COOH absorption peaks are obtained at 3550 $cm^{-1}$ to 3450 $cm^{-1}$(due to intramolecular hydrogen bonding) and 2300 $cm^{-1}$ to 2500 $cm^{-1}$ respectively. A band ~ 755 $cm^{-1}$ is obtained for C—H stretching and 1 : 2 disubstitution.

In the case of 3-hydroxy benzoic acid —OH and —COOH absorption peaks are obtained at ~ 3610 $cm^{-1}$ and 2300 $cm^{-1}$ to 2500 $cm^{-1}$ respectively. Along with these peaks at ~ 700 $cm^{-1}$, ~ 780 $cm^{-1}$ and ~ 880 $cm^{-1}$ are appeared for C—H stretching.

**Problem 19.** *Two isomers A and B of the molecular formula* $C_3H_6O$ *give IR absorption band at 1650* $cm^{-1}$ *and 1710* $cm^{-1}$ *respectively. Assign structural formula to A and B isomers.*

**Solution.** The molecular $C_3H_6O$ has the following isomers:

(*i*) $CH_3COCH_3$ (*ii*) $CH_2 = CH - CH_2OH$

(*iii*) $CH_3CH_2CHO$ (*iv*) 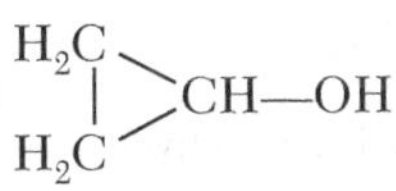

Since a band is obtained at 1710 $cm^{-1}$ it should correspond to a $>C=O$ group hence, (B) is $CH_3COCH_3$. An IR band at 1650 $cm^{-1}$ corresponds to compound (*ii*) therefore, the compound (A) is $CH_2=CH—CH_2OH$.

The compound (*iii*) will give an IR band at ~ 1740 $cm^{-1}$ and (*iv*) will not give an IR band at ~ 1650 $cm^{-1}$ so both these structures are ruled out.

**Problem 20.** *An organic compound $C_3H_8O$ shows an IR band at 2950 $cm^{-1}$ but none near 3300 $cm^{-1}$ and 1720 $cm^{-1}$. Deduce the structure of the compound.*

**Solution.** The absence of a band ~ 1720 $cm^{-1}$, ~ 3300 $cm^{-1}$ rules out the possibility of an OH group and aldehyde or ketone respectively. Thus, the compound may be ether ($CH_3—O—CH_2CH_3$). The band at 2950 $cm^{-1}$ is obviously due to C—H stretching. Therefore, the given compound is methoxy ethane (ethyl methyl ether, $CH_3OC_2H_5$).

**Problem 21.** *Using IR spectroscopy, how will you distinguish the following isomeric compounds:*

*(i) $CH_3CH_2C\equiv CH$ and $CH_3C\equiv CCH_3$*

*(ii) $CH_3CH_2OH$ and $CH_3OCH_3$*

*(iii) $(CH_3)_3N$ and $CH_3CH_2NHCH_3$*

**Solution.** (*i*) $CH_3CH_2\,C\equiv CH$ will show strong absorption bands at ~3300, 2100 and 625 $cm^{-1}$ due to $\equiv C—H$ stretching, $C\equiv C$ stretching and $\equiv C—H$ bending vibrations, respectively. All these bands will be absent in the IR spectrum of $CH_3\,C\equiv CCH_3$. It should be noted that $v_{C\equiv C}$ band will also be asbsent in the IR spectrum of $CH_3\,C\equiv CCH_3$ because the symmetrical substitution makes the $C\equiv C$ stretching frequency IR inactive.

(*ii*) $CH_3CH_2OH$ will show absorption bands in the region 3200 –3600 $cm^{-1}$ due to stretching vibration of intermolecularly hydrogen-bonded O—H group and at ~ 1050 $cm^{-1}$ due to $v_{C—O}$. These bands will be absent in the IR spectrum of $CH_3OCH_3$. It will show an absorption band around 1100 $cm^{-1}$ due to C—O—C stretching.

(*iii*) $CH_3CH_2NHCH_3$ will show a medium band in the region 3310 – 3550 $cm^{-1}$ due to N—H stretching. This band will be absent in IR spectrum of $(CH_3)_3N$.

**Problem 22.** *How will you distinguish o-hydroxybenzaldehyde (salicylaldehyde) and m-hydroxybenzaldehyde on the basis of IR spectroscopy ?*

**Solution.** In salicylaldehyde, due to intramolecular hydrogen bonding, $v_{O—H}$ and $v_{C=O}$ bands are shifted to lower wavenumbers. Since it is intramolecular, change in concentration does not cause any shift in $v_{O—H}$ and $v_{C=O}$ absorption bands. In case of *m*-hydroxybenzaldehyde, $V_{O—H}$ and $V_{C=O}$ bands occur at a still lower wavenumber due to intermolecular hydrogen bonding. In this case, $v_{O—H}$ and $v_{C=O}$ bands shift to higher wavenumbers on dilution with a nonpolar solvent.

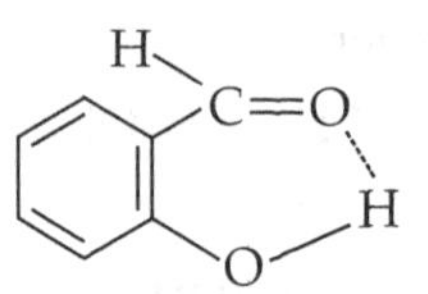

*o*-hydroxybenzaldehyde

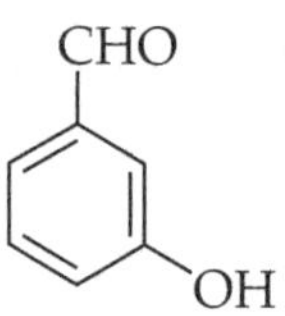

*m*-hydroxybenzaldehyde

**Problem 23.** *A hydrocarbon containing 10% hydrogen shows the following bands in its IR spectrum:*

*(i) 3295 cm$^{-1}$* *(ii) 2130 cm$^{-1}$* *(iii) 625 cm$^{-1}$*

*Deduce the structure of the hydrocarbon.*

**Solution.** Among various hydrocarbons, only alkynes show a band around 3300 cm$^{-1}$ due to ≡C—H stretching. Thus, the appearance of a band at 3295 cm$^{-1}$ clearly shows the presence of ≡ C—H group in the hydrocarbon. The presence of a band at 2130 cm$^{-1}$ due to C≡C stretching and another band at 625 cm$^{-1}$ due to ≡C—H bending confirm that the hydrocarbon is an alkyne. 10% hydrogen content shows that it is propyne ($CH_3C{\equiv}CH$).

**Problem 24.** *Give suitable explanation for the following:*

(*a*) *Stretching frequencies of C—C, C=C and C ≡ C fall in the regions which are in increasing order 800–1200 cm$^{-1}$, 1650–1670 cm$^{-1}$ and 2100–2260 cm$^{-1}$, respectively.*

(*b*) *The $v_{O—H}$ band appears near 3570 cm$^{-1}$, whereas the $v_{O—D}$ band appears near 2630 cm$^{-1}$.*

(*c*) *2-t-butylphenol has two $v_{O—H}$ bands—one at 3608 cm$^{-1}$ and the other at 3643 cm$^{-1}$, whereas 2, 6-di-t-butylphenol has only one band at 3642 cm$^{-1}$.*

**Solution.**

(*a*) According to Hooke's law, the stretching frequency of a bond increases with an increase in the force constant, *i.e.*, bond strength. Thus, the stretching frequencies of carbon-carbon single, double and triple bonds increase in the order of their increasing bond strengths, *i.e.* the increasing order of stretching frequencies is

$$C—C < C=C < C \equiv C$$

(*b*) According to Hooke's law, the stretching frequency of a bond increases as the reduced mass of the bonded atoms decreases. Since hydrogen has lesser automic mass than deuterium, the O—H stretching frequency (near 3570 cm$^{-1}$) is higher than the O—D stretching frequency (near 2630 cm$^{-1}$).

(*c*) In 2-*t*-butylphenol, two bands are present at 3608 and 3643 cm$^{-1}$ showing that some molecules are intermolecularly hydrogen bonded, whereas in others OH is not hydrogen bonded due to steric hindrance by the bulky-*t*-butyl group.

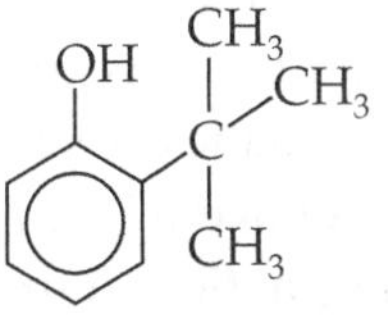

2-*t*-butylphenol

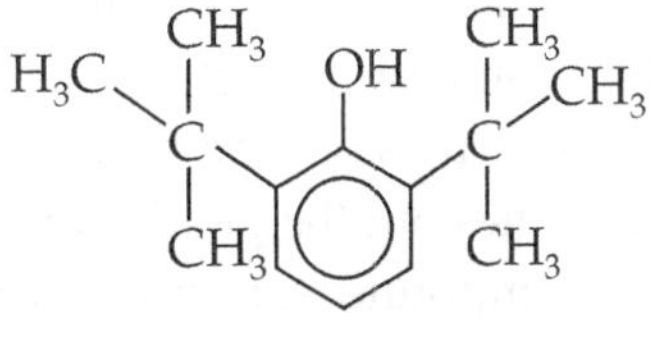

2, 6-di-t-butylphenol

In 2, 6-di-*t*-butylphenol, because of the presence of two bulky-*t*-butyl groups in the *ortho* position of the OH group, the OH groups of two different molecules are not able to approach close enough to form intermolecular hydrogen bond. Thus, 2, 6-di-*t*-butylphenol shows only one $v_{O—H}$ band at 3642 cm$^{-1}$ due to non-hydrogen-bonded OH group.

**Problem 25.** *Giving reasons arrange the following compounds in order of decreasing frequency of carbonyl absorption in their IR spectra:*

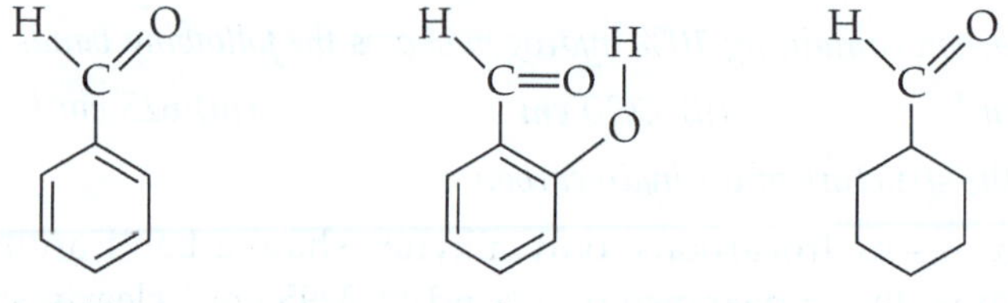

(*a*) (*b*) (*c*)

**Solution.** The following is decreasing order of frequency of carbonyl absorption

(*c*) > (*a*) > (*b*)

Cyclohexane carboxaldehyde (*c*) is a saturated aldehyde, hence will absorb around 1730 $cm^{-1}$. Due to conjugation of the C=O group with the double bonds of the benzene ring, the $v_{C=O}$ absorption of benzaldehyde (*a*) will be shifted to lower frequency (~ 1700 $cm^{-1}$). In salicylaldehyde (*b*), the lowering is due to conjugation as well as intramolecular hydrogen bonding (chelation), thus the $v_{C=O}$ absorption frequency is further lowered (~ 1665 $cm^{-1}$).

**Problem 26.** *What characteristic bands do you expect in the spectrum to toluene?*

**Solution.** Toluene 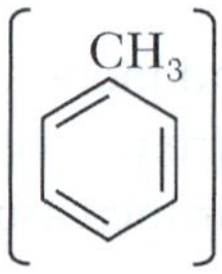 is an aromatic hydrocarbon. The expected characteristic IR bands in toluene can to due to C—H stretching of benzene ring (3100 $cm^{-1}$ to 3000 $cm^{-1}$), of methyl group (3000 $cm^{-1}$ to 2800 $cm^{-1}$) and C ⩦ C stretching of benzene ring (1600 $cm^{-1}$ to 1500 $cm^{-1}$).

**Problem 27.** *Two compounds A and B give the following important bands:*

*A: 3030 $cm^{-1}$, 1715 $cm^{-1}$, 1275 $cm^{-1}$*

*B: 3080 $cm^{-1}$, 2850 $cm^{-1}$, 2780 $cm^{-1}$, 1700 $cm^{-1}$, 1590 $cm^{-1}$, 745 $cm^{-1}$, 680 $cm^{-1}$. Deduce the probable structures of A and B.*

**Solution. A:** The band 3030 $cm^{-1}$ suggests a strongly hydrogen bonded O—H stretching possibly of a carboxylic acid. The band as 1715 $cm^{-1}$ suggests carbonyl group of carboxylic acid. The band at 1275 $cm^{-1}$ suggests C—O stretching. Therefore, the compound A is a carboxylic acid. The absence of a band at 1500 to 1600 $cm^{-1}$ rules out the possibility of aromatic carboxylic acid. The absence of a band at 1620 to 1680 $cm^{-1}$ rules out the possibility of unsaturated (C = C) carboxylic acid. Finally, we may conclude that the compound A is any **saturated aliphatic carboxylic acid.**

**B:** A possible interpretation of the IR bands in the compound follows:

| | |
|---|---|
| 3080 $cm^{-1}$: | C—H stretching of benzene ring |
| 2850 to 2780 $cm^{-1}$: | C—H stretching bands (duplet) of an aldehyde (—CHO) group. |
| 1700 $cm^{-1}$: | C = O stretching of an α, β unsaturated or an aromatic aldehyde/ketone. |

$1595\ cm^{-1}$ : C = C stretching of a benzene ring.

$745\ cm^{-1}$ to $680\ cm^{-1}$: C—H bands in a monosubstituted benzene ring.

On the basis of this discussion we may conclude that the compound B is an aromatic aldehyde. Since, there is no band at $1620\ cm^{-1}$—$1080\ cm^{-1}$ above the possibility of the C = C bond is ruled out. In other words the compounds B is not a cinnamic aldehyde ($C_6H_5CH = CH—CHO$). Therefore, the compound B is **benzaldehyde ($C_6H_5$ CHO)**

**Problem 28.** *Identify the organic compound $C_7H_7$ Br from the following IR spectrum.*

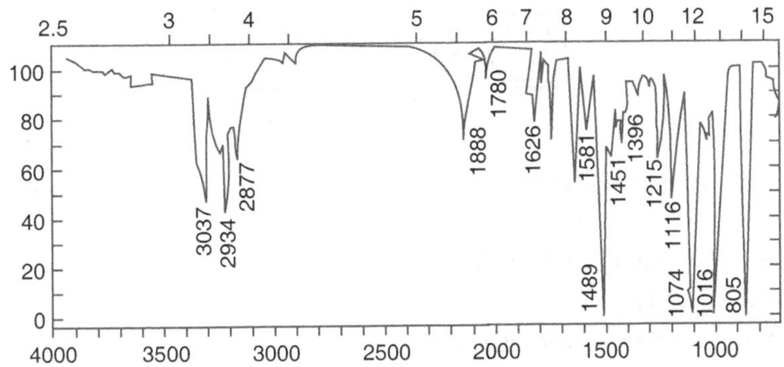


**Solution.** The infrared spectrum shows a band at $3037\ cm^{-1}$ which may be due to aromatic C—H stretching while the bands at $2934\ cm^{-1}$ are due to methyl CH stretching. The bands at $1888\ cm^{-1}$—1780 $cm^{-1}$ are due to para-substitution in benzene ring. The bands at 1626, 1581 and 1489 $cm^{-1}$ are due to phenyl ring. A band at $1396\ cm^{-1}$ is due to methyl group. The bands at 1215, 1074, 1016 $cm^{-1}$ are due to in plane C—H bonding of phenyl ring, while a band at $805\ cm^{-1}$ is due to out of plane C—H bonding of two adjacent aromatic hydrocarbons. Thus, the given compound $C_7H_7Br$ is ***p*-bromo toluene.**

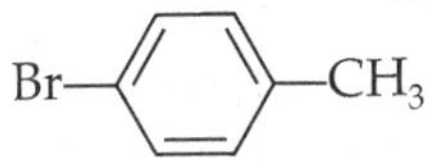


*p*-bromo toluene

**Problem 29.** *Identify the given hydrocarbon $C_{14}H_{14}$ on the basis of the following IR spectrum.*

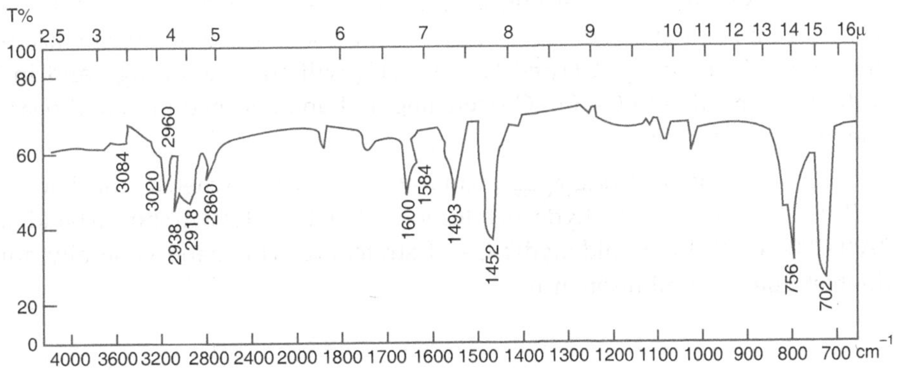

**Solution.** Different bands obtained from hydrocarbon $C_{14}H_{14}$ may be characterized as follows:

| | |
|---|---|
| 3020 cm$^{-1}$: | Aromatic CH |
| 2938, 2918, 2860 cm$^{-1}$: | Aliphatic CH |
| 1600, 1584 and 1493 cm$^{-1}$: | Aromatic C = C |
| 1452 cm$^{-1}$: | Aromatic C = C and alphatic $\delta CH_2$ |
| 756 and 702 cm$^{-1}$: | Five adjacent H–on phenyl. |

On the basis of above observation the compound $C_{14}H_{14}$ may be shown as:

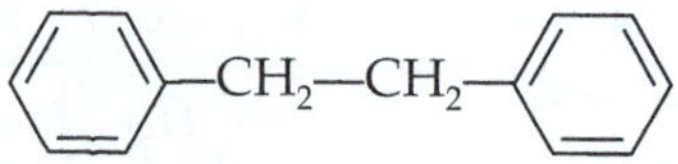


1, 2-diphenyl ethane

**Problem 30.** *How will you distinguish the two members of each of the following pairs of compound by their IR spectra:*

(*a*) $CH_3CH_2C{\equiv}CH$ *and* $CH_3CH_2\ C{\equiv}N$

(*b*) $RNH_2$ *and* $RCONH_2$

(*c*) $CH_3COCH_3$ *and* $CH_3CH{=}CHCH_2OH$

(*d*) $CH_2{=}CH{-}O{-}CH_3$ *and* $CH_3CH_2CHO$

(*e*) $CH_3CH_2CHO$ *and* $CH_3COCH_3$

**Solution.**

(*a*) $CH_3CH_2C{=}CH$ will show strong absorption band at ~ 3300 and 625 cm$^{-1}$ due to $=C{-}H$ stretching and bending vibrations, respectively. These bands will be absent in the IR spectrum of $CH_3CH_2\ C{\equiv}N$.

(*b*) $RCONH_2$ will show a strong $v_{C=O}$ band around 1690 cm$^{-1}$ (in dilute solution) which will be absent in the IR spectrum of $RNH_2$.

(*c*) $CH_3COCH_3$ will show a strong $v_{C=O}$ band at ~ 1715 cm$^{-1}$. This band will be absent in $CH_3CH{=}CHCH_2OH$. Compound $CH_3CH{=}CHCH_2OH$ will show a strong $v_{O-H}$ band in the region 3200–3650 cm$^{-1}$ and another medium $v_{C=C}$ band in the region 1620–1680 cm$^{-1}$.

(*d*) $CH_3CH_2CHO$ will show a strong $v_{C=O}$ band near 1730 cm$^{-1}$ and another weak and (generally appearing as a doublet due to Fermi resonance) in the region 2700–2900 cm$^{-1}$ due to aldehydic $v_{C-H}$. $CH_2{=}CH{-}O{-}CH_3$ will show a strong band in the range 1020–1075 cm$^{-1}$ due to C—O—C stretching and another medium band near 1650 cm$^{-1}$ due to C=C stretching.

(*e*) Acetone will show a strong $v_{C=O}$ band at ~ 1715 cm$^{-1}$, whereas this band appears at ~ 1730 cm$^{-1}$ in propionaldehyde. In addition, $CH_3CH_2CHO$ will show a band in the range 2700–2900 cm$^{-1}$ due to aldehydic C—H stretching. This band generally appears as a doublet due to Fermi resonance.

**Problem 31.** *How will you distinguish maleic acid and fumaric acid by their IR spectra?*

**Solution.**

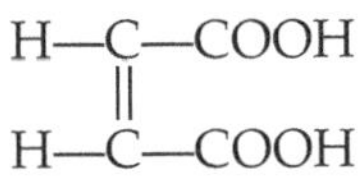

Maleic acid

$\nu_{C=O}$ 1720 $cm^{-1}$ (s)

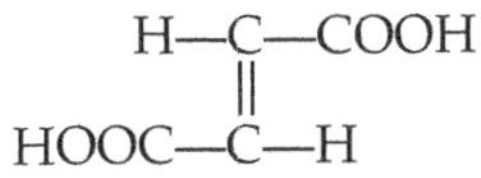

Fumaric acid

$\nu_{C=O}$ 1680 $cm^{-1}$ (s)

In maleic acid, the bulky COOH groups are on the same side of the double bond, hence due to repulsive interactions the C=O group is forced out of the plane of the C=C bond. Thus, the conjugation is diminished resulting in the appearance of $\upsilon_{C=O}$ band at a higher frequency as compared to that in fumaric acid where the C=O group is in conjugation with the C=C bond.

**Problem 32.** *Identify the compound, $C_7H_9N_2$ on the basis of the following IR spectrum.*

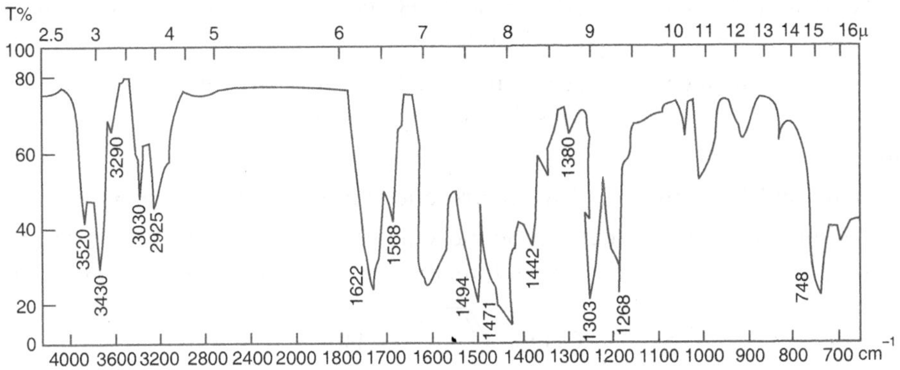


**Solution.** Different bands obtained from the IR spectrum may be characterized as follows:

| | |
|---|---|
| 3520 $cm^{-1}$ and 3420 $cm^{-1}$: | $NH_2$ (3500 and 3430 $cm^{-1}$) higher frequency of amino indicates that these bands are not involved in intermolecular hydrogen bonding due to steric hindrance. |
| 1622 $cm^{-1}$: | $NH_2$ bending |
| 1588, 1494, 1471 and 748 $cm^{-1}$: | *o*-substituted phenyl ring |
| 1442 and 1380 $cm^{-1}$: | $NH_2$ |
| 1268 $cm^{-1}$: | CN |

On the basis of this observation, the compound $C_7H_9N$ consists of $—NH_2$, $—CH_3$ and a phenyl ring. $—NH_2$ and $—CH_3$ groups are placed at *ortho* position. Therefore, the molecule is **o-toluidine.**

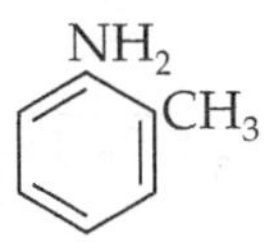

*o*-toluidine

**Problem 33.** *Identify the given compound, $C_8H_8O_2$ on the basis of the following IR spectrum.*

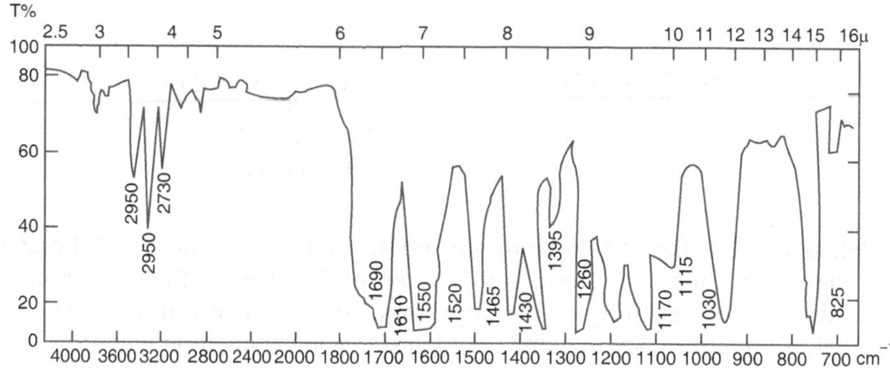


**Solution.** Different bands obtained from the compound $C_8H_8O_2$ are as follows:

| | |
|---|---|
| 2950 cm$^{-1}$ | : C—H aliphatic and aromatic stretching |
| 2820 cm$^{-1}$ and 2720 cm$^{-1}$ | : Aldehyde C—H, the —$OCH_3$ (2850 cm$^{-1}$) is overlaid by 2820 cm$^{-1}$ band. |
| 1690 cm$^{-1}$ | : Aromatic —CHO group, 1610, 1580, 1520, 1430, 1170, 115 cm$^{-1}$ *p*-substituted phenyl ring |
| 1465 cm$^{-1}$ and 1395 cm$^{-1}$ | : —$CH_3$ group |
| 1260 cm$^{-1}$ and 1030 cm$^{-1}$ | : Asymmetric and symmetric stretching of = C—O—C. |

These observations reveal that compound $C_8H_8O_2$ consists —CHO, —$OCH_3$ phenyl ring. The groups are placed at *para* position, therefore, the compound $C_8H_8O_2$ is **p-anisaldehyde.**

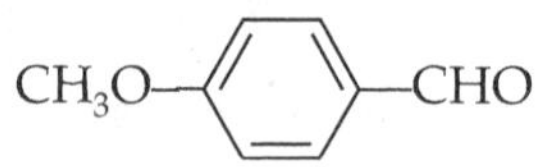


*p-anisaldehyde*

## 1.6 NUCLEAR MAGNETIC RESONANCE (NMR) SPECTROSCOPY

This spectroscopic method deals with the interaction between an oscillating magnetic field of electromagnetic radiation and the magnetic energy of some nuclei having non-zero spin. The transitions occur between magnetic sublevels of the nuclei of a molecule. The magnetic energy levels are produced by keeping the nuclei in a magnetic field and the energy required for transitions is available in radio frequency region (4 to 900 MHz).

**Spin of nuclei and angular momentum.** Magnetic field is also associated with moving electrical charges. The nuclear spin may vary from one nucleus to another. The spin of a nucleus is represented by I called the *spin quantum number*. Spin quantum number is associated with the angular momentum by the following relation:

$$I = \sqrt{I(I+1)}\,\frac{h}{2\pi}$$

where $h$ is the Planck's constant.

### 1.6.1 Rules for Nuclei Showing NMR Spectroscopy

- Nuclei with even number of protons and an even number of neutrons (*i.e.*, even atomic and mass number) have zero spin and they do not give rise to NMR signals. For example, $C^{12}$, $O^{16}$, $S^{32}$, etc.
- Nuclei with an odd number of protons and odd number of neutrons (*i.e.*, odd atomic number and even mass number) have integral spins. For example, the nuclei $H^2$, $N^{14}$ have I = 1 and for $B^{10}$, I = 3.
- Nuclei with odd mass numbers have half-integral spin. For example, $H^1$ and $N^{15}$ have I = 1/2 and $O^{17}$ have I = 5/2, etc.

### 1.6.2 Principle of $H^1$ NMR Spectroscopy

Consider the nucleus of a hydrogen atom (proton). It behaves as a tiny spinning bar magnet because it possess both electric and magnetic spin. Like any other spinning charged body, it will generate a magnetic field. If the sample contain many identical nuclei (protons) with spin quantum number I = 1/2, it can have two possible orientations (2I + 1), *i.e.*, 1/2 or –1/2 or α- and β-nuclear spins. In the absence of a magnetic field, the sample will consist of an equal number of α- and β- nuclear spins and net nuclear moment will be zero.

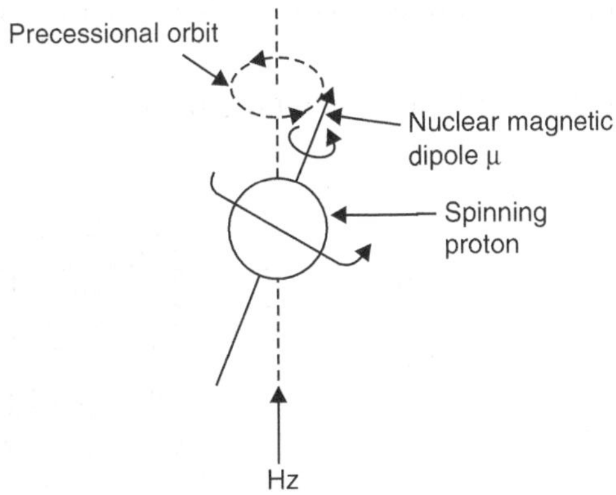


*FIGURE 1.19* **Proton precessing in a magnetic field Hz.**

When the external magnetic field is applied, the magnetic nuclei (protons), like the spinning top, will process around the axis of an applied external field. Its precessional frequency is given by:

$$\nu \propto B_o$$

The precessional frequency ν is directly proportional to the strength of the external field $B_o$. A proton exposed to an external magnetic field of 1.4 T (14,000 gauss) will precess ≈ 60 million times per second or 60 MHz. Similarly for an external field of 2.3 T, ν is 100 MHz and at 7.1 T, ν is 300 MHz. In the presence of an external magnetic field, two changes can occur:

(*i*) The precessing proton can adopt two principal orientations with respect to the applied external field.

(*a*) Aligned with the field, corresponding to low energy state, α-state.

$$E_1 = 1/2\, g\, B_n H$$

(*b*) Opposed to the field, corresponding to high energy state, β-state.

$$E_2 = -\, 1/2\, g\, B_n\, H$$

(*ii*) There will be more α-spins than β-spins. This small imbalance will result in net magnetization. When we irradiate the precessing nuclei with a beam of radio frequency energy of the correct frequency, the proton in the low energy state (α) may absorb this energy and move to the higher energy state (β). Thus, the transition takes place when the processing proton in thc aligned orientation (α-state) can absorb the energy and pass into the opposite orientation (β-state). This transition from one energy state (α) to the other energy state (β) is called the *flipping of the proton.* The energy of transition is given by:

$$\Delta E = h\nu$$

The precessing proton will absorb energy only if the precessional frequency is the same as the frequency of the radio frequency beam. When this occurs, the nuclei and radio frequency beam are said to be in resonance and hence the name nuclear magnetic resonance. The energy absorbed produces a signal at the detector and this signal is amplified and recorded as a band in the spectrum. AnMR spectrum is plotted between the absorption between signal at the detector and the strength of the magnetic field.

### 1.6.3 NMR Spectrometer

The schematic representation of the apparatus used is shown in Figure 1.20. It consists of the following parts:

(*i*) *An electromagnet* with poles M and M′ whose field strength can be varied by varying the feeding voltage.

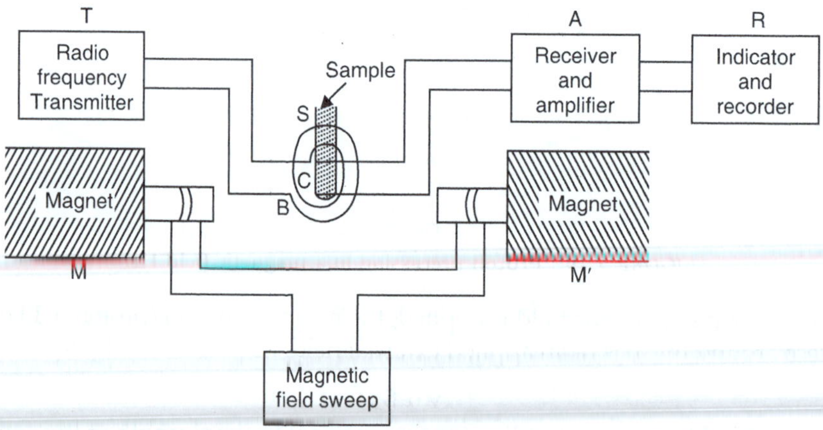


***FIGURE 1.20*** **A schematic arrangement of an NMR spectrometer.**

(*ii*) *A sample tube, S,* placed between the poles of the electromagnet containing the sample under study.

(*iii*) *A radio frequency transmitter, T,* which sends out constant frequency around the sample tube through the coil B.

(*iv*) *A receiver coil, C,* to pick up the frequency emitted by the sample.

(*v*) *A receiver and amplifier, A,* to receive and amplify the frequency picked up by the receiver coil C.

(*vi*) *An indicator, i.e., a cathode ray oscilloscope or a recorder, R,* to display or record the amplified signals.

The procedure consists of supplying some constant frequency (usually 60 megacycles/sec) to the sample from the radio frequency transmitter. A sweeping magnetic field is then applied to the sample. No absorption of energy by the sample takes place until the magnetic field strength is such that it becomes equal to the frequency being supplied by the transmitter. When this energy is absorbed, the nuclei are said to be in the excited state. They then start re-emitting the energy. The energy thus emitted is picked up by the receiver coil C, amplified by the amplifier A, and recorded on the recorder R. At this stage the sample is said to be in resonance with the applied frequency. When H is increased further no absorption of energy takes place until it again reaches a value at which the sample is in resonance with the applied frequency. In this way, a plot of the type shown in Figure 1.21 is obtained. Such a plot is called an *NMR spectrum*. Each peak corresponds to a value of H at which the sample is in resonance with the applied frequency.

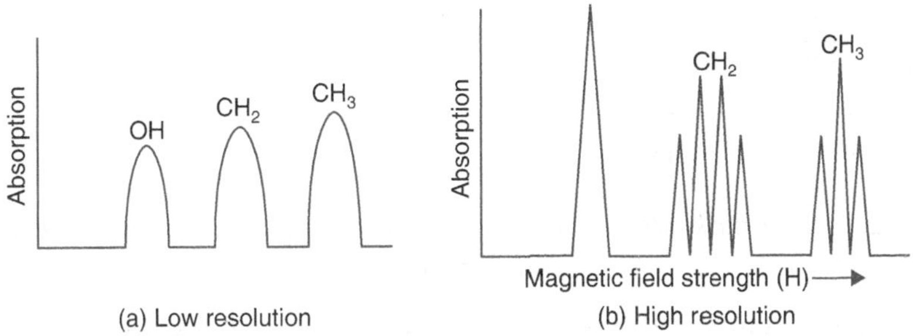


*FIGURE 1.21* **NMR spectrum of ethyl alcohol.**

### 1.6.4 Study of NMR Spectra

NMR spectra may be studied under the following headings:

#### *A. Number of Signals*

A set of protons with the same chemical environment are called equivalent protons. All equivalent protons form a *type* or a *kind* of protons and the NMR spectrum consists of as many signals as there are kinds of protons. (*a*, *b*, *c*, and *d* protons in below listed examples). For example,

| *Compound* | *Number of signals* |
|---|---|
| (*i*) H—C(H)(H)—H (methane: central C with H above, H below, H left and right) | One signal (Four equivalent protons) |
| (*ii*) $\overset{(a)}{CH_3}—\overset{(b)}{OH}$ | (Two signals) |
| (*iii*) $\overset{(a)}{CH_3}—\overset{(b)}{CH}—\overset{(a)}{CH_3}$ with OH(*c*) on the CH | (Three signals) |

(*iv*) 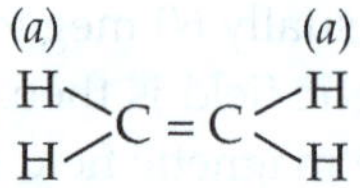 (One signal)

(*v*) 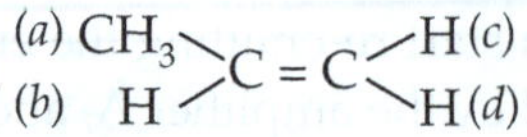 (Four signals)

(*vi*) 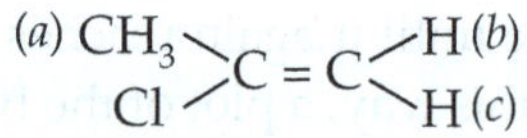 (Three signals)

(*vii*) 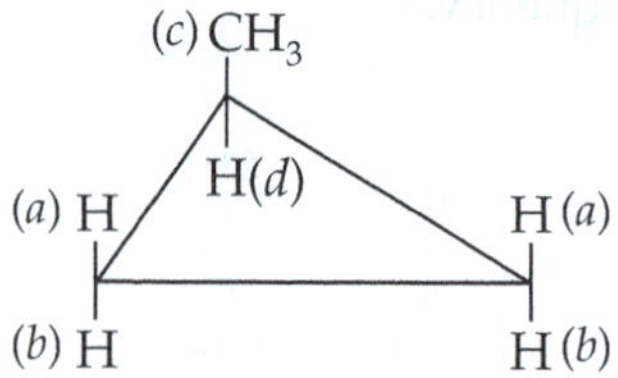 (Four signals)

**Example 22.** *Indicate the kinds of protons and number of NMR signals in the following compounds:*

(*a*) $CH_3CH_2CH_2Cl$ (*b*) $CH_3CHClCH_3$ (*c*) $CH_3COCH_3$

**Solution.** (*a*) The compound has three kinds of protons labeled as *a*, *b* and *c*, hence it will exhibit 3 NMR signals.

$C\overset{a}{H}_3C\overset{b}{H}_2C\overset{c}{H}_2Cl$ Protons *c* are enantiotopic, hence equivalent

3 NMR signals

1-Chloropropane

(*b*) $C\overset{a}{H}_3C\overset{b}{H}ClC\overset{a}{H}_3$

2 NMR signals

2-Chloropropane

(*c*) $C\overset{a}{H}_3COC\overset{a}{H}_3$

1 NMR signal

Acetone

**Example 23.** *Draw the structural formula of each of the following compounds and label all sets of equivalent protons. How may NMR signals do you expect from each of these compounds ?*

(*a*) *p-xylene,* (*b*) *Vinyl chloride,* (*c*) *Cyclobutane,*

(*d*) *Diethyl ether,* (*e*) *Two isomers of* $C_2H_4Cl_2$

**Solution.**

(*a*) 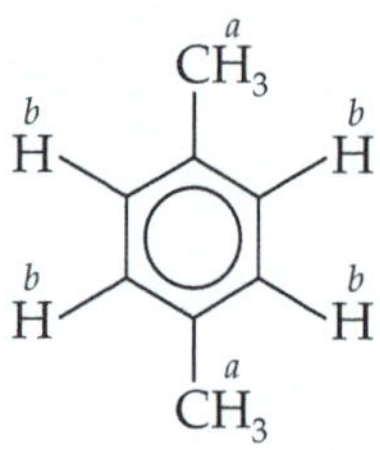

2 NMR signals *p*-xylene

(*b*) 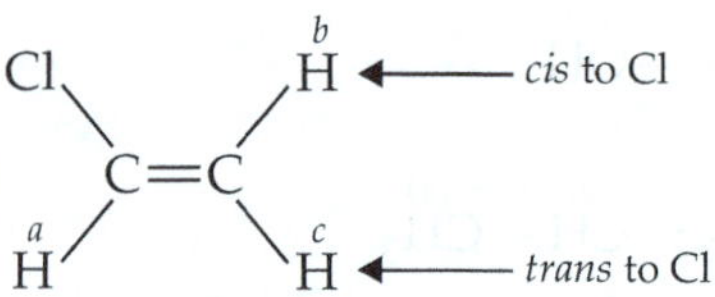


3 NMR signals vinyl chloride

(*c*) 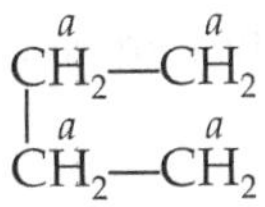

1 NMR signal
Cyclobutane

(*d*) $\overset{a}{C}H_3\overset{b}{C}H_2O\overset{b}{C}H_2\overset{a}{C}H_3$

2 NMR signals
Diethyl either

(*e*) (*i*) $\overset{a}{H_3}C—\overset{a}{C}HCl_2$

2 NMR signals
1, 1-Dichloroethane

(*ii*) $ClC\overset{a}{H_2}—C\overset{a}{H_2}Cl$

1 NMR signal
1, 2-Dichloroethane

**Example 24.** *Three isomeric dimethylcyclopropanes A, B and C give 2, 3 and 4 NMR signals, respectively. Draw the stereochemical formulae for A, B and C.*

**Solution**

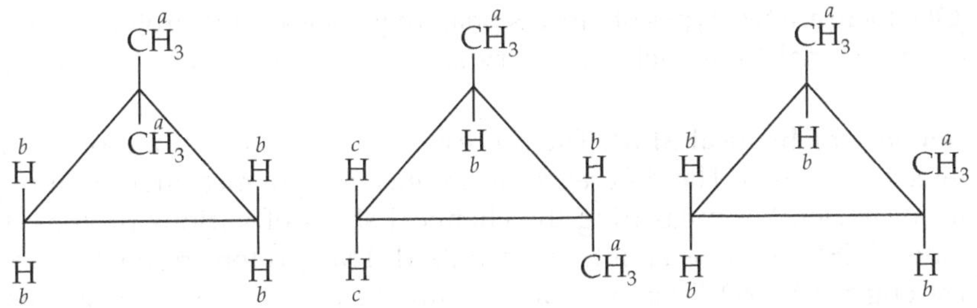

*A* *B* C

In the dimethylcyclopropane C the proton *c* is *cis* to $CH_3$ groups, whereas the proton *d* is *trans* to $CH_3$ groups. Hence these are nonequivalent.

**Example 25.** *Draw the structural formula of each of the following compounds, label the kinds of protons and indicate the expected number of NMR signals.*

(*a*) *Methylcyclopropane* (*b*) *Mesitylene*

(*c*) *Ethyl succinate* (*d*) *2, 3-Dichloropropanoic acid*

**Solution.**

(*a*) 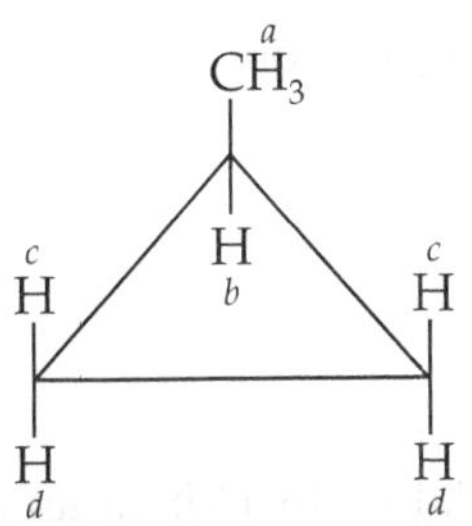

4 NMR signals
2-Methylcycloropane

(*b*) 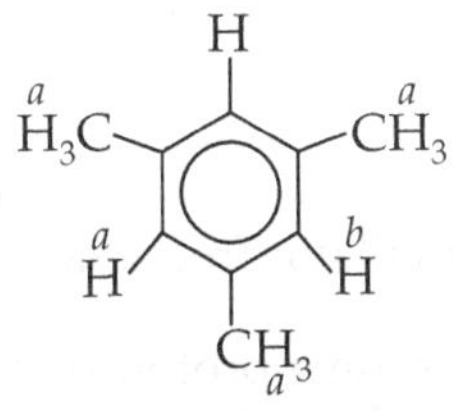

2 NMR signals
(Mesitylerse)

(*c*) 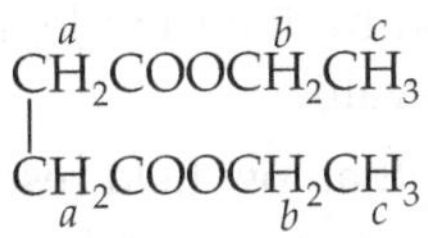

3 NMR Signals
(Ethyl succinate)

(*d*) 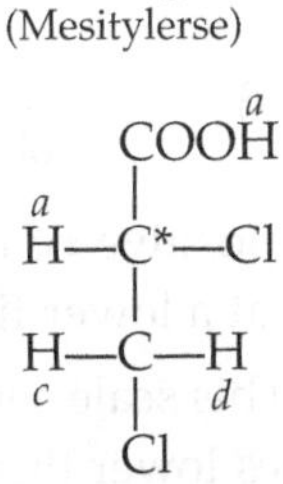

4 NMR signals
(2, 3-Dichloropropanoic acid)

Protons *c* are *cis* to the methyl group and protons *d* are *trans* to it, hence they are diastereotopic (nonequivalent).

In this compound C* is chiral. Protons $\overset{c}{H}$ and $\overset{d}{H}$ are diastereotropic, hence nonequivalent.

**B.** ***Position of Signal (Chemical Shift)***

The number of signals in an NMR spectrum tells the number of the sets of equivalent protons in the molecule. We have seen that for a field strength of 1.4T, the protons have a precessional frequency by ≈ 60 MHz. However, the precessional frequency of all the protons in the same external field is not the same as the exact value depends upon the chemical environment of the proton. Thus, different types of protons will absorb at slightly different frequencies. Since the shift in frequency depends upon the chemical environment, it is called the *chemical shift*. For example, in case of a molecule of methanol $CH_3OH$, there are two types of precessional frequencies of the proton in the spectrum. The differences in precessional frequencies are very small but can be detected by high resolution instruments.

**Measurement of Chemical Shift.** The precessional frequency for a group of nuclei cannot be measured in absolute units. The differences in frequencies are measured with respect to some reference group of nuclei. For measuring the chemical shifts of various protons, the signal for tetramethyl silane (TMS) is used as an internal standard. TMS protons resonate at exactly 60 MHz when the field strength is 1.4092 T, and we can measure the positions of other signals in relation to TMS in parts per million (ppm) at higher frequency than TMS downfield from the TMS signal. Thus, the difference from TMS signal and the signal is measured. The use of TMS has following advantages:

- It is chemically inert and has low boiling point.
- Its resonance signal is very sharp due to twelve equivalent protons.
- It is soluble in most of the organic solvents and can be added to the sample solution as an internal standard.

With this standard substance two scales are used for measuring chemical shift:

(*a*) δ *scale*. This sets the resonance of TMS arbitrarily at zero and counts the resonance at lower fields as positive.

$$\delta_{sample} = \frac{\nu_{sample} - \nu_{TMS}}{\nu_o} 10^6 \text{ ppm}$$

where $\nu_o$ is the oscillator frequency.

**Illustration.** Suppose a proton signal comes at 100 Hz relative to TMS at an oscillating frequency of 60 MHz.

$$\delta_{sample} = \frac{100}{60 \times 10^6} \times 10^6 \text{ ppm} = 1.67 \text{ ppm.}$$

In this δ scale, a positive of δ means downfield absorption by the sample, relative to TMS. Absorption at a lower field corresponds to a higher δ value.

(*b*) τ *scale (tau scale)*. This scale sets the TMS resonance at 10, so that low field resonance are measured at values lower than 10. Absorption at a lower field corresponds to a lower τ value. It can be readily seen that τ = 10 – δ.

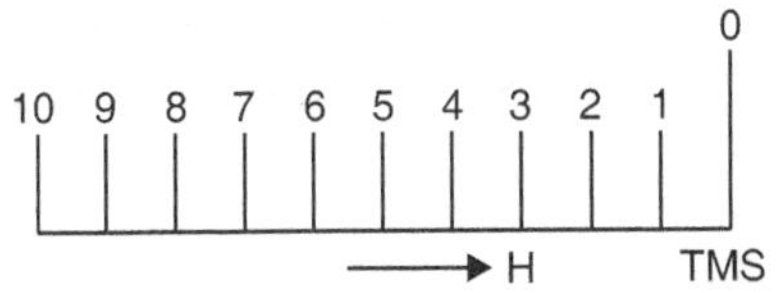


δ Scale

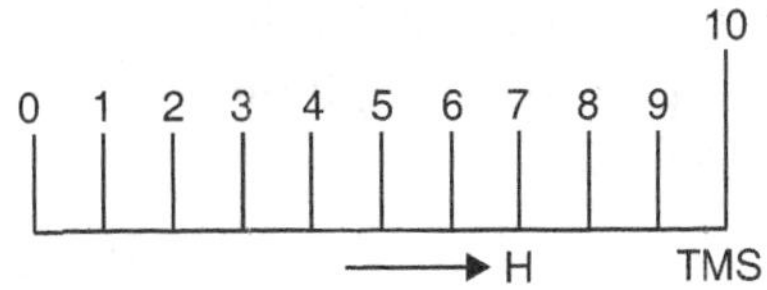


τ Scale

**Cause of Chemical Shift.** When a molecule is placed in a magnetic field, its electrons are caused to circulate and thus they produce an induced magnetic field. The induced magnetic field may oppose or reinforce with the applied field.

(*i*) When the induced magnetic field opposes the applied field the effective field felt by the proton is diminished. The shielded protons shift the absorption upfield. This is called *diamagnetic shift. Acetylenic* protons show absorption upfield.

(*ii*) When the induced magnetic field reinforce with the applied field, the effective field experienced by the nucleus is more than the applied field and the proton is said to be deshielded and the absorption position shifts to downfield in the NMR spectrum. This effect is called *paramagnetic shift.* It is seen in *ethylenic* and *aldehydic* protons.

### C. *Splitting of Signals*

Each signal in an NMR spectra represents one set of protons in a molecule. However, in certain molecules, instead of a single peak, a group of peaks (multiplet) is observed. The number of lines (multiplicity) observed in the NMR signal for a group of protons is not related to the number of protons in that group but it is related to the number of protons in the neighbouring group. The simplest rule to find the multiplicity of the signal from a group of protons is $(n + 1)$, where $n$ is the number of neighbouring protons. This can be illustrated by the following examples:

(*i*) Ethyl alcohol $\overset{(c)}{CH_3}-\overset{(b)}{CH_2}-\overset{(a)}{OH}$

It contains three types of protons, namely three methyl protons, two methylene protons, and one hydroxylic proton. Hence, three NMR signals are obtained under low resolution. However, under high resolution the proton signal from $CH_3$ splits into a triplet and the signal from $CH_2$ splits to a quartet (four peaks). This is due to the spin interaction or coupling between the adjacent sets of non-equivalent protons. The signal of the methyl protons splits into a triplet due to three types of orientations of the adjacent methylene protons as shown in the following Table 1.10. The intensities of the lines in the triplet are in the ratio of 1: 2: 1, corresponding to the number of like orientations in each case.

**TABLE 1.10**

| *Orientation of Protons* | *Number of Like Orientations (degeneracy)* | *Total Spin* |
|---|---|---|
| (*i*) αα *i.e.*, ↑↑ | 1 | $\frac{1}{2}+\frac{1}{2}=1$ |
| (*ii*) αβ, βα *i.e.*, ↑↓, ↑↓ | 2 | $\frac{1}{2}-\frac{1}{2}=1$ |
| (*iii*) ββ *i.e.*, ↓↓ | 1 | $-\frac{1}{2}-\frac{1}{2}=-1$ |

α-represents clockwise spin and β-anticlockwise spin.

Similarly, we can draw the four possible orientations of the three methyl protons, which are responsible for splitting of the signal into a quarter with the intensity ratio 1: 3: 3: 1; see Table 1.11.

**TABLE 1.11**

| *Orientation of Protons* | *Numbers of like Orientations* |
|---|---|
| (*i*) ↑↑↑ | 1 |
| (*ii*) ↑↓↓↑↓↑↑↑↓ | 3 |
| (*iii*) ↓↓↑↓↓↑↑↓↓ | 3 |
| (*iv*) ↓↓↓ | 1 |

Thus, the relative intensities of the individual lines of a multiplet correspond to the numerical coefficient of the lines in the binomial expression where $n$ = number of protons. For example,

$n = 1$; $(1 + x)^1 = 1 + x$ intensity ratio is 1: 1 (doublet)

$n = 2$; $(1 + x)^2 = 1 + 2x + x^2$ intensity ratio is 1: 2: 1 (triplet)

$n = 3$; $(1 + x)^3 = 1 + 3x^2 + 3x + x^3$ intensity ratio is 1: 3: 3: 1 (quartet)

$n = 4$; $(1 + x)^4 = 1 + 4x + 6x^2 + 4x^3 + x^4$ intensity ratio is 1: 4: 6: 4: 1 (quintet)

$n = 5$; $(1 + x)^5 = 1 + 5x + 10x^2 + 10x^3 + 5x^4 + x^5$ intensity ratio is 1: 5: 10: 10: 5: 1 (sextet)

$n = 6$; $(1 + x)^6 = 1 + 6x + 15x^2 + 20x^3 + 15x^4 + 6x^5 + x^6$ intensity ratio is 1: 6: 15: 20: 15: 6: 1 (septet)

(*ii*) 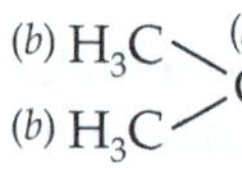

; the compound gives two signals. The signal given by the proton (*a*) split into a septet. The methyl protons '*b*' give an intense doublet.

(*iii*) 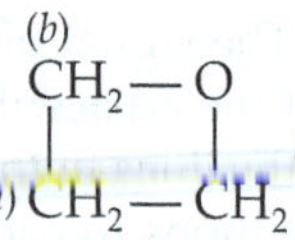

; the compound gives two signals. Signal '*a*' corresponds to four protons will appear as quintet. Signal '*b*' splits into a triplet.

(*iv*) $\overset{(b)}{CH_3}$—$\overset{(a)}{CH}$ $Br_2$; The compound gives two signals. Singal '*a*' corresponds to three proton will appear as quartet. Signal '*b*' splits into a doublet.

## Applications of NMR Spectroscopy

NMR spectroscopy is an important tool in the hands of organic chemists for getting structural information. It is also useful in stereochemical studies. An organic molecule consists of carbon-hydrogen skeleton, and thus NMR spectroscopy is especially useful in the study of this structural feature of the molecule. The important applications of NMR spectroscopy, besides obtaining routine structural information, are summarized as follows.

(*i*) **Identification of Structural Isomers**. Structural isomers can easily be distinguished by NMR spectroscopy. For example

(*i*) $CH_3OCH_3$ and $CH_3CH_2OH$

In dimethyl ether, all the six protons are equivalent. Hence, its NMR spectrum will show only one singlet. In ethanol, there are three kinds of protons. Thus, its NMR spectrum will exhibit three signals; one triplet due to the methyl protons, one quartet due to the methylene protons and one singlet due to the hydroxylic proton.

(*ii*) 1,1-dichloroethane ($CH_3CHCl_2$) and 1, 2-dichloroethane ($ClCH_2CH_2Cl$) 1,1-dichloroethane has two kinds of protons. Hence, it will exhibit two signals in its NMR spectrum; one doublet and one quartet due to $CH_3$—and—CH— groups, respectively. In 1, 2-dichloroethane all the four protons are equivalent. Hence, its NMR spectrum will show only one singlet.

(*ii*) **Detection of Aromaticity**. Aromatic protons are highly deshielded due to the circulating π electrons (ring current) and appear at very low field (δ 6-8.5). From this, the aromatic character of a compound under study can be predicted.

(*iii*) **Detection of Hydrogen Bonding.** Intermolecular and intramolecular hydrogen bondings can be detected by NMR spectroscopy because both shift the absorption position of the hydrogen-bonded proton to downfield. Besides, both types of hydrogen bonding can also be distinguished, as the intermolecular hydrogen bonding is concentration-dependent, while the intramolecular hydrogen bonding is not concentration-dependent.

(*iv*) **Distinction Between *cis-trans* Isomers and Conformers**. NMR spectroscopy can easily distinguish *cis* and *trans* isomers because the concerned protons have different values of chemical shifts as well as coupling constants. For example

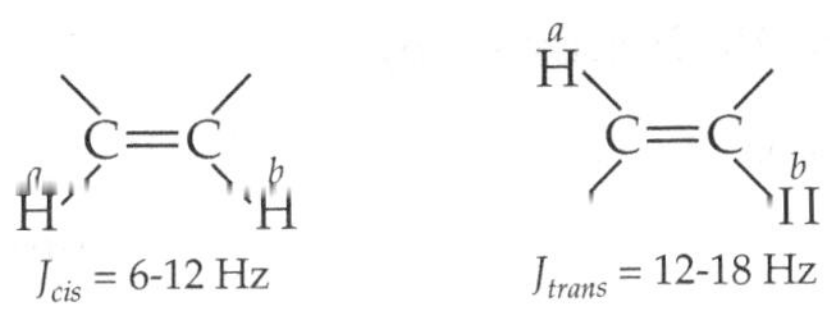

$J_{cis}$ = 6-12 Hz $J_{trans}$ = 12-18 Hz

Similarly, the axial and equatorial positions of protons or groups carrying protons and various conformations of a molecule can be distinguished on the basis of different values of their coupling constants, chemical shifts and peak areas.

(*v*) **Quantitative Analysis**. The fact that areas under the peaks are directly proportional to the number of protons causing the respective peaks is the basis for the quantitative analysis by NMR spectroscopy. For quantitative analysis, the components of the mixture must be known and each component must give at least one signal which is well separated from the other signals in the spectrum. Impure samples may be determined by the addition of a known pure compound as an internal standard. If the reactants and products are known, then the rate of the reaction may be determined.

Automatic integration of NMR signals afford an easy and rapid quantitative means for determining the ratio of compounds in a mixture provided that at least one signal from each constituent is free from overlap by other signal (*s*). The estimation of the keto-enol ratio in acetyl acetone will illustrate the quantitative analysis of a mixture.

$$\underset{\text{Keto form}}{H_3C{-}CO{-}CH_2{-}CO{-}CH_3} \rightleftharpoons \underset{\text{Enol form}}{H_3C{-}\underset{\underset{OH}{|}}{C}{=}CH{-}CO{-}CH_3}$$

Acetylacetone

In the NMR spectrum of acetylacetone, the height of the integration curve at the methylene ($-CH_2-$) signal was found to be 10 mm and that at the methine ($=CH-$) signal was 22 mm. Let us calculate the % of keto and enol forms in the sample.

The methylene group of the keto form $\equiv$ 2H$\equiv$10 mm

The methine group of the enol form $\equiv$1H$\equiv$22 mm

Therefore, 2H$\equiv$44 mm

$$\% \text{ of the keto form} = \frac{10}{44+10} \times 100 = 18.5\%$$

$$\% \text{ of the enol form} = \frac{44}{44+10} \times 100 = 81.5\%$$

NMR spectroscopy has also been used for the quantitative analysis of the mixtures of diastereomers as well as for determining the enantiomeric excess (ee), *i.e.* optical purity.

### 1.6.5 Solved Problems on NMR Spectroscopy

**Problem 34.** *How many proton signals would you expect in the NMR spectra of the following compounds?*

*(i) Cyclobutane* *(ii) Cyclohexane* *(iii) 1, 2-Dichloroethane*

*(iv) Methyl ethyl ether* *(v) 2-Chloropropane.*

**Solution.** (*i*) In cyclobutane $\begin{array}{c} CH_2-CH_2 \\ |\quad\ \ | \\ CH_2-CH_2 \end{array}$ all the hydrogen atoms are equivalent, hence, only **one NMR signal** will be observed.

Accordingly, we will get **one NMR** signal in (*ii*) cyclohexane  and

(*iii*) 1, 2; dichloroethane $\begin{array}{c} H_2C-CH_2 \\ |\quad\ \ | \\ Cl\quad Cl \end{array}$

(*iv*) Methyl ethyl ether $CH_3—O—CH_2—CH_3$. In this compound there are three types of protons so three peaks are observed.

(*v*) 2-Chloropropane $CH_3—CH(Cl)—CH_3$. In this compound there are two types of protons therefore, we will get **two NMR signals.**

**Problem 35.** *How would you identify ethane, ethene and ethyne by NMR spectroscopy?*

**Solution.** Although each of these compounds contains only one type of proton, therefore we will get one NMR signal in each case. But these protons are in different environments in different compounds, therefore, their relative position is different as given below:

| | |
|---|---|
| Ethane $CH_3—CH_3$ | $\delta$ 0.9 ($\tau$ 9.1) |
| Ethene $CH_2 = CH_2$ | 0.5 ($\tau$ 7.5) |
| Ethyne $CH \equiv CH$ | $\delta$ 2.5 ($\tau$ 7.5) |

**Problem 36.** *Predict the number of NMR signals in the following compounds:*

(*i*) *Vinyl chloride* $CH_2 = CHCl$

(*ii*) *1, 2-Dichloropropane* $CH_3—CH(Cl)—CH_2(Cl)$

**Solution.** (*i*) In vinyl chloride $\overset{(a)}{\mathrm{H}}{>}\mathrm{C{=}C}{<}\overset{(a)}{\mathrm{H}}$ (with H below each carbon) there are three types of vinylic protons marked as (*a*), (*b*) and (*c*), the structure. Therefore, we will get **three NMR signals.**

In vinylchloride, the proton *cis* to the Cl atom is in a different environment than the *trans* proton. Moreover, these protons are in different environments than the proton on the C—Cl carbon. In vinyl chloride, therefore, all three protons are non-equivalent.

(*ii*) In 1, 2-Dichloropropane we have four types of protons marked as *a*, *b*, *c*, and *d*. Therefore, we will get **four** NMR signals.

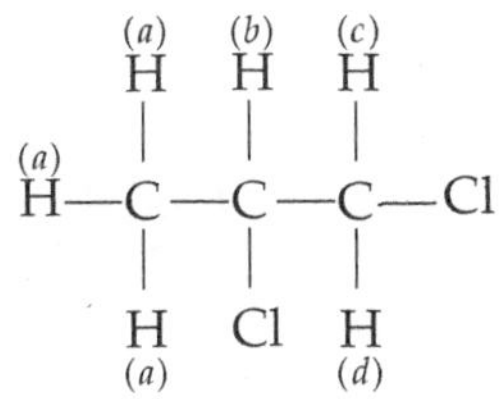

**Problem 37.** *How many NMR signals would you expect for the following compounds?*

(*i*) $H_2C = C(CH_3)H$ (ii) $(CH_3)_2{>}CH—Cl$

(iii) 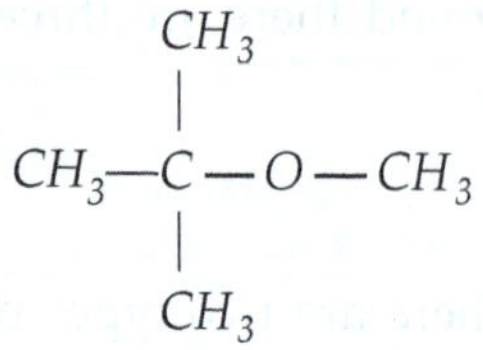

(iv) 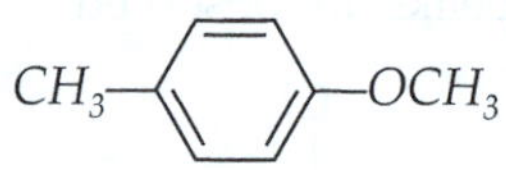

(v) 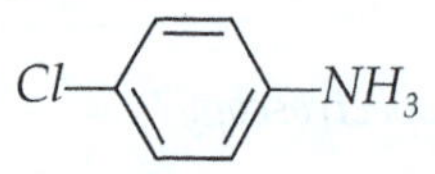

(vi) $CH_3$—CHO

(vii) $CH_3$—O—$CH_2$—$CH_3$

(viii) 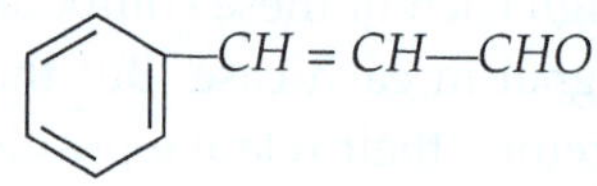

(ix) 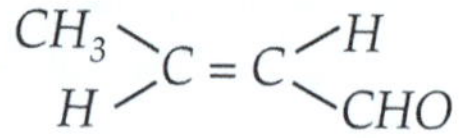

**Solution.** (*i*) 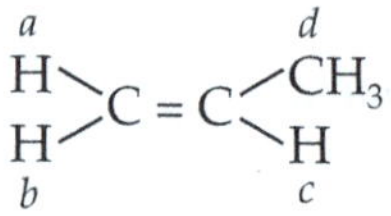

Four different protons
(four NMR signals)

(*ii*) 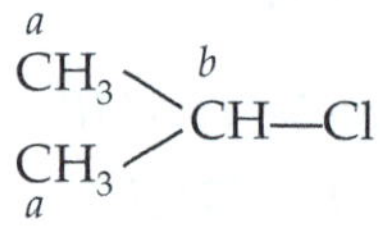

Two different protons
(two NMR signals)

(*iii*) 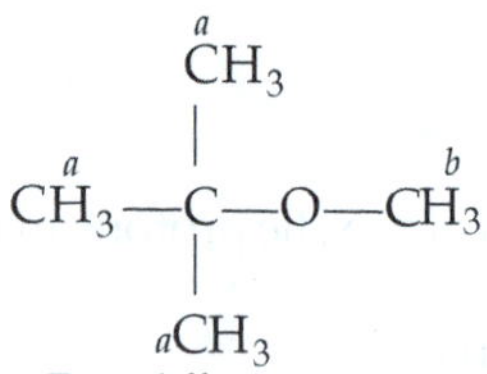

Two different protons
(two NMR signals)

(*iv*) 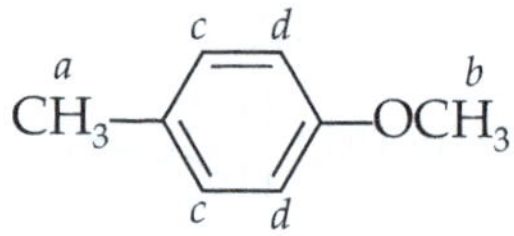

Four different protons
(four NMR signals)

(*v*) 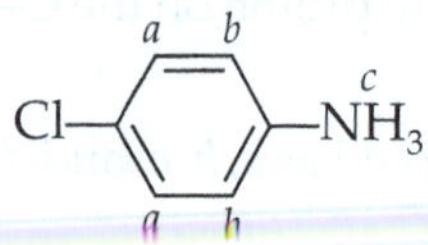

Three different protons
(three NMR signals)

(*vi*) $\overset{a}{CH_3}\overset{b}{CHO}$

Two different protons
(two NMR signals)

(*vii*) $\overset{a}{CH_3}—O—\overset{b}{CH_2}—\overset{c}{CH_3}$

Three different protons
(three NMR signals)

(*viii*) 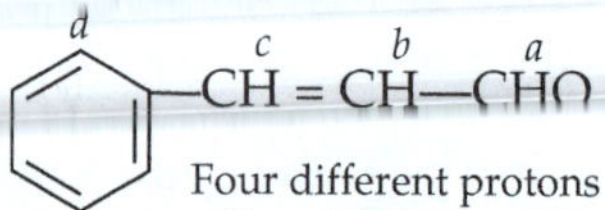

Four different protons
(four NMR signals)

(*ix*) 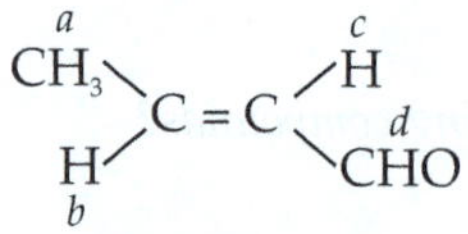

Four different protons
(four NMR signals)

**Problem 38.** *Explain the splitting pattern in the NMR spectrum of the following compounds:*

(i) $CH_3$—$CH_3$ (ii) $CH_3$— $CH_2$ Br (iii) $CH_3$—$CH_2$—$CH_2$ Cl

**Solution.** (i) **$CH_3$—$CH_3$.** Since all six hydrogens are equivalent in ethane, so no signal splitting is observed.

(ii) **$\overset{a}{CH_3}$—$\overset{b}{CH_2}$—Br.** There are two types of protons *a* and *b*. The three equivalent *a* protons of methyl group are under the influence of *b* protons (—$\overset{b}{CH_2}$—). Thus, the NMR band is spilt in $n + 1 = 2 + 1 = 3$ peaks with relative intensities 1: 2: 1. The two methylene protons (*b* type) see three neighbouring protons, their NMR bond will split into $n + 1 = 2 + 1 = 3 + 1 = 4$ peaks giving rise to a quarter with relative intensity of 1 : 3 : 3 : 1.

(iii) **$\overset{a}{CH_3}$—$\overset{b}{CH_2}$—$\overset{c}{CH_2}$—Cl.** In this molecule *a* type protons (—$CH_3$) will spilt into $n + 1 = 2 + 1 = 3$ peaks giving a triplet with relative intensities of 1 : 2 : 1. Similarly, a triplet is obtained from *c* type of proton (—$CH_2$—) with relative intensities of 1 : 2 : 1. The *b* type of protons (—$CH_2$—) are under the influence of the *a* type and *c* type of proton simultaneously. So the *b* type of protons can interact with 3 + 2 = 5 proton, so the multiplicity of the peaks is given by $n + 1$ is $5 + 1 = 6$. Thus, the *b* type of proton will exhibit a sextet with relative intensity of 1 : 5 : 10 : 10 : 5 : 1.

**Problem 39.** *Indicate the types of protons and their multiplicity in the NMR spectra of following compounds:*

(i) $ClCH_2CH_2CH_2Cl$ (ii) $CH_3CH_2COOCH\ (CH_3)_2$

(iii) $CH_2COOCH_3$
|
$CH_2COOCH_3$

(iv) 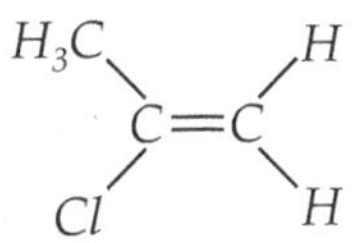

(v) 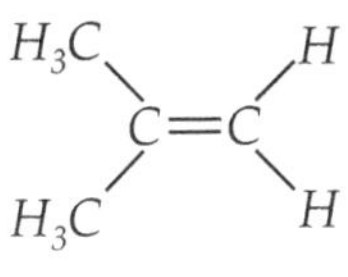

**Solution.**

(i) $Cl\overset{a}{CH_2}\overset{b}{CH_2}\overset{a}{CH_2}Cl$

There are two types of protons *a* and *d* in this compound. Their multiplicity is *a* (triplet) and *b* (quintet).

(ii) 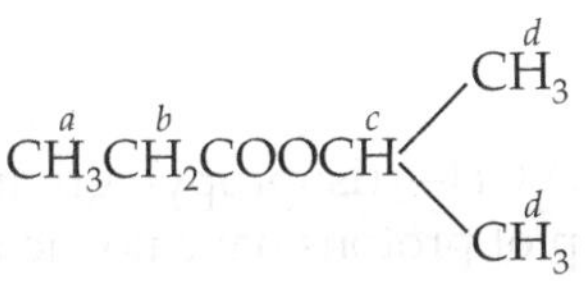

There are four kinds of protons *a*, *b*, *c* and *d*. Their multiplicity is *a* (triplet), *b* (quartet), *c* (septer) and *d* (doublet).

(*iii*) $\overset{a}{C}H_2CO\overset{b}{O}CH_3$
$|$
$\underset{a}{C}H_2COO\underset{b}{C}H_3$

There are two kinds of protons *a* and *b*. Their multiplicity is *a* (singlet) and *b* (singlet).

(*iv*) 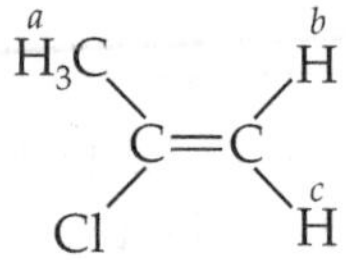

There are three kinds of protons *a*, *b* and *c*. Their multiplicity is *a* (singlet) because it has no vicinal proton, *b* (doublet) and *c* (doublet).

Protons *b* and *c* also cause splitting of signals because they are nonequivalent and are separated by only two bonds.

(*v*) 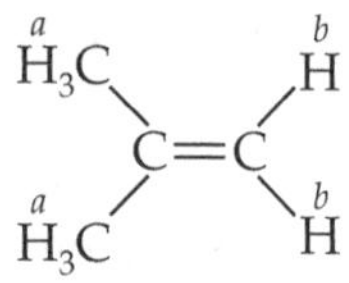

There are two kinds of protons *a* and *b*. Their multiplicity is as follows:

*a* (singlet) because there is no vicinal proton.

*b* (singlet) because there is no vicinal proton.

**Example 40.** *Draw the structure of each of the following compounds which meets the given requirements in its NMR spectrum:*

(*i*) $C_3H_3Cl_5$*; one doublet and one triplet*
(*ii*) $C_4H_{10}O$*; one singlet, one doublet and one septet*
(*iii*) $C_4H_8O_2$*; one singlet, one triplet and one quartet*
(*iv*) $C_3H_7Cl$*; one doublet and one septet*

**Solution.**

(*i*) $C_3H_3Cl_5$

The compound shows one doublet and one triplet. This indicates the presence of —CH—$CH_2$—group in the molecule. Thus, following structure with molecular formula $C_3H_3Cl_5$ fulfils this condition:

$Cl_3C\overset{a}{C}H_2\overset{b}{C}HCl_2$

*a* (doublet) and *b* (triplet).

(*ii*) $C_4H_{10}O$

The presence of a doublet and a septet indicates $(CH_3)_2CH$— (isopropyl) group. There is one singlet in the spectrum which shows that a group of protons have no vicinal proton in the molecule. Thus, following structure with the above molecular formula meets the given requirements:

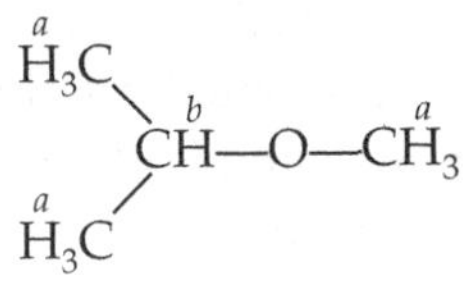

*a* (doublet) and *b* (septet).

(*iii*) $C_4H_8O_2$

The compound shows one triplet and one quartet which indicates the presence of —$CH_2CH_3$ group. The compound shows one singlet indicating that a group of three protons (—$CH_3$) has no vicinal proton. The following structures meet the above requirements:

$$\overset{a}{C}H_3COO\overset{b}{C}H_2\overset{c}{C}H_3 \text{ and } \overset{a}{C}H_3\overset{b}{C}H_2COO\overset{a}{C}H_3$$

*a* (singlet), *b* (quartet), *c* (triplet).

(*iv*) $C_3H_7Cl$

The presence of a doublet and a septet indicate $(CH_3)_2CH$— group. Thus, the structure which meets the given requirements is

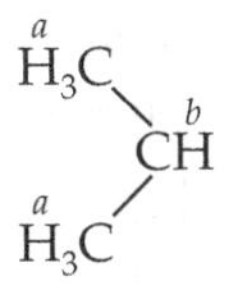

***a***(doublet) and ***b*** septet

## 1.7 MASS SPECTROSCOPY

In mass spectroscopy the substance under investigation is bombarded with an electron beam and the instrument quantitatively records the result as a spectrum of positive ion fragments. It can be studied in following steps:

### 1.7.1 Principle

The principle of mass spectroscopy is based on the following two aspects:

- The removal of a single electron from the molecule in the gas phase to form a molecular ion (*i.e.*, parent ion) occurs in the following way,

  $$M(g) + e \longrightarrow M^+ (g) + 2e,$$

  where M is the parent molecule. For example,

  $$CH_3OH + e \longrightarrow CH_3\overset{+}{O}H + 2e$$
  $$m/e = 32$$

  The parent ion produced is an ion radical. All organic molecules are even electron species and removal of one electron from such a molecule produces the ion radical.
- A mass spectrum is a presentation of mass/charge versus intensities of relative concentrations. The molecular ion peak (the parent ion peak) is usually the peak of highest $m/e$. (*i.e.*, mass number) except for isotope peaks. The largest peak, *i.e.*, maximum

height in the mass spectrum, is called *base peak* and its intensity is taken as 100. The intensities of other peaks are represented relative to the base peak. In most cases, the parent ion peak is not the base peak and is often of very small abundance. In a few cases, however, the parent ion peak may be the base peak.

### 1.7.2 The Mass Spectrum (The Base Peak)

The graphic representation of the mass spectrum of a compound is constructed by plotting mass/charge ratio (*m/e*) versus relative abundance or percentage of base peak, where the base peak is the most intense peak in the spectrum. The base peak is assigned the arbitrary intensity value of 100, while all other peaks are given their proportionate value. The common practice is to represent spectra of organic compounds, *e.g.*, of methanol (Figure 1.22) in the form of a bar graph.

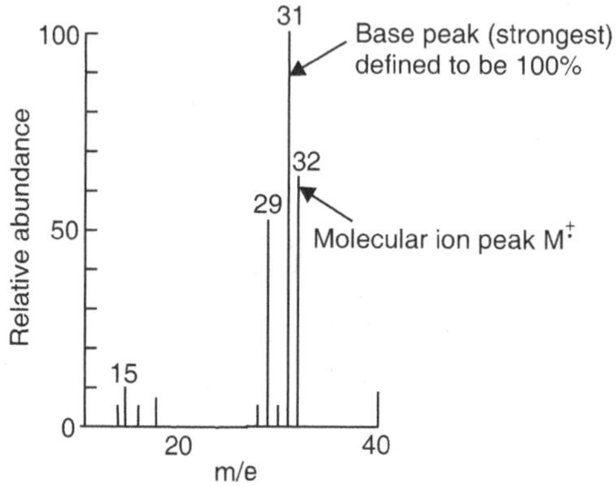


*FIGURE 1.22* **The mass spectrum of methyl alcohol.**

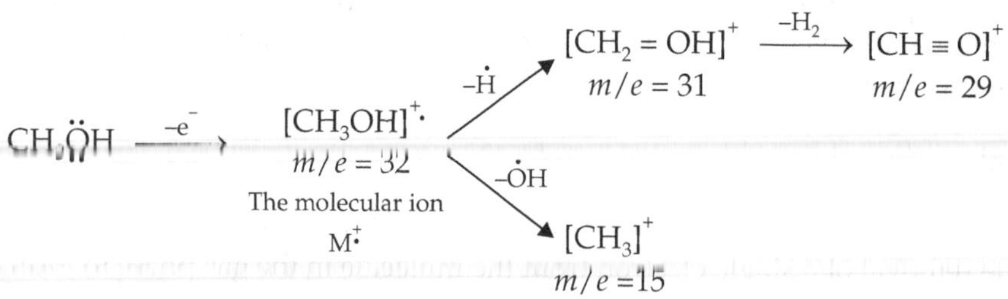


The abscissa indicates the mass, *i.e.*, the ratio of the mass to the charge (*m/e*), while the ordinate indicates the relative abundance (vertical peak intensity). In the mass spectrum of methanol (Figure 1.22), the positive ion peak at *m/e* = 32 represents the intact methanol molecule, less one electron which is knocked out by the impacting beam and is the molecular ion $M^{\overset{+}{\cdot}}$. The molecular ion subsequently produces a series of fragment ions as shown for methanol.

Thus, the parent ion (molecular ion) under further bombardment breaks down into smaller ion radicals, carbocations, and neutral molecules, and all the new positively charged ions are detected as shown below. Thus, the mass spectrum registers the masses of many different fragments, hence the name mass spectrum (literally, a spectrum of different masses).

$$M^{\overset{+}{\cdot}} \rightarrow m_1^{+} + m_2^{\cdot} \quad or \quad m_1^{\overset{+}{\cdot}} + m_2$$

Let us consider an example of Deuterated forms of ethyl methyl ketone $CH_3CH_2CO\ CH_2D$.

$$CH_3CH_2CO\ CH_2D \xrightarrow{-e} CH_2\ \overset{+}{C}H_2COCH_2D \quad m/e = 72$$

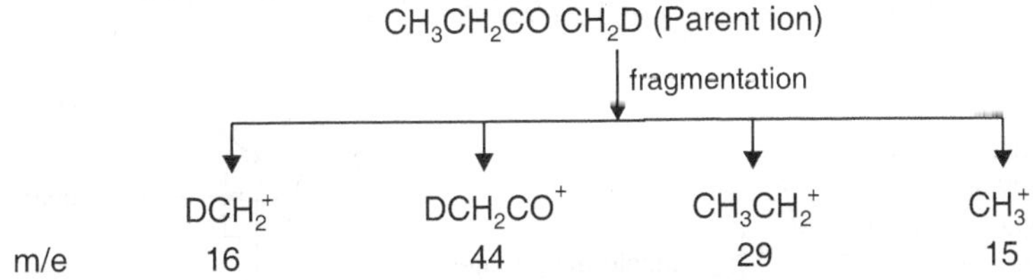


*FIGURE 1.23* **Mass spectrum of $CH_3CH_2CO\ CH_2D$.**

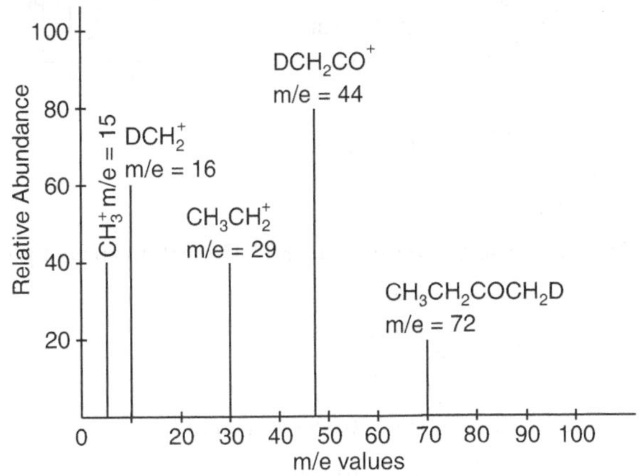


*FIGURE 1.24* **Mass spectrum of $CH_2CH_2COCH_2D$.**

### 1.7.3 Instrumentation

Most common mass spectrometer is *Magnetic Deflection Mass Spectrometer*. It consist of the following parts:

(*a*) *Ion source.* Milligram amount of the substance (pressure $\approx 10^{-6}$ mm) in the vapour phase is bombarded with electrons from a tungsten filament which have been accelerated to an energy of about 70 eV. An energy of about 10 eV is usually necessary to remove an electron from the parent molecule to give the molecular ion, while higher energy ($\approx$ 70 eV) is required for the fragmentation of the parent ion, *i.e.*, molecular ion, initially produced. The very low pressure is intended to prevent undue scattering of ions by collisions.

(*b*) *Mass analyser.* The parent ion and the fragment ions produced and accelerated in the ionization chamber enter the mass analyser. Here the fragment ions are differentiated on the basis of their $m/e$ ratios by the application of a magnetic field. The ions may be deflected in the magnetic field through any desired angle (60°, 90°, 120°, or 180°). The Figure 1.25 shows a 180° magnetic deflection mass spectrometer. The kinetic energy of the electron that enters the analyzer is expressed as

$$\frac{1}{2}mv^2 = eV \tag{1.4}$$

$$v^2 = \frac{2eV}{m}$$

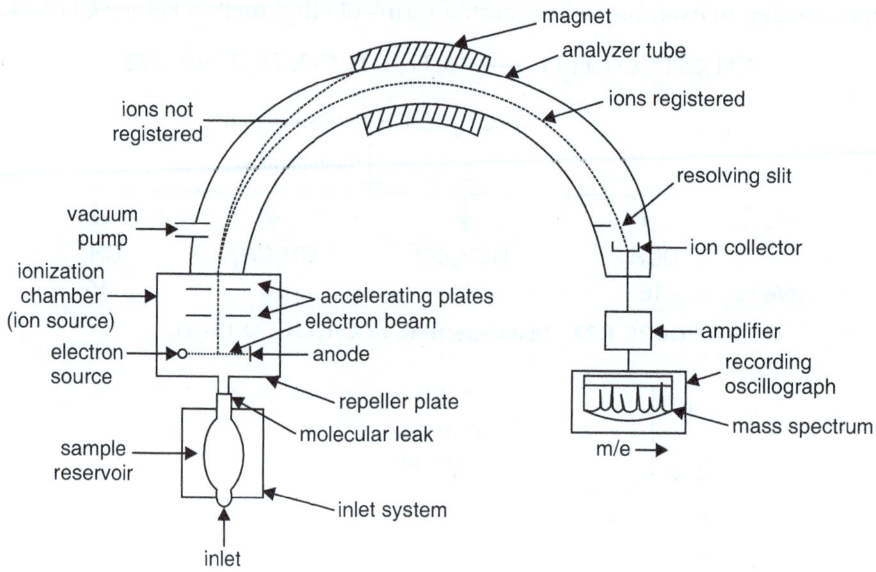


***FIGURE 1.25*** **Schematic Diagram of a 180° Deflection mass Spectrometer.**

$$v = \left(\frac{2eV}{m}\right)^{1/2} \tag{1.5}$$

where, $v$ = velocity of accelerated ions; $m$ = mass of particle

V = applied potential; $e$ = charge of an electron

Thus, KE is independent of the mass of the ion, but depends on the accelerating voltage. The magnetic field H exerts a force on the moving ions equal to H$e$V, which provides the centrifugal force for the circular deflection of ions. Thus,

$$HeV = \frac{mv^2}{r} \tag{1.6}$$

$$\frac{1}{r} = \frac{Hev}{mv^2}$$

$$r = \frac{mv}{He} \tag{1.7}$$

where, $r$ = radius of the circular ion

H = magnetic field

Now, substituting the value of $v$ from Eq. (1.5) in Eq. (1.7), we have

$$r = \frac{m}{eH}\left(\frac{2eV}{m}\right)^{1/2}$$

$$r^2 = \frac{m^2}{e^2H^2}\left(\frac{2eV}{m}\right)$$

$$r = \left(\frac{2V}{H^2} \cdot \frac{m}{e}\right)^{1/2}$$

$$\frac{m}{e} = \frac{H^2 r^2}{2V} \qquad (1.8)$$

This expression means that all ions, under a given accelerating voltage and magnetic field, will follow a circular path of radius '$r$'. Thus, ions of different $m/e$ value follow circular paths of different radii and reach the collector and are amplified and recorded.

### 1.7.4 Applications of Mass Spectroscopy

Following are the important applications of mass spectroscopy:

- It is used to determine the structure of organic molecules, *e.g.*, isomers of $n$-octane.

  $CH_3CH_2CH_2CH_2CH_2CH_2CH_2CH_3$ — $n$-octane (I)

  $CH_3CH_2CH_2CH(CH_3)CH_2CH_2CH_3$ — 4-methyl heptane (II)

  Compound (I), can be easily distinguished from (II), since the methyl group branched at C-4 in the latter compound is easily broken off as a $CH_3^+$ fragment and will give a peak of large intensity at $m/e = 15$.
- Mass spectroscopy used to find out volatile organic species pollutants in water bodies.
- For analysis of aerosol particles.
- It is used to identify drug abuse.
- This technique is used to measure molecular weights with very high accuracy; from these, exact molecular formulae can be deduced.

### 1.7.5 Solved Problems on Mass Spectroscopy

**Problem 41.** *Explain the formation of prominent ion peaks at m/e 72, 71, 57 and 43 in the mass spectrum of 2-methylbutane. Indicate the ion responsible for the base peak.*

**Solution.** The mode of formation of the prominent ions responsible for peaks at $m/e$ 72, 71, 57, 55, 43 and 41 is

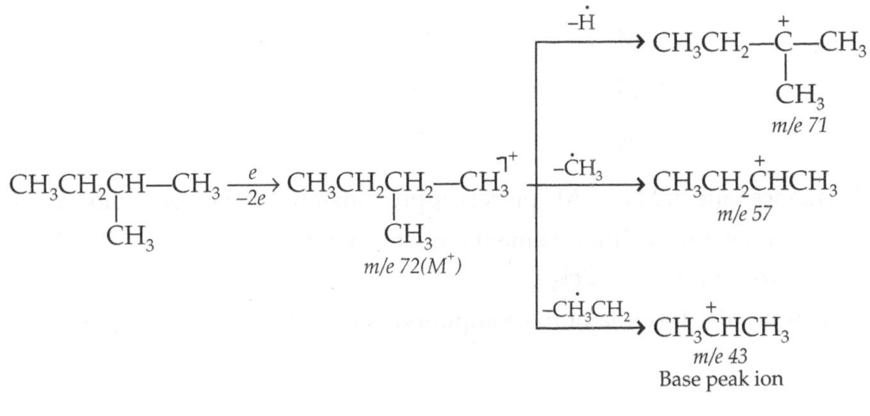

$$\underset{m/e\ 57}{CH_3CH_2CH\overset{+}{C}H_3} \xrightarrow{-H_2} \underset{m/e\ 55}{CH_3CH{=}\overset{+}{C}{-}CH_3}$$

$$\underset{m/e\ 43}{CH_3\overset{+}{C}HCH_3} \xrightarrow{-H_2} \underset{m/e\ 41}{CH_3\overset{+}{C}{=}CH_2}$$

**Problem 42.** *The mass spectrum of 2-pentene exhibits prominent peaks at m/e 70, 55, 41, 39, 29 and 27. Explain the formation of the ions corresponding to these peaks.*

**Solution.** The formation of the ions responsible for the observed peaks is explained by the following fragmentation pattern:

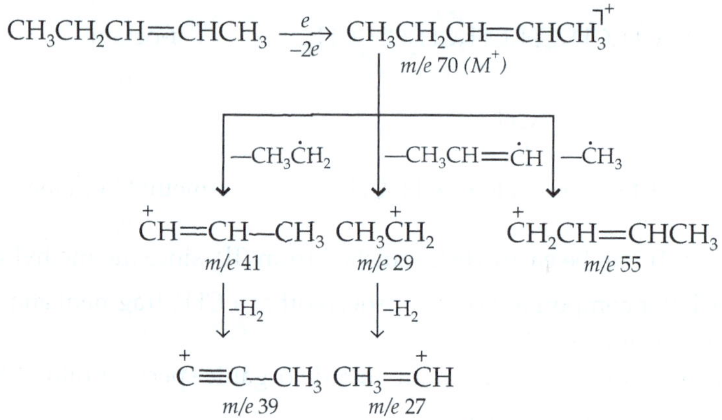

**Problem 43.** *Predict the molecular formula of an organic compound which gives the following peak in the isotopic clusters of its mass spectrum.*

| *m/e* | 94 | 95 | 96 | 97 |
|---|---|---|---|---|
| *Relative abundance* | (100) | (6.1) | (96) | (1.1) |

**Solution.** The most intense peak is at *m/e* 94 and is the molecular ion peak, so the isotopic clusters of the compounds are given by

| | | |
|---|---|---|
| M | 94 | 100 |
| $M_1^+$ | 95 | 6.1 |
| $M_2^+$ | 96 | 96 |
| $M_3^+$ | 97 | 1.1 |

The abundance ratio of M and $M_2^+$ peak is approximately 1.1 and hence indicates the presence of Br atom (mol. wt. 79 amu). Other elements must, therefore, account for 94–79 = 15 amu and correspondingly corresponds to $—CH_3$.

Thus, the molecular formula of the compound is $CH_3Br$.

**Problem 44.** *Suggest the structure of a compound with molecular formula $C_{10}H_{12}O$ whose mass spectrum shows peaks at m/e 15, 43, 57, 91, 105 and 148.*

**Solution.** The appearance of a peak at *m/e* 15 suggests a methyl ($—CH_3$) group. A peak at *m/e* ($\because$ 15 + 28 = 43) indicates that a carbonyl group (CO with *m/e* = 28) is also present as acetyl group ($CH_3CO$) in the compound. The maximum *m/e* value at 148 shows its molecular weight. The elimination of the acetyl group from the molecular given *m/e* (148 – 43 = 105) which is an observed peak in the spectrum. The peak at *m/e* 91 (105 – 14) suggest a $CH_2$ group's presence. So far we can say that the $CH_3COCH_2$ group with molecular weight 57 is present and we have to explain remaining 148 – 57 = 91. The peak corresponding to *m/e* 91 is likely to be $C_7^+H_7$, whose precursor is the stable benzyl cation $C_6H_5—CH_2^+$.

Thus, the suggested structure is $C_6H_5—CH_2—CH_2COCH_3$.

**Problem 45.** *An organic compound gave following molecular iron peak in its mass spectrum:*

| *m/e* | *124* | *123* | *93* | *77* | *65* | *51* |
|---|---|---|---|---|---|---|
| *Abundance (%)* | *2.8* | *40* | *10* | *100* | *15* | *60* |

*Suggest the structural formula of this compound.*

**Solution.** The base peak at *m/e* = 77, suggest the presence of aromatic $C_6H_5$ group. The molecular ion peak at *m/e* = 123 (*i.e.*, an odd molecular weight) suggest the presence of Nitrogen (N). The difference 123 – 77 = 46, suggest the presence of $—NO_2$ group ($\because$ 14 + 32 = 46). This is confirmed by the appearance of peak at 123 – 46 = 77. A peak is also found at *m/e* = 93 which indicates the loss of a molecule with weight 123 – 93 = 30. It can be NO group ($\because$ 14 + 16 = 30). Thus, the suggested structure is

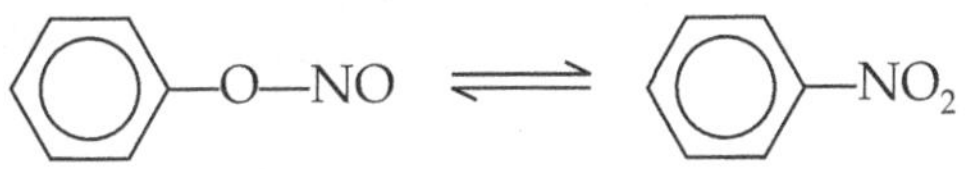

**Problem 46.** *An organic compound gave a peak at m/z 122 (M⁺) and another peak of nearly equal intensity at m/z 124 in its mass spectrum. What is the likely molecular formula of the compound ?*

**Solution.** The compound in quesiton shows the molecular ion ($M^+$) peak at *m/z* 122 and *M* + 2 peak at *m/z* 124 in the intensity ratio 1:1. This is characteristic of a monobromo compound. $M^+$ and *M* + 2 peaks are due to the contribution of $^{79}Br$ and $^{81}Br$ isotopes, respectively.

On deduction of the mass of $^{79}Br$ from the molecular mass $M^+$ of the compound, *i.e.* 122 – 79 = 43. The alkyl group corresponding to this mass is $C_3H_7$. Thus, the compound is $C_3H_7Br$.

**Problem 47.** *How will you distinguish 3-methylcyclohexene and 4-methyl-cyclohexene using mass spectroscopy ?*

**Solution.** Cyclohexene derivatives undergo fragmentation through retro-Diels-Alder reaction. Thus, 3-methylcyclohexene gives a much abundant diene fragment at *m/e* 54, whereas 4-methylcyclohexene gives the diene fragment at *m/e* 68 as

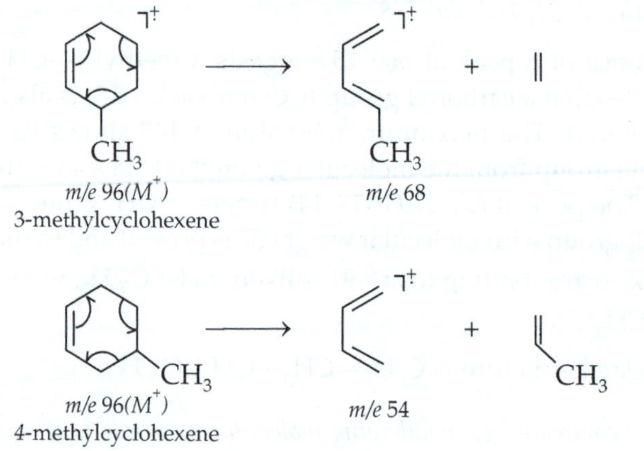


**Problem 48.** *Outline the mode of fragmentation during mass spectrometric study of the following compounds leading to the peaks at indicated m/z:*

*(i) Methylbutanoate, at m/e 74 and 59*

*(ii) Benzyl methyl ether, at m/e 91 and 65*

**Solution.**

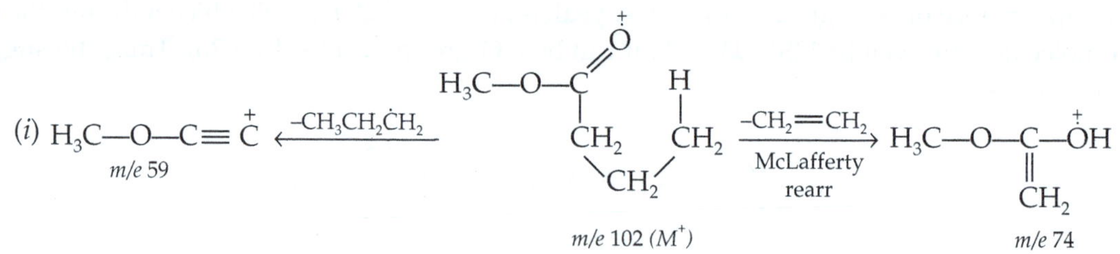


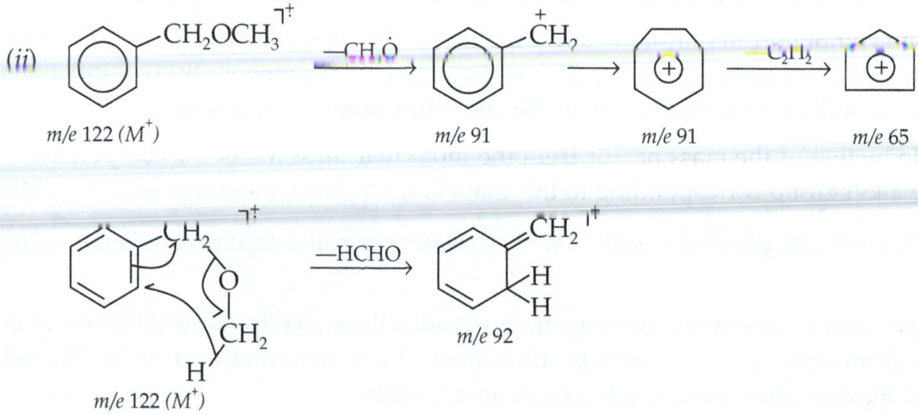

**Problem 49.** *How will you distinguish between 3-methyl and 4-methyl cyclohexene on the basis of mass spectrometry?*

**Solution.** A retro-Diels-Alder reaction of 4 methylcyclohexene will eliminate butadiene fragment at *m/e* 54 whereas a retro-Diels-Alder reaction of 3 methylcyclohexene gives a diene fragment at *m/e* 68.

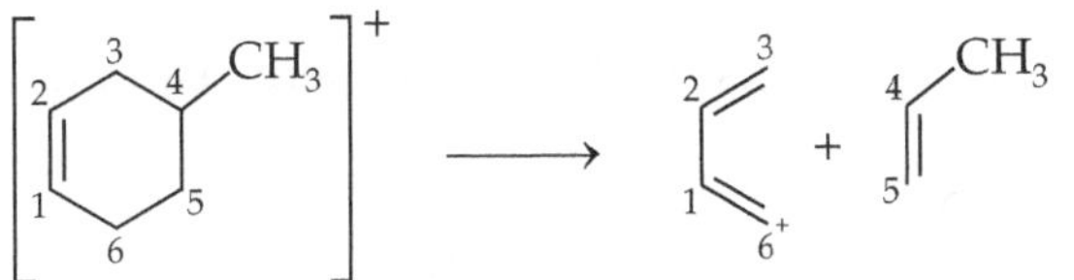

4-Methylcyclohexene $m/e = 96$ $m/e = 54$

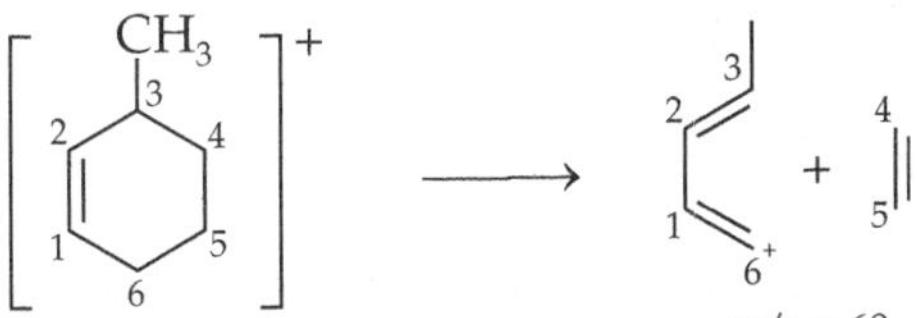

3-Methylcyclohexene $m/e = 96$ $m/e = 68$

**Problem 50.** *The parent ion of p-amino anisol appears at m/e 123. The daughter ion is formed by the loss of 15 unit mass in second field free region. Then calculate position of the metastable peak.*

**Solution.** Let the structure are like

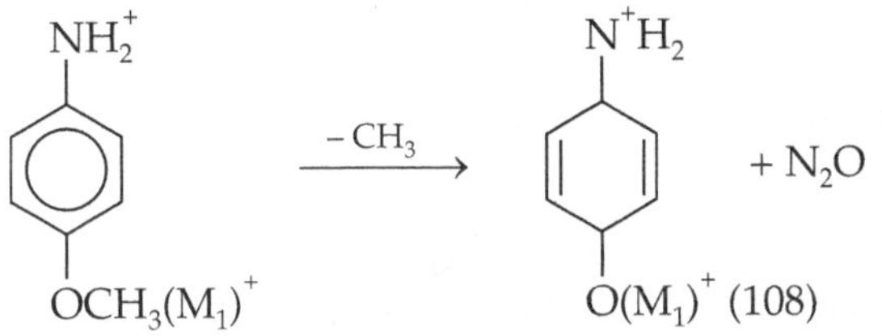

Position of metastable peak

$$M_* = \frac{M_1^2}{M_1} = \frac{108 \times 108}{123} = 94.8$$

In the 2nd field

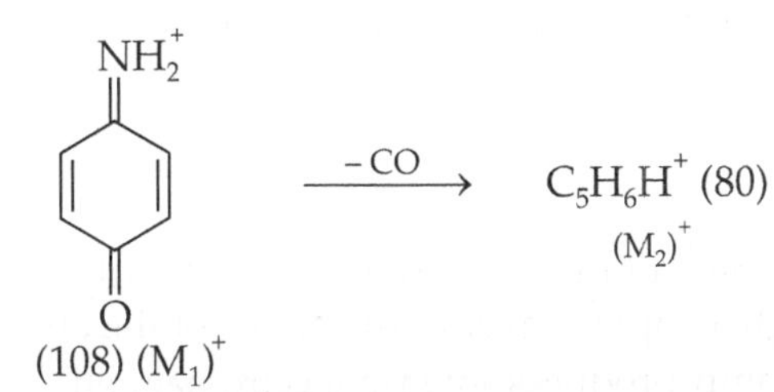

**Problem 51.** *Using mass spectrometry how will you distinguish 2-pentanone and 3-pentanone ?*

**Solution.** 2-Pentanone will show peaks at $m/e$ 71 and 43 due to acylium ions formed by the α-cleavage as shown below:

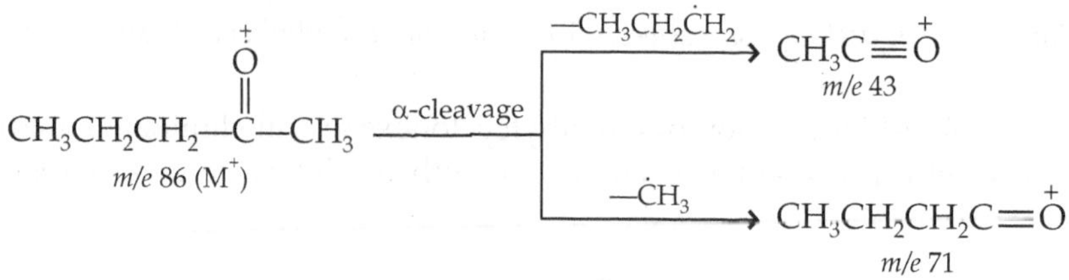

In addition, 2-pentanone will also undergo McLafferty rearrangement to give a peak at $m/e$ 58

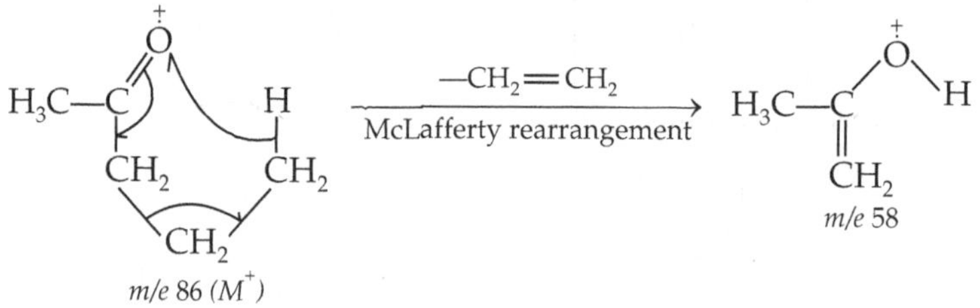

On the other hand, 3-pentanone will undergo α-cleavage to give only one acylium ion showing a peak at $m/e$ 57. Unlike 2-pentanone, 3-pentanone will not undergo McLafferty rearrangement.

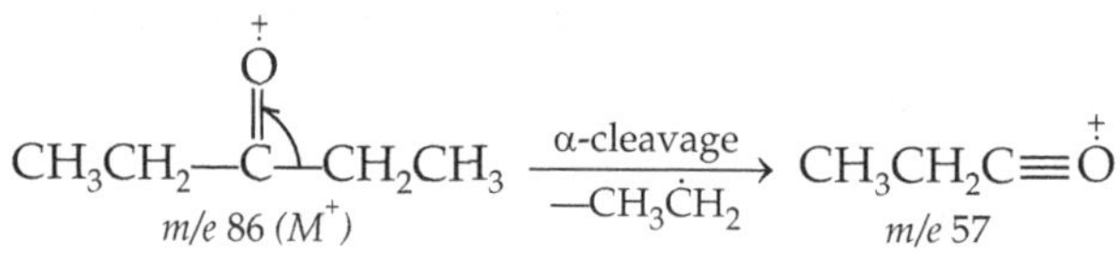

In addition, 3-pentanone will show a peak at $m/e$ 29 as follows:

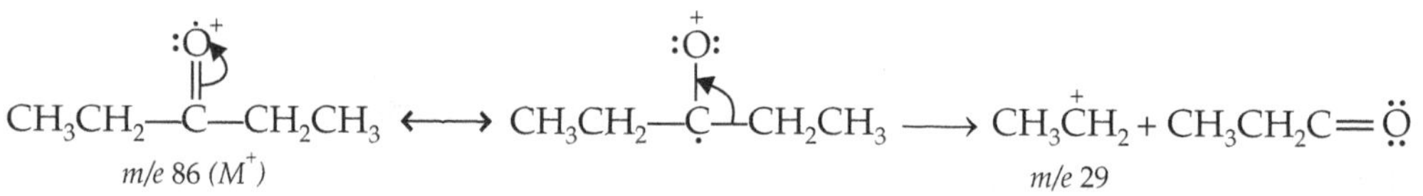

## 1.8 ATOMIC ABSORPTION SPECTROPHOTOMETER (AAS)

This is a popular method for the determination of metals in environmental samples. It was introduced by *Dr. A. Walsh* in 1955.

### 1.8.1 Principle

AAS is based on the measurement of the decrease in light intensity from a source (hollow cathode lamp), when it passes through a vapour layer of the atoms of the analytic element.

The hollow cathode lamp produces intense electromagnetic radiation with a wavelength exactly the same as that absorbed by the atoms, leading to high selectivity.

### 1.8.2 Construction and Working

The various components of AAS are shown in Figure 1.26.

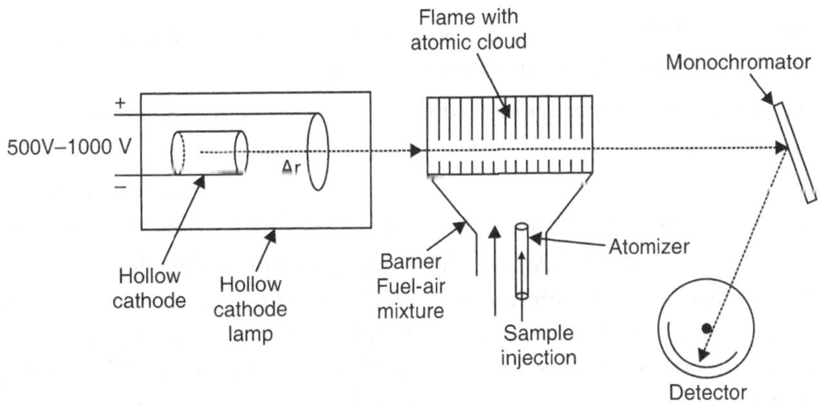


*FIGURE 1.26*

The main component is the *hollow cathode* lamp. It consists of a glass tube containing noble gases primarily argon, at several mm pressure, an anode and a hollow cathode, the inside of which is coated with the metal to be analyzed. A high potential is applied across the electrodes. An electrical current of several milliamperes flows through the lamp when it is in operation. $Ar^+$ ions produced inside the tube impinge on the cathode with a very high energy, leading to the sputtering of metal atoms from the cathode surface. These energized metal atoms emit radiation with a very narrow wavelength characteristic of the metal.

The radiation from the hollow cathode lamp passes through a flame into which the sample is aspirated. The metallic compounds are decomposed into the flame and the metal is reduced to the elemental state, forming a cloud of atoms. These atoms absorb a fraction of radiation in the flame. The decrease in the radiant energy increases with the concentration of the sought-for element in the sample, according to Beer's law. The resultant radiant energy passes through a monochromator to eliminate extraneous light resulting from the flame and finally to a detector and recorder.

### 1.8.3 Applications

Following are the imporant applications of ASS :

1. It can be used to determine about sixty trace metals in single sample.
2. It is widely used in metallurgical and general inorganic analysis.
3. It is an ideal method for analysis of many ores and minerals.
4. Determination of trace lead in ferrous alloy.
5. Determination of vanadium in lubricating oil.

## 1.9 ELECTRON SPIN RESONANCE (ESR) SPECTROSCOPY

ESR spectrocopy is an absorption spectroscopy invented by Zavoiskii in 1944. It involves the absorption of radiation in the microwave region ($10^4$ – $10^6$ MHz) by substances containing one or more unpaired electrons. This absorption of microwave radiation takes place under the influence of an applied magnetic field. The substances with one or more unpaired electrons are paramagnetic and exhibit ESR. Thus, ESR spectroscopy is also called *electron paramagnetic resonance* (EPR) spectroscopy or electron magnetic resonance spectroscopy.

Substances containing unpaired electrons, *i.e.* paramagnetic substances are of two types:

(*i*) *Stable Paramagnetic Substances.* These include simple molecules like NO, $O_2$ and $NO_2$, and the ions of transition metals and their complexes, *e.g.* $Fe^{3+}$, $[Fe(CN)_6]^{3-}$ etc. Such stable paramagnetic substances can be easily studied by ESR spectroscopy.

(*ii*) *Unstable Paramagnetic Substances.* These are generally called *free radicals* or *radical ions* and are formed either as intermediates in chemical reactions or by irradiation of a stable molecule with UV or X-ray radiation or with a beam of nuclear particles. If the lifetimes of such radicals is greater than $10^{-6}$ s, they may be studied by ESR spectroscopy. Paramagnetic substances with lifetimes shorter than $10^{-6}$ s, may also be studied by ESR spectroscopy if they are produced at low temperatures in the solid state, called *matrix technique*, as this increases their lifetimes. ESR spectroscopy is most useful in the study of free radicals.

### 1.9.1 Principle

The principle of ESR is similar to NMR, except that electron spin in involved in ESR instead of nuclear spin which is involved in NMR. An unpaired electron, like a proton, has a spin and this spin has an associated magnetic moment. An electron of spin $s = \frac{1}{2}$ can have the spin angular momentum quantum number values of $m_s = \pm\frac{1}{2}$. In the absence of an applied magnetic field, the two values of $m_s$, *i.e.* $+\frac{1}{2}$ and $-\frac{1}{2}$ will give rise to a doubly degenerate spin energy state. When a magnetic field is applied, this degeneracy disappears and two non-degenerate spin energy states result. The low energy state (more stable) has the spin magnetic moment aligned with the applied magnetic field and corresponds to the quantum number $m_s = -\frac{1}{2}$, whereas the high energy state (less stable), $m_s = +\frac{1}{2}$, has its spin magnetic moment aligned against the applied field. These energy states are illustrated in Fig. 1.27. These two states will possess energies that are split up from the original state with no applied magnetic field by the amount $-\mu_e H_0$ and $+\mu_e H_0$ for the low energy and high energy states, respectively (Fig. 1.27). Here $\mu_e$ is the magnetic moment of the spinning electron and $H_0$ the applied magnetic field acting on the unpaired electron.

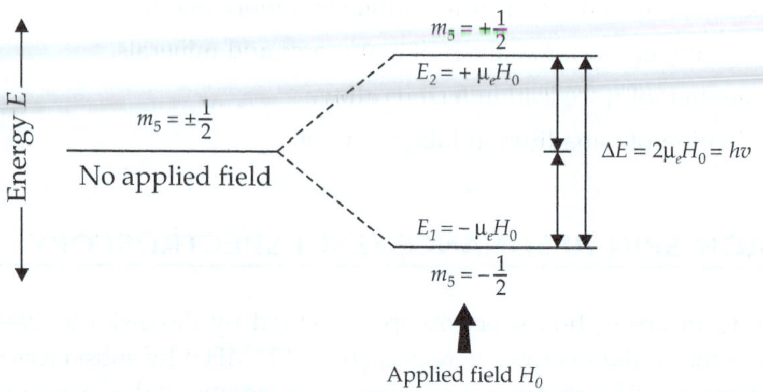


**FIGURE 1.27 Energy states of an unpaired electron in an applied magnetic field $H_0$.**

In ESR, a transition between the two different electron spin energy states takes place by absorption of a quantum of radiation of an appropriate frequency in the microwave region. When the absorption takes place, the following relation holds good:

$$2\mu_e H_0 = hv = \Delta E$$

where $2\mu_e H_0$ is the difference between the two electron spin energy states (Fig. 1.27), *i.e.*

$$\mu_e H_0 - (-\mu_e H_0) = 2\mu_e H_0 \qquad \text{...(1.1)}$$

Strictly speaking, the relation given in Eq. (1.1) holds good for a free electron.

The energy of transition $\Delta E$ in substances containing an unpaired electron is more accurately given by the relation

$$\Delta E = hv = g\beta H_0 \qquad (1.2)$$

where $h$ is the Planck's constant, $v$ the frequency in cycles per sec, $\beta$ the Bohr magneton which is a factor for converting angular momentum into magnetic moment and $g$ the spectroscopic splitting factor or Lande splitting factor. The value of $g$ ($g$ factor) is not constant but it depends on the environment of the unpaired electron. For a free electron, the value of $g$ is 2.0023 at 8388.255 MHz in a field of 0.30 T (3000 gauss). Virtually all free radicals and some ionic crystals have almost the same $g$ value, *i.e.* 2.0023 with a variation of ± 0.003 from this value. The reason for this is essentially that in free radicals the electron behaves in very much the same way as an electron in free space.

The value of $\beta$ is defined as

$$\beta = \frac{eh}{4\pi mc} = 9.273 \times 10^{-24}\ \text{JT}^{-1} \qquad (1.3)$$

where $e$ is the elctronic charge, $m$ the mass of electron and $c$ the velocity of light.

The value of $g$ depends on the orientation of the molecule containing the unpaired electron with respect to the applied magnetic field. It also depends on the physical state of the sample, *i.e.* gas, liquid or solid. In gas and liquid states, the molecules have free motion and the value of $g$ is averaged over all orientations. In case of a paramagnetic ion or radical situated in a perfectly cubic crystal site, the value of $g$ does not depend on the orientation of the crystal, *i.e.* it is the same in all directions. However, in case of a paramagnetic ion or orientation of the crystal. The values of $g$ along $x$, $y$ and $z$ axes may vary with orientation of the crystal with respect to the applied magnetic field.

The electron spin magnetic moment is about 1000 times greater than that of protons. Thus, the frequency of absorption of electrons is about 1000 times that of protons in the same magnetic field. For example, in a field of 3200 gausses, where NMR absorption would occur at about 14 MHz (in the radio-frequency region), ESR absorption occurs at 9000 MHz, *i.e.* at a much higher frequency which falls in the microwave region.

Similar to that in NMR spectroscopy, the ESR spectra are most commonly recorded by varying the strength of the applied magnetic field $H_0$ and keeping the frequency constant. However, it can also be done by varying the frequency and keeping $H_0$ constant.

### 1.9.2 ESR Absorption Positions: The *g* Factor

Eq. (1.1) shows that an ESR absorption will occur at a frequency $v = \Delta E/h$ Hz. Thus, the position of an ESR absorption may be expressed in terms of absorption frequency. It is clear from Eq. (1.2) that the absorption frequency, *i.e.* the absorption position, varies with the applied field $H_0$. Since different

ESR spectrometers operate at different magnetic fields, it is desirable to express ESR absorption positions in same form independent of the field strength. Thus, the ESR absorption positions are more conveniently expressed in terms of the observed $g$ values. Rearranging Eq. (1.2), we get

$$g = \frac{\Delta E}{\beta H_0} = \frac{hv}{\beta H_0} \qquad ...(1.4)$$

If, for example, an ESR absorption were observed at 8388.255 MHz in a field of 0.30 T, the ESR absorption position would be reported at a $g$ value of 2.0023.

For measuring the $g$ value of free radicals it is convenient to measure the field separation between the center of the spectrum of the unknown sample and that of a reference substance whose $g$ value is accurately known. The most widely used reference is 1, 1-diphenyl-2-picrylhydrazyl free radical (DPPH) which is completely in free radical state and its $g$ value is 2.0036. The reference substance is placed along with the unknown in the same dual resonant cavity.

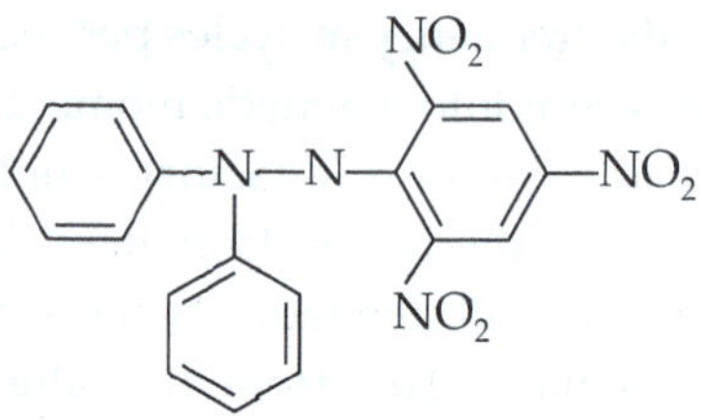

DPPH

Two signals will be observed simultaneously with a field separation of $\Delta H_0$. The $g$ factor for unknown sample is given by where $g_{ref}$ is the $g$ factor for the reference and H the resonance frequency.

$$g = g_{ref}\left(1 - \frac{\Delta H_0}{H}\right) \qquad ...(1.5)$$

$\Delta H_0$ is positive if the unknown has its center at a higher field than the reference.

### 1.9.3 Instrumentation

Since the resonance frequency for unpaired electrons is about 9,000 Hz, which lies in the microwave region, so the source of microwave radiation employed is *klystron tube*, an electronic oscillator in

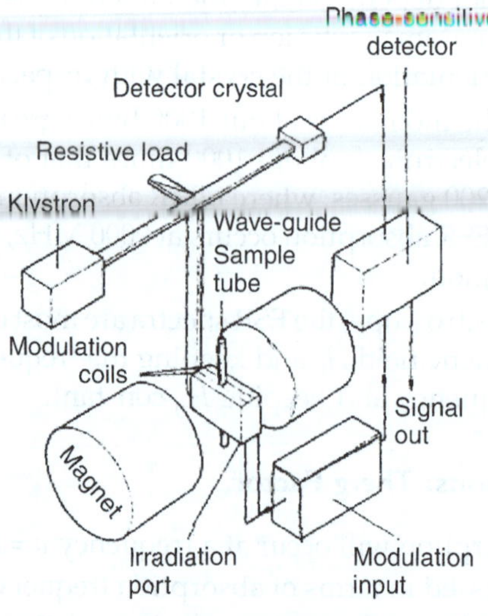


***FIGURE 1.28.*** **The layout of a typical ESR spectrometer.**

which a beam of electron is pulsed between a cathode and reflector. The oscillating output of the klystron is transmitted to a wave guide by a loop of wire, which sets up a *fluctuating magnetic field* (an electromagnetic radiation) in the guide. The wave guide transmits the microwave radiation so-produced to the sample under investigation held in a small *quartz tube* and positioned between the *poles of the permanent magnet*. With the help of *modulation* coils (Helmhotz coils), the variation of the field over the small range in which resonance occurs is achieved. With the help of *photosensitive detector*, the ESR spectra is recorded.

### 1.9.4 Applications of ESR Spectroscopy

ESR spectroscopy is a powerful tool for the study of chemical species with unpaired electrons. The important applications of ESR spectroscopy are summarized as follows.

(*i*) *Study of Free Radicals.* ESR spectroscopy is used to detect the presence of free radicals and to determine their concentration. The method is very sensitive (much more sensitive than NMR) and concentrations as low as $10^{-12}$ M can be detected and estimated under favourable conditions. For example, a signal for DPPH radical can be detected even if there is $10^{-12}$ g of material in the spectrometer. Information concerning the electron distribution (and hence, the structure) in free radicals can be obtained from the splitting pattern of the ESR spectrum (ESR signals are split by nearby protons and other magnetic nuclei).

One of the most exciting uses of ESR spectroscopy is the study of free-radical intermediates in organic reactions. For example, in the oxidation of hydroquinone in alkaline solution by oxygen, the formation of the semiquinone free radical has been conclusively proved by ESR spectroscopy. The ESR spectrum of quinone-hydroquinone redox system contains five lines with their intensities in the ratio 1 : 4 : 6 : 4 : 1. This is due to coupling between the unpaired electron and the four equivalent protons on the ring.

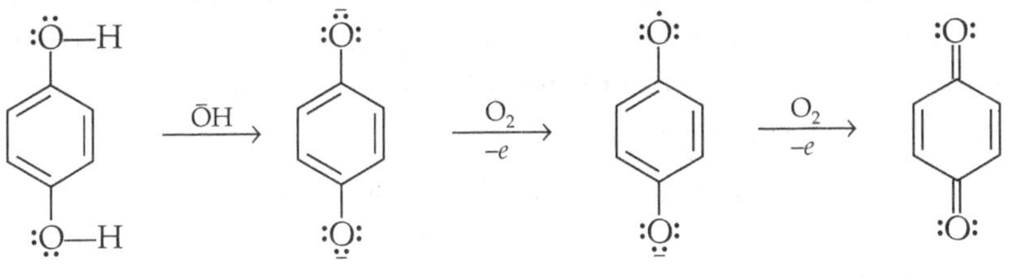


1, 4-Benzosemiquinone

(*ii*) *Reaction Velocities and Mechanisms.* A large number of organic reactions proceed through free-radical mechanism. Such reactions can be studied by ESR spectroscopy. A direct measure of variation of radical concentrations with time during these reactions would be a valuable aid for determining their mechanisms. In cases of reactions which are not extremely fast, rate studies can be simply performed by following the signal intensity of a paramagnetic reactant or product as a function of time.

Very rapid electron exchange reactions can also be studied by ESR spectroscopy. For example, when naphthalene is added to a solution of naphthalene radical anion, an electron exchange reaction occurs which causes a broadening of the hyperfine component of the ESR spectrum. The rate constant for the exchange of electron between naphthalene and naphthalene radical anion can be calculated by making use of this broadening.

(*iii*) *Structure Determination.* ESR spectroscopy is not useful for determining molecular structures because the information obtained from superfine structures is mostly about the extent of electron delocalization in the molecule. It does not tell us about the arrangement of atoms in a molecule. Sometimes, ESR gives useful information about the shape of radicals and about the symmetry of molecules. It is also used to obtain structural information about transition metals and their complexes.

(*iv*) *Effect of Ionizing Radiations on Matter.* Polymers are generally made more suitable for certain purposes by irradiation with X-rays, $\gamma$-rays, stream of electrons, $\alpha$-particles or neutrons. This is because free radicals are formed during the initial bond breaking by the radiation which subsequently react with one another to form links between the chains. Since free radicals are involved in this process, ESR spectroscopy is useful for the study of the effects of radiations on polymers. For example, polyethylene, on irradiation with $\gamma$-rays, exhibits a seven line pattern in its ESR spectrum indicating the formation of the free radical shown sideways.

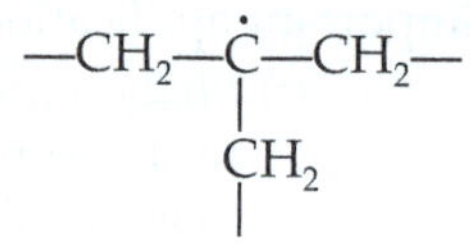

(*v*) *Study of Transition Metals and their Complexes.* The unpaired electrons of the *d* or *f* sub-shells of transition metals give ESR signals. This makes ESR spectroscopy a powerful tool for determining electron distributions in the immediate vicinity of the atom. There are several cases where evidence for electron delocalization into ligands has been thoroughly studied using ESR spectroscopy.

ESR can be used to determine the valence state of transition metals in cases where it is not obvious on chemical grounds. For example, copper (II) complexes ($d^9$ system) give a strong ESR signal, whereas copper (I) complexes ($d^{10}$ system) give a much reduced signal. This difference between the ESR signals for two oxidation states of copper has been successfully used to determine the valence state of copper in many complexes and biologically active compounds. For example, ESR studies have shown that copper protein complex has copper in divalent state, whereas it is in monovalent state in some other biologically active copper complexes.

(*vi*) *Analytical Applications.* ESR spectroscopy is used for the determination of various transition metal ions like $Mn^{2+}$, $V^{4+}$, $Cu^{2+}$, $Cr^{3+}$, $Cd^{3+}$, $Fe^{3+}$, $Ti^{3+}$ etc. For example, the ESR spectrum of a solution of $Mn^{2+}$ ions exhibits six lines. The multiplicity is given by $2I + 1$, *i.e.* $2 \times \frac{5}{2} + 1 = 6$, where $I$ for $Mn^{2+}$ ions is $\frac{5}{2}$. This accounts for the six lines in the spectrum. Thus, $Mn^{2+}$ ions can be measured and detected readily even when present in trace quantity. The method provides a very rapid measurement in aqueous solutions over the range from $10^{-8}$ to 0.1 M.

ESR technique is also used for the estimation of polynuclear hydrocarbons. These are first converted into radical cations and then absorbed in the surface of an activated silica-alumina catalyst for the estimation.

(*vii*) *Biological Applications.* The following are some examples of successful applications of ESR spectroscopy in biological systems :

(*a*) ESR spectroscopy has demonstrated the presence of free radicals in healthy and diseased tissues.

(b) ESR studies of a variety of biological systems like leaves, seeds and tissue preparations have revealed that there is correlation between the concentration of free radicals and the metabolic activity of the material.

(c) ESR studies have shown that most of the oxidative enzymes function via one-electron redox reactions involving the production of either enzyme-bound free radicals or by change in the valence state of a transition metal ion.

(d) ESR studies on photosynthesis have been generally performed with photosynthetic bacteria. These bacteria on irradiation with near infrared radiation (700-900 nm), show a single line with $g = 2.0025$ and $\Delta H = \sim 10$ G in their ESR spectra. The $g$ factor (2.0025) is characteristic of an organic free radical arising from a large conjugated system. The oxidation of bacteriochlorophyl *in vitro* exhibits an ESR signal which closely resembles that of the irradiated photosynthetic bacteria.

(e) Biological or other compounds having no unpaired electron can also be studied by ESR when they are chemically bonded to a stable free radical. This free radical (spin label) produces a sharp and simple ESR spectrum which gives detailed information about the molecular environment of the *spin label.*

### Comparison Between NMR and ESR Spectroscopy

Following is the comparison between NMR and ESR spectroscopy:

| *NMR spectroscopy* | *ESR Spectroscopy* |
|---|---|
| (*i*) Different energy states are produced due to the alignment of the nuclear magnetic moments relative to the applied magnetic field and a transition between these energy states occurs on the application of an appropriate frequency in the radio frequency region. | (*i*) Different energy states are produced due to the alignment of the electronic magnetic moments relative to the applied magnetic field and a transition between these energy states occurs on the application of an appropriate frequency in the microwave region. |
| (*ii*) NMR absorption positions are expressed in terms of chemical shifts. | (*ii*) ESR absorption positions are expressed in terms of g values. |
| (*iii*) Nuclear spin-spin coupling causes the splitting of NMR signals. | (*iii*) Coupling of the electronic spin with nuclear spins (hyperfine coupling) causes the splitting of ESR signals. |

### 1.9.5 Problems on ESR Spectroscopy

**Problem 52.** *Predict the number of lines in the ESR spectrum of each of the following radicals:*

(a) $[CF_2H]^\bullet$  (b) $[^{13}CH_3]^\bullet$  (c) $[CF_2D]$

(d) $[CClH_2]^\bullet$  (e) $[^{13}CF_2H]^\bullet$.

**Solution.** The number of lines in an ESR signal $= (2nI_i + 1)(2mI_j + 1)$ ...(1.6)

where $n$ and $m$ are the number of different kinds of coupled nuclei having spin $I_i$ and $I_j$, respectively.

(a) The spin of F $= \frac{1}{2}$, *i.e.* $I_i = \frac{1}{2}$ and H $= \frac{1}{2}$, *i.e.* $I_j = \frac{1}{2}$; $n$ and $m$ for the radical $[CF_2H]^\bullet$ are 2 and

1, respectively.

According to Eq. (1.6)

$$(2 \times 2 \times \tfrac{1}{2} + 1)\ (2 \times 1 \times \tfrac{1}{2} + 1) = 6 \text{ lines}$$

(*b*) $[^{13}CH_3]^{\bullet}$

Here $n = 1, I_i = \frac{1}{2}; m = 3, I_j = \frac{1}{2}$

According to Eq. (1.6)

$$(2 \times 1 \times \tfrac{1}{2} + 1)\ (2 \times 3 \times \tfrac{1}{2} + 1) = 8 \text{ lines}$$

(*c*) $[CF_2D]$

Here $n = 2, I_i = \frac{1}{2}; m = 1, I_j = 1$ (spin of $D = 1$)

According to Eq. (1.6)

$$(2 \times 2 \times \tfrac{1}{2} + 1)\ (2 \times 1 \times 1 + 1) = 9 \text{ lines}$$

(*d*) $[CClH_2]^{\bullet}$

Here $n = 1, I_i = \frac{3}{2}$ (spin of Cl $= \frac{3}{2}$); $m = 2, I_j = \frac{1}{2}$

According to Eq. (1.6)

$$(2 \times 1 \times \tfrac{3}{2} + 1)\ (2 \times 2 \times \tfrac{1}{2} + 1) = 12 \text{ lines}$$

(*e*) $[^{13}CF_2H]^{\bullet}$

Here $n = 1, I_i = \frac{1}{2}; m = 2, I_j = \frac{1}{2}; o = 1, I_j = \frac{1}{2}$

According to Eq. (1.6)

$$(2 \times 1 \times \tfrac{1}{2} + 1)\ (2 \times 2 \times \tfrac{1}{2} + 1)\ (2 \times 1 \times \tfrac{1}{2} + 1) = 12 \text{ lines}$$

**Problem 53.** *Which of the following will show ESR spectra and why ?*

(*a*) *Benzene,* $C_6H_6$ (*b*) *Benzene anion,* $C_6H_6^-$

(*c*) *Cyclopentadienyl cation,* $C_5H_5^+$

**Solution.** (*a*) Benzene, $C_6H_6$

From Hückel molecular-orbital (HMO) calculations on benzene, the π-electron distribution may be represented as

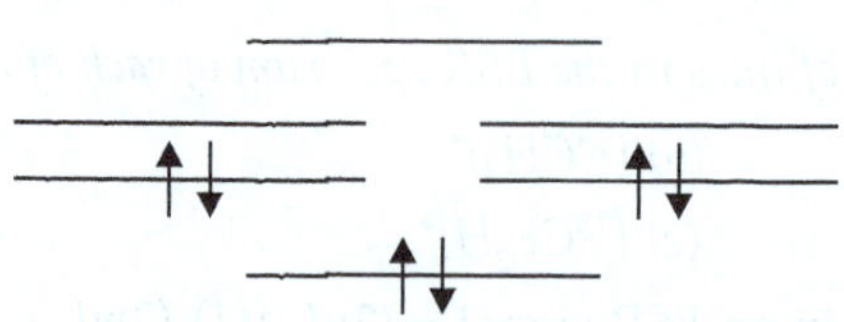

Since there is no upaired electron, benzene will not show ESR spectrum.

(*b*) Benzene anion, $C_6H_6^-$

From HMO calculations on $C_6H_6^-$, the π-electron distribution can be represented as

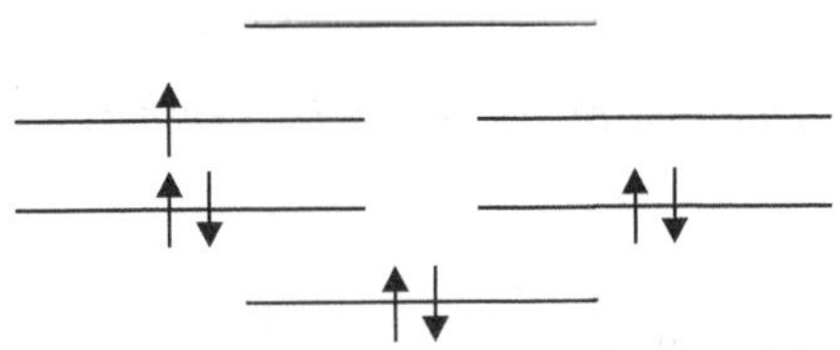

Since there is one unpaired electron, $C_6H_6^-$ will show ESR spectrum.

(*c*) Cyclopentadienyl cation, $C_5H_5^+$

The π-electron distribution in the molecular orbitals of $C_5H_5^+$ is as follows:

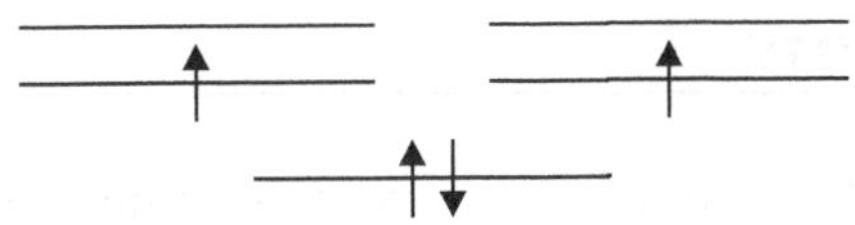

Since there are two unpaired electrons, $C_5H_5^+$ will show ESR spectrum.

**Problem 54.** *Which of the following will show electron spin resonance (ESR) spectrum ? Give reason for your choice.*

(*a*) *H* (*b*) $H_2$ (*c*) *Cl*

(*d*) $Na^+$ (*e*) $Cu^+$

**Solution.** (*a*) Hydrogen atom (H) has electronic configuration $1s^1$, *i.e.* it has one upaired electron. Thus, it will show ESR spectrum.

(*b*) Hydrogen molecule ($H_2$) has electronic configuration $(\sigma_{1s})^2$, *i.e.* it has no unpaired electron. Thus, it will not show ESR spectrum.

(*c*) Chlorine atom (Cl) has electronic configuration $1s^22s^22p^63s^23p_x{}^23p_y{}^22p_x{}^1$, *i.e.* it has one unpaired electron. Thus, it will show ESR spectrum.

(*d*) The electrronic configuration of $Na^+$ is $1s^22s^22p_x{}^22p_y{}^22p_z{}^2$. Since it has no unpaired electron, it will not show ESR spectrum.

(*e*) The electronic configuration of cuprous ions ($Cu^+$) is $3d^{10}$. Since it has no unpaired electron, it will not show ESR spectrum.

**Problem 55.** *Calculated the ESR frequency of an unpaired electron in a magnetic field of 3000 G (0.30 T).*

**Solution.** From Eq. (1.2), we have

$$\Delta E = hv = g\,\beta H_0$$

or

$$v = \frac{g\beta H_0}{h}\ \text{Hz}$$

where $g = 2.00$, $\beta = 9.273 \times 10^{-24}\ \text{JT}^{-1}$, $H_0 = 0.30$ T and $h = 6.626 \times 10^{-34}$ Js

Therefore
$$v = \frac{2.00 \times 9.273 \times 10^{-24}\ \text{JT}^{-1} \times 0.30\text{T}}{6.626 \times 10^{-34}\ \text{Js}}$$

$$= 8.397 \times 10^{9}\ \text{Hz} = 8397\ \text{MHz} = 8.397\ \text{kMHz}$$

**Problem 56.** *Calculate the g value if methyle radical shows ESR signal at 3290 G (0.3290 T) in a spectrometer operating at 9230 MHz.*

**Solution.** From Eq. (1.2)

$$hv = g\beta H_0$$

Hence,
$$g = \frac{hv}{\beta H_0} = \frac{(6.626 \times 10^{-34}\ \text{Js})\,(9.230 \times 10^{9}\ \text{Hz})}{(9.273 \times 10^{-24}\ \text{JT}^{-1})\,(0.3290\ \text{T})} = 2.004641$$

## 1.10 CHROMATOGRAPHY

The term 'Chromatography' (*chroma-a* colour ; *graphien*-to write) was coined by a Russian scientist Michael S. Tswett in 1906 to a procedure where a mixture of different coloured pigments is separated from each other. He used a column of $CaCO_3$ to separate the various components of petroleum ether chlorophyll extract into green or yellow zones of pigments. He termed such preparation as *chromatogram* and the procedure as *chromatography*. It includes all those techniques in which various components in a mixture get separated during their passage through a porous media due to differences in their migration rates. Chromatography may be defined as, *the technique of separation of substances according to their partition co-efficients in two immiscible phases.* The partition coefficient (*K*) is defined as, *the ratio of concentration of a compound in two phases (i.e., A* and *B) at equlibrium.* Hence, when both the phases are liquids.

$$K = \frac{\text{Concentration of the compound in solvent } A}{\text{Concentration of the compound in solvent } B}$$

Different compounds vary in their partition coefficients in a given biphasic system and this constitutes the underlying principle for their separation in the different types of chromatographic techniques. In this method, the separation of the components of a mixture is a function of their different affinities for a fixed or *stationary phase* (solid or liquid) and their differential solubility in a moving or *mobile phase* (liquid or gas). Separation starts when one component is held more firmly by the stationary phase than the other which tends to move on faster in the mobile phase. Thus, in chromatography the first step is to absorb the component of a mixture on an insoluble material and then to differentially remove (elute) these components one by one with suitable liquid solvents. This eventually results in the separation of various components in the mixture. The separation column is the heart of the chromatograph. It provides versatility in the types of analyses that can be performed. This versatility, due to the wide choice of materials for the stationary and mobile phases, makes it possible to separate molecules that differ only slightly in their physical and chemical properties. Broadly speaking, the distribution of a solute between two phases results from the balance of forces between solute molecule and the molecules of each phase. It reflects the relative attraction or repulsion that molecules or ions of the competing phases show for the solute and for themselves. These forces can be polar in nature, arising from permanent or induced dipole moments, or they can be due to London's dispersion forces.

### 1.10.1 Classification of Chromatographic Techniques

In chromatography, one phase is held immobile or stationary and the other one is passed over it (the mobile phase). The mobile phase can be gas or a liquid, whereas, the stationary phase can be only a liquid or solid. The designations GC and LC refers to the physical state of the mobile phase. Further classification can be made by naming the mobile and stationary phases. Thus, we have gas-solid (GS), gas-liquid (GL), liquid liquid (LL), and Liquid-solid (LS) chromatography. Recently, supercritical fluids have been used as mobile phases and these techniques are known as *supercritical fluid chromatography* (SFC) irrespective of the state of the stationary phase (Fig. 1.29). In Fig. 1.29, besides the state of the two phases, the configuration of chromatographic bed is also described. There are two popular configurations for the bed, a column and a planar surface. When the separation involves a simple partitioning between two immiscible liquid phases, one stationary and the other mobile, the process is called *liquid-liquid chromatography* (*LLC*). When physical surface forces are mainly involved in the relative ability of the stationary phase, the process is called *liquid-solid chromatography* (*LSC*). When ionic components of the sample are separated by selective exchange with counterions of the stationary phase, the process is called *ion-exchange chromatography* (*IEC*). The use of exclusion packings as the stationary phase brings about a classification of molecules based largely on molecular geometry and size, this is called *exclusion chromatography* (*EC*). EC is referred to as gel permeation chromatography by polymer chemist and as gel filtration by biochemists.

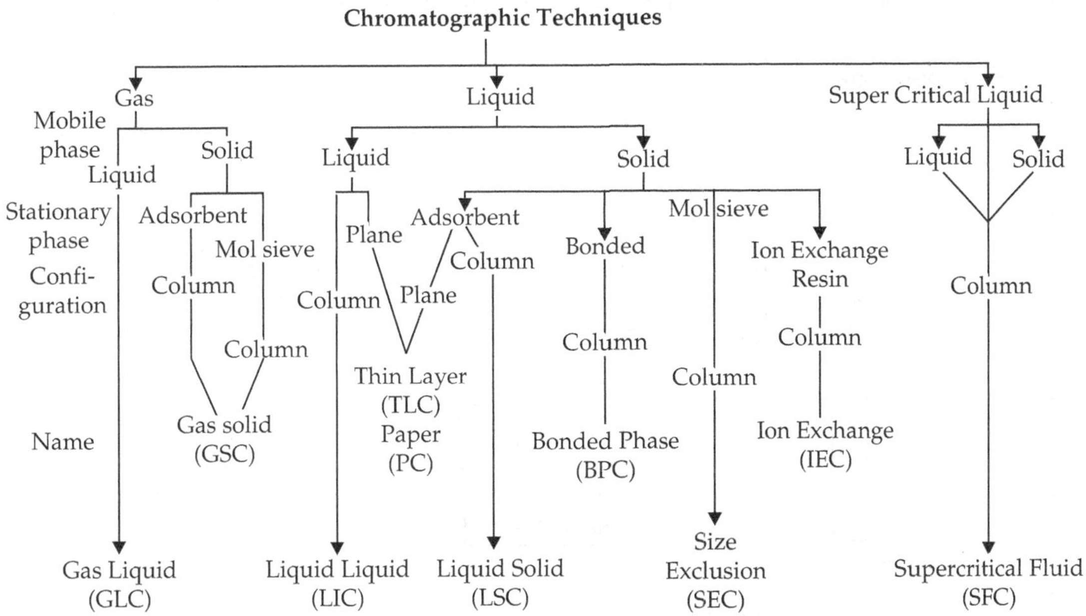


*FIGURE 1.29.* **Classification of Chromatographic Techniques.**

When the mobile phase is gas, the method is called *gas-liquid chromatography* (*GLC*) and *gas-solid chromatography* (*GSC*). These chromatographic techniques are discussed in detail in coming sections.

### 1.10.2 Gas Chromatography

#### Introduction and Principle

Gas chromatography (GC) is particularly suited for the separation of gases and volatile liquids or solids in their gaseous state. The compounds of low polarity are best separated by gas chromatography. The technique is highly sensitive, reproducible and has high speed of resolution. When the stationary phase is an active solid such as silica, the method is referred to as *gas solid chromatography* (GSC). However, if the stationary phase is a liquid such as polymers of silicone coated onto the surface of an inert granular solid then the technique is known as *gas-liquid chromatography* (GLC). The availability of versatile and specific detectors and the possibility of coupling the gas chromatograph to a mass spectrometer or an infrared spectrophotometer further enhance the usefulness of gas chromatography.

The stationary phase, whether a solid or a liquid coated as a thin film on surface of a solid support, is packed in a glass or stainless steel column which is narrow, coiled, 1 to 3 m long and with 2 – 4 mm internal diameter. An inert carrier gas (mobile phase) such as nitrogen, helium or argon is made to flow through the column. The temperature of the column is maintained high in an oven to keep the compounds to be separated in their volatile state. These volatilized compounds get partitioned between the liquid or solid stationary phase and the gaseous mobile phase and hence get separated because of differences in their partition coefficients. After leaving the column, the separated compounds pass through the detector, sensed and recorded by the recorder.

#### Gas Liquid Chromatography

The essential components of a gas chromatography equipment, are shown in Fig. 1.30 and may be described as :

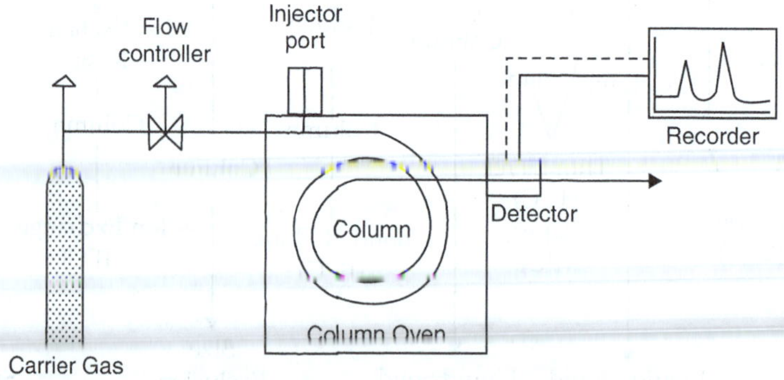


*FIGURE 1.30.* **Components of a gas chromatographic system.**

**Carrier gas.** The primary function of the carrier gas is to carry the volatile components through the column. The gas used should be inert and should not react either with sample or with stationary phase. Its secondary purpose is to carry the separated components to the detector, so it should be suitable for detector use. It should be readily available in extra pure form and should be inexpensive.

Normally, nitrogen, helium and argon are the three most commonly used carrier gases. Selecting the proper carrier gas is very important because it affects both the column and detector performance. A Van Deemter plot for the three most common carrier gases is shown in Fig. 1.31. Virtually, the same minimum plate height is achieved with each gas. The difference arises in the optimum linear velocity. Although, lower for nitrogen, the curves for helium and hydrogen are flatter. This means little loss in efficiency when linear velocities higher than the optimum are used. The use of higher velocities can shorten the analysis time significantly.

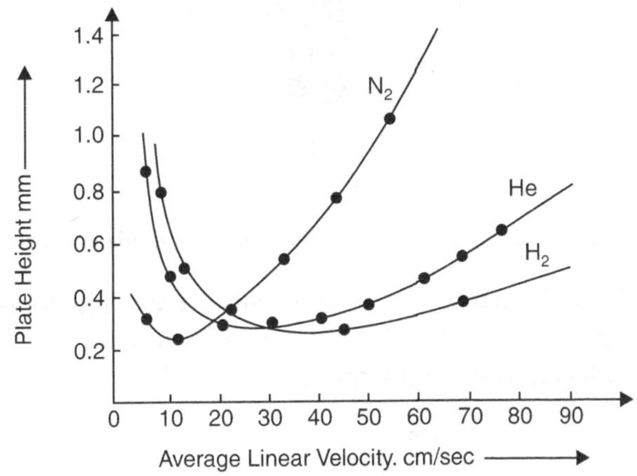


*FIGURE 1.31.* **Van Deemter graph for various carrier gases flowing through a WCOT glass column coated with OV-101. Column temperature is 175°c.**

The viscosity dictates the inlet column pressure. The ratio of carrier gas viscosity to the diffusion coefficients of the sample components should be as small as possible. In this respect, hydrogen is the best choice, followed by helium.

The purity of the carrier gas should be at least 99.995%. Impurities such as oxygen or water can cause column and detector deterioration. In temperature programming, impurities can impair the baseline stability of the chromatogram.

The column temperature must be high enough so that analysis can be accomplished in a reasonable length of time. The retention time doubles with every 30°C decrease in column temperature. Lower the temperature better is the resolution and longer analysis time. Therefore, a balance has to be struck between the peak retention time and resolution. Chromatographic separation can be achieved *isothermically*, where a constant temperature is employed or by *temperature programming*, where the temperature is increased gradually.

**Sample injection port.** The schematic diagram of a typical flash vaporizer injector port is shown in Fig. 1.32. The sample is injected by a hypodermic syringe through a rubber septum into a flash vaporizer port at the head of the column.

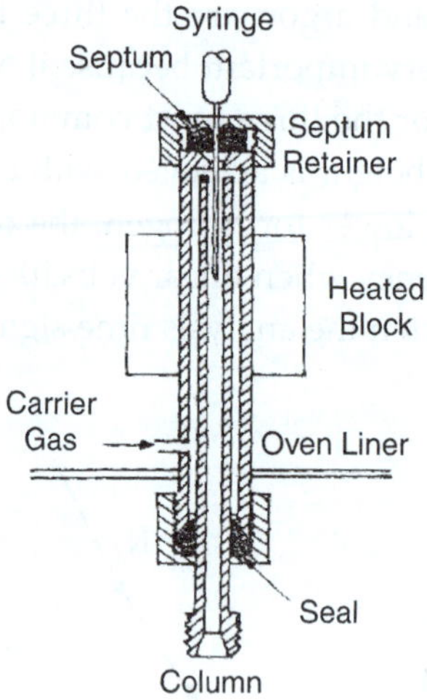


*FIGURE 1.32.* **Schematic diagram of a typical vaporizer injector port.**

The metal block is heated at a fixed temperature sufficiently high to convert virtually instantly the liquid sample into a "plug" of vapour. Samples, both liquid and gas, can be measured with a calibrated loop, then introduced into the flowing gas stream by means of a valve, (Fig. 1.33).

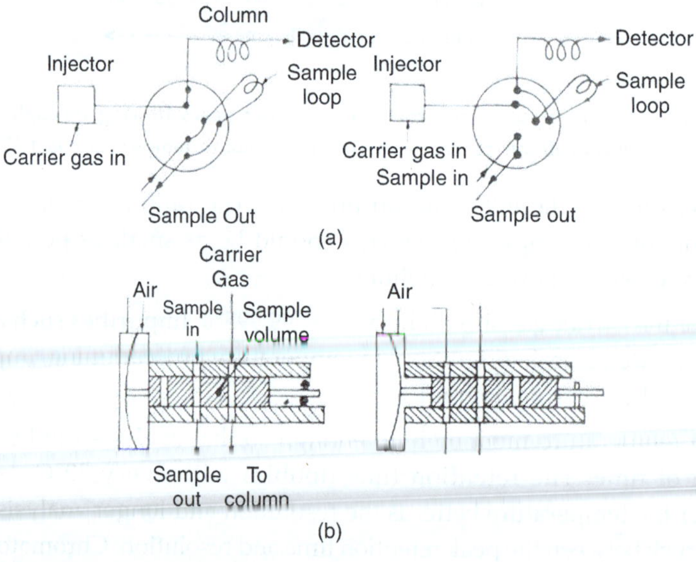


*FIGURE 1.33.* **(*a*) Six-port rotary valve and (*b*) sliding plate value.**

Reduction in sample volume is necessary when working with capillary columns. This is accomplished by an injector-splitter, where typically a 1 μL sample is injected by only 0.01 μL enters the capillary the remainder is vented. This technique prevents column overload but wastes a significant portion of the sample.

When analyses are performed on small amounts of sample with some component concentrations in the parts-per-billion range, too little material is placed on the column if splitting is used for these samples. Splitless injection is required for these samples. The entire sample, including any solvent, is injected onto the open tubular column through a modified heated flash vaporizer. A larger solvent tail avoided by venting the injection port to the atmosphere at that time (perhaps 30 second) when most of the solvent and essentially all the sample have entered the column. The proper time until venting is critical, too short a time causes loss of sample components, whereas too long a time causes a solvent peak larger than necessary which might bury some of the peaks of interest.

**Sample preparation and its application.** The sample should be prepared in such a way that it does not get retained on the column for excessive period of time. This will lead to poor resolution and peak tailing. The polar groups such as $-NH_2$, $-COOH$, $-OH$ are derivatized by methylation, silanisation and trifluoro-methylsilanisation to increase the volatile character and distribution coefficients of the compounds.

Derivatization prior to gas chromatography is often desirable to :

(*i*) Improve the thermal stability of compounds, particularly compounds that contain polar functional groups.

(*ii*) Change the separation properties of compounds by the purposeful adjustment of their volatility.

(*iii*) Introduce a detector-oriented tag into a molecule.

In an ideal derivatization, relevant functional groups should be derivatized quantitatively and quickly, preferably in less than 10 min. Then, without further processing, the reaction mixture can be injected directly into the gas chromatographic system. Separation of the derivatives should be complete and excess reagent should elute with the solvent peak. A few typical examples of derivative formation are given below :

Silylating agents, such as N, *o*-bis (trimethylsilyl) acetamide or the trifluoro analog, N, *o*-bis(trimethylsilyl) trifluoroacetamide, convert one active hydrogen in polar groups such as $-OH$, $-COOH$, $-NH_2$, $=NH$, and $-SH$ to an $-O-Si(CH_3)_3$ group.

A typical silylation reaction with an alcohol is shown :

$$H_3C\overset{\overset{\displaystyle O}{\|}}{C}N\,[Si\,(CH_3)_3]_2 + -\overset{|}{\underset{|}{C}}-OH \longrightarrow -\overset{|}{\underset{|}{C}}-OSi(CH_3)_3 + CH_3\,\overset{\overset{\displaystyle O}{\|}}{C}NHSi\,(CH_3)_3$$

The use of perdeuterated derivatives, such as N, *o*-bis (trimethylsilyl-$d_{18}$) accetamide, assists materially in interpreting the mass spectra of silylated compounds.

(*ii*) *o*-Alkylhydroxylamines are used to prepare *o*-alkyloximes of aldehydes and ketones :

$$R_1\,R_2\,C=O + H_2N-O\,(CH_2)_3CH_3 \xrightarrow{\text{Pyridine}} R_1\,R_2\,C=N\text{—}O\,(CH_2)_2CH_3 + H_2O$$

The reagents are particularly useful in the analysis of compounds that contain both hydroxyl and carbonyl groups.

(*iii*) The trifluoro group, such as trifluoroacetyl, pentafluoropropionyl and pentafluorobenzoyl, is commonly used for sensitizing substances to detection by electron capture.

(*iv*) A derivatization method makes feasible the quantitative and qualitative analysis of amino acids. After they are dry, amino acids are converted to butyl ester hydrochlorides by treatment with 1-butanol and hydrogen chloride at 100°C for 30 min :

$$\underset{\displaystyle\overset{|}{NH_2}}{RCHCOOH} + C_4H_9OH + HCl \longrightarrow \underset{\displaystyle\overset{|}{NH_3^+Cl^-}}{RCHCOOC_4H_9} + H_2O$$

The butyl esters are converted to volatile N-trifluoroacetyl butyl esters in sealed tubes :

$$(CF_3CO)_2O + \underset{\displaystyle\overset{|}{NH_3^+\,Cl^-}}{RCHCOOC_4H_9} \longrightarrow \underset{\displaystyle\overset{|}{NHCOCF_3}}{RCHCOOC_4H_9} + HCl + CF_3COOH$$

Subsequent chromatographic resolution into single peaks by temperature programming is then easily accomplished.

Solvents such as ether, heptane or methanol are used to dissolve the sample which is then injected with the help of a microsyringe onto the column through a rubber septum in the injection port. The temperature of the injection port is generally maintained higher than the temperature of the column to ensure rapid and complete volatilization of the sample. As a thumb rule, the temperature of the injection port should be 50°C higher than the boiling point of the sample. Too high temperature of the injection port may decompose the sample. Therefore, the temperature of the injection port should be such that it causes rapid vaporization of the sample without decomposing it.

**Column packing.** There are two general types of column, *packed* and *capillary* (also known as *open tubular*).

*Packed columns.* Packed columns are constructed from tubing of stainless steel, nickel or glass. Inner diameters may range from 1.6 to 9.5 mm. Length is often 3 m. These columns are packed with an inert support, usually a diatomaceous earth whose internal pore diameters range from 2 μm for material derived from firebrick to 9 μm for materials derived from filter aids. The internal column diameter should be at least eight times the diameter of the support particles. For example, the best particle size is 100/120 mesh (149 – 125 μm) for 2-mm-bore columns, 80/100 mesh (177 – 149 μm) for 4-mm-bore columns.

Special packing material may be needed for particular applications. Very lightly loaded (roughened or texturized) glass beads are used for very rapid analyses well below the boiling point of the sample components. Sieved teflon supports are used when corrosive substances are handled. In gas-solid chromatography the packing material is an adsorbent, such as silica gel, a bonded-phase support, or a molecular sieve.

*Capillary columns*. These columns are usually constructed of fused silica, having much higher degree of crosslinking within the silicon-oxygen matrix. There are generally two types of capillary columns :

(*i*) *Wall coated open tubular* (WCOT) columns, consisting of a capillary tube, whose walls are coated with liquid stationary phase.

(ii) *Support coated open tubular* (SCOT) columns, in which the inner wall of the capillary is lined with a thin layer of support material such as diatomaceous earth, onto which the stationary phase has been adsorbed.

The inner surface of the wide-bore (WCOT) capillary column (0.53 mm i.d.) is coated with the stationary liquid phase. Uniform coating of the surface was difficult to obtain until bonded phase (polymer) columns were introduced. The polymer wets the surface of a fused silica column well, with the result that a very uniform film covers the column walls. A variety of functional groups can be blended into the polysiloxane chain to provide stationary phases of different polarity or selectivity.

The result is a film that is both thermostable and non-extractable. The columns can be flushed with pure solvents to remove contaminants, non-volatiles and pyrolysis products.

The major advantage of wide-bore capillaries regardless of the separation is an increase in the speed of analysis, (Fig. 1.34). Using packed column flow rates, the wide-bore column produces separations with packed column efficiencies but with a total run time roughly three times faster. However, when the flow rate is optimized for the tubing diameter, the wide-bore column produces far superior efficiencies and therefore superior separations, with analysis time approximately equal to those for packed columns. Thus, there are six extremely good reasons for changing from packed columns to wide-bore capillaries ; shorter retention times, greater inertness, longer life, lower bleed, higher efficiencies and greater reproducibility.

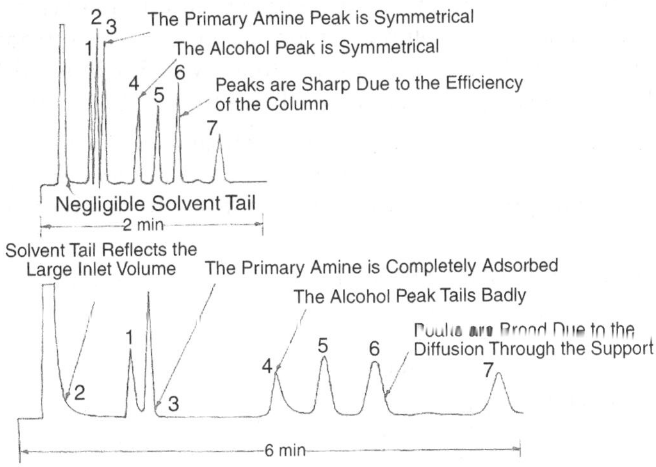


**FIGURE 1.34.** **Comparison of chromatograms obtained on a wide-bore column (upper) and a packed column. Peaks : (1) 4-chlorophenol, (2) do-decane, (3) 1-decylamine, (4) 1-undecanol, (5) tetradecane, (6) acenaphthene, and (7) pentadecane. (from J & W Scientific, Inc.)**

**Solid support and stationary phase.** The purpose of the solid support is to provide a large uniform, inert surface area for holding a thin layer of the liquid stationary phase. The support should be inert, and should have high mechanical strength, large surface area, regular shape and uniform size. The most commonly used support is celite, the OH groups of which are modified by silanization with hexamethyl disilazane to minimize interaction with the sample.

The function of the sationary liquid phase is to separate the sample components into the discrete peaks.

The correct choice of the stationary phase is perhaps the most important parameter in GC. Ideally, the stationary phase must be non-volatile and thermostable at temperature used for analysis. It should be chemically inert towards the solutes of interest at the column temperature. The high boiling point organic compounds such as polyethylene glyscols, methyl-phenyl and methyl-vinyl silicon gums, esters of adipic, succinic and phthalic acid, polyesters, polyethylene glycols are used as liquid stationary phases. The operating temperature must be compatible with the phase chosen. At very high temperature, the phase may get volatilized and cause excessive column bleeding which may contaminate the detector.

**Retention indices.** It is the measure of the degree to which polar compounds are retarded relative to their elution on a non-polar phase. This retention data can be expressed in a systematic method by *Kovats retentions indices* (RI). These indices indicate where compounds will appear on a chromatogram with respect to straight chain alkanes injected with the sample. By definition, the RI for a normal paraffin is 100 times the number of carbon atoms in the compound regardless of the columns used or the chromatographic conditions. Thus, the RI for pentane is 500, for hexane 600, for hepatne 700, and so on. RI is given by the equation, $RI = a + b \text{ In } t_R'$. The type of column and the operating conditions, such as liquid loading and any pretreatment, must be specified. For example, suppose that on a squalane column the retention times for hexane, 1-nitropropane, and octane were 15, 16 and 25 min.,respectively. On a graph of $ln\ t_R'$ of the alkanes versus their retention indices, a RI of 652 for 1-nitropropane is off the graph. The number 652 for 1-intropropane means that it elutes halfway between hexane and heptane on a logarithmic time scale. When the experiment with an OV-275 (dicyanoallylsilicone) column is repeated, the RI for 1-nitropropane is found to be 1177, between undecane and dodecane. This implies that OV-275 will retard 1-nitropropane much more than will squalane ; *i.e.*,OV-275 is more polar than squalane by 525 units.

Within a homologous series, a plot relating logarithm of the adjusted retention time (or volume) to the number of carbon atoms is linear, provided the lowest member of the series is excluded. Members of homologous series are easily identified from a plot of $ln(t_R - t_M)$ versus *the RI of the paraffins.*

**Detectors.** A detector is located at the exit of the separation column, it senses the presence of the individual components as they leave the column. The detector volume must be small to prevent

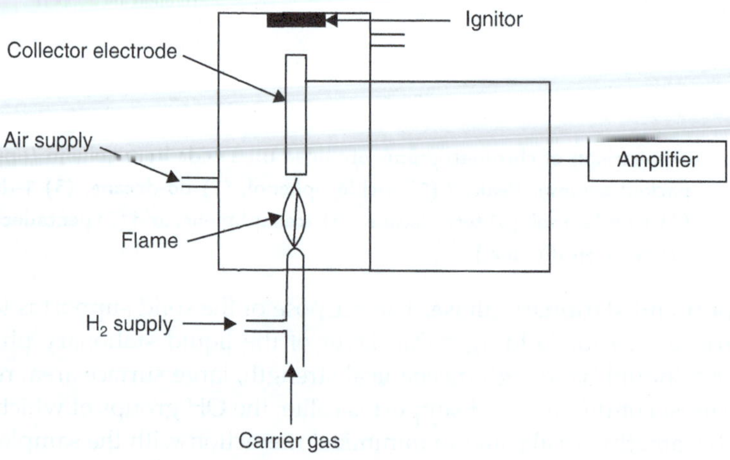


***FIGURE 1.35.*** **Flame ionization detector.**

the remixing of components separated on the columns. The choice of detector should be based on characteristics such as selectivity, sensitivity, response, noise, minimum detactable quantity and linear range. The detector should be simple to operate, inexpensive and as far as possible insensitive to changes in flowrate and temperature. Summary of common GC detectors is given in Table 1.12.

**Flame ionization detector (FID).** It is the most widely used detector as it responds to all organic compounds. It has a wide linear range ($10^6$) and can detect as low a concentration as 1 ng. The detector consists of two electrodes. One of the electrodes is the jet of the flame which is produced by introducing a mixture of hydrogen and air into the detector, while the other electrode is made of brass or platinum wire which is mounted near the tip of the flame. When the carrier gas carrying the sample components emerges from the column, the sample components are ionized in the flame, sensed by the detector giving the signal which is recorded by the recorder. An FID has an upper temperature limit of 400°C and the minimum quantity which can be detected by this detector is of the order of $5 \times 10^{-12}$ g. A typical FID is shown diagramatically in Fig. 1.35.

**TABLE 1.12. Summary of Common GC Detectors.**

| *Detector* | *Type* | *Support gases* | *Selectivity* | *Detectability* | *Dynamic range* |
|---|---|---|---|---|---|
| Flame ionization (FID) | Mass flow | Hydrogen and air | Most organic compounds | 100 pg | $10^7$ |
| Thermal conductivity (TCD) | Concentration | Reference | Universal | 1 ng | $10^7$ |
| Electron capture (ECD) | Concentration | Make-up | Halides, nitrates nitriles, peroxides, anhydrides, organometallics | 50 fg | $10^5$ |
| Nitrogen phosphorus | Mass flow | Hydrogen and air | Nitrogen, phosphorus | 10 pg | $10^6$ |
| Flame photometric (FPD) | Mass flow | Hydrogen and air possibly oxygen | Sulphur, phosphorus, tin, boron, arsenic, germanium, selenium, chromium | 100 pg | $10^3$ |
| Photo ionization (PID) | Concentration | Make-up | Aliphatics, aromatics, ketones, esters, aldehydes, amines, heterocyclics, organosulphurs, some organometallics | 2 pg | $10^7$ |
| Hall electrolytic conductivity | Mass flow | Hydrogen, oxygen | Halide, nitrogen nitrosamine, sulphur | | |

**Amplifiers and recorders.** When components leave the column and pass through the detectors discussed above, small and weak electrical signals are produced which are amplified by an amplifier before they are fed to the recorder. Recorders generally consist of two basic parts, *viz.*, a chart paper and a pen which moves on the chart paper and traces the signals being activated from the amplifier in form of peaks.

### Gas Solid Chromatography (GSC)

In GSC, the columns are packed with interactive solids such as molecular sieves or porous polymers through which the carrier gas flows. Sample injection is by means of a gas sample loop or value. Normal operations involve columns 1-4 m long and temperatures in the range 40 – 90°C. Inorganic molecular sieves are naturally occuring or synthetic zeolites. They are frameworks of aluminium and silicon with oxygen that consist of interconnected cavities with precisely uniform openings. Only molecules smaller than these openings are able to enter the cavities, Pore size is very uniform and is dependent on the cation used. Zeolites act as adsorbents but differ from other adsorbents in that they take up molecules that fit into their micropores. A specific affinity based on molecular size and shape is obtained. In addition, normal adsorption based on the physicochemical structure of the components occurs in the micropores as well as in the macropores outside the sieves. Adsorption continues until all the pores are filled. Raising the temperature will displace the molecules from the pores.

Column packings may be porous polymers analogous to the materials used in exclusion chromatography. Those made from copolymers of aromatic hydrocarbons provide packings of low to moderate polarity. Polymers made from acrylic esters provide packings of moderate to high polarity. Most gases and low-boiling point liquids can be separated on one or more of these porous polymers within the temperature range 50 – 250°C.

In GSC separation the major drawback is that almost all gas mixtures contain some components that cannot be separated on a particular column or cannot pass through a column in a reasonable time. This can be overcome by *backflushing technique*. For example, in the trace analysis of hydrocarbons in natural gas, interest is usually centred on the $C_1$ to $C_5$ hydrocarbons. Consequently, after the pentane peak has eluted from the column, the carrier gas flow is reversed so as to enter the column exit. The $C_6$ and higher hydrocarbons are back flushed from the upper end of the column through the detector now positioned at the entrance to the column. With proper programming, the column can be backflushed before any component has eluted. This provides a sharper initial combined peak of $C_6$ and higher hidrocarbons for more precise quantization (Fig.1.36).

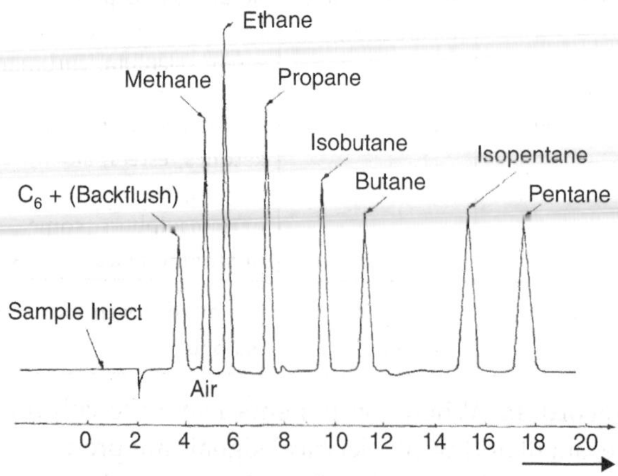


*FIGURE 1.36.* **Chromatogram of natural gas liquids. Two minutes after sample injection, the $C_6$ and higher hydrocarbons are back flushed. The forward carrier gas flow is resumed after 2 min of back flushing.**

## QUESTION BANK

### UV Spectroscopy

1. Describe Beer's law and define all the terms. What are the reasons for the deviation from Beer's law ?
2. Describe the components of a spectrophotometer and discuss the use of each component.
3. Predict the transition involved in the following compounds:

   (*a*) Methyl chloride (*b*) Methyl iodide (*c*) Methyl alcohol (*d*) Trimethyl amine.
4. State Beer's law. If the path length of a beam of light through the sample is doubled and the concentration is made half, what will be the change in the value of absorbance?
5. How can 1, 3-pentadiene and 1, 4-pentadiene be distinguished by UV spectroscopy?
6. Discuss various types of electronic transitions in UV spectroscopy.
7. Name two molecules which undergo $n \rightarrow \pi^*$ electronic transition. Discuss the effect of solvents on these transitions.
8. Assign the figures shown below to the respective isomer on the basis of the following UV spectral information:

   α-isomer shows a peak at 228 nm (ε = 1400) while β-isomer has a band at 296 nm (ε = 1100)

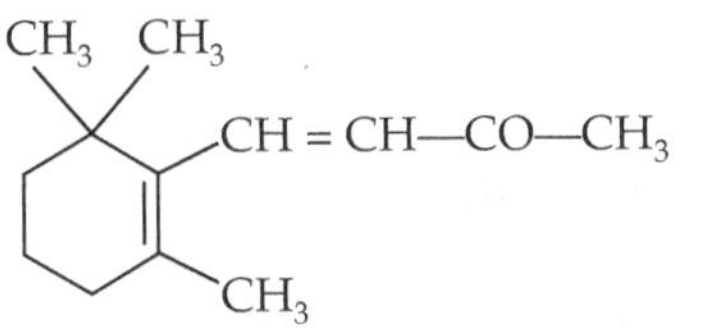

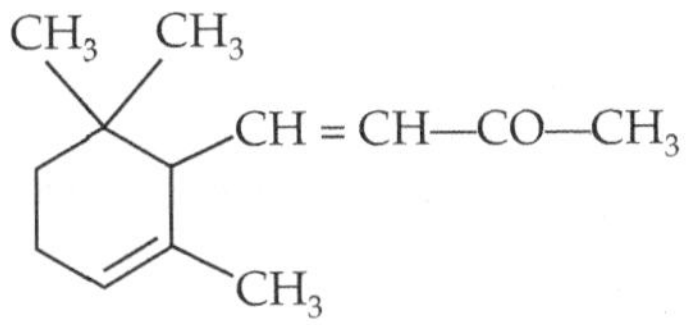

9. Explain and give examples of the types of transitions which occur in organic compounds?
10. Describe the spectral effect of adding chromophores to a molecule.
11. The wavelength of maximum absorption for methyl chloride is $\lambda_{max}$ = 173 nm, while for methyl iodide it is $\lambda_{max}$ = 253 nm; explain.
12. Why amines absorb at higher wavelengths in comparison to alcohols?
13. Why hydrogen bonding shifts the absorption to shorter wavelengths?
14. Predict the transition involved in the following compounds.

    (*a*) Alkenes (*b*) Alkynes (*c*) Carbonyl compounds (*d*) Azo compounds (*e*) Cyanides.
15. Predict $\lambda_{max}$ and $E_{max}$ of the following compounds:

    (*a*) Nitroethane (*b*) Nitrobenzene (*c*) Propionic acid (*d*) Methyl *iso*-butyl ketones (*e*) Hydrogen cyanide (*f*) Formaldehyde (*g*) Propene.
16. Predict the transition involved in saturated aldehydes and ketones.
17. Describe the term bathochromic, hypsochromic shifts, give an example of each.
18. Aniline absorb at $\lambda_{max}$ = 230 nm, but in acid solution it absorbs at 203 nm (shorter wavelength). Explain.
19. Why is ethanol a good solvent in UV spectroscopy?
20. What are essential requirement for a solvent to be used in UV spectroscopy?

21. Which of the following compounds will have higher $\lambda_{max}$?

(a) $C_6H_5CH = CH.C_6H_5$ (b) $C_2H_5CH = CH — CH = CH — C_6H_5$

22. Which of the following compounds will absorb at higher $\lambda_{max}$?

(a) $H_2\ddot{N} — C_6H_5$ (I) (b) $H_3N^+ — C_6H_5$ (II)

23. Discuss the effect of solvent polarity on $n \to \pi^*$ and $\pi \to \pi^*$transitions with suitable examples.

24. What are chromophores? Name some chromophores and explain how you can identify a particular chromophore in a compound.

25. What are auxochromes? Why and how do auxochrome increase the coloring power of a chromophore?

26. The position of absorption of acetone shifts in different solvents are : 279 nm is hexane, 272 nm is ethanal and 264.5 in water. Explain.

27. Predict the $\lambda_{max}$ of the following compounds :

(a) $CH_4$ (b) $CH_3 — CH_3$ (c) $CH_3Cl$ (d) $CH_3I$ (e) $CH_3OH$ (f) $CH_3NH_2$ (g) $(CH_3)_3N$
(h) $CH_3 — O — CH_3$.

28. The UV spectrum of acetone shows two peaks at $\lambda_{max}$ = 189 nm and $\lambda_{max}$ = 273 nm. Identify the electronic transition for each peak.

29. List the electronic transition possible when UV light is absorbed by the following :

(a) $CH_4$ (b) $CH_3Cl$ (c) HCHO.

30. Why the $\lambda_{max}$ for the diene I is observed at lower nm than II?

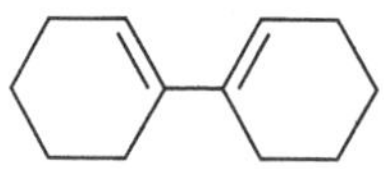

I

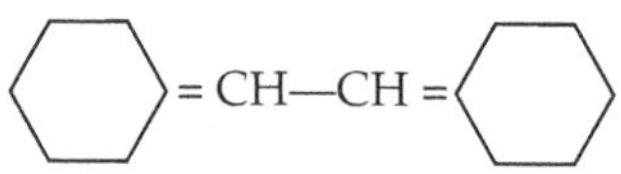

II

31. Measurement of UV spectra of four compounds (ketones) gave $\lambda_{max}$ at 221, 249, 233, and 258 nm. Assign structures to appropriate $\lambda_{max}$.

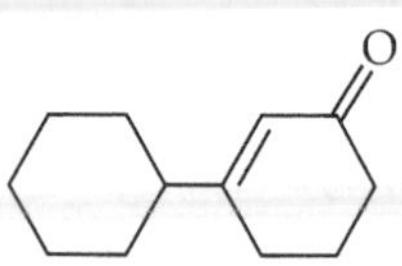

I

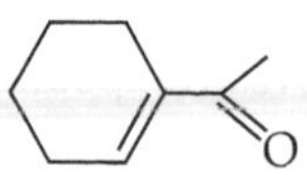

II

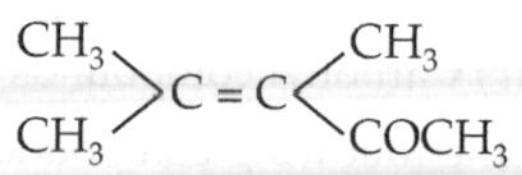

III

$CH_3CH_2\overset{O}{\overset{\|}{C}}COCH_3$

IV

32. Biphenyl exhibits intense absorption at $\lambda_{max}$ = 252 nm, but 2, 2-dimethyl derivative absorbs at $\lambda_{max}$ = 262, similar to o–xylenes. Explain.

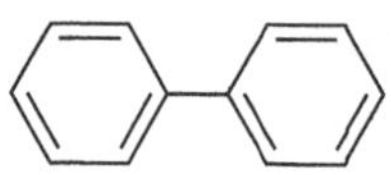

Biphenyl, $\lambda_{max}$ = 252

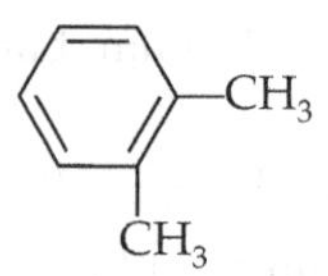

o-xylene, $\lambda_{max}$ = 252

## IR Spectroscopy

33. Why wavenumber in $cm^{-1}$ mostly used to measure the positions of a given infrared absorption?
34. Why infrared (IR) spectroscopy widely has been used for the identification of organic compounds?
35. How is the IR region of electromagnetic spectrum used for qualitative analysis?
36. What are the problems of using the IR region for quantitative analysis? How are these problems circumvented?
37. Explain clearly the origin of molecular spectra. Differentiate clearly between UV and IR spectra.
38. Write a note on modes of vibrations of polyatomic molecules. Illustrate your answer with suitable examples.
39. Discuss the basic principle of IR spectroscopy. Briefly describe the scanning of IR spectrum of an organic compound.
40. What are the different regions of IR radiations? Explain various types of stretching and bending vibrations with suitable example.
41. What are the necessary conditions required for a molecule to absorb IR radiation?
42. Why water is not a good solvent for IR spectroscopy?
43. Why conjugation increase the wavelength of absorption of a particular group?
44. Discuss the various types of stretching and bending vibrations which arise in aromatic compounds in their IR spectra.
45. Give the approximate position of the important IR absorption frequencies for the following compounds:

    (*a*) $CH_3OH$ (*b*) $C_6H_5NH_2$ (*c*) $C_6H_5CH_3$ (*d*) $CH_3COOH$ (*e*) $CH_3NH_2$
    (*f*) $CH_3CH_2OH$ (*g*) $CH_3$—CO—$CH_3$.

46. How will you distinguish between intermolecular and intramolecular hydrogen bonding by IR spectroscopy?
47. What do you understand from the number of fundamental vibrations? Calculate in the theoretical number of vibrational degrees of freedom in benzene, methane, and toluene. Why are the theoretical number of fundamental vibrations or bands not observed in IR spectra?
48. How will you distinguish the following with the help of IR spectra?

    (*a*) Aromatic and aliphatic compounds (*b*) $CH_3CHO$ and $CH_3COCH_3$ (*c*) Primary and tertiary alcohol (*d*) Aliphatic aldehyde and ketone (*e*) Primary and secondary amine (*f*) Carboxylic acid and ketone.

49. In the IR spectrum of a saturated hydrocarbon (molecular formula $C_6H_{14}$) the significant absorption bands were: stretching of saturated C–H at 2960 $cm^{-1}$, methylene (–$CH_2$–) bending at 1460 $cm^{-1}$ terminal methyl at 1380 $cm^{-1}$ and the rocking ethyl group at 775 $cm^{-1}$. Assign the structure to the compound.
50. Give the characteristic IR absorption frequencies of the following:

    (*a*) aromatic ring (*b*) alcohols, ethers, carboxylic acid, esters (*c*) aldehydes, ketones, carboxylic acid, esters (*d*) O–H monomeric alcohols, phenols (*e*) O–H hydrogen bonded alcohols, phenols.

51. Give the characteristic IR absorption frequencies of the following:

    (*a*) O–H carboxylic acids (*b*) N–H amines (*c*) C–N amines
    (*d*) C ≡ N nitriles (*e*) –$NO_2$ nitro-compounds.

52. Distinguish between ethyl alcohol and diethyl ether on the basis of IR spectroscopy.
53. How will you distinguish among aldehydes, ketones, and carboxylic acid on the basis of IR spectroscopy?

54. Why do ethene and ethyne, unlike propene and propyne, have no C to C multiple band and stretching bands?

55. What are the various types of molecular vibrations? Write the stretching frequencies of the following functional group:

$$-OH, > C = O, NH_2, \text{ether } (C - O - C), \equiv C - H, -C - H$$

## NMR Spectroscopy and Mass Spectroscopy

56. Write with suitable examples the shielding effects involved in NMR spectroscopy.
57. Describe NMR spectrum of ethanol and explain the position of each band.
58. Why do $^{12}C$, $^{16}O$, $^{18}O$, $^{32}S$ not show NMR spectra?
59. Explain the NMR spectrum of ethyl alcohol $CH_3$—$CH_2OH$.
60. Predict the NMR spectrum of ethyl chloride. $CH_3$—$CH_2$—Cl.
61. Explain clearly the chemical shift. How does it help in the elucidation of the structure of organic compounds?
62. Discuss briefly the shielding and deshielding effect involved in NMR spectroscopy, with suitable examples.
63. Explain clearly the various factors, which are responsible for effecting the magnitude of chemical shift. Illustrate your answer with suitable examples.
64. Write short notes on:

    (*a*) chemical shift (*b*) spin-spin coupling (*c*) shielding and deshielding (*d*) coupling constant.
65. Why does a signal for a particular set of protons split into a multiplet? Explain with suitable examples the term "spin-spin coupling".
66. Why is TMS used as a standard reference in NMR spectroscopy?
67. Predict the number of NMR peaks with relative areas ratio in the low resolution spectra of (*a*) Acetaldehyde (*b*) Toluene (*c*) 1, 3 dichloropropane.
68. Discuss the use of mass spectrometry in (*a*) Determination of molecular formula; (*b*) Identification of compounds from fragmentation patterns.
69. Write explanatory notes on:

    (*a*) The base peak, (*b*) Principle of mass spectrometry, (*c*) Mass spectrograph.
70. Define chemical shift. What is its significance in determination of the structure of molecules?
71. How will you distinguish between the isomers $CH_3COCH_3$ and $CH_3 - CH_2 - CHO$ on the basis of NMR spectroscopy?
72. The low resolution NMR spectrum of the compound $C_4H_8O_2$ exhibits three peaks at 8.1, 5.2, and 1.3 ppm with relative intensity ratio 1: 1: 6. Deduce the structure.
73. The low resolution NMR spectrum of the compound $C_5H_{10}O$ exhibits two peaks with relative intensities of 2 and 3 respectively. Deduce its structure.
74. What do you mean by splitting of NMR signal? How many NMR signals are expected in the NMR spectrum of acetaldehyde? Discuss their splitting.
75. What is mass spectroscopy? Give some of its important applications and advantages.
76. What is ring rule? What is its importance in the structural elucidation of unknown organic compounds?
77. In the mass spectrum of l-hexanol, a very weak molecular ion peak appears at $m/e = 102$. Some other prominent peaks appear at $m/e$ values of 100, 99, 84, 56 (Base peak) and 31. What are the most probable species responsible for these peaks?

**78.** What are metastable peaks? What are important characteristic of these peaks? How these peaks can be recognised in mass spectra and what are the importance of these peaks?

**79.** Some of the important peaks in the mass spectrum of an organic compound appear at $m/e$ = 120, 91 and 65. The molecular mass of the compound is 120 and the base peak appears at $m/e$ = 91. What are the most probable species responsible for these peaks?

**80.** Discuss the important features of the mass spectra of alkanes and alkenes with suitable examples.

**81.** In the mass spectrum of *n*-butane some of the prominent peaks appear in $m/e$ values of 58, 43, 29, and 15. What are the most probable species responsible for these peaks?

**82.** Identify the compound responsible for the following mass spectrum.

| *m/e* | *Relative Abundance* | *m/e* | *Relative Abundance* |
|---|---|---|---|
| 12 | 1.2 | 81 | 10 |
| 13 | 1.4 | 91 | 4.2 |
| 14 | 3.8 | 92 | 2.4 |
| 15 | 59 | 93 | 6.8 |
| 46 | 1.3 | 94 | 100 |
| 47 | 2.3 | 95 | 3.5 |
| 79 | 10 | 96 | 96 |

**83.** An unknown organic compound gives molecular ion peak and $m/e$ = 136. Two possible structures corresponding to $m/e$ = 136 are $C_{10}H_1$(I) and $C_{10}H_2N$(II). Which of these structures can be eliminated on the basis of nitrogen rule?

**84.** Mass spectrum of l-hexane shows some prominent peaks at $m/e$ = 27, 41, 55, 69, and 84. What are the species responsible for these peaks?

**85.** Explain various fragmentation processes which are responsible for prominent peaks, $m/e$ = 72, 57, 41, 29, and 27 in the mass spectrum of *neo*-pentane.

## ESR Spectroscopy and Chromatography

**86.** Discuss the principle, instrumental details and working of calorimetry.

**87.** Differentiate between NMR and ESR spectroscopy.

**88.** Write an explanatory note an *g-factor.*

**89.** Which of the following will show ESR spectrum. Give reason for your choice

(*i*) $Cu^{2+}$ (*ii*) $CO_2$

(*iii*) $NO_2$ (*iv*) $Cl^-$

**90.** Predict the number of lines in the ESR spectrum of each of the following radicals.

(*i*) $CH_3\,\dot{C}H_2$ (*ii*) $(CH_3)_2\dot{C}H$

(*iii*) $(CH_3)_3\dot{C}$

**91.** Describe the principle and instrumentation of ESR spectroscopy.

**92.** Write the applications of ESR spectroscopy.

**93.** Define the term chromatography. Discuss its principle and classification.

**94.** What is the principle of gas chromotagraphy? Distinguish between GSC and GLC.

**95.** Mention important applications of gas chromatography.

# Chapter 2 POLYMER CHEMISTRY

## 2.1 INTRODUCTION

The word 'polymer' is made from greek words, *poly* meaning *many* and *meros* meaning *parts* or *units*. It is defined as *a high molecular weight compound formed by the combination of a large number of one or more types of molecules of low molecular weight*. In other words polymers are giant molecules of high molecular weight, called *macromolecules*, which are built up by linking together large numbers of small molecules, called *monomers*, and the reaction by which the monomers combine to form polymers is known as *polymerization*. For example, if a simple molecule is $x$, then the molecule $x_n$, represented by the formula

$$\left( x - x - x - x - x_n \right)$$

is the polymer of $x$.

Polymers with a high degree of polymerization are called *high polymers* and those with low degree of polymerization are called *oligo polymers*. Macromolecules of biological significance, such as proteins, carbohydrates, etc., are called *biopolymers*. Silicates, silicones, etc. are termed as *inorganic polymers*.

## 2.2 CLASSIFICATION OF POLYMERS

Polymers can have different chemical structures, physical properties, mechanical behaviour, thermal characteristics, etc., and can be classified in different ways (see Figure 2.1). Some common classes of polymers based on their characteristics are:

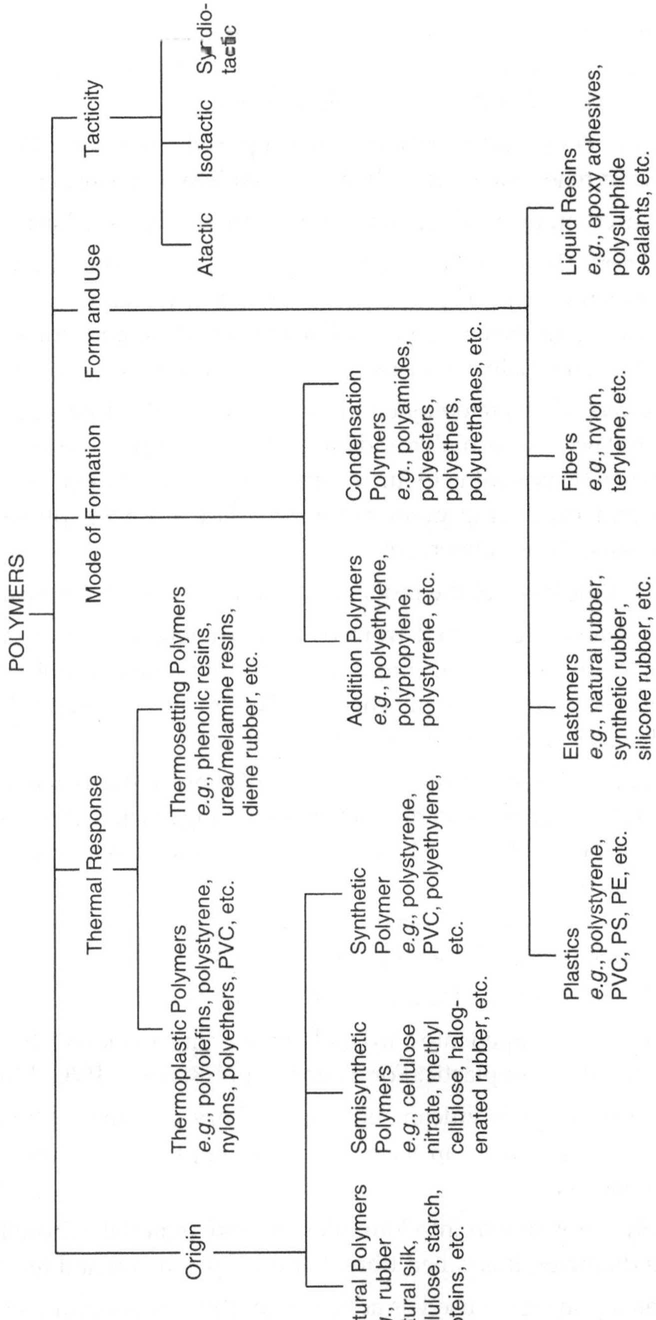


*FIGURE 2.1* **Classification of polymers based on their characteristics.**

1. **Origin.** Based on the origin, polymers are classified as:
   (*a*) *Natural Polymers.* These are available in nature and include natural rubber, natural silk, cellulose, starch, proteins, etc.
   (*b*) *Semi-synthetic Polymers.* These are chemically modified natural polymers such as hydrogenated, halogenated or hydrohalogenated natural rubber, esters and ethers of cellulose such as cellulose nitrate, methyl cellulose, etc.
   (*c*) *Synthetic Polymers.* These are synthetically prepared polymers. For example, polyethylene, polystyrene, polyesters, phenol-formaldehyde resins, etc.
2. **Thermal response.** Based on thermal response, polymers are classified as:
   (*a*) *Thermoplastic Polymers.* These can be softened or plasticized repeatedly on application of heat, without much change in properties if treated with certain precaution. For example, polyolefins, polystyrene, nylons, linear polyesters, and polyethers, PVC, etc. They are soluble, and fusible after many cycles of heating and cooling.
   (*b*) *Thermosetting Polymers.* They are obtained in soluble and fusible forms in early stages of their synthesis, but become insoluble and infusible when further heated. The process involves chemical reactions leading to further growth and cross linking of the polymer chain molecules and producing giant molecules. For example, phenolic resins, urea/melamine resins, diene rubber, etc.
3. **Mode of formation.** On the basis of the mode of formation, polymers are classified as:
   (*a*) *Addition Polymers.* These are formed from olefinic, diolefinic, vinylic, and related monomers. These are formed by the simple additions of monomer molecules to each other in a quick succession by a chain mechanism. This is called *addition polymerization.* For example, polyethylene, polypropylene, polystyrene.
   (*b*) *Condensation Polymers.* These are formed by the intermolecular reactions between bifunctional or polyfunctional monomer molecules having reactive functional groups such as –OH, –COOH, $–NH_2$, –NCO, etc. The related functional group react with each other in a stepwise manner producing condensed structural linkages, usually with the elimination of a simple molecule in each step. These include polyamides, polyesters, polyethers, polyurethanes, epoxy resins, etc.
4. **Form and use.** Based on its ultimate form and use, polymers can be classified as:
   (*a*) *Plastic.* When a polymer is shaped into hard and tough utility articles by the application of heat and pressure, it called plastic. For example, polystyrene, PVC, PMMA, etc.
   (*b*) *Elastomer.* When a polymer is vulcanized into rubbery products exhibiting good strength and elongation, it is called an elastomer. For example, natural rubber, synthetic rubber, silicone rubber, etc.
   (*c*) *Fibers.* When a polymer is drawn into long filament-like material, whose length is at least 100 times its diameter, it is called fibre. For example, nylon and terylene.
   (*d*) *Liquid Resin.* When a polymer is used as adhesive, potting compounds, sealants, etc., in liquid form, it is called liquid resin. For example, epoxy adhesives and polysulphide sealants.

5. **Tacticity.** Based on tacticity (configuration), polymers are classified as:
   (*a*) *Atactic.* When the arrangement of side substituents (R) is at random around the carbon back bone chain, such polymers are called *atactic polymers.*
   (*b*) *Isotactic.* If the substituent R, is on the same side of the carbon backbone, the polymer is called *isotactic.*
   (*c*) *Syndiotactic.* When the substituent R is arranged alternately around the carbon backbone chain, it is called *syndiotactic.*

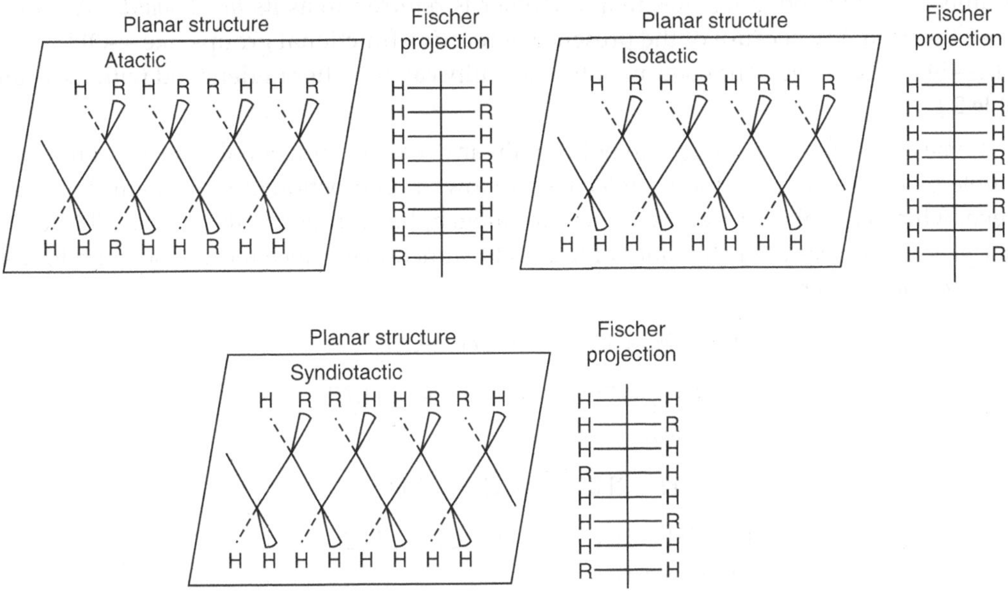


*FIGURE 2.2* **Atactic, isotactic and syndiotactic configuration of polymers.**

The planar structure and Fischer projection structure of atactic, isotactic and syndiotactic polymers are given in Figure 2.2.

## 2.3 CHARACTERISTICS AND PROPERTIES OF POLYMERS

### 2.3.1 Degree of Polymerization ($D_p$)

The subscript $n$ in $[—CH_2—]_n$ indicates the number of monomeric units contained in the polymer and is known as *degree of polymerization*. Degree of polymerization is related to the molecular weight of the polymer [M] by the equation

$$D_p = \frac{M}{m}$$

where $m$ is the molecular weight of the monomer unit.

$$\therefore \quad M = m.D_p$$

M = Molecular weight of the polymer.

$D_p$ = Degree of polymerization.

Most of the polymers have molecular weights ranging from 5,000–2,00,000.

### 2.3.2 Polymer Functionality

The number of active bonding sites in a monomer is referred to as its *functionality*. A compound assumes functionality because of the presence of reactive functional groups like —OH, —COOH, —$NH_2$, —SH, —NCO, etc. The concept of functionality can be better understood with the examples in Table 2.1.

Some compounds, however, do not contain any reactive functional group, but the presence of double or triple bonds in these molecules bestows polyfunctionality on them. For example, ethylene, which can take on two hydrogen or halogen, has a functionality of two (bifunctional). Similarly, acetylene has a functionality of four (tetra-functional), since it can react with four atoms of hydrogen or halogen.

$$CH_2 = CH_2 + H_2 \longrightarrow CH_3—CH_3$$

$$CH_2 = CH_2 + Cl_2 \longrightarrow CH_2Cl—CH_2Cl$$

(Bifunctional)

$$CH \equiv CH + 2H_2 \longrightarrow CH_3—CH_3$$

$$CH \equiv CH + 2Cl_2 \longrightarrow CHCl_2—CHCl_2$$

(Tetra-functional)

**TABLE 2.1 Examples Showing Functionality of Compounds**

| *Compound* | *Chemical Formula* | *No. of Functional Groups Present Per Molecule* | | | | *Functionality* |
|---|---|---|---|---|---|---|
| | | —OH | —COOH | —$NH_2$ | —NCO | |
| (*i*) Acetic acid | $CH_3COOH$ | - | 1 | - | - | Mono-functional |
| (*ii*) Malonic acid | $HOOC.CH_2.COOH$ | - | 2 | - | - | Bifunctional |
| (*iii*) Ethylene glycol | $HOCH_2.CH_2.OH$ | 2 | - | - | - | Bifunctional |
| (*iv*) Pentaerythritol | $C.(CH_2OH)_4$ | 4 | - | - | - | Tetra functional |
| (*v*) Hexamet—hylene diamine | $H_2N(CH_2)_6NH_2$ | - | - | 2 | - | Bifunctional |
| (*vi*) Aniline | $C_6N_5NH_2$ | - | - | 1 | - | Monofunctional |
| (*vii*) Lactic acid | $CH_3CH(OH)COOH$ | 1 | 1 | - | - | - |
| (*viii*) Tartaric acid | $CO_2H(CHOH)_2CO_2H$ | 2 | 0 | - | - | |

Depending on the functionality of the monomers used, we get linear, branched, or three dimensional cross linked polymers. For example:

(*i*) If we take bifunctional monomers, each monomer can add two more monomers to give a linear polymer as shown:

$$\underset{\text{Bifunctional polymer}}{n\cdot X\cdot} \longrightarrow \underset{\text{Linear polymer}}{\cdot X—X—X—X—X—X\cdot}$$

(*ii*) If we take trifunctional monomers, we get a cross linked polymer as:

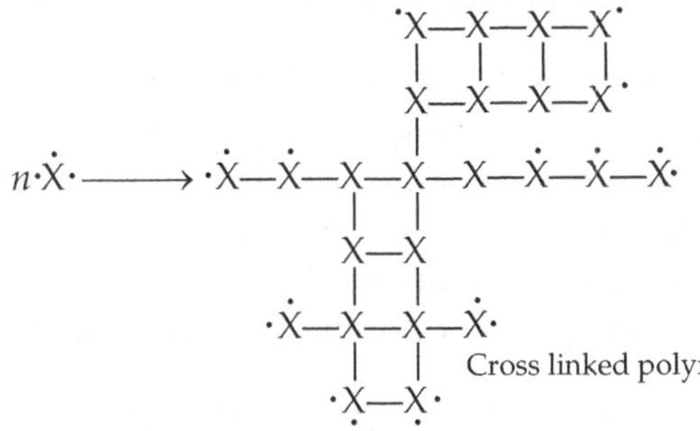

Cross linked polymer

## 2.3.3 Molecular Weight of Polymers

Molecular weight of a polymer influences various properties of a polymer such as solution viscosities, softening temperature, tensile and impact strength, and heat resistance. The low molecular weight polymers are generally soft and gum-like resins. The high molecular weight polymers are tougher and more heat resistant. Thus, control of molecular weight of a polymer is often important for industrial polymerization. Therefore, the determinations of molecular weight are of great importance in both the research and industry.

The molecular weight of a polymer depends upon the number of simple molecules joined together during polymerization reactions, *i.e.*, upon the degree of polymerization. But the polymerization chains are broken at different stages, therefore, the final product will contain the macromolecules of different weights. Most of the polymers are polydisperse, *i.e.*, the molecules of the polymers sample do not have identical molecular weights (M). Hence it is necessary in their case to take average molecular weight. There are two types of average molecular weights.

(*i*) *The Number-Average Molecular Weight, $M_n$.* It is obtained by dividing the total weight of the dispersed material with the number of molecules present, *i.e.*,

$$M_n = \frac{n_1M_1 + n_2M_2 + n_3M_3 + \ldots\ldots}{n_1 + n_2 + n_3 + \ldots\ldots}$$

where $n_1$, $n_2$, $n_3$ ...... are the number of molecules having masses $M_1$, $M_2$, $M_3$, ...... respectively. The above equation may also be written as

$$\overline{M}_n = \frac{\sum n_i M_i}{\sum n_i} \tag{2.1}$$

where, $M_i$ = simple monomer molecules

$n_i$ = number of monomer molecules

The number average molecular weight $\overline{M}_n$ assumes that each molecule makes an equal contribution to polymer properties regardless of the size or weight $\overline{M}_n$ of the polymer is obtained from measurement of colligative properties like osmotic pressure.

(*ii*) *The Weight-Average Molecular Weight, $\overline{M}_w$.* It gives representation to various molecular species in proportion to their weight in the given material. Light scattering and sedimentation equilibrium techniques, which not only depend upon the number of particles in a given mass or volume but also on the molecular size, give a weight average molecular weight of the macromolecule. Hence, in the averaging process, molecular weight of each individual species is multiplied by the weight and not by the number, *i.e.*,

$$\overline{M}_w = \frac{m_1M_1 + m_2M_2 + m_3M_3 + \ldots\ldots}{m_1 + m_2 + m_3 + \ldots\ldots}$$

If $n_1$, $n_2$, ... denote the number of molecules having masses $M_1$, $M_2$, ...... then $m_1 = n_1M_1$, $m_2 = n_2M_2$, ...... Hence the above equation may be written as,

$$\overline{M}_w = \frac{n_1M_1^2 + n_2M_2^2 + n_3M_3^2 + \ldots\ldots}{n_1M_1 + n_2M_2 + n_3M_3 + \ldots\ldots}$$

$$= \frac{\sum n_i M_i^2}{\sum n_i M_i} \tag{2.2}$$

From equations (2.1) and (2.2), it is clear that $\overline{M}_w$ is always greater than $\overline{M}_n$ for polydisperse systems. For monodisperse system $\overline{M}_w = \overline{M}_n$. Thus, the ratio $\overline{M}_w/\overline{M}_n$ is known as *distribution ratio* or an *index of polydispersity*.

For more complex averages, *z-average molecular weight* $M_z$ is obtained from the measurements of radial distribution of the refractive index gradient in sedimentation equilibrium.

$$\overline{M}_z = \frac{\sum n_i M_i^3}{\sum n_i M_i^2} \tag{2.3}$$

The *z*-average molecular weight is proportional to the mean cubic molecular weight. For a polymer having uniform distribution of molecular weight,

$$\overline{M}_z = \overline{M}_w = \overline{M}_n$$

But for real polymer samples

$$\overline{M}_z > \overline{M}_w > \overline{M}_n$$

and $$\overline{M}_n : \overline{M}_w : \overline{M}_z = 1 : 1.386 : 1.617$$

Thus when,

$$\overline{M}_w = 100{,}000$$

$$\overline{M}_n = 72{,}000$$

$$\overline{M}_z = 1{,}16{,}700$$

the molecular weight of a polymer can be determined by using following methods:

(*i*) Colligative property measurements such as Osmometry, Ebullioscopy and Cryoscopy.

(*ii*) Light scattering measurements.

(*iii*) Viscosity measurements.

(*iv*) Ultracentrifuge.

(*v*) Chemical analysis of end groups.

***Relation between molecular weight and $D_p$***

Both molecular weight and $D_p$ are related to molecular size. Hence like molecular weight, $D_p$ can also be averaged over the size of the sample. The 'number average' and 'weight average' degree of polymerization can be defined in a manner similar to that in which, $\overline{M}_n$ . $\overline{M}_w$ are defined, *viz.*,

$$(\overline{D}_p)_n = \frac{\Sigma n_i (D_p)_i}{\Sigma n_i} \tag{2.4}$$

and

$$(\overline{D}_p)_w = \frac{\Sigma n_i (D_p)_i^2}{\Sigma n_i (D_p)_i} \tag{2.5}$$

Each of these averages can be related to the corresponding molecular weight averages by the following two equations.

$$\overline{M}_n = (\overline{D}_p)_n . m \tag{2.6}$$

$$\overline{M}_w = (\overline{D}_p)_w . m \tag{2.7}$$

### 2.3.4 Crystallinity and Properties of Polymers

When X-ray studies were extended to polymers, it was seen that a majority of polymers diffract X-rays like crystalline substances, whereas many behave like amorphous solids. In fact, it was observed that they contain both crystalline and amorphous regions in their structure. In the highly ordered region, the polymer chains are aligned in a regular fashion, while in a disorder region the polymer chains are present randomly. These two regions are known as *crystalline* and *amorphous* regions respectively (Figure 2.3).

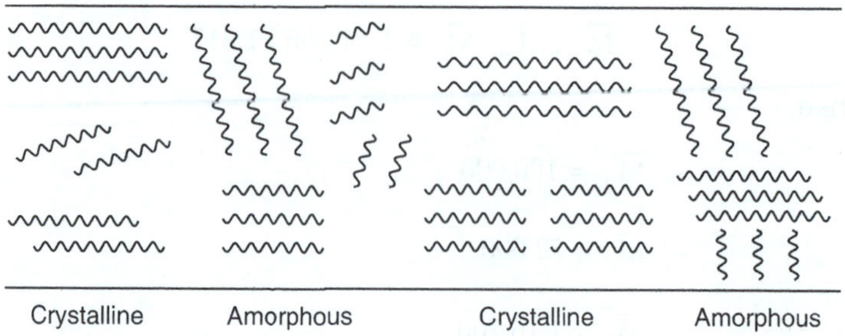


**FIGURE 2.3 Crystalline and amorphous regions in a polymer.**

Crystalline regions occur when several polymer molecules on structures of polymer molecules are oriented parallel to each other. So that the weak inter-chain forces hold the molecule together conferring strength on the polymer. On the other hand, when polymer molecules are not sufficiently close to each other for these inter-chain forces to operate, amorphous regions occurs.

In addition to the attractive forces between the polymer molecules, the relative amount of crystallinity and amorphousness of polymers also depend upon the structural regularity of the polymer chain and the presence and absence of bulk side groups. For example, polyethylene has a highly crystalline structure since it is possible for the polymer chain to line up in a zig-zag conformation. (Figure 2.4(*a*))

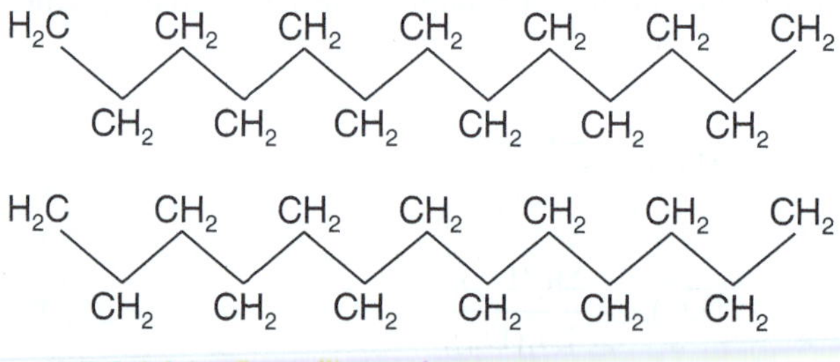

**FIGURE 2.4 (*a*) Crystalline polymer segment of polyethylene.**

On the other hand, polypropylene may be crystalline or amorphous depending upon the following geometric arrangement of the methyl groups (Figure 2.4(*b*)):

(*i*) *Isotactic polypropylene.* When the methyl groups are arranged on the same side of the polymer chain.

(*ii*) *Syndiotactic polypropylene.* When the methyl groups are arranged regularly on alternate sides of the polymer chain.

(*iii*) *Atactic polypropylene.* When the methyl groups are arranged randomly on the both sides of the polymer chain.

Since in isotactic and syndiotactic configurations the polymer chains are at a distance at which there is an interaction of attractive forces, these polymers are generally more crystalline than the atactic polymers.

Thus, it is clear that crystalline and non-crystalline (amorphous) components of a polymer coexist. Crystallinity of a polymer sample, thus is expressed in terms of that fraction of polymer which is crystalline. The degree of crystallinity of a polymer sample can be expressed in the following ways:

(*i*) In terms of density, $X_v$, the degree of crystallinity by volume is expressed as:

$$X_v = \frac{d - d_a}{d_c - d_a} \tag{2.8}$$

where, $d$, $d_a$, and $d_c$ are densities of the sample, the fully amorphous, and fully crystalline components, respectively.

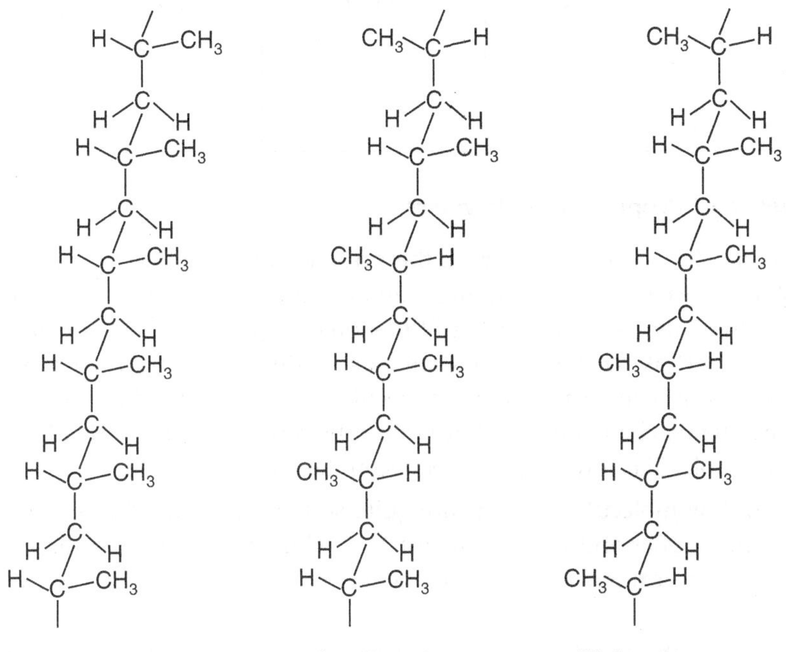

(i) Isotactic (ii) Syndiotactic (iii) Atactic

***FIGURE 2.4 (b)*** **Geometric arrangement in polypropylenes.**

(*ii*) In terms of specific volume (V), the degree of crystallinity by mass is expressed as:

$$X_m = \frac{V_a - V}{V_a - V_c} \tag{2.9}$$

where, V, $V_a$, and $V_c$ are the specific volumes of the sample, the fully amorphous and fully crystalline components, respectively.

(*iii*) In term of enthalpy ($X_{cal}$), the degree of crystallinity is expressed as:

$$X_{cal} = \frac{H_a - H}{H_a - H_c} \tag{2.10}$$

where, $H_a$, $H_c$, and H are fully amorphous, fully crystalline, and sample enthalpies, respectively.

Various properties of polymers such as density, hardness, permeability, heat capacity and modulus are affected by crystallinity. Two samples of polyethylene A (less crystalline) and polyethylene B (more crystalline), when compared of properties such as density, hardness, tensile strength, and flexural modulus gave the results given in Table 2.2.

**TABLE 2.2 Comparison of Properties of Two Polyethylene Samples having Different Crystallinities**

| *Property →* | *Density* | *Hardness (share D)* | *Tensile Strength ($10^5$ Pa)* | *Flexural Modulus ($10^5$ Pa)* |
|---|---|---|---|---|
| Sample A (less crystalline) | 0.917–0.932 | 44–50 | 83–314 | 241–331 |
| Sample B (more crystalline) | 0.952–0.965 | 66–73 | 221–310 | 1000–1551 |

### 2.3.5 Structure and Properties of Polymers

The major structural feature of polymers is the presence of large numbers of monomeric units which are combined successively. The number and chemical nature of monomeric units gives an overall picture of the polymer. The number of monomeric units depicts the macro-structure, and the mode of their interlinking indicates the micro-structure of the polymer molecule. Polymers have two types of bonds, chemical and intermolecular, which greatly differ in energy and length. The atoms in the polymer chain are joined by strong chemical bonds about 1–1.5 Å in length, while much weaker intermolecular forces operate between the chains at distances about 3–4 Å.

Polymers of low molecular weights are quite soft and gummy. On the other hand, higher chain polymers are tougher and more heat resistant. Thus, by controlling the chain length, it is possible to vary the physical properties of the polymer from soft and flexible to hard and horn-like substances. The presence of intermolecular forces can be increased by increasing polar groups such as carboxyl, hydroxyl, chlorine, fluorine, and nitrile along the chain length. For example, nylons, teflon, polyester, etc. Polymers having three dimensional networks have chemical bonds between their chains, consequently, they are more strong and tough, since the movement of intermolecular chains is totally restricted.

### 2.3.6 Polymer Degradation

Polymer degradation may be defined as *an undesirable change that occurs in a polymer affecting its physical and chemical properties.* For example, wearing of rubber tires, loss of a plasticizer from a system by evaporation, separation of a polymer from its fillers, etc. The polymer degradation is brought about by the combined action of ultraviolet radiation, heat, chemicals, and mechanical energy. A polymer can suffer degradation mainly at two stages. First during the fabrication process, and second during its daily usage.

Polymer degradation is characterized by uncontrolled change in the constitution of the polymer and reduction in its molecular weight. Polymer degradation is broadly of two types:

(*i*) *Chain end degradation.* In this type, the degradation starts from the chain ends, resulting in successive release of the monomeric units (Figure 2.5). This leads to a decrease in the molecular weight of the polymer and a large quantity of the monomer is liberated simultaneously. For example, thermal depolymerization of poly α-methyl styrene. This polymer in a solvent such as anhydrous tetra hydrofine, upon heating from – 70°C to + 60°C, undergoes degradation from the chain end and at 60°C the entire polymer is converted into monomer. This type of degradation occurs when, the backbone bonds are weaker than the bonds of the side groups. Polystyrene, polybutadiene and polyisoprene also undergo depolymerization atom to a limited extent. Chain end polymerization can be shown as:

$$M_n^\bullet \longrightarrow M_{n-1}^\bullet + M$$

$$M_{n-1}^\bullet \longrightarrow M_{n-2}^\bullet + M, \text{ and so on.}$$

where M is the monomer and $M_n^\bullet$ . $M_{n-1}^\bullet$, and $M_{n-2}^\bullet$, so on represent the polymer chain carrying an active end group.

(*ii*) *Random degradation.* This type of degradation occurs at any random points along the polymer chain instead of at the chain ends (Figure 2.6). It can be shown as:

$$M_n \longrightarrow M_x + M_y$$

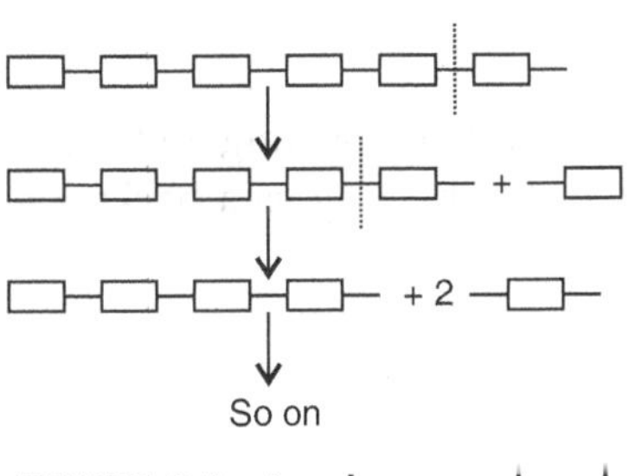


*FIGURE 2.5* **A polymer undergoing chain end degradation.**

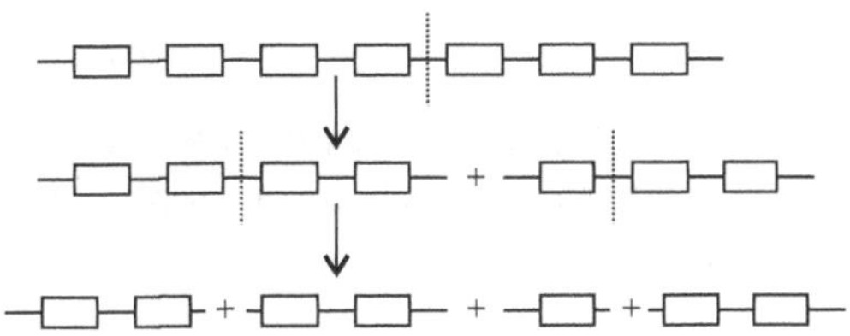

*FIGURE 2.6* **A polymer undergoing random degradation.**

The polymer degrades to lower molecular weight fragments but practically no monomer is liberated. Almost all the polymers can undergo random degradation, for example, polyesters undergo hydrolytic degradation.

## 2.4 COPOLYMERIZATION

Copolymerization is the combined polymerization of two or more monomer species. Such polymers are called *copolymers.* A macromolecule may consist of monomers of identical or different chemical structures, known as *homopolymers* and *copolymers* respectively.

... —A—A—A—A— ... Homopolymers

... —A—B —A—A— ... Copolymers

Depending upon the variations in distribution of monomer units within the polymer chains, the following types of copolymers are found:

(*a*) *Random copolymer.* In this type of copolymer, the monomer units are randomly distributed along the polymer chain. Majority of commercial copolymers of *butadiene and styrene, butadiene and acrylonitrile*, etc. are random copolymers.

··· —A—B—B—A—B—A—A—A—B—B—A—A—B—B— ···

Random Sequence

(*b*) *Alternate copolymer.* In this type, the monomer units are arranged in a regular alternating fashion. These include *polyesters, polyamides, styrene-acrylonitrile, styrene-maleic anhydride*, etc.

··· —A—B—A—B—A—B—A—B—A—B—A—B— ···

··· —AB—AB—AB—AB—AB—AB—AB—AB— ···

(*c*) *Block copolymer.* In this polymer a block of repeating unit of one kind of monomer is followed by a block of another kind of monomer, then again followed by the first kind and so on.

··· —A—A—A—A—B—B—B—B—A—A—A—A

Block Copolymer

(*d*) *Graft copolymer.* In this polymer, a chain formed by one kind of repeating unit of monomer is grafted on to a chain of monomer units of another kind acting as a backbone of the polymer.

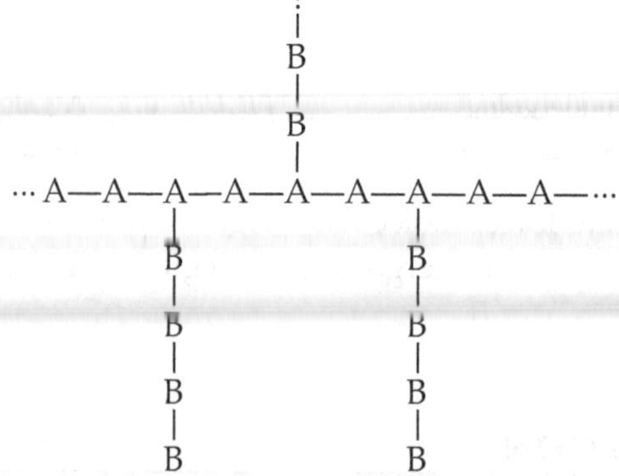

Graft Polymer

Some common copolymers include:

1. Copolymer of *acrylonitrile* and *vinyl chloride.* This polymer is readily soluble in water, whereas polyacrylonitrile and polyvinyl chloride are soluble only in high boiling and difficult available solvents.

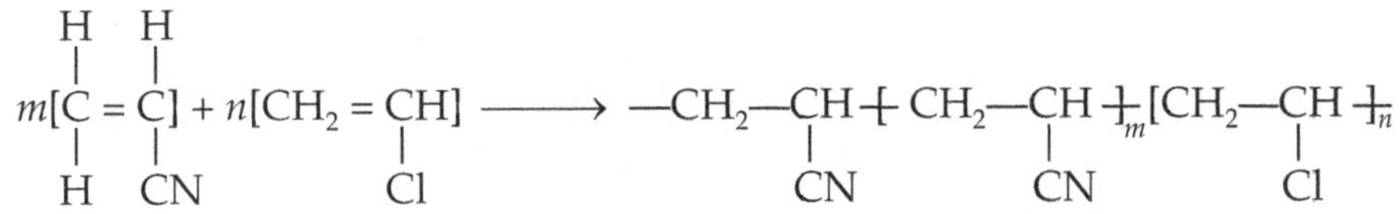

Copolymer

2. Copolymer of *isobutene* with a small amount of *isoprene* gives butyl rubber having high gas impermeability.

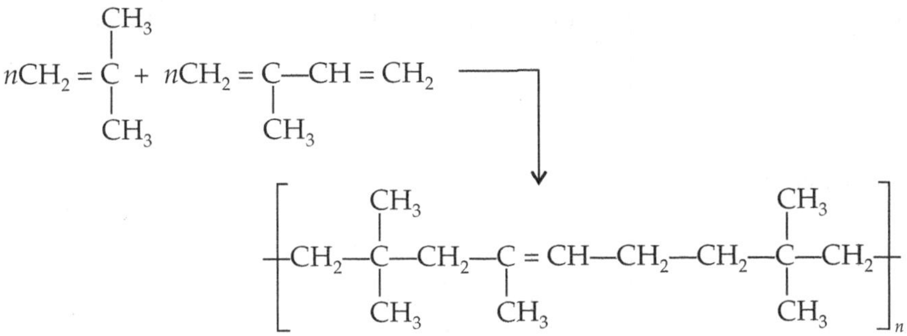

Copolymer

3. Buna-S Rubber. Refer section 2.5.4.3.
4. Buna-N Rubber. Refer section 2.5.4.3.

## 2.5 ADDITION POLYMERIZATION

**TABLE 2.3 Commonly Occurring Addition Polymers and their Repeating Units**

| *Name* | *Structural Formula of Monomer* | *Repeating Polymer Linkage* |
|---|---|---|
| Polyethylene | $CH_2 = CH_2$ | $-CH_2-CH_2-$ |
| Polypropylene | $CH_2 = CH(CH_3)$ | $-CH_2-CH(CH_3)-$ |
| Polyacrylonitrile | $CH_2 = CH-CN$ | $-CH_2-CH(CN)-$ |
| Polyvinyl chloride | $CH_2 = CH-Cl$ | $-CH_2-CH(Cl)-$ |

| | | |
|---|---|---|
| Polystyrene | 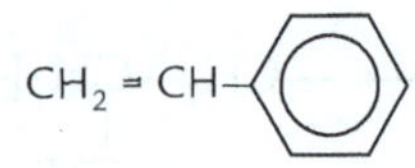 | 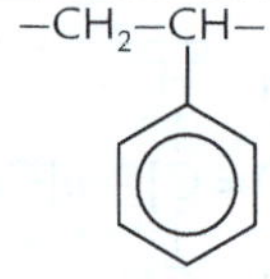 |
| Polymethylmethacrylate | 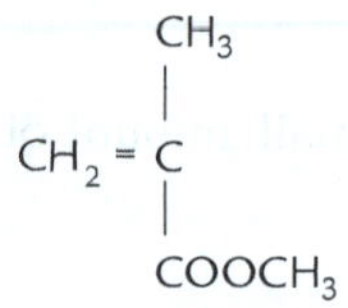 | 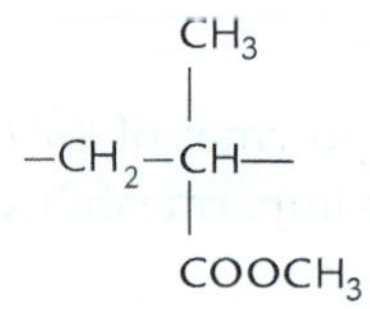 |
| Polyvinyl acetate | $CH_2 = CH{-}OCOCH_3$ | 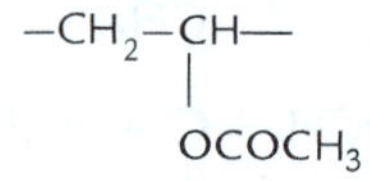 |
| Polyvinylidene chloride | 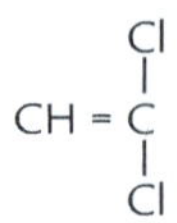 | 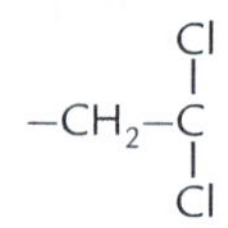 |
| Polytetrafluorethylene (Teflon) | 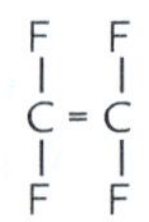 |  |
| Polyisoprene | 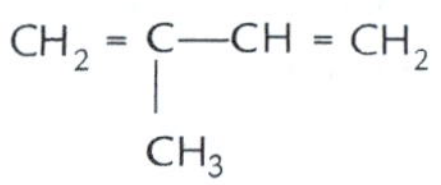 | 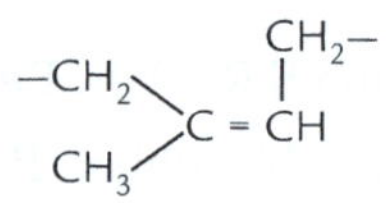 |

Addition polymerization is the most common method of synthesizing the polymers. This type of polymerization occurs among molecules containing double or triple bonds. The reaction occur through the addition of a monomer unit to a radical initiator or an initiation ion by which the active state is transferred from an initiator to the added monomer. This is followed by the addition of other monomer molecules (2,000–20,000 molecules) unit the active state is terminated. In this process no side products are formed. Some common examples of addition polymers are given in Table 2.3.

Depending upon the nature of active centre, the addition polymerization is classified as:

### 2.5.1 Free Radical and Ionic Polymerization

It involves the formation of free radicals (FR) as the active centres. The mechanism can be studied in the following steps:

1. **Chain initiation.** It involves the formation of a FR from a radical initiator. Peroxides and aliphatic azo compounds are commercially used as radical initiators. Commonly used radical initiator are:

(*i*) *Benzoyl peroxide*

$$C_6H_5-\overset{\overset{\large O}{\|}}{C}-O-O-\overset{\overset{\large O}{\|}}{C}-C_6H_5 \longrightarrow 2C_6H_5-\overset{\overset{\large O}{\|}}{C}-\dot{C} \longrightarrow \underset{\text{Phenyl free radical}}{2\dot{C}_6H_5} + 2CO_2$$

(*ii*) *Azo-bis-isobutyronitrile (AIBN)*

$$(CH_3)_2-\underset{\underset{CN}{|}}{C}-N=N-\underset{\underset{CN}{|}}{C}-(CH_3)_2 \longrightarrow 2(CH_3)_2-\underset{\underset{CN}{|}}{\dot{C}} + N_2$$

These reactions can be represented as,

$$R_2 \longrightarrow 2R^{\bullet}$$

where $R_2$ is a radical initiator and $R^{\bullet}$ is a free radical (FR)

The FR ($R^{\bullet}$) adds to a molecule of monomer producing another FR.

$$R^{\bullet} + CH_2=\underset{\underset{X}{|}}{CH} \longrightarrow R-CH_2-\underset{\underset{X}{|}}{\dot{C}H}$$

2. **Chain propagation.** The FR, $RCH_2\dot{C}HX$, further adds to a monomer molecule forming another FR, and successive addition of monomer to these radicals results in the formation of larger chain radicals.

$$R-CH_2-\underset{\underset{X}{|}}{\dot{C}H} + CH_2=\underset{\underset{X}{|}}{CH} \longrightarrow RCH_2\underset{\underset{X}{|}}{CH}-CH_2\underset{\underset{X}{|}}{\dot{C}H}$$

$$\xrightarrow{nCH_2=\underset{\underset{X}{|}}{CH}} R\left[CH_2-\underset{\underset{X}{|}}{CH}\right]_{n+1}CH_2-\underset{\underset{X}{|}}{\dot{C}H}$$

The chain propagation results in the change of $\pi$-bonds into $\sigma$-bonds. This step always occurs with the liberation of heat.

In vinyl polymerization radicals attach monomers at the methylene group, giving head-to-tail addition. This addition is also favored because this carbon atom is less sterically hindered than the other carbon atoms. This addition also leads to resonance stabilization. For example,

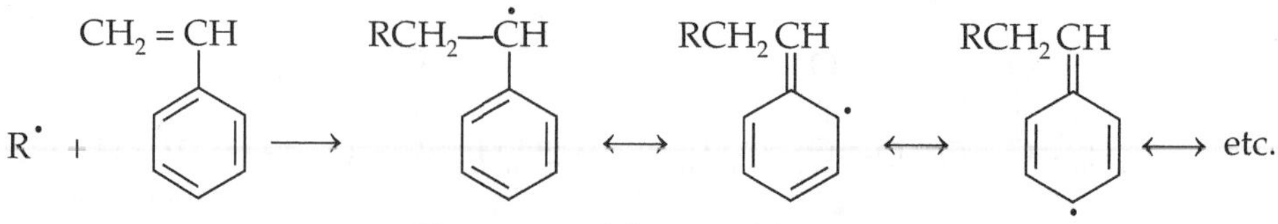

Resonance stabilization of the FR

3. **Chain termination.** The growing polymer chains can be terminated in the following two ways:

   (*i*) *Coupling or Recombination.* It involves the addition of two growing radical chains to form a saturated macromolecule.

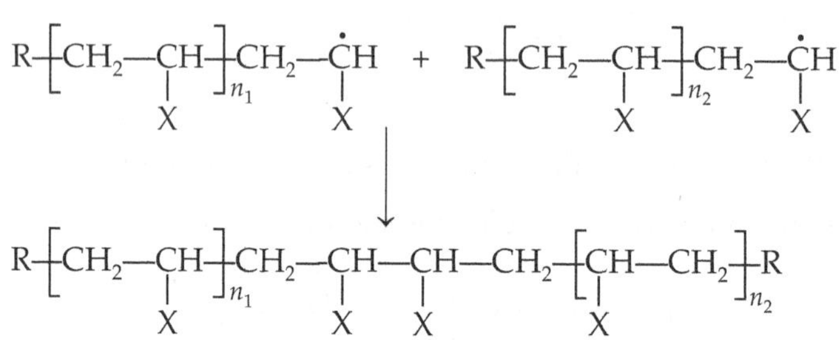

   (*ii*) *Disproportionation.* It involves the transfer of a hydrogen atom from one growing chain to another with the formation of an unsaturated end group on the chain by losing the hydrogen atom.

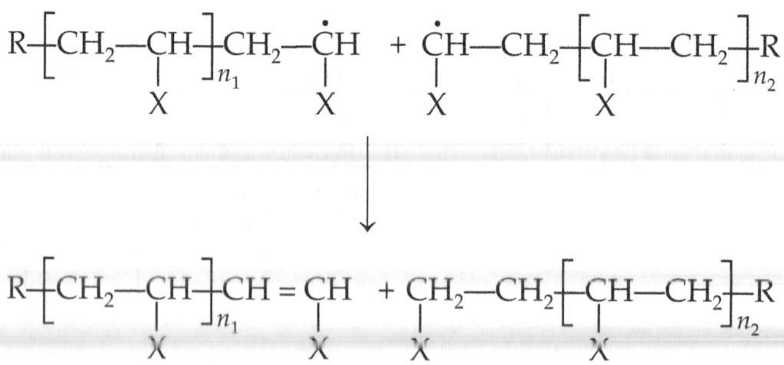

### 2.5.2 Ionic Polymerization

This type of polymerization may be anionic when electron withdrawing groups such as chlorine are present, or cationic when electron donating groups such as methyl are present. These reactions occur at very low temperatures as the initiation reaction requires only a small activation energy.

1. **Anionic polymerization.** Monomers containing electronegative substituents such as acrylonitrile, methylmethacrylate and styrene are polymerized by electropositive initiators. The initiators used in anionic polymerization are alkali metals such as sodium,

potassium, and lithium in liquid ammonia, alkali metal alkyls, triphenylmethyl sodium, and grignard reagents.

**Mechanism.** It consists of following three steps:

(*a*) *Chain initiation.* In this step the initiator produces an anion which adds to the monomer molecule to generate a carbanion as the active centre for chain growth.

$$K\text{—}NH_2 \longrightarrow K^+ + NH_2^-$$

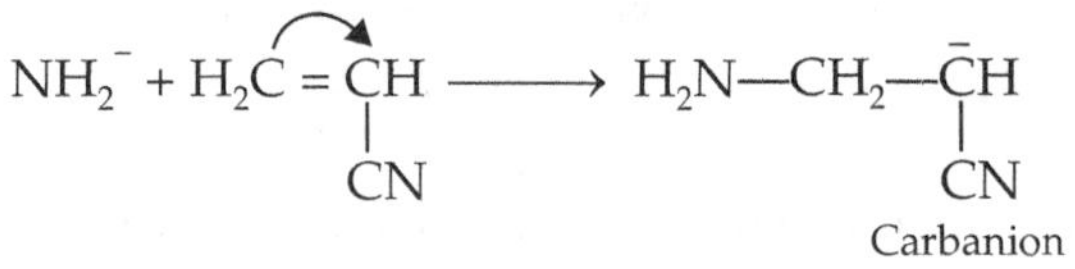

(*b*) *Chain Propagation*

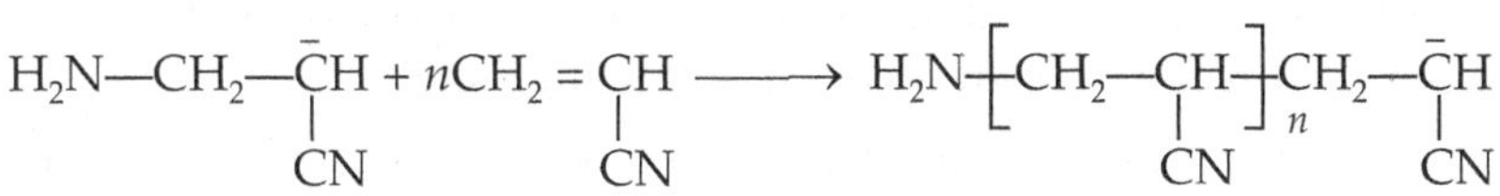

(*c*) *Chain Termination*

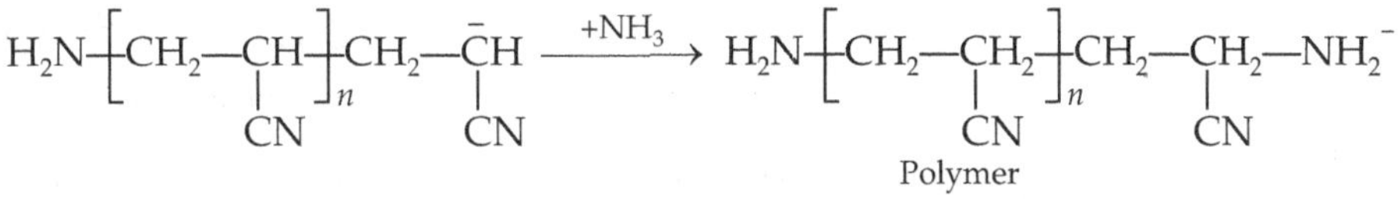

Other examples include,

(*i*) *Synthesis of Polymethylmethacrylate*

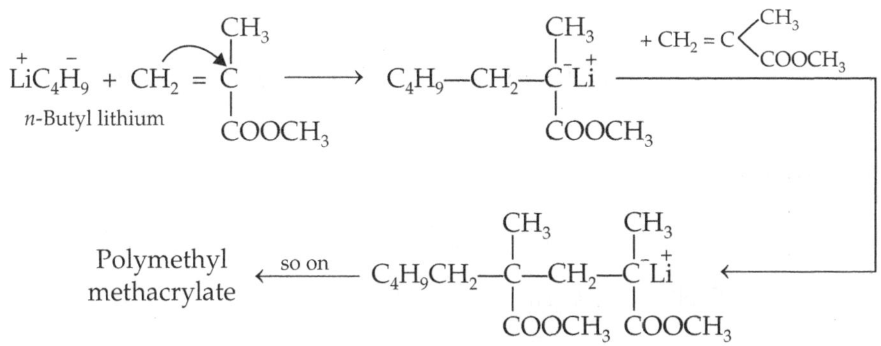

*(ii) Synthesis of Polystyrene*

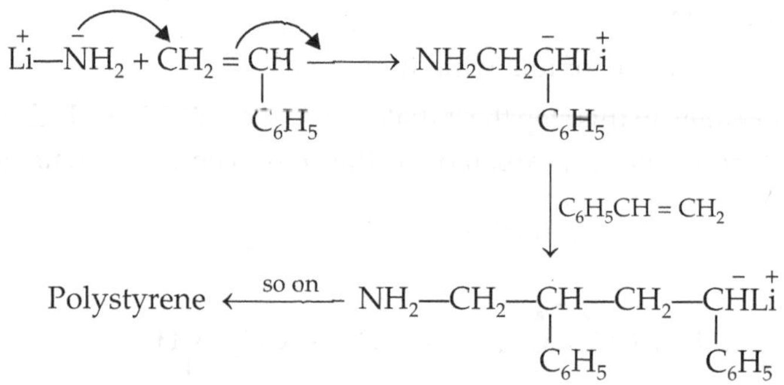

2. **Cationic polymerization.** Ethylene derivatives containing electron donating groups and other unsaturated monomers readily polymerize in the presence of very small amounts of acids and Friedal craft catalysts ($AlCl_3$, $BF_3$, $SnCl_4$, etc.). The mechanism proceeds *via* the formation of a carbonium ion. For example, polymerization of styrene in the presence of $SnCl_4$. This mechanism occurs in following steps:

   *(a) Chain Initiation.* It involves the formation of active carbonium ion centre.

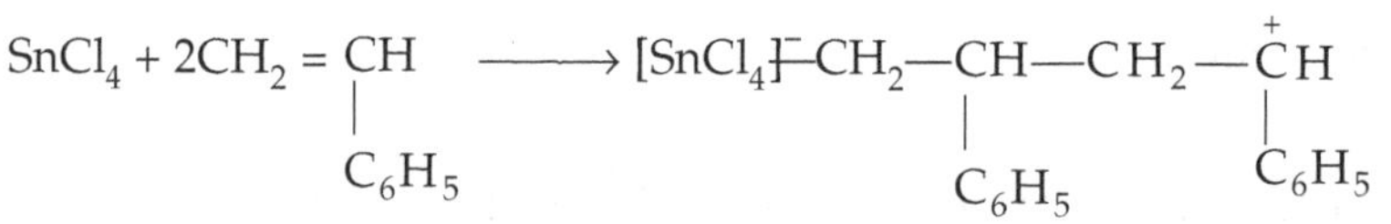

Carbonium ion active centre

   *(b) Chain Propagation.* This step involves addition of monomer molecules with regeneration of carbonium ion.

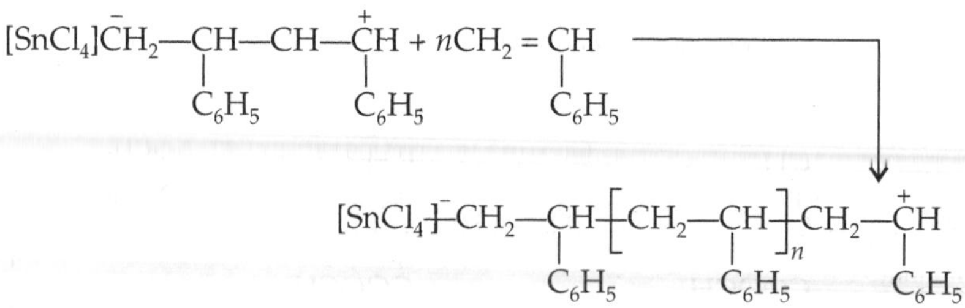

   *(c) Chain Termination*

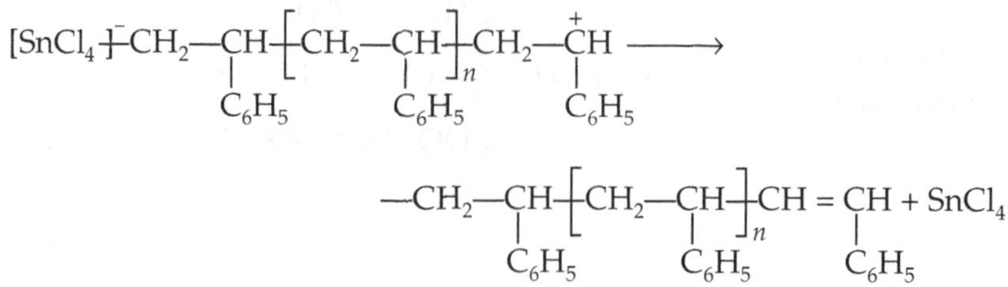

Polystyrene

Other examples include, polymerization of *ethylenederived compounds* using $BF_3$ (Lewis base) and $H_2O$ as catalyst.

(*a*) *Chain Initiation*

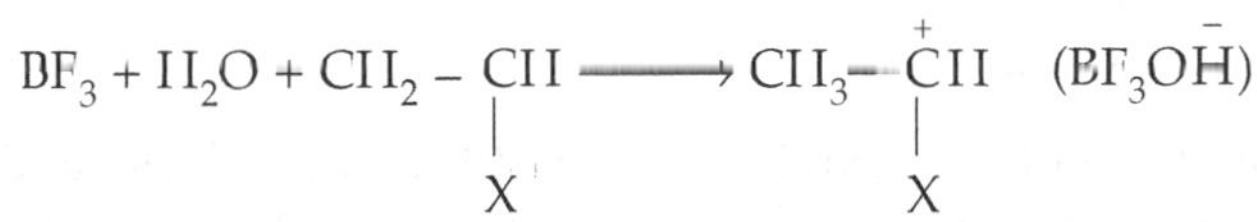

(*b*) *Chain Propagation*

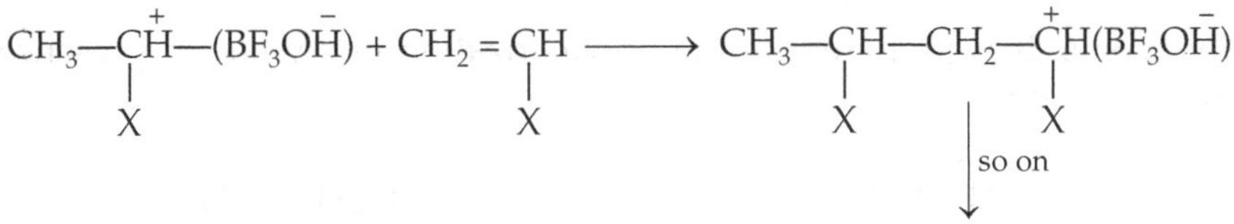

(*c*) *Chain Termination*

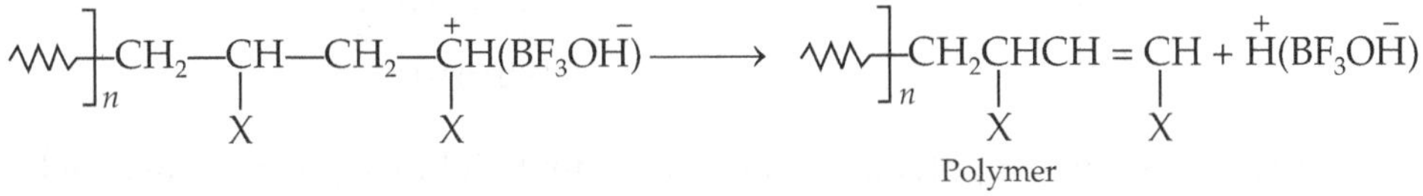


### 2.5.3 Commercially Important Addition Polymers or Vinyl Resins

1. **Polyethylene.** It is prepared by the polymerization of monomer ethylene.

$$nCH_2 = CH_2 \xrightarrow{\text{Polymerization}} -\!\!\left[CH_2—CH_2\right]\!\!-_n$$

Two varieties of polyethylene are obtained:

(*a*) *Low density polyethylene.* It is obtained by polymerizing ethylene at a pressure of 1000-3000 atmosphere and temperature 180–300°C. It is branched and used in making film and sheets, injection moulding, wire and cable, in coating and blow moulding. LDPE films are mainly used for packing frozen food, textile products, etc. Due to chemical inertness and resistance to breakage, LDPE is used in making squeeze bottles and in many attractive containers. LDPE's pipes are used for agricultural, irrigation, and domestic water line connections.

(*b*) *High density polyethylene.* It is obtained by polymerization with metal oxide as a catalyst. It is linear, possesses high crystallinity, higher melting point, greater tensile strength and hardness. It is used for blow moulding, injection moulding, and manufacture of film, sheet, wire, and cable insulation. It is also used for making toys and household goods.

2. **Polypropylene.** It is obtained by the polymerization of monomer propylene in the presence of *Ziegler Nata Catalyst.* [Al (*iso.* $C_4H_9)_3$ and $TiCl_3$].

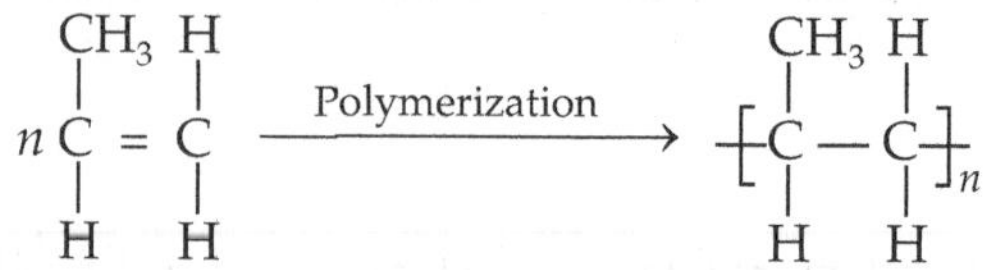


It is a high crystalline polymer exhibiting high stiffness, hardness, and tensile strength. It is used for producing packaging films, pipes, storage tanks, seat covers, monofilaments, and ropes. It is also used for making refrigerator, radio, washing machine parts and sterilizable hospital equipments.

3. **Polystyrene.** It is prepared by the polymerization of styrene in presence of benzoyl peroxide catalyst.

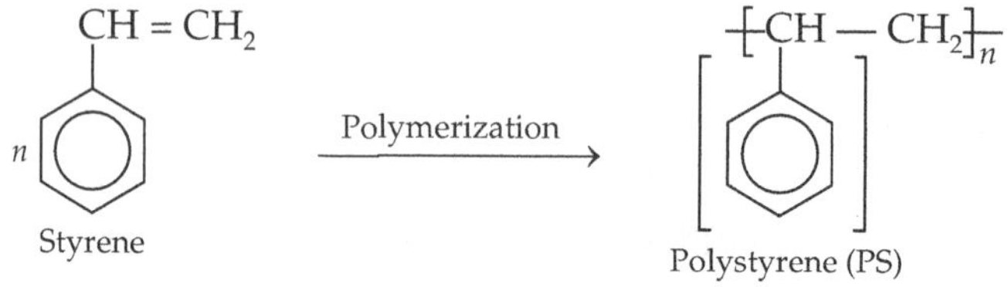


Polystyrene is a linear polymer and is transparent, light weight and have high index of refraction. It is resistant to the action of moisture, chemicals and is highly electric insulating. It is *used* for making radio and television parts, refrigerator parts, aircraft instrument panels, chemical equipments, lenses, and containers for acids and alkalies.

4. **Polyvinyl chloride.** It is produced by the polymerization of monomer, vinyl chloride, in the presence of the catalyst benzoyl peroxide.

   Structurally, the PVC molecule is partially syndiotactic and does not have a completely regular structure, so PVC has a low crystallinity. The polymer molecules are either linear or only slightly branched.

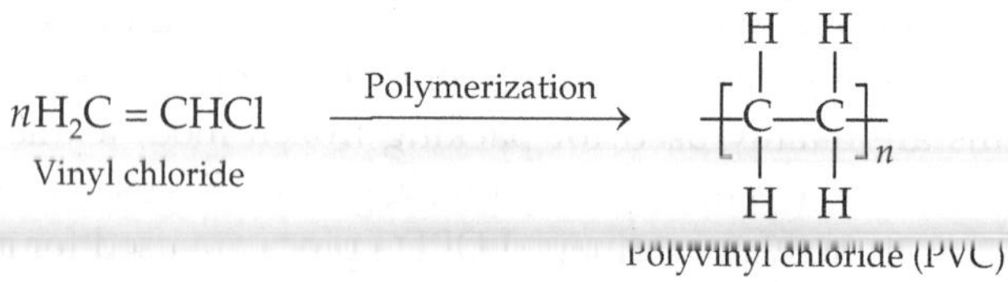


PVC, as such, is a horn-like material and difficult to process. It is, therefore, compounded with plasticizers. Depending on the plasticizer percentage, fully rigid to highly flexible finished products can be obtained.

It is insoluble in alcohols, vinyl chlorides, hydrocarbons, water, and alkalies. It is used for large scale production of cable insulations, equipment parts, pipes, laminated material and in fibre manufacture. The total chlorine content in PVC is around 56.8% and can be further increased to 60–65% by dissolving it in a suitable solvent such as chlorbenzene. This is called *chlorinated PVC*, it is mainly used in the manufacture of adhesive coatings and fibres.

5. **Polytetrafluoroethylene (PTFE).** It is also known as *teflon* and is obtained by polymerization of tetrafluoroethylene (TFE) under pressure in the presence of benzoyl peroxide as catalyst.

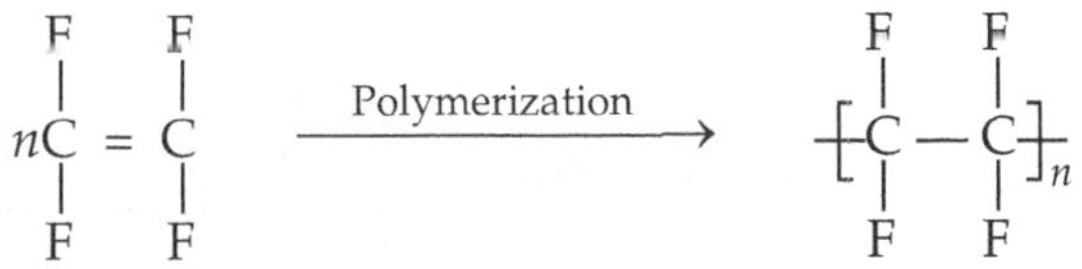

TFE Teflon

It is a linear high crystalline polymer with melting point around 330°C. It has highly regular structure and the presence of highly electronegative fluorine atoms results in strong attractive forces which give the PTFE extreme toughness, high density, extremely good electrical and mechanical properties, and high chemical resistance. Teflon is resistant to corrosion, alkalies, and organic solvents and does not dissolve in any of the strong acids. It is *used* in making articles such as pump valves and pipes where chemical resistance is required. It is used as insulation material for motors, transformers, cables, wire fitting, etc. Its fibres are used to form belts, fibre clothes, etc.

6. **Polymethylmethacrylate (PMMA).** It is prepared by polymerization of methyl-methacrylate in presence of acetyl peroxide or $H_2O_2$. It is commonly known as *orlon.* It is also known as *lucite* or, commercially, plexiglass.

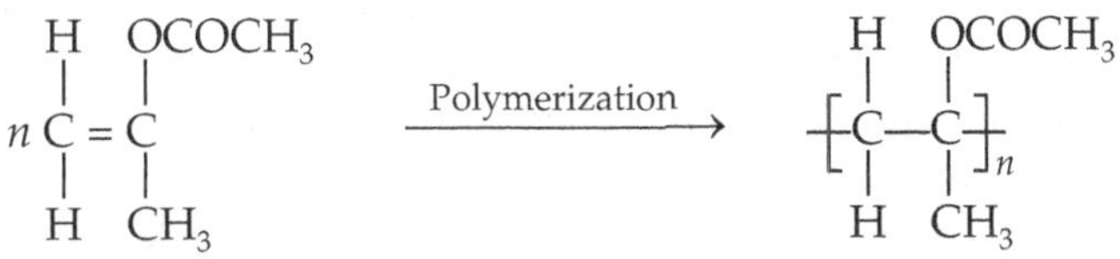

MMA PMMA

It is a colourless, transparent plastic and fairly rigid material resistant to many chemicals but soluble in organic solvents. It is used for making emulsions, adhesives, paints, TV screens, wide screen guards, durable lenses for automobile lighting, signboards, etc. It is also used in buildings for decorative purposes.

7. **Polyacrylonitrite (PAN).** Polyacrylonitrile, also known as polyvinyl cyanide.

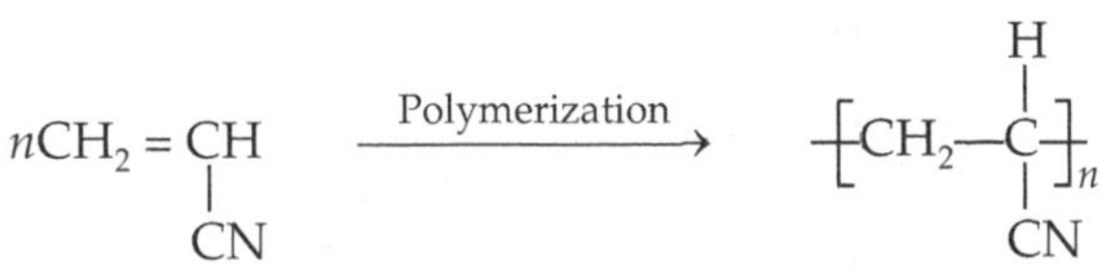

Polyacrylonitrile

PAN has remarkable resistance to heat upto around 220°C and exhibits very good mechanical properties. It is used for making warm clothes, carpets, blankets, etc.

8. **Polyvinylacetate (PVA).** It is prepared by heating vinyl acetate in the presence of benzoyl peroxide or acetyl chloride as catalyst. Vinyl acetate is obtained by treating acetylene with acetic acid in presence of $HgSO_4$ catalyst.

$$HC \equiv CH + CH_3COOH \longrightarrow \underset{\text{Vinylacetate}}{CH_2 = CHOCOCH_3}$$

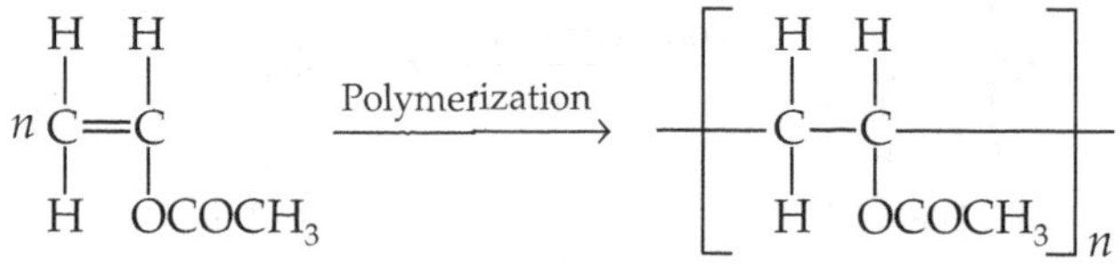

Vinylacetate Polyvinylacetate (PVA)

PVA is a colourless, transparent material having low softening point, resistant to water, atmospheric oxygen and chemicals. It is fairly soluble in organic acids and organic solvents. It is harmless if taken orally. It is used for making records, chewing gums, surgical dressings, paints, lacquors, plastic emulsions, coatings, card-boards, wrapping paper, finishing textiles and other fabrics and bonding paper, leather, textiles, etc. It is not used for moulding purposes as under the influence of compressive or textile force, they get distorted.

### 2.5.4 Rubber

Rubbers are high polymers, having elastic properties in excess of 300%. A rubber-band can be streched to 4 to 10 times its original length and as soon as stretching force is released, it returns to its original length. The elastic deformation in rubber arises from the fact that in the unstressed condition, the rubber molecule is not straight-chained, but in the form of a coil (Figure 2.7) and thus it can be streched like a spring. The unstreched rubber is amorphous. As streching is done, the macromolecules get partially aligned with respect to one another, causing crystallization and making it stiff. On releasing the deforming stress, the chains get reverted back to their original coiled state and the material again becomes amorphous.

#### 2.5.4.1 Natural Rubber

Natural rubber is mainly extracted from the tree *hevea brazilensis.* The rubber is obtained from *latex* collected from the cuts made in the bark of the tree. The latex contains 25–40% rubber hydrocarbon which coagulates to crude rubber when treated with 1% acetic acid. The crude rubber contains 90–95% of rubber hydrocarbon ($C_5H_8$).

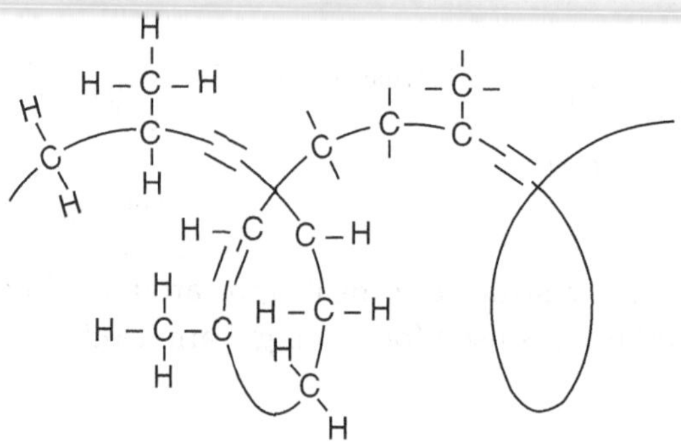

*FIGURE 2.7* **Natural rubber in coiled form.**

Rubber on destructive distillation gives isoprene ($C_5H_8$) as the main product suggesting it to be a polymer of isoprene. Hence it can be represented as $(C_5H_8)_n$. X-ray diffraction studies have shown that in natural rubber the isoprene units are arranged in the *cis*-form while in gutta-percha rubber isoprene units are arranged in *trans*-form (Figure 2.8). So, natural rubber is generally amorphous, elastic, and has a low melting point, while gutta-percha is hard, crystalline, and have high melting point.

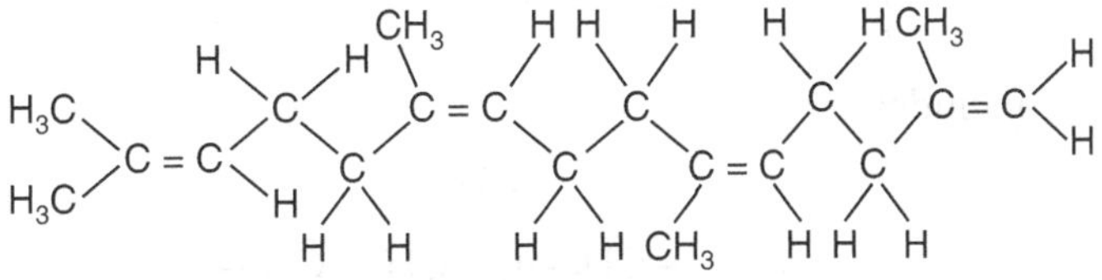

Natural rubber (all cis configuration)

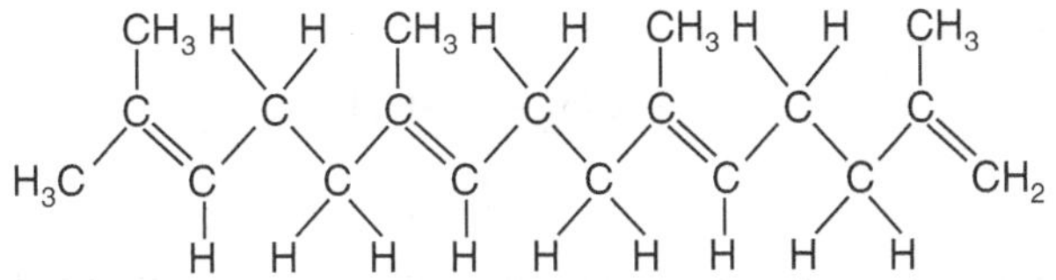

Gutta-percha (all trans configuration)

*FIGURE 2.8* **Configurations of natural rubber and Gutta-percha.**

### 2.5.4.2 Vulcanised Rubber

Natural rubber is soft, has a low tensile strength, poor resistance to abrasion, possesses elasticity only over a limited range and becomes more plastic and sticky on heating. All these undesirable properties of natural rubber may be overcome by its vulcanisation. In this process the rubber is heated with sulphur or sulphur compounds at a temperature of 150°C for a few hours. It converts rubber into a tougher product which has better elasticity over a wide range of temperature. The vulcanisation was discovered by *Charles Goodyear* in 1839.

During vulcanisation, sulphur and sulphur compounds react with rubber molecules to effect cross-linking in the following manner:

(*i*) 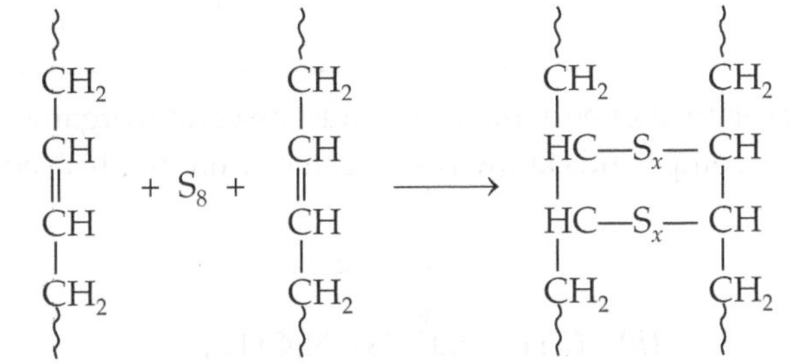

(ii) 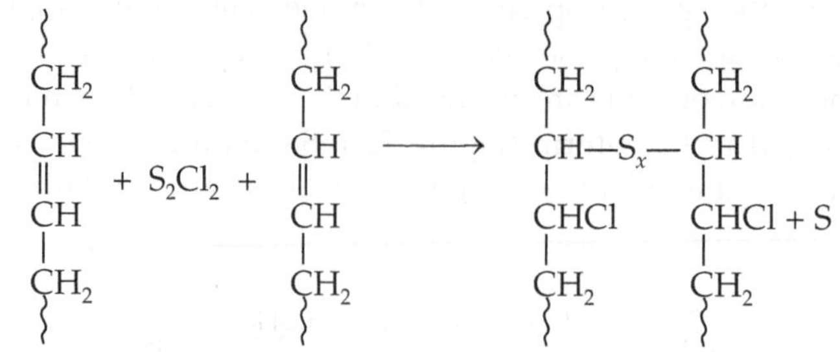

Other crosslinks include:

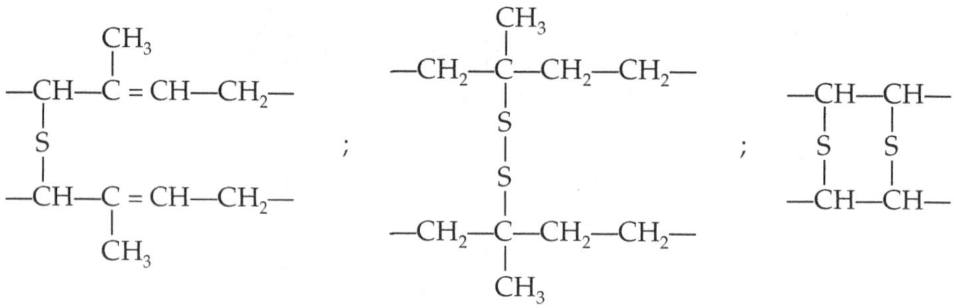

The vulcanisation process converts the natural rubber into either soft rubber (containing 5% sulphur) or hard rubber (containing 30–35% sulphur). The comparison of the properties of natural and vulcanised rubber are given in Table 2.4.

**TABLE 2.4 Comparison of Properties of Natural and Vulcanised Rubber**

| *S. No.* | *Property* | *Natural Rubber* | *Vulcanised Rubber* |
|---|---|---|---|
| 1. | Tensile strength | 200 kg/cm$^2$ | 2000 kg/cm$^2$ |
| 2. | Elasticity | very high (300–1000%) | Low |
| 3. | Water absorption | Large | Small |
| 4. | Swelling in organic solvents | Soluble | Less solubility |
| 5. | Workable temperature range | 10 to 60°C | - 40 to 100°C |
| 6. | Chemical resistance | Poor | Very good |

Properties of vulcanised rubber can further be enhanced by adding certain organic substances known as *accelators* and certain inorganic compounds known as *activators*. Commonly used accelators are:

(i) 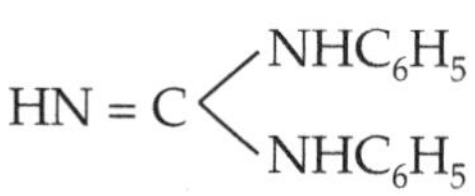

Diphenylguanidine

(ii) 

Tetramethyl thiouran disulphide

Commonly used activators are oxides and salts of zinc, calcium and lead. To prevent the oxidation of rubber a small amount (1–2%) of antioxidants such as phenyl β-naphthylamine, hydroquinone monobenzyl ether, etc. is added.

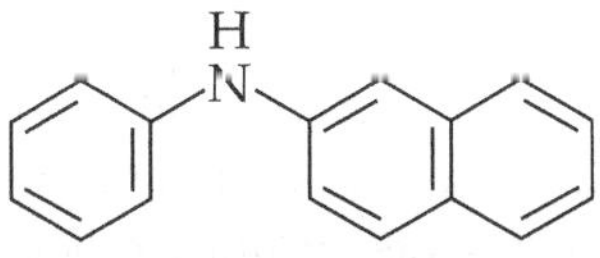

Phenyl β-naphthylamine

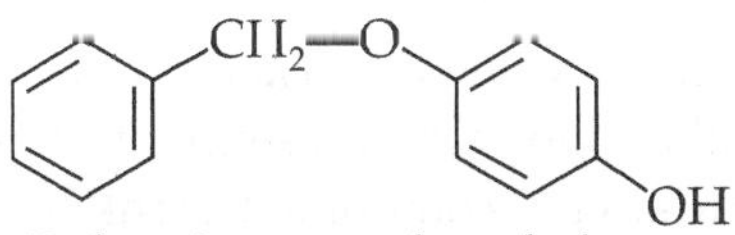

Hydroquinone monobenzyl ether

Vulcanised rubber is mainly used for making rubber-bands and gloves, but the rubber required in tyres and belting material require a greater tensile strength, stiffness, tear resistance, and abrasion resistance. These properties are achieved by adding *reinforcing materials* such as *carbon black, silica talc, zinc oxide,* etc. The reinforcing filler carbon black forms the chemical bonds with the rubber hydrocarbon during the processing thus making it stiff, and abrasion and tear resistant.

**Advantages of Vulcanised Rubber**

Vulcanisation improves the properties of natural rubber in following manner :

- Vulcanised rubber has good tensile strength and extensibility.
- It has excellent resilience, *i.e.,* articles made from it regains their original shape when the deforming load is removed.
- It possesses low water absorption tendency.
- It has higher resistance to oxidation and abrasion.
- It has higher resistance to wear and tear as compared to raw rubber.
- It is a better electrical insulator than natural rubber.
- It is resistant to organic solvents such as petrol, benzene, $CCl_4$, fats, and oils.
- It has a wide temperature range extending from – 40 to 100°C.

### 2.5.4.3 Synthetic Rubbers (Elastomers)

An elastomer is any vulcanisable man made rubber-like polymer, which can be stretched atleast twice its length, but it returns to its original shape and dimensions as soon as stretching force is released. The main types of synthetic rubbers are:

(*i*) *Buna-S (GR-S) or Styrene Rubber.* It is a copolymer of butadiene (75% by weight) and styrene (25% by weight).

$$\underset{\text{Butadiene}}{n(CH_2=CH{-}CH=CH_2)} + n\,\underset{\text{Styrene}}{\underset{\displaystyle C_6H_5}{\underset{|}{CH}}=CH_2} \xrightarrow{\text{Polymerization}}$$

$$\underset{\text{Buna-S}}{\left[ -(CH_2{-}CH=CH{-}CH_2)_x{-}\underset{\displaystyle C_6H_5}{\underset{|}{CH_2}}{-}CH_2- \right]_n}$$

Styrene rubber possesses high abrasion resistance, high load bearing capacity and resilience. It swells in oils and can be vulcanized by sulphur as $S_2Cl_2$. It is mainly used for manufacture of motor tyres and also for floor tiles, shoe soles, gaskets, wire and cable insulations, adhesives, tank linings, etc.

(*ii*) *Buna-N (GR-N) or Nitrile Rubber.* It is a copolymer of butadiene and acrylonitrile. This rubber has excellent resistance to heat, sunlight, oils, acids and salts but is less resistant to alkalies due to the presence of a CN group. It is used for making conveyor belts, high altitude aircraft components, tank-linings, hoses, gaskets, printing rollers, adhesives, etc.

$$\underset{\text{Butadiene}}{mCH_2=CH-CH=CH} + \underset{\text{Acylonitrile}}{n\underset{\substack{|\\ CN}}{C}H=CH_2} \xrightarrow{\text{Polymerization}} \underset{\text{Buna-N}}{\left(CH_2-CH=CH-CH_2\right)_m\left(CH_2-\underset{\substack{|\\ CN}}{C}H\right)_n}$$

(*iii*) *Neoprene (GR-M) or Polychloroprene.* It is obtained by the polymerization of chloroprene. It is highly resistance to vegetable and mineral oils, aging, and high temperatures but is more soluble in organic solvents. It is used for making hoses, gaskets, and tubing for carrying corrosive gases and oils. It is also used for making sponges, conveyor belts, adhesives, etc.

$$n(CH_2=\underset{\substack{|\\ Cl}}{C}-CH=CH_2) \xrightarrow{\text{Polymerization}} \underset{\text{Neoprene}}{\left(CH_2-\underset{\substack{|\\ Cl}}{C}=CH-CH_2\right)_n}$$

(*iv*) *Butylrubber (GR-I).* It is obtained by the copolymerization of isobutane with small amounts (1.5%) of isoprene.

$$\underset{\text{Isobutane}}{m\left[CH_2=\underset{\substack{|\\ CH_3}}{C}-CH_3\right]} + \underset{\text{Isoprene}}{n\left[CH_2=\underset{\substack{|\\ CH_3}}{C}-CH=CH_2\right]} \xrightarrow{\text{Polymerization}} \underset{\text{Butyl rubber}}{\left[\left[CH_2-\underset{\substack{|\\ CH_3}}{\overset{\substack{CH_3\\ |}}{C}}\right]_m\left[CH_2-\underset{\substack{|\\ CH_3}}{C}=CH-CH_2\right]\right]_n}$$

Due to small proportion of isoprene, the percentage of double bonds is very little in the copolymer. The arrangement of these monomers in the chains of the polymer is regular, so the inter-chain forces are high. So the given tensile strength is quite high in this elastomer.

It possess outstanding low permeability to air and other gases. It has excellent resistance to heat, abrasion, ageing, chemicals (such as $H_2SO_4$, $HNO_3$, HCl and HF), polar solvents (alcohol and acetone), but is soluble in benzene. It has high resistance to ozone and good electrical insulating properties.

It is used for making cycle and automobile tube, automobile parts, hoses, conveyor belts, tank linings, insulation for high voltage wires and cables.

(*v*) *Polysulphide rubber or Thiokol or GR-P.* These elastomers are also called "Thiokol". They are the condensation polymers produced by the reaction between sodium atoms of tetrasulphide and chlorine atoms of ethylene dichloride, forming sodium chloride. This reaction results in a bond being formed between the sulphur and carbon atoms and a long chain molecule is thus formed.

$$\underset{\text{Ethylene dichloride}}{Cl—CH_2—CH_2—Cl} + \underset{\text{Sodium polysulphide}}{Na—S—S—Na} \longrightarrow \underset{\text{Thiokel rubber}}{[CH_2—CH_2—S—S]_n} + 2NaCl$$

Polymer chains show sufficient regularity to crystallise on stretching, so it has sufficient tensile strength. These possess good resistance to mineral oils, fuels, solvents, oxygen, ozone and sunlight. It cannot be vulcanised and hence does not form hard rubber.

(*vi*) *Chlorosulphorated polyethylene rubber* (Hyplon). It is obtained by polymerizing polyethylene with chlorine and sulphur dioxide in following manner :

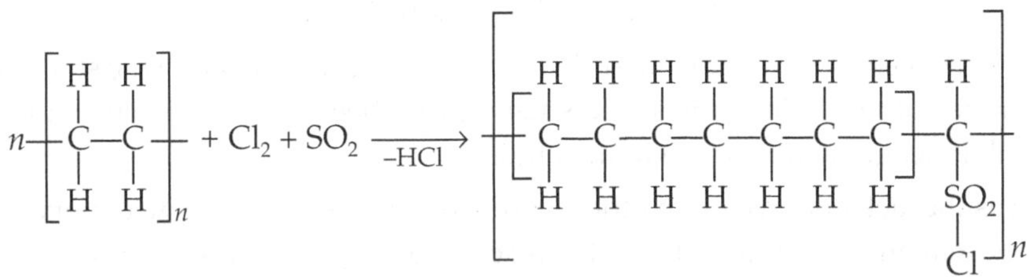

It shows high oxidation resistance Hyplen and high chemical resistance and is mainly used in chemical industry.

(*vii*) *Polyurethanes.* These are obtained by reacting polyhalcohols with di-isocyanates, *e.g.*,

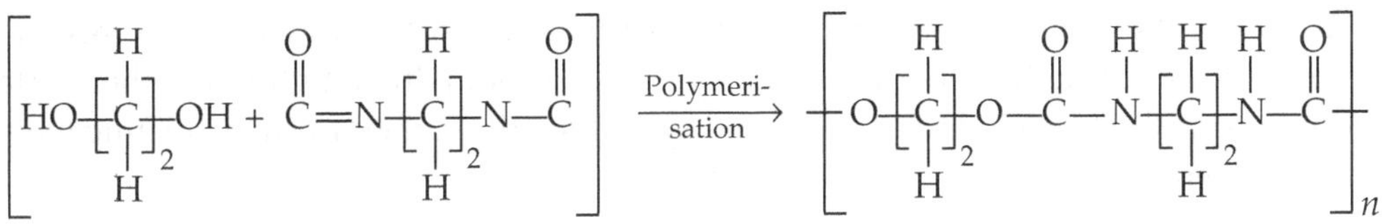

Ethylene glycol Ethylene di-isocyanate Polyurethane rubber

Di-isocyanates can also react with some of the active hydrogen atoms attached to the polymer chain, leading to a cross-linking, that is vulcanisation process. So in the manufacture of urethane polymers, casting and vulcanisation take place in one stage. This is a great advantage.

Di-isocyanates react with water vigorously to form carbon dioxide gas, as shown.

$$OCN—R—NCO + 2H_2O \longrightarrow H_2N—R—NH_2 + 2CO_2$$

Advantage is taken of this reaction is in the formation of vulcanised foamed urethane elastomers. The process involves mixing of a little water together with the short chain polymer and di-isocyanate. Simultaneously foaming, polymerization and vulcanisation take place and foamed rubber is produced.

Urethane polymers have got high gum tensile strength with the result that these elastomers crystallise on stretching.

Polyurethanes are highly resistant to oxidation, many organic solvents but are attacked by alkalies and acids. The polyurethane foams are highly tough and resistant to heat abrasion, chemicals and weathering. These are used for surface coatings and manufacture of foams and spandex fibres

**Properties of Synthetic Rubbers**

The following are the properties of synthetic rubbers which are taken into consideration before it is put to a particular use.

**1. Elasticity, Nerve and Flatness.** Elasticity is the opposite of plasticity and is measured by measuring plasticity. The plasticity is measured by plastometers such as, Goodrich plastometer. In this plastometer a cylinder of rubber of definite height is compressed by a constant load between parallel plates for 30 seconds. Then the load is released and the recovery is measured after 30 seconds. The maximum compression is a measure of softness. The proportion of compression retained by the rubber cylinder is a measure of retentivity. The two values added together give a measure of plasticity. When the value is high, the stock is plastic.

If a stock recovers considerably after distortion, the stock is called *nervy* stock. Thus a nervy stock has a low retentivity. When the stock is nervy, the sheets made of such a stock in a mill or calender, shrink in length and increase in thickness.

*Flatness* is the lack of nerve or, in other words, flatness is the opposite of nerve. Sheets made out of flat stock remain smooth with little shrinkage after coming out of the mill.

Elasticity or plasticity is a property of great importance in an elastomer in processing of the stock in factory. Rubbers with low plasticity (high elasticity) require excessive power and time on the mill for mixing them thoroughly. When the stock is nervy the operations with such a stock are very slow.

**2. Tack.** Tack is the special characteristic of rubber by virtue of which two fresh surfaces adhere or coalesce. Though rubber is not sticky and does not adhere tightly to other surfaces, it does adhere very well to itself. Tack is increased by making the surface fresh by treating the surface with some solvent.

Tack in rubber is of very great importance. By the use of this property composite rubber articles such as tyres are made. The freshened surfaces simply stuck together and then rolled into place. Synthetic rubbers generally lack this property.

**3. Stress-Strain Properties (Tensile Strength, Elongation, and Modulus).** Tensile strength, elongation and modulus are determined on a properly cured sample of the rubber.

At first the stock is cured in a mould of definite length for a definite period. A piece of dumbbell shape is cut out with a die. On the narrow portion are placed two marks one inch apart as shown below.

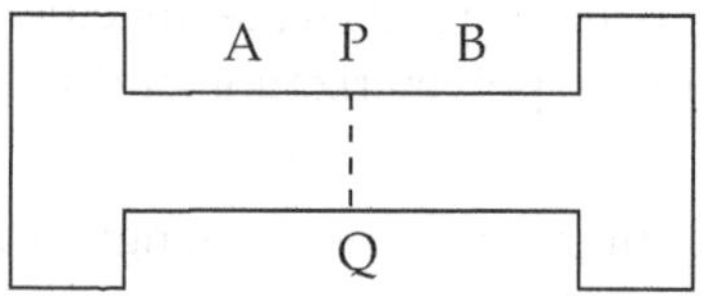

The test piece is then placed in the machine and stretched until it breaks. The load at which it breaks is the tensile strength of the stock and is expressed as pounds per square inch of the original cross-section area. (PQ is the cross-section in the original piece). Elongation at break is also noted. It is expressed as the percentage of the original length to which the test piece is stretched. If the original marks A and B are one inch apart and are then stretched to 3 inches the elongation is 200 per cent. A stress-strain curve is plotted to show the relationship between these two properties.

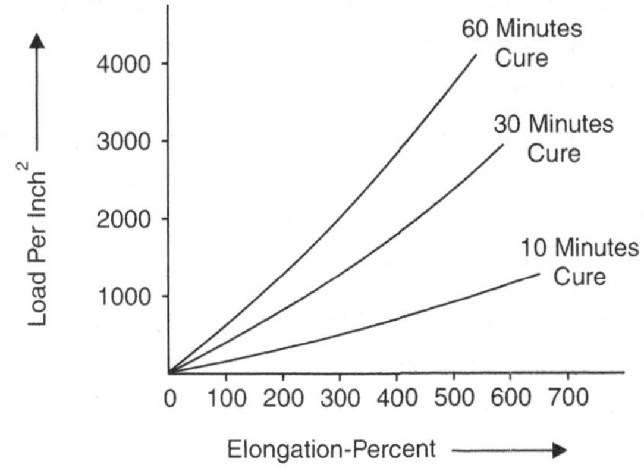


*FIGURE* 2.9

In Fig. 2.9. are given stress-strain curves for three different periods of cure.

4. **Rebound.** Rebound is a measure of resilience of a synthetic rubber. The determination of rebound is the determination of the height to which a rubber ball will bounce when dropped. In practice the test for measuring this property is carried out by allowing a weighted pendulum device to travel through a measured arc and to strike a block of the rubber. The rebound of the pendulum is read directly from a calibrated scale.

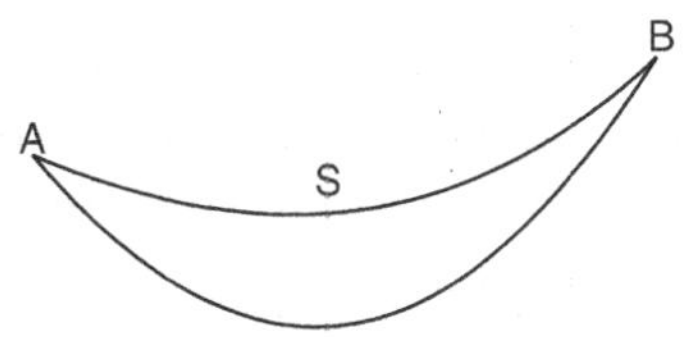


***FIGURE* 2.10. Tear test**

5. **Tear.** Resistance to tear of a stock is very important and is determined by the crescent tear test. A crescent-shaped piece AB of definite size is cut from vulcanised stock. A small slit S is made in the inner curved surface of the test specimen. It is then placed in a tensel machine and is loaded to rupture. The load required to separate the sample into two is directly proportional to the resistance to tear.

6. **Hardness.** Hardness of rubber is determined with the *Shore Durometer*. A metallic point, worked by means of a spring, is pressed against the surface of the stock, and the strain produced on the spring denotes the hardness of the rubber and the values of hardness are directly read on the scale of the instrument.

7. **Abrasion Resistance.** Abrasion is a process of wearing away of surface by friction. Abrasion resistance of a stock is measured by machines in which an abrasive material is rubbed against the surface of the stock. Volume loss is measured directly. The loss in volume in definite period denotes the loss on abrasion. The greater the loss, less is the abrasion resistance of the stock.

8. **Compression Set.** It is the extent to which distortion expressed as a percentage of the original thickness has become permanent, when a test piece is subjected to known load between plates. The resistance of vulcanised rubber to permanent distortion while subjected to compression stresses is very important.
9. **Hysteresis.** When a sample of rubber is deformed by means of a load and then it is released by releasing the load, the mechanical energy (the load applied) is converted to potential energy and then back (from potential energy to mechanical energy). A part of this energy is not recovered, that is, not converted into heat. The energy converted to heat is the hysteresis loss.

Hysteresis loss is very important as the heat generated due to hysteresis loss raises the temperature of the stock. The higher the hysteresis loss, the greater is the rise in temperature of the stock. A truck tyre formed out of a stock having high hysteresis, may reach temperatures of 200°C to 300°F in service and such a high temperature may char the tyre.

Hysteresis loss is measured by a *flexometer* in which a cylinder of vulcanised stock is compressed under constant load. On top of this constant load compression there is a very rapid cyclic defection. This leads to a rapid heating up of the sample. The hysteresis is $\Delta T$ or the heat rise during the test.

### 2.5.5 Zeigler-Nata Catalysts

These are the special type of coordination catalysts, comprising of two components, *i.e.*, the catalyst and the cocatalyst. The catalyst component consists of halides of IV–VIII group elements having transition valence and the cocatalysts are organometallic compound such as alkyls, aryls and hydrides of group I–IV metals (Zeigler-Nata Catalysts). Although these are organo aluminium compounds such as triethyl aluminium ($AlEt_3$) or diethyl aluminium chloride ($AlEt_2Cl$) in combination with titanium chlorides—both tri and tetra ($TiCl_3$ and $TiCl_4$) are most commonly used. Aluminium alkyls act as an electron acceptor and the electron donor is titanium halides and the combination, therefore, readily forms coordination complexes. The complex formed is insoluble in the solvent and is, hence, heterogeneous in nature. Some of the following structures have been proposed for these complexes:

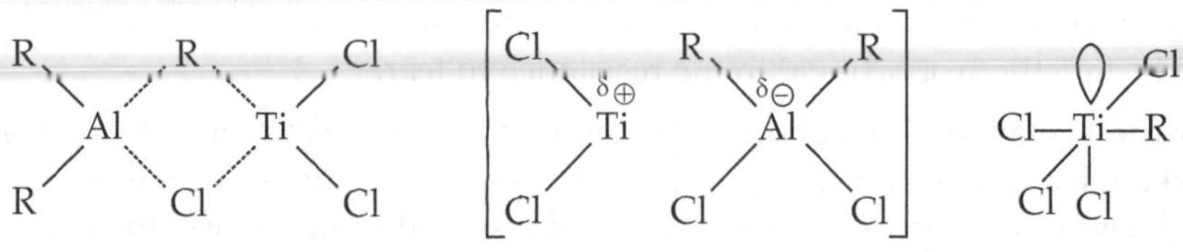

The active centres from where the polymer chain growth propagates are formed at the surface of the solid phase of the catalyst complex, and the monomer is complexed with the metal ion of the active centre before its insertion into the growing chain. The complex formed now acts as the active centre. The monomers then attach toward the Ti—C bond (C from the alkyl group R) in the active centre, when it forms a $\pi$ complex with the Ti ion.

**Formation of active centre.** The five coordinated titanium ions on the surface of the catalyst is assumed to have a vacant orbitals as shown at 6 in the following structure:

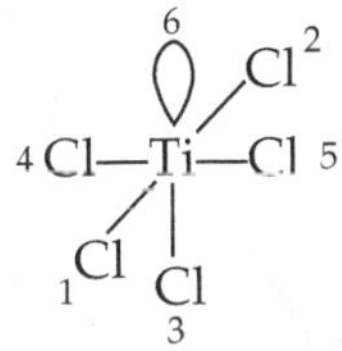

After the chemisorption of aluminium alkyl on the $TiCl_3$ crystal, $Ti^{3+}$ is alkylated by an exchange mechanism, as follows:

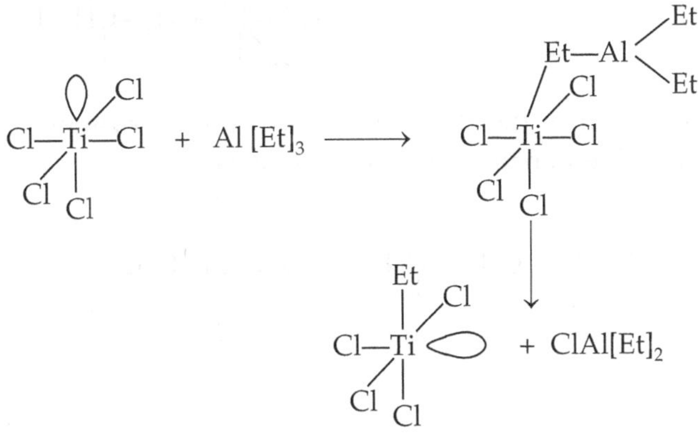

active catalyst

Once the active catalyst is formed, the monomer is attached towards the vacant orbital which then forms a transition $\pi$ complex with the Ti.

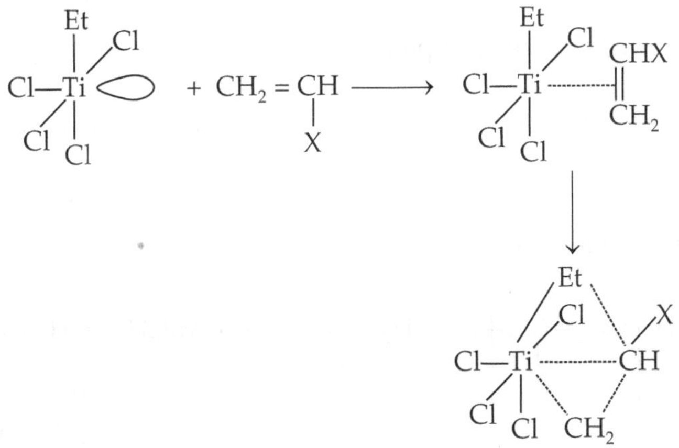

Transition state

The transition state quickly gives rise to the growth of the polymer chain by the monomeric insertion at the Ti-Et bond.

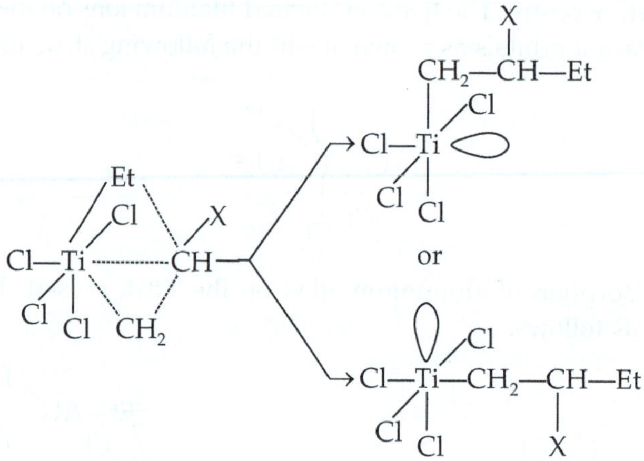

Therefore, on the basis of the above discussion, polymerization is characterised by the initiation, propagation, and termination reaction as follows:

(*i*) *Initiation*

$$\underset{\text{active centre}}{\text{Mt—R}} + CH_2 = \underset{\text{X}}{\underset{|}{CH_2}} \longrightarrow \text{MtCH}_2\underset{\text{X}}{\underset{|}{\text{C}}}\text{HR}$$

(*ii*) *Propagation*

$$\text{MtCH}_2\underset{\text{X}}{\underset{|}{\text{C}}}\text{HR} + n\text{CH}_2 = \underset{\text{X}}{\underset{|}{\text{CH}}} \longrightarrow \text{MtCH}_2\underset{\text{X}}{\underset{|}{\text{CH}}}\left[\text{CH}_2\text{—}\underset{\text{X}}{\underset{|}{\text{CH}}}\right]_n\text{R}$$

(*iii*) *Termination*

(*a*) By an active hydrogen compound

$$\text{MtCH}_2\underset{\text{X}}{\underset{|}{\text{CH}}}\left[\text{CH}_2\text{—}\underset{\text{X}}{\underset{|}{\text{CH}}}\right]_n\text{R} + \text{R}'\text{H} \longrightarrow \text{MtR}' + \text{CH}_3\underset{\text{X}}{\underset{|}{\text{CH}}}\left[\text{CH}_2\text{—}\underset{\text{X}}{\underset{|}{\text{CH}}}\right]_n\text{R}$$

(*b*) By Transfer with monomer

$$\text{Mt CH}_2\underset{\text{X}}{\underset{|}{\text{CH}}}\left[\text{CH}_2\text{—}\underset{\text{X}}{\underset{|}{\text{CH}}}\right]_n\text{R} + \text{CH}_2\text{—}\underset{\text{X}}{\underset{|}{\text{CH}}} \longrightarrow \text{MtCH}_2\text{—}\underset{\text{X}}{\underset{|}{\text{CH}_2}} + \text{CH}_2 = \underset{\text{X}}{\underset{|}{\text{CH}}}\left[\text{CH}_2\underset{\text{X}}{\underset{|}{\text{CH}}}\right]_n\text{R}$$

(*c*) By spontaneous internal transfer

$$\text{MtCH}_2\underset{\text{X}}{\underset{|}{\text{CH}}}\left[\text{CH}_2\text{—}\underset{\text{X}}{\underset{|}{\text{CH}}}\right]_n\text{R} \longrightarrow \text{MtH} + \text{CH}_2 = \underset{\text{X}}{\underset{|}{\text{C}}}\left[\text{CH}_2\text{—}\underset{\text{X}}{\underset{|}{\text{CH}}}\right]_n\text{R}$$

Here Mt denotes transition metals such as Ti, Mo, Cr, V, Ni, or Rh, Zeigler Natta polymerization is used to prepare polypropylene, polyethylene polydiene, etc.

## 2.6 CONDENSATION POLYMERIZATION

This polymerization involves the condensation of two or more monomers with the elimination of simple molecules like $H_2O$, $NH_3$, R—OH, etc. Some common condensation polymers are listed in Table 2.5.

**TABLE 2.5 Common Condensation Polymers**

| *Type* | *Characteristic Linkage* | *Polymerization Reaction* |
|---|---|---|
| Polyamide | –CO–NH– | $H_2N–R–NH_2 + HOOC–R'–COOH \longrightarrow H(NH–R–NH–CO–R'–CO)_{\overline{2}}OH + H_2O$<br>$H_2N–R–NH_2 + ClCO–R'–COCl \longrightarrow H(NH–R–NHCOR'–CO)_{\overline{n}}Cl + HCl$ |
| Polyester | –CO–O– | $HO–R–OH + HOOC–R'–COOH \longrightarrow HO(R–OCO–R'COOH)_n + H_2O$ |
| Polysiloxane | –Si–O– | $Cl–SiR_2Cl \longrightarrow HO–SiR_2–OH \longrightarrow HO–(SiR_2–O)_{\overline{n}}H$ |
| Phenolform-aldehyde | $–Ar–CH_2–$ | 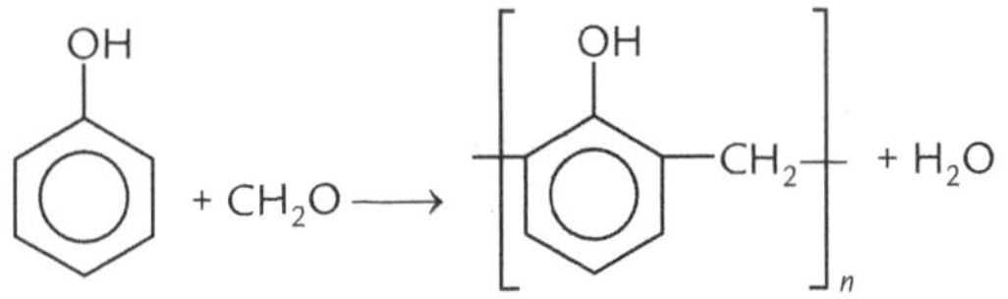 |
| Urea form-aldehyde | $–NH–CH_2–$ | $H_2N–CO–NH_2 + CH_2O \longrightarrow (NH–CO–NH–CH_2)_n + H_2O$ |
| Melamine formaldehyde | $–NH–CH_2–$ | 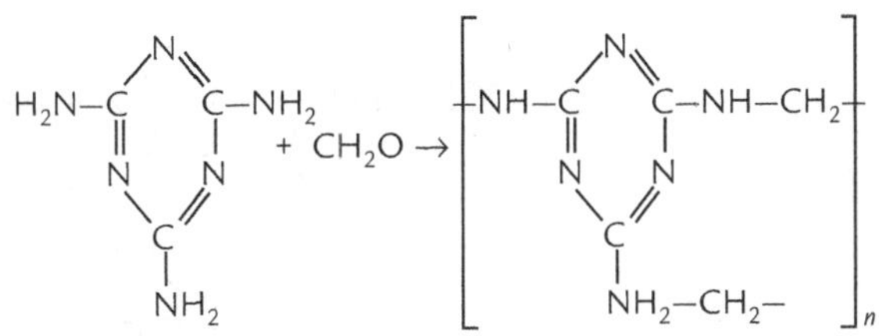 |
| Polyacetal | –O–CH(R)–O– | $RCHO + HO–R'–OH \longrightarrow (O–R'–OCHR)_n + H_2O$ |
| Polysulphide | –S– | $xClCH_2CH_2Cl + Na_2S \longrightarrow Cl\,(CH_2CH_2S)_x\,Na$ |

### 2.6.1 Commercially Important Condensation Polymers

**1. Polyesters.** These polymers have ester linkages as follows :

$$\left[ O-R-O-\underset{\underset{O}{\|}}{C}-R'-\underset{\underset{O}{\|}}{C} \right]_n$$

The most important polymer of this class is *polyethylene terephthate* (PET) also known as *terylene* or *dacron*.

**[A] Terylene or PET.** It is prepared by heating ethylene glycol and terephthalic acid. In commercial practice dimethyl terephthalate (DMT) is taken in place of terephthalic acid. The various steps of manufacture are as follows :

(*a*) Dimethyl terephthalate is prepared by esterification of terephthalic acid with methyl alcohol in the presence of an acid catalyst such as sulphuric acid.

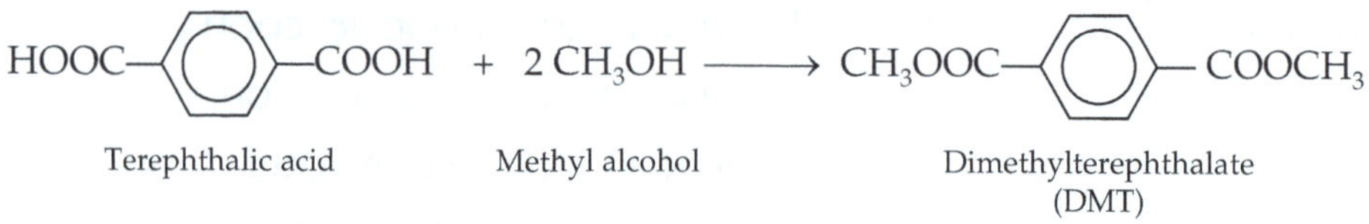


(*b*) Dimethyl terephthalate is reacted with ethylene glycol to give diethylene glycol terephthalate. The reaction mixture is heated to a temperature below the boiling point of ethylene glycol but well above the boiling point of methyl alcohol.

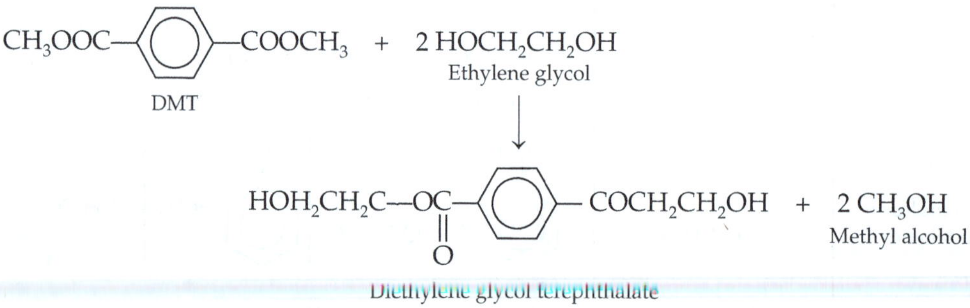


(*c*) Di ethylene glycol terephthalate is polymerised into polyethylene terephthalate by heating diethylene glycol terephthalate in vacuum for distilling off the excess of glycol.

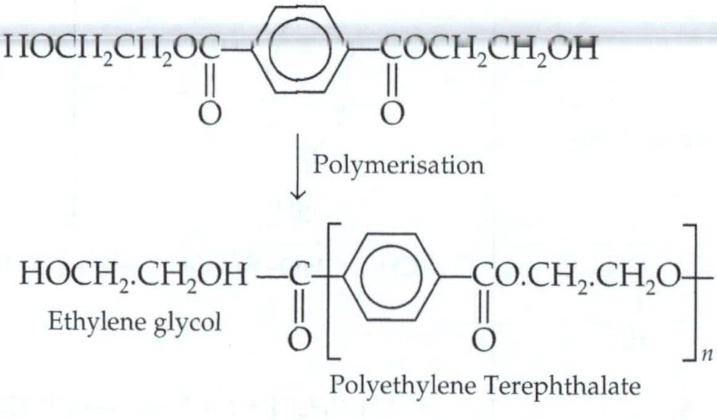

**Properties of Ethylene Terephthalate**

(*i*) Ethylene terephthalate starts softening at 70°C and melts between 257°C and 265°C. Amorphous variety has density of 1.33 and drawn yarns have density of 1.38 to 1.41.

(*ii*) It is not soluble in common types of organic solvents but dissolves in a certain number of polar solvents upon heating. Such polar solvents are phenol, *o*-chloro phenol, *m*-cresol, trifluoroacetic acid, nitrobenzene and phenol-teraclhloroethane (10 . 3 by weight).

(*iii*) Polyethylene terephthalate is highly resistant to mineral and organic acids. At 40°C in 72 hours, 30% HCl or 70% $H_2SO_4$ has hardly any effect on it. It is, however, less resistant to alkalies. At 100°C at low concentration, alkalies have no effect on the polymer but heating with concentrated alkalies or under pressures attacks it.

(*iv*) Polyethyelene glycol can be depolymerised to a lower molecular weight product by heating in the presence of an excess of glycol.

**Applications of Polyethylene Terephthalate**

(*i*) *Fibres.* Synthetic fibres known as *Terylene* or *Dacron* are prepared out of polyethylene terephthalate by melting. The hot melt is spun or extruded. Fine filaments are used as textile materials and coarse ones are used for brush bristles.

Dacron fibres have high resistance to stretch as the initial modulus of elasticity of Dacron is quite high. This high modulus of elasticity is due to the high stiffness of polymer chains. The crease resistance and high rate of recovery after deformation of Dacron fibres are also attributed to this stiffness or high strength of chains in this polymer.

Polyethylene terephthalate or Dacron fibres are superior to nylon fibres in resistance to heat, light degradation and exposures to bleaches and acids. However, nylon fibres are superior in dyeability, tenacity and resistance to alkalies.

Both nylon and Dacron fibres melt before burning and leave a hard bead after the flame is extinguished. Nylon gives an odour of burnt hair when heated in a flame and Dacron emits heavy black smoke.

(*ii*) *Films.* Extruded films of polyethylene terephthalate are transparent, tough, durable and impermeable to a number of organic and inorganic gases. They remain flexible and stable over a range of temperature from – 60 to 150°C. They exhibit high dielectric strength and relatively low power factor. So they are used for electrical insulation in addition to their uses as a fine packaging material, as a base for industrial tapes, as a lining material and as a glass replacement.

(*iii*) Polyethylene terephthalate can also be injection moulded but the products are somewhat brittle.

(*iv*) Manufacture of magnetic recording tapes.

(*v*) Manufacture of bottles for cold drinks, fruit juice and sauces.

(*vi*) Used for blending with wool.

**[B] Alkyl Resins (Glyptal).** These are prepared by condensation polymerization of polybasic acids or anhydrides with polyhydric alcohols (like glycerols) in correct proportions, in presence of heat and catalyst ($CO_2$ gas).

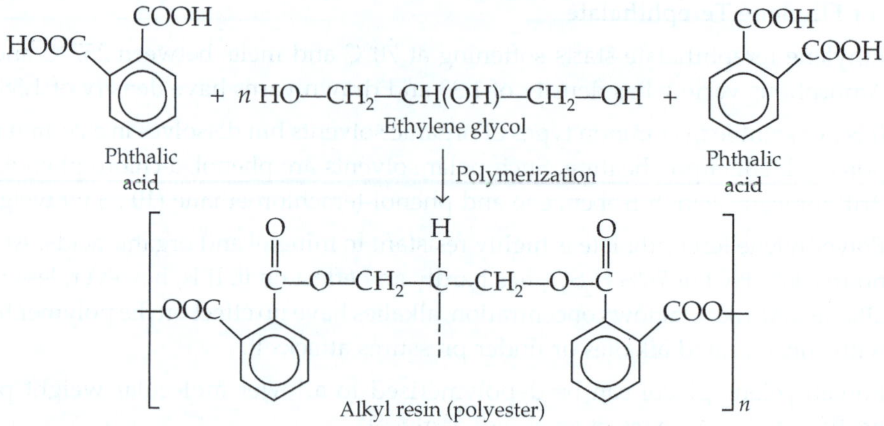


Alkyl resins are modified either by oil and/or fatty acids. These are insoluble, infusible, dimensionally stable, hard, resistant towards atmospheric oxygen and acids. Alkyl resins are used in preparation of :

(*i*) Switches and gears,

(*ii*) Aircraft and automobile parts,

(*iii*) Car panels and bodies,

(*iv*) Sheets, rods, tubes, etc. and ;

(*v*) Good insulators and circuit break insulators.

**2. Polyamides.** These have amide linkages 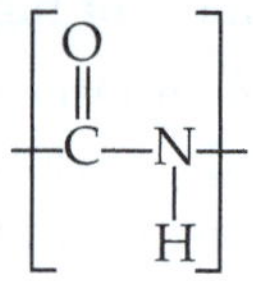 in the chain. These are popularly known as *nylons*.

**[A] Nylon-6 : 6.** Nylon–6 : 6 is manufactured from hexamethylene diamine and adipic acid.

Hexamethylene diamine is prepared by catalytic dehydration of the ammonium salt of adipic acid to give adiponitrile which is hydrogenated to the diamine.

$$\underset{\text{Adipic acid}}{HOOC(CH_2)_4COOH} + \underset{\text{Ammonia}}{2NH_3} \longrightarrow \underset{\text{Adiponitrile}}{NC(CH_2)_4CN} + 2H_2$$

$$NC(CH_2)_4CN + 4H_2 \xrightarrow[\text{Catalyst}]{\text{Heat}} \underset{\text{Hexamethylene diamine}}{H_2N(CH_2)_6NH_2}$$

Adipic acid is prepared by hydrogenating phenol to form cyclohexanol which is converted to adipic acid by oxidation with air or nitric acid.

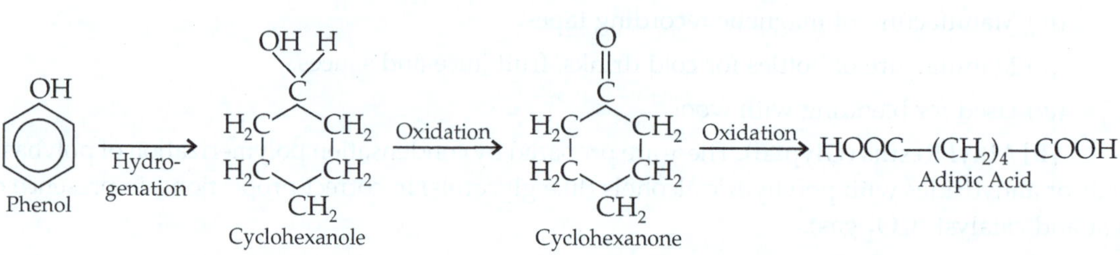

**Polymerisation.** Polymerisation of adipic acid and hexamethylene diamine involves heating the two in the absence of air so that the amine and carboxyl groups condense to form amide linkage with the evolution of water.

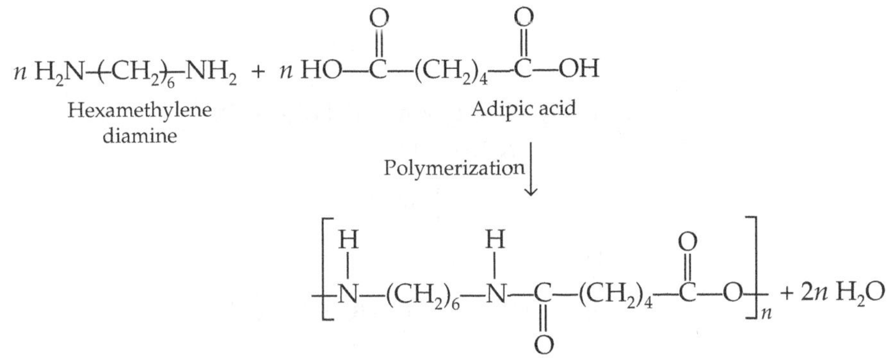


Nylon-6 : 6

The polymer is extruded into a ribbon through a heated valve. The ribbon is cut into chips which are dried in hot air. Nylon–6 : 6 has a melting point of 264°C and a specific gravity of 1.14.

**[B] Nylon-6.** It has the monomer caprolactum which is obtained from cyclohexane.

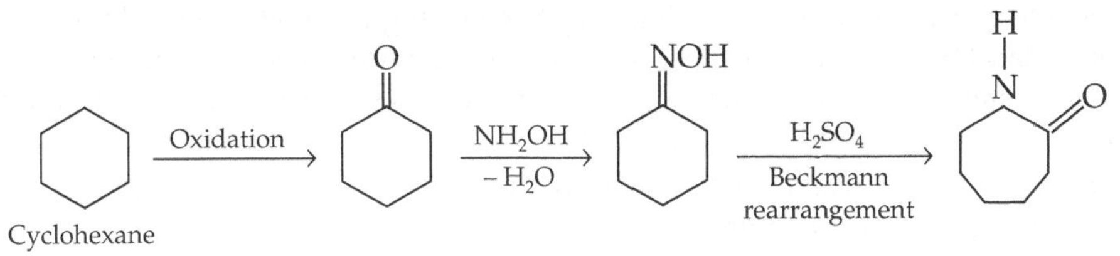


Caprolactum on heating with traces of water gives aminocaproic acid which polymerises to give Nylon-6.

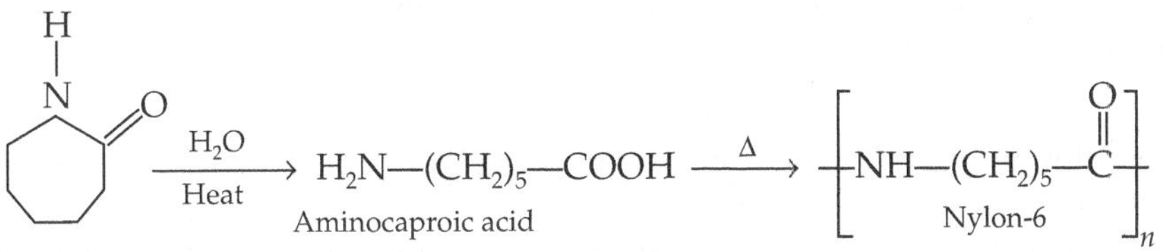


It is used in the manufacture of tire cords, fabrics, and ropes.

**[C] Nylon-11.** It is obtained by the self condensation of ω-aminoundecanoic acid.

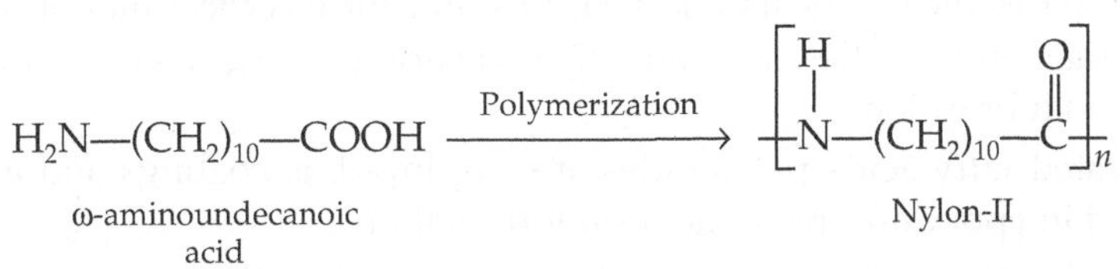


**[D] Nylon-6 : 10.** It is produced from hexamethylene diamine made from phenol and sebacic acid made from castor oil. The method employed is similar to that employed in the production of the 6 : 6 polymer. Hexamethylene diamine and sebacic acid are first reacted together to form nylon 6 : 10 salt, hexamethylene diammonium sebacate, which is polymerised further.

The polymer formed is extruded to form a ribbon. The ribbon is cut into chips.

$$nH_2N-(CH_2)_6-NH_2 + nHOOC-(CH_2)_8-COOH$$

Hexamethylene diamine        Sebasic acid

$$\downarrow \text{Polymerization}$$

$$\left[H_3\overset{+}{N}-(CH_2)_6-NH_3OOC-(CH_2)_8-COO\right]_n$$

Nylon- 6 : 10 salt

$$\downarrow \text{Polymerization}$$

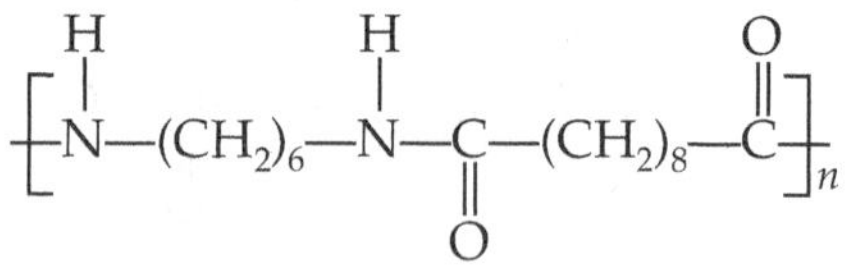

Nylon- 6 : 10

Nylon 6 : 10 has melting point of 220°C and a specific gravity of 1.0

**Properties and Uses of Polyamides.** Polyamide resins are horny, whitish, translucent, high melting polymers. Polyamide resin can be essentially transparent and amorphous when their melts are quenched. On annealing or cold drawing, they become highly crystalline and translucent. These polymers are used for fibres, bristles, bearings, gears, moulded objects, coatings and adhesives. The term nylon refers specially to those synthetic polyamides which can be spun into fibres. The following are the properties and uses of nylon :

(*i*) Nylon is a thermoplastic material. The natural colour of nylon is milky white but coloured materials are also available.

(*ii*) Nylon used for injection moulding and extrusion has high melting points, 160° to 240°C. Plasticised nylon flows easily.

(*iii*) Nylon is insoluble in common organic solvents such as methylated spirits, benzene and acetone. It is soluble in phenol and formic acid.

(*iv*) Nylon has low specific gravity, 1.1. It has high water absorption.

(*v*) Nylon mouldings and extrusions have good physical strength, especially high impact strength.

(*vi*) Nylon 6 : 6 and nylon 6 : 10 are primarily used for making fibres. However, they are also employed in moulding compositions.

(*vii*) Nylon 6 and Nylon 11 are mainly used as moulding compositions for the production of machine parts such as gears and bearings and for electrical mountings. They are also used as filaments, bristles and films. Nylon bearings and gears perform quietly and need little or no lubrication.

(*viii*) Unsaturated fatty acids polyamides are employed as coatings and adhesives and are also used in epoxy and phenolic resin formulations.

**3. Phenolic Resins.** These are formed by the condensation polymerization of phenol derivatives (like phenol, resorcinol, etc.) with aldehydes (like formaldehyde, furfural, etc). On heating these resins become highly cross linked, forming insoluble, infusible product. For example,

[A] **Phenol-formaldehyde resins or Bakelite.** It is a condensation polymer obtained from phenol and formaldehyde in the presence of a basic catalyst.

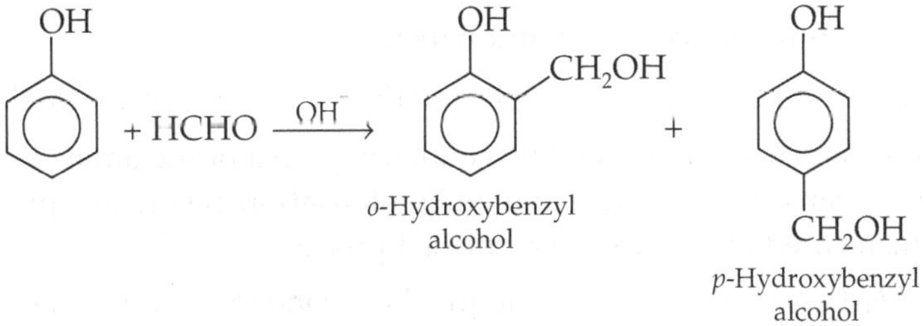


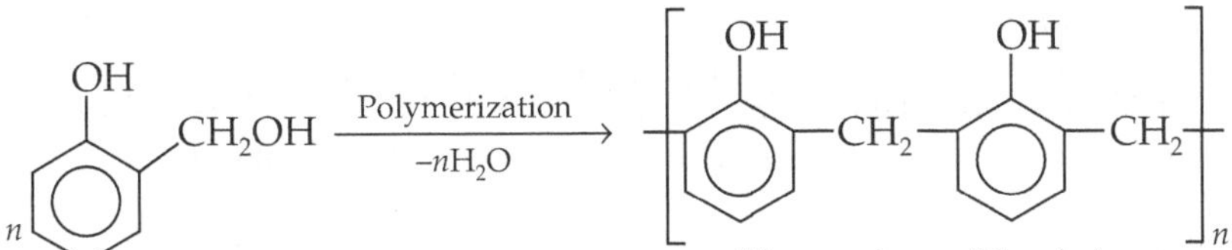


Linear polymer (Novolac)

The *o*- and *p*-substituted phenols can undergo polymerization to produce a cross-linked polymer known as *bakelite*.

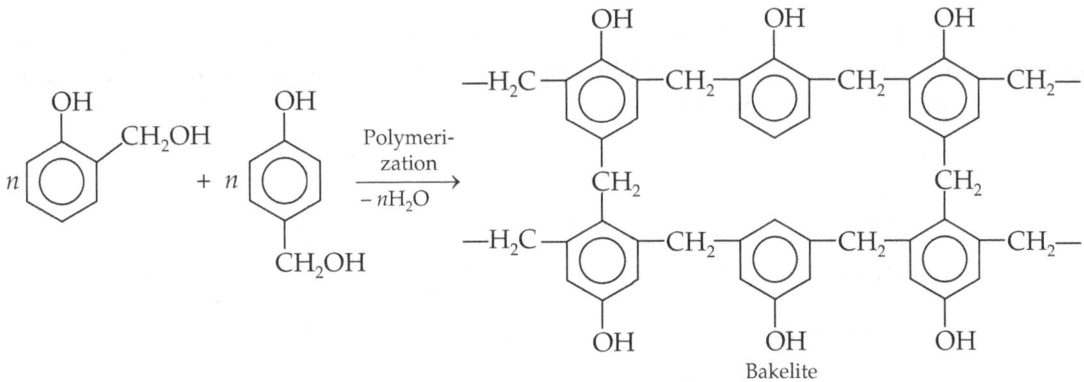


Bakelite
(low degree of polymerization)

During moulding hexamethylene tetramine $[(CH_2)_6 N_4]$ is added, this provides formaldehyde which converts the soluble and fusible *novolac* into hard infusible solid of cross linked structure.

**Properties.**

(*i*) Phenolic resins are, hard, scratch resistant, infusible, water-resistant, insoluble solids.

(*ii*) They are resistant to non-oxidizing acids, salts and many organic solvents, but are attacked by alkalis, because of the presence of free hydroxyl group.

(*iii*) They posses excellent electrical insulating character.

**Uses.**

(*i*) For making electric insulator parts like switches, plugs, switch boards, heater handles, etc.

(*ii*) Soft bakelites with low degree of polymerization are used for making glue for binding laminated wooden plantes and in varnishes.

(*iii*) For impregnating fabrics, wood and paper.

(*iv*) For making bearings, used in propellar shafts for paper industry and rolling mills.

[B] **Amino Resins.** These are formed by condensing urea or melamine with formaldehyde. In preparing amino resins, firstly intermediate condensation resins are prepared. These are compounded and then cured to give rise to a finished product.

(*a*) *Urea-formaldehyde resins* (UF). The intermediate condensation product is formed as :

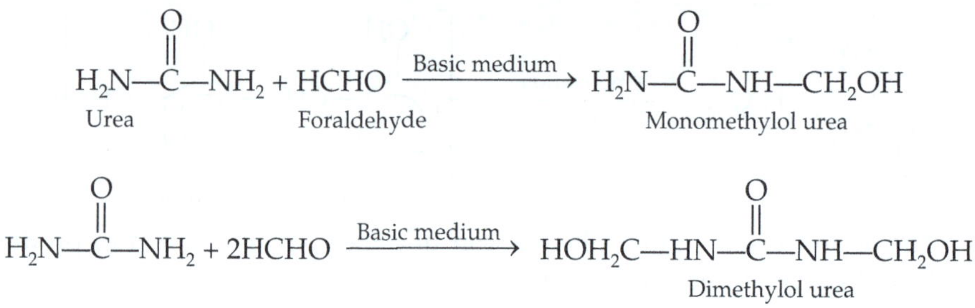


Four moulding, the methylol derivatives are compounded (with fillers, plasticizers, pigments and catalyst) and then curved by applying heat and pressure. During curring, long —C—N—C—N— chains are formed.

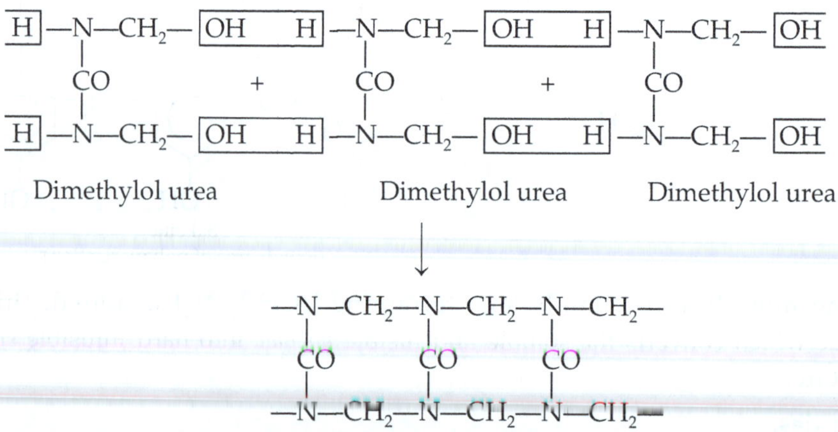


Urea formaldehyde resin

Urea-formaldehyde resins are clear, water-white products having good tensile strength, good electrical insulation, good chemical resistance, great hardness and good abrasion resistance. These are used for following purposes :

(*i*) For bonding grinding wheels.

(*ii*) Binder of glass fibres, rodewood, etc., which are used for filtration and insulation purposes.

(*iii*) Binder for foundry cores.

(*iv*) Electrical insulation.

(v) Decorative articles like plates, drinking glasses, dishes, etc.

(b) *Melamine-formaldehyde resins* (MF). It is formed by condensing melamine with formaldehyde at about 80°C. It is obtained in following steps :

(i) Melamine, a trimer of dicyanidimide is formed as,

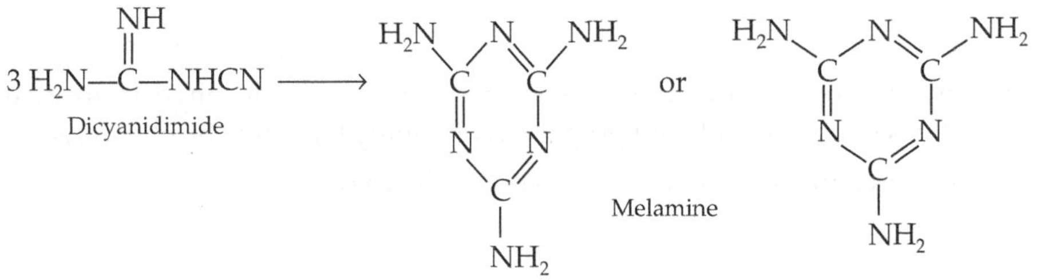


(ii) Melamine condenses with formaldehyde forming trimethylol melamine as,

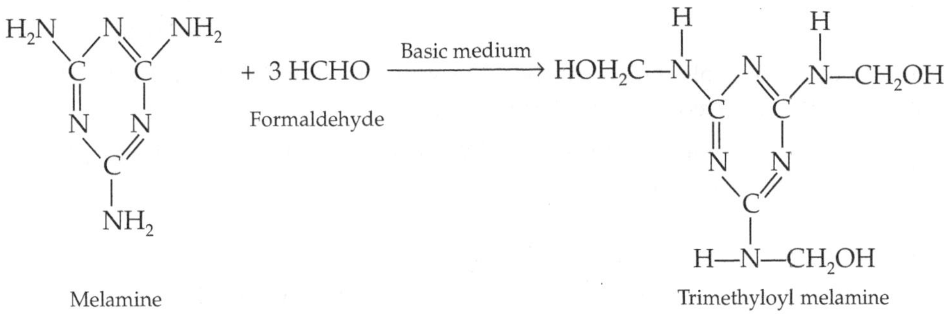


(iii) Trimethyloyl melamine polymerizes to give melamine formaldehyde resins as,

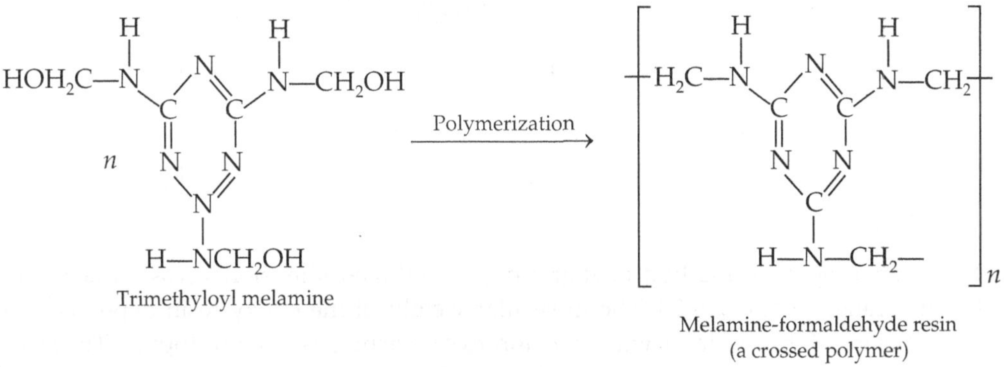


MF resin is better the UF resin *w.r.t.* water, heat, chemical resistances, hardness and tensile strength. MF resins are used for following purposes :

(i) In surface adhesives, lacquers, paper industry, plywood industry (as adhesive) etc.

(ii) In making crockery, cups and plates which are highly durable and do not break on being dropped.

**4. Epoxy Resins.** These resins are mainly formed from *epichlorohydrin*, an organic epoxide. Epichlorohydrin is a colourless toxic liquid boiling at 115.2°C. It is prepared as,

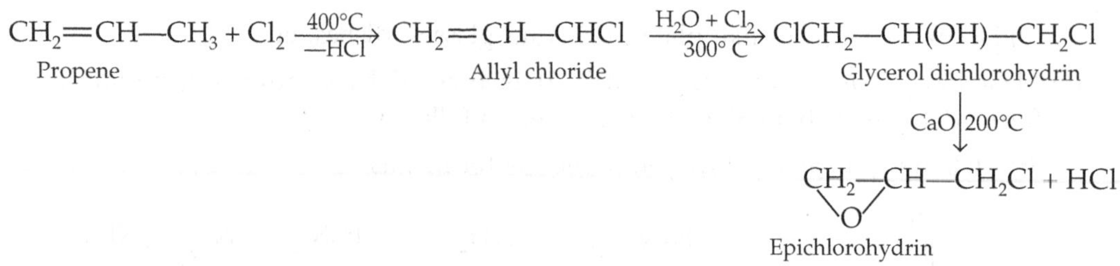

Many dihydric compounds may be condensed with epichlorohydrin to give resinous materials but 2, 2′-bis (*p*-hydroxy-phenyl propane), commonly known as bisphenol, is widely used.

Bisphenol is prepared by condensing phenol with acetone as,

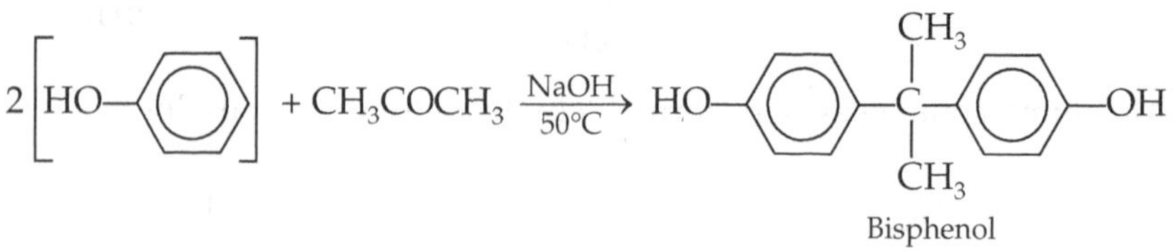

The other phenols which can be used with epichlorohydrin include trihydroxy benzene, resoricinol, ethyene glycol, dimethyl glycol, etc.

Condensation reaction between *epichlorohydrin* and *bisphenol* takes place at 60°C in the presence of alkaline catalyst.

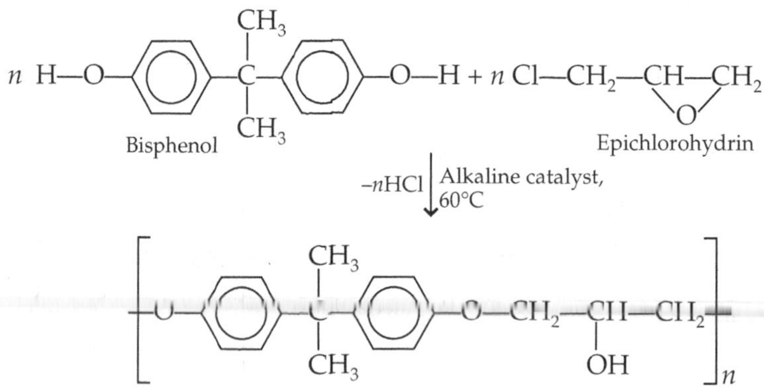

Epoxy Resin

The reactive, epoxide and hydroxyl groups give a three dimensional cross linked structure. The value of $n$ ranges from 1 to 20. The molecular weight of the epoxy resin depends upon the relative proportions of the reactants with epichlorohydrin acting as a chain stopper. The molecular weight of the epoxy resins may range from as low as 350, a mobile easy flowing liquid to as high as 8000, a solid melting at 145-155°C.

The ultimate properties and particular applications are the main factors in the selection of an epoxy resin of particular molecular weight. For example, the lower molecular weight liquid epoxy resins are used in lamination, casting and adhesives while higher molecular weight resins are more useful in surface coatings.

**Curing of Epoxy Resins.** Epoxy resins show outstanding properties of toughness, chemical inertness, flexibility and adhesion in finished products only in three-dimensional cross-linked state. Linear epoxy resins as discussed above are converted into three-dimensional polymers by reaction with chemicals such as diethylene triamine, triethylene tetramine and meta phenylene diamine. This is called curing of epoxy resins. Epoxy resins thermoset when heated with cross-linking reagents.

**Applications of Epoxy Resins**

1. Epoxy resins are used as adhesives. They stick to a number of substances including metals and glass.
2. They are used for surface coatings.
3. Moulds made with epoxy resins are employed for the production of metal components for aircraft and automobiles.
4. They are also used as laminating and casting materials, as potting compounds for electrical equipments and as stabilisers for PVC resins.
5. Epoxy resins are used for skid resistant surfaces for highways. The advantages are : (*i*) long wearing of road surface in hot or cold climate, (*ii*) permanent high traction even under wet or oily conditions, (*iii*) excellent resistance to freezing conditions, de-icing salts, solvents and water, (*iv*) non-porosity which protects the original pavement from scaling or spalling, (*v*) light weight, especially useful for surfacing bridges, (*vi*) fast curing, thereby causing minimum interruption of the flow of traffic.
6. Epoxy resins are applied over cotton, rayon and bleached fabrics to impart crease resistance and shrinkage control. Although epoxies are more expensive than urea formaldehyde and melamine formaldehyde resin, yet they have the advantage of not retaining chlorine even after strong alkaline washing and hot chlorination. So no amine odours are developed in the fabric.

## 2.7 CONDUCTING POLYMERS

Ordinary polymers, made by the usual methods of polymerization are pure insulators. Conducting polymers are long, carbon based chains composed of simple repeating units called monomers and are capable of conducting current down to the chain. To make the polymer materials conductive, they are doped with atoms that donate negative or positive charges (oxidizing or reducing agents) to each unit, enabling the current to travel down the chain. Depending on the dopant, conductive polymers exhibit either *p*-type or *n*-type conductivity.

The most extensively studied conductive-polymer systems are based on polyaniline, polythiophene, polypyrrole, and polyacetylene. The principal attractions of these polymers over conventional conducting materials are their potential ease of processing, relative robustness, and light weight. Successful commercial applications require a fine balance of conductivity, processability, and stability, but until recently material researchers could not obtain all three properties simultaneously.

Conductive polymers are much more electrically conductive than standard polymers but much less than metals such as copper. In practice, the conductivity of these materials is characterized by low-charge carrier mobility—a measure of how easily electric charge moves. This characteristic limits response speed in the case of a transistor, for example, making such a device rather inefficient.

### 2.7.1 Classification

The conducting polymers can be classified into the following five categories:

1. **Conductive element-filled polymer.** It is a resin or polymer filled with conducting elements such as carbon black, metallic fibres, metal oxides, etc. In this, the polymers acts as the binder to hold the conducting elements together in the solid entity. These polymers possess reasonably good bulk conductivity; and are, generally, low in cost, light in weight, mechanically durable and strong, and easily processable in different forms, shapes, and sizes.
2. **Conjugated π-electrons conducting polymer.** In this polymer backbones or associated groups consist of delocalized electron-pair or residual charge. Such polymers essentially contain a conjugated π-electrons backbone, which is responsible for conducting electrical charge. In an electric field, conjugated π-electrons of the polymer get excited, and thereby can be transported through the solid polymeric material. Overlapping of orbitals of conjugated π-electrons over the entire backbone results in the formation of valence bands as well as conduction bands, which extends over the entire polymer molecule. Presence of conjugated π-electron in a polymer increases its conductivity to a larger extent.
3. **Doped conducting polymer** is obtained by exposing a polymer to a charged transfer agent in either gas phase or in solution. In general, doping increases the surface conductivity of the polymer.
4. **Coordination conducting polymer (or inorganic polymer)** is a charge transfer complex containing polymer obtained by combining a metal atom with a polydentate ligand. Although the degree of polymerization in such polymers is small (≤ 18), yet they exhibit corrosion characteristics.
5. **Blended conducting polymer** is obtained by blending a conventional polymer with a conducting polymer either by physical or chemical change. Such polymers can be easily processed and possess better physical, chemical, and mechanical properties.

Important commercially produced conducting polymers include:

(*i*) Polyacetylene polymers, *e.g.*, poly-*p*-phenylene, polyquinoline, polyphenylene-co-vinylene, poly-*m*-phyenylene sulphide, etc.

(*ii*) With condensed aromatic rings, *e.g.*, polynanthrylene, polyphenanthrylene, etc.

(*iii*) With aromatic heteroaromatic and conjugated aliphatic units, *e.g.*, polypyrrole, polythiophene, polyazomethane, polybutadienylene, etc.

### 2.7.2 Structure of Conducting Polymers

The nature of a conducting polymer's intra- and intermolecular structure and the associated structural phase behavior are fundamental issues which strongly impact the physical properties manifested by this unique class of materials. Even relatively small changes in the specific chemical architecture and/or processing procedure can lead to significant variations in the resultant structural forms and in their physical properties. Ultimately a deeper understanding of the various structure-property interrelationship will, in part, form the foundation for future efforts that require even more specialized conducting polymer structures with highly specific properties.

Conducting polymers have some similarities to conventional polymeric materials, but it is clearly the extensive main chain 3-conjugation and its implicit "stiffness" with respect to chain bending and twisting that most influences the overall physical behavior. As a direct consequence, virtually all linearly unsubstituted conducting polymers, as shown in Figure 2.11, are found to be intractable and infusible. These model systems also tend to form crystalline phase structures with many common features. Hence these compounds may be conveniently lumped together to form one basic "class" of conducting polymer materials that have, loosely speaking, similar structural characteristics.

(*a*) Poly (*p*-phenylene)

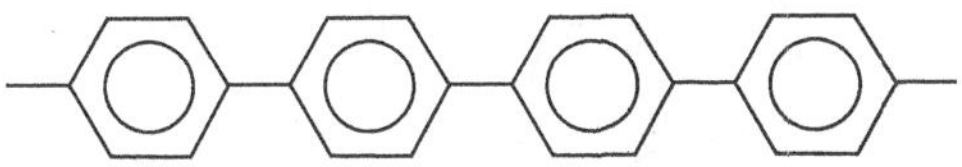

(*b*) Poly (*p*-phenylene vinylene)

(*c*) *trans*-Polyacetylene

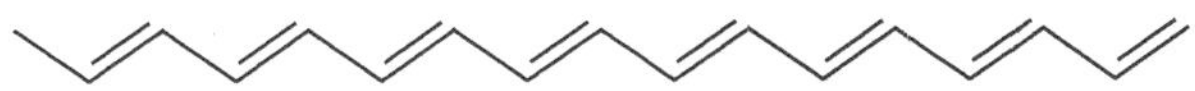

(*d*) Polythiophene

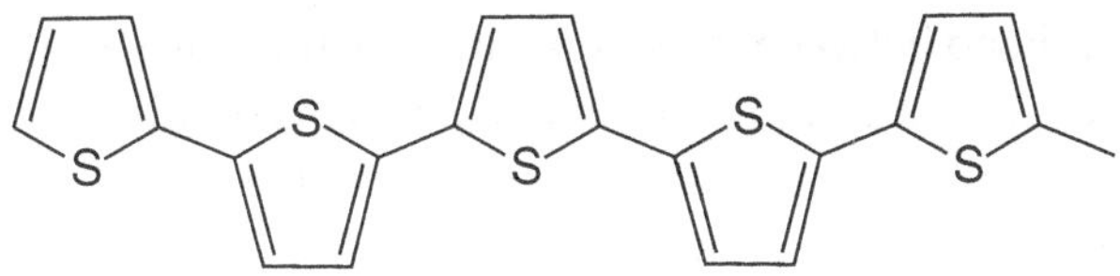


(*e*) Polypyrrole

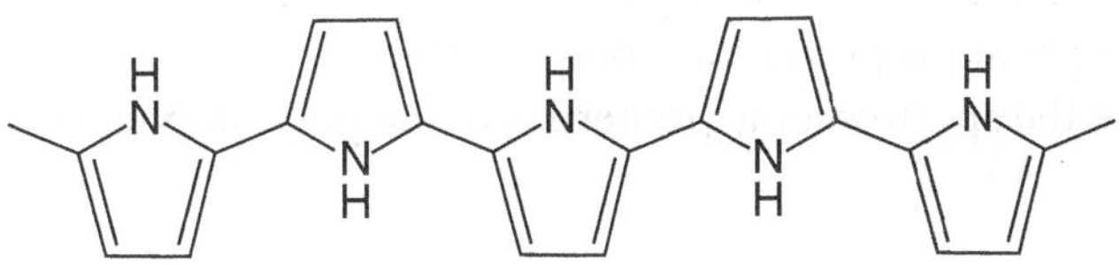


(*f*) Polyaniline

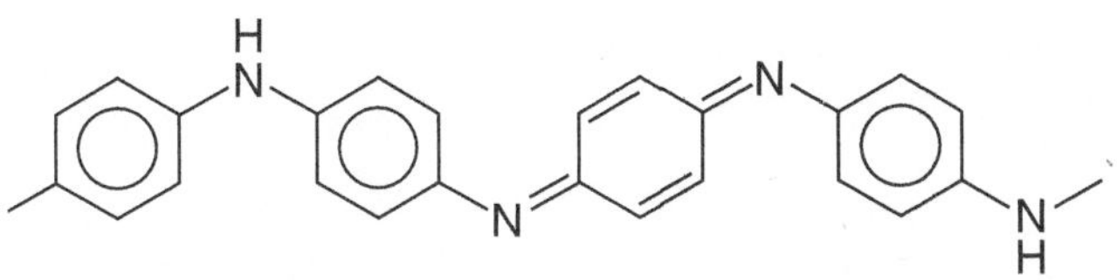


***FIGURE 2.11*** **Five stiff linearly unsubstituted conducting polymer (*a* – *e*) and are semiflexible host system (*f*).**

### 2.7.3 Polypyrrole

Polypyrrole is an inherently conductive polymer due to interchain hopping of electrons. It can be easily prepared by oxidative polymerization of the monomer pyrrole.

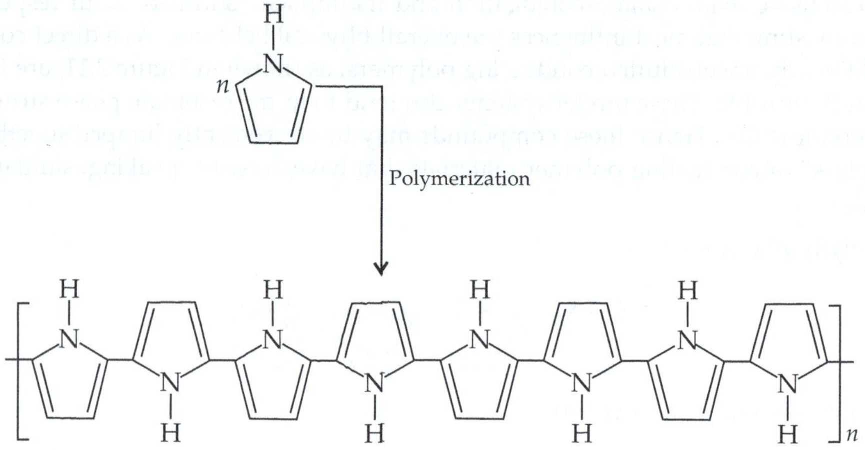


Polypyrrole

The mechanism of polymerization process, occurring at potentials above + 600 mV, is shown in Figure 2.12.

### 2.7.4 Polythiophenes

Polythiophenes consist of a chain of alternating double-and single-bonds like poly-acetylene, however, each first and fourth carbon atom are connected by a sulphur atom forming a thionyl ring (Figure 2.13). Therefore the bond between the second and the third carbon atom get more single-bond character than the other C—C bonds and consequently also the bonds connecting the thionyl rings are more of single-bond character. Due to this weaker mesomerization than in polyacetylene, the band-gaps of polythiophenes are shifted to the blue and UV. Besides of their structure-defining function, the sulphur atoms will have also a direct influence on the electronic and optical properties of a polythiophene. Nevertheless, the conjugation of the C-atoms will have also a direct influence on the electronic and optical properties of a polythiophene. Nevertheless, the conjugation of the C-atoms will dominate the spectroscopic properties of the polythiophenes.

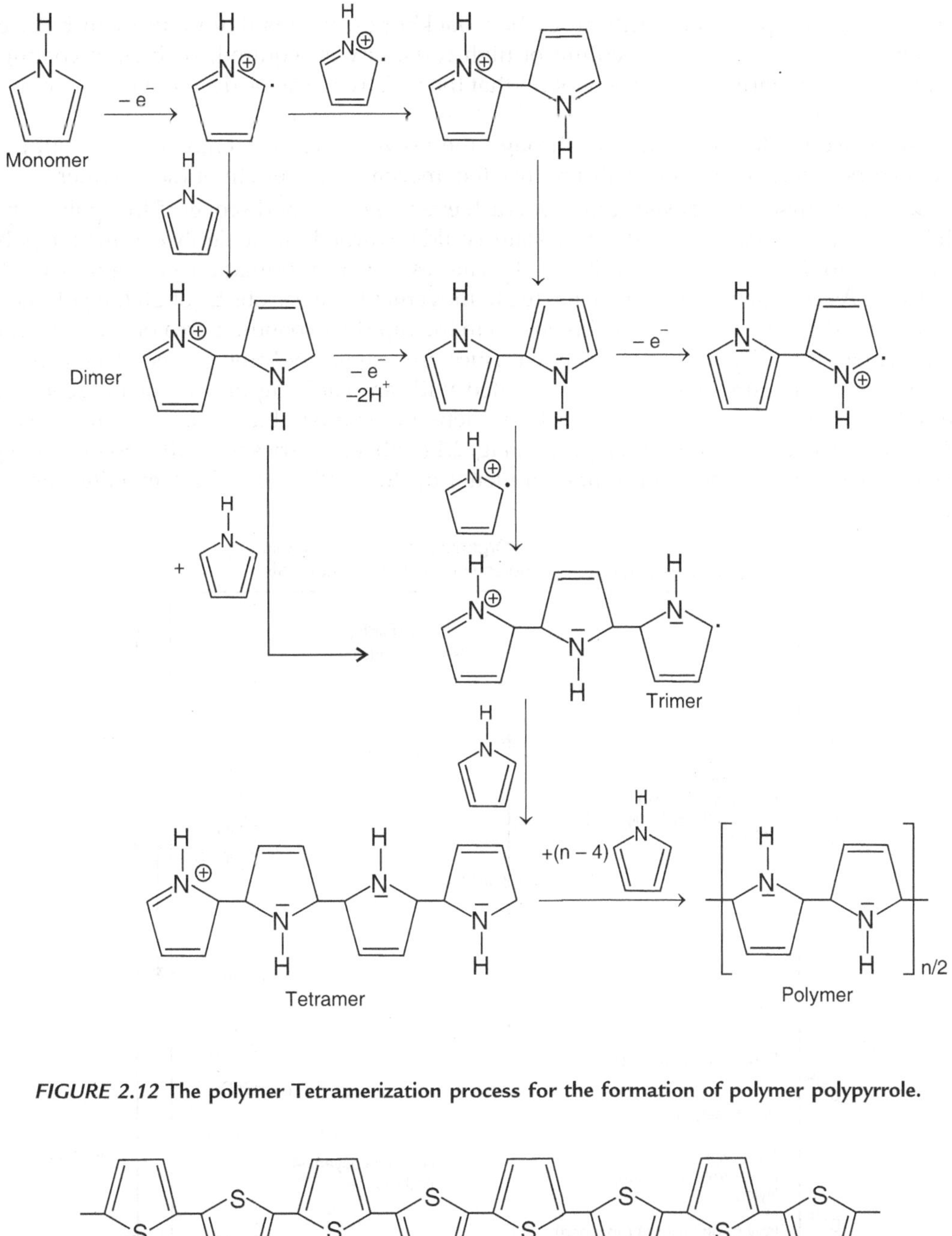


*FIGURE 2.12* **The polymer Tetramerization process for the formation of polymer polypyrrole.**

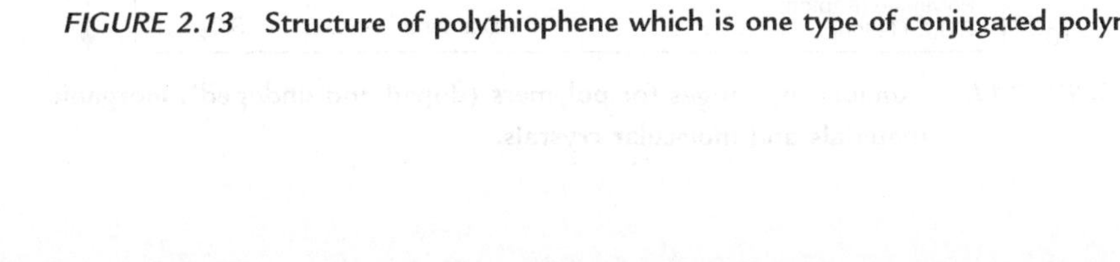

*FIGURE 2.13* **Structure of polythiophene which is one type of conjugated polymer.**

### 2.7.5 Conduction in Polymers

Inherently, organic polymers with all carbon backbones are insulators and can be used as encapsulating materials when a medium of high resistivity is required, such as in coatings for cables and electrical wiring. It has been found that the resistivity can be decreased if a composite of the polymer with carbon black or finely-divided metal is fabricated, but the conduction in these cases takes place via the filler and not through the polymer which merely acts as a supporting matrix. Incorporation of a filler can also reduce the mechanical strength of the polymer.

In 1977 the first major breakthrough was achieved when it was discovered that polyacetylene, which is a very poor conductor in the pure state, could be turned into a highly conductive polymer by conversion to the salt on reacting it with $I_2$. The result was a dramatic increase of over $10^{10}$ in conductivity. As conduction appears to be due to movement of electrons through the polymer, this discovery has added an exciting new dimension to the rapidly expanding area of synthetic metals. Other polymers that display similar characteristics are usually polyconjugated structures which are insulators in the pure state but when treated with an oxidizing or a reducing agent can be converted into polymer salts with electrical conductivities comparable to metals. Some idea of the possible range of conductivities ($\sigma$) is given in Fig. 2.14. where $\sigma$ varies from $10^{-18}$ S cm$^{-1}$ for a good polymeric insulator (*e.g.*, poly tetrafluoroethylene) up to ~ $10^6$ S cm$^{-1}$ for a metallic conductor, copper.

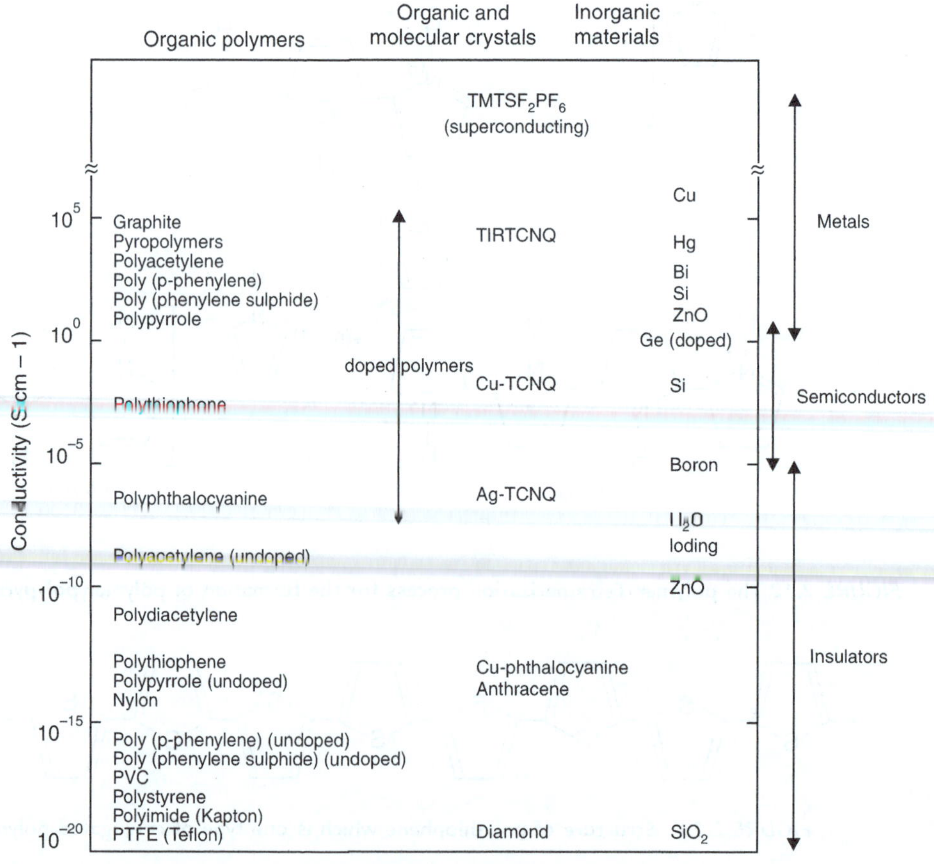


*FIGURE 2.14.* **Conductivity ranges for polymers (doped and undoped), inorganic materials and molecular crystals.**

### Conduction Mechanisms

Electrical conductivity is a function of the number of charge carriers of species '$i$' ($n_i$), the charge on each carrier ($\varepsilon_i$), and carrier mobilities ($\mu_i$) described by the relation $\sigma = \Sigma\, \mu_i\, n_i$ where the units of conductivity are S $cm^{-1}$. Conduction is solids is usually explained in terms of the band theory which postulates that when atoms or molecules are aggregated in the solid state, the outer atomic orbitals containing the valence electrons are split into bonding and antibonding orbitals, and mix to form two series of closely-spaced energy levels. These are usually called the valence band and the conduction band respectively. If the valence band is only partly filled by the available electrons, or if the two bands overlap so that no energy gap exists between them, then application of a potential will raise some of the electrons into empty levels where they will be free to move throughout the solid thereby producing a current. This is the description of a conductor. If, on the other hand, the valence band is full and is separated from the empty conduction band by an energy gap, then there can be no net flow of electrons under the influence of an external field unless electrons are elevated into the empty band and this will require a considerable expenditure of energy. Such materials are either semiconductors or insulators, depending on how large the energy gap may be, and the majority of polymers are insulators. The band model then assumes that the electrons are delocalized and can extend over the lattice (see Fig. 2.15).

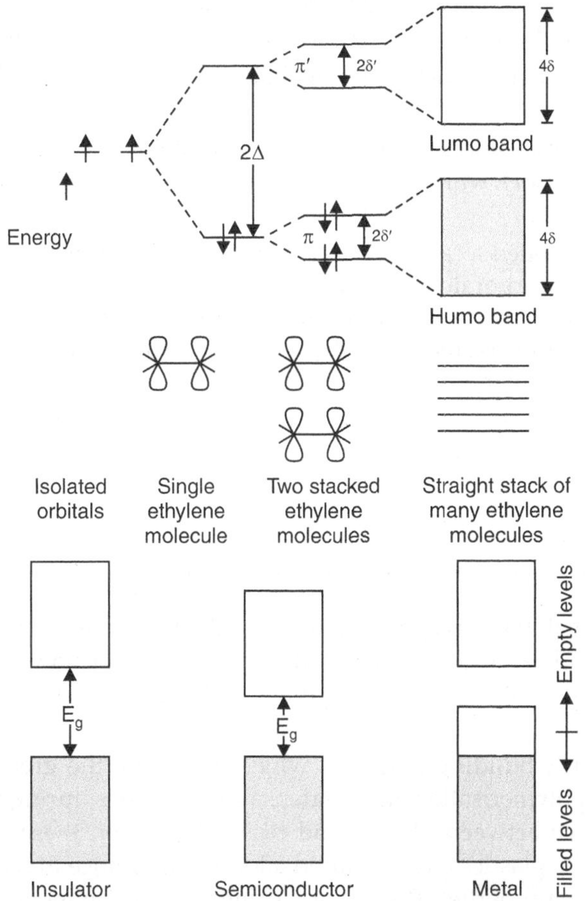


*FIGURE 2.15* **Schematic diagram showing the principles of band theory as described in the text. The dark regions represent the bands filled with electrons, and the light regions the bands that are available for conduction. The energy gap between filled and empty states is $E_g$.**

When we come to consider electronic conduction in polymers, band theory is not totally suitable because the atoms are covalently bonded to one another, forming polymeric chains that experience weak intermolecular interactions. Thus macroscopic conduction will require electron movement, not only along chains but also from one chain to another.

### 2.7.6 Applications of Conducting Polymers

1. Polythiophene is marketed under the trade name *Baytron.* It can be used to make plastics paintable by adding the conductive agent first, and also in the electrodes of small, high performance tantalum capacitors found in telecommunications, computer, and automotive products.
2. *Contex*, a fibre is coated with a conductive polymer polyprrole can be woven to create an antistatic fabric which can be used in the carpet industry.
3. The list of potential applications for conductive polymers remains a long one, and includes antiradiation coatings, batteries, catalysts, deicer panels, electrochromic windows, electromechanical actuators, embedded-array antennas, fuel cells, lithographic resists, nonlinear optics, radar dishes, and wave guides.
4. The polymers are used in making button type batteries. These batteries are longer lasting, rechargeable, and can produce current density up to 50 mA/cm$^2$.
5. These can be used in producing photovoltaic devices, *e.g.*, in Al/polymer/Au photovoltaic cells.
6. An emerging application for electrically conductive polymeric materials is biosensors and chemical sensors, which can convert chemical information into a measurable electrical response.

   *Abtech* has developed a range of enzyme biosensors. For example, immobilized glucose oxidase can be incorporated into this polymer transducer system, which acts like a glucose sensitive biosensor as the enzyme catalyzed oxidation of the glucose produces an oxidant byproduct that is measured indirectly. Levels of therapeutic drugs in patients can also be monitored in a similar way.
7. These polymers find extensive applications in telecommunication systems, micro electronic devices, biomedical uses, etc.

## 2.8 PLASTICS

Plastics are the organic materials of high molecular weight, which can be moulded into any desired form, when subjected to heat and pressure in the presence of a catalyst. In broad sense, plastic refers to a material which exhibits *plasticity i.e.*, the ability to get deformed or to undergo change of shape under pressure.

Resins are the basic building materials which constitute the greater bulk of plastics and which have undergone polymerisation and condensation reactions during their preparation. There is thus a subtle difference between plastics and resins. However, these two terms are currently used interchangeably. The plastics have acquired a lot of importance in recent years in every walk of life due to their unique properties. Plastics are formed when polymers are blended with certain external materials known as *compounding or moulding ingredients*. These ingredients provide sturdiness and other desired characteristics to polymers.

### 2.8.1 Moulding constituents of Plastics

A high polymer material like plastic is usually mixed with 4 to 10 ingredients, each of which either discharges a useful function during moulding or imparts some useful property to the finished artifact. This is called a *mix*. These ingredients are discussed below :

1. **Fillers.** The addition of *fillers* confers special properties to the polymers in addition to reducing the cost of production; sturdiness, heat resistance, insulation characteristics, opacity and increasing impact strength are imparted to the polymers. Asbestos, talc, chalk, sand, carbon black, quartz, mica, silicates, alumina, barium salts, magnesia, silicon carbide, zinc oxide are particulate fillers while nylon, polyesters, acrylics, cellulosic fibres, carbon fibres and boron filaments are some of the fibrous fillers. Bakelite when compounded with wood flour is less brittle but when compounded with asbestos has an increased heat resistance. Carbon black in rubber enhances its tensile and shear strength.
2. **Plasticisers.** They increase plastic properties and induce greater flexibility to polymers. They are usually high boiling liquids (of low volatility). The plasticiser should have a high boiling point so that it does not evaporate during the processing. As a liquid plasticiser passes into the polymer phase, a colloidal or molecular dispersion may occur. Such dispersions or solutions are known as *plastisols.* The plasticisers are subdivided into solvent (primary) and non-solvent (secondary) plasticisers. Esters, phosphate esters like tricresyl phosphate, tributyl phosphate, triphenyl phosphate, adipates, fatty acid esters, sulphonamides, camphor, non-drying vegetable oils, etc. are commonly used plasticisers. Castor oil, added to cellulose nitrate, is an example of a non-solvent plasticiser.
3. **Colourants.** The colouring agents in the form of organic dyes and opaque inorganic pigments when added to powdered plastics provide pleasing and decorative colours. A concise list of materials used as colourants to provide the appropriate shade to plastics is given in Table 2.6.

**TABLE 2.6 Materials used as colourants**

| *Name of the pigment used (as colourant)* | *Imparted colour* |
|---|---|
| Ultramarine | Blue |
| Carbon black | Black |
| Calcium carbonate | White |
| Zinc oxide | White |
| Chromium trioxide | Green |
| Ferric oxide | Red |
| Antimony sulphide | Crimson red |

4. **Lubricants.** These additives promote a good finish to the plastic materials, in addition to improving the flow characteristics and friction reduction in the processing machines. Waxes, oils, stearates, oleates and soaps are employed to make the moulding of plastics easier and help in providing a glassy finish to the products.

5. **Stabilisers.** Alkaline earth oxides, organometallic salts, epoxy compounds and amine type compounds serve as *stabilisers* which are added to polymers to prevent their degradation. Lead salts like white lead, lead chromate, litharge, red lead serve as opaque moulding compounds. Stearates of lead, cadmium and barium are transparent moulding compounds.
6. **Antioxidants.** The oxidative degradation of polymers can be prevented by incorporating certain compounds called antioxidants which are capable of interacting with reactive radicals formed. Di-*t*-butyl-*p*-cresol (TBC), phenyl-β-naphthylamine (PBNA), diphenyl *p*-phenylene diamine (DPPD) are a few common antioxidants in use.

### 2.8.2 Classification of Plastics

They are categorised as (*a*) *thermoplastics* (TP) and (*b*) *thermosetting* (TS) *plastics*. Thermoplastics are materials which can be softened by heat and reformed. Thermosetting plastics are completely hardened once formed. Addition polymerisation yields TP while condensation polymerisation give TS plastics (thermosets).

The differences between the two types of plastics are mentioned in Table 2.7.

**TABLE 2.7 Differences between thermoplastics and thermosets**

| *Thermoplastics* | *Thermosets* |
|---|---|
| Formed by addition polymerisation. | Formed by condensation polymerisation. |
| Contain long-chain with negligible crosslinks. | Contain three-dimensional array of structures. |
| Soften very readily on heating and stiffen on cooling (reversible change). | Become stiff and hard on heating, but do not soften on cooling (irreversible change). |
| Can be remoulded or reshaped and used. | Cannot be remoulded and reused. |
| Usually soluble in organic solvents; some swell up in solvents. | Insoluble in any solvent. |
| Can be reclaimed from wastes. | Cannot be reclaimed. |
| Usually soft, weak and less brittle. | Hard, strong and more brittle. |
| Neighbouring polymeric chains are held together by weak van der Waals' forces. | Neighbouring polymeric chains are held together by strong covalent bonds in the form of crosslinks. |

### 2.8.3 Fabrication Techniques

The method of fabricating plastics depends on whether they are TP or TS and also on the shape of the finished product. The industrially adopted plastic fabrication methods are of two main types: (1) Moulding process and (2) Foaming process.

**(1) Moulding Process.** This technique involves fabrication of plastics under severe heat and pressure and is applicable to both TP and TS materials. The moulding process involves different techniques:

(*a*) *Compression moulding.* In this method, a synthetic plastic material that has to be moulded is mixed with a filler and the other ingredients and then placed in the mould.

The mould is closed and heated wtih simultaneous application of pressure. As a result, the cavities in the mould get filled up by the fluidised plastic. Once the moulding is over the material is withdrawn after cooling. Finally, curing is done by either heating (for TS materials) or cooling (for TP types) (Fig. 2.16). Currently, fully automatic moulding presses are available which quicken the process. Door handles, screw caps and electricity metres are obtained by this method.

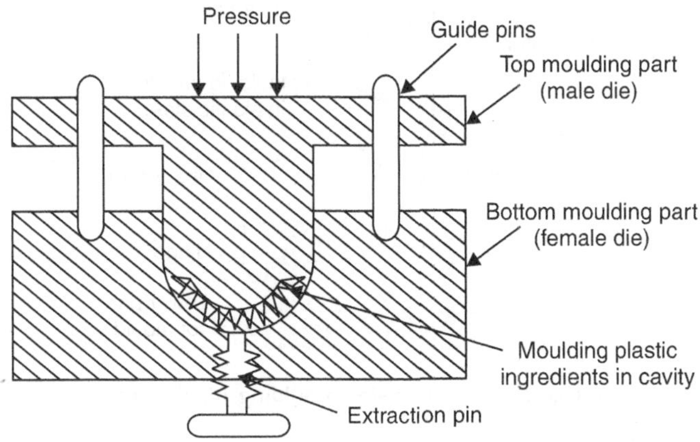


*FIGURE 2.16* **Compression moulding of plastics.**

(*b*) *Injection moulding.* The plastic in the form of powder is fed into a hot cylinder from where it is injected at a regulated rate into a tightly-locked mould by means of a screw arrangement or by a piston plunger (Fig. 2.17). The mould is kept cooled to enable the hot plastic to be cured and become rigid. Once the material is sufficiently cured, half of the mould is opened to allow the ejection of finished product without any deformation. This method is adopted to fabricate TP materials. Telephones, buckets, dustbins etc. are made by this technique.

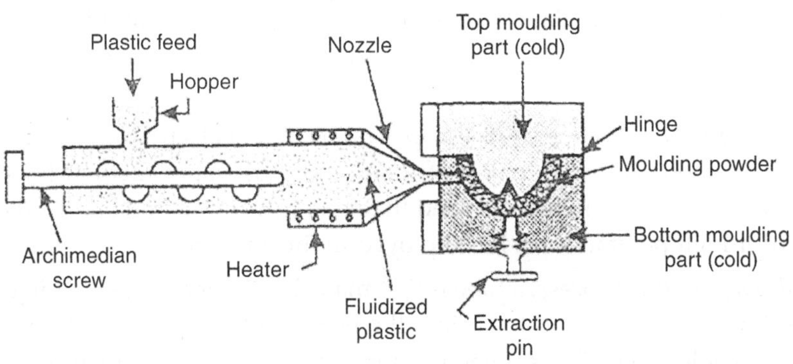


*FIGURE 2.17* **Injection moulding of plastics.**

(*c*) *Transfer moulding.* This is used for TS materials. Essentially, in this process, the principle of injection moulding is operative. The moulding powder is placed in a heated chamber maintained at the minimum temperature at which the mould powder just begins to become plastic. This plastic material is then injected through an orifice into the mould through a plunger working at high pressure. As a result of the friction that develops at the orifice, the temperature of the material at the time of ejection rises to such an extent that the moulded material becomes almost a liquid. Hence, it easily flows into the mould which is also heated up to the required curing temperature. The moulded material gets ejected mechanically (Fig. 2.18).

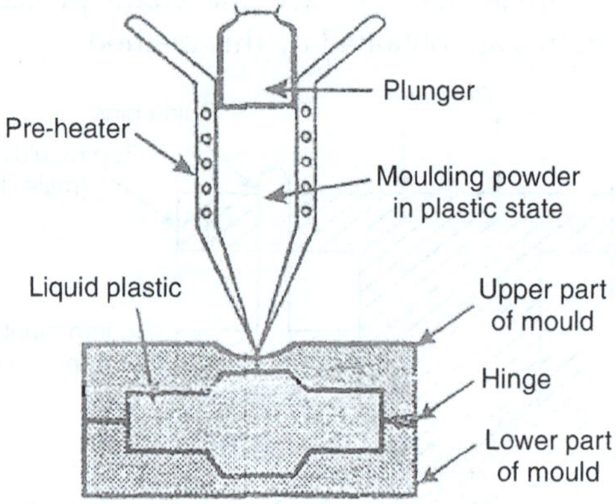


*FIGURE 2.18* **Transfer moulding of plastics.**

**Advantages**

(*i*) More complicated shapes can be fabricated by this method.

(*ii*) Shrinkage and distortion are minimum.

(*iii*) Very delicate articles can be made by this method.

(*iv*) Blistering is almost eliminated.

(*v*) Less expensive.

(*vi*) Even thick pieces can be made and cured almost completely and uniformly.

(*vii*) Fine wires and glass fibres can be inserted into the mould.

(*d*) *Extrusion moulding*. In this method a thermoplastic material is softened by heat and squeezed out by means of a screw conveyer into a die having the required shape of a desired product. Tubes, rods, strips and insulated electric cables can be produced by this method. This is the most efficient and most rapid method for fabricating long tubes, tooth-paste tubes, rods and filaments. The limitation of the process is that only linear high polymers can be employed in this method.

(*e*) *Blowing*. In this process, a tube is first made by either extrusion or injection moulding technique. Air is then blown into the heated tube kept in a split mould with easily detachable halves. The softened material then gets blown into the shape of the mould. Blow moulding can also be made fully automatic. Bottles, toys, etc. can be made by this way.

(*f*) *Casting*. This is a moulding method without application of pressure. The material is melted and poured into the moulds and subjected to curing by heating or cooling. Flexible plastics can be made by slush casting method.

(g) *Cold moulding.* This is mostly used for phenol-formaldehyde resins. The cold moulded articles can be cured/baked in an oven for several hours. Electrical fixtures are made by this method.

(h) *Vacuum forming.* A thermoplastic sheet held tightly in a frame is softened by heating and the plastic material is drawn into the mould by means of a vacuum pump. Complicated shapes like boxes can be designed by this method.

The thermoplastic material is first fed between heated rollers and finally between cold rollers, in order to get it in the form of sheets. This process is referred to as calendering. Often, it becomes necessary to coat a surface with plastics for decorative or protective purposes. For instance, PE is used to coat soap containers, steering wheels, refrigerator parts, etc. Kitchen utensils are given a PTFE (teflon) coating to provide a non-sticky surface.

2. **Foaming Process (Foamed Plastics).** According to the American Society for Testing and Materials (ASTM), a *foamed plastic* is one whose apparent density is substantially decreased due to the presence of numerous cells dispersed throughout its mass. In other words, foamed plastics are biphasic wherein a gas is dispersed in the dispersion medium (like synthetic rubber or plastic). They are prepared by expanding the fluid polymer into a cellular state and then stabilised. Polymers like PS, PVC, PE, etc. are subjected to heating with a blowing agent (pentanes, hexanes, halocarbons, etc.) which gets entrapped into the polymer structure. On heating these polymer articles containing the blowing agent in steam, they swell up. The expanded beads can then be used to get the desired structure. The foamed structures are allowed to cool and then taken out. They are also called *cellular plastics* or *expanded plastics*.

Polyurethanes are obtained by condensing a polymeric diol (HO ∼∼∼ R ∼∼∼ OH) with an excess of diisocyanates (OCN—R—NCO) or toluene diisocynate (TDI). The prepolymer OCN-R-NH-COO ∼∼∼ R ∼∼∼ OOC—NH—RNCO obtained is mixed with more of the polymeric alcohol and a halocarbon or water (blowing agent). The isocyanate on hydrolysis provides $CO_2$ gas for expanding the fluid polymer and also provides cross-linking. The product is polyurethane foam.

$$[\sim\sim\sim NCO + H_2O + OCN\sim\sim\sim \longrightarrow NHCONH \sim\sim\sim + CO_2]$$

Cellular polymers can also be obtained by the frothing process. Here, a gas is dispersed in a fluid to form a foam which is sub-sequently stabilised. Urea-formaldehyde resin can be obtained by this way. An aqueous solution of a surfactant and a catalyst is agitated to produce foam. A second aqueous suspension of partially cured urea-formaldehyde resin is mechanically stirred into the first system. In the presence of the catalyst, the resin gets further cured in the cellular state. Foam rubber is obtained by whisking suitably compounded rubber latex to obtain a froth. The foam can be set by adding sodium fluorosilicate. It can be vulcanised in a mould to give the desired structure. Foamed plastics are widely used for pillows and mattresses, thermal and electrical insulators, packing materials for delicate instruments and in sea-borne cables.

*Thermocole* is a foamed plastic obtained by blowing air/gas through molten PS or polyurethane. It is strong and light. It is ideal as a packing material due to its shock-proof characteristic and light weight. Thermocole slabs are used for false ceilings in air-conditioned rooms and for good acoustic effect. It is used as an insulating material and also for decorative purposes.

### 2.8.4 Uses of Plastics

Plastics are indispensable in daily human life. They find numerous applications, some of these are discussed below :

1. Due to their beauty, strength, oil and electric resistance, they are employed in the manufacture of aeroplanes and automotives.
2. On account of their resiliency and strength they are used for making safety glass, laminated gears, cans, pulleys, self-lubricating bearings, etc.
3. In combination with metals it is used for the manufacture of steering wheels and plastic-covered dash boards.
4. In electrical industries moulded and laminated organic plastics are used as solid insulating materials.
5. Due to their resistance to corrosive gases, liquids and chemicals etc., they are used in chemical process industries for making tubing, piping, plain or laminated piping, tanks, absorption towers etc.

#### [A] Use of Plastics in Electronic Industry

1. **Printed circuits.** Radio and television circuits are printed on resin bonded paper laminated sheet (paper having a layer of resin) which is laid on copper sheet. In this way copper sheet or foil gets an acid resistant coating in the pattern of the desired circuit.

The laminated (copper foil with resin bonded laminated sheet) is immersed in an acid etching solution which is generally ferric chloride. The exposed and unwanted copper is etched away leaving a thin conducting pattern of copper bonded to the laminate. Then miniature components are positioned automatically on the copper clad laminates and the jointing is done in one operation by dipping into a solder bath.

Usually paper based phenolic laminates (paper coated with phenolic resin) are used but sometimes glass fabric-based epoxide laminates (glass fabric coated with epoxide resin) are also applied. The laminate to be used should have the following properties :

(*i*) It should have high insulation resistance, low permitivity drift and low power factor.

(*ii*) It should have high dimensional stability (that is, it should not chip off) when components are attached to the copper bonded laminate by automatic process.

(*iii*) It should easily get punched and should not chip off by punching.

(*iv*) It should be impervious to moisture.

(*v*) It should be able to withstand high temperature which is attained during the operation of the television or radio set.

(*vi*) The bond strength between the copper and the laminate should be quite high.

(*vii*) The laminate should be able to withstand the action of acids.

(*viii*) The laminate should be able to withstand soldering conditions.

(*ix*) The surface of the laminate should be such that it should be able to take printing inks.

**2. Use of the printed circuits**

(*i*) Radio and television sets have become more compact by the adoption of miniaturised electronic circuits (that is, by adopting printed circuits and miniature components).

Before the use of printed circuits, it was very difficult to design an automatic set due to handling of maze of wires associated with the point-to-point wiring system. This also resulted in the production of big-sized sets.

(*ii*) A printed circuit can be designed for easy location, automatic component fitting and for dip soldering. This results in saving of time, wires, screws, tags and hand-soldered circuits.

(*iii*) Dip-soldered circuits can be tested automatically as they are in one plane and connections are exposed.

(*iv*) There is a lot of weight-saving by the use of printed circuit laminate as chasis.

(*v*) By using the same labour and floor space, the output can be increased by 30%.

(*vi*) The expense of training labour for assembly and writing circuits can also be eliminated.

### [B] Use of Plastics in Heavy Engineering Industry

Glass-fibre-reinforced polyester resin is used in the end sections of condensers in electricity generating stations.

Before the use of this plastic in condensers, there was always graphitic corrosion on large cast iron end casings or water boxes of main and auxiliary condensers, particularly when water has high oxygen content. Sacrificial electrodes and protective coatings were employed but they did not prove useful.

### [C] Use of Plastics in Chemical Plant

In chemical plants polyvinyl chloride resin has been used in many countries in place of stainless steel.

Polyvinyl chloride resin has got the tendency to break at higher temperature, so stabilisers like basic lead carbonate or tribasic lead sulphate (which are acid acceptors) are added to the resin. But no plasticiser is added to the resin as by adding plasticiser, the resistance of the resin to chemicals decreases.

For the fabrication of chemical plants, polyvinyl chloride with stabilisers and no plasticiser is converted into sheets. Such sheets (without any plasticisers) are called rigid sheets.

PVC sheets can be worked in the same manner as wood or metal, that is, it can be sawn, drilled and welded. The welding is carried out by means of a hot air torch and a weld rod of the same composition as the sheet used. The hot air heats the edges of the sheet to be joined to fusion point and also softens the weld rod which is fed into the joint. It is thus possible to produce self-supporting structures of rigid PVC or the same material can be used for lining metal or wooden tanks. The lining of metal tanks can be accomplished by cleaning the inside surface and applying to it an adhesive which is generally solution of PVC resin in ethylene dichloride. The solvent is allowed to evaporate and the lining is then put in place. In order to get satisfactory bond between the metal and PVC, the metal is heated from outside. In this way the adhesive is softened and then the lining is pressed into place and the two surfaces are bonded together.

Pipes etc. are also extruded from PVC resin and are used in chemical plants wherever necessary. These pipes can be welded etc. or can be joined together by adhesive containing PVC resin and solvent.

Rigid PVC is resistant to strong mineral acids at room temperature and to organic acids and larg range of organic chemicals. It is also resistant to caustic alkalies but it does not resist the attack of aromatic hydrocarbons, esters, phenols and amines.

The following are few other uses to of PVC rigid sheets :

(*i*) A fume duct as high as 125 metres and 0.5 metre in diameter has been erected for conveying nitric acid fumes to atmosphere by Bayer at Leverkusen in Germany.

(*ii*) Complete bottle-filling machine for corrosive liquids using rigid PVC sheet has been developed in Australia.

(*iii*) Corrugated sheets of PVC have lately come into great use. Corrugated sheets can be produced in opaque, clear and translucent forms. These sheets can withstand atmospheric conditions which include snow and intense sunlight. These corrugated sheets have got a great future.

### [D] Use of Poly olefines in Chemical Plants

(*a*) **Polythene (polyethylene)** has got high resistance to alkalies and acids other than oxidising acids at elevated temperatures.

Polythene pipes are used as water pipes, particularly in farms, due to its lightness, resistance to corrosion and ease with which it could be laid. Polythene pipes are also used for cold water supplies for domestic purposes. In industry these pipes have been used for carrying corrosive effluents. In Germany polythene pipes have been used for conveying telephone cables under the Rhine.

Polythene pipes are joined by hot air welding, wherein heated plates are used to soften the ends of pipes or sheets which can then be butt-jointed.

Polythene films are used for packaging and also for giving weather protection on building sites and as a water barrier in the laying of concrete.

The only disadvantages of polythene is that its softening point is 60°C and anything fabricated out of it can be used only below its softening point.

(*b*) **Polypropylene.** It is produced by the low pressure polymerisation process derived from the original Ziegler polymerisation of polyethylene. It has all the properties of polythene in addition to its high softening point of 100-120°C. So it is quite certain that polypropylene has got a great scope for use in chemical plants.

## 2.9 SILICON RESINS

Polymers which contain alternating atoms of silicon and oxygen with organic substituents (radicals) attached to silicon atoms are called silicone resins.

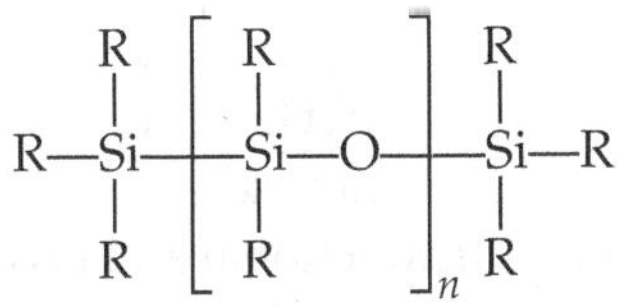

where, R = alkyl or phenyl radical

Silicone resins are also called organo polysiloxanes. Silicone resins may be liquids, viscous liquids, semi-solids like greases, solids or rubbers.

### 2.9.1 Manufacture of Silicon Resins

Silicone resins are prepared by the condensation of hydroxy organosilicon compounds which are formed by the hydrolysis of organosilicon halides. Organosilicon halides are obtained by reacting a silicon and an alkyl halide or by reacting a silicon halide with Grignard reagent.

$$\underset{\text{Methylchloride}}{CH_3Cl} + Si \longrightarrow CH_3SiCl; (CH_3)_2SiCl_2$$

$$\underset{\text{Silicon tetrachloride}}{SiCl_4} + \underset{\text{Methyl magnesium chloride}}{CH_3MgBr} \longrightarrow \underset{\text{Organosilicon chlorides}}{(CH_3)_3SiCl; (CH_3)_4Si}$$

The reaction products which is a mixture of different organosilicon halides is fractionally distilled to separate different organosilicon halides. Organosilicon halides are polymerised by careful controlled hydrolysis.

(*i*) *Dimethyl-silicon dichloride* is *bifunctional* and can yield very long chain polymers, *e.g.*,

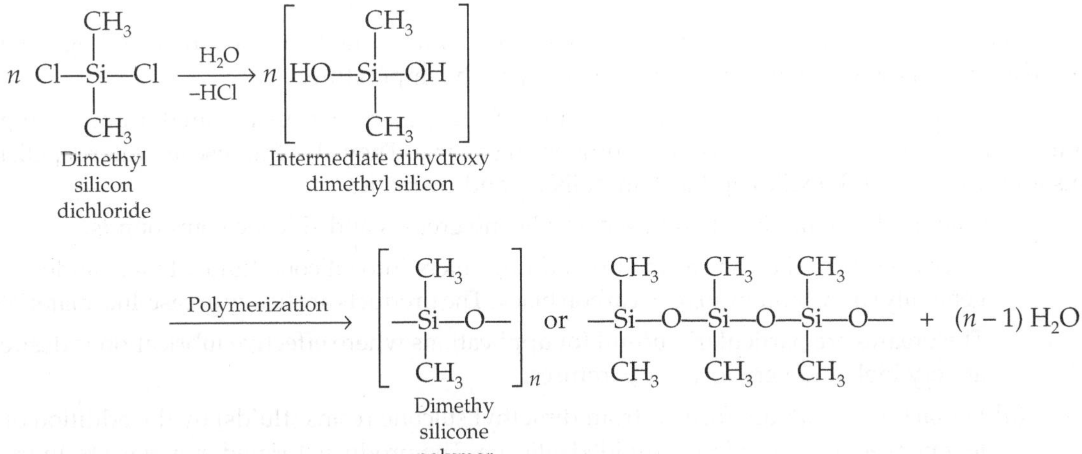

(*ii*) *Trimethyl silicon chloride* is monofunctional and hence can act as a chain stopper. So, it is used in proportions to limit the chain length.

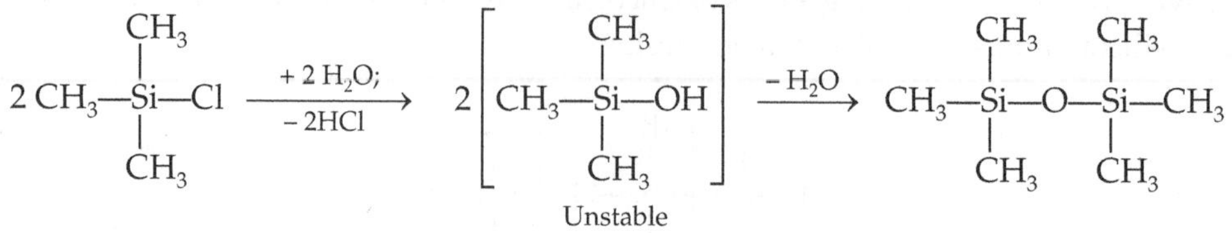

(*iii*) Monomethyl silicon chloride is trifunctional and gives cross-linking to the final polymer. Its proportion used in polymerization, determines the amount of cross-linking that can be obtained.

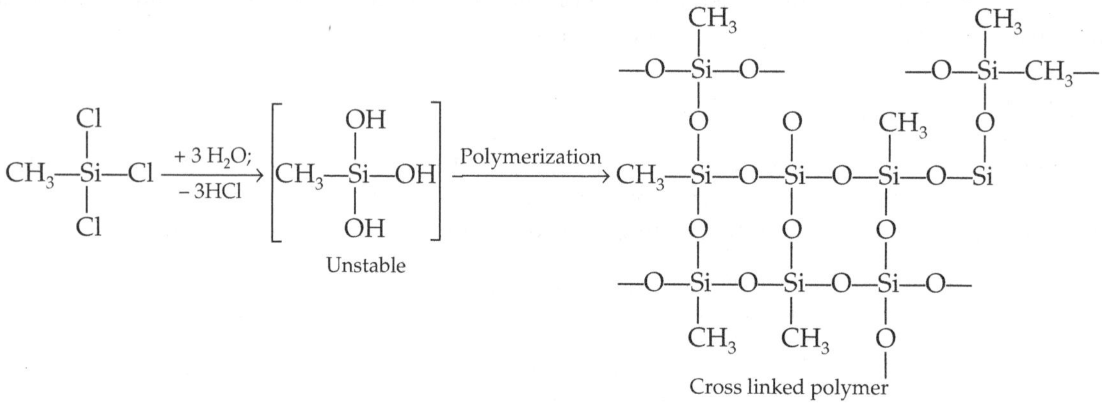

### 2.9.2 Types of Silicone Polymers and their Uses

Depending upon the proportion of various alkyl silicon halides used during their preparation, silicones may be liquids, viscous liquids, semi-solid, rubber like and solids.

**1. Liquid Silicone Polymers.** They are also called silicone fluids and also as silicone oils. They are generally dimethyl silicones or methyl and phenyl silicones (copolymers) of relatively low molecular weights. They have low molecular weight, low surface tension, great wetting power for metals and show very small change in viscosity with temperature.

They are used as damping and hydraulic fluids, as heat transfer media and as high temperature lubricants. They are used as antifoaming agents. They also find use as water repellent finishes for leather and textiles and also in polishes and cosmetics.

Silicone fluids are modified and used as silicone greases and silicone compounds.

(*i*) *Silicone greases.* They are formed by adding fillers into silicone fluids. Fillers added are generally silica, lithium soap or carbon black. The products are heavy grease-like materials.

The greases are particularly useful for applications where effective lubrication is desired at very high and very low temperatures.

(*ii*) *Silicone compounds* are formed from dimethyl silicone resins (fluids) by the addition of a few per cent of a very finely divided silicane. The products formed may vary from free-flowing liquids to heavy grease-like materials. The common uses include :

(*a*) They find use as stopcock grease in laboratory as its viscosity is unaffected by temperature and the grease continues to act as lubricant and seal for glass joints under severe conditions. Complete removal of grease with organic solvents is not possible and can only be accomplished by dipping in concentrated sulphuric acid for about 2 hours.

(*b*) They are used as a filler for high voltage conduits on aircraft.

(*c*) They are used for pump packing where ther are conditions of high temperature or corrosive atmosphere.

(*d*) They are excellent mould release agents.

(*e*) They are also used as damping medium in phonograph pick ups.

**2. Silicone Resins (Solid).** Silicone resins (solid) are obtained by hydrolysing and condensing (or polymerising) mixtures of bifunctional and trifunctional alkyl silicon halides such as $(CH_3)_2SiCl_2$ and $CH_3 \cdot SiCl_3$.

They are heat-stable resins thermosetting for use in moulding compositions for the manufacture of waterproof insulation components for high tension ignition systems and high frequency equipment. They are used for coating applications where thermal stability in the range 300–500°C is required. Due to their dielectric properties they find use in many electrical applications particularly in electrical insulation exposed to high temperatures.

**3. Silicone Rubbers.** Silicone rubbers are compositions containing high molecular weight dimethyl silicone linear polymer finely divided silicone dioxide as the filler and a peroxidic curing agent. Cross-linking takes place through reaction of the peroxide with two methyl groups on adjacent chains to form a dimethylene bridge.

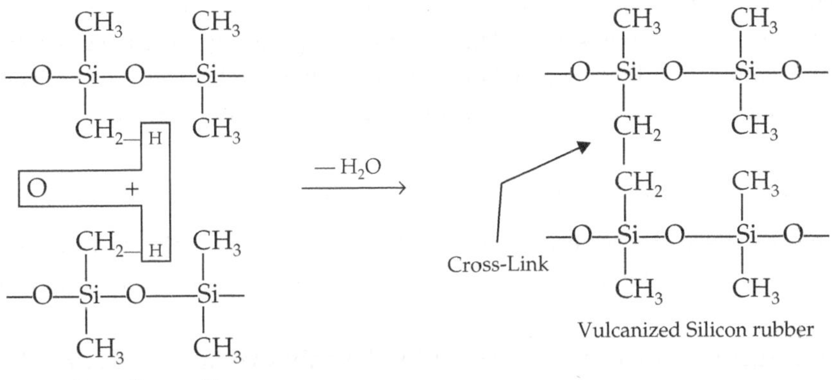


Raw silicon rubber

Vulcanized Silicon rubber

They possess exceptional resistance to prolonged exposure to sunlight, weathering, most common oils, boiling water, dilute acids and alkalies. They remain flexible in the temperature range of 90–250°C and hence, find use in making tyres of fighter aircrafts. Important applications of silicone rubbers include :

(*i*) For insulating the electrical wiring in ships.

(*ii*) For manufacture of tyres for fighter aircrafts.

(*iii*) In making lubricants, paints and protective coatings for fabric finishing and water-proofing.

(*iv*) As adhesives in electronics industry.

(*v*) For making artificial heart valves, transfusion tubings and padding for plastic surgery.

(*vi*) As a sealing material in search-lights and in aircraft engines.

## 2.10 ADHESIVES

Adhesive *is a substance capable of holding materials together by adhering strongly to the surface of both the materials.* The bodies held together by an adhesive are known as *adherends*, while the process of holding one adherend to another by adhesive is called *bonding* and the final assembly of two adherend and the adhesive is called *bond*. The bonds formed between adherends and adhesive should be strong internally.

The quality of an adhesive is judged by following factors :

(*i*) Degree of stickness

(*ii*) Rapidity of bonding

(*iii*) Strength of bond setting on drying

(*iv*) Durability

In recent years, adhesive is the general term used for bonding agents and also includes paste, glue, cement, etc. *Glues*, which are mostly used in wood work industry, are prepared from animal proteins such as hoofs, hides, mucilage and tendons. *Pastes* may be regarded as adhesive compositions having plastic consistency. These can be prepared by mixing starch and water and then boiling. *Cement* means adhesive based on rubbers or thermoplastic resins dispersed in organic solvents. These adhesives undergo setting by dehydration or loss of solvent.

### 2.10.1 Adhesive Action and Strength

The attachment of surfaces by adhesive is either due to chemical or physical forces of attraction. This is known as *specific adhesion*. The process of adhesion is believed to take place in the following way.

At first, actual wetting of the adherend by the adhesive takes place. By this, contact in molecular dimensions is established. Now if the interfacial boundary energy of the adhesive and the adherend surfaces is lower than the sum of the surface energies of the adhesive and adherends, permanent adhesion takes place.

In some cases, adhesive may fill the voids, the porous or rough surfaces and hold the surfaces by interlocking action. This is known as *mechanical adhesion*. In still some cases, the surfaces are partially dissolved in the adhesive or its solvent and bonding of surfaces takes place. This is known as *fusion adhesion*.

The adhesive develops strength is four different ways :

(*i*) *By loss of solvent.* The adhesive is dissolved or dispersed in a suitable solvent including water. This dispersion is applied to the surfaces to be bonded which are then brought together. During this operation, a part of solvent is lost by evaporation or by entering into the pores of the adherend surfaces. Loss of solvent results in hardening of the adhesive thereby resulting in the development of bond strength. Water-soluble and oil-soluble elastomers belong to this type of adhesives.

(*ii*) *By cooling of hot molten mass.* These adhesives are applied hot in liquid conditions to the bonding surfaces. The mass of fusible adhesive solidifies, on cooling, imparting adhesive strength. Fusible adhesives, like thermoplastics, belong to this type.

(*iii*) *By applying pressure.* Highly viscous and pressure-sensitive adhesives develop bond strength instantaneously, when high pressure is applied.

(*iv*) *By chemical reaction.* The ingredients of thermosetting adhesive materials react together during curing and cross-linking process, imparting bond strength.

### 2.10.2 Factors Affecting Adhesive Actions

#### [A] Physical Factors

The bond strength developed by an adhesive depends upon following physical factors.

1. **Surface tension.** A liquid adhesive is suitable for an adherend surface only if it can wet that particular surface. The wetting tendency of a liquid adhesive in turn depends upon its surface tension and viscosity. Now we know that the attraction between the liquid adhesive and the adherend surface is maximum when the interfacial tension between the two is minimum which in turn is possible when the adhesive as well as adherend surface is soluble in the same solvent.

2. **Porous and smoothness of the adherend surfaces.** Porous materials like paper, wood, leather etc. have a large number of capillaries to which more mobile portion of the adhesive flows leading to disturbance of equilibrium between the solute and solvent in the adhesive. Thus in such cases, a weak bond is formed. However, in some cases, such phenomenon may lead to stronger bond also. For example, when sodium silicate adhesive is applied to porous surfaces, loss of water by diffusion through the capillaries, causes a strong bond.

   In case of wood, smoth and properly plained surfaces are found to be more tenaciously bonded than rough surfaces.

3. **Physical characteristics of the adhesive film.** The physical characteristics of the adhesive film like tensile strength, shear strength, compressive strength, modulus of elasticity, creep rate and thermal coefficient of expansion greatly affect the bond strength of the adhesive.

   (*i*) In general, tensile, shear and compressive strengths of the adhesive film directly affect the bond strength of the film, *i.e.* greater the value of these strengths, stronger will be the bond.

(*ii*) A good adhesive film should have negligible creep under the stress of structure.

(*iii*) When the adhesive film and the contacting surfaces have quite different thermal coefficients, variation in temperature may develop stress in the film. In such case, the thermal coefficient of expansion of the adhesive can be brought to a range near to that of adherend surfaces by adding finely powdered inorganic substances like aluminium oxide to the adhesive.

4. **Thickness of the adhesive film.** Generally, a thin film is more strong than a thick one. If bonding of surfaces by adhesives is done under pressure, thickness of the adhesive film is controlled to get more bond strength. The addition of a solvent or plasticizer lowers the viscosity of a high polymer producing a thin film having increased bond strength.

   However, in case of liquid adhesive having no volatile matter, the strength of the adhesive film is independent of thickness.

5. **Factors prevailing during application of adhesive.** Factors like pressure, temperature and time have definite effect upon the final strength of the bond. Time and temperature actually determines whether full curing and drying of the adhesive film have occurred or not. Proper curing and drying of the adhesive film must be done before using.

   Perfectly flat and non-porous surfaces require very little pressure because adhesive film may squeeze out under pressure. On the other hand, porous surfaces (like wood) require pressure for keeping the joined surfaces together.

### [B] Chemical Factors Affecting Adhesive Action

1. **Polar characteristics of adhesives.** Adhesives containing polar group are strongly absorbed by the *organic* contacting surfaces. On the other hand, non-polar adhesives form weak bonds with such surfaces.

2. **Degree of polymerization of the polymeric adhesive.** In general, the degree of polymerization of an adhesive capable of giving maximum bond strength should lie between unpolymerised and highly polymerised fractions. Adhesive prepared from partially degraded cellulose derivatives are most suitable.

3. **Nature of side chain in adhesive molecules.** Chemical nature, length and complexity of the side chain in the adhesive molecule also affect the bond strength of the adhesive. In case of cellulose esters, best adhesive action is observed when the fatty acids (constituting the side chain) have 6-11 carbon chain. Similarly, resins obtained by the condensation of phenol with higher aldehydes have better adhesive action than that obtained from lower aldehyde and phenol.

4. **Effect of *p*H.** In general, strong acids and alkalies are harmful to the adhesive bond strength. Acids and alkalis affect more on cellulosic materials (*e.g.* wood) than on non-porous organic plastic and metallic surfaces. Glues derived from proteins have better adhesive strength in presence of lime (a weak alkali). Vulcanized rubber and organic plastics show better adhesive strengths at lower pH, *i.e.* in acidic medium.

### 2.10.3 Process of Bonding

Because of complex nature of adherend surfaces, adhesive and required service conditions, the exact bonding process, generally differs and supplied by the adhesive suppliers. However, the following steps are generally encountered during the process.

1. **Preparation of adherends.** Adherend surfaces must be treated properly before applying adhesive so that a strong bond is formed between adherend and adhesive. Way of treatment of adherend surface depends upon the nature of the adherend.

(*i*) *Wood or metal surfaces* are machined to provide flat and true plain surface.

(*ii*) *Wood* containing oil or resinous extract are either treated with a specific solvent or given alkaline surface treatment.

(*iii*) *Plastic surfaces* are slightly sandblasted to expose fresh surfaces which helps in better specific adhesion.

(*iv*) Metal surfaces are given chemical treatment to remove oil, grease and dust particles and metallic oxides from their surfaces.

2. **Preparation of adhesives.** This step consists of following two steps.

(*i*) Dissolution of the adhesive in proper solvent or melting of solid adhesive, or thinning of liquid adhesive.

(*ii*) Addition of additives like catalysts, fillers, extenders, plasticisers, hardeners etc.

3. **Application of adhesives.** Uniform application of adhesives over the adherend surface can be carried out by any of the following processes.

(*i*) By brushing.
(*ii*) By spraying.
(*iii*) By roller coating.
(*iv*) By laying a film or sheet of dry adhesive.

(*v*) By applying one component (*e.g.* thermosetting resin) on one surface, and the other component (such as a reactivating solvent or catalyst) on the other surface. This method is applied when the two components are chemically reactive.

(*vi*) By applying the two components (resin and catalyst) to the adherend surface by double spray directly to the adherend surfaces, so that their mixing takes place at the very moment of the application. This method is applied when the components react and harden immediately.

4. **Joining of the adhesive coated adherends.** This can be achieved in the following ways.

(*i*) Molten thermosetting adhesives dry and develop strength rapidly, hence in such cases adherends are assembled immediately.

(*ii*) Thermoplastics or rubber solution in organic solvents are first applied on the adherend surfaces. The surfaces are then exposed to air for long time for drying and volatilisation of solvent, before joining the two surfaces. In such cases, air drying is necessary to develop required tackiness and to avoid blisters or voids formation in the bond film.

(*iii*) Wood adhesives are applied on surfaces which are then assembled by laying them together in close contact, the assembled adherends are allowed to stand for sometime during which loss of solvent and completion of the reaction occur. By doing so, consistency of the adhesive layer increases with the result excess adhesive is not squeezed out during application of pressure.

(*iv*) For metal adhesives, the coated adherends are heated in an oven either before or after the assembly of the adherend surfaces and pressure is applied in the last. This process is known as *pre-curing*.

5. **Application of pressure and temperature.** When the adhesive applied on the adherend surface has reached the desired consistency, pressure is applied at the joint. The magnitude of pressure applied should be such that (*a*) the adherend surfaces come in uniform contact to each other (*b*) the excess of liquid adhesive squeezes out uniformly over the surfaces, so that no voids or air bubble remain in the adhesive film.

The applied pressure is immediately released in case of pressure-sensitive or fast-setting adhesives. On the other hand, in case of slow setting adhesives (*e.g.* thermosetting resin adhesives), the pressure is applied till chemical reactions and other allied processes are complete to achieve the full bond strength.

Temperature used during the joining of the adherend surfaces depends upon the type of mechanism involved in the development of bond strength of the adhesive film. In case of a chemically reactive or drying adhesive, heating is done under pressure, while in case of molten adhesive cooling under pressure is required.

6. **Conditioning after bonding.** Following steps are taken after bonding :

(*i*) Since the use of the thermoplastic based adhesives involves heating under pressure, adhesive film produced in this process must be cooled below a certain temperature, before bonding pressure is released, otherwise weak bond will be formed.

(*ii*) In case of thermosetting based adhesives, the joint may not have developed full strength after the release of pressure, so the joint is further conditioned by keeping it for some more time after the release of pressure so that full bond strength may develop.

(*iii*) In some joints, internal stresses may also develop after bonding. This may be due to different temperature and moisture contents of adherends and adhesive films. In such cases, stress is minimised by a number of processes.

### 2.10.4 Classification of Adhesives

There are various ways for classifying hundreds of commercially known adhesives preparations. However, the most widely used classification is based upon the chemical nature of adhesives. The important classes of adhesives along with their typical representative member are summarised below.

1. **Thermosetting synthetic resins.** These resins have three dimensional cross-linked structures and possess great adhesive properties. The bond produced by them is insoluble, infusible, and resistant to water, heat, insects and fungi. Important members of this class are discussed below.

(*i*) *Phenol-formaldehyde resin.* It is sold in the form of solid, liquid or impregnated film. The adhesive resin is coated on the two surfaces and cured by heating and pressing.

Phenol formaldehyde resin is used for bonding articles in aircraft and ship building industries. It is also used in making water proof plywoods and laminates.

(*ii*) *Urea-formaldehyde resin.* It is a transparent, syrupy liquid which can be used as such or after mixing with water. The addition of water in the urea formaldehyde resin (*a*) helps in the formation of cross-linkages, (*b*) promotes rapid setting, and (*c*) avoids development of cracks in the adhesive film. Although it can be used in cold, a little heating accelerates the setting process. The bond film formed by this resin is quite strong and resistant to moisture, insects and fungi, however, film is deteriorated by the action of acids or alkalies.

This resin is used for bonding wooden surfaces, plywoods, laminates, articles in aircraft and ship industries.

(*iii*) *Polyesters.* These on mixing with alkyd resins are used as adhesives in making laminated glass and cloth. The bond film has low heat resistance.

(*iv*) *Epoxy resins.* Epoxy resins may be used in solid or liquid form. Usually, these are modified by adding some external compounds and some solvents. The solvent evaporates, leaving a very thin film. These are used for sticking a number of substances including metals, glasses and ceramic articles. *Araldite*, the widely used adhesive, is a epoxy-resin type adhesive.

(*v*) *Silicone resins.* Silicone resins are strong adhesives and have following remarkable properties.

(*a*) These are strongly water repellent.

(*b*) These are inert against insects, fungi and other living organisms.

(*c*) These have marked stability at elevated temperatures upto 250°C.

(*d*) These have immunity to chemicals and atmospheric conditions.

Silicone resins are used for bonding metals, ceramics, plastics, rubbers etc.

2. **Thermoplastic synthetic resins.** Adhesives formed by thermoplastic synthetic resins are soluble, fusible and poor water-resistant. These become soft on heating. Important examples are given below.

(*i*) *Cellulose derivatives.* Collodion (cellulose nitrate mixed with alcohol) is the widely used adhesive. The adhesive film produced by collodion is tough and water-proof. However, it is inflammable and effected by strong acids and alkalis and deteriorates on ageing.

It is used in cloth and foot wear industries.

*Cellulose acetate.* Cellulose acetate mixed with benzene is although more resistant to heat than cellulose nitrate, it has less adhesive strength than the nitrate.

*Cellulose esters (methyl and ethyl).* Methyl cellulose when mixed with hot water has excellent adhesive property. Similarly, ethyl cellulose admixed with some modifiers and plasticisers is also used as an adhesive.

Cellulose esters are water resistant, but easily catch fire and deteriorate on ageing.

(*ii*) *Polyvinyls* like polyvinyl chloride and polyvinyl acetate are good adhesives. They produce hard film which is resistant to light, atmosphere and chemicals.

Polyvinyls are used for bonding non-porous surfaces like glass, metals etc. and as sealing material in food packings.

(*iii*) *Acrylics.* The most important acryl derivative used as adhesive is methyl methacrylate under the name of *plexiglass.* Plexiglass, when used in solution, produces adhesive film which is stable under all conditions. However, its cost restricts its use.

3. **Natural resin adhesives**

(*i*) *Shellac resin.* This is one of the oldest adhesives and must be applied while hot. It has good bond strength, good hardness and low thermal conductivity.

It is used in making belts, conveyers, etc.

(*ii*) *Asphalt.* It is a crude adhesive, used in bonding paper, cloth and metals.

4. **Starch adhesives.** Starch adhesives are prepared by heating a starch suspension in water (or dil. HCl and 150 to 175°C). Starch adhesives although have low adhesive strength and poor moisture resistant, these are widely used because of following advantages.

(*i*) These are easily prepared and cheap.

(*ii*) These spread easily and dry quickly.

(*iii*) These can be applied in cold as well as hot state.

(*iv*) These do not give any foul smell.

Starch adhesives are used in making envelopes, stamps, note-books, binding books and other paper goods.

5. **Vegetable glues or Protein glues**

(*i*) *Casein adhesives.* These are prepared from milk derived proteins, called *casein.* These adhesives have widely been used in wood work industry and in the manufacture of drinking cups, straws, and ice cream containers.

(*ii*) *Soyabean protein adhesives.* Dry soyabeans are crushed to extract the oils. The residue is mixed with milk of lime and caustic soda to form a Jelly-like adhesive paste.

(*iii*) *Zein protein.* Zein, the corn protein is also used as an adhesive to some extent in combination with other bases.

(*iv*) *Peanut protein hydrates.* These are the latest protein based adhesives. These have been found to be suitable for making gummed tape and flexible glues for boxes and books.

6. **Animal glues.** Animal glues are derived from boiling animal hides, tendons or bones which are high in collagen. Various adhesives can be prepared from animal glues by incorporating various additives such as starch, plasticisers, wetting agents, fillers and preservatives.

Animal glue obtained from hides is known as *hide glue* and that obtained from bones is known as *bone glue.*

Animal glues have good adhesive strengths, but possess poor resistance towards moisture and fungus. These are used for gluing furniture, musical instruments, pencils, labels, etc. These have also been used in paper and paper board, straw boards, sizing etc.

## SOLVED PROBLEMS

**Problem 1.** *28 g of ethane was polymerized by radical polymerization process and the average degree of polymerization of polythene was found to be 1,000.*

*Calculate: (i) the number of molecules of ethene in the original sample, (ii) the number of molecules of polythene produced.*

**Solution.** (*i*) Number of molecules of ethene in 28 g

$$= 28 \text{ g} \times (6.02 \times 10^{23} \text{ molecules}/28 \text{ g})$$
$$= 6.02 \times 10^{23} \text{ molecules}$$

(*ii*) Number of molecules of polythene formed

$$= \frac{\text{Number of ethene molecules}}{\text{Degree of polymerization}} = \frac{6.02 \times 10^{23}}{1,000}$$

$$= \mathbf{6.02 \times 10^{20} \text{ molecules.}}$$

**Problem 2.** *Calculate the maximum percentage of sulphur that can be present in vulcanized rubber.*

**Solution.** 2 monomer units of isoprene require = 2 Sulphur atoms for cross links

∴ 2 × 68 g of isoprene requires = 2 × 32 g sulphur

or 68 g of isoprene requires = 32 g of sulphur

or (68 + 32) g of vulcanized rubber contains = 32 g of sulphur

Hence, maximum percentage of sulphur in vulcanized rubber

$$= \frac{32}{100} \times 100\% = 32\%.$$

**Problem 3.** *216 g butadiene is copolymerized with 104 g of styrene. What is the molecular formula of the copolymer?*

**Solution.** 216 g of butadiene = 216/54 = 4 mol

104 g of styrene = 104/104 = 1 mol

∴ Molecular formula of copolymer is:

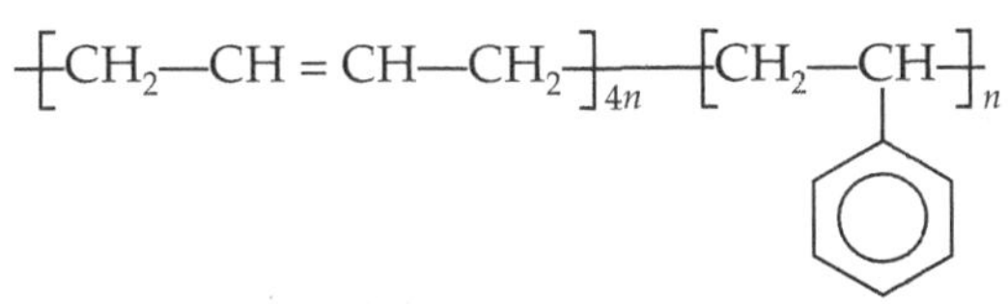

**Problem 4.** *If a polymer sample has a population as :*

*10 molecules of molecular mass each = 5,000*

*20 molecules of molecular mass each = 7,500*

*20 molecules of molecular mass each = 10,000*

*25 molecules of molecular mass each = 15,000*

*20 molecules of molecular mass each = 20,000*

*5 molecules of molecular mass each = 25,000*

*Calculate its number-average and weight-average molecular mass of polymer.*

**Solution.** Number average molecular-mass of polymers,

$$\overline{M}_n = \frac{[10 \times 5{,}000 + 20 \times 7{,}500 + 20 \times 10{,}000 + 25 \times 15{,}000 + 20 \times 20{,}000 + 5 \times 25{,}000]}{10 + 20 + 20 + 25 + 20 + 5}$$

$$= \frac{50{,}000 + 150{,}000 + 200{,}000 + 375{,}000 + 400{,}000 + 125{,}000}{100}$$

$$= \frac{1.3 \times 10^6}{100} = \mathbf{13{,}000.}$$

Weight-average molecular mass of polymer,

$$\overline{M}_w = \frac{[10 \times (5{,}000)^2 + 20 \times (7{,}500)^2 + 20 \times (10{,}000)^2 + 25 \times (15{,}000)^2 + 20 \times (20{,}000)^2 + 5 \times (25{,}000)^2]}{1{,}300{,}00}$$

$$= \frac{250 \times 10^6 + 1{,}125 \times 10^6 + 2{,}000 \times 10^6 + 5{,}625 \times 10^6 + 8{,}000 \times 10^6 + 3{,}125 \times 10^6}{1.3 \times 10^6}$$

$$= \frac{20{,}125 \times 10^6}{1.3 \times 10^6} = \mathbf{15{,}480.}$$

**Problem 5.** *(i) If two polymers of molecular weight 10,000 and 100,000 are mixed together in equal parts by weight, determine the number average and weight average molecular weights.*

*(ii) If the above polymers are mixed so that equal number of molecules are added determine* $\overline{M}_n$ *and* $\overline{M}_w$.

**Solution.** If equal parts of the two polymers of molecular weight M = 100,000 and M = 10,000 are mixed, by definition

$$\overline{M}_n = \frac{100{,}000 \times 1 + 10{,}000 \times 10}{1 + 10} = \mathbf{18181}$$

$$\overline{M}_w = \frac{1 \times 10^{10} + 10 \times 10^8}{10^5 + 10^4} = \mathbf{91819.}$$

**Problem 6.** *If 1000 g of a polymer of molecular weight 1000 g/mole and 1000 g of a polymer of molecular weight* $10^6$ *g/mole is mixed, what is the ratio* $\overline{M}_w / \overline{M}_n$?

**Solution.**

| Fraction | $M_i$(moles) | $N_i$ | $M_iN_i$ | $M_i^2N_i^2$ |
|---|---|---|---|---|
| 1 | 1000 g/mole | 1 | 1000 | $10^6$ |
| 2 | 1000 g/mole | $10^{-3}$ | 1000 | $10^9$ |

$$\overline{M}_n = \frac{\Sigma N_i M_i}{\Sigma N_i} = \frac{1000 + 1000}{1 + 10^{-3}} = \frac{2000}{1.001} = \mathbf{2000}$$

$$\overline{M}_w = \frac{\Sigma N_i M_i^2}{\Sigma N_i M_i} = \frac{10^5 + 10^9}{1000 + 1000} = \mathbf{5 \times 10^5}$$

$$\frac{\overline{M}_w}{\overline{M}_n} = \frac{5 \times 10^5}{2000} = 2.5 \times 10^2 = \mathbf{250.}$$

**Problem 7.** *100 g of novolac is cross linked by one —$CH_2$— group of each benzene ring. What weight of formaldehyde is required for achieving this?*

**Solution.** 2 molecules of 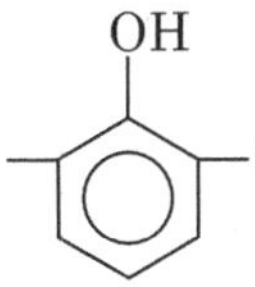 $-CH_2^-$ unit of novolac requires = 1 molecule of HCHO

$\therefore$ 2 × 96 g of novolac requires = 30 g of HCHO

or 100 g of novolac requires $= \frac{30 \times 100}{2 \times 96}$ g

$= \mathbf{15.625\ g\ HCHO.}$

## QUESTION BANK

1. What are organic polymers? Give the structure of five organic polymers.
2. What is the difference between a monomer and a polymer? Classify polymers on the basis of occurrence and chemical nature.
3. What are graft and block copolymers? Give their examples.
4. What are addition polymers? Give five examples.
5. If 20 gm of poly (ethylene) were completely burnt in the presence of excess of air, how many moles of $CO_2$ will be produced?
6. Write four units for the polymers obtained from the following monomers (assuming head to tail structure), and name the monomeric units.
7. Identify the repeating unit in the following and determine the structure of monomer.

(*i*) —CH—C(CH$_3$)$_2$—CH$_2$—C(CH$_3$)$_2$—CH$_2$—C(CH$_3$)$_2$—

(*ii*) —CH$_2$—CH(CN)—CH$_2$—CH(CN)—CH$_2$—CH(CN)—

(*iii*) 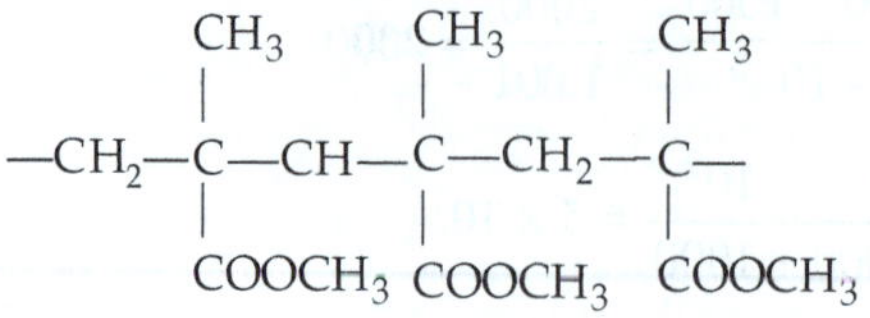

(*iv*) 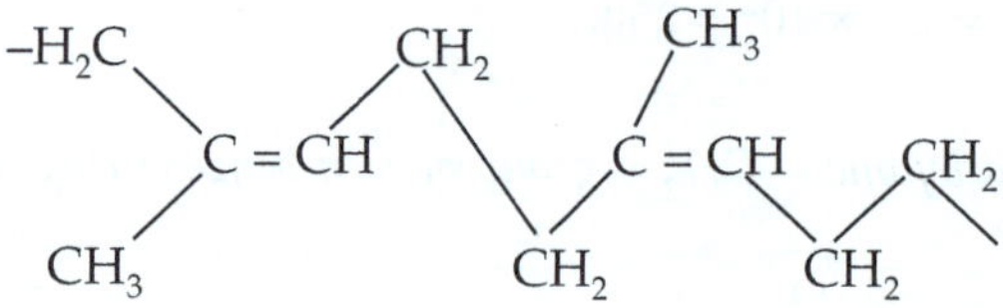

(*v*) 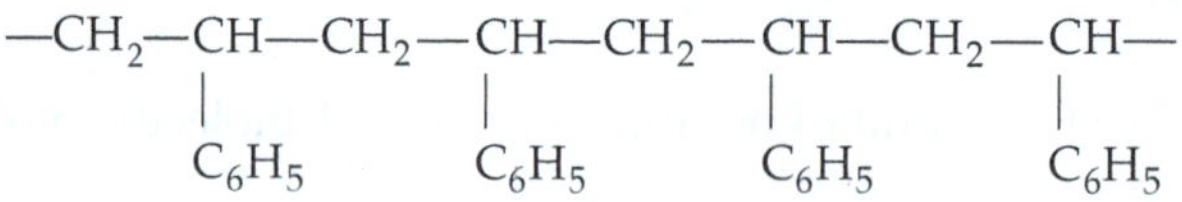

8. What are vinyl monomers? Give four examples.
9. Give the mechanism of free radical polymerization of vinyl monomers.
10. What are initiators? Name four free radical initiators.
11. What are inhibitors? Give examples.
12. (*a*) Define the terms
    (*i*) degree of polymerization and
    (*ii*) copolymerization.
    (*b*) What is the effect of polymer structure on the strength and crystalline nature of polyerms?
    (*c*) Give four uses of rubber.
13. (*a*) Explain with examples the terms: addition polymerization, condensation polymerization and copolymerization.
    (*b*) Differentiate between thermoplastics and thermosetting plastics. Give two examples for each type.
    (*c*) What are elastomers? Give the preparation and uses of SBR and neoprene.
    (*d*) Give the representative formula and two important application of each of, (*i*) plexiglas; (*ii*) teflon; (*iii*) PVC.
14. (*a*) Distinguish between (*i*) thermoplastics and thermosetting ; (*ii*) addition and condensation polymerization.
    (*b*) Explain the following with suitable examples (*i*) condensation polymerization; (*ii*) copolymerization.
15. (*a*) Outline the general method of preparing urethanes and state their applications.
    (*b*) List out the difference between LDPE and HDPE.
16. Give the structure of the following polymers:
    (*a*) Nylon 6.6 (*b*) Terylene
    (*c*) Orlon (*d*) Chloroprene.
17. What are condensation polymers? Give four examples of condensation polymers.
18. What are bio-polymers? Give three examples.
19. If average degree of polymerization of polymethyl methacrylate is $10^3$, calculate its average molecular weight.
20. State the differences between addition and condensation polymerization.

21. Write two uses of the following polymers; low density polyethylene; polystyrene, polymethylmethacrylate, Nylon 66, epoxy resins.
22. State the differences between RNA and DNA.
23. What are proteins? Write any three tests for proteins.
24. What are polysaccharides? Write examples and their building unit.
25. What are the characteristics of a polymer? Why do polymers have an average molecular weight?
26. What is a copolymer? Write structure of Buna-N, Buna-S. Classify the copolymers on the basis of arrangement of two monomer units.
27. State the differences between Free radical and ionic polymerization.
28. Write structure of phenol formaldehyde resin, urea formaldehyde resin polymides, polycarbonate, and polyurethanes.
29. (*i*) If two polymers of molecular weight 10,000 and 1,00,000 are mixed together in equal parts by weight, determine the number average and weight average molecular weights.

    (*ii*) If the above polymers are mixed so that an equal number of molecules are added, determine $\overline{M}_n$ and $\overline{M}_w$.
30. A sample of polystyrene is composed of a series of fractions of different sized molecules.

| *Fraction* | *Weight Fraction* | *Molecular Weight* |
|---|---|---|
| A | 0.10 | 12000 |
| B | 0.19 | 21000 |
| C | 0.24 | 35000 |
| D | 0.18 | 49000 |
| E | 0.11 | 73000 |
| F | 0.08 | 102000 |
| G | 0.06 | 122000 |
| H | 0.04 | 146000 |

Calculate the number average and weight average molecular weights of this polymer sample.

31. Fractions of a polymer when dissolved in an organic solvent gave the following intrinsic viscosity values at 25°C.

| M (g mol$^{-1}$): | 34000 | 61000 | 130000 |
|---|---|---|---|
| [η]: | 1.02 | 1.60 | 2.73 |

Determine K and α.

32. (*a*) What are polymers? Explain the reason for averaging molecular masses of polymers.

    (*b*) A polymer sample has the following distribution of molecular masses.

| Molecular masses: | 3000 | 4000 | 5000 |
|---|---|---|---|
| Percentage: | 20 | 30 | 50 |

Calculate the number average and weight average molecular masses of the polymer.

33. Distinguish between number average and weight average molecular weights of polymers. Under what conditions do these become identical?
34. What are the main features of coordination polymerization? Explain the mechanism of the polymerization of ethylene monomers on Ziegler-Natta catalysts.

35. (*a*) Write down the possible structural isomers of polymers produced by the mechanism of coordination polymerization.

    (*b*) Give comparative characteristics of free radical and ionic polymerization.

    (*c*) What type of monomers and catalysts participate in anionic polymerization reaction?

    (*d*) Write the chain initiation step of anionic polymerization.

36. Describe the main features of cationic polymerization reactions, the nature of monomer and catalysts. Write the mechanism of cationic polymerization.

37. What are conducting polymers? Name five of them.

38. Write the structure of five conducting polymers.

39. Write a short note on:

    (*i*) Polypyrrole (*ii*) Polythiophene.

40. Give important applications of conducting polymers.

41. What are adhesives ? Explain the bonding process by adhesive.

42. What are adhesives and give uses ?

43. (*a*) What are the advantages and limitations of adhesives ?

    (*b*) What are the various physical and chemical factors influencing the adhesive strength ?

44. Write a brief note on different types of adhesives.

45. Write note on adhesives.

46. Write short note on adhesive action.

47. Write short note on natural and synthetic adhesives.

48. Write a note on characteristics of a god adhesive.

49. Write a note on synthetic adhesives.

50. Name an adhesive used for bonding vaneers to produce plywood. What is the nature of this adhesive?

# Chapter 3 LUBRICANTS

## 3.1 INTRODUCTION

Whenever a machine works, its moving, sliding or rolling parts rub against each other resulting in the development of friction. This friction causes a lot of wear and tear of the concerned surfaces and hence such parts of machine require replacement at frequent intervals. Furthermore, due to friction large amount of energy are dissipated in the form of heat and thus causes loss in the efficiency of machine. And also, the heat produced due to friction heats up and thus causes damage to the moving part.

*A substance which is capable of reducing the friction between two surfaces which are sliding over one another is known as lubricant.* The friction developed by the motion of two contacting surfaces is reduced by lubricant. In other words the loss of energy due to friction is considerably reduced by lubricant. The lubricant acts in a number of manners *e.g.*

(*i*) It forms a thin film between the rubbing surfaces. Thus, rubbing surfaces do not come in direct contact with each other.

(*ii*) It acts as a coolant, as heat of friction is generated due to rubbing of surfaces.

(*iii*) It avoids power loss in the internal combustion engines. Since, it acts as a seal by sealing the piston and cylinder wall at the compression rings hence, there is no leakage of gases even at high pressure in combustion chamber.

Important Functions of a lubricant include :

(*i*) It reduces loss of energy in the form of heat by acting as a coolant.

(*ii*) It reduces wear and tear and surface deformation, by avoiding direct contact between the rubbing surfaces.

(*iii*) It reduces the expansion of metal by local frictional heat.

(*iv*) It reduces the maintenance as well as running cost of the machine.

(*v*) It minimizes the liberation of frictional heat and avoids seizures of moving surfaces.

(*vi*) It minimizes rough relative motion of the moving (or sliding) parts.

(*vii*) It increases the efficiency of the machine by reducing the waste of energy.

(*viii*) It acts as a seal in internal combustion engines.

## 3.2 CLASSIFICATION OF LUBRICANTS

Lubricants are classified on the basis of the states of aggregation or consistency—as solid, semi-solid, liquid lubricants and emulsions (Fig. 3.1).

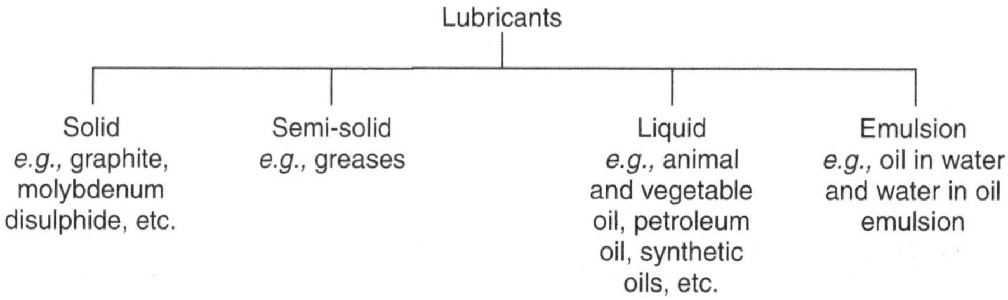


***FIGURE 3.1*** **Classification of lubricants.**

### 3.2.1 Solid Lubricants

Such lubricants are not widely used. Since, these contain grains of particles hence, may damage delicate parts of the machinery. Solid lubricants are used in the following conditions:

(*i*) Heavy machinery working on a crude job at very high load and slow speed.

(*ii*) Where parts to be lubricated are not easily accessible.

(*iii*) Where the operating temperatures and pressures are very high and liquid lubricants are combustible.

An ideal solid film lubricant should have low shear strength, good thermal conductivity, chemical inertness and stability at the operating conditions. This type of lubricants may be either inorganic or organic compounds.

*Molybdenum sulphide* and *graphite* are the widely used solid lubricants having laminar layer lattice structures. Layer lattice structure alone is not sufficient for good lubricating ability. For example, mica and boron nitride though having layer lattice are ineffective solid lubricants. Other materials used as solid lubricants are $WS_2$. $CdCl_2$, $PbI_2$, $CoCl_2$ and borax. Other soft materials without layer lattices like talc, ZnO, white lead and PbO also find use as solid lubricants.

Graphite has layered structure (Fig. 3.2). In graphite each and every carbon atom exhibits $sp^2$-hybridisation. Thus, it has a layer structure. The distance between two layers is 3.4 Å while two carbon atoms in a ring are 1.42 Å apart. There is a weak force of attraction between two layers therefore, one layer slides easily on the other and it acts as a lubricant. On the other hand molybdenum disulphide has a sandwich-like structure (Fig. 3.3) in which a layer of molybdenum atoms lies between two layers of sulphur atoms. These two layers are 3.13 Å apart while interatomic distance between two molybdenum and two sulphur atoms is 3.08 Å and 3.15 Å respectively. Again due to weak attraction forces between two layers, this acts as lubricant.

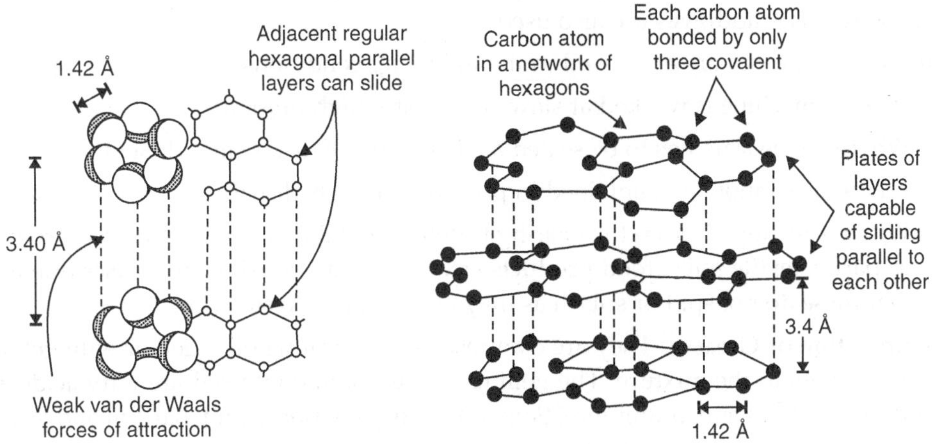


*FIGURE 3.2* **The structure of graphite.**

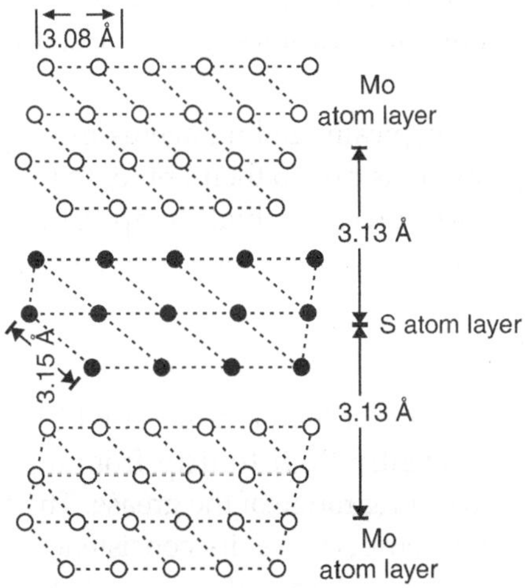


*FIGURE 3.3* **Layer structure of $MoS_2$.**

### 3.2.2 Semi-Solid Lubricants

When liquid lubricants are used, the bearings are to be sealed in order to avoid the loss of lubricant on account of the deforming effect of gravity. In order to avoid this sealing such lubricants are used that have the deforming effect of gravity. Such solidified lubricants include greases and vaslines.

Semi-solid lubricants are the petroleum products and may be defined as, *a semi-solid or solid combination of a petroleum product and a soap or a mixture of soaps, with or without filters, suitable for a certain types of lubrication is called semi-solid lubricant.*

Most of the common greases are made with soap thickness. Generally soap content is 7-18% but this limit may vary from 3-50%. The term grease has also been used for semi-solid lubricants prepared from chalk, talc, mica, graphite and also for combination of fats, waxes and petroleum residues. The properties of grease depend upon the nature and amount of the thickener and base oil used. In some cases additives are also used.

Lubricating greases are employed in the following conditions:

(*i*) When a machine is worked at slow speeds and high pressure.

(*ii*) Where the bearing has to be sealed against entry of dirt, water, dust etc.

(*iii*) Where oil cannot be maintained in position due to bad seal.

(*iv*) Where spurting oil from the bearing is deterimental to the manufactured products such as paper, textile mills, food products etc. Greases are ideal in such cases due to the fact that these do not spill or splash as they are designed to stay put.

**Composition of Greases.** They are composed of lubricating oil together with gelling agent which gives stiffness to the mixture. The gelling agents commonly used are fatty acids, soaps of lithium, sodium, calcium and aluminium. Sometimes soaps of potassium, barium and lead are also used. Fatty acids used are oleic, palmitic, stearic and other carboxylic acids obtained from tallow, hydrogenated fish oil and castor oil.

Lubricating oils employed are generally petroleum fractions. In some special greases, the oil used may be synthetic such as aliphatic diesters, polymerized ethylene oxides, siloxanes and fluorocarbons.

**Structure of Greases.** Microscopically gelling agents in soap based greases are crystallites of fibrous shape. Their gelling property is due to their ability to form a tangle in the interstices due to which the oil is held by capillary forces. At high temperature, the soap dissolves in oil, the fibrous structure disappears and the liquefication of grease takes place.

## Properties of Greases

The *dropping point* of a grease is the lowest temperature at which the grease, when heated under specified conditions, becomes sufficiently fluid to drip from a cup through a hole of specified diameter. This also indicates the operating range of the grease. The soap, besides acting as a gelling agent, also influences the operating temperature, its consistency and its resistance to water and oxidation. The viscosity of the oil used in their manufacturing dictates the lubricating characteristics.

*Consistency* is expressed in terms of penetration, which is defined as, *the distance in millimeter that a standard cone penetrates vertically into the sample under standard conditions of load, temperature and time.*

The *drop point* is the temperature at which the grease passes from semi-solid to the liquid state. It determines the upper temperature limit of the applicability of the grease. The different greases, their properties and uses are listed below in Table 3.1.

TABLE 3.1 Properties and uses of different greases

| *Type of grease* | *Properties* | *Uses* |
|---|---|---|
| Sodium-soap grease (soda-base grease) | Slightly soluble in water and can be used only in dry conditions; high dropping point | Used in high-temperature lubrication in ball bearings |
| Calcium-soap grease (lime-base grease, cup grease) | Water-resistant; beyond 70°C, the grease separates into soap and oil; cheapest; lesser dropping point than soda-base grease | Used for lubrication of water pumps and for general purposes. |
| Lithium-soap grease | Resistant to water and heat; expensive; superior to all other types | Used for lubrication of engines; for use at low temperatures. |
| Barium-soap grease | Water-resistant; good adhesiveness | Used for lubrication of automotives. |
| Aluminium-soap grease | Highly adhesive; water-resistant; expensive | Used for lubrication of chains and oscillating surfaces to which other types of greases will not adhere. |
| Axle grease (Resin grease) (talc and mica added as fillers) | Water-resistant; very cheap | Used for less delicate instruments under high loads and low speeds. |

### 3.2.3 Liquid Lubricants (Lubricating oils and Lube oils)

Majority of the lubricating oils are of petroleum origin. The lube oils are of animal or vegetable origin. These lubricants may be classified as follows:

(*i*) Vegetable oils

(*ii*) Animal oils

(*iii*) Mineral oils

(*iv*) Compounded or Blended oils

(*v*) Synthetic oils

Now, these lubricants are described as:

(*i*) **Vegetable oils.** Glyceride of higher fatty acids which are in liquid state at 20°C are known as oils. The oils which are obtained from vegetable, are known as *vegetable oils*. The oils which decomposes on heating but do not distill, they are called ***fixed oil***. Fixed oils have a property called oiliness by virture of which they are absorbed on the metal tenaciously and offer lower coefficient of friction. Such oils are also used in the preparation of greases and cutting emulsion *e.g.* ***Olive oil*** is a colourless to golden yellow oily liquid that is extracted from the olive trees. The lower grade olive oil is used as lubricant for bearing and machine parts subjected to low pressure and high speeds while high grade olive oil is used for edible purposes.

***Castor oil*** is a colourless to pale green oily liquid. It specific gravity is near to water etc., 0.96%. It has excellent oiliness. It is obtained by compression of castor seeds. It is immiscible with petroleum. It is used as lubricant for bearing and machine parts which are subjected to low pressure and high speeds.

***Palm oil*** is a pale yellow, pleasant smelling liquid, having, specific gravity of 0.91. It is widely used as lubricant for delicate articles such as watches. clocks, sewing machines, scientific instruments etc. It is obtained by compression of kernels of palm fruits.

Rape seed oil hazel nut oil etc. are some other examples of vegetable oils.

(*ii*) **Animal oils.** Chemically fats and oils are the same. Fats are solid at 20°C while oils are liquid at 20°C. Animals oils are extracted by rendering process. In this process the enclosing tissure is broken by treatment with steam or with the combined action of steam and water. Some common animal oils are as follows:

***whale oil*** is obtained by the distillation of different parts of sperm whale flesh. It is thin pale yellow oily liquid having specific gravity of 0.88. It is used as lubricant for light machines.

***lard oil*** is obtained from fats, kidneys and intestine of pigs. It is colourless oil and used in ordinary machines.

***tallow oil*** is obtained by the compression of cattle fat under hgih pressure. It is used for high pressure and low speed machines.

***neat foot oil*** is obtained by boiling foot of neat with water. It is pale yellow oily liquid of specific gravity 0.915. It is used as lubricant for delicate articles such as watches, clocks, sewing machines etc.

(*iii*) **Petroleum oils (mineral oils) ($C_{12}$—$C_{50}$).** They are the most widely used as they are abundantly available, quite stable under operating conditions, and are cheap.

The residue obtained on distilling petroleum under reduced pressure gives rise to three fractions called light, medium and heavy oils. They cannot be used as such because of the presence of wax, asphalt, coloured substances and other oxidisable impurities, and hence need to be purified before use. Wax interferes with lubricating properties by separating out at lower temperatures. Wax can be removed by chilling the oil fraction. The asphaltic material tends to leave carbonaceous deposits on the engine parts and can be removed by extraction with solvents. The coloured substances as well as micro-crystalline wax can be removed by filtration through Fuller's earth or some form of activated clay. The other oxidisable impurities can cause sludge formation. The oil can also be refined by treatment with hydrogen in the presence of nickel as catalyst to remove the sulphur compounds (desulphurisation) and to transform the unsaturated compounds into saturated ones. The solvent used for the removal of asphalt and other impurities should be capable of selectively extracting the undesired constituents and also be immiscible with oil. Phenol, *o*-cresol, nitrobenzene, furfural, liquid sulphur dioxide and dichloroethyl ether serve as solvents for extraction of impurities.

**Blended oils.** No single oil serves as the most ideal lubricant for many of the modern machineries. Specific additives can be incorporated into petroleum oils to improve their characteristics. These are called *blended oils* and confer desired properties to the lubricants required for a particular machinery.

- Vegetable oils like coconut or castor oil and fatty acids like palmitic, stearic or oleic acid can serve as additives to improve the oiliness of a lubricant.
- Aromatic, phenolic or amino compounds act as antioxidant additives to lubricants used in internal combustion engines, turbines, etc., β-Naphthol, di-*t*-butyl-*p*-cresol, *p*,*p*′-N,N-dimethylamino-diphenylmethane, etc., are some of the antioxidants. Sodium salts of sulphonic acids act as emulsifying agents.

- Addition of hexanol helps in improving the viscosity index of lubricants.
- Materials like polystyrene, polyesters, etc., serve as thickening agents and give the lubricating oils a higher viscosity.
- Additives in the form of organic compounds of phosphorus or antimony act as corrosion protectives by preventing contact between the metal surfaces and corrosive substances.
- Glycerol and methylsilicone polymers help in decreasing foam formation while the detergents in the form of salts of phenol and corboxylic acids inhibit deposit formation on internal combustion engines. Tricresyl phosphate acts as abrasive inhibitor.

**Synthetic lubricant oils.** Easy availability, low cost and desirable viscosity-temperature characteristics are responsible for preferring petroleum oil lubricants. However, under conditions of high-speed machinery, wide variations in temperature conditions, fire risk and heavy load, the petroleum oils cannot be effective as they tend to get oxidised at higher temperatures while wax separation will occur at low temperatures. Further, for the industrial and military needs of lubricants, the conditions will be too severe for petroleum oils. This has necessarily led to the development of synthetic lubricants. They perform well even below – 26°C and above 121°C. It is also possible to synthesize lubricants with molecular structures that can cope with specific operating conditions. Dibasic acid esters, polyglycol ethers, fluoro- and chlorohydrocarbons, organophosphates, silicones and silicate esters, etc. are currently in use as synthetic lubricants. Diesters of the type ROOC—$(CH_2)_n$—COOR derived from adipic ($n = 4$) or sebacic ($n = 8$) acids esterified with $C_8$ or $C_9$ branched alcohols like iso-octyl alcohol ($C_8H_{17}OH$), 2-ethylhexanol are used in jet engines and military equipments. These lubricants have low pour point and high flash points. Phosphate esters of the type $PO(OR)_3$ (R = alkyl, aryl or other groups) are fire-resistant and used as hydraulic fluids. Tricresylphosphate is an additive for anti-wear properties. Silicate esters, $Si(OR)_4$ (R = ethyl, 3-methylpentyl, 2-ethylhexyl) are used as high-temperature hydraulic fluids. They have high viscosity index.

Silicones are employed as high temperature lubricants. Fluoro- and chlorohydrocarbons are used in equipment for manufacturing and handling liquid oxygen and hydrogen peroxide due to their thermal stability, hydrolytic stability and fire resistance.

### 3.2.4 Emulsions

Emulsion is a colloidal system in which both the dispersed phase and the dispersion medium are immiscible liquids *e.g.* milk, vanishing cream, butter etc. Generally, two types of the emulsions are recognised.

(*i*) **Oil in water emulsion (aqueous emulsion).** In which oil particles (dispersed phase) are distributed in water (dispersion medium) *e.g.* when a little nitrobenzene is distributed in large excess of water an emulsion of oil in water is formed. Other examples are milk, vanishing cream, etc.

Such emulsions are very useful in the following types of lubrications:

(*a*) Cooling and lubricating fluids for cutting tools etc.

(*b*) For internal cooling and lubrication of large diesel motor pistons and other types of large internal cumbustion engines.

Such emulsion also gives rust protection as the oil and emulsifier film which separates out from the emulsion get themselves coated on the metal surfaces.

(*ii*) **Water in oil emulsion (Oily emulsion).** In which water particles (dispersed phase) are distributed in oily liquid (dispersion medium) *e.g.*, when a little water is shaken with large excess of nitro-benzene, an emulsion of water in oil is formed. Other examples are butter, cold creams etc.

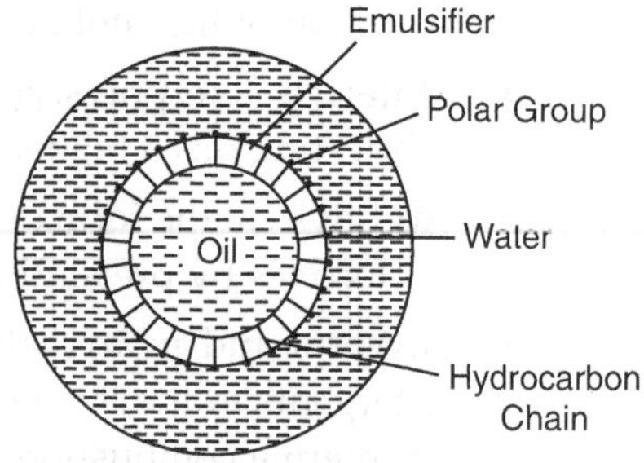

Aqueous Emulsion

The emulsions, so obtained, are not stable. In order to stabilise the emulsion, a third substance known as ***emulsifier*** or ***emulsifying agent*** is added. Soap, gelatin, gun arabic etc., are good emulsifiers. These emulsifiers form a thin film ground every drop of the dispersed phase and thus, prevents them from coming together.

These emulsions are very useful as compressor and pneumatic tool lubricants as they give internal cooling effect due to the evaporation of water.

### 3.2.5 Synthetic Lubricants

Mineral lubricants cannot be used under several abnormal conditions like extreme temperature, chemically reactive atmosphere, etc., under such conditions synthetic lubricants are used. Such synthetic lubricants are classified as follows :

**1. Hydrocarbons.** Hydrocarbons which are used as synthetic lubricants are polymerized ethylene, propylene, butylene, etc., of molecular weight range from 250 to 50,000, They do not differ in chemical and physical properties from the equivalent petroleum fractions. Their advantages over their equivalent petroleum products are the following :

(*i*) They are free from non-hydrocarbons.

(*ii*) They are light in colour.

(*iii*) They have better electrical properties and are used as electrical oil.

(*iv*) At high temperature they can be used as residue-free high temperature lubricants.

**2. Polyglycols and Related Compounds.** The compounds used under this heading are generally the following:

(*i*) Polyethylene glycol.

(*ii*) Polypropylene glycol and their ethers and esters.

(*iii*) Higher polyalkylene oxides.

(*iv*) Polyglycidyl ethers.

(*v*) Polythioglycols.

The first two are water soluble. So water diluted lubricants can be prepared from these for use in rubber bearings and joints. When these are used for metal surfaces, the advantage is that they can be easily removed by water flushing. The next two, that is, higher polyalkylene oxides and polyglycidyl ethers, though not soluble in water yet absorb lot of water. Their viscosity is not affected by temperature and their viscosity is low at sub zero. They have got low freezing point also. Polyalkylene oxides decompose into volatile parts under high temperature and oxidation of these volatile parts takes place. So they are very useful lubricants at high temperature and when it is desired that lubricants should not leave any residue. Example is roller bearing of sheet glass machinery.

**3. Esters.** Natural fats and oils are esters of glycerides but they are not useful in lubrication as their freezing point is quite high and they are easily oxidised. Esters which are used as synthetic lubricants are of the following types:

(*i*) Diester of aliphatic dibasic acids and aromatic dibasic acids.

(*ii*) Triesters of phosphoric acid.

(*iii*) Diesters of glycol.

The advantages of these are the following :

(*i*) Their viscosity is not much affected by the temperature.

(*ii*) They act as very good lubricants even at low temperature.

(*iii*) Their volatility is low.

(*iv*) Though they can be easily oxidised yet this property of susceptibility to easy oxidation can be reduced by adding inhibitors.

The disadvantages of using esters as lubricants is that they can be easily decomposed by steam and chemicals like acids and alkalies.

**4. Polycarbonates.** They are similar in behaviour and action to esters. The only difference is that they are less inflammable than esters.

**5. Amines, Imines and Amides.** Organic amines, imines and amides are used as synthetic lubricants. They have got high viscosity index and low pour point. They are used commonly due to the easiness with which they can be oxidised.

**6. Silicons.** Silicons are the products obtained by hydrolysis and polymerisation of organic chlorosilicons *e.g.*, $(CH_3)_2SiCl_2$ and $CH_3SiCl_3$.

The advantages of using silicons is that they have low viscosity temperature coefficient and they are not easily oxidisable up to 200°C. But above 200°C they are oxidised very quickly and then start cracking at 225°C.

Silicons gives good results only when one of the rubbing surfaces is non-ferrous. They are very useful for low temperature lubrication of small parts.

**7. Fluorine Derivatives.** Fluorine derivatives used are fluorocarbons and other organic compounds such as tertiary amines, ethers and esters which are completely fluorinated.

They are chemically inert and they can neither be easily decomposed by heat nor are they oxidisable. They are resistant to all chemicals excepting molten sodium. The advantage of using these as lubricants is that they have high viscosity index but they are costly.

## 3.3 THEORIES OR MECHANISMS OF LUBRICATION

The main function of the lubricant is to reduce the friction acting between two moving surfaces and acts as a coolant. There are three main types of mechanisms by which lubrication can be effected depending on the operating conditions. They are:

(*i*) Thick-Film Lubrication

(*ii*) Thin Film Lubrication

(*iii*) Extreme pressure lubrication

### 3.3.1 Thick Film Lubrication

It is also known as ***Fluid-film*** or ***Hydro-dynamic lubrication*** (Fig. 3.4). In this type of mechanism of lubrication, one of the rubbing surfaces is slightly displaced in relation to the other and a wedge-shaped film of the lubricant is formed between the two surfaces rubbing against each other. Thus, the pressure developed in the wedge-shaped film supports the load and prevents the direct contact between the rubbing surfaces. The lubricant film fills the irregularities of the sliding surfaces and lubricant forms a thick layer in between them. Due to the formation of a thick-film surfaces are not in direct contact with each other and reduces the wear considerably. The friction between two lubricated surfaces depends upon the dimensions of the film and viscosity of the lubricant.

In ***fluid-film lubrication*** the lubricant has minimum viscosity under working conditions and at the same time it is expected to remain in place and separate the surfaces. Evidently in such a system friction depends upon the viscosity thickness of the lubricant and not upon the load. The coefficient of friction in such cases is very low. It is of the order of 0.001 to 0.03.

The ***hydro-dynamic friction*** is developed in the case of a shaft running at a fair speed a film of lubricating oil covers the irregularities of shaft and that of bearing surfaces. At the same time there is no direct contact of metallic surfaces with each other. Thus, the resistance to movement is caused due to the internal resistance of the lubricant.

The hydro-dynamic lubrication is provided in delicate instruments such as watches, clocks, sewing machines, scientific instruments etc. In such cases generally hydrocarbons are used as a lubricant. To maintain proper viscosity in different seasons certain long chain polymers are added. Some ***antioxidants*** *e.g.*, aminophenols are also added to hydrocarbon petroleum fraction because these antioxidants usually contain minute quantity of unsaturated hydrocarbons which are susceptible to oxidation under operating conditions metallic surfaces. Such lubricants are good oiliness, low pour point, high resistance to oxidation and heat and high viscosity index.

In lubrication, the layer of lubricant, which is just in immediate contact with surface is held thereby forces due to physical or chemical adsorption which results because of chemical reaction between the metal surface and lubricant or by the combination of both physical and chemical type of forces. The tenacity with which this layer of the lubricant adheres to the surface depends directly on the oiliness of the lubricant. The rent layer is several molecules thick and remains stationary. This is known as *boundary layer*. The development of boundary or thin layer lubrication is due to the fact that, there is a fluid film of the lubricant, which carries the load and is continuously sheared. In case of fluid layer is squeezed out due to some reasons, then boundary layer remains in contact with the lubricated part.

Lubrication by boundary lubrication does not depend upon the viscosity and resistance to shearing of the lubricant due to the fact that lubricant layer in this type or lubrication is in immediate contact with the metal surface.

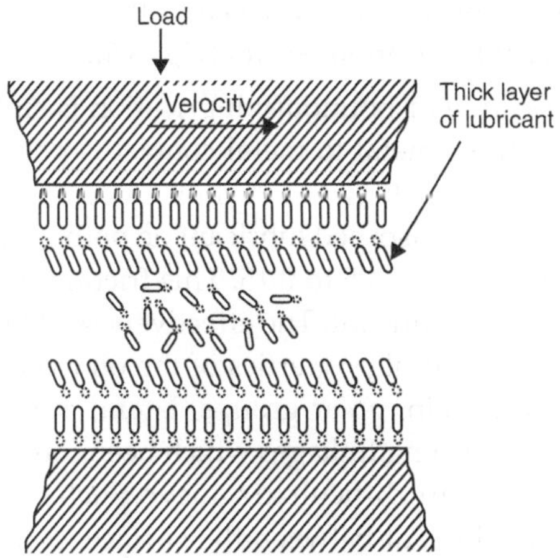


*FIGURE 3.4* **Fluid-film lubrication.**

Hydrocarbon oils are satisfactory lubricants for fluid film lubrication. To maintain the viscosity all through the seasons of the year, ordinary hydrocarbon lubricants are blended with certain long chain polymers. Some antioxidants (like aminophenols) are also used in journal bearing to avoid oxidation (under operating conditions) of unsaturated hydrocarbons present in small quantities in hydrocarbon petroleum fractions.

### 3.3.2 Thin-Film Lubrication

It is also known as ***boundary lubrication***. Such type of lubrication may take place under any one of the following conditions:

(*a*) When a heavily loaded bearing is suddenly stopped or started.

(*b*) When the load on bearing becomes greater than the pressure developed in the lubricating film.

(*c*) When the surfaces are not perfect and the load is carried on high spots.

When two metal surfaces are in contact with a hydrocarbon lubricant, a thin film of the latter gets adsorbed on the metal surface and held by van der Waals' forces. The long hydrocarbon chain gets oriented outwards in almost a perpendicular direction in the monolayer of adsorbed oil. Multilayer adsorption may then occur over the monolayer. As the distance from the metal surface increases, the oil layers become less oriented or randomly arranged.

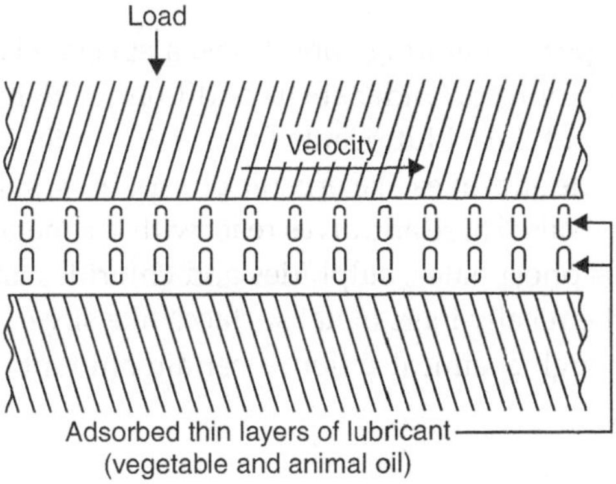


*FIGURE 3.5* **Boundary lubrication.**

Under conditions of heavy load and low speed, a thick film of lubricant cannot work an, under these conditions, boundary lubrication occurs (Fig. 3.5).

Evidently thin-film lubrication or boundary lubrication is suitable when a continuous film of lubricant cannot persist and direct metal to metal contact is possible due to reasons described above. Under all these conditions friction between the lubricated surfaces lies on absorbed lubricant. The load is carried by the layers of the adsorbed lubricant on both the metal surfaces. The coefficient of friction in such a case is usually from 0.05 to 0.15. This friction depends upon the nature of the metal and chemical properties of the lubricant. The effectiveness of boundary lubrication primarily depends upon the structure and chemical properties of the oil to be used. Such type of lubricants should contain long hydrocarbon chain, polar groups, lateral attraction between the chains and active atoms or groups capable of forming linkages which been found to affect lubrication, because true boundary lubrication conditions are difficult to achieve. Such properties of lubricant determine the formation of a wedge of the lubricant between the surface in case of fluid lubrication. In this type of lubrication, metals are not effectively separated by a lubrication film. The thin film of the lubricant is penetrated by irregularities of the metal surfaces which are never absolutely plane. When boundary lubrication has to work under vacuum, the lubricating film goes out. Thus, the friction increases due to the occurrence of the direct contact of the metals. The presence of air increases the efficiency of boundary lubrication due to the formation of protective film of iron oxide on the surface. On the other hand if fatty acids are the lubricants then the presence of moisture increases the effectiveness of boundary lubrication. Usually the friction decreases as fluid lubrication wear decreases. But in boundary lubrication decrease in friction does not decrease the wear to the same extent as decrease in friction and so minimum friction is not achieved by minimum wear.

Metals react with organic acid to form film which are chemically attached. The contact between the surfaces can be reduced effectively by these films, if these acid can undergo condensation and polymerisation. This is the reason that oleic acid is better lubricant than stearic acid. Because oleic acid can be polymerised while stearic acid cannot be. The lubricating property of oleic acid is also reduced if unsaturated impurities are removed from it.

Graphite and molybdenum disulphide serve as good boundary lubricants. High viscosity index, good oiliness, resistance to heat and oxidation and low pour points are some of the characteristics expected of boundary lubricants.

### 3.3.3 Extreme-Pressure Lubrication

Under heavy load and high speed operating conditions, a special class of lubricants called high pressure lubricants are to be employed. Under such conditions, a high temperature develops at the surface due to frictional heat and liquid lubricants fail to stick to the surface and may decompose or vaporise. To counter this, some additives like organic phosphorus, sulphur and chloro compounds need to be added to the mineral oils. These additives react with the metal surface at high temperature forming durable films such as phosphates, sulphides and chlorides and reduce the friction. These additives are ineffective on chemically inert metal surfaces like silver, chromium and titanium. As the lubrication is effected through chemical action, a certain amount of wear of the surface can be expected to occur.

## 3.4 PROPERTIES OF LUBRICANTS

### 3.4.1 Viscosity

Viscosity is the most important property determining the behaviour of lubricants. It is the property of a fluid which resists its flow or shear. It may be defined as the property of a liquid or fluid by virtue of which it offers resistance to its own flow. A liquid in a state of steady flow on a surface may be supposed to consist of a series of parallel layers moving one above the other. Any two layers will move with different velocities ; top layer moves faster than the next lower layer, due to viscous drag (*i.e.*, internal friction).

Consider two layers of a liquid separated by a distance, $d$ and moving with a relative velocity difference, $v$, then, force per unit area ($F$) required to maintain this velocity difference is given by:

$$F = \eta \frac{v}{d}$$

Where, $\eta$ (*eta*) is constant of the liquid, called *coefficient of viscosity*. If $v = 1$ unit (*e.g.*, cm/s), $d = 1$ unit (*e.g.*, cm), then $F = \eta$ . So, coefficient of viscosity may be defined as, *the force per unit area required to maintain a unit velocity gradient between two parallel layers*. The unit of viscosity is *poise* (dynes $cm^{-2}$ or $g\ cm^{-1}\ S^{-1}$). Centipoise is however the unit in practice.

**Viscosity-Temperature Relation.** Viscosity of liquids decreases with increasing temperature and, consequently, the lubricating oil becomes thinner as the operating temperature increases. Hence, viscosity of a good lubricating oil should not change much with change in temperature, so that it can be used continuously, under varying conditions of temperature.

Theoritically this decrease in viscosity at higher temperatures is due to the increase in the free volume available for the molecules to move about. In case of lubricants having larger molecules heat also provides energy required to deform the large molecules in order to make them slip part each other and thus decreasing the viscosity.

**Viscosity-Index.** The rate at which the viscosity of an oil changes with temperature is measured by an arbitrary scale, known as the *viscosity-index* (VI). If the viscosity of an oil falls rapidly as the temperature is raised, it has a low viscosity-index. On the other hand, if viscosity of an oil in only slightly affected on raising the temperature, its viscosity-index is high.

**Determination of Viscosity-Index.** For this purpose, two types of standard oils, *paraffinic-base Pennysylvanian oils* (VI = 100) and *naphthanic-base Gulf oils* (VI = 0) are used. Against each of these is marked their viscosities at 100°F and 210°F. The former are known as *H oils* and latter as *L-oils*. The VI determination is done in following steps:

(*i*) The viscosities of the oil under test at 100°F and also at 210°F are first found out, let these values be $U$ and $V$ respectively. The difference between the two values should be low, if the oil is good; and high if the oil is poor.

(*ii*) Now from the list of *H-oils* (with VI = 100), the oil which has the same viscosity at 210°F as the oil under-test is selected, and its corresponding viscosity at 100°F is read off. Let it be $H$.

(*iii*) Then, from the list of *L-oils* (*i.e.*, with VI = 0), the oil which has same viscosity at 100°F is read off. Let it be $L$. Then,

$$\text{VI} = \frac{L - U}{L - H} \times 100$$

where, $U$ = Viscosity at 100°F of the test.

$L$ = Viscosity at 100° F of the low-viscosity standard oil (*i.e.*, Gulf oil) having a VI of 0 and also having the same viscosity at 210°F as the oil under test.

$H$ = Viscosity at 100°F of the high viscosity standard oil (*i.e.*, Pennysylvanian oil) having a VI of 100 and also having the same viscosity at 210°F as the oil under test.

### Measuring Viscosity of a Lubricating Oil

Viscosity of an oil can be determined by Redwood (UK) or Engler or Saybolt (USA) or U-tube method. In UK, the time required for 50 ml of the liquid to flow through an orifice of a Redwood viscometer (Fig. 3.6) is expressed in Redwood seconds: in USA, the time required for 60 ml of the liquid to flow through an orifice in a Saybolt viscometer is termed Saybolt Universal Seconds. A brief description of Redwood viscometer is given below.

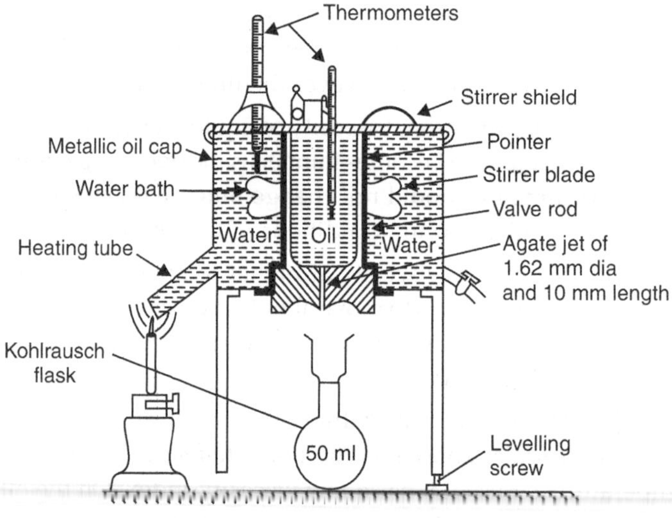


***FIGURE 3.6*** **Redwood viscometer.**

The essential parts of the experimental setup consist of *an oil cup, a heating bath, a stirrer, a spirit level, levelling screws* and *Kohlrausch flask*. The oil cup is a silver-coated brass cylinder (90 mm height and 46.5 mm in diameter), with an open upper end. The bottom part of the cylinder is fitted with an agate jet of bore diameter 1.62 mm and length 10 mm. The jet is opened by a valve rod which is a silver-plated brass ball fixed to a stout wire. The level to which the cylinder is to be filled with the oil is shown by a pointer which is a stout, tapered, upward-pointing wire fitted on the inner side of the cylinder. The lid of the cup is fitted with a thermometer to show the temperature of the oil. The oil cup is surrounded by a cylindrical copper bath containing water. It is provided with a tap for siphoning off water and a long side-tube projecting outwards for heating the bath water by means of a gas burner. A thermometer reads the temperature of the water. There is a stirrer outside the cylinder. It is provided with a circular shield at the top to avoid splashing of water into the oil cylinder. The jet can be level-monitored with a spirit level. The whole apparatus rests on three legs provided with levelling screws at the bottom. There is a 50 mL Kohlrausch flask (special shaped) meant to receive the oil form the jet outlet.

The levelled oil cup is cleaned and the ball of valve rod is placed on the agate to close it. The test oil, free from any suspension etc., is filled in the cup up to the pointer level. An empty Kohlrausch flask is placed just below the jar. Water is filled in the bath. The side-tube is slowly heated with constant stirring of the bath. When the oil reaches the desired temperature, the ball valve is lifted and suspended from the thermometer bracket. The time required for 50 mL of the oil to collect in the flask is noted and the valve is immediately replaced to prevent any overflow of the oil. The result is reported in Redwood No.1 seconds at the particular temperature.

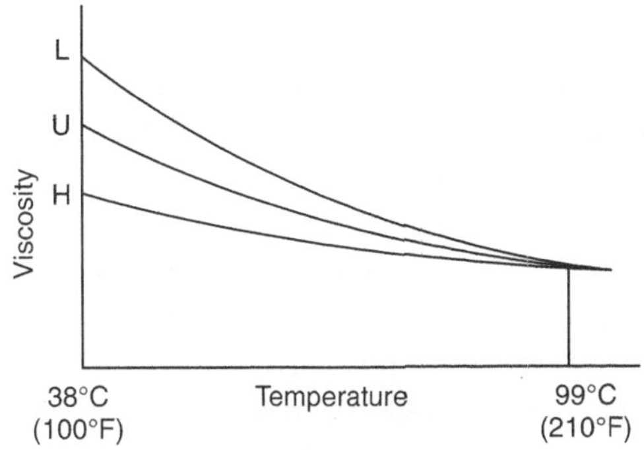


*FIGURE 3.7* **Viscosity index.**

Redwood Viscometer No.1 is commonly used for determining viscosities of thin lubricating oils. On the other hand, viscosity of viscous oils is measured by means of Redwood Viscometer No.2 where the jet bore diameter is 3.8 mm and length 15 mm).

Viscosity-index of lubricating oils can be increased by adding certain polymers, which are only partially soluble in oil.

(*i*) At low temperature, when the solubility of the added polymer in oil is slight, so the effect of the polymers on the viscosity of the oil is also slight.

(*ii*) At high temperature, when the solubility of the polymer in oil is considerable, its effect will be to increase viscosity of the oil.

Thus, by the correct addition of organic polymers, it is possible to produce oil-polymer blends which have a very slight temperature coefficient of viscosity or even none at all, *i.e., viscous static*.

### 3.4.2 Cloud Point and Pour Point

When an oil is cooled slowly, the temperature at which it becomes cloudy or hazy in appearance, is called its *cloud point*, while the temperature at which the oil ceases to flow or pour, is called its *pour point*.

Most of the petroleum-based lubricants contain dissolved paraffin wax and asphaltic impurities. When the oil is cooled, these impurities will start crystallising out giving a cloudy appearance. Further, cooling of the oil causes the oil to crystallise out and minute crystals of wax get interlocked prevents the flow of the oil. This is the reason for considering the pour point. Thus, a knowledge of the cloud point will be helpful in deciding the lowest temperature up to which it can be used without any clogging. The pour point helps to establish the lowest temperature at which the flow of the oil under gravity is reliable. Both these parameters may be useful in identifying the source of the oil. The pour point, however, is more important in the sense that it indicates the approximate minimum temperature at which the oil can be transferred by pouring, or below which lubrication by gravity flow is unreliable. The cloud and pour points can be used to assess the suitability of lubricants under cold conditions. The lubricants intended for use in machines at low temperatures should have low pour point. the presence of waxes in lubricating oils will tend to raise their pour points.

A simple way of determining the pour point is to read the maximum temperature at which the oil does not move when kept in a horizontal jar for five seconds. The pour point will be 5°F more than this temperature.

**Determination of pour point.** The apparatus employed for the determination of pour point is shown in Fig. 3.8. It consists of a flat-bottomed tube (about 3 cm diamter × 12 cm height) enclosed in an air-jacket. The jacket is placed in a freezing mixture contained in a jar. The tube is half-filled with the oil. A thermometer is inserted into the oil. There will be a gradual but continuous drop in temperature as the cooling proceeds slowly. For every degree fall of temperature of the oil, the tube is withdrawn for a moment from the air-jacket and examined. It is replaced immediately. The temperature at which cloudiness appears is recorded as the cloud point. The cooling is continued and the tube is withdrawn after every 3°C fall of temperature and tilted to observe the flow or pour of the oil. The temperature at which the oil does not flow in the test-tube even when positioned horizontally is taken as the pour point.

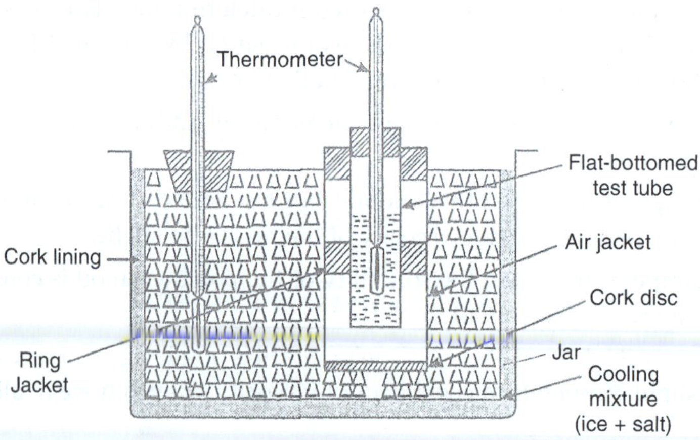


***FIGURE 3.8*** **Determination of pour point.**

*Paraflow* (commercial name) a polyalkylbenzene, is a pour point depressant and it can bring down the pour point by about 50°F. Lowering the viscosity of the oil or dewaxing can also lower the pour point.

### 3.4.3 Flash Point and Fire Point

These two parameters were originally devised to assess the volatility of an oil which affects the consumption of an oil in an IC engine. The *flash point* is the lowest temperature at which sufficient vapours are given off by an oil to cause a momentary flash as the vapours come in contact with a test flame. Flash point of burning oil (kerosene) is 73°F while that of power kerosene is 90°F. The *fire point* is the lowest temperature at which sufficient vapours are given off by an oil so that it burns continuously for at least five seconds on being lighted by a test flame.

The flash point gives a rough indication of the tendency of the oil to volatilise, and therefore is useful in process control for maintaining the uniformity of a product. In general, the fire points are 5 to 40° higher than the flash points. A good lubricant should have its flash point at least above the temperature it is to be used. This provides protection against fire hazards during use, transport and storage of the lubricant. High flash and fire points are good characteristics of a lubricant. A low flash point is indicative of evaporation losses and vapour locks in fuel passages of spark ignition engines. Oils of paraffinic base possess higher flash points as compared to naphthenic bases. Thus, the types of petroleum crude can be distinguished on the basis of flash points.

The flash and fire points do not have any bearing with the lubricating property of the oil, but there important when oil is exposed to high temperature service. The flash and fire points are usually determined by using *Pensky-Marten's* apparatus, (Fig. 3.9) which essentially consists of :

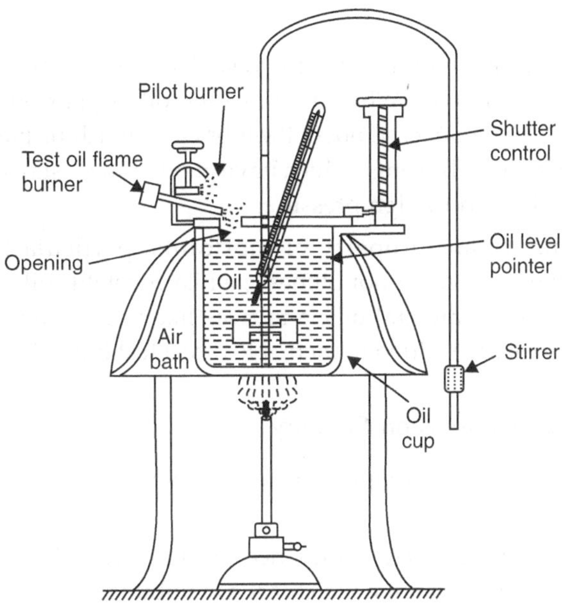


*FIGURE 3.9* **Penskey-Marten's flash-point apparatus.**

(*i*) **An oil cup.** It is about 5 cm in diameter and 5.5 cm deep. The level to which oil is to be filled is marked inside the cup. The cup lid is provided with four openings of standard sizes. Through one of these passes a thermometer; while the second opening is used for introducing test flame. Through third opening passes stirrer carrying two brass blades; while the fourth is meant for admission of air.

(*ii*) **Shutter.** It is a lever mechanism, provided at the top of the cup. By moving the shutter, opening in the lid opens and flame (carried by a flame exposure device) is dipped into this opening, thereby bringing the flame over the oil surface.

(*iii*) **Flame exposure device.** It is a tiny flame, connected to the shutter by a lever mechanism.

(*iv*) **Air bath**. Oil cup is supported by its flange over an air-bath, which is heated by a gas burner.

(*v*) **Pilot burner.** As the test-flame is introduced in the opening, it gets extinguished, but when the test flame is returned to its original position, it is automatically lighted by the pilot burner.

**Working**. Oil under-examination is filled upto the mark in the oil cup and then heated by heating the air-bath by a burner. Stirrer is worked between tests at a rate of about 1 to 2 revolutions per second. Heat is applied so as to raise the oil temperature by about 5°C per minute. At every 1°C rise of temperature, test-flame is introduced for a moment, by working the shutter. The temperature at which a distinct flash (a combination of a weak sound and light), appears inside the cup, is recorded as the flash-point. The heating is continued thereafter and the test-flame is applied as before. When the oil ignites and continues to burn for at least 5 seconds, the temperature reading is recorded as the fire-point of the oil.

### 3.4.4 Oiliness

*Oiliness* of a lubricant is a measure of its capacity to stick on the surfaces of machine parts, under conditions of heavy pressure or load. When a lubricating oil of poor oiliness is subjected to high pressure, it has a tendency to be sequeezed out of the lubricated machine parts, thereby its lubrication action stops. On the other hand, lubricants, which have good oiliness stay in-between the lubricated surfaces, when they are subjected to high pressure.

Oiliness is very important property of lubricants, particularly for extreme-pressure lubrication. Mineral oils have got very poor oiliness ; while vegetable oils have good oiliness. So, in order to improve the oiliness of mineral oils, additives like vegetable oils and higher fatty acids (such as oleic and stearic acids) are added to them. No direct test is available for measuring oiliness.

### 3.4.5 Decomposition and Corrosion Stability

Lubricating oils must be stable to decomposition at the operating temperatures. Oils are, usually, broken up by three chemical actions:

(*a*) *Oxidation* is the main destructive influence encountered. It is usually, partially prevented by adding certain antioxidants, which act as inhibitors.

(*b*) *Hydrolysis* : The presence of moisture in the oil causes hydrolysis of such components as esters in the lubricants, releasing alcohols and destructive fatty acids.

(*c*) *Pyrolysis* is the cracking of petroleum chains, due to the high temperature found within the engine bearings. Such a reaction is the main cause for the deposition of gummy and carbon sediments within the lubricant.

The harmful effects of contamination of the lubricant by these decomposition processes can be minimized by the use of an efficient system of oil filtration and periodic change of the oil.

**Corrosion stability** of a lubricating oil is estimated by carrying out corrosion test. A polished copper strip is placed in the lubricating oil for a specified time at a particular temperature. After the stipulated time, the strip is taken out and examined for corrosion effects. It the copper strip has tarnished, it shows that oil contains any chemically active substance. A good lubricant should not affect the copper strip. To retard corrosion effects of oil, certain inhibitors are added to them. Commonly used inhibitors are organic compounds containing phosphorus, arsenic, antimony, chromium, bismuth or lead.

### 3.4.6 Emulsification

The ability of an oil to get intimately mixed with water forming an emulsion is called emulsification. Certain oils form emulsions with water easily. Emulsions have a tendency to collect dirt, grit, foreign matter, etc., thereby causing abrasion and wearing out of the lubricated parts of the machinery. So, a good lubricating oil should form an emulsion with water, which breaks off quickly. The tendency of lubricant-water emulsion to break is determined by ASTM test. In this, 20 mL of oil is taken in a test-tube and steam at 100°C is bubbled through it, till the temperature is raised to 90°C. The tube is then placed in a bath maintained at 90°C and the time in seconds is noted, when the oil and water separate out in distinct layers. The time in second in which oil and water emulsion separates out in distinct layers, is called *steam emulsion number* (SEN) A good lubricant should passess a low steam emulsion number.

### 3.4.7 Aniline Point

The tendency of a lubricant to mix with aniline is a measure of its aromatic content. This can be expressed in terms of aniline point which can be defined as the lowest temperature at which the oil is completely miscible with an equal volume of freshly distilled aniline. The aniline point is inversely related to the aromatic content of the oil. Aromatic hydrocarbons have a tendency to dissolve natural rubber and certain types of synthetic rubbers. Low aromatic content is therefore desirable in the lubricants. A high aniline point suggests a higher percentage of paraffinic hydrocarbons.

The aniline point can be determined by mixing mechanically equal volumes of an oil sample and aniline in a tube. It is then heated until a homogeneous solution is formed. The tube is then allowed to cool at a controlled rate. The temperature at which the two phases separate distinctively is taken as the aniline point.

### 3.4.8 Volatility and Carbon Residue

When a lubricating oil is used in heavy machinery working at high temperature, a portion of oil may vaporize, leaving behind a residual oil, which have different lubricating properties (like increased viscosity). Good lubricant should have low volatility. The volatility of an oil is determined by an apparatus, called *vaporimeter* (see Fig. 3.10), which consists essentially of a furnace heated by some fuel gas. In the centre of the furnace passes a coiled-form of copper tube, through which air can be passed. A known weight of oil under-examination is taken in a platinum tray, which is then introduced into the copper tube. Dry air at a rate of 2 litres/minute is passed through the copper tube. After 1 hour of heating, the tray is taken out, cooled and weighed. The loss is weight is calculated as percentage of the original weight of oil taken.

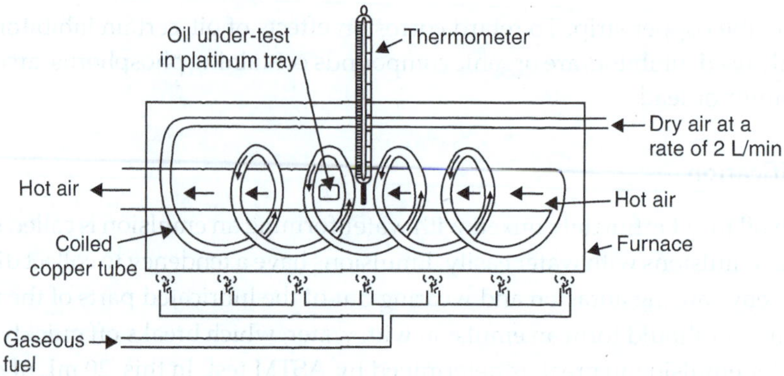


*FIGURE 3.10* **Vaporimeter.**

**Carbon residue.** Lubricating oils contain high percentage of carbon in combined form. On heating, they decompose depositing a certain amount of carbon. The deposition of such carbon in machine is introlerable, particularly in internal combustion engines and air-compressors. A good lubricant should deposit least amount of the carbon in use.

The estimation of carbon residue is, generally, carried out by *Conradson method* (Fig. 3.11). A weighed quantity of oil is taken in a silica crucible (of about 65-85 mL capacity). The skidmore crucible is provided with a lid, having a small tube-type opening for the escape of volatile matter. The combination is then placed in a wrought iron crucible (about 8 cm in dia and 6 mm high) covered with chimney-shaped iron hood (of about 10 cm diameter). The wrought iron crucible is heated slowly for 10 minutes, till flame appears. Slow heating is continued for 5 minutes more. Finally, strong heating is done for about 15 minutes, till vapours of all volatile matter are burnt completely. Apparatus is then allowed to cool and weight of residue left is determined. The result is expressed as percentage of the original weight of oil taken.

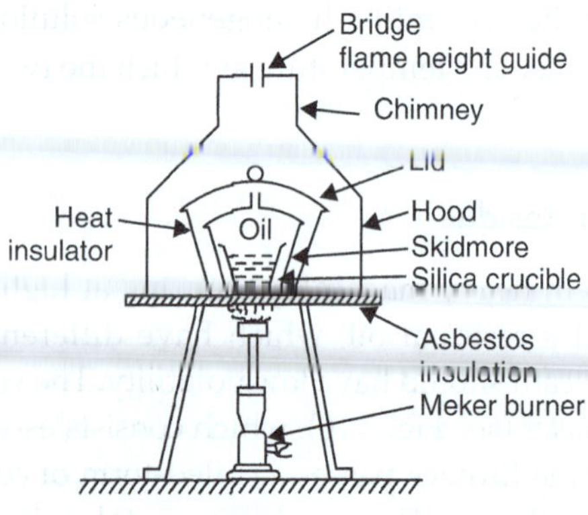


*FIGURE 3.11* **Conradson's apparatus for carbon residue estimation.**

### 3.4.9 Other Properties

Other important properties of lubricants include:

**1. Mechanical Stability.** Mechanical stability of a lubricant is judged by performin certain tests under high pressure. Commonest such test is *four-balls extreme-pressure lubricant test* (Fig. 3.12).

In this test the lubricant under-test is poured in a machine containing four balls. The lower three balls are stationary; while the upper ball is rotated. Load is gradually increased and the balls withdrawn and examined at specific intervals for scale formation, etc., on them. If the lubricant is satisfactory under the given load, the ball bearings after the test comes out clean. However, when the load is progressively increased, and if liberated heat welds the balls together, then the lubricant is said to have failed completely. Thus, this test enables us to determine the maximum load that can be carried safely by a lubricant.

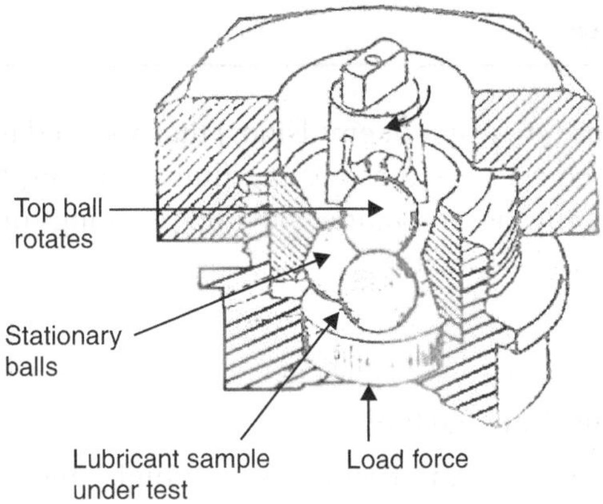


*FIGURE 3.12* **Four-balls extreme-pressure lubricant tester.**

**2. Precipitation Number.** It shows the quantity of precipitated asphalt when diluted with petroleum ether and then centrifuged. The test is carried out under standard conditions. Precipitation number differentiates between the different classes of lubricating oils.

**3. Specific Gravity.** It is a numerical value. It is the ratio of the weight of an equal volume of water at a particular temperature. Specific gravity is very helpful in identifying unknown oils, as the gravity of an oil gives the indication of the type of crude from which the lubricant has been prepared.

Specific gravity is generally, expressed in terms of API (American Petroleum Institute) degree, which is defined as :

$$\text{API gravity} = \frac{141.5}{\text{S.p. gr. at temp. } (60°\text{F})} - 131.5$$

The figure is termed as modulus of the API scale.

***API hydrometer*** adapted for oil, starts at 10, which is equivalent for 1, the specific gravity of water of 60°F. Liquids lighter than water are, therefore, assigned values higher than 10 and a heavier liquid, a value lesser than 10. Thus, heavier naphthalenic-base mineral oils have lower API gravity than those of lighter paraffin-base oils. Gravity gives no indication of quality of lubricating oil, but it is used for interconversion of weight and volume measurements of oils.

**4. Saponification Number.** It is the number of milligrams of KOH required to saponify 1 g of oil. Mineral oils do not saponify at all, but vegetable and animal oils do. Consequently, this test helps us to ascertain whether the oil under reference is animal and vegetable oil or mineral or a compounded oil containing mineral and vegetable oils.

**5. Neutralization Number.** It refers to the determination of *acidic* or *basic* constituents of an oil. Determination of acidic constituents is more common and it is referred to as *Acid number or value*, which is defined as the number milligrams of KOH required to neutralize the free acids in 1 g of the oil. Generally, free acids are not present in the lubricants, unless refined in faulty manner. A lubricating oil should possess acid value *less than 0.1*. Value greater than 0.1 indicates that oil has ben oxidized. This will, consequently, lead to corrosion, besides gum and sludge formation.

## 3.5 CUTTING FLUIDS

Cutting fluids are those liquids (oil, water or emulsions) that are used to cool as well as to lubricate. Emulsions of oil-in-water are mostly used as cutting fluids. Cutting fluids are required for tools used in the machine shop for cutting, threading, sawing, planning turning, drilling, etc. ; and the cutting fluid performs either of the functions.

(*i*) to cool the tools,

(*ii*) to lubricate the tools, or

(*iii*) to cool as well as lubricate the tools.

In such machining operations, the friction is very high, due to close contact between the work-piece and the tool ; and this generates a large amounts of local heat, thereby the tool gets overheated and it may even lose its temper and hardness. Consequently, in such a case, the cutting fluid provides cooling, besides lubrication.

In order to provide satisfactory service, the cutting fluid should possess following characteristics:

(*i*) good lubricating property,

(*ii*) low viscosity, so that the lubricant can easily fill in the cracks formed on the work-piece,

(*iii*) chemical stability,

(*iv*) non-corrosive nature towards the metals of the work-piece as well as tool, and

(*v*) high thermal conductivity.

(*i*) ***Oil as cutting fluid.*** In fine work, a stream of oil (like lard or rape or sperm oil) is directed over the work-piece. Here the cutting fluid (the oil) acts more as lubricant and less as heat-carrying agent. When the speed is low and the pressure is high (*e.g.*, in broaching), the oils are mostly used as cutting fluids. Compounded oils obtained by mixing petroleum oils with vegetable and animal oils are also used for such purposes.

(*ii*) ***Water as cutting fluid.*** In rough situation, like rough grinding and turning, only cooling action is sufficient. Water being good cooling agent and cheap, so it is mostly employed as cutting fluid. Water is primarily used, when only cooling is required, *e.g.*, in high-speed operations.

(*iii*) ***Emulsion as cutting fluid.*** In most cases, use of either water or oil alone is not satisfactory; and both a cooling agent and a lubricant is needed simultaneously. For such situations, oil-in water emulsions are employed. These are employed by mixing small amount of petroleum oil in water, and then stabilizing by adding emulsifier [like soap, tallow oil, lard oil, sperm oil, sulphonated oils, *e.g.*, $H_2HO_4$ -treated castor oil, chloro-sulphonated organic compounds, etc.]. Moreover, to prevent gelling (or thickening) of the concentrate, glycols or alcohols are added. In addition to these, certain germicide (like sodium nitrite, triethanol amine) is added.

**Cutting emulsions.** Serves as :

(*i*) lubricant to reduce frictional heat ; and

(*ii*) medium for the transfer of heat produced. In other words, use of cutting emulsions reduce power consumption as well as wear of tool. Before choosing a cutting fluid for a job it is, however, essential that is does not cause rusting of the metals involved.

## 3.6 SELECTION OF LUBRICANTS

In selecting a lubricant for a particualr job, the service condition requirements are to be related to the properties of the lubricant. The properties of a properly selected lubricant should not change under service conditions. If the properties of a chosen lubricant alters during service, it may cause certain problems. For example, if a lubricants, used at a high temperature, undergoes volatilization of a portion of it ; leaving being a residual oil, which will have different lubricating properties like higher viscosity. Such a mistake must be avoided. So a careful study of various properties and their correct interpretation is necessary for the selection of lubricants for a particular job. Selection of a lubricant for a few typical jobs are illustrated as follows:

1. **Lubricants for cutting tools.** Cutting fluids are lubricants used in machining, cutting, turning and grinding of metals. In these operations, a metal is continuously removed from the surface, and a fresh metal surface, free from any oxides, is continually exposed to the tool used. This, consequently, causes high friction, which results in excessive power consumption and liberation of large amount of heat. Heat liberated may overheat the tool and may lower its temper. Therefore, the main functions of cutting fluids are :

(*i*) to cool the tool, so as to reduce wear and prevent metal losing its harness,

(*ii*) to cool the metal work-piece, so as to prevent destrotion and demenstional inaccuracies,

(*iii*) to reduce power consumption by lubrication action, and

(*iv*) to improve surface finish.

Two situations can be there in cutting tool:

(*a*) **For heavy cutting.** For this purpose the most effective lubricants are cutting oils. The cutting oils are, essentially, mineral oils of low viscosities containing additives like fatty oils, sulphurized fatty oils and chlorinated compound, which by virtue of their polar groups attach themselves to the surface of continuously exposed fresh metals. As the shear strength of such an oil film is much less than that of the metal, a considerable reduction in friction results, thereby decreasing both the power consumption and the extent of heat generation.

(*b*) **In light cuttings.** Here the most effective lubricants are oil-emulsions. Oil-emulsions have somewhat smaller lubricating effects than cutting oils, but they are more efficient as cooling media, due to the high heat capacity of water, which is present in them as an external phase.

**2. Lubricants for internal combustion engines.** In an internal combustion engine, the lubricant is to be exposed to high temperatures. Therefore, the lubricant should possess high viscosity-index (*i.e.*, low variation of viscosity with temperature) and high thermal stability, *i.e.*, it should neither evaporate easily nor decompose at the operating high temperatures. However, as the lubricant has not to come in direct contact with water, a property of lubricant like emulsification need not be considered at all. Consequently, petroleum oils containing additives, which impart high viscosity-index and oxidation stability to them, are used as lubricants for internal combustion engines.

**3. Lubricants for gears.** Such Lubricants are subjected to extreme-pressures. So they should: (*i*) possess good oiliness, *i.e.*, stick well to the gear teeth, (*ii*) not to be removed by centrifugal force from the place of application, (*iii*) possess high resistance to oxidation, and (*iv*) have a high load-carrying capacity. Consequently, thick mineral lubricating oils, containing extreme-pressure additives (like metallic soaps and chlorine, sulphur or phosphorus compounds) are employed.

**4. Lubricants for delicate instruments** (like watches, clocks, scientific equipments, sewing machines, etc.) are not exposed to high temperature or to water or to extreme load, etc., so properties relating to these conditions are not considerd for such uses. Consequently, for such purpose, thin vegetable and animal oils (like palm oil, near oil. etc.) are employed.

**5. Lubricants for very high-pressure and low-speeds** such as for tractor rollers, concrete mixers, lathes, railways track joints, etc. Under these conditions, oil/grease films cannot be maintained. So solid lubricants (like graphite, soapstone, mica, etc.), are employed either in dry powder form or as emulsion in oil or water.

**6. Lubricants for high pressure and low-speeds** (like rail axle boxes, wire ropes, tractor rollers, etc.) are greases and blended thick oils.

**7. Lubricants for transformers.** The functions of the lubricating oil in an electrical transformer are to insulate the windings and to carry away the heat generated, when the transformer is on-load. The oil must possess good dielectric properties and, because efficient heat transfer depends on ready circulation, it must be of low viscosity. Moreover, transformer oil to some extent is exposed to air. Consequently, when the transformer is on load, such exposure coupled with elevated temperature, electrical stress and catalytic influence of copper, tend to produce chemical changes in the oil, leading to the formation of acids and sludge. Acids tend to attack the copper conductors, the transformer tank, the cotton insulation ; while sludge coats the windings and interferes with cooling, thereby increasing the temperature. High temperature accelerates chemical changes and also promotes the formation of more acids and sludge. The production of these injurious substances should be as small as possible, so highly refined mineral oils of high insulating quality, optimum oxidation resistance and chemical stability are employed.

The insulating ability of an oil may be measured by its dieletric strength, which represents the voltage which can be applied to electrodes of certain dimensions and distance apart immersed in the oil, before a spark passes. According to British Standards Institute, a minimum of 30,000 volts with electrodes 4 mm apart is required. It may be pointed here that provided an oil of suitable quality is used, the insulating ability of the oil depends almost entirely on the degree of absence of extraneous impurities such as dirt, fibres and moisture. Their presence, even in relatively small quantities, may appreciably lower the dielectric strength of the oil. If this reduction proceeds too great, the oil may cease to be effective as an insulator, with consequent damage to the transformer in which it is being used. For the reason, transformer oils should be properly filtered and dried, before being put into use.

**8. Lubricants for spindles in textile industry.** For spindles moving at very high speeds, thin oils are used. For better results, oxidation and rust inhibitors are added to the oil.

**9. Lubricants for refrigeration system.** Oil with low pour-point, low viscosity and low cloud-point is needed. So naphthalene-base oils, possessing such characteristics, are according employed mostly. The pour-point requirements are –40°F (maximum) for the lightest grade and – 13°F for the heaviest grade oils. Their viscosity range is 85 to 325 SUS (Saybold Universal Seconds) at 100°F.

## QUESTION BANK

1. What do you mean by lubricant? State the function of a lubricant.
2. What are the criteria for a good lubricant?
3. Discuss the important properties of lubricants and indicate the significance of these properties.
4. State the different mechanisms of lubrication with examples.
5. What are solid lubricants? Where they can be used? Give details about graphite.
6. Describe any four desirable properties of a lubricant oil.
7. Mention the various steps involved in the purification of mineral oil lubricants. State the advantages and disadvantages of the mineral oil as a lubricant.
8. What is extreme pressure additives? Why these additives are used?
9. Differentiate between vegetable oils and mineral oils.
10. What is meant by blended oils? Discuss the different blended oils and their uses with examples.
11. How is grease made? Give examples.
12. Differentiate between fluid film and boundary lubrication.
13. Write notes on different greases and their functions.
14. Explain the terms oildag and aquadag and state their significnce.
15. What are the characteristics features of synthetic lubricating oils?
16. What is iodine value of an oil? Mention its significance.
17. Synthetic lubricants have added advantage over natural lubricants. Explain.
18. What is lubricating emulsions? How they are prepared.
19. Differentiate between oil in water and water in oil emulsions.
20. Write notes on:
    (*a*) Lubricating grease.
    (*b*) Lubricant emulsions.
    (*c*) Semi solid lubricants.
    (*d*) Silicone synthetic lubricants.
21. Explain the properties of lubricants such as viscosity index, saponification number and neutralization number.

**22.** Write notes on:

(*i*) Cloud and pour points.

(*ii*) Flash and fire points.

(*iii*) Aniline points.

(*iv*) Carbon Residue.

**23.** Give the differences between graphite and molybdenum disulphide lubricants.

**24.** What do you mean by consisting and drop point of a grease sample?

**25.** Discuss the importance of:

(*i*) Oiliness (*ii*) Pour point

(*iii*) Flash point (*iv*) Fire point

(*v*) Oxidation stability (*vi*) Corrosion stability.

**26.** Explain the mechanism of extreme pressure lubrication.

**27.** Give short account of cutting oil.

**28.** Fatty oils are no longer used as lubricants. Why?

**29.** Explain hydrodynamic lubrications and their functions.

**30.** Under what situations greases are used? What are the main function of soap in grease?

**31.** An oil of unknown viscosity-index has a Saybolt universal viscosity of 58 seconds at 210°F and of 580 seconds at 100°F. The high viscosity-index standard (*i.e.*, Pennysylvanian) oil has Saybolt viscosity of 58 seconds at 210°F and 430 seconds at 100°F. The low viscosity-index standard (*i.e.*, Gulf oil) has a Saybolt universal viscosity of 58 seconds at 210°F and 780 seconds at 100°F. Calculate the viscosity-index of unknown oil.

**Hint :** Here $L = 780s$, H = 430s, and U = 580. So viscosity-index of oil,

$$\text{V.I.} = \frac{\text{L} - \text{U}}{\text{L} - \text{H}} \times 100 = \frac{780 - 580}{780 - 430} \times 100 = \mathbf{57.1}.$$

**32.** An oil sample under-test has a Saybolt universal viscosity of 64 s at 210°F and 564 s at 100°F. The low viscosity standard (Gulf oil) possesses a Saybolt viscosity of 64 s at 210°F and 774 s at 100°F. The high viscosity standard (Pennysylvanian oil) gave the Saybolt viscosity values of 64 s at 210°F and 414 s at 100°F. Calculate the viscosity-index of the oil sample under-test. [**Ans.** 58.33]

**33.** A lubricating oil has Saybolt universal viscosity of 59 seconds at 210°F and 564 seconds at 100°F. The low VI standard Gulf oil has Saybolt universal viscosity of 59 seconds at 210°F and 758 seconds at 100°F. The high viscosity-index standard Pennyslyvanian oil gave the Saybolt universal viscosity value of 59 seconds at 210°F and 420 seconds at 100°F. Find the VI of lubricating oil. [**Ans.** 57.9]

**34.** An oil sample under-test has a Saybolt universal viscosity same as that of standard Gulf oil (low viscosity standard) and Pennsylvanian oil (high viscosity-index standard) at 210°F. Their Saybolt universal viscosities at 100°F are 61, 758 and 420s respectively. Calculate viscosity-index of the sample oil. [**Ans.** 206.2]

Chapter 4

# PRODUCTS OF SILICA INDUSTRY

## 4.1 CEMENTING MATERIALS

Cementing materials are materials that exhibit characteristic properties of setting and hardening, when mixed to a paste with water. This makes them capable of joining rigid masses like bricks, stones and tiles, etc. into coherent structures. These materials can be classified as:

(*i*) **Hydraulic Cements.** These are the materials which set and harden after combining with water, as a result of chemical reactions with the mixing water and, after hardening, retain strength and stability even under water. The key requirement for this is that the hydrates formed on immediate reaction with water are essentially insoluble in water. Most construction cements today are hydraulic, and most of these are based upon **Portland cement,** which is made primarily from limestone, certain clay minerals, and gypsum, in a high temperature process that drives off carbon dioxide and chemically combines the primary ingredients into new compounds.

(*ii*) **Non-hydraulic Cements.** Non-hydraulic cements include such materials as (non-hydraulic) lime and gypsum plasters, which must be kept dry in order to gain strength, and oxychloride cements which have liquid components. Lime mortars, for example, "set" only by drying out, and gain strength only very slowly by absorption of carbon dioxide from the atmosphere to re-form calcium carbonate.

### 4.1.1 Cement

In the most general sense the word *cement* means *binder*, a substance which sets and hardens independently, and can bind other materials together. The name "cement" goes back to the Romans who used the term "opus caementitium" to describe masonry which resembled concrete and was made from crushed rock with burnt lime as binder. The volcanic ash and pulverized brick additives which were added to the burnt lime to obtain a hydraulic binder were later referred to as cementum, cimentum, cäment and cement.

#### *4.1.1.1 Classification*

Cement may broadly be described as, *a material possessing adhesive and cohesive properties, and capable of bonding materials like stones, bricks, building blocks, etc.* The various types of cements are discussed below.

**A.** *Portland Cements*

1. **Natural Cement.** It is made by calcining a naturally occurring argillaceous limestone (i.e. 20-40% clay-containing limestone) at a high temperature and subsequently, pulverizing the calcined mass. During calcination, silica and alumina, present in sufficient quantities, combine with the calcium oxide to form the corresponding calcium silicates and aluminates.

   Natural cement possesses *hydraulic* qualities, but is quick-setting and of relatively low strength. Mortar made from natural cement and sand is satisfactorily used in laying bricks and setting stones. It is also used in large masses of concretes such as dams and foundations. After the discovery of Portland cement, the use and production of natural cement has been nearly dropped.
2. **Portland Blastfurnace Cement** contains up to 70% ground granulated blast furnace slag, with the rest Portland clinker and a little gypsum. All compositions produce high ultimate strength, but as slag content is increased, early strength is reduced, while sulfate resistance increases and heat evolution diminishes. Used as an economic alternative to Portland sulfate-resisting and low-heat cements.
3. **Portland Flyash Cement** contains up to 30% flyash. The flyash is pozzolanic, so that ultimate strength is maintained. Because flyash addition allows a lower concrete water content, early strength can also be maintained. Where good quality cheap flyash is available, this can be an economic alternative to ordinary Portland cement.
4. **Portland Pozzolan Cement** includes flyash cement, since fly ash is a pozzolan, but also includes cements made from other natural or artificial pozzolans. In countries where volcanic ashes are available (*e.g.* Italy, Chile, Mexico, the Philippines) these cements are often the most common form in use.
5. **Portland Silica Fume Cement.** Addition of silica fume can yield exceptionally high strengths, and cements containing 5-20% silica fume are occasionally produced. However, silica fume is more usually added to Portland cement at the concrete mixer.
6. **Masonry Cements** are used for preparing bricklaying mortars and stuccos, and must not be used in concrete. They are usually complex proprietary formulations containing Portland clinker and a number of other ingredients that may include limestone, hydrated lime, air entrainers, retarders, waterproofers and coloring agents. They are formulated to yield workable mortars that allow rapid and consistent masonry work. Subtle variations of Masonry cement in the US are Plastic Cements and Stucco Cements. These are designed to produce controlled bond with masonry blocks.

7. **Expansive Cements** contain, in addition to Portland clinker, expansive clinkers (usually sulfoaluminate clinkers), and are designed to offset the effects of drying shrinkage that is normally encountered with hydraulic cements. This allows large floor slabs (up to 60 m square) to be prepared without contraction joints.
8. **White blended cements** may be made using white clinker and white supplementary materials such as high-purity metakaolin.
9. **Colored cements** are used for decorative purposes. In some standards, the addition of pigments to produce "colored Portland cement" is allowed. In other standards (*e.g.* ASTM), pigments are not allowed constituents of Portland cement, and colored cements are sold as "blended hydraulic cements".

**B.** *Non-Portland Hydraulic Cements*

1. **Pozzolan-lime Cements.** Mixtures of ground pozzolan and lime are the cements used by the Romans, and are to be found in Roman structures still standing (*e.g.* the Pantheon in Rome). They develop strength slowly, but their ultimate strength can be very high. The hydration products that produce strength are essentially the same as those produced by Portland cement.
2. **Slag-lime Cements.** Ground granulated blast furnace slag is not hydraulic on its own, but is "activated" by addition of alkalis, most economically using lime. They are similar to pozzolan lime cements in their properties. Only granulated slag (*i.e.* water-quenched, glassy slag) is effective as a cement component.
3. **Supersulfated Cements.** These contain about 80% ground granulated blast furnace slag, 15% gypsum or anhydrite and a little Portland clinker or lime as an activator. They produce strength by formation of ettringite, with strength growth similar to a slow Portland cement. They exhibit good resistance to aggressive agents, including sulfate.
4. **Calcium Aluminate Cements** are hydraulic cements made primarily from limestone and bauxite. The active ingredients are monocalcium aluminate $CaAl_2O_4$ (CA in Cement chemist notation) and Mayenite $Ca_{12}Al_{14}O_{33}$ ($C_{12}A_7$ in CCN). Strength forms by hydration to calcium aluminate hydrates. They are well-adapted for use in refractory (high-temperature resistant) concretes. *e.g.* for furnace linings.
5. **Calcium Sulfoaluminate Cements** are made from clinkers that include ye'elimite ($Ca_4(AlO_2)_6SO_4$ or $C_4A_3$ in Cement chemist's notation) as a primary phase. They are used in expansive cements, in ultra-high early strength cements, and in "low-energy cements. Hydration produces ettringite, and specialized physical properties (such as expansion or rapid reaction) are obtained by adjustment of the availability of calcium and sulfate ions. Their use as a low-energy alternative to Portland cement has been pioneered in China, where several million tonnes per year are produced. Energy requirements are lower because of the lower kiln temperatures required for reaction, and the lower amount of limestone (which must be endothermically decarbonated) in the mix. In addition, the lower limestone content and lower fuel consumption leads to a $CO_2$ emission around half that associated with Portland clinker. However, $SO_2$ emissions are usually significantly higher.

#### *4.1.1.2 Manufacture of Portland Cement*

The manufacture of portland cement requires following raw materials:

(*i*) *Calcaerous materials*, CaO [such as limestone chalk, marl, etc.];

(*ii*) *Argillaceous materials*, $Al_2O_3$ and $SiO_2$ [such as clay, shale, slate, etc.];

(*iii*) *Powdered coal or fuel oil*, and

(*iv*) *Gypsum* ($CaSO_4 . 2 H_2O$).

The various functions performed by these raw materials include:

(*i*) *Lime* is the principal constituent of cement. Its proportion must be properly regulated. However, *excess of lime reduces the strength of cement*, because it makes the cement to expand and disintegrate. On the other hand, presence of lesser amount of lime than required also reduces the strength of cement and makes it quick-setting.

(*ii*) *Silica* imparts strength to cement.

(*iii*) *Alumina* makes the cement quick-setting. Excess of alumina, however, weakens the cement.

(*iv*) *Calcium sulphate* (gypsum) helps to retard the setting action of cement. *It actually enhances the initial setting time of cement.*

(*v*) *Iron oxide* provides: (*a*) colour, (*b*) strength, and (*c*) hardness to the cement.

(*vi*) *Sulphur trioxide*, in small proportion, is desirable. When present in small amount, it imparts *soundness* to cement. However, its excess reduces the soundness of cement.

(*vii*) *Alkalis*, if present in excess, cause the cement efforescent.

**Manufacture of Portland cement** involves the following steps:

1. **Mixing of raw materials** can be done either by: (*a*) dry process or (*b*) wet process.

(*a*) **Dry Process.** The raw materials [limestone (or chalk) and clay (or shale)] are *crushed* (in gyratory crushers) into roughly 2-5 cm size pieces. Then, these are ground to *fine powder* (in ball mills/tube mills). Each separate powdered ingredient is stored in a separate hopper. Then, the powdered materials are mixed in the requited proportions to get dry '*raw mix*', which is stored in storage bins (called *silos*) and kept ready to be fed in a rotary kiln. Raw materials are mixed in calculated proportions so that the average composition of the final product is as follows:

| *Component* | *Percentage range by mass* |
|---|---|
| Lime (CaO) | 60–69 |
| Silica ($SiO_2$) | 17–25 |
| Alumina ($Al_2O_3$) | 3–8 |
| Iron oxide ($Fe_2O_3$) | 2–4 |
| Magnesium oxide (MgO) | 1–5 |
| Sulphur trioxide ($SO_3$) | 1–3 |
| Alkali oxides ($Na_2O + K_2O$) | 0.3–1.5 |

(*b*) **Wet process.** The calcareous raw materials are crushed, powdered and stored in big storage tanks (called *silos*). The argilaceous material (say clay) is thoroughly mixed with water in wash mills, to remove any adhering organic matter, etc. The basin-washed clay is also stored. Powdered limestone (from silos) and washed wet-clay (from basins) are allowed to flow in a channel, in the right proportions. From the channel, the two raw materials are led to *'grinding mills'*, where they are mixed intimately to form a paste, called *slurry* [Grinding operation is carried out in either tube mill or ball mill or both]. The slurry is led to a *'correcting basin'*, where its chemical composition may be adjusted, if necessary. *This slurry contains about 38 to 40 per cent water.* The slurry is finally stored in storage tanks and kept ready for feeding to a rotary kiln.

**TABLE 4.1 Merits and demerits of dry and we processes**

| *Dry Process* | *Wet Process* |
|---|---|
| 1. It is adopted when the raw materials are *quite hard.* | It can be used for any *type* of raw materials. |
| 2. Fuel consumption is *low.* | Fuel consumption is *higher.* |
| 3. Process is *slow.* | Process is comparatively *faster.* |
| 4. Cement produced is of *inferior* quality. | Cement produced is of *superior* quality. |
| 5. Cost of production of cement is *less.* | Cost of production of cement is somewhat *higher.* |
| 6. On the whole, the process is *costly.* | On the whole, the process is *cheaper.* |

**2. Burning.** It is usually done in *rotary kiln,* which is a steel tube, about 2.5 to 3.0 m in diameter and 90 to 120 m in length, lined inside with *refractory bricks.* The kiln is laid in *slightly inclined* position at a gradient of 1 in 25 to 1 in 30. This rests on *roller bearings,* which are supported on column of masonry or concrete. The kiln is *capable of rotating* at 1 r.p.m. (revolution per minute) about its longitudinal axis. *Burning fuel* (usually powdered coal or oil) and *air* are injected at the lower end. A long hot flame is produced, which heats the interior of the kiln upto a maximum temperature of about 1,750°C.

**Process.** The *'raw-mix'* or *'corrected-slurry'* is injected into the kiln at its upper end; while *hot flames* is forced into the kiln from the lower end. Due to slope and slow rotation of the kiln, *the materials fed in move continuously towards the hottest-end at a speed of about 15 m per hour.* As the mixture or slurry gradually descends, the temperature *rises.*

**Chemistry** . (*i*) In the upper part of kiln, where the temperature is around 400°C, most of the water in the slurry gets evaporated (*drying zone*).

(*ii*) In the central part of kiln, where the temperature is around 1,000°C, limestone of dry mix or slurry undergoes *decomposition* to form quick-lime and carbon dioxide ; and the latter escapes out. The material forms small lumps, called *nodules* [*calcination zone*].

$$\underset{\text{Limestone}}{CaCO_3} \rightleftharpoons \underset{\text{Quicklime}}{CaCO_3} + CO_2 \uparrow$$

(*iii*) In the lower part of the kiln, the temperature is between 1,500 to 1,700°C. Here lime and clay (of nodules) undergo chemical interaction or fusion, yielding *calcium aluminates and silicates* (*clinkering zone*).

$$2\,CaO + SiO_2 \longrightarrow \underset{\text{Dicalcium silicate } (C_2S)}{Ca_2SiO_4}$$

$$3\,CaO + SiO_2 \longrightarrow \underset{\text{Tricalcium silicate } (C_3S)}{Ca_3SiO_5}$$

$$3\,CaO + Al_2O_3 \longrightarrow \underset{\text{Tricalcium aluminate } (C_3A)}{Ca_3Al_2O_6}$$

$$4\,CaO + Al_2O_3 + Fe_2O_3 \longrightarrow \underset{\text{Tricalcium aluminoferrite } (C_4AE)}{Ca_4Al_2Fe_2O_{10}}$$

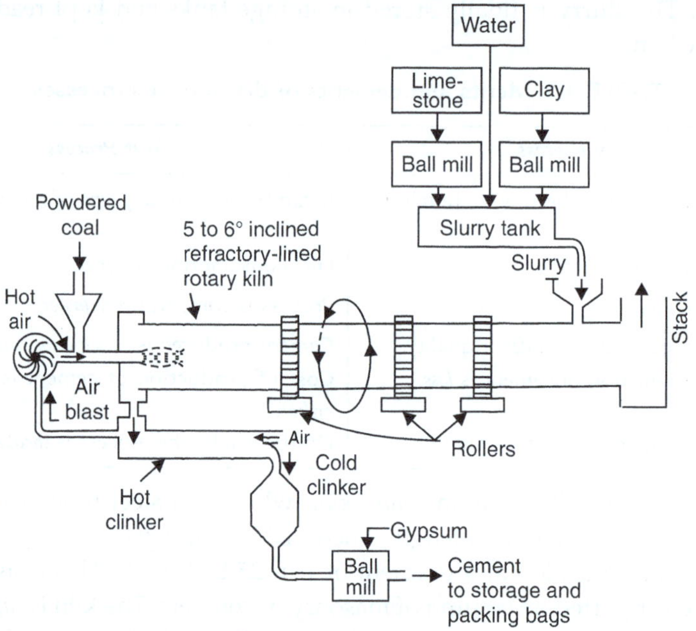


***FIGURE 4.1*** **Rotary cement kiln.**

*The aluminates and silicates of calcium then fuse together* to form small (of about 0.5 to 1 cm diameter) hard, greyish stones, called *clinkers.* These clinkers are very hot (at about 1,000°C). The rotary kiln at the base in provided with another *small rotary kiln.* In this, hot clinkers fall and cool air is admitted from opposite direction. Air counter-blast cools the clinkers. *Hot air so-produced is used for burning powdered coal oil.* The cooled clinkers are collected in small *trolleys.*

3. **Grinding.** The cooled clinkers are ground to a fine powder in ball mills or tube mills. During final grindings, a small quantity (2-3%) of powdered gypsum is added, so that the resulting cement *does not set very quickly,* when it comes in contact with water. *Gypsum, thus, acts as a retarding agent for early setting of cement.*

After the initial set, the cement-water paste becomes stiff, but *gypsum retards the dissolution of* $C_3A$ by forming tricalcium sulphoaluminate ($3\,CaO\,.\,Al_2O_3\,.\,x\,CaSO_4\,.\,7\,H_2O$), which is *insoluble.* Thus :

$$\underset{\text{After initial set}}{3\,CaO\,.\,Al_2O_3} + \underset{\text{Gypsum}}{x CaSO_4\,.7\,H_2O} \longrightarrow \underset{\substack{\text{Tricalcium sulphoaluminate}\\ (\textit{Insoluble})}}{3\,CaO\,.\,Al_2O_3\,.\,x\,CaSO_4\,.7\,H_2O}$$

Formation of insoluble tricalcium sulphoaluminate prevents too early further reactions of setting and hardening.

4. **Packing.** The ground cement is stored in *silos,* from which it is fed to automatic packing machines. Each bag, usually, contains 50 kg of cement.

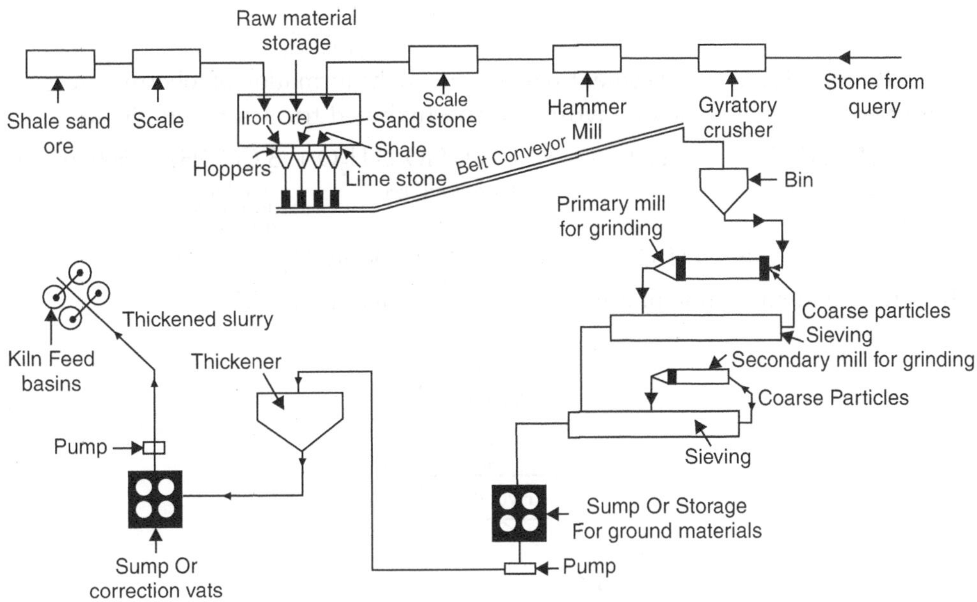


*FIGURE 4.2* **Flow sheet diagram for cement manufacture.**

#### *4.1.1.3 Setting and Hardening of Portland Cement*

Cement has the property of setting to hardening after being mixed with water which is called cement paste. The process of solidification involves setting and then hardening. *Setting* is defined as stiffening of the original cement paste, due to initial gel formation. *Hardening* is development of strength due to crystallization. The strength developed by cement paste at any time, depends upon the amount of gel formed and the extent of crystallization.

*Initial setting* of cement-paste is mainly due to the hydration of tricalcium aluminate ($C_3A$) and gel formation of tetracalcium aluminoferric.

$$3\,CaO.Al_2O_3 + 6\,H_2O \longrightarrow 3\,CaO.Al_2O_3.6H_2O + 880\text{ kJ/kg}$$

or

$$\underset{\text{Tricalcium aluminate}}{C_3A} + 6\,H_2O \longrightarrow \underset{\substack{\text{Hydrated tricalcium aluminate}\\(\textit{Crystalline})}}{C_3\,A\,.6\,H_2O} + 880\text{ kJ/kg}$$

$$\underset{\text{Tetracalcium aluminoferrite}}{4\,CaO.Al_2O_3.Fe_2O_3} + 7\,H_2O \longrightarrow \underset{(\textit{Crystalline})}{3\,CaO.Al_2O_3.6\,H_2O} + \underset{(\textit{Gel})}{CaO.Fe_2O_3.H_2O} + 420\text{ kJ/kg}$$

or

$$C_4AF + 7\,H_2O \longrightarrow C_3A.6\,H_2O\ (\textit{crystalline}) + CF.H_2O\ (\textit{gel}) + 420\text{ kJ/kg}$$

Also dicalcium silicate starts hydrolysing to tobermonite gel (which possesses a very high surface area and thus, very high adhesive property), which also contributes to initial setting

$$2\,[2\,CaO.SiO_2] + 4\,H_2O \longrightarrow 3\,CaO.2SiO_2.6\,H_2O + Ca(OH)_2 + 250\ kJ/kg$$

or

$$\underset{\text{Discalcium silicate}}{2\,C_2S} + 4\,H_2O \longrightarrow \underset{\text{Tobermonite gel}}{C_3S_2.6H_2O} + \underset{\text{Crystalline calcium hydroxide}}{Ca(OH)_2} + 250\ kJ/kg$$

*Final setting and hardening* of cement-paste is due to the formation of tobermonite gel (formed above) plus crystallization of calcium hydroxide and hydrated tricalcium aluminate.

$$\underset{\text{Tricalcium silicate}}{2\,[2\,CaO.SiO_2]} + 6\,H_2O \longrightarrow \underset{\text{Tobermonite gel}}{3\,CaO.2SiO_2.3\,H_2O} + \underset{\text{Calcium hydroxide (Crystalline)}}{3\,Ca(OH)_2} + 500\ kJ/kg$$

or

$$2\,C_3S + 6\,H_2O \longrightarrow C_3S_2.3\,H_2O + 3Ca(OH)_2 + 500\ kJ/kg$$

The setting and hardening of cement may be depicted in Fig. 4.3.

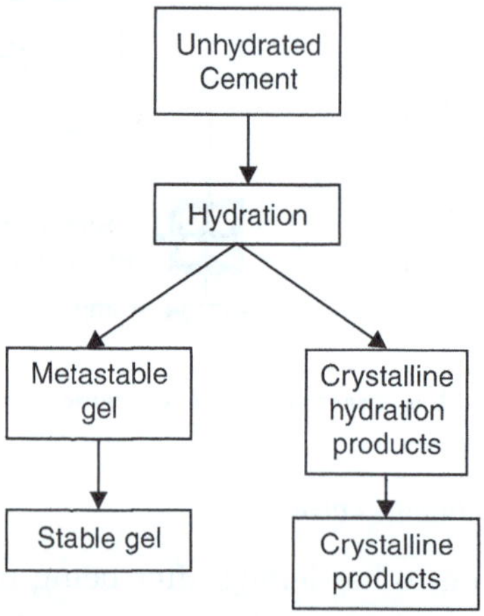


*FIGURE 4.3* **Setting and hardening of cement.**

**Sequence of chemical reactions during setting and hardening of cement.** When water is added to cement, its various constituents undergo hydration and crystallization at different rates. (*i*) At first, hydration of tricalcium aluminate ($C_3A$) and tetracalcium aluminoferrite ($C_4AF$) takes place. (*ii*) Next, the hydration of tricalcium silicate ($C_3S$) begins within 24 hours and gets completed in 7 days. (*iii*) The gel of aluminate begins to crystallize and at the same time, dicalcium silicate ($C_2S$) begins to hydrate in 7 to 28 days. Thus, the *initial set* of cement is due to the *hydration of aluminate.* The development of *early-strength,* between 1 to 7 days, is due to the *hydration of tricalcium silicate* and the further hydration of aluminate. The *increase of strength,* between 7 to 28 days, is due to *hydration of dicalcium silicate and continued hydration of tricalcium silicate.*

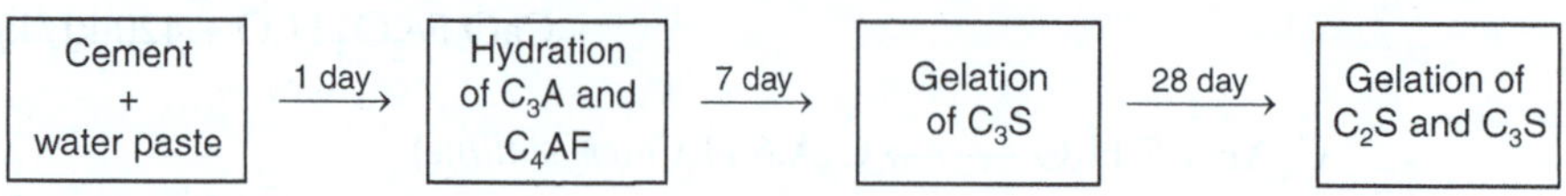


*FIGURE 4.4* **Sequence of changes during setting and hardening of cement.**

**Function of gypsum in cement.** Tricalcium aluminate ($C_3A$) combines with water very rapidly with the evolution of a large amount of heat.

$$C_3A + 6\,H_2O \longrightarrow C_3A.\,6\,H_2O + \text{Heat}$$

After the initial-set, the paste becomes somewhat stiff. However, the added gypsum retards the dissolution of $C_3A$ by forming *insoluble calcium sulpho-aluminate :*

$$3CaO.Al_2O_3.\,xCaSO_4.\,7\,H_2O$$

This reaction prevents a high concentrations of alumina in the cement solution, thereby *retarding the early initial-set of the cement.*

#### *4.1.1.4 Specifications for Portland Cement*

**(A) British Standard Specifications (BSS). The following are BSS for Portland cement:**

1. $\dfrac{CaO}{2.8\,SiO_2 + 1.2\,Al_2O_2 + 0.65\,Fe_2O_3}$ should not be greater than 1.02 and less than 0.66,

   where CaO, $SiO_2$, $Al_2O_3$ and $Fe_2O_3$ indicate their percentage in cement.

2. $\dfrac{Al_2O_3}{Fe_2O_3}$ shall not be less than 0.66, where $Al_2O_2$ and $Fe_3O_3$ indicate the percentages of alumina and iron oxide.

3. Weight of insoluble residue should not exceed 1.0%.
4. Magnesia shall not exceed 4%.
5. Total sulphur content calculated as $SO_3$ shall not exceed 2.75%.
6. Total loss on ignition shall not exceed 4%.

**Other Requirements Include:**

1. *Fineness.* Residue on B.S. 170-mesh test sieve not to exceed 10%.
2. *Setting Time.* Initial setting time, not less than 30 minutes. Final, not more than 12 hours.
3. *Tensile Strength.* Not less than 300 lbs./sq. inch after 3 days. Shall show after 7 days an increase in the tensile strength at 3 days, the value being not less than 375 lbs./sq. inch.
4. *Compressive Strength.* Not less than 1600 lbs./sq. inch in 3 days. Shall show after 7 days an increase on the compressive strength at 3 days, the value being not less than 2500 lbs. per sq. inch.

**(B) Bureau of Indian Standards (BIS) Specifications:**

1. $\dfrac{CaO}{2.8\,SiO_2 + 1.2\,Al_2O_3 + 0.65\,Fe_2O_3}$ should not be greater than 1.02 and not less than 0.66

   CaO.$SiO_2$, $Al_2O_3$ and $Fe_2O_3$ indicate their percentages in cement.

2. $\dfrac{Al_2O_3}{Fe_2O_3}$ not less than 0.66 $Al_2O_3$ and $Fe_2O_3$ indicate their percentages in cement.

3. Weight of insoluble residue not more than 1.5%.
4. Weight of magnesia, not more than 5%.
5. Total sulphur content calculated as $SO_3$ not more than 2.75%.
6. Total loss on ignition not more than 4%.

**Other Requirements Include:**

1. *Fineness.* After sieving the residue by weight on B.S. 170-mesh test sieve not to exceed 10%.
2. *Setting time.* Initial setting time 30 minutes and final setting time not more than 600 minutes.
3. *Tensile strength.* After 3 days (72 hours) not less than 300 lbs./sq. inch. After 7 days (168 hours) not less than 375 lbs./sq. inch.
4. *Compressive strength.* After 3 days not less than 1600 lbs./sq. inch. After 7 days not less than 2500 lbs./sq. inch.
5. *Heat of hydration.* In 7 days not more than 65 calories per gm. In 28 days not more than 75 calories per gm.

**(C) Heat of Hydration of Cement:**

When water is mixed with cement, hydration, hydrolysis and gelation reactions start and some heat is liberate simultaneously. On an average, 500 kJ/kg of heat is evolved during complete hydration of cement. The heats of hydrations of various constituents of cement are as follows:

| ***Constituent of cement*** | $C_3A$ | $C_3S$ | $C_4AF$ | $C_2S$ |
|---|---|---|---|---|
| ***Heat of hydration (kJ/kg)*** | 880 | 500 | 420 | 250 |

Consequently, in order to avoid the shrinkage cracks on setting and hardening, it is quite necessary to pour concrete in position in such amounts that the *heat liberated* during *hydration is dissipated as quickly as possible.*

**(D) Chemical Constitution of Portland Cement:**

*Average* compound composition of Portland cement is:

| ***Name of compound*** | ***Chemical formula*** | ***Abbreviation used*** | ***Average %*** | ***Setting time*** |
|---|---|---|---|---|
| Tricalcium silicate | $3\ CaO.SiO_2$ | $C_3S$ | 45 | 7 days |
| Dicalcium silicate | $2\ CaO.SiO_2$ | $C_2S$ | 25 | 28 days |
| Tricalcium aluminate | $3\ CaO.Al_2O_3$ | $C_3A$ | 1 | 1 day |
| Tetracalcium aluminoferrite | $4\ CaO.Al_2O_3.Fe_2O_3$ | $C_4AF$ | 9 | 1 day |
| Calcium sulphate | $CaSO_4$ | – | 5 | – |
| Calcium oxide (*free*) | CaO | – | 2 | – |
| Magnesium oxide | MgO | – | 4 | – |

**Characteristics of constituents :** (*i*) *Tricalcium silicate* ($C_3S$) has *"medium"* rate of hydration ; and *develops high ultimate strength quite rapidly*. Moreover, its ultimate-strength is also the *highest* amongst all the constituents. So it is responsible for ultimate-strength. The heat of hydration is about 880 kJ/kg.

(*ii*) *Dicalcium silicate* ($C_2S$) has quite low early-strength, but develops ultimate-strength of *slightly less than* that of $C_3S$. It hydrates *very slowly*. It is also responsible for ultimate-strength. It possesses the heat of hydration of 420 kJ/kg.

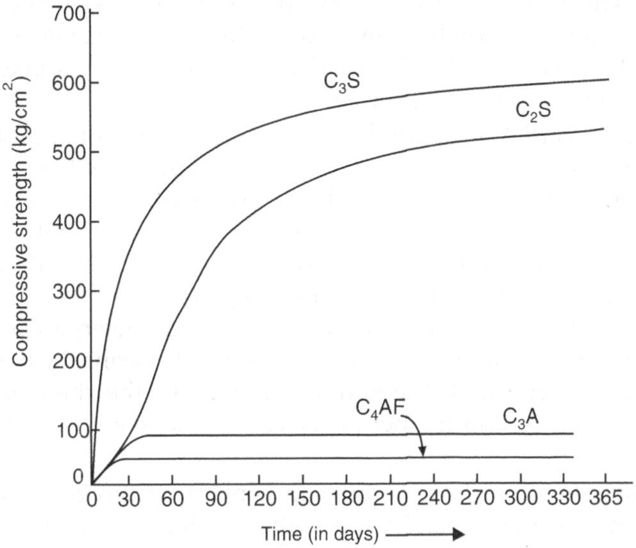


*FIGURE 4.5* **Development of strength by major constituents in cement.**

(*iii*) *Tricalcium aluminate* ($C_3A$) hydrates slowly and *does not* contribute much to the strength of cement, since its early-strength and ultimate-strength are *poorest* amongst all the constituents. Its heat of hydration is about 250 kJ/kg.

**Analysis of Indian Portland Cement.** Indian Portland cement on average had been found to have the following composition :

| | | |
|---|---|---|
| $SiO_2$ | ... | 23% |
| $Fe_2O_3$ | ... | 2.19% |
| $Al_2O_3$ | ... | 5.97% |
| CaO | ... | 64.86% |
| MgO | ... | 1.00% |
| $SO_3$ | ... | 1.75% |
| Losses | ... | 0.74% |

#### *4.1.1.5 Vertical Shaft Kiln Process of Cement Manufacture*

The use of vertical shaft kiln process for cement manufacturing dates back to the year 1824 when Portland cement was invented. However, because of its non-continuous operation, this process was found uneconomical and produced clinker of non-uniform quality.

Later after the development of continuous pan type noduliser, a significant advance was achieved and the shaft kiln performance improved considerably thus permitting its adoption and production of uniform quality of clinker on a continuous basis. Cement manufacture by using

Modern Vertical Shaft kiln is reportedly being carried out in different countries like China, India, West Germany, Australia, Italy, Indonesia, Nepal, Bhutan etc. In fact, a wide variety of processes have been adopted for the manufacture of cement through Vertical Shaft Kiln and it would be desirable to mention some of these in the following sections.

(A) *Black Meal Process*

This is the most popular type of process used for manufacture of cement through Modern Vertical Shaft Kiln in which the fuel (usually low volatile coal or coke breeze) is interground with the raw mix for efficient and uniform burning. Most of the cement plants using Vertical Shaft Kiln like CRI—Modern Vertical Shaft Kiln, Vertical Shaft Kiln supplied by M/s. Loesche of West Germany etc. are mainly based on the black meal process.

In this process, the vertical shaft kilns are fed with raw mix of appropriate composition in the form of nodules. The feed operation usually requires crushing of limestone to about—15 mm size in storage bins, proportioning of raw materials and grinding in a roller or ball mill, homogenizing and nodulizing in pan type nodulizer. The successful operation of shaft kiln, to a great extent, depends upon the size of nodules, their uniformity, porosity and thermal stability. The noduliser consists of an inclined disc or pan rotating about its axis. Raw meal is charged into noduliser by means of screw conveyor and water is sprayed, while all the parameters are maintained at optimum conditions. A specially designed rotating scraper continuously cleans the bottom and the collar of the drum of the raw meal deposited. The nodules slide down the chute and are charged in the vertical shaft kiln evenly all around the periphery continuously with the help of a rotary feeder situated on the top of the kiln.

The vertical shaft kiln in which the nodules are converted into clinker consists of a cylindrical shell with conical portion at the top and lined with refractory bricks. The sintering is normally complete within the conical portion, which is specially designed to accommodate for shrinkage of nodules. The various zones of reaction starting from the top of the kiln are ; the drying zone, the calcining zone, the sintering zone and the cooling zone. The combustion air supplied by Roots blower ascending from below in the cooling zone absorbs heat from the descending clinker. The whole kiln charge composed of unburnt nodules and clinker rests on a flat grate rotating slowly at the bottom of the kiln and mounted over the kiln shaft. The grate is driven with the help of variable speed motor in order to control the discharge rate of clinker. The clinker is finally taken out of kiln bottom with the help of a triple air lock mechanically (or hydraulically) operated discharge gates or Gamma Ray controlled Material Block Tube System used in Modern Vertical Shaft Kilns. (Fig. 4.6)

(B) *Coal Slurry Process*

A variation of the black meal process as proposed by Dr. Steven Gottlieb is 'Coal Slurry Process' the main difference being in the method of feed preparation. The method requires the raw meal to be ground 'white', *i.e.* without fuel. The fuel (coal with volatile matter up to 16–18%) is ground separately in a wet ball mill to make slurry with about 50% moisture. The 'white meal' and the 'coal slurry' are separately stored in hoppers over the noduliser platform and are pumped at controlled rates (through flow meters) into a double paddle mixer to mix the feed continuously and discharge into the standard noduliser where practically no or very little water is added to give final shape to the nodules. These nodules are charged and burnt in the vertical shaft kiln in the same manner as in the case of "Black Meal" process.

The main advantage claimed by this process is that by wetting of the coal particles before nodulising, the retention time for the volatile component of the fuel is increased, thereby allowing this to travel to a zone where adequate oxygen is available to complete the combustion of the volatile matter within the bed-thereby permitting a higher volatile content in the coal when compared to that used in the 'Black Meal' process.

(C) *Differential Heat Burning Process*

This process, which is a slight variation of the black meal process, has been reportedly tried out in a number of cement plants in China. In this process about 50% of the total fuel required is interground with the raw meal and the balance is added later prior to nodulisation. This has been based on the assumption that the heat requirement at the periphery of the bed in a VSK is usually more by about 150 kcal/kg of clinker than that in the middle and in this process some additional fuel is added at the periphery of the bed as compared to that in the middle to balance the heat distribution.

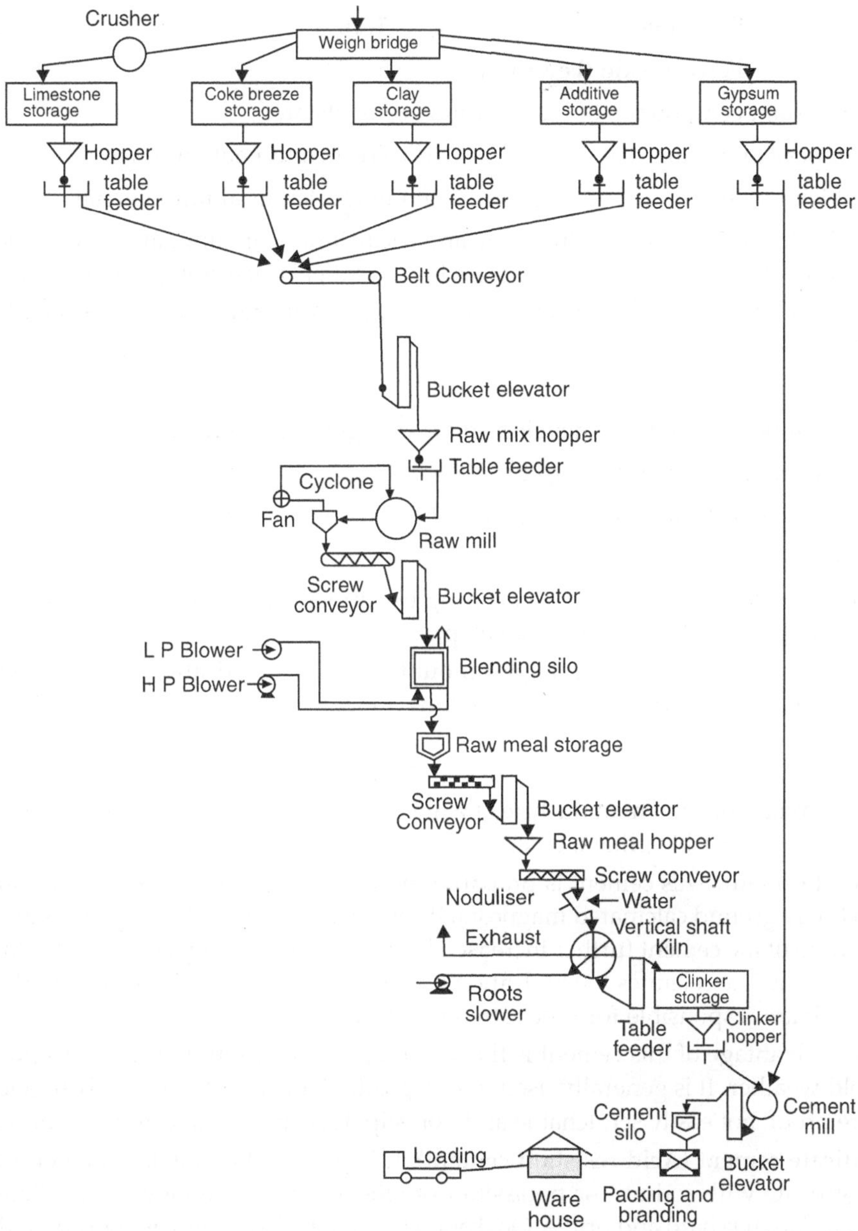


***FIGURE 4.6*** **Flow sheet of black meal vertical shaft kiln cement plant.**

#### 4.1.1.6 Special Cements

1. **High alumina cement.** The high alumina cement is obtained by fusing limestone and bauxite with small amount of $SiO_2$ and $TiO_2$ at 1500° to 1600°C in a rotary kiln and then grinding the resulting products to the finess in the same way as that of Portland Cement. The essential constituents of high alumina cement is:

| | | |
|---|---|---|
| Calcium Oxide (CaO) | = | 38–40% |
| Alumina ($Al_2O_3$) | = | 45–50% |
| Silica ($SiO_2$) | = | 5–10% |
| Ferric Oxide ($Fe_2O_3$) | = | 5–10% |

The important compounds of high alumina cements are:

(*i*) Monocalcium aluminate (*ii*) Tricalcium pentaluminate

(*iii*) dicalcium silicate (*iv*) tetracalcium aluminoferrite.

On hydration, monocalcium aluminate and tricalcium pentaluminate form stable crystalline complex $Ca_3Al_2(OH)_{12}\ 3\ H_2O$ from the hydrated colloidal gel $Ca(Al(OH)_5\ 3\ H_2O$. The setting time of this cement is about a week but its rate of hardening is very rapid and full strength is attained within 1 day.

**Advantages:**

(*i*) It sets much quicker and develops high strength within one day.

(*ii*) It withstands temperature up to 1500°C.

(*iii*) It has good resistance to sea water and dilute mineral acids.

**Disadvantages:**

(*i*) Exposure to hot wet condition or using two much water when mixing the high alimina cement and sand, cracks will develop.

(*ii*) Due to rapid hardening a large amount of heat of hydration evolved which cannot dissipated immediately so cracks will develop.

2. **White Portland cement.** When iron is not present in the cement, it will take the white color and the colorisation of this cement is possible by addition color pigments. It is used for mosaic work and manufacture of white tiles. It may be used as an adhesive for repairing and joining marble pillars, blocks etc.

3. **Sorel cement.** This cement is prepared by the addition of a strong magnesium chloride to finely ground calcinated magnesia. It sets to hard mass within 3 to 4 hrs and the rate of harden of the cement further increased by addition of calcium chloride (1.5%) or formate. Since $CaCl_2$ accelerates the hydration process of dicalcium silicate and tricalcium silicate which are responsible for inner strength of the cement.

   The advantage of this cement is the accelerated rate of setting is more useful during the cold weather. It is generally used in composite flooring. The other advantage is that it is fire proof, not easily scratchable and non slip. It is also durable and capable of oil polish.

4. **Silicate cement (Acid resistant cement).** It is prepared by mixing an iner-acid resisting aggregate with sodium or potassium or silicon ester in suitable proportion in the raw materials. It is hard and brittle. On drying the cement becomes very porous and permeable to liquids.

5. **Water proof cement.** When water proofing substances like calcium stearate, aluminium stearate and gypsum with tannic acid is added to the ordinary Portland cement at the grinding stage, water proof cement is obtained. A water proof cement has two distinct functions (*i*) Resistance to absorption of water (*ii*) impermeable of water under pressure which depends on the nature and quantity of water proofing substance added. Water repelling agents can also be employed to the portland cement to make second category water proof cement. These water repelling agents are, resins, vegetable oils, waxes, coal tar residues, soap of calcium or aluminum, and bitumen to block the pores in the cement.
6. **High early strength cement.** When in the raw materials for cement contains high lime to silica ratio, it is called high early strength cement. Since tricalcium Silicate compound is responsible for ultimate strength, its quantity should be large, so that quicker hardening with greater evolution of heat occur. For the same reason, addition of gypsum generally avoided. The setting time of this cement is within 5 to 30 minutes.
7. **Barsium and strontium cements.** When in the raw materials for cement, calcium is replaced by either barium or strontium, it is called barium or strontium cement. It is resistant to the radio active radiation. Thus, this cement has greater applications in concrete shields for atomic piles.
8. **Low heat cement.** Since it contains high percentage of $C_4AF$ and $C_2S$ the heat of hydration is low. Hence no cracks occur. This is mainly used for the construction of dams.

### *4.1.1.7 Concrete and RCC*

**Concrete** *is building and structural material, obtained by mixing a binding material (cement or lime), inert mineral aggregates (sand, crushed stone, gravel, broken brick, slag, etc.) and water in a suitable proportion and which can be readily worked or moulded into almost any desired shape and when sel, it is compact, rigid, strong and durable.*

It consists of cement, sand (below 3/16) inch mesh size) and coarse aggregates (above 3/16 inch mesh size). The mixture is made into a solid mass. The size of coarse aggregate varies with the purpose for which the concrete is required. Common maximum sizes of coarse aggregate are $\frac{3}{4}$ inch or $1\frac{1}{2}$ inches. For some purposes, the sizes of coarse aggregate may be as small as $\frac{1}{2}$ inch. For heavy mass concrete the size is as high as 6 inches and some times even more than that.

The common proportions for cement, sand and coarse aggregate are—

(*a*) 1 : 1½ : 3 (*b*) 1 : 2 : 4 and (*c*) 1 : 3 : 6.

Concrete is by far the most important of all non-metallic materials of construction. It ranks *second* only to steel in its wide and varied constructional importance. Its typical uses are in roads, buildings, floors, roofs, columns, arches, tanks, sewers, foundations, abutments, piers, reinforce-works, water-proof structures and other purposes where compressive strength is required.

**Reinforced Concrete.** Concrete has high compressive strength and relatively low tensile strength. Concrete can resist loads which tend to crush it. However it is relatively low in strength to forces which tend to bend it or pull it apart, that is, concrete has low tensile strength. So in order to impart tensile strength to concrete another form of concrete called reinforced concrete is used.

Reinforced concrete is ordinary concrete reinforced with steel rods or heavy wire mesh. Concrete is poured around steel rods or heavy wire mesh. On setting, the concrete adheres very strongly to the reinforcements. The reinforced concrete can withstand not only high tensile strength but also high compressive stresses.

**Advantage of R.C.C. over Plain Concrete Include**

(*i*) R.C.C. is easier to make and cast into any desired shapes, which can bear *all types of loads.*

(*ii*) It possesses *greater* rigidity, moisture and fire-resistances.

(*iii*) Steel reinforcement also tends to *distribute the shrinkage cracks,* thus preventing the formation of large cracks.

(*iv*) Its maintenance cost is *practically negligible.*

**Decay of Concrete.** Although cement concrete is mechanically quite strong, yet it highly susceptible to chemical attack, because concrete contains some free lime (CaO). In acidic water (*viz.*, pH < 7), lime of concrete dissolves, thereby making it *weak,* pH of most natural water is slightly greater than 7 (*i.e.* they are slightly alkaline), consequently, such waters do not have any marked effect on the strength of concrete. However, as the acidity increases (or pH decreases), the deterioration of concrete enhances. It may be pointed here that lime is more soluble in soft water than hard water. Consequently, deterioration of concrete is more quick, when in contact with soft water. Lime of concrete is also removed by sulphates and chlorides, present in water. If concrete is soaked in mineral oil for sometime, its resistance to abrasin decreases. Sugar also causes concrete failure. If even as low as 0.1% sugar is added to cement, the setting time is delayed and its strength is greatly reduced, during the first four weeks. The cause of concrete failure, due to presence of sugar has been traced in the use of old sugar sacks for carrying sand or for covering the freshly concrete used during curing.

The most serious type of damage to concrete takes place in the presence of *sulphates.* The sulphates combine with tricalcium aluminate to form sulpho-aluminates, which occupies *more volume.* This causes expansion, thereby the life of concrete is greatly reduced. Such a danger can be avoided by eliminating tricalcium aluminate from the cement composition and manufacturing cement containing tetracalcium aluminoferrite, instead of aluminate.

**Protection of concrete :**

1. *By giving a coating of bituminous material.* This prevents direct contact between concrete and water. This is a very cheap method.
2. *By coating the surface with silicon fluoride* in a soluble form together with oxides of Zn, Mg or Al. The precipitate of calcium fluoride so-formed in the capillaries prevents dissolution of lime.

#### *4.1.1.8 Cement Business*

In 2002 the world production of hydraulic cement was 1,800 million metric tons. The top three producers were China with 704, India with 100, and the United States with 91 million metric tons for a combined total of about half the world total by the world's three most populous states. For the past 18 years, China consistently has produced more cement than any other country in the world. China's cement export peaked in 1994 with 11 million tons shipped out and has been in steady decline ever since. Only 5.18 million tons were exported out of China in 2002. Offered at $34 a ton, Chinese cement is pricing itself out of the market as Thailand is asking as little as $20 for the same quality." Demand for cement in China is expected to advance 5.4% annually and exceed 9 billion metric tons in 2008, driven by slowing but healthy growth in construction expenditures. Cement consumed in China will amount to 44% of global demand, and China will remain the world's largest national consumer of cement by a large margin." In 2006 it was estimated that China manufactured 1.235 billion metric tons of cement, which is 44% of the world total cement production.

In India the raw materials required for the manufacture of cement are available in almost all the states. The first cement industry in India was established in 1904 in Tamil Nadu. Now, there are more than 120 cement industries spread throughout India.

### 4.1.2 Lime

Lime is an ancient chemical, because it has been manufactured by man since the beginning of agricultural civilization. During the Roman, Greek and Egyptian civilizations, the science of making lime and the act of using it for constructing dwellings were well-known. In the nineteenth century, in the U.K. and the USA its manufacture was scientifically developed. In the twentieth century, with a better understanding of its chemistry and a vast improvement in the knowledge of chemistry, the manufacture of lime was perfected and the cost greatly reduced. Before the process of cement making was known, it was the primary material used as a mortar.

Lime may be defined as, *a substance consisting essentially either of calcium oxide (CaO) or of calcium oxide together with a smaller proportion of magnesium oxide (MgO), formed by burning a natural rock (like limestone, $CaCO_3$) at such a temperature that it will slake, when brought in contact with water.*

#### *4.1.2.1 Classification of Lime*

Lime (CaO) is generally obtained by heating the mineral limestone ($CaCO_3$). Limes are classified as fat lime, lean lime, hydraulic lime and dolomite lime. These are discussed below.

1. **Fat Limes.** These are also known as high calcium limes and are obtained by heating pure limestone. These contain about 95-98% CaO and the rest are $SiO_2/Al_2O_3$ and MgO. These limes are rapid slaking, with evolution of large amount of heat and afford large volumes (2 to 2.5 times) of powder. These are non-hydraulic and the setting of fat lime paste (or mortar) is by drying only; and when continuously moist, they never set. Hence, they are unsuitable for use in damp situations.

   Fat lime is suitable for making mortars, used in joining thin brick-works and coarse stone masonry works. It is, generally, also used for white-washing, water softening, in metallurgy, and in glass industries.

2. **Lean Limes.** These limes contain only about 70–80% of CaO and hence are known as poor limes. The other impurities present in it are MgO, $SiO_2$ and $Fe_2O_3$.

These limes *slake very slowly.* They show very low shrinkage, better plasticity and workability than fat limes. Therefore, these limes are better suited for *plaster finishing.* These limes must be thoroughly slaked ; otherwise they will slowly expand in the finished work, producing symptoms of *unsoundness.*

They are employed mainly for mortar purposes and interior works. They are also used in places, where good fat lime is not available.

3. **Hydraulic Limes.** These limes generally contain 70–80% CaO, 5–30% silica and alumina and MgOL 2% when mixed with water, these limes undergo slow setting to a hard mass and hence known as hydraulic lime. Hydraulic limes *slake with difficulty*, so these must be finely ground before use. Such limes do not shrink or crack and so they require only a *small* quantity of sand to be added.

As hydraulic limes set slowly without cracking, they are used as a *substitute for cement.* For example, new foundations, marine works, shafts of tall chimneys and thick walls, where there is *no much access of air.*

4. **Dolomite Limes.** When dolomite ($CaCO_3.MgCO_3$) is heated in the presence of limited supply of air, the sample produced is called dolomite lime or high magnesia lime. These limes contain 30–40% MgO and less percentage of $SiO_2$ and $Al_2O_3$. Their shaking reactions are very slow, liberating only a small quantity of heat. These limes are used in plastic mortars. They are also used in metallurgy as flux and in the manufacture of basic refractories.

### 4.1.3 Manufacture of Lime

Raw material for lime manufacture is limestone which occurs in nature. Production of lime from limestone involves two processes : *Calcination* and *hydration.*

(*a*) **Calcination.** Calcining means the burning of limestone. Magnesium and calcium carbonates in limestone start dissociating at 1400°F and 1600°F respectively.

$$Ca\,CO_3 \longrightarrow CaO + CO_2 + 4{,}250{,}000 \text{ B.T.U. per ton of lime produced.}$$

Dolomite (containing $MgCO_3$) lime is easier to calcine than calcium limestone because dolomite limestone dissociates at lower temperature.

Calcination of limestone is done in kilns which are of two types, rotary kilns and vertical kilns.

**Calcination in Rotary Kiln.** The size of rotary kiln varies. The diameter of the rotary kiln is from 6 to 11.5 ft. and the length varies from 60 to 400 ft. The commonly used length is 150 ft. They have preheaters and coolers in order to save fuel. The rotary kiln rotates at a speed of one revolution in 95 seconds. The maximum temperature obtained in the furnace is 2500°F (1370°C) and lime is discharged at a temperature of 1660°F (904°C). Coal used is about 12% of the lime obtained. (*See* Fig. 4.7).

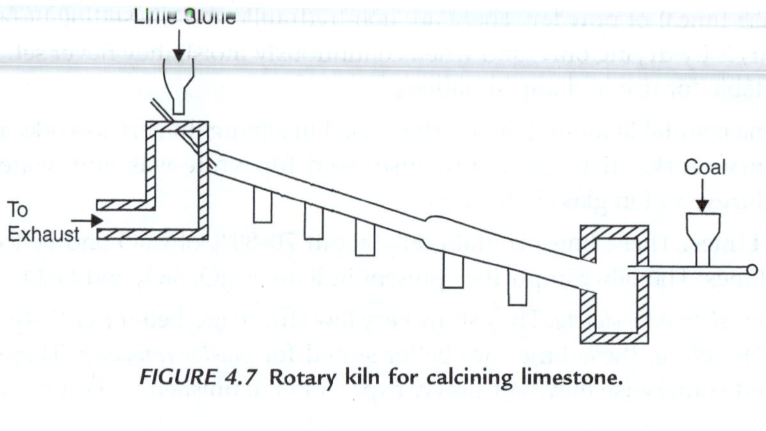


*FIGURE 4.7* **Rotary kiln for calcining limestone.**

Size of limestone which is calcined in such a kiln is $\frac{1}{4}$ to $2\frac{1}{2}$ inches.

Cooling air is let in by the draft on kiln into the discharge end. This air cools the lime and gets heated up. It is used in kiln as a secondary combustion air. A separate fan sucks hot air from the outlet. This is used as primary combustion air for drying coal. Advantage of the rotary kiln is high production per man per hour. Disadvantages are high investment cost and high cost of fuel per unit of lime produced. Production of lime is 200 tons per 24 hours by a rotary kiln.

**Calcination in Vertical Kiln.** It is a refractory cell having a casing of steel. It is 10 to 24 ft. in diameter and 35 to 75 ft. in height. Limestone is fed at the top. At the lower end there is a fire box where coal is burnt. Kiln is lined with refractory bricks. Vertical kiln burns limestone of 6 to 8 inches size. The kiln may be gas-fired or oil fired. There are four zones in the kiln.

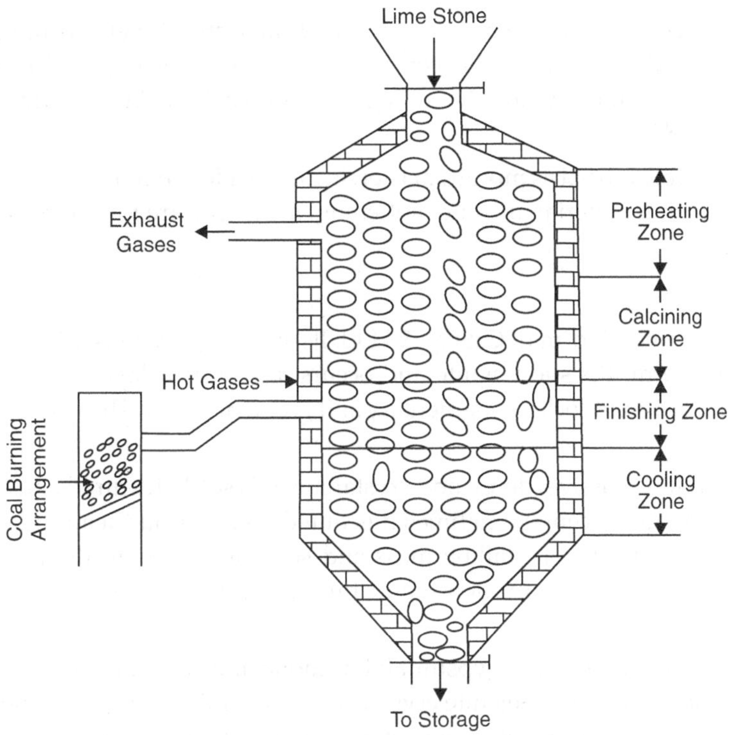


*FIGURE 4.8*

(*i*) *Cooling zone.* Air and lime pass each other in opposite directions. Heat interchanges. Lime is cooled and air is heated up.

(*ii*) *Finishing zone.* Fuel is burnt with high excess of air in this zone to finish calcination of lime in a zone of moderate temperature.

(*iii*) *Calcining zone.* In this zone most of calcination takes place (about 90%).

(*iv*) *Preheating zone.* In this zone gases are cooled and limestone is preheated.

Temperature, length of burning time and the feed of fuel are controlled in the kiln.

Advantages of vertical kiln are small investment, low fuel costs and easy operation. Vertical kiln operates continuously and produces up to 80 tons of lime per 24 hours.

**Quicklime.** By calcining limestone, quicklime is obtained. The size of quicklime lumps from vertical kiln is 6 to 8 inches, while from rotary kiln the lumps obtained are of the size from $\frac{1}{4}$ to $2\frac{1}{2}$ inches. Quicklime is then ground to such size that 100% passes through an 8-mesh sieve and 40 to 60% passes through a 100-mesh sieve.

**Hydrated Lime.** Quicklime when treated with limited quantity of water, develops heat, then swells and further disintegrates to a fine powder consisting of hydrated lime :

$$CaO + H_2O \longrightarrow Ca(OH)_2 - 15900 \text{ cal.}$$

The hydration of lime takes place with the evolution of heat as the reaction is exothermic.

For the production of hydrated lime calculated quantity of water is mixed with lime in a mechanical hydrator. More than the theoretical quantity of water required is added as there is some loss of water in steam formation due to high heat of reaction. After hydration the material is ground in a hammer mill.

**Use of Lime.** Lime is used as mortar, plaster, in insecticides, in refractory bricks, in metallurgy as flux, in paper as sizing material, in water softening, in agriculture for liming soil and fertilizing.

### 4.1.4 Gypsum

Gypsum is a natural mineral form of $CaSO_4 \cdot 2\,H_2O$. Its specific gravity is 2.31 to 2.33 and hardness is 1.5 to 2.0. In pure form it is snow white and occurs in stratified layers. This mineral occurs in various forms from hard solid rock to fragmentary and granular forms. The various forms of gypsum include :

1. **Alabaster.** It is massive, dense and crystalline. It is softly textured and is pure gypsum. It is generally translucent. Sometimes it is tinted with various beautiful colours due to the optical effects in the crystalline mass and sometimes due to the presence of traces of foreign matter. Being soft, dense and fine-textured it is used for carving images and other pieces of art.
2. **Selenite.** It is pure form of gypsum. It has monoclinic crystal structure and occurs in the form of sheets or plates. Selenite crystals up to several inches of thickness are completely transparent. Thin sheets of selenite polarize light. So they are used in laboratory equipment for polarization light.
3. **Satin Spar.** It is a form of pure gypsum and is also known as gypsum var. It has monoclinic crystals and occurs in the form of parallel threads and resembles the common form of asbestos. When it is densely formed, it is translucent. It is used for preparing small trinkets.
4. **Anhydrite.** It is the mineral form of anhydrous $CaSO_4$ ; has specific gravity 2.8 to 2.9 and hardness 3 to 3.5. It occurs in the form of dense mass and sometimes shows tints of blue and pink etc.
5. **Gypsite.** It consists of finely crystalline gypsum mixed with loams, clays, sands and humus. Gypsum content varies from 60 to 90%.
6. **Gypsum Sands.** These are deposits of pure gypsum in the form of white sands.

**Uses of Gypsum:**

1. It is used in the manufacture of Portland cement for lengthening the time of setting.
2. It is used for the treatment of soils, particularly soils for leguminous crops. It is also used for the culture of peanuts. Gypsum, when finely powdered and mixed with manures, stabilises the volatile and nitrogenous compounds. This prevents their loss by volatilisation or by leaching. In arid climates it is used for neutralising of black alkali soils by reacting with carbonates.
3. Being stable, tasteless, odourless, non-abrasive and chemically inert, it is used in paints, pharmaceuticals, paper filling, insecticides, yeast manufacture and water treatment etc.
4. In brewery it is added to water to lower the pH.

**Plaster of Paris ($CaSO_4$. 0.5 $H_2O$).** *It is produced by heating fairly pure gypsum to a temperature of about 120–160°C.* If the gypsum by heated above 200°C, the anhydrous sulphate is produced, which *loses* the power of readily combining with water. The various changes may be depicted as follows:

$$\underset{\text{(Gypsum)}}{CaSO_4 \cdot 2\,H_2O} \rightleftharpoons \underset{\substack{\text{(Orthorhombic}\\ \text{dihydrate)}}}{CaSO_4 \cdot 2\,H_2O} \overset{120°C}{\rightleftharpoons} \underset{\text{(Plaster of Paris)}}{CaSO_4 \cdot \tfrac{1}{2}\,H_2O} \xrightarrow{200°C} CaSO_4$$

The operation of manufacturing plaster of Paris from gypsum consists in first crushing and grinding of gypsum, and then calcining the ground product in kilns (heated to about 150°C) and finally, pulverizing the calcined product.

**Setting and hardening.** When plaster of Paris is mixed with water, the powder forms a plastic mass, which quickly hardens or sets, *expanding slightly* in the process and regains the closely-packed crystalline structure of gypsum ($CaSO_4 \cdot 2\,H_2O$). Setting of plaster of Paris can be accelerated by admixing it with alkali sulphates like $K_2SO_4$, $Na_2SO_4$ (or alums), which initiate as well as hasten the crystallization process.

**Uses.** The important uses of plaster of Paris include:

(*i*) **Use in room interiors.** Plaster was a common building material for wall surfaces in a process known as *lath and plaster*, whereby a series of wooden strips are covered with a semi-dry plaster and then hardened into surface. The plaster used in most lath-and-plaster construction was mainly lime plaster. Lime plaster cure time is about a month. To stabilize the lime plaster during curing, small amounts of Plaster of Paris were mixed into the putty. Because Plaster of Paris sets quickly. "retardants" were used to slow setting time enough to allow workers to mix large working quantities of lime putty plaster. A modern form of this method uses expanded metal mesh over wood or metal structures, which allows a great freedom of design as it is adaptable to both simple and compound curves. Today this building method has been partly replaced with *drywall*, also composed mostly of gypsum plaster. In both these methods a primary advantage of the material is that it is resistant to a fire within a room and so can assist in reducing or eliminating structural damage or destruction provided the fire is promptly extinguished.

(*ii*) **Use in theatrical and movie sets.** One of the skills used in movie and theatrical sets is that of "plasterer", and the material is often used to simulate the appearance of surfaces of wood, stone, or metal. Nowadays, plasterers are just as likely to use *expanded polystyrene*, although the job title remains unchanged.

(*iii*) **Use in architecture.** Plaster may also be used to create complex detailing for use in room interiors. These may be geometric (simulating wood or stone) or naturalistic (simulating leaves, vines, and flowers). These are also often used to simulate wood or stone detailing found in more substantial buildings.

(*iv*) **Use in medicine.** Plaster is widely used as a support for broken bones; a bandage impregnated with plaster is moistened and then wrapped around the damaged limb, setting into a close-fitting yet easily removed tube, known as a *cast* ; however, this is slowly being replaced by a fibreglass variety.

Plaster is also used within radiotherapy when making immobilisation casts for patients. Plaster bandages are used when constructing an impression of the patients head and neck, and liquid plaster is used to fill the impression and produce a plaster bust. Perspex is then vacuum formed over this bust creating an immobilisation shell.

Dentists use it for making impression for dentures, inlays and for the casting of metal fillings, bridges etc.

(*v*) The chief uses are as plaster for walls and in plaster board, which is made up of alternate layers of a fibrous material such as felt or paper and gypsum plaster.

## 4.2 REFRACTORIES

A refractory is a mateiral which does not melt easily, because its fusion temperature is very light. So, refractories may be defined as, *the materials which can withstand high temperatures.* In addition to *high temperature withstanding power,* a useful refractory should possess following characteristics.

1. It should contract and expand uniformly (and not suddenly) with temperature.
2. It should be chemically inert towards corrosive action of gases, metallic liquids and slags produced in the furnace.
3. It should be resistant to the cutting action of flue gases and flames.
4. It should be able to withstand the pressure which is likely to be put on it.

### 4.2.1 Classification of Refractories

Refractories are classified into three main types, on the basis of the nature of their constituents.

1. **Acid refractories.** These are made up of acidic oxides like a aluminium ($Al_2O_3$) and silica ($SiO_2$). These are not attacked by acidic substances, but are easily attacked by basic substances. Alumina, silica and fire-clay refractories belong to this class.
2. **Basic refractories.** These are made up of basic oxides like CaO, zirconia and MgO. Hence these are not attacked by basic substances, but are easily attacked by acidic substances. Dolomite and magnesite are important members of this class.
3. **Neutral refractories.** These are made up of weakly acidic/basic substances like carbon, chromite, etc. Graphite, chromite and carborundum (Sic) are important members of this class.

### 4.2.2. General Characteristics of Refractories

1. **Porosity.** Porosity of a material is defined as the ratio of its pores volume to the bulk volume. Thus porosity,

$$P = \frac{W - D}{W - A} \times 100$$

where $W$ = Wt. of saturated specimen ; $D$ = Wt. of dry specimen

$A$ = Wt. of saturated specimen submerged in water.

Thus porosity is an important property of refractories because it effects several other characteristics like strength, abrasion resistance, thermal conductivity and chemical stability. Porosity decreases the strength, thermal conductivity, resistance to abrasion and resistance to corrosion. On the other hand, it increases the penetration of slags, molten charge, gases, etc. into the refractory material and resistance to thermal spalling (thermal shock resistance). Thus, in general, *a good refractory should have lower porosity.*

2. **Refractoriness.** *It is the ability of a material to withstand the heat, without appreciable deformation or softening under particular service conditions.* Refractoriness is, generally measured as the *softening* or *melting temperature* of the material.

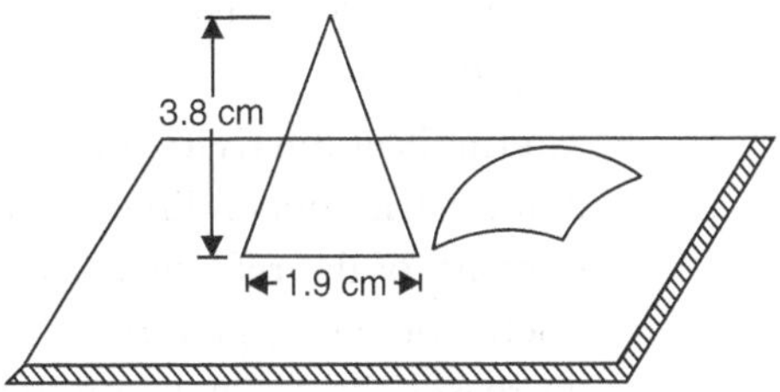

***FIGURE 4.9*** **Seger cone test for determining softening temperature of a refractory material.**

Since most of the refractory materials are mixture of several metallic oxides. They do not have a sharp fusion temperature. Hence softening temperature rather than fusion temperature is determined. The softening temperature of a refractory material is generally determined with the help of pyrometric cones test (*Seger cones test*).

The refractory material whose softening temperature is to be determined is moulded into small tetrahedral cones. The softening behaviour of these cones are compared with that of standard seger cones of known chemical compositions and softening characteristics. The seger cones are 3.8 cm (1.5″) high and have a triangular base with 1.9 cm (0.78″) long sides. The temperature at which softening of cone to be tested occurs is estimated by noting which of the standard seger cone has just bent over to the base.

3. **Spalling.** *Spalling is breaking or fracturing of a refractory due to uneven expansion and contraction of a refractory on account of heating or cooling.* Spalling may also be due to slag penetration into the refractory brick, thereby causing variation in the coefficient of expansion. Spalling may be reduced in following ways.

(*i*) By using refractory bricks having high porosity, low coefficient of expansion and good thermal conductivity. All these factors resist the development of internal stresses.

(*ii*) By avoiding sudden temperature changes.

(*iii*) By heating (overfiring) the refractories at high temperature for a sufficiently long time when mineral inversion, etc. takes place.

This makes the refractory less susceptible to uneven expansion and contraction when they are heated or cooled.

(*iv*) By modifying the furnace design.

(*v*) By mixing thoroughly the raw material during its manufacture.

(*vi*) By using binding material in appropriate proportion.

4. **Strength.** *It is the resistance of the refractory to compressive loads, tension and shear stresses in the cold as well as under high temperature.* Some refractories like fire clay and alumina brick collapse under appreciable load far below their fusion point. However, a good refractory material must possess high mechanical strength to bear the maximum possible load without breaking.

5. **Thermal expansion.** Practically all solids expand on heating and contract on cooling. Expansion of refractories may produce following two effects.

   (*i*) Expansion decreases the capacity of the furnace.

   (*ii*) Repeated expansion and contraction may lead to wear and tear and even breakage of the refractory material. Hence a *good refractory material should have least possible thermal expansion.*

6. **Thermal conductivity.** Refractories of low as well as high thermal conductivity are used in industries depending upon the type of furnace. Thermal conductivity depends upon porosity of the refractory material. It decreases when the porosity increases. In porous bricks the entrapped air, in the pores, acts as non-heat-conducting material and thus decreases the thermal conductivity. In most cases furnace is lined with refractories of low heat conductivity to reduce the heat loss to the outside by radiation. However, refractories used in muffle furnace walls, coke-oven batteries in which charge is separated from the flame should have high thermal conductivity.

7. **Resistance to abrasion or erosion.** A refractory should not be eroded (or least eroded) by descending hard charge, fine gases escaping at high speeds, carbon particles, etc.

8. **Electrical conductivity.** Refractories used for lining electric furnaces should have low electrical conductivity. All refractories except graphite are poor conductors of electricity.

9. **Texture.** Because of large porosity, coarse or light textured bricks are light in weight and hence more resistant to sudden change in temperature. However, such bricks are more susceptible to the action of abrasion and corrosion, and their crushing strength is low. On the contrary, fine or dense-textured bricks are heavier in weight and hence possess contradictory property to that of light-textured bricks.

10. **Chemical inertness.** Slags, fuel ashes, furnace gases, and products formed in refractories may react with refractory materials. Acid refractories react with basic products and basic refractories react with acidic products. Hence chemical reactions of the products formed in refractory must be considered before using a particular refractory. Hence such a refractory should be selected that the above mentioned reacting substances have the least effect on the refractory.

### 4.2.3 Manufacture of Refractories

Manufacture of refractories consists of the following steps:

1. **Grinding.** The materials are properly ground so that the ratio of coarse particles to the fine one is 55 : 45.
2. **Mixing.** The ground particles are thorougly mixed. It is done for two purposes, *viz.*, (*i*) for distributing evenly the coarse and fine particles, and (*ii*) for making moulding easy.
3. **Moulding.** Moulding may be done in two ways.
   (*i*) *By hand.* When moulding is done by hand, the density and strength of the refractory is comparatively low.
   (*ii*) *Mechanical moulding.* During mechanical moulding, de-airing of refractory is essential. De-airing increases the density of the refractory since the air with lower density than the refractory is excluded. De-airing is carried out by the following methods.
      (*a*) Air inside the voides in the refractory is allowed to escape by decreasing the rate of pressure application and release.
      (*b*) The material is first pressed and allowed to crack. It is repressed to close the voids.
      (*c*) By passing a gas such as butane to the mould in order to displace the air. When pressure is applied, the gas is absorbed by clay.
      (*d*) By applying vacuum through vents in the moulds.
4. **Drying.** The moulded refractory is dried to remove its moisture. Drying should be so regulated that neither voids are left in the refractory nor the refractory is shrunk to produce internal stresses. Generally drying is carried out very slowly and under particular set of conditions of humidity and temperature depending upon the type of refractory.
5. **Burning.** The dried refractory is burnt for vitrification and development of stable mineral forms. Burning of refractory produces following changes.
   (*i*) Removal of water of hydration.
   (*ii*) Calcination of carbonates and oxidation of $Fe^{2+}$ to $Fe^{3+}$.
   (*iii*) Shrinkage in volume may also take place which in turn produces stresses in the refractory. However, this can be avoided by the use of pre-stabilised material. Pre-stabilization consists of appropriate sizing and pressing.

Refractories are burnt in the following types of kilns.

(*i*) **Down draught kilns.** This is the most important type of firing refractories. It is either round or rectangular in shape. Refractories are stacked over the floor of the kiln and heating is commenced. The temperature is raised at a fixed rate. Flue gases and other combustion products go downward, pass over the refractories and ultimately go to chimney.

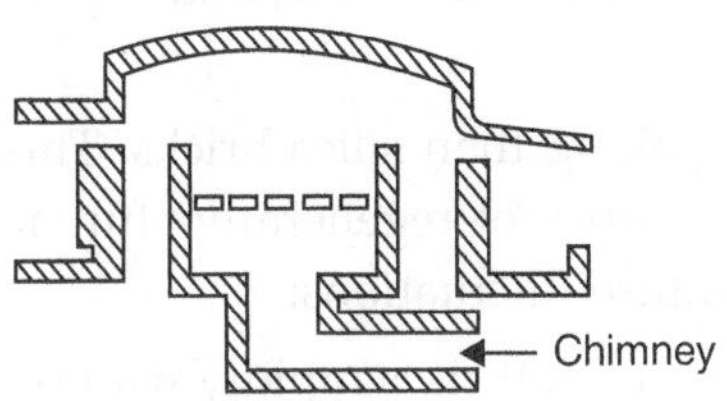


*FIGURE 4.10* **A rectangular down draught kiln.**

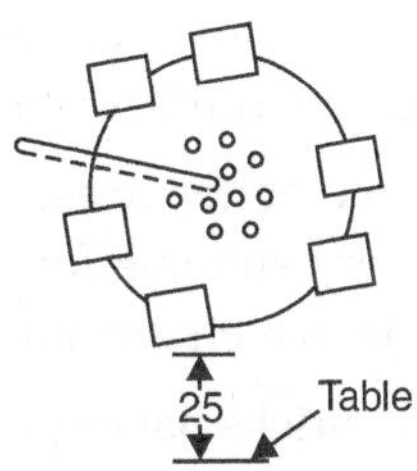


*FIGURE 4.11* **A round down draught kiln.**

(*ii*) **Tunnel kiln.** As the name represents it is a continuous kiln and consists of a long tunnel through which trolleys fitted with stacks for refractories are continuously passed. The trolleys are moved in the kiln very slowly (3 to 5 ft/hour) and generally a trolley remains in the tunnel for 50 to 100 hours depending upon the nature of the refractory. The conditions remain constant in all the parts of the tunnel and at all the times.

### 4.2.4 Acid Refractories

Acid refractories are of two types :

#### *4.2.4.1 Fire-clay Refractories*

Fire clays are the most widely used refractory materials and are well suited for a variety of applications. These contain 25 to 44% of $Al_2O_3$ and upto 70% $SiO_2$. The chief raw materials used for the manufacture of this type of alumino-silicate refractory are raw and calcined fire clays, and since these clays can withstand high temperatures, these are called fire clay refractories. Calcined fire clay, known as the "Grog", accounts for 50% or more of the batch mix. Grog is prepared by calcining fire clay at about 1450°C in rotary kilns, followed by crushing, sieving and grading. The graded fractions are suitably blended. Plastic clay is the usual bonding material used. For high grog bricks, additional bonding material like molasses, dextrin or sulphite may also be required. Addition of water may vary from 3 to 30% depending upon the method of forming proposed to be used subsequently. Moulding is done by hand or by machine. Hand moulding is done for non-standard sizes and shapes. In this the mixture containing 15 to 20% water is rammed and beaten into mould taking care to see that air pockets, voids etc., are absent. Bricks of low porosity and high slag resistance are generally produced by dry pressing the mix containing 3 to 7% water in de-airing type mechanical or hydraulic toggle presses at a pressure of 75 to 150 $MN/m^2$. Hollow bricks like sleeves, pipes and tuyers are prepared in specially designed vertical auger machines using semi-plastic mixes having 10 to 12% water.

Hand moulded shapes are first dried on cold floors for several days and then on hot floors. Machine moulded shapes are usually dried in tunnel dryers upto 1% moisture content. Then the bricks are fired at temperatures ranging from 1100 to 1400°C, lower range for casting pit refractories and higher range for the refractories used for the roofs of open hearth checkers, blast furnace hearths, etc. The firing and cooling schedule takes about 10 days.

**Properties of Fire-clay refractories:**

(*i*) Fire-clay refractories are generally light yellow to reddish brown in colour.

(*ii*) These are as hard as steel ; the exact hardness depends upon the firing temperature.

(*iii*) They possess low porosity (8 to 24%) ; the exact porosity depends upon the temperature to which the refractory is fired.

(*iv*) They have better resistance to thermal spalling than silica bricks. This makes fire-clay bricks very suitable for making a checker work of regenerative furnaces and charging doors etc. which are subjected to temperature fluctuations.

(*v*) Fire-clay bricks have specific heat as low as 0.25. It rises rapidly with rise in temperature and is 0.264 at 1300°C.

(*vi*) **Chemical properties.** Fire-clay refractories are slightly acidic in character ; hence they combine with alkalies such as soda, potash, lime, magnesia, iron oxide, etc. at high temperatures.

(*vii*) **Strength or resistance to crushing.** Crushing strength varies with temperature. An ordinary fire-clay has crushing strength as 950 kg/cm$^2$ at room temperature, but it decreases with temperature as is evident from the following data.

| *Temperature* | *Crushing load (kg/cm$^2$)* | *Temperature* | *Crushing load (kg/cm$^2$)* |
|---|---|---|---|
| 800°C | 755 | 1300°C | 360 |
| 1000°C | 575 | 1500°C | 65 |

(*viii*) **Refractoriness.** They have low refractoriness than silica bricks. On the basis of refractoriness, fire-clay refractories have been grouped in three grades:

(*a*) First grade having softening temperature between 1770° and 1790°C.

(*b*) Second grade having softenhing temperature between 1670° and 1770°C.

(*c*) Third grade having softening temperature between 1580° and 1670°C.

(*ix*) **Refractoriness under load.** Most of the fire-clay refractories have softening point of 1350°C under a load of 50 lbs/square inch.

(*x*) **Effect of heat.** Fire-clay refractories undergo following changes on heating.

(*a*) *Baking*. It involves heating of clay till it is decomposed and there is no more loss of water. Baking takes place between 500–650°C.

(*b*) *Vitrification*. It involves heating of clay till some of its ingredients are fused and fill the pores. During vitrification kaolinite, $Al_2O_3.2SiO(2\ H_2O)$ is changed to mullite, $3\ Al_2O_3.2SiO_2$. It takes place between 1250 and 1400°C.

(*c*) *Squatting or softening*. It involves heating of clay till it begins to loose its shape. It takes place between 1550 and 1770°C.

(*d*) *Fusion*. Fusion is that stage of heating during which material changes from solid to liquid state. It takes place above 1770°C.

(*xi*) **Thermal conductivity.** Thermal conductivity of fire-clay refractories varies with temperature. Fire clay refractories are although bad conductor of heat, they possess a high power of radiation. So materials placed in fire chambers mode of fire-clay may be heated to a very high temperature though the conductivity of fire-clay is very low.

(*xii*) **Durability.** The most durable and best fire-clay refractory should have high apparent density, high resistance to slags, high crushing strength and should have no volume change when heated.

**Uses.** The steel industries are the largest consumers of fire clay refractories. They are used for lining of blast furnaces, open hearths, stoves, ovens, flues, crucible furnaces etc. Fire clay refractories are also widely used in foundries, lime kilns, regenerators, pottery kilns, continuous ceramic and metallurgical kilns, glass furnaces, cupolas, brass and copper furnaces, boiler settings, gas-generators, etc.

### *4.2.4.2 Silica Refractories*

In silica brick, the two oxides of interest are $SiO_2$ and CaO. Owing to the predominance of silica as a separate phase the thermal behaviour of the refractory is mostly influenced by the performance of the various polymorphous crystalline forms of different temperature stability.

The quartz from which all silica refractories are made is converted during the firing step to a mixture of cristobalite and tridymite, which have good high temperature stability. The CaO added (about 2%) to the silica brick mixes as a "mineralizer" accelerates the formation of cristobalite and tridymite, and also helps to form a bond during firing. Cristobalite when heated to about 230°C inverts suddenly from its low-temperature-crystalline form to its high-temperature crystalline form accompanied by a large increase in volume. The reverse process occurs on cooling. In order to obviate this problem, silica refractories are generally heated and cooled slowly in this temperature range.

Silica bricks have particular ability to withstand compressive loads almost to their complete fusion point (1700°C) and hence they are particularly suitable for constructing furnace roofs. This rigidity is due to the interlocking of the crystalline silica and the fact that the amount of liquid (actually, two immiscible liquids) formed is comparatively small and quite viscous. The presence of even small percentages of alumina and alkalies eliminates the immiscibility and thus reduces the refractory properties; further, these compounds form lower-melting eutectics than what $SiO_2$ and CaO do alone. This is the reason why quartzites are beneficiated to lower the percentage of alumina and alkalies for producing silica bricks for high temperature use.

Silica refractories contain at least 94% silica. The chief raw material for silica refractories is quartzite which is used without precalcination. The grading is finer than that used for fire clay refractories. The bond additions are lime (1-2%) and $Fe_2O_3$ and/or $TiO_2$ (1%) which are so small that they require thorough blending. Sometimes, 1 to 1.5% of molasses or sulphite lye (1-1.5%) is used for green strength. For non-standard shapes and sizes, hand moulding with pneumatic ramming is used. For standard shapes, dry pressing is used. Drying is usually carried out in tunnel dryers at 90-120°C for about 24 hours until the moisture is reduced to 0.2%. Dry pressed bricks can be set directly on the kiln cars and passed through tunnel dryers.

Firing of the silica bricks is usually done in round down draught periodic kilns or continuous tunnel kilns. Firing is the most vital step in the manufacture of silica bricks. As stated earlier, on heating and cooling, quartz (which is the mineralogical constituent of quartzite undergoes a number of different crystalline allotropic transformations, some of which are stable and others unstable

$$\alpha\text{-quartz} \underset{}{\overset{573^\circ C}{\rightleftharpoons}} \underset{\text{Sp. gr.}=2.65}{\beta\text{-quartz}} \overset{867^\circ C}{\rightleftharpoons} \underset{\text{Sp. gr.}=2.20}{\text{Tridymite}} \overset{1470^\circ C}{\rightleftharpoons} \underset{\text{Sp. gr.}=2.32}{\beta\text{-cristobalite}}$$

The inversion points are 573°C for quartz, 260°C for cristobalite, ande 120°C, 160°C and 250°C for tridymide. After firing, the silica bricks must consist of cristobalite and tridymite only and quartz should be absent. If during firing of silica brick, quartzite is not converted into cristobalite and tridymite, the bricks will expand during use in the furnace, which may lead to collapse of the structure. The conversion of quartz to cristobalite is accelerated by lime, whereas the formation of tridymite is accelerated by $Fe_2O_3$ and $TiO_3$. Since these reactions are rather slow, the temperature during firing is gradually from 90°C to the final firing temperature of 1400-1500°C in about 7 days. The bricks are soaked at this temperature for 2 days. The cooling rate, particularly below 300°C, is very vital for the inversions of cristobalite and tridymite. The cooling takes about 6 days.

**Properties:**

(*i*) Silica bricks are yellowish in colour with brown speck throughout the body.

(*ii*) Its specific gravity varies from 2.2 to 2.6 depending upon the temperature to which silica is burnt.

(*iii*) Silica refractories expand on heating. However, this effect is reversible and the bricks contract to its original size when cooled. These volume changes are due to allotropic changes occurring in silica. Actually, these allotropic changes cause changes in specific gravity which in turn causes change in volume.

$$\underset{(\text{sp. gr.} = 2.65)}{\text{Quartzite}} \underset{}{\overset{870°\text{C}}{\rightleftharpoons}} \underset{(\text{sp. gr.} = 2.26)}{\text{Tridymite}} \underset{}{\overset{1470°\text{C}}{\rightleftharpoons}} \underset{(\text{sp. gr.} = 2.23)}{\text{Cristobalite}}$$

Thus at 870°C when tridymite is formed, there is increase in volume of 14 to 16%. On the other hand, heating of silica refractory upto 1470°C (*i.e.* conversion of silica to cristobalite) is accompanied by an increase in volume by 20%.

(*iv*) Silica bricks are homogeneous in texture ; these are free from air-pockets and moulding defects.

(*v*) **Mechanical strength.** Their mechanical strength varies from 900 to 3700 lbs/inch$^2$ depending upon the composition. The crushing strength of silica brick is proportional to fineness of quartz used, the proportion of binding material and water added to make paste.

(*vi*) **Refractoriness under load.** Silica brick when heated under a load of 50 lbs/inch$^2$ will disintegrate between 1350 and 1470°C. The maximum working temperature of silica brick under load is usually 1680°C.

(*vii*) **Refractoriness.** The best silica refractory has softening temperature from 1750°C to 1760°C. The refractoriness depends upon the following factors :

(*a*) Proportion of lime, soda, potash and alumina present in the refractory.

(*b*) Texture of refractory.

(*c*) Mode and extent to which refractory is burnt.

(*viii*) **Spalling.** Silica bricks have low resistance to thermal spalling and a sudden change in temperature causes fine cracks in the brick. The spalling is due to following factors.

(*a*) Spalling occurs in the refractory when the conversion from quartz to cristobalite or tridymite is either complete or occurs suddenly due to rapid rise in temperature of the refractory. However, this type of spalling can be avoided by repeatedly raising the temperature of the refractory at a slow rate upto 800°C and then cooling it slowly.

(*b*) Spalling occurs due to reversible expansion and contraction occurring too rapidly. This can also be avoided by heating the refractory slowly.

(*c*) Spalling may also occur due to the action of slags.

In general, spalling can be decreased in silica refractory by the following instructions.

(*i*) By using a coarse-grained refractory.

(*ii*) By increasing the porosity and decreasing the density.

(*iii*) By repeatedly heating the refractory to 800°C and then cooling it slowly. This causes conversion of quartz to tridymite rather than cristobalite.

(*ix*) **Thermal conductivity.** It is higher than fire-clay bricks.

(*x*) **Action of chemical agents.** Among acids, silica is effected only by hydrofluoric acid in cold and phosphoric acid at 400°C.

Alkalies and bases react with silica refractory, particularly at high temperatures, forming silicates.

Silica reacts with carbon at about 1200°C to form Si and CO or SiC in some cases.

**Uses.** Silica bricks are used for arches in large furnaces because of their high physical strength. Silica bricks are used for main arch, side walls, port arches and bulk-heads of open hearth furnaces, coproduct coke-ovens, gas retorts, glass furnaces, electric furnace, roofs, copper stove domes, acid converter linings, etc.

#### *4.2.4.3 High Alumina Refractories*

When the alumina content in fire clay brick reaches above 47.5%, then it is called high alumina refractory brick. These are made from clays rich in bauxite and diaspore. The refractoriness and the temperature of incipient vitrification increase with the alumina content, as can be seen from the phase diagram illustrated in Fig. 4.12.

The essential oxides of both fire clay and high-alumina refractories are $Al_2O_3$ and $SiO_2$. Points A and D indicate the melting points of the two essential oxides at 1723 and 2050°C. In fired refractories, at or near these points, silica is in the form of cristobalite and alumina is in the form of corundum. Point B indicates the melting point of the eutectic composition in this system at 1545°C. Point C locates the temperature of 1810°C at which mullite, the only compound in the system, melts incongruently to form corundum and a liquid. This point also shows the composition of the liquid. Points 1 to 6 represent several commercial refractory material compositions in the following order : Siliceous fire clay ; Kaolin (dehydrated); the sillimanite minerals (and alusite, kaynite and

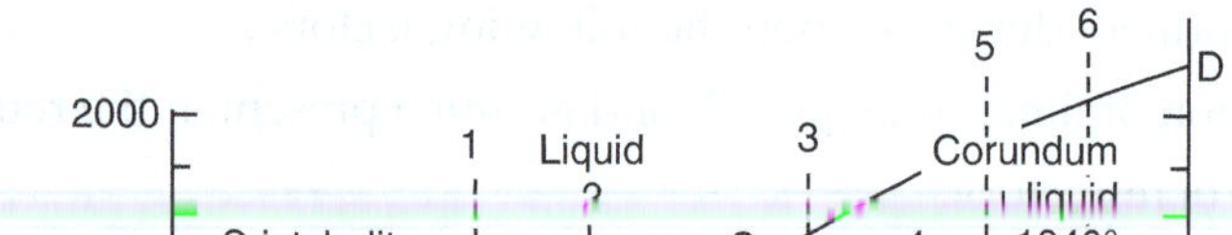


*FIGURE 4.12* **Phase diagram of MgO—SiO$_2$ System.**

sillimanite); mullite; diaspore; and bauxite (both dehydrated). Even though fire clay refractories, which are predominantly kaolin, may contain an appreciable proportion of siliceous liquid at elevated working temperatures, they are good refractories, because the liquid is so viscous. Refractories in this system containing enough alumina to be near or above mullite in composition are mostly crystalline at furnace temperatures and thus perform as super refractories. High-alumina refractory bricks are superior to fire clay bricks and silica bricks as they have more refractoriness, better slag resistance, inertness to carbon monoxide and stability to natural gas environment upto 1000°C. Special high-alumina refractories are made from silimanite and kyanite which are anhydrous aluminosilicate minerals : $Al_2O_3.SiO_2$. They are bonded with plastic clay aluminosilicate minerals : $Al_2O_3.SiO_2$. They are bonded with plastic clay and the mixture is moulded into bricks, followed by drying and firing at 1600°C. The extensively used silimanite refractories have the approximate composition of about 63% alumina and 34% silica. HIgher percentages of alumina are classed as super-refractories and almost pure alumina refractories are classed under pure oxide refractories.

**Properties:**

(*i*) **Porosity.** Alumina bricks are highly porous which decreases their durability in furnace.

(*ii*) **Shrinkage.** Bauxite refractories show great shrinkage even at high temperature. However, this drawback can be eliminated by making the refractory from fused bauxite.

(*iii*) **Resistance to abrasion.** These have great resistant to abrasion.

(*iv*) **Mechanical strength.** Bauxite refractories have high mechanical strength.

(*v*) **Refractoriness.** Bauxite refractories are heat resistant from 1790° to 1880°C. Under a pressure of 50 Ibs/inch$^2$ bauxite refractories soften at about 1500°C.

(*vi*) **Spalling.** Bauxite refractories have very little tendency to spall.

(*vii*) **Chemical reactions.** Bauxite is inert to gases like $CO_2$, $H_2$ and also to oxidizing and reducing agents. Being basic in character, bauxite is not attacked by basic slags. This property of bauxite makes it superior to fire-clay refractories which are subject to such attacks.

**Uses.** High-alumina refractory bricks are employed in cement industry, paper industry, for boiler settings, in the linings of glass furnaces, oil-fire furnaces, high-pressure oil stills, in the roofs of lead softening furnaces and in regenerator checkers of blast furnaces.

### 4.2.5 Basic Refractories

Basic bricks are usually made from magnesia, dolomite, magnesite, chromite and forsterite. In order to achieve the strength and other physical properties, the basic bricks are generally power-pressed and are either hard-burned or chemically bonded. The drawbacks of lack of bond and volume stability in unburnt basic or other bricks have been overcome by the following innovations in the manufacturing process : (*a*) Maximum interfitting of grains was achieved by using only selected particle sizes mixed in proper proportions to fill all the voids (*b*) Increasing the forming pressures to the order of 10,000 *psi* and utilizing de-airing equipment to reduce the air voids between the grains and (*c*) Use of a refractory chemical bond.

#### *4.2.5.1 Magnesite Refractories*

These are made from magnesites. These bricks do not withstand much higher load at elevated temperatures. This drawback is overcome by blending with chrome ores.

The magnesite bricks are generally made from dead-burnt magnesite grains which is prepared by burning magnesite in a rotary kiln or a shaft at about 1700°C using 3 to 4% iron oxide or mill scale powder to reduce the sintering temperature. It is then crushed, sized and graded. Molasses or sulphite lye is used as a binder. 2 to 6% of alumina is added to impart thermal shock resistance of the refractory. The ingredients are blended after adding requisite quantity of water and the mix is aged for 1 to 10 days to ensure complete hydration of any free lime present. Then the mix is moulded by dry pressing at pressures above 70 MN/m$^2$. Then the shapes are dried first in open sheds and then in controlled humidity dryers. Then they are subjected to firing which should be slow at the beginning and then at the final temperature range of 1400–1600°C. The ceramic bond is usually developed by forsterite ($2MgO.SiO_2$). However, the current trend is to prepare the bricks from purer materials and firing them at 1700°C or more so that direct bonding takes place between the periclase grains (Periclase is a stable form of magnesia). The cooling rate also should be slow. Since magnesite brick cannot bear heavy load during firing in kilns, they are boxed in load bearing silica bricks.

Fire bonded magnesite bricks are susceptible to spalling. This drawback is removed by chemical bonding or by metal casing. For preparing chemically bonded magnesite refractories, the mechanical strength is developed by adding a chemical bonding agent *e.g.*, oxychloride or oxysulphate, instead of by firing. In this case, higher forming pressures are needed to minimize voids. Curing and drying are carried out in specially designed tunnel dryers.

Metal cased magnesite refractories are prepared by ramming dead burnt magnesite into soft steel containers, which welds together at the heated end in high temperature service forming a practically monolithic structure.

**Properties:**

(*i*) Magnesia bricks are usually grey dark brown in colour with a specific gravity of 3.05 to 3.58.

(*ii*) Magnesia bricks have good crushing strength but are low resistance to abrasion.

(*iii*) They show little shrinkage.

(*iv*) They show great spalling.

(*v*) These are resistant to slags. However, being basic in nature, they react with acid refractories. They also react with iron oxide and carbon.

(*vi*) Magnesia bricks can be used upto 2000°C without load and upto 1500°C under a load of 3.5 kg/cm$^2$.

(*vii*) Heat conductivity of magnesia refractory is very low, so it is very useful for raising and maintaining high temperature in furnace.

**Uses of magnesia refractory:**

(*i*) Magnesia refractories are used mainly in the steel industry for the lining of basic convertors and open hearth furnaces.

(*ii*) Magnesia bricks are also used for constructing furnaces for melting lead, antimony and copper.

(*iii*) These are also used for gold, silver and platinum refining furnaces.

(*iv*) Bottom and side walls of electric furnaces are constructed with magnesia bricks.

(*v*) Magnesia bricks are used for covering fire-clay bricks.

#### *4.2.5.2 Forsterite Refractories*

Forsterite ($2MgO.SiO_2$) is used both as bonding material as well as a base for high-temperature refractories. When forsterite is used as a base, the refractories are usually made from olivine

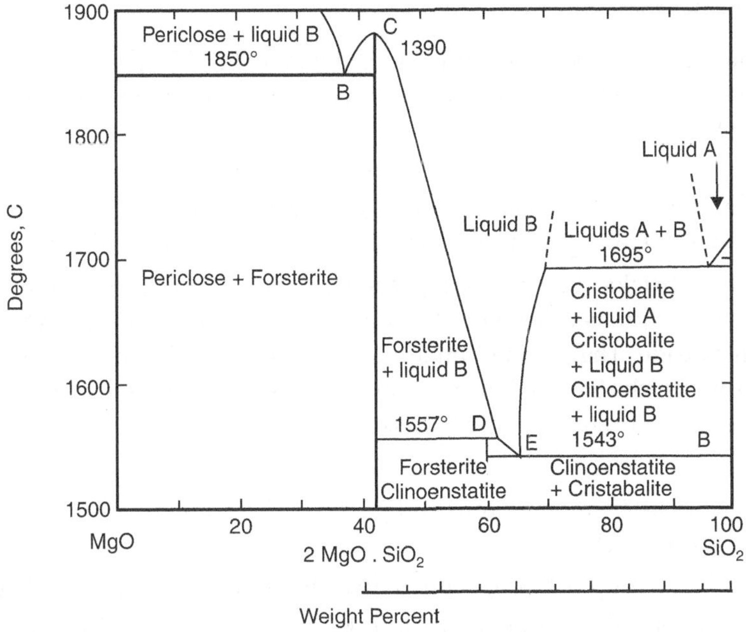


*FIGURE* 4.13

2 (Mg, Fe) $O.SiO_2$ which is characterized by its high refractoriness. In the manufacture of forsterite refractories, dead burned magnesite is usually added to convert some accessory minerals also to forsterite, which is the most stable silicate at high temperatures, as can be seen from the phase diagram of $MgO.SiO_2$ system shown in Fig. 4.13. For instance, enstatite or clinoenstatite which occurs in mined olivine rocks, is converted into forsterite:

$$\underset{\text{Enstatite}}{MgO.SiO_2} + \underset{\text{Magnesia}}{MgO} \longrightarrow \underset{\text{Forsterite}}{2MgO.SiO_2}$$

Such materials are called super-refractories which possess high melting point, good volume stability and stability against transformations during heating. Their preparation does not require calcining.

**Uses.** Forsterite is used in glass tank super-structures and checkers because they possess high chemical resistance to the fluxes used and good strength at high temperatures allow increased tank output. These refractories are also used in open-hearth end walls and copper refining furnaces.

### *4.2.5.3 Magnesite-chrome and Chrome-magnesite Refractories*

Magnesia bricks do not stand much load at elevated temperatures. In order to overcome this difficulty, they are blended with chrome ores. If the blend contains more of magnesite, it is called magnesite-chrome and if chromite predominates, it is called chrome-magnesite. These refractories are made from dead-burnt magnesite and chrome ore mixed in proper proportions. This may be fire bonded, chemically bonded or metal cased as in the case of magnesite bricks described eariler. A coarse grading provides better thermal shock resistance.

Their resistance to acid slag is low and resistance to basic slag is high. They have high resistance to abrasion, erosion and spalling. Their load-bearing capacity and volume stability at elevated temperature is high. Chrome-magnesite refractories have a tendency to bursting in contact with iron oxide.

#### *4.2.5.4 Dolomite Refractories*

Dolomite is magnesium lime stone with the composition, Ca Mg $(CO_3)_2$. It occurs in nature. Calcined dolomite particles are united with the help of magnesium silicate, iron ore, basic slags, tar, etc. (binding materials).

Dolomite is calcined, powdered and mixed with binding material and water. Mixing is performed in edge runner and then allowed to age by keeping it in wet condition. The mixture is moulded to bricks which are air-dried and burnt to a temperature ranging between 1200° to 1650°C.

**Stabilised dolomite bricks.** For preparing such bricks, dolomite is mixed with serpentine ($MgO.SiO_2$) and the mixture is calcined to form di- and tri-calcium silicates which do not absorb water and carbon dioxide. The calcined mixture is mixed with a binding agent (usually silicate) and moulded into bricks which when burnt at about 1500°C for about a day gives stabilized dolomite bricks. Such bricks are quite stable towards basic slags.

**Properties**

(*i*) Dolomite bricks are more porous and possess more softness, shrinkage and less strength than magnesia bricks.

(*ii*) They can be used upto 2300°C without load and upto 1650°C under a load of 3.5 $kg/cm^2$.

(*iii*) Dolomite bricks are corroded by silica. However, stabilized dolomite bricks are more resistant to slags and spalling than magnesia bricks because the former contains dicalcium silicate which is more resistant than magnesium silicate to dust and slag containing lime and iron compounds.

**Uses.** Dolomite is generally used as a repairing material rather than as a direct refractory because of its defects like great porosity, shrinkage and softness. However, stabilized dolomite bricks are used in electric furnaces, Bessemer convertors, open-hearth furnaces, etc. and as a cheap substitute for magnesia bricks.

### 4.2.6 Neutral Refractories

#### *4.2.6.1 Carbon Refractories*

The raw materials are low ash content (0.15%) carbon *e.g.*, petroleum coke (for electrodes) or coke derived from coal (for bricks and blocks). The crushed and finely graded coke is dried in a rotary dryer and hot blended at 50°C to 80°C with 15 to 18% tar or pitch in a dry pan mill. Forming of shapes is done by dry pressing of the hot mix at a pressure of 30 to 40 $MN/m^2$ or by pneumatic ramming in case of large shapes. Hardening takes place within 1 to 6 days, depending upon the size. Firing is carried out in a reducing atmosphere at 800 to 1000°C packing the shapes in carbon. Specially designed down draught periodic kiln or chamber kiln or tunnel kiln may be used. A tunnel kiln car takes a load upto 1600 kg in which bricks are set edgewise into muffles built of fire clay bricks, all the interstices, being filled with crushed coke and a layer of coke is laid on the top on which another layer of fire clay bricks is embedded in fire clay or refractory mortar. The heating and cooling cycle may take 4 to 6 weeks in a periodic kiln but the schedule can be minimised if tunnel kilns are used.

**Properties**

(*i*) They are light grey in colour.

(*ii*) They possess high crushing strength.

(*iii*) If not exposed to air, they can withstand high temperatures. However, when heated in presence of air, they burn.

(*iv*) They are not attacked by acids, alkalies and slags.

**Uses.** Carbon bricks are used at bottom and sides of iron and steel furnaces. They are also used in furnaces of smelting copper, lead and aluminium.

#### *4.2.6.2 Graphite Refractories*

Graphite is a natural allotropic form of carbon having dark colour, bright metallic lustre. In addition to various other industrial applications, graphite is also used as a refractory material. Graphite refractories are produced by one of the following two methods : (1) by graphatising of pear shaped carbon products by prolonged heating (3 to 5 weeks) at about 2850°C in an electric furnace (2) shaping of graphite flakes bonded with fire clay or carbon. This method is particularly used for preparing crucibles. The shapes are dried at 80 to 100°C for one day in controlled humidity driers and fired at 900 to 1200°C in down draught periodic kilns or continuous muffle kilns. In order to prevent oxidation, the crucible surface is coated with boric acid, borax, soda ash, sodium silicate, $CaCl_2$ or $MgCl_2$. The fired crucibles are slowly cooled. In India, clay or carbon bonded graphite crucibles upto 100 kg capacity are produced in small scale sector.

The properties of graphite refractories are more or less similar to the carbon refractories excepting that they have higher oxidation resistance than the latter.

Graphite crucibles are widely used in industry. Graphite bricks are used for construction of electrodes, linings of chemically resistant equipments, atomic reactors, electric furnaces and in non-ferrous metal smelting furnaces.

#### *4.2.6.3 Silicon Carbide (Carborundum) Refractories*

Silicon carbide is a very hard abrasive having a very high melting point. It is made by heating a mixture of 60% sand and 40% coke together with some saw dust and salt in an electric furnace at 1500°C. Saw dust helps in increasing the porosity and salt helps in the removal of iron etc. in the form of volatile chlorides. The SiC so formed is in the form of interlocking crystals. This is crushed, sized and suitably graded so as to get a dense packing. The bonding agents used are less than 10% of plastic fire clay, graphite, ethyl silicate or even finely divided SiC itself. Recently, silicon nitride or oxynitride is preferred as bonding agent because it imparts superior oxidation resistance to the product. The final firing of the bricks is carried out in a reducing atmosphere at about 1500°C. The silicon carbide refractories generally contain 85–90% SiC. However, for less severe service conditions, SiC content may be as low as 60%.

Self-bonding type of silicon carbide refractory bricks can be prepared by mixing SiC particles with a temporary binding agent like glue, then pressing and firing at 2000°C when inter-crystalline bond develops.

**Properties:**

(*i*) Silicon carbide bricks are dark grey to blue black in colour.

(*ii*) These can be used upto 1750°C.

(*iii*) They have high thermal conductivity and very low co-efficient of expansion.

(*iv*) When heated in air at a temperature range of 900–1000°C, they are oxidised to silica. However, this oxidation can be prevented by coating them with a thin layer of zirconium.

Self bonded silicon carbide has very useful properties like high density, greater resistance to abrasion, great strength, high refractoriness and higher resistance to chemicals.

**Uses.** Carborundum refractories are useful because of its hardness, low co-efficient of expansion, great strength and toughness. They are used for partition walls of chamber kilns, coke ovens and muffle furnaces. They are also used as heating elements in furnaces in the form of rods and bars.

#### *4.2.6.4 Chromite Refractories*

These are made by firing chromite ($Cr_2O_3.FeO$) blended with clay at a temperature of 1500 to 1700°C. These are neutral refractories having good slag resistance. They possess high density, moderate spalling resistance and moderate thermal conductivity. They can be used upto 1800°C. Their crushing strength is about 350 to 550 kg/cm$^2$ and their refractoriness under a load of 3.5 kg/cm$^2$ is 1430°C. They are used in furnace linings, sodium carbonate recovery furnaces and bottoms of soaking pits.

### 4.2.7 Special Refractories

#### *4.2.7.1 Pure Oxide Refractories*

These refractories have high refractoriness and can withstand very severe operating conditions. Common among these refractories are : alumina, magnesia, zirconia, beryllia and thoria.

Alumina, magnesia and zirconia are available in high purity (greater than 97%) and are composed of electrically fused grains. Beryllia is not used commercially in heavy wear because of its exorbitant cost and volatalizing tendency above 1610°C in presence of water vapour. Thoria has a disadvantage of its being radio-active and hence is under strict controls. Among the pure oxide refractories, sinter alumina has the widest application. It was successfully used upto 1870°C. Magnesia is a basic refractory and is reduced at high temperatures. Its applications are limited to oxidizing atmospheres upto about 2,200°C. Pure zirconia undergoes crystalline transformation from monoclinic to tetragonal form at about 980°C, accompanied by drastic volume change on inversion. Hence stabilization of the crystal structure of the cubic form (which does not undergo inversion) is necessary. This can be accomplished by adding certain metallic oxides *e.g.*, CaO and MgO. Processing temperatures with the fused stabilized zirconia are of the order of 2600°C. The pure zirconia refractories are used in kilns needed for firing barium titanate resistors.

*(i) Pure Alumina Refractories*

These are made from calcined or fused alumina having a purity greater than 97%. The usual additions are 2% of Kaolin (which reduces the sintering temperatures and improves pressing characteristics) and 2% of talc or magnesite (which controls the grain size and improves pressing characteristics). The binding agent used depends upon the method adopted for forming. For dry pressing, organic binders *e.g.*, cellulose or polyvinyl alcohol are used. Firing temperatures are in the range of 1700-1800°C depending upon the purity of alumina. Forming can also be done by hot pressing in graphite moulds. Pure alumina refractories belong to neutral type having good resistance to oxidising and reducing atmospheres even upto 1900°C. High temperature kilns furnish alumina bricks which approach pure corundum in properties which have high slag resistances. Such refractories are used where severe slagging occurs.

*(ii) Magnesia Refractories*

These are prepared from fused magnesia grains. Cellulose is normally used as binder. Firing temperatures are in the range 1900-1950°C. Care must be taken in their setting to ensure that they do not react with the kiln material at this high temperature. Forming by hot pressing is similar as in the case of pure alumina refractories. This is a basic type of refractory having refractoriness of the order of 2300°C.

#### 4.2.7.2 Mixed Oxide Refractories

*(a) Mullite ($3Al_2O_3.2SiO_2$) Refractories.*

These are prepared from high alumina minerals and fire clay in proper proportions. The grog is first prepared by fusing on sintering the raw material mix at 1820°C or by calcining it at 1600°C. After blending the fine powder with a binder such as plastic clay or sulphite lye, the shapes are formed by the usual methods followed by firing at about 1750°C. Addition of a small quantity of aluminium fluoride during blending might reduce the mullitisation temperature.

Electrocast or corhart mullite refractory blocks are prepared from electrically fused mullite. The manufacturing procedure consists of introducing a mixture of diaspore clays of high alumina content (furnishing the proportion of $3Al_2O_3.2SiO_2$) into a top of an electric furnace. Molten aluminium silicate at a temperature of about 1870°C is tapped from the furnace periodically into moulds built from sand slabs. They are annealed for 6 to 10 days before the blocks are usable. The refractory blocks so obtained have a vitreous non-porous structure which exhibits a linear coefficient of expansion of about one-half of that of good fire-brick. The electrocast has only 0.5% voids as compared to 17 to 29% for fire clay blocks. Cast refractories are used in glass furnaces, as linings of hot zones of refractory kilns, modern boiler furnaces exposed to severe duty and in metallurgical equipment such as forging furnaces. The higher initial cost of these refractories is balanced by the long life and minimum wear. A still superior type of fused cast refractories, with 83 to 95% alumina is also available now.

Mullite refractories belong to neutral type of refractories. They possess high refractoriness and high resistance to acid slag, basic slag, oxidising atmosphere, reducing atmosphere and to thermal and structural spalling.

*(b) Zircon ($ZrO_2.SiO_2$) Refractories*

These are prepared from micronised zircon mixed with finely ground clay or alumina. The grog is prepared by the partial sintering of zircon. Firing temperature is 1650-1700°C. These refractories belong to acid type and possess high resistance to acid slag. Their resistance to thermal shock or spalling is high. Their thermal expansion is low and resistance to molten metals is high. Their resistance to oxidising atmosphere is low.

#### 4.2.7.3 Non-oxide Refractories

These include silicon nitride, boron carbide, molybdenum disilicide, zirconium boride and titanium boride. They can be formed into dense shapes by hot-pressing or sintering in controlled conditions. Molybdenum disilicide exhibits oxidation resistance even at 1800°C, while others tend to oxidize from 1000°C. Non-oxide refractories are expensive and are mostly used at present in nuclear and space research programmes.

#### 4.2.7.4 *Insulating Refractories*

These are characterised by low thermal conductivity and low bulk density. They are of two types :

(*a*) Backing up insulations which are made from highly porous inorganic refractory materials like asbestos, mica, vermiculite and diatomite.

(*b*) Hot face insulations which are similar in composition to dense refractories. The raw materials used for this type of insulating refractories are chromite, magnesite, fire clay, silica and alumina. The insulating value of this type of refractories is due to the method of their preparation. For example, waste cork is ground and sized and is mixed with fire clay. This is then moulded and burnt in a kiln. The cork burns out leaving a highly porous and light weight refractory brick.

The following methods are used for rendering the insulating refractories porous :

(*i*) by adding to the mix organic materials like rice husk, saw dust, ground cork, coke or pitch, which are burnt out during firing leaving a porous mass.

(*ii*) by adding to the mix a highly porous inorganic refractory material like porous grog, expanded mica, calcined diatomite or vermiculite.

(*iii*) by adding to the mix a granular volatile substance like naphthalene flakes which sublimes at high temperature leaving behind a porous product. The fired product will be flux-free unlike in the other methods. The sublimed naphthalene can be cooled and reused.

(*iv*) by converting the mix to the consistency of a thick slip and introducing gas bubbles (by chemical effervescence or by inclusion of a preformed foam) which are so stable that they remain in it until the product is dried and fired. Chemical effervescence can be obtained by Zn or Al and an alkali which produces hydrogen bubbles or by adding finely divided lime and acid which produces $CO_2$ bubbles. Preformed foam can be obtained by stirring into the slip soap, sodium resinate or saponin in water. In both the methods, stabilizers like plaster of paris, glue or gum arabic are added which prevent coalescence of the bubbles.

## 4.3 GLASS

Ordinary glass is a homogeneous mixture of substances. When quartz or silica is heated with sodium carbonate, a compound known as sodium silicate is obtained. When it is allowed to cool, it sodifies to a glass mass called *water glass*. If sand or quartz is heated with other metal oxides or carbonates (other than sodium or potassium carbonate) such as lime, alumina or oxide of lead etc., mixture of silicates are obtained. When allowed to coal, they solidify to glasses but donot dissolve in water, and constitute what is known as *glass*. Glass, has, therefore no definite chemical composition. Physically glass may be defined as, a hard, rigid, undercooled, brittle, non-crystalline substance having no definite melting point and sufficiently high viscosity to prevent crystallization. It is usually transparent, but sometimes opaque or translucent.

Chemically, glass is a mixture of number of silicates and hence no definite chemical formula can be assigned to it. However, it is sometimes represented as $xM_2O.yM'O.6SiO_2$, where M is a monovalent alkali metal like Na, K, etc.; M′ is a bivalent metal like Ca, Pb, Zn, etc. and $x$ and $y$ are whole numbers. Thus the ordinary glass may be represented as $Na_2O.CaO.6SiO_2$. Since it is not a true solid, it has no definite melting point and melts at high temperatures. When heated, it gradually softens and can be moulded into any desired shape. It is this property of glass which makes it a highly useful material since a number of articles of different shapes and forms can be made out of it.

### 4.3.1 Manufacture of Glass

Manufacture of glass can be studied under the four headings, *viz.* raw materials, furnace, the actual manufacturing steps and annealing.

1. **Raw materials.** The following types of raw materials are used in the manufacture of glass.
   (*i*) *Silica* ($SiO_2$). Silica is generally added in the form of sand of a uniform size (too fine particles make the reaction violent while too course particles prevente reaction). It should be free from impurities like iron oxide (which acts as a colouring matter) and organic matter. Small quantities of alumina fund in sand is harmless and hence need not to be removed.

   (*ii*) *Compounds of alkali metals* (*sodium or potassium*). These are added in the form of soda ash ($Na_2CO_3$), sodium sulphate ($Na_2SO_4$), sodium nitrate ($NaNO_3$), potassium carbonate ($K_2CO_3$) or potassium nitrate ($KNO_3$).

   (*iii*) *Compounds of alkaline earth metals* (*calcium or barium*). The chief sources of calcium are lime (CaO) and limestone ($CaCO_3$). Barium is added when glass of high refractive index is needed. It is usually added as barium carbonate ($BaCO_3$).

   (*iv*) *Oxides of heavy metals.* Lead is added in the form of litharge (PbO) or red lead ($Pb_3O_4$). The latter is preferred owing to its high oxygen content. It should, however, be free from metallic lead and oxides of copper and iron.

In the modern heat-resisting glass, calcium or lead oxide is replaced by zinc oxide (ZnO) along with boric acid or borax which decreases the coefficient of expansion.

   (*v*) *Calcium phosphate.* Phosphate is added in the form of calcium phosphate, $Ca_3(PO_4)_2$. This is added only when *opalescent glasses* is to be manufactured. In such cases, arsenic and antimony oxides are also added.

   (*vi*) *Colouring materials.* Certain metallic oxides are added to the molten mass (fused silicates) during the manufacture of *coloured glasses.* For example, the following colours are produced by substances noted against each.

| *Colour* | *Substances added* |
|---|---|
| Green | Ferrous ion, chromium |
| Yellow | Ferric iron, sodium uranate |
| Lemon yellow | Cadmium sulphide |
| Blue | Cobalt oxide |
| Peacock-blue | Cupric salts |
| Purple | Manganese oxide |
| Red | Cuprous oxide ($Cu_2O$), selenium oxide |
| Ruby | Auric chloride ($AuCl_3$) |
| Amber | Organic matter and carbon |

**2. Furnace.** Two types of furnaces are used in the manufacture of glass.

(*i*) *Pot furnace.* A pot is a large monkey-shaped clay crucible (see Fig. 4.14). A number of such pots are arranged in a circle in a pot furnace. The furnace is generally fired on the regenerative principle of heat economy (Fig. 4.15). In this system, the hot burnt gases (flue gases) from the furnace are made pass through a fire-brick checker-work which, in turn, gets heated. After about 20-30 minutes, the direction of the shaft is reversed and the incoming air and fuel gas are led through this hot checker-work while the hot flue gases go to heat another checker-work. In this way the heat of the burnt waste gases (flue gases) are utilised in heating up the incoming gas and air previous to burning. In this type of furnace, the processes of charging the raw material and the removal of glass are discontinuous.

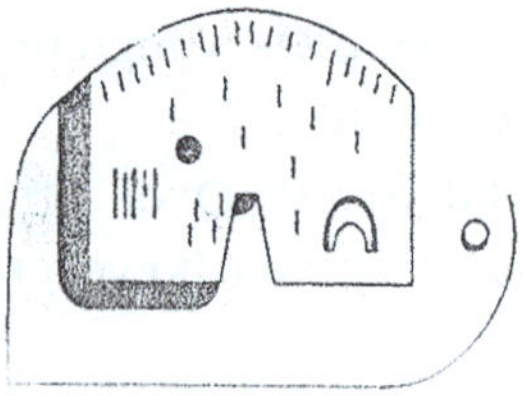

**FIGURE 4.14** A melting pot.

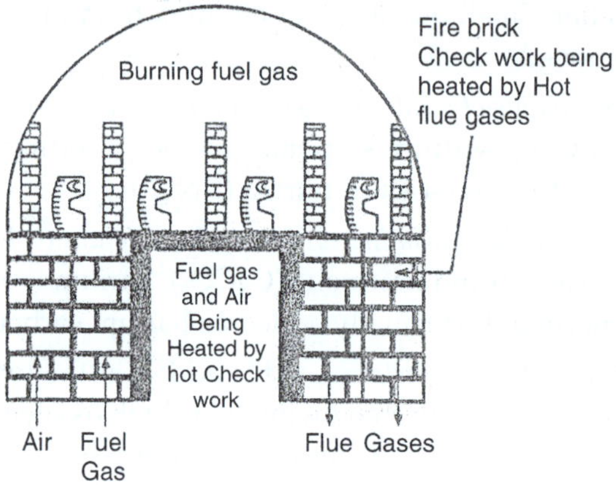


***FIGURE 4.15*** **A pot furnace of regenerative type.**

(*ii*) *Tank furnace.* A common tank furnace consists of a large rectangular clay tank in which glass is melted. It is heated by producer gas and works on the regenerative system of heat economy as explained in the case of pot furnace. In this type of furnace, the charging of the batch and the removal of glass for working are continuous. Here, a high temperature is maintained at the melting end than at the working end.

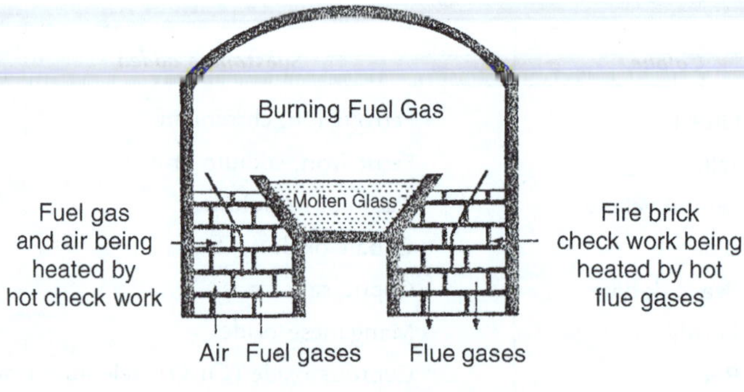


***FIGURE 4.16.*** **A tank furnace.**

*Choice of furnace.* A tank furnace is used where the output expected is large, *e.g.*, for bottle-making or for sheet-glass. On the other hand, pot furnace is used where highest quality of glass is to be manufactured.

3. **Manufacture of glass (Melting of raw materials).** The raw materials are ground and mixed. The whole charge is again ground to get thorough mixing. This homogeneous mixture is known as **batch.** It is either transferred to the tank of a tank furnace or to be pots of the pot furnace. It is mixed with a small amount of broken glass known as **cullet.** The cullet melts at a low temperature and assists in melting the rest of the charge. Now the tank or pot is filled to the desired height by adding more charge.

The mass is now fused with the hot gases. In the early stages, considerable frothing occurs on account of formation of carbon dixoide. However, on continued heating the entire amount of carbon dioxide is driven out and a clear viscous fused mass is obtained. Completeness of the fusion is tested by taking out test samples on an iron hook from time to time ; when the sample seems to be transparent and free from bubbles of $CO_2$ and $SO_2$, the fusion is said to be complete and such fused mass is known **plain.** Heating is now stopped and the furnace is allowed to cool. Undecomposed sulphates and chlorides of calcium, alkali metals and aluminium along with other impurities rise to the top in the form of a scum called **glass ball** and can be skimmed off easily.

*Chemical reactions in the furnace.* The ordinary glass, also known as winding glass, approximately represented as $Na_2.O.CaO.6SiO_2$ is formed according to the following reactions:

(*i*) Sand fuses with sodium carbonate to form a glassy mass, known as *water glass* which is soluble in water

$$Na_2CO_3 + SiO_2 \longrightarrow Na_2SiO_3 + CO_2$$

(*ii*) Lime stone or calcium carbonate reacts with sand to give a glossy substance known as *calcium silicate* which is insoluble in water and soluble in acids.

$$CaCO_3 + SiO_2 \longrightarrow CaSiO_3 + CO_2$$

This is the case when either the lime stone or the sodium carbonate is used. When both the substances are present, sodium calcium silicate is produced which is insoluble in water and acids both.

$$Na_2CO_3 + CaCO_3 + 6SiO_2 \longrightarrow Na_2O.CaO.6SiO_2 + 2CO_2$$

Proportions of soda ash, lime stone and sand are 35 : 15 : 100.

Sodium carbonate may be replaced by a mixture of salt cake ($Na_2SO_4$) and charcoal. The chemical reactions between $Na_2SO_4$, C, $CaCO_3$ and $SiO_2$ to form glass is as below.

$$2Na_2SO_4 + 2SiO_2 + C \longrightarrow 2Na_2SiO_3 + CO_2 + 2SO_2$$

$$Na_2SiO_3 + CaCO_3 + 5SiO_2 \longrightarrow Na_2O.CaO.6SiO_2 + CO_2$$

4. **Shaping (Forming).** Molten glass is worked into articles of desired shape by either blowing or moulding.

5. **Annealing of glass.** Glass manufactured above, if cooled suddenly, being a bad conductor of heat, its superficial layer cools down with the result its interior portion remains in a state of strain. Owing to this unequal expansion, the glass is likely to crack to pieces. To avoid this, these must be cooled very slowly. This process of slow cooling is known as

**annealing.** Annealing is carried out by passing the glass articles through a long tunnel-like furnace which is very hot at the mouth-end and very cold at the tail-end. The longer the annealing period, the better the quality of glass.

6. **Finishing.** After annealing all the types of glasswares are subjected to finishing processes such as cleaning, grinding, polishing, cutting, blasting, etc.

### 4.3.2 Types of Glasses

Some common varieties of commercial glasses can be obtained by changing the composition of batch.

1. **Soda-lime** or **soft glass.** The raw materials are *silica* (sand), *calcium carbonate* and *soda ash.* Their approximate composition is $Na_2O.CaO.6SiO_2$. They are low in cost, resistant to devitrification and relatively resistant to water. They *melt easily* and hence, can be not-worked easily. Such glasses are, however, *attacked by common reagents like acids.*

   They are widely used as *window glasses, electric bulbs, plate-glasses, bottles, jars, building blocks, and cheaper tablewares,* where high temperature-resistance and chemical-stability are not required.

2. **Potash-lime** or **hard glass.** *It is obtained from silica (sand), calcium carbonate, and potassium carbonate.* Their approximate composition is $K_2O.CaO.6SiO_2$. They possess *high melting-point, fuse with difficulty* and are *less acted upon by acids, alkalis and other solvents than ordinary glasses.*

   These glasses (*costlier* than soda-lime glasses) are used for *chemical apparatus, combustion tubes,* etc., which are to be used for heating operations.

3. **Lead glass** or **flint glass** *is made by using lead oxide, instead of calcium oxide, for fusing together with silica.* For *dense optical glasses,* as much as 80% of PbO is incorporated. In addition, $K_2O$ is used, instead of sodium oxide. So, its approximate composition is $K_2O.PbO.6SiO_2$.

   Lead glass is a good deal *more expensive* to make than ordinary lime-soda glass, but much *easier to shape and work with.* Lead glass has a lower softening temperature than soda-glass and also a *higher refractive-index.* It has excellent electrical properties. It is bright, lustrous and possesses high specific gravity (3 to 3.3).

**Uses.** Lead glasses are used widely for *high-quality tablewares, optical purposes* (like lenses, etc.) ; *neon sign tubings, cathode-ray tubes, electrical insulators* and *in art objects,* because of their high lustre. High lead-content glasses are used for *extra-dense optical glasses* for windows and shields to protect personnel from X-rays and gamma-rays in medical and atomic energy fields respectively.

4. **Borosilicate glass or Pyrex glass or Jena glass.** It is a mixture of sodium aluminium borosilicates. It contains boron trioxide and is very rich in silica but poor in alkali and alumina. This can withstand alterations of temperature very well, *i.e.* its coefficient of expansion is very low. Borosilicate glasses have a very high softening point, excellent resistance to shock and chemicals.

   Borosilicates are widely used in industry for pipelines for corrosive liquids, for laboratary glasswares (such as flasks, beakers, etc.), kitchen wares, television tubes, etc.

**5. Vycor glass or 96% silica glass.** High silica glass has following adavantages :

(*i*) It can resist the sudden change of temperature and does not break even when plunged in water while red hot.

(*ii*) It resists the action of acids and alkali.

High silica glasses can't be prepared directly from silica because it is vary difficult to melt pure silica to convert into glass. Hence the following method is adopted for preparing high silica glass.

Borosilicate glass (prepared above) is moulded into the required shape, but just some what bigger than the final dimensions desired. The moulded articles are subjected to heat-treatment which causes the glass to separate out into two distinct layers. One of these layers is rich in silica (it is insoluble in acid solutions) ; while the other layer is mainly alkali oxides and boron oxide (it is readily soluble in hot acid solutions). The heat-treated article is now dipped in a 10% hydrochloric acid bath at 90°C which dissolves away the alkali and boron oxides layer, leaving behind a fine porous structure consisting of about 96% silica and 3% $B_2O_3$. The acid-treated articles are now washed to remove impurities and then annealed to 1200°C during which the article is dehydrated and the cellular structure of the body is converted into a non-porous vitreous glass. The articles after undergoing all these processes undergo a shrinkage of 14 to 20% of the original size.

96% silica glasses possess very low coefficient of expansion, high softening temperature and high chemical resistance (see Table 4.2). These are corroded only by hydrofluoric acid, hot phosphoric acid and concentrated alkali solutions.

***TABLE 4.2*** **Comparison of Properties of Some Glasses**

| *Property* | *Common glass* | *Pyrex glass* | *90% silica glass* | *Fused quartz* |
|---|---|---|---|---|
| Softening point | 696°C | 820°C | 1450°C | 1670°C |
| Annealing point | 510°C | 533°C | 931°C | 1140°C |
| Strain point | 475°C | 510°C | 860°C | 1070°C |
| Max. temp. for use | — | 500-50°C | 900-1000°C | 1000°C |
| Specific gravity | 2.47 | 2.23 | 2.18 | 2.21 |
| Coefficient of linear expansion | $92 \times 10^{-7}$ | $32$ to $33 \times 10^{-7}$ | $7.8$ to $8 \times 10^{-7}$ | $5.5$ to $5.95 \times 10^{-7}$ |

96% silica glasses are used for the construction of chemical plants, laboratory crucibles, electrical insulators, etc.

**6. Vitreosil glass or 99.5% silica glass.** It is produced by heating pure sand ($SiO_2$) to its melting point (about 1750°C). However, the glass so produced, can't be shaped because it has a very high viscosity at the working temperature. Its softening temperature (1650°C) is higher, while thermal expansion is lower than that of 96% silica glass.

Vitreosil glass is used for the purposes for which 96% silica glass is used. It is especially used in the pipelines used for carrying hot concentrated acids. However, because of its opaque nature it is generally mistaken as metal pipes and thus may be broken by mistake.

Vitreosil glass when heated for a prolonged time above its melting point, becomes transparent and named as clear silica glass. The clear silica glass has considerable light transmission properties and hence used in manufacturing wares for chemical plants, laboratory wares, electrical insulating materials in electrical instruments, etc.

7. **Safety glass.** *It is made by taking two or three flat sheets of glass and in between them* ***alternate thin layer of vinyl plastic*** *is introduced and the whole is subjected to slight pressure. It is then heated, till the glass layers and plastic layers merge into one another to give a sandwich.* On cooling, the glass becomes quite *tough*. When such a glass breaks, it *does not fly into pieces*, sicne the inner plastic layer tends to hold back the broken pieces of the glass. It is commonly used in automobile and aeroplane industries as a wind shield. A similar fine layered laminated glass in used is building construction.

8. **Optical glass.** Optical glass has certain predetermined optical characteristics so that it can be used in optical instruments. The important characteristics of optical glass are summarised below :

   (*i*) Its viscosity should be very low so that it becomes free from bubbles in its molten state.

   (*ii*) The raw materials should be free from impurities like iron so that the glass produced is colourless without adding any decolourising agent.

   (*iii*) It should not be devitrified even upon long annealing.

   (*iv*) It should be able to take up the desired polish on grinding and polishing.

   (*v*) Its surface should not be spoiled even on long exposure to the atmosphere.

Optical glass is prepared in pot type of furnace. The batch free from extraneous and undesired impurities and cullet are charged into the pot which is heated to about 1450°C. The molten charge is stirred with a hollow clay tube. The stirring removes all the entrappel gases and thus ensures the removal of all seeds and bubbles. When the glass becomes too viscous to be stirred further stirring is stopped and the tube is removed. The pot is now removed from the furnace and glass is cooled rapidly till the annealing temperature is obtained. The glass is covered with an insulating powder and the pot is covered with an insulating drum. The insulating drum and insulating powder are removed after 3 and 4 days respectively. Now the pot is broken to remove glass which is further broken into small pieces. The pieces not having desired properties (about 75%) are rejected.

**Crooke's glass** is a special variety of optical glass containing cerium oxide which cuts off U.V. rays harmful to eyes.

Optical glasses are relatively soft, posses low m.p., low durability and low chemical resistance than the ordinary glasses.

9. **Glass wool or Fibre glass.** Glass wool is a fibrous wool-like material composed of intermingled fine threads or filaments of glass. Glass filaments are obtained by forcing molten glass through small offices (or holes) of average diameter of 0.005 to 0.007 mm continuously. The thread of glass so obtained is thrown over a rapidly revolving drum. In this way glass is drawn in fine fibres like fibres of wool or silk.

**Properties:**

(*i*) It is non-combustible, fire-proof and heat-roof.

(*ii*) Its density and thermal conductivity are low.

(*iii*) Its tensile strength is about eight times more than steel.

(*iv*) It does not absorb moisture

(*v*) It resists action of most of chemicals.

Uses:

(*i*) Glass wool fibres can be woven into cloth and hence used in producing decorative textiles. Such fibres are fireproof. Glass wool fibres can be blended with cotton fibres.

(*ii*) Glass wool, being resistance to chemicals, is used for filtration of corrosive liquids like acids.

(*iii*) It is used for electrical and sound insulation.

**10. Laminated glass** *is made by pressing (or bonding) together two or more sheets/plates of glass with one or more alternating layer of bonding material like plastic resin, asphalt or synthetic rubber.* The laminated glass is : (*i*) *shatter-proof* (*i.e.*, its pieces do not fly off, when suddenly broken) ; (*ii*) *shock-proof* (*i.e.*, it can withstand sudden changes of temperature and even some pressure, without breaking).

***A bullet-resistant laminated glass*** *is obtained by pressing together several layers of glass with vinyl resins in alternate layers.* Ordinarily, thickness of such glass varies from 12.7 mm to 76.5 mm. Even thicker types are made for *specific uses.*

**Uses.** As an ideal material for use as *safety glass in aircrafts, automobiles, helicopters, submarines,* etc. *Bullet-resistant laminated glass* finds application in *making automobile wind-screens, looking windows,* etc.

## 4.4 CERAMICS

The term ceramic is derived from the older Greek word "*Keramos*" meaning **pottery** and has its root in older Sanskrit word meaning **"to burn"**. Therefore, *ceramics are inorganic, non-metallic materials that are processed and/or used at high temperatures.* Chemically, ceramics include silicates, metallic oxides and their combination. Ceramics can be grouped into three broad divisions namely clay products, glasses and refractories.

**Clay.** Clay is one of the disintegration product of rocks. The term clay denotes certain earths, which are highly plastic, when wet, and when heated to redness, they lose their plasticity and converted into a hard mass, which is unaffected by water. Clays are formed by the weathering of igneous and felspathic rocks by various agencies and are composed essentially of hydrated aluminium silicates ($Al_2Si_2O_7.2H_2O$), together with other substances like mica and quartz.

$$\underset{\text{Potash feldspar}}{2KAlSi_3O_8} + 2H_2O + CO_2 \longrightarrow K_2CO_3 + \underset{\text{clay}}{Al_2O_3.2SiO_2.2H_2O} + 4SiO_2$$

If the clay remains at the original location, it is **primary clay** usually white in colour with a low iron content, thus burns white and called **kaolin** or **china clay.** If the clay is transported (by water, glacier or wind) to another location, it is a *secondary clay* and contains lime stone powder, hydrated oxides, mud and organic impurities.

Clay products can be sub-divided into three types.

(*i*) *The structural clay products.* These contain iron oxides and used for bricks, tiles and similar products.

(*ii*) *The whitewares.* These are paler substances, examples are porcelain and china.

(*iii*) *Chemical stone wares.* These are specially treated to impart hardness and non-porosity.

**Plasticity of clay.** The ***plasticity of clay*** can be defined as *its ability to form a plastic mass (dough) with water. The mass can be easily moulded to any shape, but retains sufficient rigidity to prevent deformation on standing.* Dry clay is *not plastic,* and a certain amount of *water* has necessarily to be added to produce the required plasticity. *The function of water is to form a **film** around the flaky clay particles, so that their parallel orientation and movement under pressure is facilitated. The amount of water required to make a clay plastic, depends on the size and shape of the clay particles, their surface characteristics, and presence of electrolytes* (*salts*). For every clay, there is a certain minimum water-content, below which it ceases to behave as plastic material and becomes crumbly. This is called, the ***plastic-limit*** of a clay. *As the proportion of water increases, the clay becomes more and more plastic, until the point is reached at which it starts flowing and becomes wet and sticky.* This point, is called the ***liquid-limit.*** The difference in the water-content between the liquid limit and the plastic-limit, is called the ***plasticity index,*** which represents the *plasticity-range* of the clay.

**Plasticity of clay and its effect on moulding.** Depending on the methods employed, *different plasticities* os raw mixes are required to produce a product of desired size and properties. Thus :

(*i*) ***Hand-moulding and maching-extrusion*** require a mass of *plastic* consistency, classified as *'stiff mud'*.

(*ii*) ***Machine-pressing*** is best accomplished on *'stiff plastic'*, and *semi-dry mixes.*

(*iii*) ***Slip-casting*** requires a mix of *smi-liquid consistency,* so that it can be easily poured in a mould.

The *required plasticity* can be secured by :

(*i*) ***Weathering and grinding*** produces a *uniform clay-water paste,* in which clay particles are well dispersed and surrounded by water-films. This treatment *greatly improves the plasticity and homogeneity of the mass.*

(*ii*) ***Blending,*** *i.e., mixing two or more ingredients in such proportions that a mass of required composition and plasticity is obtained.* The usual ingredients are *clay, non-plastic materials* (like finely ground quartz or flint), and *fluxes.* This is particularly used in the manufacture of *high-grade products* such as whitewares, chemicalwares and porcelain. *Blending of clay with **non-plastic materials** reduces the proportion of water required in the clay to get mass of a **desirable plasticity,** thereby preventing an excessive drying shrinkage. **Fluxes** are substances added to a mix to **lower its fusion-point,** thereby making possible more complete vitrification, even at not too high temperature.*

**Whitewares or White-Pottery.** Whiteware is the general term used for ceramic products which are usually white in colour and consist of fine texture.

The raw materials used for the manufacture of whitewares contain *little or no compounds of iron.* Whitewares are made by mixing *China clay, feldspar* ($KAlSiO_8$) and *flint* ($SiO_2$). The proportions of these components are adjusted according to the properties desired in the finished article.

Whitewares possess *good strength, translucency* and *very low porosity.* This is attained by firing the products at high temperatures (*viz.,* 1,450–1,500°C), when *partial vitrification* of the ceramic constituents takes places.

***Whiteware products*** *consist of refractory body and glossy coating, called the* ***glaze.*** They are made by *two* main processes, namely:

(*i*) *Porcelain process,* in which the glaze and the body are developed in *one* firing process.

(*ii*) *China process,* in which the glaze is applied in a *separate* state. In consequence, the glaze of China objects is definitely *better* than that of porcelain objects.

**Manufacture of white-pottery** involves the following **steps :**

**Step I. The body of ware.** The ingredients, *kaolin* and *feldspar,* are reduced to a state of fine powder, and then suspended in water in the form of a *cream.* Surplus of water is removed by means of a *filter press.* Then, article from the paste is produced by either *moulding* (pressing by hand in a plaster mould) or *throwing* (shaping on a potter's wheel) or *casting* (pouring slip into a porous mould). Next, the article so-produced is dried slowly and fired to the desired density in a *biscuit-oven* to get porous-ware, called *"bisque".*

**Step II. The glaze.** In order to get *water-tight* article, the porous article is covered with a glaze by melting it over the surface of body. The constituents of the *glaze* (quartz, feldspar and a little boric oxide ; sometime, a little lead oxide) are finely ground and then mixed with water to form a *slip* or *slurry.* The bisque is dipped into the slip and then *fired again* at a high temperature, until the material composing the glaze fuses and produces a smooth and glossy surface. *The glaze must be so-chosen as to resist the reagents to which it is to be exposed, and must have the* ***same coefficient of expansion*** *as the body ; otherwise it will crackle, when the vessel is exposed to changes of temperature.*

**Step III. Decoration.** If the article is to be decorated, the *design* may either be painted upon the body before glazing, when it is said to be *'under-glazed',* or be painted upon the glazed-article and the article *fired again* so that the pigment melts into the glaze. In the latter case, it is said to be *'over-glazed'.* In the former case, the pigments used as a rule are *metallic oxides ;* while in the latter case are often *coloured glasses.*

**Uses of whitewares.** Whitewares find applications as spark-plugs, electrical insulators, laboratory equipments, crucibles, dishes. high-class potteries, etc.

## QUESTION BANK

1. What are cementing materials? Classify them.
2. Discuss various types of cements.
3. What are the different types of Portland cement ? Indicate their general composition.
4. Explain the term setting and hardening of cement. What are the compounds, which contribute to the hardening ?
5. Write a short note on decay of cement concrete and its prevention.
6. Give ISI specification of cement.
7. Explain the setting and hardening of Portland cement with chemical reactions involved in it.
8. What are the constituents of cement ? Discuss the mechanism of setting and hardening of cement.

9. Explain how microconstituents of Portland cement contribute towards the setting and hardening of cement.
10. Write short-notes on : (*i*) cement additives, (*ii*) raw materials for cement.
11. What are the different steps involved in the manufacture of cement by the "wet process" ? Discuss various reactions taking place in rotary klin.
12. Explain : (*i*) heat of hydration, (*ii*) soundness of cement.
13. Describe wet process of Portland cement manufacture.
14. Draw a labelled diagram of rotary kiln required for the manufacture of cement and mention the reactions in each zone.
15. Discuss the merits and demerits of dry and wet process for the manufacture of Portland cement.
16. Discuss the role of gypsum in Portland cement.
17. Write an exhaustive note on properties of cement.
18. (*i*) What are clinkers ? (*ii*) Give the characteristics and uses of : (*i*) low heat cement, (*ii*) blended cement.
19. Describe the processes of 'setting' and 'hardening' of cement concrete, giving the reactions involved in such processes. What is the difference, if any, between 'setting' and 'hardening' ?
20. Write what you know about hydraulic limes. What is the difference between hydraulic lime and Portland cement ?
21. Distinguish between concrete and RCC, including their uses.
22. Give compound composition of Portland cement.
23. Explain how quicklime and slacked lime are obtained. Give applications of lime.
24. What is white cement ?
25. Discuss the bonding action of cement and the role of retarders in this action. ?
26. Describe various specifications for Portland cement.
27. Explain vertical shaft kiln process of cement manufacture. How does it differ from rotary kiln method.
28. Write a note on special cements.
29. How lime is classified? Discuss the manufacture of lime.
30. What is gypsum ? Discuss its various forms and uses.
31. How plaster of Paris is obtained ? Discuss its uses.
32. What are refractories ? How are they classified ? What are the functions and characteristics of refractories ?
33. Differentiate between a refractory and an insulator with suitable examples.
34. How do porosity, spalling strength and thermal conductivity influence the nature and utility of a refractory ?
35. Write a note on silica refractories.
36. Write what you know about different types of refractories with special reference to their chemical composition and properties.
37. Illustrate with examples the criteria for the selection of a refractory.
38. What are the causes for failure of a refractory ?
39. Define the term 'refractory'. Mention the silica refractories in common use. How the chemical composition of refractory related to its use ?

**40.** What are the binding materials used in the manufacture of silica refractories ? What are their functions ?

**41.** What are refractories ? Explain the essential requirements of a good refractory. Name the refractories used in :

(*i*) blast furnace (*ii*) regenerative furnace

(*iii*) Bessemer converter.

**42.** (*a*) Describe the manufacture, properties and uses of either silica refractories or fire-clay refractories.

(*b*) Write a note on the spalling of refractories.

**43.** (*a*) Write a note on a 'fireclay' refractories.

(*b*) Write a short note on spalling of refractories.

**44.** Discuss the physical and chemical factors which affect the use of refractories.

(*b*) Write in short the manufacture of silica refractories.

**45.** (*a*) Give the composition, preparation, properties and uses of alumina bricks and silicate refractories.

(*b*) What is meant by thermal shock resistance ?

**46.** (*a*) Discuss the properties of refractories.

(*b*) Give the composition, preparation and uses of dolomite bricks.

**47.** (*a*) Discuss the following properties of refractories.

(*i*) Fusion point (*ii*) Spalling.

(*b*) Write short note on pyrometric cones.

**48.** (*a*) What are super refractories ? Why are they employed ?

(*b*) Present a critical account of the criteria of a good refractory material.

(*c*) Mention the possible causes for the failure of a refractory.

**49.** What is glass ? Give an account of its preparation, properties and uses. How does it differ from a refractory ?

**50.** What are the raw materials used in the manufacture of glass ? Describe the manufacture of glass using pot furnace.

**51.** What is the composition of an ordinary glass ? What will be the effect of increasing the percentage of silica in glass on the properties of glass ? Explain the regenerative principle of heat with the help of a neat diagram for the tank furnace. Explain some different types of glasses.

**52.** (*a*) Bring out the difference in the chemical composition of soft glass and borosilicate glass. How is colour imparted to glass ?

(*b*) Write short note on safety glass.

**53.** (*a*) What is glass and how is it prepared ? Mention its important varieties.

(b) What is the composition of pyrex glass ? How is glass annealed and why ?

**54.** (*a*) How is glass manufactured ? What are the important constituents of

(*i*) pyrex glass and (*ii*) flint glass ?

(*b*) What are fibre-glass and high silica glass and how are they prepared ?

(*c*) How is optical glass manufactured ?

**55.** Write notes on the following :

(*i*) Types of glass and their composition (*ii*) Soda glass

(*iii*) Safety glass (*iv*) Coloured glass

(*v*) Manufacture of glass (*v*) Glass.

56. What are ceramics and how are they classified ?

57. (*a*) Give the composition, properties and uses of flint glass.

(*b*) Discuss the manufacture of ordinary glass with the help of flow-sheet-diagram. Give the reactions involved.

58. (*a*) Give the composition, properties and uses of borosilicate glass.

(*b*) What are the functions of various ingredients used in the manufacture of glass ?

59. Classify various types of glass and write short notes on :

(*i*) crookes glass, (*ii*) pyrex glass,

(*iii*) flint glass, (*iv*) glass wool.

60. (*a*) Write short note on soda-lime glass.

(*b*) Write short note on raw materials in glass making.

61. Name the different ceramic products. Discuss the theories underlying the production of ceramic products. What is whiteware ?

62. (*a*) What is meant by whitewares ? Give their *two* uses. Explain vitrification.

(*b*) Describe the chemical conversion that takes place in the manufacture of ceramic products.

(*c*) Explain liquid glazing of ceramics.

63. Write note on polymorphism in ceramic materials.

# Chapter 5 EXPLOSIVES AND ABRASIVES

## 5.1 EXPLOSIVES

A material when subjected to sudden shock releases energy in an extremely brief period which is observed in the form of an explosion. An explosive *may be defined as a material, which under the influence of thermal or mechanical shock, decomposes rapidly and spontaneously with the evolution of a great amount of heat and large volume of gases*. At the temperature of the explosion, the volume of the evolved gases is perhaps. 15,000 – 20,000 times as that of the explosive itself. The first real explosive, called *gun powder* or *black powder* was discovered by *Roger Bacon*. It is an explosive mixture of sulphur, charcoal and saltpetre and its action as explosive is at the expense of the oxygen present in the saltpetre. Generally gun powder consists of about 15% charcoal, 10% sulphur and 75% saltpetre. Various physical and mechanical improvements have now been made in gun power into modern times. *Gun cotton* was discovered in 1846 by *C.F. Schonbein*. Cotton consists of cellulose, which is a compound of carbon, hydrogen and oxygen. When cellulose is allowed to react with a mixture of Conc. $H_2SO_4$ and Conc. $HNO_3$, various compounds of cellulose with nitric acid are formed, *the composition and properties of which depend on the strength of acid mixture and the temperature, as well as on the duration of the reaction*. Such compounds are called *nitrocellulose*. Wood fibre or Wood pulp and other vegetable fibres also contain cellulose and hence they can also be used as a source of nitrocellulose. A further advance in the chemistry of explosives was made by *Alfred Noble* with the discovery of nitroglycerine in 1862. This compound was found difficult to handle on account of its great sensitiveness to shock, and was the cause of many fatal accidents. It was found that the explosive could be transported and handled with great care without any danger, if liquid nitroglycerine was mixed with Kieselguhr. The mixture was then called *dynamite*. A most powerful explosive than dynamite was also obtained by mixing 8% nitro cotton or collodion cotton with 92% nitroglycerine. The resulting tough jelly like mass was called *blasting gelatin*. Another explosive, called *cordite* was prepared by mixing a paste of gun cotton (65%) with nitroglycerine (30%) and rest acetone and vaseline.

### 5.1.1 Classification of Explosives

Explosives are classified into following three main groups :

**1. Primary or initiating explosives.** These are also called *detonators* and are very sensitive and may be exploded by means of even a slight blow. They are very dangerous to handle, and used in comparatively small amounts to initiate the explosion of large quantities of less sensitive explosives. These are usually the inorganic salts and generally used in primers, detonators and percussion caps. Examples are lead azide, murcury fulminate diazodinitrophenol, nitromannite etc.

**2. Low Explosives.** These are also called as *propellents*. They only burn and do not explode, like high explosives. The action of low explosives is less shattering and they evolve large volumes of gas on combustion in a definite and controlled manner. They are mainly used in the cartridge cases of rifles, pistols etc. Examples of low explosives are colloidal cellulose nitrate (smokeless powder, nitrocotton, dinitrotoluene, black powder etc.)

**3. High explosives.** These explosives are quite insensitive to flame and mechanical shock. They explode violently by means of a detonation, which is a rapid chemical destruction progressing directly through the mass of the explosive. Examples of high explosives are T.N.T., picric acid, dinitramine etc. These are largely the organic materials and most of high explosives, when unconfined and unshocked, merely burn, if ignited. High explosives are generally grouped into four categories. These are :

(*a*) *Single compound explosives*, which contain only one chemical compound. Examples are ammonium nitrates, 2:4:6 trinitro toluene (TNT), pentaerythritol tetranitrate (PETN), RDX or cyclonite, tetryl and picric acid.

(*b*) *Binary Explosives*, which consist of mixture of TNT with other explosives. TNT is an important ingredient of these explosives because of the fact that its melting point is low (81°C). Examples of binary explosives and *amatol* (TNT + 50 – 80% Ammonium nitrate), *pentolite* (TNT + 50% PETN), *trtrytol* (30% TNT + 70% Tetryl), *Tropex* (40% TNT + 40% RDX + 20% Al powder) and *titronal* (80% TNT + 20% Al flakes).

(*c*) *Plastic Explosives*, which are combination of explosives in plastic state and can be hand moulded/press moulded and made into various shapes without any serious risk of detonation. They contain a high explosive like PETN and available in the form of flexible sheet explosives.

(*d*) *Dynamites*, which contain nitroglycerine (NG) as the principal ingredient. *Straight dynamites* contain about 15-16% NG in wood meal with $NaNO_3$ (oxidising agent for providing oxygen). These are used generally in blasting of hard rocks, coal and other minerals. *Blasting or gelatin dynamites* are those in which nitroglycerine has been partly gelatinised by nitro cotton or collodion cotton. An approximate composition of blasting gelatin is : NG 91.5% Nitrocotton = 8.0% and $CaCO_3$ = 0.5%.

### 5.1.2 Characteristics of Explosives

**1. Power of Explosives.** The explosive powers are determined comparatively by making use of some standard methods, such as Trauzl block test or sand blast test.

In Trauzl block test, a lead block measuring 200 mm. in height and 200 mm. in diameter, with a 125 mm. deep hole of 25 mm. diameter is taken. 10 gms of the explosive are exploded and the results are reported in terms of cubic centimetres of increase in volume, caused by the detonation of the explosive. It is indicative of the explosive power. In the sand blast test, 200 gms of ottawa sand with particles in a range of 20–30 mesh are taken in a container, which remains unaffected by explosion. The explosive is placed inside the sand and detonated. The sand is blasted into finer particles. The amount of sand which passes through 30 mesh sieve represents the power of explosion.

**2. Sensitivity of Explosives.** Sensitivity of an explosive to an impact, though measurable by several methods, does not give good results. Even by the same methods, the results are different. Reasonable results are, however, obtained if determinations are repeated 15–20 times. The sensitivity of an explosive to impact can be determined by dropping a standard weight of the striker from different heights on an anvil in order to detonate the explosive (5-10 mg) placed on the anvil. The height from which the striker falls and detonates the explosive determines the sensitivity. T.Urabankski and Sikorska (1958) determined the sensitivity of TNT by piston apparatus, which gave similar results at high temperatures.

**3. Heat of Explosion.** Ordinarily fuels give more heat on combustion than the explosives. In the case of explosives, the energy is liberated instantaneously.

**4. Rate of Detonation.** The rate of detonation accounts for the time during which the explosive reaction takes place. It is measured by burning the explosive at a constant volume in an apparatus, called *time pressure bomb*. High speed photography is used to determine the rate of detonation and the measurement is carried out by peezoelectric quartz or tourmaline crystal gauges by pressure change w.r.t. time. *The rate of detonation depends upon the density of the explosive, heat of explosion, size of explosive particle and the composition of the explosive.* In many cases the rate is increased with the increase in density of the explosive. Explosives with low heat of explosion have low detonation rates. The rate of detonation becomes irregular with an increase in the size of the explosive particle.

### 5.1.3 Requirements of Good Explosives

1. The rate of decomposition of explosive is expected to be fast so that a large volume of gaseous products is formed exothermically. As a result, effective increase in volume will be large because of heating effect. It has been found that one volume of an explosive gives about 10–15 volumes of hot gases at room temperature.

2. The explosive should be cheap and stable under normal conditions.

3. It must have atleast one chemical bond (such as N—N, N—O, N—Cl, O—Cl etc.) which can be easily ruptured or broken. In other words, an explosive molecule should possess low dissociation energy. Explosives generally contain bonds given above in which electronegativity difference between the two atoms is either very small or zero.

4. The explosive must possess a positive oxygen balance. The oxygen balance shows the oxygen contained in the explosive molecule that can be used to oxidise C to $CO_2$ and H to $H_2O$.

### 5.1.4 Primary Explosives

**1. Lead Azide [$PbN_6$].** Lead azide can be prepared by treating sodium azide (prepared from sodium amide and nitrous oxide) with lead acetate or lead nitrate.

$$NaNH_2 + N_2O \longrightarrow \underset{\text{Sodium azide}}{NaN_3} + H_2O$$

$$2NaN_3 + Pb(CH_3COO)_2 \longrightarrow \underset{\text{Lead azide}}{Pb(N_3)_2} + 2CH_3COONa$$

Lead azide has partially replaced mercury fulminate as an initiating or primary explosive for blasting caps. Mercury fulminate is somewhat less stable than desired and needs costly raw material Hg. Lead azide and mercury fulminate decompose exothermically on being struck in the following manner.

$$Pb(N_3)_2 \longrightarrow Pb + 3N_2 + 0.6 \text{ k. cals.}$$

$$Hg(ONC)_2 \longrightarrow Hg + N_2 + CO + 117 \text{ k. cals.}$$

Lead azide is very popular for military purposes due to its

(*i*) low cost,

(*ii*) excellent initiating action, and

(*iii*) stability in storage.

Lead azide reacts with brass, so the caps loaded with it are made of aluminium.

**2. Mercury fulminate [$Hg(CNO)_2$].** It is prepared by dissolving mercury in excess of nitric acid and subsequently adding the solution in ethyl alcohol. The solution starts boiling spontaneously and mercury fulminate is precipitated. It is more sensitive and more expensive than lead azide. However, it is slightly toxic, consequently, it is used to a lesser extent than lead azide.

**3. Tetracene.** It can be prepared by first reducing nitro guanidine to aminoguanidine. The latter produces nitroso aminoguanidine on treatment with sodium nitrite and an acid. Some of the nitroso aminoguanidine loses water molecule and forms guanyl azide, which combines with remaining nitroso aminoguanidine to form tetracene.

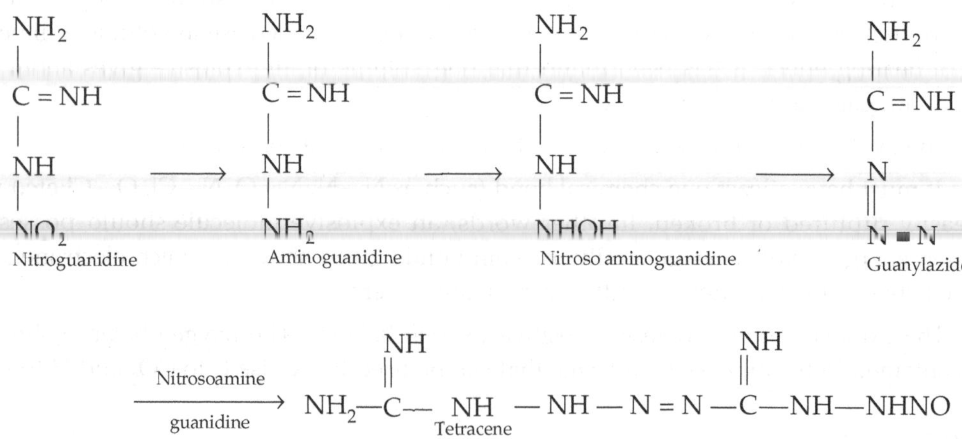


Tetracene is used as a substitute of lead azide or mercury fulminate. It ignites easily, has light heat of explosion and produces a large volume of gases. However, it is not used by military and is mainly used as detonator.

**4. Dinol.** Dinol is diazodinitrophenol or 4, 6-dinitrobenzene. 1, 2 diazoxide. It was first diazo compound ever discovered by Griess (1858). It is generally used as a primary explosive and prepared by diazotizing picramic acid.

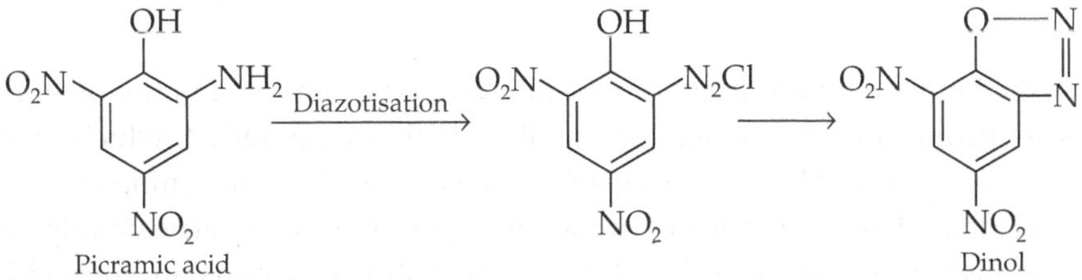


The dinol is formed as a result of internal coupling. It is widely used in commercial blasting caps.

### 5.1.5 Low Explosives

**1. Nitrocellulose (or smokeless powder).** It is prepared from *cotton linters* or the short fibres which are left after the cotton used textile purposes, had been removed from the cotton seed. Cotton linters, after being freed from dust are purified by boiling in Kiers (vats) with a dilute solution of caustic soda, in order to remove natural oil and non-cellulosic matter and then bleached with $CaOCl_2$ or NaOCl. The bleached linters are dried and weighed. The bleached and dried cotton linters are then treated with the mixed acid having 21% $HNO_3$, 63% $H_2SO_4$, 0.5% $N_2O_4$, and 5.5% water under carefully controlled conditions for about 2–2.5 hours in a mechanical dipper nitrator or shallow earthenware pans. One nitrator charge is formed by 32 lbs of purified and dried cotton linters. The linters are agitated with about 1500 lbs of mixed acid at 30°C. After nitration, the entire charge is dropped into a centrifuge, where the spent acid is centrifuged from the nitrated cellulose. The waste or spent acid is partly sold or otherwise disposed off. The nitrated cotton is then throughly washed and boiled with water to render it more stable and again washed in a beater.

In order to produce smokeless powder, more stable on storage, it is necessary to destroy unstable sulphate esters and to remove free acid completely.

$$\underset{\text{Cellulose}}{C_6H_7O_2(OH)_3} + 3HONO_2 + H_2SO_4 \longrightarrow \underset{\text{Nitrocellulose}}{C_6H_7O_2(ONO)_3} + 3H_2O + H_2SO_4$$

*If the three hydroxyl groups in the cellulose are esterified, the cellulose nitrate should contain 14–16 percent of nitrogen, but actually the cellulose nitrate contains only 12.6 percent of nitrogen.* Nitration is facilitated by the presence of sulphuric acid, which contracts the fibre. Complete nitration is not possible as the cellulose fibre swells and makes the diffusion of nitric acid difficult for nitration. Thus only 96–99% of the hydroxy groups are nitrated. During nitration in presence of sulphuric acid, some sulphate esters are also formed. *These esters render the explosive unstable with great explosion risks. Hence sulphate esters are removed completely before storing.*

The nitrated cotton is washed several times with water after about 40 hours of boiling and then fibre is made in the form of a pulp by means of a beater. The washed nitrated cotton is then treated with $Na_2CO_3$ solution to remove sulphate esters. 5 lbs. of $Na_2CO_3$ per tonne of the cellulose are required. The cotton is again washed several times with boiling water. *The process of making nitrated cotton free from sulphate esters is called poaching.* The poached nitrocellulose is centrifuged to remove most of its water. The water contents are further reduced by alocohol percolation under pressure dehydration.

The finished nitrocellulose should not be allowed to become acid in use or storage, because this catalyses its further decomposition. A stabilizer is therefore added, which reacts with any traces of nitrous, nitric, or sulphuric acid. A stabilizer, such as diphenyl amine or diphenyl urea is added to the finished nitrocellulose to check the formation of nitrous, nitric or sulphuric acid that may be released because of the decomposition of nitrocellulose. Grains are formed by extrusion through dies and these are dried and blended to form smokeless powder. For smokeless powder diphenyl amine or diphenyl urea is used as stabilizer and for celluloid, urea is used as stabilizer. The diphenyl amine forms a series of innocuous compounds with the evolved gases. Examples of such compounds are

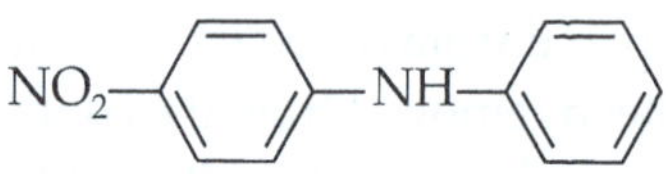


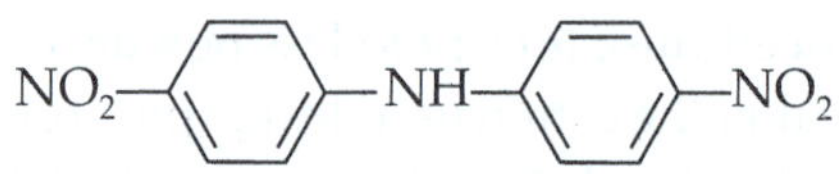


**Smokeless powder** is colloidal nitrocellulose containing about 1% diphenyl amine to improve its storage life and a small amount of plasticizer such a dibutyl phthalate. *The nitrocellulose formed in this manner contains about 12.6% nitrogen and is called pyrocotton.* The nitrogen content may be increased to 13.6% by using a stronger acid. Cotton nitrated to contain 13.2% or more nitrogen is known as *gun cotton*. The nitrocellulose may be colloidized by mixing with alcohol, ether and diphenyl amine. Modern military smokeless powders are prepared from a blend of pyrocotton and gun cotton and contain about 13.15% nitrogen.

**2. Black Powder (or gun powder).** It is a mixture of 75% $KNO_3$, 15% charcoal and 10% sulphur. It contains sufficient oxygen to combine with the other elements present, when it is ignited. The decomposition reaction is

$$10\,KNO_3 + 3S + 8C \longrightarrow 3K_2SO_4 + 2K_2CO_3 + 6CO_2\uparrow + 5N_2\uparrow$$

The excess of carton and sulphur take part in the slower reduction processes, leading to the evolution of more gases.

$$4K_2CO_3 + 7S \longrightarrow K_2SO_4 + 3K_2S_2 + 4CO_2\uparrow$$

$$4K_2SO_4 + 7C \longrightarrow 2K_2CO_3 + 2K_2S_2 + 5CO_2\uparrow$$

large volume of gaseous products account for its explosive effect.

In the manufacture of black powder, the raw materials are finely ground, mixed together under a wheel mill and then passed through a fine mesh sieve of copper or brass wire. Thorough mixing can be achieved by grounding the mixture in a moist state in a mill under heavy edge runners. The cake so obtained is broken up into small pieces and then subjected to a pressure of about 400 psi. As a result, a hard mass is obtained, which is granulated by passing between rollers.

The resulting grains are polished or glazed in rotating wooden drums into which a small amount of graphite is sometimes also added. The gun powder or black powder thus obtained in finally dired at about 40°C.

It is an excellent and cheap explosive for blasting down of coal in mines. Its other uses include for blasting in shells, igniters and primer assemblies for propellants, practice bombs, and saluting changes.

### 5.1.6 High Explosives

These fall in following four categories.

**[A] Single Compound Explosives**

**1. Ammonium Nitrate.** It is very stable, non-toxic, cheap and has low brisance value. It is about half as powerful as TNT and is mainly employed in making binary explosives.

**2. Trinitrotoluene (TNT).** 2, 4, 6-trinitro toluene, commonly known as TNT has become the principal military explosive since 1920. It is most widely used explosive because of its economic and safe preparation, low sensitivity to impact, high explosive power, comparatively low toxic effects and its low melting point (80.8°C). It is manufactured by three stage nitration of toluene with mixed acids. The three stage nitrogen gives better yields and is more safe than one stage nitration. *In one stage nitration, large volume of strong mixed acids at high temperature would be required which will lead to the loss of TNT because the later is soluble in acids and undergoes partial oxidation in presence of acids. Moreover, in three stage nitration process, waste acid from the final stage may be recycled to be used in the initital stage.*

The crude TNT or trioil is then washed with water and soda ash to neutralise the excess of acid and then with acidulated 16% $Na_2SO_3$ to remove the unsymmetrical TNT (2, 3, 4, ; 2, 3, 6 and 2, 4, 5, trinitro toluenes) as water soluble sodium dinitro sulphonates. The decomposition of TNT gives the following gases.

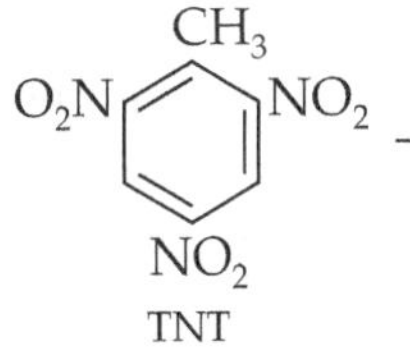


TNT

$$\longrightarrow 3.5\,CO + 3.5\,C + 2.5\,H_2O + 1.5\,N_2 + 190\text{ K cal}$$

The TNT is superior to picric acid because of its low melting point and the fact that it has no tendency to form sensitive salts with metals. It has two major disadvantages (reactivity with alkali and toxicity towards humans). These disadvantages have largely been removed by using proper technical methods.

**3. Picric Acid.** 2, 4, 6-trinitrophenol or picric acid can not be prepared directly by the nitration of phenol, because of the formation of various oxidation by products. It is therefore manufactured by the nitration of phenol-2, 4-disulphonic acid as follows.

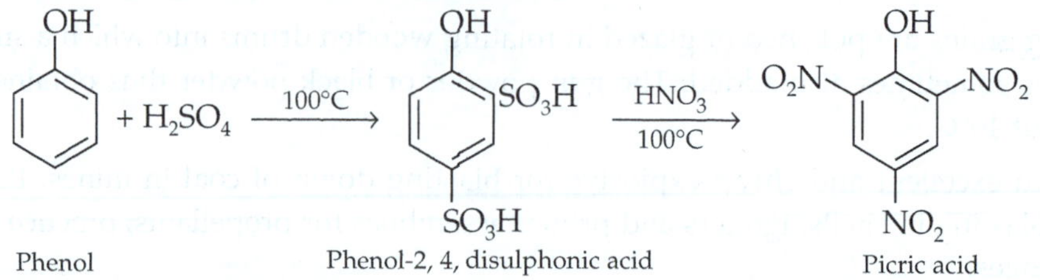


Picric acid can also be prepared from chlorobenzene. The latter is not easily oxidised by nitric acid. When it is nitrated with conc. $HNO_3$ and conc $H_2SO_4$, it form 2, 4-dinitrochlorobenzene. The resulting dinitrochlorobenzene is then boiled with aqueous sodium carbonate solution, as a result of which active chlorine atom is replaced by hydroxy group and 2, 4-dinitrophenol is obtained. The latter is again nitrated with conc. $HNO_3$ and conc. $H_2SO_4$ to picric acid.

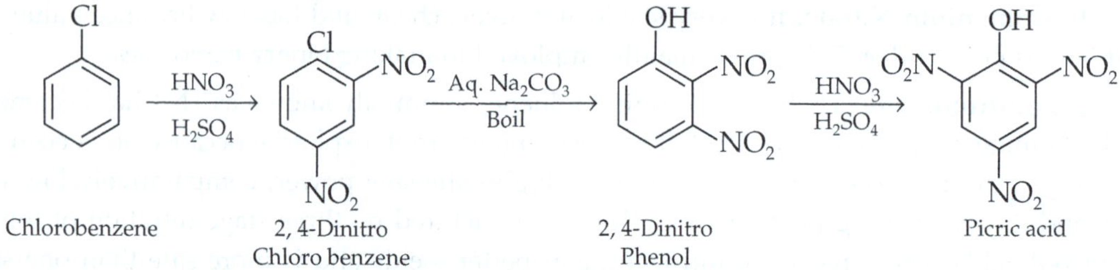


Picric acid is a high explosive, exploding violently on detonation. TNT, however, has replaced picric acid as an explosive.

**4. Cyclonite (or RDX).** RDX or cyclonite is sym-trimethylene trinitramine $[(CH_2)_3N_3(NO_2)_3]$. It is one of the most powerful explosives known at present time. It can be prepared by the following methods.

(*a*) **Hale Method**. A British chemist G.C. Hale (1925) prepared RDX by destructive nitration of hexamethylene tetramine with conc. $HNO_3$.

$$\underset{\text{Hexamethylene tetramine}}{(CH_2)_6N_4 \text{ or } C_6H_{12}N_4} + 3HNO_3 \longrightarrow \underset{\text{RDX or Cyclonite}}{C_3H_6O_6N_6} + 3HCHO + NH_3$$

(*b*) **Ebele-Schlessler-Ross Method.** This method consists in addition of paraformaldehyde and ammonium nitrate to acetic anhydride as 70°C.

$$3HCHO + 3NH_4NO_3 + 6(CH_3CO)_2O \longrightarrow \underset{\text{RDX or Cyclonite (60\% yield)}}{[(CH_2)_3N_3(NO_2)_3] \text{ or } C_3H_6O_6N_6} + 12CH_3COOH$$

The process takes place in two stages.

(*i*) $6HCHO + 4NH_4NO_3 + (CH_3CO)_2O \longrightarrow (CH_2)_6N_4 + 4HNO_3 + 6CH_3COOH$

(*ii*) $\underset{\text{Hexamethylene tetraime}}{(CH_2)_6N_4} + 4HNO_3 \longrightarrow \underset{\text{RDX}}{(CH_2N.NO_2)_3} \text{ or } C_3H_6O_6N_6 + NH_4NO_3 + 3HCHO$

The formaldehyde and ammonium nitrate from the step (*ii*) are further utilized alongwith more ammonium nitrate and acetic anhydride and the process is thus repeated. But the yield of cyclonite is about 60%, because some HCHO gets converted into unreactive methylene and polymethylene diacetates.

(*c*) **Bachmann Method.** A combination process was developed by *Bachmann (1949)*, who utilized the by product to get a second mole of RDX. This is the combination of above two methods.

$$(CH_2)_6N_4 + 2NH_4NO_3 + 4HNO_3 + 6(CH_3CO)_2O \longrightarrow \underset{\text{Cyclonite (80\% yield)}}{2C_3H_6O_6N_6} \text{ or } 2(CH_2N.NO_2)_3 + 12CH_3COOH$$

(*d*) **Wolfram Method.** This method was developed in Germany and involves the following steps:

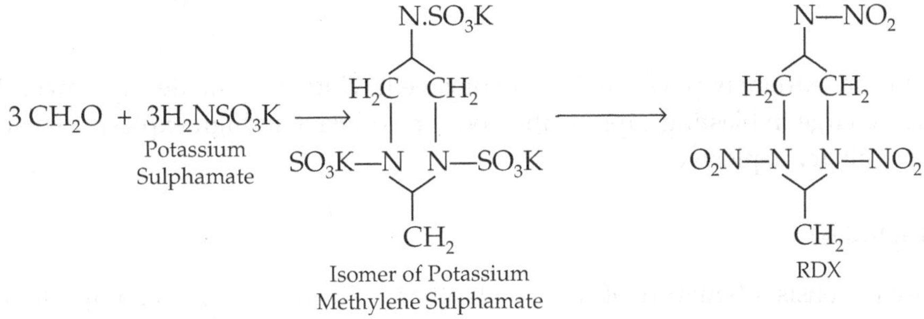


RDX is used in a mixture with TNT and aluminium, known as *Torpex*, for mines, depth charges and torpedo warheads. It is also used as an ingredient for explosive shells and bombs and it is desensitized by wax or oily material.

**5. PETN.** Pentaerythritol tetra nitrate $[C(CH_2ONO_2)_4]$ is one of the most brisant and sensitive military high explosive. For use as a booster explosive, a bursting charge or a plastic demolition explosive, it is desensitized by admixture with TNT (the mixture is called pentrolite) or by the addition of wax. PETN is prepared by the nitration of pentaerythritol with strong 96% $HNO_3$ at about 50°C.

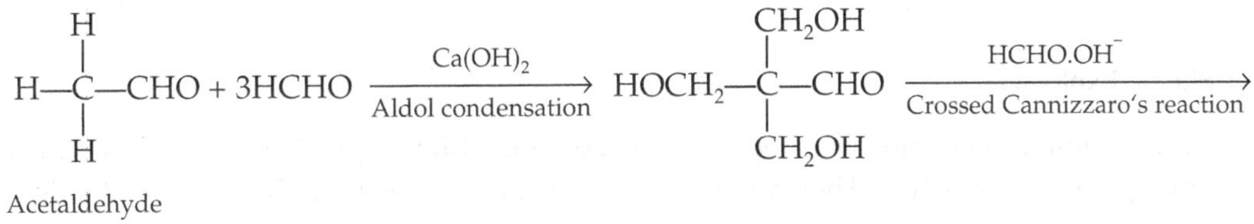


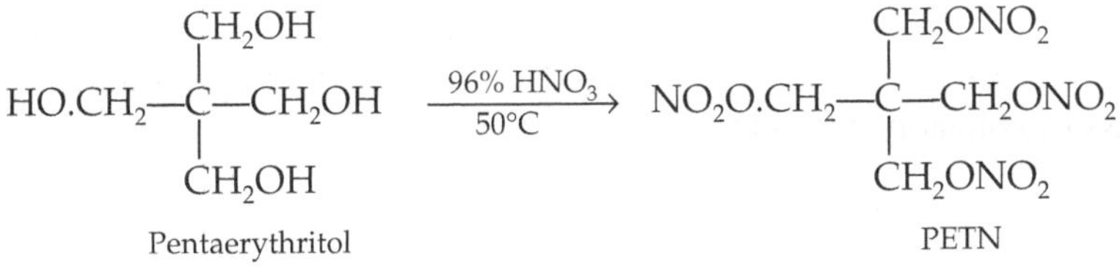


The decomposition of PETN probably takes place in the following manner.

$$C(CH_2ONO_2)_4 \longrightarrow 4H_2O + 3CO_2 + 2CO + 2N_2$$

The high sensitivity of PETN allows it to be detonated by the impact of a rifle bullet.

**6. Tetryl.** Tetryl (2, 4, 6-trinitrophenyl) methylnitramine, $[C_6H_2(NO_2)_3NCH_3NO_2]$ is generally prepared by the action of mixed sulphuric acid and nitric acid on dimethyl aniline in a multistage nitration.

$$C_6H_5N(CH_3)_2 + 2HNO_3 \xrightarrow{H_2SO_4} C_6H_5N(CH_3)_2.NO_2NO_2 + 2H_2O$$

Tetryl can also be prepared by the alkylation of 2.4 dinitrochlorobenzene with methyl amine and then nitrating the product with conc $H_2SO_4$ and conc. $HNO_3$.

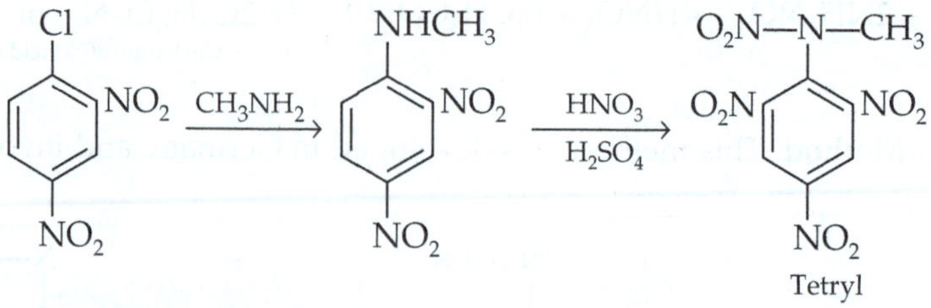

Tetryl is an extremely powerful high explosive having great shattering power. It is chiefly used as a base charge in blasting caps, as the booster explosive in high explosive shells and as an ingredient of a binary explosive.

### [B] Binary Explosives

These explosives consist of mixture of, TNT with other explosives. TNT is an important ingredient in binary explosives, because it has low melting point. Important binary explosives include :

**1. Amatal.** It is an explosive mixture of TNT with ammonium nitrate (50 to 80%). It is white smoke giving explosive with strength equal to TNT. It is hygroscopic and is used as blasting charges in large caliber shells.

**2. Pentolite.** It is a mixture of TNT and PETN, 50% each.

**3. Tetrytol.** It is a mixture of 70% tetryl and 30% TNT.

**4. Troplex.** It is a mixture of 40% RDX, 40% TNT, and 20% Al powder.

**5. Titronal.** It consist of 80% TNT, and 20% Al flakes.

### [C] Plastic Explosives

They are combination explosives because they contain a high explosive admixed with wax or oil, to render them less sensitive. They are in plastic state and can be moulded or press loaded into the desired shape without any serious risk. They are used for industrial or military purposes and are available as flexible sheet explosives. Usually, RDX or PETN is the high explosive involved. These sheets can be cut into required shapes and can be easily fixed to a metal part by an adhesive. Special caps are needed to trigger explosions in them.

### [D] Dynamites

These are the explosives containing nitroglycerine (NG) as the principal ingredient. Nitroglycerine $C_3H_5(ONO_2)_3$ is an ester, not a nitro compound and should be called as ***glycerol trinitrate***. It was the first high explosive to be employed on large scale. It was manufactured by slowly adding very pure glycerine to mixed acid having 59.5% $H_2SO_4$. 40% $NHO_3$ and 0.5% water at 10–25°C. The nitration is carried out in agitated nitrators, fitted with steel cooling coils carrying brine at 5°C to maintain the temperature at about 10°C. The nitration is complete within one to one and a half hours. After nitration, the mixture of nitroglycerine and spent acid is allowed to flow through a trough into setting tanks, where nitroglyceine forms the upper yellow layer being lighter and is separated from the spent acid or nitrating mixture. The yellow layer of nitroglycerine is washed with cold water in wash tank and then with dilute sodium carbonate solution. It is again washed with warm water to remove even traces of alkali. It should be noted that after nitration of nitroglycerine, spent acid is allowed to flow a into separating and settling tanks, at some distance from the nitrator and nitroglycerine is carefully separated from the acid and sent to wash tank, where it is washed twice with warm water and with 2% $NaHCO_3$ solution to ensure complete removal of any remaining acid. Additional washes with warm water are also necessary for the removal of last traces of alkalinity that remains.

Nitroglycerine is a poisonous, heavy oil (Sp. gravity 1.592) and is highly sensitive to shock. It explodes very violently on detonation.

$$\begin{array}{l} CH_2OH \\ | \\ CHOH \\ | \\ CH_2OH \end{array} + 3HNO_3 \xrightarrow{H_2SO_4} \underset{\text{Nitroglycerine}}{\begin{array}{l} CH_2ONO_2 \\ | \\ CHONO_2 \\ | \\ CH_2ONO_2 \end{array}} + 3H_2O + (H_2SO_4)$$

Nitroglycerine is quite sensitive to blow and freezes at 56°F. It is less sensitive when solid, but it has a strong tendency to explode incompletely in this state. To make nitroglycerine easier and safe to handle, it is usually manufactured into dynamite.

***Original dynamite*** was a mixture of 75% nitroglycerine and 25% kieselguhr, a porous, silicious earth. ***Modern dynamites*** use wood flour, ammonium nitrate or sodium nitrate as the absorbing agent for nitroglycerine. An oxidiser is also added sometimes. Dynamite is easy to handle. It is a high explosive, used for rock blasting and gold mining. Because of the demand for a non-freezing dynamite for work in cold weather, several dynamites having other materials have also been designed to lower the freezing point of nitroglycerine. For example, *ethylene glycol dinitrate* **(DEGN)** was developed in Germany as a replacement of nitroglycerine in double base powders. Diethyl glycol is nitrated to the dinitrate in practically the same manner as glycerine nitration and such non-freezing dynamites have potentials almost as great as *straight dynamites*. A jelly like mixture of 7–8 percent nitrocellulose and 82–93 percent nitroglycerine is an exceptionally powerful high explosive, commonly known as *gelatin dynamite* or *blasting gelatin*. It can also be handled safely. Almost all nitro compounds and nitric acid esters used as explosives are toxic.

***Cordite.*** Cordite is prepared by mixing a paste of nitrocellulose (65%) and nitroglycerine (30%) with acetone, and adding a quantity of vaseline (5%). The mixture is then forced by hydraulic pressure through a die in the form of a thread or cord. Hence the name cordite. The latter forms a horn like material after the removal of acetone. This material is very sensitive to shock and easy to handle. The acetone may be recovered by making use of activated characoal, which possesses the property of adsorbing different gases and vapors on its surface.

### 5.1.7 Other Explosives

1. **EDNA.** EDNA or haleite is ethylenedinitramine, $[CH_2NHNO_2)_2]$, was first developed as a military explosive. It was later on also found to be useful as bursting charge explosive and as an ingredient of the binary explosive *ednatol*. It can be economically prepared by nitrating ethylene urea (2-imidazolidine), either with conc. $HNO_3$ or a mixture of nitric acid and conc. sulphuric acid to the dinitro compounds, which is hydrolysed to form EDNA or haleite. The latter is separated as shining crystals. It is more powerful than TNT, but not as stable as RDX.

2. **HMX.** HMX is 1, 3, 5, 7-tetranitro-1,3,5,7 tetra-azacyclo octane and always associated with the RDX preparation. However, it can be prepared by the nitrolysis of hexamine tetranitrate.

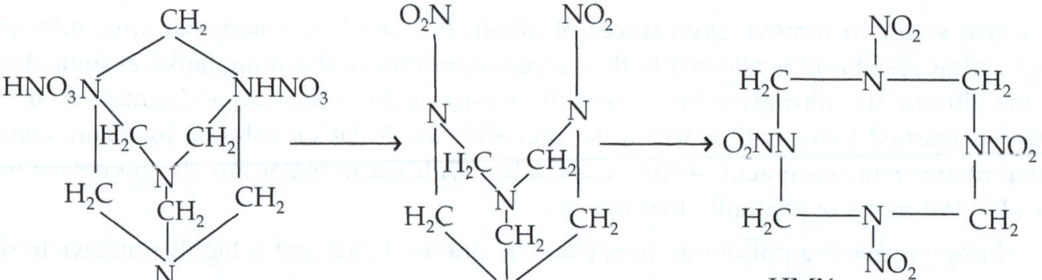

HMX

HMX has the same explosive power as cyclonite. Wax or oily material desensitised RDX and HMX are used as ingredients of explosive for shells and bombs. It should be noted that RDX (cyclonite), HMX, Tetryl, Pentryl, Hexyl, EDNA (haleite) etc are classified as *nitramine explosives.*

3. **Pentryl.** Pentryl is sym-trinitrophenylnitraminomethyl nitrate and can be prepared by the following methods.

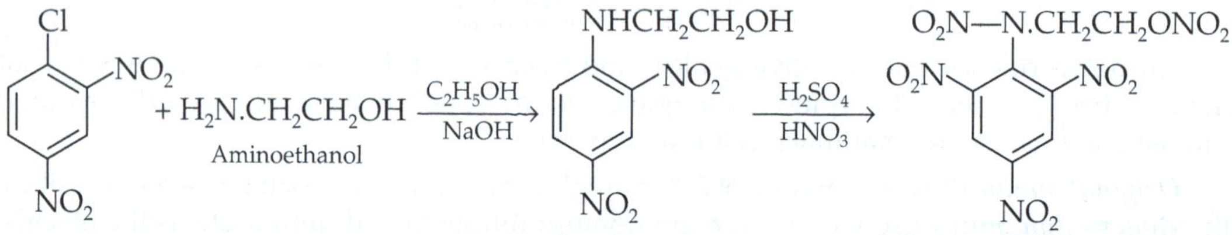

2, 4–Dinitrochloro benzene; Aminoethanol; 2, 4–Dinitrophenol aminoethane; Pentryl

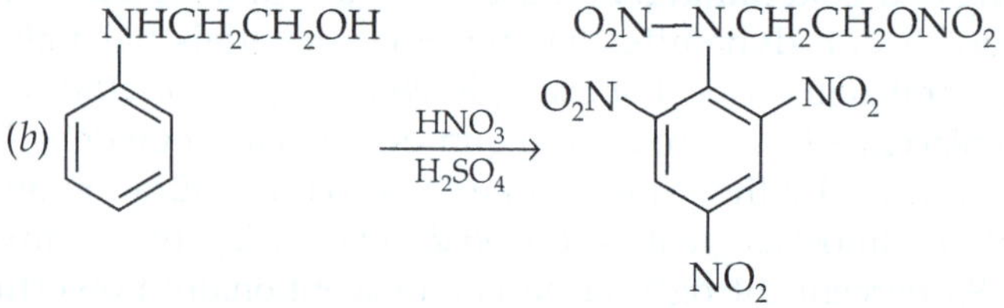

Pentryl

Pentryl is less sensitive than tetryl.

4. **Hexyl.** Hexyl is hexanitrodiphenylamine and can prepared in the following manner.

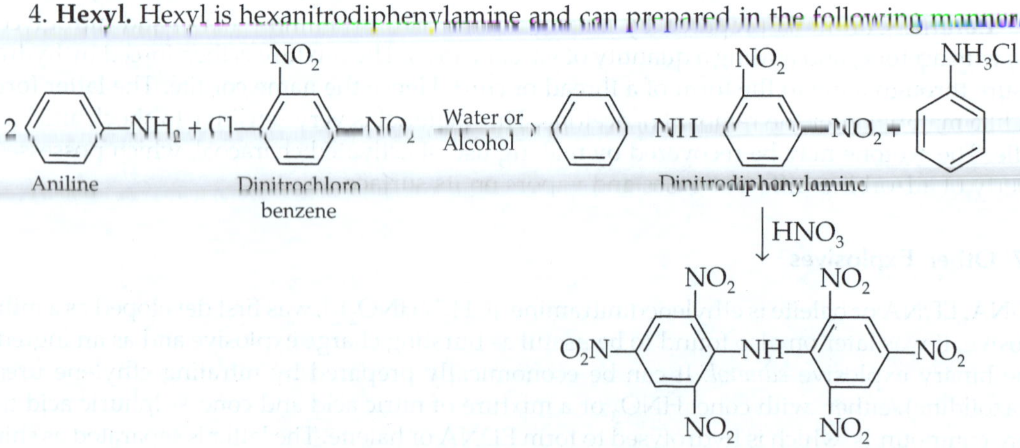

Aniline; Dinitrochloro benzene; Dinitrodiphenylamine; Hexyl

Hexyl is generally used with other explosives as a filler for bombs, mines and warheads for torpedoes.

5. ***Dinitro Benzene (DNB).*** Dinitro benzene is prepared by the nitration of benzene with the nitrating mixture consisting of 40 parts $HNO_3$ and 60 parts of $H_2SO_4$ at about 100°C. A mixture of all the three isomeric dinitrobenzene is thus obtained, but the m-dinitrobenzene predominates in the mixture.

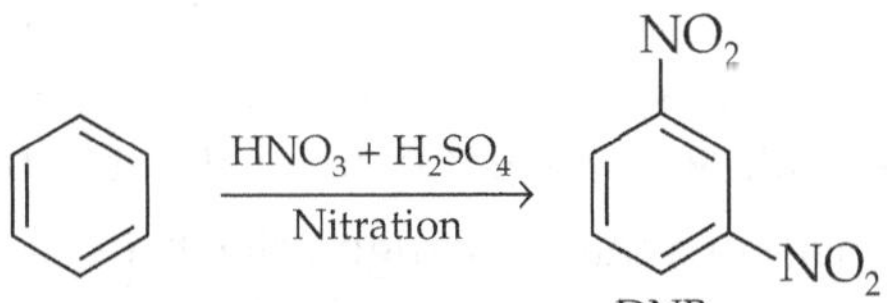

DNB

It is used as a **blasting explosive,** mixed with chlorates. The commercial explosives containing dinitrobenzene are *belite No. 1* (dinitrobenzene = 15–18%, Ammonium nitrate = 85–82%), *tonite No. 3* (Dinitrobenzene 13%, Guncotton 19% and Barium nitrate 68%) etc.

**6. Trinitrobenzene (TNB).** TNB exists in three isomeric forms, but the sym-trinitrobenzene (1,3,5-trinitrobenzene) is most stable and more powerful explosive. It is prepared by drastic nitration of benzene with conc. $HNO_3$ and conc $H_2SO_4$ (mixed acids).

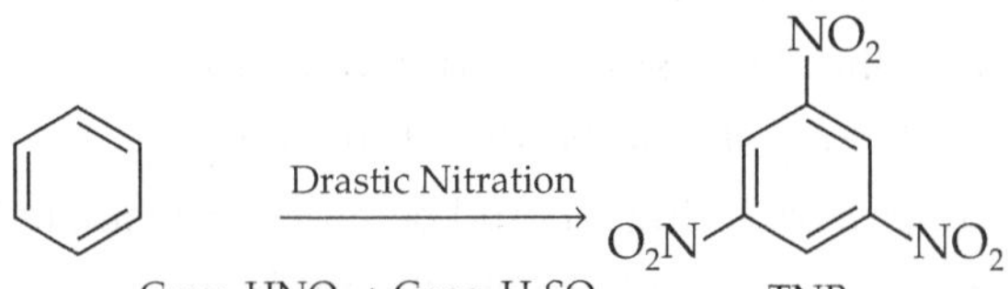

TNB

TNB is more powerful than TNT and proposed as a substitute for picric acid. Its use is, however restricted because of its high melting point and difficulty in its preparation.

Characteristics of some explosives is shown in Table 5.1

**TABLE 5.1 Characteristics of some explosives**

| *Name* | *Formula* | *Products Per Formula Weight* | *Heat of Explosion at Constant Volume, cal/kg* | *Explosion Temp-erature, 0°C* | *Potential × 10⁵ kg.m* |
|---|---|---|---|---|---|
| Gun Powder | $2KNO_3 + 3C + S$ | $N_2 + 3CO_2 + K_2S$ | 501 | 2090 | 2.1 |
| Nitro Cellulose | $C_{24}H_{29}O_9(NO_3)_{11}$ | $20.5CO + 3.5CO_2 +$ $14.5H_2O + 5.5N_2$ | 1250 | 2800 | 5.3 |
| Nitroglycerine | $C_3H_5(NO_3)_3$ | $3CO_2 + 2.5H_2O + 1.5N_2$ | 1526 | 3360 | 6.5 |
| Ammonium Nitrate | $NH_4NO_3$ | $2H_2O + N_2 + 0.5O_2$ | 384 | 1100 | 1.6 |
| TNT | $C_7H_5(NO_2)_3$ | $6CO + CO + 2.5H_2 + 1.5N_2$ | 656 | 2200 | 2.8 |
| Picric acid | $C_6H_2OH(NO_2)_3$ | $6CO + H_2O + 0.5H_2 + 1.5N_2$ | 847 | 2717 | 3.6 |
| Ammonium Picrate | $C_6H_2(NO_2)_3ONH_4$ | $6CO + H_2O + 2H_2 + 2N_2$ | 622 | 1979 | 2.6 |
| Tetryl | $C_7H_5N_5O_8$ | $7CO + H_2O + 1.5\ H_2 + 2.5N_2$ | 908 | 2781 | 3.9 |
| Mercury Fulminate | $Hg(ONC)_2$ | $Hg + 2CO + N_2$ | 420 | 4105 | 1.8 |
| Lead azide | $Pb(N_3)_2$ or $PbN_6$ | $Pb + 3N_2$ | 684 | 3180 | 2.9 |

### 5.1.8 Blasting Fuses

A fuse may be defined as, a thin water-proof canvas length of tube containing gun powder or TNT arranged to burn at a given speed, for setting off charges of explosive. These are of two types :

**1. Safety Fuse.** It consist of a small diameter core of black powder enclosed in water-proof fabrics. It has an approximate burning speed of 30–40 seconds per foot while blasting, a sufficient length of the fuse is used so as to allow sufficient time for the shot firer to reach a point of safety.

**2. Detonating Fuse.** It consist of a charge of high velocity explosives (6000 meters per second), *e.g.*, TNT contained in a small-diameter bent tube. The line of fuse is in contact with the charge throughout its length and this is responsible for instantaneous detonation. These are mainly used for exploding charges of explosives in deep holes.

### 5.1.9 Safety Measures in Storage of Explosives

Proper storage of explosives is a big challenge, following safety measures must be taken into consideration while storing the explosives.

(*i*) Different explosives should be stored in different boxes.

(*ii*) Detonators and explosives should be stored separately.

(*iii*) All the electric fittings and wirings should be properly insulated and frequently checked.

(*iv*) Smoke or firing of any sort should be strictly prohibited within the radius of 50 metres from the explosive store.

(*v*) Only authorized persons should be allowed to the explosive store.

(*vi*) The boundary of explosive stove should be protected, by a high barbed wire fencing and proper 'Caution Boards' should be installed around it.

(*vii*) Lightning conductors should be provided to safeguard the magazine.

### 5.1.10 Rocket Propellants

A number of satellites have been launched by different countries including India for remote sensing and communication purposes. India launched its first successful satellite launch vehicle SLV-3 from Sriharikota. In order to push these rocket satellites into the space the lift off these space vehicles is achieved by having a propulsion system. It comprises of rocket engines that are powered with chemicals called *propellants*. The propellant consists of a fuel and an oxidizer. When propellant is iginited, the combustion of fuel occurs with the release of gases which are passed through the nozzle of the rocket motor. This provides the necessary thrust for the rocket uplift.

The main force responsible for the push *i.e.*, propulsion or thrust ($w$) is proportional to the rate ($v$) at which the fuel is burnt and also the velocity of the exhaust gases ($v_e$). In other words, the performance of the rocket engine depends on the exhaust velocity. The crucial parameter for the functioning of a rokect is the specific thrust or impulse ($I_{sp}$), which is *the thrust per unit mass of the fuel consumed per second*. $I_{sp}$ depends on $\sqrt{T/m}$, where $T$ is the temperature at which the fuel burns and $m$ is the mean molecular weight of the gaseous products of combustion. A higher T and $m$ will contribute to more efficiency of propellants. Some propellants in common use include :

(*i*) $O_2$—$C_2H_5OH$

(*ii*) $O_2$—$H_2$

(*iii*) $N_2O_4$—$N_2H_2$—$(CH_3)_2N_2H_2$

(*iv*) $N_2O_4$—$CH_3N_2H_3$

(*v*) $H_2O_2$ monopropellant

(*vi*) $N_2H_4$ monopropellant

(*vii*) Fuming $HNO_3$–Mixed amine

(*viii*) Polybutadiene-aluminium perchlorate

(*ix*) $NH_4ClO_4$—Al

**Classification of Propellants.** Depending upon the physical state, propellants are classified as

1. **Solid Propellants.** They involve the use of solid fuel and solid oxidizer. They are furthen classified as:

(*i*) **Composite Propellants.** These consist of polymeric binder, like polyurethane or polybutadiene, which is used as a fuel and ammonium perchlorate which is used as an oxidiser. The performance of these propellants can be enhanced by using additives like alumimium or magnesuim in the finely divided form with the fuel. For example, gun powder.

(*ii*) **Double base Propellants.** These contain two materials. For example, *ballisite,* containing nitroglycerine and nitrocellulose mixture. *Cordite,* composed of 65% nitrocellulose, 30%. nitroglycerine, and 5% of petroleum jelly. In this, petroleum jelly works as a plasticizer.

2. **Liquid Propellants.** These comprise fuels such as kerosene, alcohol, hydrazine or liquid hydrogen mixed with an oxidizer such as liquid oxygen, dinitrogen tetroxide ($N_2O_4$) or mistric acid. The main advantage of liquid propellants is that,

(*i*) They give a better thrust than salid propellants.

(*ii*) Their thrust can be controlled by regulating the flow of propellants.

Liquid propellants may further be classified as :

(*a Monopropellant.* A monopropellant has a fuel as well as oxidiser in the same molecule or in a solution conaining both these. For example, hydrogen peroxide, nitromethane, ethylene oxide, hydrazine propyl nitrate and a mixture of 21.4% methanol and 78.6% hydrogen peroxide, etc.

(*b*) *Bipropellants.* These consist of two liquids. In these, liquid fuel plus oxidiser, kept separately, are injected in the combustion chamber separately. Commonly used fuels are liquid hydrogen, hydrazine, ethyl alcohol, aniline, and kerosene oil. The common oxidisers employed are liquid oxygen, ozone, hydrogen peroxide, fuming nitric acid and liquid fluorine.

3. **Hybrid Propellants.** These consist of a solid fuel and a liquid oxidiser. For example, a mixture of acrylic rubber and liquid dinitrogen tetraoxide.

Distinction between solid and liquid propellant is described below:

| Solid Propellant | Liquid Propellant |
|---|---|
| (*i*) They have low specific impulse. | They have high specific impulse. |
| (*ii*) They can easily be handled, stored and transported. | They have storing, handling and transporting problems. |
| (*iii*) The engines used for them are of simple design. | The engines used for them are delicate and cannot withstand any rough handling |
| (*iv*) They are more economical in use. | They are comparatively less economical in use. |
| (*v*) Manipulations with solid propellants is difficult. | They are more versatile. |
| (*vi*) Common examples include, gun powder, ballistite, cordite, etc. | Common examples include, hydrogen peroxide, nitro methane, hydrazine, ethyl alcohol, kerosene etc. |

**Propellants used in Different Space Vehicles**

(*i*) **Saturn booster rocket** of American space programme used a combination of *liquid oxygen* and *kerosene* as the propellant for initial stage whereas *liquid hydrogen* and *liquid oxygen* were used for the upper stage.

(*ii*) Russion rockets such as **PROTON** used kerosene and liquid oxygen.

(*iii*) **SLV–3** and **ASLV,** the Indian satellites used composite solid propellants.

(*iv*) For **PSLV-C4 rocket**, which was launched or ar 12th sep. 2002, solid propellants, *hydroxyterminated polybutadiene* (HTPB) were used for first and third stage while liquid propellants, $N_2O_4$ and unsymmetrical dimethyl hydrazine (UDMH), and $N_2O_4$ and monomethyl hydrazine (MMH) were used for the second and fourth stages respectively.

## 5.2 ABRASIVES

Abrasives are the substances which are employed to rub down objects, in order to provide them desired size, shape or finish. The capacity of an abrasive to rub-down other objects is called **abrasive power** and mainly depends upon *hardness, toughness* and *refractoriness. The capacity of a substance to scratch other substances is usually known as hardness. The greater the hardness, the greater the capacity of the abrasive to scratch or rub off the other objects. In general, the hardness of an abrasive should always be greater than the substance to be rubbed off by that abrasive. Mohr, a mineralogist has suggested a reference scale for determining the hardness of abrasive substances. The scale is : Talc, Gypsum, Calcite, Fluorite, Apatite, Feldspar, Quartz, Topaz, Corundum, Dimond. This indicates that each member of the scale can scratch those below it and in turn are capable of being scratched by those above. Brittleness is the opposite of toughness. The more an abrasive is brittle or friable, the more sharp it is and hence capable of removing more materials from the object which is ground.* An alteration in the manufacturing procedure or a reduction in the size of the crystals may cause brittleness or friability in an abrasive. The resistance of an abrasive to breaking of its grain when the temperature is increased is called refractoriness of the abrasive.

### 5.2.1 Classification of Abrasives

Abrasives are generally classified as natural abrasives and artificial or synthetic abrasives. The examples of natural abrasives are diamond, emery, garnet, pumice, quartz, corundum, flinit, sandstone, sand etc. Silicon carbide, calcium carbide etc are the examples of arifticial or synthetic abrasives.

#### *5.2.1.1 Natural Abrasives*

**1. Diamond** is the best known abrasive and is the hardest material in scratching power. It has been used for drill points for drilling rocks, for polishing gems, for cutting, turning and dressing grinding wheels and for lapping. Diamond wheels made with an inner zone or core of some resinoid or metallic alloy and a very much less thick abrasive face containing diamond have also been used in industry in recent years.

**2. Corundum** is the crystalline form of aluminium oxide ($Al_2O_3$) having scratching hardness next to diamond. It is used for the manufacture of abrasive wheels which are very important in metal industry.

**3. Emery** is a dark coloured mixture of corundum and black magnetite ($Fe_2O_3$). It has a number of important properties such as high scratch hardness, reasonable toughness, gradual break down of crystal under pressure etc. These properties make emery a good abrasive.

**4. Granet** consists of a number of similar silicate minerals having different tint as well as hardness ranging from 6.5 to 7.5 in *Mohr's scale of hardness*. It is used in the manufacture of coated papers and cloths and for polishing hard woods.

**5. Pumice** is generally found in porous blocks, pale grey in colour as a combination of silicates of Al, K and Na. Pumice is used directly without any processing as cleaning and polishing stone in *lithography*.

**6. Quartz** is crystallised silica which is brittle and has shell like fracture. Its hardness on *Mohr's scale* is about 7 and it is used mostly for making sand papers.

**7. Sand stone** is used for making grind stone, pulp stones and sharpening stones etc.

**8. Flint** is a modified silica. It is light in colour and has got conchoidal fracture. For making mill stones and sharpening stones, it is usually crushed and then bonded together.

**9. Diatomite,** also known as diatomaceous earth, infusorial earth or kieselguhr, is a deposit of the beautiful siliceous skeletons of unicellular organisms called diatoms.

**10. Tripoli** is similar to diatomite but has minute spongy appearance and does not contain diatom structure. Diatomite at first floats on water but tripoli sinks at once in water. Diatomite as well as tripoli have been used in metal polishing creams.

#### *5.2.1.2 Artificial or Synthetic Abrasives*

**1. Silicon carbide** is manufactured by heating sand and coke with some saw dust in an electric furnace. Silicon carbide (SiC) is ordinary grey or black and green. The green variety is translucent.

**2. Fused alumina or alundum** ($Al_2O_3$) is not as hard as silicon carbide.

**3. Calcium carbide ($CaC_2$)** can be manufactured from carbon and quick lime at 2000-2200°C in a specially designed electric furnace.

**4. Boron carbide ($B_4C$)** is the hardest material ever synthesised. It is very resistant to abrasion.

**5. Steel shot** is chilled cast iron and classified as metallic abrassive. Steel shots can be prepared by exposing streams of molten cast iron to air and about 80lb steam pressure. As a result, metal separates into drops of various sizes which are chilled into a water tank. They are further treated and conditioned to make them tougher and harder and finally screened according to sizes. Finer sizes are used in metal blasting and shots having sizes greater than 15 mesh are used in sawing, grinding, and polishing stone 'granite' marble etc.

**6. Steel grit** is another metal abrasive that is prepared by crushing high carbon steel and further treating for hardening. Then it is screened into various sizes. Working stones are made of grits coarser than 90 mesh and finer grits are used in steel cement and for grinding optical lenses. Steel wool in the long fibres is manufactured by shaving steel wire of high manganese open hearth steel wire, by making use of stationary knife. It is used in papering and scraping.

In addition to these abrasives, a number of other substances such as rouge ($Fe_2O_3$). black rouge (black magetite), green rouge (chromium oxide) etc are used as abrasives.

## 5.2.2 Artificial Abrasives

### *5.2.2.1 Silicon Carbide or Carborundum*

Silicon carbide (SiC) is now manufactured by making use of sand or silica and carbon. The sources of carbon are coke, pitch, petroleum cokes and anthracite. The sand contains pure silica (98–99.5%) $SiO_2$. For the manufacture of silicon carbide a special type of open top indirect heating *resistance furnace* is used. The furnace consists of a permanent bed of curved cast iron pieces lined with fire brick on which the charge is placed. The charge is built up in the furnace around a heating core of granular carbon. The 30–50 ft long furnace is provided with about 60 carbon electrodes which are one metre long and 8 cm in diameter and the central core of iron connects these electrodes. The ends of the furnace are permanent and the sides are built up every time with the charge. Excessive heat loss does not take place, because the outside unreacted charge acts as an insulator. The sides are pulled down after the completion of the process to take out the product. The charge, which is a mixture of 53.5% $SiO_2$, 40% coke, 5% saw dust and 1.5% salt, is added to furnace until it is full. Saw dust increases the porosity of the charge to permit the circulation of vapours and the escape of carbon monoxide formed. The CO burns at the top of the furnace. Salt is added as a flux. A typical initial current between the electrodes is 6000 amps. at 250 volts, but its resistance gradually decreases as the reaction proceeds and the final current becomes 20000 amperes at 75 volts. After about 2 hours, the current increases rapidly from 6000 to 20000 amps and remains steady during the whole run of 36 hours. The temperature at the core is 2200°C and the energy efficiency is 50%. After about 36 hours, the reaction is complete. The product is then cooled for 24 hours and crystals of silicon carbide are removed from the furnance. The yield is about 6-8 tonnes per furnace. The black crystals are broken and then treated successively with sulphuric acid and sodium hydroxide solution. It is dried in a kiln and graded through a screening or sieving system into the powders of various degrees of fineness.

*It should be noted that the temperature of the core should not exceed 2200°C. Higher temperature leads to the decomposition of silicon carbide* into graphite with the volatilization of silicon. The graphite formed in this manner is called *artifical graphite.* The outer unreacted part of the charge is combined with the next charge for the furnace and next to the core some graphite formed as a result of decomposition of SiC is obtained. After the complete run, which takes about 36 : 24 = 60 hours, the graphite can be separated as a by product from silicon carbide and converted to desired shapes. A part of the core used in the furnace can be made of coke suitable for graphite manufacture. A good yield of carborundum can be obtained by taking the following points into consideration.

*(i) Silica or sand used should be of about 98% purity. (ii) The carbon used may be petroleum coke, metallurgical coke, anthracite etc, having low ash and sulphur contents. (iii) $Fe_2O_3$ and $Al_2O_3$ in silica should be as low as possible as they catalyse the decomposition of SiC. The carbide undergoes decomposition at 2830°C. (iv) The temperature of formation of silicon carbide is about 1840 ± 30°C (v) 100 parts by weight of finest grade sand, 60 parts by weight of coke and 19 parts by weight of saw dust is the charge composition for a good yield of SiC. Saw dust increases the porosity of the charge or escape of CO and other volatile matter during SiC formation.*

**Reactions.** The various reactions involved in the manufacture of silicon carbide are,

$$SiO_2 + 2C\text{ (amorph)} \rightarrow Si + 2CO + 144.8 \text{ K. cals}$$

$$Si + C\text{ (amorph)} \rightarrow SiC - 30.5 \text{ K. cals}$$

$$SiO + 3C\text{ (amorph)} \rightarrow SiC + 2CO + 114.3 \text{ K. cals}$$

The temperature is maintained at 2000°C. Higher temperature is prevented as SiC is decomposed into graphite at high temperature.

## USES

**1.** The most notable property of silicon carbide is its hardness. It is almost as hard as diamond and is used for making grinding machines, grinding stones, crushers for giving edge to cutting tools (razors, knives, swords etc.).

**2.** On account of its refractory nature and higher heat conductivity, it is also used for making crucibles.

**3.** It has also been used as a refractory material, both in the form of bricks and as loose material.

**4.** Silicon carbide has also been used as an abrasive for making the modern grinding tools, which are used as in fabricating the metal parts of modern automobilies, tanks, engines, riflex, aeroplanes and other machines. It is also used for carbon rods in resistance heaters.

Silicon carbide has high thermal conductivity due to which it is resistant to shock because of temperature fluctuations. Its conductivity lies between those of metals and non-metals or insulators and so it acts as semiconductor. Its radiation resistance is much higher than pure silicon. It is exceptionally stable to most chemicals at low temperatures but reacts at high temperatures.

### *5.2.2.2 Calcium Carbide*

**Raw Materials.** Calcium carbide is commercially manufactured from carbon and quick lime at 2000 – 2200°C in a specially designed calcium carbide electric furnace.

$$CaO + 3C\text{ (amorph)} \rightarrow CaC_2 + CO\ ;\ \Delta H = +\ 103 \text{ K. cals}$$

The carbon is obtained from coke, anthracite or petroleum coke. Coke has, however, widely been used. The coke must be compact and have a low ash content, a low ignition point and high electrical resistivity (or low electrical conductivity) so that the bulk of the furnace charge may be highly resistant to the flow of energy. As a result, the energy becomes concentrated in the charge rapidly producing high temperature required for the reaction to take place and to go for completion. *The coke should also be free from phosphorus, because it forms a phosphide, which is converted to poisonous phosphine ($PH_3$), when acetylene is formed by the action of water on carbide.* The presence of phosphine may even cause serious explosion. The coke must contain low ash content (about 3% or less), otherwise the carbide formed will be viscous and, therefore, will create difficulty during tapping from the furnace.

The quicklime is obtained by buring limestone containing about 97 percent $CaCO_3$.

$$CaCO_3 \rightarrow CaO + CO_2$$

Magnesia, silica and iron oxides present in the form of impurities hamper production and give a less pure carbide. These impurities should, therefore, be properly controlled before the charge is fed to the furnace.

**Method of Manufacture.** Carbide furnaces fall in the class of combined resistance and direct arc heating, since some of the heat is due to the resistance of the raw materials, but most part of it is due to sparking across the charge. The furnace consists of a steel shell 10 × 3 metre in area or more with the side walls lined with refractory material and the bottom covered with carbon blocks or anthracite to withstand extremely hot alkaline conditions. Most of the recent large furnaces have closed top (where almost all the CO evolved during the reaction is collected and utilized) and three phase electric current, *i.e.,* they have three vertical electrodes suspended in the shell and each is connected to one of the low tension bars. The electrodes are made of graphite, fitted and bound. They may be as long as 10 × 2 ft and may be round or oval in cross section. They are also provided with arrangement for lowering because they are consumed.

Recently *Soderberg Continuous Self Baking Electrodes* have been used in place of prebacked electrodes, since the former permit large capacity furnaces. The electordes are dipped in the charge which almost fills the furnace. The layer of carbon or anthracite in the bottom acts as the bottom electrode. The charge consisting of 60% calcium oxide and 40% coke is fed from the top. Before it reaches the region of high temperature (2000°C) provided by the furnace, it gets heated by the burning of carbon monoxide (liberated in the reaction) in air. The latter is pumped through the side tubes in the middle of the furnace. Calcium carbide is formed at the bottom as a result of reaction of calcium oxide and coke. The molten calcium carbide is periodically tapped out of the furnace and fresh charge is added into cast iron chilled pots of about 5 tonnes capacity each. The carbide is cooled, crushed and sized.

Commercial calcium carbide thus obtained is generally black, reddish black or reddish brown, depending upon the presence of iron or other constituents. A good yield of calcium carbide is obtained when following points are taken into consideration.

*(i) Lime is prepared from lime stone of about 96–97% purity. (ii) MgO content of lime should not exceed 1%. Higher amount may cause reduction of fluidity of molten $CaC_2$ creating difficulty in tapping. (iii) $Al_2O_3$ should be as small as possible, because it also creates the same problem as created by MgO. (iv) $SiO_2$ should not exceed 1.25%. (v) $Fe_2O_3$ + $Al_2O_3$ should not exceed 0.5% because $Fe_2O_3$, if present, reacts with*

*$SiO_2$ and produce ferrosilicon. The latter settles down below $CaC_2$ and attacks the refractory material of the furnace. (vi) Impurities of phosphorus and sulphur in lime should not be more than 0.1%, because they would produce $Ca_3P_2$ and CaS which in their turn would produce $PH_3$ and $H_2S$ during water treatmeat of the carbide. (vii) Metallurgical coke, petroleum coke, and soft coke with low ash and volatile matter should be used. Petroleum coke is most commonly used because of its low ash content and high resistivity. (viii) The moisture content of coke used should not exceed 2% (ix) 60 parts of CaO and 40 parts by weight of coke are used. (x) The electrodes may be prebaked, square or rectangular made from amorphous carbon, graphite or continuous self baking type made from calcined anthracite and coal, and pitch as binder.*

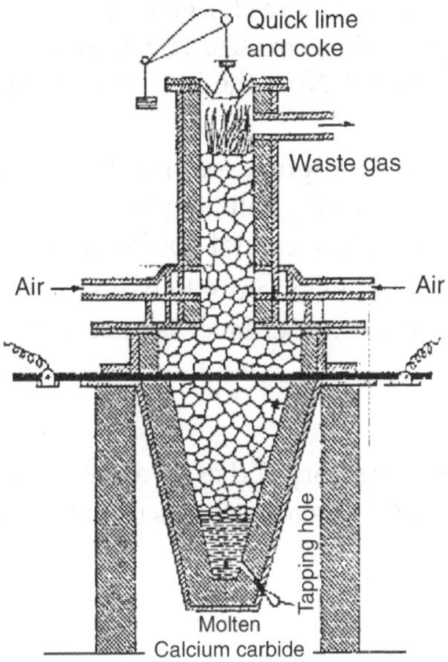


*FIGURE 5.1* **Manufacture of $CaC_2$.**

USES

Today calcium carbide is used as an abrasive and for the manufacture of acetylene which is used in oxyacetylene blow pipes acetylene lamps and as starting material for the preparation of many compounds, such as $CH_3CHO$, $CH_3COCH_3$, $CH_3COOH$, $CH_3COOC_2H_5$. It is also used in the manufacture of calcium. A small amount of calcium carbide is used as a reducing agent in some metallurgical processes and as a drying agent.

### *5.2.2.3 Alundum*

Alundum is fused aluminium oxide ($Al_2O_3$) and is prepared by impure bauxite often called *red bauxite*. It is also prepared from pure alumina and then it is usually called as *corundum* or *sapphire*. In other words, *fused aluminium oxide made from impure bauxite or red bauxite is called alundum and obtained from pure alumina is called corundum or sapphire.* Iron and silicon oxides present in the form of impurities in the bauxite must be carefully controlled as they have a marked effect on the structure and properties of the product. Alumina is purified by the methods described in the chapter of electrochemical industries.

**Raw Materials.** Red bauxite is first calcined and a charge is prepared by mixing 89% calcined bauxite, 2% coke and 9% iron scraps. The vertical arc resistance furnace consists of a circular steel shell about 7–8 ft high and standing on a carbon base of about 5 ft diameter. Now some charge is placed at the base and round carbon electrodes are lowered into the furnace, then arcs are drawn between them and the fused furnace charge. When the furnace is in operation, the outer walls of it are water cooled so that the unfused alumina around the inner for the furnace. The fusion is usually started by forming a trench in the top of the starting batch and packing it with coarse coke. The charge is added as gradually as it fuses between the arcs and this molten alumina carries the current. Iron oxide and silica are reduced by coke and an iron silicon alloy is formed which sinks to the bottom. By the passage of electric current, the temperature becomes 2100–2200°C and at this temperature alumina melts. As the level of fused alumina rises in the furnace, the electrodes are raised by an automatic arrangement and more charge is added simultaneously inorder to maintain the arc between the molten charge and electrodes. The addition of charge and melting of alumina continues until the furance becomes full. Now the current is stopped and the entire mass is cooled, solidified, crushed and graded.

**Manufacture or Corundum or Sapphire**—Corundum or sapphire is white in colour and is made from pure alumina. It is prepared by crystallizing through fusion of pure alumina in a hydrogen oxygen upside down flame at 3500°C. The crystal boules are cut and polished as desired.

It can also be manufactured by fusing pure bauxite in an arc furnace at about 3000°C. The impurities are allowed to settle and the upper layer is collected, cooled and crushed to a powder.

#### USES

Alundum has been used as a basic refractory and as an abrasive. It is also used for making grinding wheels, polishing jelly and paper. Artificial corundum is used for hard bearings necessary for watches and modern instruments.

#### *5.2.2.4 Boron Carbide*

It is another very hard abrasive, also known as *norbide*. It is manufactured by heating boric oxide with coke in a carbon resistance furnace at 2600°C. The product is about 99% boron carbide ($B_4C$).

$$2B_2O_3 + 7C \longrightarrow B_4C + .6CO$$

It is used in the nozzles for sand blasting.

#### *5.2.2.5 Boron Nitride*

Its hardness is comparable with that of diamond, with the advantage of resisting oxidation much better. It is also known as *borazon* and is prepared by converting hexagonal boron nitride into cubic form at 3000°F temperature and I million p.s.i. pressure in an electric furnace.

### 5.2.3 Grinding Wheels

The abrasives are widely used for grinding of different materials such as metal, glass, stone wood, etc. in the form of wheels for preparing grinding wheels, the lumps of the proper abrasive materials is first crushed in jaw crushers and roller mills, and then graded, according to the grain size. Then, abrasive material of proper size is mixed with a binding material and shaped into moulds by pressing. The production of grinding wheels is governed by the following factors :

**1. Type of abrasive.** Generally artificial abrasives like carborundum, fused alumina, silica, etc. are used due to their hardness purity and uniformity in chemical composition.

**2. Grain size.** Abrasive materials usually exist in the form of big lumps, so these are first broken up in jaw crushers and then pulverized in heavy roller mills. Then, different sizes of grains called *grit* are separated by sieving, through different size mesh screens. These are usually graded according to the mesh numbers as 4, 8, 20, 12, 64, 96, 180, and 220. The use of particular size of grain depends mainly on the nature of job to be performed by the grinding wheel.

**3. Types of binding material.** There are six types of bonds or bonding materials, which are mainly used in the manufacture of grinding wheels. These are discussed below :

**Silicate bond** consists of water glass or sodium silicate solution and some clay. The abrasive is mixed with the bond and pressed into the mould. Silicate bonded wheels are soft and in such wheels, silicate bond releases the grains more quickly than any other bond and used to grind edged tools (*e.g.*, sharpening of knives). These are also used where large wheels are needed or where there is a need of keeping down the heat generated.

**Vitrified bonds** consists of a mixture of clay, feldspar, flint and frit in varying proportions according to the toughness needed. Wheels with this type of bond are prepared either by **pudding or by pressing.** In *pudding method,* wet slurry of the mixture is poured into moulds of plaster of Paris and the wheels so formed are shaved to proper size on a potters wheel. In pressed method, the stiff mixture of abrasive and bond is pressed in moulds by making use of hydraulic pressure and then taken out of the moulds for drying. The dried wheels are then fired in a kiln upto a temperature of about 1300°C, depending upon the toughness required. At such a high temperature firing, vitrification takes place as a result of chemical combination of abrasive grain and the bond. The vitrified bond is somewhat brittle and thus used for heavy work. The vitrified grinding wheels are resistant to moisture, mild acids and oils etc.

**Shellac bonds** are prepared by mixing shellac with abrasive grain in a mixer heated by steam. The mixture is then hydraulic pressed into moulds. The hydraulic pressure during mouling is heated by steam. After few hours, the wheels are taken out. Shellac wheels are used for shaping and sharpening knife blades, for giving high finishes on camshafts and steel rolls and for sharpening the teeth of saws in saw mills. Fused alumina (500–600 mesh) is mixed with shellac to obtain wheels for edge of safety razor blades.

Rubber bonds are produced by mixing abrasive grains with natural or synthetic rubber, sulphur (as vulcanising agent) and other ingredients (for modifying the properties of rubber) between the heavy steel rollers and then whole mixture is formed into sheet having suitable thickness by passing through calenderrolls. Wheels are finally cut from the sheet using a round cutter and then heat cured. Rubber bonded wheels have widely been used for high speed grinding as in sawing. Thin rubber bonded wheels have also been used for slitting pin points.

**Resin bonds** are obtained by mixing phenol formaldehyde resins or alkyd resins with the abrasive grains in presence of some ingredients such as catalysts, plasticizers and fillers. In actual practice, plasticizer is mixed with abrasive grains till the grains are wetted. Then a mixture of powdered resin and filler is added and the mixture is fed to a mould which is then compressed in a hydraulic press, either cold or hot according to the nature of the resin employed in the bond. The green wheel is laid on a flat, circular, ceramic disc and placed in the oven for curing under strict temperature control.

Metallic bonds are obtained by mixing alloys of copper, tin and steel with abrasive grains and these bonds are mainly used in the manufacture of diamond surfaced wheels, where diamonds are embedded by powder metallurgy in a solid metal alloy. The latter reinforces the diamonds against mechanical injury and also prevents under heating.

### 5.2.4 Abrasive Paper and Abrasive Cloth

Abrasives in the form of abrasive paper and abrasive cloths have also been used for cleaning and polishing of different materials. These are generally manufactured by passing a roll or strip of paper or cloth through a series of rollers and a thin coating of soft glue is applied on its upper side. This glued paper or cloth is then allowed to pass under a hopper, from which the grit of abrasive is dropped which spread evenly on the glued paper or cloth. The paper or cloth is again given a thin coating of glue and then dried in warm drying rooms. Finally, the paper or cloth is allowed to age for about a week inorder to set the glue firmly. The cloth or paper rolls are then cut into sheets of desired sizes. Emery, garnet, flint, sand, fused alumina ($Al_2O_3$) carborundum (SiC) etc are generally used as abrasives for making abrasive papers or cloths. The abrasive papers or cloths are of different numbers, depending upon the hardness of abrasive, grit and grain size. Abrasive cloths are more durable than abrasive papers because they are more resistant to friction. Abrasive papers and cloths have widely been used in polishing furnitures, automobiles, and in leather industries.

Now artificial abrasives such as carborundum, silica, fused alumina are generally employed because of their hardness, purity and uniformity in chemical composition in preference to natural abrasives such as emery, corundum etc. For grinding of materials like copper, aluminium alloys, brass, marble and other low tensile strength materials, silicon carbide or carborundum has widely been used. Fused alumina has been used for materials of high tensile strengths such as high carbon steels, alloy steels etc.

Since abrasive materials are generally found in the form of lumps, the latter are broken up in jaw crushers and then pulverised in heavy roll mills. Then different sized grains, called **grit,** are separated by sieving, through various types of sieving screens. These are graded according to the mesh numbers such as 4, 8, 12, 20, 64, 96, 180, 220. The use of a particular size of grit depends mainly on the nature of work to be performed on the grinding wheel. Fine grains cut slowly but give better and smooth finish in comparison to coarse grains, which cut quickly.

### 5.2.5 Uses of Abrasives

**1. Coated Abrasives.** Sand papers and its different variations are generally termed as *coated abrasives* and consist of three parts : *Backings*, *Abrasives* **and** *Bond*. **(a)** *Backings* consisting of paper, cloth and combinations of paper and cloth. **(b)** *Abrasives* such as fused alumina, emery, flint, garnet and silicon carbide etc. **(c)** *Bond* such as hide glue. For water proof coated abrasives, resins are generally used instead of hide glue.

The size of grit varies from 12 to 600 mesh. In the process of coating grains of abrasive are electrostatically attracted to the glued backing end and these are well anchored in upright position. The process is known as *electrostatic process*. Electrostatically coated abrasives have a longer life and a sharper surface. The coated abrasives are generally used for wood sanding, metal polishing, painting, tanning, varnishing, shoe making etc.

**2. Polishing Abrasives.** This type of abrasives make the surface smooth as well as clean, remove tool marks, scratches and other irregularities on the surface and provide good appearance to the object. The finer the polish required, the finer would be the abrasive needed. For complete polishing of the surface, various polishes are applied on the surface with abrasive powders by making use of finer mesh for every subsequent polishing. For example, auto bumpers are polished with emery powder completely through four polishes with 60–90 mesh, 120 mesh, 150–180 mesh and 220 mesh polishing respectively.

Mechanical polishing with flexible polishing wheel provides rapid as well as efficient polishing. In such type of polishing, discs of cloth stitched together or leather covered wooden wheels have been employed. Fused alumina, emery and other hard and tough abrasives, which have uniform chemical composition and crystal structure, have been employed in mechanical polishing. High grade adhesive glue is applied on the polishing surface and the abrasive is then sprinkled on the glue.

**3. Buffing.** Buffing is refined polishing in which 500–600 mesh abrasive is used without any adhesive to carry the abrasive grain on buffing wheels. The process consists in pouring the adhesive to the buffing wheels by making use of buffing sticks made by combining abrasive and hard grease or bees wax. The finer type of buffering is called *colouring*, which imparts true colour as well as fine lustre.

**4. Lapping.** Lapping is used to improve accuracy and finish and is carried out on parts of instruments and where two surfaces must fit each other tightly. A careful grinding is carried out before lapping and silicon carbide, fused alumina, emery and garnet have widely been used abrasives for the purpose. In lapping, abrasive is carried by a wooden wheel, grooved cast iron wheel or an abrasive wheel by a carrier known as *lap*. It should be noted that lap must not be harder than the material being lapped. The composition of lapping consists of an abrasive in a vehicle such as kerosene oil which can spread the abrasive and keeps it in contact with the lapping surface.

**5. Sharpening Sticks.** A number of articles such as oil stones, whetstones, hones etc are used as sharpening sticks, but artificial or synthetic abrasives have now been much used for sharpening instruments than these sharpening sticks.

**6. Grinding Stones.** A number of stones such as Grind stones, Burr stones, Pulpstones, Mill stones etc. have widely been used as grinding stones. These stones are now made of bonded silicon carbide or fused alumina. *Burr stones* are more efficient than *millstones* and used for the grinding of pigments, cement and grains. *Steel rolls* are, however, used in flour mills. *Pulpstones* are used for the grinding of logs into pulp for paper manufacture.

## QUESTION BANK

1. How are explosives classified ? Give a brief account of the preparation, properties and uses of TNT.
2. What are the essential requirements of an explosive ? Classify explosives with suitable examples.
3. What are explosives ? How are they classified ? Write not on TNT :
4. Write note on dynamite. Write short note on modern explosives.
5. Write notes on (*i*) Dynamites (*ii*) Blasting gelatin (*iii*) Cordite.

6. Detonators are used in conjugation with high explosives ? Give reasons.
7. What are explosives ? Write short note on nitroglycerine.
8. What are the requirements of a good explosive?
9. Distinguish between primary and secondary explosives.
10. Write a shart note an TNT.
11. What are rocket propellants? How they are classified? What are the requirements for the selection of a propellants?
12. Give two examples of monopropellants and bipropellants used in rocket.
13. Explain the term, detonation and discuss the nature of the process. Give examples of a few detonators and high explosives.
14. Write short note on explosives and dynamites.
15. Write short note on moderm explosives.
16. What are blasting fuses?
17. Write a note on handling and storage of explosives.
18. Describe the characteristics of explosives.
19. Write the preparation and uses of following explosives
    (*a*) Lead azide
    (*b*) Tetracene
    (*c*) Dinol
    (*d*) Smokeless gun powder
    (*e*) Gum powder
    (*f*) TNT
    (*g*) PETN
    (*h*) Tetryl and Hexyl
    (*i*) Plastic explosives.
20. What are abrasives ? How they are classified ? Name some natural and artificial abrasives.
21. What are abrasives ? Describe briefly the manufacture of silicon carbide. What are the important uses of silicon carbide? What is the function of saw dust in the manufacture of carborundum?
22. How calcium carbide is prepared ? Mention some of the uses of calcium carbide. What are corundum and alundum ?
23. What are abrsives ? Give important uses of abrasives. Give some important uses of graphite.
24. What are natural and artificial graphites.
25. Write short notes on (*i*) Bonded grinding wheels (*ii*) Grinding stones. Write a note on various forms of industrial carbon.
26. How are abrasives classified? Describe the manufacture of alundum. Describe the manufacture of abrasive paper and cloth.
27. Give two examples each of hard and soft abrasives and discuss their applications. What are abrasives ? How are they classified ? How are artificial abrasives produced?
28. Write a note on synthetic abrasives. Give two examples of artificial abrasives. Describe the manufacture and uses of synthetic abrasives.
29. What are abrasives? Name two natural and two artificial abrasives. How are abrasive papers made? What are their uses?
30. Mention the names of common types of bonds used in making grinding wheels. Describe the merits and demarits of each bond. What are abrasives? How are carbides prepaired? How are they classified? Give examples ? Give the preparation and technical applications of silicon carbide.
31. Write not on norbide and abrasive paper. Write briefly about structure of commonly used abrasives.

Chapter 6

# New Engineering Materials

The advances in various branches of science have helped humans in greatly enhancing the quality of life. Among such sciences, chemistry has played a significant role in processing raw materials and converting them into articles of great utility for the man. The production of metals and alloys, plastics, drugs, ceramics, industrial catalysts, etc. has largely been the outcome of chemists. Applying the principles of engineering, those raw materials are exploited to benefit mankind. Such materials are called *new engineering materials.* The main such materials include: super conductors, optical fibres, organic electronic materials, fullerenes, industrial catalysts, etc. These materials are described in this chapter.

## 6.1 SUPERCONDUCTORS

The phenomenon of superconductivity was discovered by Kamerlingh Onnes, in 1911, when he was measuring the resistivity of mercury at low temperature. He observed that the electrical resistivity of pure mercury drops abruptly to zero at about the boiling point of helium *i.e.,* 4.2 K. He concluded that mercury has passed into a new state which is called the *superconducting state*. The temperature at which the resistance disappears is called the transition temperature or critical temperature. The critical temperature of mercury is shown in Fig. 6.1. *Thus when a substance loses its electrical resistance i.e., a current can continue through it without changing its value, the phenomenon is called superconductivity*. The subtance showing this property is called a *superconductor*. Other substances showing super-conductivity are silver, lead, gallium, iridium etc. Table 6.1 lists the critical temperature $T_c$ for some elements and compounds.

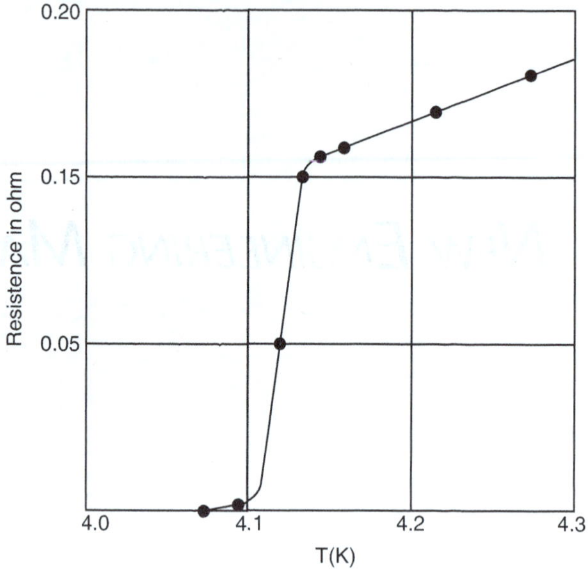


*FIGURE 6.1* **Resistance of mercury as a function of temperature.**

### 6.1.1 Experimental Study

In the superconducting state the dc electrical resistivity is zero, or so close to zero that persistent electrical currents have been observed to flow without attenuation in superconducting rings for more than a year, until at last the experimentalist wearied of the experiment. The decay of *supercurrent* in a solenoid was studied by File and Mills using precision nuclear magnetic resonance methods to magnetic field associated with the supercurrent. They concluded that the decay time of the supercurrent is not less than 100,000 years. In some superconducting materials, particularly those used for superconducting magnets, finite decay times are observed because of an irreversible redistribution of magnetic flux in the material.

The magnetic properties exhibited by superconductors are as dramatic as their electrical properties. The magnetic properties cannot be accounted for by the assumption that the superconducting state is characterized properly by zero electrical resistivity. It is an experimental fact that a bulk superconductor in a weak magnetic field will act as a perfect diamagnet, with zero

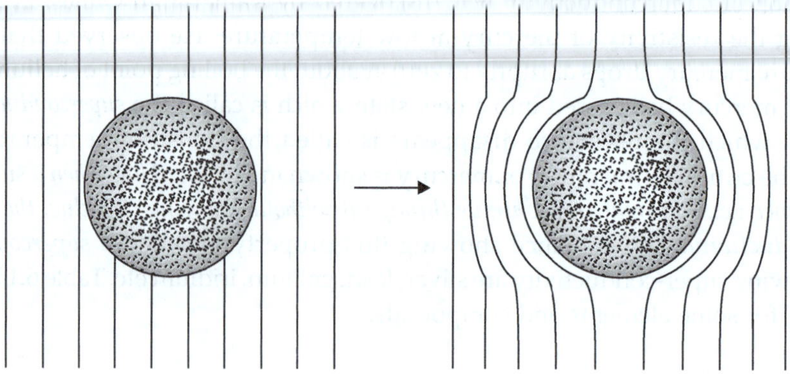

*FIGURE* **6.2**

magnetic induction in the interior. When a specimen is placed in a magnetic field and is the cooled through the transition temperature for superconductivity, the magnetic flux originally present is ejected from the specimen. This is called the *Meissner effect*. The sequence of events is shown in Fig. 6.2. The unique magnetic properties of superconductors are of central importance to the characterization of the superconducting state.

*The superconducting state is known to be an ordered state of the conduction electrons of the metal.* The order in the formation of loosely associated pairs of electrons. The electrons are ordered at temperatures below the transition temperature, and they are disordered above the transition temperature. The nature and origin of the ordering was explained by Bardeen, Cooper, and Schrieffer.

### 6.1.2 Occurrence of Superconductivity

Superconductivity occurs in many metallic elements of the periodic system and also in alloys, intermetallic compounds, and semiconductors. The range of transition temperatures at present extends from 23.2 K for the alloy $Nb_3Ge$ to 0.01 K for some semiconductors. In many metals superconductivity has not been found down to the lowest temperatures at which the metal was examined, usually well below 1 K. Thus Li, Na, and K have been investigated for superconductivity down to 0.08 K. 0.09 K. and 0.08 K, respectively, where they were still normal conductors. Similarly, Cu, Ag and Au have been investigated down to 0.05 K, 0.35 K, and 0.05 K, where they are still normal conductors. It has been predicted by theoretical calculation that if sodium and potassium superconduct at all, the transition temperature will be much less than $10^{-5}$ K. These statements refer to atmospheric pressure. Cesium becomes a superconductor ($T_c$ = 1.5 K) at a pressure of 110 kbar, after several phase transformations. The elements known to be superconducting are displayed in Table 6.1.

Will every nonmagnetic metallic element become a superconductor at sufficiently low temperatures? We do not know. In experimental searches for superconductors with very low transition temperatures it is important to eliminate from the specimen even trace quantities of foreign paramagnetic elements, because they can lower the temperature severely. A few parts per millilon of Fe will destroy the superconductivity of Mo, which when pure has $T_c$ = 0.92 K; and 1 at percent of gadolinium lowers the transition temperature of lanthanum from 5.6 K to 0.6. Nonmagnetic impurities have not very marked effect on the transition temperature.

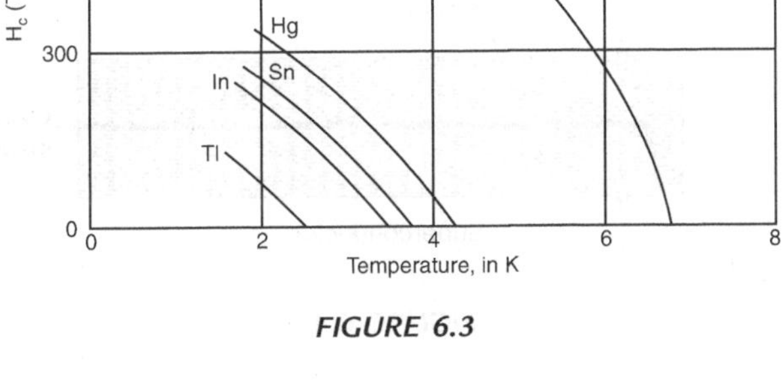


***FIGURE 6.3***

### 6.1.3 Destruction of Superconductivity by Magnetic Fields

A sufficiently strong magnetic field will destroy superconductivity. The threshold or critical value of the applied magnetic field for the destruction of superconductivity is denoted by $H_c(T)$ and is a function of the temperature. The variation of the critical field with temperature for several superconducting elements is shown in Fig. 6.3. The threshold curves separate the superconducting state in the lower left of the figure from the normal state in the upper right.

**Note:** We should denote the critical value of the applied magnetic field as $B_{ac}$, but this is not common practice among workers in superconductivity. In the CGS system we shall always understand that $H_c \equiv B_{ac}$ and in the SI we have $H_c \equiv B_{ac}/\mu_0$. Where the symbol Ba denotes the applied magnetic field.

### 6.1.4 BCS Theory of Superconductivity

**Bardeen, Cooper and Scruffier** offered an explanation for the superconductors with a theory called **BCS theory** (after the name of the inventors).

They proposed that by using the knowledge of solids and applying the quantum mechanics, it is possible to explain the superconducting state in which the motions of all the electrons in conduction band are locked together. In this state the electrons are coupled together in pairs at very low temperatures. One electron interacts with a crystal lattice and perturbs it and this perturbed lattice then interacts with the second electron in the pair in such a way that there is an attraction between the two electrons and this attraction overcomes the coulomb repulsion between them. The electron interaction via lattice produces an energy gap between the superconducting state in which electrons act collectively while in non-superconducting (normal) state these electrons act individually.

The energy required to excite the system to the higher non-superconducting state and destroy superconductivity is not available at lower temperature. Thus the material becomes superconductor at such a low temperature.

BCS theory has been successfully able to explain all the experimental facts of superconductivity. In the latest research efforts are being made to have underground power lines to transport electricity from generating plants to consumers. These lines will have negligible resistance and so energy losses and heating effects will altogether be prevented. However, for this we will require liquid helium to make the wires act as superconductors at low temperatures and it is quite a difficult job.

**Josephson Effect.** In 1962, Brian Josephson, an English Physicist predicted that if a very thin layer (about 2 nm) or an insulator (normally an oxide) is sandwiched between two superconductors

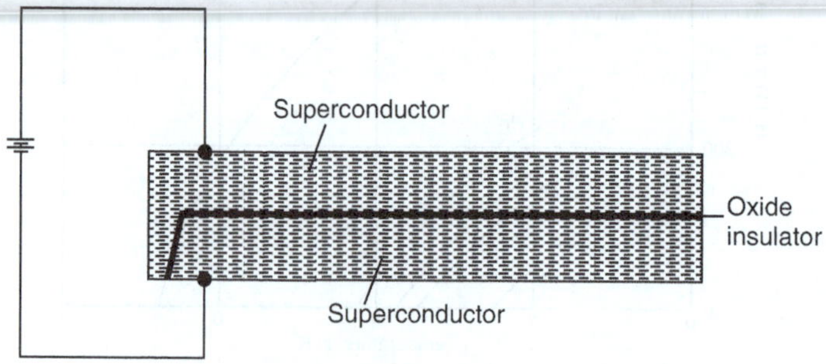


*FIGURE* **6.4**

and a voltage $V$ is applied across the two superconductors, a current starts flowing (Fig. 6.4). This current has two components (*i*) continuous dc component which persists even when the voltage is removed (*ii*) ac component which is present only when the voltage $V$ is present.

An important thing to note is that the ac component is always having frequency exactly $v = \frac{2eV}{h}$, no matter which materials are used for the so called Josephson function.

So this phenomenon is useful for finding accurately the ratio $e/h$ using the equation.

$$\frac{e}{h} = \frac{v}{2V} \qquad ...(6.1)$$

Since frequency can be measured very accurately, the value of $\frac{e}{h}$ can be measured precisely.

The results obtained are so accurate that volt can now be defined in terms of requency using equation 6.1. So instead of standard call or other measuring instruments, this method gives very good *calibration of voltmeter*.

Recently, Josephson junctions have led to the development of very sensitive detectors and fast and efficient memory computers. SQUID (superconducting quantum interference device) detectors have been successful used to measure poison levels of iron of liver, so that iron build up can be treated before much harm is done to the body.

### 6.1.5 Types of Superconductors

There are two types of super conductors :

(*i*) **Type I superconductor.** It is that type of superconductor which exhibits a *complete meissner effect or perfect diamagnetism i.e.,* above the critical field $H_C$, the specimen is a normal conductor. Below critical field, the specimen excludes all the magnetic lines of force inside the specimen or it will become a *diamagnetic material*. Such type of superconductors are also called soft superconductors. The values of $H_C$ are quite low for these materials. *Examples*: Al, Zn, Ga.

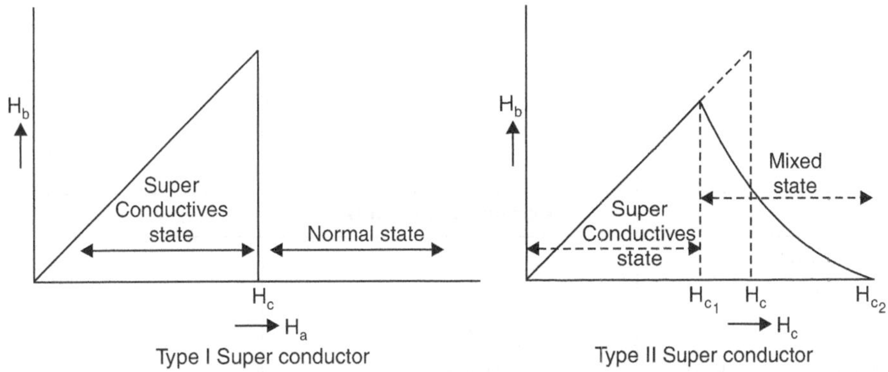


*FIGURE 6.5*

$H_a$ ⟶ Applied external magnetic field

$H_b$ ⟶ Magnetic field produced by the super conducting currents induced in the presence of an external field.

(*ii*) **Type II superconductors.** These are completely diamagnetic at low applied magnetic fields and field exclusion is total. However, their transition from the superconducting state to normal state is *gradual* and occurs between *lower critical field* ($H_{C1}$ and *upper critical field* $H_{C2}$). Consequently, the magnetic flux lines start to penetrate into the body of the material at $H_{C1}$ and with increasing magnetic field, this penetration is complete. In other words, for fields between $H_{C1}$ and $H_{C2}$, the material is in *mixed state, i.e.*, both normal and superconducting regions are present. By virtue of their higher critical temperatures and critical magnetic fields, type II suuperconductors are preferred over type I for most practical applications. At present, most commonly utilized superconductors are: (*i*) niobium-titanium (Nb-Ti) alloys, (*ii*) niobium-titanium (Nb-Ti) alloys, (*iii*) niobium-tin intermediate compound ($Nb_3Sn$). Table 6.1 lists several type of I and II superconductors, their critical temperature and critical magnetic flux densities.

**TABLE 6.1 Critical temperatures and magnetic fluxes for selected superconducting materials**

| *Material* | *Critical temperature $T_C$ (K)* | *Critical magnetic flux density, $B_C$ (teslas)* |
|---|---|---|
| | **Elements** | |
| Tungsten | 0.01 | 0.0001 |
| Titanium | 0.40 | 0.0056 |
| Aluminium | 1.19 | 0.0105 |
| Tin | 3.72 | 0.0305 |
| Mercury (α) | 4.15 | 0.0411 |
| Lead | 7.19 | 0.0803 |
| | **Compounds and alloys** | |
| Nb-Ti alloy | 10.2 | 12 |
| Nb-Zr alloy | 10.8 | 11 |
| $PbMo_6S_8$ | 14.0 | 45 |
| $V_3Ga$ | 16.5 | 22 |
| $Nb_3Sn$ | 18.3 | 22 |
| $Nb_3Al$ | 18.9 | 32 |
| $Nb_3Ge$ | 23.0 | 30 |
| | **Ceramic compounds** | |
| $Ba_{0.7}K_{0.3}BiO_3$ | 30 | |
| $Li_{1.8}Sr_{0.2}CuO_4$ | 44 | |
| $Ba_2Cu_3O_7$ | 92 | |
| y $Ba_2Cu_3O_{7-y}$ ($y > 0.5$) | 100 | |
| $Bi_2Sr_2Ca_2Cu_3O_{10}$ | 110 | |
| $Tl_2Ba_2Ca_2Cu_3O_{10}$ | 125 | |
| $HgBa_2Ca_2Cu_2O_8$ | 153 | |

**Note :** The symbols $T_c$ and $T_c$ represent the curie temperature and superconducting critical temperature respectively and these are totally different entities and should not be confused.

**Applications of superconductivity.** The phenomenon of superconductivity has many practical applications such as:

1. *Superconducting magnets* capable of generating high fields with low power consumptions are being employed in scientific tests and research equipments.
2. They are also used for *magnetic resonance imaging* (MRI) in the medical field as a diagnostic tool. On the basis of the production of cross-sectional images, any abnormalities in body tissues and organs can be detected. Chemical analysis of body tissues in also possible by *Magnetic Resonance Spectroscopy* (MRS).
3. Other numerous fields in which superconductivity is used are:
   (*i*) Electrical power transmission through superconducting materials—power loss is extremely low and equipment operates at low voltage levels.
   (*ii*) Magnets for high-energy particle accelerators.
   (*iii*) High-speed switching and signal transmission for computers.
   (*iv*) High-speed magnetically levigated trains, wherein the levigation results from magnetic repulsion.
   (*v*) Memory or storage elements on computers.
   (*vi*) For amplifying very small direct current and voltages.
   (*vii*) *In magneto-hydrodynamic (MHD) power generators* to maintain plasma. These superconducting materials are smaller in size and consume less energy.
4. They form a new generation of energy saving power systems. Superconducting generators are very small and compact in size and weight as compared to the conventional generators. In U.S.A. many companies have built superconducting generators.
5. In Japan and France, superconducting magnets have been used to raise the trains above the railway tracks and attain very high speeds as frictional losses are minimized. A similar magnetic propelsion system is being used to launch the satellites into orbits directly from the earth without the use of rockets.
6. High efficiency ore-separating machine may be built using super conducting magnets. Also super conducting magnets may be used to separate tumour cells from the healthy cells and is finding great use in medical sciences.
7. Superconductors can act as a relay or switching system in a computer. The basic principle can be understood from the Fig. 6.6. The fact that when we apply a magnic field whose value is more than the *critical field* of the superconductor, the *superconducting properties disappear* is used in gating circuits or in cyclotrons. Consider a wire *A* of superconducting material, inside a superconducting material *B*. Let the temperature of the two be *lower than the transition temperature* of the two materials. Current in central wire *A* can be controlled by a current in coil *B* as the magnetic field produced by *B* may exceed the critical field of the core material

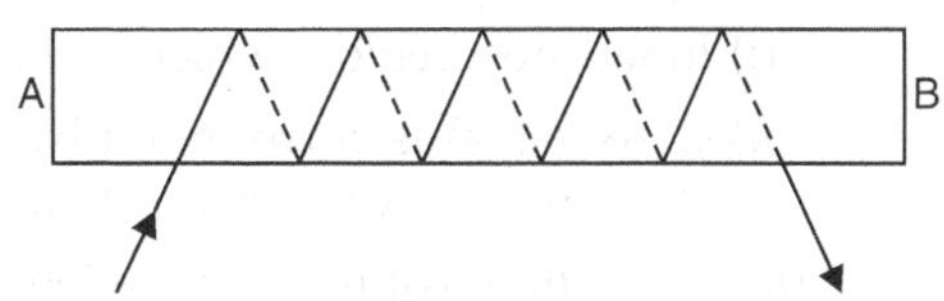


*FIGURE* 6.6

(wire) *A* at the operating temperature. It means that A will become a normal conductor and system can be used as *relay or a switch* in computers.

8. Very fast and accurate computers can be constructed using superconductors and the power consumption is also very low ~ 0.5 watt.

## 6.2 OPTICAL FIBRES

In case of electronic communication, the information is normally carried over in the form of radio waves and microwaves, through copper wires and co-axial cables. However, the information carrying capacity of these wires is restricted, due to their limited band-width and is not sufficient, as per the needs of the modern communication techniques.

If light waves are used instead of radio waves, or microwaves, the number of signals transmitted can be increased enormously. Technically, Grahm Bell in 1880 invented Photophone, which transmitted speech on a beam of light. By 1960, it was established that light could be guided by glass fibre. However, the glass available at that time *heavily attenuated light propagating through them*.

It was due to the invention of solid state lasers in 1970 and fabrication of low-loss glass fibres, that made the optical communication practical. With this the era of photonics started.

In addition to the use of optical fibre in communication, these have been extensively used as sensors to detect electrical, mechanical and thermal energies. Also they are being used in copying machines and in medical diagnostics (Endoscopy).

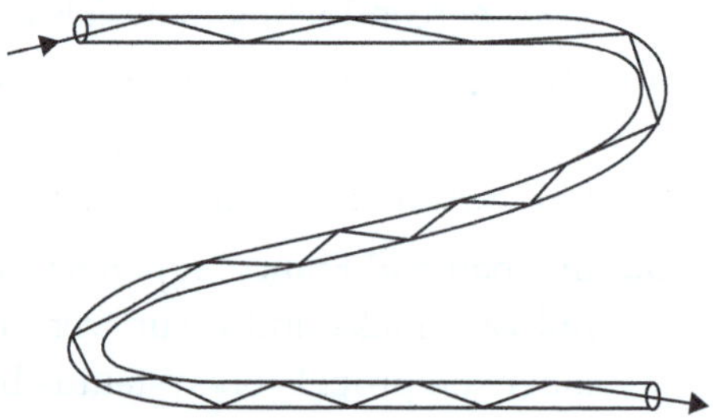

FIGURE 6.7

Optical fibres are made up of glass or plastic conduicts as thin as a human hair which is designed to guide light waves along the length of the fibre with the help of successive total internal reflections from side walls of the fibre. The movement of light is in a *zig-zag* path and some fraction of light may escape through side walls but major fraction comes out from the other end of the fibre. Fig. 6.7.

An optical fibre consists, in general, the following three regions:

(*i*) Inner-most region, which is the light guiding region is called *core*.

(*ii*) Core region is surrounded by a middle region, called *cladding*. The refractive index of cladding is always lower than that of the core.

(*iii*) The outermost region is called *sheath/coating*. The sheath protects the core and cladding from moisture, abrasion contamination and to give mechanical strength to fibre.

Optical fibres are either made as a single fibre or a flexible bundle of cable. A bundle fibre is a number of fibres in a single jacket.

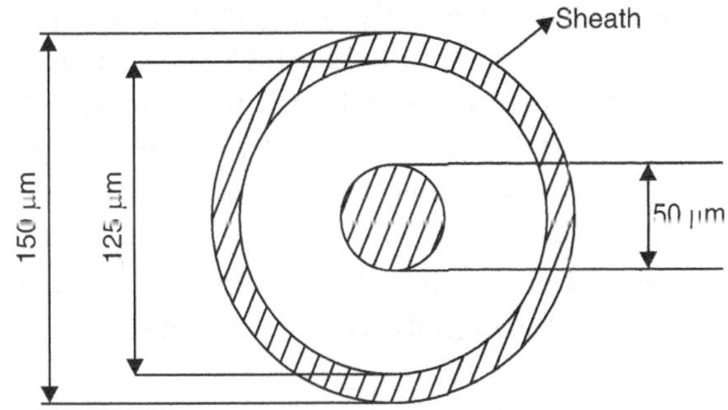


*FIGURE* 6.8

### 6.2.1 Light Transmission in an Optical Fibre

The principal of the process is the *total internal reflection* at the interface between the core of optical fibre and the outer sheath or cladding. Core is usually vitreous silica doped with germenia ($GeO_2$). Any light rays travelling at oblique angles to the fibre axis are reflected back into the core. Internal reflection is accomplished by varying the refractive index of the core and cladding glass material. In this regard, *two* designs are employed:

(*i*) **Step-index optical fibre design.** The index of refraction of the cladding is *slightly lower* than that of the core. The index profile and manner of internal reflection are shown in Fig. 6.9 (*b*) and (*d*). For this design, the output pulse will be *broader* than the input one [*see* 6.9 (*c*) and (*e*)].

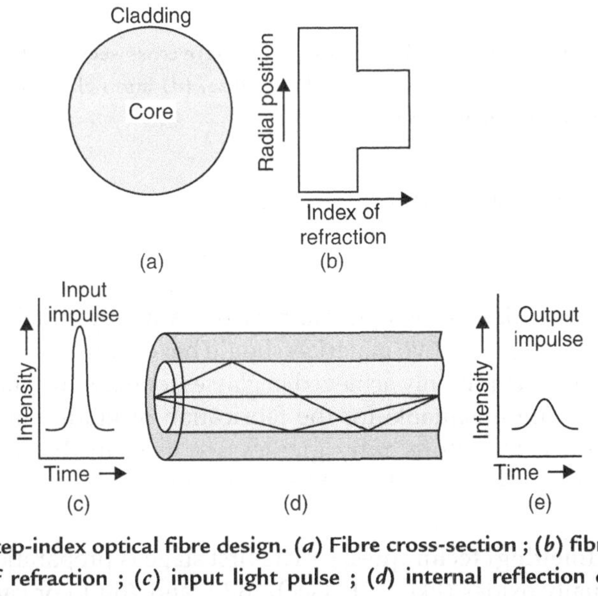


*FIGURE* 6.9 **Step-index optical fibre design. (*a*) Fibre cross-section ; (*b*) fibre radial index of refraction ; (*c*) input light pulse ; (*d*) internal reflection of light rays ; (*e*) output light pulse.**

This phenomenon is, however, undesirable because it limits the rate of transmission. Broadening of pulse results because various light rays, although being injected at approximately the same instant, arrive at the output at different times and they traverse different trajectories, thereby they have a variety of path lengths.

(*ii*) **Graded-index optical fibre design.** By using this design, pulse broadening is largely avoided. Here, impurities like boron oxide ($B_2O_3$) or germenia ($GeO_2$) are added to the silica glass such that the index of refraction is made to vary "parabolically" across the cross-section [*see* Fig. 6.10 (*b*)]. Consequently, *the velocity* of *light within the core varies with radial position*, being greater at the periphery than at the centre. Consequently, light rays that traverse longer path lengths through the outer periphery of the core travel faster in this lower index material and arrive at the output at approximately the same time as undeviated rays that pass through the centre portion of the core.

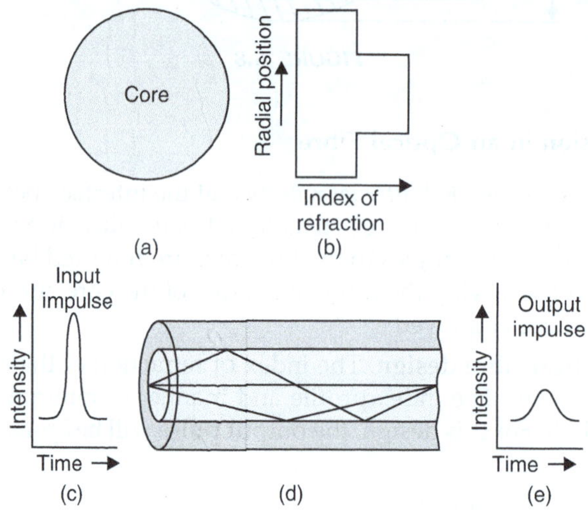


*FIGURE 6.10* **Graded-index optical fibre design. (a) Fibre cross-section, (b) fibre radial index of refracted profile, (c) input light pulse, (d) internal reflection of a light ray, (e) output light pulse.**

### 6.2.2 Preparation of Optical Fibres

The choice of suitable materials for fabrication of optical fibres is limited to either glasses or glass-like materials and monocrystalline structures (certain plastics). In case of graded index fibres, it is essential that the refractive index of the material may be varied by suitable doping with another compatible material. Moreover, these two materials should have mutually solubility over a relatively wide range of concentrations. This is only achieved in glasses (or glass-like materials). Consequently, monocrystalline materials are unsuitable for the fabrication of graded index fibres, but may be used for step-index fibres. Therefore, only glasses are used in the fabrication of fibres for telecommunication applications, while plastic fibres find some use in short-haul, low-band width applications.

**1. Liquid-phase (melting) techniques.** (*i*) The first stage is preparation of *ultrapure* material powders, which are usually oxides (like $SiO_2$, $GeO_2$, $B_2O_3$ and $Al_2O_3$) or carbonates (like $Na_2CO_3$, $K_2CO_3$, $CaCO_3$ and $BaCO_3$) of the required constituents. Very high initial purity is very essential and purification itself accounts for a major proportion of the material cost. The purification involves combined techniques like fine filtration and coprecipitation, followed by solvent extraction by recrystallization and final drying in a vacuum to remove any resident OH ions.

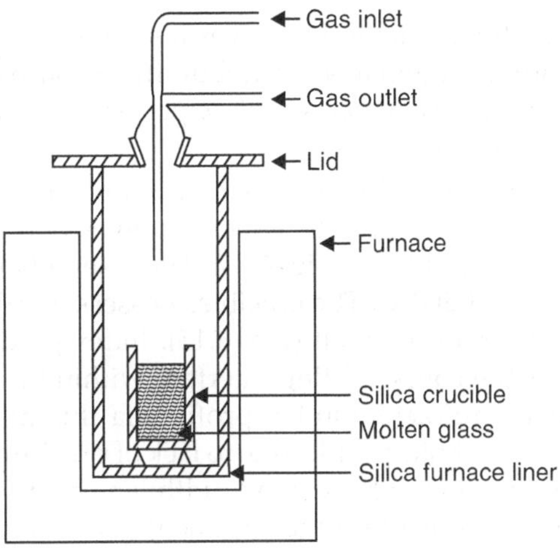


***FIGURE 6.11*** **Glass making furnace for the production of high purity glasses.**

The next stage is to melt these high purity powdered low melting glass materials to form a homogeneous bubble-free multicomponent glass. A refractive index variation may be achieved by either a *change in composition* of constituents or by *ion exchange*, when the materials are in the molten phase. The melting of such multi-components glass occurs at about 900 to 1,300°C and this is usually carried out in a *silica crucible* (*see* Fig. 6.11). However, use of silica crucible can give dissolution into the melt, thereby introducing inhomogeneities to the glass at high melting temperatures. So for getting homogeneous glass, melting is carried out directly by using *radio frequency* (RF approximating 5 MHz) *induction furnace* ; while cooling the silica by gas or water flow (Fig. 6.12). The materials are preheated to around 1,000°C, where they exhibit sufficient ionic conductivity to enable coupling between the melt and the RF field. In both the above techniques, the glass is homogenised and dried by bubbling pure gases through the melt. After the melt has been processed, it is cooled and formed into long rods (canes) of the multicomponent glass.

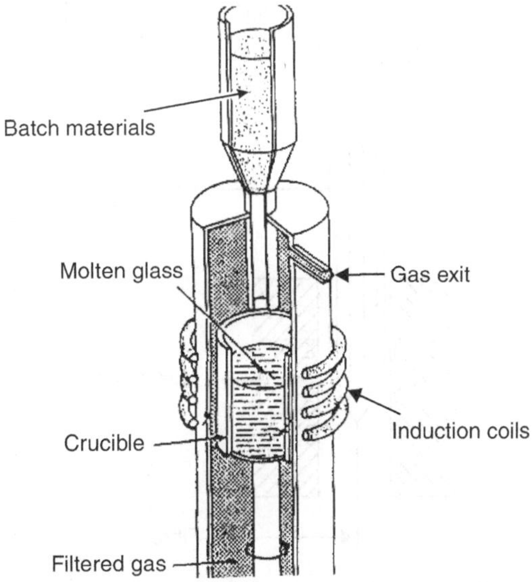


***FIGURE 6.12*** **High purity melting using a radio frequency induction furnace.**

(*ii*) **Fibre drawing.** Traditional method involves inserting a rod of core glass into a tube of cladding glass and then preform is drawn in a vertical muffle furnace (*see* Fig. 6.13). This technique is used for the production of *step-index fibres with large core and cladding diameters*, where the achievement of low attenuation is not critical.

Another technique for the production of large core diameter step-index fibres, with reduced core-cladding problems is *stratified melt-process*, which involves pouring a layer of cladding glass over the core glass in a platinum crucible (*see* Fig. 6.14). The assembly is located in a muffle furnace capable of heating between 800 to 1,200°C. The crucibles possess nozzles in their bases from which the clad fibre is drawn directly from the melt (Fig. 6.15). Index grading is achieved through the diffusion of mobile ions across the core-cladding interface within the molten mass. Although it is possible to achieve a reasonable refractive index profile via this diffusion process, yet it is not possible to obtain the optimum parabolic profile, due to lack of precise control. Hence, graded-index fibres produced by this technique do not have bandwidth length of optimum profile fibres. Some of the materials used in the production of multicomponent glass are given in Table 6.2.

**TABLE 6.2 Materials used in the fabrication of multicomponent glass fibres by the double crucible technique**

| *Type of fibres* | *Core glass* | *Cladding glass* |
|---|---|---|
| Step-index | $Na_2 - B_2O_3 - SiO_2$ | $Na_2O - B_2O_3 - SiO_2$ |
| | $Na_2 - LiO - CaO - SiO_2$ | $Na_2O - Li_2O - CaO - SiO_2$ |
| | $Na_2 - CaO - GeO_2$ | $Na_2O - CaO - SiO_2$ |
| | $Na_2O - BaO - GeO_2 - B_2O_3 - SiO_2$ | $Na_2O - B_2O_5 - SiO_2$ |
| | $P_2O_5 - Ga_2O_3 - GeO_2$ | $P_2O_5 - Ga_2O_3 - SiO_2$ |
| Graded-index | Base glass | Diffusion mechanisms |
| | $R_2O - B_2O_3 - CaO - SiO_2$ | $Na^+ \rightleftharpoons K^+$ |
| | $R_2O - B_2O_3 - SiO_2$ | $Tl^+ \rightleftharpoons Na^+$ |
| | $Na_2O - B_2O_3 - SiO_2$ | $Na_2O$ diffusion |
| | $Na_2O - B_2O_3 - SiO_2$ | CaO, BaO diffusion |

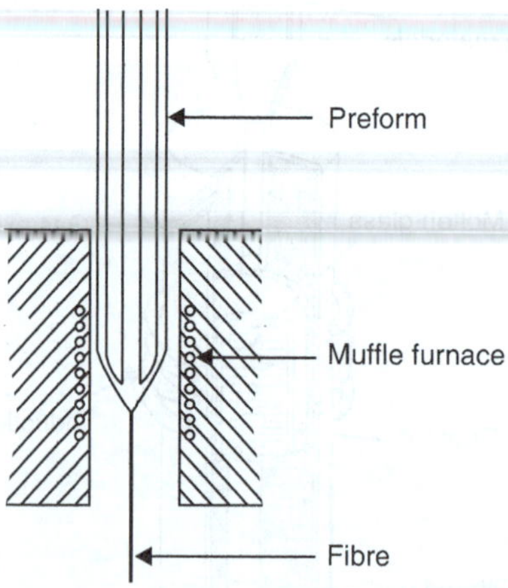


*FIGURE 6.13* **Optical fibre from a preform.**

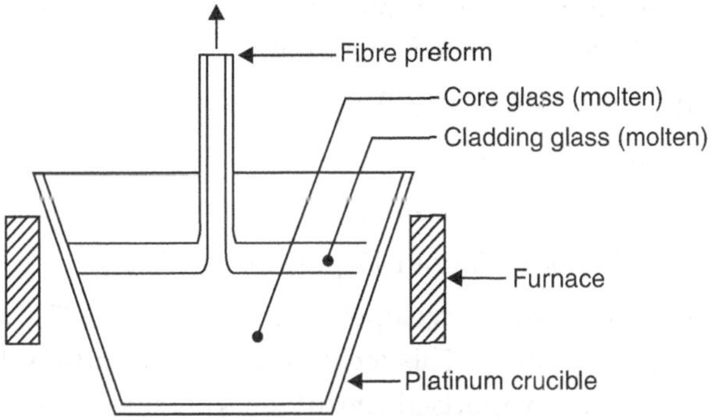


*FIGURE 6.14* **Stratified melt process for producing glass-clad rods or preforms.**

By using very high purity melting technique and double crucible drawing methods, step-index and graded-index fibres with attenuations of the order 3.4 dB $km^{-1}$ and 1.1 dB $km^{-1}$ respectively have been produced. Further, these methods give continuous production of optical glass fibres.

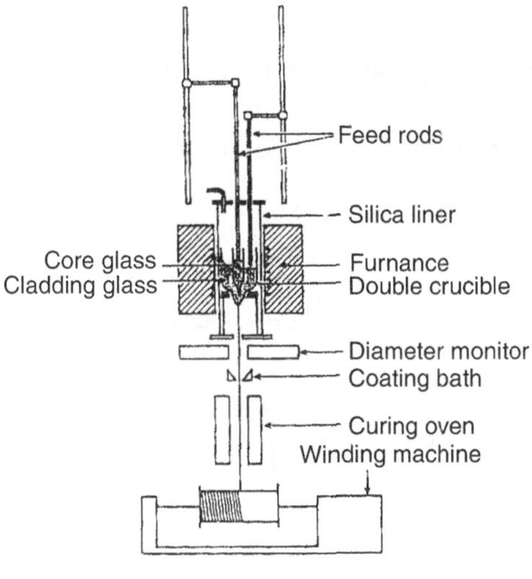


*FIGURE 6.15* **The double cruscible method for optical fibre drawing.**

**2. Vapour-phase deposition techniques.** These techniques are employed for producing *silica-rich glasses* of high transparency and with optimal optical properties. The starting materials are volatile components like $SiCl_4$, $GeCl_4$, $SiF_4$, $BCl_3$, $BBr_3$, $O_2$ and $POCl_3$, which may be distilled to reduce the concentration of transition metal impurities below 1 part per $10^9$, thereby giving negligible absorption losses. Modification of refractive index is achieved through the formation vapour phase dopant like $TiO_2$, $GeO_2$, $P_2O_5$, $B_2O_3$ and $F_2$. Gaseous mixtures of the silica-containing compound, the doping material and oxygen are combined in the vapour phase on to a substrate or within a hollow tube and is thus built up as a stack of successive layers. Consequently, the dopant concentration can be varied gradually to produce a graded-index profile or may be maintained to give a step—index profile.

In case of the *substrate*, this directly gives a solid rod (or preform) ; while the *hollow tube* is collapsed to give a solid preform from which the fibre is drawn.

There are a number of variations of vapour-phase deposition, which have been successfully utilized to produce low loss fibres. Major techniques are:

(*i*) Outside Vapour Phase Oxidation (OVPO) process.

(*ii*) Vapour Axial Deposition (VAD).

(*iii*) Modified Chemical Vapour Deposition (MCVD).

(*iv*) Plasma-Activated Chemical Vapour Deposition (PCVD).

**Vapour Axial Deposition (VAD)** developed by Izaw et al in 1977 is a continuous process for the production of low loss optical fibres. This technique uses an end-on deposition onto rotating fused silica target (*see* Fig. 6.16). The vaporised constituents are injected from the burners and react to form *'silica soot'* by flame hydrolysis. This is deposited on the end of the starting target in the axial direction forming a solid porous glass preform in the shape of a *boule*. The preform which is growing in the axial direction is pulled upwards at a rate which corresponds to the growth rate. It is initially dehydrated by heating with $SOCl_2$ and is then sintered into a solid preform in a graphite

$$\underset{\text{(Vapour)}}{H_2O} + \underset{\text{(Vapour)}}{SOCl_2} \longrightarrow \underset{\text{(Gas)}}{2HCl} + \underset{\text{(Gas)}}{SO_2}$$

resistance furnace at an elevated temperature of about 1,500°C. Consequently, this method may be adapted to draw fibre continuously. However, at present it is also operated as a batch process, since the resultant preforms can yield more them 100 km of fibre.

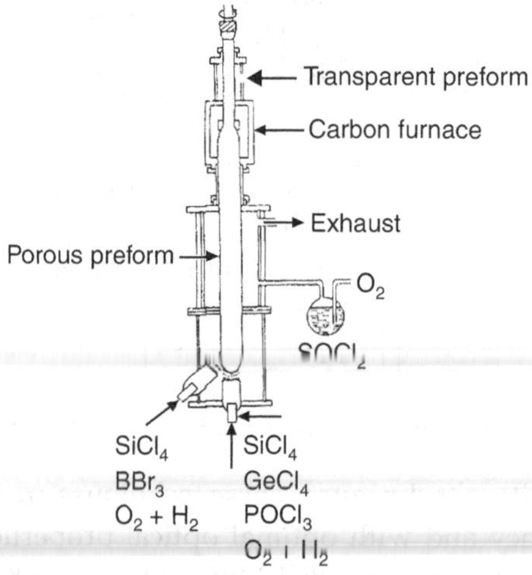


***FIGURE 6.16*** **The VAD process of continous process of production of optical fibres.**

A spatial reflective index profile is achieved by using the deposition properties of $SiO_2$–$GeO_2$ particles within the oxygen-hydrogen flame. The concentration of these constituents deposited on the porous preform is controlled by the substrate temperature distribution, which can be varied by altering the gas flow condition. Fibres produced by VAD possess attenuation in the range of 0.7 to 2.0 dB km$^{-1}$ at a wavelength of 1.181 μm.

**3. Plastic-clad optical fibres.** Plastic-clad optical fibres are multimode and have either a step-index or a graded-index. They have a plastic cladding of *silicone rubber* and a *glass core,* which is often silica (*i.e.,* plastic-clad silica – PCS). They exhibit lower radiation-induced losses than silica-clad silica fibres and consequently, possess an improved performance in certain environments. They are, usually, slightly cheaper than the corresponding silica fibres, but usually have more limited performance characteristics. A typical structure of more common step-index plastic-clad optical fibre is shown in Fig. 6.17.

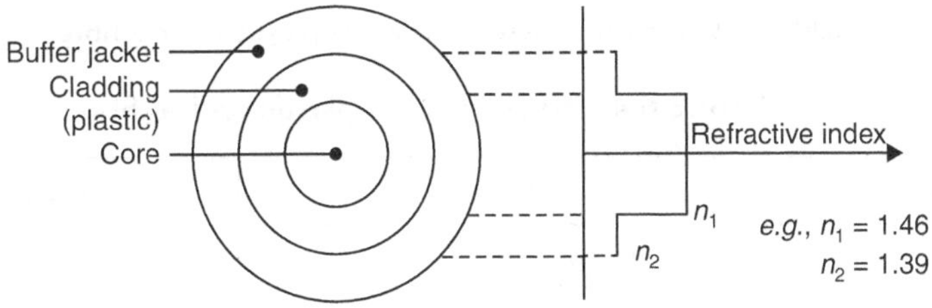


*FIGURE 6.17* **Typical structure of a plastic-clad silica multimode step-index optical fibre.**

**Table 6.3 Plastic-clad optical fibres**

| *Structure* | *Step-index* | *Graded-index* |
|---|---|---|
| Core diameter | 100 to 500 μm | 50 to 100 μm |
| Cladding diameter | 300 to 800 μm | 125 to 150 μm |
| Buffer jacket diameter | 500 to 1,000 μm | 250 to 100 μm |
| Performance characteristics (attenuation) | 5 to 50 dB km$^{-1}$ | 4 to 15 dB km$^{-1}$ |

**4. All-plastic optical fibres.** All-plastic (or polymeric) optical fibres are exclusively of the multimode step-index type with large core and cladding diameters. Consequently, these do not require any buffer jacket for protection and strength. These fibres are usually cheaper to produce and easier to handle than the corresponding silica-based glass optical fibres. However, their performance (particularly optical transmission in IR) is restricted. However, generally these have large numerical apertures, which permit easier coupling of light into the fibre from a multimode source.

Early plastic optical fibres fabricated with a polymethyl methacrylate (PMMA) and a fluorinated acrylic cladding exhibited losses around 500 dB km$^{-1}$. With the subsequent development of a continuous casting process for PMMA, losses as low as 100 dB km$^{-1}$ were achieved in the visible region. The loss mechanisms in PMMA and PS (polystyrene) core fibres are similar to those of glass fibres. These fibres exhibit losses including absorption and Rayleigh scattering which results from density fluctuations and the anisotropic structure of the polymers.

Subsequent progress in the fabrication of higher performance plastic fibres in recent years led to the development PMMA core fibres with losses as low as 20 dB km$^{-1}$ in operating wavelength of 0.68 μm. A typical all plastic fibre structure is illsutrated in Fig. 6.18.

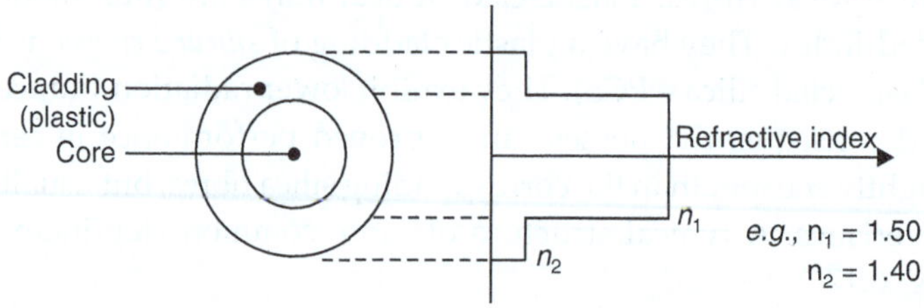


*FIGURE 6.18* **Typical structure for as all-plastic optical fibre.**

**TABLE 6.4 Structure of all-plastic optical fibres**

| *Structure* | *Core diameter* | *Cladding diameter* |
|---|---|---|
| | 200 to 600 μm | 450 to 1000 μm |
| **Performance characteristics** | **Attenuation** | **Bandwidth** |
| | 50 to 1000 dB $km^{-1}$ at a wavelength of 0.65 μm | Limited to tens of metres |

**TABLE 6.5 Important all-plastic optical fibres**

| *Core material* | *Cladding materials* |
|---|---|
| 1. Polymethyl methacrylate (PMMA) | Fluorinated acrylic polymer |
| 2. Polystyrene (PS) | PMMA |
| 3. Deuterrated PMMA (PMMA-d8) | PMMA-df |
| 4. Polymethyl methyacrylate -co- vinyl phenyl acetate *p* (MMA-co-VPAC) | PMMA |
| 5. Polycarbonate (PC) | PMMA |
| 6. Ploymethyl methyacrylate -co- vinyl benzoate *p* (MMA-co-VB) | PMMA |

**5. Fabrication of All plastic optical fibres.** Core materials used for the fabrication of all-plastic optical fibres are PMMA and PS, which are usually synthesized by emulsion polymerization. This method involves the dispersal of monomer in another liquid so as to produce granular polymeric products. However, removal of contaminants in this process is quite difficults, so a special bulk polymerization technique in a closed system is adopted for the preparation of all-plastic optical fibres. Fig. 6.19 shows the essential features of the *continuous extrution method*. The desired monomer containing traces of polymerization initiator and chain transfer agent is continuously fed to the heated reactor ; while the clad fibre is continuously withdrawn from the die. At the reactor temperature of about 150°C, the polymer-monomer concentrated fluid is pumped to the devolatilizing extruder by the gear pump. In devolatilizing cum core polymer extruder, excess monomer evaporates (due to reduction in pressure), which is recycled to reactor. As the polymer emerges from the die, it is coated by means of another cladding polymer extruder.

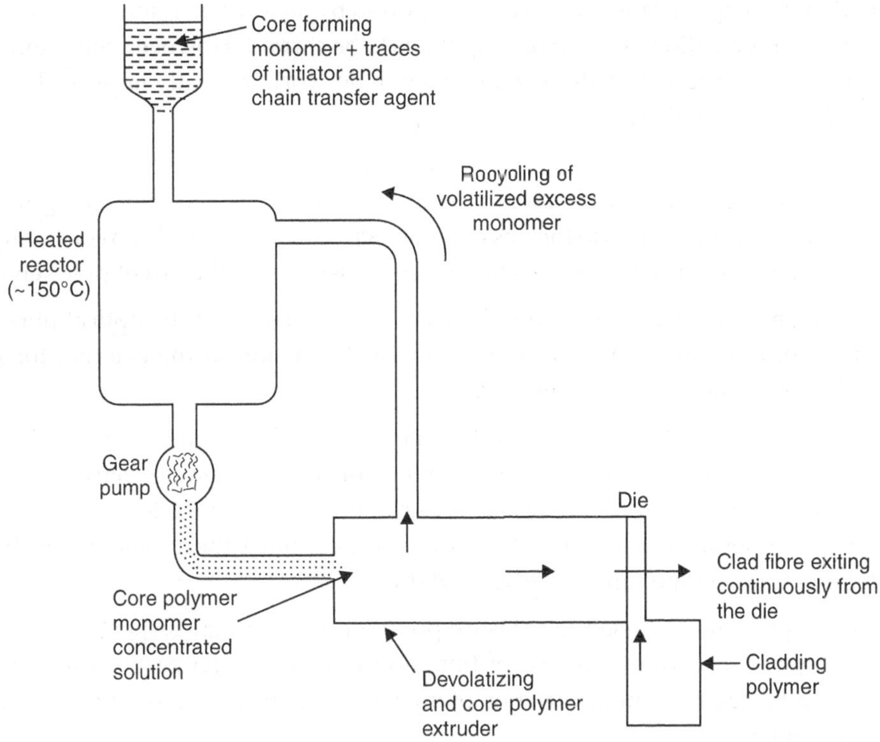


*FIGURE 6.19* **Schematic representation of continuous extrusion process for obtaining all-plastic optical fibre.**

### 6.2.3 Functions of Optical Cable

(*i*) *Fibre protection.* The major function of the optical cable is to protect against fibre damage and breakage, both during installation and throughout the life of the fibre.

(*ii*) *Stability of the fibre transmission characteristics.* The cabled fibre must possess stable transmission characteristics, comparable with the uncabled fibre. Increases in optical attenuation due to cable, if any, must be minimized within the cable design.

(*iii*) *Cable strength.* Optical cables must possess similar properties to electrical transmission cables so that they may be handled in the same manner. These mechanical properties include tension, tortion, compression, bending, vibration, etc. The cable strength may, therefore, be improved by incorporating a suitable strength member and by giving the cable a property designed thick outer sheath.

(*iv*) *Identification and joining of the fibres with the cable.* This is particularly important for cables including a large number of optical fibres. If the fibres are arranged in a suitable geometry, then it may be possible to use multiple jointing techniques rather than jointing each fibre individually.

### 6.2.4 Applications of Optical Fibres

**1. Military applications.** Optical fibres are finding lot of applications in various warfares and military operations.

If conventional copper cables are to be transported by an aircraft, a ship or a tank for control mechanisms of communication, then lot of weight has to be carried. However, with fibre optics, the weight is enormously reduced and also communication network is greatly improved, as with fibre optics; secrecy is also maintained.

During recent wars, fibre guided missiles have been extensively used. Sensors are mounted on missiles to collect the video information. This information is passed on to the ground control van which sends further commands to the missiles through sensors. Control van constantly monitors the course of missiles and corrects the direction of the missile to hit the target precisely.

**2. Optical fibre Sensors.** In this case the basic characteristics of the optical fibres are used, *viz*, the variation of refractive index of the fibre under the action of the external forces may be utilized for using the optical fibre as transducer.

In case the fibre is subjected to heat, the refractive index changes because of change of temperature. As temperature increases, the difference of refractive index between the core and cladding is reduced, thereby some light leaks into cladding. This characteristics can be used for measurement of temperature using a LED as light source. Such thermometer can be used for measuring temperatures in the range of 80°C to 700°C.

Also fibre optics may be used to measure pollution and foreign suspended particles in the air. A beam of light is send from one end of fibre and in case foreign particles are present, they scatter the light and measurement of variation of intensity of light will account on the extent of the presence of foreign particles.

Fibre optics may also be used to measure the level of a liquid in a container. A part of fibre is suspended in the liquid and light is directed to pass through it and its intensity is measured. A bare core loses more light when it is immersed in liquid than when it is placed in air. A sudden change in intensity of light will indicate the liquid level. A LED source, a photodector and a MMF is used to monitor the liquid level and in finding many applications in the filling of petroleum tanks.

In addition, fibre optics can be used in the fabrication of various sensors using the phenomenon of double refraction and birefringes. Due to the application of external forces, the optical properties of the fibre are varied. For example, we can have a pressure sensor using the induced birefringes, current sensor using the Faradays rotation, magnetic sensor using the variation of magnetic fields and voltage sensor to account for the variation of voltages.

**3. Entertainment Applications.** A coherent optical fibre bundle can greatly enhance the size of the image on a TV screen. In case we have to use the conventional method of getting larger images then the projection system becomes quite bulky and unmanageable.

**4. Medical Applications.** Fibre optics are finding large number of applications in medical fields. Many parts of the human body which are otherwise inaccessible can be studied by illumination with fibre optics. A bundle of MMF is used to illuminate a part of human organ and the other part to collect the reflected light. This fibrescopic technique is employed for endoscopic applications. Fibre optics can be employed to attach a detached retina or rectify other eye defects using lasers.

In heart problems also lasers angioplasty has replaced the ballon angioplasty and bypass surgery. A special catheter is used which consists of three channels, viz. (*i*) to visualize the various veins (*ii*) to supply power for lasers (*iii*) an open tube to collect and suck the debris and other blocking materials.

**5. In Communications.** A fibre optic communication system is similar to the traditional communication system with the exception that this system is highly efficient.

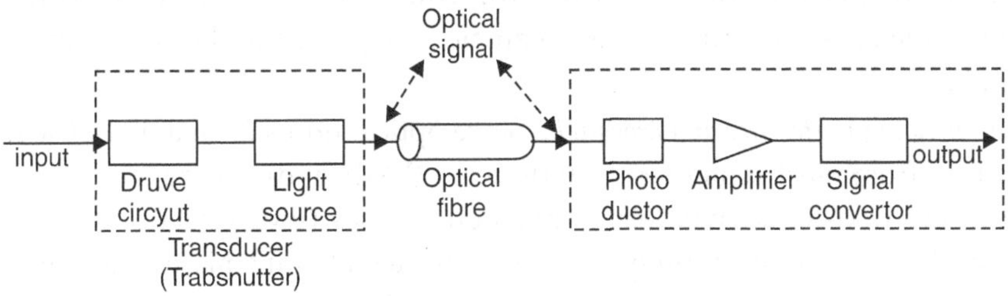


*FIGURE 6.20*

A block diagram of the typical communication system is shown in Fig. 6.20. A transducer changes a non-electrical message to an electrical signals and in case of fibre communication light waves act as carrier waves. Light waves are generally generated by a laser diode or a light emitting diode (LED) and the intensity of light given by these sources is proportional to the current passing through them.

As a message in the form of an electrical signal is received by the optical source, light output will follow the path of variation of signal with the message digital modulation is preferred in optical communication over the analogue modulation as it can pass through a larger distance with the same power.

However, one basic difficulty arises due to the attenuation of optical signal over large distances because of various losses and dispersion in the optical fibre. So to overcome it various repeaters are used in the transmission line to amplify and reshape the signal.

At the receiver end an optical coupler is used to direct the light coming from the fibre into a photoduetor or, which converts light into electric current, which will contain the message fed into the optical fibre. In the next step, the message is extracted from this current and is amplified. Then this message is put into a transducer to get it either in video or radio form depending upon the nature of the original message.

**Advantages of optical fibre communication**

(*i*) *Enormous potential bandwidth.* The optical carrier frequency in the range $10^{13}$ to $10^{16}$ Hz (generally in the near infrared around $10^{14}$ Hz or $10^5$ GHz) yields a far greater potential transmission band width than metallic cable system (*i.e.*, coaxial cable bandwidth upto around 500 MHz).

(*ii*) *Small size and weight.* Optical fibres have very small diameters, usually not greater than the diameter of a human hair. Hence, even when such fibres are covered with protective coatings, they are far small and much lighter than corresponding copper cables. For example, it would required 30,000 kg (30 tonnes) of copper to transmit the same amount of information as only 1 kg of the optical fibre materials.

(*iii*) *Electrical isolation*. Since optical fibres are fabricated from glass (or sometimes a plastic polymer), so they are electrical insulators and therefore, unlike their metallic counterparts, do not exhibit earth loop and interface properties. Moreover, this property makes optical fibre transmission ideally suited for communication in electrically hazardous environments, since optical fibres create no arcing or spark hazard at abrasions or short circuits.

(*iv*) *Immunity to interference and cross-talk*. Since optical fibres form a dielectric waveguide, so are free from electromagnetic interference (EMI), radio frequency interference (RFI), switching transients given electromagnetic pulses (EMP). Hence, the operation of an optical fibre communication system is not affected by transmission through an electrically noisy environment. Moreover, optical fibre cable is not susceptible to lightning strikes, if used overhead rather than underground.

(*v*) *Signal security*. Optical fibres provide a high degree of signal security, since the light from optical fibres does not radiate significantly.

(*vi*) *Low transmission loss*. Optical fibres have been fabricated with losses as low as 0.2 dB $km^{-1}$ and this feature has become a major advantage of optical fibre communication.

(*vii*) *Ruggedness and flexibility*. Optical fibre cables are, generally, superior in terms of storage, transportation, handling and installation to corresponding copper cables, whilst exhibiting comparable strength and durability.

(*viii*) *System reliability and ease of maintenance*. The lifetime of optical fibres is predicted as 20 to 30 years. This in-turn have led to reduction in their maintenance costs.

(*ix*) *Improved speed of transmission, information density, transmission distance and reduction in error rate*. With regard to speed, optical fibres can transmit, in one second information equivalent to three episodes of your favourite television program. Or relative to information density, two small optical fibres can transmit the equivalent of 24,000 telephone calls simultaneously.

## 6.3 ORGANIC ELECTRONIC MATERIALS

Certain newly produced organic materials are found to be electrical conductors, these behaving like metals in this respect a few of them are superconductors. Such materials are called organic electronic materials. For details of organic electronic materials **please refer to section 2.**

## 6.4 FULLERENES

Pleases refer to **section 15.8** for details.

## QUESTION BANK

1. Define superconductivity. Discuss its origin.
2. Describe BCS theory of superconductivity.
3. Discuss different types of superconductors.
4. What is superconductivity? Describe its main applications.
5. What are optical fibres? Explain their structure.
6. Explain the process of light transmission in optical fibres.
7. Write short notes on following:
   (*a*) Vapour phase deposition technique
   (*b*) Plastic-clad optical fibres
   (*c*) All plastic optical fibres
8. Explain the function of an optical cable.
9. Describe applications of optical fibres.
10. Enumerate various advantages of optical fibre communication.
11. Write short notes on following:
    (*a*) Organic electronic materials
    (*b*) Fullerenes
12. Elaborate how new engineering materials have revolutionized today's life.

Chapter 7

# CHEMICAL KINETICS

## 7.1 INTRODUCTION

Chemical kinetics deals with the study of the rate of chemical reactions. Such studies also enable us to elucidate a mechanism by which a reaction occurs. In chemical kinetics the rate of the reaction is followed and time as a variable is introduced.

Experimentally, it is found that the rate of a chemical reaction is dependent on factors like temperature, pressure and the concentration of the reacting species. The presence of catalyst can increase or decrease the rate of a reaction by many powers of ten.

From the kinetics point of view, the reactions are classified into two groups, viz., (*a*) homogeneous reactions which occur in one phase only; it may be a gaseous phase or a liquid phase, and (*b*) heterogeneous reactions which take place in two or more phases, *e.g.*, gaseous reactions taking place on the surface of a solid catalyst or on the walls of the container.

## 7.2 RATE OF A CHEMICAL REACTION

*Rate, speed or velocity* of a reaction has the same meaning in the chemical kinetics. The rate of a reaction tells us to what speed the reaction occurs. It may be defined as *the change in concentration of any one of the reactants or prodacts per unit time.* Consider a simple reaction

$$\underset{\text{Reactants}}{A} \longrightarrow \underset{\text{Products}}{B}$$

The rate of the reaction may be expressed as :

Rate of reaction = rate of disappearance of A

= rate of appearance of B

$$= \frac{-\Delta[A]}{\Delta t} \qquad \text{...(7.1)}$$

$$= +\Delta[B]/\Delta t \qquad \text{...(7.2)}$$

where $\Delta[A\}$ represents decrease in the concentration of A in $molL^{-1}$, $\Delta[B]$ represents increase in concentration of B in $molL^{-1}$.

The negative sign along with the change in concentration of the reactant shows that the concentration of the reactant decreases with time while the plus sign along with the concentration of the product denotes that the concentration of the product increases with time.

The decrease in concentration of reactant(s) and increase in concentration of product(s) can graphically be represented in the Fig. 7.1.

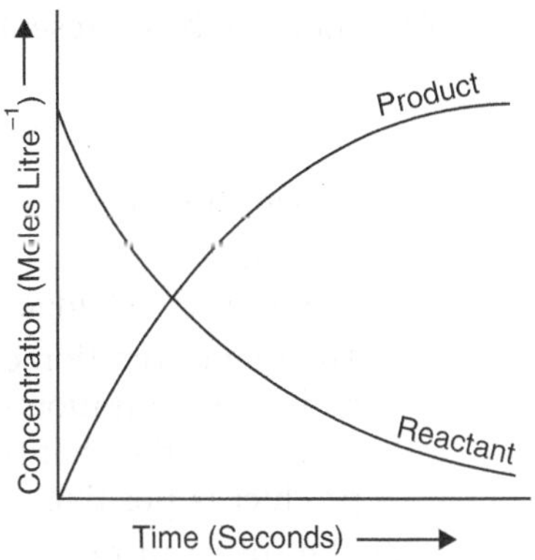


***FIGURE 7.1*** **Variation of concentration with time.**

Now let us extend this concept of expressing the rate of reaction to different kind of chemical reactions.

(*a*) **Reactions having same stoichiometric coefficients of reactants and products.**

For the reaction, $PCl_5 \rightleftharpoons PCl_3 + Cl_2$

Suppose during a time interval, $\Delta t$,

Decrease in concentration of $PCl_5$ is $\Delta [PCl_5]$

Increase in concentration of $PCl_3$ is $\Delta [PCl_3]$

Increase in concentration of $Cl_2$ is $\Delta [Cl_2]$

Where square brackets show molar concentration, *i.e.*, moles litre$^{-1}$ of the concerned substance.

Thus, rate of reaction is given by

$$\text{Rate of reaction} = -\frac{\Delta [PCl_5]}{\Delta t} = +\frac{\Delta [PCl_3]}{\Delta t} = +\frac{\Delta [Cl_2]}{\Delta t}$$

Likewise, for the reaction

$$NO_2(g) + CO(g) \longrightarrow CO_2(g) + NO(g)$$

$$\text{Rate of reaction} = -\frac{\Delta [NO_2]}{\Delta t} = -\frac{\Delta [CO]}{\Delta t} = +\frac{\Delta [CO_2]}{\Delta t} = +\frac{\Delta [NO]}{\Delta t}$$

(*b*) **Reactions having different stoichiometric coefficients of reactants and products**

Let us consider the reaction involving thermal decomposition of nitrogen pentaoxide ($N_2O_5$) as $2N_2O_5(g) \rightleftharpoons 4NO_2(g) + O_2(g)$, then

$$\text{The rate of disappearance of } N_2O_5 = -\frac{\Delta [N_2O_5]}{\Delta t}$$

$$\text{The rate of appearance of } NO_2 = +\frac{\Delta [NO_2]}{\Delta t}$$

$$\text{The rate of appearance of } O_2 = +\frac{\Delta [O_2]}{\Delta t}$$

Now, during the period in which one mole of $O_2$ is formed, four moles of $NO_2$ are formed and two moles of $N_2O_5$ have disappeared. Therefore, the rate of appearance of $NO_2$ is four times the rate of appearance of $O_2$. The rate of disappearance of $N_2O_5$ is twice the rate of appearance of $O_2$.

So, the rate of reaction irrespective of the species selected is expressed in the following form.

$$\text{Rate of reaction} = -\frac{1}{2}\frac{\Delta [N_2O_5]}{\Delta t} = +\frac{1}{4}\frac{\Delta [NO_2]}{\Delta t} = +\frac{\Delta (O_2)}{\Delta t}$$

In brief, the rate of reaction can be equated for any species (either reactant or product) by just dividing the rate of change of concentration by the stoichiometric coefficient of reactant or product expressed by the reaction.

Similarly, for the reaction

$$N_2(g) + 3H_2(g) \rightleftharpoons 2NH_3(g)$$

rate of reaction can be represented by

$$-\frac{\Delta[N_2]}{\Delta t} = -\frac{1}{3}\frac{\Delta[H_2]}{\Delta t} = +\frac{1}{2}\frac{\Delta(NH_3)}{\Delta t}$$

So, we can sum up the important points to express the rate of reaction as follows.

(*a*) The rate of reaction is expressed by the change in concentration divided by the time taken for the change.

(*b*) The rate of reaction can be given by utilizing any of the reactant or the product. Since molar concentration of reactant decreases with time and that of the product increases with time, the change in molar concentration of reactant is negative whereas that of product is positive.

(*c*) The different expressions for the rate can be made equivalent by dividing the rate expression by the stoichiometric coefficient shown by the balanced chemical equation.

**Units of the Rate of Reaction**

The units of the rate of reaction depend on the unit used to express the concentration of the reactant or product.

(*i*) The concentration is usually expressed in moles litre$^{-1}$ and the time in seconds or minutes. Therefore, the unit of the rate of reaction is

Moles litre$^{-1}$ sec$^{-1}$ or Moles litre$^{-1}$ min$^{-1}$

(*ii*) When the concentration of either gaseous reactant or product or both is expressed in atmosphere then the unit of the rate of reactions is given as atmosphere per second (atm sec$^{-1}$).

## 7.3 AVERAGE AND INSTANTANEOUS RATE OF A REACTION

The rate expression given in preceding equations (7.1) and (7.2) are average rates of reaction over the time interval ($\Delta t$). Average rate may be expressed as:

$$\text{Rate} = \frac{\text{Change in concentration}}{\text{Time interval}} = \frac{\Delta x}{\Delta t}$$

As the reaction proceeds, the concentration of the reactants keep on decreasing rapidly in the initial stages and more slowly later on. Thus the rate of reaction may not be constant in the time interval and therefore it cannot be determined simply by dividing the total change in the concentration by the time taken. The rate of reaction may be expressed at a particular moment of time as *the rate of change of concentration of any one of the reactants or products at a given time*. Such reaction rate is called *instantaneous rate.* For this purpose the time interval $\Delta t$ is made as small as possible so that the rate of reaction remains almost constant during that time interval.

Mathematically, instantaneous rates may be represented as $\frac{-d[A]}{dt}$ or $\frac{d[B]}{dt}$ instead of $\frac{-\Delta[A]}{\Delta t}$ or $\frac{\Delta[B]}{\Delta t}$ respectively. $d$[A] and $d$[B] represents infinitesimally small changes in concentration of A or B in infinitesimally small interval of time, $dt$. Thus, average rate of reaction approaches instantaneous rate of reaction as $\Delta t$ becomes smaller and smaller, *i.e.*,

$$-\left[\frac{\Delta(A)}{\Delta t}\right]_{\Delta t \to 0} = \frac{-d[A]}{dt} \text{ and } \left[\frac{\Delta[B]}{\Delta t}\right]_{\Delta t \to 0} = \frac{d[B]}{dt}$$

For a general reaction,

$$aA + bB \longrightarrow cC + dD$$

the instantaneous rate of the reaction is defined as,

$$\text{Rate of reaction} = -\frac{1}{a}\frac{d[A]}{dt} = -\frac{1}{b}\frac{d[B]}{dt} = \frac{1}{c}\frac{d[C]}{dt} = \frac{1}{d}\frac{d[D]}{dt}.$$

**Experimental Determination of Reaction Rate**

The rate of a reaction is determined by determining the concentration changes of either reactant or product. As per the usual practice a small portion of the reaction mixture is withdrawn (2.0 m L or 5.0 mL) at a suitable interval of time. The samples are immediately cooled to about 0°C to stop the reaction. This is called freezing the reaction. The concentration of reactant or product is determined by using a suitable method *e.g.*, by titration against a suitable reagent, by measuring electrical conductivity, optical rotation or refractive index, etc.

A curve is then plotted between concentration of the reactant and time as shown in Fig. 7.2. The rate of reaction is calculated by drawing a tangent to the curve at the point. The tangent meet,

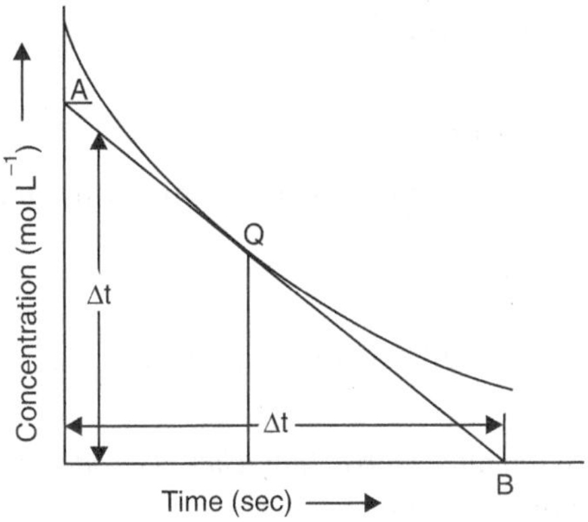


***FIGURE 7.2*** **Determination of instantaneous rate of reaction.**

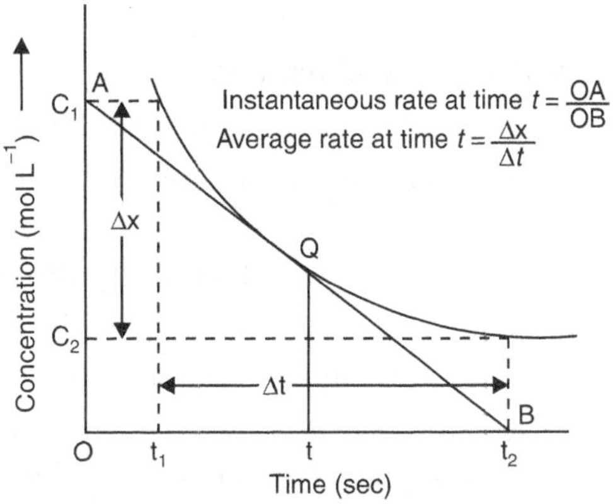


***FIGURE 7.3*** **Calculation of average rate of reaction.**

the axis at points A and B, then,

$$\text{Rate of reaction} = \frac{\text{Change in Concentration}}{\text{time}}$$

Therefore,

$$= \frac{OA}{OB} = \frac{\Delta x}{\Delta t}$$

$$\text{Instantaneous rate} = \left(\frac{\Delta x}{\Delta t}\right)_{\Delta t \to 0} \text{ or } \frac{d[B]}{dt}$$

$$= \text{Slope of the tangent}$$

Hence, slope of the tangent gives the instantaneous rate of the reaction.

However, to calculate the average rate of reaction, the change in concentration during the time interval from $t_1$ to $t_2$ is measured as represented in the curve of Fig. 7.3.

$$\text{Average rate of reaction} = \frac{C_1 - C_2}{t_2 - t_1} = \frac{\Delta C}{\Delta t}$$

## 7.4 RATE LAW AND RATE CONSTANT

In 1864 *Gulberg* and *Waage* established that the rate of a homogeneous reaction is proportional to some power of the concentration of the reactants. Thus, for a substance A undergoing reaction

$$\text{rate} \propto [A]^n$$

or

$$\text{rate} = k\,[A]^n \tag{7.3}$$

For a chemical reaction

$$2A + B \longrightarrow \text{products}$$

$$\text{rate} = k\,[A]^m\,[B]^n \tag{7.4}$$

*An expression which shows how the reaction rate is related to concentrations is called rate law or rate equation.* The proportionality constant $k$ is called the *rate constant.* For example,

| *Reaction* | *Rate law* |
|---|---|
| (*i*) $H_2 + I_2 \longrightarrow 2HI$ | rate = $k\,[H_2]\,[I_2]$ |
| (*ii*) $2NO_2 \longrightarrow 2NO + O_2$ | rate = $k\,[NO_2]^2$ |

Consider a general reaction,

$$aA + bB + cC \longrightarrow \text{Products}$$

where, A, B and C are the reactants and $a$, $b$, and $c$ are stoichiometric coefficients of the balanced equation,

$$\text{Rate} = [A]^p\,[B]^q\,[C]^r \tag{7.5}$$

The values of $p$, $q$, and $r$ are to be determined experimentally and may or may not be equal to $a$, $b$, and $c$ coefficients in the reaction respectively.

*Rate Constant or Specific Reaction Rate* (*k*) *is equal to the rate of a reaction when the concentration of each of the reactants is unity.* For example, consider a general equation

$$aA + bB + cC \longrightarrow \text{products}$$

$$r = k\,[A]^a\,[B]^b\,[C]^c$$

if

$$[A] = [B] = [C] = 1, \text{ then,}$$

$$rate = k.$$

### Characteristics of Rate Constant

Following are the important characteristics of rate constant:

(*i*) The value of $k$ is constant and characteristic of a reaction at a given temperature.

(*ii*) The value of $k$ differs from reaction to reaction.

(*iii*) Larger value of $k$ reflects a faster reaction whereas small value of $k$ refers to a slower reaction.

(*iv*) For a specific reaction $k$ depends on temperature but is independent of concentration.

### Comparison between Rate of Reaction and Reaction Rate Constant

The comparison between the rate of reaction and reaction rate constant is shown in Table 7.1.

**TABLE 7.1 Comparison between Rate of Reaction and Reaction Rate Constant**

| *Rate of Reaction* | *Reaction Rate Constant* |
|---|---|
| (*i*) Rate of reaction is defined as the change in concentration of the reactant or product per unit time. | (*i*) It is the proportionality constant in the rate law equation when molar concentration of reactants is taken as unity, its value is equal to rate of reaction. |
| (*ii*) Rate of reaction varies with the concentration of the reactant. | (*ii*) Since it is proportionality constant, it is independent of the concentration of the reactant. |
| (*iii*) Rate of reaction increases with increase in temperature. | (*iii*) Its value also varies directly with the temperature. |
| (*iv*) The units of rate of reaction is moles $liter^{-1}$ $tim^{-1}$. | (*iv*) The units of reaction rate constant depends upon the order of reaction, *i.e.*, its unit differs according to order of reaction. |

## 7.5 ORDER OF A REACTION

It is defined as *the sum of the powers to which the concentration terms are raised in the rate law expression.* For example, consider a general equation,

$$aA + bB + cC \longrightarrow \text{Products}$$

The rate of the reaction is given by the rate law as:

$$\text{Rate} = k\,[A]^p\,[B]^q\,[C]^r$$

then, the order of the reaction $n$, is given by,

$$n = p + q + r \qquad (7.6)$$

where $p$, $q$ and $r$ are the orders with respect to individual reactants and overall order of the reaction is the sum of these exponents.

When $n = 1$, the reaction is of first order, if $n = 2$ the reaction is of second order, and so on.

Common examples of reaction orders are given below:

**1. First order reactions**

(*i*)
$$NH_4NO_2 \longrightarrow N_2 + 2H_2O$$
$$\text{Rate} = k\,[NH_4NO_2]$$

(*ii*)
$$\underset{(g)}{N_2O_5} \longrightarrow \underset{(g)}{2NO_2} + \frac{1}{2}\underset{(g)}{O_2}$$

$$\text{Rate} = k\,[N_2O_5]$$

**2. Second order reactions**

(*i*) $$2NO_2 \longrightarrow 2NO + O_2$$

$$\text{Rate} = k\,[NO_2]^2$$

(*ii*) $$H_2 + I_2 \longrightarrow 2HI$$

$$\text{Rate} = k\,[H_2]\,[I_2]$$

**3. Third order reactions**

(*i*) $$2NO + O_2 \longrightarrow 2NO_2$$

$$\text{Rate} = k\,[NO]^2\,[O_2]$$

(*ii*) $$2NO + Cl_2 \longrightarrow 2NOCl$$

$$\text{Rate} = k\,[NO]^2\,[CI_2]$$

**4. Fractional order reactions**

(*i*) $$CH_3CHO \longrightarrow CH_4 + CO$$

$$\text{Rate} = k\,[CH_3CHO]^{3/2} \qquad \text{order} = \frac{3}{2}$$

(*ii*) $$\underset{(g)}{H_2} + \underset{(g)}{Br_2} \longrightarrow 2HBr$$

$$\text{Rate} = k\,[H_2]\,[Br_2]^{1/2} \qquad \text{order} = 1 + \frac{1}{2} = \frac{3}{2}.$$

## 7.6 ZERO ORDER REACTIONS

*A zero order reaction is one where the reaction rate does not depend on the concentration of the reactant or the rate is proportional to zeroth power of the concentration of reactants.* For example,

(*i*) Decomposition of $N_2O$ on a hot platinum surface

$$N_2O \longrightarrow N_2 + \frac{1}{2}O_2$$

$$\text{Rate} = k\,[N_2O]^0 \qquad \text{order} = 0$$

(*ii*) Decomposition of $NH_3$ at the surface of metals like gold, platinum, etc.

$$2NH_3 \xrightarrow{\text{Platinum}} N_2 + 3H_2$$

$$\text{Rate} = k\,[NH_3]^0 \qquad \text{order} = 0$$

(*iii*) $$NO_2 + CO \longrightarrow NO + CO_2 \text{ at } 200°C$$

$$\text{Rate} = k\,[NO_2]^2\,[CO]^0$$

The order of the reaction is zero with respect to CO and two with respect to $NO_2$.

## 7.7 MOLECULARITY OF A REACTION

*The number of reacting species (molecules, atom or ions) which collide simultaneously to bring about a chemical reaction* is called molecularity of a reaction. Thus in the reaction,

$$aA + bB \longrightarrow \text{Product}$$

$$\text{Molecularity} = a + b$$

The molecularity of the reaction may be 1, 2, 3, etc., according to participation of one, two, or three reactant molecules in the reaction. On the basis of molecularity, reactions are classified as:

(1) *Unimolecular reactions.* If a reaction involves the decomposition of only a single species, it is called unimolecular reaction. For example,

(*i*) $$H_2O_2 \longrightarrow H_2O + \frac{1}{2}O_2$$

(*ii*) 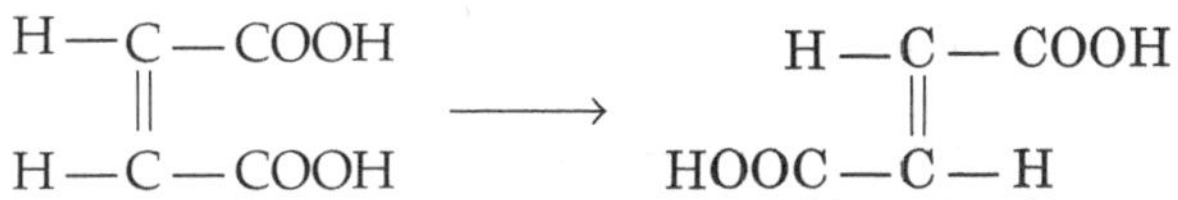

Maleic acid → Fumaric acid

(2) *Bimolecular reactions.* Reaction involving collision of two species.

For example,

(*i*) $$2HI \longrightarrow H_2 + I_2$$

(*ii*) $$CH_3COOCH_3 + H_2O \longrightarrow \underset{\text{Acetic acid}}{CH_3COOH} + \underset{\text{Methyl alcohol}}{CH_3OH}$$

(3) *Termolecular reaction.* For example,

(*i*) $2NO + O_2 \longrightarrow 2NO_2$ (*ii*) $2NO + Cl_2 \longrightarrow 2NOCl$

All the reactions discussed above are simple reactions, however, in complex reactions involving more than three molecules or ions in the balanced equation, the chances of all the reacting molecules to came close together and collide are very rare. All such complex reactions take place in number of steps and the slowest step is the rate determining step. The molecularity of the complex reactions, therefore, does not exceed beyond three and the molecularity of slowest step is called the molecularity of the reaction. Consider the following examples,

(*a*) **The decomposition of $N_2O_5$**

$$2N_2O_5 \longrightarrow 4NO_2 + O_2$$

This is a complex reaction and occurs in following three simple reactions, each having its own molecularity equal to the number of participating molecules.

(i) $2N_2O_5 \longrightarrow 2NO_2 + 2NO_3$ (slow, molecularity = 2)

(ii) $NO_2 + NO_3 \longrightarrow NO + NO_2 + O_2$ (slow, molecularity = 2)

(iii) $NO + NO_3 \longrightarrow 2NO_2$ (fast, molecularity = 2)

As is clear from the above simple reactions, the step (ii) is the rate determining step and hence the molecularity of the reaction is 2.

(b) **Reaction between $NO_2$ and $F_2$**

The reaction between $NO_2$ and $F_2$ gives $NO_2F$ which can be represented as

$$2NO_2 + F_2 \longrightarrow 2NO_2F$$

On the basis of experimental observation, the rate equation is as follows

$$\text{Rate} = k[NO_2][F_2]$$

On the basis of kinetic studies, the proposed mechanism for the reaction involves following two steps:

(i) $NO_2 + F_2 \longrightarrow NO_2F + F$ (slow step)

(ii) $NO_2 + F \longrightarrow NO_2F$ (fast step)

$2NO_2 + F_2 \longrightarrow 2NO_2F$ (overall reaction)

∴ Molecularity of reaction = 3

Order of the reaction = 2

(c) **Decomposition of nitrous oxide**

The decomposition of nitrous oxide is given by the following reaction:

$$2N_2O \longrightarrow 2N_2 + O_2$$

The experimentally observed rate law equation is found to be

$$\text{Rate} = k\,[N_2O]$$

*i.e.*, the rate of reaction depends only upon single power of $N_2O$. So concentration of only one molecule of $N_2O$ is altered in the slow step. Thus, the mechanism for this reaction may be written as

(i) $N_2O \longrightarrow N_2 + O$ (slow step)

(ii) $N_2O + O \longrightarrow N_2 + O$ (fast step)

Hence, the above reaction is a first order reaction and its molecularity is two.

TABLE 7.2 **Differences between Order and Molecularity of a Reaction**

| *Order of Reaction* | *Molecularity of Reaction* |
|---|---|
| (*i*) It is the sum of the power of concentration terms in the rate law. | (*i*) It is the number of reacting species taking part in a single step chemical reaction. |
| (*ii*) It is always determined experimentally. | (*ii*) It is a theoretical concept. |
| (*iii*) It may be a whole number, zero or fractional. | (*iii*) It is always a whole number. |
| (*iv*) It can change with conditions such as pressure and temperature. | (*iv*) It is invariant for a chemical equation. |
| (*v*) It gives information about the slowest step and hence provides clue to the mechanism of the reaction. | (*v*) It does not give any clue about the mechanism of the reaction. |
| (*vi*) Order of reaction is based on the overall reaction. | (*vi*) The overall molecularity of the complex reaction has no significance, only the molecularity of the slowest step is the molecularity of the reaction. |

## 7.8 PSEUDO-UNIMOLECULAR REACTION

Acid catalyzed hydrolysis of ethyl acetate is represented by the equation

$$\underset{\text{Ethyl acetate}}{CH_3COOC_2H_5} + H_2O \xrightarrow{H^+} \underset{\text{Acetic acid}}{CH_3COOH} + \underset{\text{Ethyl alcohol}}{C_2H_5OH}$$

The apparent molecularity of this reaction is two but the kinetic studies have shown that the rate of reaction is dependent only on the concentration of ethyl acetate, *i.e.*,

$$-\frac{d\,[CH_3COOC_2H_5]}{dt} = k\,[CH_3COOC_2H_5]$$

It shows that the concentration of water does not change appreciably during hydrolysis. Water is present to such a large extent that its concentration remains almost constant during the reaction. $H^+$ or $H_3O^+$ functions as a catalyst, which is regenerated unchanged after the reaction. So it does not participate in the reaction.

A similar reaction is the acid catalyzed inversion of cane sugar

$$\underset{\text{Sucrose}}{C_{12}H_{22}O_{11}} + H_2O \xrightarrow{H^+} \underset{\text{Glucose}}{C_6H_{12}O_6} + \underset{\text{Fructose}}{C_6H_{12}O_6}$$

Here also the rate of reaction is proportional to the concentration of sucrose

$$-\frac{d\,[\text{sucrose}]}{dt} = k\,[\text{sucrose}]$$

Therefore, such reactions which are not truly first order reactions but show first order kinetics under certain conditions are termed as pseudo-first order reactions or pseudo-unimolecular reaction.

## 7.9 INTEGRATED RATE EQUATIONS

### 7.9.1 First Order Rate Equation

Consider the general first order reaction,

$$A \longrightarrow \text{Products}$$

Let $a$ mol $L^{-1}$ be the concentration of A at the beginning of the reaction and after time $t$ let $x$ mol $L^{-1}$ of A have changed into products. So, concentration of A at $t$ will be $(a - x)$ mol $L^{-1}$. For the first order reaction, the rate of the reaction is directly proportional to the concentration of the reactant A. Therefore,

$$\text{Rate} = \frac{dx}{dt} = k(a - x)$$

or

$$\frac{dx}{(a - x)} = kdt \qquad (7.7)$$

On integrating,

$$\int \frac{dx}{(a - x)} = k \int dt$$

$$-\ln(a - x) = kt + I \qquad (7.8)$$

where I is the constant of integration. At $t = 0$, $x = 0$

$$-\ln(a - 0) = I$$

or

$$I = -\ln a$$

Putting in equation (7.8)

$$-\ln(a - x) = kt - \ln a$$

$$kt = \ln \frac{a}{(a - x)}$$

$$k = \frac{1}{t} \ln \frac{a}{(a - x)}$$

Changing into common logarithm, $\ln x = 2.303 \log x$, we get

$$\boxed{k = \frac{2.303}{t} \log \frac{a}{(a - x)}} \qquad (7.9)$$

This equation (7.9) is the integrated rate expression for the first order reaction. The value of $k$ can be found by substituting the values of $a$ and $(a - x)$ determined experimentally at time interval $t$ during the course of the reaction.

This equation may be written in alternate form. Let the initial concentration of A be $[A_0]$ mol $L^{-1}$ at $t = 0$ and [A] mol $L^{-1}$ at time '$t$'. Therefore,

$$a = [A_0] \text{ and } (a - x) = [A]$$

thus, equation (7.9) becomes

$$\boxed{k = \frac{2.303}{t} \log \frac{[A_0]}{[A]}} \tag{7.10}$$

Equation (7.9) may also be written in following form,

$$\boxed{k = \frac{2.303}{t_2 - t_1} \log \frac{(a - x_1)}{(a - x_2)}} \tag{7.11}$$

where $x_1$ and $x_2$ are the amounts decomposed in time $t_1$ and $t_2$ respectively from the start of the reaction.

### Common Examples of First Order Reaction

1. **Decomposition of $N_2O_5$ in $CCl_4$**

$$2N_2O_5 \longrightarrow 4NO_2 + O_2$$

The progress of the reaction is monitored by measuring the volume of $O_2$ evolved from time to time. Let $V_t$ be the volume of $O_2$ at any time $t$ and $V_\infty$ be the final volume of $O_2$. When the reaction is completed, then $V_\infty$ is the measure of the initial concentration of $N_2O_5$ and $(V_\infty - V_t)$ is the measure of undecomposed $N_2O_5(a - x)$ remaining at time $t$, thus,

$$k = \frac{2.303}{t} \log \frac{V_\infty}{(V_\infty - V_t)} \tag{7.12}$$

2. **Decomposition of $H_2O_2$ in aqueous solution with platinum as catalyst**

$$H_2O_2 \xrightarrow{Pt} H_2O + O$$

The progress of reaction is monitored by titrating with known solution of $KMnO_4$ at different time intervals.

3. **Hydrolysis of an ester.** The hydrolysis of an ester catalyzed by $H^+$ ions is given by the equation:

$$CH_3COOC_2H_5 + H_2O \xrightleftharpoons{H^+} CH_3COOH + C_2H_5OH$$

For studying the kinetics of the reaction, a known volume of ethyl acetate is mixed with a large excess of acid HCl (N/2). At various intervals of time a known volume of the reaction mixture is titrated against a standard alkali solution N/50 NaOH. As the reaction proceeds, the volume NaOH used goes on increasing.

The amount of NaOH used is equivalent to the amount of HCl present originally and the amount of acetic acid produced in the reaction. The amount of HCl present can be determined by titrating against the same alkali before the start of the reaction. The amount of acetic acid produced after different intervals of time $t$ can be determined.

The amount of acetic acid formed at the end of the reaction is equivalent to the initial amount of $a$ of the ester. Suppose $V_o$, $V_t$ and $V_\infty$ are the volumes of N/50 NaOH solution used at zero, $t$ and infinite time, respectively.

The amount of acetic acid produced after time $t$, *i.e.*, the value of $x$ is proportional to $(V_t - V_o)$. The initial concentration of methyl acetate, *i.e.*, the value of $a$ is proportional to $(V_\infty - V_o)$. Therefore, the amount of ester present of time $t$,

$$(a - x) \propto (V_\infty - V_o) - (V_t - V_o)$$

$$\propto (V_\infty - V_t)$$

Hence from first order rate expression

$$k = \frac{2.303}{t} \log \frac{V_\infty - V_o}{V_\infty - V_t} \qquad (7.13)$$

4. **Inversion of cane sugar catalyzed with dil. HCl.** The reaction can be represented by the equation

$$C_{12}H_{22}O_{11} + H_2O \longrightarrow \underset{\text{D-Glucose}}{C_6H_{12}O_6} + \underset{\text{D-Fructose}}{C_6H_{12}O_6}$$

The reaction is bimolecular but as the concentration of water remains almost constant, the reaction obeys the first order rate expression. The progress of reaction is monitored by noting the optical rotation of reaction mixture with polarimeter at different time intervals. The optical rotation goes on changing as D-glucose rotates plane polarized light (PPL) to right and D-fructose rotates it to the left. The change in rotation is directly proportional to the amount of sugar decomposed.

Let the intitial rotation be $r_o$, at time $t$ the rotation be $r_t$ and final rotation be $r_\infty$ then,

initial concentration, $a \propto (r_o - r_\infty)$

concentration at time $t$, $(a - x) \propto (r_t - r_\infty)$

Substituting in the first order rate equation

$$\therefore \qquad k = \frac{2.303}{t} \log \frac{(r_o - r_\infty)}{(r_t - r_\infty)} \qquad (7.14)$$

### 7.9.2 Second Order Rate Equation

Reactions where rate is determined by the change of two concentration terms are known as second order reactions. In case of second order reactions, there are two possibilities:

**Case (i)** When both the reactants are same, the reaction may be represented schematically as,

$$2A \longrightarrow \text{Products}$$

and the rate may be expressed

$$\frac{dx}{dt} = k(a - x)^2 \tag{7.15}$$

where, $a$ mol $L^{-1}$ is the initial concentration $x$ moles is the amount of A that has reacted in time $t$, and $(a - x)$ mol $L^{-1}$ is the concentration of A after time $t$.

On rearranging we have

$$\frac{dx}{(a - x)^2} = kdt \tag{7.16}$$

On integrating the above equation, we have

$$\int \frac{dx}{(a - x)^2} = k \int dt$$

$$\frac{1}{(a - x)} = kt + I \tag{7.17}$$

where, I is the constant of integration.

At the start of the reaction, when $t = 0$, $x = 0$, putting in above equation

$$I = \frac{1}{a}$$

Substituting in equation (7.17)

$$kt = \frac{1}{a - x} - \frac{1}{a}$$

$$k = \frac{1}{t}\left(\frac{1}{a - x} - \frac{1}{a}\right)$$

$$\boxed{k = \frac{1}{t}\left(\frac{x}{a(a - x)}\right)} \tag{7.18}$$

**Case (ii)** When both the reactants are different, the reaction may be represented schematically as

$$A + B \longrightarrow \text{Products}$$

Let the initial concentration of A and B be $a$ mol $L^{-1}$ and $b$ mol $L^{-1}$, respectively, and after time $t$, let $x$ moles of A and B react. Thus, the reactant concentrations at time $t$ are $(a - x)$ and $(b - x)$ in mol $L^{-1}$, respectively.

The rate may be expressed as

$$\frac{dx}{dt} = k\,(a - x)\,(b - x) \tag{7.19}$$

On rearranging we have

$$\frac{dx}{(a - x)\,(b - x)} = k\,dt \tag{7.20}$$

where $k$ is the second order rate constant.

On resolving into partial fraction (assuming $a > b$) we have

$$\frac{1}{(a - x)\,(b - x)} = \frac{1}{(a - b)}\left[\frac{1}{b - x} - \frac{1}{a - x}\right]$$

On substituting in equation (7.20) and integrating the equation

$$\frac{1}{(a - b)}\left[\int \frac{dx}{b - x} - \int \frac{dx}{a - x}\right] = k \int dt \tag{7.21}$$

$$\frac{1}{(a - b)}\,[-\ln\,(b - x) - \{-\ln\,(a - x)\}] = kt + \text{I} \tag{7.22}$$

$$\frac{1}{(a - b)} \ln\left[\frac{(a - x)}{(b - x)}\right] = kt + \text{I} \tag{7.23}$$

where I is the constant of integration. At the start of the reaction when $t = 0$, $x = 0$

$$\frac{1}{(a - b)} \ln\left[\frac{a}{b}\right] = \text{I} \tag{7.24}$$

substituting in equation (7.23)

$$\frac{1}{(a - b)} \ln\left(\frac{a - x}{b - x}\right) = kt + \frac{1}{(a - b)} \ln\left(\frac{a}{b}\right) \tag{7.25}$$

$$k = \frac{1}{(a - b)\,t}\left[\ln\frac{(a - x)}{(b - x)} - \ln\left(\frac{a}{b}\right)\right]$$

$$k = \frac{1}{(a - b)\,t}\left[\ln\frac{b\,(a - x)}{a\,(b - x)}\right] \tag{7.26}$$

When $b > a$, the result may be written as

$$k = \frac{1}{(b - a)t}\left[\ln\frac{a\,(b - x)}{b\,(a - x)}\right] \tag{7.27}$$

## 7.10 CHARACTERISTICS OF 1ST AND 2ND ORDER REACTIONS

### 7.10.1 Half-life of a Reaction

Half-life of a reaction is the time required to convert the concentration of the reactant to half and is denoted by $t_{1/2}$ or $t_{0.5}$.

**Half-life period of first order reaction.** The first order rate equation is given by

$$k = \frac{2.303}{t} \log \frac{a}{a-x}$$

where $a$ mol L$^{-1}$ is the initial concentration and $x$ is the number of moles of reactant that have reacted in time $t$. At time $t_{1/2}$, $x = \frac{a}{2}$ mol L$^{-1}$. Putting these values in first order rate equation, we have

$$k = \frac{2.303}{t_{1/2}} \log \frac{a}{(a - a/2)}$$

$$t_{1/2} = \frac{2.303}{k} \log \frac{a}{a/2} = \frac{2.303}{k} \log 2 = \frac{2.303}{k} \times 0.3010$$

$$\boxed{t_{1/2} = \frac{0.693}{k}} \qquad (7.28)$$

Therefore a first order reaction half-life period is independent of the initial concentration of reactants.

**Half-life period of second order reaction.** The second order rate equation is given by

$$k = \frac{1}{t}\left[\frac{x}{a(a-x)}\right]$$

Substituting, $t = t_{1/2}$ and $x = \frac{a}{2}$, we have

$$k = \frac{1}{t_{1/2}}\left[\frac{a/2}{a\left(a - \frac{a}{2}\right)}\right]$$

$$t_{1/2} = \frac{1}{k}\left[\frac{a/2}{a^2/2}\right] = \frac{1}{k}\left[\frac{1}{a}\right]$$

$$\boxed{t_{1/2} = \frac{1}{ka}} \qquad (7.29)$$

Thus half-life period for a second order reaction is inversely proportional to the initial concentration of reactants.

**Half-life period of nth order reaction.** The rate equation of $n$th order reaction is given by

$$k = \frac{1}{t\,(n-1)}\left[\frac{1}{(a-x)^{n-1}} - \frac{1}{a^{n} - 1}\right]$$

Substituting, $x = a/2$ and $t = t_{1/2}$, we have

$$t_{1/2} = \frac{1}{k\,(n-1)}\left[\frac{1}{(a-a/2)^{n-1}} - \frac{1}{a^{n} - 1}\right]$$

$$\boxed{t_{1/2} = \frac{1}{k\,(n-1)}\left[\frac{2^{n-1}-1}{a^{n} - 1}\right]} \qquad (7.30)$$

Thus the half-life of the $n$th order reaction varies inversely to $(n - 1)$th power of the initial concentration of reactants.

### 7.10.2 Units of Rate Constant

Units of rate constants for various reactions are summarized in Table 7.3.

**TABLE 7.3 Units of Rate Constants**

| *Order of Reaction* | *Rate Law Equation* | *Units for k* |
|---|---|---|
| First | $k = \frac{2.303}{t} \log \frac{a}{a-x}$ | $k$ = time$^{-1}$ $\log_{10} \frac{\text{mol L}^{-1}}{\text{mol L}^{-1}}$ = **time$^{-1}$** |
| Second | $k = \frac{1}{t}\left[\frac{x}{a\,(a-x)}\right]$ | $k = \frac{1}{\text{time}} \times \frac{\text{mol L}^{-1}}{\text{mol L}^{-1} \times \text{mol L}^{-1}}$<br>= **Liter mol$^{-1}$ time$^{-1}$** |
| Third | $k = \frac{1}{2t}\left[\frac{x(2a-x)}{a^2(a-x)^2}\right]$ | $k$ = **Liter$^2$ mol$^{-2}$ time$^{-1}$** |
| Zero | $k = \frac{x}{t}$ | $k$ = **Mol Liter$^{-1}$ time$^{-1}$** |

## 7.11 DETERMINATION OF ORDER OF REACTION

(*i*) **Integration method.** This method consists in performing the reaction between required amounts of the reactants. The products formed are analyzed from time to time and the concentration of reactants is determined. These values are substituted in equations for various order reactions, *e.g.*,

$$k = \frac{2.303}{t}\log_{10}\frac{a}{a-x} \qquad \text{(for first order reaction)}$$

$$k = \frac{1}{t}\times\frac{x}{a(a-x)} \qquad \text{(for second order reaction)}$$

$$k = \frac{1}{2t}\times\frac{x(2a-x)}{a^2(a-x)^2} \qquad \text{(for third order reaction)}$$

and the order of reaction is then known by that equation which gives constant value of velocity constant $k$. This method is the most common method for determinating the order of the reaction.

(*ii*) **Graphical method.** The reaction velocity in a first order reaction varies as one concentration term; while in second order reactions, the rate is determined by variation of two concentration terms and so on. Mathematically,

For a first order reaction: $\frac{dx}{dt} = k(a-x)$

For a second order reaction: $\frac{dx}{dt} = k(a-x)^2$

For a third order reaction: $\frac{dx}{dt} = k(a-x)^3$

In general for a $n$th order reaction: $\frac{dx}{dt} = k(a-x)^n$

Here, it is assumed that the initial concentrations of reactants are same.

The various values of $\frac{dx}{dt}$ are plotted against $(a-x)$ and if the graph so obtained is a straight line then the reaction will be of first order (Figure 7.4.). But if straight line is obtained by plotting $\frac{dx}{dt}$ and $(a-x)^2$, then the reaction is of second order (Figure 7.5).

Similarly, if a straight line is obtained between $\frac{dx}{dt}$ and $(a-x)^3$, it will be a third order reactions.

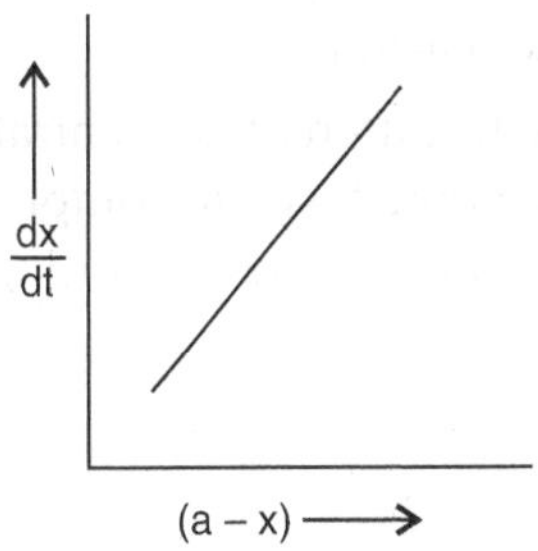


*FIGURE 7.4* **Plot of $(a - x)$ against $\frac{dx}{dt}$ za for a first order reaction.**

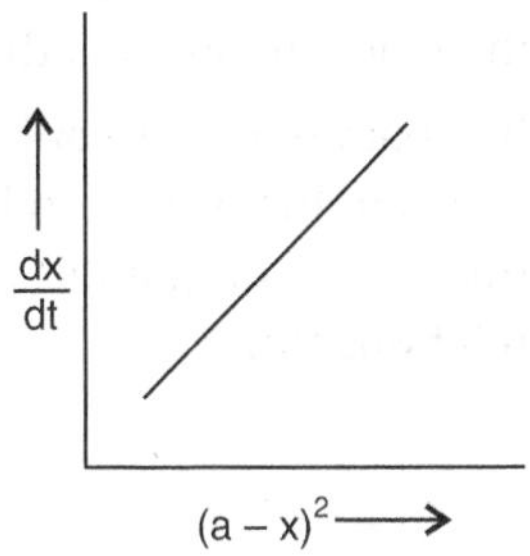


*FIGURE 7.5* **Plot of $(a - x)^2$ against $\frac{dx}{dt}$ for a second order reaction.**

(*iii*) **Isolation method.** In this method all the reactants are taken in very large extent except one and the order of the reaction is determined by first method. The concentrations of the substances taken in excess are considered to be constant. The sum of the orders, obtained by the process of isolation of each reactant is the order of the reaction. The method may be illustrated by the reaction between sodium formate and silver acetate.

$$\underset{\text{Silver acetate}}{2CH_3COOAg} + \underset{\text{Sodium formate}}{HCOONa} \longrightarrow Ag\downarrow + CO_2\uparrow + CH_3COOH + CH_3COONa$$

When silver acetate is taken in excess the reaction is first order, but when sodium formate is taken in excess the reaction is second order. Hence, overall, the reaction is third order.

(*iv*) **Half life period method.** In this method, the time at which one-half of the substance has disappeared is determined and the experiment is repeated with different initial concentrations of the reactants. The order is then known with the help of first order.

(*a*) If half decomposition period is constant and is independent of initial concentration, the reaction is first order.

For first order reaction, $t_{1/2} = \dfrac{0.693}{k}$

(*b*) If half decomposition period is inversely proportional to the initial concentration, the reaction is second order.

For second order reaction, $t_{1/2} = \dfrac{1}{ka}$

(*c*) If half decomposition period is inversely proportional to the square of the initial concentrations, the reaction is third order.

For third order reaction, $t_{1/2} = \dfrac{3}{2ka^2}$

Thus in general, $$t \propto \left[\frac{1}{a}\right]^{n-1} \quad (7.31)$$

where $n$ is the order of reaction, $t$ is the time required to complete a definite fraction of the reactant, and $a$ is the initial concentration of the reactant.

If separate experiments are carried out, starting with different concentrations of the reactant, the order can be determined by finding the time taken to complete the same fraction of the change with initial concentrations $a_1$ and $a_2$. Then according to general relationship,

$$t_1 \propto \frac{1}{a_1^{\,n-1}}$$

and $$t_2 \propto \frac{1}{a_2^{\,n-1}}$$

$$\frac{t_1}{t_2} = \frac{a_2^{\,n-1}}{a_1^{\,n-1}} = \left[\frac{a_2}{a_1}\right]^{n-1} \tag{7.32}$$

Taking logarithms,

$$\log \frac{t_1}{t_2} = (n-1) \log \frac{a_2}{a_1}$$

$$n = 1 + \frac{\log \frac{t_1}{t_2}}{\log \frac{a_2}{a_1}}$$

$$n = 1 + \frac{\log t_1 - \log t_2}{\log a_2 - \log a_1} \tag{7.33}$$

where $n$ is the order of reaction and thus can be calculated by knowing the values $a_1$, $a_2$, $t_1$ and $t_2$.

(*v*) **Vant Hoff's differential method.** We know that the rate of a reaction varies as the $n$th power of the concentration of the reactant where $n$ is the order of reaction. Thus, in general

$$r = -\frac{dC}{dt} = kC^n \tag{7.34}$$

For two different concentrations $C_1$ and $C_2$, the rates $r_1$ and $r_2$ will be given by

$$r_1 = -\frac{dC_1}{dt} = kC_1^{\,n}$$

$$r_2 = -\frac{dC_2}{dt} = kC_2^{\,n}$$

On taking logarithm, we get

$$\log r_1 = \log k + n \log C_1 \tag{7.35}$$

and

$$\log r_2 = \log k + n \log C_2 \tag{7.36}$$

On subtracting equation (7.35) from equation (7.36), we get

$$\log r_2 - \log r_1 = n \log C_2 - n \log C_1$$

$$\log r_2 - \log r_1 = n (\log C_2 - \log C_1)$$

$$n = \frac{\log r_2 - \log r_1}{\log C_2 - \log C_1} \tag{7.37}$$

Thus, in short, the method involves the determination of concentration of reactant(s) at different intervals of time.

The values of $r_1$ and $r_2$ are evaluated by plotting concentration-time graphs corresponding to $C_1$ and $C_2$ concentrations. Substituting these values in equation (7.37), the value of $n$ can be evaluated.

## 7.12 EFFECT OF TEMPERATURE ON RATE OF REACTION

It has been observed that rate of all reactions increases with the increase in temperature. It is observed that 10°C rise in temperature generally doubles or triple the reaction rate. *The ratio of the velocity constants of a reaction at two temperatures separated by 10°C is known as the temperature coefficient.* The temperatures usually selected for this purpose are 25°C and 35°C. Thus mathematically,

$$\text{Temperature coefficient} = \frac{\text{Velocity constant at } 35°\text{C}}{\text{Velocity constant at } 25°\text{C}} = \frac{k_{35}\,°\text{C}}{k_{25}\,°\text{C}}$$

In most of reactions the temperature coefficient is nearly 2 although in some cases it approaches even 3. For example, the temperature coefficients for dissociation of HI, and for the reaction between methyl iodine and sodium ethoxide are 1.8 and 2.9 respectively.

$$2HI \rightleftharpoons H_2 + I_2$$

$$CH_3I + C_2H_5ONa \longrightarrow CH_3OC_2H_5 + NaI$$

Further, the greater the difference in temperature, the greater will be the difference in reaction rates. For example, increase of 160°C in temperature during the decomposition of acetaldehyde increases the reaction rate by 453 times. Similarly, the rate of decomposition of acetone dicarboxylic acid goes up by 220-folds for a 60° rise in temperatures.

$$\underset{\text{Acetone dicarboxylic acid}}{COOHCH_2.CO.CH_2COOH} \longrightarrow \underset{\text{Acetone}}{CH_3.CO.CH_3} + 2CO_2$$

However, in some reversible reactions the reaction rate decreases with an increase in temperature. This is so because in such cases the rate of backward step increases more rapidly with a rise in temperature than the rate of the forward step.

**Explanation of temperature coefficient.** The increase in reaction rate with the increase in temperature is explained on the basis of collision theory, which says that although a reactant molecule collides with another reactant molecule a billion times ($10^6$ times) per second, only those collisions which possess a certain minimum amount of energy result in a chemical reaction. This minimum amount of energy which must be associated with molecules so that their mutual collisions result in chemical reaction is called *threshold energy.*

In a given sample of a gas all the molecules do not possess similar velocity and kinetic energy. In fact, at a given temperature (say $T_1$) there is a definite distribution of velocities or kinetic energies among the various molecules of gas. At low temperature ($T_1$) only a small part of the molecules have very low kinetic energy, another small part has very high kinetic energy, and a larger part has intermediate kinetic energy. Further, as the temperature is increased from $T_1$ to $T_2$, the energy distribution undergoes a change and more number of the molecules have acquired high kinetic energy as evidenced by the increase in shaded area of the curve in Figure 7.6.

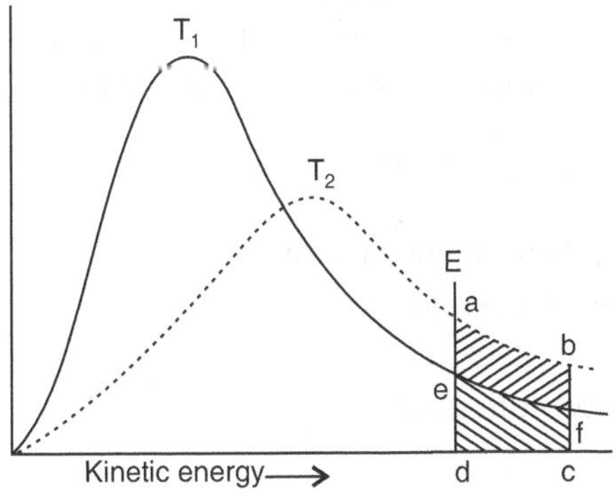


***FIGURE 7.6*** **Energy distribution in molecules at two different temperatures ($T_2 > T_1$); shaded portions of the curves represent molecules having energy more than threshold energy (E).**

The excess energy acquired by the real reacting molecules is known as *activation energy, i.e.,*

Activation energy = Threshold energy – Actual energy possessed by molecules.

Thus according to activation energy concept, no-nactive molecules (having energy less than the threshold energy) can be activated by absorption of extra energy known as the *activation energy*. In other words, the reactant species have to acquire a certain amount of energy before they can react to yield products, *i.e.*, there is an energy barrier which must be crossed over by the reactants to be converted into products. This energy barrier determines the magnitude of threshold energy which reactant molecules must acquire before they can yield products (see Figure 7.7).

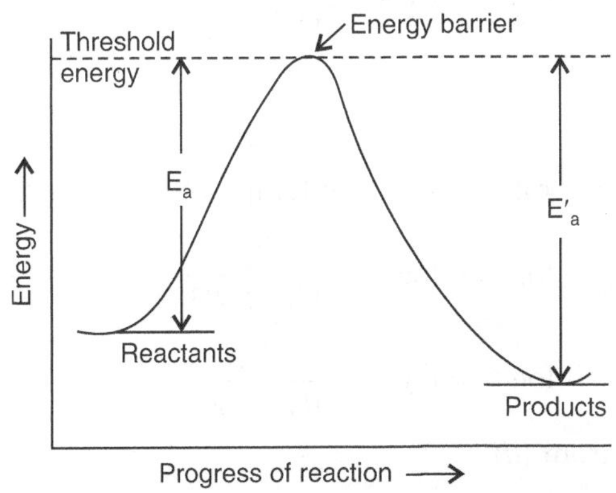


***FIGURE 7.7*** **Concept of energy barrier in chemical reactions, where $E_a$ is the activation energy of reactants and $E_a'$ is the activation energy of products.**

From Figure 7.7 it is obvious that as the energy of products is less than that of the reactants, the reaction is exothermic (*i.e.*, heat will be evolved during the reaction) and thus $\Delta H$ will be negative.

Different reactions have different amounts of activation energy. A reaction which has a higher activation energy is slow at ordinary temperatures, while the reaction having lower activation energy will proceed at a faster rate at a given temperature. Thus the difference in activation energy is mainly responsible for observed difference in rates of reactions.

**Calculation of activation energy.** According to Arrhenius, activation energy of a reaction with $k$ as the rate constant and at temperature T may be given by the following emperical equation:

$$k = Ae^{-E_a/RT} \qquad (7.38)$$

where, A = Constant, known as frequency factor

$E_a$ = Activation energy of the reaction

R = Gas constant

T = Temperature on absolute scale

$e$ = 2.718

The values $k_1$ and $k_2$ of a reaction at temperatures $T_1$ and $T_2$ are related by the following relation.

$$\log \frac{k_2}{k_1} = -\frac{E_a}{2.303\,R}\left[\frac{1}{T_2} - \frac{1}{T_1}\right] \qquad (7.39)$$

**Derivation of the above relation**

$$k = Ae^{-E_a/RT}$$

Taking log of both the sides,

$$\log k = \log A - \frac{E_a}{RT}\log e$$

or
$$\log k = \log A - \frac{E_a}{RT}\cdot\frac{1}{2.303}$$

or
$$\log k = \log A - \frac{E_a}{RT\,2.303}$$

Now, let $k_1$ and $k_2$ are the rate constants at temperatures $T_1$ and $T_2$ respectively.

$$\therefore \quad \log k_1 = \log A - \frac{E_a}{RT_1\,2.303} \qquad (i)$$

$$\log k_2 = \log A - \frac{E_a}{RT_2\,2.303} \qquad (ii)$$

Subtract, equation (*i*) from (*ii*)

$$\log k_2 - \log k_1 = -\frac{E_a}{RT_2\,2.303} + \frac{E_a}{RT_1\,2.303}$$

or
$$\log \frac{k_2}{k_1} = \frac{E_a}{R\,2.303}\left[\frac{1}{T_1} - \frac{1}{T_2}\right]$$

Thus, activation energy of a reaction can be determined from the above equation by knowing the values of rate constants $k_1$ and $k_2$ at temperatures $T_1$ and $T_2$ respectively.

Alternatively, the value of $E_a$ can be evaluated by plotting a graph between $\log k$ and $\frac{1}{T}$ (Figure 7.8). The graph will give a straight line whose slope is equal to $-\frac{E_a}{2.303\,R}$ or $-\frac{E_a}{4.576}$, from which the value of $E_a$ can be determined.

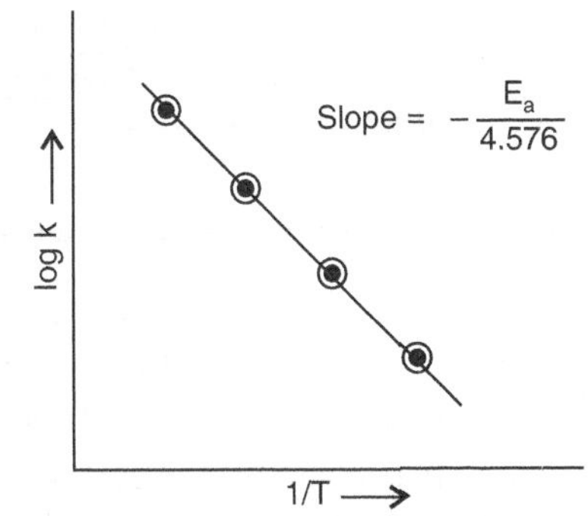


*FIGURE 7.8* **Arrhenius plot for log k vs $\frac{1}{T}$.**

## 7.13 THEORIES OF REACTION RATE

1. **Collision Theory.** According to this theory the molecules of the reactant(s) collide to carry out the reaction. Only those collisions are effective which are capable to overcome activation energy. Such collisions are called *effective collisions.* The two main postulates of collision theory are:

   (*i*) *The molecules must collide with sufficient kinetic energy.* Consider a general reaction,

$$A-A+B-B \longrightarrow 2A-B$$

   A chemical reaction requires breaking of bonds between the atoms of reacting molecules and formation of new bonds in the products. The energy required in breaking the bonds comes from the kinetic energy possessed by the reacting molecules before the collision. Thus, the colliding molecules must possess certain minimum energy (threshold energy) to make the collisions effective.

   Figure 7.9 shows activation energy ($E_a$), that is the minimum energy necessary to cause a reaction between the colliding molecules. Only the molecules that collide with kinetic energy greater than $E_a$, react and get converted into products. The collisions between molecules with KE less than $E_a$ fail to surpass the energy barrier and are non-productive.

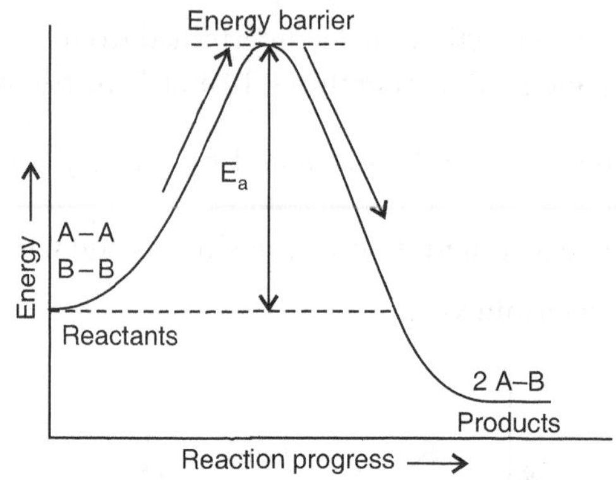


***FIGURE 7.9*** **The activation energy $E_a$ acting as an energy barrier.**

(*ii*) *The molecules must collide with correct orientation.* The reacting molecules must collide with correct orientation. The correct orientation is one that ensures direct contact between the atoms involved in the breaking and forming of bonds (Figure 7.10).

Thus, from above two postulates it is clear that only the molecules colliding with kinetic energy greater than $E_a$ and with correct orientation results in the formation of products.

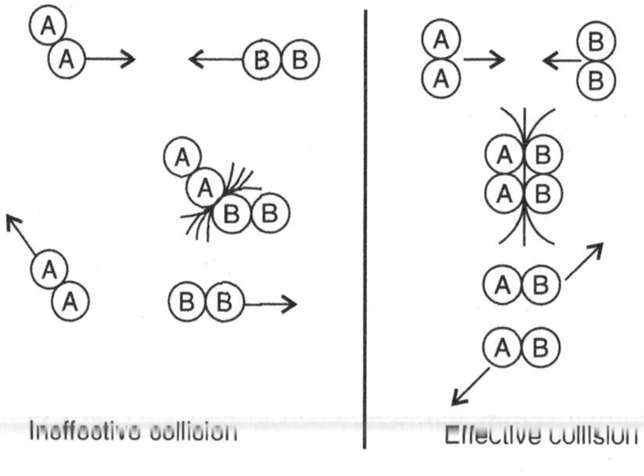


*FIGURE 7.10*

## Drawbacks of Collision Theory

Following are the major drawbacks of collision theory:

(*i*) This theory does not advance any explanation for the abnormally high rates that are observed sometimes.

(*ii*) This theory cannot be applied to reversible reactions with complete success.

2. **Transition State Theory.** This is also known as *activated complex theory.* According to this theory the molecules of reactants combine to form activated complex or an intermediate product (Figure 7.11). Since the energy of the activated complex is greater than that of the reactants and products, the activated complex changes to the product:

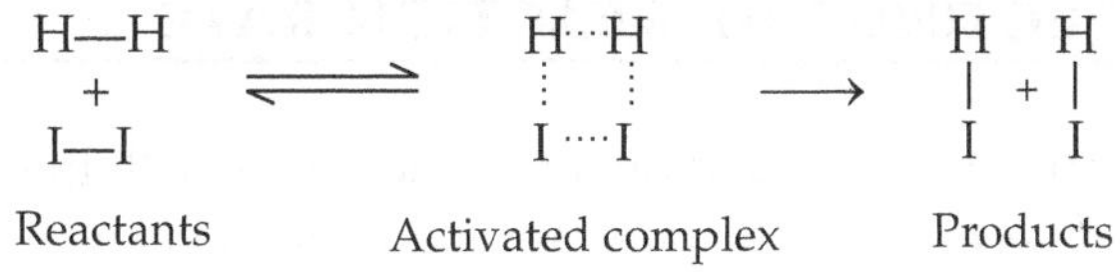


or $$H_2 + I_2 \longrightarrow 2HI$$

where $E_t$ = Threshold energy

$E_a$ = Activation energy forward reaction

$E_a'$ = Activation energy backward reaction

$P_1$ = Potentials energy of reactants

$P_2$ = Potential energy of products

$E_t - P_1 = E_a$ = Activation energy of forward reaction

$E_t - P_2 = E_a'$ = Activation energy of backward reaction

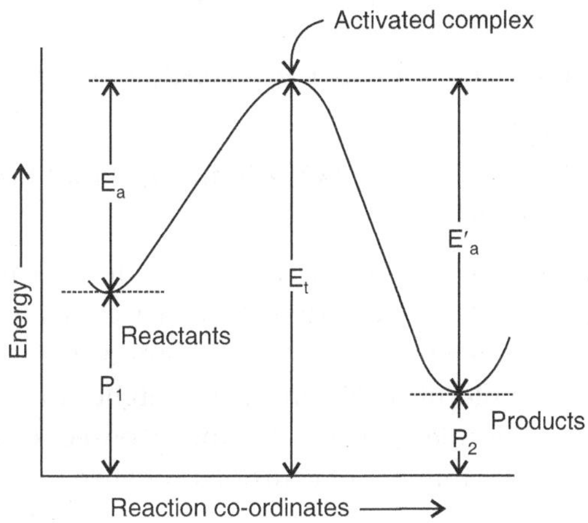


*FIGURE 7.11*

### Characteristics of an Activated Complex

Following are the important characteristics of an activated complex:

(*i*) The formation of the activated complex is regarded as the characteristic of all chemical changes.

(*ii*) The activated complex may be treated as a molecule in which one of the vibrations is stiff.

(*iii*) The activated complex has a transient existence only and breaks up at a definite rate to form the products of the reaction.

(*iv*) The potential energy of the activated complex is maximum.

## 7.14 FACTORS AFFECTING THE REACTION RATE

The rate of a chemical reaction is not a fixed quantity and varies with the experimental conditions. A brief account of factors affecting the reaction rate is discussed below.

1. **Nature of the reactant.** It has been observed that reactants having ionic bonds are relatively fast to form product than the reactant possessing covalent bond. Probably it is due to the fact that in aqueous conditions, the ionic reactants are ionized resulting in the rapid formation of product while in case of covalent reactants, part of the energy is used up for dissociation of bonds. So, these observations may be generalised as below.
   (*a*) Reactions that do not involve cleavage of bonds usually tend to be fast at room temperature *e.g.*, ionic reactions are fast.
   (*b*) More the number of covalent bonds cleaved during a chemical reaction, slower will be the product formation at room temperature.

For example, decolourization of acidified $KMnO_4$ with ferrous ions is very fast because it involves ionic reactions or cleavage of few chemical bonds. However, decolourization of acidified $KMnO_4$ oxalate ions is quite slow and requires heating because it involves cleavage of several covalent bonds.

$$2KMnO_4 + 8H_2SO_4 + 10FeSO_4 \longrightarrow K_2SO_4 + 2MnSO_4 + 5Fe_2(SO_4)_3 + 8H_2O \text{ (fast reaction)}$$

$$2KMnO_4 + 3H_2SO_4 + 5\begin{array}{c}COOH\\|\\COOH\end{array} \longrightarrow K_2SO_4 + 2MnSO_4 + 10CO_2 + 8H_2O \text{ (slow reaction)}$$

2. **Concentration of the reactant.** During the course of a chemical reaction, concentration of reactants decrease while that of products increases *i.e.*, reactants change to products. The rate of change of reactants to products is proportional to the number of collisions between the reacting molecules. We know that higher the concentration, greater is the number of reacting molecules per unit volume. Consequently, higher concentration of reactant will have greater number of collisions between reacting molecules. So, it results in quick acquiring of activation energy by reactant to change over to product.

$$\text{Rate of reaction} \propto [\text{Reactants}]$$

This means that the rate of reaction becomes double, when the concentration is doubled.

This relationship which express the rate of a reaction to the concentration of the reactant is termed as Rate law.

3. **Effect of catalyst in a chemical reaction.** *A catalyst is a substance which alters the rate of a chemical reaction, without itself being permanently changed.* The catalysts usually enhance the rate of a reaction and are termed as *positive catalysts*, while in some reactions catalyst reduces the rate of reaction and termed as *negative catalyst*.

$$\underset{\text{Pot. perchlorate}}{2KClO_3} \xrightarrow{MnO_2} 2KCl + 3O_2 \text{ (Positive catalyst)}$$

$$N_2 + 3H_2 \xrightarrow{Fe} 2NH_3 \text{ (Positive catalyst)}$$

$$2H_2O_2 \xrightarrow{H_3PO_4} 2H_2O + O_2 \text{ (Negative catalyst)}$$

The catalytic action of the catalyst in accelerating the rate of reaction can be explained by two main theories.

(*a*) Intermediate compound formation theory

(*b*) The adsorption theory.

(*a*) **Intermediate compound formation theory.** This theory applies to homogeneous catalytic reactions where the catalyst forms an intermediate compound with the reactant. In the intermediate compound, a temporary physical or chemical bond is formed with one of the reactants and catalyst. These highly reactive intermediate compounds have lower activation energy and thus provide a new pathway (mechanism) for the reaction. Because of the large number of effective collisions with the second reactant in the presence of catalyst, the reactions go faster to form the product and releasing the catalyst.

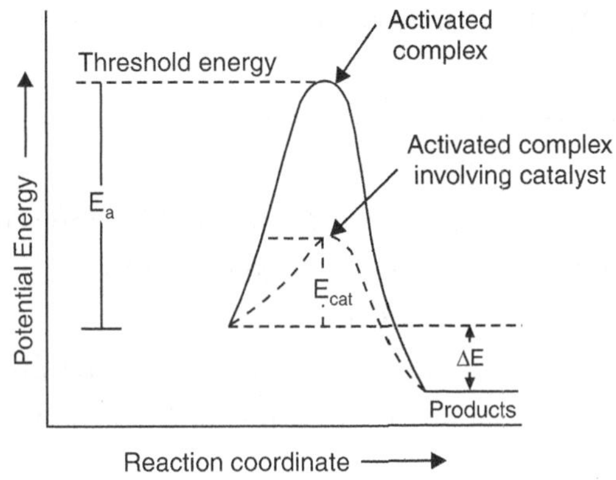


*FIGURE 7.12*

The mechanism involving formation of intermediate compound may be illustrated by taking the general reaction.

$$A + B \longrightarrow AB \qquad ...(i)$$

If C is a catalyst, then the reaction proceeds as,

$$A + C \longrightarrow AC \text{ (intermediate compound)} \qquad ...(ii)$$

$$AC + B \longrightarrow AB + C \qquad ...(iii)$$

The activation energies of reactions (*ii*) and (*iii*) are lower than that of uncatalysed reaction (*i*). Hence involvement of the catalyst in intermediate compound formation and its subsequent decomposition accelerate the rate of the reaction (*i*) which is otherwise very slow.

Many catalytic industrial processes are based on this theory. For example,

Lead chamber process for manufacturing sulphuric acid from sulphur dioxide and oxygen in presence of nitric oxide as a catalyst.

$$O_2 + \underset{\text{Catalyst}}{2NO} \longrightarrow \underset{\text{(intermediate compound)}}{2NO_2}$$

$$2NO_2 + 2SO_2 \longrightarrow 2SO_3 + \underset{\text{(Catalyst regenerated)}}{2NO}$$

$$2SO_3 + 2H_2O \longrightarrow \underset{\text{Sulphuric acid}}{2H_2SO_4}$$

In the absence of catalyst, the activation energy of the reaction is high. Therefore, the reaction is slow and takes place only between small fraction of the reacting molecules with high energy. However, in the presence of catalyst the *activation energy of the reaction is comparatively lower* and even molecules with lower activation energy undergo reaction resulting in faster reaction.

4. **Surface area of the catalyst.** Greater the surface area of the catalyst, faster is the rate of reaction. This observation may be explained by the mechanism of *adsorption theory* which is applicable to *heterogeneous catalysis*. Although the reactant molecules are adsorbed on the surface of the catalyst and are held by London forces of attraction, the reaction takes place through *chemisorption* (chemical adsorption). In chemisorption the adsorbed molecules are held to the surface of the catalyst by bodns which are similar is strength to those in chemical compounds. In the chemisorbed molecules, some bonds are stretched, weakened, and even broken. Let us illustrate the mechanism of the decomposition of dinitrogen oxide ($N_2O$) in presence of gold (catalyst). The reaction proceeds in the following steps.

(*i*) Dinitrogen oxide is chemisorbed on the surface of gold

$$N_2O \longrightarrow N_2O \qquad \text{(Chemisorbed on gold)}$$

(*ii*) The bond between O atom and the N atom of $N_2O$ molecule is weakened and finally broken, resulting in the escape of nitrogen gas ($N_2$).

$$N_2O\text{ (on gold)} \longrightarrow N_2(g) + O\text{ (on gold)}$$

(*iii*) Two oxygen atoms on the surface of gold combine to form $O_2$ molecule, which leaves the catalyst surface, enabling catalyst for further action.

$$O\text{ (on gold)} + O\text{ (on gold)} \longrightarrow O_2(g) + \text{Gold catalyst}$$

So, it is evident that smaller is the size of the particle of the catalyst, more will the surface area of the catalyst available for adsorption consequently faster will be the rate of reaction.

5. **Effect of radiation or photochemical reactions.** There are various reactions which either do not proceed or are quite slow in the dark but take place in the presence of radiations like sunlight or ultraviolet light. Such reactions are termed as *photochemical reactions*. Some of the common examples of such reactions are photosynthesis of carbohydrates, photography and various other reactions involving decomposition, polymerization, isomerization, oxidation and reduction.

The glucose formed in photosynthesis is subsequently converted into starch, cellulose, fats and proteins by the action of enzymes.

### Study of Fast Reactions

Fast reactions are those which occur instantaneously, *i.e.*, these reactions are completed within a span of few picoseconds ($10^{-12}$ second). The kinetic studies of these reactions cannot be carried out by usual methods of measuring the change in concentration of the reactant. The special technique to study the fast reactions (half-life upto one picosecond) is *flash photolysis* developed by Norrish, Porter and Thrush in 1953.

**Flash photolysis :** *Flash photolysis is a method of identifying the free radicals formed when the vapours of a compound at low pressure are exposed to an intense, but very brief, flash of radiation or a laser* *beam.* A second flash, followed shortly after the first, is used to photograph the absorption spectrum of the gases which record the free radical present. Subsequent series of flashes at regular intervals after the initial flash may be used to calculate the life time of the radical so formed. The change in absorption spectra with time gives the rate of various processes occuring during fast reaction.

Using this technique, the kinetic studies of a number of fast reactions have been made. The mechanism of photosynthetic reaction is discussed below.

**Photosynthesis.** As discussed earlier, photosynthesis is a fast reaction whcih converts carbon dioxide and water into carbohydrate by chlorophyll of the plant as sensitizer in presence of sunlight. The process of photosynthesis involves several steps that follow the absorption of sunlight by chlorophyll. A chlorophyll molecule absorbs a photon of red light and gets excited.

$$\text{Chl} \xrightarrow{\text{Photon}} \underset{\text{Excited molecule}}{\text{Chl}^*}$$

Within a few picoseconds, the excited chlorophyll molecule transfers the extra energy to another nearby molecule A, either $CO_2$ or $H_2O$, called *electron acceptor.*

$$\text{Chl}^* + \text{A} \longrightarrow \text{Chl}^+ + \text{A}^-$$

Nearly 150 picoseconds later, the electron acceptor ($A^-$) transfers the electron to another molecule and so on.

$$\text{A}^- + \text{B} \longrightarrow \text{A} + \text{B}^-$$

$$\text{B}^- + \text{C} \longrightarrow \text{B} + \text{C}^- + \text{Energy}$$

## SOLVED PROBLEMS

### TYPE-I General Problems Based on Kinetics

**Problem 1.** *For the assumed reaction $X_2 + 2Y_2 \longrightarrow 2XY_2$. Write the rate equation in terms of the rate of disappearance of $Y_2$.*

**Solution.** The rate of reaction in terms of disappearance of $Y_2 = -\frac{1}{2}\frac{\Delta[Y_2]}{\Delta t}$.

**Problem 2.** *For the reaction $N_2(g) + 3H_2(g) \longrightarrow 2NH_3(g)$. How is the rate of formation of ammonia related to the rate of disappearance of hydrogen.*

**Solution.** Rate of formation of ammonia $= +\frac{1}{2}\frac{\Delta[NH_3]}{\Delta t}$

Rate of disappearance of $H_2 = -\frac{1}{3}\frac{\Delta[H_2]}{\Delta t}$

So, the relationship is given as

$$+\frac{1}{2}\frac{\Delta[NH_3]}{\Delta t} = -\frac{1}{3}\frac{\Delta[H_2]}{\Delta t}$$

$$\frac{\Delta[NH_3]}{\Delta t} = -\frac{2}{3}\frac{\Delta[H_2]}{\Delta t}.$$

**Problem 3.** *The following reaction was carried out at 300 K.*

$$2SO_2(g) + O_2(g) \longrightarrow 2SO_3(g).$$

*The concentration of $SO_3$ gas is $5 \times 10^{-3}$ mole after 7.5 minutes of the start of the reaction. Calculate the average rate of formation of $SO_3$ during this time.*

**Solution.** Here, $SO_3$ is the product of the reaction, therefore, its concentration at $t = 0$ is zero.

We know that,

$$\text{Average rate} = \frac{\text{Change in concentration}}{\text{Time taken}}$$

$$\text{So, average rate of formation of } SO_3 = \frac{5 \times 10^{-3} \text{ moles}}{7.5 \text{ minutes}}$$

$$= \mathbf{6.66 \times 10^{-4} \text{ moles min}^{-1}}.$$

**Problem 4.** *For the chemical reaction $4NH_3(g) + 5O_2(g) \longrightarrow 4NO(g) + 6H_2O(g)$. If the rate expression for the disappearance of $O_2(g)$ is $-\frac{\Delta [O_2]}{\Delta t}$, what will be the rate expression for disappearance of $NH_3$ and appearance of $H_2O$. If $\frac{\Delta NO}{\Delta t} = 6.0 \times 10^{-4}$ atm sec$^{-1}$, determine the value of $\frac{\Delta H_2O}{\Delta t}$.*

**Solution.** (*i*) For the given reaction,

$$\text{Rate of disappearance of } NH_3 = -\frac{1}{4}\frac{\Delta[NH_3]}{\Delta t}$$

$$\text{Rate of disappearance of } O_2 = -\frac{1}{5}\frac{\Delta [O_2]}{\Delta t}$$

$$\text{Rate of appearance of NO} = +\frac{1}{4}\frac{\Delta [NO]}{\Delta t}$$

$$\text{Rate of appearance of } H_2O = +\frac{1}{6}\frac{\Delta[H_2O]}{\Delta t}$$

$$\text{So, rate of reaction } -\frac{1}{5}\frac{\Delta[O_2]}{\Delta t} = -\frac{1}{4}\frac{\Delta[NH_3]}{\Delta t} = +\frac{1}{4}\frac{\Delta[NO]}{\Delta t} = +\frac{1}{6}\frac{\Delta[H_2O]}{\Delta t}$$

or

$$-\frac{\Delta[O_2]}{\Delta t} = -\frac{5}{4}\frac{\Delta[NH_3]}{\Delta t} = +\frac{5}{4}\frac{\Delta[NO]}{\Delta t} = +\frac{5}{6}\frac{\Delta[H_2O]}{\Delta t}$$

(*ii*)

$$+\frac{5}{4}\frac{\Delta[NO]}{\Delta t} = \frac{5}{6}\frac{\Delta[H_2O]}{\Delta t}$$

So,
$$\frac{\Delta[H_2O]}{\Delta t} = \frac{6}{5} \times \frac{5}{4} \frac{\Delta[NO]}{\Delta t}$$

$$\frac{\Delta[H_2O]}{\Delta t} = \frac{3}{2} \frac{\Delta[NO]}{\Delta t}$$

Substituting the value of $\frac{\Delta[NO]}{\Delta t}$ in this expression, we get

$$\frac{\Delta[H_2O]}{\Delta t} = \frac{3}{2} \times 6.0 \times 10^{-4} \text{ atm sec}^{-1} = \mathbf{9.0 \times 10^{-4} \text{ atm sec}^{-1}}.$$

**Problem 5.** *Write the equivalent rate expressions for the following reactions.*

(*i*) $2NO_2 + F_2 \longrightarrow 2NO_2F$ (*ii*) $2ICl + H_2 \longrightarrow I_2 + 2HCl$

(*iii*) $N_2O_4(g) \rightleftharpoons 2NO_2(g)$.

**Solution.** (*i*)
$$-\frac{1}{2}\frac{\Delta[NO_2]}{\Delta t} = -\frac{\Delta[F_2]}{\Delta t} = +\frac{1}{2}\frac{\Delta[NO_2F]}{\Delta t}$$

(*ii*)
$$-\frac{1}{2}\frac{\Delta[ICl]}{\Delta t} = -\frac{\Delta[H_2]}{\Delta t} = +\frac{\Delta[I_2]}{\Delta t} = +\frac{1}{2}\frac{\Delta[HCl]}{\Delta t}$$

(*iii*)
$$-\frac{\Delta[N_2O_4]}{\Delta t} = +\frac{1}{2}\frac{\Delta[NO_2]}{\Delta t}.$$

**Problem 6.** *For the reaction: $2H_2O_2(aq) \longrightarrow 2H_2O(l) + O_2(g)$, the initial concentration of $H_2O_2$ is 0.5 m and its concentration after 5 minutes is 0.3 m. Calculate:*

(*i*) *the rate of decomposition of $H_2O_2$*

(*ii*) *the relative rate of formation of $O_2$.*

**Solution.** (*i*)
$$-\frac{\Delta[H_2O_2]}{\Delta t} = -\frac{\text{Final }[H_2O_2] - \text{Initial }[H_2O_2]}{\text{Elapsed Time}}$$

$$= \frac{-[0.3 - 0.5]}{5} = \mathbf{0.04 \text{ M min}^{-1}}.$$

(*ii*)
$$-\frac{1}{2} \cdot \frac{\Delta[H_2O_2]}{\Delta t} = +\frac{\Delta[O_2]}{\Delta t}$$

$$\frac{\Delta[O_2]}{\Delta t} = -\frac{1}{2}\frac{\Delta[H_2O_2]}{\Delta t}$$

$$= \frac{1}{2} \times 0.04 = \mathbf{0.02 \text{ M min}^{-1}}.$$

**TYPE-II Problems Based on Half-life of A Reaction**

**Problem 7.** *The rate constant of a first order reaction is 3.5 × $10^{-2}$ minute$^{-1}$. Calculate the time taken for half the initial concentration to react (half-life).*

**Solution.** Half-life of a first order reaction $t_{1/2} = 0.693/k$

Thus, $$t_{1/2} = \frac{0.693}{3.5 \times 10^{-2}} = \mathbf{19.8\ minutes.}$$

**Problem 8.** *The time for half change for a first order reaction is 60 minutes. Calculate the rate constant.*

**Solution.** $$t_{1/2} = \frac{0.693}{k} \text{ or } k = \frac{0.693}{t_{1/2}} = \frac{0.693}{60 \times 60} = \mathbf{1.925 \times 10^{-4}\ sec^{-1}.}$$

**Problem 9.** *A first order reaction takes 69.3 minutes for 50% completion. How much time will be needed for 80% completion?*

**Solution.** $$k_1 = \frac{2.303}{t} \log \frac{a}{(a-x)} \quad \text{or} \quad t_{50\%} = \frac{2.303}{k_1} \log \frac{100}{100-50}$$

or $$t_{50\%} = \frac{2.303}{k_1} \log 2$$

Now, $$t_{80\%} = \frac{2.303}{k_1} \log \frac{100}{100-80} = \frac{2.303}{k_1} \log 5$$

Thus, $$\frac{t_{80\%}}{t_{50\%}} = \frac{2.303 \log 5}{k_1} \times \frac{k_1}{2.303 \log 2} = \frac{\log 5}{\log 2}$$

$$\text{Time for 80\% change} = 69.3 \times \frac{\log 5}{\log 2} = \frac{69.3 \times 0.6990}{0.3010} = \mathbf{160.9\ min.}$$

**Problem 10.** *In a first order reaction, it takes the reactant 40.5 minutes to be 25% decomposed. Calculate the rate constant of the reaction.*

**Solution.** Since the reactant is 25% decomposed after 40.5 minutes, the remaining reactant is 75%, thus at $t$ = 40.5 minutes, $(a - x) = 0.75\ a$

Now, $$k_1 = \frac{2.303}{t} \log \frac{a}{a-x} = \frac{2.303}{40.5} \log \frac{a}{0.75a}$$

$$= \mathbf{7.11 \times 10^{-3}\ minutes.}$$

**Problem 11.** *Find the two third life, i.e., $t_{2/3}$ of the first order reaction in which $k_1 = 5.48 \times 10^{-14}$ sec$^{-1}$.*

**Solution.** If the initial concentration is $a$, then $(a - x)$ will be $(a - 2/3a)$

$$= a/3. \ (a - 2/3a) = a/3.$$

Now,

$$t_{2/3} = \frac{2.303}{k_1} \log \frac{a}{(a-x)} = \frac{2.303}{5.48 \times 10^{-14}} \log \frac{a}{a/3}$$

$$= \frac{2.303}{5.48 \times 10^{-14}} \log 3 = \mathbf{2.01 \times 10^{13}\ sec.}$$

**Problem 12.** *The half-life period of first order reaction is 60 minutes. Calculate the time for 87.5% completion of the reaction.*

**Solution.**

$$k_1 = \frac{0.693}{t_{1/2}}.$$

Thus,

$$k_1 = \frac{0.693}{60} = \mathbf{11.55 \times 10^{-3}\ min^{-1}}.$$

Now we know,

$$k_1 = \frac{2.303}{t} \log \frac{a}{(a-x)}.$$

Here $a = 100$ and $x = 87.5$

Thus,

$$11.55 \times 10^{-3} = \frac{2.303}{t} \log \frac{100}{100 - 87.5} = \frac{2.303}{t} \log \frac{100}{12.5}$$

$$= \frac{2.303}{t} \log 8 = \frac{2.303}{t} \times 0.9030$$

or

$$\mathbf{t = 180\ minutes.}$$

**Problem 13.** *A first order reaction is 15% complete in 20 minutes. How long will it take to be 60% complete?*

**Solution.** We know

$$k_1 = \frac{2.303}{t} \log \frac{a}{(a-x)}.$$

Thus,

$$k_1 = \frac{2.303}{t_1} \log \frac{a_1}{(a_1 - x_1)}$$

$$= \frac{2.303}{t_2} \log \frac{a_2}{(a_2 - x_2)}$$

In the present problem $x_1 = \frac{15a_1}{100} = 0.15a_1$ and $t_1 = 20$ min.

Similarly, $x_2 = \frac{60a_2}{100} = 0.60, a_2.t_2 = ?$

So, $$\frac{2.303}{20}\log\frac{a_1}{(a_1-0.15a_1)}=\frac{2.303}{t_2}\log\frac{a_2}{(a_2-0.60a_2)}$$

or $$\frac{1}{20}\log\frac{a_1}{0.85a_1}=\frac{1}{t_2}\log\frac{a_2}{0.40a_2} \quad \text{or} \quad t_2 = \textbf{112.7 minutes.}$$

**Problem 14.** *In the reduction of nitric oxide by hydrogen, it was found that equimolecular of gases at 340.5 mm was half changed in 102 seconds, while in another experiment with a pressure of 288 mm, the reaction was half completed in 140 seconds. Find the order of reaction.*

**Solution.** We know that $$t_{1/2} \propto \frac{1}{a^{n-1}}.$$

Thus, $$\frac{t_1}{t_2}=\left(\frac{a_2}{a_1}\right)^{n-1}=\left(\frac{p_2}{p_1}\right)^{n-1}$$

Here $t_1$ = 102 sec, $t_2$ = 140 sec, $p_1$ = 340.5 mm and $p_2$ = 288 mm. Hence,

$$\frac{102}{140}=\left(\frac{288}{340.5}\right)^{n-1} \quad \text{or } n = 2.89 = \mathbf{3}$$

Hence the order of reaction is three.

**Problem 15.** *The rate expression for a chemical reaction*

$$(CH_3)_3C\text{–}Br + OH^- \longrightarrow (CH_3)_3\,C\text{–}OH + Br^- \text{ is}$$

$$Rate = k\,[(CH_3)_3\,C\text{–}Br]$$

*Write the probable mechanism.*

**Solution.** For the given reaction, following mechanism may be written.

(*i*) $(CH_3)_3\,C - Br \longrightarrow (CH_3)_3\,C^+ + Br^-$ (slow step)

(*ii*) $(CH_3)_3\,C^+ + OH^- \longrightarrow (CH_3)_3\,C\text{–}OH$ (fast step)

**Problem 16.** *What is the rate of reaction and order of a reaction, if the mechanism is*

$$2NO + H_2 \longrightarrow N_2 + H_2O_2 \quad \textit{(slow step)}$$

$$H_2O_2 + H_2 \longrightarrow 2H_2O \quad \textit{(fast step)}$$

**Solution.** Rate = $k\,[NO]^2\,[H_2]$; order = 3.

**Problem 17.** *Nitric oxide, NO, reacts with oxygen to produce nitrogen dioxide $2NO(g) + O_2(g) \longrightarrow 2NO_2(g)$. What is the predicted rate law, if the mechanism is*

$$NO + O_2 \xrightarrow{k_1} NO_3 \quad \textit{(fast step)}$$

$$NO_3 + NO \xrightarrow{k_2} NO_2 + NO_2 \quad \textit{(slow step)}$$

**Solution.** Reaction rate of slow step

$$\text{Rate} = k\,[NO_3]\,[NO] \qquad (i)$$

Reaction rate constant from fast step

$$k_1 = \frac{[NO_3]}{[NO]\,[O_2]} \qquad (ii)$$

Substituting the value of $[NO_3]$ from equation (*ii*) in equation (*i*)

$$\text{Rate} = k_1\,k_2\,[NO]\,[O_2]\,[NO]$$

$$\text{Rate} \propto [NO]^2\,[O_2].$$

**Problem 18.** *In a second order reaction the half-life was found to be 30 minutes when the concentration was 0.1 M. Find the rate constant.*

**Solution.** For a second order reaction $t_{1/2} = 1/k_2 a$. In the present problem $t_{1/2}$ = 30 minute, $a$ = 0.1 M. So,

$$30 = \frac{1}{k_2 \times 0.1} \text{ or } k_2 = \frac{1}{3.0} = \mathbf{0.333\ moles\ lit^{-1}\ min^{-1}.}$$

**Problem 19.** *Show that in case of a first order reaction, the time required for 99.9% of the reaction to take place is ten times that required for half of the reaction.*

**Solution.**

$$t = \frac{2.303}{k} \log \frac{a}{a-x}$$

$$= \frac{2.303}{k} \log \frac{1}{1-f}$$

where $f$ is the fraction of reaction completed in time, $t$

$$t_{50\%} = t_{1/2} = \frac{2.303}{k} \log \frac{1}{1-0.5}$$

$$= \frac{2.303}{k} \log 2 \qquad (i)$$

$$t_{99.9\%} = \frac{2.303}{k} \log \frac{1}{1-0.999}$$

$$= \frac{2.303}{k} \log \frac{1}{0.001}$$

$$= \frac{2.303}{k} \log \frac{1}{10^{-3}}$$

$$= \frac{2.303}{k} \log 10^3 \qquad (ii)$$

Divide equation (*ii*) by equation (*i*)

$$\frac{t_{99.9\%}}{t_{50\%}} = \frac{2.303}{k} \times \frac{k}{2.303} \times \frac{\log 10^3}{\log 2}$$

$$= \frac{3.000}{0.3010} = 9.967 = \mathbf{10.}$$

**Problem 20.** *Show that for a first order reaction, the time required for 99.99 per cent of the reaction to take place is twice as long as the time required for 99.0 per cent of the reaction to take place.*

**Solution.** Using,
$$t = \frac{2.303}{3} \log \frac{1}{1-f}$$

where $f$ is the fraction of reaction completed in time, $t$.

$$t_{99\%} = \frac{2.303}{k} \log \frac{1}{1-0.99} = \frac{2.303}{k} \log \frac{1}{10^{-2}} \quad (i)$$

$$= \frac{2.303}{k} \log 10^2 = \frac{2.303}{k} \times 2$$

$$t_{99.99\%} = \frac{2.303}{k} \log \frac{1}{1-0.9999}$$

$$= \frac{2.303}{k} \log \frac{1}{0.0001} = \frac{2.303}{k} \log \frac{1}{10^{-4}}$$

$$= \frac{2.303}{k} \log 10^4 = \frac{2.303}{k} \times 4 \quad (ii)$$

Dividing equation (*ii*) by equation (*i*), we get,

$$\frac{t_{99.99\%}}{t_{99\%}} = \left(\frac{2.303}{k} \times 4\right) \times \left(\frac{k}{2.303 \times 2}\right) = \mathbf{2.}$$

**Problem 21.** *The initial concentration of reactant, A, in the following first order reaction, $2A \longrightarrow B + C$ is 0.5 M at 25° C. The concentration of reactant after 20 minutes was 0.05 M. Calculate the rate constant of the reaction.*

**Solution.**
$$k = \frac{2.303}{t} \log \frac{\text{Initial concentration of A}}{\text{Concentration of A at time } t}$$

$$= \frac{2.303}{t} \log \frac{0.5}{0.005}$$

$$= \frac{2.303}{20} \log 100$$

$$= \frac{2.303}{20} \times 2$$

$$= \mathbf{0.2303\ min^{-1}}.$$

**Problem 22.** *The time required to decompose $AO_2B_2$ to half its initial amount is $3.5 \times 10^4$ seconds. If the decomposition of $AO_2B_2$ follows first order kinetics, calculate the rate constant of the reaction.*

**Solution.** Using $$t_{1/2} = \frac{0.693}{k}$$

$$= \frac{0.693}{t_{1/2}}$$

$$= \frac{0.693}{3.5 \times 10^4} = \mathbf{1.98 \times 10^{-5}\ s}.$$

**Problem 23.** *For the decomposition of $2N_2O_5(g) \longrightarrow 4NO_2(aq) + O_2(g)$ first order kinetics is followed. The concentration of $N_2O_5$ was 1.32 M after 175 seconds of the start of the reaction and 1.07 M after 506 seconds of the start of the reaction. Calculate the specific reaction rate of the reaction.*

**Solution.** Using $$\log \frac{[A]_1}{[A]_2} = \frac{k(t_2 - t_1)}{2.303}$$

$$k = \frac{2.303}{t_2 - t_1} \log \frac{[A]_1}{[A]_2}$$

$$= \frac{2.303}{506 - 175} \log \frac{1.32}{1.07}$$

$$= \frac{2.303}{331} (\log 1.32 - \log 1.07)$$

$$= \frac{2.303}{331} (0.1206 - 0.0294)$$

$$= \frac{2.303}{331} \times 0.0912 = \mathbf{6.4 \times 10^{-4}\ s^{-1}}.$$

**Problem 24.** *The specific reaction rate of a first order reaction is $10^{-3}\ s^{-1}$. Calculate half-life of the reaction. If the initial concentration of the reactant is 0.1 M, what is the value of reaction rate at t = 0 sec. What is the reaction rate after one hundred seconds?*

**Solution.** $$t_{1/2} = \frac{0.693}{k}$$

$$= \frac{0.693}{10^{-3}} = \mathbf{6.93 \times 10^2\ sec}$$

Initial rate $= k[A]$

$= 10^{-3}s^{-1} \times 0.1\ M = 10^{-4}\ M\ s^{-1}$

$$t = \frac{2.303}{k} \log \frac{\text{Initial concentration}}{\text{Concentration at time } t}$$

$$\log \frac{\text{Initial concentration}}{\text{Concentration at time } t} = \frac{kt}{2.303}$$

$$= \frac{10^{-3} \times 100}{2.303} 0.0434$$

$$= \frac{[A_o]}{[A_t]} = \text{Antilog } 0.0434$$

$$= 1.105$$

$$[A_o] = \frac{[A_o]}{1.105} = \frac{0.1}{1.105}$$

$$= 0.09\ M$$

Rate of reaction after $100 = k \times [A_t]$

$= 10^{-3} \times 0.09\ M$

$= \mathbf{9 \times 10^{-5}\ M\ s^{-1}}$**.**

**Problem 25.** *The half-life for a first order reaction is* $5 \times 10^4$ *s. What percentage of the initial reactant will react in 2 hours?*

**Solution.**

$$k = \frac{0.693}{t_{1/2}}$$

$$= \frac{0.693}{5 \times 10^4}$$

Time = 2 hours = 2 × 3600 sec = 7200 s

Let initial concentration of the reactant, $a = 100\ \text{mol L}^{-1}$

The concentration that reacts in 2 hours = $x$ mol L$^{-1}$

The concentration of the reactant that remains after 2 hours, $(a - x) = (100 - x)$ mol L$^{-1}$

The first order equation is

$$k = \frac{2.303}{t} \log \frac{[A_o]}{[A_t]} = \frac{2.303}{t} \log \frac{a}{a-x}$$

$$\log \frac{a}{a-x} = \frac{kt}{2.303}$$

$$= \frac{0.693 \times 7200}{5 \times 10^4 \times 2.303} = 0.0433$$

$$\frac{a}{a - x} = \text{Antilog } 0.0433 = 1.105$$

$$\frac{100}{100 - x} = 1.105$$

$$100 - x = \frac{100}{1.105} = 90.5$$

$$x = 100 - 90.5 = \mathbf{9.5\%}.$$

**Problem 26.** *A first order reaction is found to have a rate constant $k = 7.39 \times 10^{-5}$ sec$^{-1}$. Calculate $t_{1/2}$, $t_{1/3}$, $t_{90\%}$, $t_{80\%}$, $t_{75\%}$.*

**Solution.** Let $a$ be the initial concentration of the reactant and $x$ the concentration that has reacted in time $t$, then fraction of the reactant that has reacted in time $t$ is

$$\boxed{f = \frac{x}{a} \qquad \therefore \quad x = fa}$$

or
$$t_f = \frac{2.303}{k} \log \frac{a}{a - fa}$$

$$= \frac{2.303}{k} \log \frac{1}{1 - f}$$

(*i*)
$$t_{1/2} \text{ or } t_{50\%} = \frac{2.303}{k} \log \frac{1}{1 - 0.5} = \frac{2.303}{k} \log 2$$

$$= \frac{0.693}{k} = \frac{0.693}{7.39 \times 10^{-5}} = \mathbf{0.0937 \times 10^{-5}\ s}$$

(*ii*)
$$t_{1/3} = \frac{2.303}{k} \log \frac{1}{1 - \frac{1}{3}} = \frac{2.303}{k} \log \frac{2}{3}$$

$$= \frac{0.4055}{k} = \frac{0.4055}{7.39 \times 10^{-5}} = \mathbf{0.0548 \times 10^{-5}\ s}$$

(*iii*)
$$t_{2/3} = \frac{2.303}{k} \log \frac{1}{1 - \frac{2}{3}} = \frac{2.303}{k} \log 3 = \frac{1.098}{k}$$

$$= \frac{1.098}{7.39 \times 10^{-5}} = \mathbf{0.1485 \times 10^{-5}\ s}$$

(iv) $$t_{90\%}\ (f = 0.9) = \frac{2.303}{k}\log\frac{1}{1-0.9} = \frac{2.303}{k}\log 10$$

$$= \frac{2.303}{k} = \frac{2.303}{7.39 \times 10^{-5}}$$

$$= \mathbf{0.3116 \times 10^{-5}\ s}$$

(v) $$t_{80\%}\ (f = 0.8) = \frac{2.303}{k}\log\frac{1}{1-0.9} = \frac{2.303}{k}\log 10$$

$$= \frac{1.609}{k} = \frac{1.609}{7.39 \times 10^{-5}\ s^{-1}}$$

$$= \mathbf{0.2177 \times 10^{-5}\ s}$$

(vi) $$t_{75\%}\ (f = 0.75) = \frac{2.303}{k}\log\frac{1}{1-0.75}$$

$$= \frac{2.303}{k}\log 4 = \frac{1.386}{k}$$

$$= \frac{1.386}{7.39 \times 10^{-5}\ s^{-1}}$$

$$= \mathbf{0.1875 \times 10^{-5}\ s.}$$

### TYPE-III Problems Based on First Order Rate Kinetics

#### *A. Decomposition of $N_2O_5$*

**Problem 27.** *The following results were obtained in the decomposition of nitrogen in carbon tetrachloride at 40°C.*

| *t (sec)* | *1200* | *1800* | *2400* | $\infty$ |
|---|---|---|---|---|
| *x (ml) s* | *11.40* | *15.53* | *18.90* | *34.75* |

*where x denotes volume of oxygen evolved in t seconds. Show by calculation that the reaction is first order. How long would it take half the original material to decompose?*

**Solution.** The decomposition of nitrogen peroxide in the gaseous state or in a solvent like carbon tetrachloride is a first order reaction. The reaction rate can be measured by measuring the volume of oxygen gas evolved at different time intervals.

$$2N_2O_5 \longrightarrow 4NO_2 + O_2$$

we take,

Volume of oxygen at time $t$ ($V_t$) $\propto$ Amount of $N_2O_5$ decomposed ($x$)

Volume of oxygen at infinity ($V_\infty$) $\propto$ Amount of $N_2O_5$ present ($a$)

$\therefore$ Amount of $N_2O_5$ that remain unreacted at time $t = a - x = V_\infty - V_t$

or
$$\frac{a}{a-x} = \frac{V_\infty}{V_\infty - V_t}$$

The rate constant for the first order reaction is given by the following first order rate expression.

$$k = \frac{2.303}{t} \log \frac{a}{a-x}$$

$$= \frac{2.303}{t} \log \frac{V_\infty}{V_\infty - V_t}$$

(*i*) When $t = 1200$ seconds

$$k = \frac{2.303}{1200} \log \frac{34.75}{(34.75 - 11.40)}$$

$$= \frac{2.303}{1200} \log \frac{34.75}{23.35} = \mathbf{3.31 \times 10^{-4}\ sec^{-1}.}$$

(*ii*) When $t = 1800$ seconds

$$k = \frac{2.303}{1800} \log \frac{34.75}{34.75 - 15.53}$$

$$= \frac{2.303}{1800} \log \frac{34.75}{19.22} = \mathbf{3.29 \times 10^{-4}\ sec^{-1}.}$$

(*iii*) When $t = 2400$ seconds

$$k = \frac{2.303}{2400} \log \frac{34.75}{(34.75 - 18.90)}$$

$$k = \frac{2.303}{2400} \log \frac{34.75}{15.85} = \mathbf{3.3 \times 10^{-4}\ sec^{-1}.}$$

Since all the calculated values of $k$ are nearly the same, $k$ is a constant and therefore the reaction is first order.

Average value of
$$k = \frac{(3.31 + 3.29 + 3.30) \times 10^{-4}\ sec^{-1}}{3}$$

$$= \frac{9.90 \times 10^{-4}}{3} = 3.3 \times 10^{-4}\ sec^{-1}$$

$$t_{1/2} = \frac{0.693}{k} = \frac{0.693}{3.3 \times 10^{-4} sec^{-1}}$$

$$= \mathbf{2100\ seconds.}$$

**Problem 28.** *From the following data for the decomposition of $N_2O_5$ in $CCl_4$ solution at 48°C, show that the reaction is first order*

| *t (min)* | *10* | *15* | *20* | $\infty$ |
|---|---|---|---|---|
| *Vol. of $O_2$ evolved* | *6.30* | *8.95* | *11.40* | *34.75* |

**Solution.** For a first order reaction the integrated rate equation is

$$\frac{1}{t}\log\frac{V_\infty}{(V_\infty - V_t)} = k$$

In this example, $V_\infty = 34.75$

| $t$ | $(V_\infty - V_t)$ | $\frac{1}{t}\log\frac{V_\infty}{(V_\infty - V_t)} = k$ |
|---|---|---|
| 10 | 28.45 | $\frac{1}{10}\log\frac{34.75}{28.45} = \mathbf{0.00868}$ |
| 15 | 25.80 | $\frac{1}{15}\log\frac{34.75}{25.80} = \mathbf{0.00862}$ |
| 20 | 23.35 | $\frac{1}{20}\log\frac{34.75}{23.35} = \mathbf{0.00863}$ |

Since the value of $k$ is fairly constant, it is a reaction of **first order.**

***B. Decomposition of $H_2O_2$ in Aqueous Solution***

The decomposition of $H_2O_2$ in the presence of platinum as catalyst is a first order reaction

$$H_2O_2 \xrightarrow{Pt} H_2O + O$$

The progress of the reaction is followed by titrating equal volumes of the reaction mixture against standard $KMnO_4$ solution at different time intervals.

**Problem 29.** *Solution of $H_2O_2$ when titrated against $KMnO_4$ solution at different time intervals gave the following results:*

| *t (minutes)* | *0* | *10* | *20* |
|---|---|---|---|
| *Vol. $KMnO_4$ used* | *23.8 ml* | *14.7 ml* | *9.1 ml* |

*Show that the decomposition of $H_2O_2$ is a first order reaction.*

**Solution.** The integrated rate equation for first order reaction is

$$k = \frac{2.303}{t}\log\frac{a}{a-x}$$

Since volume $KMnO_4$ used in the titration is a measure of concentration of $H_2O_2$ in solution.

$a = 23.8$ ml

$(a - x) = 14.7$ when $t = 10$ mts

$(a - x) = 9.1$ when $t = 20$ mts

Substituting these values in the rate equation above, we have

$$k = \frac{2.303}{10} \log \frac{23.8}{14.7}$$

$$= 0.2303 (\log 23.8 - \log 14.7)$$

$$= 0.2303 (1.3766 - 1.1673) = \mathbf{0.04794}$$

and

$$k = \frac{2.303}{20} \log \frac{23.8}{9.1}$$

$$= 0.10165 (\log 23.8 - \log 9.1)$$

$$= 0.10165 (1.3766 - 0.9595) = \mathbf{0.04806.}$$

Since the value of $k$ is almost constant, the decomposition of $H_2O_2$ is a first order reaction.

*C. Hydrolysis of an Ester*

The hydrolysis of ethyl acetate or methyl acetate in the presence of a mineral acid as catalyst, is a first order reaction.

$$\underset{\text{ethyl acetate}}{CH_3COOC_2H_5} + H_2O \xrightarrow{H^+} \underset{\text{acetic acid}}{CH_3COOH} + C_2H_5OH$$

For studying the kinetic of the reaction, a known volume of ethyl acetate is mixed with a relatively large quantity of acid solution, say N/2 HCl at various intervals of time, a known volume of the reaction mixture is titrated against a standard alkali solution. Hydrolysis of the ester produces acetic acid. Therefore, as the reaction proceeds, the volume of alkali required for titration goes on increasing.

**Problem 30.** *The following data was obtained on hydrolysis of methyl acetate at 25°C in 0.35 N hydrochloric acid. Establish that it is a first order reaction.*

| *t (sec)* | *0* | *4500* | *7140* | *∞* |
|---|---|---|---|---|
| *mL alkali used* | *24.36* | *29.32* | *31.72* | *47.15* |

**Solution.** For a first order reaction

$$k = \frac{2.303}{t} \log \frac{a}{a - x}$$

At any time, the volume of alkali used is needed for the acid present as catalyst and the acid produced by hydrolysis.

The volume of alkali used for total change from $t_0$ to $t_\infty$ gives the initial concentration of ester. Thus,

$$a = 47.15 - 24.36 = 22.79 \text{ ml}$$

$$(a - x) \text{ after 4500 sec} = 47.15 - 29.32 = 17.83 \text{ ml}$$

$$(a - x) \text{ after 7140 sec} = 47.15 - 31.72 = 15.43 \text{ ml}$$

Substituting values in the above equation, we have

$$k = \frac{2.303}{4500} \log \frac{22.79}{17.83} = \mathbf{0.000654}$$

$$k = \frac{2.303}{7140} \log \frac{22.79}{15.43} = \mathbf{0.000546}$$

Since the values of $k$ in the two experiments are fairly constant, the reaction is first order.

**Problem 31.** *10 ml of ethyl acetate was added to a flask containing V mL of 0.1N HCl placed in a thermostat at 60°C. Ethyl acetate hydrolyzes according to the following equation:*

$$CH_3COOC_2H_5 + H_2O \longrightarrow CH_3COOH + C_2H_5OH$$

*10 mL of the reaction mixture was withdrawn at different intervals of time, chilled to stop the hydrolysis and then titrated against a standard alkali. The following data were obtained:*

| *Time (Minutes)* | 0 | 10 | 20 | 100 | 150 | ∞ |
|---|---|---|---|---|---|---|
| *mL of alkali used* | 10 | 10.62 | 11.21 | 15.20 | 17.11 | 25 |

*Show that the hydrolysis of ethyl acetate is a first order reaction:*

**Solution.** Using first order rate equation:

$$k = \frac{2.303}{t} \log \frac{\text{Initial concentration of ethyl acetate}}{\text{Concentration of ethyl acetate at time } t}$$

$$k = \frac{2.303}{t} \log \frac{V_\infty - V_0}{V_\infty - V_t}$$

(*i*) When $t = 10$ min

$$k = \frac{2.303}{10} \log \frac{25 - 10}{25 - 10.62}$$

$$= \frac{2.303}{10} \log \frac{15}{14.38} = \mathbf{0.0042\ min^{-1}}$$

(*ii*) When $t = 20$ min

$$k = \frac{2.303}{20} \log \frac{25 - 10}{25 - 11.21}$$

$$= \frac{2.303}{20} \log \frac{15}{13.79} = \mathbf{0.00416\ min^{-1}}$$

(*iii*) When $t = 100$ min

$$k = \frac{2.303}{100} \log \frac{25 - 10}{25 - 15.20}$$

$$= \frac{2.303}{100} \log \frac{15}{9.80} = \mathbf{0.00425\ min^{-1}}$$

(*iv*) When $t$ = 150 min

$$k = \frac{2.303}{150} \log \frac{25 - 10}{25 - 17.11}$$

$$= \frac{2.303}{150} \log \frac{15}{7.89} = \mathbf{0.00428\ min^{-1}}$$

As the value of $k$ is nearly constant, hydrolysis of ethyl acetate in presence of excess of acid is a first order reaction.

***D. Inversion of Cane Sugar***

**Problem 32.** *Optical rotation of sucrose in 1N HCl at various times was found as shown below:*

| *Time (sec)* | *0* | *7.18* | *18.0* | *27.05* | $\infty$ |
|---|---|---|---|---|---|
| *Rotation (degree)* | *+ 24.09* | *+ 21.7* | *+ 17.7* | *+ 15.0* | *– 10.74* |

**Solution.** Let $r_0$, $r_1$ and $r_\infty$ be the optical rotations at time $t = 0$, $t = t$ and $t = t_\infty$ respectively, then

Initial concentration of sucrose = $a = r_0 - r_\infty$

Decrease in concentration of sucrose = $x = r_0 - r_t$

$\therefore$ Concentration of sucrose at any time $t = a - x$

$$= (r_0 - r_\infty) - (r_0 - r_t)$$

$$= r_t - r_\infty$$

Initial concentration of sucrose = $r_0 - r_\infty$

$$= +\,24.09 - (-\,10.74)$$

$$= +\,34.83$$

$$k = \frac{2.303}{t} \log \frac{a}{a - x}$$

$$\therefore \qquad k = \frac{2.303}{t} \log \frac{r_0 - r_\infty}{r_t - r_\infty}$$

(*i*) At $t$ = 7.18 sec,

$$k = \frac{2.303}{7.18\ \text{sec}} \log \frac{34.83}{21.6 - (-\,10.74)}$$

$$= \frac{2.303}{7.18} \log \frac{34.83}{32.34} = \mathbf{11.8 \times 10^{-3}\ sec^{-1}}$$

(*ii*) At $t$ = 180 sec,

$$k = \frac{2.303}{180} \log \frac{34.83}{17.7 - (-\,10.74)}$$

$$= \frac{2.303}{180} \log \frac{34.83}{28.44} = \mathbf{11.86 \times 10^{-3}\ sec^{-1}}$$

(*iii*) At $t = 27.05$ sec, $k = \frac{2.303}{27.05} \log \frac{34.83}{15-(-10.74)}$

$$= \frac{2.303}{27.05} \log \frac{34.83}{25.74} = \mathbf{11.83 \times 10^{-3}\ sec^{-1}.}$$

As the value of $k$ is nearly constant, the reaction is first order.

### TYPE-IV Problems Based on 2nd Order Kinetics

**Problem 33.** *A second order reaction is 20% complete in 500 seconds. How long will it take for the reaction to go to 60% completion?*

**Solution.** For a second order reaction $k_2 = \frac{1}{t} \cdot \frac{x}{a(a-x)}$

When the reaction is 20% complete, $x$ is 20% of the initial concentration $a$, *i.e.*, 20% of $x$

$$= \frac{20a}{100} = 0.2\,a. \text{ So } k_2 = \frac{1}{500} \times \frac{0.2a}{a(a-0.2a)} = \frac{1}{2000a}$$

Now suppose it takes $t_1$ seconds for the reaction to go for 60% completion. Then

$$x = \frac{60a}{100} = 0.6a \text{ so. Thus } k_2 = \frac{1}{t_1} \times \frac{0.6a}{a(a-0.6a)}$$

or $$\frac{1}{2000a} = \frac{1}{t_1} \times \frac{3}{2a} \quad \text{or} \quad \mathbf{t_1 = 3000\ seconds.}$$

**Problem 34.** *Hydrolysis of ethyl acetate by NaOH using equal concentration of the reactants, was studied by titrating 25 ml of the reaction mixture at different time intervals against standard acid. From the data given below, establish that this is a second order reaction.*

| *t (min)* | *0* | *5* | *15* | *25* |
|---|---|---|---|---|
| *mL acid used* | *16.00* | *10.24* | *6.13* | *4.32* |

**Solution.** Initial concentration $a = 16.00$

Concentration after 5 mts $(a-x) = 10.24$ and $x = 5.76$

Concentration after 15 mts $(a-x) = 6.13$ and $x = 9.85$

Concentration after 25 mts $(a-x) = 4.32$ and $x = 11.68$

Substituting values in the second order rate equation, we have

$$k = \frac{1}{16 \times 5} \cdot \frac{5.76}{10.24} = \mathbf{0.0070}$$

$$k = \frac{1}{16 \times 5} \cdot \frac{9.85}{6.13} = \mathbf{0.0067}$$

$$k = \frac{1}{16 \times 5} \cdot \frac{11.88}{4.32} = \mathbf{0.00675}$$

The value of $k$ being fairly constant, this reaction is second order.

### TYPE-V Problems Based on Activation Energy

**Problem 35.** *The activation energy for the reaction:*

$$C_4H_8(g) \longrightarrow 2C_2H_4(g)$$

*is 262 kJ mol$^{-1}$ at 600 K. Calculate the fraction of molecules having energy equal or greater than activation energy at 600 K.*

**Solution.** Fraction of molecules with energy $= x = e^{-E_a/RT}$ equal or greater than activation energy

$$\log x = \frac{E_a}{2.303\ RT}$$

$$= \frac{-(262 \times 10^3 J)}{2.303 \times 8.301 J \times 600 K}$$

$$= -22.8058 = \overline{23}.1942$$

$$x = \text{Antilog } \overline{23}.1942$$

$$= \mathbf{1.564 \times 10^{-23}}$$

**Problem 36.** *The rate constant for the reaction:*

$$2N_2O_5(g) \longrightarrow 4NO_2(g) + O_2(g)$$

*is $5.0 \times 10^{-4}$ s$^{-1}$ at 45°C. If energy of activation of this reaction is 102 kJ mol$^{-1}$ at 45°C, calculate the collision frequency, A, of the reaction at 45°C.*

**Solution.**

$$T = 273 + 45°C = 318 K$$

$$\log k = \log A - \frac{E_a}{2.303\ RT}$$

$$\log A = \log k + \frac{E_a}{2.303\ RT}$$

$$= \log (5.0 \times 10^{-4}) + \frac{102000}{2.303 \times 8.314 \times 318}$$

$$= -3.3010 + 16.75 = 13.449$$

$$A = \text{Antilog } 13.449$$

$$= \mathbf{2.812 \times 10^{13}\ s^{-1}}.$$

**Problem 37.** *The following values for the first order rate constant were obtained for a certain reaction.*

| *Temp. (K)* | *k* ($sec^{-1}$) |
|---|---|
| *298* | $3.46 \times 10^{-5}$ |
| *308* | $13.50 \times 10^{-5}$ |

*Calculate the activation energy.* $R = 8.3\ JK^{-1}\ mol^{-1}$.

**Solution.** Using $E_a = 2.303\ R\left[\dfrac{T_1T_2}{T_2 - T_1}\right]\left[\log\dfrac{k_2}{k_1}\right]$

$T_1 = 298\ K \qquad k_1 = 3.46 \times 10^{-5}\ s^{-1}$

$T_2 = 308\ K \qquad k_2 = 13.50 \times 10^{-5}\ s^{-1}$

$$\left[\frac{T_1T_2}{T_2 - T_1}\right] = \frac{298 \times 308}{308 - 298} = \frac{298 \times 308}{10}$$

$$\log\frac{k_2}{k_1} = \log\frac{13.50 \times 10^{-5}}{3.46 \times 10^{-5}}$$

$$= \log\frac{13.50}{3.46} = \log 3.902 = 0.5913$$

$$= 103740\ J\ mol^{-1} = \mathbf{103.74\ kJ\ mol^{-1}}.$$

**Problem 38.** *The rate of a reaction quadruples when the temperature changes from 293 to 313 K. Calculate the energy of activation of the reaction assuming that it does not change with temperature.*

**Solution.** $T_1 = 293\ K;\ T_2 = 313\ K$

$$\log\frac{k_2}{k_1} = \frac{E_a}{2.303\ R}\left(\frac{T_2 - T_1}{T_1T_2}\right)$$

$$E_a = 2.303R\left(\frac{T_1T_2}{T_2 - T_1}\right)\left[\log\frac{k_2}{k_1}\right]$$

$$\frac{T_1T_2}{T_2 - T_1} = \frac{293 \times 313}{313 - 293}$$

$$= \mathbf{4585.45\ K}$$

$$\log\frac{k_2}{k_1} = \log\frac{4}{1} = \log 4 = 0.6021$$

$$E_a = 2.303 \times 8.314\ JK^{-1}mol^{-1} \times 4585.45\ K \times 0.6021$$

$$= 52848\ J\ mol^{-1}$$

$$= \mathbf{52.8\ KJ\ mol^{-1}}.$$

**Problem 39.** *The activation energy for the reaction* $2N_2O_5(aq) \longrightarrow 2N_2O_4(aq) + O_2(g)$ *is 100 kJ mol*$^{-2}$*. The rate constant of the reaction is* $2.35 \times 10^{-4} s^{-1}$ *at 293 K. What is the rate constant of the reaction at 303 K?*

**Solution.**

$$\log \frac{k_2}{k_1} = \frac{E_a}{2.303\ R}\left(\frac{T_2 - T_1}{T_1T_2}\right)$$

$$= \frac{100 \times 10^3}{2.303 \times 8.314}\left(\frac{303 - 293}{293 \times 303}\right)$$

$$= 5222.7 \times (0.0001126)$$

$$= 0.5881$$

$$\frac{k_2}{k_1} = \text{Antilog } 0.5881 = 3.874$$

$$k_2 = k_1 \times 3.874$$

$$= 2.35 \times 10^{-1}\ s^{-1} \times 3.874$$

$$= \mathbf{9.11 \times 10^{-4}\ s^{-1}}.$$

**Problem 40.** *In the Arrhenius equation for a reaction, the value of A and* $E_a$ *(activation energy) are* $4 \times 10^{13}$ *sec*$^{-1}$ *and 98.6 KJ mol*$^{-2}$ *respectively. If the reaction is of first order, at what temperature will its half-life period be ten minutes?*

**Solution.**

$$k = \frac{0.693}{t_{1/2}} = \frac{0.693}{10 \text{ min}}$$

$$= \frac{0.693}{10 \times 60} = 0.0012 \text{ sec}^{-1}$$

$$\log k = \log 0.0012 = -2.9208$$

$$\log A = \log 4 \times 10^{13} = 13.6021$$

$$\log k = \log A - \frac{E_a}{2.303\ RT}$$

$$\frac{E_a}{2.303\ RT} = \log A - \log k$$

$$T = \frac{E_a}{2.303\ R(\log A - \log k)}$$

$$= \frac{98.6 \times 10^3}{2.303 \times 8.314\ (13.6021) - (-2.9208)}$$

$$= \frac{98.6 \times 10^3}{2.303 \times 8.314 \times 16.5229} = \mathbf{311.7\ K.}$$

**Problem 41.** *The activation energy for a certain first order reaction is 500 kJ mol$^{-2}$. The half-life of the reaction at 27°C is 5 h. Calculate the half-life at 37°C.*

**Solution.**

$$k_1 = \frac{0.693}{t_{1/2}} = \frac{0.693}{5h}$$

$$\log \frac{k_2}{k_1} = \frac{E_a}{2.303\ RT} \times \left(\frac{T_2 - T_1}{T_1T_2}\right)$$

$$= \frac{500000}{2.303 \times 8.314}\left(\frac{310 - 300}{300 - 310}\right)$$

$$= 2.8079$$

$$\frac{k_2}{k_1} = \text{Antilog } 2.8079 = 6.425 \times 10^2$$

$$k_2 = k_1 \times 6.425 \times 10^2$$

$$= \frac{0.693}{5} \times 6.425 \times 10^2$$

$$t_{1/2} = \frac{0.693}{k}$$

$$= \frac{0.693 \times 5}{0.693 \times 6.425 \times 10^2} = \mathbf{7.78 \times 10^{-3}\ h.}$$

## QUESTION BANK

1. (*a*) Explain unimolecular, biomolecular, and pseudomolecular reactions. Give appropriate examples.
   (*b*) Write down Arrhenius equation. Explain the significance of all the terms in it. How would you explain the deviation of reaction rate from Arrhenius equation?
   (*c*) What are pseudo unimolecular reaction? Give two examples.
2. (*a*) What do you understand by the order and molecularity of a reaction? Derive an expression for the velocity constant of first order reaction.
   (*b*) Write a note on order of reaction.
3. What is meant by molecularity and order of reaction? What are the unimolecular and pseudomolecular reactions? Give examples. Describe half-life period method for the determination of the order of reaction.
4. (*a*) Outline any three methods for determining the order of a reaction.
   (*b*) Discuss two different methods used for determining order of reactions.
   (*c*) Write short notes on: (*i*) Temperature coefficient of reaction, (*ii*) Energy of activation.

5. (*a*) Write brief note on (*i*) Chain reaction, (*ii*) Zero order reactions.
   (*b*) Write explanatory notes on (*i*) Activated complex, (*ii*) Temperature coefficient of a reaction.
   (*c*) Show how the rate constant of saponification of ethyl acetate by caustic soda is determined experimentally.
6. (*a*) Deduce the kinetic equation for a second order reaction where the reacting moles are same.
   (*b*) In a particular reaction, the time required to complete half of the reaction was found to increase nine times when the initial concentration of the reactants was reduced to 1/3. What is the order of the reaction?
7. Differentiate between:
   (*i*) Order and molecularity
   (*ii*) First order and Pseudo first order reaction.
8. Derive the equation for determining $k_1$ and $k_2$ for the following reaction.

$$A \underset{k_2}{\overset{k_1}{\rightleftharpoons}} B$$

9. How can the rate constant $k_1$ and $k_2$ be determined for the reaction?

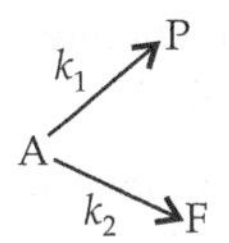

10. What do you mean by parallel and reversible reaction? Give an example of each.
11. For reaction: $A + 2B \longrightarrow C + D$
    The following data are given:

| *[A]* *mol l⁻¹* | *[B]* *mol l⁻¹* | *Rate* *mol l⁻¹s⁻¹* |
|---|---|---|
| 0.01 | 0.10 | 0.0032 |
| 0.03 | 0.10 | 0.0096 |
| 0.01 | 0.20 | 0.0064 |

    Find the order with respect to A, B and total of reaction.
12. For reaction: $2NO + Cl_2 \longrightarrow 2NOCl$
    It was found that on doubling the concentration of both reactants, the rate increases eight fold. But on doubling the concentration of $Cl_2$ alone the rate only double itself. What is the order of reaction with respect to NO? Justify your answer.
13. Integrate the equation $-\frac{dc}{dt} = k[C]^{3/2}$. If CO is the initial concentration find $k$ and calculate half-life time.
14. The following results were obtained for the saponification of ethyl acetate using equal concentrations of ester and alkali:

| *t* (minutes) | 0 | 4 | 6 | 10 | 20 |
|---|---|---|---|---|---|
| Volume (mL) of acid used | 8.04 | 5.30 | 4.58 | 3.50 | 2.22 |

    Show that the reaction is second order.

15. At a certain temperature, the half-life periods for the catalytic decomposition of ammonia were as follows:

| Pressure (mm Hg) | 50 | 100 | 200 |
|---|---|---|---|
| Relative half-life period | 3.52 | 1.82 | 0.96 |

16. From the following data, show that the determination of $H_2O_2$ in aqueous solution is a first order reaction.

| Time (minutes) | 0 | 15 | 30 |
|---|---|---|---|
| N | 25.4 | 9.83 | 3.81 |

where N is the number of ML of $KMnO_4$ required to decompose a definite volume of hydrogen peroxide.

17. The optical rotations of sucrose in 0.5 N HCl at 30°C are given below. Show that the reaction is first order.

| Time (minutes) | 0 | 15 | 30 | 45 | 60 | $\infty$ |
|---|---|---|---|---|---|---|
| Rotation (degrees) | + 32.4 | + 28.8 | + 22.5 | + 22.4 | + 16.6 | + 11.0 |

18. The rate of a reaction triples when temperature changes from 20 to 50°C. Calculate the energy of activation for such a reaction (R = 8.314 J $K^{-1}$ $mole^{-1}$). **[Ans.** 28.19 kJ $mol^{-1}$]

19. The activation energy of reaction is 52.86 kJ $mole^{-1}$ and the value of rate constant at 298 K is $3.46 \times 10^{-5}$ $sec^{-1}$. Calculate the frequency factor A.

20. The reaction $C_2H_5I \longrightarrow C_2H_4 + HI$ is of first order and its rate constants are $3.20 \times 10^{-4}$ $sec^{-1}$ at 600 K and $1.60 \times 10^{-2}$ $sec^{-1}$ at 1200 K. Calculate the energy of activation for the reaction. (Given R = 8.3 $JK^{-1}$ $mole^{-1}$).

Chapter 8

# PHASE EQUILIBRIUM—THE PHASE RULE

## 8.1 INTRODUCTION

The phase rule is an important tool in the study of heterogeneous equilibrium. It relates the conditions which must be specified to describe the state of a system at equilibrium.

A substance can exist in various states depending on the external conditions. For example, water can exist in the form of a solid, liquid, and vapor. *A change in the state of a substance is called phase transition.* Equilibrium in a system in which phase transitions occur in addition to a chemical process is called *phase equilibrium.*

Before we proceed to describe and derive phase rule, let us define various terms used in phase equilibrium.

## 8.2 PHASE

A phase is defined as *the part of a system which is homogeneous throughout and is separated from other homogeneous parts of the system by means of a definite boundary. A phase is a physically distinct and mechanically separable part of a system.*

Common examples that describe phase are:

(*i*) A system consisting of water and ice; all the pieces of ice form one phase, and the water another phase. Thus there are *two phases.*

(*ii*) A mixture of gases such as helium, hydrogen, and argon constitutes a single phase, since gases are completely miscible.

(*iii*) Liquids may or may not form a single phase. It depends upon their miscibility. Completely miscible liquids constitute *one phase* system. For example, water and alcohol, benzene and chloroform.

(*iv*) Immiscible liquids constitute different phases. For example, carbon tetrachloride and water do not mix with each other and form *two phases*.

(*v*) Each solid makes up a separate phase except in the case of a solid solution. For example, many forms of sulphur can exist together and all are separate phases.

## 8.3 COMPONENT

The components of a system can be described in terms of the species present in the system. The component of a system are defined as *the minimum number of independent variable constituents by means of which the composition of any phase of the system can be expressed either directly or in the form of a chemical equation.* For example,

(*i*) Water exists in three phases;

$$\text{Ice } (s) \rightleftharpoons \text{liquid } (l) \rightleftharpoons \text{Vapor } (g)$$

However, the composition of each phase can be expressed in terms of $H_2O$. Hence, it is a *one component system.*

(*ii*) The *sulphur system* consists of four phases; rhombic, monoclinic, liquid, and vapor. The chemical composition of each phase can be expressed in terms of S. Hence, it is a *one component system*.

(*iii*) Consider an equilibrium system containing sodium sulphate and water. Such a system consists of various phases—water, vapor, ice, and various hydrates of sodium sulphate. However the composition of each phase can be expressed by a chemical equation which involves only anhydrous $Na_2SO_4$ and water.

Consider that three phases $Na_2SO_4.10H_2O$, $Na_2SO_4$ and solution are present. Then

$$Na_2SO_4.10H_2O \rightleftharpoons Na_2SO_4 + 10H_2O$$

$$Na_2SO_4 \rightleftharpoons Na_2SO_4 + 0H_2O$$

$$\text{Solution} \rightleftharpoons xNa_2SO_4 + yH_2O$$

The smallest number of independent variable constituents required to express the composition of each phase is $Na_2SO_4$ and $H_2O$. Thus it is a *two component system*.

(*iv*) *Decomposition of $CaCO_3$*. There are three different molecular species—$CaCO_3$, CaO, and $CO_2$, but the composition of each phase can be expressed in terms of any two of the species. If any two of the constituents are chosen, the composition of all the phases can be expressed. For example, if CaO and $CO_2$ are two constituents, then:

$$CaO = CaO + 0CO_2$$

$$CO_2 = 0CaO + CO_2$$

$$CaCO_3 = CaO + CO_2$$

If $CaCO_3$ and $CO_2$ are chosen, then:

$$CaO = CaCO_3 - CO_2$$

$$CaCO_3 = CaCO_3 + 0CO_2$$

$$CO_2 = 0CaCO_3 + CO_2$$

Thus the composition of the phases present at the equilibrium can be fixed by two variables. Therefore it is a *two component system.*

## 8.4 DEGREE OF FREEDOM

The degree of freedom or variance is defined as *the smallest number of independent variables such as pressure, temperature, and concentration that must be specified in order to describe completely the state of the system.*

Systems having degrees of freedom three, two, one, or zero are known as *trivariant, bivariant, univariant* (or monovariant), and *nonvariant* systems, respectively. Common examples include:

1. **For ice-water-vapor system, F = 0.** In the system, ice $\rightleftharpoons$ water $\rightleftharpoons$ vapor, the three phases co-exist at the freezing point of water. As the freezing temperature of water has a fixed value, the vapor pressure of water also has a definite value. The system has two variables (T and P), both of which are already fixed. The system is completely defined automatically and there is no need to specify any variable. It has no degree of freedom, *i.e.*, F = 0.
2. **For saturated NaCl solution, F = 1.** The saturated solution of sodium chloride in equilibrium with solid NaCl and water vapor, *i.e.*,

   $$NaCl(s) \rightleftharpoons NaCl(solution) \rightleftharpoons \text{Water vapor}$$

   is completely defined if we specify temperature only. The other two variables, *i.e.*, composition of NaCl solution and vapor pressure have a definite value at a fixed temperature. So, the system has one degree of freedom.
3. **For a pure gas, F = 2.** For a sample of pure gas, PV = RT. If the values of P and T are specified, volume (V) can have only one definite value or that volume, *i.e.*, the third variable, is fixed automatically. Any other sample of the gas under the same pressure and temperature as specified above will be identical to the first one. So, the system containing a pure gas has two degrees of freedom.

## 8.5 THE PHASE RULE

The phase rule was deduced from thermodynamic considerations by *Willard Gibbs* in 1876. This rule is stated in terms of the number of phases (P), the number of components (C) and the degrees of freedom (F) of a heterogeneous equilibria. According to this rule, *for any equilibrium thermodynamic system, the sum of the number of degrees of freedom (F) and the number of phases (P) are equal to the sum of number of independent components C and the number of external factors n (temperature and pressure) affecting the physical state of the system.*

$$F + P = C + n$$

or $$F + P = C + 2 \qquad (\because \; n = 2)$$

or $$F = C - P + 2$$

where, $n$ = Number of external factors such as temperature and pressure.

### 8.5.1 Derivation of the Phase Rule

Consider a system in equilibrium containing **P** phases and **C** components. The composition of each phase can be specified if the amount of each component is known. So, we have to describe the mole fraction of each component. If **C** is the number of components required to describe the composition of a phase, then the total number of composition variables for the **P** phases are **PC.** Besides, there are two more variables, temperature and pressure, which have to be considered. The total number of independent variables is **PC** + 2.

In each phase there is a relationship between the mole fractions:

$$X_1 + X_2 + X_3 + \ldots\ldots + X_c = 1$$

$$\sum_i X_i = 1 \qquad (8.1)$$

From equation (8.1), the mole fraction of any component in a phase can be calculated if the mole fractions of all other components are known. For each phase there is one such equation and for **P** phases there are **P** equations similar to equation (8.1). Therefore, the total number of independent variables to be specified are:

$$\mathbf{CP + 2 - P}$$

or $$\mathbf{P(C - 1) + 2}$$

When a heterogeneous system is in equilibrium at a constant temperature and pressure, the chemical potential of a particular component must be the same in all the phases in which it appears. For a one component system having two phases $\alpha$ and $\beta$, the equality of chemical potential implies

$$\mu(\alpha) = \mu(\beta)$$

Hence, there is one equilibrium relation that gives the distribution of the component between two phases. Similarly, in a one component system having three phases $\alpha$, $\beta$, and $\gamma$, we have

$$\mu(\alpha) = \mu(\beta) = \mu(\gamma)$$

Thus, two additional equilibrium relations between the variables exist. In general, for each component in **P** phases, (**P** – 1) relations are possible. Since there are **C** components, the total number of these equations are **C(P** – 1).

Therefore the degrees of freedom or the number of independent variables **F** is given by

$$\begin{aligned} F &= \text{Total number of variables} \\ &\quad - \text{Total number of relations among the variables} \\ &= PC + 2 - P - C(P - 1) \\ &= C - P + 2 \end{aligned}$$

or

$$F + P = C + 2 \tag{8.2}$$

Equation (8.2) is the phase rule given by Willard Gibbs, the fundamental relation controlling the equilibria in heterogeneous systems.

### 8.5.2 Advantages of the Phase Rule

Following are the advantages of the Phase Rule:

(*i*) It gives a simple method of classifying equilibrium states of systems.

(*ii*) It confirms that the different systems having the same number of degrees of freedom behave in a like manner.

(*iii*) It predicts the behavior of systems when subjected to changes in the variables such as pressure, temperature, and volume.

(*iv*) The phase rule is applicable to macroscopic systems.

(*v*) It is applicable to physical as well as to chemical phase reactions.

(*vi*) Phase rule predicts that a number of substances will remain in equilibrium or not in equilibrium if some of the substances have been transformed into the new substances.

### 8.5.3 Limitations of the Phase Rule

Following are the limitations of the Phase Rule:

(*i*) It is applicable to heterogeneous systems in equilibrium; it is, therefore, irrelevant for such systems which are slow in reaching the equilibrium state.

(*ii*) In phase rule, various variables are temperature, pressure, and composition. It does not consider electric and magnetic influences.

(*iii*) All the phases in the system must be present under the same pressure, temperature, and gravitational force.

(*iv*) No liquid or solid phases should be finely divided, otherwise their vapor pressure will change their normal values.

## 8.6 ONE COMPONENT SYSTEM

### 8.6.1 The Water System

Water is the most common example of the one component system, it can exist in three phases; solid ice, liquid water, and water vapors. The three phases may occur in four possible combinations in equilibrium as follows:

(*i*) Liquid $\rightleftharpoons$ Vapor

(*ii*) Liquid $\rightleftharpoons$ Solid

(*iii*) Solid $\rightleftharpoons$ Vapor

(*iv*) Solid $\rightleftharpoons$ Liquid $\rightleftharpoons$ Vapor

The existence of phases in equilibrium at any time depends upon the conditions of temperature and pressure. The phase diagram for a water system is shown in Figure 8.1. This phase diagram has the following parts:

(*i*) Stable Curves OA, OB and OC (represented by solid lines)

(*ii*) Metastable Curve OA′ (represented by dotted line)

(*iii*) Areas AOC, BOC and AOB

(*iv*) Triple Point O.

**Curves**

(*i*) *OA (Vaporization Curve)*: This curve is known as the *vaporization curve* as it represents the equilibrium between liquid water and vapor at different temperatures.

$$\text{Liquid water} \rightleftharpoons \text{Vapor}$$

At every point of the curve two phases are present. The starting point O of the curve, corresponds to the freezing point of water (0.0075°C at 4.58 mm). The curve ends at A, the critical temperature (374°C at 218 atm). Beyond this temperature, liquid water cannot exist. At 100°C, the vapor pressure of water equals the atmospheric pressure (760 mm), and therefore is the boiling point of the water. Applying the phase rule,

$$F = C - P + 2$$

$$C = 1, \quad P = 2$$

$$F = 1 - 2 + 2 = 1$$

So, this system has a degree of freedom 1 and is univariant.

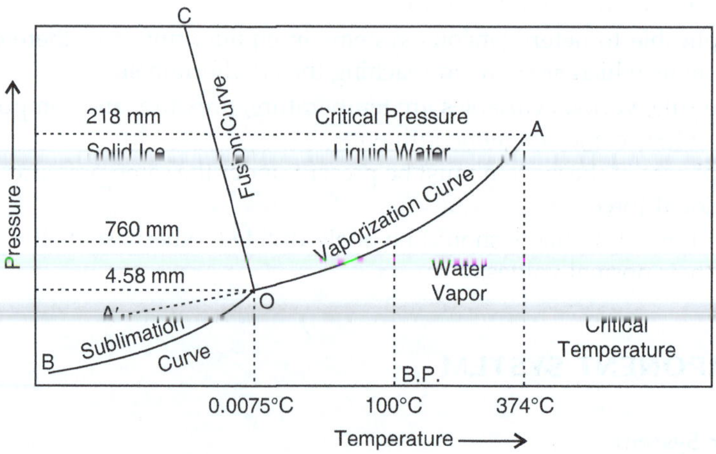


***FIGURE 8.1*** **Phase diagram for water system (Not to scale).**

(*ii*) *OB* (*Sublimation Curve*). This curve represents the equilibrium between ice and vapor and is called *vapor pressure curve* of ice or sublimation curve of ice. The curve starts from point O and ends at B (absolute zero). This curve shows the effect of pressure on the freezing point of water. This system has a degree of freedom 1.

(*iii*) *OC* (*Fusion Curve*). This curve represents the equilibrium between ice and liquid water at various pressures. The curve shows the effect of pressure on the melting point of ice. As can be seen, the curve OC inclines toward the pressure axis which shows that expansion takes place on the freezing of water and the melting point of ice is decreased by an increase of pressure. The upper point of this curve is OC (2000 atm, 20°C); beyond this point, one type of ice changes into another solid but the solid-liquid equilibrium still remains there. This system is a univariant system and the degree of freedom is 1.

(*iv*) *OA′* (*Metastable Curve*). The dotted curve OA′ is a continuation of the OA curve and represents the vapor pressure of supercooled water. With due care, it is possible to cool water (any liquid) below its freezing point without separation of ice and the water is then said to be supercooled. The liquid $\rightleftharpoons$ vapor system along the curve OA′ is said to be in metastable equilibrium as even the slight disturbance in this phase changes the supercooled water to ice and the curve merges into OB. This curve OA′ lies above the curve OB and shows that the metastable system has a higher vapor pressure than the stable one at the same temperature.

**Areas or Regions between Curves**

The systems represented by areas have two degrees of freedom and are bivariant. Pressure and temperature are the two quantities required to define the system completely at any point in the three areas. From phase rule

$$F = C - P + 2$$

$$C = 1, P = 1$$

$$F = 1 - 1 + 2 = 2$$

So the systems are bivariant.

As discussed in a water system there are three areas:

(*i*) AOB — water vapors

(*ii*) BOC — Solid Ice

(*iii*) AOC — Liquid water

**Triple Point**

At this point, three curves OA, OB, and OC meet together and all the three phases co-exist. The point O represents the triple point. Temperature and pressure are fixed at 0.0075°C and 4.58 mm, respectively. The effect of heat will simply result in more of the solid melting but there will be an increase in temperature or pressure until all of the solid has completely changed into liquid. Then the system will change from nonvariant to univariant. Similar is the effect of pressure. Thus the system has zero degrees of freedom at the point O and can be shown by the phase rule equation:

$$F = C - P + 2$$

$$C = 1, P = 3$$

$$F = 1 - 3 + 2 = 0.$$

The main features of the phase diagram of water can be summed as:

| *System* | *Phase* | *Degree of Freedom* |
|---|---|---|
| **Curves (OA, OB, OC, OA′)** | 2 | 1 |
| OA: Vaporization curve | | |
| OB: Sublimation curve | | |
| OC: Fusion curve | | |
| OA′: Metastable curve | | |
| **Areas (AOB, BOC, AOC)** | 1 | 2 |
| AOB: Vapors only | | |
| BOC: Ice only | | |
| AOC: Liquid water | | |
| **Triple point (O)** | 3 | 0 |

## 8.6.2 The Carbon Dioxide System

The phase diagram of carbon dioxide resembles that of water and has three distinct areas for solid, liquid, and gaseous phases. The phase diagram for a carbon dioxide system is shown in Figure 8.2.

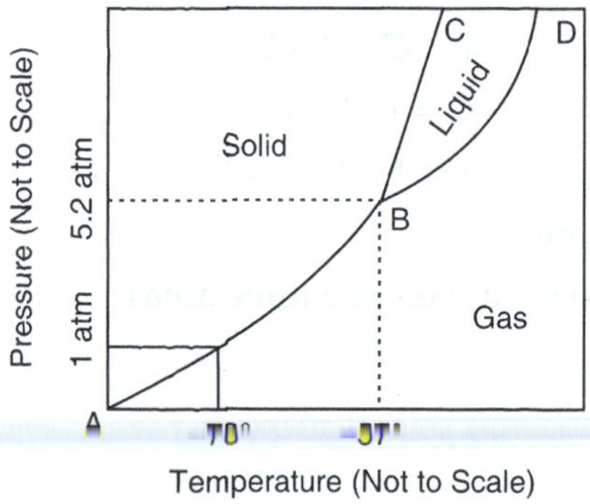


*FIGURE 8.2* **Phase diagram for $CO_2$ system.**

### Curves

The system is univariant and has two phases all along these curves.

(*i*) AB is the sublimation curve along which solid carbon dioxide is in equilibrium with gas.

(*ii*) BD is the vaporization curve and represents the liquid carbon dioxide in equilibrium with gas.

(*iii*) BC is the fusion curve and represents the solid and liquid carbon dioxide in equilibrium with each other.

### Areas

The system is bivariant in the areas.

(*i*) ABC — Solid carbon dioxide
(*ii*) BCD — Liquid carbon dioxide
(*iii*) ABD — Gaseous carbon dioxide

**Triple point**

B is the triple point at which all three phases of carbon dioxide exist in equilibrium. The temperature of the system at this point is – 57°C and the pressure is 5.2 atm. The system is nonvariant at this point.

**Difference from Water System**

(*a*) The fusion curve slopes away from the pressure axis.

(*b*) The solid carbon dioxide can exist in equilibrium with liquid carbon dioxide at very high pressure (5.2 atm), while in the case of a water system, ice and water can co-exist even at a very low pressure (4.58 mm).

(*c*) The vapor pressure of solid carbon dioxide, even at extremely low temperatures, is very high and many times higher than that of ice.

As evident from the diagram, carbon dioxide gas can be directly solidified at (1 atmospheric pressure) without the appearance of the liquid phase merely on cooling to – 78°C. It is for this reason that solid carbon dioxide is commonly known as *dry ice*. The main features of the phase diagram can be summed as:

| *System* | *Phase* | *Degree of Freedom* |
|---|---|---|
| **Curves** | 2 | 1 |
| AB: Sublimation Curve | | |
| BD: Vaporization Curve | | |
| BC: Fusion Curve | | |
| **Areas** | | |
| ABC: Solid $CO_2$ | 1 | 2 |
| BCD: Liquid $CO_2$ | | |
| ABD: Gaseous $CO_2$ | | |
| **Triple point O** | 3 | 0 |

## 8.7 SULPHUR SYSTEM

It is a one component system having more than one solid phase. It consist of following four phases :

(*i*) Rhombic sulphur — $S_R$ Solid phase
(*ii*) Monoclinic sulphur — $S_M$ Solid phase
(*iii*) Liquid sulphur — $S_L$ Liquid phase
(*iv*) Vapor sulphur — $S_V$ Vapor phase

It is a typical example of one component system exhibiting polymorphism and solid transformation. At 95.6°C, the two crystalline forms rhombic and monoclinic sulphur are in equilibrium with each other *i.e.* $S_R \rightleftharpoons S_M$. This is known as the *transition temperature or transition point*. At transition temperature both rhombic sulphur and monoclinic sulphur exists. Rhombic sulphur has octahedral crystal form and monoclinic sulphur has prismatic crystal form. Below the transition temperature rhombic sulphur is stable and monoclinic sulphur slowly changes into rhombic sulphur. Sulphur melts around 120°C which is higher than the transition temperature. Any such polymorphic form of a substance whose transition temperature is lower than the fusion temperature is said to be enantiotropic and the phenomenon is called *enantiotropy*.

**Characteristics of the transition temperature (Transition point).** A transition point is definite likewise a melting point. Many transition points have been determined correctly to the third place of decimal. It is affected by pressure just like melting point. By the application of Le Chatelier's principle, effect of pressure on transition point can be ascertained quantitatively. Transformation of one form into another form at the transition point takes place much more slowly than melting point. If this transformation (*i.e.*, $S_R \rightleftharpoons S_M$) does not get sufficient time *i.e.* $S_R$ is heated rapidly, it is possible to heat it to a temperature well above its transition point without getting $S_M$. Thus it is made possible to study the properties of metastable $S_R$ up to 115°C (its melting point). Super heating is not possible at melting point because at this point it changes into liquid state. The complete equilibrium diagram can be shown on the graph paper. It is shown in the figure 8.3.

According to phase rule all the four phases cannot exist in equilibrium at a time. It can be shown by phase rule equation.

$$F = C - P + 2 = 1 - 4 + 2 = -1 \qquad [\because \; C = 1, P = 4]$$

Because we get a negative degree of freedom which is impossible in a system consisting of single component. However, it is possible to have three or lesser number of phases to exist together. These conditions have been determined experimentally and are summarised here in Fig. 8.3.

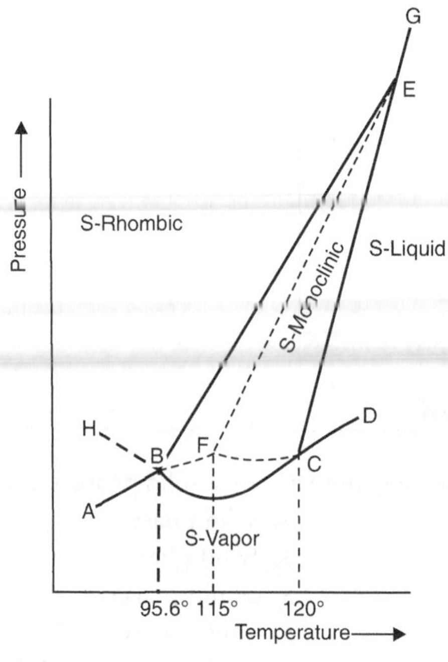


*FIGURE 8.3* **Phase diagram for sulphur.**

The diagram consists of six stable curves *AB, BC, CD, CE, BE,* and *EG* and four metastable curves. These curves divide diagram into four areas *ABCD, GECD, ABEG,* and *BEC.*

**Areas.** The areas are bivariant because to locate any point in the areas, both temperature and pressure are to be fixed up. Just like water system, area represents conditions for the stable existence of a single phase.

| *Area* | *Phase* |
|---|---|
| ABCD | $S_V$ |
| GECD | $S_L$ |
| ABEG | $S_R$ |
| BEC | $S_M$ |

In areas, there are two degrees of freedom which can be shown by phase rule equation.

$$F = C - P + 2 = 1 - 1 + 2 = 2 \qquad [\because \ C = 1; P = 1]$$

**2. Study of Curves.**

(*i*) *Vapour pressure curve of rhombic sulphur, Curve AB (BF).* $S_R$ and $S_V$ exist in equilibrium along this curve and degree of freedom is one. The curve starts from the point B, the transition temperature of $S_R$ to *SM* and ends at A (50°C) below which vapour pressure of $S_R$ cannot be measured.

Point B (95.6°C) is the transition point. At this point $S_R$ changes to $S_M$ but it requires some time. If sufficient time is not allowed *i.e.* heating is rapid the transition from one crystalline form to another crystalline form is not possible and we have vapor pressure curve of $S_R$ extending even beyond B, and it reaches the point F (115°C), the melting point of $S_R$. It is already stated that $S_R$ is metastable above the transition point, dotted curve BF is metastable curve and point *F* is the melting point of metastable $S_R$.

(*ii*) *BC (Vapor pressure curve of monoclinic sulphur).* The two different phases, $S_M$ and $S_V$, are present along this curve and degree of freedom is one. This curve meets the vapor pressure curve of $S_R$ at *B*. Thus point *B* is common to both the curves : vapor pressure curve of $S_R$ and vapor pressure curve of monoclinic sulphur. At this point $S_R$ and $S_M$ can coexist with their vapors. It is a triple point and is nonovariant. The point *C* (120°C), the melting point of $S_M$, is the other end of the vapor pressure curve of $S_M$.

If the temperature of monoclinic sulphur is lowered, then *ordinarily* it would change to $S_R$ at 95.5°C but it is possible to prevent this conversion by carefully cooling it. At that time we will obtain a curve *BH*, a metastable vapor pressure curve of monoclinic sulphur. Here, it should be noted, that the metastable phase has the higher vapor pressure than stable phase at the same temperature, which is clear from diagram.

(*iii*) *CD (CF) vapor pressure curve of liquid sulphur.* We have two phases $S_L$ and $S_V$ in equilibrium along this curve and degree of freedom is one. This curve starts at *C*, the melting point of $S_M$ and ends at *D*, the critical point after which only $S_V$ can exist.

The dotted curve *CF* is a prolongation of *CD* and represents the metastable vaporisation curve of $S_L$. It corresponds to the metastable curve of super cooled water. *F* is the melting point of the two metastable curves *CF* and *BF*. *F* is a metastable triple point ; at this point metastable $S_R$, metastable $S_M$ and vapor co-exists.

(*iv*) *BE, (Transition curve of $S_R$ to $S_M$).* This curve shows the effect of pressure on the transition temperature of $S_R$. The slope of the curve is away from the pressure axis which indicates that the transition point is raised by the rise of pressure which may be explained as under.

The rhombic sulphur changes to monoclinic sulphur accompanying the absorption of heat and increase of volume.

$$\text{Density of } S_R = 2.04 \text{ gm/ml}$$

$$\text{Density of } S_M = 1.9 \text{ gm/ml}$$

By the application of Le Chatelier's principle we know that increase of pressure will shift the equilibrium in a direction in which decrease of volume takes place. So, increase of pressure displaces the equilibrium of the system $S_R + Q$ (heat energy) $\rightleftharpoons S_M$ towards left hand side, forming more and more of $S_R$. Since the formation of $S_R$ is accompanied by evolution of heat, the transition temperature will increase by the increase of pressure.

The transition curve *BE* ends at point *E*, after which $S_M$ disappears.

(*v*) *CE (The fusion or melting point curve for $S_M$).* This curve shows the equilibrium between monoclinic sulphur and liquid sulphur. It is deduced by the study of effect of pressure on the melting point of sulphur. This curve also inclinds away from the pressure axis showing that the melting point is raised by the increase of pressure. This is due to the fact that $S_M$ melts by taking its heat of fusion and thus expansion in volume takes place, *i.e.*

$$S_M + Q \text{ (heat energy)} \rightleftharpoons S_L$$

According to Le Chatelier's Principle, increase in pressure will push the reaction in backward direction and since this would result in the liberation of heat, melting point would be raised.

(*vi*) *Curve EG (FE) (the melting point curve of rhombic sulphur).* Any point on the curve *FE* shows the equilibrium between the metastable phase $S_L$ and $S_R$. At point *E*, for which conditions are 151°C, 1288 atm pressure, the two curves *BE* and *CE* meet together. It follows that if liquid sulphur were cooled at a pressure higher than 1288 atm then $S_R$, rhombic sulphur, crystallises taking no time, and no $S_M$ will be formed. The curve *FE*, beyond the point *E*, becomes a stable melting point curve *EG* of rhombic sulphur.

**3. Triple Points**

| | *Triple points* | *Temperature* | *Phase in equilibrium* |
|---|---|---|---|
| (*i*) | B | 95.6°C | $S_R - S_M - S_V$ |
| (*ii*) | C | 120°C | $S_M - S_L - S_V$ |
| (*iii*) | E | 151°C | $S_R - S_M - S_L$ |
| (*iv*) | F | 115°C | $S_R - S_L - S_V$ |

(*F*) is a metastable triple point.

There are three stable triple points (*B, C, E*) and a metastable triple point (*F*) in the phase diagram. If we change either temperature or pressure, one of the phase disappears. Therefore, the system has no degree of freedom at these triple points. This can be shown from the phase rule equation.

$$F = C - P + 2 = 1 - 3 + 2 = 0 \qquad (\because \; C = 1, P = 3)$$

TABLE 8.1 The sulphur system (*Component* = 1)
(Metastable systems are given in brackets)

| *Name of the system represented in the diagram* | *Phases in the equilibrium* | *Degree of freedom (Variance)* $F = C - P + 2$ |
|---|---|---|
| **Areas :** | | |
| | | Two variant |
| (*i*) ABEC | $S_R$ | as $F = 1 - 1 + 2 = 2$ |
| (*ii*) DCEG | $S_L$ | —do— |
| (*iii*) ABCD | $S_V$ | —do— |
| (*iv*) BEC | $S_M$ | —do— |
| **Curves :** | | |
| | | One variant |
| (*i*) AB (BF) | $(S_R - S_V)$ | as $F = 1 - 2 + 2 = 1$ |
| (*ii*) BC (BH) | $(S_M - S_V)$ | —do— |
| (*iii*) CD (CF) | $(S_L - S_V)$ | —do— |
| (*iv*) FE (EG) | $S_R - S_L$ | —do— |
| (*v*) BE | $S_R - S_M$ | —do— |
| (*vi*) CE | $S_M - S_L$ | —do— |
| **Triple points :** | | |
| | | Zero (Invariant) |
| (*i*) B | $S_R - S_M - S_V$ | $F = 1 - 3 + 2 = 0$ |
| (*ii*) C | $S_M - S_L - S_V$ | —do— |
| (*iii*) E | $S_R - S_M - S_L$ | —do— |
| (*iv*) (F) | $S_R - S_L - S_V$ | —do— |

## 8.8 TWO COMPONENT SYSTEMS

A two component system can be defined as that in which it is possible to express the composition of all the phases, in terms of two substances. In two component system, when $P = 1$, $F$ has the highest value (*i.e.* 3)

$$F = C - P + 2$$
$$= 2 - 1 + 2 = 3$$

Thus three variables (pressure, temperature and concentration) of one of the components must be specified in order to describe the system. Thus a set of three coordinate axes at right angles to one another would be required to represent these relations graphically. However, in actual practice a simple plain diagram only with two variables is considered, the third variable is considered as constant. Thus following three types of diagrams are used.

(*i*) Pressure-temperature (*p-t*) diagram
(*ii*) Composition-pressure (*c-p*) diagram
(*iii*) Temperature-composition (*t-c*) diagram.

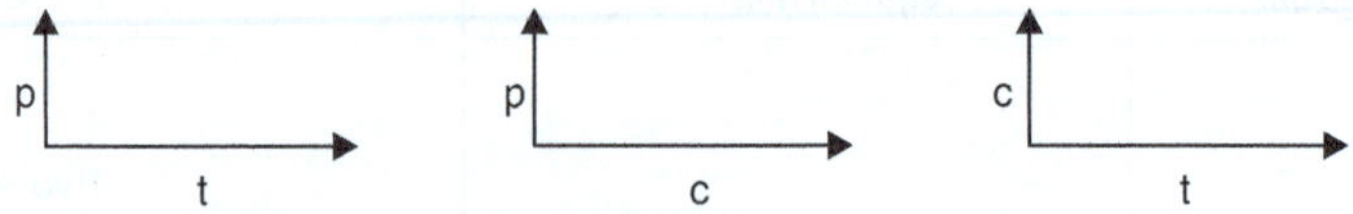


*FIGURE 8.4*

However, in most cases pressure is kept constant at atmospheric value, *i.e.* $t - c$ diagram is used. However, for solid-liquid equilibria the effect of pressure is negligible since the vapor pressures of solids are small. Thus the results obtained are substantially the same as if the system were in true equilibrium with its vapor. Such systems are called **condensed systems.** Since in such case the degree of freedom is reduced by one, the **reduced phase rule equation** can be written as

$$F' = C - P + 1$$

Where $F'$ represents the number of degrees of freedom which the system can possess in addition to pressure.

**Types of two component systems involving solid-liquid equilibria.** The two component solid liquid systems are divided into several types based on their miscibility and on the nature of the solids separating from the solution. However, here only one type is discussed in which the two components do not react with each other but simply mix into each other in the molten state or in the solution, such system is known as eutectic system; *e.g.*, silver-lead system.

## 8.8.1 Phase Diagram of Lead-Silver System

It is a two component system with four possible phases, namely solid *Ag*, solid *Pb*, solution of *Ag* + *Pb* and vapor. As the gaseous phase is practically absent and one variable **(pressure) in neglected, the condensed phase rule (reduced phase rule equation) will be applicable.**

$$F' = C - P + 1$$

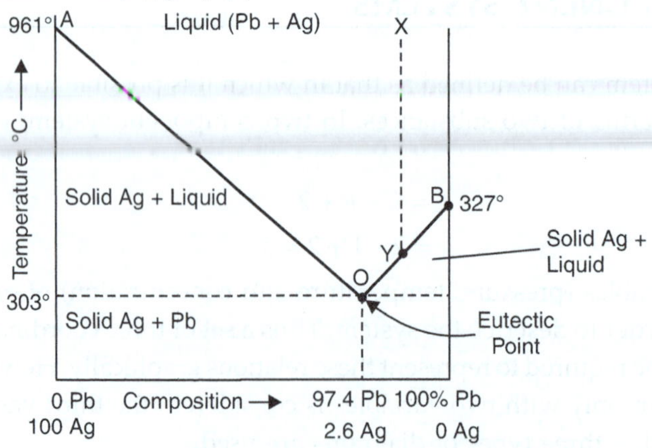


*FIGURE 8.5* **Phase diagram of lead-silver system.**

Thus when $F' = 0$, $P$ will be 3, *i.e.* the maximum number of phases (other than vapor) that can coexit would be three.

The phase diagram of the system is drawn taking temperature and composition along the axes. The equilibrium diagram consists of two curves $OA$ and $OB$, intersecting at point $O$.

**1. Curve AO.** It represents the freezing point curve of Ag on the successive addition of small quantities of lead. The starting point $A$ represents the melting point (961°C) of pure silver, thus at this point pure silver coexists as solid and liquid (vapor being neglected). It is observed that the melting point of silver is lowered gradually by the addition of lead into it. All along this curve the added lead goes into the solution and the separation of silver occurs till the point $O$, is reached where no more lead goes into solution and hence the point $O$ represents the lowest possible temperature (303°C) in the system which corresponds to fixed composition, *viz.* 2.6% Ag and 97.4% Pb. The point $O$ is called the *eutectic point*.

Thus the curve $AO$ represents the effect of addition of lead on the melting point of pure silver. All along the curve $AO$, two phases (solid silver and liquid) are present in equilibrium neglecting the vapor phase. Thus the system is univariant

$$F' = C - P + 1 = 2 - 2 + 1 = 1$$

**2. Curve BO.** It represents the freezing point curve of lead to which successive amounts of silver are added. The point $B$ represents the melting point of pure lead (327°C). It is observed that the addition of silver to lead lowers the melting point of lead. All along this curve, the added silver goes into the solution and separation of lead occurs till the point $O$ is reached. At this point, no more silver goes into the solution, *i.e.*, the soution is saturated with respect to silver. Hence the melting point of lead does not fall any more.

All along this curve, two phases (solid lead and liquid melt) are present in equilibrium. Therefore, as in the curve $AO$, here also the system is univariant.

$$F' = C - P + 1 = 2 - 2 + 1 = 1$$

**3. Eutectic point O.** This is the point where the two freezing point curves $AO$ and $BO$ intersect each other. The point $O$ represents a fixed composition of the two constituents (2.6% Ag and 97.4% lead) and is called *eutectic point*; the composition at this point is called *eutectic composition* and the temperature at this point is called *eutectic temperature* (303°C). NO mixture of lead and silver has a melting point lower than the eutectic temperature. At this point, the temperature remains constant until the whole of the melt solidified en-block to become solid of eutectic composition. However, further cooling leads to simultaneous crystallization of a mixture of Ag and Pb in relative amounts corresponding to eutectic point, $O$.

Since at eutectic point three phase (solid lead, solid silver and their solution) are present in equilibrium, the system will be invariant *i.e.* it has no degree of freedom as evident from the **reduced phase rule equation**.

$$F' = C - P + 1 = 2 - 3 + 1 = 0$$

Below the eutectic line $DC$, we have two regions in the diagram.

(*i*) Region having eutectic + solid Ag. In this region, crystalline silver and solid eutectic are stable.

(*ii*) Region having eutectic + solid Pb. In this region, crystalline lead and solid eutectic are stable.

**Application of the lead-silver system to Pattinson's process.** The above principle is applied in the Pattinson's process for the desilverisation of lead. If a sample of argentiferrous lead, containing less than 2.6% Ag is heated to a temperature well above its melting point so that the system consists only of liquid phase. Let the composition of the alloy (liquid melt) corresponds to point X in the Fig. 8.5. The melt is now allowed to cool gradually along the line XY till the point Y is reached where lead starts separating out as solid and thus the residual solution (melt) will comparatively be richer in silver. Further cooling will shift the sytem along the line *YO* with the result lead continues to separate out and the melt continues to be richer and richer in silver, till the point *O* is reached where an alloy containing about 2.6% Ag is obtained.

**Eutectic System.** *A binary system of two substances which are miscible in all proportions in the liquid phase but do not react chemically is known as* **eutectic** (easy to melt) **system.** On the other hand, *eutectic mixture is that solid solution of fixed proportions of two or more substances which has the lowest freezing point.*

**TABLE 8.2 Eutectic compositions and temperatures of some alloys**

| *Component metals with M.P.* | | *Eutectic temp.* | *Eutectic composition* |
|---|---|---|---|
| Ag (960°C) | Cu (1083°C) | 788°C | 71.8% Ag + 28.2% Cu |
| Bi (273°C) | Cd (321°C) | 140°C | 60% Bi + 40% Cd |
| Pb (327°C) | Zn (419°C) | 303°C | 97.6% Pb + 2.4% Zn |
| Zn (419°C) | Al (658.7°C) | 280.5°C | 95.6% Zn + 4.4% Al |
| Cd (321°C) | Zn (419°C) | 284.5°C | 82.6% Cd + 17.4% Zn |
| Pb (327°C) | Sb (631°C) | 245°C | 87% Pb + 13% Sb |

**Applications of eutectics.** (*i*) Low melting alloys are used in safety devices, *e.g.* as plugs in automobiles, fire sprinklers, etc. Very low melting alloys are obtained by suitable choice of metals, *e.g.*, wood's metal (alloy containing 50% Bi, 25% Pb, 12.5% Cd and 12.5% Sn) melts at 65°C.

(*ii*) ***Freezing mixtures.*** A system at the eutectic point can maintain its temperature over long periods and such systems are used as constant temperature baths. The lowest possible temperature attainable with a certain freezing mixture depends upon the eutectic temperature of the system. Some common freezing mixtures are given in the table 8.3.

**TABLE 8.3 Some common freezing mixtures**

| *Eutectic system* | *% of Salt* | *Eutectic temperature* |
|---|---|---|
| $NaCl \cdot 2H_2O$ + Ice | 23 | - 22°C |
| KI + Ice | 52 | - 23°C |
| $CaCl_2 \cdot 6H_2O$ + Ice | 15.2 | - 55.9°C |
| $NaNO_3$ + Ice | 33.3 | - 18.1°C |
| $NH_4Cl$ + Ice | 20.1 | - 16°C |
| $NH_4NO_3$ + Ice | 42.98 | - 18°C |

## SOLVED PROBLEMS

**Problem 1.** *Write the number of components and number of phases and evaluate the degree of freedom for the following:*

(*i*) $N_2O_4(g) \rightleftharpoons 2NO_2(g)$

(*ii*) $NH_4Cl(g) \rightleftharpoons NH_3(g) + HCl(g)$ *when (a)* $pNH_3 \neq pHCl$ *and (b)* $pNH_3 = pHCl$

(*iii*) *Solid carbon in equilibrium with gaseous CO,* $CO_2$, *and* $O_2$ *at 100°C.*

(*iv*) *A dilute solution of sulphuric acid in water.*

(*v*) *An aqueous solution saturated with respect to both sodium chloride and potassium chloride and in equilibrium with the vapor phase.*

(*vi*) *An aqueous solution of a mixture of sodium chloride and potassium bromide and in equilibrium with the vapor phase.*

(*vii*) *A mixture of nitrogen, hydrogen, and ammonia at temperatures at which the equilibrium* $N_2(g) + 3H_2(g) \rightleftharpoons 2NH_3(g)$ *is readily established.*

(*viii*) *Reduction of* $Fe_2O_3(s)$ *by CO in the presence of (a) air and (b) oxygen alone.*

**Solution.**

(*i*) Number of components,

C = Total number of components – The relations between them

$= 2 - 1 = 1$

Number of phases = 1

Hence the degree of freedom, $F = C + 2 - P = 1 + 2 - 1 = 2$

(*ii*) (*a*) Number of components, $C = 3 - 1 = 2$

$P = 2$

$F = 2 + 2 - 2 = 2$

(*b*) Number of components is reduced by one because $pHCl = pNH_3$

or $C = 3 - 2 = 1$

$P = 2$

$F = 1 + 2 - 2 = 1$

(*iii*) The different equilibria are:

$$C(s) + O_2(g) \rightleftharpoons CO_2(g)$$

$$C(s) + \frac{1}{2}O_2(g) \rightleftharpoons CO(g)$$

Hence, $C = 4 - 2 = 2$

$$P = 2$$
$$F = 2$$

(*iv*) The ionization equilibria in the aqueous sulphuric acid solution are:

$$H_2SO_4 + H_2O \rightleftharpoons H_3O^+ + HSO_4^-$$
$$HSO_4^- + H_2O \rightleftharpoons H_3O^+ + SO_4^{2-}$$

There are five different species in the solution. In addition to these two relations between these species, the solution is electrically neutral. Hence the number of components is C = 5 – 3 = 2, the number of phases is P = 1, and the degrees of freedom is F = 3.

(*v*) The total number of chemical species are seven [$NaCl(s)$, $KCl(s)$, $K^+(aq)$, $Na^+(aq)$, $Cl^-(aq)$, $H_2O(l)$ and $H_2O(g)$]. In addition to the condition of electroneutrality, the following three equilibria exist between the components.

(*a*) $NaCl(s) \xrightleftharpoons{H_2O} Na^+(aq) + Cl^-(aq)$

(*b*) $KCl(s) \xrightleftharpoons{H_2O} K^+(aq) + Cl^-(aq)$

(*c*) $H_2O(l) \rightleftharpoons H_2O(g)$

Therefore the number of components is C = 7 – 4 = 3

P = 4 (two solid phases, one liquid phase and one gas phase)

F = 1

(*vi*) The following equilibria exists between the components:

(*a*) $NaCl(s) \xrightleftharpoons{H_2O} Na^+(aq) + Cl^-(aq)$

(*b*) $KBr(s) \xrightleftharpoons{H_2O} K^+(aq) + Br^-(aq)$

(*c*) $NaBr(s) \xrightleftharpoons{H_2O} Na^+(aq) + Br^-(aq)$

(*d*) $NaCl + KBr \rightleftharpoons NaBr + KCl$

(*e*) $H_2O(l) \rightleftharpoons H_2O(g)$

Including the condition of electroneutrality, there are six relations between the different species. The number of components is thus 10 – 6 = 4 and the number of phases is six. Hence the degree of freedom is zero.

(*vii*) There are three chemical species and one equilibrium condition

$$N_2(g) + 3H_2(g) \rightleftharpoons 2NH_3(g)$$

Hence the number of components is two. The number of phases is 1 and hence the degree of freedom is 3.

(*viii*) (*a*) Here in the presence of air (containing mainly nitrogen and oxygen), the different species which could exist are FeO(*s*), $FeO_3$(*s*), Fe(*s*), CO(*g*), $CO_2$(*g*), $N_2$(*g*), and $O_2$(*g*). The total number of phases is four and the following reactions may occur

$$Fe_2O_3(s) + CO(g) \rightleftharpoons 2FeO(s) + CO_2(g)$$

$$FeO(s) + CO(g) \rightleftharpoons Fe(s) + CO_2(g)$$

$$CO(g) + \frac{1}{2}O_2(g) \rightleftharpoons CO_2(g)$$

Since the number of components is 7 – 3 = 4 and the degree of freedom F is

$$4 + 2 - 4 = 2$$

(*b*) If the reduction is carried out in the presence of oxygen, the number of components is given by 6 – 3 = 3 and the degree of freedom is 3 + 2 – 4 = 1.

**Problem 2.** *Determine the number of degree of freedom in each of the following systems. Suggest the variables that could correspond to these degrees of freedom.*

(*a*) *Liquid water and water vapor in equilibrium.*

(*b*) *Liquid water and water vapor in equilibrium at a pressure of 1 atm.*

**Solution.** From the Gibbs phase rule, F = C – P + 2

(*a*) In this case $\quad C = 1$

$\therefore \quad F = 1 - 2 + 2 = 1$

That is, only one variable T or P needs to be specified.

(*b*) Since the pressure is held constant, the degree of freedom reduces by one. Hence, the phase rule equation becomesF = C – P + 1

In this case, $\quad C = 1$ and $P = 2$

$\therefore \quad F = 1 - 2 + 1 = 0$

The system under the given conditions is thus **invariant.**

**Problem 3.** *Is it possible to have a quadruple point in a phase diagram for a one component system?*

**Solution.** From the Gibbs phase rule F = C – P + 2.

For a quadruple point, the number of phases in equilibrium with one another must be 4. This corresponds to F = 1 – 4 + 2 = –1

*i.e.,* F is negative hence, the answer is **'no'**.

**Problem 4.** *An alloy AB of 10g weight contained A at 25%. The molten AB on cooling gave out B and an eutectic alloy containing equal percentage of A and B. What is the amount of B formed ?*

(*Anna BEI, 1996*)

**Solution.** Wt. of alloy = 10 g

Wt. of *A* in 10 g of alloy = 2.5 g

Wt. of *B* in 10 g of alloy = 7.5 g

Since the eutectic alloy contains equal amount of *A* and *B*, and the amount of *A* is 2.5 g, amount of *B* in the eutectic alloy should also be 2.5.

Hence the amount of *B* formed = Total amount of *B* – Amount of *B* in eutectic

$$= 7.5 - 2.5 = \mathbf{5.0\ g.}$$

## QUESTION BANK

1. Define the terms phase, component, and degree of freedom. State the phase rule and discuss its application to the system of water, vapor, and ice.
2. State the phase diagram of water and label the various parts.
3. Define the following terms:

   Degree of freedom, Transition temperature, Heterogeneous system.
4. (*a*) State the phase rule and explain the terms involved.

   (*b*) Draw the diagram of the water system and explain the term triple point.
5. Define phase rule, explaining the terms: phase, component, and degree of freedom.
6. Discuss the applications of the phase rule to an ice-water-water-vapor system.
7. Define and explain the terms: phase, component, and the degree of freedom.
8. Explain the phase diagram of a one component system defining the terms involved.
9. Explain how phase rule is helpful in studying heterogeneous equilibria.
10. What do you understand by the term degree of freedom?
11. Explain the terms: phase, components, and degree of freedom as applied in the phase rule studies. Illustrate, giving suitable examples.
12. Explain the term chemical potential. Show that in a heterogeneous system in equilibrium, the chemical potential of any given constituent is the same in all the phases in equilibrium.
13. State the phase rule. Explain the various terms involved. Discuss the derivation of the phase rule from thermodynamic considerations.
14. Discuss the applicability of the phase rule to sulphur system.
15. Explain the term eutectic temperature.
16. Write a neat sketch, explain the salient features of S-system.
17. Draw the diagram of Pb-Ag system forming eutectic alloy and label all the phases in the diagram.
18. Discuss the application of phase rule to water system or Ag-Pb system.

Chapter 9

# ELECTROCHEMISTRY AND CORROSION

## 9.1 INTRODUCTION TO ELECTROCHEMISTRY

Electrochemistry studies the relationship between chemical and electrical phenomena and the laws of their interaction. Such interactions are observed at the boundaries between electrolytes and electrodes dipping in them.

Electrochemistry is of great significance, since the laws of electrochemistry forms the basis of technical processes like electrolysis and electrosynthesis. The branch holds importance for a number of reasons:

(*i*) It is of huge economical importance because of the costly destruction caused by corrosion and the possibilities of fuel cells which generate electricity directly from fuel.

(*ii*) It provides techniques for measuring the thermodynamical state functions and for predicting equilibrium concentrations of dissolving and reacting ions.

Thus the branch of electrochemistry is of major technological importance.

## 9.2 ELECTROCHEMICAL CELLS OR GALVANIC CELLS

An electrochemical cell is a single arrangement of two electrodes and an electrolyte for producing an electric current due to chemical action within the cell, or for producing chemical action due to passage of electricity. Thus electrochemical cells may be used for two purposes.

(*i*) to convert chemical energy into electrical energy.

(*ii*) to convert electrical energy into chemical energy.

### 9.2.1 Galvanic Cell

A galvanic cell, also known as voltaic cell, is a device in which electrical current is generated by a spontaneous redox reaction. A simple voltaic cell is shown in Figure 9.1 and the reaction taking place is:

$$Zn(s) + CuSO_4(aq) \longrightarrow ZnSO_4(aq) + Cu(s)$$

A bar of zinc metal (anode) is placed in zinc sulphate solution in the left container. A bar of copper (cathode) is immersed in a copper sulphate solution in the right container. The zinc and copper electrodes are joined by a copper wire. A salt bridge containing a potassium chloride solution interconnects the solutions in the anode and cathode compartment.

The oxidation occurs in the anode compartment.

$$Zn \longrightarrow Zn^{2+} + 2e^-$$

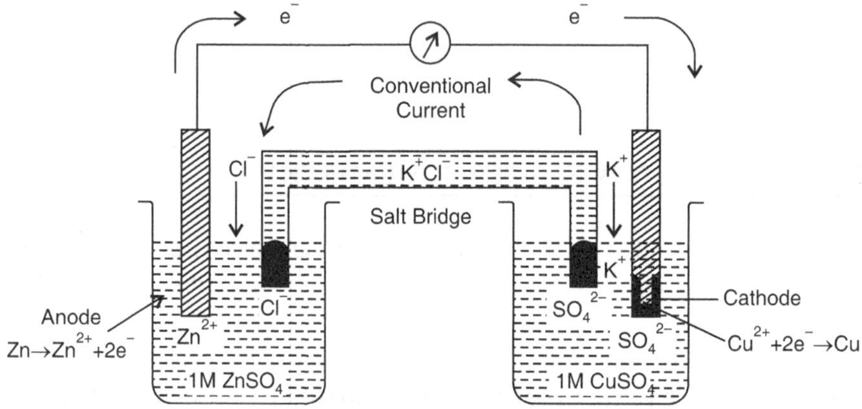


***FIGURE 9.1*** **A simple galvanic cell.**

The reduction takes place in the cathode compartment.

$$Cu^{2+} + 2e^- \longrightarrow Cu$$

When the cell is set up, electrons flow from the zinc electrode through the wire to the copper cathode. As a result, zinc dissolves in the anode solution to form $Zn^{2+}$ ions. The $Cu^{2+}$ ions in the cathode half-cell pick up electrons and are converted to Cu atoms on the cathode. At the same time, sulphate ions from the cathode half-cell migrate to the anode half-cell through the salt bridge. Likewise, $Zn^{2+}$ ions from the anode half-cell move into the cathode half-cell. This flow of ions from one half-cell to the other completes the electrical circuit which ensures continuous supply of current.

As the cell operates, the weight of the copper rod will increase while that of the zinc rod will decrease. Consequently, the cell will function until either the zinc metal or copper ion is completely used up.

A typical type of voltaic cell (galvanic cell) in which the salt bridge is replaced by a porous pot is known as *Daniell cell* (Figure 9.2).

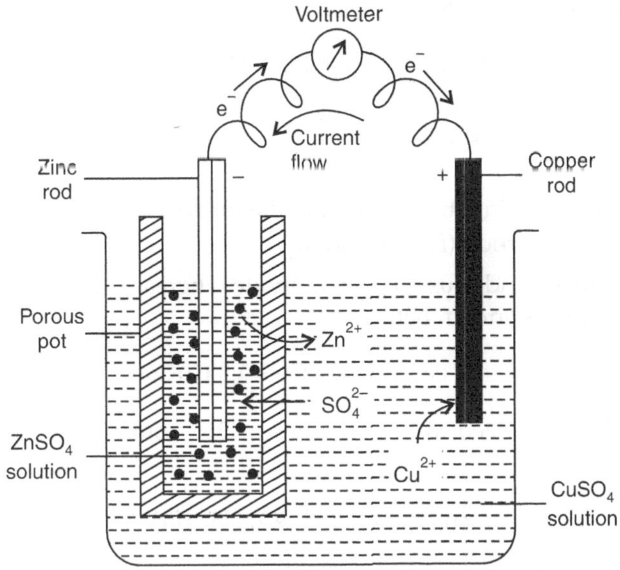


*FIGURE 9.2* **Daniell cell.**

Daniell cell resembles the above voltaic cell in all details except that $Zn^{2+}$ ions and $SO_4^{2-}$ ion flow to the cathode and the anode respectively through the porous pot instead of through the salt bridge. In spite of this difference, the cell diagram remains the same.

Porous Pot

$$\underset{\text{Anode}}{Zn \mid Zn^{2+}} \parallel \underset{\text{Cathode}}{Cu^{2+} \mid Cu}$$

### Salt Bridge and its Functions

A salt bridge consists of a U shaped glass tube filled with a semisolid paste prepared by adding agar-agar or gelatine to the aqueous solution of a strong electrolyte, *e.g.*, NaCl, KCl, $NaNO_3$, $KNO_3$, $NH_4NO_3$, $K_2SO_4$, etc. The strong electrolyte should be chemically inert. The main functions of a salt bridge are as follows:

(*i*) It provides a low resistance electrical connection between the two half-cells and allows the flow of ions from one half-cell to the other half-cell. Thus, the salt bridge completes the circuit.

(*ii*) It keeps the solutions in the two half-cells electrically neutral. Due to oxidation in the anodic half cell, $Zn^{2+}$ ions are accumulated near the zinc electrode. An equal number of $SO_4^{2-}$ ions will move from the salt bridge toward the zinc electrode and thus the positive charge around the zinc electrode is neutralized. Similarly, $K^+$ ions will move from the salt bridge toward the copper electrode, thereby neutralizing the negative charge. Therefore, the solutions in the two half-cells are electrically neutral.

**Example 1.** *Construct the cells in which the following reactions are taking place.*

(*i*) $Fe + CuSO_4 \longrightarrow FeSO_4 + Cu$ (*ii*) $Zn + H_2SO_4 \longrightarrow ZnSO_4 + H_2$

(*iii*) $Fe + SnCl_2 \longrightarrow FeCl_2 + Sn$ (*iv*) $\frac{1}{2} H_2 + \frac{1}{2} Cl_2 \longrightarrow HCl.$

**Solution.** The metal which goes into the solution in the form of its salt is undergoing oxidation and thus will act as a negative electrode (*i.e.*, cathode).

(*i*) Here, Fe is oxidized and thus will act as anode whereas $Cu^{2+}$ ions are reduced. Hence the cell is represented as shown below.

$$Fe \mid FeSO_4 \parallel CuSO_4 \mid Cu$$

(*ii*) In this cell, Zn is oxidized to $ZnSO_4$ whereas $2H^+$ ions are reduced to $H_2(g)$. Hence the cell is shown as:

$$Zn \mid ZnSO_4 \parallel H_2SO_4 \mid H_2 \text{ (Pt)}$$

(*iii*) In this cell, Fe is oxidized to $FeCl_2$ and $Sn^{2+}$ are reduced on Sn. Hence the cell is shown as:

$$Fe \mid FeCl_2 \parallel SnCl_2 \mid Sn$$

(*iv*) In this cell, $H_2$ is oxidized to $H^+$ ions and $Cl_2$ is reduced to $Cl^-$ ions. Therefore, the cell is written as:

$$(Pt)H_2 \mid HCl \parallel Cl_2(Pt)$$

**Example 2.** *Write the cell reaction for the following cells.*

(*i*) $Zn \mid Zn^{2+} \parallel Cu^{2+} \mid Cu$ (*ii*) $Fe \mid Fe^{2+} \parallel H_2SO_4 \mid H_2 \text{ (Pt)}$

(*iii*) $(Pt)H_2 \mid HCl \parallel Cl_2 \text{ (Pt)}$ (*iv*) $Fe \mid FeSO_4 \parallel SnCl_2 \mid Sn.$

**Solution.** (*i*) The reaction of oxidation half-cell is

$$Zn \longrightarrow Zn^{2+} + 2e^-$$

The reaction of reduction half-cell is

$$Cu^{2+} + 2e^- \longrightarrow 2u$$

Thus, the net cell reaction is obtained by adding the two half-cell reactions, *i.e.*,

$$Zn + Cu^{2+} \longrightarrow Zn^{2+} + Cu$$

(*ii*) For the cell, $Fe \mid F^{2+} \parallel H_2SO_4 \mid H_2(Pt),$

Oxidation reaction: $Fe \longrightarrow Fe^{2+} + 2e^-$

Reduction reaction: $2H^+ + 2e^- \longrightarrow H_2$

Cell reaction: $Fe + 2H^+ \longrightarrow Fe^{2+} + H_2$

(*iii*) For (Pt)$H_2$ | HCl || $Cl_2$(Pt) cell,

Oxidation reaction: $\frac{1}{2}H_2 \longrightarrow H^+ + e^-$

Reduction reaction: $\frac{1}{2}Cl_2 + e^- \longrightarrow Cl^-$

Cell reaction: $\frac{1}{2}H_2 + \frac{1}{2}Cl_2 \longrightarrow H^+ + Cl^-$

(*iv*) For the cell, Fe | $FeSO_4$ || $SnCl_2$ | Sn,

Oxidation reaction $Fe \longrightarrow Fe^{2+} + 2e^-$

Reduction reaction: $Sn^{2+} + 2e^- \longrightarrow Sn$

Cell reaction: $Fe + Sn^{2+} \longrightarrow Fe^{2+} + Sn$

**Example 3.** *Write the anode, cathode, and the net cell reactions for the following cells:*

(*i*) $Zn(s) \mid Zn^{2+}(aq) \parallel Br^-, Br_2 \mid Pt(s)$

(*ii*) $Cu(s) \mid Cu^{2+}(aq) \parallel Au^+(aq) \mid Au(s)$

(*iii*) $Cr(s) \mid Cr^{3+}(aq) \parallel I^-, I_2 \mid Pt(s)$.

**Solution.**

(*i*) Anode reaction: $Zn(s) \longrightarrow Zn^{2+}(aq) + 2e^-$

Cathode reaction: $Br_2 + 2e^- \longrightarrow 2Br^-(aq)$

Net cell reaction: $Zn(s) + Br_2 \longrightarrow Zn^{2+} + 2Br^-(aq)$

(*ii*) Anode reaction: $Cu \longrightarrow Cu^{2+}(aq) + 2e^-$

Cathode reaction: $2Au^+ + 2e^- \longrightarrow 2Au(s)$

Net cell reaction: $Cu(s) + 2Au^+(aq) \longrightarrow Cu^{2+}(aq) + 2Au(s)$

(*iii*) Anode reaction: $[Cr(s) \longrightarrow Cr^{3+}(aq) + 3e^-] \times 2$

Cathode reaction: $[I_2 + 2e^- \longrightarrow 2I^-(aq)] \times 3$

Net cell reaction: $2Cr(s) + 3I_2 \longrightarrow 2Cr^{3+}(aq) + 6I^-(aq)$

### 9.2.2 Half-cell Reactions and Electrodes

A redox reaction is a reaction in which there is a transfer of electrons from one substance to another. Any redox reaction may be expressed as the sum of the two half reactions. One of the half reactions shows a loss of electrons and the other shows a gain of electrons. For example, the reaction occurring in a Galvanic cell,

$$Zn(s) + Cu^{2+}(aq) \longrightarrow Zn^{2+}(aq) + Cu(s)$$

can be expressed as the sum of the following two half-cell reactions:

$$Cu^{2+}(aq) + 2e^- \longrightarrow Cu(s) \qquad \text{(Reduction of } Cu^{2+}\text{)}$$

$$Zn(s) \longrightarrow Zn^{2+}(aq) + 2e \qquad \text{(Oxidation of Zn)}$$

The half-cell in which oxidation takes place is called *oxidation half-cell* while the other in which reduction takes place is called *reduction half-cell.* In common practice, both the half-cell reactions are written as reduction reactions and then the overall reaction is the difference of the two:

$$Cu^{2+}(aq) + 2e^- \longrightarrow Cu(s) \qquad \text{(Reduction of } Cu^{2+}\text{)}$$

$$Zn^{2+}(aq) + 2e^- \longrightarrow Zn(s) \qquad \text{(Reduction of } Zn^{2+}\text{)}$$

$$Cu^{2+}(aq) + Zn(s) \longrightarrow Zn(s) + Cu^{2+}(aq) \qquad \text{(Overall)}$$

The reduced and oxidized substances in a half reaction form a redox couple, denoted as Ox/Red. In this reaction $Cu^{2+}/Cu$ and $Zn^{2+}/Zn$ are the redox couple.

As shown in Figure 9.1 in an electrochemical cell the two half reactions take place in two different compartments. When the reaction takes place, the electron released in one compartment travels through the external circuit and enters the cell through the other electrode. For example:

$$Zn(s) \longrightarrow Zn^{2+}(aq) + 2e^- \qquad \text{(oxidation)}$$

Zn will lose the electrons which are travelled through the circuit. The electrode where the oxidation half reaction takes place is called the **anode**. The anode is assigned a negative charge, because oxidation is accompanied by liberation of negatively charged electrons. These electrons are given to the electrode, which thus acquires a negative charge.

The electrons given at the anode are travelled through the circuit to the other electrode where the reduction half reaction will take place by gain of electrons. In a galvanic cell, the electrons lost by anode are accepted by the copper ions to form copper metal.

$$Cu^{2+}(aq) + 2e^- \longrightarrow Cu(s)$$

The electrode at which reduction takes place is called the **cathode**. Since the copper electrode requires the electrons for reduction, the electrode is electron deficient and pulls the electrons from the external circuit. Thus the cathode is assigned a positive charge.

When the electrons flow from the anode (zinc rod) toward the cathode (copper rod), the electrical energy is produced. Thus, an electrochemical cell converts the chemical energy into electrical energy.

### 9.2.3 Representation of a Cell or Cell Diagram

A cell diagram is an abbreviated symbolic depiction of an electrochemical cell. For this purpose, consider that a cell consists of two half-cells as discussed in earlier sections. Each half-cell is again made of a metal electrode in contact with metal ions in solution. In 1953, IUPAC recommended the following conventions for writing cell diagrams. We will illustrate these with reference to a Zinc-Copper cell.

(*i*) A single vertical line | represents a phase boundary between a metal electrode and an ion solution (electrolyte). Thus, the two half-cells in a voltaic cell are indicated as:

Phase boundary

$Zn \mid Zn^{2+}$ (Anode Half-Cell)    $Cu^{2+} \mid Cu$ (Cathode Half-Cell)

The metal electrode in the anode half-cell is on the left, while in the cathode half-cell it is on the right of the metal ion.

(*ii*) A double vertical line represents the salt bridge, porous partition, or any other means of permitting ion flow while preventing the electrolyte from mixing.

(*iii*) The anode half-cell is written on the left and the cathode half-cell on the right.

(*iv*) In the complete cell diagram, the two half-cells are separated by a double vertical line (salt bridge) in between. The zinc-copper cell can now be written as:

Salt Bridge

$Zn \mid Zn^{2+} \parallel Cu^{2+} \mid Cu$

Anode Half-Cell    Cathode Half-Cell

(*v*) The symbol for an inert electrode like the platinum electrode is often enclosed in a bracket. For example,

Inert Electrode

$Mg \mid Mg^{2+} \parallel H^{+} \mid H_2(Pt)$

(*vi*) The value of Emf of a cell is written on the right of the cell diagram. Thus, a zinc-copper cell has Emf 1.1 V and is represented as:

$$Zn \mid ZnSO_4 \parallel CuSO_4 \mid Cu \qquad E = +1.1\ V$$

### 9.2.4 EMF of the Cell or Cell Potential

An electrochemical cell is made up of two electrodes. One of these electrodes must have a higher electrode potential (higher tendency to lose electrons) than the other electrode. As a result of this potential difference, the electrons flow from the electrode at higher potential to the electrode at lower potential. A cell that has not reached equilibrium can do the electrical work until a given transfer of electrons is accomplished. The electron transfer depends upon the potential difference between two electrodes. This potential difference is called the *cell potential* and is measured in volts. The potential is also known as *electromotive force* (Emf) of the cell.

The magnitude of the Emf of a cell reflects the tendency of electrons to flow externally from one electrode to another. The electrons are transported through the cell solution by ions present and pass from the positive electrode (Cu in case of Daniell cell) to the negative electrode. This corresponds to a clockwise flow of electrons through the external circuit. Thus the Emf of the cell is given the + sign. If the Emf acts in the opposite direction through the cell circuit, it is quoted as – ve value. For example, Daniell cell has an Emf of 1.1 V and the copper electrode is positive. This can be expressed in two ways:

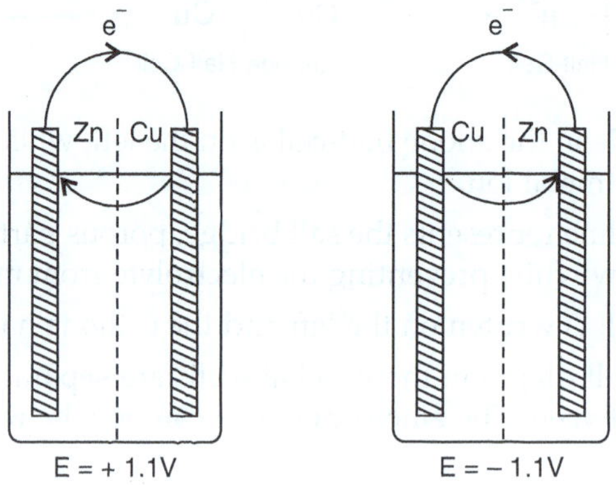


*FIGURE 9.3* **Illustration of emf sign in Daniell cell.**

$Zn \mid ZnSO_4 \parallel CuSO_4 \mid Cu$ $\quad E = + 1.1$ V

$Cu \mid CuSO_4 \parallel ZnSO_4 \mid Zn$ $\quad E = - 1.1$ V

The negative sign indicates that the cell is not feasible in the given direction. The reaction will takes place in the reverse direction.

### Calculation of Emf

The Emf of a cell can be calculated from the half-cell potentials of the two cells (anode and cathode) by using the following formula:

$$E_{cell} = E_{cathode} - E_{anode}$$

$$\boxed{E_{cell} = E_R - E_L}$$

where $E_R$ and $E_L$ are the reduction potentials of the right-hand and left-hand electrodes, respectively. It may be noted that absolute values of these reduction potentials cannot be determined. These are found by connecting the half-cell with a standard hydrogen electrode whose reduction potential has been arbitrarily fixed as zero.

### Difference between EMF and Cell Potential

The points of difference between the Emf of a cell and the cell potential are listed in Table 9.1.

**TABLE 9.1 Difference between Emf and cell potential**

| EMF | Cell Potential |
|---|---|
| 1. Emf is the potential difference between two electrodes when no current is flowing in the circuit. | Cell potential measures the difference in the potentials of the two half-cells when electric current flows through the cell. |
| 2. Emf is the maximum voltage that can be obtained from the cell. | Cell potential is always less than the maximum voltage obtainable from the cell. |
| 3. Emf corresponds to the maximum useful work obtainable from the galvanic cell. | Cell potential does not correspond to the maximum useful work. |
| 4. Emf is measured by a potentiometer. | Cell potential is measured using a voltmeter. |

## 9.3 ELECTRODE POTENTIAL (REDUCTION POTENTIAL)

When a piece of metal is immersed in a solution of its own ions as shown in Figure 9.4, a potential difference is created at the interface of the metal and the solution. The magnitude of the potential difference is a measure of the tendency of the electrode to undergo oxidation or reduction or the tendency to lose or gain electrons. The metal and ion (M and $M^{2+}$) represent the half-cell and the reaction is half reaction. The immersed metal is an electrode and the potential due to reaction at the interface of the electrode and the solution is called the *electrode potential*. Thus, electrode potential is the tendency of an electrode to lose or gain electrons. If the reduction takes place at the electrode, it is termed as *reduction potential.*

If the oxidation takes place at the electrode, it is called the *oxidation potential.*

$$M \longrightarrow M^{2+} + 2e^-$$

More recently, the reduction potential has been adopted by the International Union of Pure and Applied Chemistry (IUPAC) for the designation of electrode potential.

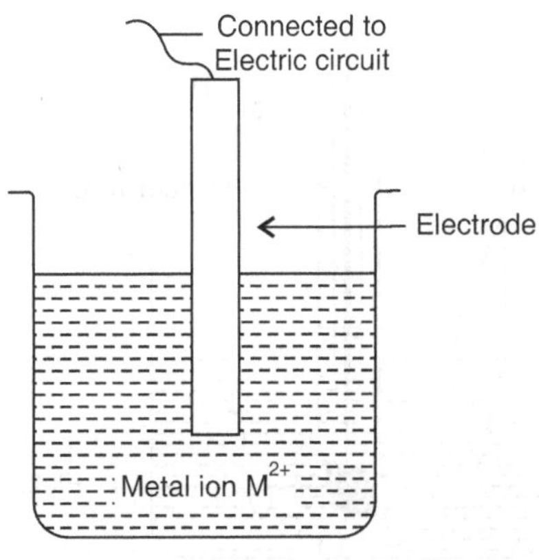


*FIGURE 9.4* **Electrochemical half-cell.**

When the half-cell reaction is carried out at a temperature of 298 K and the electrode is suspended in a solution of one molar concentration, the electrode potential is termed as the *standard electrode potential* and is represented by $E_o$. The Standard electrode potential $E_o$ enables one to assess the thermodynamic activity of various chemical substances. But there are no methods available by which we can measure its absolute value. The electrode potential of an electrode is measured with respect to a standard hydrogen electrode.

The electrode potential of an electrode depends upon the concentration of ions in solution in contact with metal. Simply, oxidation potential of an electrode is inversely proportional to the concentration of ions and reduction potential is directly proportional to the concentration of ions.

### 9.3.1 Measurement of Electrode Potential

A galvanic cell consists of the conjunction of two electrodes, with each making a characteristic contribution to the cell potential. Although it is not possible to determine the absolute value of the electrode potential of a single electrode, it can be determined with respect to a reference electrode. The reference electrode used is the standard hydrogen electrode (SHE) which is assigned the value zero.

A standard hydrogen electrode contains hydrogen gas in contact with $H^+$ ions of 1 M concentration at 298 K and at 1 atmospheric pressure (Figure 9.5).

In a hydrogen electrode, hydrogen gas is bubbled through a solution of hydrogen ions. The inert metal (which is often Pt) acts as a source or sink of electrons, but takes no other part in the reaction other than acting as a catalyst for it. This electrode is denoted as

$$\mathrm{Pt} \mid \mathrm{H_2}(g) \mid \mathrm{H^+}(aq)$$

The hydrogen electrode may act as either a cathode or an anode, depending on the other electrode in the cell and the spontaneous direction of the overall reaction. When the electrode is acting as a cathode, the reaction is:

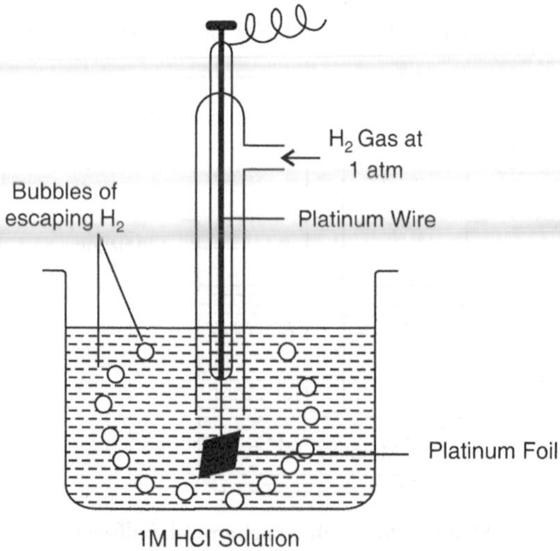


*FIGURE 9.5* **Standard hydrogen electrode.**

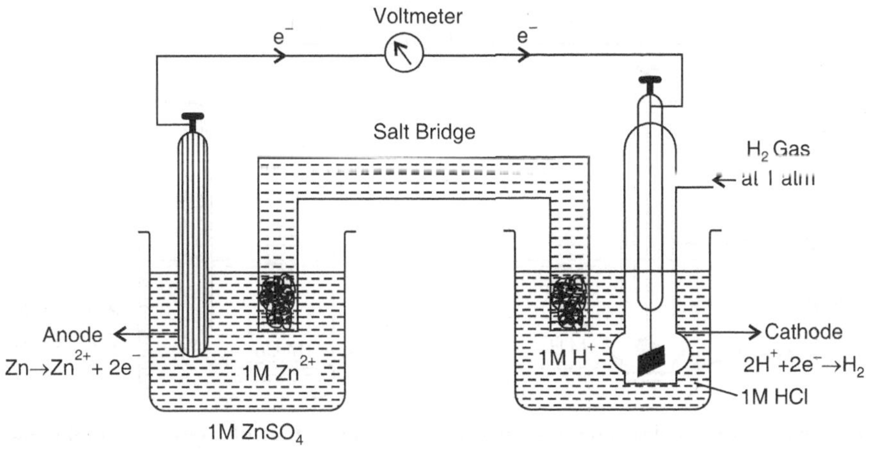


***FIGURE 9.6*** **Standard electrode potential (reduction potential) of zinc by combining with SHE.**

$$2H^+(aq) + 2e^- \longrightarrow H_2(g) \qquad \text{(Reduction)}$$

when the hydrogen electrode is acting as an anode

$$H_2(g) \longrightarrow 2H^+(aq) + 2e^- \qquad \text{(Oxidation)}$$

To determine the electrode potential of the given electrode, the electrode is coupled with SHE. The Emf of the cell is then determined with the help of a potentiometer. The electrode potential of SHE is arbitarily taken as zero, the Emf of the cell gives the value of electrode potential of the given electrode.

For example, when the zinc electrode (Zn rod dipped in an electrolyte) is coupled with SHE, oxidation takes place at the zinc electrode (Figure 9.6). The two half-cell reactions are represented as:

Zinc electrode : $Zn(s) \longrightarrow Zn^{2+} + 2e^-$ (Oxidation)

Hydrogen electrode : $2H^+ + 2e^- \longrightarrow H_2$ (Reduction)

The Emf of the cell is 0.76 V. Therefore, the oxidation potential of the zinc electrode is taken as + 0.76 V and the reduction potential is taken as – 0.76 V.

Similarly, when a copper electrode (copper rod dipped in an electrolyte) is coupled with SHE, reduction takes place at copper electrode (Figure 9.7)

Hydrogen electrode : $H_2(g) \longrightarrow 2H^+ + 2e^-$ (Oxidation)

Copper electrode : $Cu^{2+} + 2e^- \longrightarrow Cu(s)$ (Reduction)

Emf of the cell = 0.34 V

Hence, the reduction potential of copper electrode is taken as + 0.34 V and oxidation potential is – 0.34 V. The negative sign of oxidation potential indicates that the oxidation at the electrode will not be spontaneous.

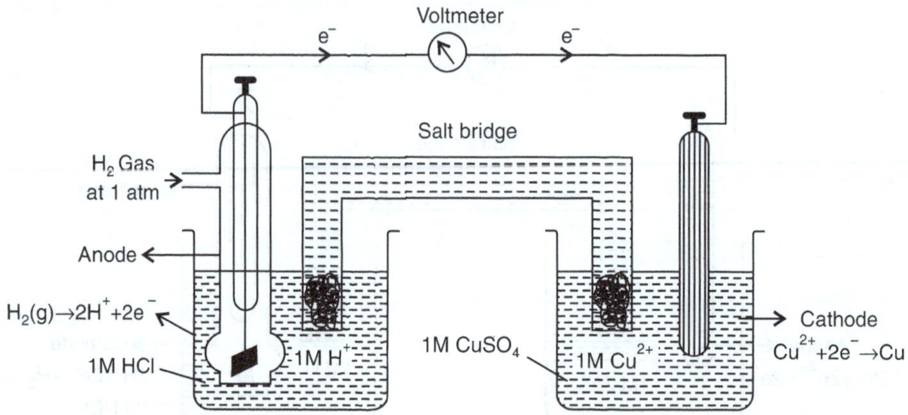


*FIGURE 9.7* **Standard electrode potential (reduction potential) of copper by combining with SHE.**

## 9.4 ELECTROCHEMICAL SERIES

The standard potential of an electrode can be measured with reference to a standard hydrogen electrode, which has been assigned the value zero. These electrodes can be arranged in order of their increasing reduction potentials. *Electrochemical series* is the arrangement of various electrodes in the increasing order of their standard electrode potential values. The values reported in Table 9.2 for standard electrode potential correspond to the half-cell reaction in which reduction is taking place. Therefore, E° values are also known as standard reduction potentials. The electrochemical series is also called the *activity series.*

### Important features of electrochemical series

(*i*) Metals near the top of the series are strongly electropositive (or weakly electronegative). They lose electrons readily to give cations.

(*ii*) The lower the position of metal in the series, the greater is its tendency to be reduced. Thus, metals having a higher position in the series are strong reducing agents and their ions are stable. On the other hand, those near the bottom of the series are stable metals and their ions are easily reduced to the metals. Accordingly, any metal will reduce any cation below it in the electrochemical series, if one molar solution is used.

(*iii*) Weakly electronegative metals in the upper part can displace more electro- negative metals below them from their salts. For example, iron displaces copper from $CuSO_4$ solution, Cu displaces silver from silver salt solution, silver displaces gold from gold salt solution as in photography during gold toning process.

(*iv*) Metals above hydrogen displace hydrogen from dilute acids.

(*v*) An oxidizing agent with a higher reduction potential will oxidize any reducing agent with a lower reduction potential, *e.g.*, silver (reducing potential + 0.80 V) will be deposited from its solution when treated with copper metal (reduction potential + 0.34 V).

(*vi*) A reducing agent with a lower reduction potential will reduce any oxidizing agent with a higher reduction potential, *e.g.*, $Li^+/Li$ has the lowest reduction potential. This means that all ions below it can be reduced by lithium.

**TABLE 9.2 Standard electrode potentials**

| *Elements* | *Electrode Reaction* | *E°$_{red}$ (volts)* |
|---|---|---|
| Li | $Li^+(aq) + e^- \longrightarrow Li(s)$ | - 3.05 |
| K | $K^+(aq) + e^- \longrightarrow K(s)$ | - 2.93 |
| Ba | $Ba^{2+}(aq) + 2e^- \longrightarrow Ba(s)$ | - 2.90 |
| Ca | $Ca^{2+}(aq) + 2e^- \longrightarrow Ca(s)$ | - 2.87 |
| Na | $Na^+(aq) + e^- \longrightarrow Na(s)$ | - 2.71 |
| Mg | $Mg^{2+}(aq) + 2e^- \longrightarrow Mg(s)$ | - 2.37 |
| Al | $Al^{3+}(aq) + 3e^- \longrightarrow Al(s)$ | - 1.66 |
| Zn | $Zn^{2+}(aq) + 2e^- \longrightarrow Zn(s)$ | - 0.76 |
| Cr | $Cr^{3+}(aq) + 3e^- \longrightarrow Cr(s)$ | - 0.74 |
| Fe | $Fe^{2+}(aq) + 2e^- \longrightarrow Fe(s)$ | - 0.44 |
| Cd | $Cd^{2+}(aq) + 2e^- \longrightarrow Cd(s)$ | - 0.40 |
| Pb | $PbSO_4(s) + 2e^- \longrightarrow Pb(s) + SO_4^{2-}(aq)$ | - 0.31 |
| Co | $Co^{2+}(aq) + 2e^- \longrightarrow Co(s)$ | - 0.28 |
| Ni | $Ni^{2+}(aq) + 2e^- \longrightarrow Ni(s)$ | - 0.25 |
| Sn | $Sn^{2+}(aq) + 2e^- \longrightarrow Sn(s)$ | - 0.14 |
| Pb | $Pb^+(aq) + e^- \longrightarrow Pb(s)$ | - 0.13 |
| **$H_2$** | **$2H^+ + 2e^- \longrightarrow H_2(g)$** | **0.00** |
| Cu | $Cu^{2+}(aq) + 2e^- \longrightarrow Cu(s)$ | + 0.34 |
| $I_2$ | $I_2(s) + 2e^- \longrightarrow 2I^-(s)$ | + 0.54 |
| Fe | $Fe^{3+}(aq) + e^- \longrightarrow Fe^{2+}(aq)$ | + 0.77 |
| Hg | $Hg_2^{2+}(aq) + 2e^- \longrightarrow 2Hg(l)$ | + 0.79 |
| Ag | $Ag^+(aq) + e^- \longrightarrow Ag(s)$ | + 0.80 |
| Hg | $Hg^{2+}(aq) + 2e^- \longrightarrow Hg(l)$ | + 0.85 |
| $Br_2$ | $Br_2(aq) + 2e^- \longrightarrow 2Br^-(aq)$ | + 1.08 |
| $O_2$ | $\frac{1}{2}O_2(g) + 2H_3O^+(aq) + 2e^- \longrightarrow 3H_2O$ | + 1.23 |
| Cr | $Cr_2O_7^- + 14H^+ + 6e^- \longrightarrow 2Cr^{3+} + 7H_2O$ | + 1.33 |
| $Cl_2$ | $Cl_2(g) + 2e^- \longrightarrow 2Cl^-(aq)$ | + 1.36 |
| Au | $Au^{3+}(aq) + 3e^- \longrightarrow Au(s)$ | + 1.42 |
| Mn | $MnO_4^-(aq) + 8H_3O^+(aq) + 5e^- \longrightarrow Mn^{2+}(aq) + 12H_2O(l)$ | + 1.51 |
| $F_2$ | $F_2(g) + 2e^- \longrightarrow 2F^-(g)$ | + 2.87 |

### 9.4.1 Importance of Electrochemical Series

1. **Calculation of standard Emf of cell (E°$_{cell}$).** In order to calculate the standard Emf of the cell, the electrochemical reaction is to be split into two half reactions :

   (*i*) Oxidation half reaction (*ii*) Reduction half reaction.

From the values of standard reduction potentials of the two half-cells, the standard Emf can be calculated as follows.

Standard Emf of the cell

$$= \begin{bmatrix} \text{Standard reduction} \\ \text{potential of the reduction} \\ \text{half reaction} \end{bmatrix} - \begin{bmatrix} \text{Standard reduction} \\ \text{potential of the oxidation} \\ \text{half reaction} \end{bmatrix}$$

or $$(E^\circ_{cell}) = (E^\circ_{cathode}) - (E^\circ_{anode}) \quad (9.1)$$

By convention, the cathode is written on the right and anode on the left in an electrochemical cell. Thus, $(E^\circ_{cell})$ may be written as

$$(E^\circ_{cell}) = (E^\circ_{right}) - (E^\circ_{left}) \quad (9.2)$$

**Example 4.** *A cell is prepared by dipping a copper rod in 1M $CuSO_4$ solution and a Zn rod in 1M $ZnSO_4$ solution. The standard reduction potentials of Cu and Zn are 0.34 and – 0.76 respectively.*

(*i*) *Which electrode will be positive?*

(*ii*) *How will the cell be represented?*

(*iii*) *What is the cell reaction?*

(*iv*) *What will be the Emf of the cell?*

**Solution.**

(*i*) The cathode is the positive electrode and has greater reduction potential than the second electrode of the cell. Therefore, the positive electrode is the copper electrode.

(*ii*) Oxidation reaction : $Zn(s) \longrightarrow Zn^{2+}(aq) + 2e^-$

Reduction reaction : $Cu^{2+}(aq) + 2e^- \longrightarrow Cu(s)$

∴ Cell is represented as follows

$$\underset{}{Zn} \mid \underset{(1M)}{ZnSO_4} \parallel \underset{(1M)}{CuSO_4} \mid Cu$$

(*iii*) Cell reaction is the sum of oxidation half-cell reaction plus reduction half-cell reaction. It is written as follows

$$Cu(s) + ZnSO_4(aq) \longrightarrow CuSO_4(aq) + Zn(s)$$

(*iv*)
$$\begin{aligned}(E^\circ_{cell}) &= (E^\circ_{right}) - (E^\circ_{left}) \\ &= (E^\circ_{cathode}) - (E^\circ_{anode}) \\ &= 0.34 - (-0.76) = \mathbf{1.1\ V}\end{aligned}$$

**Example 5.** *For the cells shown below, write cell reactions*

(*a*) $Cd(s) \mid Cd^{2+} \parallel Ag^+ \mid Ag(s)$ (*b*) $Zn(s) \mid ZnSO_4(aq) \parallel CdSO_4(aq) \mid Cd(s)$

(*c*) $Fe(s) \mid FeCl_2(aq) \parallel AuCl_3(aq) \mid Au(s)$

*Calculate standard cell potential, if standard reduction potentials are*

$$E^\circ_{\frac{Zn^{2+}}{Zn}} = -0.76\ V;\quad E^\circ_{\frac{Cd^{2+}}{Cd}} = -0.40\ V;\quad E^\circ_{\frac{Fe^{2+}}{Fe}} = -0.44\ V$$

$$E^\circ_{\frac{Au^{+3}}{Au}} = +1.42\ V;\quad E^\circ_{\frac{Ag^{+}}{Ag}} = +0.80\ V.$$

**Solution.**

(*a*) Anodic reaction : $Cd(s) \longrightarrow Cd^{2+} + 2e^-$

Cathodic reaction : $2Ag^+ + 2e^- \longrightarrow 2Ag(s)$

∴ Cell reaction : $Cd(s) + 2Ag^+(aq) \longrightarrow Cd^{2+}(aq) + 2Ag(s)$

Standard cell potential:

$$E^\circ_{cell} = E^\circ_{cathode} - E^\circ_{anode}$$

$$= 0.80 - (-0.40) = \mathbf{1.2\ V}$$

(*b*) Balanced cell reaction:

$$Zn(s) + CdSO_4(aq) \longrightarrow ZnSO_4(aq) + Cd(s)$$

Standard cell potential:

$$E^\circ_{cell} = E^\circ_{right} - E^\circ_{left}$$

$$= 0.40 - (-0.76) = \mathbf{1.16\ V}$$

(*c*) Balanced cell reaction:

$$3Fe(s) + 2AuCl_3(aq) \longrightarrow 3FeCl_2(aq) + 2Au(s)$$

Standard cell potential:

$$E^\circ_{cell} = E^\circ_{right} - E^\circ_{left}$$

$$= 1.42 - (-0.44) = \mathbf{1.86\ V}$$

**Example 6.** *Calculate the standard reduction electrode potential of* $\frac{Zn^{2+}}{Zn}$ *electrode when the cell potential* $E^\circ$ of the cell $Zn(s) \mid Zn^{2+}(1M) \parallel Cu^{2+}(1M) \mid Cu(s)$ *is 1.10 V*

$$E^\circ_{\frac{Cu^{2+}}{Cu}} = +0.34\ V.$$

**Solution.**

Oxidation reaction: $Zn(s) \longrightarrow Zn^{2+}(aq) + 2e^-$

Reduction reaction: $Cu^{2+}(aq) + 2e^- \longrightarrow Cu(s)$

Cell reaction: $Zn(s) + Cu^{2+}(aq) \longrightarrow Cu(s) + Zn^{2+}(aq)$

$$E^\circ_{cell} = E^\circ_{cathode} - E^\circ_{anode}$$

$$1.10 = 0.34 - E^\circ_{\frac{Zn^{2+}}{Zn}}$$

$$E^\circ_{\frac{Zn^{2+}}{Zn}} = \mathbf{-\,0.76\ V}$$

**Example 7.** *Two half-cell reactions with their oxidation potentials are*

(*a*) $Pb(s) - 2e^- \longrightarrow Pb^{2+}(aq);\ E^\circ = -\,0.13\ V$

(*b*) $Ag(s) - e^- \longrightarrow Ag^+(aq);\ E^\circ = +\,0.80\ V$

*Write the cell reaction and calculate its Emf.*

**Solution.** Here E° oxidation values are given, in reduction potentials E° values are

$$E^\circ_{Anode} \text{ or } E^\circ_{\frac{Pb^{2+}}{Pb}} = -\,0.13\ V \quad E^\circ_{\frac{Ag^+}{Ag}} = +\,0.80\ V$$

$$E^\circ_{cell} = E^\circ_{cathode} - E^\circ_{anode}$$

$$= 0.80 - (-\,0.13) = \mathbf{0.93\ V}$$

Oxidation reaction: $Pb(s) \longrightarrow Pb^{2+}(aq) + 2e^-$

Reduction reaction: $2Ag^+(aq) + 2e^- \longrightarrow 2Ag(s)$

∴ Cell reaction: $Pb(s) + 2Ag^+(aq) \longrightarrow Pb^{2+}(aq) + 2Ag(s)$

**Example 8.** *The Emf of the following cells are*

$Ag \mid Ag^+(1M) \parallel Cu^{2+}(1M) \mid Cu \quad E^\circ = -\,0.46\ V$

$Zn \mid Zn^{2+}(1M) \parallel Cu^{2+}(1M) \mid Cu \quad E^\circ = +\,1.10\ V$

*Calculate the Emf of the cell*

$Zn \mid Zn^{2+}(1M) \parallel Ag^+(1M) \mid Ag$

**Solution.** (*i*) E° of first cell, $E^\circ_{cell} = E^\circ_{\frac{Cu^{2+}}{Cu}} - E^\circ_{\frac{Ag^+}{Ag}} = -\,0.46\ V$

(*ii*) E° of second cell, $E^\circ_{cell} = E^\circ_{\frac{Cu^{2+}}{Cu}} - E^\circ_{\frac{Zn^{2+}}{Zn}} = 1.10\ V$

subtracting (*i*) from (*ii*)

$$\left[E^\circ_{\frac{Cu^{2+}}{Cu}} - E^\circ_{\frac{Zn^{2+}}{Zn}}\right] - \left[E^\circ_{\frac{Cu^{2+}}{Cu}} - E^\circ_{\frac{Ag^+}{Ag}}\right] = 1.10 - (-\,0.46)$$

$$E^\circ_{\frac{Ag^+}{Ag}} - E^\circ_{\frac{Zn^{2+}}{Zn}} = \mathbf{1.56\ V.}$$

2. **Predicting whether a metal can liberate hydrogen from the acid solution or not.** The metals having lesser reduction potential compared to SHE can liberate hydrogen from the acids (HCl, $H_2SO_4$, etc.). For example Zn, Fe, Mg and Ni release hydrogen when placed in HCl or $H_2SO_4$. However, metals which have positive reduction potential cannot liberate hydrogen from acids, *e.g.*, copper, silver, gold, etc. In other words, hydrogen will be liberated when metal has a tendency to lose electrons (get oxidized) and $H^+$ ions have a tendency to get reduced (by accepting electrons).

**Example 9.** *Predict reaction of 1 N $H_2SO_4$ with the following metals.*

(*i*) Zn (*ii*) Fe (*iii*) Cu (*iv*) Cd

*Given* $E^\circ_{\frac{Zn^{2+}}{Zn}} = -0.76\ V,\ E^\circ_{\frac{Fe^{2+}}{Fe}} = -0.44\ V,\ E^\circ_{\frac{Cu^{2+}}{Cu}} = +0.34\ V,\ E^\circ_{\frac{Cd^{2+}}{Cd}} = -0.40\ V.$

**Solution.** The reduction potential of $H^+$ is $E^\circ_{\frac{H^+}{H}} = 0.0$ V.

(*i*) The reduction potential of Zn is lower than reduction potential of $H^+$ ions. Therefore, zinc will be oxidized and $H^+$ will undergo reduction. Hence, the reaction will be spontaneous and $H_2$ gas will be released.

(*ii*) Since $E^\circ_{\frac{Fe^{2+}}{Fe}}$ is lower than $E^\circ_{\frac{H^+}{H}}$, iron will be oxidized to $Fe^{2+}$.

Hence the reaction is spontaneous and $H_2$ gas is formed.

(*iii*) The reduction potential of Cu is greater than the reduction potential of $H^+$ ions; thus Cu should undergo reduction. Therefore, no reaction occurs between copper and $H_2SO_4$.

(*iv*) Since $E^\circ_{\frac{Cd^{2+}}{Cd^2}}$ is lower than $E^\circ_{\frac{H^+}{H}}$, the reaction between the two is spontaneous or $H_2$ gas will be released.

3. **Predicting feasibility of a redox reaction.** A redox reaction is feasible (or occurs) spontaneously when it fulfills the following conditions:

(*a*) The species which has higher reduction potential undergoes reduction and the one with lower reduction potential is oxidized.

(*b*) If value of the $E^\circ_{cell}$ is positive, the reaction is spontaneous. However, if $E^\circ_{cell}$ comes out to be negative, the forward reaction is not feasible.

**Example 10.** *Predict the feasibility of the reaction between $Ag^+(aq)$ and $Cu(s)$. Given*

$$E^\circ_{\frac{Ag^+}{Ag}} = +0.80\ V,\ E^\circ_{\frac{Cu^{2+}}{Cu}} = +0.34\ V.$$

**Solution.** In this cell, the copper electrode will function as an anode and silver as a cathode. Thus, the cell will be

$$Cu(s)\ |\ Cu^{2+}(aq)\ ||\ Ag^+(aq)\ |\ Ag(s)$$

and the Emf of the cell is $E^\circ_{cell} = E^\circ_{\frac{Ag^+}{Ag}} - E^\circ_{\frac{Cu^{2+}}{Cu}} = (0.80 - 0.34)\ V = \mathbf{0.46\ V}$

Since E° cell is positive, the reaction occurs between the reacting species. The cell reaction is

$$Cu(s) + 2Ag^+(aq) \longrightarrow 2Ag(s)\ Cu^{2+}(aq).$$

**Example 11.** *Standard oxidation potential of $Cd/Cd^{2+}$ and $Pb/Pb^{2+}$ electrodes is – 0.40 and – 0.13 V respectively. State whether the following cell is feasible.*

**Solution.** The standard reduction potentials of the electrodes will have signs opposite to their oxidation potentials

$$E^\circ_{(Pb^{2+}/Pb)} = 0.13\ V \text{ and } E^\circ_{(Cd^{2+}/Cd)} = +\ 0.40\ V$$

$$E^\circ_{cell} = E^\circ_{red} - E^\circ_{oxid}\ 0.40 - 0.13 = \mathbf{0.27\ V}$$

Since calculated value of $E^\circ_{cell}$ is positive, the cell is feasible.

**Example 12.** *Can we use a copper vessel to store 1 M $AgNO_3$ solution? Given*

$$E^\circ_{Cu^{2+}/Cu} = 0.34\ V;\ E^\circ_{Ag^+/Ag} = 0.80\ V.$$

**Solution.** According to the standard reduction potential values, copper will function as an anode (*i.e.*, Cu metal will be oxidized to $Cu^{2+}$ and silver ions will be reduced to silver metal at the cathode). Therefore, the copper vessel will dissolve in $AgNO_3$ if kept in it. Hence, we cannot store $AgNO_3$ in a copper vessel.

**Example 13.** *A copper wire is dipped in $AgNO_3$ solution kept in beaker A and a silver wire is dipped in a solution of $CuSO_4$ kept in beaker B. Predict in which breaker the ions present will get reduced. Given*

$$Cu^{2+} + 2e^- \longrightarrow Cu \qquad E^\circ = 0.34\ V$$
$$Ag^+ + e^- \longrightarrow Ag \qquad E^\circ = 0.80\ V$$

**Solution.** Silver ions will get reduced in beaker A, while no reaction will occur in beaker B.

**Example 14.** *Answer the following using E° values from Table 9.2. Give reasons.*

(*a*) *Can a solution of 1M $ZnSO_4$ be stored in a vessel made of copper?*

(*b*) *In which will iron pieces dissolve. $ZnSO_4(aq)$ or $NiSO_4(aq)$?*

(*c*) *Why does the blue color of $CuSO_4$ solution get discharged by adding iron pieces or zinc powder to it?*

(*d*) *Is it safe to store 1M $AgNO_3$ solution with a copper spoon?*

(*e*) *Can a solution of a 1M $CuSO_4$ be stored in a vessel made of Ni metal?*

**Solution.** (*a*) Yes, (*b*) Fe will dissolve in $NiSO_4$ (*c*) Blue colored $Cu^{2+}$ ion undergoes reduction to Cu. (*d*) No (*e*) No.

4. **Predicting the anode and cathode.** The values of the electrode potential of the electrodes in a cell help in predicting the anode and cathode. An electrode with lower reduction

potential (lying in the lower part of the series) can easily lose the electrons and will act as an anode. The electrode with higher electrode potential will act as a cathode.

**Example 15.** *Consider the Zn-Cu Cell*

$$Zn^{2+} + 2e^- \longrightarrow Zn; \qquad E^\circ = -0.76\ V$$

$$Cu^{2+} + 2e^- \longrightarrow Cu; \qquad E^\circ = +0.335$$

*Which of these will act as anode and which as cathode?*

**Solution.** Standard reduction potential (E°) of Cu is more than the standard reduction potential of Zn. Thus Zn will act as anode and Cu will act as cathode.

5. **Predicting the oxidizing or reducing ability.** Let us consider a series of elements Cu, $H_2$, Ni, Zn, and their ions. These four elements could act as reducing agents. On the other hand, their ions $Cu^{2+}$, $H^+$, $Ni^{2+}$, and $Zn^{2+}$ can act as electron acceptors or oxidizing agents. If we list the respective half-reactions (or electrodes) in order of descending E° values, we will have placed the oxidizing agents in descending order of their ability to attract electrons.

| Increasing strength as oxidizing agent (↑) | Half-reaction | Increasing strength as reducing agent (↓) | E° |
|---|---|---|---|
| | $Cu^{2+} + 2e^- \longrightarrow Cu$ | | E° = + 0.34 V |
| | $2H^+ + 2e^- \longrightarrow H_2$ | | E° = 0.00 |
| | $Ni^{2+} + 2e^- \longrightarrow Ni$ | | E° = – 0.25 |
| | $Zn^{2+} + 2e^- \longrightarrow Zn$ | | E° = – 0.76 |

The value of E° becomes more negative down the series. This means that $Cu^{2+}$ is the best oxidizing agent (most electron attracting ion) of those in the list. That is, $Cu^{2+}$ shows the greatest tendency to be reduced. Conversely, $Zn^{2+}$ is the worst oxidizing agent, being the least electron-attracting ion. Of the elements Cu, $H_2$, Ni, and Zn; Zn is the best reducing agent (best electron donor) since E° for the half-reaction,

$$Zn \longrightarrow Zn^{2+} + 2e \quad E^\circ = +0.76\ V$$

has the most positive value. By the same reasoning, Cu is the worst reducing agent.

The previous table of half reaction potentials tells us that at standard conditions the following reactions occur spontaneously:

| Increasing oxidizing ability (↑) | | Increasing reducing ability (↓) | |
|---|---|---|---|
| | $Cu^{2+}$ can oxidize $H_2$, Ni, Zn | | $H_2$ can reduce $Cu^{2+}$ |
| | $H^+$ can oxidize Ni, Zn | | Ni can reduce $H^+$ and $Cu^{2+}$ |
| | $Ni^{2+}$ can oxidize Zn | | Zn can reduce $Ni^{2+}$, $H^+$ and $Cu^{2+}$ |

Some important points concerning the Table of Standard Reduction Potentials (Table 9.2) are:

1. The more positive the value of E°, the better the oxidizing ability (the greater the tendency to be reduced) of the ion or compound, on moving upward in the table.

2. The more negative the value of E°, the better the reducing ability of the ions, elements, or compounds on moving downward in the table.
3. Under standard conditions, any substance in this table will spontaneously oxidize any other substance placed lower than it in the table.

## 9.5 TYPES OF CELLS

When various electrodes are combined, two general classes of cells are formed.

(*i*) Chemical cells, in which the Emf is due to a chemical reaction occurring within the cell.

(*ii*) Concentration cells, in which the Emf is due to transfer of material from one electrode to the other due to species involved. Each type of the cells may or may not involve a junction potential and it is usually said, the cell may or may not have transference.

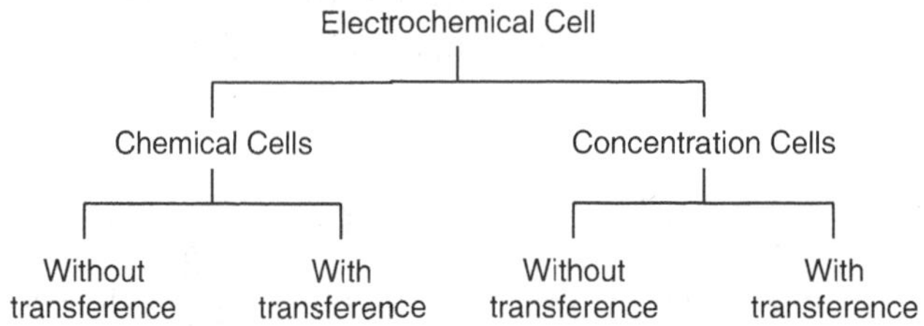


*FIGURE 9.8*

### 9.5.1 Chemical Cells

(*i*) *Chemical cells without transference.* The simplest type of chemical cell without transference, or with no liquid junction potential, consists of two electrodes and an electrolyte common to both the electrodes. A typical example of chemical cells without transference is the cell consisting of the hydrogen electrode and the silver electrode dipped in a common electrolyte, *i.e.*, hydrochloric acid.

$$Pt \mid H_2(g) \mid HCl(aq) \mid AgCl(s) \mid Ag$$

In the cell, oxidation takes place at the hydrogen electrode and reduction at the silver silver electrode.

$$\frac{1}{2}H_2 \longrightarrow H^+ + e^- \qquad \text{(Oxidation)}$$

$$Ag^+ + e^- \longrightarrow Ag(s) \qquad \text{(Reduction)}$$

(*ii*) *Chemical cell with transference.* In chemical cells with transference, the Emf results from a chemical reaction occurring within cell, but the assembly of electrode contains liquid junction between solution of different electrolytes. Cells of this type are very common. The typical example is the Daniell Cell (Figure 9.2) in which the redox couple at one electrode is $Cu^{2+} \mid Cu$ and the other is $Zn^{2+} \mid Zn$.

$$Zn \mid Zn^{2+} \parallel Cu^{2+} \mid Cu$$

### 9.5.2 Concentration Cells

In concentration cells, Emf results from the transfer of a material from one electrode to another due to a concentration difference between the two. The difference in concentration arises due to:

(*i*) Both electrodes dipping in the same electrolyte have different concentrations. These type of cells are known as *electrolyte concentration cells*. For example,

(*a*) $H_2(g, 1\text{ atm}) \mid H^+_{(a_1)} \mid H^+_{(a_2)} \mid A_2(g, 1\text{ atm})$

(*b*) $Ag \mid Ag^+_{(a_1)} \mid Ag^+_{(a_2)} \mid Ag$

(*c*) $Pt \mid H_2(P_1) \mid HCl \mid H_2(P_2) \mid Pt$

(*d*) $Cd(Hg)(C_{Cd} = C_1) \mid Cd^{++} \parallel Cd(Hg)(C_{Cd} = C_2)$

(*ii*) The electrodes may have different concentrations either because they are gas electrodes operating at different pressure or because they are amalgams with different concentrations. A typical concentration cell is shown in Figure 9.9. It consists of two silver electrodes; one immersed in 0.1 M silver nitrate solution and the other in 1 M solution of the same electrolyte. The two solutions are in contact through a membrane (or salt bridge). When the electrodes are connected by a wire, it is found experimentally that electrons flow from the electrode in the more dilute (0.1 M) solution to that in the more concentrated (1 M) solution.

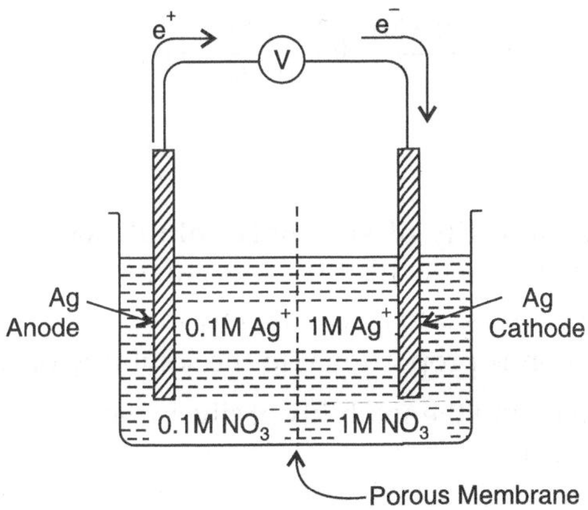


*FIGURE 9.9* **A typical concentration cell.**

The concentration of $Ag^+$ ions in the left compartment is lower (0.1 M) and in the right compartment it is higher (1 M). There is a natural tendency to equalize the concentration of $Ag^+$ ions in the two compartments. This can be done if the electrons are transferred from the left compartment to the right compartment. This electron transfer will produce $Ag^+$ ions in the right compartment by the half-cell reactions:

$$Ag - e^- \longrightarrow Ag^+ \qquad \text{(left compartment)}$$

$$Ag^+ + e^- \longrightarrow Ag \qquad \text{(right compartment)}$$

Thus in a bid to equalize concentration of $Ag^+$ ions in the two compartments, the cell will develop an Emf (Le Chatelier's Principle) to cause the transfer of electrons. Eventually, the solutions in the two compartments will have equal $Ag^+$ ion concentrations and there will be no Emf recorded thereafter.

### Emf of Concentration Cell

Suppose the concentrations in the two half-cells are $C_1$ and $C_2$ at 25°C, $C_2$ being greater than $C_1$. Then Emf, E, of the concentration cell will be given by the difference between the two electrode potentials. In terms of the Nernst equation:

$$E = \left(E_M + \frac{0.0591}{n}\log C_2\right) - \left(E_M + \frac{0.0591}{n}\log C_1\right)$$

$$\boxed{E = \frac{0.0591}{n}\log\left[\frac{C_2}{C_1}\right]} \qquad (9.3)$$

where $E_M$ is the standard electrode potential of the metal M and $n$ is the valence of the ions in contact with it. For example, the Emf, E, of the concentration cell

$$Ag \mid Ag^+(1M) \parallel (0.1\ M)Ag^+ \mid Ag$$

can be found by substituting the values in the above equation

$$E = \frac{0.0591}{1}\log\frac{1\,M}{0.1\,M} = 0.0591.$$

### Applications

1. **Determination of solubility of sparingly soluble salts.** The ionic concentration of a solution be calculated from the Emf of a concentration cell. In case of a sparingly soluble salt, the salt can be supposed to be completely ionized even in saturated solution. Hence the ionic concentration is proportional to the solubility of the salt.

   Suppose we want to find the solubility of silver chloride. This can be done by measuring the Emf, E, of the cell.

   $$Ag \mid N/100\ AgNO_3 \parallel \text{Saturated } AgCl \mid Ag$$

   The Emf of the cell at 25°C is given by the relation

   $$\boxed{E = \frac{0.0591}{n}\log\frac{C_2}{C_1}}$$

   Here, $n$, the valence of $Ag^+$ ion is 1 and the concentration of $Ag^+$ in N/100 $AgNO_3$ solution is 0.01 gram ion per liter.

Hence, $$E = \frac{0.0591}{1} \log \frac{0.01}{C_1}$$

From the above expression, the concentration of AgCl in gram ions of silver per liter can be calculated. Multiplying this by 143.5, the molecular weight of silver chloride, we get the solubility of AgCl in grams per liter.

**Example 16.** *The Emf of the cell*

$$Ag \mid AgI \text{ in } 0.045\ M\ KI \parallel 0.045\ M\ AgNO_3 \mid Ag$$

*is 0.788 at 25°C. Calculate (i) the solubility product of AgI and (ii) the solubility of AgI in water at 25°C.*

**Solution.** *(i) Calculation of* $K_{sp}$. At 25°C the concentration of $Ag^+$ in the cathodic half-cell is 0.045 M and the concentration of $I^-$ ion in anodic half-cell is 0.045 M KI. Let the concentration of $Ag^+$ ion in the anodic half-cell due to solubility of AgI be $C_1$.

$$E = \frac{0.0591}{1} \log \frac{0.045}{C_1}$$

or $$0.788 = \frac{0.0591}{1} \log \frac{0.045}{C_1}$$

or $$\log \frac{0.045}{C_1} = \frac{0.788}{0.0591} = 13.33$$

or $$\frac{0.045}{C_1} = 2.138 \times 10^{13}$$

$$C_1 = \frac{0.045}{2.138 \times 10^{13}} = \mathbf{2.105 \times 10^{-15}}$$

*(ii) Calculation of Solubility*

$$\boxed{\text{Solubility of AgI} = \sqrt{K_{sp}}} \qquad (9.4)$$

$$= \sqrt{9.472 \times 10^{-17}}$$

$$= 9.732 \times 10^{-9} \text{ g mol liter}^{-1}$$

$$= 9.732 \times 10^{-9} \times 143.5 \text{ g liter}^{-1}$$

$$= \mathbf{1.396 \times 10^{-6} \text{ g liter}^{-1}}$$

2. **Determination of valence.** The expression for the Emf, E, of a concentration cell is

$$E = \frac{0.059}{n} \log_{10} \frac{C_2}{C_1}$$

where $n$ is the valence of the metallic ion in solution, while $C_1$ and $C_2$ are the concentrations of the ions in the two half-cells. Knowing the experimental values of E, $C_1$, and $C_2$, $n$ can be calculated.

For example, the valence of mercury in mercurous nitrate, $Hg_2(NO_2)_2$, can be determined by the concentration cell method. The following cell is constructed:

$$Hg \mid 0.05\ N\ Hg_2(NO_3)_2 \parallel 0.5\ N\ Hg_2(NO_3)_2 \mid Hg$$

and its Emf, found experimentally, is 0.029 volts.

Let $C_1$ be the concentration of mercurous ion in 0.05 N $Hg_2(NO_3)_2$ solution in the left half-cell and $C_2$ be the concentration of mercurous ion in 0. 5 N $Hg_2(NO_3)_2$ solution in the right half-cell.

Substituting the values in the equation (9.3), we have,

$$0.29 = \frac{0.0591}{n} \log \frac{0.5}{0.05}$$

$$= \frac{0.0591}{n} \log 10 = \frac{0.0591}{n}$$

Therefore,
$$n = \frac{0.0591}{0.029} = 2$$

## 9.6 CELL POTENTIAL MEASUREMENT AND THERMODYNAMIC FUNCTIONS

### 9.6.1 Relation between ΔG° and the Cell Potential

In an electrochemical cell, chemical energy is converted into electrical energy, so it is a source of electrical energy. The net electrical work performed by a reaction is the product of electromotive force E and the quantity of electricity Q.

$$\text{Net electrical work done} = QE$$

$$\text{Quantity of electricity} = nF$$

where $n$ is the number of electrons taking part in the reaction. F is the charge of 1 mole of electrons and is equal to 96,485 C.

Thus, $\text{Net electrical work done} = nFE_{cell}$

where, $E_{cell}$ = Emf of the cell

We know that any work performed by the cell can be accomplished only at the expense of a decrease in free energy occurring within the cell. Further, when the cell is operating reversibly, the electrical work done is maximum; the decrease in free energy–ΔG, is equal to the electrical work done.

Therefore, $$- \Delta G = nFE \qquad (9.5)$$

Thus the reversible Emf of any cell is determined by the free energy change of the reaction occurring in the cell. In the cell reaction, the substances involved are in their standard states (one atm pressure and 298 K) then,

$$\Delta G^{\circ} = nFE^{\circ}_{cell}$$

The significance of the equation (9.5) is:

(*i*) The various thermodynamic properties of reaction can be calculated from Emf measurement by using this equation.

(*ii*) This equation helps to know about the spontaneity of the reaction at constant temperature and pressure.

When $\Delta G = -$ ve The reaction is spontaneous

$\Delta G = +$ ve The reaction is nonspontaneous

$\Delta G = 0$ For equilibrium

From equation (9.5), it may be concluded that for any spontaneous reaction the cell potential will be positive, and for any nonspontaneous reaction E will be negative, while for any reaction in equilibrium E will have to be zero. The relation between spontaneity of a reaction, $\Delta G^{\circ}$ and $E^{\circ}$ values is summarized in Table 9.3.

**TABLE 9.3**

| Reaction | ΔG° | E° |
|---|---|---|
| Spontaneous | – ve | + ve |
| Nonspontaneous | + ve | – ve |
| Equilibrium | 0 | 0 |

### 9.6.2 Entropy of the Reaction

The standard cell potential is related to standard free energy or cell function through equation

$$\Delta G = - nFE$$

The thermodynamic relation

$$\left(\frac{\partial(\Delta G)}{\partial T}\right)_P = - \Delta S$$

By substituting the value of $- \Delta G$ from equation 9.5,

$$\left(\frac{\partial(- nFE)}{\partial T}\right)_P = - \Delta S$$

$$- nF\left(\frac{\partial E}{\partial T}\right)_P = - \Delta S$$

$$\Delta S = nF\left(\frac{\partial E}{\partial T}\right)_P \tag{9.6}$$

The temperature coefficient of cell potential gives the energy of the cell reaction.

### 9.6.3 Enthalpy of the Reaction

The enthalpy of the reaction is related at ΔG and ΔS as

$$\Delta H = \Delta G + T\Delta S$$

Substituting the values of ΔG and ΔS

$$\Delta H = -\,nFE + nFT\left(\frac{\partial E}{\partial T}\right)_P$$

$$\Delta H = -\,nF\left[E - T\left(\frac{\partial E}{\partial T}\right)_P\right]$$

$$E_{cell} = -\frac{\Delta H}{nF} + T\left(\frac{\partial E}{\partial T}\right) \tag{9.7}$$

Whether the electrical work done is greater or less than the heat of the reaction depends upon the temperature coefficient of the cell; that is, $E_{cell}$ varies with T.

If $\frac{\partial E}{\partial T}$ is negative, then ΔG > ΔH and heat is absorbed from the surroundings during the working of the cell.

If $\frac{\partial E}{\partial T}$ is positive, then ΔH > ΔG and heat is given out to the surroundings of the cell during working.

Thus ΔG, ΔH, and ΔS can be calculated for the cell reaction from measurements of Emf and temperature coefficient of Emf.

## 9.7 EMF OF THE CELL AND EQUILIBRIUM CONSTANT OF A CELL REACTION

Consider the general electrochemical cell reaction

$$aA + bB \rightleftharpoons cC + dD$$

The decrease in free energy – ΔG for the process is given by the thermodynamic equation

$$-\,\Delta G = -\,\Delta G^\circ - RT\ln Q \tag{9.8}$$

where $-\Delta G^\circ$ = decrease in free energy of system when the substances are in their standard states of unit activity,

$$Q = \text{Reaction quotient}$$

$$Q = \frac{[C]^c\,[D]^d}{[A]^a\,[B]^b}$$

The electrical energy produced during the reaction by the cell is $n$EF volt.

$$-\Delta G = nFE$$

$$nFE = -\Delta G^\circ - RT\ln Q \qquad (9.9)$$

We know that $\quad -\Delta G^\circ = RT\ln K$

where K is the equilibrium constant for the cell reaction.

Substituting in equation (9.9), we get

$$nFE = RT\ln K - RT\ln Q$$

$$nFE = RT\ln K - RT\ln\frac{[C]^c\,[D]^d}{[A]^a\,[B]^b}$$

Changing ln to log 10, we get

$$E_{cell} = \frac{2.303\,RT}{nF}\log K - \frac{2.303\,RT}{nF}\log\frac{[C]^c\,[D]^d}{[A]^a\,[B]^b} \qquad (9.10)$$

At 25°C, the value of $\frac{2.303\,RT}{nF}$ is

$$= \frac{2.303 \times 8.313\ JK^{-1}\ mol^{-1} \times 298\ K}{(n) \times 96500\ C\ mol^{-1}} = \frac{0.0591}{n}\ \text{volt}$$

Putting value in equation (9.10)

$$E_{cell} = \frac{0.0591}{n}\log K - \frac{0.0591}{n}\log\frac{[C]^c\,[D]^d}{[A]^a\,[B]^b} \qquad (9.11)$$

The equilibrium constant of the cell can be calculated from this equation if we know the Emf of the cell and the concentration of reactants and products of the cell reaction.

## 9.8 THE NERNST EQUATION

At any instant, the value of $\Delta G$ for a reacting system depends on the concentration of the species in solution and the reaction quotient Q. If a system is not at equilibrium, $\Delta G$ is not zero and as the reaction proceeds $\Delta G$ continually changes until the system reaches the equilibrium, at which $\Delta G = 0$.

$$\Delta G = \Delta G^\circ + RT \ln Q$$

We know $\Delta G = -nEF$

$$-nFE_{cell} = -nE^\circ_{cell}F + RT \ln Q$$

Divide this equation by $n$F

$$E_{cell} = E^\circ_{cell} - \frac{RT}{nF} \ln Q$$

where E = Electrode potential, T = Temperature,
E° = Standard electrode potential, F = Faraday of electricity,
R = Gas constant, $n$ = number of electrons.

$$E_{cell} = E^\circ_{cell} - \frac{2.303\ RT}{nF} \log Q$$

The value of 2.303 RT/$n$F at 25°C is $= \frac{0.0591}{n}$ volts

$$E_{cell} = E^\circ_{cell} - \frac{0.0591}{n} \log Q$$

Consider for an electrode reaction

$$M^{n+}(aq) + e^- \longrightarrow M(s)$$

$$Q = \frac{[M(s)]}{[M^{n+}(aq)]}$$

$$E_{cell} = E^\circ_{cell} - \frac{0.0591}{n} \log \frac{[M(s)]}{[M^{n+}(aq)]} \tag{9.12}$$

For a reaction

$$Zn^{2+}(aq) + 2e^- \longrightarrow Zn(s)$$

$$E_{cell} = E^\circ_{cell} - \frac{0.0591}{n} \log \frac{[Zn(s)]}{[Zn^{2+}(aq)]}$$

The activity of solids is taken as unity. For solids, ($a$ = 1)

$$E_{cell} = E^\circ_{cell} - \frac{0.0591}{n} \log \frac{1}{[Zn^{2+}(aq)]}$$

$$E_{cell} = E^\circ_{cell} + \frac{0.0591}{n} \log [Zn^{2+}(aq)]$$

For a general reaction (9.12), [M(*s*)] is taken as unity

$$E_{cell} = E^\circ_{cell} + \frac{0.0591}{n} \log [M^{n+}(aq)] \quad (9.13)$$

This equation gives the effect of concentration of $M^{n+}$ ions on the potential of $M^{n+}$, M electrode is known as the ***Nernst equation.*** It can be modified as:

(*i*) Consider the reaction

$$Fe^{3+}(aq) + e^- \longrightarrow Fe^{2+}(aq)$$

Hence,
$$E_{cell} = E^\circ_{cell} - \frac{0.0591}{n} \log \frac{[Fe^{2+}]}{[Fe^{3+}]}$$

where $n = 1$

$$E_{cell} = E^\circ_{cell} + \frac{0.0591}{n} \log \frac{[Fe^{3+}]}{[Fe^{2+}]}$$

From this, for an electrode reaction we can write

$$\text{Oxidized state} + ne^- \rightleftharpoons \text{Reduced State}$$

Correspondingly the electrode potential is given by

$$E_{cell} = E^\circ_{cell} - \frac{0.0591}{n} \log \frac{[\text{Reduced State}]}{[\text{Oxidized State}]}$$

$$E_{cell} = E^\circ_{cell} + \frac{0.0591}{n} \log \frac{[\text{Oxidized State}]}{[\text{Reduced State}]}$$

(*ii*) Suppose in a chlorine electrode, chlorine gas is passed at a pressure of 1 atmosphere through a solution containing $Cl^-$ ions,

$$\frac{1}{2}Cl_2(g)(1\text{ atom}) + e^- \rightleftharpoons Cl^-(aq)$$

The cell potential is

$$E_{cell} = E^\circ_{cell} - \frac{0.0591}{n} - \left[\frac{Cl-}{Cl_2}\right]$$

Since activity of chlorine gas at 1 atm pressure is taken as unity, therefore

$$E_{cell} = E^\circ_{cell} - \frac{0.0591}{n} \log [Cl^-]$$

(*iii*) *Calculation of equilibrium constant (K) from Nernst equation.* Consider the galvanic cell,

$$Zn(s) \mid Zn^{2+}(aq) \parallel Cu^{2+}(aq) \mid Cu(s)$$

The state of equilibrium may be written by the reaction,

$$Zn(s) + Cu^{2+}(aq) \rightleftharpoons Zn^{2+}(aq) + Cu(s)$$

In equilibrium state, the concentration of $Zn^{2+}(aq)$ and $Cu^{2+}(aq)$ will be the equilibrium concentrations and EMF of the cell will be zero. Therefore the Nernst equation is modified as,

$$O = E^\circ_{cell} - \frac{2.303\,RT}{nF} \log \frac{[Zn^{2+}(aq)]_{equilib}}{[Cu^{2+}(aq)]_{equilib}}$$

$$E^\circ_{cell} = \frac{2.303\,RT}{nF} \log K \qquad (9.14) \qquad \left( \because \; K = \frac{[Zn^{2+}(aq)]_{equilib}}{[Cu^{2+}(aq)]_{equilib}} \right)$$

where K = equilibrium constant at 298 K.

$$\therefore \qquad E^\circ_{cell} = \frac{0.0591}{n} \log K \quad \text{or} \quad \log K = \frac{nE^\circ_{cell}}{0.0591}$$

$$\boxed{K = \text{Antilog}\, \frac{nE^\circ_{cell}}{0.0591}} \qquad (9.15)$$

## SOLVED PROBLEMS

### TYPE-I Problems on Galvanic Cell

**Problem 1.** *Calculate the Emf of the cell: Ni/Ni²⁺(1 M) || Pb²⁺ (1 M) | Pb at 25°C. Write down its cell reactions. Standard electrode potential of nickel and lead are – 0.24 V and – 0.13 V respectively at 25°C.*

**Solution.** $$E^\circ_{cell} = E^\circ_{right} - E^\circ_{left} = E^\circ_{Pb^{2+}/Pb} - E^\circ_{Ni^{2+}/Ni}$$

$$0.13\text{ V} - (-0.24\text{ V}) = \mathbf{+\,0.11\ V}$$

Cell reactions are:

Anode: $Ni(s) \longrightarrow Ni^{2+}(aq) + 2e^-$ (Oxidation)

Cathode: $Pb^{2+}(aq) + 2e^- \longrightarrow Pb(s)$ (Reduction)

Net reaction: $Ni(s) + Pb^{2+}(aq) \longrightarrow Ni^{2+}(aq) + Pb(s)$

**Problem 2.** *What is the Emf of the following cell at 25°C?*

$$Zn(s)\, /\, Zn^{2+}(0.2\ M)\, //\, Ag^{+}(0.002\ M)\, /\, Ag\ (s)$$

*The standard Emf of the cell is 1.54 V.*

**Solution.**

$$E_{cell} = E^\circ_{cell} + \frac{0.0591}{2}\log\frac{[Ag^+]^2}{[Zn^{2+}]}$$

$$= 1.54 + 0.0296 \log\frac{(0.002)^2}{(0.2)}$$

$$= 1.54 + 0.0296 \log 2 \times 10^{-5}$$

$$= 1.54 + 0.0296\,[0.301 - 5]$$

$$= 1.54 + 0.0296\,(-4.699)$$

$$= 1.45 - 0.139 = \mathbf{1.401\ V}$$

**Problem 3.** *A cell is formed by dipping Zn-rod in 0.01 M $Zn^{2+}$ solution and Ni-rod in 0.5 M $Ni^{2+}$ solution. The standard electrode potentials of Zn and Ni are – 0.76 V and – 0.25 V respectively. Write the cell representation, cell reaction, and calculate the Emf of the cell.*

**Solution.** Since E°(= – 0.25 V) for nickel electrode is higher than that of zinc electrode (– 0.76 V), the nickel electrode is the cathode; while the zinc electrode is the anode. Hence, the cell is represented as:

$$Zn(s)\,|\,Zn^{2+}(0.01\ M)\,||\,Ni^{2+}\,(0.5\ M)\,|\,Ni(s)$$

$$\therefore \quad E_{cell} = \left[E^\circ\,(Ni^{2+}/Ni) - E^\circ(Zn^{2+}/Zn)\right] + \frac{0.0592}{2}\log\frac{[Ni^{2+}]}{[Zn^{2+}]}$$

$$= [-0.25 + 0.76] + 0.0296 \log\,[0.5/0.01]$$

$$= 0.51 + 0.0296 \log 50$$

$$= 0.51 + 0.0296 \times 1.6987$$

$$= 0.51 + 0.0503 = \mathbf{0.5603\ V}$$

**Problem 4.** *Sketch the electrochemical cell and write down the cell reaction, if Zn and Ag electrodes are dipped in respective solutions. Also find out the standard emf of the cell, given that $E^\circ_{Zn/Zn^{2+}}$ = 0.76 V and*

*$E^\circ_{Ag/Ag^+}$ = – 0.8 V at 25°C.*

**Solution.** Standard redox potential, $E^\circ_{Zn^{2+}/Zn} = -E^\circ_{Zn/Zn^{2+}}$ = – 0.76 V and $E^\circ_{Ag^+/Ag} = -E^\circ_{Ag/Ag^+}$ = – (– 0.8 V) = 0.8 V. Since SRP of Ag electrode is higher than that of Zn electrode, so Ag electrode is cathodic with respect to Zn electrode. Therefore, the electrochemical cell is:

$$\underset{\text{Anode}}{Zn(s)\,|\,Zn^{2+}\,(aq)}\,||\,\underset{\text{Cathode}}{Ag^+\,(aq)\,|\,Ag(s)}$$

| | | |
|---|---|---|
| Reaction at anode: | $Zn(s) \longrightarrow Zn^{2+}(aq) + 2e^-$ | (Oxidation) |
| Reaction at cathode: | $2Ag^+(aq) + 2e^- \longrightarrow 2Ag(s)$ | (Reduction) |
| Cell reaction: | $2Ag^+(aq) + Zn(s) \longrightarrow 2Ag(s) + Zn^{2+}(aq)$ | |

Standard Emf of cell: $E^\circ_{cell} = E^\circ_{cathode} - E^\circ_{anode} = E^\circ_{right} - E^\circ_{left}$

$$= 0.8 \text{ V} - (-0.76 \text{ V}) = \mathbf{1.56 \text{ V}}$$

**Problem 5.** *Predict whether the reaction:*

$$2Ag(s) + Zn^{2+}(aq) \longrightarrow 2Ag^{+}(aq) + Zn(s) \text{ is feasible or not.}$$

**Solution.** The cell half reactions are

Anode: $2Ag(s) \longrightarrow 2Ag^{+}(aq) + 2e^{-}$ $E^\circ = 0.80$ V

Cathode: $Zn^{2+}(aq) + 2e^{-} \longrightarrow Zn(s)$ $E^\circ = -0.763$ V

$$E^\circ_{cell} = E^\circ_{cathode} - E^\circ_{anode}$$

$$E^\circ_{cell} = -0.763 - 0.80 = \mathbf{-1.563}$$

Since $E^\circ_{cell}$ is negative, the given reaction is not feasible.

**Problem 6.** *Determine the feasibility of the reaction:*

$$2Al(s) + 3Sn^{4+}(aq) \longrightarrow 2Al^{3+} + 3Sn^{2+}(aq).$$

**Solution.** The given reaction consists of the following half reaction

Anode: $2Al(s) \longrightarrow 2Al^{3+} + 6e^{-}$ $E^\circ = -1.66$ V

Cathode: $3Sn^{4+} + 6e^{-} \longrightarrow 3Sn^{2+}$ $E^\circ = +0.15$ V

$$E^\circ_{cell} = 0.15 - (-1.66) = 1.81 \text{ V}$$

Since $E^\circ_{cell}$ is positive, the reaction is feasible.

**TYPE-II Problems on Concentration Cells**

**Problem 7.** *Calculate the valency of mercurous ions with the help of the following cell:*

$$Hg \left| \begin{array}{c} 0.001 \text{ N mercurous} \\ \text{nitrate in } 0.1 \text{ N } HNO_3 \end{array} \right| \begin{array}{c} 0.01 \text{ N mercurous} \\ \text{nitrate in } 0.1 \text{ N } HNO_3 \end{array} \left| Hg^{+} \right.$$

*when the Emf observed at 18°C is 0.029 volt (F = 96,500).*

**Solution.** Emf of the given concentration cell is given by:

$$E_{cell} = \frac{2.303 \text{ RT}}{nF} \log \frac{C_2}{C_1} = \frac{0.058 \text{ V}}{n} \log \frac{C_2}{C_1} \qquad \text{(At 18°C)}$$

Here $E_{cell} = 0.029$ V; $C_2 = 0.01$ N and $C_1 = 0.001$ N.

$$\therefore \quad 0.029 = \frac{2.303 \text{ RT}}{n} \log \frac{0.01}{0.001} = \frac{0.058}{n} \times 1 = \frac{0.058}{n}$$

or $$n = \frac{0.058}{0.029} = \mathbf{2}$$

Hence, valency of mercurous ion is 2 and its formula is $\mathbf{Hg_2^{2+}}$.

**Problem 8.** *Design a reversible cell in which the reaction occurs.*

$$AgCl(s) + \frac{1}{2}H_2(g) \rightleftharpoons Ag(s) + HCl(aq)$$

*Write down the Emf expression of this cell. Use this cell to determine the activity of HCl, when Emf of this cell is 0.15 V at 25°C.*

**Solution.** The cell can be represented as:

$$Ag(s)\ Ag^+(aq)\ ||\ H^+(aq)\ H_2(g)$$

∴ Redox reactions are:

RHE (red): $$H^+(aq) + e^- \longrightarrow \frac{1}{2}H_2(g); \qquad E° = 0.0\ V$$

LHE(ox): $$Ag(s) \longrightarrow Ag^+ + e^-; \qquad E° = -0.2224\ V$$

Overall reaction: $$H^+(aq) + Ag(s) \longrightarrow \frac{1}{2}H_2(s) + Ag^+(aq); \quad E°_{cell} = 0.2224\ V$$

Now, $$E_{cell} = E_{hydrogen} - E_{Ag^+/Ag}$$

∴ $$0.15 = -0.0592\ pH + 0.2224$$

$$pH = \frac{0.0724}{0.0592} = 1.223$$

or $$-\log [H^+] = 1.223$$

∴ $$[H^+] = \text{Antilog}(-1.223)$$

$$= \text{Antilog}\ \bar{2}.777 = \mathbf{0.06\ mol\ L^{-1}}$$

Hence, activity (or effective concentration) or HCl is 0.06 mol $L^{-1}$.

**Problem 9.** *The Emf of the cell: Ag | 0.0093 N $AgNO_3$ || satd. $NH_4NO_3$ | $AgNO_3$ (x) | Ag is 0.086 volt at 25°C. Find the concentration (x), (2.303 RT/F = 0.059).*

**Solution.** Emf of the above concentration cell is given by

$$E_{cell} = \frac{0.059}{1}\log\frac{x}{0.0093} \qquad (\because\ n = 1)$$

∴ $$0.086 = 0.059 \log\frac{x}{0.0093}$$

or $$\log\frac{x}{0.0093} = \frac{0.086}{0.059}$$

whence, $\frac{x}{0.0093} = 1.4576 = 28.68$

Hence, $x = 28.68 \times 0.0093 = \mathbf{0.267\ M}$

**Problem 10.** *Find the Emf of the following at 25°C:*

$$Ag \mid Ag^+ \ (0.01\ M) \parallel Ag^+ \ (0.1\ M) \mid Ag$$

**Solution.** Emf of concentration cell is given by:

$$E_{cell} = \frac{0.0592\ V}{n} \log \frac{C_2}{C_1} = \frac{0.0592}{1} \log \frac{0.1}{0.01} \qquad (n = 1 \text{ for } Ag^+ + e^- \longrightarrow Ag)$$

$$= 0.0592 \log 10 = \mathbf{0.0592\ V}$$

**Problem 11.** *Calculate the equilibrium constant for the Daniell cell at 25°C from the following data: R = 8.316 joules; F = 96500 coulombs, SOP for zinc electrode 0.765 V, SOP for copper electrode = – 0.337 V.*

**Solution.** $E_{cell}$ = SOP of (Cu-Zn) electrode – SOP of (Zn-Cu) electrode

$$= [0.765 - (-0.337)] = 1.102$$

Now $$\log K_{eq} = \frac{nE°}{0.0592} = \frac{2 \times 1.102}{0.0592} = 37.23 \qquad \text{(At 25°C)}$$

Hence, $K_{eq}$ = Antilog 37.23 = $\mathbf{1.697 \times 10^{37}}$

**Problem 12.** *Calculate the potential of a silver electrode in a solution that is saturated with AgI and has an iodide concentration of exactly unity. ($K_{sp}$ for AgI = $8.3 \times 10^{-17}$, $E°_{Ag^+/Ag}$ = + 0.799 V).*

**Solution.** $$[Ag^+] = \frac{K_{sp}}{[I^-]} = \frac{8.3 \times 10^{-17}}{1} = \mathbf{8.3 \times 10^{-17}\ M}$$

**Problem 13.** *The potential of a hydrogen gas electrode in a solution of an acid of unknown strength is 0.29 V at 298 K, as measured against normal hydrogen electrode. Calculate the pH of acid solution.*

**Solution.** $$E_{cell} = E°_{hydrogen} - E_{hydrogen} = 0\ V - (-\ 0.0592\ V\ pH)$$

$$= 0.0592\ pH$$

$$pH = \frac{E_{cell}}{0.0592} = \frac{0.29}{0.0592} = \mathbf{4.9}$$

**Problem 14.** *Find out the pH of a solution in a quinhydrone half-cell, which is coupled with standard calomel electrode. The Emf of the combined cell was determined to be 0.123 V = (0.6990 V – 0.0592 V pH) – 0.2415 V.*

**Solution.** $$pH = \frac{0.6990 - 0.2415 - 0.123}{0.0592} = \mathbf{5.65}$$

**Problem 15.** *A hydrogen electrode is immersed in a solution of pH = 3.5 and the other half-cell is SHE. Calculate the Emf of the cell, if the two half-cells are connected by a salt bridge.*

**Solution.** The cell is:

$$\text{Pt, H}_2(g) \mid \text{H}^+(p\text{H}) = 3.5 \parallel \text{H}^+(1\text{ M}) \mid \text{H}_2(g), \text{Pt}$$

$$E_{cell} = E° \text{ of SHE} - E° \text{ of hydrogen electrode}$$

$$\therefore \quad = 0 - (-0.592 \text{ V } p\text{H}) = 0.0592 \text{ V} \times 3.50 = \mathbf{0.207\ V}$$

**Problem 16.** *Emf of the cell:*

*Pt, $H_2(g)$ $H^+$ ($10^{-6}$ M) || $H^+$ (C = ?) | $H_2(g)$, Pt is 0.118 V at 25°C. Calculate the value of C.*

**Solution.**
$$E_{cell} = 0.0592 \text{ V} \log \frac{C}{10^{-6}}$$

$$\therefore \quad 0.118 = 0.0592 \log 10^6 C$$

or
$$\log 10^6 C = 0.118/0.0592 = 1.9932$$

or
$$C = \frac{\text{Antilog } 1.9932}{10^6} = \frac{98.5}{10^6} = \mathbf{9.85 \times 10^{-5}\ M}$$

**TYPE-III Problems On Nernst Equation**

**Problem 17.** *Calculate the half-cell potential at 298 K for the reaction*

$$Zn^{2+}(aq) + 2e^- \longrightarrow Zn(s) \text{ if } [Zn^{2+}] \text{ is 5.0 M and } E^\circ_{Zn^{2+}/Zn} = -0.76 \text{ V.}$$

**Solution.** For the electrode reaction $Zn^{2+}(aq) + 2e^- \longrightarrow Zn(s)$

The Nernst equation is

$$E_{cell} = E^\circ_{\frac{Zn^{2+}}{Zn}} - \frac{0.059}{2} \log \frac{[Zn]}{[Zn^{2+}]} = -0.76 - \frac{0.059}{2} \log \frac{1}{5}$$

$$= -0.76 + \frac{0.059}{2} \times 0.669 = 0.76 + 0.0206 = \mathbf{-0.739\ V}$$

**Problem 18.** *Calculate the Emf of the cell:*

$$Zn \mid Zn^{2+} (0.001 \text{ M}) \parallel Ag^+ (0.1 \text{ M}) \mid Ag$$

*The standard potential of $Ag/Ag^+$ half-cell is + 0.80 V and $Zn/Zn^{2+}$ is – 0.76 V.*

**Solution. Step 1.** Write the half-cell reactions of the anode and the cathode. Then add the anode and cathode half reactions to obtain the cell reaction and the value of $E^\circ_{cell}$.

| | | |
|---|---|---|
| Cathode: | $2Ag^+ + 2e \longrightarrow 2Ag$ | $E° = +0.80$ V |
| Anode: | $Zn \longrightarrow Zn^{2+} + 2e^-$ | $E° = -0.76$ V |
| Cell: | $Zn + 2Ag^+ \rightleftharpoons Zn^{2+} + 2Ag$ | $E° = 1.56$ V |

**Step 2.** K for the cell reaction = $\frac{[Zn^{2}]}{Ag^{2}}$

Substitute the given values in the Nernst equation and, solving for $E_{cell}$, we have

$$E_{cell} = E^{\circ}_{cell} - \frac{0.0591}{n} \log K$$

$$= 1.56 - \frac{0.0591}{2} \log \frac{[Zn^{2+}]}{[Ag^{+}]^{2}}$$

$$= 1.56 - \frac{0.0591}{2} \log \frac{[10^{-3}]}{[10^{-1}]^{2}}$$

$$= 1.56 - \frac{0.0591}{2} (\log 10^{-1})$$

$$= 1.56 + 0.02955 = \mathbf{1.58955\ V}$$

**Problem 19.** *Calculate the half-cell potential of the following half-cells.*

(a) $Fe \left| \begin{array}{l} FeSO_4 \\ 0.1\,M; \end{array} \right. E^{\circ}_{\frac{Fe^{2+}}{Fe}} = 0.44\ V$ (b) $\begin{array}{r} Pt\ Cl_2 \\ 10\ atm \end{array} \left| \begin{array}{l} HCl \\ 0.1\,M; \end{array} \right. E^{\circ}_{\frac{1/2\,Cl_2}{Cl^-}} = +1.36\ V.$

**Solution.** (*a*) Since change at the electrode is oxidation, the reaction is

$$Fe \longrightarrow Fe^{2+} + 2e^{-}$$

$$\therefore \qquad E_{cell} = E^{\circ}_{\frac{Fe}{Fe^{2+}}} - \frac{0.059}{2} \log \frac{[Fe^{2+}]}{Fe} = 0.44 - \frac{0.059}{2} \log [0.1]$$

$$= 0.44 + \frac{0.059}{2} = \mathbf{0.4695\ V}$$

(*b*) The oxidation reaction is $2Cl^{-} \longrightarrow Cl_2 + 2e$

$$E_{cell} = E^{\circ}_{\frac{Cl^-}{Cl_2}} - \frac{0.059}{2} \log \frac{[Cl_2]}{[Cl^{-}]^{2}} = -1.36 - \frac{0.059}{2} \log \frac{10}{0.1 \times 0.1}$$

$$= -1.36 - 0.085 = \mathbf{-1.4485\ V}$$

**Problem 20.** *What is the potential of a half-cell consisting of zinc electrode in 0.01 M $ZnSO_4$ solution at 25°C. E° = 0.763 V.*

**Solution.** The half-cell reaction is

$$Zn \longrightarrow Zn^{2+} + 2e^{-}$$

The Nernst equation for the oxidation half-cell reactions

$$E = E^{\circ} - \frac{0.0591}{n} \log [Zn^{2+}]$$

The number of electrons transferred $n = 2$ and E° = 0.763 V. Substituting these values in the Nernst equation we have

$$E = 0.763 - \frac{0.0591}{2} \log (0.01)$$

$$= 0.763 - \frac{0.0591}{2}(-2) = 0.763 + 0.0591 = \mathbf{0.8221\ V}$$

**Problem 21.** *Calculate the equilibrium constant for the following reaction.*

$$Cu(s) + 2Ag^+(aq) \longrightarrow Cu^{2+}(aq) + 2Ag(s)$$

*Given* $E^\circ_{cell} = 0.46\ V.$

**Solution.** $E^\circ_{cell} = \frac{0.0591}{n} \log K$

Substituting the given data

$$0.46 = \frac{0.0591}{2} \log K = \frac{2 \times 0.46}{0.0591} = 15.567$$

$$K = \text{Antilog } 15.567 = \mathbf{3.69 \times 10^{15}}$$

**Problem 22.** *For the cell reaction: $Ni(s) \mid Ni^{2+}(aq) \parallel Ag^+(aq) \mid Ag(s)$*

*Calculate the equilibrium constant at 25°C.*

$$E^\circ_{Ni^{2+}/Ni} = -0.25\ V;\ E^\circ_{Ag^+/Ag} = 0.80\ V.$$

**Solution.** $E^\circ_{cell} = E^\circ_R - R^\circ_L = 0.80 - (-0.25) = 0.80 + 0.25 = 1.05\ V$

Substituting the data in Nernst equation

$$1.05 = \frac{0.0591}{2} \log K$$

$$\log K = \frac{1.05 \times 2}{0.0591} = 35.533$$

$$K = \text{Antilog } 35.533 = \mathbf{3.41 \times 10^{35}}$$

**Problem 23.** *The standard reduction potential of the reaction*

$$2H_2O + 2e^- \rightleftharpoons H_2(g) + 2OH^- - 0.8227\ V.$$

*Calculate equilibrium constant for the reaction*

$$2H_2O \rightleftharpoons H_3O^+ + OH^- \text{ at } 25°C.$$

**Solution.** Let us split the desired reaction into two half-cell reactions

Oxidation half reaction: $H_2O + \frac{1}{2}H_2(g) \longrightarrow H_3O^+ + e^-$ $\quad E° = 0.00$ V

Reduction half reaction: $H_2O + e^- \longrightarrow \frac{1}{2}H_2 + OH^-$ $\quad E° = -0.8277$ V

Net reaction: $2H_2O \rightleftharpoons H_3O^+ + OH^-$ $\quad E°_{cell} = -0.8277$ V

So, the number of electrons involved in redox reaction, $(n) = 1$

We know that
$$E°_{cell} = \frac{0.0591}{n} \log K$$

$$\log K = \frac{C°_{cell} \times n}{0.0591} = \frac{(-0.8277)\times 1}{0.0591} = -14.005$$

$$K = \text{Antilog } [\overline{15}.995] = \mathbf{9.88 \times 10^{-15}}$$

**Problem 24.** *Determine the equilibrium constant of the following reaction at 298 K.*

$$2Fe^{3+} + Sn^{2+} \longrightarrow 2Fe^{2+} + Sn^{4+}$$

*Also predict whether $Sn^{2+}$ ions can reduce $Fe^{3+}$ ions to $Fe^{2+}$ quantitatively or not.*

$$E°_{\frac{Fe^{3+}}{Fe^{2+}}} = 0.771\ V,\ E°_{\frac{Sn^{4+}}{Sn^{2+}}} = 0.150\ V.$$

**Solution.**
$$E°_{cell} = E°_R - E_L = 0.771 - 0.150 = 0.621 \text{ V}$$

We know,
$$E°_{cell} = \frac{0.0591}{2} \log K$$

$$0.621 = \frac{0.0591}{2} \log K \quad \text{or} \quad \log K = \frac{2 \times 0.621}{0.0591} = 21.015$$

$$K = \text{Antilog } 21.015 = \mathbf{1.035 \times 10^{21}}$$

**Problem 25.** *Calculate the equilibrium constant for the reaction between silver nitrate and metallic zinc.*

**Solution. Step 1.** Write the equation for the reaction

$$2Ag^+ + Zn \rightleftharpoons Zn^{2+} + 2Ag \qquad E°_{cell} = 1.56 \text{ V}$$

**Step 2.** Substitute values in the Nernst equation at equilibrium

$$\log K = \frac{nE°_{cell}}{0.0591}$$

$$0 = 1.56 - 0.03 \log K$$

$$-1.56 = -0.03 \log K$$

$$\log K = \frac{-1.56}{-0.03} = 52$$

$$K = 1 \times 10^{52}$$

**TYPE-IV Problems on Work Done by the Cell**

**Problem 26.** *Calculate the maximum work that can be obtained from the cell:*

$$Zn \mid Zn^{2+} (1\ M) \ \| \ Ag^{+}(1\ M) \mid Ag.$$

*Given that* $E^\circ_{\frac{Zn^{2+}}{Zn}} = -0.76\ V;\ E^\circ_{\frac{Ag^{+}}{Ag}} = +0.80\ V.$

**Solution.** $E^\circ_{cell} = E_R - E_L = 0.8 - (-0.76) = 1.56\ V$

Here $n = 2$, F = 96500 coulombs

$\because$ $E^\circ_{cell} = 1.56\ V$

Substituting the given values in the equation

$$W_{max} = nF\,E^\circ_{cell} = 2 \times 96500 \times 1.56\ V = \mathbf{301.08\ kJ}$$

**Problem 27.** *For the cell reaction: Mg(s) | $Mg^{2+}$ (aq) || $Ag^{+}$(aq) | Ag(s). Calculate the equilibrium constant at 25°C and also the maximum work that can be obtained by operating the cell.*

$$E^\circ_{\frac{Mg^{2+}}{Mg}} = -2.37\ V \text{ and } E^\circ_{\frac{Ag^{+}}{Ag}} = 0.80\ V.$$

**Solution.** $E^\circ_{cell} = E^\circ_R - E^\circ_L = 0.80 - (-2.37) = 3.17\ V$

*Calculation of* $K_c$

$$E^\circ_{cell} = \frac{2.303\ RT}{nF} \log K_c$$

$$3.17 = \frac{2.303 \times 8.314 \times 298}{2 \times 96500} \log K_c$$

$$3.17 = \frac{0.0591}{2} \log K_c$$

or $$\log K_c = \frac{2 \times 3.17}{0.0591} = 107.276$$

$$K_c = \text{Antilog } 107.276 = 189 \times 10^{107}$$

*Calculation of maximum work* ($W_{max}$)

$$W_{max} = -\Delta G^\circ = nF\,E^\circ_{cell}$$

$$= 2 \times 96500 \times 3.17 = 611810\ J = \mathbf{611.810\ kJ}$$

**Problem 28.** *Given the cell: Al | $Al^{3+}$ (0.1 M) || $Fe^{2+}$ (0.2 M) | Fe;*

$$E^\circ_{\frac{Al^{3+}}{Al}} = -1.66 \text{ V and } E^\circ_{\frac{Fe^{2+}}{Fe}} = -0.44 \text{ V}$$

*Calculate the maximum work that can be obtained by the cell.*

**Solution.** Cell reaction

$$2Al(s) + 3Fe^{2+}(aq) \longrightarrow 2Al^{3+}(aq) + 3Fe(s).$$

We know that,

$$E_{cell} = E^\circ_{cell} - \frac{0.0591}{n} \log Q = (E^\circ_R - E^\circ_L) - \frac{0.0591}{n} \log Q$$

$$= (-0.44 + 1.66) - \frac{0.0591}{6} \log \frac{[Al^{3+}]^2}{[Fe^{2+}]^3}$$

$$= 1.22 - \frac{0.0591}{6} \log \frac{[0.1]^2}{[0.2]^3} = 1.22 - \frac{0.0591}{6} \log \frac{[0.01]}{[0.08]}$$

$$= 1.22 - 0.00095 = 1.219 \text{ V}$$

*Calculation of $W_{max}$*

$$W_{max} = -\Delta G = nF\,E_{cell} = 6 \times 96500 \times 1.219 = 705801 \text{ J} = \mathbf{705.801\ kJ}$$

## 9.9 INTRODUCTION TO CORROSION

When metals are exposed to atmospheric conditions they react with air or water in the environment to form undesirable compounds, usually oxides. This process is called *corrosion.* Almost all metals except the less reactive metals such as gold, platinum, and palladium, undergo corrosion. For example, silver tarnishes, copper develops a green coating, lead, and stainless steel lose their lustre to corrosion. Thus, corrosion may be defined as *the gradual eating away, disintegration, decaying, or deterioration of a metal by a chemical or electrochemical reaction with its environment.*

## 9.10 THEORIES OF CORROSION

### 9.10.1 The Electrochemical or Wet or Immersed Theory

This theory is based on Nernst theory, according to which all the metals have a tendency to pass into solution. The tendency of a metal to pass into solution when immersed in a solution of its salt is measured in terms of electrode potential. If a metal having higher electrode potential comes into contact with another metal having a lower electrode potential, a galvanic cell is set up and the metal having the higher electrode potential becomes anodic and goes into solution to a measurable extent. The extent of galvanic corrosion depends upon the difference in the electrode potential of the two electrodes and their respective areas. The greater the difference in the potentials of the cathode and anode, the greater will be the corrosion. Also the smaller the area of the anode as compared to the cathode, the more will be the attack.

**Corrosion reactions.** The corrosion process involves the formation of galvanic cells. The various reactions that occur during corrosion are:

(*i*) **Anode reactions.** The anode is the metal which undergoes oxidation and loses electrons to the environment and passes into solution in the form of positive ions. For example,

$$M(s) \longrightarrow M^{n+}(aq) + ne^-$$

$$Fe(s) \longrightarrow Fe^{2+}(aq) + 2e^-$$

$$Fe^{++}(aq) + 2OH^-(aq) \longrightarrow Fe(OH)_2$$

This continues as long as the electrons and ions are removed from the environment. If they are not removed, the corrosion will not proceed further.

(*ii*) **Cathode reactions.** The electrons released at the anode are carried to the cathode and are responsible for following reactions:

(*a*) *Hydrogen electrode or hydrogen type corrosion.* In acidic solution (in the absence of oxygen), the common reaction is the reduction of $H^+$ ions to form $H_2$ gas. $H^+$ ions are present due to the ionization of water

$$H_2O \longrightarrow 2H^+ + O^{2-}$$

$$2H^+ + 2e^- \longrightarrow H_2$$

Such types of corrosion in which hydrogen is evolved is called *hydrogen type corrosion.* (Figure 9.10). At the anode, the metal ions pass into the solution as

$$M \longrightarrow M^{2+} + 2e^-$$

The *overall reaction* can be written as

$$M + 2H^+ \longrightarrow H_2 + M^{2+}$$

For example, *iron corrosion* (Figure 9.10) in an acidic solution consists of two half-cell reactions.

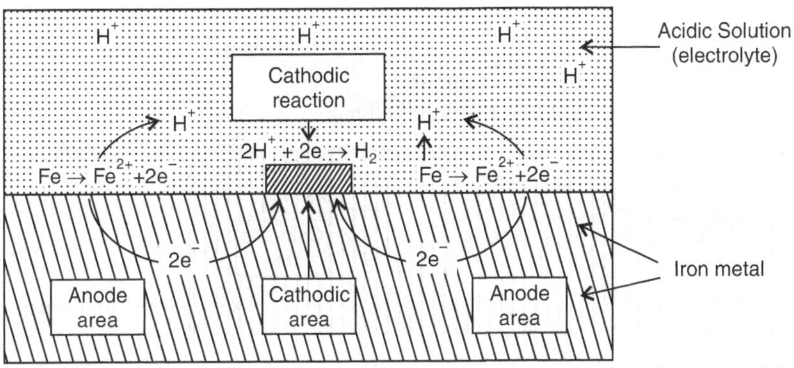


***FIGURE 9.10*** **Diagrammatic representation of the mechanism of wet corrosion by hydrogen evolution.**

$$Fe \longrightarrow Fe^{2+} + 2e^- \qquad \textit{(Anode Reaction)}$$

$$2H^+ + 2e^- \longrightarrow H_2 \qquad \textit{(Cathode reaction)}$$

(*b*) *The water electrode.* In oxidizing acids, the cathodic reaction is the formation of water as a byproduct.

$$O_2 + 4H^+ + 4e^- \longrightarrow 2H_2O$$

(*c*) *The oxygen electrode.* Hydrogen ion concentration is very low in neutral or basic solutions, thus evolution of hydrogen gas is not favorable. In such cases reduction of oxygen to hydroxyl ion occurs.

$$O_2 + 2H_2O + 4e^- \longrightarrow 4OH^-$$

For example,

The *corrosion of iron* occurs by oxygen in the presence of aqueous solution. At anode, iron dissolves to form ions as

$$Fe \longrightarrow Fe^{2+} + 2e \qquad \textit{(Anode reaction)}$$

At cathode, the electrons evolved by the above reaction will be intercepted by oxygen in the presence of water.

$$2e^- + \frac{1}{2}O_2 + H_2O \longrightarrow 2OH^- \qquad \textit{(Cathode reaction)}$$

*The overall reaction* will be

$$Fe + \frac{1}{2}O_2 + H_2O \longrightarrow Fe^{2+} + 2OH^-$$

The $OH^-$ ion produced reacts with the positively charged $Fe^{2+}$ ion forming $Fe(OH)_2$ or *rust*. This process is known as *rusting of iron* (Figure 9.11).

$$Fe^{2+} + 2OH^- \longrightarrow \underset{\text{Rust}}{Fe(OH)_2} \downarrow$$

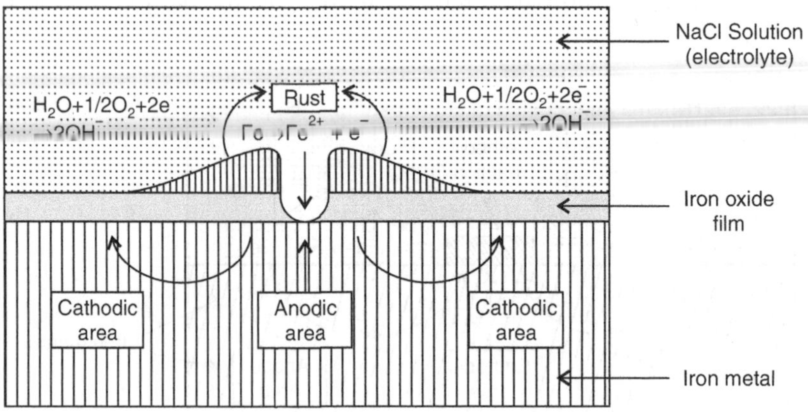


***FIGURE 9.11*** **Diagrammatic representation of the wet corrosion mechanism by absorption of oxygen type corrosion.**

(*i*) If enough oxygen is present, ferrous hydroxide is easily oxidized to ferric hydroxide

$$4Fe(OH)_2 + O_2 + 2H_2O \longrightarrow 4Fe(OH)_3$$

This product, called *yellow rust*, actually corresponds to $Fe_2O_3.H_2O$.

(*ii*) If the supply of oxygen is limited, the corrosion product may be black anhydrous magnetite, $Fe_3O_4$.

The corrosion proceed as long as oxygen is available. Such type of corrosion is called *oxygen type corrosion.*

Thus the essential requirements of electrochemical corrosion are:

(*a*) formation of anodic and cathodic areas

(*b*) electrical contact between the cathodic and anodic parts

(*c*) an electrolyte through which the ions can diffuse or migrate. This is provided by moisture.

**Rate of rusting of iron.** During the rusting of iron, the weight of the product (rust, $Fe_3O_4.xH_2O$) is higher than that of the parent metal, Fe. So the rate of rusting can be measured by placing a weighed sample of iron in the rusting environment, followed by weighing it (after drying) at an interval of 1, 2, 3, 4 ... etc. weeks. From the corresponding increase in weight, the rate of rusting can be evaluated.

### 9.10.2 Chemical Corrosion or Dry Corrosion or Direct Chemical Attack Theory

Direct chemical attack includes all kinds of corrosion in which there is no appreciable flow of current through the metal for percipitable distances. This type of corrosion occurs mainly through the action of atmospheric gases such as oxygen, halogens, sulphur dioxide, hydrogen sulphide, nitrogen, etc. on the metal surfaces. In this type of corrosion, generally, a solid film of the corrosion product is formed on the surface of the metal which resists further corrosion. However, if a soluble or liquid corrosion product is formed, then the metal is exposed to further attack. For example, chlorine and iodine attack silver, generating a protective film of silver halide. During detining of tinned low-carbon steel cans using $Cl_2$ gas at high temperatures (over 120°C), tin is converted into volatile $SnCl_4$ and hence all tin is readily removed, however dry $Cl_2$ attacks the base metal only superficially because the $FeCl_3$ formed on the surface protects the rest of the metal. Dry or chemical corrosion is of three main types:

1. **Oxidation corrosion.** This type of corrosion is brought about by the direct chemical action of oxygen at low or high temperatures on metals, generally in the presence of moisture. Alkali and alkaline metals (*e.g.*, Na, Ca, Mg, etc.) suffer extensive oxidation even at low temperatures, whereas at high temperatures, practically all metals except Ag, Au, and Pt are oxidized. The reactions involved are:

$$2M \longrightarrow \underset{\text{metal ion}}{2M^{n+}} + 2ne^- \qquad \text{(Oxidation)}$$

$$nO_2 + 2ne^- \longrightarrow \underset{\text{oxide ion}}{2nO^{2-}} \qquad \text{(Reduction)}$$

$$2M + nO_2 \longrightarrow \underbrace{2M^{n+} + 2nO^{2-}}_{\text{Metal oxide}}$$

In the case of zinc and aluminium, the outward diffusion of metal is very great. As the temperature increases, the rate of diffusion of oxygen into the metal through oxide film increases. As a result, the temperature of the metal is increased and the metal undergoes melting or volatilization.

In high temperature oxidation of iron in dry air, three different oxide layers are formed on iron. $Fe_2O_3$ is the outermost layer, FeO is the innermost, and $Fe_3O_4$ is the intermediate layer. The reactions may be written as

$$4Fe + 3O_2 \longrightarrow 2Fe_2O_3$$

$$3Fe + 2O_2 \longrightarrow Fe_3O_4$$

$$2Fe + O_2 \longrightarrow 2FeO$$

The nature of the oxide film formed is very important in oxidation corrosion.

$$\text{Metal + Oxygen} \longrightarrow \text{Metal oxide (corrosion product)}$$

When oxidation starts, a thin layer of oxide is formed on the metal surface and the nature of this film decides the further action. The film may be

(*i*) *Stable.* A stable layer is fine-grained in structure and can adhere tightly to the parent metal surface. Hence, such a layer can cutoff penetration of the attacking oxygen to the underlying metal. Such a film is protective in nature, thereby shielding the metal surface. The oxide films on Al, Sn, Pb, Cu, Pt, etc., are stable, tightly adhering and impervious in nature. Consequently, further oxidation corrosion is prevented.

(*ii*) *Unstable.* The oxide layer formed, decomposes back into the metal and oxygen.

$$\text{Metal oxide} \rightleftharpoons \text{Metal + Oxygen}$$

Consequently, oxidation corrosion is not possible in such a case. Thus, Ag, Au, and Pt do not undergo oxidation corrosion.

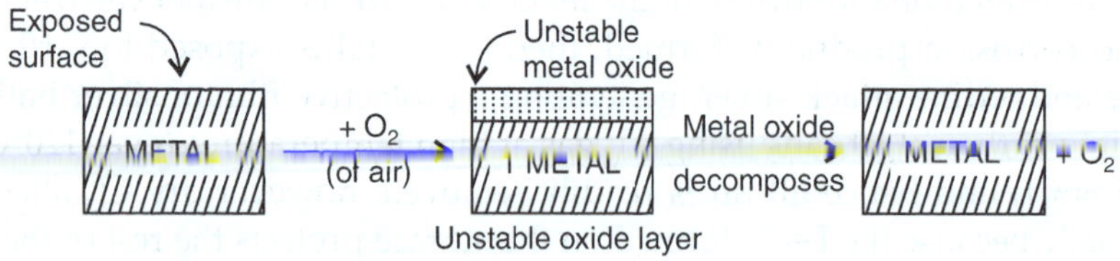


*FIGURE 9.12*

(*iii*) *Volatile.* Here the oxide layer volatilizes as soon as it is formed, thereby leaving the underlying metal surface exposed for further attack. This causes rapid and continuous corrosion, leading to excessive corrosion, *e.g.*, molybdenum oxide ($MoO_3$) is volatile.

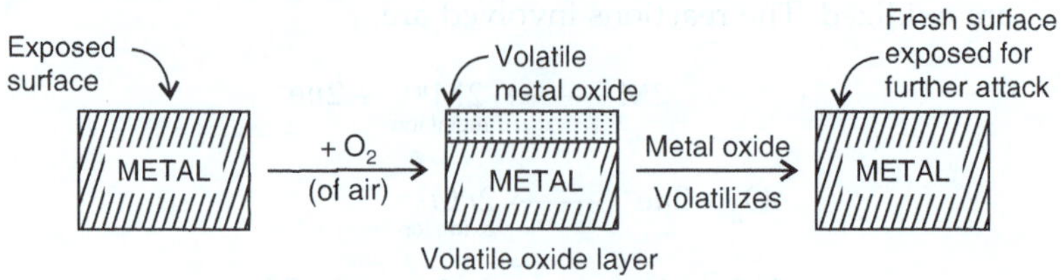


*FIGURE 9.13*

(*iv*) *Porous.* The oxide film formed has pores or cracks. In this case, the atmospheric oxygen has access to the underlying surface of metal, through the pores or cracks of the layer, thereby the corrosion continues unobstructed until the entire metal is completely converted into its oxide.

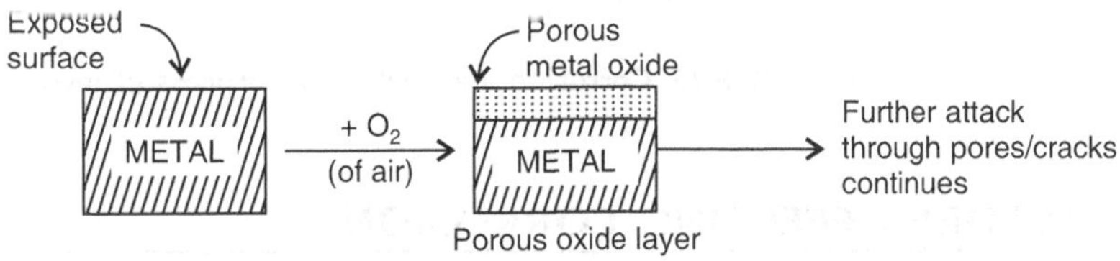


*FIGURE 9.14*

2. **Corrosion by other gases.** Gases like $CO_2$, $SO_2$, $Cl_2$, $H_2S$, $F_2$, etc. also cause corrosion of metals. The extent of corrosion depends mainly upon the chemical affinity between the metal and the gas involved. The degree of attack depends upon the formation of *protective* (*nonporous*) or *nonprotective* (*porous*) films on the metal surface.

   For example, the *AgCl Layer* formed from the attack of chlorine on silver metal is protective and nonporous. Therefore, the intensity of the attack decreases and protects the silver metal from further attack. When $Cl_2$ attacks tin it forms a *$SnCl_4$ layer* which gets volatilized from the surface of tin and leaves the fresh metal for further attack. In the petroleum industry, steel is attacked by $H_2S$ gas at high temperature and the *FeS scale* is formed. It interferes with the normal operation of the industry. The silver metal also gets tarnished by $H_2S$ gas in the presence of moisture.

$$Fe + H_2S \longrightarrow FeS + H_2$$

$$2Ag + H_2S \xrightarrow{H_2O} Ag_2S + H_2$$

3. **Liquid metal corrosion.** This type of corrosion is due to the chemical action of flowing liquid metal at high temperature on solid metal or alloy. It occurs in devices used for nuclear power. The corrosion involves dissolution of a solid metal or internal penetration of the liquid metal into the solid metal.

## 9.11 CAUSE OF CORROSION

Most metals in nature exist as compounds (ores) such as sulphides, sulphates, oxides, etc. because these compounds represent their thermodynamically stable state. The metals are extracted from these ores, and unless the nature of the metal is substantially changed (*e.g.*, by alloying), the metal will have a natural tendency to revert back to its natural thermodynamically stable state. However, metals existing in an elemental state (gold) naturally have excellent corrosion resistance.

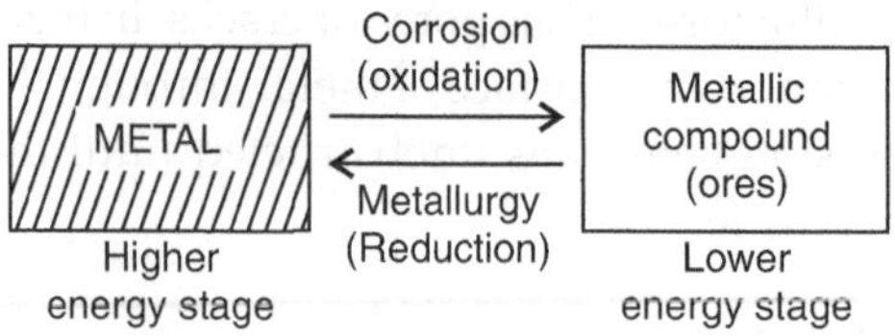


*FIGURE 9.15* **Corrosion and metallurgy process of metals.**

## 9.12 FACTORS AFFECTING CORROSION

The rate and extent of corrosion depends upon the following factors:

### 9.12.1 Nature of the Metal

(*i*) *Position in the galvanic series.* When two metals are in direct contact in presence of an electrolyte, the metal higher up in the galvanic series suffers corrosion. The rate and extent of corrosion depends upon the difference in the position of metals in the electrochemical series. The greater the difference, the faster is the rate of corrosion of the anodic metal.

(*ii*) *Over-voltage.* When a metal having a high position in a galvanic series (*e.g.*, zinc) is placed in 1N $H_2SO_4$, it undergoes corrosion, but the initial rate of reaction is quite slow because of high over-voltage (0.70 V) of zinc metal, which reduces the effective potential to a small value. However, if a few drops of $CuSO_4$ are added, the corrosion rate of zinc metal is accelerated because some copper gets deposited on the zinc metal, forming minute cathodes, where the hydrogen over-voltage is only 0.33 V. Thus, reduction in over-voltages of the corroding metal accelerates the corrosion rate.

(*iii*) *Relative areas of anodic and cathodic parts.* When two dissimilar metals are in contact, the corrosion of anodic part is directly proportional to the ratio of the areas of the cathodic part and the anodic part. Corrosion is more rapid and severe when the anodic area is small (*e.g.*, a small steel pipe fitted in a large copper tank), because the current density at a smaller anodic area is much greater and the demand for electrons (for cathode) can be met by smaller anodic only through a rapid corrosion rate.

(*iv*) *Purity of metal.* Impurities in a metal cause heterogeneity and form minute galvanic cells at exposed parts, thereby the anodic parts get corroded. For example, zinc metal having impurities Pb or Fe undergoes corrosion due to the formation of local galvanic cells. The rate and extent of corrosion increases with the increasing exposure and extent of impurities.

(*v*) *Solubility of the corrosion products.* In electrochemical corrosion, if the corrosion product is soluble in the corroding medium, then corrosion proceeds at a faster rate and *vice versa*.

(*vi*) *Volatility of the corrosion product exposes* the metal surface to further attack, and thus leads to excessive corrosion.

(*vii*) *Nature of the surface film.* All metals in the atmosphere get covered with a thin surface film of metal oxide. The ratio of the volumes of the metal oxide to the metal is known as *specific volume ratio.* The greater the specific volume ratio, the less is the rate of oxidation corrosion. The specific volume ratios of Ni, Cr, and W are 1.6, 2.0, and 5.6 respectively, thus tungsten get least corroded.

(*viii*) *Metals like Ti, Al, Mg, Ni, and Co* get corrosion passive and they exhibit a much higher position in the galvanic series, due to the formation of a highly protective but very thin film of oxide on the metal.

### 9.12.2 Nature of the Corroding Environment

(*i*) *Temperature.* The rate of chemical reaction and diffusion rate increases with the increase of temperature, thus corrosion rate is greatly influenced.

(*ii*) *Humidity of air.* Humidity is the deciding factor for atmospheric corrosion. Critical humidity is defined as *the relative humidity above which the atmospheric corrosion rate of metal increases sharply.* The value of critical humidity depends on the physical characteristics of the metal and the nature of corrosion products. Rate of corrosion increases in humid conditions because the gases $CO_2$, $O_2$ etc. and vapors present in the atmosphere furnish water to the electrolyte essential for setting up an electrochemical corrosion cell.

(*iii*) *Effect of pH.* Acidic media are generally more corrosive then alkaline and neutral media. Amphoteric metals like Al, Zn, and Pb cause corrosion of iron in oxygen free water at slower rate until *p*H falls below 5.0.

(*iv*) *Conductance of the medium.* This is an important factor for corrosion in under-ground or submerged structures because corrosion current depends on this factor. Conductance of clay and mineralized soils is higher than that of dry sandy soils, hence stray currents from power leakage will damage the metal structures burried under soils of higher conductance to a greater extent than those under dry sandy soils having higher resistance.

(*v*) Other factors influencing corrosion include, nature of anions and cations present, flow velocity of process stream, formation of oxygen concentration cells.

## 9.13 TYPES OF CORROSION

1. **General or Uniform corrosion.** This is the most common form of corrosion and accounts for the greater proportion of metal deterioration in terms of mass of metal corroded and cost. In this corrosion, there is a uniform decrease in the volume of metal and soluble corrosion products are formed. For example, oxidation and tarnishing of metals, active dissolution of metals in acids, passivity and chemical and electrochemical polishing.
2. **Pitting corrosion.** It is a localized attack resulting in the formation of holes in the otherwise relatively unattacked surface (Figure 9.16). The pitting corrosion may be due to the following reasons:

   (*i*) Heterogeneity in the metallic surface.

   (*ii*) Cracking of protective surface film on a metal at some points in the presence of an appropriate environment leading to concentrated attack at those points forming pits or holes.

(*iii*) Environment is not homogeneous.

Pitting is usually the result of the breakdown or cracking of the protective film on a metal at specific points by various mechanical factors or by chemical attack. For example, stainless steel and aluminium are affected by halide ions and produce characteristic pitting. It is a very destructive type of corrosion, which frequently ruins tubes or pipes or vessels. As a result of pitting, small anodic and large cathodic areas are formed and this results in specific corrosion. Pitting occurs mostly when chloride ion concentration is high.

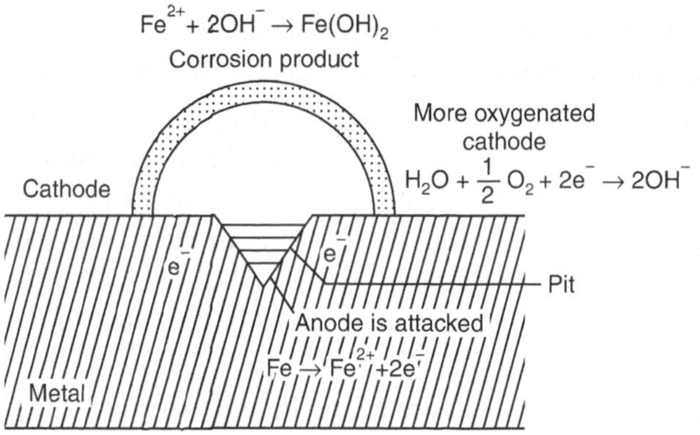


*FIGURE 9.16* **Pitting corrosion at the surface of metal.**

3. **Crevice corrosion.** It is a local corrosion and is created by dirt deposits, corrosion products, crack in paint coatings, etc. It ranges from small pits to extensive corrosion over the whole surface. The crevices may be formed due to:

   (*i*) The geometry of the structure.

   (*ii*) Contact of the metal with nonmetallic solids, *e.g.*, plastics, rubber, and glass.

   (*iii*) Deposits of sand, dirt or permeable corrosion products on the metal surface.

   Maintaining clean surfaces, selection of resistant materials, and proper design are the measures to control this type of corrosion.

4. **Intergranular corrosion.** When a molten metal is cast, the solidification starts at many randomly distributed nuclei, each of which grows in a regular atomic array to form grains. This type of corrosion takes place because of loss of coherence between the grains. It occurs along grain boundaries and leaves the grain interiors unattacked or only very slightly attacked. Grain boundaries are more susceptible for attack because these might have lost an element necessary for adequate corrosion resistance. Alloys are generally more susceptible to intergranular corrosion. During welding of stainless steel, chromium carbide gets precipitated at the grain boundaries, thus depleting the adjacent region of chromium which therefore become more anodic to the alloy composition and also to the precipitated chromium carbide. This renders the regions near the granular boundaries more susceptible to corrosion.

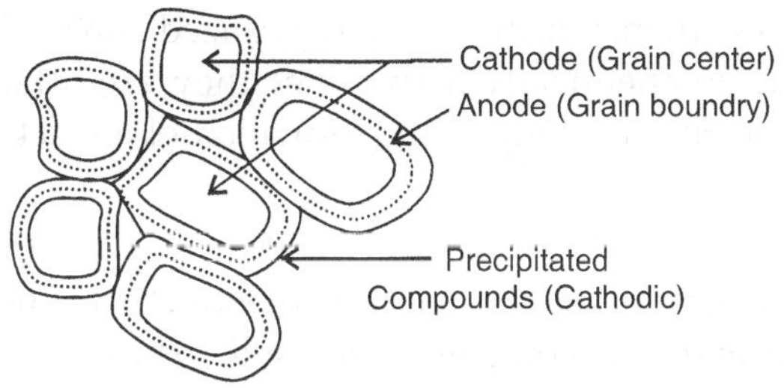


*FIGURE 9.17* **Intergranular corrosion.**

5. **Filiform corrosion.** It is characterized by the formation of a network of thread-like filaments of corrosion products on the surface of the metal coated with paint film when exposed to humid atmosphere. It is observed in steel, zinc, magnesium, and aluminium coated with lacquers and paints. It generally occurs when the relative humidity is above 60%. Between 65–80% relative humidity the filaments are very thin, at 80–95% they are much wider and at 95% they broaden to form blisters.
6. **Exfoliation and Selective leaching.** This type of corrosion involves the selective removal of one metal (electrochemically active) from certain alloys under certain environmental conditions. This results in either localized attack, perforation, or uniform attack and causes weakening of the strength of component. For example, selective leaching of zinc from brasses. Other metals such as Al, Fe, Ni and Cr can also be removed from their alloys. This type of corrosion is highly undesirable as it yields metal with poor mechanical qualities. This can be reduced by using nonsusceptible alloys.
7. **Waterline corrosion.** When water is stored in a steel tank, it is generally found that the maximum amount of corrosion takes place along a line just beneath the level of water meniscus. The area above the waterline is highly oxygenated and acts as the cathode and is completely unaffected by corrosion. This corrosion is of more importance to marine engineers. Use of special antifouling paints resists this kind of corrosion.
8. **Stress corrosion.** This corrosion is seen in fabricated articles of some alloys such as zinc brasses and nickel brasses. This is a combined effect of static tensile stress and corrosive environment on an alloy. It is characterized by highly localized attack and takes place when overall corrosion is almost negligible. The corrosion agents are highly specific and selective in cases of stress corrosion such as caustic alkalies and strong nitrate solution for mild steel, traces of ammonia for brass, and acid chloride solution for stainless steel. Stress corrosion can be divided into two types:
   (*i*) *Season cracking* is applied to stress corrosion of copper alloys, mainly brasses.
   (*ii*) *Caustic embrittlement* is a very dangerous form of stress corrosion occurring in mild steel exposed to alkaline solutions at high temperatures. It is generally seen in water boilers employed for industrial use. It can be prevented by addition of sodium sulphate to the boiler water and also using tannin and lignin. Stress corrosion is a localized electrochemical corrosion occurring along narrow paths, forming anodic areas with respect to the large cathodic areas at the metal surface. Presence of stress corrosion

produces strain, which results in localized zones of higher electrode potential. They become so chemically active that they are attacked even by a mild corrosive environment, resulting in the formation of a crack which grows and propagates in plants.

9. **Underground or Soil corrosion.** The corrosiveness of the soil depends upon the acidity of the soil, degree of aeration, electrical conductivity, moisture and salt contents, presence of bacteria, and micro-organisms and soil texture.

   Soil corrosion is purely electrochemical in character. In case of *gravel* or *sandy* soil corrosive conditions, and the corrosion rate is governed by the amount of moisture content of the soil. In *water logged* soil, various bacteria and micro-organisms can grow, leading to micro-biological corrosion. Fine gravel, sand, and salt containing soils may produce a localized and intense corrosion.

## 9.14 CORROSION CONTROL

Corrosion control methods may be classified into the following categories:

### 9.14.1 Material Selection and Design

1. Choice of the right type of material is the main factor for corrosion control. The selection of metal should be based on its chemical properties as well as its cost and structure.
2. Noble metals are most immune to corrosion but they cannot be used for economical reasons. The metal used should be as pure as possible. Presence of even minute amounts of some impurities may lead to severe corrosion. For example, minute quantities of iron in magnesium or lead in zinc in die-casting alloys are detrimental.
3. Strength and corrosion resistance may be improved by alloying. For example, stainless steel containing chromium forms an oxide film protecting it from further attack. Nimonic alloys (Ni-Cr-Mo) used in gas turbines are resistant to hot gases.
4. Heat treatment like annealing reduces internal stress and corrosion.
5. When two metals have to be in contact, they should be selected so that their oxidation potentials are as near as possible. Further, the area of more active metal, *i.e.*, the anode should be larger than that of the cathode or less active metal. A protective coating over the two metals reduces pitting.
6. If an active metal is used, it should be insulated from the more cathodic area.
7. Avoid crevices between assembled or joined materials. Welding is a better joining technique than soldering, for mechanical fastness. Galvanic cells develop in case of soldering or brazing.
8. The components should be designed in such a way that a stagnant pool of liquid should not be collected.
9. Moisture should be excluded wherever practicable; no corrosion occurs without the presence of moisture. The assembly should be designed such that the moisture should be as low as possible.

10. Sometimes it is not possible to control the corrosion, then the assembly should be designed in such a manner that the corroded part can easily and economically be replaced.

### 9.14.2 Protective Coatings

Protective coatings are used to isolate the anode and cathode regions. These protective coatings may be metallic, inorganic, nonmetallic, or of organic substances. For applying any type of coating,

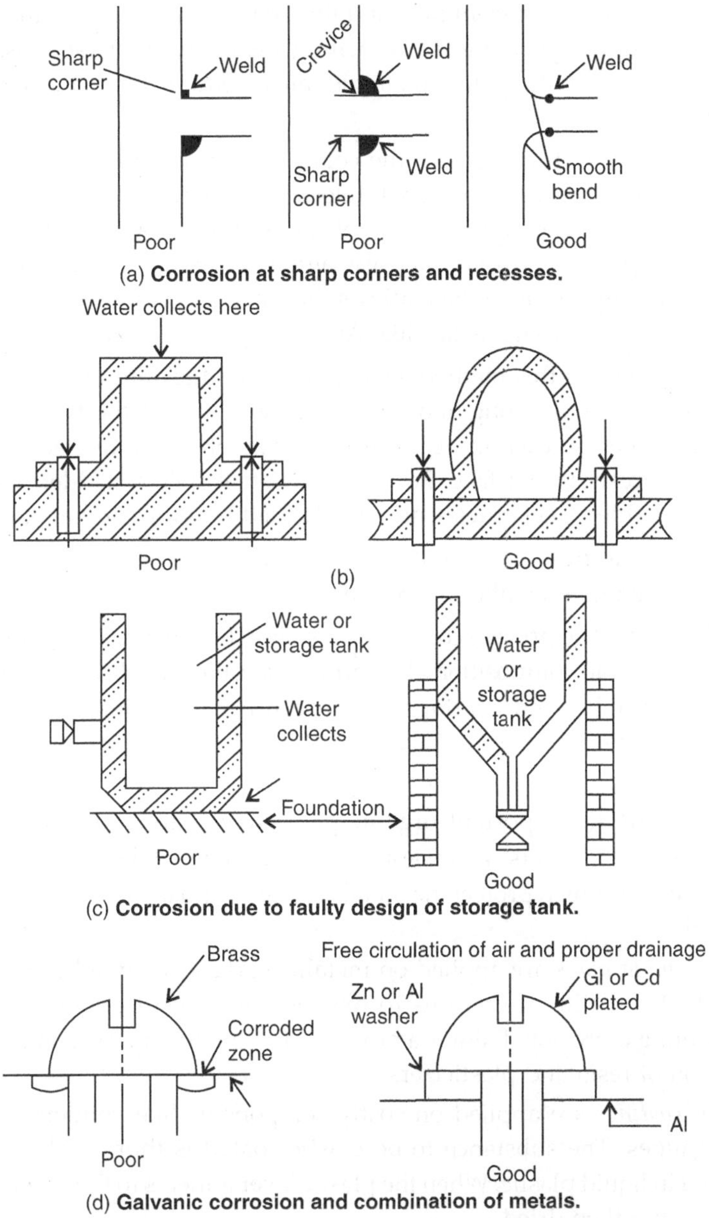


*FIGURE 9.18* **Corrosion control by material design and selection.**

the metal surface is prepared in the following manner: (*a*) grease and other surface contamination is removed; (*b*) oxide scales, rust, and corrosion products are removed; (*c*) etching treatment to aid in proper adhesion or buffing or polishing is given. Protective coatings are of the following two types:

1. **Metallic coatings.** These are the coatings of metals applied as a protective coating on center base metals. The metallic coatings are usually imparted by the following methods:
   (*i*) *Hot dipping.* In this method, the metal to be coated is dipped in the molten bath of the coating metal for sufficient time and then removed along with the adhering film. For example, applying coating of low melting metals and alloys such as, Zn, Pb, Sn, solder. The process providing zinc coating is called *galvanizing* and providing tin coating is called *tinning*.

   (*ii*) *Electroplating.* In this method, the coating metal is deposited on the base metal by passing a direct current through the electrolyte solution containing a salt of the coating metal. The coating material is made of anode and the base metal is the cathode. Plating depends upon the current intensity, agitation, temperature of the solution, and the chemical composition of the plating solution. For example, in steel, tin plating or zinc plating is applied. The metals like Au, Ag, Cr, Ni, Cu, Sn, etc. may be electroplated.

   (*iii*) *Metal cladding.* In this method a comparatively thick lining is placed upon the metal surface forming a strong alloy between the plates of the two metals. This band is formed either by casting or hot rolling. The objective of cladding is to produce a corrosion resistant surface. For example, steel sheets cladded with stainless steel, copper-cladded steel articles, and tin-cladded lead foils are prepared by this method.

2. **Nonmetallic coatings.** Nonmetallic coatings may be of organic or inorganic nature. The commonly used nonmetallic coatings are:
   (*i*) *Oxide coating.* The process consists of dipping the metallic compound in an electrolyte bath of suitable composition. The process is called *anodizing*. ZnO and MgO are the two commonly used coatings for preventing corrosion. In nuclear reactors, the zirconium alloy containing Sn, Fe, Cr, Ni is often used for protection of uranium fuel rods. The alloy is protected by $ZrO_2$ formed on it by controlled oxidation.

   (*ii*) *Organic coating.* Important organic protective coatings include paints, varnishes, enamels, and lacquers. When applied on cleaned metal surfaces, they act as effective inert barriers which protect the metal corrosion and also give decorative and aesthetic appeal.

   (*iii*) *Paints and lacquers* are applied on metallic surfaces to avoid contact of air with base metals. Paint film are porous to oxygen and moisture, but are fairly effective in preventing corrosion. Paint is a mixture of pigment, dry oil, and thinner. Lacquer is a solution of resin and plasticizers.

   (*iv*) *Plastic coatings* are applied on costly components like verniers, micrometers, tools, and gauges. The substance to be plastic coated is thoroughly cleaned, dried, and dipped in liquid plastic. When the plastic layer adheres to the substance, it is removed, cooled, and then dried.

(*v*) *Vitreous enamelling* is carried out by applying a dispersion of vitreous enamel in water containing silicate and clay on the base metal surface. The enamel coating has good resistance to dilute alkalies, acids, boiling water, and can withstand high temperature.

### 9.14.3 Cathodic and Anodic Protection

When it is impossible or impractical to alter the nature of the corrosion medium, corrosion control may be achieved by cathodic or anodic protection.

**Cathodic Protection.** The principle involved in cathodic protection is to force the metal to behave like a cathode. Since there will be no anode area on the metal, corrosion will not occur. It is applied in two ways:

1. *Sacrificial anode or galvanic protection.* In this method a more active metal is connected to the metal structure to be protected so that all the corrosion is concentrated at the more active metal, thus saving the metal structure (Figure 9.19). Generally, sacrificial anodes are made of magnesium or zinc or aluminium alloys. The sacrificial anode, typically zinc or magnesium, is consumed requires replacement. Important applications of cathodic protection include:

   (*i*) Protection from soil corrosion of underground cables and pipelines.

   (*ii*) Protection from marine corrosion of cables, ship hulls, etc.; offshore drilling platforms, and water heaters.

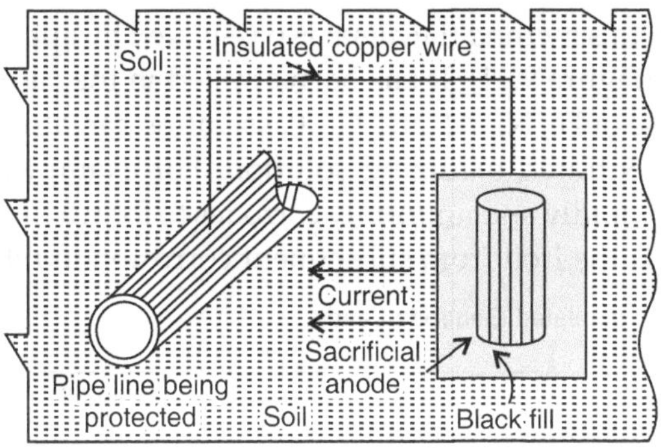


*FIGURE 9.19* **Galvanic protection using sacrificial anode.**

2. *Impressed current cathodic protection.* In this method an impressed current is applied to convert the corroding metal from anode to cathode. This is done by applying a sufficient amount of direct current (DC) from a DC source to an anode (*e.g.*, graphite or high silicon iron) buried in the soil or immersed in the corrosion medium and connected to the corroding metal structure which is to be protected (Figure 9.20). The electrons flow to the metals, as a result the metal acts as a cathode and the auxiliary anode, such as scrap iron, corrodes. This type of protection is of particular value in case of buried structures such as tanks and pipelines, transmission line towers, marine piers, laid-up ships, etc.

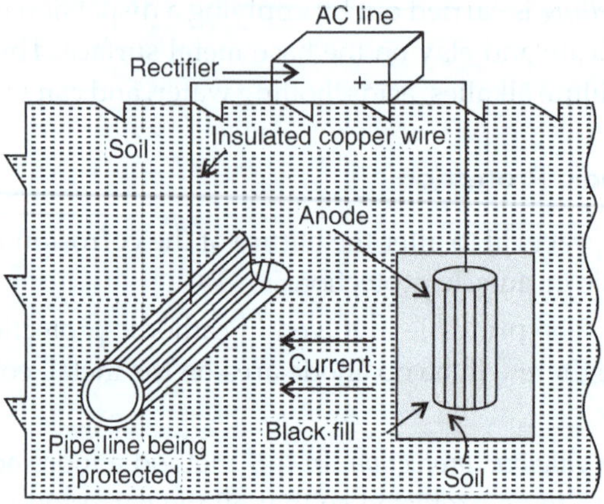


*FIGURE 9.20* **Impressed current cathodic protection.**

**Limitations with Cathodic Protection**

(*i*) A cathodic protection system may increase the corrosion of an adjacent pipeline or an adjacent metal structure because of strong currents.

(*ii*) Capital investment and maintenance costs are quite high.

(*iii*) The structure should not be over protected as it may lead to evolution of $H_2$ or formation and accumulation of $OH^-$ ions.

**Anodic Protection.** In this method the metal is passivated by applying a current in a direction that renders it more anodic (Figure 9.21). It has been applied in the case of steel and stainless steel and to some extent in the case of Fe, Ni, Al, and Cr. Passivation can also be introduced by exposing the metals to a highly concentrated oxidizing solution. For example, if iron is dipped in very concentrated $HNO_3$, the iron rapidly and uniformly corrodes to form a thin, protective iron hydroxide coating. The coating prevents the iron from subsequent corrosion in nitric acid.

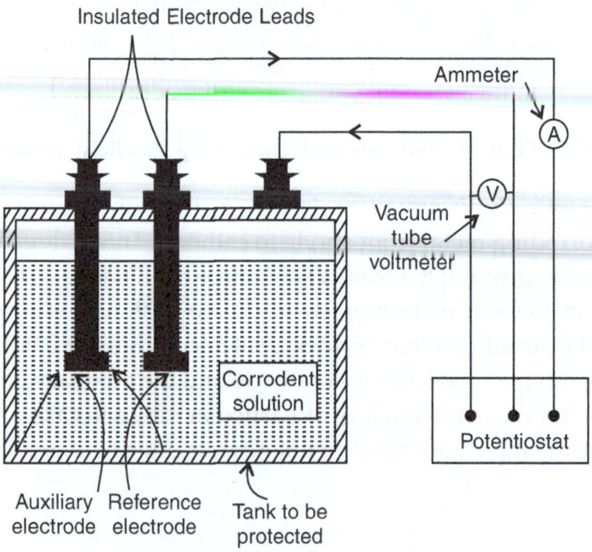


*FIGURE 9.21* **Anodic protection system for tanks.**

### 9.14.4 Corrosion Inhibitors

Inhibitors are the substances which, when added to the electrolyte solution, reduce the rate of corrosion. These are of two major classes:

1. *Inorganic inhibitors.* Inorganic inhibitors such as silicates, chromates, borates, etc. suppress the rate of corrosion by acting on the anode. Alkaline sodium nitrite alone or in combination with other inhibitors such as phosphate has been used to control the corrosion of tankers and pipelines. Sodium benzoate has been used as an inhibitor for mild steel and in preventing corrosion in cooling systems such as automobile radiators. A mixture containing 0.1% $NaNO_2$ and 1.5% sodium benzoate has been used as an inhibitor in anti-freeze solutions. Lime acts as a cathodic inhibitor by precipitating $CaCO_3$ in water containing temporary hardness or dissolved $CO_2$. Colloidal particles of $CaCO_3$ having a positive charge are attracted to the cathodic areas and get deposited there, thus reducing the corrosion current.
2. *Organic inhibitors.* Organic inhibitors act by different mechanisms, *e.g.*, organic colloidal inhibitors form protective layers on the metal surface by adsorption. Surface active reagents known as surfactants containing polar groups promote spreading and oriented adhesion to the surface, thus forming a protective film. Organic bases, *e.g.*, amines, pyridines, quinoline, and their derivatives contain a hydrophobic group or radicals. These positively charged cationic groups attach themselves to the cathodic areas through the nitrogen and provide inhibition. The efficiency of inhibition depends upon the size and number of the alkyl groups, *e.g.*, primary amyl amine ($C_5H_{11}NH_2$) has been found to be more effective an inhibitor than primary ethyl amine ($C_2H_5NH_2$).

   High molecular weight amines derived from rosin have been used as corrosion inhibitors. The product obtained by rosin amine and pentachlorophenol was used in castings for underground pipelines to protect them from corrosion by soil bacteria. Organic inhibitors such as soluble salts of rosin, *e.g.*, stearate or naphthenate, have been used as inhibitors in the various metal cleaning operations like acid pickling of metals, cleaning of boilers, condensers, heat exchangers, chemical equipment, pipelines, etc., to prevent undue and excess corrosion of the metal while removing rust, scales, and deposits.

   Some vapor phase inhibitors (VPI) such as dicyclohexylammonium nitrite and cyclohexylamine carbonate are used to inhibit the corrosion in storing aircraft engines as well as during storage, packing, shipping, etc.

## 9.15 PASSIVITY

It is defined as a *phenomenon in which metal or an alloy exhibits a much higher corrosion resistance than its position in the electrochemical series would indicate.* This is due to the formation of a thin (0.0004 mm thick) but highly protective film of oxide on the surface of the metal or alloy. The passivated metal can be rendered active by a change in the environmental conditions.

Metals which exhibit passivity include iron, chromium, nickel, titanium, and alloys of these metals. Passivation is usually associated with an oxidizing environment and is characterized by the formation of a very thin protective oxide film. However, it should be emphasized that a metal which is passive in one environment may not necessarily be so in a different environment because passivity results from a continuing reaction at a very slow rate. Hence, passivity offers an excellent method of corrosion control.

The passivation is not static in nature but is quite stable under certain environmental conditions such as oxidizing conditions, *e.g.*, the metals or alloys of Al, Cr, Ti, etc., exhibit high corrosion resistance in oxidizing environments, but are quite unstable in reducing environments. The passivity of some metals are in the order of:

$$\text{Ti} \longrightarrow \text{Al} \longrightarrow \text{Cr} \longrightarrow \text{Be} \longrightarrow \text{Mo} \longrightarrow \text{Mg} \longrightarrow \text{Ni} \longrightarrow \text{Co} \longrightarrow$$
$$\text{Fe} \longrightarrow \text{Mn} \longrightarrow \text{Zn} \longrightarrow \text{Cd} \longrightarrow \text{Sn} \longrightarrow \text{Pb} \longrightarrow \text{Cu}.$$

The presence of oxide films may affect the initial rate of attack by acids. Iron dissolves readily in very dilute $HNO_3$, but as the concentration of the acid is raised, its oxidizing power rises and a point is reached at which the acid directly oxidizes the metal to its oxide. The layer of the oxide formed on the surface makes the iron passive. The passive iron has a high, positive electrode potential and is not easily corroded. The oxidized film is easily damaged by scratching, heating, or because of the presence of chloride ions. Under such conditions, the metal continues to corrode. However some passive films are not so easily damaged as that of iron, and even if there are cracks in the film, these films can readily heal themselves in an oxidizing atmosphere, for example stainless steel composed of 18% Cr and 8% Ni. The highly protective coating of $Cr_2O_3$ is believed to be responsible for the corrosion resistance. Other examples of increased resistance to oxidation by alloying are silicon containing cast iron and silicon containing molybdenum.

### 9.15.1 Oxide or Protective Layer Theory

According to this theory, the passivity of metals is due to formation of an invisible oxide film on the surface. The oxide layer so formed is not soluble in acids. The thickness of this film depends on the treatment used to produce it. This theory is supported by the following facts:

(*i*) Removal of this film either mechanically or chemically makes the metal active again.

(*ii*) When passive iron is heated in nitric acid at 75°C, it becomes active again. This is because at 75°C the $Fe_2O_3$ layer begins to dissolve.

(*iii*) Certain metals like lead derive protection from atmospheric corrosion in the presence of a film of oxide.

It is the oxide film that is responsible for the occurrence of passivity. But in some cases films other than those of oxides are responsible for the passivity. Therefore, any protective film produced during the chemical reaction is sufficient to cause passivity and thus the theory is termed as *protective layer theory.*

The oxide film is extremely thin, transparent, and invisible so long as it is in contact with the bright reflecting metallic surface. During the formation, it is assumed that oxygen reacts extremely fast with the metal and the rate of film growth is inversely proportional to the thickness of the film. Therefore,

$$\frac{dy}{dt} = \frac{k}{y}$$

where, $y$ = film thickness, $t$ = time, and $k$ = constant.

On integration, we get

$$y^2 = 2kt + I$$

where, I = integration constant.

Experimentally it is observed that the rate of film growth does not obey the above equation. In other words the film is not necessarily uniform throughout its thickness.

## QUESTION BANK

### Electrochemistry

1. Describe the construction of a simple electrochemical cell. Indicate the positive and negative electrodes? What are the reactions taking place at these electrodes? Give the overall reaction and cell diagram.
2. (*i*) What is the basic difference between the voltage and Emf of the cell? Explain.
   (*ii*) Define electrode potential. How is it different from liquid junction potential? How and why is the later eliminated?
   (*iii*) What will be the reaction in the cell which is made by combining standard Cd half-cell with standard Cu half-cell? Also calculate the Emf of the cell. The standard oxidation potential of half-cells are 0.403 and – 0.337 volt.
3. (*i*) Give an example each for gas/ion and redox electrodes. Write also the reversible reactions occurring in them.
   (*ii*) To which type of reversible electrode does the calomel electrode belong? Write the reduction reaction occurring in it and the expression for its reduction potential.
4. (*i*) Differentiate between the following giving suitable examples.
   (*a*) Electrochemical and electrolytic cells.
   (*b*) Reversible and irreversible cells.
   (*ii*) What are fuel cells? Describe construction and working of a hydrogen oxygen fuel cell with diagram.
5. (*i*) What is a reference electrode? Describe the construction of normal hydrogen electrode.
   (*ii*) Write the (*a*) electrode reaction: (*b*) cell reaction and calculate the Emf of the given cell:

$$Ag(s) \mid Ag^+ (a = 0.1) \parallel Zn^{2+} (a = 0.1) \mid Zn$$

$$E^\circ_{Ag/Ag^+} = 0.799\text{ V};\ E^\circ_{Zn^{2+}/Zn} = 0.763\text{ V}$$

   Is the reaction spontaneous? Give reasons.
6. Represent the cell according to conventions which is obtained by connecting a standard Zn half-cell with a standard hydrogen electrode.
7. (*i*) Define the construction and working of a quinhydrone electrode.
   (*ii*) Write cell reactions for the following electrochemical cells:
   (*a*) $Cd \mid Cd^{2+} \parallel KCl \mid Hg_2Cl_2 \mid Hg$
   (*b*) $Pt \mid Tl^+, Tl^{3+} \parallel Cu^{2+} \mid Cu$

(*c*) Pb | $PbSO_4$(*s*) $SO_4^{2-}$ || $Cu^{2+}$ | Cu

(*d*) Zn | $Zn^{2+}$ || $Fe^{3+}$, $Fe^{2+}$ | Pt.

8. (*i*) Why is a salt bridge used? What is the parameter of utmost importance when an electrolyte is chosen or construction of a salt bridge?

   (*ii*) Explain what is implied by reversibility of an electrode with respect to same ion? Illustrate your answer with reference to $H^+$ and $Cl^-$ reversible electrodes.

9. (*i*) What do you understand by single electrode potential? How can it be measured experimentally?

   (*ii*) Write the cell reaction for the following cell:

$$Zn \mid Z^{++}(a = 0.1) \parallel Ag^+ (a = 0.01) \mid Ag.$$

10. (*i*) What are different types of electrodes? Give examples of each type. Discuss construction, working and use of hydrogen electrode.

    (*ii*) How will you determine activity and activity coefficient from cell potential measure- ments?

11. (*i*) Write a concentration cell with transference. Derive the mathematical expression for its Emf.

    (*ii*) How will you decide whether the mercurous ion is to be represented as $Hg_2^{++}$ for $Hg^+$ using concentration cell?

    (*iii*) How can you find the *p*H of a solution using

    (*a*) hydrogen electrode and (*b*) quinhydrone electrode?

    Explain the merits and demerits of these methods.

12. (*i*) Explain in terms: (*a*) diffusion current and (*b*) half-wave potential.

    (*ii*) Draw and discuss a diagramatic sketch of the electrode assembly for determining half-wave potential.

    (*iii*) Write briefly a quinone-hydroquinone and saturated KCl calomel electrode. How can *p*H of a buffer solution be determined using the above mentioned electrode? Give complete experimental details.

    (*iv*) Give details of a titration method involving open electrode by which dissociation constant of a weak acid can be determined.

13. (*i*) Describe the concentration cell method of determining the solubility product of a sparingly soluble salt in water.

    (*ii*) What is meant by hydrogen over voltage? Mention two of its applications.

14. Calculate the Emf of voltaic cell: Fe | $Fe^{2+}$(*aq*) || $Cu^{2+}$(*aq*) | Cu,

    given the electrode potentials of copper and ion are 0.34 volt and – 0.44 volt respectively.

15. Calculate the standard Emf of the following cell at 25°C, writing its half-cell reaction and net cell reaction:

$$Zn \mid ZnSO_4 \parallel CuSO_4 \mid Cu$$

    Standard potentials of copper and zinc electrodes are + 0.34 volt and – 0.76 volt respectively.

16. What will be the Emf of the cell at 25°C?

$$Zn \mid Zn^{2+}\ (0.001\ M) \parallel Fe^{2+}\ (0.005\ M) \mid Fe,$$

    given that E°($Zn^{2+}$. Zn) = – 0.76 V and E° ($Fe^{2+}$. Fe) = – 0.44 V and mention the polarity of the electrodes.

17. The standard Emf of the cell:

$$Pt \mid H_2(g)\ 1\ atm \mid HCl(aq) \parallel Hg_2Cl_2(s) \mid Hg(l)$$

is found to be 0.26699 V at 293 K and 0.2669 V at 303 K. Calculate the standard Gibbs function, enthalpy and entropy of the reaction at 293 K.

18. (*i*) Calculate the free energy change of the following cell at 25°C

$$Sn \mid Sn^{++}\ (0.5\ N) \parallel Pb^{++}\ (0.03\ N) \mid Pb$$

The standard Emf of the cell is 0.14 V.

(*ii*) Define electrode potential. Describe calomel electrode.

19. For the following cell:

$$Cr(s) \mid Cr^{3+}(aq)\ (0.5\ M) \parallel Fe^{2+}(aq)\ (0.\ 8\ M) \mid Fe$$

Given $E^\circ_{Fe^{2+}/Fe} = -\ 0.71$ V at 298 K. Calculate $\Delta G^\circ$ for the cell reaction at 298 K.

20. Write the half-cell and net cell reactions of the following cell:

$$Zn \mid Zn^{2+}\ (1\ M) \parallel Cu^{2+}\ (1\ M) \mid Cu$$

Find the Emf of the above cell, given: $E^\circ_{Zn^{2+}/Zn} = -\ 0.76$ volt and $E^\circ_{Cu^{2+}/Cu} = 0.34$ volt.

21. Will a reaction take place, if a silver bar is placed in a solution of $FeSO_4$? Explain using the following data:

$$E^\circ_{Ag^+/Ag} = 0.799\ V, \text{ and } E^\circ_{Fe^{2+}/Fe} = -\ 0.44\ V.$$

22. Calculate the single electrode potential of cadmium electrode dipped in 0.025 M $CdSO_4$ solution at 25°C. Standard reduction potential of cadmium electrode is – 0.40 V at 25°C.

23. What is the single electrode potential of a half-cell for zinc electrode dipping in a 0.01 M zinc sulphate solution at 28°C? Standard electrode potential of zinc = – 0.763 V, R = 8.313 $JK^{-1}\ mol^{-1}$, F = 96,500 C $mol^{-1}$.

24. Zinc rod is dipped in 0.01 M $ZnSO_4$ solution at 298 K. What is single electrode potential of Zn half-cell, if $E^\circ_{cell} = +\ 0.763$ V?

25. Calculate the potential of a copper electrode dipped in a 0.1 M solution of copper sulphate at 298 K. The standard electrode potential for $Cu^{2+}/Cu$ is + 0.34 V at 298 K.

26. Calculate the Emf of the cell at 25°C:

$$Cu \mid Cu^{2+}\ (0.01\ M) \mid Cu^{2+}\ (0.1\ M) \mid Cu.$$

27. Consider the cell: Fe | $Fe^{2+}$ (0.01 M) || $Cu^{2+}$ (0.5 M) | Cu. The standard electrode potential of iron and copper are – 0.44 V and + 0.34 V respectively. Write the cell reaction and calculate the Emf of the cell.

## Corrosion

28. What is corrosion of metals? Explain the basic reason of metallic corrosion.
29. Describe briefly the consequences of corrosion.
30. Define metallic corrosion. Explain electrochemical theory of corrosion.
31. State and explain Pilling-Bedworth rule.
32. Discuss in detail chemical corrosion.
33. Differentiate between chemical and electrochemical corrosion with suitable examples.
34. What is stress corrosion? Give two examples. How can it be controlled?
35. Discuss any two factors which influence the corrosion rate in detail.

36. Why does corrosion occur in steel pipes connected to copper plumbing? Explain.
37. Discuss briefly (*a*) Soil corrosion (*b*) Galvanic corrosion (*c*) Pitting corrosion (*d*) Crevice corrosion.
38. Explain briefly the various factors influencing corrosion.
39. Silver and copper metals do not undergo corrosion like iron in a moist atmosphere. Explain.
40. Explain methods of prevention of corrosion.
41. Discuss briefly corrosion inhibitors.
42. What is corrosion? Briefly discuss the various methods employed for protection of metals from corrosion.
43. How are metals protected against corrosion by modifying the environment?
44. How will you protect an underground pipe line from corrosion by sacrificial anodic and impressed current cathodic protection methods?
45. What are cathodic and anodic protection methods for controlling corrosion? Discuss their merits and demerits.
46. Discuss any four methods of corrosion control.
47. Distinguish between anodic and cathodic inhibitors.
48. What are corrosion inhibitors? Explain with examples how anodic and cathodic inhibitors provide protection against corrosion.
49. Explain rusting of iron with the help of electrochemical theory of corrosion.
50. Discuss briefly passivity and its significance.
51. Discuss briefly various types of corrosion.
52. Distinguish between anodic and cathodic coating.
53. Explain the preliminary treatment given to a metallic surface before coating.
54. Discuss the importance of design and material selection in controlling corrosion.
55. What is meant by dry corrosion?
56. Discuss briefly the nature of dry corrosion product.
57. Discuss liquid metal corrosion briefly.
58. How does the physical state of the metal influence the corrosion rate?
59. Discuss in detail the nature of a corroding environment.
60. Write short notes on: (*a*) corrosion inhibitors; (*b*) passivity.

Chapter 10 
# WATER AND WATER TREATMENT

## 10.1 HARDNESS OF WATER

*The soap consuming capacity of water is known as the hardness of water.* It is generally caused by calcium and magnesium ions present in water. Polyvalent ions of metals like strontium, iron, aluminium, zinc, and manganese etc., are also capable of precipitating the soap and thus contribute to hardness. However, the concentration of these ions is very low in natural waters, therefore hardness is generally measured as a concentration of only calcium and magnesium ions as calcium carbonate.

### 10.1.1 Types of Water Hardness

The water hardness may be classified as:

1. *Temporary hardness.* The hardness that can be removed by boiling the water is called *temporary hardness.* It is caused by the presence of bicarbonates of calcium and magnesium. On boiling, bicarbonates are decomposed into insoluble carbonates.

$$Ca(HCO_3)_2 \xrightarrow{\text{Heat}} \underset{\text{Insoluble}}{CaCO_3} \downarrow + H_2O + CO_2$$

$$Mg(HCO_3)_2 \xrightarrow{\text{Heat}} \underset{\text{Insoluble}}{MgCO_3} \downarrow + H_2O + CO_2$$

   The insoluble carbonates can be easily removed by filtration. Thus, boiling not only removes $Ca^{2+}$ and $Mg^{2+}$ ions but also decreases the total ionic content of the water.

2. *Permanent hardness.* This is due to the presence of sulphates, chlorides, and nitrates of calcium, magnesium, and other metals and cannot be removed by boiling.

The hardness is determined by *EDTA method.* Calcium and magnesium form a complex of wine red color with EBT (Eriochrome black-T) at $pH\,10 \pm 0.1$. The EDTA (Ethylene diamine tetracetic acid) has stronger affinity for $Ca^{2+}$ and $Mg^{2+}$ ions, and therefore by addition of EDTA, the blue color of the indicator (EBT) is restored to give the end point. The reactions involved are:

$$Ca^{2+} + \underset{\text{EDTA}}{H_2Y^{2-}} \longrightarrow CaY^{2-} + 2H^+$$

$$Mg^{2+} + H_2Y^{2-} \longrightarrow MgY^{2-} + 2H^+$$

$$\text{Hardness as mg/L } CaCO_3 = \frac{\text{ml EDTA used} \times 1000 \times E}{\text{ml sample taken for titration}}$$

where E is mg $CaCO_3$ equivalent to 1.0 ml EDTA.

### 10.1.2 Units of Water Hardness

The hardness of water is expressed in the following units:

(*i*) *Milligram per liter (mg/l).* It is the number of milligrams of $CaCO_3$ dissolved in the 1L of water.

$$1 \text{ mg/liter} = 1 \text{ mg } CaCO_3 \text{ equivalent in 1L of water.}$$

(*ii*) *Parts per million.* It is the number of parts by weight of calcium carbonate present per million parts by weight of water.

$$1 \text{ ppm} = 1 \text{ part of } CaCO_3 \text{ in } 10^6 \text{ parts of water.}$$

(*iii*) *Degree Clark (°Cl).* It is the number of grains of $CaCO_3$ present per gallon of water.

$$1° \text{ Clark} = 1 \text{ grains of } CaCO_3 \text{ per gallon of water}$$

$$= 1 \text{ part of } CaCO_3 \text{ per 70 000 parts of water.}$$

(*iv*) *Degree French (°Fr).* It is the part of $CaCO_3$ equivalent hardness per lac parts of water.

$$1° \text{ Fr} = 1 \text{ part of } CaCO_3 \text{ equivalent per } 10^5 \text{ parts of water.}$$

### 10.1.3 Effects of Water Hardness

Following are the harmful effects of hard water:

(*i*) Hard water is harmful for potable purposes. It causes deposition of calcium in the bone joints and the possibility of forming calcium oxalate crystals in urinary tracts is increased.

(*ii*) It causes many problems in boilers when it is directly fed for steam generation. The troubles include scale and sludge formation, corrosion, priming and foaming and caustic embrittlement.

(*iii*) It does not form lather with soaps or detergents. On the other hand it produces precipitates of calcium and magnesium soaps. The formation of such insoluble, sticky precipitates continues until all calcium and magnesium salt present in water are precipitated. After that, the soap (sodium stearate) gives lather with water. The reaction involved is:

$$\underset{\text{soap}}{C_{17}H_{35}COONa} + H_2O \rightleftharpoons C_{17}H_{35}COOH + NaOH$$

$$\underset{\text{stearic acid}}{C_{17}H_{35}COOH} + \underset{\text{soap}}{C_{17}H_{35}COONa} \longrightarrow \text{Lather}$$

This causes wastage as a lot of soap is being used.

(*iv*) Due to the presence of dissolved hardness producing salts, the boiling point of water is elevated. Consequently, more fuel and time are required for cooking.

### 10.1.4 Equivalents of $CaCO_3$ Concept

The concentration of hardness and nonhardness constituting impurities are usually expressed in terms of equivalent amount of $CaCO_3$. Calcium carbonate is taken as a standard substance due to the following two reasons:

(*i*) The molecular weight of $CaCO_3$ is 100 (equivalent weight = 50) which makes all the calculations simple and quick.

(*ii*) It is the most insoluble salt that can be precipitated in water treatment.

All the substances causing hardness should be converted into their respective $CaCO_3$ equivalent, as a matter of convention and convenience.

$$CaCO_3 \text{ equivalent of a hardness causing impurity} = \frac{\text{Wt. of the impurity}}{\begin{array}{c}\text{Chemical equivalent}\\ \text{of the impurity}\end{array}} \times 50$$

(since chemical equivalent weight of $CaCO_3$ = 50).

For instance, 136 parts by weight of $CaSO_4$ would contain the same amount of Ca as that of 100 parts by weight $CaCO_3$. Hence, in order to convert the weight of $CaSO_4$ as its $CaCO_3$ equivalent, the weight of $CaSO_4$ should be multiplied by a factor of $\frac{100}{136}$ or $\frac{50}{68}$.

Conversion factors for some of the commonly occurring impurities in water are given in Table 10.1.

**TABLE 10.1 Calculation of Equivalents of $CaCO_3$**

| *Salt/Ion* | *Molar Mass* | *Chemical Equivalent* | *Multiplication Factor to Convert into its $CaCO_3$ Equivalent* |
|---|---|---|---|
| $Ca(HCO_3)_2$ | 162 | 81 | 100/162 |
| $Mg(HCO_3)_2$ | 146 | 73 | 100/146 |
| $CaSO_4$ | 136 | 68 | 100/136 |
| $CaCl_2$ | 111 | 55.5 | 100/111 |
| $MgSO_4$ | 120 | 60 | 100/120 |
| $MgCl_2$ | 95 | 47.5 | 100/95 |

| | | | |
|---|---|---|---|
| $Mg(NO_3)_2$ | 148 | 74 | 100/148 |
| $Ca^{2+}$ | 40 | 40 | 100/40 |
| $Mg^{2+}$ | 24 | 24 | 100/24 |
| $HCO_3^-$ | 61 | 61 | $\frac{100}{61\times 2}$ |
| HCl | 36.5 | 36.5 | $\frac{100}{36.5\times 2}$ |
| $H_2SO_4$ | 98 | 49 | 100/98 |
| $CO_2$ | 44 | 22 | 100/44 |
| $Al_2(SO_4)_2$ | 342 | 57 | 100/114 |
| $FeSO_4.7H_2O$ | 278 | 139 | 100/278 |
| $CaCO_3$ | 100 | 50 | 100/100 |
| $MgCO_3$ | 84 | 42 | 100/84 |
| $NaAlO_2$ | 82 | 82 | $\frac{100}{82\times 2}$ |
| $OH^-$ | 17 | 17 | 100/34 |
| $CO_3^{2-}$ | 60 | 30 | 100/60 |
| $H^+$ | 1 | 1 | 100/2 |

**Example 1.** *20 mL of 0.1 $Na_2CO_3$ solution was added to 100 mL of a sample of hard water. The filtrate from the solution required 30 mL of 0.05N $H_2SO_4$ for complete neutralization. Calculate the hardness of the water sample.*

**Solution.** The filtrate ≡ 30 mL of 0.05 $H_2SO_4$ = $\frac{30 \times 0.05}{0.1}$ mL of 0.1 N $H_2SO_4$.

15 mL of 0.1N $Na_2CO_3$ solution consumed in precipitating hardness in 100 mL of hard water = 20 – 15 = 5 mL. So,

100 mL of water sample ≡ 5 mL of 0.1N $Na_2CO_3$ solution

or $$100 \times N_{\text{Hardness}} = 5 \times 0.1$$

or $$N_{\text{Hardness}} = \frac{5 \times 0.1}{100} = 0.005$$

Strength of hardness as $CaCO_3$ equivalents = 0.005 × 50 g. per liter

$$= \frac{0.005 \times 50 \times 10^6}{1000} \text{ mg/liter} = \mathbf{250\ ppm.}$$

**Example 2.** *100 ml of tap water was titrated with N/50 HCl with methyl orange as an indicator. If 30 ml of HCl were required, calculate the hardness as parts of $CaCO_3$ per 100 000 parts of water. The hardness is temporary.*

**Solution.** For the temporary hardness,

$$Ca(HCO_3) + 2HCl \longrightarrow CaCl_2 + 2CO_2 + 2H_2O$$

Thus 2 equivalents of HCl ≡ $Ca(HCO_3)_2$ ≡ $CaCO_3$ (since 100 parts)

$$CaCO_3 + H_2O + CO_2 = Ca(HCO_3)_2$$

Therefore, 1 equivalent of HCl = 50 gm $CaCO_3$

Now 100 ml of tap water = 30 ml of N/50 HCl = $\frac{30}{50}$ ml of N HCl

$$= \frac{30 \times 0.05}{50} = 0.03 \text{ gm } CaCO_3$$

Therefore, 10,000 ml tap water contains **30 gm of $CaCO_3$.**

**Example 3.** *How many grams of $FeSO_4$ dissolved per liter gives 210.5 ppm of hardness? (Fe = 56, S = 32, O = 16, Ca = 40, C = 12).*

**Solution.** $FeSO_4$ ≡ $CaCO_3$

56 + 32 + 64 = 152 mg  100 g

∴ 100 ppm of hardness= 152 ppm of $FeSO_4$

$$210.5 \text{ ppm of hardness} = \frac{152 \times 210.5}{100} = 319.96 \text{ ppm of } FeSO_4$$

= **319.96 mg/L** or 0.3199 g/L of $FeSO_4$.

**Example 4.** *A sample of water is found to contain 40.5 mg/L $Ca(HCO_3)_2$ 45.5 mg/L $Mg(HCO_3)_2$; 27.6 Mg/L $MgSO_4$, 32.1 mg/L $CaSO_4$ and 22.45 mg/L $CaCl_2$. Calculate the total hardness of water (At. wts. Ca = 40, Mg = 35.5, C = 12, S = 32, O = 16, H = 1).*

**Solution.** Conversion into $CaCO_3$ equivalents:

| *Constituent* | *Amount (mg/L)* | *Multiplication Factor* | *$CaCO_3$ Equivalent (mg/L)* |
|---|---|---|---|
| $Ca(HCO_3)_2$ | 40.5 | 100/162 | 40.5 × 100/162 = 25.0 |
| $Mg(HCO_3)_2$ | 46.5 | 100/146 | 46.5 × 100/146 = 31.85 |
| $MgSO_4$ | 27.6 | 100/120 | 27.6 × 100/120 = 23.00 |
| $CaSO_4$ | 32.1 | 100/136 | 32.1 × 100/136 = 23.60 |
| $CaCl_2$ | 22.45 | 100/111 | 22.45 × 100/111 = 20.26 |

∴ Total hardness of water = [25.0 + 31.85 + 23.00 + 23.60 + 20.26] = **123.71 mg/L.**

**Example 5.** *100 mL of water sample has a hardness equivalent of 12.5 mL of 0.05 of 0.08 N $MgSO_4$. What is its hardness in ppm?*

**Solution.** 100 mL of water sample = 12.5 mL of 0.08 N $MgSO_4$

$$= 12.5 \times 0.08 \text{ mL } 1N\ MgSO_4$$

$$= 1 \text{ mL of } 1N\ MgSO_4 = 1 \text{ mL of } 1N\ CaCO_3 \text{ eq.}$$

$$= 0.001 \text{ of } 1N\ CaCO_3 \text{ eq.}$$

$$= 0.05\text{g } CaCO_3 \text{ eq. (or 50 mg } CaCO_3)$$

$$\therefore \quad 1{,}000 \text{ mL (or/L) of water sample} = \frac{50 \text{ mg } CaCO_3 \text{ eq} \times 1000 \text{ mL}}{100 \text{ mL}}$$

$$= 500 \text{ mg } CaCO_3 \text{ eq.}$$

Hence, the hardness of water is 500 mg $CaCO_3$ eq. per liter or **500 ppm.**

## 10.2 CHARACTERISTICS IMPARTED BY IMPURITIES IN WATER

**A. Physical Contaminants of Water.** Important physical contaminants include:

(*i*) *Color.* Even pure water is not colorless. It has a pale green blue tinge in large volumes. Color in natural waters may occur due to the presence of humic acids, fulvic acids, metallic ions such as iron and manganese, peat, tannins, suspended matter, algae, weeds, etc. and industrial wastes. Color change is harmful only when it is associated with a toxic chemical, but it affects the penetration power of sunlight beneath water bodies. Generally, in a surface water, yellow color indicates the presence of organic matter and chromium; red brown color indicates peaty matter, and yellowish-red color indicates presence of iron.

(*ii*) *Turbidity.* Turbidity in water is mainly due to colloidal matter, fine suspended particles, and soil erosion. Turbidity makes water unfit for domestic purposes, food and beverage industries, and many other industrial uses. It is measured by Jackson candle turbidimeter and is expressed in terms of *Jackson turbidity unit (JTU)*. It is removed by sedimentation followed by coagulation, filteration, and settling.

(*iii*) *Taste and odor.* The taste and odors are present in water mainly due to dissolved impurities of organic nature. Presence of manganese, oils, hydrocarbons, chlorophenols, petroleum products, synthetic detergents, pesticides, etc., produce characteristic tastes in water. Decomposed algae, organic matter, fungi, etc., produce peculiar tastes. These pollutants can be removed by chlorination, chemical precipitation, and using activated charcoal.

Odor is generally caused by chemicals such as $H_2S$, free $Cl_2$, $NH_3$, phenols, alcohols, esters, etc., and biological organisms such as algae, fungi, etc. It can be eliminated by using chlorine, ozone, activated carbon, aeration, and chlorination.

**B. Chemical Contaminants of Water.** Chemical contamination of water is due to the presence of inorganic and organic chemicals such as acids, alkalies, toxic inorganic compounds, dissolved inorganic compounds, suspended inorganic compounds, suspended organic compounds, and dissolved organic compounds. The industrial wastes of certain industries contain acids, organic chemicals such as proteins, fats, carbohydrates, etc., these contaminants find access to water bodies. Important toxic inorganic pollutants include free chlorine, chloramines, $NH_3$, $H_2S$, metals such as Pb, Ni, Cu, Cr, Hg, As, U, etc. Chemical pollution of water causes changes in alkalinity, acidity, dissolved oxygen, and other gases in water.

C. **Biological Contamination of Water.** Biological contamination of water is due to the presence of pathogenic bacteria, certain fungi, pathogenic protozoa, virus, parasitic worms, etc. The main sources of biological contamination are domestic sewage and industrial wastes.

   Bacterial contamination in water is caused by the excretory products of warm blooded mammals including man, and wild and domestic animals. The main pollutants belong to the coliform group, for example faecal streptococci, etc. Contaminated water creates infections in the intestinal tract like dysentry, cholera, typhoid and gastroenterititis, polio, and infectious hepatitis.

## 10.3 SOFTENING OF WATER OR WATER TREATMENT

Water can be softened either by the removal of calcium and magnesium salts or by their conversion into corresponding salts of sodium. There are two main methods of removing hardness of water: Lime-soda process and zeolite process. These are discussed in the following sections.

### 10.3.1 Lime-Soda (L-S) Process of Water Softening

In this process, lime and soda ash ($Na_2CO_3$) are used for the precipitation process of softening. The lime is required for the precipitation of compounds causing temporary or carbonate hardness and the soda-ash is used for the precipitation of compounds causing permanent hardness. The reactions involved in the process are:

(*i*) **For carbonate or temporary hardness**

$$CO_2 + Ca(OH)_2 \longrightarrow CaCO_3\downarrow + H_2O$$

$$Ca(HCO_3)_2 + Ca(OH)_2 \longrightarrow 2CaCO_3\downarrow + 2H_2O$$

$$Mg(HCO_3)_2 + Ca(OH)_2 \longrightarrow MgCO_3 + CaCO_3\downarrow + 2H_2O$$

$MgCO_3$ formed further reacts in following manner

$$MgCO_3 + Ca(OH)_2 \longrightarrow Mg(OH)_2\downarrow + CaCO_3\downarrow$$

(*ii*) **For noncarbonate or permanent hardness**

$$CaSO_4 + Na_2CO_3 \longrightarrow CaCO_3\downarrow + Na_2SO_4$$

$$MgSO_4 + Ca(OH)_2 + Na_2CO_3 \longrightarrow Mg(OH)_2\downarrow + CaCO_3\downarrow + Na_2SO_4$$

$$MgCl_2 + Ca(OH)_2 \longrightarrow Mg(OH)_2\downarrow + CaCl_2$$

$$CaCl_2 + Na_2CO_3 \longrightarrow CaCO_3\downarrow + Na_2SO_4$$

It is clear from the above reactions that in case of *temporary hardness*, each unit of calcium bicarbonate requires one mole of lime, whereas each unit of $Mg(HCO_3)_2$ requires two moles of lime. In case of *permanent hardness*, magnesium salts require one mole each of lime and soda ash, whereas calcium salts require one mole of soda ash only. So, it can be summarized that for the removal of permanent hardness by $Na_2CO_3$, no lime is necessary to remove permanent hardness caused by calcium salts, but some lime is necessary to remove such hardness caused by the presence of magnesium salts. If there is excess lime in softened water, the water will become hard due to soluble $Ca(OH)_2$. Hence, it is most essential to add the requisite amount of $Ca(OH)_2$ and $Na_2CO_3$ to the hard water for softening. The L-S process may be broadly classified as:

**Cold lime-soda process.** In this process the required amount of lime and soda ash are added with the hard water in the chemical tank and the contents are thoroughly mixed. The $CaCO_3$ formed during the reaction does not precipitate completely but forms supersaturated solution. This $CaCO_3$ may separate during the use of the soft water later on. To prevent supersaturation, some previously formed sludge is thoroughly mixed with the treated water. Coagulants or flocculants, *e.g.*, alum, aluminium sulphate, sodium aluminate, etc. are added to help the formation of coarse precipitates. Use of sodium aluminate also helps with the removal of silica as well as oil, if present in water.

$$\underset{\text{Sod. aluminate}}{NaAlO_2} + 2H_2O \longrightarrow NaOH + Al(OH)_3$$

$$\underset{\text{Coagulant}}{Al_2(SO_4)_3} + \underset{\text{Hardness in water}}{3Ca(HCO_3)_2} \longrightarrow 2Al(OH)_3\downarrow + 3CaSO_4 + 6CO_2\uparrow$$

This process uses two kinds of softeners:

(*a*) *Intermittent type softeners.* In this type the requisite amounts of lime and soda ash are mixed with raw water simultaneously in the tank with continuous stirring. A small amount of precipitate from the previous filtration is added to prevent supersaturation. When the tank fills up, the reaction is completed and the precipitate is allowed to settle down. The sludge is taken out from the bottom and the clear softened water is taken out through a pipe into another tank.

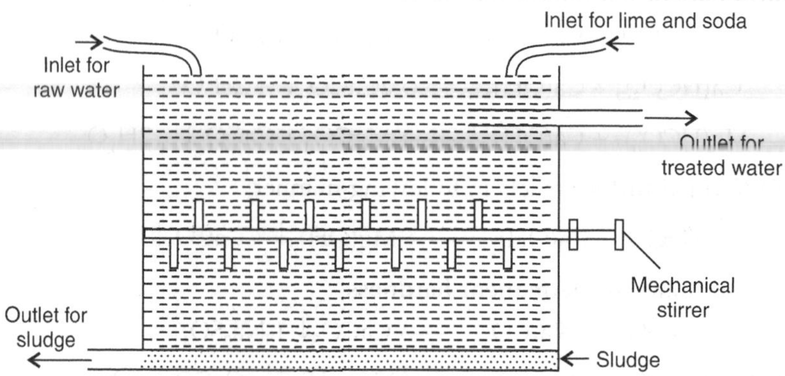


*FIGURE 10.1* **Intermittent type softener.**

(*b*) *Continuous type softener.* It consists of a big steel tank with inner and outer chambers as shown in Figure 10.2. Raw water and requisite chemicals are let into the inner chamber at room temperature with continuous stirring. Softening of water occurs in this chamber and the softened water goes upwards in the outer chamber where it removes all the sludge after complete filtration. Filtered water comes out through the water outlet and the sludge is removed from the bottom.

Cold lime-soda ash process is not used for softening the boiler feed water for the following reasons:

(*i*) Softened water does not have zero hardness.

(*ii*) The residual hardness is harmful for water tubes in the boiler.

**Hot Lime-Soda Process.** It is similar to the cold lime-soda process except that the mixture of raw water, softening chemicals, and a small amount of sludge are heated at a temperature of 80° to 150°C (Figure 10.3). Thus the reactions are much faster, the precipitation becomes complete in a short time, the sludge formed settles down quickly, no coagulants are required, and one tank is sufficient. The softened water has the hardness of 20–25 ppm.

**Advantages of hot L-S process over cold L-S process**

Following are the advantages of hot L-S process over cold L-S process:

(*i*) The reaction is completed within a short period of time.

(*ii*) Silica also get precipitated by adding MgO and lime at 50°C.

(*iii*) The quantity of chemicals used is reduced.

(*iv*) The precipitate formed is granular and more dense and can easily be removed.

(*v*) The temporary hardness, free $CO_2$, dissolved oxygen (DO), and other dissolved gases are removed by pre-heating.

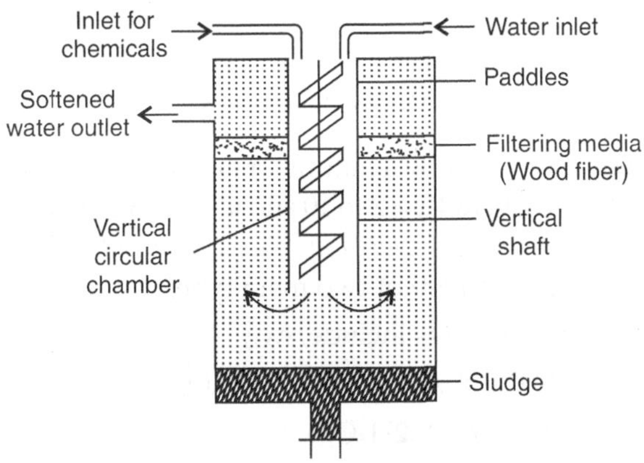


*FIGURE 10.2* **Continuous type softener.**

**Disadvantages of L-S Process**

Following are the disadvantages of L-S process:

(*i*) It cannot be used for household purposes as the requisite amount of chemicals is not known.

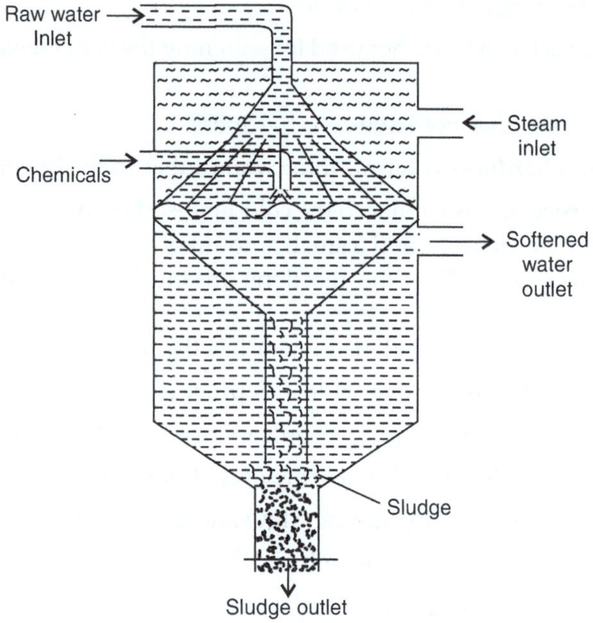


***FIGURE 10.3*** **Continuous type hot lime soda softener.**

(*ii*) It is also not fit for many industrial purposes as the water obtained is supersaturated with $CaCO_3$.

## 10.3.2 Zeolite or Permutit Process of Water Softening

This process involves softening of water by a natural or artificial *zeolite*. Zeolite is hydrated sodium aluminosilicate, $Na_2O.Al_2O_3.xSiO_2.yH_2O$($x = 2$ to 10 and $y = 2$ to 6). *Zeolites are the substances insoluble in water and can act as base exchangers when placed in contact with water containing cations.* Zeolites are commercially known as permutits and are of two main types:

(*a*) *Natural zeolites.* Natural zeolites include, *e.g.*,

(*i*) Thomsonite, $(Na_2O.CaO)\ Al_2O_3.2SiO_2.2.5H_2O$;

(*ii*) Natralite, $Na_2O.Al_2O_3.3SiO_2.2H_2O$.

(*iii*) Laumonite, $CaO.Al_2O_3.4SiO_2.4H_2O$, etc.

These zeolites are derived from green sands by washing, heating and treatment with NaOH.

(*b*) *Synthetic zeolites* are prepared from sodium silicate and sodium aluminate.

Natural zeolites are more durable whereas synthetic zeolites have higher exchange capacity per unit weight. Zeolites are now being replaced slowly by high capacity resins having greater stability.

**Process.** The zeolite softener consists of a steel tank packed with a thick layer of loosely packed zeolite over a layer of coarse sand and gravel (Figure 10.4). The water enters at the top and passes through the bed of zeolite, where the following reactions take place:

$$Na_2Z + Ca(HCO_3)_2 \longrightarrow CaZ + 2NaHCO_3$$

$$Na_2Z + Mg(HCO_3)_2 \longrightarrow MgZ + 2NaHCO_3$$

$$Na_2Z + CaCl_2 \longrightarrow CaZ + 2NaCl$$

$$Na_2Z + MgCl_2 \longrightarrow MgZ + 2NaCl$$

$$Na_2Z + CaSO_4 \longrightarrow CaZ + Na_2SO_4$$

$$Na_2Z + MgSO_4 \longrightarrow MgZ + Na_2SO_4$$

where $Z = Al_2Si_2O_8.xH_2O$

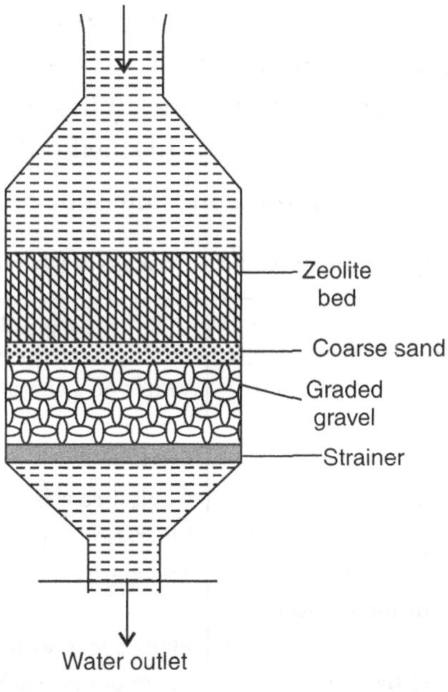


*FIGURE 10.4* **Zeolite softener.**

The process removes both temporary and permanent hardness. The utility of the process lies in the fact that when the zeolite bed gets exhausted, *i.e.*, it loses its softening capacity, it can be *regenerated* simply by passing 10% solution of sodium chloride or brine through it. The following reaction takes place:

$$\underset{\text{exhausted zeolite}}{CaZ} + 2NaCl \longrightarrow \underset{\text{regenerated zeolite}}{Na_2Z} + CaCl_2$$

$$MgZ + 2NaCl \longrightarrow Na_2Z + MgCl_2$$

**Limitations**

Following are the limitations of Zeolite or Permutit Process of Water Softening:

(*i*) The feed water must be free from suspended particles, otherwise the permutit will get clogged.

(*ii*) Acid present in water must be neutralized to prevent burning of zeolite.

(*iii*) Colored ions should be removed by pretreatment.

(*iv*) This method cannot remove lead.

(*v*) Water treated by zeolite process contains about 25% more dissolved solids than that treated by L-S process.

**Advantages**

Following are the advantages of Zeolite or Permutit Process of Water Softening:

(*i*) The equipment occupies a small space and is easy to operate.

(*ii*) The method can even produce zero hardness water.

(*iii*) No sludge formation occurs during the process.

(*iv*) The method is cheap as the permutit can be regenerated.

(*v*) It automatically adjusts to waters of different hardness.

**Disadvantages**

Following are the disadvantages of Zeolite or Permutit Process of Water Softening:

(*i*) Treated water contains more sodium salt.

(*ii*) All acidic ions such as $HCO_3^-$ and $CO_3^{2-}$ are left in softened water.

(*iii*) Highly turbid water cannot be treated by this method.

Comparison between the zeolite process and lime-soda process is summarized in Table10.2.

**TABLE 10.2 Comparison of Zeolite Process with Lime-soda Process**

| *Zeolite Process* | *Lime-soda Process* |
|---|---|
| 1. Water of 10–15 ppm residual hardness is obtained. | Water of, generally, 15–50 ppm hardness is obtained |
| 2. Treated water contains larger amount of sodium salts than in original raw water. | Treated water contains lesser amount of sodium salts. |
| 3. Cost of plant and material is higher. | Capital cost is lower. |
| 4. Operation expenses are lower. | Operation expenses are higher. |
| 5. It cannot be used for treating acidic water, because the permutit material undergoes disintegration. | There are no such limitations. |
| 6. The plant occupies less space. | Plant occupies more space. |
| 7. The raw water to be softened must be free from suspended matter otherwise the pores of permutit material are blocked and the bed loses its exchange capacity. | There are no such limitations. |
| 8. It involves no problem of settling, coagulation, filtration, and removal of the sludges and precipitates. | It involves difficulty in settling, coagulation, filteration and removal of the precipitates. |
| 9. Treated water contains more dissolved solids. | Treated water contains lesser amount of dissolved solids. |

### 10.3.3 Solving Problems on Water Treatment by Lime-soda Process

Based on reactions taking place in the lime-soda process given earlier, the following inferences can be made:

(*i*) One equivalent of calcium temporary hardness requires one equivalent of lime

$$Ca(HCO_3)_2 + Ca(OH)_2 \longrightarrow 2CaCO_3\downarrow + 2H_2O$$

(*ii*) One equivalent of magnesium temporary hardness requires two equivalents of lime

$$Mg(HCO_3)_2 + 2Ca(OH)_2 \longrightarrow Mg(OH)_2\downarrow + 2CaCO_3 + 2H_2O$$

(*iii*) One equivalent of calcium permanent hardness requires one equivalent of soda.

$$CaSO_4 + Na_2CO_3 \longrightarrow CaCO_3 + Na_2SO_4$$

$$CaCl_2 + Na_2CO_3 \longrightarrow CaCO_3 + 2NaCl$$

(*iv*) One equivalent of magnesium permanent hardness requires one equivalent of lime and one equivalent of soda.

$$MgSO_4 + Ca(OH)_2 + Na_2CO_3 \longrightarrow Mg(OH)_2 + CaCO_3 + Na_2SO_4$$

$$MgCl_2 + Ca(OH)_2 + Na_2CO_3 \longrightarrow Mg(OH)_2 + CaCO_3 + 2NaCl$$

(*v*) Lime reacts with HCl, $H_2SO_4$, $CO_2$, $H_2S$, salts of iron, aluminium, etc. Accordingly their respective equivalents must be considered for calculating the lime requirement

$$Ca(OH)_2 + 2HCl \longrightarrow CaCl_2 + 2H_2O$$

$$Ca(OH)_2 + H_2SO_4 \longrightarrow CaSO_4 + 2H_2O$$

$$Ca(OH)_2 + CO_2 \longrightarrow CaCO_3 + H_2O$$

$$Ca(OH)_2 + H_2S \longrightarrow CaS + 2H_2O$$

$$Ca(OH)_2 + FeSO_4 \longrightarrow CaSO_4 + Fe(OH)_2$$

$$2Fe(OH)_2 + H_2O + \frac{1}{2}O_2 \longrightarrow 2Fe(OH)_3$$

$$2Ca(OH)_2 + Al_2(SO)_3 \longrightarrow 2Al(OH)_3 + 3CaSO_4$$

(*vi*) Lime, while reacting with HCl, $H_2SO_4$, $MgSO_4$, $MgCl_2$, $Mg(NO_3)_2$, salts of Fe, Al etc., generates the corresponding quantities of calcium permanent hardness. Accordingly, these constituents should also be considered while calculating the soda requirement.

(*vii*) Two equivalents of $HCO_3^-$ react with two equivalents of lime as follows:

$$\underset{\text{(2 equivalents)}}{2HCO_3^-} + \underset{\text{(2 equivalents)}}{Ca(OH)_2} \longrightarrow CaCO_3 + H_2O + \underset{\text{(2 equivalents)}}{CO_3^{2-}}$$

It is evident that in the above reaction, 2 equivalents of $CO_3^{2-}$ are generated. Thus for every one equivalent of $HCO_3^-$ present, the corresponding reduction in the dose of soda has to be made in the calculations for soda requirement.

It may be noted that the chemical reactions taking place during lime-soda process are slow, and more over the precepitates formed are fine and they tend to produce a supersaturated solution. This eventually results in *after deposition,* later in the pipes and boiler tubes, producing sludges. These in turn decrease the effective diameters of pipes, clog tubes and valves, cause their corrosion etc. These drawbacks are overcome by adding *accelators* (activated charcoal) that bring down the fine particles and *coagulants* or *flocculants, e.g.,* alums, 10.3 that help in the formation of coarse precipitates.

For *solving numerical problems on lime-soda requirements* for softening of hard water, the following steps may be followed:

1. The units in which the impurities analyzed are expressed, *i.e.*, ppm (or mg/l), grains per gallon (or degrees Clark), etc., are to be noted.
2. Substances which do not contribute toward hardness (*i.e.*, KCl, NaCl, $SiO_2$, $Na_2SO_4$, $Fe_2O_3$, $K_2SO_4$, etc.) should be ignored while calculating lime and soda requirements.
3. Convert ion of all the hardness causing into their respective $CaCO_3$ equivalents. (Please refer to section 10.1.4)
4. If the impurities are given as $CaCO_3$ or $MgCO_3$, these should be considered to be due to $Ca(HCO_3)_2$ or $Mg(HCO_3)_2$ respectively and they are only expressed in terms of $CaCO_3$ and $MgCO_3$.
5. The amount expressed as $CaCO_3$ does not require any further convertion. However, the amount expressed as $MgCO_3$ should be converted into its $CaCO_3$ equivalent by multiplying with 100/84.
6. Calculate the lime and soda requirements as follows:

   Lime required for softening

$$= \frac{74}{100} \times \left\{ \begin{array}{l} \text{Temporary calcium hardness} + (2 \times \text{Temporary magnesium} \\ \text{hardness}) + \text{Permanent Mg hardness} + CO_2 + HCl + H_2SO_4 \\ + HCO_3^- + \text{salts of } Fe^{2+}, Al^{3+} \text{ etc.} - NaAlO_2 \text{ ; all expressed} \\ \text{in terms of their } CaCO_3 \text{ equivalents.} \end{array} \right\} \times \text{Vol. of water}$$

   Soda required for softening

$$= \frac{106}{100} \times \left\{ \begin{array}{l} \text{Permanent calcium hardness + Permanent Mg} \\ \text{hardness + Salts} \\ \text{of } Fe^{2+}, Al^{3+} \text{ etc.,} + HCl + H_2SO_4 - HCO_3^- \text{ ;} \\ \text{all expressed in terms of their } CaCO_3 \text{ equivalents.} \end{array} \right\} \times \text{Vol. of water}$$

7. If the analytical report shows the quantities of $Ca^{2+}$ and $Mg^{2+}$, then one equivalent of soda is required for $Ca^{2+}$; whereas one equivalent of lime and one equivalent of soda is required for $Mg^{2+}$.
8. If the lime and soda used are impure and if the % purity given, then the actual requirements of the chemicals should be calculated accordingly. Thus, if lime is 90%

pure, then the value obtained under step 6 above should be multiplied by $\frac{100}{90}$ to get the actual requirement of lime. Similarly, if soda is 95% pure, then the value obtained under 6 above should be multiplied by $\frac{100}{95}$ to get the actual soda requirement.

**Example 6.** *Calculate the amount of lime (84% pure) required for treatment of 20 000 liters of water whose analysis is as follows:*

| | | | | | |
|---|---|---|---|---|---|
| $Ca(HCO_3)_2$ | — | *40.5 ppm* | $CaSO_4$ | — | *34.0 ppm* |
| $Mg(HCO_3)_2$ | — | *36.5 ppm* | $CaCl_2$ | — | *27.75 ppm* |
| $MgSO_4$ | — | *30.0 ppm* | $NaCl$ | — | *10.0 ppm* |

*Also, calculate the temporary and permanent hardness of the water sample.*

**Solution.**

| *Salt* | | *$CaCO_3$ equivalent* |
|---|---|---|
| $Ca(HCO_3)_2$ | — | $40.5 \times \frac{100}{162} = 25$ ppm |
| $Mg(HCO_3)_2$ | — | $36.5 \times \frac{100}{146} = 25$ ppm |
| $MgSO_4$ | — | $30.0 \times \frac{100}{120} = 25$ ppm |
| $CaSO_4$ | — | $34.0 \times \frac{100}{136} = 25$ ppm |
| $CaCl_2$ | — | $27.75 \times \frac{100}{111} = 25$ ppm |
| NaCl | — | Ignored as it does not contribute to hardness. |

Temporary hardness $= [Ca(HCO_3)_2] + [Mg(HCO_3)_2] = 25 + 25 = 50$ ppm

Permanent hardness $= [MgSO_4] + [CaSO_4] + [CaCl_2] = 25 + 25 + 25 = 75$ ppm

Lime is required for $Ca(HCO_3)_2$, $Mg(HCO_3)_2$ and $MgSO_4$.

Hence, 84% pure lime is required for treating 20 000 liters of water

$$= \frac{74}{100}\,[\underset{Ca(HCO_3)_2}{25} + \underset{Mg(HCO_3)_2}{(2 \times 25)} + \underset{MgSO_4}{25}] \times \underset{\text{Purity factor}}{\frac{100}{84}} \times \underset{\text{grams for 20 000 liters}}{\frac{20\,000}{1\,000}}$$

$= 1761.905$ g $=$ **1.7619 kg.**

**Example 7.** *Calculate the amount of lime and soda required for softening of 50,000 liters of water, which on analysis gave the following results.*

$$Mg^{2+} = 12 \text{ ppm}, Ca^{2+} = 40 \text{ ppm}, HCO_3 = 164.7 \text{ ppm}, CO_2 = 30.8 \text{ ppm}.$$

**Solution.**

**Calculation of $CaCO_3$ Equivalents**

| *Constituent* | *Amount (ppm)* | *Multiplication Factor* | *$CaCO_3$ Equivalent (ppm)* |
|---|---|---|---|
| $Mg^{2+}$ | 12.0 | 100/24 | 12 × 100/24 = 50 |
| $Ca^{2+}$ | 40.0 | 100/40 | 40 × 100/40 = 100 |
| $HCO_3^-$ | 164.7 | 100/61 | 164.7 × 100/122 = 135 |
| $CO_2$ | 30.8 | 100/44 | 30.8 × 100/44 = 70 |

Lime requirement $= \frac{74}{100}$ (Bicarbonate hardness + $Mg^{2+}$ + $CO_2$)

$$= \frac{74}{100}(135 + 50 + 70) \text{ ppm}$$

$$= \frac{74}{100} \times 255 \text{ ppm} = 74 \times 2.55 \text{ mg/liter}$$

1 liter of water requires 74 × 2.55 mg/liter

50,000 liters of water $= 74 \times 2.55 \times 50{,}000$ mg.

$$= \frac{74 \times 2.55 \times 50{,}000}{1000 \times 1000} \text{ kg} = \mathbf{9.435 \text{ kg}}$$

Soda requirement $= \frac{106}{100}(Ca^{2+} + Mg^{2+} - HCO_{3^-}) = \frac{106}{100}(100 + 50 - 135)$ ppm

$$= \frac{106}{100} \times 15 \text{ ppm} = 15.9 \text{ mg/liter}$$

Soda required for 50,000 liters of water

$$= 15.9 \times 50{,}000 \text{ mg}$$

$$= \frac{15.9 \times 50{,}000}{1000 \times 1000} \text{ kg}$$

$$= \mathbf{0.795 \text{ kg.}}$$

**Example 8.** *10 000 liters of hard water was made soft with zeolite; the zeolite required a total amount of 8 liters of NaCl solution containing 150 grams/liter of sodium chloride for regeneration. Calculate the hardness of the water.*

**Solution.** 1 liter of NaCl solution contains 150 g of NaCl

8 liters of NaCl solution contains = 150 × 8 g = 1200 g of NaCl

10 000 liters of hard water = 1200 g of NaCl

$$= 1200 \text{ g of NaCl}$$

$$= \frac{1200 \times 50}{58.5} \text{ g of } CaCO_3 \text{ equivalent hardness}$$

1 liter of hard water $= \frac{1200 \times 50}{58.5 \times 10\,000} \text{ g of } CaCO_3$ equivalent hardness

$$= \frac{1200 \times 50 \times 1000}{58.5 \times 10\,000} \text{ mg of } CaCO_3 \text{ equivalent hardness}$$

$$= 102.5 \text{ mg of } CaCO_3 \text{ equivalent hardness}$$

Hence the hardness of water = 102.5 mg/liter

= **102.5 ppm.**

**Example 9.** *A completely exhausted zeolite requires 120 liters of sodium chloride solution having 100 g/liter of NaCl. How many liters of a water having hardness 500 ppm be softened by the zeolite?*

**Solution.** 1 liter of NaCl solution = 100 × 120 = 12000 g of NaCl

$$= \frac{12\,000 \times 50}{58.5} \text{ g of } CaCO_3 \text{ eq. hardness}$$

Hardness of water = 500 ppm or 500 mg/liter

= 0.5 g/liter.

Thus, 0.5 of hardness = 1 liter of hard water

$$= \frac{12000 \times 50}{58.5} \text{ g of hardness}$$

$$= \frac{1 \times 12000 \times 50}{58.5 \times 0.5} \text{ liters}$$

= **20518 liters of water.**

### 10.3.4 Ion Exchange Resins for Water Softening

In an ion exchange process, a reversible exchange of ions take place between the stationary ion-exchange phase and the external liquid mobile phase. Ion exchange resins are *insoluble, cross linked, long chain organic polymers with a microporous structure and the functional groups attached to them are responsible for ion exchanging properties.*

The purely synthetic organic exchangers are made by:

1. *Polycondensation, e.g.,* between a substituted phenol and formaldehyde, with some unsubstituted phenol to get the desired cross-linking between the linear chains.

2. *Polymerization, e.g.,* of styrene in presence of divinyl benzene, to give the desired cross-linking.

The functional groups are then introduced into the cross-linked resin network, either by subsequent treatment of the resin or by introducing the functional groups into the starting material itself. It is these functional groups which decide the nature of the exchanger (*i.e.,* cationic or anionic). Based on the acidity or basicity of the functional group, exchangers are further classified as follows:

| ***Type of Ion-exchanger*** | ***Functional Groups*** |
|---|---|
| Strongly acidic cation exchangers | $-SO_3H$ |
| Moderately strong cation exchangers | $-PO(OH)_2$ |
| Weakly acidic cation exchangers | $-COOH$ or $-OH$ |
| Strongly basic anion exchanger | $-NR_3^+$; $\equiv P^+-CH_3$; etc. |
| Weakly basic anion exchanger | $-NH_2$; $(C_2H_4)_x(NH)_y^-$ |

Ion exchange resins may be classified as:

(*i*) *Cation exchange resins* (*RH*$^+$) are mainly styrene-divinyl benzene copolymers, which on sulphonation or carboxylation, become capable to exchange their hydrogen ions with cations in the water.

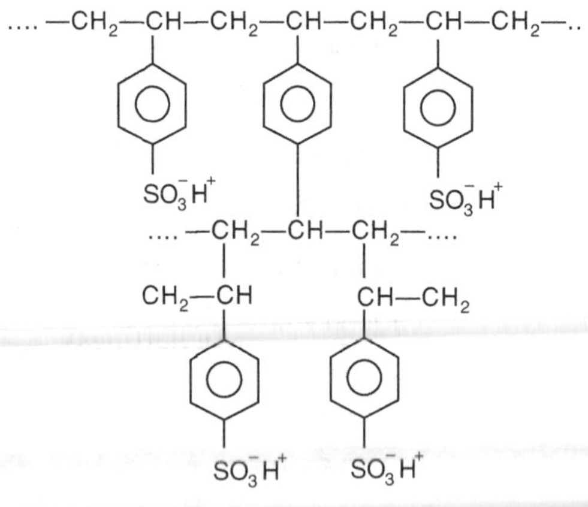


***FIGURE 10.5*** **Acidic or cation exchange resin (sulphonate form).**

(*ii*) *Anion exchange resins* ($R'OH^-$) are styrene-divinyl benzene or amine-formaldehyde copolymers, which contain amino or quaternary ammonium or quarternary phosphonium or tertiary sulphonium groups as an integral part of the resin matrix. These, after treatment with diluted NaOH solution, become capable of exchanging their $OH^-$ anions with anions in water.

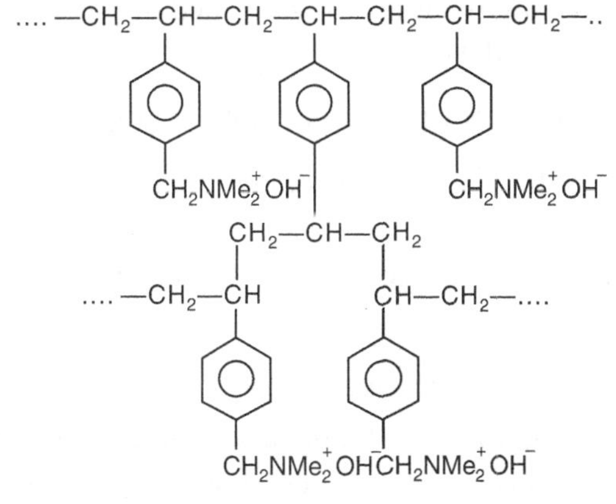

*FIGURE 10.6* **Basic or anion exchange resin (hydroxide form).**

**Process.** The hard water is passed first through a *cation exchange column,* which removes all the cations like $Ca^{2+}$. $Mg^{2+}$, etc. from it, and the equivalent amount of $H^+$ ions are released from this column to water. Thus:

$$2RH^+ + Ca^{2+} \longrightarrow R_2Ca^{2+} + 2H^+$$

$$2RH^+ + Mg^{2+} \longrightarrow R_2Mg^{2+} + 2H^+$$

After the cation exchange column, the hard water is passed through *anion exchange column,* which removes all the anions like $SO_4^{2-}$, $Cl^-$, etc. present in the water and the equivalent amount of $OH^-$ ions are released from this column to water. Thus:

$$R'OH^- + Cl^- \longrightarrow R'Cl^- + OH^-$$

$$2R'OH^- + SO_4^{2-} \longrightarrow R_2'SO_4^{2-} + 2OH^-$$

$$2R'OH^- + CO_3^{2-} \longrightarrow R_2'CO_3^{2-} + 2OH^-$$

$H^+$ and $OH^-$ ions (released from cation exchange and anion exchange columns respectively) get combined to produce water molecule.

$$H^- + OH^+ \longrightarrow H_2O$$

Thus, the water coming out from the exchanger is free from cations as well as anions. Ion-free water known as *deionized* or *demineralized water.*

**Regeneration.** When capacities of cation and anion exchangers to exchange $H^+$ and $OH^-$ ions respectively are lost, they are then said to be exhausted.

The exhausted cation exchange column is regenerated by passing a solution of diluted HCl or diluted $H_2SO_4$. The regeneration can be represented as:

$$R_2Ca^{2+} + 2H^+ \longrightarrow 2RH^+ + Ca^{2+} \quad \text{(washing)}$$

The column is washed with deionized water and washing (which contains $Ca^{2+}$, $Mg^{2+}$, etc. ions) is passed to sink or drain.

The exhausted anion exchange column is regenerated by passing a solution of diluted NaOH. The regeneration can be represented as:

$$R_2'SO_4^{2-} + 2OH^- \longrightarrow 2ROH^- + SO_4^{2-} \quad \text{(washing)}$$

The column is washed with deionized water and washing (which contains $Na^+$ and $SO_4^{2-}$ or $Cl^-$ ions) is passed to sink or drain.

The regenerated ion exchange resins are then used again.

**Advantages**

Following are the advantages of Ion-Exchange Process:

(*i*) The process can be used to soften highly acidic or alkaline water.

(*ii*) It produces water of very low hardness (about 2 ppm). So it is very good for treating water for use in high-pressure boilers.

**Disadvantages**

Following are the disadvantages of Ion-Exchange Process:

(*i*) The equipment is costly and more expensive chemicals are needed.

(*ii*) If water contains turbidity, then the output of the process is reduced. The turbidity must be below 10 ppm. If it is more, it has to be removed first by coagulation and filtration.

For effective water treatment, ion exchangers should possess the following properties:

(*i*) They should be nontoxic.

(*ii*) They should not discolor the water being treated.

(*iii*) They should possess a high ion-exchange capacity. (It depends upon the total number of ion active groups per unit weight of the exchanger and is expressed as mil-equivalents per gram of the exchanger).

(*iv*) They should be physically durable.

(*v*) They should be resistant to chemical attack.

(*vi*) They should be cheap and commonly available.

(*vii*) They must be capable of being regenerated and back-washed easily and economically.

(*viii*) They should have a large surface area since ion-exchange is a surface phenomenon. At the same time, their resistance to flow must be compatible with hydraulic requirements.

## 10.4 BOILER FEED WATER

For steam generation, boilers are almost invariably employed in industries. The boiler feed water requires that the following conditions are met before being used as feed water.

(*i*) Hardness of water should be below 0.2 ppm

(*ii*) Caustic alkalinity should be between 0.15–0.45

(*iii*) Soda alkalinity should lie between 0.45–1.00 ppm.

(*iv*) Excess soda ash should be 0.3–0.5 ppm.

When the natural water without any pretreatment is directly fed into the boilers, various physical as well as chemical reactions in boiler water take place with the action of heat. This results in the following defects:

### 10.4.1 Scale or Sludge Formation

In a boiler, water is continuously converted into steam, increasing the concentration of salts. A stage is reached when these change into a loose and slimy precipitate called *sludge* and if there are hard adhering coatings, these are called *scales*. Thus scales are hard deposits, which stick very firmly to the inner surfaces of the boiler. On the basis of chemical composition, scales formed in the boiler units can be divided into the following groups:

(*a*) *Alkaline earth metal scales.* These consists the compounds such as $CaCO_3$, $CaSO_4$, $CaSiO_3.(PO_4)_2$, $Mg(OH)_2$, $Mg_3(PO_4)_2$, $3MgO.2SiO_2.H_2O$. These substances are precipitated from the solution and get deposited on a hot or cooled surface. These are formed as:

$$Ca(HCO_3)_2 \longrightarrow \underset{\text{Scale}}{CaCO_3} \downarrow + H_2O + CO_2\uparrow$$

$$MgCl_2 + 2H_2O \longrightarrow \underset{\text{Scale}}{Mg(OH)_2} \downarrow + 2HCl\uparrow$$

(*b*) *Iron-oxide scales.* These consists of iron oxides, iron silicates and iron phosphates. These are formed due to the presence of iron impurities in feed water.

(*c*) *Copper scales.* These contain 20–30% of copper, iron oxides and calcium and magnesium compounds.

(*d*) *Silicate scales.* Silica present in feed water forms scale with calcium forming calcium silicate and with aluminium forming sodium aluminium silicate.

#### Effect of Scale or Sludge Formation

Following are the effects of scale or sludge formation:

(*i*) These are poor conductors of heat so they hinder the flow of heat from source to water.

(*ii*) Decreases the efficiency of the boiler by clogging the boiler tubes.

(*iii*) Due to scales, boiler plates are overheated, reducing the life of the plates.

(*iv*) Scales of magnesium chloride corrode the plates.

(*v*) Boilers sometimes crack with the deposition of scales. The water enters through these cracks to the very hot metal and gets changed into steam. This sudden change sometimes bursts the boiler.

#### Prevention of Scale or Sludge Formation

Following are the methods to prevent scale or sludge formation:

(*i*) Scale or sludge formation is prevented by introducing kerosene, tannins, and other colloidal substances. These colloids get coated over the scale, forming particles, and are suspended in water. These can be easily blown off.

(*ii*) In ***Carbonate conditioning process,*** sodium carbonate is added to boiler water. Scales react with it and form insoluble $CaCO_3$.

$$CaSO_4 + Na_2CO_3 \longrightarrow CaCO_3\downarrow + Na_2SO_4$$

Since the solution is saturated with respect to calcium carbonate, deposition of calcium sulphate does not takes place.

For a salt to be precipitated, sufficient amount of the ions forming the salt must be present so that the product of their concentrations (*i.e.*, ionic product) exceed a limiting value known as the solubility product. Thus, for a salt like $CaCO_3$ to be precipitated, the product of the concentration of $Ca^{2+}$ and $CO_3^{2-}$ must exceed the solubility product of $CaCO_3$, represented as $K_{CaCO_3}$.

The formation of $CaCO_3$ may be represented by

$$Ca^{2+} + CO_3^{2-} \rightleftharpoons \underset{\text{solution}}{CaCO_3} \rightleftharpoons \underset{\text{solid}}{CaCO_3}$$

When the equilibrium is established,

$$\underset{\substack{\text{(Concentration}\\ \text{of } Ca^{2+})}}{[Ca^{2+}]} \quad \underset{\substack{\text{(Concentration}\\ \text{of } CO_3^{--})}}{[CO_3^{2-}]} = \underset{\substack{\text{(Solubility product}\\ \text{of } CaCO_3)}}{K_{CaCO_3}} \tag{10.1}$$

at any given temperature.

If, under these conditions, some $Na_2CO_3$ is added, the $CO_3^{2-}$ concentration increases. Then, in order to maintain the $K_{CaCO_3}$ constant, some $Ca^{2+}$ will get precipitated.

Similarly, in a saturated $CaSO_4$ solution,

$$Ca^{2+} + SO_4^{2-} \rightleftharpoons \underset{\text{solution}}{CaSO_4} \rightleftharpoons \underset{\text{solid}}{CaSO_4}$$

and

$$\underset{\substack{\text{(Concentration}\\ \text{of } Ca^{2+})}}{[Ca^{2+}]} + \underset{\substack{\text{(Concentration}\\ \text{of } SO_2^{2-})}}{[SO_4^{2-}]} = \underset{\substack{\text{(Solubility product}\\ \text{of } CaSO_4)}}{K_{CaSO_4}} \tag{10.2}$$

Thus, if a solution is saturated with $CaCO_3$ and $CaSO_4$ (as occurring in boiler water conditions), both the relations (10.1) and (10.2) apply, and by dividing (10.1) by (10.2), we get

$$\frac{[CO_3^{2-}]}{[SO_4^{2-}]} >= \frac{K_{CaCO_3}}{K_{CaSO_4}} = K \tag{10.3}$$

From this, it is clear that if

$$\frac{[CO_3^{2-}]}{[SO_4^{2-}]} > \frac{K_{CaCO_3}}{K_{CaSO_4}} \quad \text{or} \quad K' \tag{10.4}$$

or in other words, if $[CO_3^{2-}] > K'[SO_4^{2-}]$, then only $CaCO_3$ will be precipitated in preference to $CaSO_4$ (because the solubility product of $CaSO_4$ cannot be attained under these conditions).

These principles are used in the carbonate conditioning. When sodium carbonate solution is added to boiler water, the $[CO_3^{2-}]$ increases and when it becomes greater than

$K' \times [SO_4^{2-}]$, only $CaCO_3$ gets precipitated and $CaSO_4$ remains in solution. Thus the deposition of scale-forming $CaSO_4$ is prevented.

$$Na_2CO_3 + CaSO_4 \longrightarrow CaCO_3\downarrow + Na_2SO_4$$

Carbonate conditioning is used only for low pressure boilers. In high pressure boilers the excess $Na_2CO_3$ might be converted into NaOH due to hydrolysis as follows:

$$Na_2CO_3 + 2H_2O \rightleftharpoons 2NaOH + H_2CO_3$$

$$H_2CO_3 \rightleftharpoons H_2O + CO_2\uparrow$$

NaOH causes caustic embrittlement in high pressure boilers.

(*iii*) *Phosphate conditioning*. As in the case of carbonate conditioning, if we consider a solution saturated with both $Ca_3(PO_4)_2$ and $CaSO_4$ the solubility product equations for both salts must be satisfied.

$$Ca_3(PO_4)_2 \rightleftharpoons 3Ca^{2+} + 2PO_4^{3-}$$

$$\therefore \quad [Ca^{2+}]^3[PO_4^{3-}]^2 = \underset{\text{solubility product of } Ca_2(PO_4)_2}{K_{Ca_3(PO_4)_2}} \qquad (10.4a)$$

$$\therefore \quad [Ca^{2+}][PO_4^{3-}]^{2/3} = [K_{Ca_3(PO_4)_2}]^{1/3} \qquad (10.5)$$

Similarly, for $CaSO_4$

$$CaSO_4 = Ca^{2+} + SO_4^{2-}$$

$$\therefore \quad [Ca^{2+}][SO_4^{2-}] = \underset{\text{solubility product of } CaSO_4}{K_{CaSO_4}} \qquad (10.6)$$

By dividing equation (10.5) by equation (10.6) we have

$$\frac{[PO_4^{3-}]^{2/3}}{[SO_4^{2-}]} = \frac{[K_{Ca_3(PO_4)_2}]^{1/3}}{[K_{CaSO_4}]} \quad \text{or} \quad K' \qquad (10.7)$$

or $$[PO_4^{3-}]^{2/3} = K'[SO_4^{2-}] \qquad (10.8)$$

Therefore, as long as $[PO_4^{3-}]^{2/3} > K'[SO_4^{2-}]$, only $Ca_3(PO_4)_2$ will be precipitated in preference to $CaSO_4$ (because the solubility product of $CaSO_4$ cannot be exceeded under these conditions) and hence $CaSO_4$ scale will not form.

These principles are made use in the phosphate conditioning. In this, an excess of a soluble phosphate is added to the boiler water to precipitate the residual calcium ions in the form of a nonadherent precipitate of calcium phosphate and thus the scale formation is prevented.

The three sodium orthophosphates $Na_3PO_4$, $Na_2HPO_4$, and $NaH_2PO_4$ have been used for phosphate conditioning as have sodium pyrophosphate ($Na_4P_2O_7$) and sodium metaphosphate ($NaPO_3$). The typical reactions of the various phosphates with the hardness represented as $CaCO_3$, may be summarized as follows:

$$2Na_3PO_4 + 3CaCO_3 \longrightarrow Ca_3(PO_4)_2 + 3Na_2CO_3$$
$$2Na_2HPO_4 + 3CaCO_3 \longrightarrow Ca_3(PO_4)_2 + 2Na_2CO_3 + CO_2 + H_2O$$
$$2NaH_2PO_4 + 3CaCO_3 \longrightarrow Ca_3(PO_4)_2 + Na_2CO_3 + 2CO_2 + 2H_2O$$
$$2NaPO_3 + 3CaCO_3 \longrightarrow Ca_3(PO_4)_2 + Na_2CO_3 + 2CO_2$$

The quality of the feed water dictates the choice of a particular phosphate to be used. For instance, if the feed water tends to produce an acidic condition in the boiler, the alkaline $Na_3PO_4$ should be chosen. This treatment could be supplemented with NaOH if the required alkalinity desired is in the boiler, it is preferable to use $Na_2HPO_4$ which is practically neutral. If the boiler water becomes too alkaline, the acidic $NaH_2PO_4$ would be selected.

Both sodium pyrophosphate and metaphosphate are rapidly hydrolyzed under boiler water temperatures to orthophosphate.

$$NaPO_3 + H_2O = NaH_2PO_4$$
$$Na_4P_2O_7 + H_2O = 2Na_2HPO_4$$

(*iv*) *Calgon Conditioning.* Calgon or sodium hexameta phosphate $(NaPO_3)_6$ forms soluble compounds with calcium salts at a higher temperature, *calgon* is converted into sodium orthophosphate which forms loose sludge which can be removed by blowing it off.

$$\underset{\text{Calgon}}{Na_2[Na_4(PO_3)_6]} \longrightarrow 2Na^+ + \underset{\text{Loose sludge}}{[Na_4P_6O_{18}]^{2-}}$$

$$[Na_4P_6O_{18}]^{2-} + 2CaSO_4 \longrightarrow \underset{\text{soluble complex ion}}{[Ca_2P_6O_{18}]^{2-}} + 2Na_2SO_4$$

### 10.4.2 Boiler Corrosion

Boiler corrosion is the *decay of boiler material by a chemical or electrochemical attack of its environment.* Main reasons for boiler corrosion are:

1. **Dissolved oxygen.** Water usually contains about 8 mL of dissolved oxygen per liter at room temperature. Dissolved oxygen in water, in presence of prevailing high temperature, attacks boiler material.

$$2Fe + 2H_2O + O_2 \longrightarrow 2Fe(OH)_2\downarrow$$

$$\underset{\text{Ferrous hydroxide}}{4Fe(OH)_2\downarrow} + O_2 \longrightarrow \underset{\text{(Rust)}}{2[Fe_2O_3.2H_2O]\downarrow}$$

*Removal of dissolved oxygen*

(*i*) By adding a calculated quantity of sodium sulphite or hydrazine or sodium sulphide, the following reaction occurs:

$$2Na_2SO_3 + O_2 \longrightarrow 2Na_2SO_4$$

$$\underset{\text{Hydrazine}}{N_2H_4} + O_2 \longrightarrow N_2 + 2H_2O$$

$$Na_2S + 2O_2 \longrightarrow Na_2SO_4$$

*Hydrazine* is an ideal internal treatment chemical for the removal of dissolved oxygen. It reacts with oxygen, forming nitrogen and water. Nitrogen is harmless. Consequently, hydrazine removes oxygen without increasing the concentration of dissolved solids/salts. With sodium sulphite or sodium sulphide, the sodium sulphate so formed is liable to decompose, giving $SO_2$, particularly in high pressure boilers. The $SO_2$ enters the steam pipes and appears as sulphurous acid ($H_2SO_3$) in the steam condensate.

(*ii*) By mechanical deaeration, *i.e.*, spraying water in a perforated plate-fitted tower, heated from sides and connected to vacuum pump (see Figure 10.7).

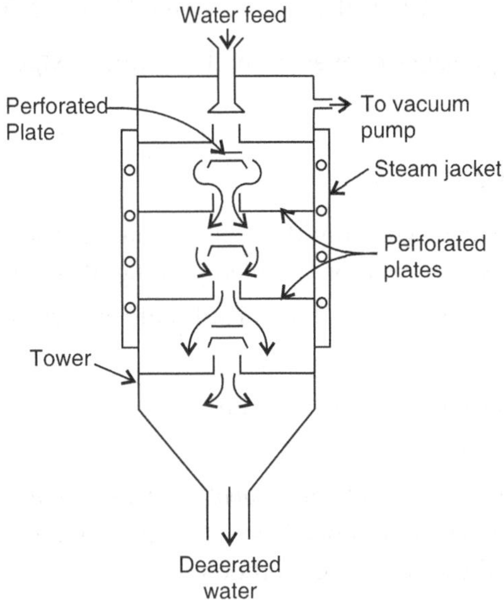


*FIGURE 10.7* **Mechanical deaeration of water.**

High temperature, low pressure and larger exposed surface (provided by perforated plates) reduces the dissolved oxygen in water.

2. **Dissolved carbon dioxide.** $CO_2$ forms carbonic acid,

$$CO_2 + H_2O \longrightarrow H_2CO_3$$

which has a slow corrosive effect on the boiler material. Carbon dioxide is also released inside the boiler, if water used for steam generation contains bicarbonate, *e.g.*,

$$Mg(HCO_3)_2 \xrightarrow{\text{heat}} MgCO_3 + H_2O + CO_2$$

*Removal of $CO_2$*

(*i*) By adding calculated quantity of *ammonia*. Thus:

$$2NH_4OH + CO_2 \longrightarrow (NH_4)_2CO_3 + H_2O$$

(*ii*) By *mechanical deaeration* process along with oxygen.

3. **Acids from dissolved salts.** Water containing dissolved magnesium salts liberate acids on hydrolysis, *e.g.*,

$$MgCl_2 + 2H_2O \longrightarrow Mg(OH)_2\downarrow + 2HCl$$

The liberated acid reacts with iron (of the boiler) in chain-like reactions producing HCl again and again. Thus:

$$Fe + 2HCl \longrightarrow FeCl_2 + H_2\uparrow$$

$$FeCl_2 + 2H_2O \longrightarrow Fe(OH)_2\downarrow + 2HCl$$

Consequently, the presence of even a small amount of $MgCl_2$ will cause corrosion of iron to a large extent.

### 10.4.3 Priming and Foaming

When a boiler is producing steam very rapidly, some water particles are carried along with the steam. This process of wet steam formation is called *priming*. It is caused by:

(*i*) the presence of large quantities of alkali sulphate and chloride in water

(*ii*) sudden boiling

(*iii*) sudden increase in steam production rate

(*iv*) improper boiler design.

*Foaming is the persistent formation of foam or bubbles in the boiler, which do not break easily.* Foaming is due to the presence of oil which generally reduces the surface tension of water in boiler. Priming and foaming can be prevented by following methods:

(*i*) Addition of antifoaming agents like castor oil.

(*ii*) Mechanical purifiers easily check priming.

(*iii*) Addition of compounds like sodium aluminate.

(*iv*) Avoiding rapid change in steaming rate.

### 10.4.4 Caustic Embrittlement

It is the intercrystalline cracking of the boiler metal from the caustic attack and is caused by using highly alkaline water in boiler. During water softening by L-S process, free sodium carbonate is generally left in small proportion in soften water which decomposes to NaOH and $CO_2$

$$Na_2CO_3 + H_2O \longrightarrow 2NaOH + CO_2$$

The NaOH containing water flows into the small cracks of boiler by capillary action dissolving the iron of surrounding area inside the boiler. At these areas corrosion takes place in such a manner that intergranular cracks occur is an irregular fashion, this is called *caustic embrittlement*. It can be prevented by following methods:

(*i*) Tannin or lignin should be added in boiler feed water.

(*ii*) During softening process sodium phosphate should be used in place of sodium carbonate.

(*iii*) Sodium sulphate should be added to water. This blocks hair cracks in the boiler.

## 10.5 MUNICIPAL WATER SUPPLY

Water is an important and most abundant substance. All animals and plants possess water as an essential constituent. It is widely available in air as water vapour and can be obtained from variety of sources. Some common sources of water are river, lake, sea, well, etc. The purest form of water is probably the rain water, because it is obtained from the evaporation of water from natural resources leaving behind many impurities.

Water is used for industrial purposes and for municipal supply. To ensure the right quality and quantity of water for these purposes it is necessary to study water supply thoroughly taking all the aspects into consideration. Water supply for any purpose requires following factors.

(*i*) The quantity of water available and the seasonal variation in quantity and quality.

(*ii*) Water analysis with regards to its chemicals, physical, microscopic and bacteriological characteristics.

(*iii*) The influence on quality of water due to industrial wastes and sewage etc.

(*iv*) Cost of getting continuous water supply in required quantity and quality.

### 10.5.1 Requisites of Potable Water

Water which is safe to drink is known as potable water. Municipalities have to supply potable water, fit for human consumption. Potable water should have following characteristics :

(*i*) Water should be obtained from a source that is free from pollution.

(*ii*) Water should be obtained from a source purified by natural agencies and municipalities.

(*iii*) Water should be protected by artificial treatment.

(*iv*) Water supply system should be free from health hazards and sanitary effects.

(*v*) All kinds of bacteria and microorganisms should be absent.

(*vi*) It should not be turbid. It should be colorless. Turbidity should not exceed 10 ppm on silica scale.

(*vii*) It should be odourless and must have an agreeable taste.

(*viii*) Soluble mineral salts should not be present beyond the prescribed limits. The limit for some of the minerals is as follows :

$Pb < 0.1$ ppm ; $Cu < 3.0$ ppm ; $Fe + Mn < 0.3$ ppm ; $Cl < 250$ ppm Sulphate $< 250$ ppm ; Phenolic compounds $< 0.01$ ppm interms of phenol and ; Total solids $< 500$ ppm.

(*ix*) Its alkalinity should not be high. Its $p$H should be about 8.0.

### 10.5.2 Purification of Water for Municipal Supply

Natural water from rivers, canals, etc., does not confirm to all the required specifications of drinking water as described above. For removing various types of impurities, the following treatment processes are employed :

### [A] Removal of Suspended Impurities

Suspended impurities (organic or inorganic) in water can be removed by the following steps.

1. **Screening.** *Screening is the process of removing floating materials from water.* Raw water is allowed to pass through a screen having a large number of perforations which removes the large and small floating matter.
2. **Sedimentation (settling).** *Sedimentation is the process of removing insoluble impurities by allowing the water to stay undisturbed for sometime.* Water is allowed to settle in big tanks or reservoirs for a number of hours with the result the suspended materials, other than living organisms which float due to the gases they produce, settle down at the bottom due to the force of gravity. The process of sedimentation is generally carried out in *continuous flow type* tanks in which water flows continuously in a horizontal, radial or vertical direction at a uniform rate. Thus there are three types of continuous flow type tanks, *viz.* horizontal, radial and vertical flow tanks.

   (*i*) *Horizontal flow tank* (Fig. 10.8.). It consists of a rectangular tank with channel type inlet and outlet extending throughout the width of the tank. It is provided with baffle walls to reduce the velocity of the incoming water. The sedimented sludge is taken out through the outlet provided at the bottom.

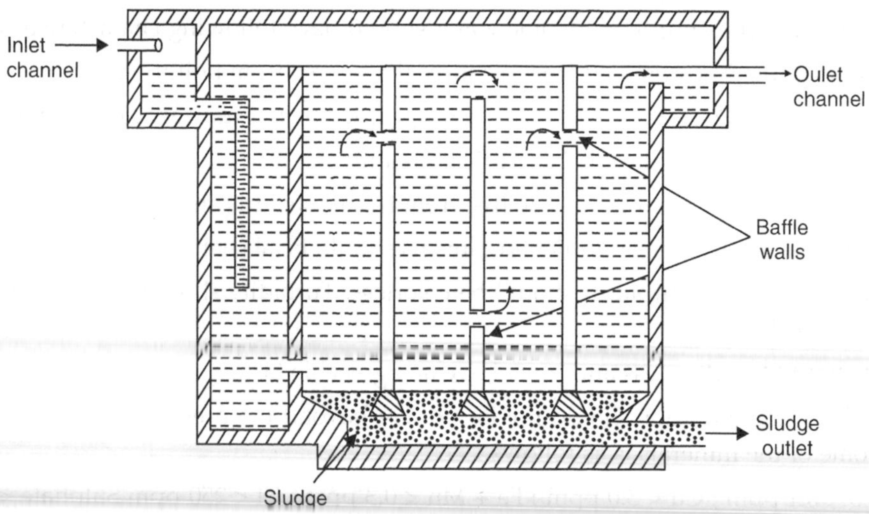


***FIGURE 10.8*** **Horizontal flow rectangular sedimentation tank.**

   (*ii*) *Radial flow circular tank* (Fig. 10.9). In this type of tank water first enters at the centre through an inlet pipe provided at the bottom, then it flows downwards in the deflector box, from where it flows upwards and ultimately leaves the tank through the outlet provided at the top. The suspended impurities settle down and are led to the sludge outlet with the help of scrapers attached to the rotating racking arm.

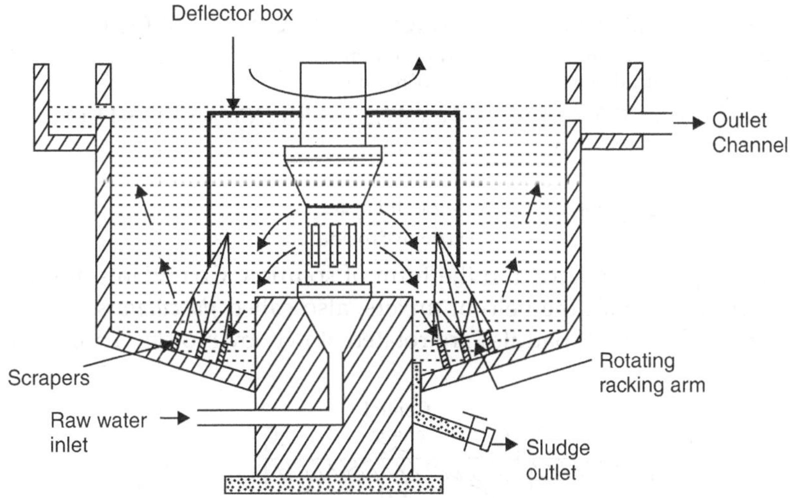


*FIGURE 10.9* **Radial flow circular sedimentation tank.**

(*iii*) *Vertical flow hopper bottom tank* (Fig. 10.10). In such a tank, water enters from the top into the deflector box. The sludge is removed from time to time by a sludge outlet pipe which also opens at the top.

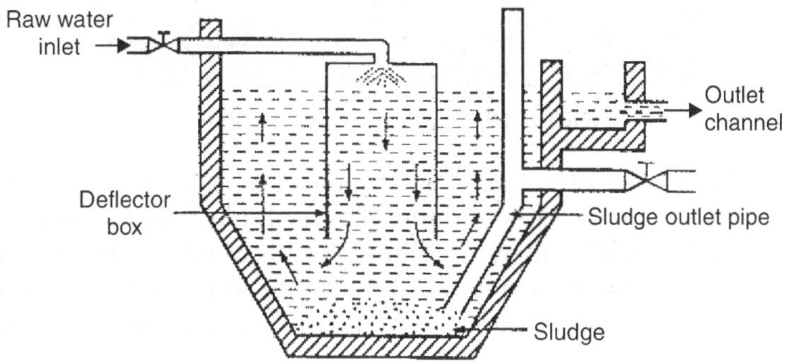


*FIGURE 10.10* **Vertical flow hopper bottom tank.**

**3. Coagulation.** *Coagulation is the process of removing fine sized particles from water with the addition of certain chemicals known as coagulants.*

Actually the fine sized particles present in water either do not settle down at all or take a long time. In order to facilitate quick settling of these particles, some chemicals agents known as coagulants are used. The commonly used coagulants are the salts of iron and aluminium, *e.g.*, alum [$K_2SO_4$ $Al_2(SO_4)_3.24H_2O$], aluminium sulphate, $Al_2(SO_4)_3.18H_2O$, ferrous sulphate, $FeSO_4.7H_2O$, ferric chloride, $FeCl_3$, ferric sulphate, $Fe_2(SO_4)_3$, etc. These salts react with carbonate and bicarbonate radicals present in water and form coagulable precipitates (*flocks*) of hydroxides of these metals. The precipitated hydroxide absorbs the suspended impurities, bacteria and other micro-organisms and causes them to settle down. Generally, coagulation occurs in about two hours.

$$\underset{\substack{\text{Aluminium}\\\text{sulphate}}}{Al_2(SO_4)_3} + \underset{\substack{\text{Calcium}\\\text{bicarbonate}}}{3Ca(HCO_3)_2} \longrightarrow \underset{\substack{\text{Aluminium}\\\text{hydroxide}}}{2Al(OH_3)} \downarrow + 3CaSO_4 + 6CO_2 \uparrow$$

$$\underset{\substack{\text{Ferrous}\\\text{sulphate}}}{FeSO_4} + \underset{\substack{\text{Magnesium}\\\text{bicarbonate}}}{Mg(HCO_3)_2} \longrightarrow \underset{\substack{\text{Ferrous}\\\text{hydroxide}}}{Fe(OH)_2} \downarrow + MgCO_3 + CO_2 + H_2O$$

$$4Fe(OH)_2 + 2H_2O + O_2 \longrightarrow \underset{\text{Ferric hydroxide}}{4Fe(OH)_3} \downarrow$$

Coagulants are generally added in the form of solution. Coagulants are thoroughly mixed with water (to be purified) with the help of mixers, also called *flocculators*. A typical mechanical flocculator is shown in Fig. 10.11, paddles are rotated slowly by motors. Thoroughly mixed water is then sedimented.

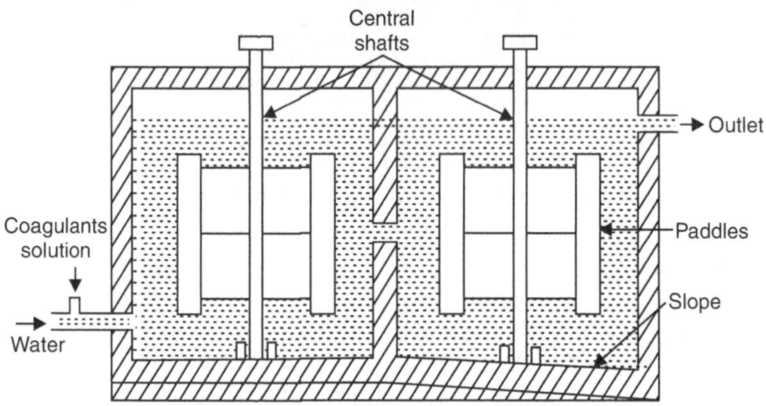


*FIGURE 10.11* **Mechanical flocculator.**

**4. Filtration.** When sludge etc. after coagulation has settled down, filtration is carried out by means of filters. Filtration is the process of removing insoluble, colloidal and bacterial impurities from water by means of filters. The filters used in water filtration consist of several layers of sand particles of different size. Following two types of filters are commonly used in water treatment.

(*i*) *Gravity sand filter.* It consists of a large rectangular tank made of concrete. The filtering medium in it consists of 3 feet of fine sand below which is 1 foot thick coarser sand which in turn is supported by 8 inches thick bed of graded gravel. Below the graded gravels are drains through which filtered water goes out.

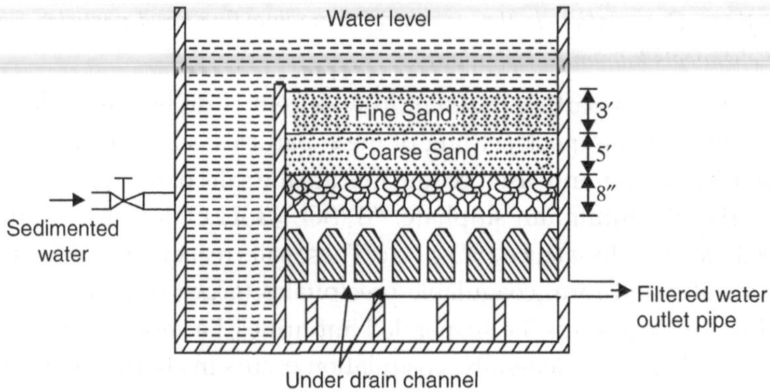


*FIGURE 10.12* **Gravity sand water.**

As the water percolates (due to gravity) through the fine sand bed, most of the objectionable materials are absorbed by it and clear (filtered) water collects in the underdrain channel, from where it is drawn out.

After some time (nearly 24 hours), the rate of filtration becomes very slow because of clogging of pores of the top layer by the impurities. Under such circumstances, fine sand bed is removed, washed and is then again replaced.

After filtration if the pH of water is below 7, it is made to 7 by adding lime, soda ash or sodium silicate.

Gravity sand filters are used where a large quantity of water is to be filtered. Since the rate of filtration is slow and very fine sand particles are used, a large portion of even the bacteriological impurities are removed, thus such sand filters are best suited for municipal water.

(*ii*) *Pressure filters.* Pressure filters are used for the filtration of water where a small amount of water is required because these are not economical for large scale supplies. Thus pressure filters are used for filtering water for industrial plants and for small scale water supply schemes.

A typical pressure filter (Fig. 10.13) consists of a cylindrical steel shell containing three layers of filtering media.

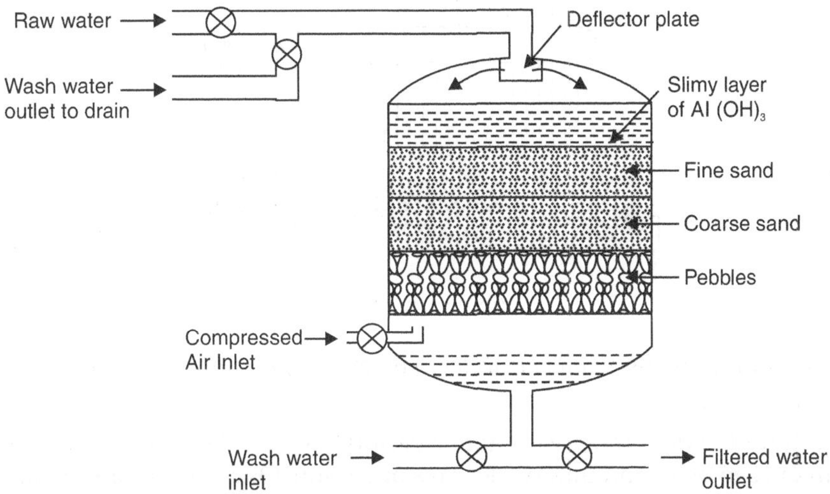


*FIGURE 10.13* **Vertical pressure filter.**

(*i*) Pebbles (10 – 25 mm grain size) layer,

(*ii*) Coarse sand (5 – 7 mm grain size) layer,

(*iii*) Fine sand (1 – 2 mm grain size) layer,

The sand used is generally graded and cleaned before it is filled in the filter.

Often alum is added to the coming stream of water from the alum tank in order to form an artificial slimy layer on the filter bed so that the colloidal and bacteriological impurities are removed. The function of the deflector plate provided at the top is to distribute the slimy layer uniformly over the top of the filter bed.

Like the sand gravity filter, the pressure filter gets clogged after some time and thus it becomes necessary to clean it periodically. This is done by *back-washing*. Clear water is forced up from the bottom, it agitates the sand and carries with it the clogging material. The wash water is removed from the outlet, provided at the top.

Now-a-days compressed air is passed to agitate sand before back washing.

**[B] Sterilization of Water (Removal of Microorganisms)**

Although most of bacteria and micro-organisms present in water are removed by coagulation followed by settling and filtration through sand beds, yet water contains a small percentage of pathogenic bacteria. These bacteria must be completely removed particularly from water supplied by municipality for drinking purposes, otherwise water-borne diseases may be produced. *The process of destroying these pathogenic (disease causing) bacteria and microorganisms is known as sterilization or disinfection.* The chemicals used for sterilization are known as *sterilizers*. Water can be sterilized in the following ways.

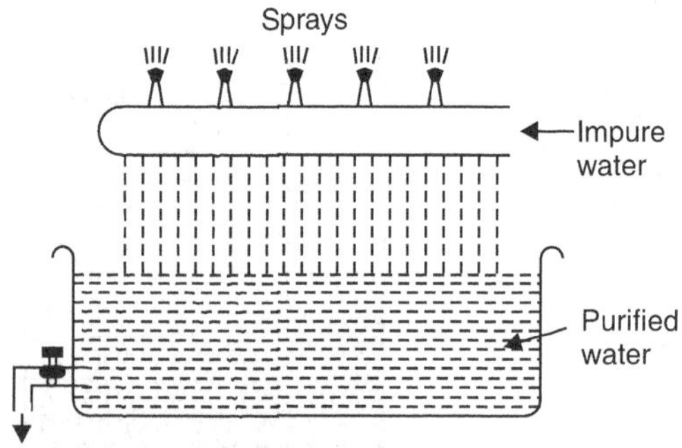


*FIGURE 10.14* Aeration.

(*i*) **Aeration.** *The process of spraying water in the form of fine droplets into the atmosphere is known as aeration.*

Water is forced under pressure through a perforated pipe. As water sprays into air, it comes in contact with oxygen of air and ultra-violet rays of the sun. Oxygen and ultra-violet rays destroy organic matter and bacteria.

Natural aeration takes place in streams and rivers when the water flows slowly in its bed or when it forms a fall.

(*ii*) **Sterilization by ultraviolet rays.** Although sunlight is also helpful in destroying bacteria, it cannot penetrate large depths of water and hence it is not an efficient method for the sterilization of water. On the other hand, ultra-violet rays destroy bacteria quite efficetively. Ultra-violet rays have been used for cleaning swimming pool water. In this process, UV light (from mercury lamp enclosed in a quartz globe and placed under water) in focussed on flowing water. However, the method is very costly and cannot be applied for water works.

(*iii*) **Boiling.** Boiling is a good method of sterilizing the water since boiling of water for 15—20 minutes destroys all the disease causing bacteria. But the method is useful only for household purposes during epidemic like cholera, etc. because the process is very expensive for town supply water.

(*iv*) **Sterilization by bleaching powder.** Bleaching powder is a good sterilizer for small water works. In practice about 1000 liters of water is mixed well with about 1 kg of the powder and the resulting solution allowed to stand for several hours. Hypochlorous acid (HOCl) and nascent oxygen produced by the action of water on bleaching powder are powerful germicides.

$$CaOCl_2 + H_2O \longrightarrow Ca(OH)_2 + Cl_2$$
$$Cl_2 + H_2O \longrightarrow HOCl + HCl$$
$$\underset{\text{hypochlorous acid}}{HOCl} \longrightarrow HCl + \underset{\text{(nascent oxygen)}}{O}$$

Bleaching powder should be used only in calculated amount because excess of it will give a bad taste and disagreeable odour, while lesser amount of it will not sterilize the water completely. The quantity of bleaching powder required for complete sterilization depends upon the following factors.

(*a*) The quantity of oxidisable matter like bacteria and microorganisms present in water.

(*b*) Turbidity of water.

(*c*) The temperature of reaction.

(*d*) Time allowed for the reaction.

However, the use of bleaching powder has the following disadvantages.

(*i*) It should always be analysed for its chlorine content before use, because it deteriorates due to its continuous decomposition during storage.

(*ii*) It introduces calcium in water which in turn makes water hard.

(*iii*) If used in excess amount, it imparts a bad taste and disagreeable odour to water.

(*iv*) **Sterilization by chlorine.** Chlorine is a commonly used sterilizer in municipal water supply. It can be used directly as a gas or as chlorine water. Its germicidal action is based on its reaction with water. It reacts with water to form hypochlorous acid and nascent oxygen both of which are powerful germicides.

$$Cl_2 + H_2O \longrightarrow HOCl + HCl \qquad HOCl \longrightarrow HCl + [O]$$

Liquid chlorine has several advantages over bleaching powder as a sterilizer.

(*a*) It does not decompose on standing and can be stored in a very little space.

(*b*) It can be obtained in pure form.

(*c*) It is more effective and economical.

(*d*) It can be used at high as well as low temperature.

(*e*) No impurity is introduced into water.

However, excess of chlorine should be avoided because it produces characteristic unpleasant taste and odour and irritating effect on mucous membrane. For filtered water about 0.3–0.5, ppm of chlorine in sufficient. The treated water should not contain more than 0.1–0.2 ppm of free chlorine.

*Process.* Water is treated with chlorine in a tower known as *chlorinator*, provided with a number of buffle trays (Fig. 10.15). Chlorine and water are introduced at the top. During their passage through the tower, they get thoroughly mixed. The treated water is taken out from the bottom.

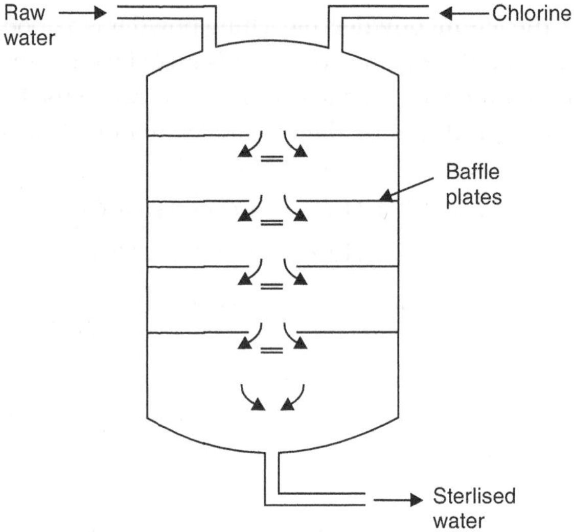


***FIGURE 10.16* Chlorinator.**

*Break-point chlorination.* When chlorine is added to a sample water and its amount is estimated after a few minutes, it will be found that the available chlorine is less than the added. The difference is due to the chlorine used up in oxidising some oxidisable matter (*e.g.*, bacteria) present in water. If we go on adding increasing amounts of chlorine in equal quantities of different samples of water and analyse the samples after a few seconds for the residual chlorine, we find that the quantity of available chlorine increases with the increase in quantity of chlorine added.

If we plot a graph between chlorine added *versus* residual chlorine, a straight line curve is obtained until a point (known as *break-point*) is reached where the addition of a small amount of chlorine decreases abruptly the amount of residual chlorine.

After this point, the quantity of residual chlorine after each increment of added chlorine agrees fairly well with the quantity of chlorine added. This is due to the fact that certain organic

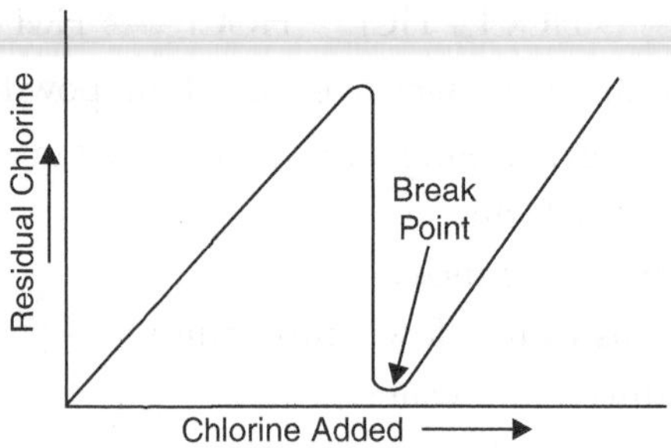


***FIGURE 10.16* Break-point chlorination curve.**

compounds which resist oxidation at a lower concentration of chlorine are oxidised after the break-point concentration of chlorine. Such compounds which are oxidised only after break point chlorination cause bad taste and/or odour in water. So chlorination upto break point removes these undesirable taste and odour producing substances from water.

Thus we may define the *break point chlorination of water as the chlorination of water to such an extent that living organism as well as other organic impurities in water are destroyed.*

*Dechlorination.* Over-chlorination after the break point produces unpleasant taste and odour of chlorine in water. These objectionable properties may be removed by the process known as ***dechlorination***. Dechlorination is achieved by filtering the overchlorinated water through a bed of granular carbon or activated carbon which removes decomposition products as well as excess of chlorine.

Alternatively, the effect of over chlorination may be removed *i.e.*, dechlorination may be done by the addition of sulphur dioxide or sodium sulphite to the super chlorinated water.

$$H_2O + SO_2 \longrightarrow H_2SO_3$$
$$H_2O + Cl_2 \longrightarrow HClO + HCl$$
$$HClO + H_2SO_3 \longrightarrow HCl + H_2SO_4$$
$$\overline{SO_2 + Cl_2 + 2H_2O \longrightarrow 2HCl + H_2SO_4}$$
$$Na_2SO_3 + Cl_2 + H_2O \longrightarrow Na_2SO_4 + 2HCl$$

(*v*) **Sterilization with chloramine.** Ammonia and chlorine when mixed in the ratio of 1 : 2 by volume produce a compound, known as chloramine.

$$Cl_2 + NH_3 \longrightarrow \underset{\text{chloramine}}{ClNH_2} + HCl$$

Chloramine is a quite stable compound and does not impart any disagreeable smell or bad taste to water, hence it is considered as better germicidal than chlorine alone.

$$ClNH_2 + H_2O \longrightarrow HOCl + NH_4OH$$
$$HOCl \longrightarrow HCl + [O]$$

(*vi*) **Sterilization with ozone.** Ozone is an unstable compound and decomposes to molecular oxygen and nascent oxygen.

$$O_3 \longrightarrow O_2 + [O]$$

The nascent oxygen thus released is very effective and kills all the germs and bacteria.

During ozone sterilization, ozone is injected through an injector into water in a contact tank. Sterilized water is collected from the top.

*Advantages of ozone as sterilizing agent*

(*i*) Ozone acts not only as a sterilizing agent but also as a bleaching, decolorising and deodourising agent.

(*ii*) It improves the taste of water and actually produces a very palatable water.

(*iii*) Its excess is not harmful since it is unstable and decomposes into oxygen.

**Removal of slime and algae.** Algae in water is produced by a type of air-borne bacteria (*Bacillus subtilis*). It can cause a lot of problem in cooling systems such as air-conditioning. In air-conditioning, spray nozzles and even screens of circulating pumps are clogged by algae and thus efficiency of plant is decreased.

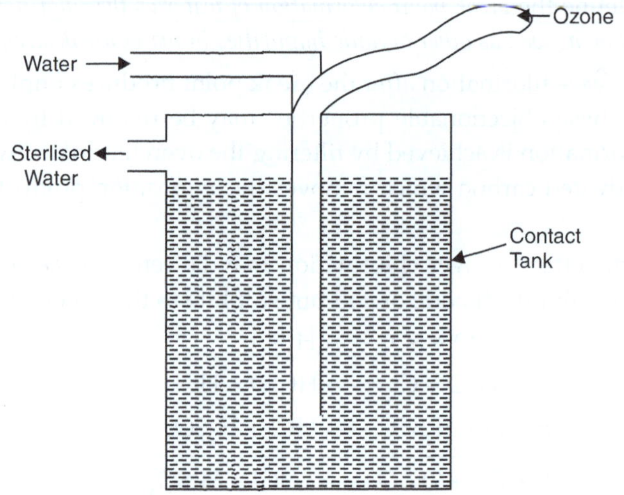


*FIGURE 10.17* **Sterilization with ozone.**

Growth of algae can be prevented by the use of certain chemicals, *viz.*, salts of mercury, copper and silver, sodium penta-chlorophenate and quaternary ammonium compounds.

## 10.6 METHODS OF ANALYSIS OF WATER

1. **Estimation of Hardness.** As described earlier there are two types of hardness of water, *viz.* temporary and permanent. Temporary hardness is mainly due to the bicarbonates of calcium and magnesium, while the permanent hardness is due to chlorides and sulphates of calcium and magnesium.

   Hardness of water can be estimated by the following three methods.

   (*i*) **O. Hehner's method.** Determination of *temporary hardness* is based upon the fact that it can be removed by boiling.

$$Ca(HCO_3)_2 \xrightarrow{\text{boil}} CaCO_3 \downarrow + H_2O + CO_2$$

$$Mg(HCO_3)_2 \xrightarrow{\text{boil}} MgCO_3 \downarrow + H_2O + CO_2$$

Thus it can be determined by estimating the alkalinity of the water before and after boiling.

*Procedure.* Take 50 ml of the sample water in a conical flask and add to it 2–3 drops of methyl orange. Titrate it against N/50 HCl solution and note the volume of HCl consumed.

Now take 100 ml of the same sample water in a pyrex beaker and evaporate it to dryness. To the residue, add 50 ml of distilled water, warm and then stir to dissolve the soluble portion. Filter off the insoluble material (carbonates), wash the residue with distilled water and then make up the filtrate to 100 ml in a measuring flask by adding more distilled water. Titrate 50 ml of this solution against N/50 HCl using methyl orange as indicator and note the volume of acid used.

*Calculations :* Let vol. of N/50 acid used by 50 ml of water before boiling = $V_1$ ml.

and vol. of N/50 acid used by 50 ml of water after boiling = $V_2$ ml.

∴ Vol. of N/50 HCl used by temporary hardness present in 50 ml. of water = $(V_1 + V_2)$ ml.

According to normality formula

$$1000 \text{ ml N/50 HCl} \equiv 1000 \text{ ml of N/50 } CaCO_3 \text{ (Hardness in terms of } CaCO_3)$$

$$\therefore \quad 1000 \text{ ml of N/50 } CaCO_3 \equiv \frac{50 \times 1}{50} = 1 \text{ gm of } CaCO_3 \qquad [\because \text{ Eq. wt. of } CaCO_3 = 50]$$

$$\therefore \quad 1000 \text{ ml of N/50 HCl} \equiv 1 \text{ gm of } CaCO_3$$

$$(V_1 - V_2) \text{ ml of N/50 HCl} = \frac{1 \times (V_1 - V_2)}{1000} \text{ gm of } CaCO_3$$

$$\text{Now 50 ml of water} = \frac{(V_1 - V_2)}{1000} \text{ gm of } CaCO_3$$

$$10^6 \text{ ml of water} = \frac{(V_1 - V_2) \times 10^6}{1000 \times 50} \text{ gm of } CaCO_3$$

$$= 20\,(V_1 - V_2) \text{ gm of } CaCO_3$$

∴ Temporary hardness = **20** $(V_1 - V_2)$ **ppm.**

Determination of *permanent hardness* is based on the fact that it can be removed by boiling with sodium carbonate with which chlorides and sulphates of calcium and magnesium form insoluble carbonates.

$$CaCl_2 + Na_2CO_3 \longrightarrow CaCO_3 \downarrow + 2NaCl$$

$$MgSO_4 + Na_2CO_3 \longrightarrow MgCO_3 \downarrow + Na_2SO_4$$

In practice, an excess but known amount of standard sodium carbonate is used and the residual sodium carbonate is then determined by titrating against a standard acid solution. The amount of $Na_2CO_3$ used is equivalent to permanent hardness of water.

*Procedure.* Boil 50 ml of the water sample with 50 ml of N/50 sodium carbonate solution in a beaker. Cool the resulting solution, filter off the precipitate and then wash it with distilled water, collecting all the filtrate and washings in the conical flask. Titrate the contents of the conical flask against N/50 HCl using methyl orange as indicator.

*Calculations.* Let the vol. of N/50 HCl used during titration = $V$ ml.

∴ Vol. of N/50 $Na_2CO_3$ used for removing permanent hardness of the water sample

$$= (50 - V) \text{ ml.}$$

According to normality formula

$$1000 \text{ ml of N/50 } Na_2CO_3 \equiv 1000 \text{ ml. of N/50 } CaCO_3$$

But 1000 ml of N/50 $CaCO_3$ contains $50 \times \frac{1}{50} = 1$ gm of $CaCO_3$ [$\because$ Eq. wt. of $CaCO_3 = 50$]

$\therefore$ 1000 ml of N/50 $Na_2CO_3 \equiv 1$ gm of $CaCO_3$

$$(50 - V) \text{ ml of N/50 } Na_2CO_3 \equiv \frac{1 \times (50 - V)}{1000} \text{ gm of } CaCO_3$$

$$\text{Now 50 ml of hard water} \equiv \frac{(50 - V)}{10^3} \text{ gm of } CaCO_3$$

$$10^6 \text{ ml of hard water} \equiv \frac{(50 - V) \times 10^6}{10^3 \times 50} \text{ gm of } CaCO_3 = \frac{(50 - V) \times 10^6}{10^3 \times 50} \text{ ppm}$$

$\therefore$ Permanent hardness $\equiv 20\,(50 - V)$ ppm.

*Magnesium hardness* of water can be estimated by treating sample water with an excess but known amount of standard caustic soda solution, when magnesium chloride and sulphate are precipitated as magnesium hydroxide.

$$MgCl_2 + 2NaOH \longrightarrow Mg(OH_2) \downarrow + 2NaCl$$
$$MgSO_4 + 2NaOH \longrightarrow Mg(OH_2) \downarrow + Na_2SO_4$$

The amount of unreacted NaOH is then determined by titration against a standard acid. Thus the amount of NaOH consumed which is equivalent to the magnesium hardness can be determined.

It is important to note that caustic soda decomposes bicarbonates (temporary hardness) to form sodium carbonate which in turn will react with calcium salts to form insoluble calcium carbonate.

$$Ca(HCO_3)_2 + 2NaOH \longrightarrow CaCO_3 \downarrow + Na_2CO_3 + 2H_2O$$
$$Mg(HCO_3)_2 + 2NaOH \longrightarrow Mg(OH)_2 \downarrow + 2Na_2CO_3 + 2H_2O$$

To avoid this, it is therefore necessary that prior to boiling with caustic soda the sample water is neutralised with HCl (using methyl orange as indicator) to convert bicarbonate into chloride.

$$Ca(HCO_3)_2 + 2HCl \longrightarrow CaCl_2 + 2H_2O + 2CO_2$$
$$Mg(HCO_3)_2 + 2HCl \longrightarrow MgCl_2 + 2H_2O + 2CO_2$$

*Procedure.* Take 50 ml of sample water in a beaker. Neutralise it to methyl orange end point with N/10 HCl. Boil off carbon dioxide gas. Add 50 ml of N/50 NaOH solution. Boil the contents to precipitate out magnesium hardness. Cool and filter the solution. Titrate the filtrate against N/50 HCl solution using methyl orange as indicator.

*Calculations.* Let the vol. of N/50 HCl used = $V$ ml.

$\therefore$ Vol. of N/50 NaOH used for removing magnesium hardness in 50 ml of water

$$= (50 - V) \text{ ml.}$$

$\therefore$ Strength of magnesium hardness in terms of $CaCO_3$ equivalent

$$= \frac{(50 - V) \times 50}{50 \times 50} \text{ gm/litre}$$

$$\therefore \quad \text{Magnesium hardness} = \frac{(50 - V) \times 50 \times 10^6}{50 \times 50 \times 10^3} \text{ ppm.} = 20(50 - V) \text{ ppm.}$$

(*ii*) **Soap titration method.** This method is based on the fact that soluble soap (sodium or potassium salt of a higher fatty acid) when added to hard water first forms insoluble compounds with the metal ions producing hardness and thus does not produce any lather. But as soon as all such metal ions are precipitated out, addition of further drop of soap solution will produce lather.

$$\underset{\text{Sod. stearate (soap)}}{2C_{17}H_{35}COONa} + Ca(HCO_3)_2 \longrightarrow \underset{\text{Cal. stearate}}{(C_{17}H_{35}COO)_2\,Ca} \downarrow + 2NaHCO_3$$

$$2C_{17}H_{35}COONa + MgCl_2 \longrightarrow (C_{17}H_{35}COO)_2Mg \downarrow + 2NaCl$$

$$2C_{17}H_{35}COONa + CaSO_4 \longrightarrow (C_{17}H_{35}COO)_2Ca \downarrow + Na_2SO_4$$

Thus the total hardness can be determined by titrating the water sample against a standard soap solution. On the other hand, permanent hardness can be determined by first precipitating the temporary hardness by boiling and then titrating the filtrate against standard soap solution. The end point is the appearance of a stable lather at least for two minutes.

#### Procedure

1. *Preparation of standard hard water.* Dissolve 1.0 gm of pure and dry calcium carbonate in the minimum amount of dil. HCl. Evaporate the solution to dryness. Dissolve the residue in one liter of distilled water. The hardness of this solution of 1g/liter or 1000 mg/liter or 1000 ppm or 1 mg/ml.
2. *Preparation and standardization of soap solution.* Dissolve 10 gm of pure and dry soap in one liter of 80% alcohol. Take this solution in the burette. Take 50 ml of hard water in a stoppered bottle and add to it soap solution from the burette dropwise and with constant shaking till a permanent (at least for a minute) lather is formed. Let the volume of soap used be $V_1$ ml.
3. *Lather factor of distilled water.* Titrate 50 ml. of distilled water against the above solution soap till permanent lather is formed. Let the volume of soap used is $V$ ml. This is lather factor and must be subtracted from all the titre values.
4. *Total hardness.* Titrate 50 ml of hard water against the soap solution as above. Let the vol. of soap used is $V_2$ ml.
5. *Permanent hardness.* Boil 250 ml of the water sample in a pyrex beaker till the volume is reduced to 50 ml to convert all bicarbonates to carbonates. Filter off the precipitate, wash the ppt. with the distilled water and collect all the filtrate and washings in a 250 ml. measuring flask. Make up the solution with distilled water. Titrate 50 ml. of this solution against soap solution and let $V_3$ ml of soap solution is used.

#### *Calculations*

(*i*) *Strength of soap solution.* Since 1 ml of standard hard water contains 1 mg of hardness causing substances in terms of $CaCO_3$,

50 ml of standard hard water = 50 mg of $CaCO_3$

Also 50 ml of standard water = $(V_1 - V)$ ml. of soap solution

$(V_1 - V)$ ml of soap solution = 50 mg of $CaCO_3$

or 1 ml of soap solution = $\frac{50}{(V_1 - V)}$ mg of $CaCO_3$

*(ii) Calculation of total hardness*

Since 1 ml of soap solution = $\frac{50}{(V_1 - V)}$ mg of $CaCO_3$

$(V_2 - V)$ ml of soap solution = $\frac{(V_2 - V) \times 50}{(V_1 - V)}$ mg $CaCO_3$

Thus 50 ml of the sample water contains = $\frac{(V_2 - V) \times 50}{(V_1 - V)}$ mg of $CaCO_3$

$\therefore$ 1000 ml (1 litre) of water sample = $\frac{(V_2 - V) \times 1000 \times 50}{(V_1 - V) \times 50}$ mg of $CaCO_3$

$\therefore$ Total hardness of water sample = $\frac{(V_2 - V) \times 1000}{(V_1 - V)}$ mg/liter

$= \frac{(V_2 - V) \times 1000}{(V_1 - V)}$ ppm. [$\because$ 1000 mg/liter = 1000 ppm]

*(iii) Calculation of permanent hardness*

Again, since 1 ml. of soap solution = $\frac{50}{(V_1 - V)}$ mg of $CaCO_3$

$(V_3 - V)$ ml. of soap solution = $\frac{(V_3 - V) \times 50}{(V_1 - V)}$ mg of $CaCO_3$

Thus 50 ml of the given water contains = $\frac{(V_3 - V) \times 50}{(V_1 - V)}$ mg of $CaCO_3$

$\therefore$ 1000 ml. of boiled water = $\frac{(V_3 - V) \times 1000 \times 50}{(V_1 - V) \times 50}$ mg of $CaCO_3$

$\therefore$ Permanent hardness $= \frac{(V_3 - V) \times 1000}{(V_1 - V)}$ mg liter

$= \frac{(V_3 - V) \times 1000}{(V_1 - V)}$ ppm

(*iv*) $\therefore$ *Temporary hardness* = Total hardness – Permanent hardness

$$= \frac{(V_2 - V) \times 1000}{(V_1 - V)} - \frac{(V_3 - V) \times 1000}{(V_1 - V)}$$

$$= \frac{1000}{V_1 - V} [(V_2 - V) - (V_3 - V)] \text{ ppm}$$

$$= \frac{1000 \times (V_2 - V_3)}{(V_1 - V)} \text{ ppm}$$

In short, **Total hardness** $= x \times \frac{V_2 - V}{V_1 - V} \times 1000$ **ppm**

**Permanent hardness** $= x \times \frac{(V_3 - V)}{(V_1 - V)} \times 1000$ **ppm**

**Temporary hardness = Total hardness – Permanent hardness**

where, $x$ = Amount of $CaCO_3$ in gm per liter or mg per ml in standard hard water.

$V$ = Volume of soap solution required by 50 ml of distilled water.

$V_1$ = Volume of soap solution used by 50 ml of standard hard water.

$V_2$ = Volume of soap solution used by 50 ml of the given hard water solution.

$V_3$ = Volume of soap solution used by 50 ml of the boiled hard water.

(*iii*) **EDTA method.** This method is based on the following facts.

(*i*) Erichrome black-T (a blue colored dye) forms wine red colored unstable complex with calcium and magnesium ions (present in water).

$$\underset{\text{(blue colored dye)}}{Ca^{2+} + \text{Erichrome black-T}} \longrightarrow \underset{\text{(unstable wine red complex)}}{Ca^{2+} \text{ chrome black-T complex}}$$

(*ii*) EDTA (ethylene diamine tetra-acetic acid) takes calcium and magnesium ions from the above unstable complex and forms a colorless stable complex with the regeneration of blue colored dye (erichrome black-T).

$$Ca^{2+} \text{ chrome black-T complex} + \text{EDTA} \longrightarrow \underset{\text{(colorless)}}{Ca^{2+} \text{ EDTA}} + \underset{\text{(blue)}}{\text{Erichrome black}}$$

Thus the amount of used EDTA corresponds the hardness of water ; the end point of the titration will be change of color from wine red to blue. The important steps of the method are as follows.

1. *Preparation of standard water.* Described already.
2. *Preparation of buffer solution.* Take about 575 ml of concentrated ammonia solution in 1 liter flask and add to this 67.5 ml of $NH_4Cl$. Mix the solution well and then dilute to 1 liter by adding distilled water.
3. *Preparation and standardisation of EDTA solution.* Dissolve about 4 gm EDTA and 0.1 gm of $MgCl_2$ in one liter of water. Rinse and fill the burette with this EDTA solution. Pipette out

50 ml of standard hard water in a conical flask. Add to it 10–15 ml of buffer solution and 4–5 drops of eriochrome black-T (indicator). Titrate the solution with EDTA solution till wine red colour changes to blue. Let $V_1$ ml of EDTA solution is used during this titration.

4. *Titration of sample water.* Titrate 50 ml of sample water using erichrome black-T as indicator exactly in the above manner and let $V_2$ ml of EDTA is used for this.
5. *Permanent hardness.* Boil 250 ml of the water sample in a pyrex beaker till the volume is reduced to 50 ml to convert all bicarbonates to carbonates. Filter off the precipitate, wash the precipitate with distilled water and collect all the filtrate and washings in a 250 ml measuring flask. Make up the solution with distilled water. Titrate 50 ml of this solution against EDTA solution using eriochrome black-T as indicator, and let $V_3$ ml of EDTA is used.
6. *Calcium hardness.* Calcium hardness alone can also be determined by this method using murexide indicator in place of erichrome T. Take 50 ml. of the boiled filtrate in a conical flask. Add to it 50 ml of N/10 caustic soda solution and two drops of murexide indicator. Titrate the pink colored solution against EDTA solution till violet color is obtained. This is the end point and is obtained as soon as whole of the calcium hardness is removed. Let the volume of EDTA solution used in $V_4$ ml.

*Calculations :*

(*i*) *Strength of EDTA solution,*

Since 1 ml. of the standard hard water contains 1 mg of hardness causing substance in terms of $CaCO_3$.

50 ml. of standard hard water contains = 50 mg of $CaCO_3$

Also 50 ml. of standard hard water = $V_1$ ml. of EDTA solution

$$\therefore \quad V_1 \text{ ml. of EDTA solution} = 50 \text{ mg of } CaCO_3$$

$$1 \text{ ml. of EDTA solution} = \frac{50}{V_1} \text{ mg of } CaCO_3$$

(*ii*) *Calculation of total hardness*

$$\text{Since 1 ml. of EDTA solution} = \frac{50}{V_1} \text{ mg of } CaCO_3$$

$$V_2 \text{ ml. of EDTA solution} = \frac{50 \times V_2}{V_1} \text{ mg of } CaCO_3$$

$$\text{Thus 50 ml. of sample water contains} = \frac{50 \times V_2}{V_1} \text{ mg of } CaCO_3$$

$$\text{1000 ml. of sample water contains} = \frac{50 \times V_2 \times 1000}{V_1 \times 50} \text{ mg} = \frac{V_2 \times 1000}{V_1} \text{ mg/liter}$$

$$\therefore \quad \text{Total hardness} = \frac{V_2 \times 1000}{V_1} \text{ppm} \qquad [\because \ 1 \text{ mg/liter} = 1 \text{ ppm}]$$

(*iii*) *Calculation of permanent hardness*

Applying the above steps and replacing $V_2$ (vol. of EDTA sol. used by total hardness) by $V_3$ (vol. of EDTA sol. used by permanent hardness), we get

$$\text{Permanent hardness} = \frac{V_3 \times 1000}{V_1} \text{ ppm}$$

(*iv*) *Calculation of temporary hardness*

Temp. hardness = Total hardness – Permanent hardness

(*v*) *Calculation of calcium hardness.* Applying the steps as in (*ii*) and replacing $V_2$ by $V_4$ (volume of EDTA used by calcium hardness), we get

$$\text{Calcium hardness} = \frac{V_4 \times 1000}{V_1} \text{ ppm}$$

(*vi*) *Calculation of magnessium hardness*

Mg hardness = Permanent hardness – Calcium hardness

**2. Alkalinity**

$Mg(HCO_3)_2$. These can be estimated by titrating against a standard solution of acid using phenolphthalein and methyl orange as indicators.

*Procedure.* Take 100 ml of the water to be analysed in a titration flask. Add to it two or three drops of phenolphthalein indicator and titrate the sample by adding N/50 sulphuric acid from a burette till the pink color is just discharged. To this solution now add 2 or 3 drops of methyl orange and continue adding N/50 $H_2SO_4$ till pink color reappears.

*Calculations :* Let vol. of acid used to phenolphthalein end-point = $V_1$ ml.

and total vol. of acid used to methyl orange end-point = $V_2$ ml.

∴ Phenolphthalein alkalinity (P) in terms of $CaCO_3$ equivalent

$$= \frac{V_1 \times 50 \times 10{,}00{,}000}{50 \times 100 \times 1000} = 10\, V_1 \text{ ppm}$$

and methyl orange alkalinity (M) in terms of $CaCO_3$ equivalent

$$= \frac{V_2 \times 50 \times 10{,}00{,}000}{50 \times 100 \times 1000} = 10\, V_2 \text{ ppm}$$

Now since phenolphthalein alkalinity indicates half the carbonates (if any) plus all the hydroxides (if any), while the methyl orange alkalinity indicates total alkalinity (due to hydroxides, bicarbonates and carbonates); the calculation of hydroxides, carbonates and bicarbonates in the various samples can be made with the help of the following table.

| *Alkalinity* | *$OH^-$ (ppm)* | *$CO_3^{2-}$ (ppm)* | *$HCO_3^-$ (ppm)* |
|---|---|---|---|
| P = 0 | 0 | 0 | M |
| P = M | M | 0 | 0 |
| $P = \frac{1}{2}M$ | 0 | 2P | 0 |
| $P < \frac{1}{2}M$ | 0 | 2P | M−2P |
| $P > \frac{1}{2}M$ | (2P−M) | 2(M−P) | 0 |

### 3. Sulphate

Sulphate ions in water are usually present in the form of compounds like $Na_2SO_4$, $CaSO_4$, $MgSO_4$, etc. The estimation of sulphate ions in water is based on the fact that sulphate ions react with benzidine hydrochloride to form benzidine sulphate which is insoluble in cold and soluble in hot to form sulphuric acid.

$$\underset{\text{Benzidine hydrochloride}}{C_{12}H_8(NH_2)_2.2HCl} + Na_2SO_4 \longrightarrow \underset{\substack{\text{Benzidine sulphate}\\ \text{(insoluble in cold)}}}{C_{12}H_8(NH_2)_2.H_2SO_4} \downarrow + 2NaCl$$

$$C_{12}H_8(NH_2)_2.H_2SO_4 + 2H_2O \xrightarrow{\text{heat}} C_{12}H_8(NH_2)_2.2H_2O + H_2SO_4$$

The free sulphuric acid can be titrated against a standard caustic soda solution using phenolphthalein as indicator.

*Procedure.* Take 100 c.c. of the water sample and add to it 10 c.c. of benzidine hydrochloride solution. Mix the contents thoroughly and allow the precipitate to settle. Filter off the precipitate through Whatmann filter paper No. 4. Wash the precipitate on filter paper to free of acid with a minimum amount of cold distilled water. Transfer the filter paper along with precipitate to a conical flask, add to 50 ml of distilled water and warm the contents to about 50°C to dissolve the precipitate. Add 2 to 3 drops of phenolphthalein and titrate the liberated $H_2SO_4$ against N/50 caustic soda solution till a permanent pink color is obtained.

*Calculations:*

$$\text{Amount of } SO_4^{2-} = \frac{V_4 \times 48 \times 10{,}00{,}000}{50 \times 100 \times 1000} = 9.6\ V_4 \text{ p.p.m.}$$

$$\text{Amount of } SO_4^{2-} \text{ as } CaCO_3 \text{ equivalent} = \frac{V_4 \times 50 \times 10{,}00{,}000}{50 \times 100 \times 1000} = 10\ V_4 \text{ p.p.m.}$$

where $V_4$ is the vol. of N/50 NaOH solution used during the titration.

### 4. Chlorides

In water, chloride ions are present in the form of compounds like $CaCl_2$, $MgCl_2$, NaCl, etc. The chloride ions are estimated by titrating the water sample against a standard solution of silver nitrate using potassium chromate as indicator.

When all the chloride ions are precipitated out as AgCl, potassium chromate (indicator) reacts with the excess drops of $AgNO_3$ to form red coloured silver chromate (end point).

$$NaCl + AgNO_3 \longrightarrow AgCl \uparrow + NaNO_3 \; ; K_2CrO_4 + 2AgNO_3 \longrightarrow Ag_2CrO_4 \uparrow + 2KNO_3$$

The end-point of the titration will thus be the change in color from bright yellow (due to $K_2CrO_4$) to red (due to $Ag_2CrO_4$).

*Procedure.* Pippette out 100 ml of the water sample in a clean titration flask. Add to it 2–3 drops of potassium chromate. Run N/50 $AgNO_3$ solution from the burrette to the flask till the color changes from yellow to brick-red.

*Calculations:*

$$\text{Chloride ions} = \frac{V_3 \times 35.5 \times 10{,}00{,}000}{50 \times 100 \times 1000} = 7.1\ V_3 \text{ ppm}$$

$$\text{Chloride ions as } CaCO_3 \text{ equivalent} = \frac{V_3 \times 50 \times 10{,}00{,}000}{50 \times 100 \times 1000} = 10\ V_3 \text{ ppm}$$

where $V_3$ is the vol. of N/50 $AgNO_3$ used during titration.

5. **Total solids**

Since high concentration of dissolved solids in water used in boilers leads to problems like priming and foaming, it becomes necessary to know its concentration. The amount of total solids in a water sample can be determined very easily.

For this, weigh accurately a thoroughly cleaned and completely dry evaporating dish. Add to it 100 ml. of the water sample and evaporate the contents carefully to dryness. Place the dish in an oven at about 110 – 120°C, and when no water seems to be present in the dish, cool it in desiccator and weigh.

*Calculations.*

Let the wt. of empty dry dish $= W_1$ gm

and the wt. of dish + residue $= W_2$ gm

$\therefore$ wt. of residue $= W_2 - W_1 = W$ gm

Now 100 ml. of sample water contains *W* gm of total solids

$$10^6 \text{ ml. of sample water will contain} = \frac{W \times 10^6}{100} \text{ ppm} = \mathbf{10^4\ W\ ppm}$$

6. **Dissolved carbon dioxide**

Take 250 ml. of sample water in a large beaker or conical flask. Add to it 2–3 drops of phenolphthalein indicator. Titrate it against N/10 sodium carbonate solution till a permanent pink colour is obtained. Let *V* ml. of $Na_2CO_3$ solution is used during the titration.

Thus dissolved $CO_2$ in the sample will be **8.8 V ppm.**

7. **Dissolved oxygen**

This determination is based on the fact that oxygen (dissolved in water) oxidises potassium iodide to iodine. The liberated iodine is titrated against a standard hypo solution as usual. However, since dissolved oxygen in water is in molecular state, it as such can't oxidise potassium iodide. Therefore, an oxygen carrier such as manganese hydroxide is used to bring about the reaction between KI and oxygen. Manganese hydroxide, in turn, is obtained by the action of KOH on manganese sulphate.

$$MnSO_4 + 2KOH \longrightarrow Mn(OH)_2 + K_2SO_4$$
$$2Mn(OH)_2 + O_2 \longrightarrow 2MnO(OH)_2$$
$$MnO(OH)_2 + H_2SO_4 \longrightarrow MnSO_4 + 2H_2O + [O]$$
$$2KI + H_2SO_4 + [O] \longrightarrow K_2SO_4 + H_2O + I_2$$
$$2Na_2S_2O_3 + I_2 \longrightarrow Na_2S_4O_6 + 2NaI$$

However, some samples of water like rain water contains ions like nitrites and sulphites which can oxidise potassium iodide to iodine with the result, determination of dissolved oxygen in their presence is liable to be wrong. To overcome this problem, sodium azide is used in alkaline iodide solution to decompose the nitrite or sulphite.

$$2NaN_3 + H_2SO_4 \longrightarrow Na_2SO_4 + 2HN_3$$
$$HNO_2 + HN_3 \longrightarrow N_2O + N_2 + H_2O$$

*Procedure.* Take 250 ml of sample water, 2 ml. of manganous sulphate solution and 2 ml.of alkaline potassium iodide solution in a bottle. Stopper the bottle and then shake thoroughly. When the precipitate is settled, add 2 ml. of conc. HCl and shake the bottle until the precipitate has completely dissolved. Take 100 ml. of this solution and titrate against N/100 hypo, solution using starch as indicator till the blue color disappears. If *V* ml. of hypo is used, then

$$\text{Amount of dissolved oxygen} = \frac{0.0008\ V \times 10{,}00{,}000}{1000} = \mathbf{0.8\ V\ ppm.}$$

8. **Chemical oxygen demand (COD)**

The chemical oxygen demand is defined as the amount of oxygen used while oxidizing the organic matter content of a sample with a strong chemical oxidant under acidic conditions.

A known amount of $K_2Cr_2O_7$ is added to the measured amount of the sample and mixture is boiled with conc. $H_2SO_4$. The organic matter gets completely oxidized to produce $CO_2$ and $H_2O$

$$C_xH_yO_z + \left(x + \frac{y}{4} - \frac{z}{2}\right) O_2 \rightarrow x\,CO_2 + \frac{y}{2}\,H_2O$$

The excess of $K_2Cr_2O_7$ remained after the reaction is titrated against standard Mohr's salt solution using ferrion as indicator. The dichromate consumed gives the oxygen required for the oxidation of organic matter.

*Procedure.* Take 50 ml of the sample in a round bottomed flask and add 1 g of $HgSO_4$ and some broken porcelain pieces. Immerse the flask in cold water and slowly add 75 ml silver sulphate reagents with continuous shaking. (As this is an exothermic process that is why the flask is immersed in cold water). Now add 25 ml of $K_2Cr_2O_7$ (0.25N) to this soluition and mix the contents of the flask. Attach the reflux condenser and reflux for 2 hours. Wash the condenser with distilled water into the flask. Cool and dilute to about 300 ml by distilled water. Add 2–3 drops of ferroin as indicator and titrate against 0.25N Mohr's salt solution till the end point *i.e., from blue to wine red.* Record the volume of Mohr's salt solution used. Let it be X ml. Perform a blank titration using distilled water in place of sample solution. For this, take 50 ml of distilled water in a round bottomed flask, add he same amounts of reagents and reflux for two hours. Titrate it the same way as is done with the sample water. Record the volume of Mohr's salt solution used. Let it be Y ml.

*Calculations:*

$$\begin{bmatrix}\text{Volume of } K_2Cr_2O_7 \\ \text{to oxidize the organic} \\ \text{matter present in water sample}\end{bmatrix} = \begin{bmatrix}\text{Volume of Mohr's salt} \\ \text{solution equivalent to} \\ K_2Cr_2O_7\end{bmatrix} = [Y - X] \text{ ml}$$

$$\underset{\text{(sample)}}{N_1V_1} = \underset{\text{(Mohr's salt solution)}}{N_2V_2}$$

$$N_1 \times 50 = \frac{1}{4}\,(Y - X)$$

$$N_1 = \frac{1}{4 \times 50}\,(Y - X)$$

Therefore COD = $N_1$ × Eq. wt of oxygen = $N_1 \times 8$ g/L

The COD of the given sample is = $N_1 \times 8 \times 1000$ mg/L

**9. Biological oxygen demand**

The BOD in a water sample is directly related to the extent of pollution by sewage or other oxygen demanding waste. If the oxygen in water is *less than* Biological Oxygen Demand, the decomposition of microorganisms takes place anaerobically (in absence of oxygen) and hence produce highly objectionable end products including $H_2S$, $NH_3$ and $CH_4$.

Therefore sufficient oxygen must be present or supplied externally to maintain aerobic conditions so that organic matter decomposes into acceptable products. To know the amount to oxygen required by the given sample, the determination of BOD is necessary.

Polluted waters will continue to absorb oxygen for many months, till the oxidation gets completed and hence it is not practical to determine this ultimate oxygen demand which would require an extended period of time. As a result, it has become standard practice to simply measure the oxygen over a shorter period of 5 days. This is known as 5-day BOD test. $BOD_5$ is the total amount of oxygen consumed by micro organisms during the first five days of biodegradation.

In its simplest form $BOD_5$ would involve diluting a known volume of sample water with a known volume of pure water, whose oxygen content is already known. Let it be $D_1$. This is kept for 5 days at 20°C in a stoppered bottle, away from light. The dissolved oxygen is measured after the period of incubation. Let it be $D_2$. The difference between the original oxygen content and residual oxygen content will indicate the oxygen consumed by the water sample in five days in oxidizing the organic matter present in water sample. Thus $BOD_5$ is given by :

$$BOD_5 = D_1 - D_2 \times \frac{\text{Volume of sample after dilution}}{\text{Volume of sample before dilution}}$$

where $D_1$ = D.O of the sample in mg/L at the start of the experiment

$D_2$ = D.O of the sample in mg/L after 5 days.

= Loss of oxygen in mg/L × Dilution Factor.

The dilution of the waste water sample with pure water is necessary because waste water contains more of oxygen demanding material and hence the dissolved oxygen may not be sufficient to decompose it aerobically. *Procedure.* Take 1000 ml of distilled water in a bottle. Add 1 ml each of phosphate buffer (pH = 7.2), $CaCl_2$ (27.5 g/L), $MgSO_4$ (22.5 g/L) and $FeCl_3$ (0.25 g/L).

Water is aerated by passing compressed air through a diffusion tube untill it is completely saturated. This is known as dilution water. Take about 5ml of the sample and dilute it to 300 ml with dilution water prepared above. The diluted sample is taken in two bottles filled upto the neck.

The dissolved oxygen in one bottle is determined immediately and in the other bottle after five days of incubation. The procedure for determining the DO is the same as described earlier.

*Calculations.* Volume of the sample before dilution = 5 ml.

Volume of the sample after dilution = 300 ml.

* (*i*) D.O in bottle no. – 1.

$$\underset{\text{(Hypo)}}{N_1V_1} = \underset{(O_2 \text{ in sample})}{N_2V_2}$$

$$\frac{1}{100} \times V_1 = N_2 \times 100$$

$$N_2 = \frac{V_1}{10000}$$

Strength of dissolved oxygen = $N_2 \times$ Eq. wt

$$(D_1) = \frac{V_1}{10{,}000} \times 0.8 \text{ g/L.}$$

$$= 0.8\ V_1 \text{ ppm.}$$

(*ii*) D.O in bottle–2.

Similarly find D.O in bottle-No-2

Let it $D_2$

$$(iii)\ \text{BOD} = D_1 - D_2 \times \frac{\text{Volume of sample after dilution}}{\text{Volume of sample before dilution}} \text{ mg/L}$$

BOD of the given water sample = ...ppm

## 10.7 WASTEWATER TREATMENT

Wastewater, both domestic or industrial water have several undesirable components that are potentially harmful to the environment and human health. Thus, treatment of wastewater and its proper management has become a necessity. The main aim of the wastewater treatment is the removal of contaminants from wastewater so that it may be reused for beneficial purposes. Wastewater treatment is carried out in following stages:

### 10.7.1 Primary Treatment of Wastewater

This treatment is of a general nature and removes suspended solids, odor, color and neutralization of high or low *p*H. This stage is further divided into the following stages:

(*i*) **Screening.** In this stage coarse solids consisting of sticks, rags, brands, and other large objects are removed from the wastewater. The purpose of screening is to protect pumps and other mechanical equipments and to prevent clogging of valves and other apparatus in the wastewater plant.

(*ii*) **Comminuting.** This process consists of shredding of coarse solids to a size of approximately 8 mm. For this purpose a comminutor is located across the flow path of wastewater.

(*iii*) **Grit removal.** A grit in wastewater include inorganic solids such as pebbles, sand, silt, egg shell, glass, metal fragments, and organics such as bone chips, seeds, etc. Most of the substances in grit are abrasive in nature and cause accelerated wear on pumps and sludge handling equipments with which it comes in contact.

Grit is removed in grit channels or tanks. Grit being heavier than organic solids, can be separated from organic solids by careful regulation of flow velocity in the grit tanks. The grit settling chambers are periodically disconnected from the main system to remove the grit manually, for possible use of land filling and road making.

(*iv*) **Skimming tanks.** A skimming tank is so arranged that the floating matter like oil, fat, grease etc. rise and remain on the surface of wastewater until removed, while the liquid flows out continuously under partitions or baffles. The floating matter can be either hand or mechanically removed. This process can be made more efficient by aeration, chlorination or vacuum floatation.

(*v*) **Sedimentation.** In this step, settleable solids are removed by gravitational settling under quiescent conditions. The sludge formed at the bottom of the tank is removed either by vacuum suction or by taking it to a discharge point at the bottom of the tank for withdrawal.

The sedimentation tanks are designed to enable smaller and lighter particles to settle under gravity. Commonly used equipment include horizontal flow sedimentation tanks and center feed circular classifiers. The settled sludge is removed by mechanical scrapping into hoppers and pumping it out subsequently. Vertical flow tanks are often used in small treatment plants. An efficient sedimentation system is expected to remove about 90% of the suspended solids and 40% of organic matter.

Finely divided suspended solids and colloidal particles cannot be efficiently removed by simple sedimentation by gravity, in such cases, mechanical flocculation or chemical coagulation is employed.

(*a*) *Mechanical flocculation.* In this process wastewater is passed through a tank with a detention time of 30 minutes and fitted with paddles rotating at an optimum speed of 0.4 m/s. Under this gentle stirring, the finely divided suspended solids coalesc into larger particle and settle out.

(*b*) *Chemical coagulation.* In this process water is treated with certain chemicals such as hydrated lime, alum, $Al_2(SO_4)_3.18H_2O$, copperas, $FeSO_4.7H_2O$, ferric chloride and chlorinated copperas, $FeSO_4.Cl$ etc. which form a floc that absorbs and entrains the suspended and colloidal particles present. Coagulation is the most effective and economical means to remove impurities.

(*vi*) **Neutralization.** When *p*H of a wastewater is too high or too low then it is neutralized by acid or alkali. *Acidic effluents* are neutralized by treatment with limestone or lime slurry or caustic soda, depending upon the type and quantity of the waste. Alkaline wastes are neutralized by treatment with sulphuric acid or $CO_2$ or waste boiler flue gases.

### 10.7.2 Secondary or Biological Treatment of Wastewater

This process involves removing, stabilizing, and rendering harmless the very fine suspended matter and solids of the wastewater that remain even after the primary treatment. The dissolved and colloidal organic matter present in the waste water is removed by bacteria and other micro-organisms bringing about the following changes:

(*i*) Coagulation and flocculation of colloidal matter.

(*ii*) Oxidation of dissolved organic matter to $CO_2$.

(*iii*) Degradation of nitrogenous organic matter to ammonia.

Thus, this process removes BOD and appreciable amounts of oils and phenols. Two of the most commonly used systems for secondary waste water treatment are:

1. **Activated sludge process.** In this process the wastewater after primary treatment enters an areation tank where it comes in contact of sludges which is heavily laden with micro-organisms. Air is introduced into the tank and the bacterial flora bring about the biological degradation of waste into stabilized low energy compounds such as nitrates, sulphates, $CO_2$ and synthesize new bacterial cells. The bacterial flora grow and remains in the form of a floc called *activated sludge.* The effluent from the reaction tank is separated from the sludge by settling and is discharged. A part of the sludge is recycled to same tank for a fresh treatment cycle. The surplus sludge is digested in a sludge digester. Flow diagram of this process is given in Figure 10.18.

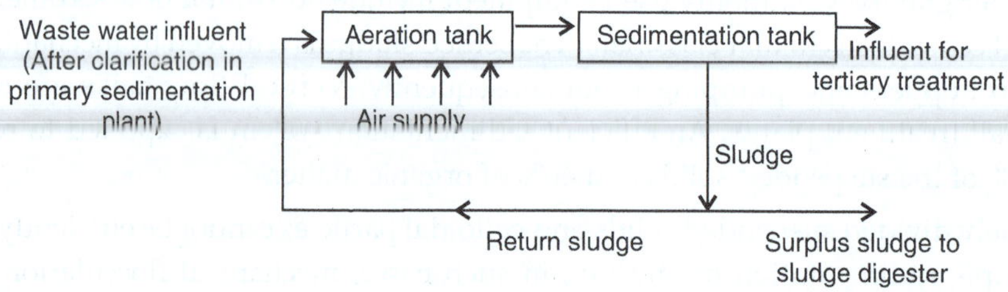


***FIGURE 10.18*** **Activated sludge process.**

This process produces a high quality effluent, with efficient systems, about 90–95% BOD removal is possible. It can be satisfactorily used for effluents from food processing sugar industries, etc. The disadvantage include high cost of operation and maintenance.

2. **Trickling filters.** These filters are also known as percolating filters and consist of circular or rectangular beds 1 to 3 m deep, made of broken stones, PVC, coal, coke, synthetic resins, gravel, etc. of size 40 mm to 150 mm, over which wastewater is sprinkled uniformly with the help of a slowly rotating distributor equipped with nozzles. The influent is sprinkled over the bed packing which is coated with a biological lime. As the liquid trickles over the bed, oxygen and dissolved organic matter diffuse into the film to be metabolized by the micro-organisms in this lime layer. As the microorganisms utilize the organic matter, the thickness of this lime film increases largely so that it cannot be supported by the solid media and thus gets detached from the surface, this is known as *sloughing* and is carried away along with the effluent. Effluent from the trickling filters settle in the settling tank and is then discharged. The sludge is then pumped to the sludge digestion unit. A schematic representation of the process is given in Figure 10.19.

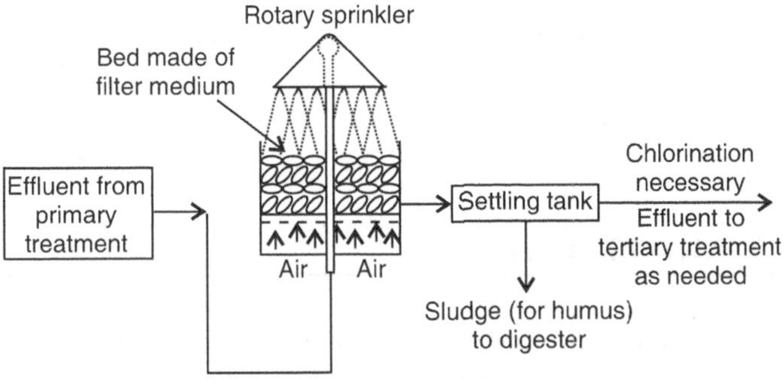


*FIGURE 10.19* **Trickling filter process.**

This process can reduce BOD to 65–85%. This process does not require manual attention and produce effluent of consistent and better quality. It has greater resistance to toxic waste as compared to activated sludge process. The disadvantages of the process include the cost of construction, the efficiency decreases with increasing load of the wastewater. The process is effectively used for the effluents from dairy, distillery, brewery, cannery, food processing, pulp and paper mills, pharmaceuticals, petrochemicals, etc.

### 10.7.3 Tertiary Wastewater Treatment

Primary and secondary treatments are sufficient to meet wastewater effluent standards. To improve its quality further, the following treatments are given:

(*i*) *Removal of fine suspended solids* is achieved with the help of micro-strainers and sand filters.

(*ii*) *Removal of bacteria* of faecal origin is generally achieved by retaining the effluents from the secondary treatment plant in maturation ponds for specified periods of time. Three or four such ponds are arranged in series to give an effluent with very low BOD and low suspended solids.

(*iii*) *Removal of dissolved inorganic solids* is achieved by following methods depending upon the given situation:

(*a*) *Evaporation* process is used only when the recovered solids or the concentrated solutions are reused or when the volume of the wastewater to be treated is less.

(*b*) *Ion-exchange* is widely used for obtaining deionized water for the use in high pressure boilers, special ion exchangers are available from the removal of toxic metal ions from industrial wastewater.

(*c*) *Adsorption* uses activated carbon to remove small quantities of organic contaminants from wastewater. It is particularly useful for the removal of pesticides (DDT) and carbonate insecticides.

(*d*) *Reverse osmosis* process is suitable and effective for the removal of dissolved solids from wastewater using a semipermeable membrane. The cost of the membranes and the fouling of the membranes are the major limitations.

(*e*) *Chemical precipitation* involves the removal of dissolved solids, particulates, the heavy metal ions by precipitating as their hydroxides with chelating agents like lime.

Several other processes such as electrodialysis, solvent extraction, freeze purification etc. can also be used.

## 10.8 REVERSE OSMOSIS (RO)

When two solutions of unequal concentrations are separated by a semipermeable membrane, flow of solvent takes place from dilute to concentrated sides due to osmosis. If, however, a hydrostatic pressure in excess of osmotic pressure is applied on the concentrated side, the flow of solvent reverses, *i.e.*, the solvent is forced to move from concentrated side to dilute side across the membrane. This process is called *reverse osmosis* or *super-filteration.*

Reverse osmosis (RO) systems can often improve the quality of water. The reverse osmosis water treatment method has been used extensively to convert brackish or seawater to drinking water, to clean up wastewater, and to recover dissolved salts from industrial processes. It is becoming more popular in the home market as home owners are increasingly concerned about contaminants that affect their health, as well as about nonhazardous chemicals that effect the taste, odor, or color of their drinking water.

### 10.8.1 Reducing Contaminants through RO

Reverse osmosis treatment reduces the concentration of dissolved solids, including a variety of ions and metals and very fine suspended particles such as asbestos, that may be found in water. RO also removes certain organic contaminants, some detergents, and specific pesticides. The various impurities that can be reduced by RO are listed below:

| **Ions and Metals** | **Organic Chemicals** |
|---|---|
| Arsenic (As) | Benzene |
| Barium (Ba) | Carbon tetrachloride |
| Bicarbonate ($HCO_3^-$) | Dichlorobenzene |

Cadmium (Cd)
Calcium (Ca)
Carbonate ($CO_3^{2-}$)
Chloride ($Cl^-$)
Chromium (Cr)
Copper (Cu)
Fluoride ($F^-$)
Iron (Fe)
Lead (Pb)
Magnesium (Mg)
Manganese (Mn)
Mercury (Hg)
Nitrate ($NO_3^-$)
Potassium (K)
Radium (Ra)
Selenium (Se)
Sodium (Na)
Sulphate ($SO_4^{2-}$)

Toluene
Trichloroethylene

**Particles**
Asbestos
Protozoan cysts

**Pesticides**
1, 2, 4-trichlorobenzene
Atrazine
Endrin
Heptachlor
Lindane
Pentachlorophenol

### 10.8.2 The Reverse Osmosis Process

In the reverse osmosis process, a cellophane-like membrane separates purified water from contaminated water. An understanding of osmosis is needed before further describing RO. Osmosis occurs when two solutions containing different quantities of dissolved chemicals are separated by a semipermeable membrane (allowing only some compounds to pass through). Osmotic pressure of the dissolved chemical causes pure water to pass through the membrane from the dilute to the more concentrated solution (Figure 10.20). There is a natural tendency for chemicals to reach equal concentrations on both sides of the membrane.

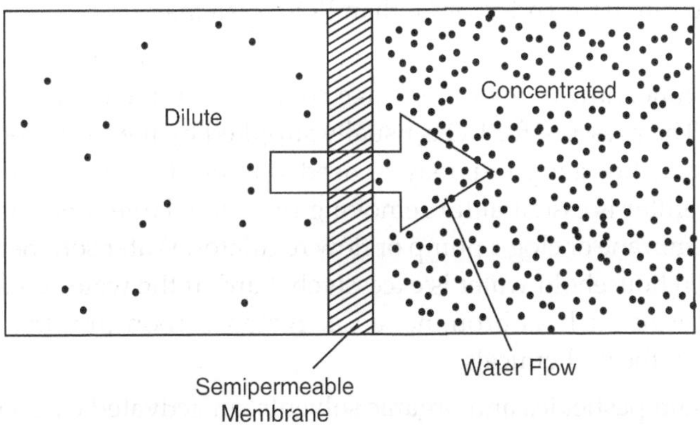


*FIGURE 10.20* **In osmosis, water flows across the membrane from the dilute to the concentrated solution.**

In reverse osmosis, water pressure applied to the concentrated side forces the process of osmosis into reverse. Under enough pressure, pure water is "squeezed" through the membrane from the concentrated to the dilute side (Figure 10.21). Salts dissolved in water as charged ions are repelled by the RO membrane. Treated water is collected in a storage container. The rejected impurities on the concentrated side of the membranes are washed away in a stream of wastewater, not accumulated as on a traditional filter.

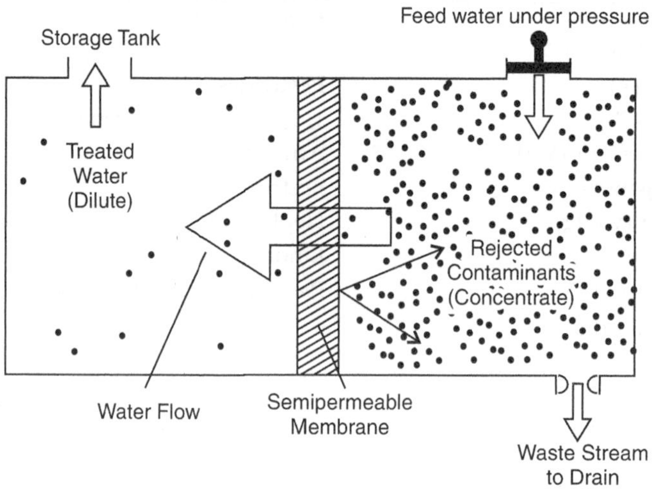


***FIGURE 10.21*** **The reverse osmosis, pressure is applied to the concentrated solution reversing the natural flow of water.**

### 10.8.3 Design of an RO System

Although the reverse osmosis process is simple, a complete water treatment system is often complex, depending on the quality of the incoming water before treatment and the consumer's needs. Most home RO systems are point-of-use (POU) units placed beneath the kitchen sink to treat water used for cooking and drinking. Point-of-entry (POE) systems that treat all the water entering the household are more expensive to purchase and operate than POU systems.

A typical home reverse osmosis system consists of pretreatment and post-treatment filter as well as the RO membrane, flow regulator, storage container for the treated water, and dispensing faucet (Figure 10.22). The pressure for RO is usually supplied by the feed line pressure of the water system in the home, but a booster pump may be needed to produce an adequate volume of treated water. A sediment prefilter is essential for removing relatively large sand grains and silt that may tear or clog the RO membrane or clog a pump or flow regulator. Water softeners are used in advance of the RO system when household water is excessively hard. If the water is chlorinated or contains other oxidizing chemicals such as bromine, an activated carbon prefilter is needed to protect membranes sensitive to these chemicals.

To remove certain pesticides and organic solvents, an activated carbon (AC) post-treatment must be included in the system. A standard AC filter positioned after the storage tank removes compounds that cause unpleasant taste and odours, including those from the tank or plastic tubing, just before water is dispersed. To remove high levels of organic chemicals such as trihalomethanes, volatile organic chemicals, and chloramines, an additional prolonged contact carbon filter (PCCF) is placed between the RO membrane and the storage tank. Combining an activated carbon filter with RO expands the range of chemicals the system can remove. Furthermore, AC treatment is improved because RO removes compounds that adversely affect AC absorption.

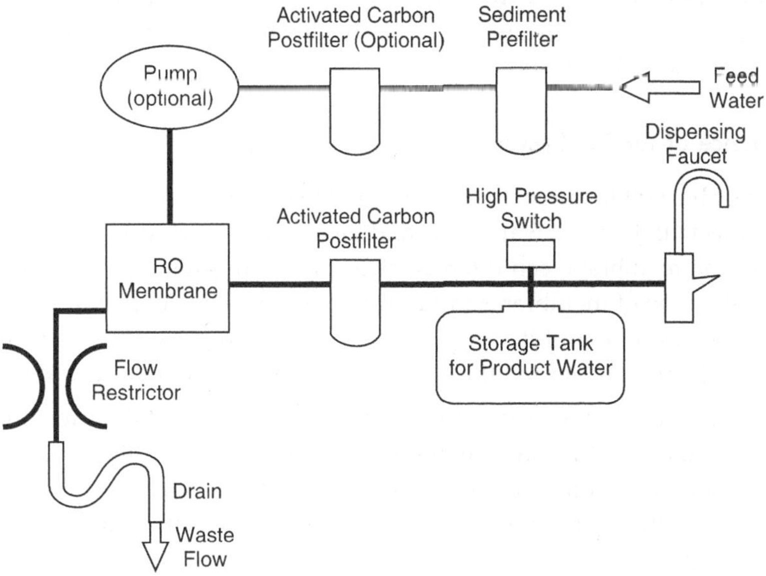


*FIGURE 10.22* **A schematic diagram of a typical RO system.**

The storage tank, tubing, and dispensing faucet should be made of plastic, stainless steel, or other nontoxic materials. The low *p*H and mineral content of RO-treated water may corrode copper pipes and allow lead to leach into the drinking water from brass components.

### 10.8.4 RO Membrane Materials

The most common RO membrane materials are polyamide thin film composites (TFC) or cellulosic types (cellulose acetate (CA), cellulose triacetate (CTA), or blends). Very thin membranes are made

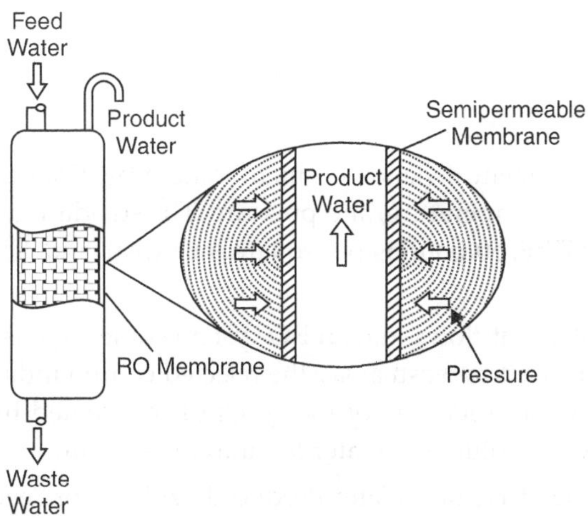


*FIGURE 10.23*

from these synthetic fibers. Membrane material can be spiral-wound around a tube, or hollow fibers can be bundled together, providing at tremendous surface area for water treatment inside a compact cylindrical element (Figure 10.23). Hollow fiber membranes have greater surface area (and therefore greater capacity) but are more easily clogged than the spiral-wound membranes commonly used in home RO systems.

### 10.8.5 Maintenance of an RO System

An RO system must be well maintained to ensure reliable performance. Clogged RO membranes, filter or flow controls will decrease water flow and the system's performance. If fouling is detected in the early stages the membrane can often be cleaned and regenerated. The cleaning procedure varies depending the type of membrane and fouling. Completely clogged or torn RO membranes must be replaced. In addition, prefilter or postfilters must be replaced once a year or more often, depending on the volume of water fed through the system and the quality of the feed water.

Damage to RO membranes cannot be seen easily. The treated water must be analyzed periodically to determine whether the membrane is intact and doing its job. Many systems now have a built-in continuous monitor that indicates a high TDS level, a sign that the system is not operating properly. It may also be necessary to test regularly for specific health-related contaminants such as nitrates or lead.

Microorganisms, dead or alive, can clog RO membranes. To prevent biofouling, RO units must be disinfected periodically with chlorine or other biocides provided by the manufacturer. Continuous chlorination can be used with cellulosic membranes to protect the system from biofouling and eliminate the particle-trapping slime that worsens other forms of fouling such as scaling.

Chlorine and other oxidizing disinfectants are harmful to thin film composite membranes. If the feed water is chlorinated, an activated carbon unit must be in place to remove the oxidizing chemicals before they reach the TFC membrane. Activated carbon (AC) prefilters should not be used on nonchlorinated water supplies because they provide a place for micro-organisms to multiply and lead to increased biofouling of the RO membrane surface. It is important to replace AC filters periodically following the manufacturer's instructions, especially after an extended shutdown period during which micro-organism can flourish.

### 10.8.6 Efficiency of RO Systems

The performance of an RO system depends on membrane type, flow control, feed water quality (*e.g.*, turbidity, TDS, and *p*H), temperature, and pressure. The standard at which manufactures rate RO system performance is 77°F, 60 pounds per square inch (psi), and TDS at 500 parts per million (ppm).

Only part of the water that flows into an RO system comes out as treated water. Part of the water fed into the system is used to wash away the rejected compounds and goes down the drain as waste. The recovery rate, or efficiency, of the system is calculated by dividing the volume of treated water produced by the volume of water fed into the system.

% Recovery = Volume of Treated Water Produced/Volume of Feed Water Used

The RO device must also overcome osmotic pressure, the bonding between water molecules and the dissolved impurities; the higher the TDS level, the greater the osmotic pressure. The net water pressure at the RO membrane can be calculated by *subtracting* back pressure and osmotic pressure from the feed line pressure. If the net water pressure can result in low rejection rates.

In some systems, once the storage tank is filled, surplus treated water is discarded, water loss from successes units is frequently excessive. A system that automatically shunts off when the pressure on the tank reach a given level saves water.

### 10.8.7 Summary of RO Process

RO removes many inorganic impurities from drinking water. Its effectiveness depends not only on the type of membrane but on feed water quality, temperature, pressure, and flow control, as well as the type and concentration of specific contaminant to be removed.

- RO is not effective for removing dissolved gases, some pesticides and solvents, hydrogen sulphide gas, THMS, VOCs, and chloramines.
- A typical RO system consists of a sediment filter, pump, reverse osmosis membrane, flow regulator, storage tank, final activated carbon filter (for taste and odors), and dispensing faucet. An AC prefilter is sometimes needed for dechlorination.
- RO is commonly used to treat only the water used for drinking and cooking at the point of use rather than at the point of entry for all household use.
- RO membrane types vary in their ability to reject contaminants and differ in capacity (the volume of treated water produced per day).
- Water pressure is an important factor in determining the RO system's rejection rate, capacity, and recovery rate (amount of treated water produced per amount of feed water used).
- Maintenance of an RO system is essential for reliable performance. High levels of TDS and micro-organisms in the system are common the cause of fouled membranes.
- The treated water should be monitored for TDS and the level of any specific contaminants that may affect human health.

## SOLVED PROBLEMS

### TYPE-I Problems on Water hardness

**Problem 1.** *A precipitate of 0.110 g of calcium oxalate ($CaC_2O_4.H_2O$) was obtained from 250 mL of water sample. Express the content of calcium in ppm.*

**Solution.**

$$CaC_2O_4.H_2O \equiv Ca$$

$$40 + 24 + 64 + 18 = 146 \text{ parts} \equiv 40 \text{ parts}$$

Now 146 parts of calcium oxalate ≡ 40 parts calcium

$$\therefore \quad 0.110 \text{ parts of calcium oxalate} = \frac{40 \times 0.11}{146} \text{ g of calcium}$$

$$= 0.03 \text{ g of Ca.}$$

$$250 \text{ mL of water contains } = 0.03 \text{ g of Ca.}$$

$$\therefore \quad 1000 \text{ mL of water contains } = \frac{0.03 \times 1000}{250} = 0.12 \text{ g of Ca.}$$

Calcium contents of water = 0.12 g per liter = 120 mg per liter = **120 ppm.**

**Problem 2.** *In an experiment for the determination of hardness of a sample of water, the following results were obtained. Volume of sample of water = 100 c.c.; N/50 $Na_2CO_3$ solution added to it = 20 c.c.; N/50 $H_2SO_4$ used to back titrate the unreacted $Na_2CO_3$ = 10 c.c. Calculate the hardness of water in grains per gallon.*

**Solution.** N/50 $Na_2CO_3$ consumed by 50 c.c. of water = 20 – 10 = 10 c.c.

$$10 \text{ c.c. of N/50 } Na_2CO_3 \text{ contain } \frac{53}{50} \times \frac{10}{1000} = 0.0106 \text{ g of } Na_2CO_3 \qquad (\text{Eq. wt of } Na_2CO_3 = 53)$$

So 100 c.c. of water has hardness equivalent to 0.0106 g of $Na_2CO_3$ or

$$100 \text{ c.c. of water has } \frac{100 \times 0.0106}{106} = 0.01 \text{ g of } CaCO_3.$$

$$\text{One gallon of water has hardness equivalent to } \frac{0.01 \times 1000}{100 \times 0.22}$$

$$= 0.454 \text{ g of } CaCO_3$$

$$= 0.454 \times 15.4 \text{ grain of } CaCO_3 = \mathbf{6.99 \text{ grain of }} CaCO_3.$$

**Problem 3.** *200 ml of a sample of water required 5 mL of N/50 $Na_2CO_3$ solution for complete precipitation of $Ca^{2+}$ and $CaCO_3$. Calculate the hardness in ppm.*

**Solution.** $5 \text{ ml of N/20 } Na_2CO_3 = \frac{5 \times 50}{20 \times 1000} \text{ g of } CaCO_3$

$$\text{So 200 parts of water contain } \frac{5 \times 50}{20 \times 1000} \text{ parts of } CaCO_3 \text{ equivalents.}$$

$$10^6 \text{ parts of water contain } \frac{5 \times 50 \times 10^6}{20 \times 1000 \times 200} = 62.5 \text{ ppm.}$$

Therefore hardness = **62.5 ppm.**

**Problem 4.** *A water sample contain 204 mg of $CaSO_4$ per liter. Calculate the hardness in terms of $CaCO_3$ equivalent.*

**Solution.**

$$\underset{136\ g\ mol^{-1}}{CaSO_4} \equiv \underset{100\ g\ mol^{-1}}{CaCO_3}$$

∴ 136 mg/L of $CaSO_4$ ≡ 100 mg/L of $CaCO_3$ equivalents.

or

$$204\ mg/L\ of\ CaCO_3 = \frac{100\ mg/L\ of\ CaCO_3\ eq. \times 204}{136}$$

$$= 150\ mg/L\ of\ CaCO_3\ equivalents.$$

Hence, hardness = **150 mg/L or ppm.**

**Problem 5.** *A sample of hard water has a hardness of 510 ppm. Express the hardness in °French and °Clark.*

**Solution.** 1 ppm = 0.07° Clark = **0.1° French.**

Thus, 510 ppm = 510 × 0.07° Clack = **35.7° Clark**

Again 1 ppm = 0.1° French, so 510 ppm = 510 × 0.1° French

= **51° French.**

**Problem 6.** *Calculate temporary hardness and total hardness of a sample of water containing $Mg(HCO_3)_2$ = 7.3 mg/L; $Ca(HCO_3)_2$ = 16.2 mg/L; $MgCl_2$ = 9.5 mg/L; $CaSO_4$ = 13.6 mg/L.*

**Solution.** Temporary hardness = [7.3 × 100/146 + 16.2 × 100/162] mg/L

= (5 + 10) mg/L = **15 mg/L**

Permanent hardness = [6.5 × 100/95 + 13.6 × 100/136] mg/L

= [10 + 10] mg/L = **20 mg/L.**

**Problem 7.** *100 mL of a sample of hard water neutralizes exactly 12 mL of 0.12 N HCl methyl orange as indicator. What kind of hardness is present? Express the same in terms of an equivalent of $CaCO_3$.*

**Solution.** Hardness of water is temporary, since methyl orange indicator does not give the value for permanent hardness.

Now 100 mL of sample = 12 mL of 0.12 N HCl

= 12 × 0.12 mL of 1N HCl

= 1.44 mL of 1N HCl or 1N $CaCO_3$ eq.

= $1.44 \times 10^{-3}$ L × 50 g $CaCO_3$ eq $L^{-1}$

= 0.072 g $CaCO_3$ eq. or 72 mg $CaCO_3$ eq.

$$\therefore\quad 1.000\ mL\ (or\ 1\ L)\ of\ water = \frac{72\ mg\ CaCO_3\ eq. \times 1000\ mL}{100\ mL}$$

$$= 720\ mg\ CaCO_3\ eq.$$

Hence, the temporary hardness of water is **720 ppm.**

**Problem 8.** *A sample of water on analysis has been found to contain the following in ppm: $Ca(HCO_3)_2 = 10.5$; $Mg(HCO_3)_2 = 12.5$; $CaSO_4 = 7.5$; $CaCl_2 = 8.2$; $MgSO_4 = 2.6$. Calculate the temporary and permanent hardness in degree French.*

**Solution.** Conversion into $CaCO_3$ equivalent:

| *Constituent* | *Amount (ppm)* | *Multiplication Factor* | *$CaCO_3$ Equivalent (ppm)* |
|---|---|---|---|
| $Ca(HCO_3)_2$ | 10.5 | 100/162 | 10.5 × 100/162 = 6.481 |
| $Mg(HCO_3)_2$ | 12.5 | 100/146 | 12.5 × 100/146 = 8.562 |
| $CaSO_4$ | 7.5 | 100/136 | 7.5 × 100/136 = 5.515 |
| $CaCl_2$ | 8.2 | 100/111 | 8.2 × 100/111 = 7.387 |
| $MgSO_4$ | 2.6 | 100/120 | 2.6 × 100/120 = 2.167 |

∴ Temporary hardness = (6.481 + 8.582) = 15.043 ppm

= 15.043 × 0.1° Fr = **1.54°Fr.**

Permanent hardness = (5.515 + 7.387 + 2.167) ppm = 15.069 ppm

= 15.089 × 0.1° Fr = **1.507°Fr.**

**Problem 9.** *100 mL of a water sample required 15.5 mL of 0.02 N sulphuric acid for complete neutralization. 100 mL of the sample was boiled with 8 mL of 0.1 N NaOH, filtered and again made to 100 mL; 25 mL of this solution required 7.1 mL of 0.02 N $H_2SO_4$ for neutralization. Describe the hardness present.*

**Solution.** From the problem it is obvious that water has only temporary hardness and magnesium hardness without any permanent hardness.

Now since 100 ml of water ≡ 15.5 ml or 0.02 N $H_2SO_4$

Strength of temporary hardness in terms of $CaCO_3$

$$= \frac{15.5 \times 0.02 \times 50}{100} \text{ g/liter}$$

$$= \frac{15.5 \times 0.02 \times 50 \times 10^6}{100 \times 10^3} \text{ ppm} = 155 \text{ ppm}$$

*Magnesium hardness*

25 mL of filtrate ≡ 7.1 mL of 0.02 N $H_2SO_4$

$$100 \text{ mL of filtrate} = \frac{7.1 \times 100}{25} \text{ mL of 0.02 N } H_2SO_4$$

$$= \frac{7.1 \times 100}{25 \times 0.1} \text{ mL of 0.01 N } H_2SO_4$$

= 5.68 mL of 0.01 N $H_2SO_4$

$\therefore$ Volume of 0.1 N, NaOH used for removing magnesium hardness

$$= 8 - 5.68 = 2.32 \text{ mL}$$

Strength of magnesium hardness in terms of $CaCO_3$

$$= \frac{2.32 \times 50 \times 0.1 \times 10^6}{100 \times 10^3} \text{ ppm} = \mathbf{116 \text{ ppm.}}$$

## TYPE-II Problems on Water Softening

**Problem 10.** *Calculate the amount of lime and soda required for softening of 15 000 liters of water, which analyzed as follows: temporary hardness = 20 ppm; permanent hardness = 15 ppm; permanent mg hardness = 10 ppm.*

**Solution.** Lime requirement

$$= 74/100 \text{ [Temporary hardness + Permanent hardness]} \times \text{Volume of water}$$

$$= 74/100 \text{ [(20 + 10) ppm or mg L}^{-1}] \times 15\,000 \text{ L}$$

$$= 22.2 \text{ mg L}^{-1} \times 15\,000 \text{ L} = 330\,000 \text{ mg} = \mathbf{330 \text{ g}}$$

Soda requirement

$$= 106/100 \text{ [Permanent hardness]} \times \text{Volume of water}$$

$$= 106/100 \text{ [15 ppm of mg L}^{-1}] \times 15\,000 \text{ L}$$

$$= 15.9 \text{ mg L}^{-1} \times 15\,000 \text{ L} = 238.500 \text{ mg} = \mathbf{238.5 \text{ g.}}$$

**Problem 11.** *An exhausted zeolite softener was regenerated by passing 150 liters of NaCl solution, having a strength of 150 g/L of NaCl. If the hardness of water is 600 ppm, calculate the total volume of water that is softened by this softener.*

**Solution.** 150 L of NaCl contains = 150 × (150 g/L) of NaCl

= 22 500 g of NaCl

= 22 500 g × (100/117) $CaCO_3$ eq.

= $1.932 \times 10^4$ g $CaCO_3$ eq.

= $1.932 \times 10^7$ mg $CaCO_3$ eq.

Let V liters of 600 ppm (or 600 mg/L) water contains hardness

= $1.932 \times 10^7$ mg $CaCO_3$ eq.

$$\therefore \text{ Number of liters of hard water, } V = \frac{1.932 \times 10^7}{600} = \mathbf{3.219 \times 10^4 \text{ L.}}$$

**Problem 12.** *A zeolite softener was 90% exhausted, when 10 000 L of hard water was passed through it. The softener required 200 L of NaCl solution of strength 50 g NaCl/L of solution. What is the hardness of water?*

**Solution.** 10 000 L of hard water = 200 L of NaCl solution

= 200 L × (50 g/L) of NaCl

= 10 000 g of NaCl

= 10 000 g × (100/117) of $CaCO_3$ eq.

= 8547 g of $CaCO_3$ eq.

∴ 1 L of hard water = 8547 g of $CaCO_3$ eq./10 000

= 0.8457 g of $CaCO_3$ eq.

= 845.7 mg of $CaCO_3$ eq.

Hence, hardness of water = **845.7 mg/L (or ppm).**

**Problem 13.** *A zeolite softener was completely exhausted and then regenerated by passing 200 liters of NaCl solution containing 100 g per liter of NaCl. How many liters of a sample of water of hardness 500 ppm can be softened by the softener?*

**Solution.** 200 liters of NaCl contains = 200 × 100 = 2 000 g of NaCl

$$= 20\,000 \times \frac{50}{58.5} \text{ g } CaCO_3 \text{ eq.}$$

= 17 094 g $CaCO_3$ eq.

Let V liters of 500 ppm (on 500 mg/l) water contains hardness

= 1709 × $10^7$ mg $CaCO_3$ eq.

$$\therefore \quad \text{Number of liters of hard water, } V = \frac{1.7094 \times 10^7}{500} = \textbf{34188 liters.}$$

**Problem 14.** *The hardness of 150 000 liters of sample of water was completely removed by passing it through a zeolite softener. This softener then required 550 liters of NaCl solution containing 100 g/liter of NaCl for regeneration. Calculate the hardness of the sample of water.*

**Solution.** 550 liter of NaCl solution contains 550 × 100 g = 55 000 g of NaCl. So 150 000 liters of hard water ≡ 55 000 g of NaCl.

$$\text{Thus, 150 000 liters of hard water} \equiv \frac{55\,000 \times 50}{58.5} \text{ g of } CaCO_3 \text{ eq.}$$

$$\text{One liter of hard water} \equiv \frac{55\,000 \times 50}{58.5 \times 150\,000} \text{ g of } CaCO_3 \text{ eq.}$$

= 0.313 g of $CaCO_3$ eq.

Thus hardness of water sample = 313 mg/liter = **313 ppm.**

## TYPE-III Problems on Analysis of Water

**Problem 15.** *Calculate the temporary hardness, permanent hardness and magnesium hardness of a water sample which gave the following analysis :*

(*i*) *50 ml of water required 13 ml of N/50 HCl for neutralization to methyl orange end-point.*

(*ii*) *50 ml of water after boiling required 8 ml of N/50 HCl for neutralization to methyl orange end point.*

(*iii*) *50 ml of water sample was boiled with 55 ml of N/50 sodium carbonate solution and the resulting solution required 35 ml of N/50 HCl for neutralization to methyl orange end-point.*

(*iv*) *Another 50 ml of water was first neutralised with hydrochloric acid in presence of methyl orange as indicator. The solution was then boiled with 40 ml of N/50 NaOH solution and filtered. The filtrate required 30 ml of N/50 HCl for neutralization to methyl orange end-point.*

**Solution.** (*i*) *Temporary hardness*

Vol. of N/50 HCl required for removing temporary hardness of 50 ml of water sample

$$= 13 - 8 = 5 \text{ ml}$$

According to normality formula,

$$1\,000 \text{ ml of N/50 HCl} = 1000 \text{ ml of } \frac{N}{50} CaCO_3 \qquad \text{(Hardness in terms of } CaCO_3\text{)}$$

$$\therefore \quad 1000 \text{ ml of } \frac{N}{50} CaCO_3 \text{ contains} = \frac{50 \times 1}{50} \text{ gm of } CaCO_3 \quad [\because \text{ Eq. wt. of } CaCO_3 = 50]$$

$$\therefore \quad 1000 \text{ ml of } \frac{N}{50} \text{ HCl} = \frac{50}{50} \text{ gm of } CaCO_3$$

$$5 \text{ ml of } \frac{N}{50} \text{ HCl} = \frac{50 \times 5}{50 \times 1000} \text{ gm of } CaCO_3$$

Since $\frac{50 \times 5}{50 \times 1000}$ gm of $CaCO_3$ hardness is present in 50 ml of water, we can say that

$$50 \text{ ml of water contains} = \frac{50 \times 5}{50 \times 1000} \text{ gm of } CaCO_3$$

$$10^6 \text{ ml of water contains} = \frac{50 \times 5 \times 10^6}{50 \times 1000 \times 50} \text{ gm of } CaCO_3$$

$$= 100 \text{ gm or } CaCO_3$$

$\therefore$ Temporary hardness = **100 ppm**

(*ii*) *Permanent hardness*

Vol. of N/50 $Na_2CO_3$ required for removing permanent hardness of 50 ml of water

$$= 55 - 35 = 20 \text{ ml}$$

Applying the above steps, permanent hardness as $CaCO_3$ equivalent,

$$= \frac{20 \times 50 \times 10,00,000}{50 \times 50 \times 1000} \text{ ppm} = \mathbf{400 \text{ ppm}}$$

*(iii) Magnesium hardness*

Vol. of N/50 NaOH used for removing magnesium hardness of 50 ml of water

$$= 40 - 30 = 10 \text{ ml.}$$

Applying the steps as in temporary hardness calculations,

$$\text{magnesium hardness in terms of } CaCO_3 = \frac{10 \times 50 \times 10{,}00{,}000}{50 \times 50 \times 1000} = \mathbf{200\ ppm}$$

**Problem 16.** *100 ml of a water sample required 15.5 ml of 0.02 N sulphuric acid for complete neutralization. 100 ml. of the same sample was boiled with 8 ml of 0.1 N NaOH, filtered and again made to 100 ml ; 25 ml of this solution required 7.1 ml of 0.02 $NH_2SO_4$ for neutralization. Describe the hardness present.*

**Solution.** From the problem it is obvious that water has only temporary hardness and magnesium hardness without any permanent hardness.

Now since 100 ml of water = 15.5 ml of 0.02 N $H_2SO_4$

$$\text{Strength of temporary hardness in terms of } CaCO_3 = \frac{15.5 \times 0.02 \times 50}{100} \text{ g/liter}$$

$$= \frac{15.5 \times 0.02 \times 50 \times 10^6}{100 \times 10^3} \text{ ppm} = \mathbf{155\ ppm}$$

*Magnesium hardness*

$$25 \text{ ml of filtrate} \equiv 7.1 \text{ ml of } 0.02 \text{ N } H_2SO_4$$

$$100 \text{ ml of filtrate} \equiv \frac{7.1 \times 100}{25} \text{ ml of } 0.02\text{N } H_2SO_4$$

$$= \frac{7.1 \times 100}{25 \times 0.1} \text{ ml of } 0.1\text{N } H_2SO_4$$

$$= 5.68 \text{ ml of } 0.1 \text{ N } H_2SO_4$$

$\therefore$ Vol. of 0.1N NaOH used for removing magnesium hardness = 8 – 5.68 = 2.32 ml.

$$\text{Strength of magnesium hardness in terms of } CaCO_3 = \frac{2.32 \times 50 \times 0.1 \times 10^6}{100 \times 10^3} \text{ ppm} = \mathbf{116\ ppm}$$

**Problem 17.** *Calculate the degree of permanent and temporary hardness in ppm of a water with the following data.*

*(i) 50 ml of the distilled water ≡ 0.5 ml of soap soluiton.*

*(ii) 50 ml of $CaCl_2$ solution ≡ 40.5 ml of soap solution.*

*(iii) 50 ml of the given water ≡ 15.5 ml of soap solution.*

*(iv) 50 ml of the boiled water ≡ 12.5 ml of soap solution.*

*[1 ml of $CaCl_2$ solution ≡ 1.0 mg of $CaCO_3$]*

**Solution.** Since 1 ml of $CaCl_2$ solution (standard water) ≡ 1 mg of $CaCO_3$

1 liter of $CaCl_2$ solution (standard water) = 1 gm of $CaCO_3$

Hence $x = 1$ gm V = 0.5 ml $V_1$ = 40.5 ml $V_2$ = 15.5 ml $V_3$ = 12.5 ml

$$\text{Now Total hardness} = x \times \frac{(V_2 \times V) \times 1000}{(V_1 - V)} \text{ ppm}$$

$$= \frac{1 \times 15 \times 1000}{40} \text{ ppm} = \mathbf{375\ ppm}$$

$$\text{Permanent hardness} = x \times \frac{(V_3 - V) \times 1000}{(V_1 - V)} \text{ ppm}$$

$$= 1 \times \frac{12 \times 1000}{40} \text{ ppm} = \mathbf{300\ ppm}$$

Hence, Temporary hardness = Total hardness – Permanent hardness

= 375 – 300 ppm = **75 ppm.**

**Problem 18.** *Titration of 50 ml of water at each time with standard soap solution gave the following results :*

*(i) Lather factor = 0.4 ml* *(ii) Total hardness = 8.2 ml*

*(iii) Permanent hardness = 2.5 ml*

*(iv) Standard hard water containing 0.2 gm $CaCO_3$/liter = 19.9 ml.*

*Calculate each type of hardness present in water.*

**Solution.** $x$ = 0.2 gm V = 0.4 ml $V_1$ = 19.9 ml $V_2$ = 8.2 ml $V_3$ = 2.5 ml

$$\therefore \quad \text{Total hardness} = x \times \frac{(V_2 - V) \times 1000}{(V_1 - V)} \text{ ppm}$$

$$= \frac{0.2 \times 7.8 \times 1000}{19.5} = \mathbf{80\ ppm}$$

$$\text{Similarly, Permanent hardness} = x \times \frac{(V_3 - V) \times 1000}{(V_1 - V)} \text{ ppm}$$

$$\therefore \quad \text{Permanent hardness in sample water} = \frac{0.2 \times 2.1 \times 1000}{19.5} \text{ ppm} = \mathbf{21.5\ ppm}$$

Hence, Temporary hardness = Total hardness – Permanent hardness

(80 – 21.5) ppm = **58.5 ppm**

**Problem 19.** *Calculate the hardness of a sample of water on the 'Clarke scale' whose 70 ml required 15 ml of the soap solution to form a permanent lather.* [*1 ml of the soap solution = 1 mg of $CaCO_3$*].

**Soluiton.**

70 ml of sample water ≡ 15 ml of soap solution

or 70 ml of sample water ≡ 15 mg of $CaCO_3$

$$\therefore \quad 70{,}000 \text{ ml sample water} = \frac{15 \times 70{,}000}{70} \text{ mg of } CaCO_3$$

$$= 15{,}000 \text{ mg of } CaCO_3 = 15 \text{ gm of } CaCO_3$$

or 70,000 gm of sample water = 15 gm of $CaCO_3$ [∵ Wt of 1 ml of water = 1 g]

Hardness of water = **15° Clarke**

**Problem 20.** *One gm of $CaCO_3$ was dissolved in dil. HCl and the solution diluted to one liter. 50 ml of this solution required 45 ml of EDTA solution, while 50 ml of the sample water required 18 ml of EDTA solution. On the other hand, 50 ml of the boiled sample water when titrated against EDTA consumed 9 ml of solution. Calculate each type of hardness in ppm.*

**Solution.**

1000 ml of standard water = 1 gm or 1000 mg of $CaCO_3$

1 ml of standard water = 1 mg of $CaCO_3$

Now, 45 ml of EDTA sol. = 50 ml of standard water

or 45 ml EDTA sol. = 50 × 1 mg of $CaCO_3$

$$\therefore \quad 1 \text{ ml of EDTA sol.} = \frac{50}{45} \text{ mg of } CaCO_3$$

Further, 50 ml of sample water = 18 ml of EDTA solution

$$= \frac{18 \times 50}{45} \text{ mg of } CaCO_3$$

$$1000 \text{ ml of sample water} = \frac{18 \times 50 \times 1000}{45 \times 50} \text{ mg of } CaCO_3 = 400 \text{ mg of } CaCO_3$$

∴ Total hardness = 400 mg/liter = **400 ppm**

$$\text{Also, 50 ml of boiled sample water} = 9 \text{ ml EDTA solution} = \frac{9 \times 50}{45} \text{ mg of } CaCO_3$$

$$\therefore \quad 1000 \text{ ml of boiled sample water} = \frac{9 \times 50 \times 1000}{45 \times 50} \text{ mg of } CaCO_3 = 200 \text{ mg of } CaCO_3$$

∴ Permanent hardness = **200 mg/liter or 200 ppm**

Hence, Temporary hardness = 400 – 200 = **200 ppm**

**Problem 21.** *100 ml of a sample of water required 7.5 ml of N/50 sulphuric acid for neutralization to phenolphthalein end point. At this stage two drops of methyl orange were added and solution again titrated against N/50 sulphuric acid and 12.5 ml of acid was further required to methyl orange end point. Calculate the alkalinity of water as calcium carbonate in ppm.*

**Solution.** Vol. of $H_2SO_4$ used for phenolphathalein end point, $V_1$ = 7.5 ml

and Vol. of $H_2SO_4$ used for methyl orange end point, $V_2$ = 7.5 + 12.5 = 20 ml

∴ Phenolphthalein alkalinity, P = 10 $V_1$ ppm = 10 × 7.5 = 75 ppm

Similarly, methyl orange alkalinity, M = 10 $V_2$ ppm = 10 × 20 = 200 ppm

Now here since $P < \frac{1}{2} M$

Alkalinity due to hydroxide ion = Nil

Alkalinity due to carbonate ion = 2P = **2 × 75 = 150 ppm.**

and Alkalinity due to bicarbonate ion = M – 2P = 200 – 150 = **50 ppm.**

**Problem 22.** *A water sample is alkaline both to phenolphthalein and methyl orange. 100 ml of water on titration with 0.02 N sulphuric acid gave phenolphthalein end point with 20 ml of the acid and methyl orange end-point with another 5 ml of the acid. Describe the character of alkalinity in ppm.*

**Solution.** Vol. of 0.02 N (N/50) $H_2SO_4$ used to phenolphthalein end point, $V_1$ = 20 ml

Vol. of 0.02 N $H_2SO_4$ used to methyl orange end point, $V_2$ = 20 + 5 = 25 ml

∴ Phenolphthalein alkalinity = 10 $V_1$ ppm = 10 × 20 = 200 ppm

and methyl orange alkalinity = 10 $V_2$ ppm = 10 × 25 = 250 ppm

Now here since $P > \frac{1}{2} M$

Alkalinity due to $OH^-$ = 2P – M = 400 – 250 = **150 ppm**

Alkalinity due to $CO_3^{2-}$ = 2(M – P) = 2(250 – 200) = 2 × 50 = **100 ppm**

**Problem 23.** *Calculate the amount of chloride ions in a water whose 100 c.c. required 10.5 ml of N/40 $AgNO_3$ solution upto potassium chloromate end-point.*

**Solution.** 100 ml of water sample = 10.5 ml of $\frac{N}{40}$ $K_2CrO_4$

$$100 \times \text{Normality of } Cl^- = 10.5 \times \frac{1}{40}$$

$$\text{Normality of } Cl^- = \frac{10.5}{100} \times \frac{1}{40}$$

$$\text{Strength of } Cl^- = \frac{10.5 \times 35.5}{40 \times 100} \text{ gm/liter}$$

$$= \frac{10.5 \times 35.5 \times 10{,}00{,}000}{40 \times 100 \times 1000} = \mathbf{93.125\ ppm}$$

In short, strength of $Cl^- = \frac{10.5 \times 35.5 \times 10{,}00{,}000}{40 \times 100 \times 1{,}000} = 93.125$ ppm

## QUESTION BANK

1. (*a*) What is hardness of water? How is it determined by EDTA method? (*b*) What is soft water? Describe zeolite process of water softening.

2. Define degree of hardness and discuss its units. Which of the following substances will
(*a*) increase hardness (*b*) decrease hardness (*c*) have no effect on hard water containing $MgCl_2$, $MgSO_4$, $Ca(HCO_3)_2$, $CaSO_4$ and $CaCl_2$.
3. (*a*) Why should water be softened before use in boilers? (*b*) Describe how water is softened by ion exchange process? (*c*) Give an account of conditioning of water used in boilers.
4. What are the causes and effect of (*i*) boiler scales (*ii*) caustic embrittlement?
5. What is scale? How is it formed? What are the disadvantage of scale formation in boilers?
6. (*a*) How water is softened before use in boilers? (*b*) Describe how water is softened by ion exchange process? (*c*) Give an account of conditioning of water used in boilers.
7. (*a*) Describe the zeolite process for treating water. (*b*) What is desalination? How is it achieved? (*c*) What is the method of treating sewage?
8. (*a*) Describe the lime soda process of water softening and compare it with cation exchange method,
(*b*) Describe in detail scale and sludge formation.
9. Write a note on caustic embrittlement.
10. How are boiler scales formed? Discuss four disadvantages of scale formation in boilers. Give the difference between a sludge and a scale.
11. Write a short note on priming and foaming?
12. 100 mL of a sample of water required 20 mL of N/40 $H_2SO_4$ for complete neutralization. Calculate the temporary hardness. [**Ans.** 250 ppm]
13. 20 mL of 0.1 N sodium carbonate solution was added to 100 mL of sample of hard water. The filtrate from this required 30 mL of 0.05 N $H_2SO_4$ for complete neutralization. Calculate the hardness of water sample. [**Ans.** 250 ppm]
14. Calculate the hardness of a water sample whose 10 mL required 20 mL of EDTA. 20 mL of calcium chloride solution, whose strength is equivalent to 1.5 g. of $CaCO_3$ per liter, required 30 mL EDTA solution. [**Ans.** 200 ppm]
15. 100 mL of a sample of water is treated with 20 mL of N/20 $NaCO_3$ solution. The solution is then gently boiled and filtered. The filtrate required 15 mL of N/20 HCl for neutralization. Calculate the hardness of the sample of water.
16. A standard hard water contains 15 g of $CaCO_3$ per liter. 20 mL of this required 25 mL of EDTA solution. 100 mL of sample of water required 18 mL EDTA solution. The same sample after boiling required 12 mL EDTA solution. Calculate the temporary hardness of the given sample of water, in terms of ppm. [**Ans.** 720 ppm]
17. What is the carbonate and noncarbonate hardness of a sample of water in containing:
$Ca(HCO_3)_2$ = 16.2 mg/L; $Mg(HCO_3)_2$ = 7.3 mg/L; $MgCl_2$ = 9.5 mg/L; $CaSO_4$ = 13.6 mg/L?
[**Ans.** Carbonate hardness = 15.0 ppm; non-carbonate hardness = 20.0 ppm]
18. Describe the methods to determine temporary, permanent, and total hardness of hard water.
19. What is the main purpose of secondary wastewater treatment? Discuss the sludge treatment method.
20. Discuss the characteristics impared by impurities in water.
21. A sample of ground water has 150 mg/l of $Ca^{+2}$ and 60 mg/l of $Mg^{+2}$. Find the total hardness express in milli equivalent per liter (meq/l) and mg/l as $CaCO_3$.
22. State the lime-soda process for the removal of hardness of water.
23. What do you understand by temporary and permanent hardness of water? Which unit is used to express hardness of water?

24. Describe the methods to determine color and odor in wastewater.

25. Write a short note on the wastewater treatment methods.

26. Calculate the amount of lime and soda required to soften 25 000 liters of water having the following analysis:
$Ca(HCO_3)_2$ = 4.86 ppm; $Mg(HCO_3)_2$ = 7.3 ppm; $CaSO_4$ = 6.8 ppm; $MgCl_2$ = 5.7 ppm; $MgSO_4$ = 9.0 ppm; $SiO_2$ = 3.5 ppm; NaCl = 5.85 ppm. [**Ans.** $L$ = 4.90; $S$ = 490 g]

27. Calculate the quantity of lime and soda required for softening 50 000 liters of water containing the following salts per liter: $Ca(HCO_3)_2$ = 8.1 mg; $Mg(HCO_3)_2$ = 5 mg; $CaSO_4$ = 13.6 mg; $MgSO_4$ = 120 mg; $MgCl_2$ = 2.0 mg; and NaCl = 4.7 mg. [**Ans.** $L$ = 1.01 kg; $S$ = 1.17 kg]

28. Calculate the amount of lime (84% pure) and soda (92% pure) required for treatment of 20 000 liters of water, whose analysis is as follows:
$Ca(HCO_3)_2$ = 40.5 ppm; $Mg(HCO_3)_2$ = 36.5 ppm; $MgSO_4$ = 30.0 ppm;
$CaSO_4$ = 34.0 ppm; $CaCl_2$ = 27.75 ppm; and NaCl = 10.00 ppm.
Also calculate the temporary and permanent hardness of water sample.

29. Calculate the amounts of lime and soda needed for softening 100 000 liters of water containing HCl = 7.3 m/L; $Al_2(SO_4)_3$ = 34.2 mg/L; $MgCl_2$ = 9.5 mg/L:: NaCl = 29.25 mg/L; (Al = 27; Mg = 24; Na = 23; H = 1; Cl = 35.5; S = 32; O = 16). [**Ans.** $L$ = 3.7 kg; $S$ = 5.3 kg]

30. Explain with equations and calculate the quantity of quick lime and soda ash required to soften 10 000 liters of water containing:
(*i*) 219 ppm of magnesium bicarbonate and 234 ppm of sodium chloride.
(*ii*) 16 ppm of $Mg^{2+}$ and 18.3 of $HCO_3^-$
(*iii*) 1.5 ppm of the free acids, 144 ppm of sulphate ions, and 71 ppm of chloride ions.
[**Ans.** $L$ = 40 kg; $S$ = 2.226 kg]

31. A water sample, using $FeSO_4 \cdot H_2O$ as coagulant at the rate of 278 ppm, gave the following data on analysis for raw water:
$Ca^{2+}$ = 240 ppm; $Mg^{2+}$ = 96 ppm; $CO_2$ = 44 ppm; $HCO_3^-$ = 732 ppm;
Calculate the lime and soda required to soften 250 000 liters of water.
[**Ans.** $L$ = 27.2 kg; $S$ = 132.5 kg]

32. Calculate the quantities of lime and soda required for softening 300 000 liters of water, using 20 ppm of sodium aluminate as a coagulant. Impurities in water are as follows: $Ca^{2+}$ = 160 ppm; $Mg^{2+}$ = 96 ppm; $HCO_3^-$ = 403 ppm; dissolved $CO_2$ = 34 ppm. [**Ans.** $L$ = 176.6 kg; $S$ = 145.4 kg]

33. Calculate the quantities of lime ($Ca(OH)_2$) and soda (anhyd. $Na_2CO_3$) required for cold softening of 125,000 L of water with the following analysis, using 10 ppm of sodium aluminate as coagulant:
Analysis of raw water: $Ca^{2+}$ = 95 ppm; $Mg^{2+}$ = 36 ppm; $CO_2$ = 66 ppm; $HCO_3^-$ = 264 ppm; $H^+$ = 2 ppm
Analysis of treated water: $CO_3^{2-}$ = 45 ppm; $OH^-$ = 34 ppm. Write the chemical equations involved.
[**Ans.** $L$ = 63 kg; $S$ = 55 kg]

34. Calculate the amount of lime and soda required to soften 10 000 liters of water containing the following ions per liter:
$Mg^{2+}$ = 4.8 mg; $Ca^{2+}$ = 16.0 mg; $HCO_3^-$ = 73.2 mg. [**Ans.** $L$ = 592 g; $S$ = Nil]

# Chapter 11 FUELS

## 11.1 INTRODUCTION

Whenever a substance is burnt in air or oxygen with the evolution of heat, the process is called *combustion* and the substance which is burnt is known as *combustible substance.* The heat evolved during a combustion can be used economically for industrial or other use. All combustible substances which contain carbon as the main constituent are called **fuels.** Thus, a fuel may be defined as, *any combustible substance, chemical or reactant containing carbon as the main constituent which on proper burning produces heat that can be used economically for domestic and industrial purposes and in generation of power.*

$$\text{Fuel} + O_2 \longrightarrow \text{Products} + \text{Heat}$$

## 11.2 CLASSIFICATION OF FUELS

The fuels are broadly classified into two types depending upon the state of matter:

1. **Primary Fuels.** These include the naturally occurring fuels found freely on earth's crust. These are further classified as:
   (*i*) *Solid fuels,* for example, wood, peat, lignite, coal, etc.
   (*ii*) *Liquid fuels,* for example, petroleum, crude oil, etc.
   (*iii*) *Gaseous fuels,* for example, natural gas, etc.
2. **Secondary or Derived Fuels.** These are artificially manufactured or are derived from primary fuels. These are further classified as:

(*i*) *Solid fuels,* for example, coke, charcoal, petroleum coke, pulverised coal, and colloidal fuels. Solid rocket fuels such as thiokol, hydrazine, nitrocellulose, etc.

(*ii*) *Liquid fuels,* for example, gasoline, diesel oil, kerosene, coaltar and its fractions, alcohol, LPG, and synthetic spirit.

(*iii*) *Gaseous fuels,* for example, coal gas, water gas, oil gas, biogas, blast furnace gas, acetylenes, etc.

The flowchart for the classification of fuels is given below:

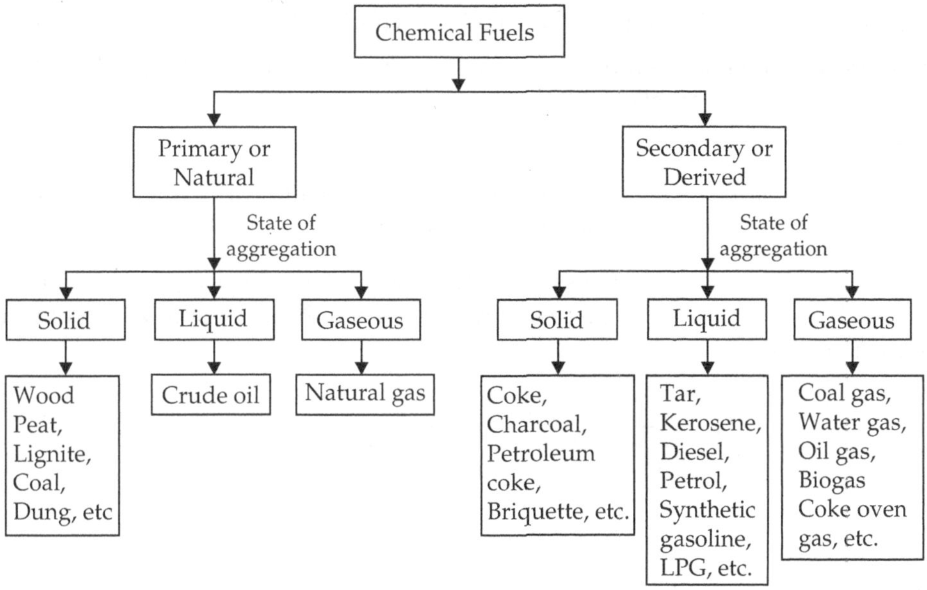


## 11.3 CHARACTERISTICS OF A GOOD FUEL

A good fuel should possess the following characteristics:

(*i*) The fuel should have high calorific value, *i.e.*, it should produce a large amount of heat on burning.

(*ii*) It should have a low moisture content as moisture reduces the heating value.

(*iii*) It should have a low noncombustible matter content as it reduces heating value.

(*iv*) The ignition temperature should be moderate as the low ignition temperature is harmful for storage and transport while high ignition temperature causes difficulty in igniting the fuel.

(*v*) It should be easy to handle, transport, and store.

(*vi*) There should be no spontaneous combustion as it causes fire hazard.

(*vii*) It should burn in air with more efficiency and without smoke.

(*viii*) It should be readily available in bulk and at low cost.

(*ix*) The ash content of the fuel should be as low as possible.

(*x*) Its combustion should not pollute the environment.

## 11.4 CALORIFIC VALUE OF A FUEL

The calorific value of a fuel is defined as *the amount of heat obtained by the complete combustion of a unit mass of the fuel.* Calorific value is of two types:

**Gross Calorific Value (GCV).** *It is the total amount of heat generated when a unit mass of fuel is completely burnt and the products of combustion are cooled down to 60°F, 288°K or 15°C.*

When a fuel containing hydrogen is burnt, the hydrogen present in the fuel produces steam. If the products of combustion are condensed to 15°C, the latent heat of condensation of steam also gets included in the measured heat. This is called *higher or gross calorific value.*

**Net Calorific Value (NCV).** *It is the net heat produced when one unit mass of fuel is completely burnt and the products are allowed to escape.*

Thus,

*Net calorific value* = Gross calorific value – Latent heat of condensation of the water vapor produced

= Gross calorific value – (Mass of hydrogen per unit weight of the fuel burnt × 9 × latent heat of vaporization of water)

For the reaction,

$$H_2 + \frac{1}{2}O_2 \longrightarrow H_2O$$

| | |
|---|---|
| 2 g | 18 g |
| 1 g | 9 g |

One part by weight of hydrogen gives nine parts by weight of water. The latent heat of steam is 587 cal/g of water vapor produced.

Thus,

$$NCV = GCV - 9 \times \frac{H}{100} \times 587$$

$$\boxed{\textbf{NCV = GCV – 0.09} \times \textbf{H} \times \textbf{587}}$$

where, H = mass of hydrogen in the fuel.

**Units of Calorific Value.** Calorific values of solid and liquid fuels are usually expressed in calories per gram, *cal/g* or kilocalories per kilogram, *kcal/kg* or British Thermal Units per pound, *B.Th.U/lb*. Calorific values of gaseous fuels are expressed as kilocalories per cubic meter (*kcal/m*$^3$) or British Thermal units per Cubic feet (*B.Th.U./Ft*$^3$).

### 11.4.1 Determination of Calorific Value

**[A] Bomb Calorimeter**

Calorific value of a solid and liquid fuel is determined by Bomb Calorimeter.

**Principle.** A known unit mass of the fuel is burnt and the quantity of heat produced is absorbed in water and measured. Then the quantity of heat produced by burning a unit mass of the fuel is calculated.

**Construction.** It consists of the following parts (Fig. 11.1):

(*i*) A strong *cylindrical bomb* made up of chromium nickel-molybdenum steel, resistant to corrosion and capable of withstanding high pressures up to 50 atmospheres. It is provided with a lid which can be screwed firmly on the bomb. The lid in turn is provided with two electrodes and an oxygen inlet valve. A small ring is attached to one of the electrode which acts as a support for the crucible.

(*ii*) A *copper calorimeter vessel* with a known weight of water and in which the bomb stands.

(*iii*) The calorimeter is surrounded by an *air jacket* and a *water jacket* to prevent the loss of heat due to radiation.

(*iv*) The calorimeter is provided with an *electrical stirrer* for stirring water and a *Beckmann thermometer which can read accurately temperature difference up to 0.01 of a degree.*

(*v*) The *crucible* used is made up of nickel, stainless steel or fused silica.

**Working.** A weighed amount of the fuel is placed in the silica crucible supported over the ring. A fine magnesium wire touching the sample of the fuel is stretched across the electrodes. About 10 mL of distilled water is introduced into the bomb to absorb vapors of sulphuric acid and nitric acid formed during the cumbustion. Oxygen supply is forced into the bomb to a pressure of 25–30 atmosphere. The bomb is then carefully placed in the calorimetric vessel containing a known amount of water. The stirrer is driven and the initial temperature of water is noted. The electrodes are then connected to a battery and circuit is completed. The cumbustion of the fuel takes place with the evolution of heat. The heat produced by burning is transferred to the water which is stirred throughout the experiment by the electric stirrer. Maximum temperature shown by the thermometer is recorded. The calorific value of fuel can be calculated as follows:

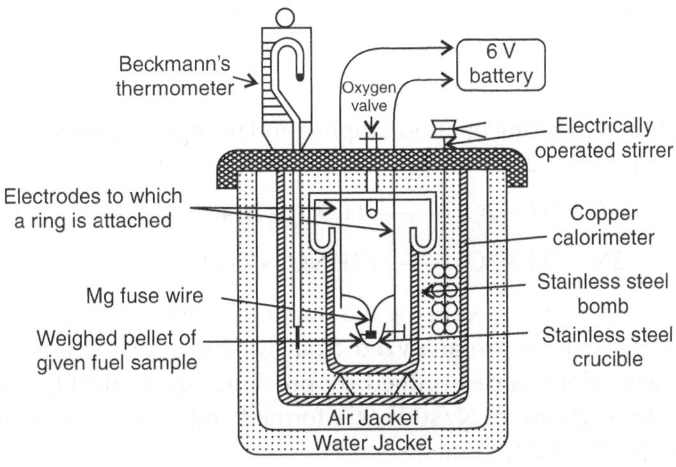


*FIGURE 11.1* **Bomb calorimeter.**

**Observations**

Weight of the fuel taken in crucible $= x$ g

Weight of water in the calorimeter = W g

Water equivalent of the calorimeter, stirrer, thermometer and bomb $= w$ g

Initial temperature of water in calorimeter $= t_1$ °C

Final temperature of water in calorimeter $= t_2$ °C

Let the higher or gross calorific value of the fuel = C cal/g

**Calculations**

Heat gained by water $= W \times (t_2 - t_1)$ cal

Heat gained by calorimeter $= w \times (t_2 - t_1)$ cal

Total heat gained $= W(t_2 - t_1) + w(t_2 - t_1)$ cal

$= (W + w)(t_2 - t_1)$ cal

Heat liberated by fuel $= x \times C$

Now, Heat liberated by fuel = Heat gained by water and calorimeter

$$x \times C = (W + w)(t_2 - t_1)$$

$$\therefore \qquad C = \frac{(W + w)(t_2 - t_1)}{x} \text{ cal/g}$$

**Corrections**

For more accurate results, the following corrections should be incorporated in the experiment:

(*i*) *Cooling correction* ($t_c$). It is the time taken for the water in the calorimeter to cool from the maximum temperature attained to room temperature. Let the actual time taken for cooling is $x$ minutes and rate of cooling is $dt°$/minute, then the cooling correction $= x \times dt$. This should be added to the observed raise in temperature.

(*ii*) *Acid correction* ($t_a$). During ignition, sulphur and nitrogen if present in the fuel are oxidized to the corresponding acids along with the evolution of heat.

$$S + 2H + 2O_2 \longrightarrow H_2SO_4 + \text{heat}$$

$$2N + 2H + 3O_2 \longrightarrow 2HNO_3 + \text{heat}$$

So, this heat is also included in the measured heat and hence it must be subtracted. The amount of these acids are analyzed from washings of the bomb by titration, while sulphuric acid alone is determined by precipitation as $BaSO_4$. *3.6 calories* should be subtracted for each mL of N/10 $H_2SO_4$ formed and *1.43 calories* must be subtracted for each mL of N/10 $HNO_3$ formed.

(*iii*) *Fuse wire correction* ($t_f$). The measured heat also includes the heat given by the ignition of the fuse wire and hence must be subtracted from the total value as per the instructions of fuse wire supplier.

Thus, higher or Gross calorific value,

$$C = \frac{(W + w)(t_2 - t_1 + t_c) - (t_a + t_f + t_t)}{x} \text{ cal/g}$$

$t_t$-cotton thread correction if used.

**[B] Junker's Or Boy's Calorimeter**

The calorific value of gaseous and volatile liquid fuels is measured by Boy's or Junker's calorimeter.

**Principle.** The gas is burnt at a known constant rate in a vessel such that the entire amount of heat produced is absorbed by water which is also flowing at a constant rate. From the volume of gas burnt, the volume of water collected and then mean rise of temperature of the water, the calorific value of the gaseous fuel is calculated.

**Construction.** The Boy's non-recording gas calorimeter (Fig. 11.1*a*) consists of two flat-flame burners B situated in a chimney C which forms the centre of the annular vessel V. The lower portion of V is provided with a trough where water condensed from the products of combustion is collected and can be remvoed through the side tube T for measurement of its volume. The products of combustion pass up to the chimney and are deflected downwards by the watercooled head H over a spiral of copper tubing through which cold water flows. The gas is again deflected upwards by an insulating baffle over another spiral of similar copper tubing and finally passes out through a number of holes in the lid of the calorimeter. The inlet water passes through the outer

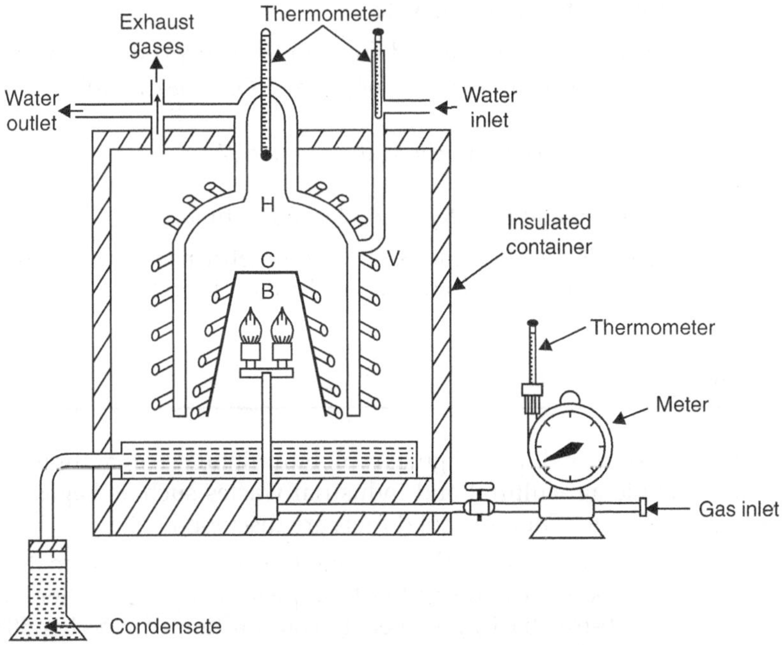


*FIGURE 11.1a* **Boy's calorimeter.**

coils downwards to return upwards through the inner coils. Finally, it flows around suitable channels on the exterior of the metal casting immediately above the chimney and passes into a mixer. The temperatures of the inlet and outlet water are measured with different thermometers. The gas flows into the burners through an accurate gas-meter.

### Procedure

The apparatus is assembled as shown in the diagram. After ensuring that the assembly is leak-proof, the gas is turned on and lighted. The water is turned on and the rate of flow is so adjusted that the rise in temperature of the water in passing through the calorimeter is as nearly as possible to 20°C. After the conditions are allowed to stabilize for about 45 minutes, the following readings are noted : (*a*) the volume of the gas burnt at a given temperature and pressure during a certain time interval (*b*) the amount of water passed through the cooling during the above time interval (*c*) the steady difference in temperature of the outlet and inlet water and (*d*) the amount of the water condensed from the products of combustion during the experiment. Also, the atmospheric pressure and the gas pressure are recorded from the barometer and manometer respectively. Further, the temperatures of the effluent gas, the ambient air and the gas meter are also noted to the nearest degree.

Then the high calorific value of the gaseous fuel can be calculated from the following relation :

$$\text{Gross calorific value,} \quad \theta = \frac{w\,(t_2 - t_1)}{V}$$

where

$w$ = wt. of the cooling water passed in time $t$.

V = volume of gas burnt at STP in time $t$.

$t_1$ = temperature of the incoming water.

$t_2$ = temperature of the outgoing water.

Further, the amount of water condensed from the steam produced by burning 1 $m^3$ of gas = m/V ; and the latent heat of steam per 1 $m^3$ of gas at 15°C = 587 Kcal)

$$\therefore \quad \text{Net calorific value} = \left(\text{Gross calorific value} - \frac{m}{V} \times 587\right)$$

where $m$ = weight of steam condensed in time, $t$.

Since the calorific value of gases is expressed on volumetric basis, it is essential to define the conditions of temperature and pressure for the volume being referred.

## 11.5 SOLID FUELS

The major solid fuels include wood, peat, lignite coal, bituminous coal, anthracite, charcoal, and coke. Solid fuels also include agricultural and industrial wastes such as bagasse, spent tar, rice husk, coconut and nut shells, etc. Common solid fuels are discussed below:

1. **Wood.** It consists of cellulose $(C_6H_{10}O_5)_n$, lignin, resins, inorganic materials and water. Water content of wood varies from 25 to 50% depending upon whether it is dry or freshly cut. On dry and ash-free basis, the average composition of wood is 50% carbon, 43% oxygen, 6% hydrogen and 1% nitrogenous and resinous material.

   Dry wood is very combustible, easily kindled and gives maximum intensity of heat very quickly. It burns with long and nonsmoky flame. Its calorific value ranges between 3000–4000 kcal/kg. It has a very low ash content ranging between 0.3 to 0.6%. Due to low calorific value it is very rarely used in industry and is largely used as a domestic fuel mainly in rural areas.

2. **Wood Charcoal.** It is obtained by the destructive distillation of wood. The carbonization is performed usually in closed retorts. During the process, gas, tar and an aqueous liquid are obtained in addition to charcoal. The yield of charcoal and gas is about 32% of wood on dry basis. The yield of gas is about 23% of wood and that of tars and oils are about 6% of wood. The gas obtained has generally the following composition:

   | | | |
   |---|---|---|
   | Hydrogen—11% | Carbon dioxide—31% | Nitrogen—2% |
   | Methane—12% | Carbon monoxide—35% | Higher hydrocarbons—8% |

   Charcoal is not pure carbon as it rapidly absorb about 9% moisture and also the gas produced during carbonization.

   Charcoal was previously used for metallurgical purposes but it has now been replaced by coke. The major use of charcoal today is for producing activated carbon which is extensively used for decolorization (*e.g.*, in sugar industry) and absorption of gases. It is also used in the production of $CaC_2$, ferro alloys, and special quality pig iron in small furnaces.

3. **Coal.** It is the most important solid fuel formed by the combined action of high temperature and high pressure on vegetable matter over geological periods of time. During conversion to coal the proportion of hydrogen and oxygen decreases while that of carbon increases. The vegetable matter, like wood, passes through various stages before getting transformed into coal. These transformation stages are peat, lignite, bituminous, and anthracite.

   (*i*) *Peat.* It is the first stage of formation of coal, it involves the gradual decaying of vegetable matter in moist places such as slightly sloping places in upland country and shallow basins in low lands. Peat is generally found in high altitudes. In upper layers of peat bog, structures of vegetable matter from which it has been formed are quite apparent. The color of these upper layers is brown and deeper layers are darker and contain fully formed peat. The fully formed peat has jelly like appearance. Composition of a fresh sample of peat is as: Moisture content—85%; Volatile matter—10.4%; Fixed carbon—4.6%.

   Peat is used only as a local fuel and is not an economic fuel due to the cost involved. The percentage composition of dried peat is:

   | | | |
   |---|---|---|
   | Carbon—57% | Sulphur—0.7% | Oxygen—35% |
   | Hydrogen—5% | Nitrogen—1.0% | Calorific value—4000–5000 kcal/kg |

   It is sometimes used as a fertilizer and also as a packaging material. In India peat deposits are found in Nilgiri Hills of South India, on both sides of river Hoogly and around Kolkata and also in Kashmir.

   (*ii*) *Lignite.* It is the second stage of transformation of wood into coal. It occurs in two forms—brown and black. In the black sample, the carbon content is comparatively greater than in the brown sample. Fresh sample of lignite has about 65% moisture content which reduces to 15% on drying. The ash content after burning is about 4% and the calorific value is 6500 kcal/kg. The percentage composition of air dried lignite is:

   | | | |
   |---|---|---|
   | Carbon—60–70% | Nitrogen—1.4% | Hydrogen—5% |
   | Sulphur—1.2% | Oxygen—20.0% | |

   Lignites are not considered as good fuel but are better than peat. Lignites are commonly made into briquettes after dehydration and marketed as such or after carbonization.

They are used for the production of producer gas and in thermal plants for steam raising. These are also used as briquettes for domestic use. Lignite deposits occur in Assam, Kashmir, Rajasthan, Tamil Nadu, (Neyveli), Malabar coast and Madhya Pradesh in India.

(*iii*) *Bituminous coal.* It is the third stage of formation of coal from vegetable matter. These are so called because they burn with smoky yellow flame like bitumen and their distillation product is coal tar which is bituminous in nature. These coals are dark and opaque. These occur in two forms:

(*a*) *Sub-bituminous coals* are black in color having waxy lustre. It is harder and denser than lignite and ignite easily having calorific value around 7000 kcal/kg.

(*b*) *Bituminous coals* are black in color with a banded appearance, laminated structure, and cubical fracture. Comparatively, it has a higher percentage of carbon than sub-bituminous coal. Moisture content is about 2% and ash content varies from 3 to 6.5%. The calorific value is around 8000–8500 kcal/kg. The approximate composition is:

Carbon—83% Hydrogen—6% Oxygen—10%

These coals are easy to handle and good in heating qualities. They are most widely used coals in the world and are used for domestic and industrial purposes, mainly for steam raising, coke, and gas production.

Major bituminous coal deposits in India are at, Gondwana, Raniganj, Bihar, West Bengal, Orissa and Madhya Pradesh.

(*iv*) *Anthracite.* It is the final stage of coal formation from vegetable matter. It is the highest ranking coal and contains maximum percentage of carbon. It is hard, lustrous, black and brittle. It contains least amount of volatile matter. Ash content is about 2.5 to 4% and calorific value is about 8600 kcal/kg. The average composition is:

Carbon—90–95% Oxygen—3% Sulphur—1%
Hydrogen—0.3% Nitrogen—1%

These coals are used in metallurgical operations, naval purposes, slow combustion stoves, central heating furnaces, generation of producer gas, etc. Special uses include manufacture of cathodes for industry and as a filter in water treatment. In India anthracites are found in Kashmir, Darjeeling and Rajhara.

## 11.6 SELECTION OF COAL

For the selection of coal for different uses, the following factors are taken into consideration:

1. **Calorific value** should be high, so that larger quantities of heat can be obtained from a smaller quantity of coal. This also reduces the cost of storage and handling.
2. **Moisture content** should be low, because moisture reduces the heating value. Moreover, moisture involves in a loss of money, because it is paid for at the same rate as the coal.
3. **Ash content** should be low, because it is noncombustible and reduces the heating value of coal; besides added cost of storage, handling and disposal. Each percent of ash in coal means a heat loss of about 1.5%. Moreover, the composition and melting point of ash are important factors:

(*i*) *Composition of ash* is of great importance in metallurgical operation, because it affects the metal and slag composition. Also, composition of ash is an important point in selecting proper flux in metallurgical operations.

(*ii*) *Fusion temperature of ash* is of great importance when coal is used in a boiler. If the constituents of ash fuse at the working temperature, then the lumps of ash (called '*clinkers*') begin to form on the fire-grates, thereby the primary air supply to fuel is restricted. Moreover, the coal particles may also get embedded in the fused lumps of ash and thus cause loss of fuel. The fused particles of ash may stick to the tube of the boiler, thereby reducing heat transfer. Low: melting ash forms molten slag, which penetrates into the pores of the refractory lining of boiler. Since the coefficient of expansion and contraction of penetrated ash is quite different from that of refractory lining material, so such absorbed low melting ash causes spoiling and reduction in the life of refractory lining of boiler. Hence, in common boilers, it is desirable that the fusion temperature of the ash should be high (above 1430°C).

4. **Calorific intensity** of coal is defined as *the maximum temperature reached when the coal is completely burnt in the theoretical amount of air.* It depends upon: (*i*) quantity, (*ii*) nature, and (*iii*) specific heat of the gaseous products of coal combustion. In most cases, the heat liberated due to combustion, preheats the air and thus affects the calorific intensity. Calorific intensity can be calculated theoretically as follows:

$$\text{Flame temperature} = \frac{\text{Heat of combustion} + \text{Sensible heat of air}}{\Sigma\,(\text{combustion products} \times \text{specific heat})}$$

## 11.7 ANALYSIS OF COAL

Coal, the most important solid fuel, is a mixture of various hydrocarbons formed by the combined action of high temperature and high pressure on vegetable matter over a geological period of time. Its composition varies widely and hence it is necessary to analyze it and interpret the results for classification, price fixation and proper industrial utilization. The analysis of the coal is carried out in the following two ways:

### 11.7.1 Proximate Analysis

It includes the determination of moisture, volatile matter, ash and fixed carbon in a given sample of coal. It gives quick and valuable information regarding commercial classification and suitability for a particular industrial use. This method involves the following procedure:

(*i*) *Determination of percentage of moisture.* A known weight of coal is taken in a preweighed silica crucible and is maintained at 105–115°C for about an hour in an electrical oven. It is then taken out, cooled, and weighed. The process of heating, cooling, and weighing is repeated many times until the constant weight of the crucible is obtained. Knowing the loss in weight of the coal, percentage of moisture can be calculated as:

Let

Weight of empty crucible $= x$ g
Weight of crucible + coal sample $= y$ g
Weight of crucible + residue (coal without moisture) $= z$ g
Weight of the coal sample $= (y - x)$ g
Weight of the moisture $= (y - z)$ g

$$\text{Percentage of moisture} = \frac{\text{Weight of moisture}}{\text{Weight of coal}} \times 100 = \frac{(y - z)}{(y - x)} \times 100$$

(*ii*) *Determination of percentage of volatile matter*. A known weight of moisture free coal is taken in a weighed crucible covered with the lid and is kept in a muffle furnace at 950°C ± 20°C for about 7 minutes. The hot crucible is taken out, cooled in a dessicator, and weighed again. The amount of weight loss is reported as volatile matter and is calculated as:

Let

| | |
|---|---|
| Weight of empty crucible | $= x$ g |
| Weight of crucible + moisture free coal | $= y$ g |
| Weight of crucible + residue (coal without volatile matter) | $= z$ g |
| Weight of moisture free coal | $= (y - x)$ g |
| Weight of volatile matter | $= (y - z)$ g |

$$\text{Percentage of volatile matter} = \frac{\text{Weight of volatile matter}}{\text{weight of moisture free coal}} \times 100$$

$$= \frac{(y-z)}{(y-x)} \times 100$$

(*iii*) *Determination of percentage of ash*. A known weight of dry coal sample obtained in the above process is taken in a weighed silica crucible. The crucible is then placed in an electric furnace maintained at 750° ± 20°C for about half an hour. The crucible is taken out, cooled and weighed again. Heating, cooling and weighing is repeated till a constant weight is obtained. The percentage of ash is calculated as:

Let

| | |
|---|---|
| Weight of empty crucible | $= x$ g |
| Weight of crucible + coal sample | $= y$ g |
| Weight of crucible + residue (ash) | $= z$ g |
| Weight of coal sample | $= (y - x)$ g |
| Weight of ash | $= (z - x)$ g |

$$\therefore \quad \text{Percentage of ash} = \frac{\text{Weight of ash}}{\text{weight of coal sample}} \times 100 = \frac{(z-x)}{(y-x)} \times 100$$

(*iv*) *Percentage of fixed carbon*. The percentage of fixed carbon is determined as,

percentage of fixed carbon = 100% of (Moisture + volatile matter + Ash)

**Significance of Proximate Analysis.** The proximate analysis plays an important role in assessing the quality of the coal in following ways:

(*i*) *Moisture*. Presence of moisture in the coal sample reduces its calorific value and a considerable amount of heat is wasted in evaporating the moisture during combustion. Hence, a high percentage of moisture is undesirable.

(*ii*) *Volatile matter*. The presence of excess volatile matter lowers the calorific value of fuel, because in this case, the combustion will be incomplete. Also high volatile matter containing coal gives long flames, high smoke and relatively low heating values. The volatile matter content is more in bituminous coal than in anthracite coals.

(*iii*) *Ash*. Ash is a non-combustible matter and it reduces the calorific value of coal. It also cause hindrance in the flow of heat. It lowers the temperature and gives rise to clinker formation. Disposal of ash is also a big problem.

(*iv*) *Fixed carbon.* The presence of fixed carbon increases the calorific value of the coal. It is this carbon that is burnt in the solid state.

### 11.7.2 Ultimate Analysis of Coal

This analysis gives us information about the percentage of carbon, hydrogen, sulphur, nitrogen, and oxygen present in the sample of coal. It involves the following determinations:

(*i*) *Determination of percentage of carbon and hydrogen.* A known weight of coal is taken in a combustion tube and is burnt in excess supply of oxygen. The $CO_2$ and $H_2O$ formed are passed through two preweighed bulbs containing anhydrous calcium chloride and a second bulb containing potassium hydroxide. The increase in the weight of calcium chloride bulb represents the weight of water formed, whereas the increase in the weight of KOH bulb represents the weight of carbon dioxide formed. The percentage of carbon and hydrogen is calculated as:

Let

The weight of the coal sample taken $= x$ g

Increase in weight of $CaCl_2$ bulb (wt. of $H_2O$) $= a$ g

Increase in weight of KOH bulb (wt. of $CO_2$) $= b$ g

Percentage of carbon is calculated as:

$$\underset{12}{C} + O_2 \longrightarrow \underset{44}{CO_2}$$

44 g of $CO_2$ contains carbon $= 12$ g

$b$ g of $CO_2$ contains carbon $= \frac{12}{44} \times \frac{b}{x}$ g

$$\therefore \quad \text{Percentage of carbon} = \frac{12}{44} \times \frac{b}{x} \times 100$$

Percentage of hydrogen is calculated as:

$$\underset{2}{H_2} + \frac{1}{2} O_2 \longrightarrow \underset{18}{H_2O}$$

18 g of $H_2O$ contains hydrogen = 2 g

$a$ g of $H_2O$ contains hydrogen $= \frac{2}{18} \times \frac{a}{x}$ g

$$\therefore \quad \text{Percentage of hydrogen} = \frac{2}{18} \times \frac{a}{x} \times 100$$

(*ii*) *Percentage of nitrogen.* Nitrogen in the sample is estimated by Kjeldahl's method. A known weight of coal sample is heated with concentrated $H_2SO_4$, $K_2SO_4$, and $HgSO_4$ for 2–3 hours so that nitrogen is converted into ammonium sulphate.

$$\text{Nitrogen} + H_2SO_4 \xrightarrow{\text{Heat}} (NH_4)_2SO_4$$

The contents are then transferred to a round bottom flask and diluted with water. The solution is then made alkaline with NaOH and liberated ammonia is absorbed in a known volume of standard acid solution. The unreacted acid is determined by back titration with the standard solution of NaOH. The percentage of nitrogen in the coal can be determined by knowing the volume of acid used and ammonia liberated as

$$\text{Percentage of nitrogen} = \frac{\text{Volume of acid used} \times \text{Normality} \times 1.4}{\text{weight of coal taken}}$$

(*iii*) *Percentage of sulphur.* It is determined from the washings obtained from the known mass of coal used in a bomb calorimeter for the determination of calorific value. During this process, sulphur is converted into sulphuric acid. The washings are treated with $BaCl_2$ solution when barium sulphate is precipitated. The precipitate is filtered, washed and heated to constant weight. The percentage of sulphur is calculated as,

$$\text{Percentage of sulphur} = \frac{\text{wt. of } BaSO_4 \text{ ppt}}{\text{wt. of coal taken in bomb}} \times \frac{32}{233} \times 100$$

(*iv*) *Percentage of oxygen.* It is calculated as:

$$\text{Percentage of oxygen} = 100 - (\text{Percentage of all other elements})$$

**Significance of ultimate analysis.** The calorific value of a fuel depends upon the carbon content, therefore, the greater the percentage of carbon the higher will be the quality of the coal. Nitrogen has no calorific value as it is inert in nature. The presence of sulphur is unwanted as it produces acid which corrodes the equipment and also causes atmospheric pollution. Hydrogen is present in combination with oxygen as water and lowers the calorific value of fuel.

## 11.8 COKE

Coke is formed by the destructive distillation of coal. It has white lustre and has got porous structure and metallic ring when struck. It has all the ash-forming mineral matter, but no volatile matter. It burns without smoke and little flame. Coke generally contains besides carbon and ash, small quantities of hydrogen (0.2 to 1.2%), oxygen traces up to 1 or 2%, organic sulphur and total sulphur from 0.8 to 1.8%.

The best cokes generally contain 90% of carbon, and less than 8% of ash. Coke is slightly hygroscopic but does not absorb more than 1 to 2% of moisture. The calorific value of coke containing 90% of carbon and 0.5% of hydrogen is about 13410 BTU per lb.

For metallurgical purposes best coke is compact, heavy, homogeneous with bright light grey lustre.

**Coking coals and non-coking coals.** When coal is distilled destructively, that is out of contact of air, it loses volatile matter. During the process some coals fuse. A coal is coking if it has tendency to soften, swell and stick together during combustion. Only bituminous coals have coking properties. Anthracite, sub-bituminous and lignitic coals are non-coking.

Coking coals are of two types. There are firm non-swelling coking coal which yield compact hard coke without any increase in the original volume. Such cokes are very useful in blast furnaces. In India coking coal of Giridih, Jharia and Raniganj coal-fields are of this type. There are other coking coals which swell excessively when coke is formed out of these. These are the coals of Raniganj series in Jharia and Raniganj coal-fields.

Coals containing low volatile matter (below 17%) high moisture (above 5%) and high volatile matter (above 40%) are not of good coking quality. All coals having H : O ratio above 58 are good coking coals.

Firmness of coke can be increased by using finely crushed coal and heating it suddenly to high temperature. For so doing finely powdered coal is charged into the coking oven in the compressed condition.

### 11.8.1 Carbonization of Coal (Manufacture of Metallurgical Coke)

Coke is manufactured from coal by heating coal without bringing it in contact with air. This process is called *carbonization*. Carbonization of coal is carried in coking ovens which are of following types :

**1. Beehive oven.** It is the earliest and cheapest process of manufacturing metallurgical coke. The oven is a dome shaped structure of bricks about 4 m wide and 2.5 m high. There are two openings (Fig. 11.2), one at the top (A) for the charging of coal and other (B) in one side for air supply as well as for coke discharge. These openings can be opened or closed as desired.

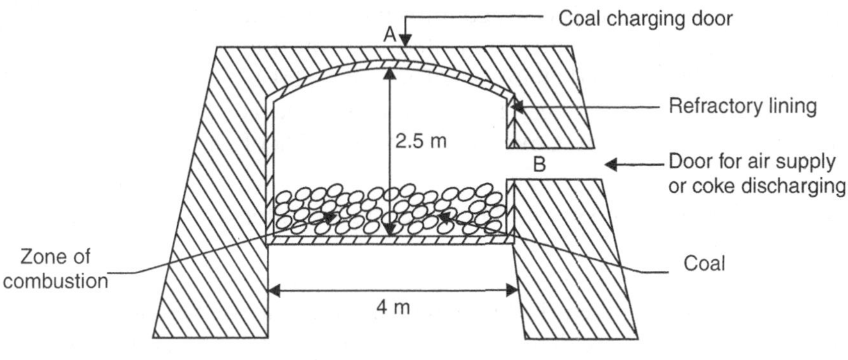


*FIGURE 11.2 Beehive Coke oven.*

Coal is charged through A and some air is supplied in and the coal is ignated. Combustion is allowed to proceed in a gradually diminishing supply of air so that slow carbonization, from the top layer to the bottom, takes place, completion of which takes about 3–4 days. When the carbonization is complete, the hot cake is quenched with water and raked out through the side door ; leaving the oven hot enough to start the carbonization of next charge batch. The yield of coke is about 80% of the coal discharged and averages about 5 to 6 tonnes coke per firing.

In this furnace the volatile products are allowed to go waste.

**2. Otto Hoffman's Oven.** This oven consists of a number of chambers (Fig. 11.3). Each chamber is constructed 14″ to 22″ wide, 12 to 14 ft. high and about 30 to 40 ft. long. At the top of each chamber there are three holes for charging (Fig 11.4). In interspaces between coking chambers, gas is burnt. The heating chambers and coking chambers alternate each other.

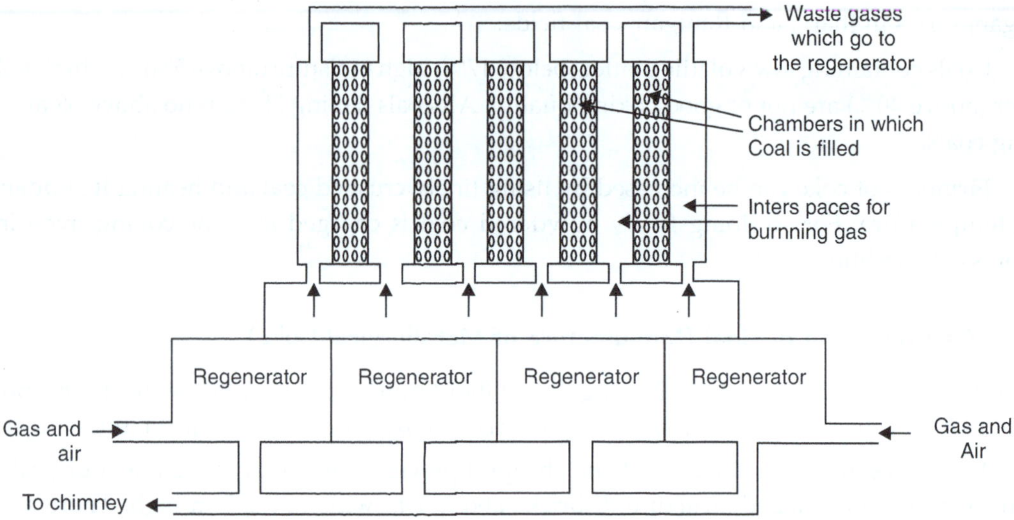


*FIGRE 11.3* **Otto Hoffman Oven.**

Every chamber at top is connected with the pipe which carries out the volatile matter. Coal is charged into chambers and the chambers are closed. Fuel gas and pre-heated air from the regenerators burn in the interspaced chambers. The coking takes palce in about 24 hrs. When coke has been formed, red-hot coke is pushed outside by means of a ram. Coke yield is about 75% of charge.

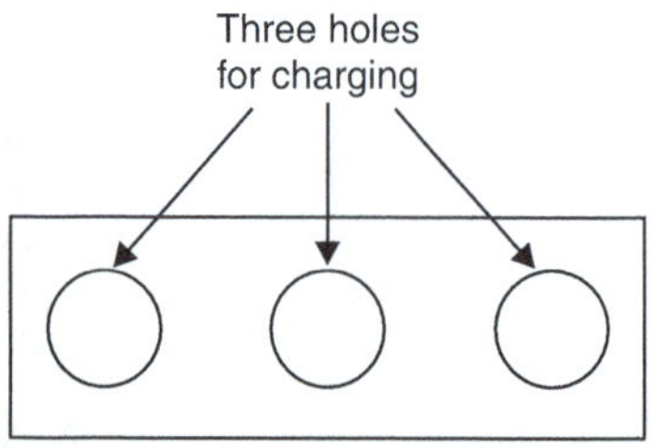


*FIGRE 11.4* **Top of Chamber.**

Heat saving is done by regenerators. Hot exhaust gases after combustion are allowed to pass through chambers which are stacked by brick work. When hot exhaust gases pass through this brick work, they heat up the brick work. After some interval, say half an hour, when brick work is heated up, the hot exhaust gases are diverted to another regenerator and through the first pass cold air and gas for pre-heating before these are burnt in combustion chamber. When it cools down, the cold air and gas are stopped and diverted through regenerator which is hot and the hot exhaust gases are passed through this first one in order to heat it up. So this process of alternately passing of hot exhaust gases and cold air and gas through regenerators continues, and this process is called *regeneration*.

**Separation of By-products from Volatile Matter.** Different by-products are separated out from the volatile matter.

**Flushing.** The volatile matter (coke oven gases) which is 25 to 30% of coal is collected into a collecting main where it is flushed with flushing liquor. This flushing liquor consists of dilute ammonia. By such flushing most of the tar separates out. Tar is collected in a separating tank where it is heated by steam coils. By so doing tar is separated from water. Dry tar is collected in a storage tank. Flushing liquor goes into flushing liquor tank.

$$2NH_4OH + H_2SO_4 \rightarrow (NH_4)_2SO_4 + 2H_2O$$

As this reaction is exothermic so the gas gets heated up. This gas is again cooled by passing it through a cooling tower.

**Removal of Naphthalene.** The gas is scrubbed with oil. Oil is sprayed in a tower where gas passes upwards.

**Removal of $H_2S$.** The gas is passed through purifiers which contains iron oxide, $Fe_2O_3$. The reactions which take place are

$$Fe_2O_3 + 3H_2S \rightarrow Fe_2S_3 + 3H_2O$$

$$Fe_2O_3 + 3H_2S \rightarrow 2FeS_3 + S + 3H_2O.$$

When oxide is saturated, it is taken out from the purifier and is exposed to air. There by the action of oxygen in air, ferric oxide is formed and free sulphur is set free.

In Thylox Process $H_2S$ is removed by employing thioarsenate as the absorbing agent. The chemical action involved is

$$Na_4As_2O_2 + H_2S \rightarrow Na_4As_2S_6O + H_2O$$

$$Na_4As_2S_6O + \frac{1}{2}O_2 \rightarrow Na_4As_2S_5O_2 + S$$

**Yields in Carbonisation.** Yields from one tonne of typical coal having 30 to 35% volatiles at 900°C is as follows :

1. **Gas.** 3500 to 6000 litre of gas of calorific value 340–452 cals/gm. Gas composition is as follows :

   $CO_2$—4 to 5% , $C_nH_n$—5% $\quad$ $C_nH_{2n+2}$—30 to 5%

   CO—4 to 11% $\quad$ $H_2$—15 to 30%

   $N_2$—4 to 5%

2. **Tar and oil.** 15 to 20 gallons. Tar and oils consist of neutral oils, tar acids and bases.
3. **Light spirit**. About 3 gallons.
4. **Ammonia**. 2 to 3 kg.
5. **Semi-coke**. 15 cwt.

**Types of Carbonization.** There are three types of carbonization :

1. **Low Temperature Carbonisation.** In this method carbonization is carried out at about 538°C. The objectives of low temperature carbonisation are as :

   (*i*) It is resorted to when low temperature semi-coke as an improved domestic fuel is required.

   (*ii*) Low carbonisation is done as a preliminary step to coals when complete gasification of coal is required.

(*iii*) Sometimes to avoid smoke etc. coal is replaced in boilers by low temperature carbonised coke.

(*iv*) Low temperature carbonisation is resorted to for coals for mixing with high volatile coal for charging into high temperature carbonisation ovens.

In short the object of low carbonisation is to get a smokeless and free burning solid fuel semi-coke. The portion causing smoke is recovered as tar oils, gases etc.

*Main difficulties of low carbonisation and their removal include :*

(*i*) Heat cannot penetrate through the whole charge which is put in a retort heated externally.

(*ii*) After carbonisation, discharging is difficult as coal after carbonisation swells a lot.

The yield of By-products in low temperature carbonisation method.

(*i*) Gas yield is less but heating value of gas is more.

(*ii*) Tar quantity is more but it is more highly paraffinic in nature.

(*iii*) Ammonia yields are low.

(*iv*) Aromatic content of tar is low.

(*v*) Tar acid content is high and consists mainly of high acids.

2. **High temperature carbonisation.** In this the carbonisation takes place at 1500° – 2000°C. Solid coke is produced from high temperature carbonisation. High temperature carbonisation ovens are Beehive Coke Oven and Otto Hoffman Oven. These have been described earlier. The by-products are collected and separated as already described.

3. **Medium temperature carbonisation.** Carbonisation temperature is between 1200°C and 1300°C. The ovens used resemble high temperature ovens. The yields in medium temperature carbonisation are as :

(*i*) Ammonia yields are same as in high temperature carbonisation.

(*ii*) Tar yields are greater than high temperature carbonisation, out tar properties differ from tar obtained by high temperature.

(*iii*) There are increases in yield of total light oils and phenols, particularly cresols. However, light oil will contain a substantial quantity of paraffin and diolefinic hydrocarbons.

(*iv*) The gas yield will be about half of the high temperature carbonisation.

## 11.9 LIQUID FUELS

Liquid fuels are becoming more and more important these days due to their usefulness and economical nature. The important liquid fuels include petroleum, petroleum products, tar, alcohols, and colloidal fuels.

### 11.9.1 Petroleum

It is dark viscous oily liquid which occurs at various depths below the surface of the earth. It is also called *rock oil* and is often associated with large quantities of natural gas and some water. When the water and natural gas are removed, the remaining liquid is called *crude oil*. Petroleum occurs mainly in middle East countries, USA, USSR, Canada, Indonesia, etc. The approximate composition of petroleum is:

| | | |
|---|---|---|
| Carbon—79.5 to 87.1% | Oxygen—0.1 to 0.9% | |
| Hydrogen—11.1 to 15.0% | Nitrogen—0.4 to 0.9% | Sulphur—0.1 to 3.5% |

### 11.9.2 Origin of Petroleum

There are two main theories regarding the origin of petroleum:

1. **Carbide or Mendeef's Theory.** According to this theory the formation of petroleum occurs by the action of steam or water on metallic carbides below the earth's crust. The following reactions take place:

   (*i*) The metals come in contact with coal under high temperature and pressure forming carbides.

   $$Ca + 2C \longrightarrow \underset{\text{Calcium carbide}}{CaC_2}$$

   $$4Al + 3C \longrightarrow \underset{\text{Aluminium carbide}}{Al_4C_3}$$

   (*ii*) These carbides react with steam to give lower hydrocarbons.

   $$Al_4C_3 + 12H_2O \longrightarrow 4Al(OH)_3 + \underset{\text{Methane}}{3CH_4}$$

   $$CaC_2 + 2H_2O \longrightarrow Ca(OH)_2 + \underset{\text{Acetylene}}{C_2H_2}$$

   (*iii*) These lower hydrocarbons undergo hydrogenation and polymerization to give paraffins, cycloparaffins, and aromatic hydrocarbons.

   $$C_2H_2 + H_2 \longrightarrow \underset{\text{Ethylene}}{C_2H_4} \qquad 3C_2H_2 \longrightarrow \underset{\text{Benzene}}{C_6H_6}$$

   $$C_2H_4 + H_2 \longrightarrow \underset{\text{Ethane}}{C_2H_6} \qquad 3C_2H_4 \longrightarrow \underset{\text{Hexane}}{C_6H_{12}}$$

   This theory fails to explain the presence of optically active compounds, nitrogen and sulphur compound, chlorophyll, heamin, etc.

2. **Organic or Engler's Theory.** It postulates that petroleum is formed from animal and vegetable debris covered up during some volcanic eruption. These, in due course of time, get converted into petroleum under high temperature and pressure. However, this theory fails to explain the presence of chlorophyll and radioactive compounds in petroleum.

### 11.9.3 Mining of Petroleum

The crude oil is found well deep below the earth's crust. Mining (Figures 11.5 and 11.6) of oil is done by drilling holes in the earth's crust and sinking pipes. Sometimes due to the pressure of natural gas above it, it rushes out as soon as it finds some natural hole. When the gas pressure becomes too low then it is mechanically pumped up with the help of lift pumps and is conveyed to a distant place for refining with the help of pipes.

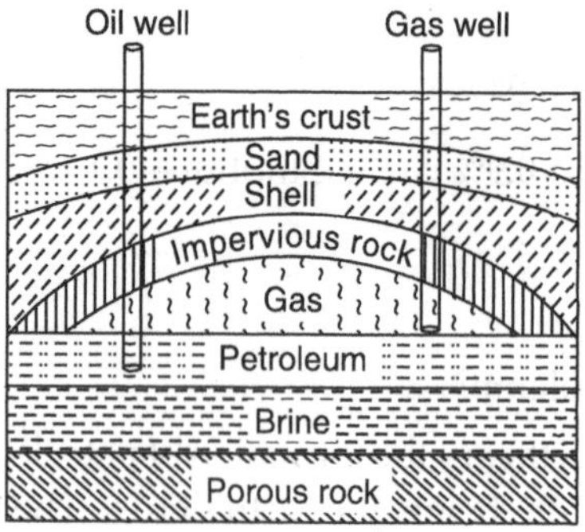


*FIGURE 11.5* **Mining of petroleum.**

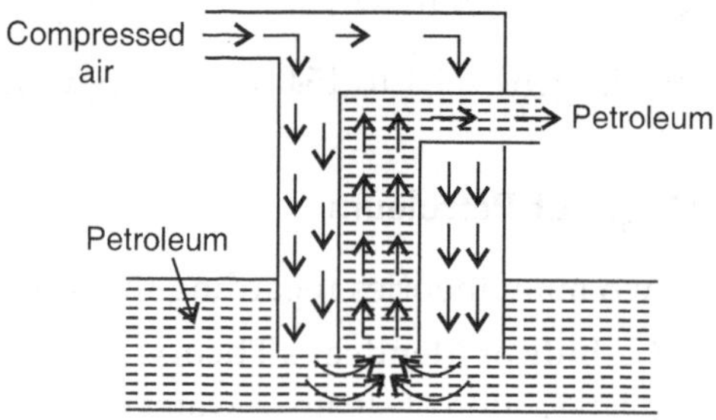


*FIGURE 11.6* **Mechanical pumping of oil.**

### 11.9.4 Refining of Petroleum

The crude oil, after removing dirt, water, sulphur and other impurities is subjected to fractional distillation as shown in Figure 11.7. This process of separating crude oil into various fractions

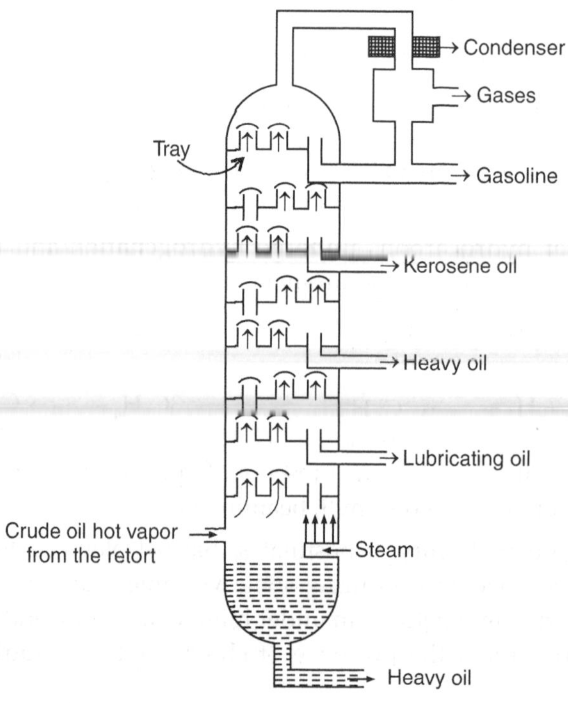


*FIGURE 11.7* **Refining of petroleum.**

having different boiling points and further treated to remove undesirable tendencies and to impart specific properties in them is broadly known as *refining of petroleum.*

The refining is done in big refineries situated at large distances from the oil fields. The oil is heated in coiled pipes in the gas heated furnace at about 400°C. The vapors obtained are introduced into the fractionating tower fitted with a large number of horizontal steel trays at short distances. Each tray is provided with small chimneys which are covered with loose caps. As the vapors go up, they become cooler and get condensed. High boiling point fractions condense first while the low boiling point fractions condense afterwards. Thus the crude oil gets fractioned into different fractions in order of their boiling ranges and are collected at different heights in the columns with the help of condensers. The different fractions of petroleum and their uses are given in Table 11.1.

**TABLE 11.1 Various Crude oil Fractions and Their Uses**

| *S. No.* | *Name of the Fraction* | *Boiling Range* | *Approximate Composition* | *Uses* |
|---|---|---|---|---|
| 1. | Uncondensed gases | Below 30°C | $C_1$ to $C_4$ | Used as domestic or industrial fuel |
| 2. | Petroleum ether | 30–70°C | $C_5$ to $C_7$ | As a solvent |
| 3. | Gasoline or petrol | 40–120°C | $C_5$ to $C_9$ | As motor fuel, solvent and in dry cleaning |
| 4. | Benzene | 120–180°C | $C_9$ to $C_{11}$ | For dry cleaning and as a solvent |
| 5. | Kerosene oil | 180–250°C | $C_{10}$ to $C_{16}$ | As an illuminant fuel and in manufacturing of oil gas |
| 6. | Diesel oil or fuel oil | 250–320°C | $C_{10}$ to $C_{18}$ | As diesel engine fuel |
| 7. | Heavy oil | 320–400°C | $C_{17}$ to $C_{30}$ | For getting gasoline by cracking |
| 8. | Residual oil on refractionation gives: | Above 400°C | $C_{30}$ onwards | |
| | (*a*) Lubricating oil | — | | As lubricant |
| | (*b*) Paraffin wax | — | | For making candles, ointment, toilet soaps, shoe polishes and vaseline. |
| | (*c*) Asphalt | — | | In manufacturing of paints, for water proofing and road making. |

The above fractions are collected separately. Higher boiling fractions are subjected to cracking to get lower boiling fractions which further are subjected to purification to make them more suitable for specific purposes.

The three most important liquid fuels obtained from crude petroleum are:

1. *Petrol.* It is obtained between the temperature range 40–120°C and consists of a mixture of hydrocarbons ranging from $C_5$ to $C_9$. Its calorific value is about 11 250 kcal/kg. It is highly volatile and highly inflammable. The approximate composition is:

   Carbon—84% Hydrogen—15% Nitrogen, Sulphur and Oxygen—1%

   It is used as a fuel for internal combustion engines of automobiles and airplanes.

2. *Kerosene oil.* It is obtained between 180–250°C and consists of a mixture of hydrocrabons ranging from $C_{11}$ to $C_{16}$. The oil is colourless to light green. Its specific gravity is 0.75–0.85. Its calorific value of about 11,100 kcal/kg and approximate composition is:

   Carbon—84% Hydrogen—16% Sulphur— $< 0.1\%$

   It is used as a domestic fuel in stove, aviation turbine fuel, illuminant, as jet engine fuel, and for making oil gas.

3. *Diesel oil.* This fraction is obtained between 250–320°C and is a mixture of hydrocarbons ranging from $C_{10}$ to $C_{18}$. Its density is 0.86–0.95. Its calorific value is about 11 000 kcal/kg. It is used as a diesel engine fuel.

### 11.9.5 Cracking

Cracking is the process by virture of which crude petroleum or their fractions are decomposed by heat to produce products which have lower boiling points. The main objective of cracking is generally the production of gasoline. Of all the fractions obtained by fractionation of petroleum, gasoline has the largest demand as a motor fuel, but its yield is only 20% of the crude. Also the quality of *Straight run* gasoline is not high. It has to be properly blended. Moreover, there is a surplus of heavier petroleum fractions. To overcome these difficulties, the middle and heavy fractions are cracked to get petrol. The petrol made by cracking has far better characteristics than *straight run* petrol. In general there are two types of cracking:

1. **Thermal Cracking.** In thermal cracking charge is heated to 970°F to 1020°F in a pipe heater under a pressure of about 1000 psi at the outlet of the heating coil so that dense phase and longer time are provided during reaction.

   $$\underset{\text{paraffin}}{C_{10}H_{22} \rightarrow C_5H_{12}} + \underset{\text{olefin}}{C_5H_{10}}$$

   The lower molecules of paraffin and olefin thus formed may be further decomposed. They can even be fully cracked to carbon and hydrogen at a certain high temperature.

   $$CH_4 \rightarrow C + 2H_2$$

The charge is fed in the fractionator in which it is fractionated alongwith the cracked products from the reaction chamber. The vapours of gas and gasoline go out from the fractionator at the top. These vapours are condensed in a condenser to obtain gas and gasoline.

The liquid products from the fractionator passs on to the hot oil pump which transfers these products to cracking tubes from where the heated products go to the reaction chamber. Here cracking reactions are allowed to complete. Vapour products from the cracked products are also separated. The vapour products go the fractionator, where they are fractionated into mixture of gas and gasoline and liquids. The coke left or deposited in the reaction chamber is periodically withdrawn.

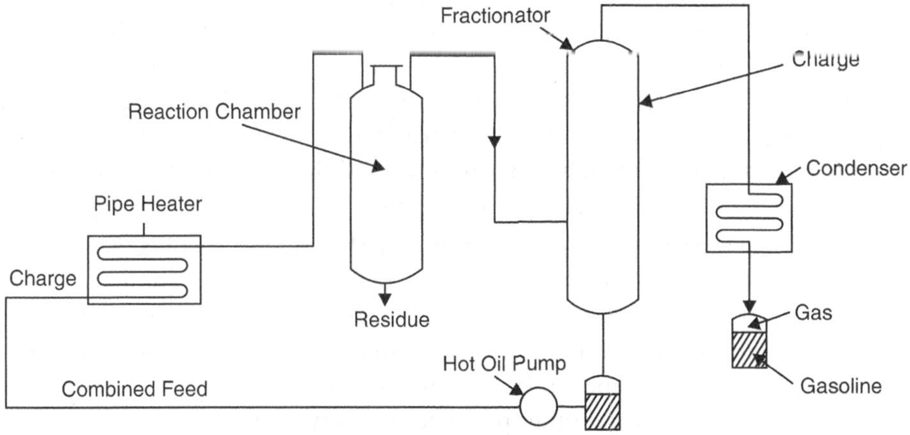


*FIGURE 11.8* **Thermal Cracking.**

It should be noted that thermal cracking may be of various types. For example,

(*a*) **Viscosity Breaking.** In this type of cracking, viscous oils and residues obtained after thermal cracking are further subjected to cracking in order to produce *fuel oils* of lower viscosity. This type of cracking, known as *viscosity breaking,* is carried out at temperatures ranging from 415°C to 490°C under a pressure range of 100-500 lbs per sq. inch. The decrease in viscosity is due to the decomposition of heavy components into intermediate products such as gas oils. During this process about 10–15% of gasoline is also obtained.

(*b*) **Vapour Pressure Cracking.** This type of cracking is carried out in such a manner that only vapour phase is obtained after cracking. As a result, only aromatic hydrocarbons and gaseous products are obtained. This type of cracking is usually performed between 550–650°C and high pressure up to 5 atmospheres.

(*c*) **Thermal Reforming.** This type of cracking consists in converting heavy gasoline of lower octane number into higher gasoline of higher octane number. The temperature range for such type of cracking is 510–540°C and pressure applied is upto 750 lbs per square inch.

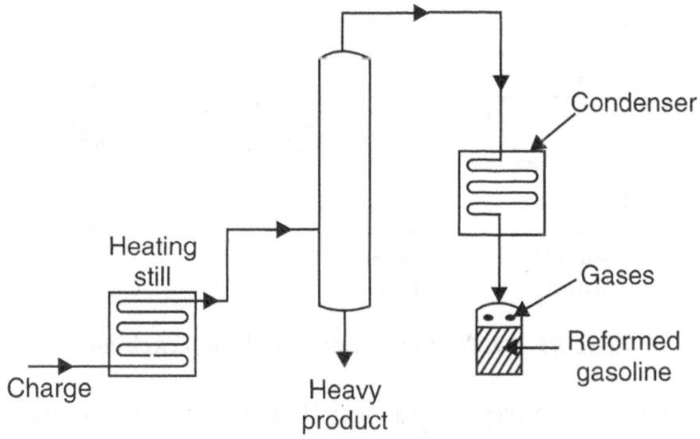


*FIGURE 11.9* **Thermal Reforming.**

**2. Catalytic Cracking.** The quality and yield of gasoline produced by cracking can be greatly improved by using suitable catalyst. The use of catalyst in cracking increases the rate of decomposition of different types of hydrocarbons in crude oils. The gasoline obtained by catalytic cracking contains low percentage of olefins and high percentage of isoparrafin and aromatic hydrocarbons. The octane number of gasolines obtained by catalytic cracking is generally about 80, while gasolines obtained by thermal cracking have octane number up to 72. If octane number of gasoline is increased by thermal cracking, the yield decreases to a much extent. So gasolines obtained by catalytic cracking are generally used as aviation gasolines because of their high octane number and stable as well as unsaturated character. The main advantages of catalytic cracking are:

(*i*) The octane number of gasoline is high. (*ii*) The yield of gasoline is also high. (*iii*) No fuel from outside is needed for catalytic cracking. The heat required is obtained from burning of coke held on the catalyst. (*iv*) The pressure required for the process is neither too low nor too high. (*v*) The gasoline formed contains much less gum and gum forming compounds. (*vi*) Sulphur is eliminated as $H_2S$ and so sulphur contents present in gasolines and other products are very small.

The Catalytic cracking is also of two types:

(*a*) **Fixed Bed Catalytic Cracking.** In this type of cracking the catalyst is in the form of granules or pellets and the beds of these catalysts are fixed in catalyst towers. Oil vapours to be cracked are passed through the catalytic beds at the cracking temperature until the catalyst becomes carbonised. The catalyst is then regenrated by burning off the deposited carbon. During reactivation of catalyst, oil vapours are transferred through the second catalyst chamber.

(*b*) **Moving Bed Catalytic Cracking.** In this type of cracking the catalyst in the form of fine powder flows down through a hopper into the reactors used for cracking and a steam of pre heated coils go up. The carbonised particles of catalysts while coming down come in contact with the rising air and the carbon deposited oven them is burnt off.

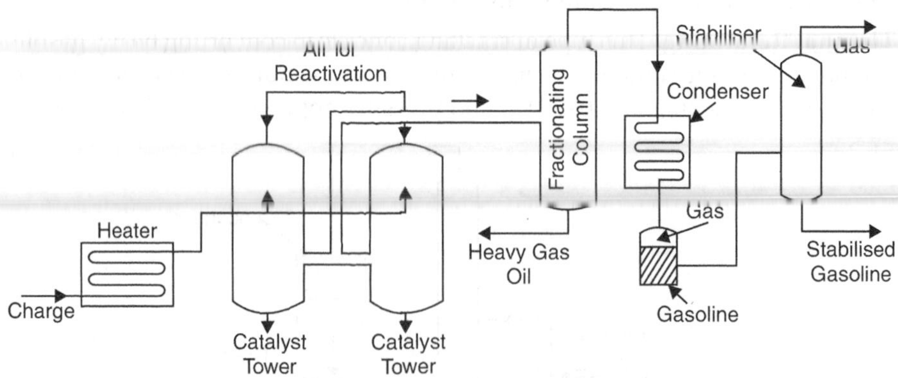


***FIGURE 11.10*** **Fixed bed catalytic cracking.**

*In fixed bed catalytic process* charge is passed through a heater and heated there to cracking temperature. The charge is then passed to the catalyst towers which contain catalyst tubes. Around catalyst tubes (catalyst used are bauxite pellets or $SiN_2.Al_2O_3$, 6 mesh pellets) molten salt mixture of sodium nitrate and sodium nitrite is allowed to circulate which distributes heat and keeps the temperature uniform in the reactors. The cracked vapours from these catalyst towers are fractionated in fractionating column to get gases and gasoline vapours from the top and heavy gas oil is collected from the bottom of the fractionating column. Gasoline vapours are cooled and condensed in condenser and gasoline is sent to stabilizer, where certain gases are removed. In this way, gasoline having the desired boiling range and vapour pressure is obtained.

*In moving bed catalytic process,* heated oil vapours go up in the reactor and powdered catalyst moves down through the hopper. When catalytic cracking of vapours is complete, the catalyst is drown off from the bottom. It is regenerated again to be dropped from the catalyst hopper through the elevator. The cracked vapours are separated from catalyst dust in dust separator and then passed through fractionator, where gas oil is separated from vapours of gas and gasoline, gas and gasoline vapours are condensed in the condenser and separated. The gas oil is withdrawn from the bottom. It should be noted that:

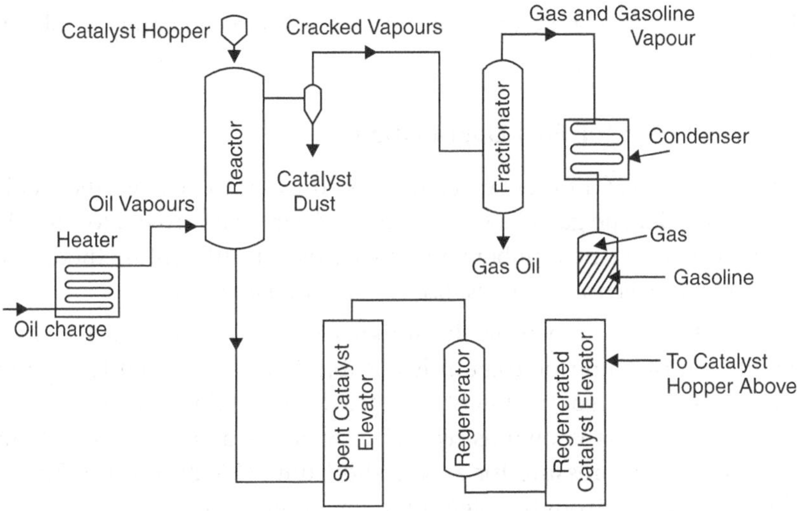


*FIGURE 11.11* **Moving bed catalytic cracking.**

(*i*) Catalysts are selective in their action and they crack only high boiling materials.

(*ii*) They decompose naphthalenic materials more than paraffins. As a result, product formed by catalytic cracking are more paraffinic than products formed from thermal cracking.

(*iii*) Isomerisation takes place in catalytic cracking and thus the gasoline is rich in iso-paraffins.

(*iv*) Polymerisable or coke forming materials are absorbed by the catalysts as soon as they are formed.

Cracking processes thus are conducted in the temperature range 400–700°C and at pressures which vary from atmospheric (vapour phase cracking) to as much as 1200 lbs per sq inch (liquid phase cracking). Thus cracking is of two types :

1. **Liquid Phase Cracking.** This cracking is carried out by heating heavy oil ($C_8H_{38}$ — $C_{48}H_{88}$) at a high temperature (475–530°C) continuously through the cracking zone in the presence of some catalyst such as silica and alumina (4 : 1) with about 1% manganese dioxide

under a very high pressure of 1000–1200 lbs per sq. inch of the surface area. High pressure is most essential to maintain the oil in the liquid state during cracking inspite of the high temperature. The large fractions are thus converted into smaller ($C_4 - C_{10}$) hydrocarbons together with some olefins and aromatics. The conversion is about 70%.

2. **Vapour Phase Cracking.** In this method generally kerosene oil or gas oil is first vapourised and then cracked using high temperature (600–800°C) and low pressure (50-150 lbs per sq. in). Heavy oil can not be cracked by this method, since these can not be completely vapourised under such conditions. The cracking may further be facilitated by using catalyst.

Advantages of catalytic cracking over thermal cracking include:

(*i*) More selective cracking and less end products. (*ii*) More olefin isomerisation, both of olefin bonds and carbon skeletons. (*iii*) More controllable saturation of double bonds. (*iv*) Greater production of aromatic compounds. (*v*) Less production of diolefins. (*vi*) Relatively more economically salable coke. (*vii*) Greater ability to tolerate high sulphur feedstocks.

### 11.9.6 Chemical Treatment for Fuel Upgradation

The most important purpose of different cracking procedures is to increase the yield of high octane gasoline from a given crude. The gasolines thus prepared can be improved to a much greater extent in their qualities as well as octane numbers by making use of different methods such as *reforming polymerisation, alkylation, isomerisation, aromatisation, hydrogenation etc.*

1. **Reforming**. In order to increase the antiknock properties of petrol or gasoline obtained by distillation of crude petroleum it is subjected to a special type of cracking called reforming. Thus thermal treatment of straight run gasoline for the purpose of increasing the octane number is known as *reforming,* which is a special type of cracking process carried out at high pressure for a very short time (10–20 second). Thus if straight run gasoline is passed through a steel pipe containing a catalyst, anhydrous aluminium chloride and promoter as anhydrous HCl at about 600°C under 400–750 pounds per square inch (psi) for 10–20 seconds, 85–95 percent gasoline is obtained with a higher octane number (around 100). This is called *aviation gasoline*. The reaction is endothermic, vapours are preheated and the catalyst bed is maintained at the optimum temperature by external heating. Under these conditions, normal paraffins, (such as *n*-butane) undergo isomeric change producing isoparaffins, (such as iso-butane).

$$\underset{n\text{-butane}}{CH_3CH_2CH_2CH_3} \rightleftharpoons \underset{\text{Iso-butane}}{CH_3-\overset{\overset{CH_3}{|}}{\underset{\underset{H}{|}}{C}}-CH_3}$$

Iso-butane is better fuel because it has greater resistance to knocking than the *n*-butane. After separation, from HCl gas, the iso-butane is refrigerated and stored. The liquid iso-butane contains some unconverted butane, which is removed by fractionation in the alkylation plant. The liquid iso-butane is alkylated as below;

$$(CH_3)_3C.H + CH_2 = C(CH_3)_2 \xrightarrow{H_2SO_4} CH_3-C(CH_3)_2-CH_2-CH(CH_3)-CH_3$$

Iso-butane    Iso-butylene    Iso-octane (2, 2, 4 trimethyl pentane)

Alkylation can be induced with catalysts such as sulphuric acid, boron fluoride, anhydrous hydrogen fluoride, aluminium chloride etc. It may also be induced by thermal treatment.

In addition to formation of branched chain alkanes (*e.g.*, iso-butane from *n*-alkanes (*e.g.*, *n*-butane), the process of reforming also involves aromatisation, cyclisation and dehydrogenation. For example, cyclohexane gives benzene as a result of dehydrogenation. Methyl cyclopentane yields cycloparaffins as a result of isomerisation and *n*-alkanes form cycloparaffins as a result of cyclisation.

The various reactions involved in above processes are :

(*i*) *Dehydrogenation*

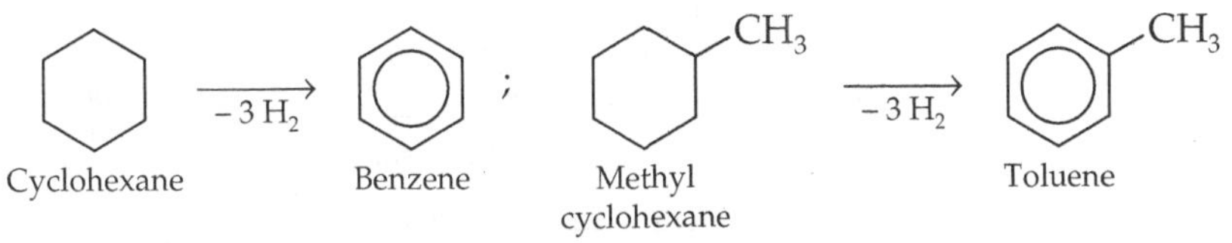


(*ii*) *Dehydrocyclization*

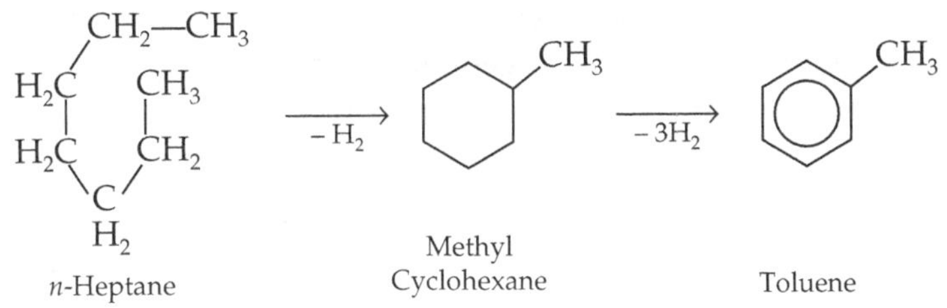


(*iii*) *Hydrocracking.*

$$CH_3-(CH_2)_8-CH_3 + H_2 \xrightarrow{[Pt]} 2\,CH_3-(CH_2)_3-CH_3$$

*n*-Decane    *n*-pentane

(*iv*) *Isomerization*

$$CH_3-(CH_2)_4-CH_3 \longrightarrow CH_3-CH(CH_3)-CH_2-CH_2-CH_3$$

2-Methyl pentane

**2. Polymerisation.** Large quantities of olefins or unsaturated hydrocarbons are obtained as a by product during cracking. These hydrocarbons can be converted into gasoline by polymerisation.

The gases obtained from the cracking of petroleum or other heavier oils are olefins such as ethylene, butenes, propene and saturated hydrocarbons such as methane, ethane, propane and butanes. When these unsaturated hydrocarbons are subjected to high pressure and high temperature in presence or absence of a catalyst, they undergo polymerisation to form higher hydrocarbons. Saturated hydrocarbons may also undergo polymerisation reactions.

$$\underset{\text{Propene}}{CH_3{-}CH = CH_2} + \underset{\text{1-Butene}}{CH_3{-}CH_2{-}CH = CH_2} \rightarrow \underset{\text{5-methyl-1-hexene}}{CH_2 = CH{-}CH_2{-}CH(CH_3)_2}$$

The unsaturated hydrocarbons, such as ethylene, propylene, butylene etc. produced in cracking are valuable starting materials for the production of higher octane motor fuels. They undergo polymerisation in the presence of catalyst like sulphuric acid and the product obtained

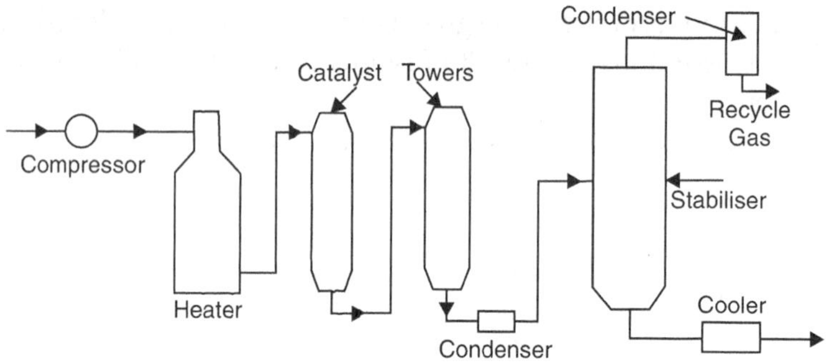


*FIGURE 11.12* **Catalytic polymerisation.**

is called *polymer gasoline*, of octane number 78–83. When cracked gases are passed in 70% $H_2SO_4$ at 10–20°C, two molecules of iso-butene polymerise leaving behind other olefins unaffected. The di-iso- butylene thus formed undergoes hydrogenation in presence of suitable catalyst to yield iso-octane, which in pure form has an octane number of 100.

$$\underset{\text{Iso-butene}}{2CH_3{-}\underset{\substack{|\\ CH_3}}{C}{=}CH_2} \rightarrow \underset{\text{(Di-isobutylene)}}{CH_3{-}\overset{\substack{CH_3\\ |}}{\underset{\substack{|\\ CH_3}}{C}}{-}CH_2{-}\underset{\substack{|\\ CH_3}}{C}{=}CH_2} \xrightarrow{H_2} CH_3{-}\overset{\substack{CH_3\\ |}}{\underset{\substack{|\\ CH_3}}{C}}{-}CH_2{-}\underset{\substack{|\\ CH_3}}{CH}{-}CH_3$$

Iso-octane (2, 2, 4 trimethyl pentane)

Polymerisation is of two types:

(*a*) *Thermal Polymerisation*. It may be *unitary thermal polymerisation*, in which polymerisation is carried out with cracked gases containing propane, propene, butanes and butenes (This polymerisation is carried out between 950–1100°F and pressure applied is 1000-1400 lbs per sq. inch) or *multiple coil thermal polymerisation*, which involves polymerisation of cracked gases and paraffinic gases at 900–1300°F at a pressure of 45–800 lbs per sq-inch.

In unitary thermal polymerisation the liquids formed are gasoline and gas oil. In multi coil thermal polymerisation, the products obtained are gasoline, fuel oil and tar. The multicoil thermal polymerisation takes place in three stages:

(*i*) Olefinic hydrocarbons undergo polymerisation (*ii*) Polymerised olefinic hydrocarbons and the residue containing paraffinic gas (after polymerisation) undergo cracking (*iii*) Olefinic hydrocarbons obtained from cracking get polymerised.

(*b*) *Catalytic Polymerisation.* The polymerisation carried out in the presence of a catalyst is known as catalytic polymerisation. In presence of a catalyst, the rate of reaction is increased and the reaction can be carried out at low temperature and low pressure. Moreover, catalytic polymerisation unit may be made small or big and can be operated without any change in efficiency. Some molecular combinations, which are not possible by thermal polymerisation are possible by catalytic polymerisation.

3. **Alkylation.** It is a process in which the iso-octane can be directly obtained in a simple step. This is another important method of converting olefins into gasoline. In this process, an olefin reacts with some isoparaffin at about 40°C in presence of a catalyst. The catalysts used are sulphuric acid, boron fluoride, anhydrous hydrogen fluoride, aluminium chloride etc. For example, the alkylation of iso-butane with iso-butylene gives iso-octane.

$$\underset{\text{Isobutane}}{(CH_3)_3CH} + \underset{\text{Isobutylene}}{CH_2 = C(CH_3)_2} \xrightarrow{H_2SO_4} CH_3-\overset{\overset{CH_3}{|}}{\underset{\underset{CH_3}{|}}{C}}-CH_2-\underset{\underset{CH_3}{|}}{CH}-CH_3$$

Iso-octane (2, 2, 4 trimethyl pentane)

Iso-butane is alkylated with butylenes in the presence of 90–100% concentrated sulphuric acid at temperatures 0–10°C in-order to reduce the oxidising property of sulphuric acid.

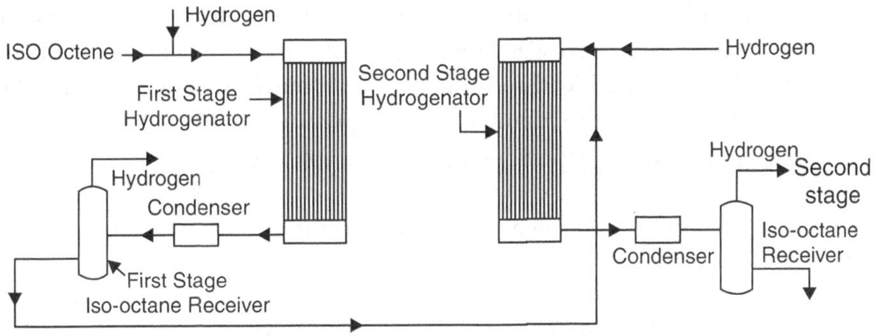


*FIGURE 11.13* **Catalytic alkylation of iso-butane.**

The mixture of iso-butane and butylene is charged in the reactor. Sulphuric acid and hydrocarbons are thoroughly mixed in the reactor to form an emulsion. The mixture is passed on to a settler, from where spent acid is removed. The hydrocarbon layer from the settler is transferred to a fractionator, where isobutane is separated. Then the alkylate goes to **de-**butaniser, where butane is separated and collected. The alkylate is then passed to a **re-**run tower for the recovery of aviation alkylate fraction from the top. High boiling alkylated products one withdrawn as bottoms.

### 11.9.7 Knocking

A mixture of petrol vapours and air acts as a fuel in an internal combustion engine. In the down stroke of the pistons a mixture of gasoline or petrol vapour and air is drawn from the carburettor into the cylinder, and on the stroke the mixture is compressed. *The ratio of the original volume of the fuel mixture to that of the final volume obtained after compression is called compression ratio.* The efficiency of the engine depends upon this ratio and higher the compression, ratio higher is the efficiency of the engine. The compression ratio infact, depends on the nature of the constituents in the gasoline used. In certain circumstances (due to the presence of some constituents in the gasoline used), the rate of oxidation becomes so great that the last portion of the fuel-air mixture gets ignited instantaneously, producing an explosive violence, known as *knocking*. The knocking results in loss of efficiency.

*Knocking* is thus *a sharp metallic sound similar to rattling of hammer which is produced in the internal conbustion engine owing to immature ignition of the air-gosoline mixture*. There is thus a huge loss of energy and it causes damage to piston and cylinder.

The phenomenon of knocking is not yet fully understood and is known to be closely related with the chemical composition of the fuel. Knocking has been found to be maximum in the fuels containing straight chain paraffins and minimum in fuels containing aromatic and branched chain hydrocarbons. Hence fuels containing straight chain paraffins are worst while those containing aromatic and branched chain hydrocarbons are best.

The tendency of fuel constituents to knock is in the following order:

*Straight-chain paraffins > branched chain paraffins > olefins > Cycloparaffins > aromatics.*

**Antiknock Compounds.** Knocking can be decreased by adding certain compounds to the fuel and these compounds are known as *antiknock compounds*. Thus, *those compounds which when added to gasoline increase the octane number and decrease the knocking are known as antiknock compounds.*

Tetraethyl lead (TEL), is a very important alkylated product and is the main antiknock compound for the gasolines. This process is called *doping*. It is prepared commercially by the action of ethyl chloride on a finely divided lead sodium alloy at about 70°C. The reaction is heated in the beginning and then must be cooled to maintain the temperature at 40–60°C.

$$4Pb\text{-}Na + 4C_2H_5\,Cl \longrightarrow \underset{\text{TEL}}{Pb(C_2H_5)_4} + 3Pb + 4NaCl$$

TEL is a colourless liquid with sweet odour. Its specific gravity is 1.62. It boils at 200°C with decomposition. It is highly poisonous and care should be taken in its storage. About 0.5 mL of TEL is added for motor spirits and about 1 mL of TEL per litre is generally added for aviation petrol.

As already mentioned, knocking in a petrol engine is due to the spontaneous ignition of the last portions of the compressed mixture of petrol and air. The process of ignition of the explosive mixture comprises of a relatively slow induction period characterized by the production of hydroxyl and other free radicals, which is followed by explosive combustion. The addition of TEL increases the induction period by 200–300% by interaction with the free radicals whereby lead oxide is produced. As soon as explosure combustion occurs, the lead oxide is rapidly reduced to metallic lead which is deposited on the spark plug joints, the cylinder walls and the piston head.

To avoid this difficulty, ethylene dibromide ($C_2H_4Br_2$) is also added to the dope to convert the lead formed by the combustion into the more volatile lead bromide, which is swept out with the exhaust gases.

If the compression ratio is suitably adjusted, an increase in octane number from 73 to 100 increases the power output by 26.7% and reduces the fuel consumption by 29%.

**Octane Rating of Fuels.** The knocking characteristics of a fuel are usually expressed by *octane number*. It was found that the hydrocarbon *iso*-octane ($C_8H_{18}$), having the structural formula

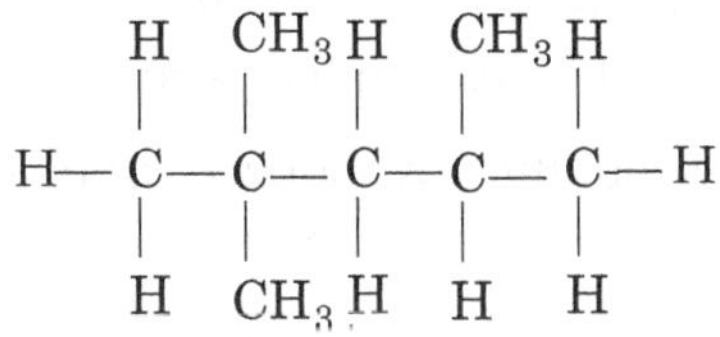

has very good combustion characteristics and exhibits very little tendency to detonate when mixed with air and ignited at high temperatures. Hence the octane number is taken as 100. On the contrary, the straight chain hydrocarbon, *n*-heptane ($C_7H_{16}$), having the structure knocks very readily and

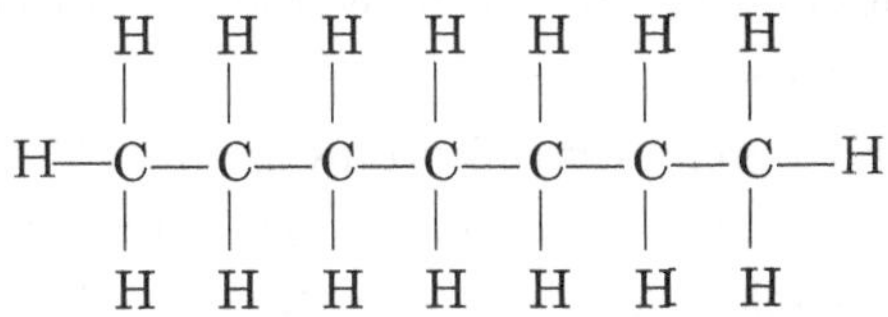

hence its octane number is taken as zero. The octane number of a given fuel is numerically taken as the percentage of *iso*-octane in a mixture of *iso*-octane and *n*-heptane which just matches with the knocking characteristics of the fuel under consideration. Thus, if a sample of petrol gives as much of knocking as a mixture of 75 parts of *iso*-octane and 25 parts of *n*-heptane, then its octane number is taken as 75. The octane ratings of some common hydrocarbons are given below :

| *Hydrocarbon* | *Octane number* |
|---|---|
| Benzene | 100+ |
| Isopentane | 90 |
| Cyclohexane | 77 |
| 2-methyl pentane | 71 |
| *n*-bentane | 62 |
| *n*-Hexane | 26 |

Fuels with octane rating greater than 100 are quite common now-a-days and they are rated by comparison with a blend of *iso*-octane with TEL which greatly diminishes the knocking tendency of any hydrocarbon with which it is mixed.

### 11.9.8 Diesel Engine Fuels

In a compression ignition engine (diesel engine), air only is first forced into the cylinder and compressed to about 30 to 50 kg/cm$^2$. As a result of the compression, the air temperature is raised to 500° to 600°C. At this stage, the fuel is injected in the form of spray into the very hot air. The fuel droplets vaporize and get heated to the temperature at which spontaneous ignition takes place and the pressure is still further increased to about 70 kg/cm$^2$. The whole process of combustion of the fuel in a diesel engine may be considered to take place in the following four steps:

(*i*) The oil droplets are shot into the very hot compressed air in the cylinder. Heat is absorbed and the first and smallest of succeeding droplets are completely vaporized.

(*ii*) Ignition of the hot vapour occurs with consequent increase in temperature and pressure within the cylinder.

(*iii*) The oil spray, still entering, burns like a jet flame, supplying heat and hot gaseous combustion products to keep up the push on the piston.

(*iv*) The injection of fuel terminates, but the unvaporized droplets of oil on cylinder surfaces continue to burn. This after-burn is increased by carbon deposits.

All these steps are completed within only a small fraction of a second.

The fuels used must have self-ignition temperatures at least 30°C below the temperatures of the compressed air. Middle or high boiling petroleum fractions are generally used as diesel fuels.

Compression ignition engines are classified as Slow, Medium or High speed, and they may operate on the two-or four-stroke cycles.

In *high speed engines*, the combustion process takes only about 1/500th of a second to complete. It is obvious that the time lag in getting the oil droplets heated to the ignition point (*i.e.*, the step 2 given above) must be as brief as possible. The diesel knock is caused by a retarded ignition (also called as ignition lag). When a fuel has a long ignition lag, a large portion of the oil gets injected and accumulated into the cylinder even before the ignition is initiated. This results in a violent combustion and a sudden increase in pressure. The nearly explosive combustion as a result of the long ignition lag causes rough and bumpy running and is responsible for the diesel knock. If the operation is ideal, rapid but steady increase of pressure reaching a maximum at the end is obtained. The greater the ignition lag, the higher is the diesel knock. The ignition lag and the consequent diesel knock depend upon (*i*) the engine design (*ii*) the type of the injector (*iii*) the size of oil droplets (*iv*) the manner in which the spray is mixed with the air, and above all (*v*) the chemical nature of the fuel.

High speed engines are very sensitive to fuel characteristics. They need the lightest diesel or gas oils having viscosities lesser than 45 seconds Redwood No.1 at 38°C. further, the fuels used for these engines should have low self-ignition temperatures. They should have only a short ignition lag. At least 65% of the oil should distil below 300°C.

Paraffins give less diesel knock than the aromatic hydrocarbons. Hence a diesel fuel should contain mostly straight chain hydrocarbons with minimum percentage of aromatic and branched chain hydrocarbon molecules.

### 11.9.9 Cetane Number

The knocking characteristics of a diesel oil are expressed in terms of *cetane number*. Cetane, $C_{16}H_{34}$, is a saturated hydrocarbon which has a very short ignition lag as compared to any commercial diesel fuel. Hence, its cetane number is taken as 100. On the contrary, α-methyl naphthalene, $C_{11}H_{10}$ (an aromatic hydrocarbon) has a very long ignition lag as compared to any commercial diesel oil. Hence its cetane number is taken as zero. Then the cetane number of a diesel oil is defined as *the percentage by volume of cetane in a mixture of cetane and α-methyl naphthalene which exactly matches in its knocking characteristics with the oil under test*.

The cetane numbers of high speed, medium speed, and low speed diesel engines should be at least 45, 35 and 25 respectively.

The cetane number of an oil can be improved by adding substances (called dopes) like ethyl nitrite, ethyl nitrate, isoamyl nitrate and acetone peroxide. The dopes are added only in small amounts (about 2%) and they are not effective on low cetane number fuels. Using of dopes for diesel oil is not as common as the use of TEL for gasoline. Other additives used for diesel fuels include inhibitors to delay or prevent gum formation chemical additions to reduce surface tension and addition of big polyhydrocarbon molecule to preserve the fluid properties during temperature variations.

In petrol engines, knocking is due to the sudden spontaneous combustion of last portions of the fuel, whereas knocking in a diesel engine is due to delay in spontaneous combustion of the first part of the fuel. Thus, an oil of high octane number has a low cetane number and vice-versa.

## 11.10 GASEOUS FUELS

### 11.10.1 Natural Gas

It is the primary gaseous fuel obtained from underground strata either as such or associated with crude oil. When it occurs along with petroleum in oil wells, it is called *wet gas* and contains appreciable amounts of propane, butane and other liquid hydrocarbons like pentanes, hexanes, etc. The wet gas is suitably treated to remove these hydrocarbons which are used as LPG. When the gas is associated with crude oil, it is called *dry gas* and consists entirely of methane and ethane along with small amounts of impurities such as $CO_2$, CO, $HO_2$, $H_2S$, $N_2$ and inert gases. The approximate composition of natural gas is:

Methane ($CH_4$)—70–90%   Ethane ($C_2H_6$)—5–10%   Hydrogen ($H_2$)—3%

Besides these, CO and $CO_2$ also occur in small amounts. It is an excellent gaseous fuel having calorific value varying from 12000 to 14000 kcal/m$^3$. It is widely used as a domestic fuel as well as industrial fuel. It also serve as raw material for the manufacture of carbon black and hydrogen, used for ammonia synthesis.

### 11.10.2 Coal Gas

It is obtained by heating good quality coal in the absence of air to a high temperature at pressure little greater than atmospheric pressure. The coal is fed into the top of large, vertical, silica retorts heated externally to about 1350°C. The flow sheet of the process is shown in Figure 11.14.

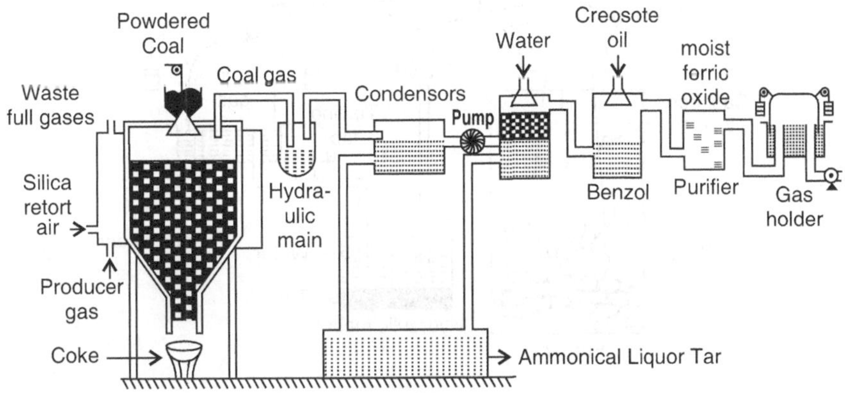


*FIGURE 11.14* **Manufacture of coal gas.**

$$\text{Coal} \xrightarrow{1350°\text{C}} \text{Coke} + \text{Coal gas} \uparrow$$

The gas obtained is associated with various impurities like tarry vapors, ammonia, and organic nitrogen containing compounds and sulphur compounds. These are removed because of their utility value and to improve the quality of gas. The various reactions involved in the removal of impurities are:

$$NH_3 + H_2O \longrightarrow NH_4OH$$

$$CO_2 + H_2O \longrightarrow H_2CO_3$$

$$6HCN + Fe_2O_3 \longrightarrow 2Fe(CN)_3 + 3H_2O$$

$$3H_2S + Fe_2O_3 \longrightarrow Fe_2S_3 + 3H_2O$$

The purified coal gas is finally stored over water in gas holder. Coal gas is a colorless gas having a characteristic odor. It is lighter than air and burns with a nonsmoky flame. Its calorific value is about 4900 kcal/m$^3$. Its average composition is:

| | | |
|---|---|---|
| Methane—32% | Hydrogen—40% | Ethylene—3% |
| Other hydrocarbons—4% | Acetylene—2% | Nitrogen—4% |
| Carbon monoxide—7% | Carbon dioxide—1% | |

**Uses**

(*i*) It is used as a heating fuel and illuminant.

(*ii*) To provide reducing atmosphere in metallurgical operations.

### 11.10.3 Oil Gas

It is prepared by cracking of kerosene oil. The equipment used in the process is shown in Figure 11.15.

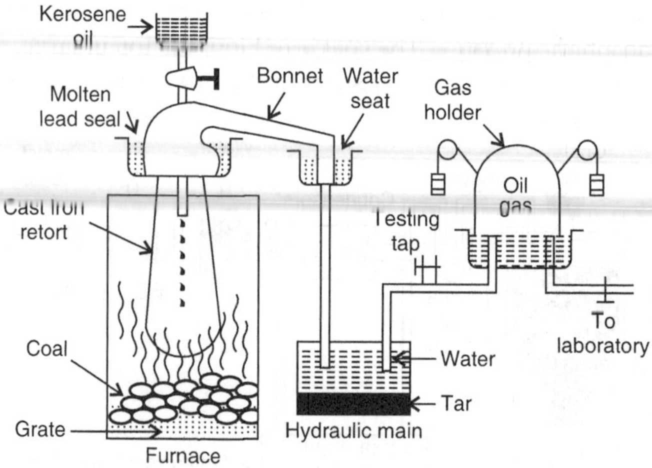


*FIGURE 11.15* **Oil gas production.**

Cracking of kerosene oil is done in a strong iron retort enclosed in a coal-fired furnace. At the mouth of the retort, a bonnet is fitted with a molten lead seal, which is further attached to a hydraulic main. The retort is made red hot and a thin continuous stream of kerosene oil is allowed to fall on the red hot bottom of the retort, resulting in cracking of oil into a number of lower hydrocarbons. The gas rises and goes to the hydraulic main through the bonnet, where tar gets condensed. Pure gas is then lead to the gas holder where it is mixed with the requisite amount of air and supplied to the place of burning.

$$C_{12}H_{26} \xrightarrow{\text{Cracking}} CH_4 + C_2H_6 + C_2H_4 + C_2H_2 + C_3H_6 + \underset{(\text{Tar})}{C}$$

The approximate composition of oil gas is:

Methane—25–30% Carbon monoxide—10–15%

Hydrogen—50–55% Carbon dioxide—3%

Oil gas burns with a smoky flame and has characteristic color. Its calorific value is about 4500–5400 kcal/$m^3$. It is mainly used as a *laboratory gas.*

### 11.10.4 Producer Gas

It is essentially a mixture of carbon monoxide and nitrogen. Its is prepared by passing a mixture of air and steam through red hot coal or coke maintained at about 1100°C in a special reactor called *gas producer* (Figure 11.16). It has an airtight cylindrical furnace lined inside with refractory bricks with an outer mild steel casting. It is provided with an inlet at the bottom through which a mixture of air and steam is blown. Another outlet is also provided at the bottom for the disposal of ash. The fuel is feed from the hopper at the top.

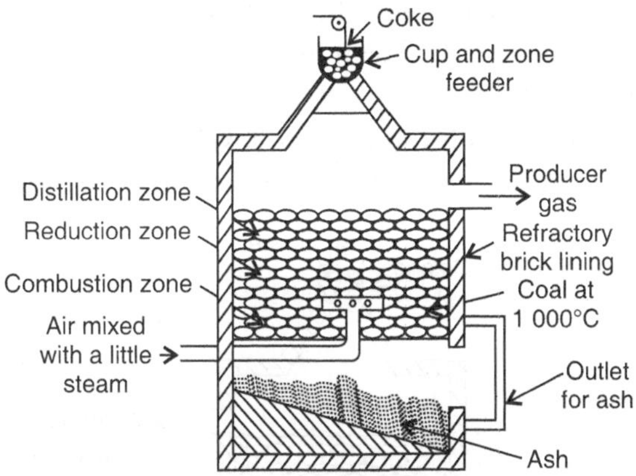


*FIGURE 11.16* **Producer gas production.**

When a mixture of air and steam is passed, the following reactions occur in different cones of the bed:

(*i*) In *oxidation zone* the coke is oxidized to CO and $CO_2$ at a temperature of about 1100°C.

$$C + O_2 \longrightarrow CO_2 + 97 \text{ kcal}$$

$$C + 0.5O_2 \longrightarrow CO + 29.5 \text{ kcal}$$

(*ii*) In *reduction zone*, $CO_2$ and steam combines with red hot coke liberating free hydrogen and CO. All these reactions are endothermic and the temperature in this region falls to 1000°C.

$$CO_2 + C \longrightarrow 2CO - 36 \text{ kcal}$$

$$C + H_2O \longrightarrow CO + H_2 - 29 \text{ kcal}$$

$$C + 2H_2O \longrightarrow CO_2 + 2H_2 - 19 \text{ kcal}$$

(*iii*) In *distillation zone*, the incoming coal is heated by the outgoing gases. The heat given by the gases and the heat radiated from the reduction zone helps to distill the fuel.

The approximate composition of producer gas is:

| | |
|---|---|
| Carbon monoxide—22–30% | Nitrogen—52–55% |
| Hydrogen—8–12% | Carbon dioxide—3% |

The producer gas is cheap, clean, insoluble in water, and poisonous in nature. Its calorific value is about 900–1300 kcal/m$^3$.

**Uses**

(*i*) It is mainly used as a fuel for furnaces like open hearth furnaces, muffle furnaces, etc.

(*ii*) It is used as a reducing agent in metallurgical operations.

### 11.10.5 Water Gas

It is essentially a mixture of CO and $H_2$. It is prepared by passing steam over red hot coke in following two steps:

(*i*) Air is passed through the coke bed when carbon burns to form $CO_2$ and the temperature of the fuel bed is about 1000°C. The outgoing gases $CO_2$ and $N_2$ are used to preheat the incoming air and steam.

(*ii*) Steam is now blown through the red hot coke when CO and $H_2$ are produced. These reactions are endothermic the temperature of the fuel bed falls below 1000°C.

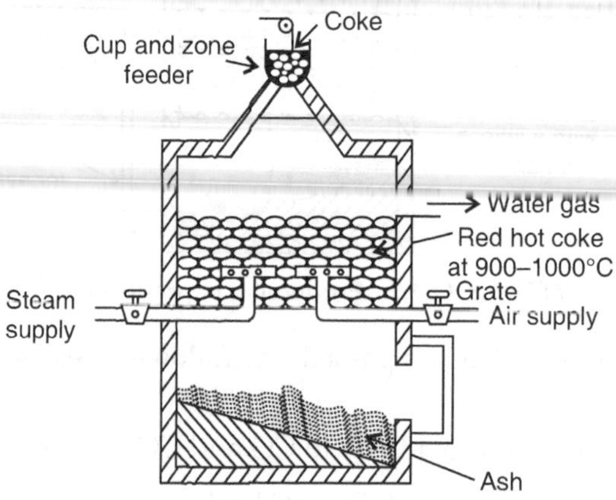


*FIGURE 11.17* **Water gas production.**

The calorific value of water gas is about 2580–2670 kcal/$m^3$. The approximate composition of water gas is:

Carbon monoxide—43%  Hydrogen—47%

Carbon dioxide—4%  Nitrogen—4%

Water gas burns with a blue flame and gives a short but hot flame. It is used for heating purposes in industrial operations.

## 11.11 COMPARATIVE ASSESSMENT OF FUELS

### SOLID FUELS

**Advantages**

*Following are the advantages of solid fuels:*

(*i*) They are easily available in abundance.
(*ii*) Their cost of production is low.
(*iii*) They possess moderate ignition temperature.
(*iv*) They are easy to transport.
(*v*) They are convenient to store.

**Disadvantages**

*Following are the disadvantages of solid fuels:*

(*i*) Their thermal efficiency is low.
(*ii*) They burn with large amounts of smoke.
(*iii*) The ash content is high and burns with clinker formation.
(*iv*) Their calorific value is quite low as compared to liquid and gaseous fuels.
(*v*) They cannot be used in internal combustion engines.
(*vi*) They require large excess of air for complete combustion.

### LIQUID FUELS

**Advantages**

*Following are the advantages of liquid fuels:*

(*i*) They have higher calorific values than solid fuels.
(*ii*) They burn without forming ash, clinkers, etc.
(*iii*) These are easy to handle, store and transport.
(*iv*) Liquid fuels can easily be kindled, the combustion can be started or stopped at once.
(*v*) They can be easily used in internal combustion engines.
(*vi*) They can easily be transported through pipe and fuel supply can be regulated to many furnaces by a single person.

### Disadvantages

*Following are the disadvantages of liquid fuels:*

(*i*) They give peculiar odors particularly when combustion is incomplete.

(*ii*) They are more expensive than solid fuels.

(*iii*) Special type of burners and storage tanks are required for their burning and storage, respectively.

(*iv*) Risk of fire hazards is great in case of inflammable and volatile liquid fuels.

(*v*) Possibility of losses during evaporation and leakage in containers is more.

## GASEOUS FUELS

### Advantages

*Following are the advantages of gaseous fuels:*

(*i*) They can be manufactured at a central location and distributed over a wide area through pipelines.

(*ii*) These are very clean to operate and are ashless and smokeless.

(*iii*) Their calorific value is quite high and burning can easily be controlled.

(*iv*) They can be ignited spontaneously and excess air needed is much less than solid and liquid fuels.

(*v*) They can be used in internal combustion engines.

### Disadvantages

*Following are the disadvantages of gaseous fuels:*

(*i*) Large storage tanks are required for their storage.

(*ii*) They are highly inflammable, hence risk of hazards are quite high.

(*iii*) Their cost of production is quite high.

## 11.12 NON-CONVENTIONAL ENERGY SOURCES

Energy may be defined as the capacity for doing useful work. Energy resources are broadly classified as primary and secondary:

A. **Primary Energy Sources.** These sources are either mined or otherwise obtained from the environment. For example,

(*i*) *Fossil fuels:* Coal, lignite, crude oil, natural gas, etc.

(*ii*) *Nuclear fuels:* Uranium, thorium, deuterium, etc.

(*iii*) *Hydroenergy:* The energy produced by falling water.

(*iv*) *Solar energy:* Electromagnetic radiations from the sun.

(*v*) *Geothermal energy:* The heat stored in the hot rocks beneath the earth's surface.

(*vi*) *Wind energy.*

(*vii*) *Tidal energy.*

The primary energy resources can be further classified as:

*Renewable resources, i.e.,* the sources which are not exhaustible and hence can provide continuous supply, *e.g.*, wood, tidal energy, solar energy, wind energy, geothermal energy, etc.

*Non-renewable resources, i.e.,* the sources which are finite and exhaustible, *e.g.*, fossil fuels, nuclear fuels, etc.

B. **Secondary Energy Resources.** These resources do not occur in nature but are derived from primary energy resources. For example, petrol, electrical energy derived from burning of coal, hydrogen obtained by electrolysis of water, etc.

## 11.13 SOLAR ENERGY

The sun provides an ideal source of energy. The solar energy is unlimited in supply which does not add to the Earth's total heat burden and does not add to air and water pollution. It is a powerful alternative to fossil and nuclear fuel. Solar energy originates from the thermonuclear fusion reactions taking place in the sun. The earth continuously receives this solar energy, a part of which is absorbed while the remaining is emitted back into space. 92% of the solar radiations reaching earth fall in the range 315 nm to 1400 nm. The earth absorbs radiations mainly in the visible region (45%) and emits radiation in the IR region. The nature of solar flux reaching the earth's upper atmosphere is about 1400 watts $m^{-2}$ $min^{-1}$. The heat equivalent of the solar radiation reaching the earth is estimated to be about $2.68 \times 10^{24}$ J per year.

Sunlight is the major energy source for all ecosystems, through the process of photosynthesis. About 0.2 to 0.5% of solar energy is trapped by photosynthesis, so only a tiny fraction of the solar energy reaching the earth drives all ecosystems. Similarly, a small percentage of solar energy reaching the earth could sustain all the energy needs for human society without altering the biosphere in any way.

### Applications of Solar Energy

Following are the important applications of Solar Energy:

1. *Space and Water Heating by Solar Energy.* The most convenient application of solar energy is in heating buildings and to provide hot water which, in developed countries like the USA, consumes 20–25% of the fuel supply. Temperature of about 20–22°C are used for space heating and 50–60°C are used for hot water. Solar energy is ideally suited for providing low temperature heat for such purposes because sunlight falling on a black surface is readily absorbed and converted to heat in the desired temperature range. For this purpose a simple flat plate collector is sufficient. Flat plate collectors basically consist of a black surface covered by a clear plastic or glass window (Fig. 11.18). The black surface absorbs sunlight and converts it to heat, and the window prevents the heat from escaping out. Air can be heated by passing it between the window and the black surface whereas water heating is done by passing it through tubes embedded in the surface.

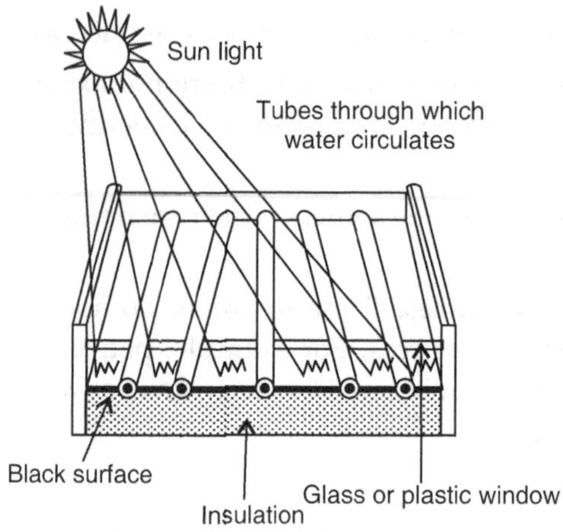


*FIGURE 11.18* **Flat plate Solar collector.**

2. *Conversion of Solar Energy into Electrical Energy.* Sunlight may be directly converted into electricity through photovoltaic effect. A *photovoltaic cell* or *solar cell* consist of two very thin layers of material. The lower layer has atoms with simple electrons in the outer orbital, which are easily lost. The upper layer has atoms lacking one electron from their outer orbital and hence can readily gain electrons. The kinetic energy of light striking this 'sandwich' dislodges electrons from the lower layer, which are trapped into the upper layer, thus creating an electrode potential between the two layers. This potential provides the electric current through the rest of circuit, which connects the upper side through an electrical device back to the lower side. Thus, without any moving parts, solar cells convert light energy directly to electrical power. These cells do not wear out because no moving parts are involved.

   The solar cells made of silicon have been used as source of power in space craft and satellites. Other solar cells developed are CdS(*n*-type)/$Cu_2S$(*p*-type), gallium arsenide and indium phosphate.

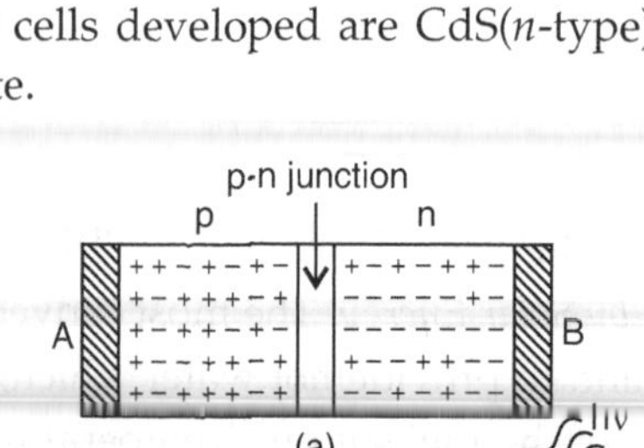


*FIGURE 11.19* **Use of semiconductors as solar cells.**

3. *Solar Energy for Driving Vehicles.* Solar energy can be used to produce hydrogen by electrolysis of water which in turn can be used as fuel for running vehicles. Hydrogen is a clean burning gas causing no pollution because the product of combustion is only water vapor.

   Present vehicles could easily be adapted to run on hydrogen with some modifications in the fuel tank, fuel intake system, and the type of service stations.

## 11.14 BIOMASS AND BIOMASS ENERGY

Biomass may be defined as *the waste material of living beings and the dead parts of the living beings, i.e., plants, trees and animals.* It includes cattle dung, wood, sewage, agricultural wastes, or crop residues such as bagasse and rice husk. Biomass contains carbon compounds and works as a source of heat energy for domestic works. Since biomass is obtained through the process of photosynthesis, so biomass energy is considered to be another form of indirect use of solar energy.

Over 90% of the biomass energy contribution come from combustion of wood, wood wastes, and municipal plus industrial solid wastes. Biomass and various waste products can be used to produce liquid fuels. Simple wood distillation can yield liquid fuels. Biogas or methane can be generated by anaerobic digesters, this methane can be used for heating steam generation and to generate electricity.

Molasses, corn, or other biomass can be fermented to yield acetone and butanol by *Clastridium acetobutylicum.* Acetone is widely used as solvent in chemical synthesis. Butanol is used in lacquer solvents, hydraulic fluids, and as intermediate in other fluids.

Various crops such as sugar and starch crops can be used as raw materials for the production of ethanol. Average yields of ethanol in liters per hectare of crop can vary from about 4000 for sugarcane to 2500 for millet. Some other crops or their residues that can be used as substrates are banana, sweet sorghum, maize, and paddy.

Cellulosic and lignocellulosic materials are abundant all over the world; they may be used for large scale fuel production. Cellulose is cheap but requires mechanical or chemical pretreatment before it is hydrolyzed enzymatically. About 60% of cellulose can be converted into fermentable sugars.

Biomass fermentations also yield valuable byproducts. Yeast biomass is sold as animal feed. Some useful byproducts are fuel oil, $CO_2$ and sugarcane bagasse. In production of ethanol from corns, animal feed, corn oil, and gluten meal are valuable byproducts.

Biomass may be used for diesel fuel production, which is gaining popularity over petroleum in developing countries. Some biomass sources for diesel fuel production are woody biomass from which turpentine is recovered, oilseeds from which fuels may be extracted, selected aquatic biomass such as water hycinth, and oil producing microorganisms (*e.g., Botryococcus*). Some vegetable oils have the potential for being used as fuel in tractors which run on diesel.

The most important factors in biomass conversions are the microorganisms that produce high levels of celluloses, hemicelluloses or ligninases. The best producer of cellulose is the fungus *Trichoderma reesei.*

## 11.15 BIOGAS

*Biogas is produced by the degradation of biological matter by the bacterial action (anaerobic bacterias) in the absence of free oxygen*. For example, *natural gas* is a biogas, which results after a long period decay of animal and vegetable matter with proper temperature and radioactive rays, etc. Probably the cheapest and easily obtainable biogas is *gobar gas* (or dung gas), which is produced by the *anaerobic fermentation* of cattle dung. Biogas can also be produced from the sewage waste and other organic wastes.

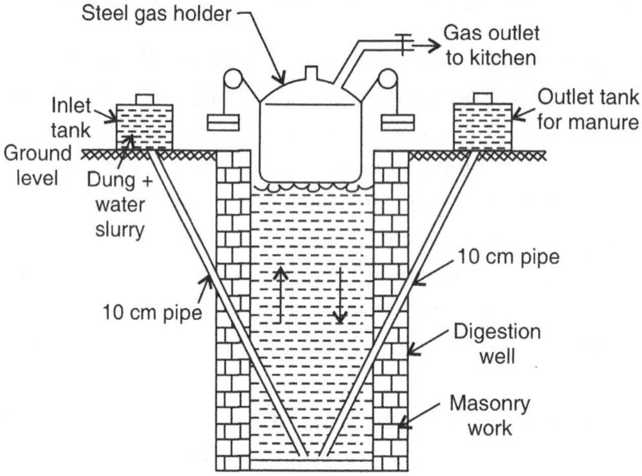


*FIGURE 11.20* **Cross-section of a gobar gas plant.**

The raw material for the biogas is mainly *dung*, which is subjected to *anaerobic fermentation* (*i.e.*, fermentation in the *absence* of free air, caused by anaerobic bacterias). The process is carried out in *closed steel* or *concrete tank*, made above ground or underground, into which a *slurry*, made by mixing equal parts of fresh cattle dung and water, is poured. *Anaerobic bacterias* (*i.e.*, which can survive and function in the *absence* of free air), present in the dung, digest this slurry forming mainly *methane* and *carbon dioxide*. The optimum temperature for this fermentation is 34–48°C. The gas generated is collected in a *steel gas holder*, placed on the top of digestion tank. The average composition of gobar gas is: $CH_4$ = 55%; $H_2$ = 7.4%; $CO_2$ = 35.0%; $N_2$ = 2.6%; and traces of $H_2S$. Its average gross calorific value is 1 200 kcal/m$^3$.

The essentials of gobar gas plant (Figure 11.20) are:

(*i*) a *digester*, a well constructed of masonry work, dug and built usually below the ground level

(*ii*) a *gas holder* made up of welded steel sheets, which cover digester

(*iii*) a *pipeline*.

**Advantages**

Following are the advantages of gobar gas:

(*i*) It has been found that 4.25 kg of fresh cattle dung (or 1 kg dry cattle dung) gives approximately 160 liters of gobar gas, which can supply 188 kcal of heat. On other hand,

1 kg of dry dung, on direct burning, gives only 23.4 kcal of effective heat. Thus, gobar gas production is highly economical.

(*ii*) The gas has all the advantages of gaseous fuel such as flexibility, optimum utilization of waste, cleanliness of utensils and surrounding areas, absence of smoke, dust, dirt, etc.

(*iii*) It does not contain any poisonous gas, carbon monoxide, as an ingredient.

(*iv*) It can provide the flame temperature of 540 °C, with proper burners.

(*v*) Gobar gas also gives simultaneously excellent yield of good manure.

The nitrogen content of the manure is 2% as against 0.75% in farm yard manure.

**Limitation.** It is necessary to have the gas lamp or stove of burner within 10 meters of the plant.

## 11.16 ANALYSIS OF FLUE GASES

Analysis of flue gases is carried out in order to have a correct idea about the combustion of the fuel. In other words, we can come to know from the flue gas analysis, whether the fuel has burnt completely or not and whether the supply of air given for combustion is in excess or not. If there is high percentage of carbon monoxide in flue gases, it shows incomplete combustion and that the oxygen supply is less than required. If there is high percentage of oxygen in flue gases, the combustion may be complete and the supply of oxygen may be excessive.

In the analysis of flue gases the percentages of carbon dioxide, carbon monoxide and oxygen are determined. In some rare cases percentages of hydrogen is also determined. Analysis of flue gases is carried out by Orsat's Apparatus.

**Orsat's Apparatus.** It consists of a horizontal tube. At one end of this tube is a three-way stopcock and the other end is connected with a burette, [Fig. 11.21].

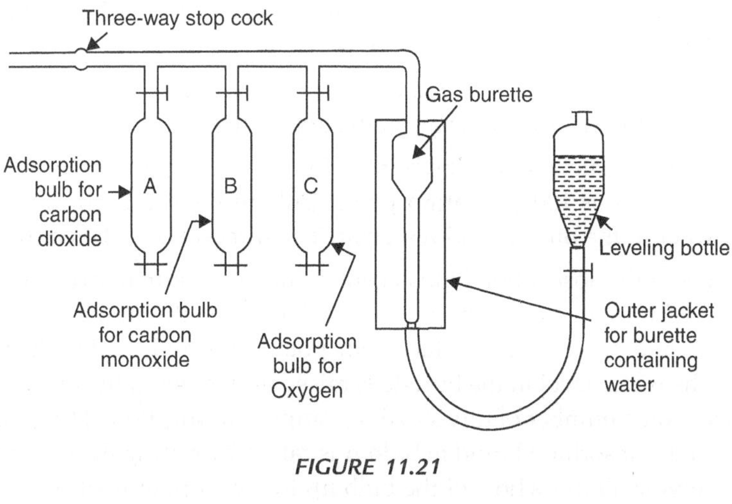


*FIGURE 11.21*

The burette has graduations up to 100 cc and it is surrounded by a water jacket in order to keep the burette at a uniform temperature. This burette is further connected by means of a rubber tube to the separating funnel containing water (water reservoir). The level of the water in separating funnel can be raised or lowered by raising or lowering the separating funnel. By so changing the level of water the flue gas can be moved into various parts of the apparatus and the pressure can also be regulated.

The horizontal tube is also connected with three different absorption bulbs, A, B and C, for absorbing carbon dioxide, carbon monoxide and oxygen. Bulb A contains potassium hydroxide to absorb $CO_2$, bulb B contains alkaline pyrogallol to absorb oxygen and bulb C contains ammonical cuprous chloride to absorb carbon monoxide.

**Working of Orsat Apparatus.** Working of Orsat apparatus consists essentially in measuring the flue gas in the burette of the apparatus and then leading the gas in turn to the different bulbs for absorption of $CO_2$, $O_2$ and CO. After each absorption the gas is measured in the burette in order to note the contraction which has taken place.

The following steps are necessary for the working of the Orsat apparatus for the analysis of a given sample of flue gas :

1. The apparatus is thoroughly cleaned. The stoppers are cleaned and greased and then are tested for air-tightness. The jacket and the separating funnel are filled with water. The necessary fresh solutions are taken into the reservoirs for bulbs A, B and C.
2. Water level in the burette is raised by raising the level of water in the reservoir. Then the stoppers of the bulbs A, B and C and the three-way stopper of the horizontal tube are closed. The stopper of bulb A connecting it with the horizontal tube is opened and the water reservoir is lowered. The absorption solution is sucked into bulb A. When the required quantity is sucked in, the stopper of bulb A is closed. Similarly, required quantities of absorption solution are sucked into bulbs B and C.
3. The level of water in the burette is raised and the free end of the horizontal tube is connected to the flue gases. By slowly lowering the water level in the burette the gas is sucked in. The sucked in gas is let out by means of the three-way stopcock and by raising the level of the water in the burette. This gas is sent out in order to exclude all the air in the apparatus prior to the sucking in of flue gas.

   This process of sucking in the flue gas and then letting it out is repeated three to four times in order to exclude all the air in the apparatus.
4. Finally the gas is sucked into the burette by connecting the three-way cock to the gas. The volume taken is 100 cc at atmospheric pressure which is ensured by equalising the level of water in the burette and reservoir. The three way cock is now closed.
5. The stopper of the absorption bulb A containing potassium hydroxide is opened and the gas is passed into the bulb A by raising the level of water in the burette. The gas enters the bulb A and comes into contact with potassium hydroxide which absorbs carbon dioxide. The water level in the burette is raised and lowered by lowering and raising the water reservoir a number of times so that complete absorption of $CO_2$ takes place. Finally the level of the absorbing liquid in bulb A is raised by raising the liquid in the absorption liquid reservoir till the whole of the bulb up to the stopper is filled. Then the stopper is closed. In this way all the residual gas is turned out of the bulb A.

The reading of the volume of residual gas is taken in the burette by equalising the water level both in the burette and the water reservoir.

The absorpiton process is again repeated in bulb A and the burette reading of residual gas again taken. If the first and second readings of the volume of residual gas are constant, it shows the completion of the absorption. If there is difference in the two readings, the absorpiton process is continued for a number of times until the two readings are constant.

The difference between the original volume (100 c.c.) and the volume of gas after absorption shows the volume of $CO_2$ absorbed. This difference is the percentage volume of $CO_2$ in the gas.

6. The stopper of bulb A is closed and the stopper of bulb B containing alkaline pyrogallol is opened. The absorption process is repeated as in step (5) and the percentage volume of oxygen in gas is determined.

## 11.17 COMBUSTION REACTIONS

Combustion is an exothermic chemical reaction, which is accompanied by developement of heat and light at a rapid rate, so that temperature rises considerably. For proper combustion, the substances must be brought to its *ignition temperature,* which is defined as, *the minimum temperature at which the substance ignites and burns without further addition of heat from outside.*

In all the combustion reactions, definite relationships exist between the masses of the fuel fired, the air used and the flue gases formed, as well as of the amounts of heat evolved or absorbed during the reactions. All these reactions are governed by well known laws such as the law of conservation of mass, the law of definite proportion, the gas laws (Boyle's law and Charle's law), the law of conservation of energy, etc.

Combustion is an exothermic reaction during which combustible elements or gases when brought to their ignition temperature burn in presence of air instantly with the evolution of heat.

The amount (volume or weight) of oxygen and air required for the complete combustion of a given quantity of the fuel can be calculated by taking the following points in consideration:

1. Substances always combine in definite proportions by their molecular weights or volume. For example, the combustion reaction of carbon,

$$\begin{array}{lcccc} & C & + \; O_2 & \longrightarrow & CO_2 \\ \text{(by wt.)} & 12 & 32 & & 44 \\ \text{(by vol.)} & \text{(1 vol.)} & \text{(1 vol.)} & & \text{(1 vol.)} \end{array}$$

Indicates that,

12 parts by weight of carbon require 32 parts by wt. of $O_2$ and 44 parts by wt. of $CO_2$ is produced by 12 parts by wt. of carbon.

1 vol. of carbon require 1 vol. of oxygen and 1 vol. of $CO_2$ is produced by 1 vol. of carbon.

In the same way, when hydrogen combines oxygen to form steam, the reaction occurs as,

$$\begin{array}{lcccc} & 2H_2(g) & + \; O_2(g) & \longrightarrow & 2H_2O \\ \text{(by wt.)} & 4 & 32 & & 36 \end{array}$$

and it indicates that mass proportions of hydrogen, oxygen, and steam ($H_2O$) formed are 4 : 32 : 36 respectively.

2. 22.4 L (or 22 400 mL) of any gas at STP (*i.e.*, 0°C and 760 mm pressure) has a mass equal to its 1 mol. Thus, 22.4 L of $CO_2$ at STP will have a mass of 44 g, because 44 g is its molar mass.
3. Air contains 21% of oxygen by volume and the mass percent of oxygen is 23. This means that 1 kg of oxygen is supplied by

$$\frac{1 \times 100}{23} = 4.35 \text{ kg of air.}$$

Similarly, 1 $m^3$ of oxygen is supplied by

$$\frac{1 \times 100}{21} = 4.76 \text{ m}^3 \text{ of air.}$$

4. Molecular mass of air is taken as 28.94 g $mol^{-1}$.
5. Minimum $O_2$ required = Theoretical $O_2$ present in the fuel.
6. Minimum $O_2$ required should be calculated on the basis of complete combustion. If the combustion products contain CO and $O_2$, then excess $O_2$ is found by subtracting the amount of $O_2$ required to burn CO to $CO_2$.
7. The mass of dry flue gases formed should be calculated by balancing the carbon in the fuel and carbon in the flue gases.
8. The mass of any gas can be converted to its volume at a particular temperature and pressure by using the *gas equation*

$$PV = nRT$$

where P = Pressure of gas in atmospheres;

V = Volume of gas in liters;

$n$ = Number of molecules of the gas

= Number of g of the gas/molar mass of the gas;

and T = Temperature on Kelvin scale.

9. Some amount of hydrogen may be present in the combined from (*i.e.*, as $H_2O$). As this combined hydrogen is a non-combustible substance, so it does not take part in combustion. The rest of hydrogen called available hydrogen only takes part in the combustion reaction:

$$4H + O_2 \longrightarrow 2H_2O + \text{Heat}$$

Mass proportions 4 32

## SOLVED PROBLEMS

### TYPE-I Determination of Calorific Values

The calorific value of coal from analysis is determined by Dulong's formula, expressed as,

$$HCV = \frac{1}{100}\left[8080\,C + 34500\left[H - \frac{O}{8}\right] + 2240\,S\right]$$

In this formula, oxygen is assumed to be present in combination with hydrogen as water.

$$LCV = \left[HCV - \frac{9H}{100} \times 587\right] cal/g$$

**Problem 1.** *The ultimate analysis of a coal gave the following results: C = 84%, S = 1.5%, N = 0.6%, H = 5.5%, and O = 8.4%. Calculate the gross and net calorific value of the coal using Dulong's formula.*

**Solution.** Given that,

$$C = 84\%,\ H = 5.5\%,\ O = 8.4\% \text{ and } S = 1.5\%$$

So,

$$GCV = \frac{1}{100}\left[8080 \times 84 + 34500\left(5.5 - \frac{8.4}{8}\right) + 2240 \times 1.5\right]$$

$$= 8356 \text{ kcal/kg}$$

$$NCV = \left[GCV - \frac{9H}{100} \times 587\right]$$

$$= 8356 - \frac{9 \times 5.5 \times 587}{100}$$

$$= 8365 - 290.56$$

$$= \mathbf{8065.44\ kcal/kg}$$

**Problem 2.** *A producer gas has the following composition by volume:*

*$CO_2$ = 8%, C = 27.69%, $H_2$ = 10%, $CH_4$ = 1.2%, $N_2$ = 52.6%, and $O_2$ = 0.6%. Calculate the GCV of the gas. Calorific values of $H_2$, CO, and $CH_4$ are 3100, 2970, and 9260 k cal/m$^3$ respectively.*

**Solution.** In the given producer gas the combustible constituents are $H_2$, CO, and $CH_4$. Their volumes per m$^3$ of the producer gas are:

$$H_2 = 0.10 \text{ m}^3,\ CO = 0.276 \text{ m}^3, \text{ and } CH_4 = 0.012 \text{ m}^3.$$

Thus,

$$GCV = 0.10 \times 3100 + 0.276 \times 2970 + 0.012 \times 9260$$

$$= 310.0 + 819.7 + 11.1$$

$$= \mathbf{1240.8\ kcal/m^3}$$

**Problem 3.** *0.72 g of a fuel containing 80% carbon, when burnt in a bomb calorimeter increased the temperature of water from 27.3° to 29.1°C. If the calorimeter contains 250 g of water and its water equivalent is 150 g. Calculate the HCV of the fuel.*

**Solution.** Given that, $x = 0.72$ g, W = 250 g, $w = 150$ g, $t_1 = 27.3$°C, and $t_2 = 29.1$°C

$$\text{HCV of fuel} = \frac{(W + w)(t_2 - t_1)}{x} \text{ kcal/kg}$$

$$= \frac{(250 + 150) \times (29.1 - 27.3)}{0.72} \text{ kcal/kg}$$

$$= \mathbf{1000 \text{ kcal/kg}}$$

### TYPE-II Problems on Bomb Calorimeter Experiment

**Problem 4.** *On burning 0.3 g of a solid fuel in a bomb calorimeter, the temperature of 3500 g of water increased from 26.5° to 29.2°C. Water equivalent of calorimeter and latent heat of steam are 385.0 g and 587.0 cal/g respectively. If the fuel contains 0.7% hydrogen, calculate its gross and net calorific value.*

**Solution.** Given that, $x = 0.83$ g, W = 3500 g, $w = 385.0$ g, $t_1 = 26.5$°C, $t_2 = 29.2$°C, percentage of hydrogen = 0.7%, and latent heat of steam = 587 cal/g.

$$\therefore \quad \text{GCV} = \frac{(W + w)(t_2 - t_1)}{x} = \frac{(3500 + 385)(29.2 - 26.5)}{0.83}$$

$$= \mathbf{12\ 638 \text{ cal/g}}$$

$$\text{NCV} = [\text{GCV} - 0.09 \times 587]$$

$$= [12638 - 0.09 \times 0.7 \times 587] \text{ cal/g}$$

$$= [12638 - 38] = \mathbf{12,600 \text{ cal/g}}$$

**Problem 5.** *The following data is obtained in a bomb calorimeter experiment:*

| | |
|---|---|
| *Weight of the crucible* | = 3.649 g |
| *Weight of the crucible + fuel* | = 4.678 g |
| *Water equivalent of the calorimeter* ($w$) | = 570 g |
| *Water taken in the calorimeter* (W) | = 2200 g |
| *Observed rise in temperature* | = 2.3°C |
| *Cooling correction* ($t_c$) | = 0.047°C |
| *Acids correction* ($T_A$) | = 62.6 cal |
| *Fuse wire correction* ($T_F$) | = 3.8 cal |
| *Cotton thread correction* ($T_t$) | = 1.6 cal |

*Calculate the GCV of the fuel sample. If the fuel contain 6.5% hydrogen, determine the NCV.*

**Solution.** We know,

$$\text{GCV} = \frac{(W + w)(t_2 - t_1 + t_c)(T_A + T_F + T_t)}{x}$$

$$= \frac{(2200 + 570)(2.3 + 0.047) - (62.6 + 3.8 + 1.6)}{(4.678 - 3.649)}$$

$$= \frac{(2770 \times 2.347) - 68}{1.029}$$

$$= \frac{6510 - 68}{1.029} = \mathbf{6261\ cal/g}$$

$$\text{NCV} = [\text{GCV} - 0.09 \times H \times 587]\ \text{cal/g}$$

$$= 6261 - 0.09 \times 6.5 \times 587$$

$$= 6261 - 343.4 = \mathbf{5917.6\ cal/g.}$$

**Problem 6.** *During the determination of calorific value of a gaseous fuel by Boy's calorimeter, the following results were recorded* **:**

| | |
|---|---|
| *Volume of the gaseous fuel burnt at normal temperature and pressure (N.T.P. or S.T.P.)* | *= 0.093 $m^3$* |
| *Weight of the water used for cooling the combustion products* | *= 30.5 kg* |
| *Weight of steam condensed* | *= 0.031 kg* |
| *Temperature of inlet water* | *= 26.1°C* |
| *Temperature of outlet water* | *= 36.5°C* |

*Determine the gross and net calorific values of the gaseous fuel per cubic meter at N.T.P, provided that the heat liberated in condensation of water vapour and cooling the condensate at 587 k cal/kg.*

**Solution.**

Gross calorific value, $$A = \frac{w(t_2 - t_1)}{V}$$

$$= \frac{30.5(36.5 - 26.1)}{0.093}$$

$$= \mathbf{3410\ Kcal/m^3}$$

Net calorific value $$= \text{Gross C.V.} - \left(\frac{m}{V} \times 587\right)$$

$$= 3410 - \frac{0.031}{0.093} \times 587$$

$$= 3410 - 195.7$$

$$= \mathbf{3214.3\ kcal/m^3}$$

## TYPE-III Coal Analysis

**Problem 7.** *A sample of coal contains C = 93%, H = 6%, and ash = 1%. The following data were obtained when the above coal was tested in Bomb Calorimeter:*

| | |
|---|---|
| *Weight of coal burnt (x)* | *= 0.92 g* |
| *Weight of water taken (W)* | *= 550 g* |
| *Water equivalent of bomb and calorimeter* | *= 2200 g* |
| *Rise in temperature* | *= 2.42°C* |
| *Fuse wire correction* | *= 10.0 cal* |
| *Acid correction* | *= 50.0 cal* |

*Calculate the gross and net calorific values of coal, assuming that the latent heat of condensation of steam in 580 cal/g.*

**Solution.** Given that, $x = 0.92$ g, W = 550 g, $w = 2200$ g

$(t_2 - t_1) = 2.42°C$, $T_A = 50.0$ cal, $T_F = 10.0$ cal

latent heat of fusion = 580 cal/g and % of hydrogen = 6%

$$\therefore \quad GCV = \frac{(W + w)(t_2 - t_1) - [T_A + T_F]}{x}$$

$$= \frac{(550 + 2200)(2.42) - (50 + 10)}{0.92} \text{ cal/g}$$

$$= \mathbf{7\,168.5\ cal/g}$$

$$NCV = [GCV - 0.09\ H \times \text{latent heat of steam}]$$

$$= (7168.5 - 0.09 \times 580) \text{ cal/g} = \mathbf{6\,855.3\ cal/g.}$$

**Problem 8.** *0.5 g of a sample of coal was used in a bomb calorimeter for the determination of calorific value. Calorific value of coal was found to be 8 600 cal/g. The ash formed in the bomb calorimeter was extracted with acid and the acid extract was heated with barium nitrate solution and a precipitate of barium sulphate was obtained. The precipitate was filtered, dried, and weighed. The weight of precipitate was found to be 0.05 g. Calculate the percentage of sulphur in the coal sample.*

**Solution.** Percentage of sulphur in the coal sample

$$= \frac{\text{Weight of } BaSO_4 \text{ precipitate obtained } \times 32 \times 100}{233 \times \text{Weight of coal sample taken in bomb}}$$

$$= \frac{0.05 \text{ g} \times 32 \times 100}{233 \times 0.5 \text{ g}} = \mathbf{1.3734\%.}$$

**Problem 9.** *1.0 g of a sample of coal was used for nitrogen estimation by Kjeldhal method. The evolved ammonia was collected in 25 mL N/10 sulphuric acid. To neutralize excess acid, 15 mL of 0.1 N sodium hydroxide was required. Determine the percentage of nitrogen in the given sample of coal.*

**Solution.** 15 mL of 0.1 N NaOH solution = 15 mL of 0.1 N $H_2SO_4$

∴ Volume of $H_2SO_4$ used to neutralize the evolved ammonia

$$= 25 \text{ mL of } 0.1 \text{ N} - 15 \text{ mL of } 0.1 \text{ N} = 10 \text{ mL of } 0.1 \text{ N}$$

$$\therefore \quad \text{Percentage of N} = \frac{\text{Volume of } H_2SO_4 \text{ used} \times \text{Normality} \times 1.4}{\text{Weight of coal sample taken}}$$

$$= \frac{10 \times 0.1 \times 1.4}{1.0} = \mathbf{1.4\%}.$$

**Problem 10.** *A sample of coal was analyzed as follows:*

*Exactly 2.500 g was weighed into a silica crucible. After heating for one hour at 110°C, the residue weighed 2.415 g. The crucible next was covered with a vented lid and strongly heated for exactly seven minutes at 950 ± 20°C. The residue weighed 1.528 g. The crucible was then heated without the cover, until a constant weight was obtained. The last residue was found to weight 0.245 g. Calculate the percentage results of the above analysis.*

**Solution.** Mass of moisture in coal sample = 2.500 – 2.415 = 0.085 g

Mass of volatile matter = 2.415 – 1.528 = 0.887 g

Mass of ash = 0.245 g

Percent of moisture $= \dfrac{0.085 \times 100}{2.500} = \mathbf{3.40\%}$

Percent of volatile matter $= \dfrac{0.887 \times 100}{2.500} = \mathbf{35.48\%}$

Percent of ash $= \dfrac{0.245 \times 100}{2500} = \mathbf{9.80\%}$

Percent of fixed carbon = 100 – (3.40 + 35.48 – 9.80) = **51.32%.**

### TYPE-IV Problems on Combustion Reactions

**Problem 11.** *Calculate the weight and volume of air required for the combustion of one kg of carbon.*

**Solution.** Carbon undergoes combustion according to the equation

$$\underset{12}{C} + \underset{32}{O_2} \longrightarrow CO_2$$

Thus, weight of $O_2$ required for combustion of 12 gm of C = 32 gm

Hence weight of oxygen required by 1 kg of carbon $= \dfrac{32}{12} \times 1 = 2.667$ kg

weight of air (containing 23% oxygen) required $= \dfrac{100}{23} \times 2.667 = \mathbf{11.59\ kg}$

Now since 32 gm of oxygen occupies 22.4 liters at NTP

$$\therefore \quad 2.667 \times 1000 \text{ gm } O_2 \text{ will occupy} = \frac{22.4}{32} \times 2.667 \times 1000 = 1866.9 \text{ L}$$

so volume of air (containing 21% oxygen) required

$$= \frac{100}{21} \times 1866.9 = 8890 \text{ Liters} = \mathbf{8.89\ m^3}$$

**Problem 12.** *A gas has the following composition by volume:*

*$H_2$ = 30%; $CH_4$ = 5%; CO = 20%; $CO_2$ = 6%; $O_2$ = 5%; and $N_2$ = 34%. If 50% excess air is used, find the weight of air actually supplied per $m^3$ of this gas, given molecular weight of air = 28.97.*

**Solution.** In one $m^3$ of the gas

| *Composition of Components* | *Volume of $O_2$ Needed* |
|---|---|
| $H_2$ (30%) = 0.3 $m^3$ | 0.3 × 0.5 = 0.15 $m^3$ |
| $CH_4$ (5%) = 0.05 $m^3$ | 0.05 × 2 = 0.1 $m^3$ |
| CO (20%) = 0.2 $m^3$ | 0.2 × 0.5 = 0.1 $m^3$ |
| | Total = 0.35 $m^3$ |
| $O_2$ (5%) = 0.05 $m^3$ | Less $O_2$ in fuel gas = − 0.05 $m^3$ |
| | Net $O_2$ needed = 0.3 $m^3$ = 300 L |

Volume of air required for 1 $m^3$ of gas using 50% excess air

$$= 300 \times \frac{100}{21} \times \frac{150}{100} = 2142.8 \text{ L}$$

Hence, weight of air actually supplied per $m^3$ of the gas

$$= 2142.8 \text{ L} \times \left(\frac{1 \text{ mol}}{22.4 \text{ L}}\right) \times \left(\frac{28.97 \text{ gm}}{\text{mol}}\right) = \mathbf{2771\ gm}$$

**Problem 13.** *A gaseous fuel containing $H_2$ = 10%, $CH_4$ = 30%, $C_2H_2$ = 15%, $C_4H_{10}$ = 25%, and $C_2H_6$ = 20% by volume was burnt in 10% excess of theoretical air by mass required for complete combustion. Find out the analysis of dry products of combustion.*

**Solution.** 100 L of gaseous fuel contains: $H_2$ = 10 L; $CH_4$ = 30 L; $C_2H_2$ = 15 L; $C_4H_{10}$ = 25 L and $C_2H_6$ = 20 L.

| *Combustion Reaction* | *Volume of $O_2$ Needed* | *Dry Products of Combustion* |
|---|---|---|
| $H_2 + 0.5\ O_2 \longrightarrow H_2O$ | 10 × 0.5 = 5 L | — — — |
| $CH_4 + 2\ O_2 \longrightarrow CO_2 + 2\ H_2O$ | 30 × 2 = 60 L | $CO_2$ = 30 × 1 = 30 L |
| $C_2H_2 + 2.5\ O_2 \longrightarrow 2\ CO_2 + H_2O$ | 15 × 2.5 = 37.5 L | $CO_2$ = 15 × 2 = 30 L |
| $C_4H_{10} + 6.5\ O_2 \longrightarrow 4\ CO_2 + 5\ H_2O$ | 25 × 6.5 = 162.5 L | $CO_2$ = 25 × 4 = 100 L |
| $C_2H_6 + 3.5\ O_2 \longrightarrow 2\ CO_2 + 3\ H_2O$ | 20 × 3.5 = 70 L | $CO_2$ = 20 × 2 = 40 L |
| | Total $O_2$ = 335 L | Total $CO_2$ = 200 L |

Theoretical air required for combustion of 100 L of gaseous fuel

$$= 335 \text{ L} \times (100/21) = 1.595 \text{ L}.$$

*Calculation of dry products of combustion using 10% excess air by mass ( or volume)*

$CO_2$ [as calculated above] = 200 L

$N_2$ [110% theoretical air × (79/100)] = 335 L × (110/100) × (79/100) = 291 L

$O_2$ [10% theoretical air × (21/100)] = 335 L × (10/100) × (21/100) = 7 L

Total volume of dry products of combustion = 498 L

*Calculation of percentage composition of dry products of combustion*

$$CO_2 = \frac{200 \times 100}{498} = \mathbf{40.16\%};\ N_2 = \frac{291 \times 100}{498} = \mathbf{58.43\%};\ O_2 = \frac{7 \times 100}{498} = \mathbf{1.41\%}$$

(**Note.** 10% excess of air by volume of air is same thing as 10% excess of air by mass.)

**Problem 14.** *A sample of coal was found to have the following percentage composition: C = 75%, H = 5.2%, O = 12.1%, N = 3.2%, and ash = 4.5%*

*(i) Calculate the minimum air required for complete combustion of 1 kg of coal.*

*(ii) Also calculate the higher calorific value and lower calorific value of coal sample. (Gross calorific value in kcal/kg, C = 8 080, H = 34 500, S = 2 240).*

**Solution.** (*i*) 1 kg of coal contains: C = 750 g, H = 52 g, O = 121 g, N = 32 g, ash = 45 g.

Minimum weight of air required for complete combustion of 1 kg of coal

$$= \text{Weight of } [C \times (32/12) + H \times (16/2) - O] \times (100/23)$$

$$= [750 \text{ g} \times (32/12) + 52 \text{ g} \times (16/2) - 121 \text{ g}] \times (100/23)$$

$$= [2\,000 \text{ g} + 416 \text{ g} - 121 \text{ g}] \times (100/23)$$

$$= 2\,295 \text{ g} \times (100/23) = \mathbf{9\,978\ g\ or\ 9.978\ kg}$$

(*ii*)

$$\text{HCV} = \frac{1}{100}\left[8\,080 \times \text{C} + 34\,500\left(\text{H} - \frac{\text{O}}{8}\right) + 2\,240 \times \text{S}\right] \text{kcal/kg}$$

$$= \frac{1}{100}\left[8\,080 \times 75 + 34\,500\left(5.2 - \frac{12.1}{8}\right) + 2\,240 \times 0\right] \text{kcal/kg}$$

$$= \frac{1}{100}\,[606\,000 + 127\,219 + 0]\ \text{kcal/kg}$$

$$= [6\,060 + 1\,272]\ \text{kcal/kg} = \mathbf{7\,332\ kcal/kg}$$

$$\text{LCV} = (\text{HCV} - 0.09\ \text{H} \times 587)\ \text{kcal/kg}$$

$$= (7\,332 - 0.09 \times 5.2 \times 587)\ \text{kcal/kg} = \mathbf{7057\ kcal/kg}$$

**Problem 15.** *Calculate the mass of air needed for complete combustion of 5.0 kg of coal containing 80% carbon, 15% hydrogen, and the rest oxygen.*

**Solution.** 5 kg of coal contains: C = 4 000 gms, H = 750 gms, and O = 250 gms.

∴ Net weight of oxygen required for combustion

$$= \left[\frac{32}{12} \times \text{C} + \frac{16}{2} \times \text{H}\right] - \text{oxygen present in fuel}$$

$$\Rightarrow \text{Net Oxygen} = \left[\frac{32}{12} \times 4000 + \frac{16}{2} \times 750\right] - 250 = 16\,416.7\ \text{gms.}$$

Now, mass of air needed for complete combustion

$$= \text{Net Oxygen} \times \frac{100}{23} = 71\,376.8\ \text{gms.}$$

$$= \mathbf{71.38\ kg}$$

**Problem 16.** *The percentage composition of coal sample is*

*C = 80%, $H_2$ = 4%, $O_2$ = 3%, $N_2$ = 3%, S = 2%, Ash = 5%, Moisture = 3%.*

*Calculate the quantity of air needed for complete combustion of 1 kg of coal, if 60% excess of air is supplied.*

**Solution.** 1 kg of coal contains: C = 800 gm, H = 40 gm, S = 20 gm, $O_2$ = 30 gm.

$$\text{Net } O_2 = \left[\frac{32}{12} \times 800 + \frac{16}{2} \times 40 + \frac{32}{32} \times 20\right] - 30 = 2443.3\ \text{gm.}$$

$$\therefore \text{ Quantity of air needed} = \text{Net } O_2 \times \frac{100}{23} \times \frac{160}{100} = 16996.9 = \mathbf{16.99\ kg}$$

**Problem 17.** *A sample of fuel having following percentage composition: C = 72%, H = 5%, O = 4%, N = 2%, S = 3%, ash = 8%, and moisture = 6%. Calculate the quantity of air required for the complete combustion of 1 kg of fuel.*

**Solution.** 1 kg of fuel contains: C = 720 gm, H = 50 gm, S = 30 gm, $O_2$ = 40 gm.

$$\therefore \qquad \text{Net } O_2 = \left[\frac{32}{12} \times 720 + \frac{16}{2} \times 50 + \frac{32}{32} \times 30\right] - 40 = 2310 \text{ gm.}$$

$$\text{so the quantity of air required} = \text{Net } O_2 \times \frac{100}{23} = 10043.5 \text{ gm} = \mathbf{10.04 \text{ kg}}$$

**Problem 18.** *Analysis of coal used in a boiler is: C = 82%, H = 4.2%, O = 4.8%, ash = rest. When this coal was used in a boiler, the analysis of flue gas was: $CO_2$ = 10%, CO = 1.5%, $O_2$ = 3.0%, $N_2$ = 80.5%. Determine:*

*(i) the amount of excess air per kg of coal burnt,*

*(ii) total mass of air supplied per kg of coal burnt,*

*(iii) percentage of excess air supplied for combustion.*

**Solution.** (*i*) 1 $m^3$ of flue gas contains: $CO_2$ = 0.100 $m^3$, CO = 0.015 $m^3$, $O_2$ = 0.080 $m^3$, and $N_2$ = 0.805 $m^3$.

*Conversion of volumetric analysis to mass analysis:*

| *Gas* | *Volume in $m^3$ (v)* | *Molecular mass (m)* | *v × m Product* | *Mass/Kg of Flue Gas* |
|---|---|---|---|---|
| $CO_2$ | 0.100 | 44 | 4.40 | 4.4/29.92 = 0.1470 kg |
| CO | 0.015 | 28 | 0.42 | 0.42/29.92 = 0.0140 kg |
| $O_2$ | 0.080 | 32 | 2.56 | 2.56/29.92 = 0.0855 kg |
| $N_2$ | 0.805 | 28 | 22.54 | 22.54/29.92 = 0.7533 kg |
| | | | Total = 29.92 | Total = 0.9998 kg |

Mass of excess $O_2$ per kg of flue gas

= Mass of $O_2$ (in flue gas – required to convert 0.014 kg CO to $CO_2$)

= [0.085 – 0.014 × (16/28)] kg = (0.085 – 0.008) kg = 0.077 kg

Mass of C present in 1 kg of flue gas

= Mass of [$CO_2$ × (12/44) + CO (12/28)]

= [0.147 × (12/44) + 0.014 × (12/28)] kg = (0.040 + 0.006) kg = 0.046 kg

∴ Mass of flue gas formed by burning 1 kg of coal

$$= \frac{\text{Weight of C/kg of coal}}{\text{Weight of C/kg of flue gas}} = \frac{0.82 \text{ kg}}{0.046} = 17.83 \text{ kg}$$

Excess $O_2$/kg of coal burnt

= [Mass of excess $O_2$/kg flue gas] × [Mass of flue gas/kg of coal burnt]

= 0.077 × 17.83 kg = 1.373 kg

∴ Excess air/kg of coal burnt = 1.373 kg × (100/23) = **5.970 kg**

(*ii*) Theoretical amount of air required for complete combustion of 1 kg coal (0.82 kg C + 0.042 kg H + 0.048 kg O)

= [0.82 kg × (32/12) + 0.042 kg × (16/12) – 0.048 kg] × (100/23)

= [(2.187 + 0.056 – 0.048) kg] × (100/23) = 9.543 kg

Hence, total mass of air supplied/kg of coal burnt

= Mass of air (for combustion + excess in flue gas)

= (9.543 + 5.970) kg = **15.513 kg**

(*iii*) Percentage of excess air supplied for combustion

$$= \frac{\text{Excess air}}{\text{Theoretical air}} \times 100 = \frac{5.970 \times 100}{9.543} = \mathbf{62.6\%}$$

## QUESTION BANK

1. (*a*) How are chemical fuels classified?

   (*b*) How have cracking processes improved the output of liquid fuels? Name the various cracking processes employed in refinery with the help of a neat sketch.

2. (*a*) How are fuels classified? What is meant by lower calorific value of a fuel? How is it experimentally determined for a gaseous fuel?

   (*b*) What is meant by cracking of oil? Describe the method used for oil cracking? Why is cracked gasoline better than straight run gasoline?

3. (*a*) How is synthetic gasoline obtained?

   (*b*) Discuss the characteristics of cracked and polymer gasolines.

   (*c*) Define net and gross calorific value of a fuel. How are these determined experimentally for solid fuels?

4. (*a*) Define calorific value of a fuel.

   (*b*) Draw a neat diagram and explain the working of a bomb calorimeter for determining the heating values of solid fuels?

5. (*a*) What is the difference between gross and net calorific value of a fuel?

   (*b*) How is the calorific value of a solid fuel determined with the help of bomb calorimeter?

6. Write notes on: (*a*) Proximate and ultimate analysis of coal, (*b*) Analysis of flue gases.

7. Why and how coal is analyzed? What is the significance of the constituents determined in the analysis of coal?

8. (*a*) Explain the significance of following constituents present in coal:

(*i*) Moisture (*ii*) Volatile matter

(*iii*) Carbon (*iv*) Ash.

(*b*) Write a short note on synthetic gasoline.

9. Calculate the gross and net calorific values of coal sample having the following composition: C = 80%, H = 7%, O = 3%, S = 3.5%, N = 2.1%, and ash = 4.4%.

[**Ans.** GCV = 8828 kcal/kg; NCV = 8458 kcal/kg]

10. The determination of calorific value of a coal sample gave the following results. Weight of sample = 0.9 g.Water equivalent of calorimeter = 440 g. Weight of water = 2560 g. Increase in temperature = 2.42°C. Cooling correction = 0.052°C and fuse wire correction = 10.0 calories. Calculate the gross calorific value and net calorific value, if the coal contains 6% hydrogen and assume latent heat of steam as 600 cal/g. [**Ans.** GCV = 8228.9 cal/g; NCV = 7904 kcal/kg]

11. A coal has the following ultimate analysis: Carbon = 84%, Sulphur = 1.5%, Nitrogen = 0.6%, Hydrogen = 5.5%, and Oxygen = 8.4%. Find the gross calorific value using Dulong's formula.

12. When 0.935 g of a fuel underwent complete combustion in excess of oxygen, the increase of temperature of water in a calorimeter containing 1.365 g of water was 2.40°C? Calculate the higher calorific value of the fuel, if the water equivalent of calorimeter, etc. is 135 g. [**Ans.** HCV = 3850.3 cal/g]

13. Liquid fuel weighing 0.98 g and containing 90% of carbon and 8% of hydrogen gave the following results in a bomb calorimeter experiment.

Amount of water taken in the calorimeter = 1.45 g; water equivalent of the calorimeter = 450 g; rise in temperature of water 1.8°C. If the latent heat of steam is 587 cal/g. Calculate the gross and net calorific value of the fuel. [**Ans.** GCV = 3490 cal/g; NCV = 3097 cal/g]

14. A gaseous fuel has the following composition by volume: $H_2$ = 22%, $CH_4$ = 4%, CO = 22%, $CO_2$ = 6%, $N_2$ = 45%. If 20% excess of air is used, find the weight of air actually supplied per $m^3$ of the gas.

[**Ans.** 2215 g]

15. A sample of coal was found to have the following percentage composition: C = 75%, H = 5.2%, O = 12.1%, N = 3.2%, and ash = 4.5%.

(*i*) Calculate the minimum amount of air necessary for complete cumbustion of 1 kg coal.

(*ii*) Also calculate HCV and LCV of the coal sample given: GCV of carbon = 8.080 kcal/kg, hydrogen = 34500 kcal/kg, and sulphur = 2240 kcal/kg. [**Ans.** (*i*) 9.978 kg (*ii*) 7332 kcal/kg; 7057 kcal/kg]

16. The temperature of 950.0 g of water increased from 25.5°C to 28.5°C on burning 0.75g of a solid fuel in a bomb calorimeter. Water equivalent of calorimeter and latent heat of steam are 400.0 g and 587.0 cal/g respectively. If the fuel contains 0.65% of hydrogen, calculate its net calorific value.

17. Calculate the gross and net calorific value of coal sample having the following composition: C = 80%, H = 7%, O = 3%, S = 3.5%, N = 2.1% and ash = 4.4%.

18. A sample of coal containing 92% C, 5% H, 3% ash. When this coal was tested in the laboratory for its calorific value in the bomb calorimeter, the following data were obtained:

| | |
|---|---|
| Weight of coal burnt | = 0.95 g |
| Weight of water taken | = 700 g |
| Water equivalent of bomb calorimeter | = 2000 g |
| Rise in temperature | = 2.48°C |
| Cooling correction | = 0.02°C |

Fuse wire correction = 10.0 cal

Acid correction = 60.0 cal

Calculate the net and gross calorific values of the coal in cal/g. (Assume the latent heat of condensation of steam as 580 cal/g). **[Ans.** GCV = 7031.6 cal/g; NCV = 6770.6 cal/g]

**19.** A fuel gas has the following composition by volume: hydrogen 24%, methane 30%, ethane 11%, ethene 4.5%, butane 2.5%, carbon monoxide 6.0%, carbon dioxide 8%, oxygen 2%, and nitrogen 12%. Calculate the theoretical amount of air required at 25°C and 750 mm pressure for complete combustion of 1 $m^3$ of fuel. **[Ans.** 7.324 $m^3$]

**20.** The percentage composition of a sample of bituminous coal by weight was found to be: C = 76%, H = 5.25%, O = 12.8%, N = 2.7%, S = 1.2%, ash = 2.1%. Calculate the minimum (*a*) weight, and (*b*) volume at NTP of air necessary for complete combustion of 1 kg of coal, and (*c*) the percentage composition by weight of dry products. **[Ans.** $CO_2$ = 26.22%; $SO_2$ = 0.226%; $N_2$ = 73.553%]

**21.** The percentage composition of a sample of bituminous coal was found to be as: C = 75.4, H = 5.3, O = 12.6, N = 3.2, S = 1.3; the rest being ash.

Calculate the minimum weight of air necessary for complete combustion of 1 kg of coal and percentage composition of the dry products of combustion by weight.

**22.** What is ultimate and proximate analysis of coal? Discuss their significance. How are ultimate and proximate analysis done?

**23.** (*a*) What is meant by net calorific value of a fuel? How can it be determined experimentally in the case of gaseous fuels?

(*b*) Give the units of calorific value of the gaseous fuels.

(*c*) Calculate the weight and volume of air needed for the combustion of 1 kg of carbon.

**24.** What do you understand by the net calorific value of a fuel? How do you find it for a liquid fuel?

**25.** What are the merits and demerits of solid, liquid and gaeous fuels?

**26.** 0.05 gm of the fuel of complete combustion in oxygen increased in temperature of water in the calorimeter from 15°C to 18.5°C. The weight of water in calorimeter was found to be 1630 gm. Calculate the calorific value of the fuel if the water equivalent of the calorimeter be 110 gm.

**27.** The determination of the calorific value of a coal sample gave the following data:

Weight of coal sample = 0.9 gm

Water equivalent of the calorimeter = 440 gm

Weight of water = 2560 gm

Rise of temperature = 2.42°C

Cooling correction = 0.052°C

Fuse wire correction = 10.0 calories

Calculate the gross and net calorific values if the coal contains 6% hydrogen and assume latent heat of steam equal to 600 cal/gm. **[Ans.** GCV = 8228.9 cal/g; NCV = 7904.9 cal/g]

**28.** A fuel has 80% carbon, 10% hydrogen and remaining 10% incombustible matter. What volume of air at N.T.P. is required to burn completely 1 kg of this fuel?

**29.** A fuel has the following percentage composition by weight:

$H_2$ = 20.24, $CH_4$ = 35.6, $C_2H_6$ = 6.15, $C_2H_4$ = 4.78, $C_4H_8$ = 2.50, CO = 7.60, $O_2$ = 0.60, and rest nitrogen. Calculate: (*i*) the theoretical amount of oxygen and air required at 30°C and 1 atmospheric pressure for complete combustion per kg of fuel burnt (*ii*) the volumetric analysis of fuel gases.

**30.** Calculate the weight and volume of air containing 20% oxygen and 80% nitrogen by volume, needed for the combustion of 1 kg of carbon. Give the composition of the combustion products.

31. Calculate the mass and volume of air needed for the combustion of 1 kg of carbon.
32. (*a*) What is the difference between the gross and net calorific values of a fuel?
    (*b*) Give a neat sketch of the Bomb calorimeter and describe how you would determine the calorific value of coal with it.
33. What is crude petroleum? What are the products obtained from petroleum? Give in detail the method employed in the manufacture and purification of gasoline and kerosene oil.
34. Describe biomass and biomass energy.
35. Justify that solar energy is a very good source of nonconventional energy. What do you understand by renewable and non-renewable fuels? Categorize the following: Coal, biomass, solar energy, electrical energy.
36. Explain the process involved in the conversion of solar energy into electrical energy and heat energy in solar plants.
37. Describe the reasons for not using solar energy at domestic and industrial levels economically, in spite of the free availability of sunlight.
38. List the raw materials which can be utilized for biogas manufacture. How is biogas obtained from cattle dung?
39. Write the constituents of the amorphous conductors used in solar energy unit. Also explain the role of impurities.

Chapter 12 ENVIRONMENTAL POLLUTION

## 12.1 INTRODUCTION

Pollution of the environment is one of the most serious ecological crises to which we are subjected today. The three basic amenities for living organisms are air, land or soil, and water. In the past, these amenities were pure, virgin, undisturbed, uncontaminated, and basically most hospitable for living organisms. But the situation is just the reverse today, because progress in science and technology is also leading to pollution of the environment and serious ecological imbalance which, in the long run, may prove disastrous for mankind. Environmental pollution is the result of urban-industrial technological revolution and speedy exploitation of every bit of natural resources. The craze of progress in agriculture, industry, transportation and technology is taken as the general criterion of development of any nation. Such activities of man have created adverse effects on all living organisms in the biosphere. Rapid industrialization has left with us polluted rivers, contaminated soil, depleted wild life, and exhausted natural resources. Today the environment has become foul, contaminated, undesirable, and therefore, harmful for the health of living organisms, including man. The unlimited rapacious exploitation of nature by man has disturbed the delicate ecological balance existing between living and non-living components on the planet Earth. The root cause of environmental pollution has been man's misbehavior with nature under the false impression that he is the master of nature. This undesirable situation created by man has threatened the survival of man himself and other living biota on the earth. The term 'environment', as far as pollution is concerned, includes the air, the water, the soil, the noise, the buildings, the landscapes, the oceans, the lakes, the rivers, the parks, the vehicles, and many other things. Thus pollution is generally defined as the addition of the constituents to water, air or land, which adversely alter the natural quality of the environment. In some cases, pollution may involve removal, rather than the addition of constituents from the environment.

Today India, which occupies seventh place among the industrialized developing countries of the world, is provided with a good industrial infrastructure in several industries like chemicals, power, nuclear energy, food, petroleum, pesticides, plastics, etc. Rapid progress in atomic and nuclear energy has added a huge amount of radioactive substances to the atmosphere. A number of industrial effluents and emissions, especially toxic gases, are spewed into the air daily. Thus the environment is deteriorated to such an extent that it has crossed the critical limit and has become lethal to all the organisms, including man.

Environmental pollution is thus a necessary evil of all man-made progress. It is not only the industrialized countries which are threatened with environmental pollution, but the menace is growing in India too. The developed countries dump a lot of effluents into the environment, polluting the whole globe. The United States is the most polluted area of the world today and New Delhi is the third most polluted metropolis in India.

*Environmental Pollution is thus the unfavorable alteration of our surroundings, wholly or largely as a byproduct of man's actions, through direct or indirect effects of changes in the energy pattern, radiation levels, chemical and physical constitution, and abundance of organisms.*

## 12.2 ORIGIN OF POLLUTION

The following are different views regarding the origin of the pollution crisis on the planet Earth.

(*i*) According to *Lynn White* (1967) and *Ion McHarg* (1969), The *Judeo—Christian* ethic is responsible for pollution, because this ethic taught man to believe that the Earth was made for man to do with as he wished, and thereby encouraged exploitation.

(*ii*) According to *Write* (1967) human greed and ignorance have allowed our culture to develop an ecological crisis like pollution. *Tuan* (1970) described an ancient oriental culture which despoiled their forests and faced ecological crises. According to *Southwick* (1976), human population explosion is the main cause of pollution. He pointed out that with more people there is more sewage, more solid waste, more fuel being burned, and more fertilizers and insecticides being used to produce more food for hungry mouths.

(*iii*) Some other authors feel that it is the wasteful aspects of our science and technology which strive always to produce more disposable items that are responsible for the pollution of our environment.

(*iv*) According to some authors, profit motives and capitalism of the modern economic system of countries like the USA, USSR, France, Britain, etc. are the main causes of pollution.

(*v*) *Benton* and *Werner* (1974), however, say that economic systems are not the real cause of pollution, because the USSR has problems very similar to western countries, *i.e.*, severe oil pollution in the Caspian sea from offshore oil drilling and flushing the holds of ships, industrial pollution from milk, fish, and linen factories and industrial complexes, etc.

(*vi*) Modern ecologists, however, point out that there are various factors such as human population explosion, rapid industrialization, deforestation, unplanned urbanization, scientific and technological advancement, etc., which are mainly responsible for the pollution crisis on Earth. It is well evident that the United States, with its vast infrastructure of industries, is one of the leading polluters of the world. The worst pollution occurs in countries where there have been the greatest scientific and technological advances.

(*vii*) Rapid population growth, combined with modern technology, has caused a serious environmental crisis. This is because of the fact that in well-developed countries, citizens consume more food, use more fertilizers and pesticides, fuel, minerals, automobiles (cars, scooters, vans, trucks, etc.), and other manufactured products of all kinds. Almost all of these products are manufactured in one or another kind of industry. These industries are responsible for adding pollutants to our environment (land, soil, water, and air) and hence cause pollution.

## 12.3 POLLUTANTS

Pollution is usually brought about by the addition of waste products of human activity to the environment. When the waste products are not efficiently assimilated, decomposed, or otherwise removed by the natural, biological, and physical processes of the biosphere, adverse effects may results as the pollutants accumulate or get converted into more toxic substances. Thus the materials which cause pollution of environment are called *pollutants*. Some other definitions of pollutant are as follows:

1. A pollutant is a physical or biotic component which adversely alters the environment by altering the growth rate of species, interfering with the food chain, health, amenities, or the property value of man.
2. A pollutant is a harmful solid, liquid, or gaseous substance present in such great concentration in the environment that it tends to be injurious for the whole living biota.
3. A pollutant is the residue of substances which are present in the wrong place, at the wrong time, and in the wrong quantity.

A pollutant may belong to any one of the following:

(*a*) A chemical substance, *e.g.*, an organophosphorous compound.

(*b*) A geochemical substance, *e.g.*, dust, sediment, etc.

(*c*) A biological organism and product.

(*d*) A physical property, *e.g.*, heat, noise, etc.

For example, demand for more food increased as a result of rise in human population. Hence, fertilizers were used to increase food production. Fertilizers actually boosted the production of food grains and vegetables three to four times and an agricultural miracle or green revolution seemed in the making all over the earth. Very soon, the world discovered that the dazzling revolution had another, costly side. The fertilizers could destroy the environment. The fertilizers did not stay entirely in the fields where they were utilized for the increase in production of food grains and vegetables. They were also washed down into the streams, rivers, lakes, seas, and oceans. The fertilizers in the water stimulated the algae to luxuriant growth. The algae consumed oxygen dissolved in water much more than that returned to water in a dissolved state. Short supply of dissolved oxygen made it impossible for the fishes and other water animals to live in water. Lakes that had been clear for more than hundreds of years, began to fill with the slimy green of the luxuriant growth of the algae. The Nainital lake in the U.P. thus turned into a thriving sewer though biologically not dead. As a result, it has become not even difficult but more costly to derive pure drinking water from the lake. In addition to fertilizers, the pollution caused by insecticides and pesticides is also alarming. DDT, for example, also washed down to the rivers and lakes and became a part of the fish body. Even a small amount of DDT in the fish (10 ppm), may cause a marked fall in the population of birds that feed on DDT-polluted fish.

Everything today is polluted by pollutants. The rivers, seas, and lakes are polluted by wastes coming from various industries, the air by toxic effluents from thermal power plants and exhausts of automobiles, and the land or soil by chemicals, fertilizers, pesticides, acid rains, insecticides, etc.

## 12.4 CLASSIFICATION OF POLLUTANTS

Pollutants can be classified in a number of ways. For example:

1. On the basis of the forms they exist in the environment after their release. On this basis, pollutants can be classified as primary and secondary pollutants.
   (*i*) *Primary Pollutants. Those substances emitted directly from an identifiable source.* These pollutants exist as such after being added or released into the environment. Examples are sulphur dioxide, nitrogen oxides, etc.
   (*ii*) *Secondary Pollutants. These are substances, derived from primary pollutants by chemical reactions.* For example, primary pollutants such as hydrocarbons and nitrogen oxides, particularly in the environment, react in presence of sunlight to form a group of nitrous compounds like peroxyacetyl nitrate (PAN) as the secondary pollutant.
2. **Biodegradable Pollutants.** Biodegradable pollutants are those which can be decomposed easily in their natural environment; that is, by the natural biological process and also in engineered systems. Plant debris, domestic sewage, etc, which on decomposition give out nutrients like carbonates, phosphates, etc., that are released in the biosphere are biodegradable pollutants. Heat or thermal pollution also belongs to biodegradable pollutants because heat can be dispersed by natural means. If organic wastes drained out from residential buildings are easily degraded completely by microorganisms, these become useful for the ecosystem. Domestic sewage can be readily decomposed by natural processes or by engineered systems (such as municipal sewage treatment plant) which enhance the capacity of nature to decompose and recycle. However, if these pollutants enter the environment in such large quantities that complete degradation of all *cannot* take place, then these become biodegradable pollutants, and thus pollute the environment.
3. **Non-biodegradable Pollutants.** Non-biodegradable pollutants are those which either do not degrade or degrade very slowly in their natural environment. Such pollutants are mostly inorganic compounds like mercury and lead salts, aluminium cans, iron, compounds like phenolic compounds, DDT, etc. Such pollutants are harmful even in low concentrations. These pollutants not only accumulate, but are often biologically magnified as they move in biochemical cycles and along food chains. They may also react with

other compounds present in the environment to produce even more toxic additional products. Their high concentration may even prove fatal.

## 12.5 TYPES OF POLLUTION

Pollutants need not be material substances. Noise can be a pollutant and electromagnetic waves can be pollutants. However, today, man is the principal source of pollution, because there is no way for nature to decompose many man-made materials and return their elements to the cycle of nature. These substances will remain and will cause whatever harmful effects they can, until they are somehow dispersed or diluted so that their action is no longer harmful.

Thus pollution may be natural (which originates from natural processes such as forest fires, natural organic and inorganic decays, volcanic eruptions, earthquakes, etc.) or artificial which originate due to the activities of man.

It is better to classify pollution according to the environment, such as air, water, soil, noise, radiation, thermal, etc. in which it occurs, or according to the pollutants by which pollution is caused.

**Classification according to environment.** When different spheres of the environment are affected by pollution, they can be categorized as,

(*i*) Air pollution. (*ii*) Water Pollution. (*iii*) Soil or Land Pollution.

## 12.6 AIR POLLUTION

Air is never found clean in nature due to natural and anthropogenic reasons. Population explosion in the last few decades has led to enormous urban growth, extensive industrialization, and the automobile revolution, all of which are threatening the Earth's environment and making it unsuitable for future generations. The release of poisonous gases like CO, $N_2O$, and gases from industries and automobiles has disturbed the balance among the air components, affecting both the living and as non-living components of Earth. Air pollution may be defined in the following ways:

(*a*) *Any atmospheric condition in which certain substances are present in such concentrations that they can produce undesirable effects on man and his environment.*

(*b*) *Air pollution is the excessive concentration of foreign matter in the air which adversely affects the well being of individuals or causes damage to property.*

### 12.6.1 Air Pollutants and their Classification

Air pollutants are the substances causing damage to target or receptor. The target may be man, animal, plant, tree, building, or material which is adversely affected by pollutants. On the basis of origin, pollutants are classified as primary and secondary pollutants:

*Primary pollutants.* These are emitted directly from the sources and are found in the atmospheres in the form in which they were emitted. For example, sulphur oxides, nitrogen oxides, hydrocarbons, ash, smoke, dust, fumes, mist, sprays, radioactive compounds, etc.

*Secondary pollutants.* These are formed in the atmosphere by chemical interactions between primary pollutants and atmospheric constituents. For example, ozone, sulphur trioxide, peroxyacyl nitrate (PAN), aldehydes, ketones, etc.

According to the states of matter, air pollutants are classified as:

*Gaseous pollutants.* These pollutants are in a gaseous state at normal temperature and pressure. Common gas pollutants are CO, $CO_2$, $NO_x$, $SO_x$, hydrocarbons, and photochemical oxidants.

*Particulate air pollutants.* These include suspended droplets, solid particles, or their mixtures in the atmosphere, commonly referred as particulates. For example, aerosols (all air suspensions smaller than 1 μm), dust, smoke, fumes, mist, fog, flyash, soot, and natural particulates such as pollen grains, protozoa, fungal spores, and volcanic dust.

### 12.6.2 Air Pollutants–Effects and Control

There are five major air pollutants which together contribute more than 90% of the world's air pollution. These include:

1. **Carbon monoxide (CO).** It is colorless, odorless, tasteless gas, chemically inert under normal conditions of temperature and pressure. It has no effect at normal concentration (0.1 ppm) but at higher concentrations it seriously affects the human metabolism.

**Sources**

(*a*) Natural processes such as volcanoes, natural gas emissions, electrical discharge during and forming, seed germination, marsh gas production, etc. contribute to CO in the atmosphere, though in small amounts.

(*b*) Transportation sources contribute about 64% of CO in the atmosphere.

(*c*) Forest fires and agricultural burning contribute about 17% of the CO in the atmosphere.

(*d*) Industrial processes such as electric furnace and blast furnaces in iron and steel industry, petroleum refining, paper industry, gas manufacture, coal mining, etc. cause about 9.6% of CO in the atmosphere.

**Effects**

(*a*) CO reduces the oxygen-carrying capacity of the blood by selectively combining with haemoglobin (Hb), forming carboxy haemoglobin (COHb). This causes giddiness, laziness, and exhaustion.

$$O_2Hb + CO \rightleftharpoons \underset{\text{Carboxy haemoglobin}}{COHb} + O_2$$

(*b*) CO reduces vision and causes cardiovascular disorders.

(*c*) CO is a very dangerous asphyxiant and, in high levels is fatal to human beings.

**Sinks**

Soil microorganisms act as the major sinks for CO. A potting soil sample weighing 2.8 kg completely removed 120 ppm CO from ambient air in just three years. The same sample when sterilized failed to remove CO from air. Since CO-producing areas have the least amount of soil sinks, CO exists in significant concentrations in the environment.

### Control

Transportation sources are responsible for most of the CO emissions, so considerable efforts for CO pollution control are directed mainly at checking automobile emissions. These include:

(*a*) Designing more efficient internal combustion engines to reduce the amounts of pollutants formed during fuel combustion.

(*b*) Developing an exhaust system that converts pollutants into more acceptable forms.

(*c*) Developing a substitute fuel for gasoline. Both compressed natural gas (CNG) and liquified natural gas (LNG) can be used as fuel.

(*d*) Developing pollution-free power sources.

2. **Nitrogen oxides ($NO_x$).** The $NO_x$ group includes eight possible oxides of nitrogen. NO, $NO_2$, and $N_2O$ are primarily involved in air pollution. NO is a colorless, odorless gas, but $NO_2$ is reddish-brown and has a suffocating odor. NO and $NO_2$ are formed as:

$$N_2 + O_2 \xrightleftharpoons{1210\text{–}1765°C} 2NO$$

$$2NO + O_2 \rightleftharpoons 2NO_2$$

The formation of NO occurs at high temperatures. The second reaction is also favored at high temperatures about 1100°C. $NO_2$ is also formed by photolytic reactions.

### Effects

(*a*) Like CO, nitric oxide can also combine with haemoglobin and reduce the oxygen-carrying capacity of blood.

(*b*) NO is moderately toxic as its concentration in the air is not high.

(*c*) $NO_2$ is more toxic; it irritates alveoli of the lungs and its high concentration may cause acute bronchitis.

### Control

(*a*) Use of catalytic converters in automobiles reduces $NO_x$ emissions.

(*b*) 90% emissions of $NO_x$ from power plants can be used in the following two ways:

(*i*) The fuel is fired at high temperatures with lower concentrations (90–95%) of air required in the process. Formation of NO is limited in the absence of excess of air.

(*ii*) Fuel is burned completely at a relatively low temperature in excess air, reducing the formation of NO.

(*c*) $NO_x$ can be removed from stack gas by chemical absorption using $H_2SO_4$ or alkaline solution in the following steps:

(*i*) Fuel gas and $NO_2$ are introduced into an oxidizer.

$$N_2 + SO_2 + H_2O \longrightarrow H_2SO_4 + NO$$

(*ii*) NO and $NO_2$ react to form $N_2O_3$ which reacts with $H_2SO_4$ in a scrubber.

$$NO_2 + NO \longrightarrow N_2O_3$$

$$N_2O_3 + 2H_2SO_4 \longrightarrow 2NOHSO_4 + H_2O$$

The cleaned fuel gas is released into the atmosphere.

(*iii*) $NOHSO_4$ formed is decomposed as:

$$2NOHSO_4 + \frac{1}{2}O_2 + H_2O \longrightarrow 2H_2SO_4 + 2NO_2$$

(*iv*) $NO_2$ forms $HNO_3$ in the $HNO_3$ reactor

$$3NO_2 + H_2O \longrightarrow 2HNO_3 + NO$$

excess NO and $NO_2$ are recirculated through step (*i*).

3. **Sulphur oxides ($SO_x$).** $SO_x$ include sulphur dioxide and sulphur trioxide. $SO_2$ is a colorless gas having a pungent and suffocating odor.

**Sources and Sinks**

Most of the $SO_x$ pollution (67%) is due to volcanic activities and other natural sources. The remaining (33%) $SO_x$ emissions are due to industrialization and automobile pollution.

$SO_2$ is partially oxidized to $SO_3$ in air by a photolytic and catalytic process involving $O_3$, hydrocarbons, and $NO_x$ in a photochemical smog.

$$2SO_2 + O_2 \rightleftharpoons 2SO_3;\ SO_3 + H_2O \longrightarrow H_2SO_4$$

$$SO_3 + H_2O \longrightarrow H_2SO_4 \longrightarrow (H_2SO_4)_n$$

$$SO_2 + \frac{1}{2}O_2 \xrightarrow[\text{metal oxides}]{\text{soot dust}} SO_3 \xrightarrow{+H_2O} H_2SO_4 \longrightarrow \underset{\text{aerosol droplet}}{(H_2SO_4)_n}$$

This gives rise to acid rain.

**Effects**

(*a*) These tend to irritate the mucous membranes of the respiratory tract.

(*b*) Higher concentrations of sulphur oxides may cause bronchitis.

(*c*) Marbles and mortar statues are affected.

(*d*) Clothes, leather, and paper are also affected.

(*e*) Plants are sensitive to high concentrations of $SO_2$ and suffer chlorosis, metabolic inhibition, plasmolysis, and even death.

**Control**

(*a*) $SO_x$ from flue gases can be removed by using chemical scrubbers such as $CaCO_3$

$$2CaCO_3 + 2SO_2 + O_2 \longrightarrow 2CaSO_4 + 2CO_2$$

(*b*) Sulphur can be removed from the fuels used for combustion; for example, sulphur in coal can be removed by grinding and washing in coal washeries.

(*c*) Using low sulphur fuels.

4. **Hydrocarbons (HC).** The gaseous and volatile hydrocarbons are mainly responsible for air pollution. Common hydrocarbons include methane, ethylene, acetylene, terpenes, etc.

**Sources**

(*a*) Trees emit large quantities of hydrocarbons in the atmosphere.

(*b*) Anaerobic decomposition of organic matter by bacteria produces hydrocarbons as:

$$2[CH_2O] \xrightarrow{\text{Bacteria}} CO_2 + CH_4$$

Domesticated animals also release $CH_4$ to atmosphere.

(*c*) Anthropogenic sources contribute about 15% of the hydrocarbons to the atmosphere, which can be broken up as:

| | | |
|---|---|---|
| Petroleum burning | 55% | Incinerators and refuse burning 28.3% |
| Coal burning | 3.3% | Solvent evaporation 11% |
| Wood burning | 2.2% | |

Hydrocarbons are removed by several chemical and photochemical reactions (refer to photochemical smog). They are oxidized through a series of steps forming $CO_2$, solid organic particulate matter or water soluble products, which are washed by rain.

Harmful effects of hydrocarbons are due to the photochemical reactions in which they are involved. They form photochemical smog as

$$\text{Unburnt Hydrocarbon} + NO_x \xrightarrow[\text{Stagnant air}]{\text{Abundant sun light}} \text{Photochemical smog}$$

Some hydrocarbons have carcinogenic and lachrymatory effects.

5. **Particulates.** These are small, solid particles and liquid droplets present in the atmosphere in fairly large numbers and sometimes pose serious air pollution problem. In size, particulates ranges from 0.002 μ in diameter to a diameter of 500 μ, with lifetimes varying from a few seconds to several months.

**Sources**

(*a*) Numerous natural processes such as volcanic eruptions, blowing of dust and soil by the wind, spraying of salt and other solid particles by the seas and oceans, etc. inject particulate matter into the atmosphere.

(*b*) Anthropogenic sources include flyash from power plants, meltors, and mining operations, and smoke from incomplete combustion processes.

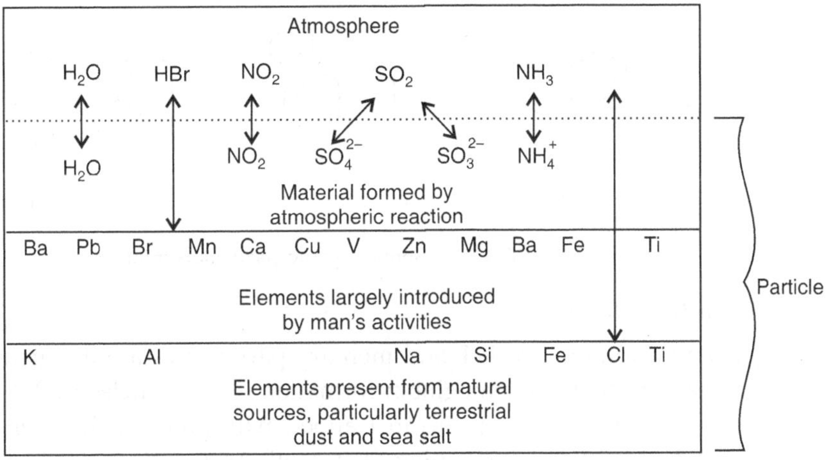


***FIGURE 12.1*** **Inorganic particulates and their origins.**

Particulate matter can be classified as:

(*i*) *Inorganic particulate matter.* These mainly include metal oxides and are produced whenever fuels containing metals are burnt. For example,

$$3FeS + 8O_2 \longrightarrow Fe_3O_4 + 6SO_2$$

$$CaCO_3 \xrightarrow{\text{Heat}} CaO + CO_2$$

Combustion of leaded gasoline produces lead halides which are volatile and on condensation form particles.

$$Pb(C_2H_5)_4 + O_2 + C_2H_4Cl_2 + C_2H_4Br_2 \longrightarrow CO_2 + H_2O + PbCl_2 + PbBrCl + PbBr_2$$

Whenever $H_2SO_4$ droplets are formed, aerosol mists appear and with basic pollutants. Salts are also formed.

$$2SO_2 + O_2 + 2H_2O \longrightarrow \underset{\text{aerosol droplet}}{2H_2SO_4}$$

$$\underset{\text{droplet}}{H_2SO_4} + \underset{\text{gas}}{2NH_3} \longrightarrow \underset{\text{droplet}}{(NH_4)_2SO_4}$$

$$\underset{\text{droplet}}{H_2SO_4} + \underset{\text{particle}}{CaO} \longrightarrow \underset{\text{droplet}}{CaSO_4.H_2O}$$

(*ii*) *Organic particulate matter.* These generally include polycyclic aromatic hydrocarbons (PAH) such as benzo-α-pyrene, chrysene, benzofluoro-anthene, etc.

Soot is also a highly condensed product PAH and most PAH compounds are absorbed in soot particle.

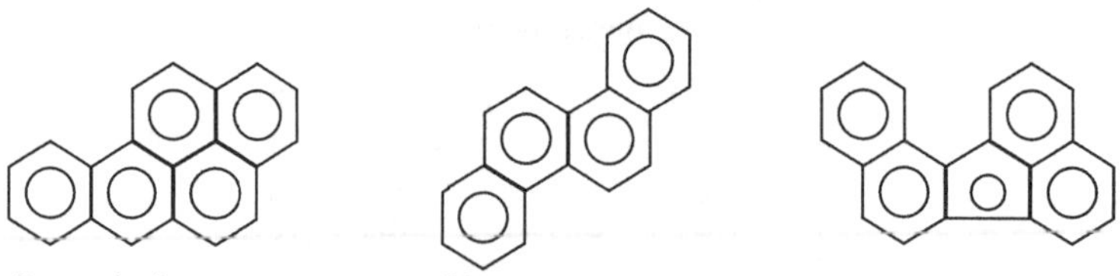

Benzo (α–) pyrene    Chrysene    Benzofluoranthene

*FIGURE 12.2* **Some organic particulate matter.**

**Harmful Effects**

(*a*) Lead, cadmium, nickel, and mercury particulates are the greatest health hazards. Lead affects the haemoglobin formation. Fine particles ($< 3\mu$) cause lung damage while larger particles ($> 3\mu$) are trapped in the nose and throat and may be eliminated easily. Metal dust containing Pb, Cr, Ni, As, Cd, Zn, and Hg affect the respiratory system.

(*b*) Particulates accelerate corrosion of metals, damage buildings, sculptures, paints, etc.

(*c*) Aerosols (fluorocarbons, $NO_x$, $SO_x$), released from emissions from jet and supersonic planes deplete the ozone layer in the higher atmosphere.

(*d*) Dust coating on leaves affects photosynthesis and reduces plant growth.

(*e*) Animals feeding on particulate-coated plant containing fluorides, arsenic, or lead suffer from toxic effects.

(*f*) Particulates in the atmosphere reduce visibility and influence the climate through formation of clouds, rain, and snow by acting as nuclei upon which water condenses.

**Control**

(*a*) **Gravity settling chamber** decreases the velocity of effluent gases resulting in settling of dust or droplets. This method is not suitable for settling of fine particles.

(*b*) **Wet scrubber** uses a liquid (usually water) for removing solid, liquid, or gaseous contaminants.

(*c*) **In electrostatic precipitators,** aerosols acquire charges when subjected to an electrical field which is attracted to a ground surface from which they are removed. The electrostatic precipitator (Fig. 12.3) works on the principle of electrostatic precipitation, *i.e.*, electrically charged particulates present in the polluted gas are separated from the gas stream under the influence of the electrical field.

A typical wire and pipe precipitator consists of:

(*i*) a positively charged collecting surface (grounded).

(*ii*) a high voltage (50 KV) discharge electrode wire.

(*iii*) an insulator to suspend the electrode wire from the top.

(*iv*) a weight at the bottom of the electrode wire to keep the wire in position.

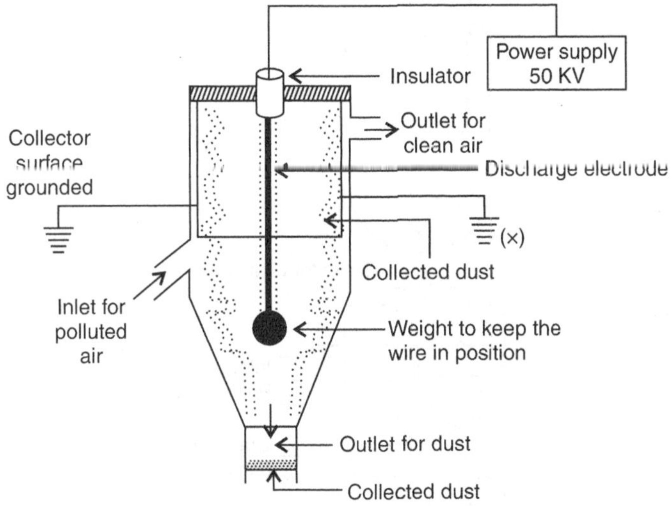


*FIGURE 12.3* **Electrostatic precipitator.**

**Mechanism of Action.** The polluted gas enters from the bottom, flows upwards (*i.e.*, between the high voltage wire and grounded collecting surface). The high voltage in the wire ionizes the gas. The negative ions migrate toward the grounded surface and pass on their negative charge to the dust particles. Then these negatively charged dust particles are electrostatically drawn toward the positively charged collector surface, where they finally get deposited.

The collecting surface is rapped or vibrated to periodically remove the collected dust particles so that the thickness of the dust layer deposited does not exceed 6 mm, otherwise the electrical attraction becomes weak and efficiency of the electrostatic precipitator get reduced.

As the electrostatic precipitation has 99 percent efficiency and can be operated at high temperatures (600°C) and pressure at less power requirement, it is economical and simple to operate compared to other devices.

(*d*) **Cyclone Separators (Reverse flow Cyclone).** In this method centrifugal force is utilized by cyclone separators to separate the particulate matter from the polluted gas. Centrifugal force several times greater than gravitational force can be generated by a spinning gas stream and this quality makes cyclone separators more effective in removing much smaller particulates than can possibly by removed by gravitational settling chambers.

A simple cyclone separator (Fig. 12.4) consists of a cylinder with a conical base. A tangential inlet discharging near the top and an outlet for discharging the particulates is present at the base of the cone.

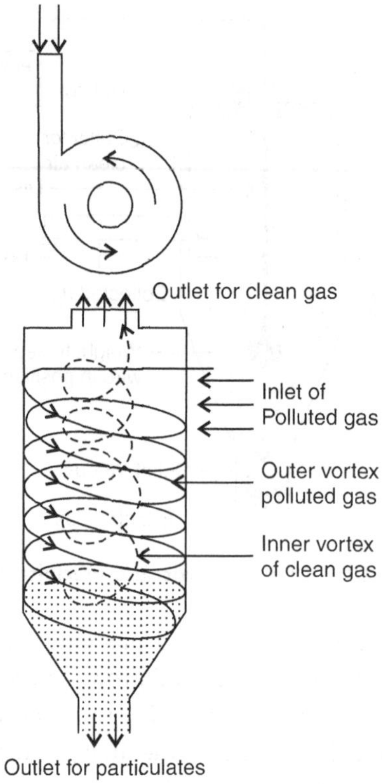


*FIGURE 12.4* **Cyclone separator (Reverse Flow Cyclone).**

**Mechanism of action.** The dust laden gas enters tangentially, receives a rotating motion, and generates a centrifugal force due to which the particulates are thrown to the cyclone walls as the gas spirals upward inside the cone (*i.e.*, flow reverses to form an inner vortex which leaves flow through the outlet). The particulates slide down the walls of the cone and are discharged from the outlet.

## 12.7 ACID RAIN

Rain water is slightly acidic in nature due to the presence of carbon dioxide in the atmosphere, forming carbonic acid. However, the presence of $SO_x$ and $NO_x$ gases as pollutants in the atmosphere further lowers the $pH$ of the rain water to as low as 2.4 due to the formation of $H_2SO_4$ and $HNO_3$, respectively. This type of precipitation of lower $pH$ is called *acid rain*.

Acid rain is a man-made global ecological problem where $H_2SO_4$ is the major contributor (60–70%) followed by $HNO_3$ (30–40%), and HCl. The various photochemical reactions occurring in the atmosphere are:

$$NO + O_3 \longrightarrow NO_2 + O_2$$

$$NO_2 + O_3 \longrightarrow NO_3 + O_2$$

$$NO_2 + NO_3 \longrightarrow N_2O_5$$

$$N_2O_5 + H_2O \longrightarrow 2HNO_3$$

some $HNO_2$ is also formed as:

$$N_2O_3 + H_2O \longrightarrow 2HNO_2$$

$HNO_3$ and $HNO_2$ then return to the Earth's surface. $HNO_3$ is removed as a precipitate or as particulate nitrates after reaction with bases ($NH_3$, particulate lime).

Sulphuric acid is formed as:

$$SO_2 + O_2 + H_2 \xrightarrow[\text{soot particles (metal oxide)}]{(HC, NO_x)} H_2SO_4$$

The presence of hydrocarbons and $NO_x$ step up the oxidation rate of the reaction. In water droplets, ions such as $Mn^{2+}$, $Fe^{2+}$, $Ni^{2+}$, and $Cu^{2+}$ catalyze the oxidation reaction. Soot particles also catalyze the oxidation of $SO_2$.

$HNO_3$ and $H_2SO_4$ combine with HCl to generate acidic precipitation, known as acid rain.

**Harmful Effects of Acid Rain**

(*a*) Acid rain causes extensive damage to building and sculptural materials of marble, limestone, slate, mortar, etc. These materials become pitted and weakened mechanically as the soluble sulphates are leached out by rain water.

$$\underset{\text{Lime stone}}{CaCO_3} + H_2SO_4 \longrightarrow CaSO_4 + CO_2 + H_2O$$

In Greece and Italy, invaluable statues have been partially dissolved by acid rain. The Tajmahal at Agra, faces the same problem.

(*b*) The acid in the rain damages the leaves of trees, plants, and retards the growth of forests which are important natural resources for production of wood pulp, paper, and board.

(*c*) Besides damaging the flora and fauna, acid rain also affects fish mortality. These pollutants ($H_2SO_4$, Cd, Pb, etc.) enters rivers and lakes, killing the fish eggs. In Sweden, 85000 lakes were slowly dying by the deluge and 4000 lakes are completely dead. In Canada, trees and lakes are also being killed by acid rain, 60% of which originates in the USA.

(*d*) Lowering of $pH$ of rain water, due to acid rain, changes the rate of metabolism of organisms.

(*e*) Acid rain causes irritation to eyes and mucous membrane.

(*f*) Acid rain accelerates the rate of corrosion.

## 12.8 SMOG FORMATION

Smog is an odd combination of smoke and fog in air which partially reduces the atmospheric visibility. Two types of smogs are recognized:

1. **Classical or London Smog.** This type of smog occurred in London in 1952. It prevailed for five days and caused death to 4000–5000 people and made thousands ill. This smog is associated with the use of the traditional fuel coal. It is characterized by high concentration of unburnt carbon soot and high levels of $SO_2$ in the atmosphere. The most important source of this type of smog is the combustion of coal. It has reducing as well as acidic properties due to the presence of $SO_2$, a mild reducing agent and an precursor of a weak acid, thus it is also known as *reducing smog*. The various reactions involved are:

$$S + O_2 \longrightarrow SO_2$$

$$SO_2 + \frac{1}{2}O_2 \longrightarrow SO_3$$

This reaction takes place in presence of sunlight and particulate matter. $SO_2$ combines with moisture from fog to form sulphurous acid.

$$SO_2 + H_2O \longrightarrow H_2SO_3$$

$$H_2SO_3 + \frac{1}{2}O_2 \xrightarrow{\text{Metal salts}} H_2SO_4$$

$$H_2SO_4 + 2NH_3 \longrightarrow (NH_4)_2SO_4$$

or

$$SO_2 + 2NH_3 + H_2O + \frac{1}{2}O_2 \longrightarrow (NH_4)_2SO_4$$

2. **Photochemical Smog.** It is an oxidizing smog having high concentration of oxidants. Photochemical smog is characterized by brown and hazy fumes which irritate the eyes and lungs and lead to the cracking of rubber and extensive damage to plant life. Photochemical smog formation occurs in the following steps:

(*i*) Reactive hydrocarbons interact with $O_2$ to form a hydrocarbon free radical $RCH_2^{\bullet}$.

(*ii*) $RCH_2^{\bullet}$ reacts with $O_2$ to form $RCH_2O_2^{\bullet}$.

(*iii*) $RCH_2O_2^{\bullet}$ reacts with NO to produce $NO_2$ and the free radical $RCH_2O^{\bullet}$.

(*iv*) $RCH_2O^{\bullet}$ interacts with $O_2$ to yield a stable aldehyde, RCHO, and hydroperoxyl radical $HO_2^{\bullet}$.

(*v*) $HO_2^{\bullet}$ then reacts with another molecule of NO forming $NO_2$ and $HO^{\bullet}$.

(*vi*) $HO^{\bullet}$ reacts rapidly with a stable hydrocarbon to give $H_2O$ and regenerated $RCH_2^{\bullet}$, thereby completing the cycle. This cycle goes on as a chain reaction. One complete cycle yields two molecules of $NO_2$, one molecule of RCHO, and regenerates $RCH_2^{\bullet}$. Very soon there is a rapid buildup of smog products.

(*vii*) RCHO formed may interact with the $HO^{\bullet}$ radical, leading to the formation of an acyl radical $RC = O^{\bullet}$, peroxyacyl radical $RCOO_2^{\bullet}$ (by reaction with $O_2$), and finally peroxyacyl nitrate (PAN) by reaction with $NO_2$. PAN is one of the most potent eye irritants found in smog.

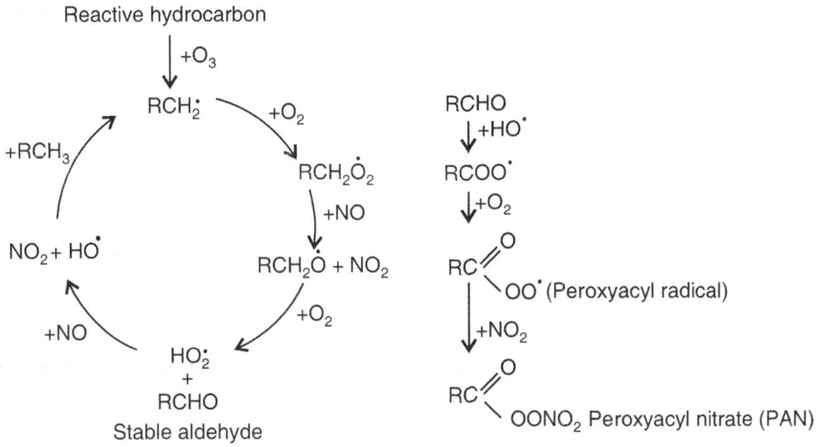


*FIGURE 12.5* **Formation of photochemical smog.**

## 12.9 EFFECTS OF AIR POLLUTION

Air pollutants adversely affect the health of human beings and animals and damage plants and properties. Effects of air pollutants can be studied under following headings:

1. **Effect on Human Health.** Air pollutants may have acute as well as chronic effects for human health. For example,
   (*a*) Irritation of eyes, nose, and throat.
   (*b*) Irritation of the respiratory tract.
   (*c*) Lead particulates cause lead poisoning leading to convulsions, coma, and even death.
   (*d*) Cadmium particulates (through cigarette smoking) cause cardiovascular diseases, kidney and liver damage, and even death.
   (*e*) Nickel particulates (in tobaco smoke) result in respiratory damages.
   (*f*) Mercury results in nerve, brain, and kidney damage.
2. **Effect on Animals.** Animals feeding on particulate coated plants get affected with arsenic and lead poisoning leading to bronchitis and lack of appetite in pet animals.
3. **Effect on Plants.** Air pollutants have the following effects on plants:
   (*a*) $NO_2$ causes suppressed growth of plants and premature fall of leaves.
   (*b*) Ozone causes necrosis (dead areas on leaves) and damages leaves.
   (*c*) $SO_2$ bleaches the leaf surface and causes chlorosis (*i.e.*, loss of chlorophyll and yellowing of leaf) in leafy vegetables.
   (*d*) Peroxyacyl nitrate (PAN) damages leafy vegetables causing premature falling, discoloration, and curling of sepals.
4. **Effect on Climate.** The concentration of $CO_2$ is expected to double by 2020. This will increase the atmospheric temperature of earth by the greenhouse effect, resulting in

melting of the polar ice caps, glaciers, etc. which will flood the coastal towns. Patterns of rainfall may also change, affecting agricultural output.

5. **Effect on Materials.** The damages caused to various materials are:
   (*a*) Acid rains, aerosols, and $SO_2$ damage the building materials.
   (*b*) $SO_2$, $H_2S$, and particulates discolorizes paints.
   (*c*) Metals undergo tarnishing and corrosion by $SO_2$ and acid gases.
   (*d*) $SO_2$ and acid gases make paper brittle and disintegrate leather.
   (*e*) $O_2$, $SO_2$, $NO_2$, and acid gases discolor, deteriorate, and reduce the tensile strength of textiles.

   The intensity of damage depends upon factors such as moisture, temperature, sunlight, air movement, and the nature and concentration of the pollutant.
6. Heavy smoke and dust storms may cause sky darkening. Visibility may be reduced due to smoke, fog, and industrial fumes containing particulates that scatter light.

## 12.10 OZONE CHEMISTRY

In the upper atmosphere the species of oxygen are $O_2$, O, $O^{\bullet}$, $O^+$, $O_2^{\bullet}$ and $O_3$, *i.e.*, molecules, atoms, ions, and excited atoms and molecules, respectively. UV radiation causes their photochemical dissociation, ionization, etc., for example,

$$O_2 + h\nu \xrightarrow{240-260\text{ nm}} O + O$$

$$O_3 + h\nu \xrightarrow{308\text{ nm}} O^{\bullet} + O_2$$

$$O + h\nu \longrightarrow O^+ + e$$

$$O_2 + h\nu \longrightarrow O_2^+ + e$$

Ozone is an important gas present in the stratosphere, with maximum concentration around 10 ppm at an altitude of 25–30 km. The ozone layer in the stratosphere acts as a protective shield for life on Earth, absorbing harmful solar radiations such as UV rays, which are harmful to life, causing DNA mutation and skin cancer. In the absence of this layer, the temperature of the lower atmosphere will rise to such an extent that the biosphere will turn into a blast furnace.

### 12.10.1 Formation of the Ozone Layer

It is formed in the stratosphere by a photochemical reaction, followed by a third body reaction

$$O_2 + h\nu \xrightarrow{242\text{ nm}} O + O$$

$$O + O_2 + M(N_2 \text{ or } O_2) \longrightarrow O_3 + M$$

The third body (M) absorbs the excess energy liberated by the above reaction and thereby stabilizes the $O_3$ molecule. The thickness of the ozone layer is measured in Dobson units (DU), where 1 DU = 0.01 mm of the compressed gas at 0°C and 760 mm Hg pressure. Average thickness varies marginally with latitudes because of different atmospheric conditions. The thickness is comparatively low in the polar regions.

### 12.10.2 Depletion of the Ozone Layer

According to the World Metrological Organization (WMO), Geneva, ozone levels have been depleting at an average annual rate of 4.3% every decade in the Northern Hemisphere and 4.1% in the Southern Hemisphere since 1997. The ozone layer is believed to be depleted by reactions with atomic oxygen, reactive hydroxyl radicals, and nitric oxide.

(*a*) $$O_3 + O \longrightarrow O_2 + O_2$$

(*b*) $$O_3 + HO^\bullet \longrightarrow O_2 + HOO^\bullet$$

$$HOO^\bullet + O \longrightarrow HO^\bullet + O_2$$

(*c*) $$O_3 + NO \longrightarrow NO_2 + O_2$$

$$NO_2 + O \longrightarrow NO + O_2$$

Apart from the above mentioned elements, *chlorine* also plays an important role in depleting ozone in the stratosphere in the following manner:

$$Cl^\bullet + O_3 \longrightarrow Cl\text{—}O^\bullet + O_2$$

$$Cl\text{—}O^\bullet + O \longrightarrow Cl^\bullet + O_2$$

$Cl^\bullet$ atoms are regenerated, so a long chain process is involved.

The following reaction helps in decreasing the concentration of $Cl^\bullet$ by forming chlorinenitrate, a stable molecule.

$$Cl-\dot{O} + NO_2 \longrightarrow Cl-O-N\begin{matrix}\nearrow O\\ \searrow O\end{matrix}$$

Chlorinenitrate

The various sources and sinks of chlorine free radical are as follows:

**Sources**

1. CFCs (chlorofluorocarbons) which are used as refrigerants and aerosol spray can propellants and in plastic foams like thermocole are the main sources. These molecules are inert in the troposphere but slowly escape into the stratosphere, where they decompose to form $Cl^\bullet$ free radicals, which immediately react with ozone.

$$CFCl_3 \xrightarrow[200\text{ nm}]{h\nu} CFCl_2 + Cl^\bullet$$

$$Cl^\bullet + O \longrightarrow ClO^\bullet + O_2$$

$$ClO^\bullet + O \longrightarrow Cl^\bullet + O_2$$

Each $ClO^{\bullet}$ radical can initiate a series of reactions leading to destruction of up to 100 000 molecules of $O_3$. These $Cl^{\bullet}$ atoms remain active for more than 100 years and are capable of breaking thousands of ozone molecules before the released chlorine gas gets converted into dilute HCl and comes down as acid rain.

2. Volcanic eruptions inject $Cl_2$ and HCl directly into the stratosphere. $Cl_2$, on exposure to UV radiation (300–400 nm), forms $Cl^{\bullet}$ while HCl reacts with ${}^{\bullet}OH$ to give $Cl^{\bullet}$.

**Sinks**

$$Cl^{\bullet} + O_3 \longrightarrow ClO^{\bullet} + O_2$$

$$ClO^{\bullet} + O \longrightarrow Cl^{\bullet} + O_2$$

$$\left.\begin{array}{l} Cl^{\bullet} + CH_4 \longrightarrow CH_3^{\bullet} + HCl \\ ClO^{\bullet} + NO_2 \longrightarrow ClONO_2 \end{array}\right] \text{rained out}$$

With $CH_4$, $Cl^{\bullet}$ forms HCl which is rained out, while the reaction of $ClO^{\bullet}$ with $NO_2$ forms chlorine nitrate, a stable product, and thus lowers the overall $O_3$ depletion rates.

The depletion of the ozone layer has been found to be more acute in polar regions, particularly Antarctica, leading to the formation of an ozone hole, detected during September to November 1985. This may be due to two reasons:

(*a*) Cold climatic conditions and other complex atmospheric parameters including turbulence.

(*b*) Absence of $NO_2$ in these areas which destroys $ClO^{\bullet}$ and checks ozone depletion. $NO_2$ at sub-zero temperature freezes into ice droplets.

### 12.10.3 Consequences of Ozone Layer Depletion

Ozone layer destruction leads to more UV radiation reaching Earth and causing similar climate changes as in the greenhouse effect. These rays are harmful to man, causing skin cancer, swelling of skin, sun burns, skin ageing, leukemia, breast cancer, cataracts, edema, haemorrhage, lung injury, lung cancer, visual impairment, DNA mutation, and many more adverse effects.

## 12.11 GREENHOUSE EFFECT AND GLOBAL WARMING

The greenhouse effect is the trapping of heat from below by the atmosphere. Earth's atmosphere admits the visible rays from the sun, but traps the infrared rays emanating from Earth's surface. Some of Earth's heat is carried from the surface by air currents or by the evaporation of water, but most of it is radiated to the atmosphere, which traps it and radiates much of it back, releasing the rest to space. But how is this trapping accomplished ? Why do we worry about $CO_2$ and other minor constituents of the atmosphere, when it is made up almost entirely of $N_2$ and $O_2$ (See Table 12.1).

**TABLE 12.1 Composition of Dry Air at Ground Level in Remote Continental Areas**

| *Constituent* | *Formula* | *Concentration (by volume, ppm)* |
|---|---|---|
| Nitrogen | $N_2$ | 780,900 |
| Oxygen | $O_2$ | 209,400 |
| Argon | Ar | 9,300 |
| Carbon dioxide | $CO_2$ | 370 |
| Neon | Ne | 18 |
| Helium | He | 5.2 |
| Methane | $CH_4$ | 1.7 |
| Krypton | Kr | 1.1 |
| Hydrogen | $H_2$ | 0.5 |
| Nitrous oxide | $N_2O$ | 0.3 |
| Xenon | Xe | 0.08 |
| Carbon monoxide | CO | 0.04–0.08 |
| Organic vapours | | 0.02 |
| Ozone | $O_3$ | 0.01–0.04 |

The answer is that the major atmospheric gases are unable to absorb infrared light. They do not meet the two fundamental requirements for the absorption of electromagnetic radiation :

(1) When radiation is absorbed by a molecule, the molecule undergoes a quantum transition, involving the movement either of its electrons or its nuclei; the energy of the radiation must therefore match the energy of the molecular transition. In the infrared region of the spectrum, the available transitions involve movement of the nuclei in molecular vibrations. That is why argon, the third most abundant atmospheric constituent (0.9 percent) is transparent to infrared radiation. Since argon is monatomic, it has no vibrations.

(2) Because radiation is electromagnetic, its absorption requires that the transition change the electric field within the molecule, that is, the transition must alter the molecule's dipole moment (the vector sum of atomic charges times their distances from the molecule's center of mass). This second requirement is the reason that $N_2$ and $O_2$ are unable to absorb Earth's infrared radiation. Although their nuclei do vibrate along the bond joining them, and the energy of the vibration is in the infrared region, the vibration does not change the dipole moment. Because the molecule is symmetrical, the dipole moment remains zero no matter how much the bond is stretched. The vibration is infrared-*inactive*. This is true for all *homonuclear* diatomic molecules. The dipole moment is altered by vibrations of *heteronuclear* diatomic molecules, such as CO, NO, and HCl, since their atoms have different partial charges. However, these molecules do not contribute significantly to the greenhouse effect because their concentration in the atmosphere is too low and their infrared absorption too weak.

In contrast, polyatomic molecules have numerous vibrations ($3n - 6$) for non-linear molecules, where $n$ is the number of atoms, or ($3n - 5$) for linear molecules; at least some of these vibrations change the dipole moment and are infrared-active. All the gases that contribute significantly to the greenhouse effect are polyatomic. The two most important greenhouse molecules are water and carbon dioxide. Their vibrations are illustrated in Figure 12.6. All three vibrations of water change its dipole moment. For carbon dioxide, the symmetric stretching motion of the two O atoms leaves the dipole moment unchanged : the dipoles that each O atom generates relative to the carbon atom cancel one another due to the linear geometry. However, the net dipole moment is altered by the asymmetric stretch and the bending vibration.

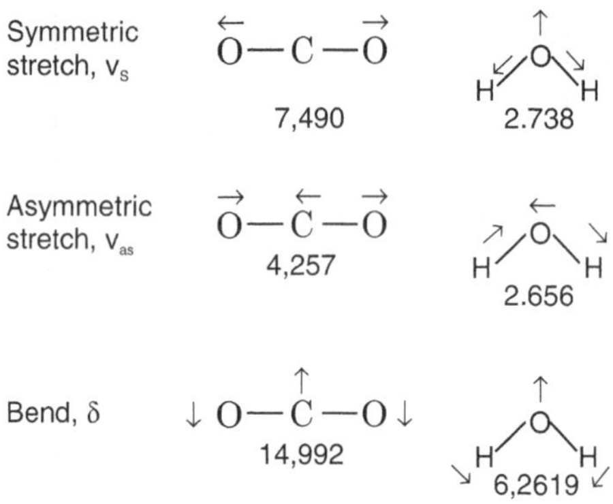


***FIGURE 12.6*** **Molecular vibrations of $CO_2$ and $H_2O$ in units of wavelengths (nm).**

Both water and carbon dioxide contribute to the greenhouse effect, but water is not listed in Table 12.1 because its content in the atmosphere varies greatly from place to place and time to time; on average, water molecules make up 0.4 percent of the atmosphere. The total amount of water is an order of magnitude greater than the amount of carbon dioxide. If Earth's temperature rises as a result of increasing carbon dioxide and other greenhouse gases, the water vapour pressure will also rise, and the amount of water in the atmosphere is expected to increase. This is an example of *positive feedback:* the greater the concentration of greenhouse gases, the higher the surface temperature; higher temperatures lead to greater amounts of atmospheric water, which in turn amplifies the greenhouse effect.

When greenhouse molecules absorb the terrestrial infrared radiation, they re-radiate it in all directions as illustrated in Figure. 12.7. Absorption and re-radiation continue repeatedly with increasing altitude, leaving less than 10 percent of the absorbed infrared radiation available near the top of the atmosphere. As a result, the lower atmosphere is warmer, but the higher atmosphere is colder than it would be in the absence of infrared absorbers.

It might appear that water and carbon dioxide would not be effective in heat trapping because Earth's emissions cover a wide spectrum of wavelengths, whereas the molecular vibrations correspond to specific energies. However, the molecules can undergo not only vibrations but also rotations. For each molecular vibration, infrared photons can induce transitions to many different rotational levels (rates of rotation). Consequently, each vibration has a broad absorption band. These bands are shown for water and carbon dioxide in Figure. 12.8. In the top panel of the figure, the absorptions for these two molecules are added up and superimposed on Earth's emission spectrum. The combined absorption bands can be seen to block most of the terrestrial radiation. There is, however, a relatively unobstructed region of the spectrum between 8,000 and 12,000 nm through which radiation can escape. This region is called the *atmospheric window.*

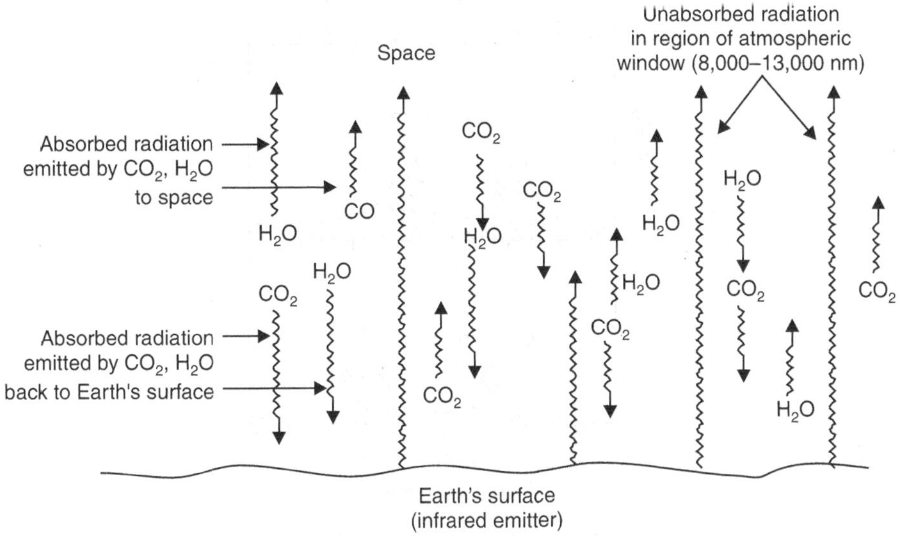


***FIGURE 12.7*** **Re-radiation of terrestrial infrared radiation by greenhouse gases.**

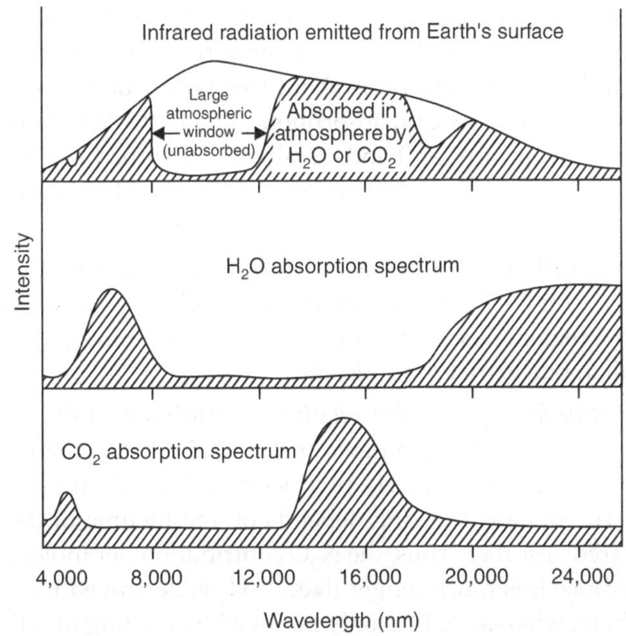


***FIGURE 12.8*** **Absorption of terrestrial radiation by water and carbon dioxide.**

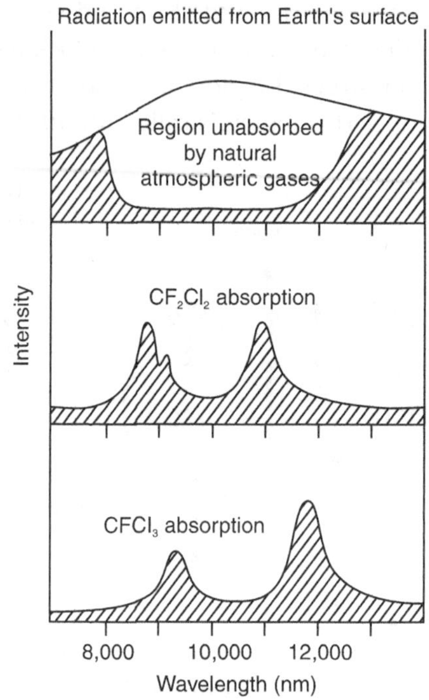


***FIGURE 12.9*** **Absorption spectra of chlorofluoromethanes ($CF_2Cl_2$ and $CFCl_3$) and their coincidence with the atmospheric window (8,000 to 13,000 nm).**

This window can be filled by other polyatomic molecules such as the chlorofluorocarbons (CFCs) (Fig. 12.9), methane ($CH_4$) and nitrous oxide ($N_2O$). The CFCs are of major concern as destroyers of stratospheric ozone, but they are also important greenhouse gases. Their impact is significant even though their concentration is about five orders of magnitude lower than that of carbon dioxide. The reason is that the $CO_2$ absorptions are nearly "saturated," that is, most of the radiation emitted within the absorption bands is already absorbed. As a result, each extra $CO_2$ molecule contributes only a relatively small amount to the total absorption. (The same is true for each additional water molecules.)

In contrast, the contribution to overall absorption is relatively high for each extra CFC molecule precisely because the CFCs are very dilute and absorb only a small fraction of the radiation, but do so in the window region. One extra CFC molecule contributes thousands of times more to the greenhouse effect than one extra $CO_2$ molecule.

The relative *radiative forcing* (*i.e.*, the relative contribution to the infrared absorption per molecule added to the atmosphere) is compared in Table 12.2 for methane, nitrous oxide CFC-11 ($CF_2Cl_2$), historically one of the most widely produced CFCs, and HFC-23 ($HCF_3$), currently the major CFC substitute. The radiative forcing depends on the lifetime of the gas, as well as on its effectiveness as an infrared absorber. Thus, the $N_2O$ contribution per molecule is much larger than that of $CH_4$, mainly because it is much longer lived. $CH_4$ is destroyed by reaction with hydroxyl radicals in the atmosphere, whereas $N_2O$ is only removed by drifting into the stratosphere, where it is photolyzed by ultraviolet rays.

**TABLE 12.2 Summary of Properties of Greenhouse Gases Affected by Human Activities**

| | $CO_2$ | $CH_4$ | $N_2O$ | CFC-11 | HCF-23 |
|---|---|---|---|---|---|
| Atmospheric concentration | ppmv | ppbv | ppbv | pptv | pptv |
| Preindustrial (1750  1800) | 200 | 700 | ~270 | zero | zero |
| Current | 370 | 1745 | 314 | 268 | 14 |
| Current rate of change/year* | 1.5/yr+ | 7.0/yr+ | 0.8/yr | - 1.4/yr | 0.55 |
| (% increase/year) | 0.41 | 0.40 | 0.25 | - 0.52 | 3.92 |
| Atmospheric lifetime (years) | 5 to 200‡ | 12 | 114 | 45 | 260 |
| Per molecule ratio of radiative forcing | 1 | 23 | 296 | 4,000 | 11,700 |
| [$\Delta F(GHG)/\Delta F(GHG\ CO_2)$] | | | | | |
| Major removal mechanism | I | II | III | III | III |

ppmv, ppbv, and pptv are parts per million, parts per billion, and parts per trillion by volume, respectively.

* Rate is calculated over the period 1990–1999.

† Rate has fluctuated between 0.9 ppm/yr and 2.8 ppm/yr for $CO_2$, and between 0 and 13 ppb/yr for $CH_4$, over the period 1990–1999.

‡ No single lifetime can be defined for $CO_2$ because of the different rates of uptake by different removal processes.

§ Quoted forcings are for a 100-year time horizon.

I — slow exchange of carbon between surface waters and deeper layers of ocean; uptake into biomass.

II—reaction with hydroxyl radical in troposphere.

III—photolysis in stratosphere.

***Sources :*** *(a) Working Group I, IPCC (2001) Technical Summary of the Working Group I Report. Intergovernmental Panel on Climate Change : Geneva, Switzerland. (b) Carbon Dioxide Information Center (2000). Current Greenhouse Gas Concentrations. Oak Ridge National Laboratory : Oakridge, Tennessee.*

Thus Principal Greenhouse Gases are Carbon Dioxide $CO_2$, Methane $CH_4$, Nitrous Oxide $N_2O$, Halocarbons. The gases differ in terms of atmospheric concentration rates of increase of concentration warming potential per molecule atmospheric lifetime. Their relative contributions to total potential warming depend on the time horizon considered and the emissions scenarios used. $CO_2$ is the major contributor-over a 100 year horizon, it contributes 70% + of the total greenhouse gas radiative forcing.

### 12.11.1 Current Trends in the Growth of Greenhouses Gases

The first assessments of the carbon dioxide content of the atmosphere at the beginning of the Industrial area indicate levels of $CO_2$ in the lower atmosphere (troposphere) are rising year by year. During the last ice age, they were below 300 ppm, and they are still around 280 ppm before the industrial revolution, but they have risen to 360 ppm by 1990. Over the period 1860–1960, the rise in $CO_2$ was 9% $a^{-1}$ and by the year 2000, the increase over the century as a whole was been 25%$a^{-1}$ (Fig. 12.10).

According to present information, the overall average temperature is rising by 0.5°C (0.9°F) for each increase in the carbon dioxide content by 18% with a net increase of 1°C by the end of 2005.

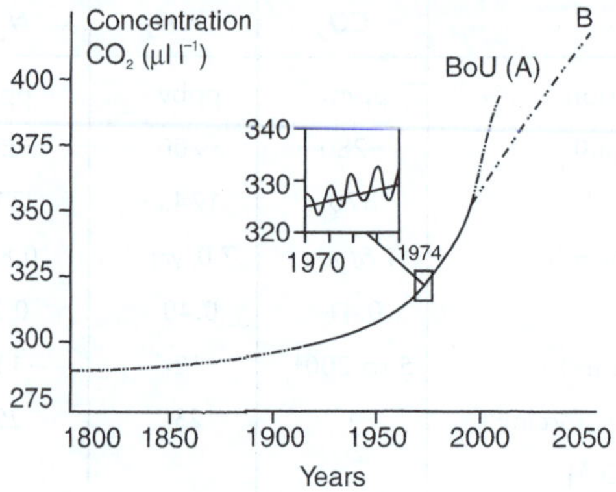


*FIGURE 12.10* **Global changes in atmospheric $CO_2$ with time. The solid part of the graph represents actual measured values and the dashed regions on the right are those most likely to occur in the future either with 'business as usual'-BaU (A) -or if the IPCC predications of Scenario B(B) are brought into reality by international implementation. Typical seasonal fluctuations are shown in the inset Figure Values in the northern summer are lowest because of the higher photosynthetic activity of northern forest.**

The current global trend in deforestation along with increased combustion of fossil fuels have a cumulative effect on the net increase in carbon dioxide content. Forests are the areas where a great deal of photosynthesis occurs through which they act as sinks for $CO_2$. Moreover, they maintain vast reservoirs of fixed but readily oxidizable carbon in the form of wood and humus. The overall result is that they serve to maintain a balance in the atmospheric $CO_2$ level.

The major sink for $CO_2$ is the ocean which contains the bulk of dissolved carbon dioxide as bicarbonate. Covering some 70 percent of the planet, the oceans act as gigantic fly-wheel in the Earth's machinery. As a huge reservoir of carbon, the oceans regulate the atmospheric $CO_2$ level and hence global climate. Another important sink is the biomass, *viz.*, living green plants, in which the photosynthesis rate is known to accelerate with increase in the carbon dioxide level.

Most scientists now agree that the global temperature increases are due to various human activities (Fig. 12.11). By burning large quantities of fossil coal or oil, and destroying tropical rain forests at the rate of 0.5 ha s$^{-1}$ and replacing them with grasslands and waste scrub, human activity

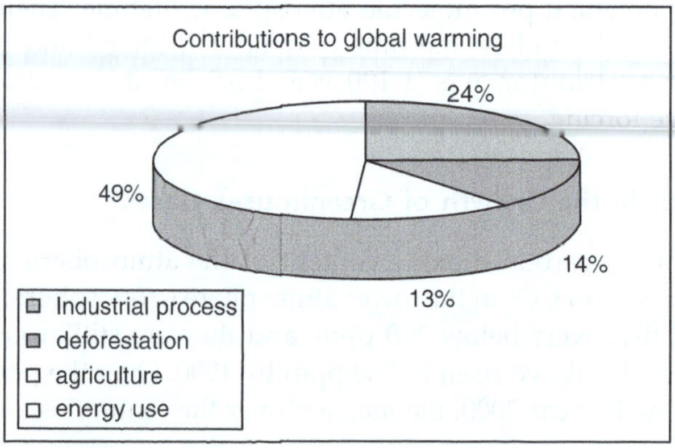


*FIGURE 12.11* **Contributions to global warming by different types of human activities in 1991.**
***Source : Data from World Resources Institute.***

has tipped the net balance of the natural carbon cycle in favour of more $CO_2$ being returned to the atmosphere than at any time since the end of the last ice age. When expressed in global quantities of carbon exchanged over one year the full extent of this imbalance can be appreciated. The 5 Tt. of fossil fuels burned in 1976, for example, contributed an extra 2.3 ppm. $CO_2$ to the atmosphere. Of this total, 1.6 ppm was either absorbed by the oceans or added to biomas, leaving a net increase of 0.7 ppm $CO_2$ in the atmosphere by the end of 1976.

If current trends continue, preindustrial $CO_2$ concentrations will be doubled by the year 2075 (Fig. 12.12). Computer models predict that doubling atmospheric $CO_2$ could cause global temperatures to rise 1.5°C to 4.5°C based on contribution of different human activities for global warming (Fig. 12.12). This may not sound like much change, but the difference between the temperature now and the last ice age about 10,000 years ago when glaciers covered much of North America was only about 5°C.

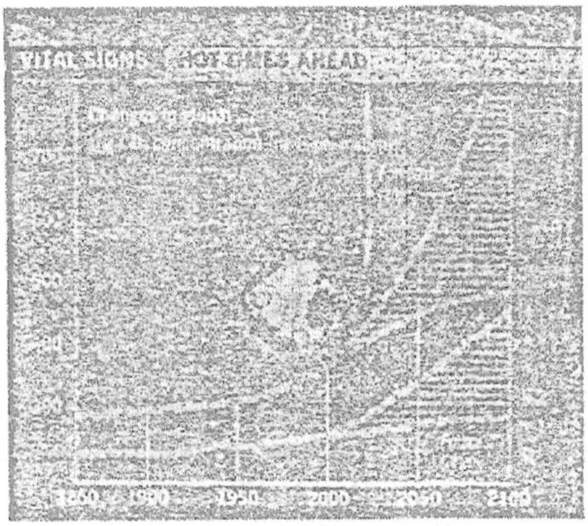

*FIGURE 12.12* **Correlation between $CO_2$ levels and global temperature.**

## 12.11.2 Impacts of Global Warming

The potential effects of global climate change can be considered in terms of ecological systems, sea-level rise, water resources, agriculture, electricity demand, air quality, and health effects. Since climate influences the location and composition of plants and animals in the natural environment, changes in climate have numerous consequences on ecological systems. Shifts in forests in geographic range and composition are likely to occur.

1. **Ecological Impact.** Biodiversity is also impacted by climate change, with *diversity* defined as the variety of species in ecosystems and the variability within each species. Examples of impact include an increased extinction of many species; growth or losses in freshwater fish populations depending on geographic location; and mixed effects on migratory birds, with some arctic-nesting herbivores benefiting and continental nesters and shorebirds suffering.

   Sea-level rises cause increased losses of coastal wetlands, inundation of coastal lowlands, increased erosion of beaches, and increased salinity in estuaries. Coastal wetlands currently total 13,145 sq. mi; with a 1-m rise in sea level, 26 to 66% of these wetlands may be lost.

2. **Impact on Water Resources.** The main consequence of climatic changes to inland waters include the following; (*a*) changes in the global amount of water resources and in the spatial and temporal distribution of these resources; (*b*) changes in soil moisture; (*c*) changes in extreme phenomena related to water resource, *i.e.*, floods and droughts; (*d*) changes in water quality; (*e*) changes in sedimentation processes; and (*f*) changes in water demand.

The consequences of climatic change on water quality include possible changes in the precipitation regime and the occurrence of acid rain. Direct consequence of climatic change on water quality occur. For example, temperature increases can decrease levels of dissolved oxygen in the water. Second, the biochemical oxygen demand (BOD) also increase with temperature. These two effects can decrease the dissolved oxygen concentration in a surface water system. Also, climatic changes can have indirect consequence on water quality since a decrease of river discharges, particularly during the dry season, can increase the concentration of pollutants in water bodies.

Climatic changes influence not only water available but also water demand. For example, water demand irrigation is largely affected by climatic change which conditions evapotranspiration. Water demands for domestic or industrial use are also affected by climatic change, hence water consumption for cooling systems, bathing, washing and gardening.

The simplest way to view the implications of global change on water resource is to consider the relationship between increasing atmospheric $CO_2$ and the hydrologic cycle; which may result in the following:

When the atmosphere reaches a new and warmer equilibrium, more precipitation balances faster evaporation. This faster hydrologic cycle is predicted to raise global averages of the up-and-down arrows of precipitation and evaporation by 7 to 10 time. Soil moisture predicted to fall in some places. Also, precipitation is predicted to increase some seasons and decrease in others. Thus Global climate change caused by the greenhouse effect appears to be a reality although scientific opinion differs as to the rate of change and its impact on water resources.

3. **Impact of Agriculture.** Global warming exhibits direct and indirect geographical effects on agricultural productivity. Direct effects occur through changes in the length of the growing season, the frequency of heat waves, and altered patterns of rainfall; while indirect effects result from changes in topsoil management practices. Dryland yields of corn, wheat, and soybeans could decrease in many regions as a result of higher temperatures which shorten a crop's life cycle. In contrast, dry land yields of corn and soybeans could double as warmer temperatures could decrease by 5 to 25%. Irrigation would probably increase in many areas because irrigated yields are more stable than dry-land yields under conditions of increased heat stress and reduced precipitation.

   Climate-sensitive electric uses include space heating, cooling to a lesser degree, water heating and refrigeration. Summer cooling electricity demands would increase, while winter heating demands would decrease. As a result of climatic change, annual electricity demands are expected to increase by 4 to 6% by the year 2055, additional power plants will be required to meet peak demands.

4. **Impact on Air Quality.** Global climate change also has implications for ambient air quality. Increased emissions of $SO_x$, $NO_x$ and CO are associated with power plants meeting increased electric demands. Plume rises from stacks would decrease due to higher ambient air temperatures and reduced buyoyancy effects; these decreases would be manifested

in higher ground level concentrations of air pollutants located closer to their stacks. Air pollutant dispersion is also affected by weather variables such as wind speed and direction, temperature, precipitation patterns, cloud cover, atmospheric water vapour, and global circulation patterns. Ozone pollution problems in many urban areas would worsen due to higher reaction rates at higher temperatures and lengthened summer seasons.

5. **Impact on Human Health.** Human Health effects are manifested by changes in morbidity and increases in mortality, particularly for the elderly during hotter and extended summer periods. Geographical patterns in relation to health effects are also expected.

### 12.11.3 Predicted Effects of Global Warming

- According to greenhouse theory, any warming on Earth should be accompanied by cooling in the stratosphere (the second layer of Earth's atmosphere) and the mesosphere (the layer of the atmosphere above the stratosphere). Some cooling of the stratosphere has been observed, but most of this is probably the result of ozone depletion, recent measurements indicate a 2°C drop in the temperature of the lower mesosphere.
- To date, any temperature changes possibly caused by an enhanced greenhouse effect have been too small to exceed normal short-term variations in mean atmospheric temperature caused by volcanic eruptions, air pollution, and other climatic factors.
- Mean global temperatures never vary more than a degree-either side of 12°C. Should this change, even by a degree, then the consequences are likely to be considerable. For example, between 6000 and 8000 years ago, a temperature rise of 2°C was responsible for melting the ice sheets spread across both hemisphere and brought about the end of the last ice age.
- Warming or cooling by more than 1°C over a few decades would be disastrous for Earth's ecosystems and for human economic and social systems. Such rapid climate change would alter conditions faster than some species, especially plants, could adapt or migrate. These changes might also shift areas in which it is possible for people to grow food. Many areas might become uninhabitable because of drought, or floods following a rise in average sea levels.
- We don't yet know enough about how Earth works to accurately project the possible effects of our emissions of greenhouse gases on global and regional climates or on the biosphere.

A warmer troposphere would have different consequences for different peoples and species. Some places would get drier, some wetter. Some would get hotter, others cooler. Here are some possible effects of a warmer global climate.

Mean global temperatures are likely to rise by 1.4°C (± 0.7°) by year 2030, and by 2.1°C (± 0.8°C) by year 2050 (Figure 12.13). Furthermore, global mean sea level which has risen 10–15 cm in the last century may raise by another 10–20 cm by 2025 and 50–200 cm by 2100. Most of this increase in sea level comes about from an increase in sea volume due to thermal expansion and rather less from the melting of polar ice. These predictions of sea level rise are also less than those of a few years ago but will still cause extensive flooding of low-lying areas in the world, especially during storms. The countries most at risk are the river delta areas of Bangladesh, India, Egypt, Thailand, and China (most vulnerable first) but many small islands will also suffer. The immediate effect of inundation will be crop destruction but more importantly, the surrounding groundwaters will become increasingly saline and large areas of productive land rendered permanently useless.

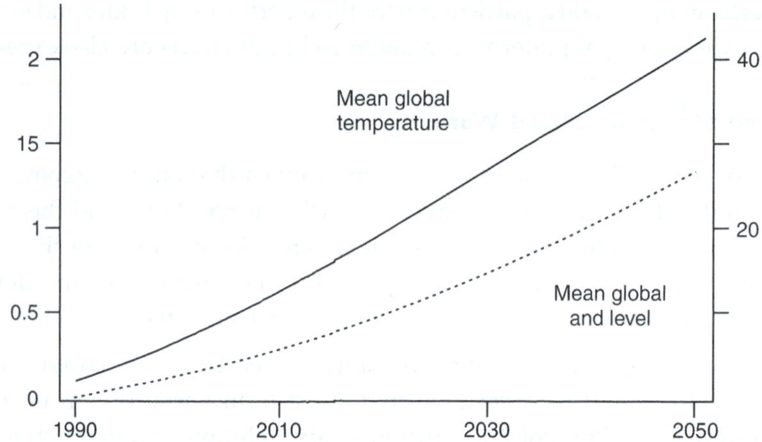


*FIGURE 12.13* **Likely changes in mean global temperatures and sea levels (IPCC predictions).**

### 12.11.4 Global Warming Effects in India

According to Dr. G.B. Pant, working at the Indian Institute of Tropical Meteorology, Pune, India has also warmed up somewhat in the last 3 decades. While the average temperature had risen by 0.4°C for the better part of the 20th century, the rate has climbed to 0.6% 100 years in the last 30 years. Days are hotter and the nights are the same. This has increased the difference between the maximum and minimum temperatures. So the weather is changing.

Despite the repeated warnings of scientists that the greenhouse effect could eventually have highly adverse effects on human society, the general public, as well as policy makers, were generally unconcerned with the threat until 1998. In 1998 Americans experienced unusual drought, searing heat, floods, and hurricanes—the precise events scientists predicted would occur with increasing frequency and severity as the levels of greenhouse gases rise.

### 12.11.5 International Agreements on Greenhouse Gases

Global warming is a problem of unprecedented dimension for human society, and it is not surprising that debate on what to do about it is lengthy and rancorous. To stem the tide of rising greenhouse gas (GHG) levels will require profound changes in many countries, with uncertain consequences for their economics. The scale of the problem is illustrated in Figure 12.14, which projects the trajectories of atmospheric $CO_2$ concentrations (which is the dominant anthropogenic greenhouse gas) for various levels of carbon emissions. Even if annual $CO_2$ emissions were capped in the next 20 years at 9 Gt, and then declined to 2 Gt (the level in the year 1950) over the next two centuries, the stabilized $CO_2$ concentration would still reach 450 ppm and stay there for hundreds of more years (mixing time of the oceans). Stabilizing $CO_2$ concentrations at 550 ppm, about double their pre-industrial level of 280 ppm and a common benchmark in descussions about global warming, would eventually require a reduction by all nations to 60 percent of the 1990 level. This is far more than cuts currently contemplated in international negotiations.

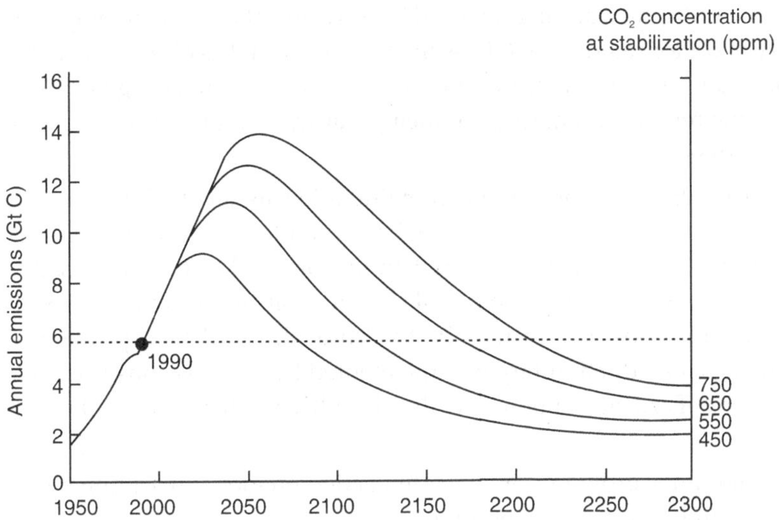


*FIGURE* 12.14 **Scenarios of greenhouse gas emissions; dependence of long-term stabilization levels of atmospheric $CO_2$ upon timing and magnitude of peak emissions of carbon. Dashed line shows emissions in 1990, the benchmark year for emissions reductions in the Kyoto Protocol on climate change.**

***Source :*** *Adapted from Wigley et. al., (1996). Economic and environmental choices in the stabilization of atmospheric $CO_2$ emissions. Nature 379 : 240–243. Copyright © 1996 by Nature, reprinted with permission from Nature and authors of journal article.*

Recognizing the seriousness of the problem, the industrialized countries signed a Framework Convention on Climate change at the 1992 Earth Summit in Rio de Janeiro. The Convention called for reducing emissions of GHGs to 1990 levels by the year 2000, but was non-binding and had no impact on subsequent emissions, which rose steadily through the 1990s (see Figure 12.14). It did, however, set the stage for subsequent international negotiations, culmiating in the 1997 Kyoto protocol. The goal was now to reduce GHG emissions to 5 percent below 1990 levels by 2008–2012, and specific targets were set for each country. The highest emitters—the U.S., Japan, and Europe—were assigned the greates reductions, 6–8 percent and less-developed countries (LDCs) were exempted.

The Protocol allowed that targets could to some extent be accomplished by emissions trading, although a trading system has yet to be specified. Emissions trading has worked well in lowering the cost of controlling $SO_2$ from U.S. power plants, but GHG emissions trading across international borders will be a much more complex task. The Protocol also established a Clean Development Mechanism, by which developed countries could gain GHG credits by investing in GHG-reducing projects (*e.g.*, tree planting) in LDCs. There remains significant disagreement on the extent to which targets can be achieved by these "market-based" mechanisms, as opposed to actual emission reductions.

The most contentious issue is the counting of biosphere uptake as a GHG credit. The U.S., with a large land area and evidence for substantial $CO_2$ uptake by its vegetation, claimed acredit worth about 20 percent of its target, a claim rejected by European countries, who want the Protocol to concentrate exclusively on emission reductions.

The fate of the Kyoto Protocol is currently in doubt, since the U.S. President, George W. Bush, has rejected it outright, claiming it would unduly hurt the U.S. economy. Nevertheless, there is considerable support for reigning in GHGs in the U.S., as well as elsewhere. For example, the city of Seattle has announced that it will exceed the Kyoto $CO_2$ reduction goals by imroving energy efficiency, promoting renewable energies, reducing traffic, recycling industrial waste materials and heat, and planting trees.

The European Union is committed to maintaining its Kyoto Protocol reduction target (8 percent by 2008–2012). The U.K., historically a high $CO_2$ emitter, already reports a 7 percent reduction in 1999, relative to 1990, thanks largely to switching from coal to natural gas. A number of large corporations have established energy and/or GHG reduction targets. Even in the oil industry, which spearheaded political opposition to the Kyoto Protocol, the BP Corp. has broken ranks and announced its own GHG reduction target, implemented by an emissions trading program among units of the company. Ford has announced that it will voluntarily upgrade fuel economy of its SUVs.

Thus, momentum is growing for concerted action on global warming. However, the task is huge, and will be with us for a very long time.

## CFC Substitutes

CFCs have been found to contribute in ozone layer depletion and greenhouse effect. So a great deal of effort has been applied to find the substitutes for CFCs main strategy has been to explore the suitability of hydrochlorofluorocarbons (HCFCs) and hydroflourocarbons (HFCs). These molecules have hydrogen, as well as chlorine and/or fluorine substituents on the carbon. The presence of C—H bonds allows the HCFCs and HFCs to be attacked by hydroxyl radicals and thereby be destroyed in the troposphere. At the same time, the Cl and/or F substituents lend these chemicals some of the desirable properties of CFCs, such as low reactivity and fire suppression, good insulating and solvent characteristics, and boiling points suitable for use in refrigerator cycles.

Some of the CFCs and currently available substitutes are listed in Table 5.3. $CHF_2Cl$ (HCFC-22) is a refrigerant that can substitute for $CCl_2F_2$ (CFC-12) in compressors of home air conditioners or refrigerators. The polyurethane foam insulation in refrigerator walls can be blown with $CH_3CFCl_2$ (HCFC 141b) or $CF_3CHCl_2$ (HCFC-123) instead of $CCl_3F$ (CFC-11). They are good insulators and have low flammability.

However, while HCFCs have significantly shorter atmospheric lifetimes than CFCs, some of the molecules survive and drift into the stratosphere, where they contribute to ozone depletion. The three HCFCs shown in the table have ozone depletion potentials (ODP) ranging from 0.02 to 0.11 (relative to CFC-11). If enough of these HCFCs were to be produced, they could still deplete the ozone layer significantly and contribute to climate change. For this reason, HCFCs are viewed as transitional CFC substitutes; the amended Montreal Protocol calls for the eventual elimination of HCFCs and for their permanent replacement by substances without chlorine. HFCs, having no chlorine, are not ozone destroyers, and are not covered by the Montreal Protocol. They have been used as blowing agents, refrigerants, and fire extinguishers.

**TABLE 12.3 Chlorofluorocarbons (CFCs) and Their Substitutes—Hydrochlorofluorocarbons (HCFCs) and Hydrofluorocarbons (HFCs)**

| *Trade-name* | *Chemical formula* | *Market* | *Atmospheric lifetime (yr)* | *100 year GWP** | *ODP+* |
|---|---|---|---|---|---|
| CFC-11 | $CCl_3F$ | Blowing agent | 50 | 4,000 | 1.0 |
| CFC-12 | $CCl_2F_2$ | Refrigerant | 102 | 8,500 | 1.0 |
| CFC-113 | $CCl_2FCClF_2$ | Cleaning agent | 85 | 5,000 | 0.8 |
| HCFC-22 | $CHF_2Cl$ | Refrigerant<br>Blowing agent | 12.1 | 1,700 | 0.055 |
| HCFC-141b | $CH_3CFCl_2$ | Blowing agent | 9.4 | 630 | 0.11 |
| HCFC-123 | $CF_3CHCl_2$ | Blowing agent | 1.4 | 93 | 0.02 |
| HFC-134a | $CH_2FCF_3$ | Refrigerant<br>Blowing agent | 14.6 | 650 | 0.0 |
| HFC-23 | $CHF_3$ | Fire extinguisher | 260 | 11,700 | 0.0 |
| HFC-227ea | $C_3HF_7$ | Fire extinguisher | 36.5 | 2.900 | 0.0 |
| HFC-245fa | $C_3H_3F_5$ | Blowing agent | 6.6 | 790 | 0.0 |

*GWP signifies "global warming potential." It is a measure of the degree of radiative forcing of a given molecule compared to a molecule of $CO_2$, which is assigned a GWP value of 1.

+ODP signifies "ozone depletion potential." It is the ratio of the impact on ozone of a chemical compared to the impact of a similar mass of $CFCl_3$ (CFC-11), which is assigned the value of 1.0.

***Sources :*** *IPCC Special Report on Emission Scenarios. Halocarbons and other halogenated compounds. (htpp://www.grida.no/climate/ipcc/emission/123.htm). Montreal Protocol on Substances that Deplete the Ozone Layer (http://www.unep.org/ozone/Montreal-Protocol/Montreal-Protocol 12000.shtml).*

However, both HCFCs and HFCs are potent greenhouse gases. Their global warming potentials (GWP) are many times larger on a per molecule basis than $CO_2$ (see Table 12.1). Although their lifetimes (with the exception of $CHF_3$) are shorter than those of the CFCs, they are longer than unsubstituted hydrocarbons, because the fluorine atoms stabilize the C—H bonds, making them slower to react with hydroxyl radicals. Concern for their potential impact on global climate has led to their inclusion on the list of greenhouse gases targeted for emissions reductions under the Kyoto Protocol. Current market trends indicate that relatively few CFCs will be replaced by HFCs. To date only HFC-134a has penetrated the market to any significant extent. The low percentage of replacement reflects the introduction of non-HFC substitutes, and increased efficiency, containment, and recycling of halocarbons currently in use.

New technologies that rely on neither HCFCs nor HFCs to replace CFCs are being developed. Aerosol propellants, for example, can be isobutane or dimethyl ether (mixed with water to suppress flammability). Similarly, CFCs have been replaced by hydrocarbons as blowing agents in styrofoam production. The foam insulation in refrigerator walls, first blown with CFC-11 and now with HCFC-141b, may soon be replaced with vacuum panels that are filled with a solid filler material and sealed under vacuum in a gas-tight envelope. The electronics industry, which relied heavily on CFC solvents for cleaning circuit boards, has switched to aqueous detergent cleaners and new imprinting methods that reduce the amount of cleaning needed.

The working fluids of refrigerators and air conditioners are the hardest to replace. Here, too, alternatives are coming into view. There has been much interest in older materials such as ammonia and hydrocarbons, but deterrents include the toxicity and corrosiveness of ammonia and the flammability of hydrocarbons. Flammability, at least, can be managed with proper engineering (we are exposed to flammability hazards all the time, for example, in handling automobiles and gas furnaces); on the market now is a German-made domestic refrigerator (called ''Greenfreeze'') that uses a propane-butane mixture. Moreover, there are air conditioners in development that do not have compressors, but that rely instead on evaporative cooling combined with desiccant drying of the cooled air. In the longer term, there is also interest in using sound waves for cooling.

There are currently no adequate substitutes for the halons, which are used to flood enclosed spaces such as offices, airplanes, and military tanks, for example, in case of fire. Since 1994 when production ceased, halons are being carefully banked, pending the development of substitutes. The halons exhibit a combination of low reactivity and effective fire suppression that is hard to match. The most promising candidate appears to be $CF_3I$, which, like $CF_3Br$ (halon-1301) is heavy enough to blanket and smother fires. The C—I bond is rapidly photolyzed by solar photons, even at ground level, so the molecule's atmospheric lifetime is short. However, toxicity and corrosiveness have not been fully evaluated.

Overall, the shift away from massive reliance on CFCs is happening more quickly than anyone thought possible a few years ago. As usual, necessity is the mother of invention.

## QUESTION BANK

1. What are the pollutants responsible for causing acid rain? What are its detrimental effects on material and terrestrial ecosystems?
2. What are the different methods adopted to limit the levels of various air pollutants in the atmosphere?
3. Name important air pollutants and mention their effects on human life.
4. What is meant by the terms pollutant, contaminant, receptor, sink, reservoir?
5. Discuss the sources, harmful effects, and methods of control of the following pollutants (*a*) CO (*b*) particulate matter (*c*) photochemical oxidants.
6. Explain the harmful effects of air pollution with respect to
   (*a*) Depletion of ozone layer (*b*) Photochemical smog
   (*c*) Greenhouse effect.
7. Define pollution. List various sources of air pollution and discuss their effects on the environment.
8. Explain briefly (*a*) Acid rain (*b*) Photochemical smog (*c*) Greenhouse effect (*d*) Ozone hole (*e*) Radiation balance of Earth.
9. Describe the sources and constituents of air pollution.
10. What is meant by the greenhouse effect? Describe two monitoring techniques used for the greenhouse effect.

11. What is the greenhouse effect? How do pollutants affect this?
12. What is the effect of ozone depletion? How does it occur?
13. How does automobile exhaust contribute to air pollution?
14. How do you account for Ozone depletion? Indicate its consequences.
15. Explain the sources and harmful effect of the following air pollutants.
    (*a*) CO (*b*) $SO_2$
    (*c*) Smoke (*d*) Particulate matter.
16. Give a detailed account of the various environmental segments.
17. Describe characteristics of major atmospheric regions.
18. Write a detailed and informative note on the following: (*a*) Earth's radiation balance (*b*) Lapse rate and temperature inversion.
19. How are air pollutants classified? What are the major effects of CO, $SO_2$, NO, hydrocarbon, and particulates on human health?
20. What are the sources of NO and $SO_2$ in the global atmosphere? Give an account of the global efforts of control the release of these two pollutants into the atmosphere by anthropogenic sources.
21. How would you broadly divide the major regions of the atmosphere? State their respective altitudes and temperature ranges. What are the important chemical elements in each region?
22. "Carbon dioxide, a nonpollutant, is perhaps the single most important environmental question facing us at present." Discuss in terms of the greenhouse effect.
23. "Oxygen plays a key role in the troposphere, while ozone, in the stratosphere." Elucidate.
24. Discuss briefly ozone hole and its effect on human health.

Chapter 13

# ADVANCED THEORY OF CHEMICAL BONDING

## 13.1 VALENCE BOND THEORY OF BONDING

This theory was proposed by *W. Heitler* and *F. London* in 1927 and later developed by *Pauling* and *Slater* in 1933. It is based on the Linear Combination of Atomic Orbitals (LCAO). Its salient features are:

1. A covalent bond is formed by the overlap of half filled atomic orbitals of the different atoms.
2. The overlapping atomic orbitals must have electrons with opposite spins.
3. The bonded electron pair is localized between the two linked atoms.
4. The stability of the covalent bond is due to the exchange of the valence electrons between the participating atoms, which lowers the potential energy of the bonded atoms.
5. Each atom of the covalent compound tends to acquire a noble gas configuration by sharing of electrons.

The valence bond treatment of bonding begins with the independent atoms and then considers the interaction between them. When the atoms are brought nearer to each other there is a change in energy due to the rearrangement of the electron cloud of one atom under the influence of the other atom. This theory is based on the following two principles of quantum mechanics:

1. If $\psi_A$ and $\psi_B$ are the wave functions for two independent atoms A and B, then the total wave function for the system of two atoms will be the product of these two wave functions, *i.e.*,

$$\psi = \psi_A \psi_B \tag{13.1}$$

The energy of the system is given by:

$$E = E_A + E_B$$

2. If there are several wave functions $\psi_1, \psi_2, \psi_3 \ldots\ldots \psi_n$ for the same system, then the wave function $\psi$ for the system is obtained by combining the several wave functions linearly, *i.e.*,

$$\psi = C_1\psi_1 + C_2\psi_2 + C_3\psi_3 + \ldots\ldots + C_n\psi_n \quad (13.2)$$

where $C_1, C_2, C_3 \ldots\ldots$ are the various coefficients indicating the contribution of each wave function. These coefficients are so adjusted as to give a state of lowest energy, *i.e.*, maximum stability.

### 13.1.1 Formation of $H_2$ Molecule

Let there be two hydrogen atoms A and B associated with electrons 1 and 2, respectively. Such hydrogen atoms may be represented as $H_A(e_1)$ and $H_B(e_2)$, respectively. If their starting wave functions are $\psi_A(1)$ and $\psi_B(2)$, then the wave function of the separated atom can be given as:

$$\psi = \psi_A(1)\,\psi_B(2) \quad (13.3)$$

Equation (13.3) can be used to calculate the energy of an $H_2$ molecule. The energy changes that take place are shown in Figure 13.1.

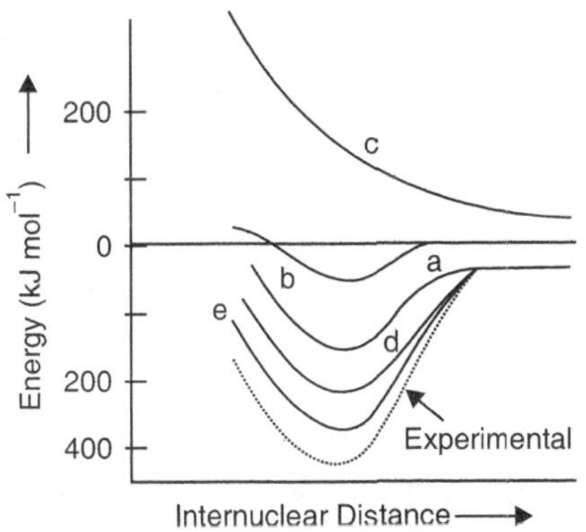


*FIGURE 13.1* **Change of energy with internuclear distance in the formation of $H_2$ molecule.**

It is clear that when the two atoms of hydrogen are at an infinite distance, the total energy of the system is zero. It increases as the internuclear distance decreases, as shown in the figure. It is known that in the formation of $H_2$ molecule, internuclear distance ($r_o$) is 0.74 Å while decrease in energy is 458 kJ/mol. Curve *a* gives the value of ($r_o$) as 0.90 Å and decreases in energy 24 kJ/mol, which is quite different from the experimental values. Hence equation (13.3) doesn't show the wave function of an $H_2$ molecule. Certain improvements have been introduced to make the energy value comparable to the experimental value. These improvements are:

1. **Exchange of electrons.** When two hydrogen atoms come close to each other then their electrons get exchanged and a bond is formed between the two hydrogen atoms and the

two shared electrons become indistinguishable from each other. This state of two H atoms may be represented as:

$$H_A(e_1)\ H_B(e_2) \qquad\qquad H_A(e_2)\ H_B(e_1)$$
$$\text{(I)} \qquad\qquad\qquad\qquad \text{(II)}$$

If the wave functions of the two states (I) and (II) are represented by $\psi_I$ and $\psi_{II}$ respectively, the wave functions may be written as:

$$\psi_I = \psi_A(1)\ \psi_B(2)$$
$$\psi_{II} = \psi_A\ (2)\ \psi_B(1)$$

The law of linear combination may be used for calculating the total wave function depending upon the mode of interactions as symmetrical ($\psi_s$) and antisymmetrical ($\psi_a$). These functions may be given as:

$$\psi_s = \psi_I + \psi_{II} = \psi_A\ (1)\psi_B(2) + \psi_A(2)\ \psi_B(1) \qquad (13.4)$$

$$\psi_a = \psi_I - \psi_{II} = \psi_A(1)\psi_B(2)\ - \psi_A(2)\ \psi_B(1) \qquad (13.5)$$

The energy of the systems as a function of internuclear distance can be calculated for $\psi_s$ and $\psi_a$ as shown by the curves *b* and *c*, respectively.

The energy of the system is higher for $\psi_a$, hence an $H_2$ molecule will not be formed and it is known as *nonbonding or repulsive state* while the energy of the system is lower for $\psi_s$ hence an $H_2$ molecule will be formed and is known as *bonding state*. The value of $r_0$ and the decrease in energy calculated from equation (13.4) comes out to be 0.80 Å and 303 kJ/mol respectively. So, on considering exchange of electrons, the energy is decreased from 24 to 303, *i.e.*, 279 kJ/mol. This 279 kJ/mol is known as the exchange energy for the $H_2$ molecule.

2. **Screening effect of electrons.** The energy value corresponding to the curve *d* is still far short of the experimental value of 458 kJ/mol. The wave function may further be improved by considering the fact that the electrons shield each other and the effective nuclear charge will be less than the actual nuclear charge. This gives the value of $r_0$ and the decrease in energy as 0.76 Å and 365 kJ/mol, respectively, shown by the curve *d*.

3. **Ionic character of H-H bond.** There is a possibility that the electron of one H atom is totally transferred to the other hydrogen atom, thus the system may be represented as:

$$(e_1)H_AH_B \text{ or } H_A^-H_B^+ \qquad\qquad H_A^+H_B^- \text{ or } H_AH_B(e_1)$$
$$\text{III} \qquad\qquad\qquad\qquad \text{IV}$$

Due to the ionic nature of the $H_2$ molecule, there will be a decrease in the interionic distance and an increase in the attractive force between the H atoms. If $\psi_{III}$ and $\psi_{IV}$ are the wave functions for the systems (III) and (IV), respectively, then,

$$\psi_{III} = \psi_A(1)\ \psi_A(2) \qquad (13.6)$$

$$\psi_{IV} = \psi_B(1)\ \psi_B(2) \qquad (13.7)$$

The total ionic function ($\psi_{ionic}$) is given as:

$$\psi_{ionic} = \psi_{III} + \psi_{IV} = \psi_A(1)\,\psi_A(2) + \psi_B(1)\,\psi_B(2) \quad (13.8)$$

Thus, the combined wave function for the hydrogen molecule may be written as

$$\psi - \psi_{covalent} + \psi_{ionic}$$
$$= [\psi_A(1)\,\psi_B(2) + \psi_A(2)\,\psi_B(1)] + \lambda[\psi_A(1)\,\psi_A(2) + \psi_B(1)\,\psi_B(2)] \quad (13.9)$$

The constant '$\lambda$' is the measure of the degree to which the ionic structure contributes to the bonding.

With the modifications, the calculated bond energy of the $H_2$ molecule becomes 388 kJ/mol and the value of '$r_o$' becomes 0.75 Å. The differences in value still persist due to some complicated electronic interactions.

### 13.1.2 Pauling and Slater Theory

This theory deals with the directional nature of the bond formed and is an extension of Heitler-London theory. According to this theory:

1. The strength of the covalent bond depends upon the extent of overlapping. The greater the overlapping, the stronger is the bond formed.
2. Only the electrons present in the valence shell of the combining atoms participate in bonding. The inner orbitals from each atom remain undisturbed.
3. A spherically symmetrical orbital (*s*-orbital) does not show any preference in direction, whereas nonspherical orbitals *p*- or *d*- tend to form a bond in the direction of maximum electron density with the orbital.

### 13.1.3 Types of Overlapping

1. ***s*-*s*-Overlapping.** Overlapping between *s*-*s* electrons of two similar or dissimilar atoms is known as *s*-*s* overlapping and forms a single covalent bond. The overlapping of orbitals takes place almost along the same axis and the resulting bond is very strong. The bond formed as a result of overlapping along the same axis is known as *sigma bond*, thus *s*-*s* is a sigma bond. The formation of $H_2$, $Li_2$, and LiH molecules can be explained by this overlapping.
2. ***s*-*p*-Overlapping.** Overlapping between *s*-*p* electrons is known as *s*-*p* overlapping. Ammonia $NH_3$ is formed by *s*-*p* overlapping between three *p*-electrons of nitrogen ($1s^2, 2s^2, 2p^3$) and three *s*-electrons of hydrogen ($1s^1$). The overlapping takes place along the same axis so the resulting bond is strong and is thus a σ-bond (Figure 13.2).

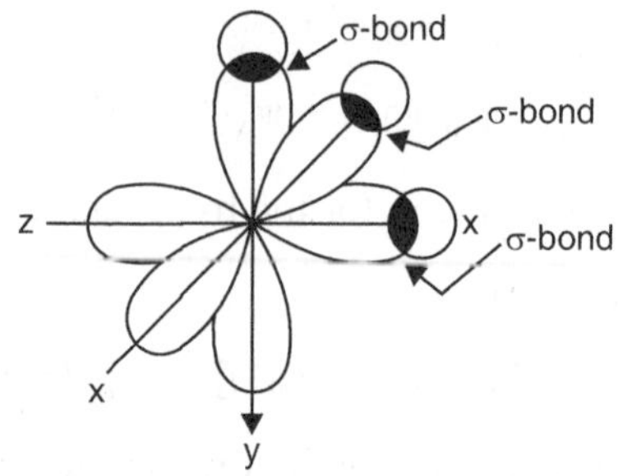


*FIGURE 13.2* **Formation of ammonia molecule.**

3. ***p-p*-Overlapping.** This is produced by overlapping of two *p*-orbitals. For example,

   *Formation of chlorine molecule.* The electronic configuration of chlorine is $1s^2\ 2s^2\ 2p^6\ 3s^2\ 3p_x^2\ 3p_y^2\ 3p_z^1$, with one electron $3p_z$, being unpaired. In the formation of the $Cl_2$ molecule, 3*p*-orbital of one chlorine atom may overlap with 3*p*-orbital of another chlorine atom (Figure 13.3). The bond formed is again the σ-bond as the overlapping takes place along the same axis.

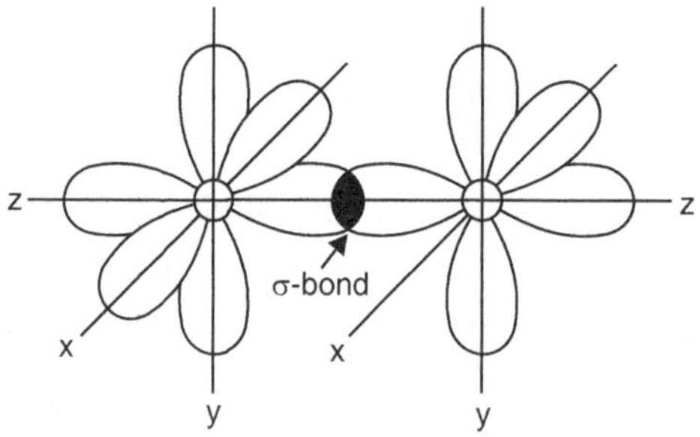


*FIGURE 13.3* **Formation of chlorine molecule by the overlapping of their $3p_z$ orbitals.**

*Formation of $N_2$ molecule.* The electronic configuration of a nitrogen atom is $1s^2\ 2s^2\ 2p_x^1\ 2p_y^1\ 2p_z^1$. Thus, there are three bonding orbitals $2p_x^1 2p_y^1 2p_z^1$ directed along the three axes at right angles to one another. When a nitrogen molecule is formed, $2p_x$ orbitals of two nitrogen atoms overlap along the same axis forming a σ-bond. The other 2*p*-orbitals ($2p_y^1$, $2p_z^1$), each containing an electron, are at right angles to each other and so the electron cloud in these orbitals interacts to give a π-bond by sideways overlapping. This overlapping is partial and the bond produced is a weak bond. Thus, a nitrogen molecule contains one sigma and two π-bonds.

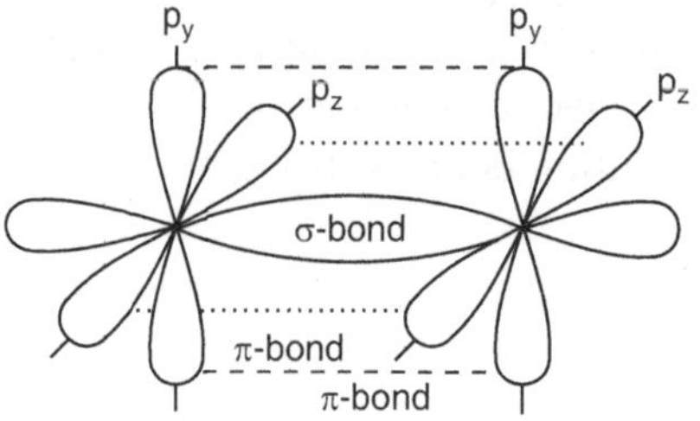


*FIGURE 13.4* **Formation of nitrogen molecule by overlapping of $2p_x$, $2p_y$ and $2p_z$ orbitals.**

## 13.2 SHAPES OF MOLECULES

The different atoms in a molecule have definite geometric arrangements in space around the central atom. This definite relative arrangement of the bonded atoms in a molecule is known as the geometry or shape of the molecule. The shape of a molecule decides many of its physical and chemical properties. For example, the unique properties of a water molecule are due to its angular shape.

Ionic solids generally arrange themselves in a definite three-dimensional pattern known as *lattice*. Covalent bonds, on the other hand, are directional. In a covalent compound, the bonded atoms adopt that particular arrangement in space around the central atom, which keeps them on the average as far apart as possible. This state will have minimum energy and maximum stability.

### 13.2.1 Valence Shell Electron Pair Repulsion (VSEPR) Theory

This theory was proposed by *Sidwick* and *Powell* in 1940 and is primarily based on the repulsive interactions of the electron pairs in the valance shell of the atoms. The main postulates of VSEPR theory are:

1. The geometry of a molecule depends upon the number of valence shell electron pairs (whether bonded or not) around the central atom.
2. Electron pairs tend to repel one another because their electron clouds are negatively charged. Thus, the electron pairs try to stay as far apart as possible so as to acquire a state of minimum energy or maximum stability.
3. Repulsion between the lone pair and lone pair of electrons is different than that between two bond pairs or one lone pair and one bond pair. The repulsive interactions decrease in the order:

   *Lone pair-Lone pair > Lone pair-Bond pair > Bond pair-Bond pair.*

   Thus, the presence of lone pairs in addition to bond pairs, will result in certain distortions in the regular geometries of molecules.
4. Repulsive forces decrease sharply with increasing angle between the electron pairs. They are strong at 90°, weaker at 120°, and weakest at 180°.

This theory is very simple to use. In this theory no distinction is made between *s*- and *p*-electrons and only the number of electron pairs present in the valence shell of the central atom are considered. For example, if there are only two electron pairs around the central atom, the only way to keep them as far apart as possible is to arrange them at an angle of 180° to each other. The molecule in such a case will adopt a linear geometry. Similarly, for three, four, five, and six electron pairs around the central atom, the molecule adopts trigonal, tetrahedral, trigonal bipyramidal, and octahedral geometries respectively. The geometries of the molecules based on VSEPR theory are given in Table 13.1.

TABLE 13.1 Types of Hybridization and Shape of Molecules

| *Total Number of Electrons* | *Type of Hybridization* | *Shape (geometry) of Molecule* | *Bond Angle* | *Example* |
|---|---|---|---|---|
| 2 | *sp* | Linear | 180° | $BeF_2$, $CH \equiv CH$ |
| 3 | $sp^2$ | Plane triangle (trigonal) | 120° | $BCl_3$, $NO_3^-$ |
| 4 | $sp^3$ | Tetrahedral | 109°28′ | $SiF_4$, $SO_4^{2-}$ |
| 4 | $sp^2d$ | Square planar | 90° | $[Ni(CN)_4]^{2-}$ |
| 6 | $sp^3d^2$ | Trigonal bipyramidal | 120° and 90° | $PCl_5$, $XeF_2$ |
| 6 | $sp^3d^2$ | Octahedral | 90° | $SF_6$, $XeF_4$ |
| 7 | $sp^3d^3$ | Pentagonal bipyramidal | 72° and 90° | $IF_7$, $XeF_6$ |

### 13.2.2 Structure of Molecules

1. **Water ($H_2O$) molecule.** In water the central oxygen atom (Z = 8, $1s^2$, $2s^2$, $2p^4$) has two half-filled orbitals ($2p_y^1$, $2p_z^1$) which overlap 1*s*-orbital of two hydrogen atoms to form two single bonds between the oxygen atom and two hydrogen atoms.

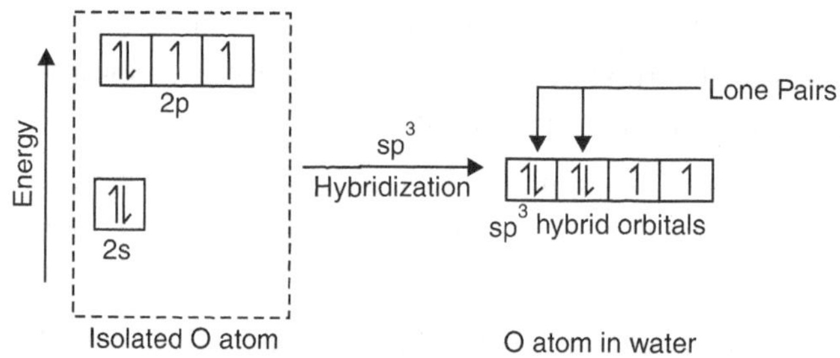


*FIGURE 13.5 (a)* **$sp^3$ Hybridization of orbitals of oxygen atom.**

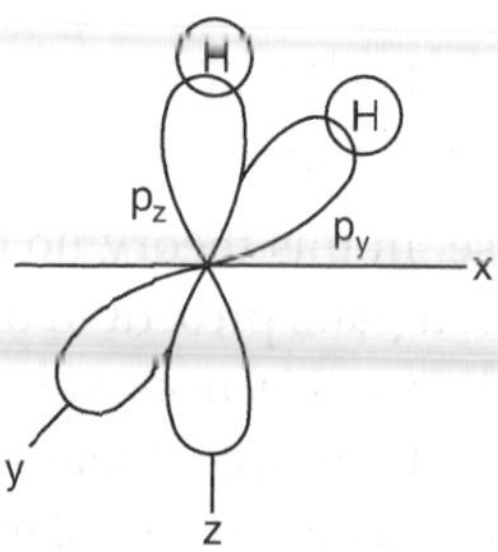


*FIGURE 13.5* **(*b*) Schematic and pictorial overlap of $2p_y$ and $2p_z$ orbitals of O-atom with 1*s* orbitals of two H-atoms.**

The discrepancy between the expected and the experimental values of the bond angle is best explained with the concept of hybridization. The valence orbitals of oxygen atoms hybridize to give four tetrahedrally dispersed $sp^3$ hybrid orbitals. The overlapping of 1*s* orbital of H-atoms with the orbitals of the oxygen atom will occur along $y$ and $z$. Therefore, the bond angle should be 90°, but experiments show that the $H_2O$ molecule is angular and the bond angle is 104.5°. According to VSEPR theory, in order to secure maximum separation between them, the four hybrid orbitals are directed to the corners of a tetrahedron and the predicted bond angle is 109.5°. But due to the presence of two lone pair of electrons in addition to the bond pairs, the force of repulsion between the lone pairs is larger than the force of repulsion between the two bond pairs, thus the two one-pair electrons move away from each other while the two O—H bonds are forced to come closer to each other and the bond angle decreases to 104.5°.

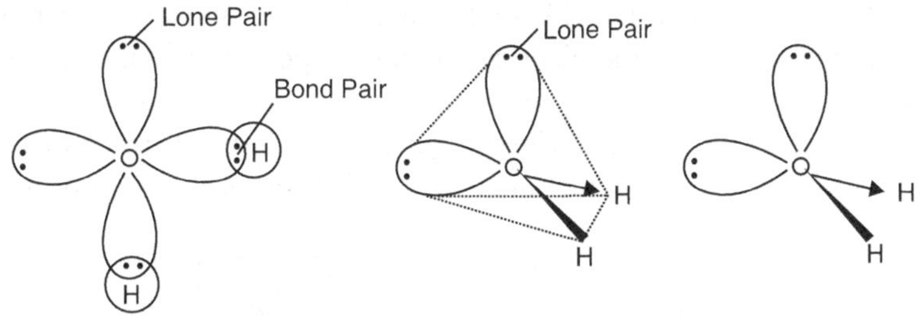


***FIGURE 13.5*** **(*c*) Tetrahedral model of water molecule showing two lone pairs at O-atom and *bp-bp* and *lp-lp* repulsions.**

2. **Ammonia ($NH_3$) molecule.** Nitrogen is the central atom in ammonia and has three half-filled ($2p_x^1$, $2p_y^1$, and $2p_z^1$) orbitals which overlap with 1$s$-orbital of the three H-atoms to form three single bonds. Two electrons do not take part in orbital overlap and remain as

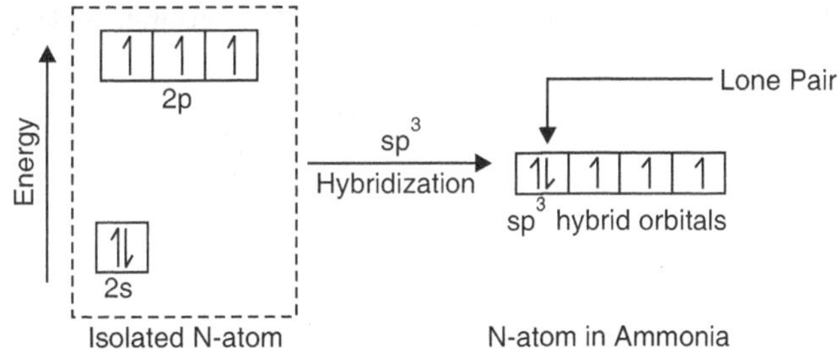


***FIGURE 13.6*** **(*a*) Hybridization of valence orbitals of N-atom.**

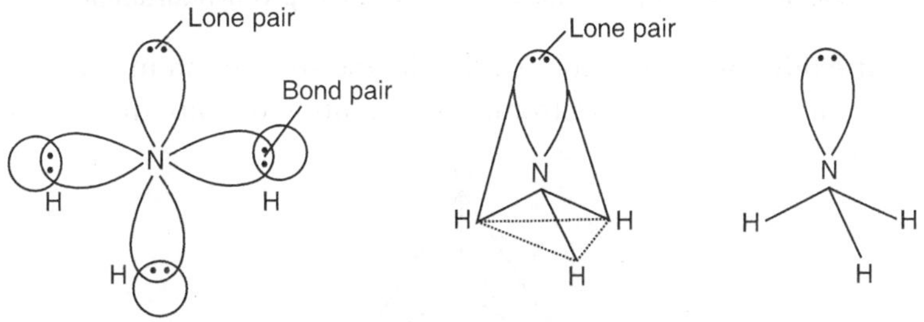


***FIGURE 13.6*** **(*b*) Tetrahedral $NH_3$ molecule with a lone-pair orbital concentrated in the vicinity on N-atom.**

a lone pair. The outer shell then has a share in eight electrons. Four electron pairs give rise to a tetrahedral structure with a bond angle of 109°28′, but in ammonia the bond angle is 107°48′. This slight distortion is due to the presence of lone pair on nitrogen atom. According to VSEPR theory, a lone pair exerts greater repulsion than the bonding pairs do on each other. As a result, the bonds of the $NH_3$ molecule are pushed slightly closer and the bond angle becomes 107°28′.

3. **Phosphorus pentachloride ($PCl_5$).** The electronic structure of phosphorus is $1s^2, 2s^2, 2p^6, 3s^2, 3p^3$. Thus it has five valence electrons which hybridize to form five $sp^3d$ hybrid orbitals (Figure 13.7(*a*)). Three of these five hybrid orbitals lie in a plane inclined at an angle of 120°, while the other two hybrid orbitals are directed above and below the plane in a direction perpendicular to the plane giving the shape of a trigonal bipyramidal. These five orbitals of phosphorus atoms overlap with those of five chlorine atoms, forming $PCl_5$ molecule (Figure 13.7(*b*)). The trigonal bipyramidal structure of $PCl_5$ is not a regular structure, since some bond angles are of 90° and others are 120°. Thus it is highly reactive.

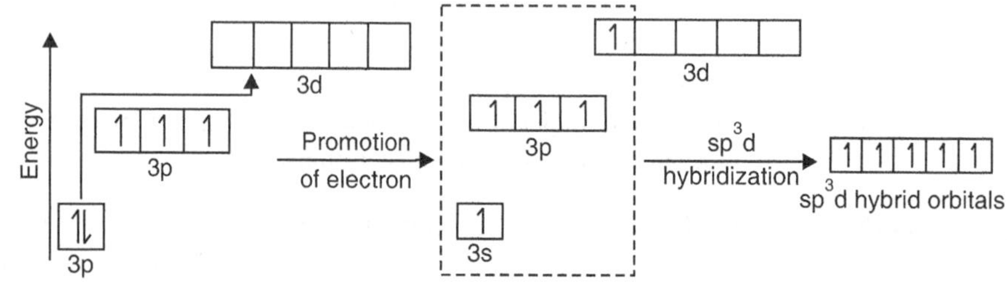


*FIGURE 13.7* **(*a*) Various states of phosphorus atom with its orbitals undergoing $sp^3d$ hybridization.**

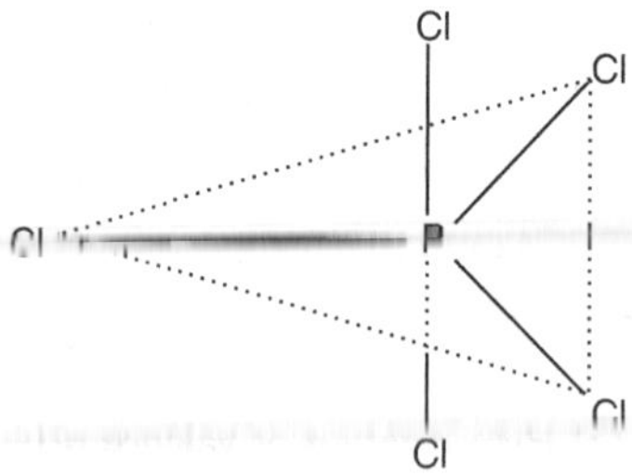


*FIGURE 13.7* **(*b*) $PCl_5$ molecule formed by $sp^3d$ hybridization.**

4. **Sulphur trioxide ($SO_3$).** In sulphur trioxide, the sulphur atom is in an excited state and has six vacant orbitals. The electronic configuration of sulphur is $1s^22s^22p^63s^23p^4$.

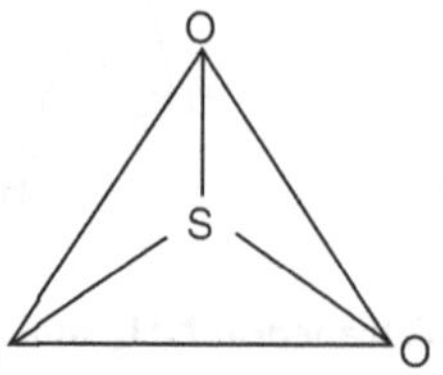


*FIGURE 13.8* **Shape of $SO_3$ Molecule.**

Sulphur atom in ground state

Sulphur atom in 1st excited state

Sulphur atom in 2nd excited state

$S_{excited}$

Sulphur atom having gained 6 electrons to form six bonds to oxygen atom.

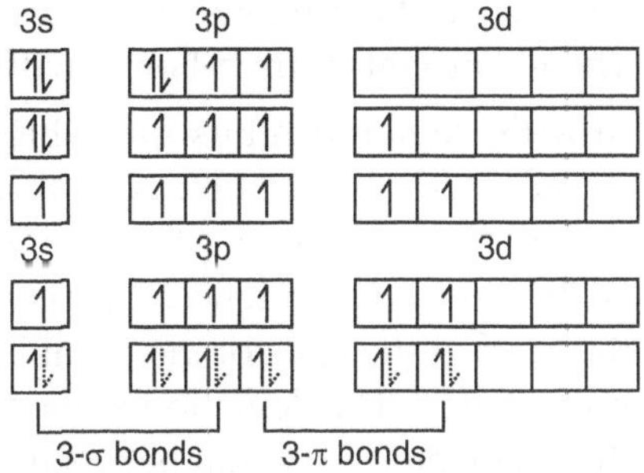


The three π-bonds are ignored in determining the shape of the molecule. The three σ-bonds orbitals are directed toward the corners of an equilateral triangle, and thus the $SO_3$ molecule is a regular plane triangle. The π-bond shortens the bond length but does not affect the shape of the molecule.

5. **$SO_2$ molecule.** In an $SO_2$ molecule, the S-atom is bonded to one O-atom by a double bond and to the other O-atom by a single bond. The double bond arises from $p\pi\text{–}d\pi$ bonding due to lateral overlap of *p*-orbitals of oxygen with *d*-orbitals of sulphur. The sulphur atom also has an unshared electron pair. A double bond in the VSEPR model is counted as a single electron pair. Thus, the S-atom is surrounded by three electron pairs. For maximum repulsion, the three electron pairs are directed to the corners of an equilateral triangle. The predicted bond angle is of 120° (Figure 13.9(*a*)). But with one lone pair repelling the bond pair, the bond angle is actually reduced to 119.5°.

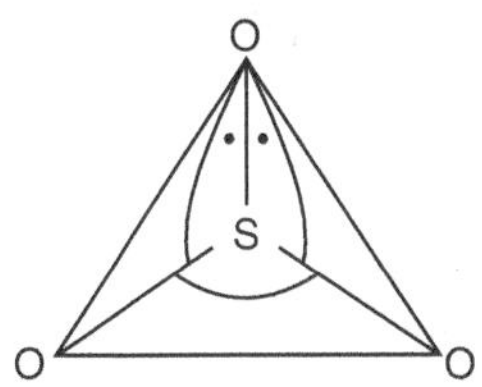


***FIGURE 13.9*** **(*a*) VSEPR model of $SO_2$ molecule.**

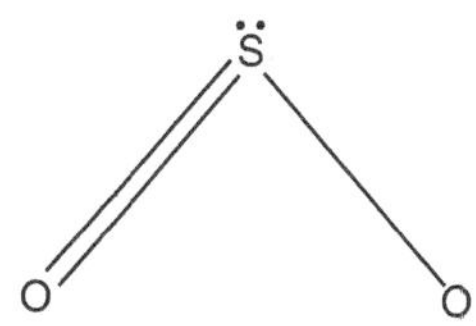


***FIGURE 13.9*** **(*b*) Bent molecule of $SO_2$ with bond angle of 119.5°.**

6. **$XeO_3$ molecule.** The electronic configuration of Xenon is:

Xe-atom in ground state

Xe-atom in their excited state

Xe-atom in $XeO_3$ molecule

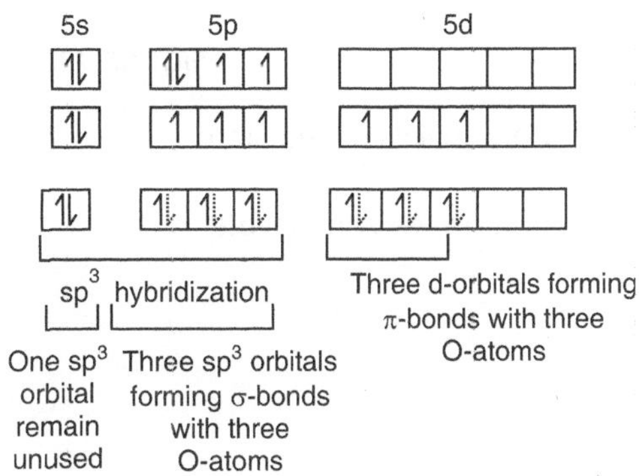

One $sp^3$ orbital remains unused, three $sp^3$ orbitals form σ-bonds with three O-atoms. Three *d*-orbitals form π-bonds with three O-atoms.

In $XeO_3$, xenon exhibits $sp^3$ hybridization. In this way the three 5*d*-orbitals are not included in hybridization. Now three $2p_y$ orbitals of three oxygen atoms form 3σ-bonds with hybrid orbitals and the other three $2p_z$ orbitals of 3 O-atoms form 3π-bonds with 5*d* hybrid orbitals. Hence, the $XeO_3$ molecule has a trigonal pyramidal structure with the atom at the apex of the pyramid (Figure 13.10). In the $XeO_3$ molecule the O–Xe–O bond is 103°. The deviation in the bond angle from a normal tetrahedral bond angle is due to greater repulsion between the lone pair and the bond pair when compared to bond pair-bond pair repulsion.

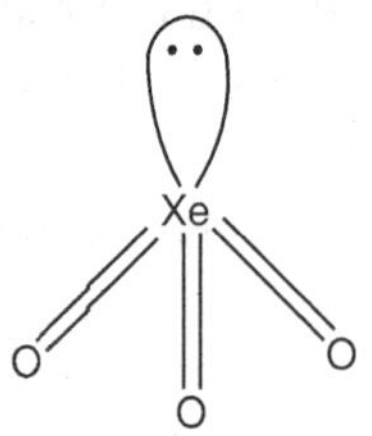

*FIGURE 13.10* **Structure of $XeO_3$.**

7. **$XeO_4$ molecule.** In $XeO_4$, all the eight valence electrons are unpaired and xenon exhibits $sp^3$ hybridization. The four 5*d*-orbitals do not take part in hybridization. Now four electrons of four oxygen atoms form four σ-bonds with hybrid orbitals and the other four electrons of four oxygen atoms form four π-bonds with 5*d*-orbitals. The shape of $XeO_4$ is thus tetrahedral.

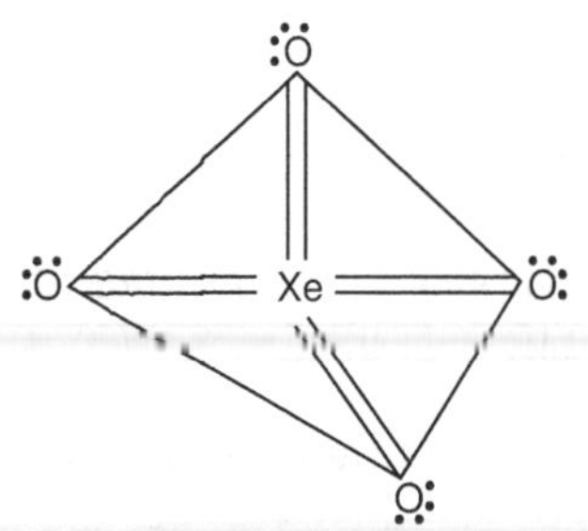

*FIGURE 13.11* **Structure of $XeO_4$.**

Xe-atom in 4th excited state

Xe-atom in $XeO_4$ molecule

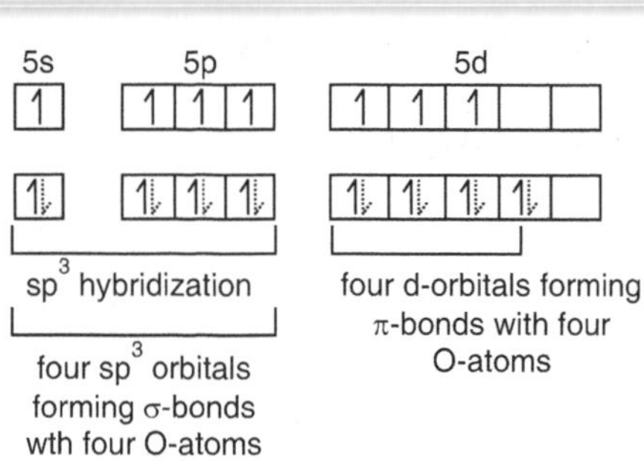

8. **$XeOF_4$ molecule.** The lewis structure (electron dot structure) of $XeOF_4$ molecule suggests that the Xe-atom is surrounded by five σ-bond pairs and one lone pair. This Xe-atom in its excited state [Xe = $5s^2 5p^3 5d^3$] is $sp^3d^2$ hybridized [Xe = $(sp^3d^2)^2$ $(sp^3d^2)^1$ $(sp^3d^2)^1$ $(sp^3d^2)^1$ $5d^1$]. One 5*d*-orbital of the Xe-atom remains excluded from hybridization. This orbital is used in forming Xe-O π-bond. Four Xe-F σ-bonds are formed by the overlap between singly-filled four $sp^3d^2$ hybrid orbitals and four singly-filled $2p_x$ orbitals on four F-atoms (F = $2s^2$, $2p_x^1$ $2p_z^2$) while Xe-O σ-bonds results by the overlap between the singly-filled $sp^3d^2$ hybrid orbital and singly-filled 2*p* orbitals on the O-atom (O = $2s^2$, $2p_x^1$ $2p_y^1$ $2p_z^1$). The Xe-O π-bond results due to the overlap between the singly-filled $5d_{xy}$ orbital on the Xe-atom and the singly-filled 2*p* orbital on the O-atom. Thus the Xe-O π-bond is a (*d*-*p*) π-bond. Although the spatial arrangement of the six electron pairs round the central Xe-atom is octahedral due to the presence of one lone pair of electrons in the axial position of the octahedron, the shape of the $XeOF_4$ molecule is distorted and becomes square pyramidal (Figure 13.12(*b*)) in which four F-atoms are co-planar, since they lie in the same plane (square plane). The lone pair of electrons lies opposite to the O-atom, which is bonded to Xe-atom by a double bond. This arrangement of the lone pair of electrons minimizes the repulsion between electron pairs and a stable molecule is obtained.

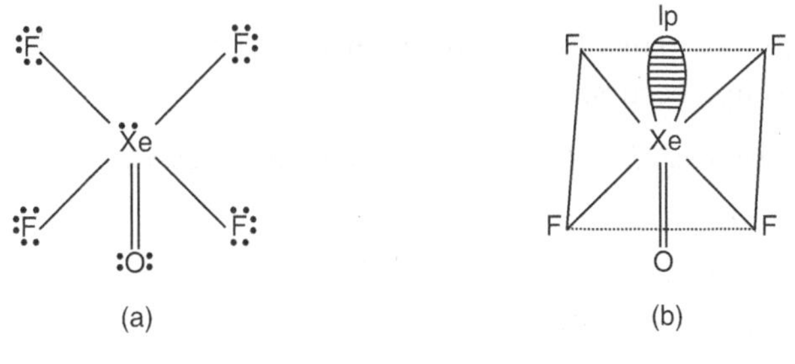


*FIGURE 13.12* **Electron dot and square pyramidal shape of $XeOF_4$ molecule.**

### 13.2.3 Limitations of VSEPR Theory

Following are the limitations of VSEPR theory:

(*i*) It is applicable to simple covalent molecules and does not cover ionic compounds.

(*ii*) It does not make any distinction between the electrons present in the different types of orbitals (*i.e.*, *s*, *p*, *d*, or *f*) of the valence shell.

(*iii*) It does not predict the exact bond angles in molecules such as $H_2O$, $NH_3$, etc., having distorted geometries.

(*iv*) It cannot explain the difference in bond angle in a given series of compounds (*i.e.*, $NH_3$, $PH_3$, $AsH_3$, $SbH_3$) in which the central atom is different but is attached to similar atoms and surrounded by the same number of valence electron pairs in each case.

### 13.2.4 Calculation of Hybridization in a Molecule

The following rules are observed in calculation of type of hybridization in a molecule:

(*i*) Calculate the total number of valence electrons.

(*ii*) Calculate the number of duplet or octet

$$= \frac{\text{Total valence electrons}}{2}$$

$$= \frac{\text{Total valence electrons}}{8}$$

(*iii*) Calculate the number of lone pairs of electrons

$$= \frac{\text{Total number of electrons} - 2 \times \text{Number of duplet}}{2}$$

$$= \frac{\text{Total number of electrons} - 8 \times \text{Number of octet}}{2}$$

(*iv*) Calculate the number of orbitals used = Number of duplet or octet + Number of lone pairs of electrons.

(*v*) If there are no lone pair of electrons, then structure and shape are identical, otherwise distortion in shape occurs.

**Example 1.** *Calculate the type of hybridization geometry of the following molecules:* $H_2O$, $SO_4^{2-}$.

**Solution. (*a*) $H_2O$ molecule**

(*i*) Total valence of electrons = 1 × 2 + 6 = 8

(*ii*) Required number of orbitals = 2

(*iii*) Required electrons for duplet = 2 × 2 = 4

(*iv*) Number of lone pairs of electrons $= \frac{(i) - (iii)}{2} + \frac{8-4}{2} = 2$

(*v*) Number of orbitals = (*ii*) + (*iv*) = 2 + 2 = 4.

Therefore, oxygen exhibits $sp^3$- hybridization and its structure is tetrahedral with distorted bond angle due to the presence of two lone pairs.

**(*b*) $SO_4^{2-}$ ion**

(*i*) Total number of valence electrons = 6 + 6 × 4 + 2 = 32

(*ii*) Required number of orbitals = 4

(*iii*) Required '*e*' for octet = 4 × 8 = 32

(*iv*) Number of lone pairs $= \frac{(i) - (iii)}{2} = \frac{32-32}{2} = 0$

(*v*) Number of orbitals = (*ii*) + (*iv*) = 4 + 0 = 4

∴ The ion involves $sp^3$-hybridization and is a regular tetrahedron.

### 13.2.5 Calculation of Percentage *s*- and *p*-Character

It is given by the equation:

$$\cos\theta = \frac{s}{s-1} = \frac{p-1}{s}, \text{ where } \theta \text{ is the bond angle.}$$

### 13.2.6 Calculation of Number of σ- and π- Bonds

The number of σ- and π-bonds in a covalent molecule are calculated as:

(*i*) The number of π-electrons in a noncyclic molecule which doesn't contain any hydrogen atom is given by $6n + 2 - v$, where $n$ is the total number of atoms of molecule and $v$ is the sum of the valence electrons of the molecule.

(*ii*) The number of π-electrons in a noncyclic molecule containing hydrogen atoms is given by $6n - 6q + 2 - v$, where $q$ is the number of hydrogen atoms of the molecule.

(*iii*) If the molecule is of A × *n* type and none of them is a hydrogen atom, then,

$$\text{Number of } \sigma\text{-bonds} = \frac{\text{Total number of valence electrons}}{8}$$

(*iv*) If the molecule contains a hydrogen atom, then

$$\text{Number of } \sigma\text{-bonds} = \frac{\text{Total number of valence electrons}}{2}$$

**Example 2.** *Calculate the percentage s-orbital and p-orbital character of a hybrid orbital if the bond angle between the hybrid orbital is 105°.*

**Solution.** We know that

$$\cos\theta = \frac{s}{s-1} = \frac{p-1}{s}, \text{ where } \theta \text{ is the bond angle.}$$

$$\therefore \quad \frac{s}{s-1} = \cos 105° = \cos(90+15) = -\sin 15° = -0.2588$$

or $$s = \frac{0.2588}{1.2588} = 0.2056$$

Similarly $$\frac{p-1}{p} = \cos 105° = -0.2588$$

$$p = 0.7944$$

**Example 3.** *Calculate the number of π and σ-bonds in the following molecules.*

| (*i*) $CO_2$ | (*ii*) $CO_3^{2-}$ | (*iii*) $H_2O$ |
|---|---|---|
| (*iv*) $NH_3$ | (*v*) $C_2H_2$ | (*vi*) $C_2H_4$ |

**Solution.** (*i*) $\mathbf{CO_2}$ Number of π-electrons

$$= 6 \times 3 + 2 - (4 + 2 \times 6) = 4$$

$\therefore$ Number of $\pi$-bonds $= \dfrac{4}{2} = 2$

Number of $\sigma$-bonds $= \dfrac{4 + 2 \times 6}{8} = 2$

(*ii*) $\mathbf{CO_3^{2-}}$ Number of $\pi$-electrons $= 6 \times 4 + 2 - (4 + 3 \times 6 + 2) = 2$

$\therefore$ Number of $\pi$-bonds $= 1$

Number of $\sigma$-bonds $= \dfrac{4 + 3 \times 6 + 2}{8} = 3$

(*iii*) $\mathbf{H_2O}$ Number of $\pi$-electrons $= 6 \times 3 - 6 \times 2 + 2 - 8 = 0$

Number of $\pi$-bonds $= 0$

Number of $\sigma$-bonds $= \dfrac{2 + 6}{2} = 4$

(*iv*) $\mathbf{NH_3}$ Number of $\pi$-electrons $= 6 \times 4 - 6 \times 3 + 2 - 8 = 0$

Number of $\sigma$-bonds $= \dfrac{5 + 3}{2} = 4$

(*v*) $\mathbf{C_2H_2}$ Number of $\pi$-electrons $= 6 \times 4 - 6 \times 2 + 2 - (2 \times 4 + 2) = 4$

Number of $\pi$-bonds $= 2$

Number of $\sigma$-bonds $= \dfrac{2 \times 4 + 2 \times 1}{2} = 5$

(*vi*) $\mathbf{C_2H_4}$ Number of $\pi$-electrons $= 6 \times 6 - 6 \times 4 + 2 - (2 \times 4 + 4) = 2$

Number of $\pi$-bonds $= 1$

Number of $\sigma$-bonds $= \dfrac{2 \times 4 + 4 \times 1}{2} = 6$

## 13.3 MOLECULAR ORBITAL THEORY

This theory was proposed by *Hund* and *Mulliken* in 1932. According to this theory, the atomic orbitals combine and form a resultant orbital known as the *molecular orbital* in which the identity of both the atomic orbitals is lost. All the electrons belonging to both atoms move along the entire molecule under the influence of all the nuclei.

Molecular orbitals are formed by the combination of atomic orbitals of bonded atoms. As atomic orbitals can be expressed in terms of wave function ($\psi$) obtained from the solutions of Schrodinger wave equations, similarly wave equations can be written for whole molecules. However, it is very difficult to solve the wave equation for a molecule, so an approximate method known as *Linear Combination of Atomic Orbitals (LCAO)* is used.

### 13.3.1 Linear Combination of Atomic Orbitals (LCAO) Method

According to this method, the molecular orbitals are formed by the linear combination (addition or subtraction) of the atomic orbitals (AOs) of the atoms which form the molecule. Consider that A and B are the two atoms of hydrogen and $\psi_A$ and $\psi_B$ are the wave functions of the electrons of atoms A and B, respectively. When these two hydrogen atoms approach each other, they combine to form molecular orbitals. The linear combination of AOs takes place by addition and subtraction of wave functions of AOs to give two molecular orbitals (MOs) in the following manner:

1. MOs obtained by addition of wave function of atoms

$$\psi\ (MO) = \psi_A + \psi_B \tag{13.10}$$

This MO is called *bonding molecular orbital* (BMO).

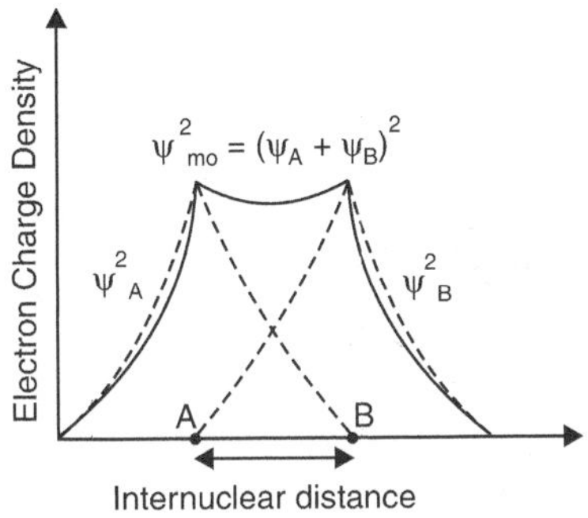


*FIGURE 13.13* **(*a*) Bonding molecular orbital.**

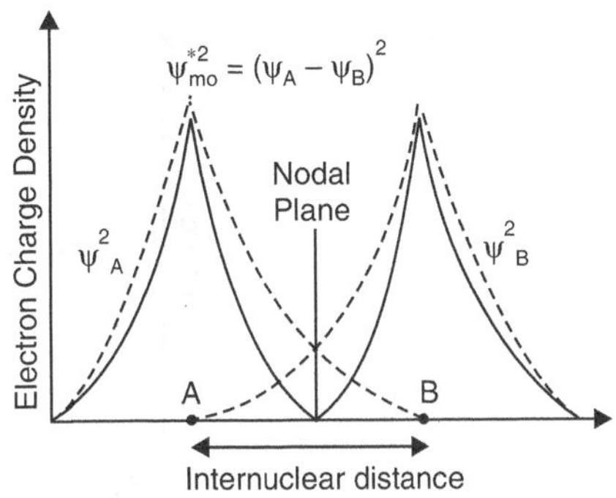


*FIGURE 13.13* **(*b*) Antibonding molecular orbital.**

2. MOs obtained by subtraction of wave functions of atoms

$$\psi^*\ (MO) = \psi_A - \psi_B \tag{13.11}$$

This molecular orbital is called *antibonding molecular orbital* (ABMO).

The pictoral representations of these MOs formations are shown in Figure 13.3. In the case of BMO the orbital electron density is concentrated between the two nuclei of two atoms. In the case of ABMO the nuclei of the two atoms come close to each other but due to the same charge and absence of electron density, the nuclei repel each other.

We know that the square of the wave function ($\psi^2$) is known as the probability of finding the electron, hence, squaring equations (13.10) and (13.11) we get:

$$(\psi_{mo})^2 = (\psi_A + \psi_B)^2$$

$$\psi^2_{mo} = \psi_A{}^2 + \psi_B{}^2 + 2\psi_A\psi_B \tag{13.12}$$

$$(\psi^*_{mo})^2 = \psi_A - \psi_B$$

$$\psi^{*2}_{mo} = \psi_A{}^2 + \psi_B{}^2 - 2\psi_A\psi_B \tag{13.13}$$

Thus $\psi_{mo}^2$ is greater than the sum of the probability densities of individual atoms $(\psi_A{}^2 + \psi_B{}^2)$ by a factor of $2\psi_A\psi_B$, therefore, the probability of finding the electron in BMO is greater than in either of the atomic orbitals ($\psi_A$ or $\psi_B$). Whereas, in ABMO, $\psi^{*2}$ is less than $(\psi_A{}^2 + \psi_B{}^2)$ by a factor of $2\psi_A\psi_B$, so the probability of finding the electron in ABMO is less than in either of the atomic orbitals ($\psi_A$ or $\psi_B$).

### Relative Energies of BMO and ABMO

In the case of BMO, the attraction of both the nuclei for both the electrons is increased, resulting in lowering of energy, while in ABMO the electrons try to go away from the nuclei, resulting in higher energy.

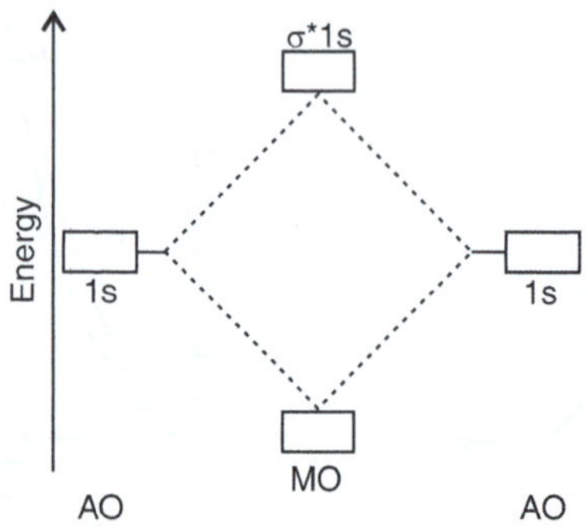


*FIGURE 13.14* **Relative energies of BMO and ABMO.**

**TABLE 13.2 Differences between the Bonding and Antibonding Molecular Orbital**

| *BMO* | *ABMO* |
|---|---|
| 1. BMO is formed by the addition of overlapping atomic orbitals $\psi_{mo} = \psi_A + \psi_B$. | 1. ABMO is formed by the subtraction of overlapping atomic orbitals $\psi^*_{mo} = \psi_A - \psi_B$ |
| 2. It has greater electron density in the region between the two nuclei of bonded atoms. | 2. It has lower electron density in the region between the two nuclei of the atoms. |
| 3. Electrons in BMO contribute to attraction between the two atoms. | 3. Electrons in ABMO contribute to repulsion between the two atoms. |
| 4. It possesses lower energy than isolated atomic orbitals. | 4. It possesses higher energy than isolated atomic orbitals. |
| 5. These are stabilized due to less energy contents. | 5. These are stabilized due to higher energy contents. |

## 13.4 SHAPES OF MOLECULAR ORBITALS

On the basis of symmetry of orbitals, there are two types of molecular orbitals.

1. **Sigma (σ) molecular orbitals.** These are formed by the overlap of AOs that have a cylindrical symmetry around the internuclear axis. Depending upon the magnitude of overlap, there are two types of σ–MOs:

   **A.** *Bonding (σ) MOs,* formed by the overlapping or combination of the following AOs:

   (*i*) (1*s* + 1*s*) or (2*s* + 2*s*) overlap

   (*ii*) (2*s* + $2p_x$) overlap

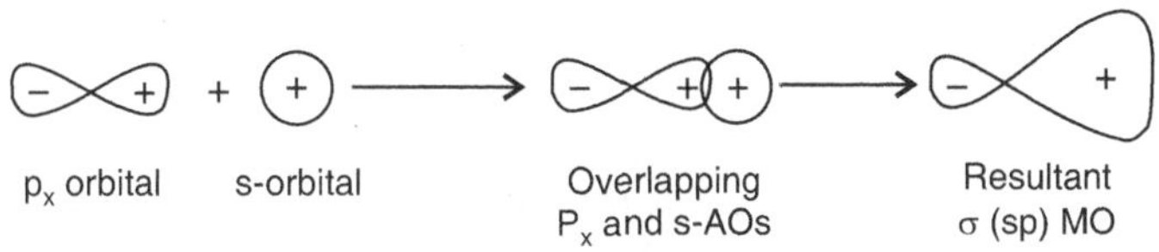


   (*iii*) ($2p_x$ + $2p_x$) overlap

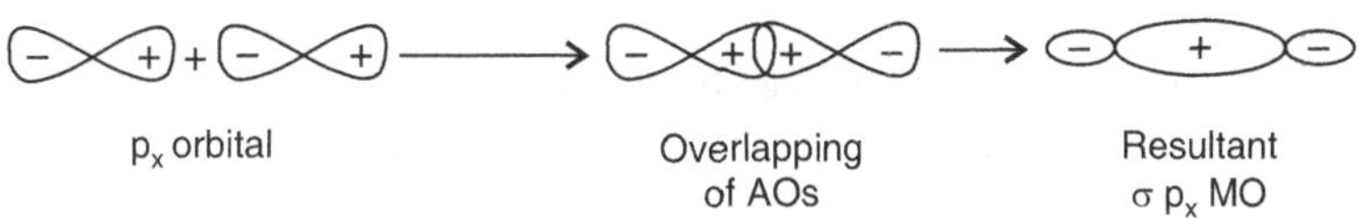


   **B.** *Antibonding (σ*) MO,* formed by the subtraction overlap of the following AOs:

   (*i*) (1*s*-1*s*) overlap

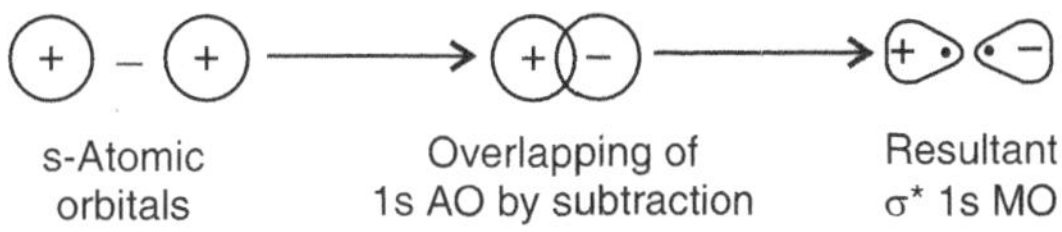


   (*ii*) (*s*-*p*) overlap

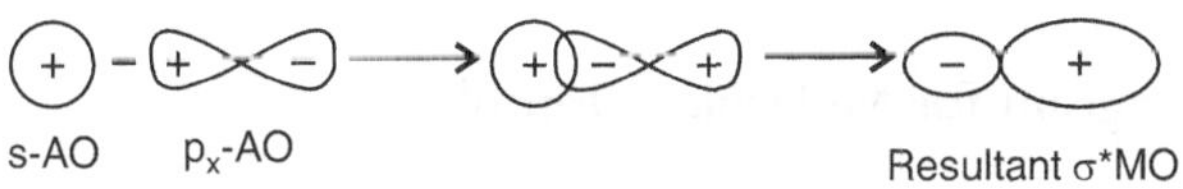


   (*iii*) (*p*-*p*) overlap

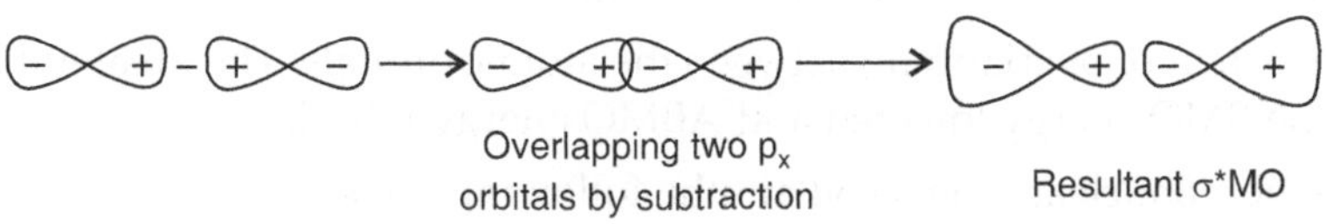


2. **π-molecular orbitals.** When $p_y$ and $p_z$-orbitals of the atoms overlap, they do so laterally on side to side and the resulting molecular orbital formed is called π-orbital. These are of two types:

**A.** *π-Bonding Molecular Orbitals,* formed by the following combination:

(*i*) $p_y + p_y$ (*ii*) $p_z + p_z$

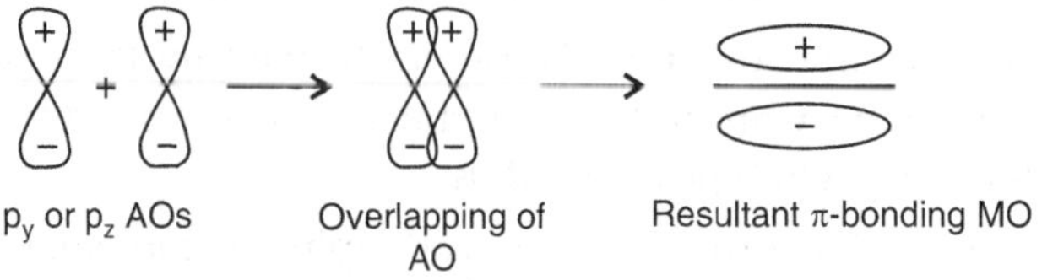


**B.** *π-Antibonding Molecular Orbitals,* formed by the negative overlap in the following manner:

(*i*) $p_y - p_y$ (*ii*) $p_z - p_z$

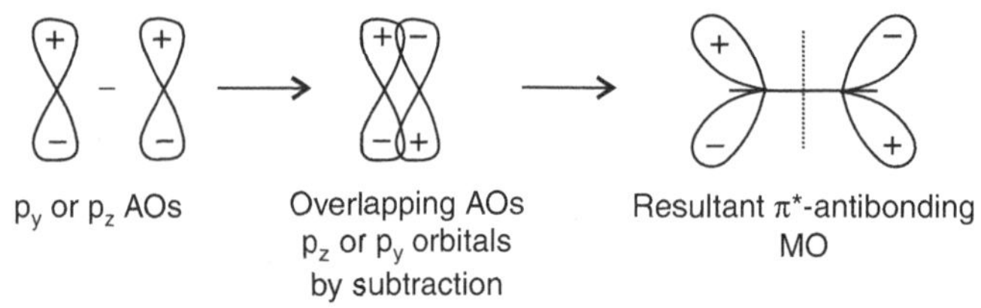


**TABLE 13.3 Differences between σ- and π-Molecular Orbitals**

| *σ-MO* | *π-MO* |
|---|---|
| 1. It is formed by head on overlap of AOs along the internuclear axis. | 1. It is formed by sideways overlapping of AOs perpendicular to the internuclear axis. |
| 2. The overlapping region is maximum. | 2. The overlapping region is minimum. |
| 3. It leads to the formation of a strong bond. | 3. It leads to the formation of a weak bond. |
| 4. The orbital is symmetrical to the rotation about the internuclear axis. | 4. The orbital is not symmetrical to rotation about internuclear axis. |

### 13.4.1 Energy Level Diagram for Molecular Orbital

The relative energies of MOs depend upon:

1. The energies of AOs combining to form MOs.
2. The extent of overlapping between the atomic orbitals. The greater the overlap, the more is the BMO energy lowered and ABMO energy raised.

Simple homonuclear atoms contain the following MOs:

*Bonding MOs:* $\sigma 1s$, $\sigma 2s$, $\sigma 2p_x$, $\pi 2p_y$, $\pi 2p_z$

*Antibonding MOs:* $\sigma*1s, \sigma*2s, \sigma*2p_x, \pi*2p_y, \pi*2p_z$

The energy levels of these MOs have been determined experimentally by spectroscopic methods and these follow the order:

$$\underrightarrow{\sigma 1s, \sigma*1s, \sigma 2s, \sigma*2s, \sigma 2p_x, [\pi 2p_y \equiv \pi 2p_z], [\pi*2p_y \equiv \pi*2p_z], \sigma*2p_x}$$

increasing energy

However, experimental evidences show that the above order is not correct for all the molecules. In second row molecules such as $B_2$, $C_2$, $N_2$ the $\sigma 2p_x$ MO is higher in energy than $\pi 2p_y$ and $\pi 2p_z$ MOs. The order of MOs for these molcules is:

$$\underrightarrow{\sigma 1s, \sigma*1s, \sigma 2s, \sigma*2s, [\pi 2p_y \equiv \pi 2p_z], \sigma 2p_x, [\pi*2p_y \equiv \pi*2p_z], \sigma*2p_x}$$

increasing energy

Thus, the molecular orbital diagram (MOD) for molecules $Li_2$, $Be_2$, $B_2$, $C_2$, $N_2$ is shown in Figure 13.15(*b*) and for molecules $O_2$, $F_2$, $Ne_2$, MOD is shown in Figure 13.15(*a*).

The change in energies of MOs is due to the mixing of orbitals. The combination of 2*s* AOs of two atoms forms $\sigma 2s$ and $\sigma^* 2s$ MOs.

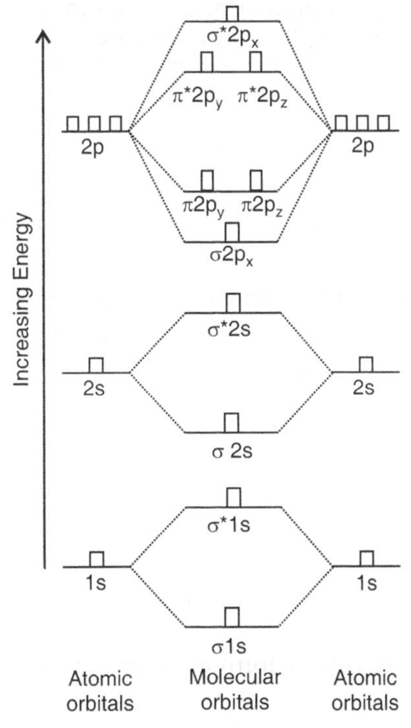


*FIGURE 13.15* **(*a*) $O_2$ type MOD.**

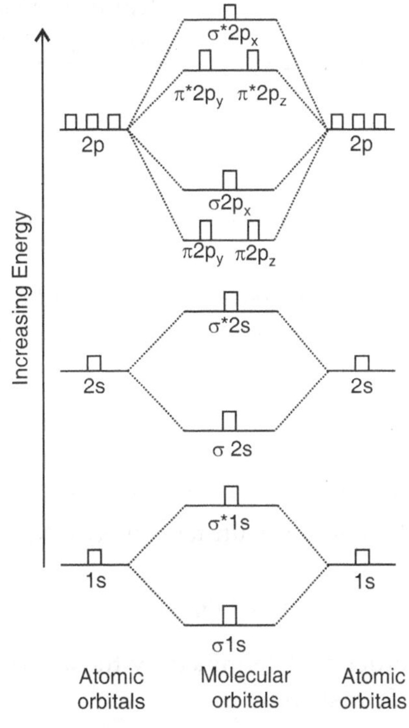


*FIGURE 13.15* **(*b*) $N_2$ type MOD.**

Similarly, the combination of $2p_x$ AOs of two atoms forms $\sigma 2p_x$ and $\sigma^*2p_x$. But if the difference between 2*s* and 2*p* AOs is smaller, then there is a possibility of mixing of 2*s* and $\sigma\ 2p_x$ AOs to acquire *sp* character. Under these conditions, neither of the MOs retain pure *s*-character ($\sigma$ 2*s* and $\sigma$*2*s*) nor *p* character ($\sigma 2p_x$ and $\sigma{*}2p_x$). Due to this, the energies of the four MOs change in such a way that ($\sigma$2*s* and $\sigma$*2*s*) become more stable (attain lower energy level) and $\sigma 2p_x$ and $\sigma{*}2p_x$ becomes less stable (attain higher energy level). Since $\pi$ (pi) orbitals are not involved in mixing, their energy levels remain unchanged. So a new sequence of energy of MOs is obtained for $Li_2$, $B_2$, $C_2$, $N_2$ molecules, since the difference in energy levels between 2*s* and $2p_x$ AOs of these atoms is lower, while that of O and F is very large as shown below:

| ***Element*** | Li | Be | B | C | N | O | F |
|---|---|---|---|---|---|---|---|
| ***kJ/mol*** | 178 | 262 | 449 | 510 | 570 | 1430 | 1970 |

### 13.4.2 Bond Order

Bond order is correctly predicted by the molecular orbital theory. *The number of bonds in a molecule is one-half of the difference of the number of electrons in the bonding molecular orbitals and the number of electrons in the antibonding molecular orbitals.* Mathematically,

$$\text{Bonding order} = \frac{\begin{pmatrix}\text{Number of electrons in} \\ \text{bonding orbitals}\end{pmatrix} - \begin{pmatrix}\text{Number of electrons in} \\ \text{antibonding orbitals}\end{pmatrix}}{2}$$

$$= \frac{N_b - N_a}{2}$$

In common practice, only MOs formed from valence orbitals are considered for determining bond order.

The predicated bond order is found to be in conformity with other methods of determining it. Thus, molecular orbital theory helps in establishing whether the existence of a bond is feasible or not. For instance, the formation of $He_2$ molecule is completely ruled out as the bond order predicated is zero.

In general

(*i*) A molecule is stable if $N_b > N_a$

(*ii*) A molecule is unstable if $N_b < N_a$

**Information given by Bond Order**

The bond order as determined by the above formula is very helpful as it gives us the following valuable information:

(*i*) *Stability of the molecule/ion.* A molecule/ion is stable if $N_b > N_a$

(*ii*) *Bond dissociation energy.* It depends upon the bond order. The greater the bond order, the greater is the bond dissociation energy.

(*iii*) *Bond length.* Bond order is inversely proportional to the bond length. The higher the bond order, the smaller the bond length.

(*iv*) *Magnetic properties.* The presence of unpaired electrons of molecular orbitals of a species makes it paramagnetic in nature. The greater the number of unpaired electrons, the more will be its paramagnetic character. On the other hand, if there are no unpaired electrons in MOs, the species will be diamagnetic in nature. The magnetic moment shown by any atom, molecule, or ion is calculated by the formula $\mu = \sqrt{n(n+2)}$ where, $\mu$ = magnetic moment, $n$ = number of unpaired electron.

### 13.4.3 Homonuclear Diatomic Molecules

1. **Hydrogen ($H_2$) molecule.** The hydrogen molecule is formed from $1s^1$ atomic orbitals of two hydrogen atoms. They give rise to two molecular orbitals $\sigma$ (1s) and $\sigma$*(1 s). The molecule has two electrons to be accommodated in these two molecular orbitals and are filled as desired by Aufbau order of filling. Both these electrons go to the lower energy bonding molecular orbital and the antibonding orbital remains vacant (Figure 13.16). The electronic configuration of the molecule is represented by the equation:

$$2H\,(1s^1) \longrightarrow H_2[\sigma\,(1s^2)]$$

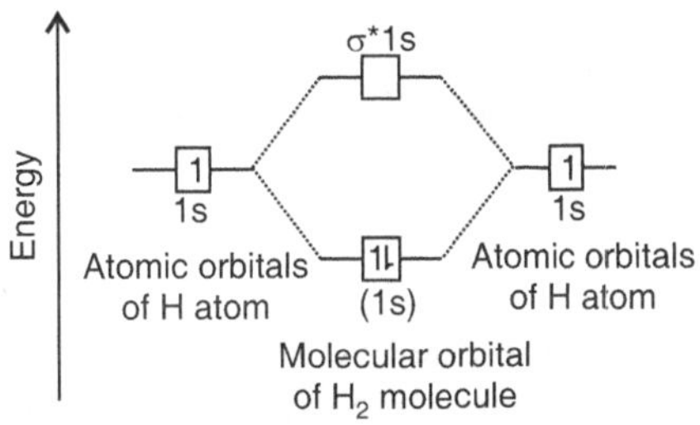


*FIGURE 13.16*

The number of bonds in the molecule, accordingly, is $\frac{1}{2}(2-0) = 1$. Thus two hydrogen atoms are bonded through only one bond in the molecule.

2. **Helium ($He_2$) molecule.** Helium atoms have the configuration $1s^2$. For the formation of $He_2$, the molecular orbitals required are $\sigma(1s)$ and $\sigma^*(1s)$ which have to accommodate four electrons (two each contributed by the atoms). Each one of these molecular orbitals would accommodate two electrons as shown below. The molecule of helium is represented by the equation

$$2He(1s^2) \longrightarrow He_2\,[\sigma(1s^2)\,\sigma^*(1s^2)].$$

There are two electrons in each of the bonding and antibonding orbitals (Figure 13.17) giving zero bonds. $(2-2) = 0$, *i.e.*, this molecule does not exist.

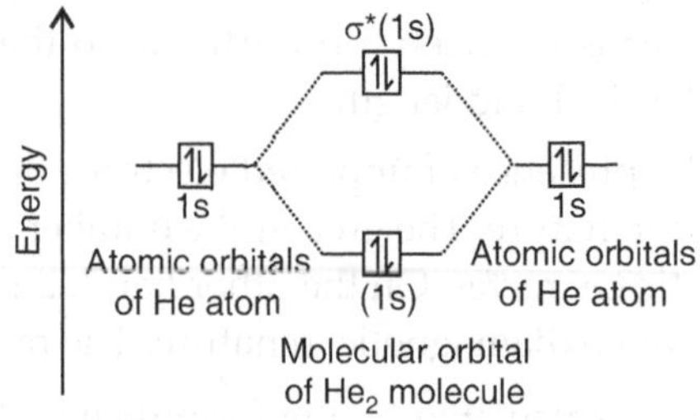


*FIGURE 13.17*

3. **Lithium ($Li_2$) molecule.** The lithium atom has one electron in the valence shell $2s^1$, the other electrons are nonbonding. Two such atoms give a molecule having two molecular orbitals $\sigma(2s)$ and $\sigma*(2s)$ and only two electrons are to be filled in them. These are accommodated in the molecular bonding orbital while the antibonding molecular orbital remains vacant. The number of bonds $= \frac{(2-0)}{2} = 1$. Thus, the atoms form a molecule involving a single covalent bond. The bond energy is about 25 kcal/mole which is much less than of $H_2$ molecule (103) kcal/mole.
4. **Beryllium ($Be_2$) molecule.** Diatomic beryllium molecule is not formed because two valence electrons present give two molecular orbitals $\sigma(2s)$ and $\sigma*(2s)$, each having two electrons. This formation can be represented as

$$2Be(1s^2\ 2s^2) \longrightarrow Be_2\ [\sigma(1s^2)\sigma*(1s^2)\ \sigma(2s^2)\ \sigma*(2s^2)]$$

$$\therefore \quad \text{Number of bonds} = \frac{1}{2}(2-2) = 0.$$

5. **Nitrogen ($N_2$) molecule.** Each of the two N-atoms $1s^2, 2s^2, 2p_x^{\ 1}, 2p_y^{\ 1}, 2p_z^{\ 1}$ contributes five valence electrons and four atomic orbitals toward the molecule. The molecule would have

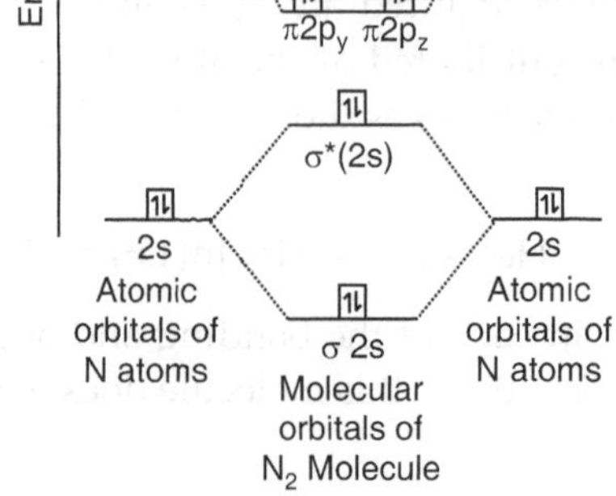


*FIGURE 13.18* **Molecular orbital diagram of nitrogen molecule.**

eight molecular orbitals and ten electrons to be accommodated in them. The electronic configuration in the molecule is represented in Figure 13.18.

There are eight electrons in the bonding molecular orbitals and only two in the antibonding orbitals. This gives the number of bonds $\frac{1}{2}(8-2)=3$. The two N-atoms are bonded with three covalent bonds, *i.e.*, a triple bond. The $N_2$ molecule is diamagnetic as it has no unpaired electron.

6. **$N_2^+$ cation.** $N_2^+$ ion shows that it is formed by the loss of one electron from the $N_2$ molecule. Thus the MO electronic configuration of $N_2^+$ ion is

$$N_2^+ = KK\,(\sigma\, 2s)^2\,(\sigma{*}2s)^2\,(\pi 2p_y)^2\,(\pi 2p_z)^2\,(\sigma 2p_x)^1$$

The characteristic properties of the cation are:

(*a*) Bond order $= \frac{1}{2}\,[N_b - N_a] = \frac{1}{2}\,[7-2] = 2\frac{1}{2}$.

(*b*) The bond order of the cation is smaller than that of the $N_2$ molecule. Therefore, it is less stable than the $N_2$ molecule.

(*c*) It has one unpaired electron in its MO electronic configuration. So, it is paramagnetic in nature.

7. **$N_2^-$ anion.** It is formed when the $N_2$ molecule gains one electron. This additional electron will enter the empty $\pi{*}2p_y$ or $\pi{*}2p_z$ orbital. The MO electronic configuration for $N_2^-$ anion is

$$N_2^- = KK(\sigma\, 2s)^2\,(\sigma{*}2s)^2\,(\pi 2p_y)^2\,(\pi 2p_z)^2\,(\sigma 2p_x)^2\,(\pi{*}2p_y)^1$$

$$\text{Bond order} = \frac{1}{2}[N_b - N_a] = \frac{1}{2}\,[8-3] = 2.5$$

So this ion is less stable than the $N_2$ molecule. Further, the $N_2^-$ anion has one unpaired electron ($\pi{*}2p_y$). So, it is paramagnetic in behavior.

8. **$N_2^{2-}$ anion.** This anion is formed when two electrons are added to the $N_2$ molecule. These two additional electrons will be taken, one each by two nitrogen atoms. According to Hund's rule, these two electrons will occupy the empty $\pi{*}2p_y$ and $\pi{*}2p_z$ orbitals. So the MO electronic configuration is

$$N_2^{2-} = KK\,(\sigma\, 2s)^2\,(\sigma{*}2s)^2\,(\pi 2p_y)^2\,(\pi 2p_z)^2\,(\sigma 2p_x)^2\,(\pi{*}2p_y)^1\,(\pi{*}2p_z)^1$$

The consequences of this MO electronic configuration are:

(*a*) *Bond order* $= \frac{1}{2}[N_b - N_a] = \frac{1}{2}\,[8-4] = 2$

(*b*) *Stability.* Since the bond order for $N_2^{2-}$ is 2, which is less than $N_2^+$ and $N_2^-$ (bond order = 2.5 each) as well as $N_2$ molecule (bond order = 3), its stability is least.

Thus, among the various species of nitrogen the stability order is as:

$$N_2^{2-} < N_2^- \equiv N_2^+ < N_2$$

(*c*) *Bond length.* It is inversely proportional to the bond order so the relative bond length will be:

$$N_2^{2-} > N_2^- \equiv N_2^+ > N_2$$

9. **Oxygen ($O_2$) molecule.** The two participating oxygen atoms ($1s^2$, $2s^2$, $2p_x^{\,2}\ 2p_y^{\,1}\ 2p_z^{\,1}$) contribute a total of 12 valence electrons and 2 × 4 = 8 atomic orbital toward the oxygen molecule. There are eight molecular orbitals and 12 electrons are accommodated as shown in Figure 13.19. The bond order is calculated if $BO = \frac{8-4}{2} = 2$, *i.e.*, a double bond. It has two unpaired electrons so it is paramagnetic in nature.

10. **$O_2^+$ cation.** This cation is formed when one electron is ionized from the oxygen molecule. From the MO diagram of the $O_2$ molecule, if one electron is ionized from ($\pi{*}2p_y$) or ($\pi{*}2p_z$), then the MO electronic configuration is

$$O_2^+ = KK(\sigma 2s)^2\ (\sigma^* 2s)^2\ (\sigma 2p_x)^2\ (\pi 2p_y)^2\ (\pi 2p_z)^2\ (\pi^* 2p_y)^1$$

(*a*) *Bond order* $= \frac{1}{2}(N_b - N_a) = \frac{1}{2}\,[8-3] = 2\frac{1}{2}$

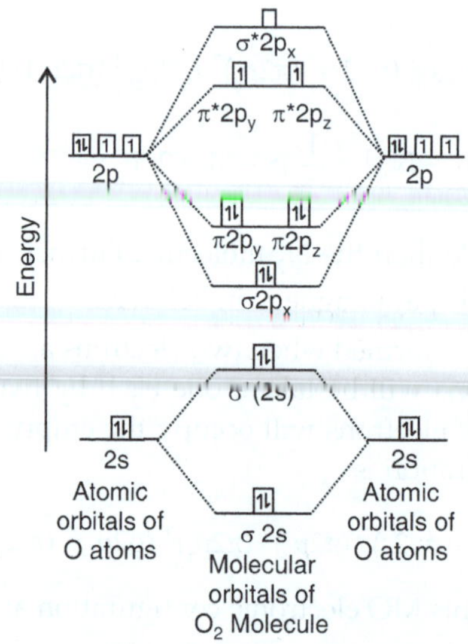


*FIGURE 13.19* **Molecular orbitals of $O_2$ molecule, having one full, two half-full and one vacant antibonding orbitals; and all the bonding molecular orbitals are full.**

(*b*) *Stability.* The bond order of the $O_2^+$ cation is higher than that of the $O_2$ molecule. Therefore, its stability will be more than that of the $O_2$ molecule [*i.e.*, $O_2^+ > O_2$].

(*c*) *Bond length.* Bond length being inversely proportional to bond order, it decreases with an increase of bond order. Therefore, $O_2^+$ has a lower bond length than the $O_2$ molecule.

(*d*) *Magnetic parameter.* The MO configuration has one unpaired electron ($\pi*2p_y$). This is paramagnetic in nature.

**11. $O_2^-$ ion (Superoxide ion).** This ion is formed when the $O_2$ molecule gains one electron. This new electron pairs with either of the two half-filled antibonding orbitals ($\pi*2p_y$) or ($\pi*2p_z$). The MO electronic configuration is

$$O_2^- = KK\ (\sigma 2s)^2\ (\sigma^* 2s)^2\ (\sigma\ 2p_x)^2\ (\pi 2p_y)^2\ (\pi 2p_x)^2\ (\pi* 2p_y)^2\ (\pi* 2p_z)^1$$

(*a*) *Bond order* $= \frac{1}{2}[N_b - N_a] = \frac{1}{2}[8-5] = 1\frac{1}{2}$

(*b*) *Stability.* As the bond order of the superoxide ion is less than that of the molecule, it is less stable than the $O_2$ molecule ($O_2^- < O_2$).

The higher the bond order, the shorter is the bond length. Thus, the bond length of the $O_2^-$ ion is more than the $O_2$ molecule ($O_2^- > O_2$).

(*c*) *Bond length.* If the bond length is inversely proportional to the bond order, its bond length will be greater than the $O_2$ molecule, *i.e.*, $O_2^{2-} > O_2$.

**12. Fluorine molecule ($F_2$).** The two participating F atoms ($1s^2 2s^2 2p_x^2 2p_y^2 2p_z^1$) contribute a total of 14 electrons. There are eight molecular orbitals formed. In these MOs, 14 electrons are accommodated. The electronics configuration of the $F_2$ molecule is

$$F_2 = KK\ \sigma(2s)^2\ \sigma*(2s)^2\ \sigma(2p_x)^2\ \pi(2p_y)^2\ \pi(2p_z)^2 \pi*(2p_y)^2 \pi*(2p_z)^2$$

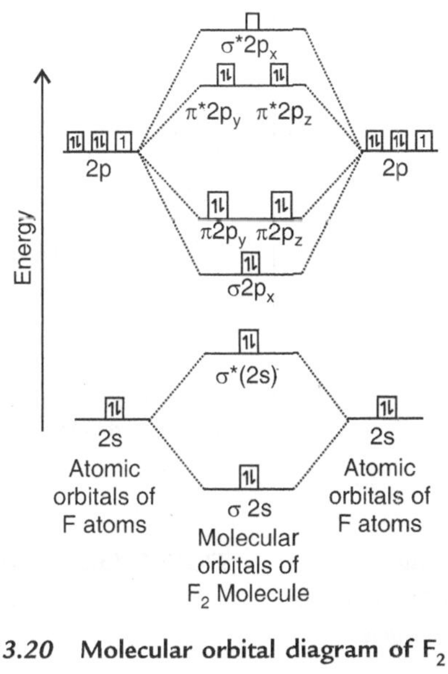


*FIGURE 13.20* **Molecular orbital diagram of $F_2$ molecule.**

The bond order in $F_2$ molecule is

$$= \frac{N_b - N_a}{2} = \frac{8-6}{2} = 1$$

Since the $F_2$ molecule has no unpaired electron, it is diamagnetic in nature. The diagram of the $F_2$ molecule is shown in Figure 13.20.

### 13.4.4 Heteronuclear Diatomic Molecules

In heteronuclear diatomic molecules, the electronegativity of the two atoms is different. Therefore, the MOD will not be symmetrical like homonuclear diatomic molecules. But the method of electron distribution is the same. In heteronuclear diatomic molecules, the electrons in the bonding MOs are nearer to the more electronegative atom, whereas the electrons in the antibonding MOs are nearer to the less electronegative atom. The formation of $\psi_b$ and $\psi_a$ molecular orbitals from two AOs having different energies is shown in Figure 13.21.

1. **Carbon monoxide (CO) molecule.** The electronic configurations of participating C and O atoms are C = ($1s^2\ 2s^2\ 2p^2$) and O = ($1s^2\ 2s^2\ 2p^4$).

   The total number of valence electrons is 10 and the electronic configuration of the CO molecule may be written as

$$CO = KK\ \sigma(2s)^2\ \sigma^*(2s)^2\ \pi(2p_y)^2\pi(2p_z)^2\ \sigma(2p_x)^2$$

$$\text{The bond order} = \frac{N_b - N_a}{2} = \frac{1}{2}(8-2) = 3$$

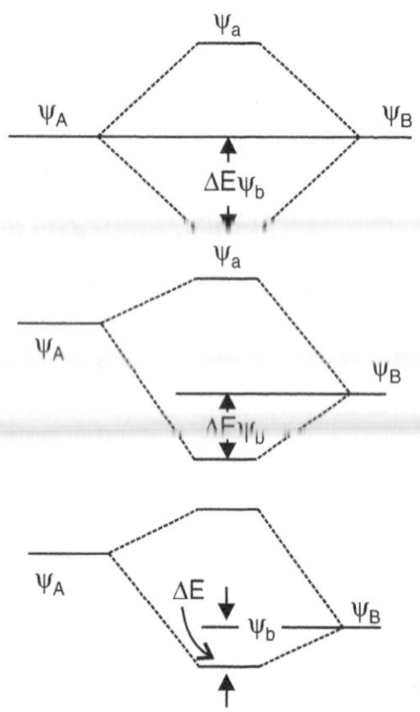


*FIGURE 13.21* **Formation of bonding and antibonding MO from two AOs of different energies.**

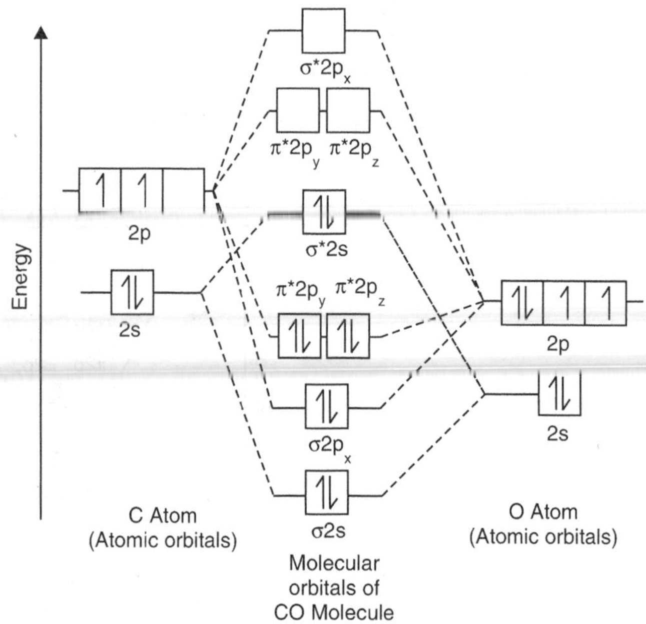


*FIGURE 13.22* **Molecular orbital diagram of CO Molecule.**

This shows that the carbon and oxygen atoms in the CO are bonded to each other by a triple bond (one σ- and two π- bonds). There is no unpaired electron in the CO molecule and hence it is diamagnetic in nature. The molecular orbital diagram for the CO molecule is shown in Figure 13.22.

2. **CN molecule.** The electronic configuration of participating C and N atoms is: C = ($1s^2$ $2s^2$ $2p^2$) and N = ($1s^2$ $2s^2$ $2p^3$).

   The total number of valence electrons is 9 and the electronic configuration of the CN molecule may be written as:

$$CN = KK\ \sigma(2s)^2\ \sigma*(2s)^2\ \sigma(2p_x)^2\ \pi(2p_y)^2\ \pi(2p_z)^1$$

   The bond order $= \frac{1}{2}(N_b - N_a) = \frac{1}{2}(7 - 2) = 2\frac{1}{2}.$

   Since the bond order in the CN molecule is lesser than CO molecule, the former is less stable than the latter. Also, there is one unpaired electron in the CN molecule. It is paramagnetic in nature. The molecular orbital diagram of CN molecules is shown in Figure 13.23.

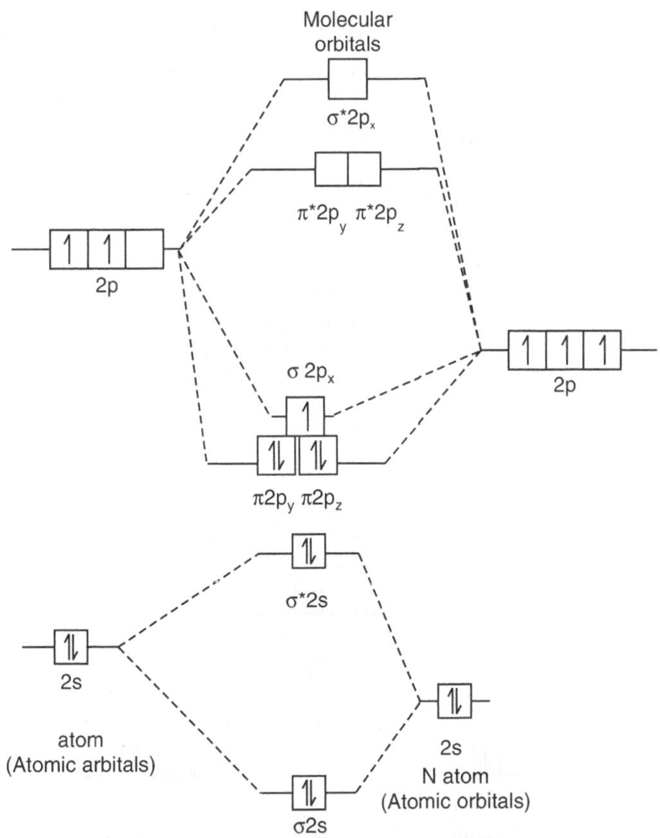


*FIGURE 13.23* **Molecular orbital diagram of CN molecule.**

3. **Nitric oxide (NO) molecule.** The electronic configuration of participating N and O atoms are N = ($1s^2 2s^2$ $2p^3$) and O = ($1s^2$ $2s^2$ $2p^4$).

The total number of valence electrons is 11 and the electronic configuration of the NO molecule may be written as

$$NO = KK\ \sigma(2s)^2\ \sigma*(2s)^2\ \sigma(2p_x)^2\ \pi(2p_y)^2\ \pi(2p_z)^2\pi*(2p_y)^1$$

The bond order $= \frac{1}{2}(N_b - N_a) = \frac{1}{2}(8-3) = 2\frac{1}{2}.$

It is clear that one σ-bond and two π-bonds with an unpaired electron in an antibonding $\pi*(2p_y)^1$ molecular orbital are formed. This molecule is less stable than the $N_2$ molecule (bond order = 3). Due to the presence of one unpaired electron in the NO molecule, it is paramagnetic in nature. The MO diagram for the NO molecule is shown in Figure 13.24.

The electronic configurations bond order and the magnetic character of $NO^+$ and $NO^-$ ions are summarized in Table 13.4.

**TABLE 13.4 Characteristics of $NO^+$ and $NO^-$ ions**

| *Species* | *Number of Electrons* | *Electronic Configuration* | *Bond Order* | *Magnetic Character* |
|---|---|---|---|---|
| $NO^+$ | 14 | $KK\sigma(2s)^2\ \sigma*(2s)^2\ \sigma(2p_x)^2\ \pi(2p_y)^2\ \pi(2p_z)^2$ | $\frac{8-2}{2} = 3$ | Diamagnetic |
| $NO^-$ | 16 | $KK\sigma(2s)^2\ \sigma*(2s)^2\sigma(2p_x)^2\pi(2p_y)^2\ \pi(2p_z)^2\pi*(2p_y)^1\ \pi*(2p_z)^1$ | $\frac{8-4}{2} = 2$ | Paramagnetic |

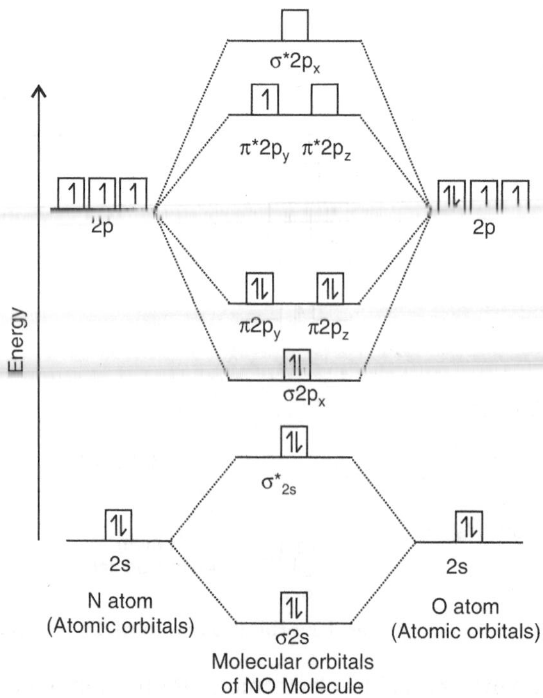


*FIGURE 13.24* **Molecular orbital diagram of NO molecule.**

**4. Hydrogen fluoride (HF) molecule.** The electronic structure of participating H and F atoms H = $1s^1$ and F = $1s^2 2s^2 2p^5$.

Thus, in HF molecule, there are ten electrons that need to be accomodated in the molecular orbitals of HF molecule. The formation of HF molecule occurs by linear combination of 1*s* AO of H-atom with $2p_x$ orbital of F-atom. Combination with 1*s* and 2*s* AOs of F-atom is ruled out because their energies are too low. If overlap occured between the fluorine $2p_y$ and $2p_z$ orbitals it would be non-bonding. This is because the positive lobe of H-atom will overlap equally with the positive and negative labes of $2p_y$ and $2p_z$ AOs of F-atom. Thus, the only effective overlap is with the fluorine $2p_x$ orbital.

The combination of H($1s^1$) and F($2p_{x^1}$) gives both bonding ($\sigma_{spx}$) and antibonding ($\sigma^*_{spx}$) orbitals and the two electrons occupy the bonding molecular orbital leaving the antibonding molecular orbital empty. The MO diagram of HF molecule is shown in Fig. 13.24*a* and the electronic configuration of HF molecule is $1s^2\ 2s^2\ \sigma^2_{spx}$. $[2p_{y^2} = 2p_{z^2}]\ \sigma^*_{spx}$. The bond order of HF molecule can be calculated as

$$\frac{1}{2}(N_b - N_a) = \frac{1}{2}(2 - 0) = 1$$

Since all the MOs (bonding or non-bonding) are completed by filled, so HF molecule is diamagnetic.

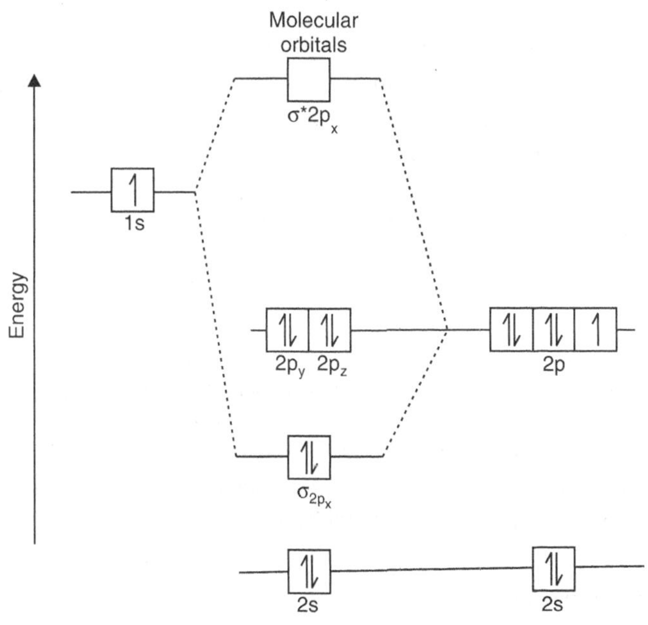


*FIGURE 13.24(a)* **Molecular orbital diagram of HF.**

## 13.5 THEORIES OF METALLIC BONDING

More than 80 elements in the periodic table are metals and possess characteristic properties such as high thermal and electrical conductivity, bright luster, malleability, ductility, high melting and boiling points, and crystallize with high coordination numbers of 8, 12, or 14. These properties of metals cannot be explained by normal ionic or covalent bonding. The ionic bond is formed between atoms of different electronegativities, while metals cannot form 12 or 14 covalent bonds due to an insufficient number of valence electrons and van der Waal forces are too weak to account for the high melting and boiling point of metals. Thus, metals are neither held by ionic nor by covalent bonds. These properties are explained by a special type of bond called *metallic bond*. It may be defined as:

1. The force that binds a metal ion to a number of electrons within its sphere of influence.
2. A bond which is formed between electropositive elements.
3. The attractive force which holds the atoms of two or more metals together in a metal crystal or in an alloy.

The following theories have been put forth to explain bonding in metals:

### 13.5.1 Electron Sea Theory

*Drude* (1900) proposed metal as a lattice with electrons moving through it in the same way as the molecules of gases. *Lorentz* (1923) suggested that metals comprised of lattice of rigid spheres (positive ions) embedded in a gas of free valence electrons which could move in the interstices. The important features of this theory are:

1. The positively charged kernels of metal atoms are arranged in a regular fashion in a metallic lattice.
2. Each kernel in metallic lattice is surrounded by the loosely held valence electrons.
3. The valence electrons of each atom are loosely held to its kernel and there is a complete freedom of electrons in the metallic lattice to move. The valence electrons in metals are thus regarded as mobile electrons.

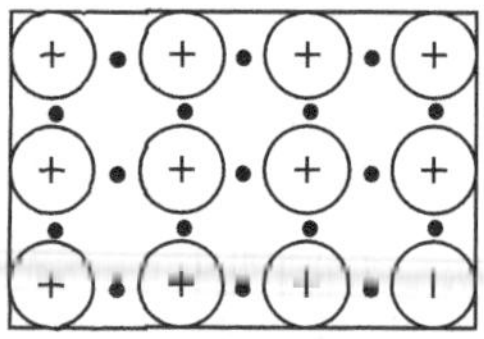

*FIGURE 13.25* **Representation of sea of mobile electrons.**

Thus, according to this theory, *metallic solids may be supposed to be collection of positive atomic cores immersed in a fluid of moving electrons or a sea of mobile electrons*. Thus, the force that binds a metal ion to the mobile electrons is called *metallic bond*.

#### Explanation of Metallic Properties

1. *Metallic luster.* As the beam of light falls on the surface of a metal, the mobile electrons move to some excited state. On coming back from the excited state to the ground state in one or more jump, the light of all wavelengths in the visible region is emitted and thus it appears as if light is being reflected.

2. *Electrical conductivity.* This property is also due to the presence of mobile electrons. When a potential difference is applied across the metal sheet, the free mobile electrons start moving toward the positive electrode constituting electric current through the metal. (Figure 13.26).

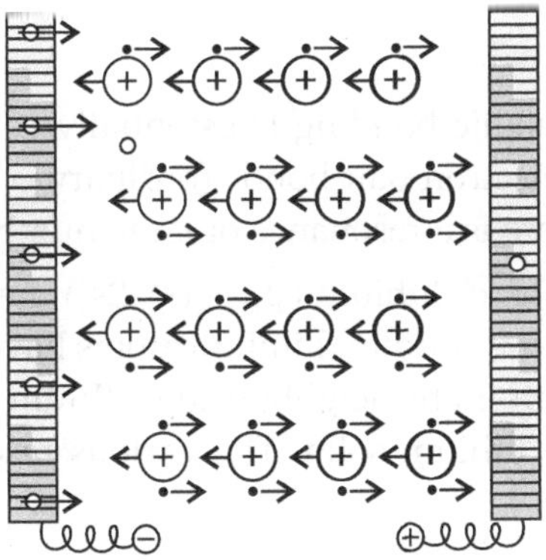

*FIGURE 13.26* **Electrical conductivity in metal crystals.**

3. *Thermal conductivity.* When a part of a metal is heated, the kinetic energy of the electrons in that region is increased and they start moving rapidly through the crystal and conduct heat to other parts of the metal. Hence, the heat energy is transferred to all corners of the metal. (Figure 13.27)

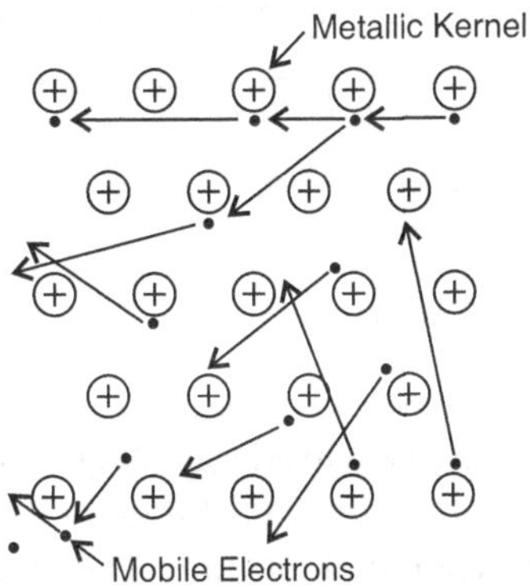


*FIGURE 13.27* **Thermal conductivity in metallic crystals.**

4. *Malleability and ductility.* Metals can be beaten into sheets (malleability) and drawn into wires (ductility) due to the nondirectional nature of the metallic bond. Whenever any stress is applied to metals, the position of the metallic kernels is altered without destroying the crystal. Due to slippage of the adjacent layers of kernels from one part to another, everywhere the metallic lattice gets deformed and the environment of kernels does not change as delocalized electrons are available.

5. *Hardness.* The hardness of metals is due to the strength of the metallic bond which depends upon the number of valence electrons and the smaller size of the kernel of the metal atom, *e.g.*, alkali metals have only one valence electron and larger atomic kernels, thus they form weak metallic bonds and are soft metals.

### 13.5.2 Valence Bond Theory

According to this theory, the metallic bonding is essentially covalent in origin and the metallic structure involves resonance of electron pair bonds (ordinary covalent bond) between each atom and its nearest neighbours, so there is a resonance of large number of canonical forms.

Consider the metallic crystal of lithium ($_3Li - 1s^2, 2s^1$) metal in which each atom has eight nearest neighbours as shown in X-ray studies. Lithium atoms have only one valence electron which is insufficient to form bonds with eight neighbouring lithium atoms. Thus, it is assumed that resonance occurs throughout the lithium solid. The various resonating structures of the lithium solid may be shown as:

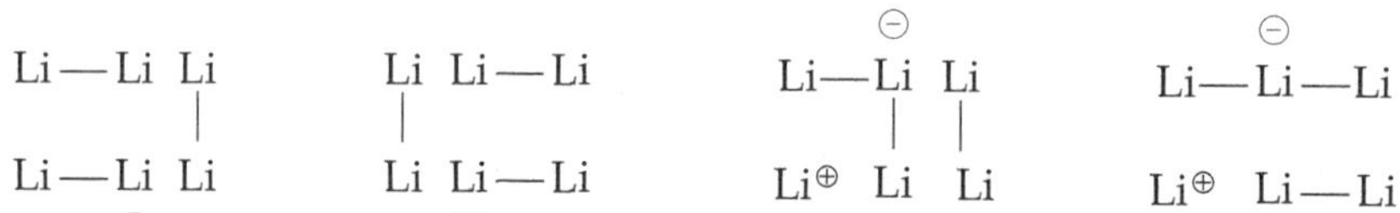

I II III IV

Here, only four atoms are shown but actual bonding includes all the atoms of the crystal and is three dimensional, the contributing structures III, IV contain lithium atom with a negative charge bonded to other two lithium atoms by covalent bonds. Covalent bonds are formed as:

$$Li_a(1s^2, 2s^1)—Li_b(1s^2, sp^1, sp^1\ 2p_y{}^0, 2p_z{}^0) \text{ or } 1s^2, 2s^1, 2s^1, 2p^1$$

$$Li_d\ (1s^2)—Li_c(1s^2, 2s^1)$$

### 13.5.3 Molecular Orbital or Band Theory

When a diatomic molecule is formed, two AOs of equal energies combines to give two MOs, one BMO, and one ABMO each with its own discrete energy. In general, *n* atomic orbitals on combination will yield *n* MOs. In metals, where many atoms exist together very closely, electron delocalization occurs. Consider the case of the lightest metal, lithium. The electronic structure of the Li atom is

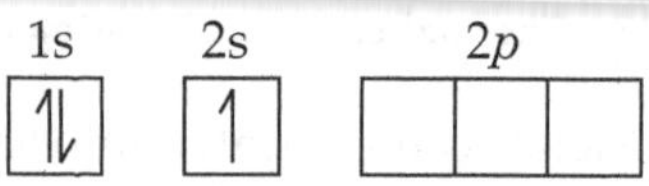

The $Li_2$ molecule exists in vapor state, and the bonding occurs using the 2*s* AOs. The 2*s* AOs on each of the Li atoms combine to give two MOs—one bonding and one antibonding. The valence electrons occupy the bonding MOs.

Suppose three Li atoms join to form $Li_3$. Three 2*s* AOs would combine to form three MOs—one bonding, one nonbonding, and one antibonding. The energy of the nonbonding MO is between bonding and antibonding. The three valence electrons from the three atoms occupy the two lowest energy MOs, which are both bonding orbitals.

In $Li_4$, the four AOs form four MOs—two bonding, two antibonding. The presence of two nonbonding MOs between the bonding and antibonding orbitals reduces the energy gap between the orbitals. The four valence electrons will occupy the two lowest energy MOs, which are both bonding orbitals.

As the number of electrons in the cluster increases, the spacing between the energy levels of various orbitals decreases further, and when there are large numbers of atoms, the energy levels of the orbitals are so close together that they almost form a continuum. For $n$ centers (where $n$ is very large), we may have $n$ discrete energy levels. If $n$ is of the order of $10^{23}$ in a metallic crystal, the discrete level which corresponds to $ns$-orbitals in a band of levels which appear to be continuous is called *quasi continuous*. Within the band the electrons will occupy the stage of lowest energy and greatest stability.

In Lithium, the inner 1*s* bond is full, and since there is one valence electron in the 2*s* orbital, only half the MOs in the 2*s* valence band are filled, *i.e.*, the bonding MOs. The energy gap between the two bands is so large that it prevent promotions of electrons from the lower to the higher band. Such an energy gap is known as a *forbidden zone.* The filled part of the band is called *nonconduction band* and the empty band is called *conduction band.*

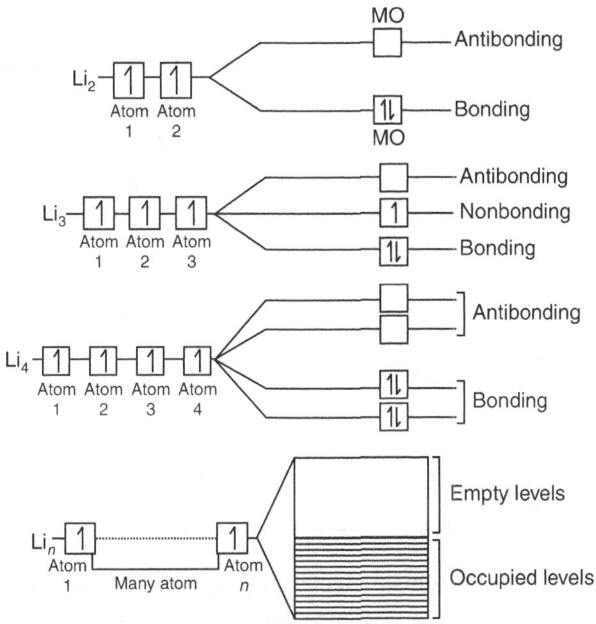


*FIGURE 13.28* **Formation of 1s band in lithium metal.**

The bandwidth increases with the increase in overlapping of orbitals, which in turn increases with decrease in interatomic distance. At a certain distance between atoms, the two adjoining bands may be wider to such an extent that they start overlapping each other and thus form a single band (Figure 13.29).

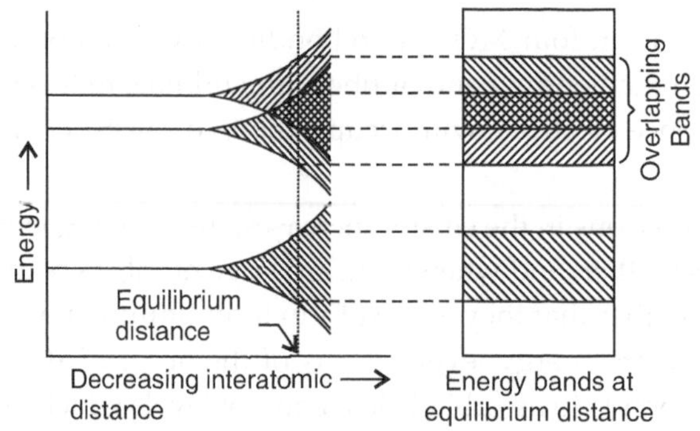


*FIGURE 13.29* **The broadening and overlapping of bands as a function of interatomic distance.**

It is clear from Figure 13.29 that if the internuclear distance increases, there is lesser overlapping or the orbitals remain essentially atomic. Similarly, if the internuclear distance decreases, there is more overlapping or the orbitals become molecular orbitals. The filling of these bands and the width of forbidden zones gives the idea of conductance in metals, nonmetals, and semiconductors (Figure 13.30) as discussed below:

(*a*) The true metal (conductor) has either a partially filled band or an unoccupied band overlapping an empty band. As the potential difference is applied, the electrons in the partially filled energy bands get extra energy and jump to a higher energy level of the same band. Hence all the energy levels are delocalized and the metal acts as a conductor. In the same way, the metal behaves as a conductor at high temperature due to the vibratory motion of the atoms.

(*b*) When the bands are completely filled, the metal acts as an insulator (*i.e.*, the substance has extremely low or no electrical conductivity) because electrons fail to jump in the filled bands.

(*c*) When the bands are completely filled but the next energy band is nearer, such metals act as a semiconductor, *e.g.*, Si, Ge, Se, etc.

## 13.6 SEMICONDUCTORS

Semiconductors are those materials whose electrical properties lie between that of insulators and good conductors. These substances have a valence band (completely filled) and an empty conduction band. The energy gap between these two bands is very small, thus less energy is required to promote the valence band electrons to the empty conduction band. As the temperature increases, the electron can jump to the conduction band, *i.e.*, conductivity increases with temperature.

When the energy gap between the valence band and the empty conduction band is less than 3 eV, then it is called a *semiconductor.*

### Classification of Semiconductors

Semiconductors are of two types:

1. **Intrinsic semiconductors.** When semiconductors are in their extremely pure form, they are called intrinsic semiconductors. The pure elements like silicon (Si), germanium (Ge), selenium (Se), etc. have four valence electrons in their atoms and their forbidden energy gap is very small (3 eV), so at ordinary temperature the electrons from the valence band can be promoted to the empty conduction band. For example, in the pure form, the crystal structure of Ge each atom is surrounded by four neighbouring atoms and bonded covalently in a two dimensional network. When a small energy is given, an electron from the valence band is shifted to the conduction band, leaving behind a positively charged hole Figure 13.31 (*a*). This +ve hole can travel to an adjacent atom by accepting an electron '*e*' from the latter one in a repeated manner. Consequently, the hole can move from one site of crystal lattice to another. Thus '*e*' and holes in equal numbers are the charge carriers is intrinsic semiconductors.

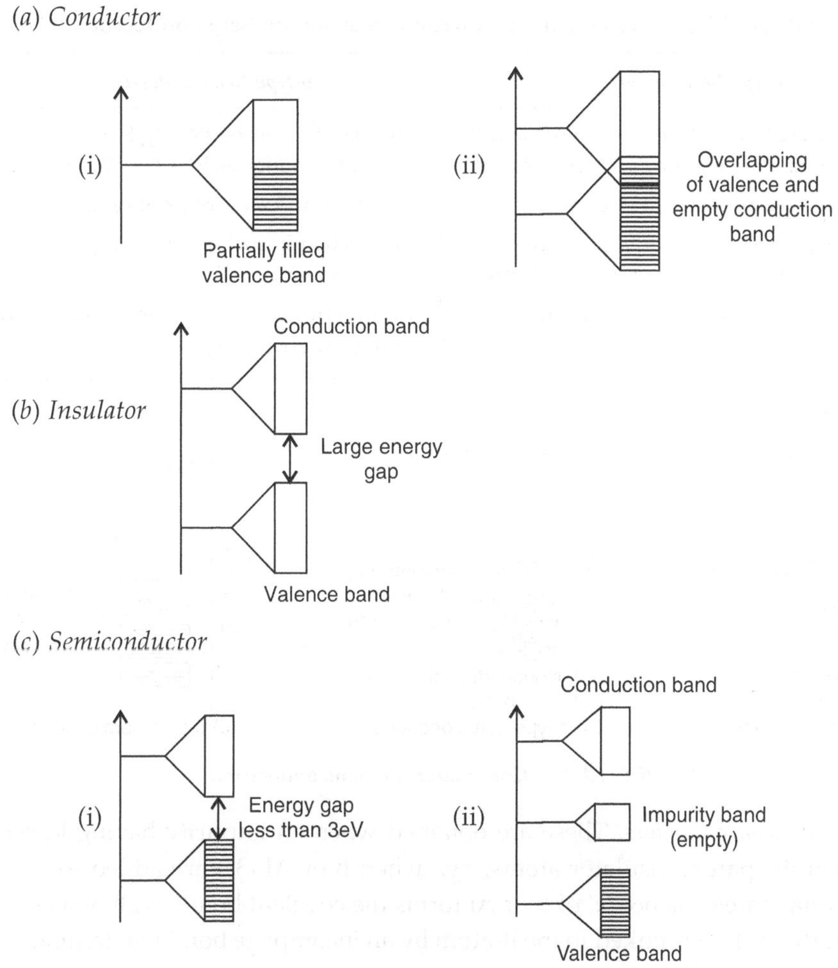


***FIGURE 13.30*** **Band structure for conductors, insulators and semiconductors.**

2. **Extrinsic semiconductors.** Such types of substances have very low electrical conductance in the pure state (*e.g.*, Si, Se, Ge, etc.) but their conductance increases by the addition of traces of an element belonging to another group of the periodic table. These are of two types: *n*-type and *p*-type semiconductors.

   (*a*) *n-type semiconductors (n-stands for negative).* The addition of a small amount of impurity to a semiconductor before it crystallizes is known as *doping*. When a small amount of phosphorus (with five valence electrons) is added to silicon (4 valence electrons), the fifth valence electron remains unshared and is a surplus electron which moves through the crystal under the influence of the electric field. This gives an *n*-type semiconductor (Figure 13.32(*a*)). These surplus electrons occupy a delocalized level called the donor impurity level which is just below the empty conduction band of Si crystal. These '*e*' can very easily be excited to empty *c* and action bands by application of electric or thermal energy so the crystal of Si becomes a semiconductor.

**TABLE 13.5** ***n*-type Semiconductor versus *p*-type Semiconductor**

| *n-type Semiconductor* | *p-type Semiconductor* |
|---|---|
| (*i*) They are formed by doping Si or Ge with a pentavalent impurity such as P, As, etc. | They are formed by doping Si or Ge with a trivalent impurity such as Ga, Al, Be, etc. |
| (*ii*) They have an excess of electrons. | They have an excess of positive holes. |
| (*iii*) Conduction in such a semiconductor is due to the movement of the free electron. | Conduction in such a semiconductor is due to the movement of holes. |
| (*iv*) Conduction is due to movement of the negative electric charge. | Conduction is due to movement of the positive electric charge. |

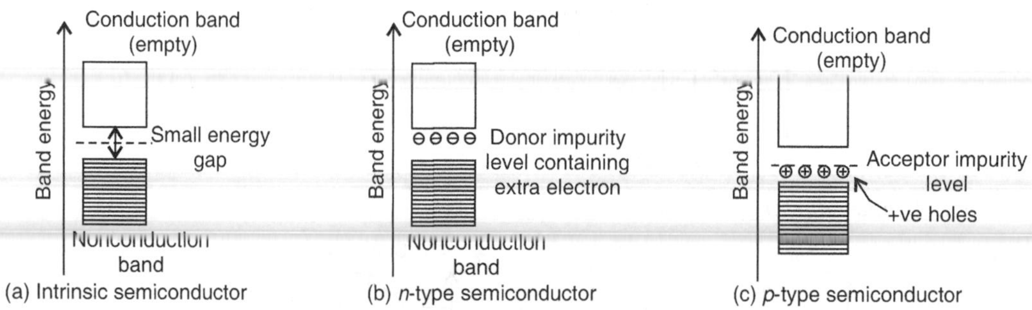


*FIGURE 13.31* **Conduction in semiconductors.**

   (*b*) *p-type semiconductors.* These are obtained when an impurity having lesser electrons than the parent insulator atoms, *e.g.*, when B or Al (3*e*) are added to pure Si or Ge crystal, three valence '*e*' of B or Al forms the covalent bonds with 3-Si atoms and the fourth Si atom is linked to the B-atom by an incomplete bond (containing 1 – *e*). Thus, in this bond in lattice there is an electron deficiency which creates a +ve hole in the valence band of Si. There are as many +ve holes as boron atoms. These holes occupy

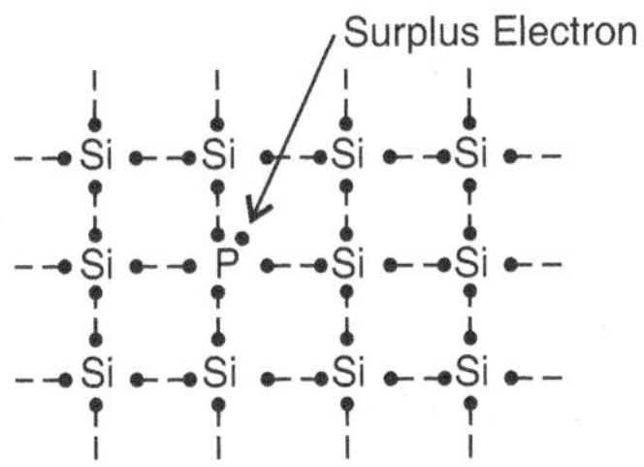


**FIGURE 13.32** (*a*) ***n*-type extrinsic semiconductor.**

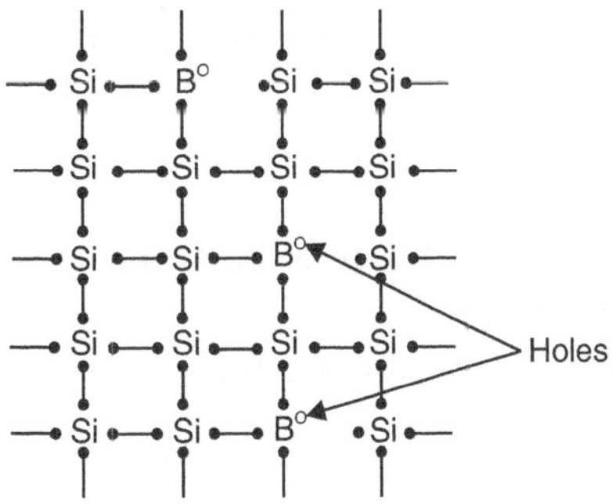


**FIGURE 13.32** (*b*) **Si with boron impurity (*p*-type semiconductor).**

the level close to the filled valence band called *acceptor impurity level.* Electrons can jump to this empty acceptor impurity level of the holes as shown in Figure 13.32 (*b*). The positive holes so formed in the crystal lattice conduct current, when an electric field is applied. The movement of holes through the crystal of a *p*-type semiconductor is nothing but movement of a positive electric charge.

### Applications of Semiconductors

Following are the important applications of semiconductors:

1. The combination of *n*-type and *p*-type semiconductors, known as *n-p* functions, is used in the manufacture of transistors. This device can conduct electric current more easily in one particular direction than in the reverse direction and so can be used as rectifier for converting AC into DC.
2. Semiconductors are used in transistors and in exposure meters as photo electric devices.

## 13.7 THE BORN-HABER CYCLE

**Lattice Energy.** The lattice energy is a measure of the energetic stability of the crystal. It is defined as *the energy change when one mole of a crystal is formed from its component ions in the gaseous state.* For example, in the case of magnesium oxide,

$$Mg^{2+}(g) + O^{2-}(g) \longrightarrow MgO; \qquad \Delta H_{LE} = -3933 \text{ kJ/mol.}$$

For sodium chloride,

$$Na^{+}(g) + Cl^{-}(g) \longrightarrow NaCl(s); \qquad \Delta H_{LE} = -788.5 \text{ kJ/mol.}$$

The *Born* and *Haber* (1919) cycle relates the lattice energy of an ionic crystal to other thermodynamic data. The values of lattice energy can be obtained indirectly by means of an energy cycle, connecting the ionic solids, the gaseous ions, and the elements in their standard states. Reactions may be considered to be made up of several component steps and the heat effect of the entire process may be calculated.

Consider the formation of sodium chloride by a reaction between sodium and chlorine.

$$Na(s) + \frac{1}{2}Cl_2(g) \longrightarrow Na^+Cl^-(s); \qquad \Delta H^\circ_f = -411 \text{ kJ/mol.}$$

This reaction can be broken into a number of hypothetical steps with each step involving an enthalpy change:

(*i*) Conversion of solid sodium into gaseous sodium atoms:

$$Na(s) \longrightarrow Na(g); \qquad \Delta H_s = 108.4 \text{ kJ/mol}$$

where $\Delta H_s$ is the sublimation energy. The reaction is endothermic in nature.

(*ii*) Conversion of sodium atom into sodium ions:

$$Na(g) \longrightarrow Na^+(g); \qquad \Delta H_i = 496 \text{ kJ/mol}$$

where $\Delta H_i$ is the ionization energy of sodium. The reaction is endothermic.

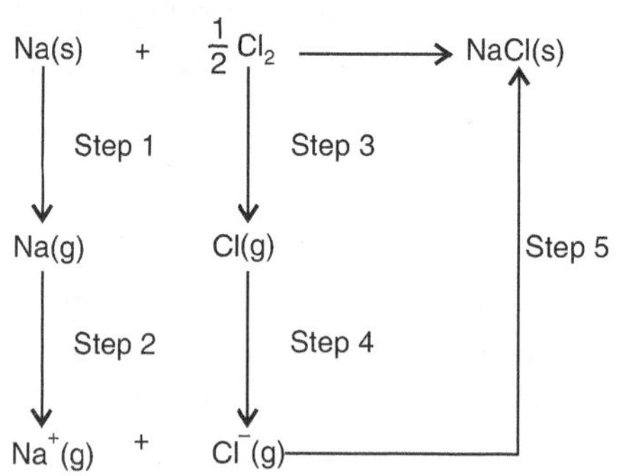


FIGURE 13.11 Born-Haber cycle for the formation of NaCl crystal from its elements.

(*iii*) Conversion of chlorine molecule into chloride atoms:

$$\frac{1}{2}Cl_2(g) \longrightarrow Cl(g); \qquad \Delta H = 121.1 \text{ kJ/mol.}$$

where $\Delta H$ is the dissociation energy. It is an endothermic reaction.

(*iv*) Conversion of the chloride atom into the chloride ion:

$$Cl(g) \longrightarrow Cl^-(g); \qquad \Delta H_e = 348 \text{ kJ/mol}$$

where $\Delta H_e$ is the electron affinity of chlorine. This change is an exothermic reaction.

(*v*) Formation of solid sodium chloride:

$$Na^+(g) + Cl^-(g) \longrightarrow NaCl(s); \quad \Delta H_{LE} = ?$$

The process of bringing together sodium and chloride ions results in the liberation of energy known as *lattice energy*. The formation of NaCl(s) by the above two processes is called Born-Haber cycle. According to Hess law, the heat of formation of NaCl(s) must be the sum of both the processes. Therefore,

$$\Delta H_f = \Delta H_s + \Delta H_i + \frac{1}{2}\Delta H + \Delta H_e + \Delta H_{LE}$$

$$-411 = 108.4 + 496 + 121.1 - 348 + \Delta H_{LE}$$

$$\Delta H_{LE} = -788.5 \text{ kJ/mol}$$

### Applications of Born-Haber Cycle

Following are the important applications of Born-Haber Cycle:

1. The cycle can be used to calculate electron affinities by using known crystal structures.
2. By knowing the values of electron affinity, the cycle is used to calculate the lattice energy of unknown crystals.
3. Lattice energy may provide some information about ionic covalent bonding.
4. It is helpful in analyzing and correlating the stabilities of various ionic solids.

**Example 4.** *Calculate the lattice energy of a crystal of NaCl. The heat of sublimation is 108.8 kJ mol$^{-1}$, the heat of dissociation of chlorine gas is 242 kJ mol$^{-1}$, the ionization energy of sodium is 498.3 kJ mol$^{-1}$, the electron affinity of chlorine is – 380.7 kJ mol$^{-1}$, and the heat of formation of solid sodium chloride from its elements is – 410.9 kJ mol$^{-1}$.*

**Solution.** We know that

$$\Delta H_f = \Delta H_s + \Delta H_i + \frac{1}{2}\Delta H + \Delta H_e + \Delta H_{LE}$$

$$\Delta H_{LE} = \Delta H_s + \Delta H_i + \frac{1}{2}\Delta H + \Delta H_e + \Delta H_f$$

$$= 108.8 + 498.3 + \frac{1}{2}(242.7) - 380.7 + 410.9$$

$$\mathbf{\Delta H_{LE} = -758.65 \text{ kJ mol}^{-1}}$$

**Example 5.** *Formation of MgO(s) takes place as follows:*

$$Mg(s) + \frac{1}{2}O_2(g) \longrightarrow Mg^{2+}O^{2-}(s)$$

*Different values (kJ mol$^{-1}$) for Mg, $O_2$, and MgO are:*

*$\Delta H_s = 152.7$; $\Delta H_i = 2178$; $\Delta H = 495.4$; $H_e = 630$; $\Delta H_{LE} = -3933$*

*Calculate the heat of formation of MgO.*

**Solution.** Since

$$\Delta H_f = \Delta H_s + \Delta H_i + \frac{1}{2}\Delta H + \Delta H_e + \Delta H_{LE}$$

$$= 152.7 + 2178 + 247.7 + 630 - 3933 = \mathbf{-\ 724.6\ kJ\ mol^{-1}}$$

## 13.8 HYDROGEN BONDING

When a hydrogen atom linked to a highly electronegative atom (like F, O, or N) comes under the influence of another strongly electronegative atom, then a weak bond is developed between them. This type of bond is called a *hydrogen bond* and is designated by dotted lines as shown below:

$$\text{------}\overset{\delta+}{H} - \overset{\delta-}{F}\text{------}\overset{\delta+}{H} - \overset{\delta-}{F}\text{------}\overset{\delta+}{H} - \overset{\delta-}{F}\text{------}\overset{\delta+}{H} - \overset{\delta-}{F}\text{------}$$

Thus we see that hydrogen in hydrogen fluoride is bonded to two highly electronegative fluorine atoms. The linkage between hydrogen and one of the fluorine atoms is covalent while the linkage between hydrogen and the other fluorine is simply due to electrostatic forces of attraction which results in a weak bond, commonly known as a hydrogen bond or hydrogen bonding. Thus a hydrogen bond may also be defined as *an electrostatic force of attraction between a covalently bonded hydrogen atom of one molecule and an electronegative atom (such as F, O, and N) of another molecule.*

A hydrogen bond is merely a strong electrostatic attractive force and not a normal chemical bond. It is very weak (strength about 2–10 k cal $mol^{-1}$ or 8.4–42 kJ $mol^{-1}$).

### Conditions for Hydrogen Bonding

Following are the conditions for Hydrogen Bonding:

(*i*) The molecule must possess a high electronegative atom bonded to a hydrogen atom. The greater is the electronegativity difference, the more is the polarization of the bond of the molecule.

**Differences between a Hydrogen Bond and a Covalent Bond**

| *Hydrogen Bond* | *Covalent Bond* |
|---|---|
| (*i*) It is formed between a hydrogen atom and a highly electronegative atom such as F, O, or N | It is formed between two electronegative atoms of the same element or of different elements. |
| (*ii*) Hydrogen bond is the result of dipole-dipole attractive interaction. | It is formed by sharing of electrons. |
| (*iii*) The strength of H-bond is very small, *e.g.*, bond strength of H----O bond is 29.3 kJ $mole^{-1}$ | The bond strength of covalent bond is sufficiently high, *e.g.*, bond strength of H—H bond is 433.0 kJ $mole^{-1}$. |

(*ii*) The size of the electronegative atom should be small. The smaller the size of the electronegative atom the more is the electrostatic attraction.

The previous two conditions are met only by F, O, and N atoms and thus only these atoms form effective hydrogen bonding.

### 13.8.1 Examples of Hydrogen Bonding

Following are some important examples of Hydrogen bonding:

(*i*) **Water ($H_2O$).** The water molecule has a highly electronegative oxygen atom linked to a hydrogen atom. Thus the bond between oxygen and hydrogen becomes polar, giving hydrogen an electropositive and oxygen an electronegative nature. Electrostatic attraction between the negative end of one molecule and the positive end of the other molecule results in the formation of the hydrogen bond. As a result of hydrogen bonds, each oxygen is tetrahedrally surrounded by four hydrogen atoms of which two are covalent bonds and two are hydrogen bonds as shown in Figure 13.34. This cluster of water molecules, also known as associated molecule, may be described as $(H_2O)_n$.

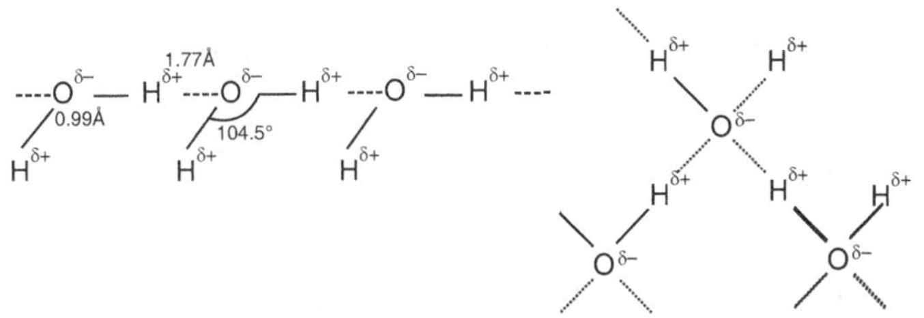


*FIGURE 13.34* **Hydrogen bonding in water.**

(*ii*) **Alcohols (R—OH).** Similar to water molecules, alcohols also have an electronegative oxygen atom linked to a hydrogen atom and thus form the following associated molecules.

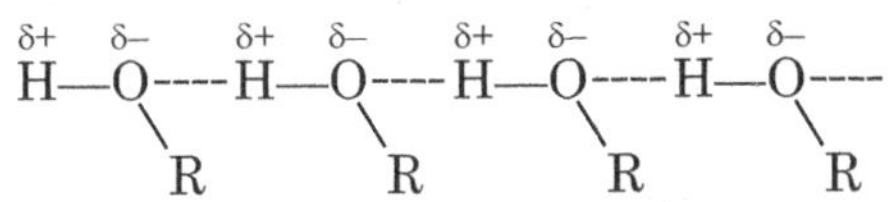

(*iii*) **Ammonia ($NH_3$).** An ammonia molecule has an electronegative nitrogen atom covalently linked to three hydrogen atoms. Due to their electronegativity difference, each hydrogen acquires a positive charge and the nitrogen becomes negatively charged. Hence, electrostatic attraction between ammonia molecules takes place, resulting in the formation of an associated molecule as shown below.

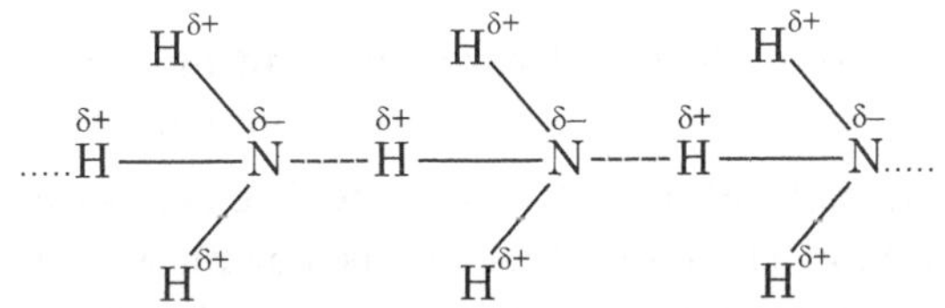

(*iv*) **Carboxylic acids (R–COOH).** Unlike hydrogen bonding in HF, $H_2O$, R—OH, $NH_3$ etc., where a large number of molecules associate, hydrogen bonding in carboxylic acids results in dimer formation only,

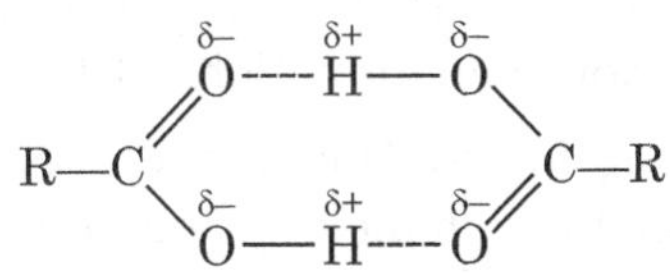

Dimer of carboxylic acids

### 13.8.2 Strength of Hydrogen Bond

A hydrogen bond is a weak bond whose bond strength is intermediate between a covalent bond and van der Waal's forces of attraction. The bond dissociation energy of a hydrogen bond ranges from 12.6 to 41.8 kJ $mol^{-1}$ depending upon the electronegativity of the other atom. The higher the electronegativity of the atom linked to the hydrogen, the greater is the strength of hydrogen bond and the higher is the bond dissociation energy.

H----F bond dissociation energy is 41.8 kJ $mol^{-1}$

H----O bond dissociation energy is 29.3 kJ $mol^{-1}$

H----N bond dissociation energy is 12.6 kJ $mol^{-1}$

However, bond dissociation energy of a covalent bond is 209.0 to 418.0 kJ $mol^{-1}$.

### 13.8.3 Types of Hydrogen Bond

Hydrogen bonding is of two types:

1. **Intermolecular hydrogen bonding.** When hydrogen bonding takes place between various molecules of the same or different compounds, it is called intermolecular hydrogen bonding, *e.g.*,

   (*i*) Various molecules of the same compound like HF, $H_2O$, alcohol, carboxylic acid, etc.

   (*ii*) Various molecules of different compounds like water and alcohol or water and ammonia.

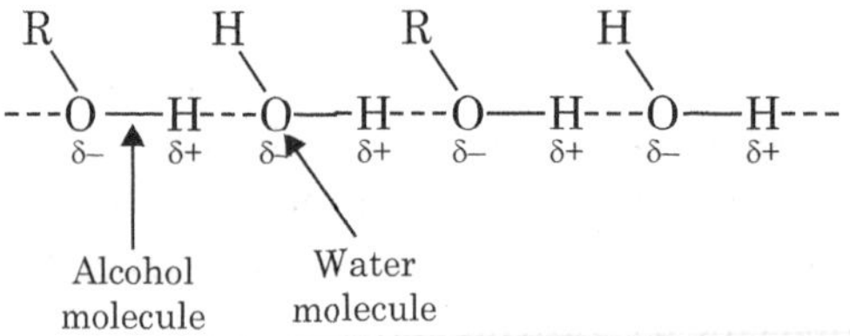


**Consequences of intermolecular hydrogen bonding.** Formation of an intermolecular hydrogen bond influences the properties of various substances in the following manner:

(*i*) *Association.* The molecules of carboxylic acids exist as dimers due to intermolecular hydrogen bonding. Therefore, their molecular masses are found to be double than calculated from their simple formula. For example, the molecular mass of acetic acid is found to be 120.

(*ii*) *Dissociation.* Among halogens, only fluorine forms hydrogen bonding because of its small size and high electronegativity. In aqueous solution, hydrogen fluoride dissociates and gives the difluoride ion ($HF_2^-$) instead of fluoride ion ($F^-$). This supports the existence of $KHF_2$; however, formation of $KHCl_2$, $KHBr_2$, and $KHI_2$ is not possible.

(*iii*) *High melting and boiling points.* The compounds having intermolecular hydrogen bonding show abnormally high melting and boiling points. For example:

(*a*) According to a general trend the boiling point of halogen acids should decrease with a decrease in molecular mass which is found to be true in case of HCl, HBr, and HI. HCl BP. 198K < HBr BP 206K < HI BP. 237K. The boiling point of HF (292.4 K) should be less than HCl but it is abnormally high. The unusually high boiling point of HF (292.4K) is due to the existence of hydrogen bonding in HF, which requires some extra energy for dissociation of intermolecular hydrogen bonding.

(*b*) $H_2O$ is a liquid whereas hydrides of other elements of this group *i.e.*, $H_2S$, $H_2Se$, and $H_2Te$ are gases. This is due to the presence of intermolecular hydrogen bonding in $H_2O$ molecules. The associated water molecules have higher boiling points and hence exist as liquid. On the other hand, there is no hydrogen bonding in $H_2S$, $H_2Se$, and $H_2Te$ molecules.

(*c*) A similar explanation can be given for the higher boiling point of $NH_3$ among the hydrides of group 15 elements.

Relative melting and boiling points of different hydrides of group 15, 16, and 17 have been shown graphically in Figure 13.35.

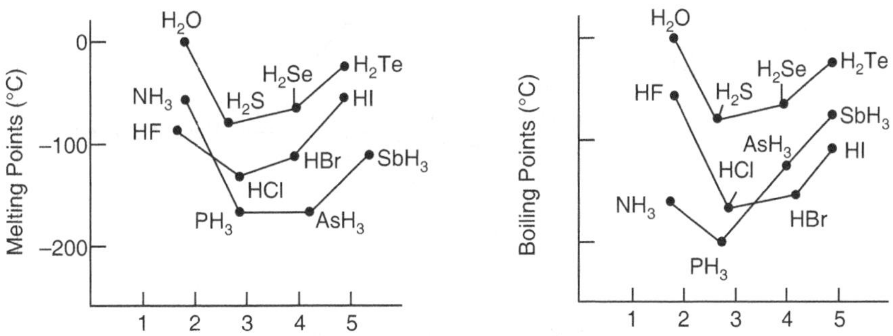


***FIGURE 13.35*** **Abnormally high mp and bp of $NH_3$, $H_2O$ and HF.**

(*iv*) *Ethanol has a higher boiling point than diethyl ether.* Due to intermolecular hydrogen bonding in ethanol, it exists as associated molecule whereas no hydrogen bonding occurs in diethyl ether. Therefore, ethanol has a higher boiling point than diethyl ether.

(*v*) *Solubility in water.* Compounds which can form hydrogen bonding with water will dissolve in water. For example, lower alcohols are soluble in water because their molecules can form H-bonds with water.

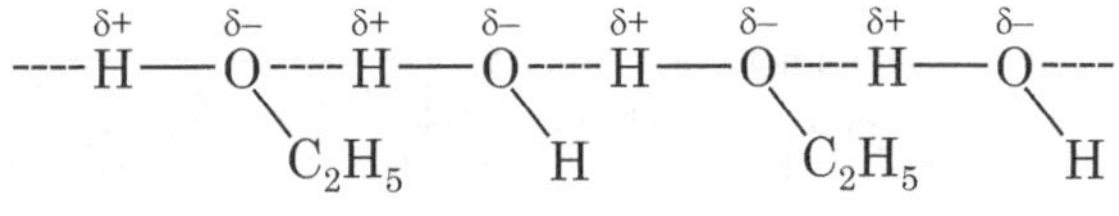


Similarly, ammonia ($NH_3$) and amines (R—$NH_2$) are soluble in water because of their ability to form H-bonds with water.

(*vi*) *Higher viscosity and surface tension.* Substances which can give more extensive hydrogen bonding will have higher viscosity and high surface tension. Glycerol forms more extensive hydrogen bonding than glycol and glycol gives more extensive hydrogen bonding than ethanol. Therefore, the viscosity and surface tension follow the order.

$$\underleftarrow{\text{Glycerol} > \text{Glycol} > \text{Ethanol}}$$

Higher viscosity and surface tension

(*vii*) *Lower density of ice than water.* Water in solid state (*i.e.*, ice) has a cage-like structure because of intermolecular hydrogen bonding as shown below. When ice melts, the cage-like structure collapses and water molecules come closer to each other. So in ice, $H_2O$ molecules are not so closely packed as they are in liquid state. Hence in the liquid state its volume decreases for the same mass of water, *i.e.*, its density increases. Evidently, the ice (or water at 273 K) will have lower density than water. So the ice floats on water.

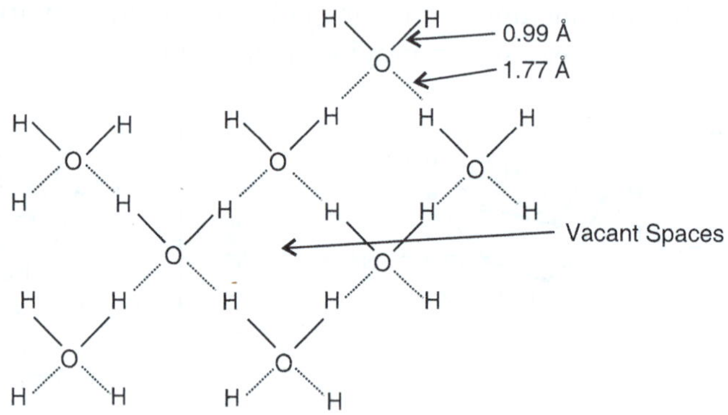


Cage-like structure of $H_2O$ in ice

(*viii*) *Volatility.* The compounds involving intermolecular hydrogen bonding have a higher boiling point, hence such compounds will have low volatility, *e.g.*, ethanol has lower volatility than *n*-butane ($C_4H_{10}$).

2. **Intramolecular hydrogen bonding (Chelation).** This type of hydrogen bonding involves the electrostatic forces of attraction between hydrogen and an electronegative element of the same molecule. When hydrogen bonding occurs within the same molecule, it is termed intramolecular hydrogen bonding. Common examples of intramolecular hydrogen bonding are given below :

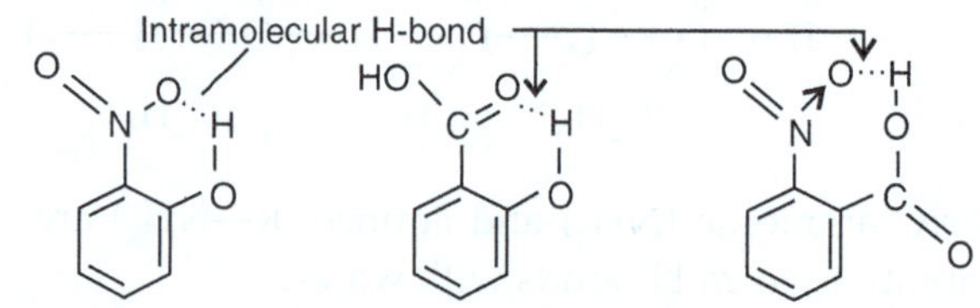


*o*-Nitrophenol Salicylic acid *o*-Nitro benzoic acid

**Consequences of intramolecular hydrogen bonding.** Effect of intramolecular hydrogen bonding on physical properties of compounds is described below:

(*i*) *Low melting point.* Compounds having an intramolecular hydrogen bond (*i.e.*, *o*-isomers) have a lower melting point than the other isomers (*meta* or *para* isomers). This is due to formation of an intramolecular hydrogen bond in the *o*-isomer whereas the *m*-and *p*-isomers have intermolecular hydrogen bonding.

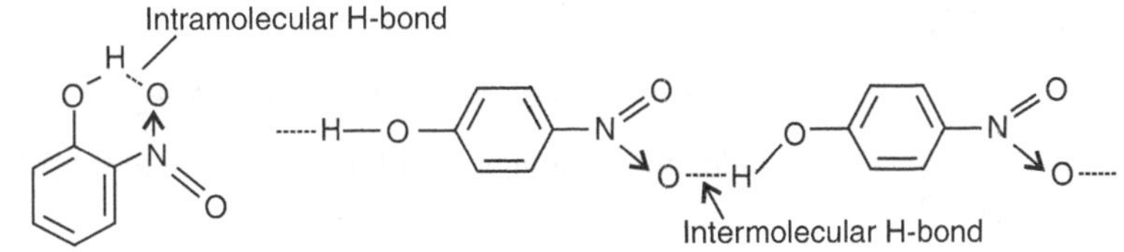


*o*-Nitrophenol *p*-Nitrophenol

(*ii*) *Solubility in non-polar solvents.* Compounds involved in chelation become nonpolar, *i.e.*, salicylaldehyde. Consequently, such compounds are soluble in nonpolar solvents like ether, benzene, etc. and are only sparingly soluble in water, whereas the corresponding *meta* and *para* isomers are more soluble in water and less soluble in nonpolar solvents.

(*iii*) *More volatility.* Since compounds having intramolecular hydrogen bonds have low melting or boiling points, these will be more volatile than the corresponding *p*-isomers.

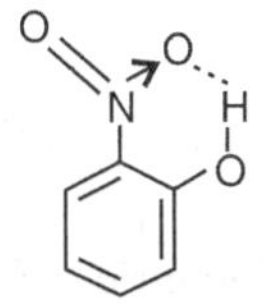

*o*-Nitrophenol (More volatile)

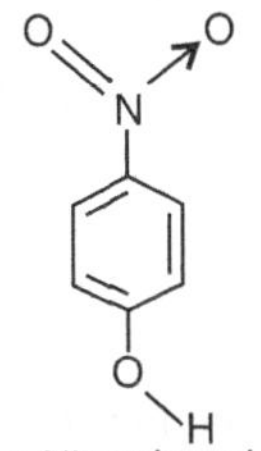

*p*-Nitrophenol (Less volatile)

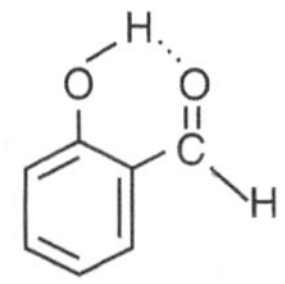

Salicylaldehyde

(*iv*) *Ionization of carboxylic acids.* When the anion of an acid is stabilized by intramolecular hydrogen bonding, there is marked increase in the dissociation of the corresponding carboxylic acid, *e.g.*, salicylate ion is stabilized by intramolecular hydrogen bonding. Therefore, salicylic acid is 17 times more acidic than benzoic acid.

However, if the dissociating H atom is involved in the intramolecular H-bond, the acidic character becomes less, *e.g.*, *o*-halophenols are weaker acids than the *m*- and *p*-isomers.

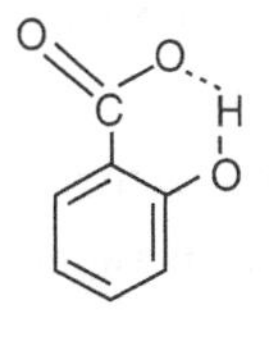

Salicylate ion

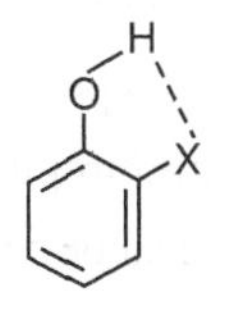

*o*-Halophenol (Weaker acid)

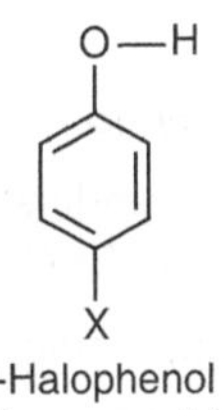

*p*-Halophenol (Stronger acid)

## QUESTION BANK

1. What is hybridization? Calculate percentage *s* orbital character of a hybrid orbital if the bond angle between the hybrid orbital is 105°.
2. State the difference between σ (Sigma) and π (Pi) bonds.
3. Calculate the number of σ (Sigma) and π (Pi) bonds in $C_2H_2$, $C_2H_4$, $CO_3^{2-}$, $NH_3$, $PCl_5$.
4. Predict the shape, bond angle, dipole moment, and hybridization of the following compound on the basis of VSEPR theory:

   $$Ag(CN)_2^+,\ XeO_3,\ H_2O_3,\ H_2O,\ PCl_5,\ XeF_4$$

5. Predict the shape, bond angle, dipole moment, and hybridization of the following compounds on the basis of VSEPR theory:

   $$SO_4^{2-},\ NH_4^+,\ BF_4^-,\ XeO_4,\ NH_3$$

6. Explain valence shell electron pair repulsion theory on the basis of bond angle of methane, ammonia, and water.
7. Discuss the salient features of the molecular orbital theory of bonding. How does it differ from valence bond theory?
8. Predict the type of hybridization and geometry of the following:

   (*i*) $XeO_3$ (*ii*) $NH_3$ (*iii*) $H_2O$
   (*iv*) $PCl_5$ (*v*) $SO_2$ (*vi*) $XeO_4$

9. With the help of the molecular orbital diagram, calculate the bond order of:

   (*i*) $O_2$ (*ii*) $O_2^{2}$ (*iii*) $He_2^+$ (*iv*) $NO^-$

10. What do you understand by the term Metallic bond? Explain it on the basis of molecular orbital theory.
11. What do you understand by stoichiometric and nonstoichiometric defects?
12. Write information notes on:

    (*i*) Frenkel defect (*ii*) F-centers (*iii*) Metal excess defects
    (*iv*) Dislocation (*v*) Schottky defects

13. Calculate the bond order in $B_2$, $N_2$, $O_2$, and NO.
14. What do you understand by covalent bond? Give three characteristics of a covalent bond.
15. Explain conductivity in conductors, semiconductors, and insulators.
16. Differentiate between *p*-type and *n*-type semiconductors.
17. What do you understand by extrinsic and intrinsic semiconductors?
18. Explain conduction in *p*-type and *n*-type semiconductors.
19. What is the effect of temperature and impurities on conductivity of a semiconductor?
20. What do you understand by Born-Haber cycle? What is its significance? Draw Born-Haber cycle for the formation of KCl.
21. Derive Braggs equation for diffraction of X-rays by crystals.
22. What is meant by 'Bond order'? Give a formula to calculate it. Calculate bond order for NO.

23. Explain paramagnetic nature of $O_2$ and its lesser stability in comparison to that of $N_2$.
24. Explain the formation of bonding and antibonding molecular orbitals on the basis of wave function.
25. State rules for LCAO method for formation of molecular orbitals.
26. Which of the following is paramagnetic:

$$O_2^-, O_2^{2-}, N_2, C_6H_6, CO_3^{2-}$$

27. What do you understand by overlap of atomic orbitals? Explain the overlap criterion in explaining bond strength.
28. Explain the formation of the hydrogen molecule with VB and MO theories.
29. Write the hybridization involved in (*i*) Beryllium dichloride and (*ii*) Boron trifluoride.
30. Explain the following on the basis of molecular orbital theory:
    (*i*) $Br_2$ molecule is not stable
    (*ii*) $N_2$ has a stronger bond than $B_2$
    (*iii*) $F_2$ is diamagnetic while $O_2$ is paramagnetic.
31. (*a*) Write down the rules for hybridization of orbitals.
    (*b*) $CH_4$, $NH_3$ and $H_2O$ have tetrahedral geometry yet their bond angles are different. Why?
32. (*a*) Explain the terms bonding and molecular orbitals.
    (*b*) Explain, with the help of MO theory, the paramagnetic character of oxygen.
    (*c*) What are σ and π molecular orbitals and give their characteristics?
    (*d*) Explain the formation of HF molecule with MO diagram.
33. Explain the state of hybridization of oxygen atom in water and indicate its shape.
34. Discuss, with the help of MO theory, the formation of $N_2$, $O_2$, and $F_2$ molecules. How does this theory explain the difference in reactivity of $N_2$, $O_2$, and $F_2$ molecules?
35. (*a*) Fill up the blanks:
    (*i*) HF molecules are bonded through ...... bond.
    (*ii*) The geometrical shape of $CH_4$ molecule is ...... .
    (*b*) Explain the type of hybridization and geometrical shape of $PCl_5$ molecule with the help of an orbital diagram.
    (*c*) Explain the shape of $NH_3$ and $H_2O$ with the help of VSEPR theory.
36. What are different types of overlap between *s* and *p* orbitals? Give one example of each type.
37. (*a*) What are the characteristics of hybrid orbitals?
    (*b*) What is tetrahedral hybridization? Explain the formation of C-H bond in $CH_4$.
    (*c*) What are sigma bond (σ) and pi (π) bonds? Distinguish between them.
38. (*a*) Draw the shape of $ClO_4^-$ ion on the basis of hybridization.
    (*b*) Write short note on LCAO.
    (*c*) Draw the shapes of $\sigma 2p_x$ and $\pi 2p_z$ molecular orbitals.
    (*d*) Compare the bond length of $N_2$, $O_2$, $F_2$ on the basis of MO Theory.
39. (*a*) What is meant by hybrid orbital? On the basis of the concept of hybridization how do you account for the observed H-O-H bond angle in water molecule?
    (*b*) Compare the polarity character of the following pairs:
    (*i*) $NH_3$ and $NF_3$ (*ii*) $CO_2$ and $SO_2$

**40.** (*a*) Compare VBT and MOT.
(*b*) Draw the molecular orbital picture of NO.
(*c*) What are the differences between bonding and antibonding molecular orbitals?

**41.** (*a*) Why is the bond angle in $H_2S$ less than $H_2O$?
(*b*) Explain the structure and hybridization of $SO_4^{2-}$ ion.
(*c*) Explain the structure and various bond angles in $IF_7$.
(*d*) Why is the bond angle of $OF_2$ smaller than $Cl_2O$?

**42.** Explain:
(*i*) Anhydrous HCl is bad conductor of electricity but aqueous HCl is a good conductor.
(*ii*) Glycol is more viscous than methanol.
(*iii*) $\sigma$ bond is stronger than $\pi$ bond.

**43.** Explain the structure of $O_2$ and $N_2$ molecules by MO theory. Show that $O_2$ is paramagnetic according to MO theory.

**44.** Account for the following with reasons:
(*i*) Melting point of $MgCl_2$ is much higher than that of $AlCl_3$
(*ii*) KI is soluble in alcohol but KCl is not.

**45.** (*a*) Discuss molecular orbital theory.
(*b*) Give the electron density diagram for bonding and nonbonding orbitals of hydrogen molecule.
(*c*) Predict the bond order in $C_2$ molecule with the help of MO diagram.

**46.** Explain the following on the basis of MO theory.
(*a*) Oxygen molecule is paramagnetic while nitrogen molecule is diamagnetic.
(*b*) Hydrogen forms diatomic molecule while helium remains monoatomic.
(*c*) The bond order in $O_2$ is less than that in $O_2$ which, in turn, is less than that in $O_2$.
(*d*) $Ne_2$ does not exist.

**47.** Draw the MO diagram for the following species:

$Be_2$, $O_2^-$, CO, BN

**48.** (*a*) Calculate the bond orders in the following molecules and ions:

$He_2$, $O_2$, $O_2^+$, and $O_2^-$

(*b*) Describe the hybridization of each C atom in the propylene molecule:

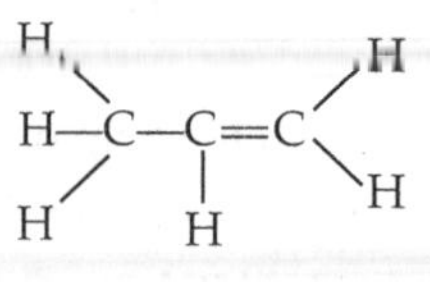

**49.** Draw the energy level diagrams for the molecular orbital in $F_2$ and CO molecules and obtain the bond order in each case.

**50.** Use molecular orbital method to write down the electronic configuration of NO molecules and discuss the bonding. Will NO be paramagnetic or diamagnetic?

**51.** Use valence bond method to discuss the electronic structure of hydrogen molecules. Use this example to discuss the relative merits and demerits of valence bond and molecular orbital (MO) methods.

**52.** (*a*) Explain, giving examples, bonding, antibonding, and nonbonding orbitals.
(*b*) Why is it not possible to visualize a stable $He_2$ molecule?

**53.** Give the essential features of valence bond and molecular orbital theories.

**54.** What are sigma ($\sigma$) and pi ($\pi$) bonds? Explain the formation of all $\sigma$ and $\pi$ bonds in the following molecules: HCN, $SO_3$, and $N_2$.

Chapter 14

# COORDINATION CHEMISTRY

## 14.1 INTRODUCTION AND TERMINOLOGY

It is well known that when an acid reacts with an alkali, a neutralization reaction occurs and a simple salt is formed as:

$$NaOH + HCl \longrightarrow \underset{\text{Sodium Chloride}}{NaCl} + H_2O$$

Also, when solutions containing two or more salts in simple proportions are allowed to evaporate, two types of *addition compounds* are formed. For example,

(*i*) *Double salt.* These salts give test of all constituents *i.e.*, individual components do not lose their identity. For example,

$$\underset{\text{Mohr's salt}}{FeSO_4(NH_4)_2SO_4 6H_2O} \longrightarrow \underset{\text{Constituent ions}}{Fe^{2+} + 2NH_4^+ + 2SO_4^{2-}} + 6H_2O$$

(*ii*) *Coordination compounds.* In these compounds individual components lose their identity and a complex ion is formed. For example, when solutions of $Fe(CN)_2$ and KCN are mixed and evaporated, potassium ferrocyanide is formed as:

$$Fe(CN)_2 + 4KCN \longrightarrow Fe(CN)_2 4KCN \text{ or } \underset{\text{Potassium ferrocyanide}}{K_4[Fe(CN)_6]}$$

$$K_4[Fe(CN)_6] \rightleftharpoons 4K^+ + \underset{\text{complex ion}}{[Fe(CN)_6]^{4-}}$$

The other examples of such compounds are

$$K_2[PtCl_6] \rightleftharpoons 2K^+ + [PtCl_6]^{2-}$$

$$[Pt(NH_3)_4Br]\ Br_2 \rightleftharpoons [Pt(NH_3)_4Br]^{2+} + 2Br^-$$

Thus, a complex ion is defined as *an electrically charged radical formed by the union of a metal cation with one or more neutral molecules or anions.*

**Coordination number (CN).** *The total number of the ligands that can coordinate to the central metal ion is the coordination number of the metal cation.* Numerically, it represents the total number of chemical bonds formed between the central metal ion and donor atoms of the ligand. For example, in ferrocyanide $[Fe(CN)_6]^{4-}$ ion, the coordination number is 6.

Some common coordination numbers are 2, 3, 4, and 6. The lighter transition metals generally exhibit 4 and 6 coordination numbers while the heavier transition elements generally exhibit coordination number greater than 6, *i.e.*, 7 or 9. Certain metals have been found to possess more than one coordination number (see Table 14.1).

**TABLE 14.1 Oxidation States and Coordination Numbers Exhibited by Some Metals**

| *Monovalent Ions* | *Divalent Ions* | *CN* | *Trivalent Ions* | *CN* | *Tetravalent Ions* | *CN* |
|---|---|---|---|---|---|---|
| $Ag^+$ | $Ag^{2+}$ | 4 | $Au^{3+}$ | 4 | $Pd^{4+}$ | 6 |
| $Cu^+$ | $Cu^{2+}$ | 4, 6 | $Fe^{3+}$ | 6 | $Pt^{4+}$ | 6 |
| $Pb^+$ | $Fe^{2+}$ | 6 | $Co^{3+}$ | 6 | | |
| $Tl^+$ | $Co^{2+}$ | 4, 6 | $Os^{3+}$ | 6 | | |
| | $Ni^{2+}$ | 4, 6 | $Cr^{3+}$ | 6 | | |
| | $Sn^{2+}$ | 4 | $Ir^{3+}$ | 6 | | |
| | $V^{2+}$ | 6 | $Sc^{3+}$ | 6 | | |
| | $Pd^{2+}$ | 4 | | | | |
| | $Pt^{2+}$ | 4 | | | | |

**Coordination sphere.** The central metal ion and the ligands that are directly attached to it are enclosed in a square bracket together and are called *coordination sphere.*

**Ligands.** The neutral molecules or ionic species which are attached with the central metal ion are called *ligands.*

## 14.2 CLASSIFICATION OF LIGANDS

Ligands are generally classified on the basis of their mode of attachment to the central metal ion. These are classified as:

(*i*) *Monodentate or unidentate ligands.* These are attached to the central metal ion through one point. These may be neutral or anionic. For examples, please refer to Table 14.2.

**TABLE 14.2 Some Common Unidentate Ligands**

| *Formula* | *Name* | *Formula* | *Name* |
|---|---|---|---|
| **Neutral** | | | |
| $H_2O$ | Aquo | $Br^-$ | Bromo |
| $NH_3$ | Ammine | I | Iodo |
| CO | Carbonyl | $CN^-$ | Cyano |
| **Anionic** | | | |
| $F^-$ | Fluoro | $OH^-$ | Hydroxo |
| $Cl^-$ | Chloro | | |

(*ii*) *Bidentate ligands.* These are attached to the central metal ion through two positions. These may be neutral or anionic in nature. For examples, see Table 14.3.

**TABLE 14.3 Some Common Bidentate Ligands**

| *Formula* | *Name* | *Formula* | *Name* |
|---|---|---|---|
| **Anionic** | | **Neutral** | |
| 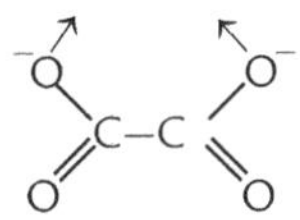 | Oxalato | 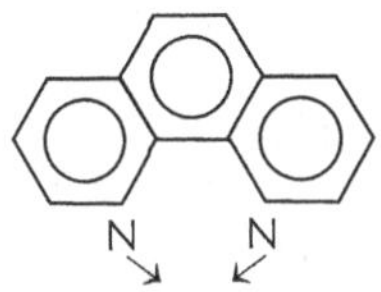 | 1,10-Phenanthroline (Phen) |
| $\ddot{N}H_2CH_2CO—O^-$ | Glycinato | $CH_2-CH_2$<br>$\ddot{N}H_2$ $\ddot{N}H_2$ | Ethylene diamine (en) |
| 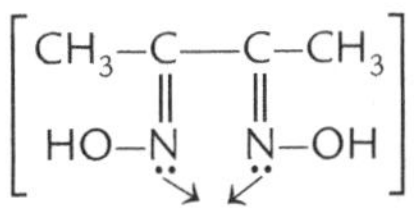 | Dimethyl-glyoximato | 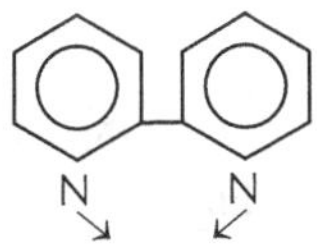 | 2, 2′ Dipyrdyl (dipy) |

(*iii*) *Tridentate ligands.* These ligands are attached to the central metal ion through three points. These may also be neutral or anionic. For examples, see Table 14.4.

**TABLE 14.4 Some Common Tridentate Ligands**

| *Formula* | *Name* | *Formula* | *Name* |
|---|---|---|---|
| **Anionic** | | **Neutral** | |
| $^-OOC-CH_2-CH-COO^-$<br>$\ddot{N}H_2$ | Asparate ion, ($Asp^{2-}$) | $CH_2—CH—CH_2$<br>$\ddot{N}H_2$ $\ddot{N}H_2$ $\ddot{N}H_2$ | Triaminopropane |
| $CH_2-CH-COO^-$<br>$\ddot{N}H_2$ $\ddot{N}H_2$ | Diamino propanoate | 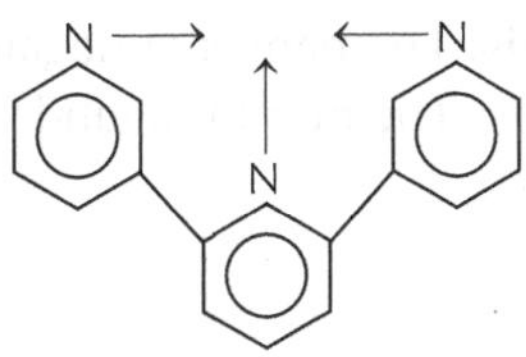 | Terpyridine (terpy) |

(*iv*) *Tetradentate ligands.* These ligands have four points of attachment to the central metal ion. For example,

(*a*) *Neutral.* Triethylenetetramine (trien)

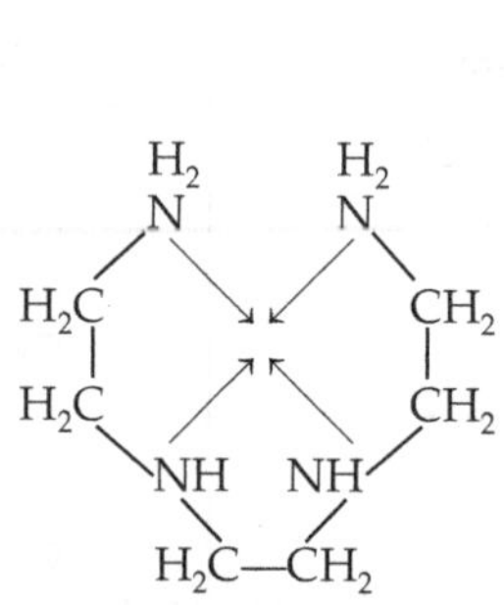

(*b*) *Anionic.* Nitrito triacetate (nta$^{3}$)$^{-}$

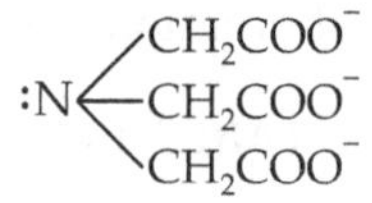

(*v*) *Pentadentate ligands.* The ligands which coordinate the central metal ion through five points are called *pentadentate ligands.* For example, ethylenediaminetriacetic acid.

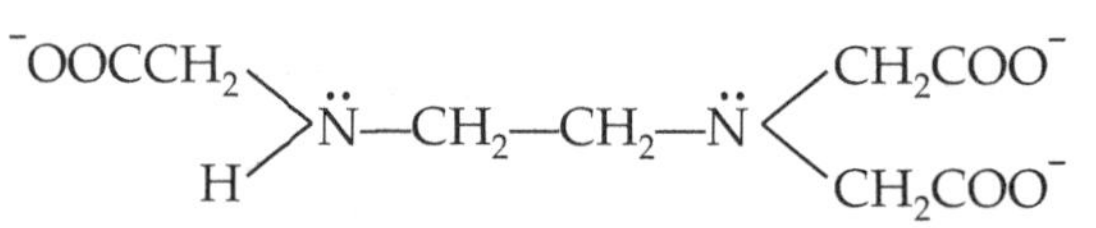

(*vi*) *Hexadentate ligands.* The ligand having six sites of attachment to the metal ion is termed as *hexadentate ligand*. For example, ethylenediaminetra acetato (EDTA).

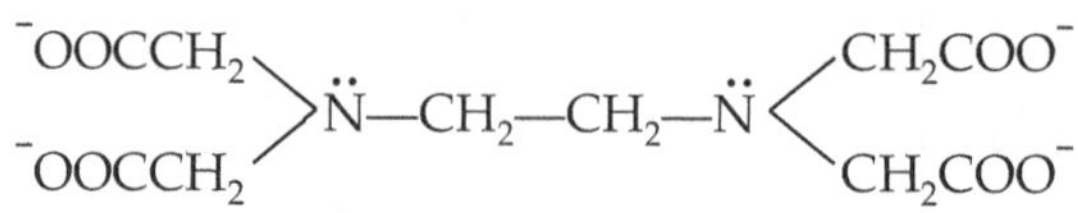

## 14.3 RULES FOR NAMING OF COORDINATION COMPOUNDS

The following rules have been given by IUPAC (International Union of Pure and Applied Chemistry) in naming the coordination compounds:

1. In ionic complexes, the cation is named first followed by the anion. Non-ionic complexes are given one-word names.
2. In naming the coordination sphere, the ligands are named first and then the central metal ion.
3. **Naming of ligands.** The names of negative ligands end in *-o* and the names of positive ligands end in *-ium.* The neutral ligands are named as such. For example,

(*i*) *Negative ligands*

| | | | | | | | | |
|---|---|---|---|---|---|---|---|---|
| $F^-$ | — | Fluoro | $NO_2^-$ | — | Nitro | $Cl^-$ | — | Chloro |
| $ONO^-$ | — | Nitrito | $Br^-$ | — | Bromo | $SO_4^{2-}$ | — | Sulphato |
| $OH^-$ | — | Hydroxo | $(C_2O_4)^{2-}$ | — | Oxalato | $CN^-$ | — | Cyano |
| $CH_3COO^-$ | — | Acetato | $O^{2-}$ | — | Oxo | SCN | — | Isothiocyanato |
| $N_3^-$ | — | Azido | $O_2^{2-}$ | — | Peroxo | $NCS^-$ | — | Thiocyanato |
| $H^-$ | — | Hydrido | $CO_3^{2-}$ | — | Carbonato | $NO_3^-$ | — | Nitrato |
| $NH^{2-}$ | — | Imido | $NH_2^-$ | — | Amido | $S_2O_3^{2-}$ | — | Thiosulphato |
| $ClO_3^-$ | — | Chlorato | | | | | | |

(*ii*) *Positive ligands*

$NO^+$ — Nitrosonium

$NH_2NH_3^+$ — Hydrazinium

$NO_2^+$ — Nitronium

$C_5H_5N^+$ — Pyridinium

(*iii*) *Neutral ligands*

$NH_2-CH_2-CH_2-NH_2$ — Ethylenediamine

$C_6H_5N$ — Pyridine

$(C_6N_5)_3P$ — Triphenylphosphine

Exceptions in naming neutral ligands

$H_2O$ — Aquo $NH_3$ — Ammine

NO — Nitrosyl CO — Carbonyl

(*iv*) *Order of naming ligands.* When more than one type of ligands are present, they are named in alphabetical order regardless of their charge without separation by hyphen.

(*v*) When there are several ligands of the same kind, the prefixes di-, tri-, tetra-, *etc.* are used to indicate their number. When the name of the ligand includes the numerical prefixes (di, tri, etc.), then the prefixes *bis-*, *tirs-*, *tetrakis-* are used, such ligands are called *complex ligands.*

(*vi*) *Ending of names.* When the complex is anionic, the name of the central metal atom ends in *-ate*. For cationic and neutral ligands, the name of the metal is written without any characteristic ending.

The names of some metals in anionic complex ion are given below:

| | | |
|---|---|---|
| Ga — Gallate | B — Borate | Pd — Palladate |
| Cd — Cadmate | Pt — Platinate | Re — Rhenate |
| Zn — Zincate | Cr — Chromate | Ir — Iridinate |
| Si — Silicate | Co — Cobaltate | Os — Osmate |
| Ni — Nickelate | Al — Aluminate | |

For certain metals latin names are given:

| | | |
|---|---|---|
| Fe — Ferrate | Au— Aurate | Pb — Plumbate |
| Sn — Stannate | Ag— Argentate | |

For example,

$[Co(NH_3)_6]Cl_3$ — Hexaamminecobalt (III) chloride

$K[Pt(NH_3)Cl_5]$ — Potassium amminepentachloroplatinate (V)

(*vii*) *Oxidation state of the central metal atom.* The oxidation state of the metal ion is designated by Roman numerals such as, I, II, III, etc. in the brackets at the end of the name of the complex without a gap between the two.

(*viii*) *Naming of bridging ligands.* Ligands which act as bridges between two metal atoms, the Greek letter μ (mu) is written before their names. For example,

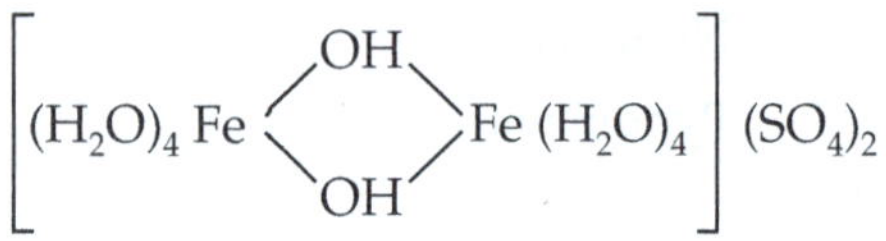

Tetraaquoiron (III)-μ-dihydroxotetraaquoiron (III) sulphate
or octaaquo-μ-dihydroxodiron (III) sulphate

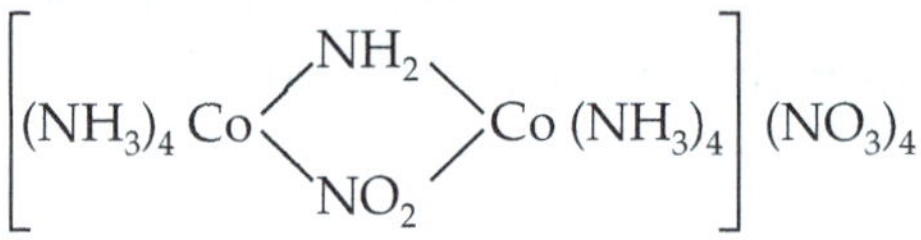

Tetraamminecobalt (III)-μ-amido-μ-nitritotetraamminecobalt (III) nitrate
or (octaammine-μ-amido-μ-nitritodicobalt (III) nitrate)

If any lattice components such as water or solvent of crystallization are present, these follow the name, and are preceded by the number of these groups is Arabic numberals.

(*ix*) The optically active complexes are designated by (+) and (–) or by *d*– and *l*– respectively.

Common examples describing these rules are given in Table 14.5.

**TABLE 14.5**

| Formula | IUPAC Name |
|---|---|
| **A. Anionic complexes** | |
| $Na_2[ZnCl_4]$ | Sodium tetrachlorozincate (II) |
| $K_3[Fe(CN)_3NO]$ | Potassium pentacyanonitrosylferrate (II) |
| $Na_3[Ag(S_2O_3)_2]$ | Sodium bis-(thiosulphato) argentate (I) |
| $K_2[Cr(CN)_2O_2(O)_2NH_3]$ | Potassium amminedicyanodioxoperoxochromate (VI) |
| $Na_2[SiF_6]$ | Sodium hexafluorosilicate (IV) |
| $K_3[Al(C_2O_4)_3]$ | Potassium trioxalatoaluminate (III) |

| | |
|---|---|
| $(NH_4)_3[Cr(NCS)_6]$ | Ammonium hexaisothiocyanatochromate (III) |
| $(NH_4)[Cr(NH_3)_2(NCS)_4]$ | Ammonium diamminetetraisothiocyanatochromate (III) |
| $K[SbCl_5C_6H_5]$ | Potassium pentachlorophenylantimonate (V) |
| $Na_2[CrOF_4]$ | Sodium tetrafluoroxochromate (IV) |
| **B. Cationic complexes** | |
| $[Co(NH_3)Cl_3]$ | Hexaamminecobalt (III) chloride |
| $[CoSO_4(NH_3)_4]NO_3$ | Tetraamminesulphatocobalt (III) nitrate |
| $[CoCl.CN.NO_2(NH_3)_3]$ | Triamminecyanonitrocobalt (III) |
| $[Cd(SCN)_4]^{2+}$ | Tetraisothiocyanatocadmium (II) |
| $[Cr(en)_3]Cl_3$ | Tris-(ethylenediamine) chromium (III) chloride |
| $[Pt(py)_4]Cl_2$ | Tetrakis-(pyridine)platinum (II) chloride |
| $[Pt(en)(NH_3)_2(NO_2)Cl]SO_4$ | Diamminechloroethylenediamine nitroplatinum (IV) sulphate |
| $[Ni(H_2O)_6](ClO_4)_2$ | Hexaaquonickel(II) perchlorate |
| $[Co(en)_2Cl_2]^+$ | Dichloro bis-(ethylenediammine) cobalt (III) ion |
| $[Fe(C_5H_5)_2]$ | Bis-(cyclopentadienyl) iron (II) |
| **C. Molecular complexes** | |
| $[Pt(py)(NH_3)(NO_2)(Cl)(I)]$ | Amminebromochloroiodonitro (pyridine) platinum (III) |
| $[Pd(H_2O)_2(ONO)_2I_2]$ | Diaquodiiododinitritopalladium (IV) |
| $[Co(NH_3)_3(NO_2)_2]$ | Triamminetrinitrocobalt (II) |
| $[CuCl_2(CH_3NH_2)_2]$ | Dichloro bis-(methylamine) copper (II) |
| **D. Complexes with both cation and anion** | |
| $[Cr(NH_3)_6][Co(CN)_6]$ | Hexaamminechromium (III) hexacyanocobaltate (III) |
| $[Pt(NH_3)_4Cl_2]\ [PtCl_4]$ | Tetraamminechloroplatinum (IV) tetrachloroplatinate (II) |
| $[Pt(py)_4]\ [PtF_4]$ | Tetrapyridineplatinum (II) tetrafluoroplatinate (II) |
| $[CO(NH_3)_6]\ [Cr(C_2O_4)_3]$ | Hexaamminecobalt (III) trioxalatochromate (III) |
| **E. Bridged polynuclear complexes** | |
| $[(NH_3)_5\ CoO_2{-}Co(NH_3)_5]^{4+}$ | Decaammine-μ-peroxodicobalt (III) ion |
| 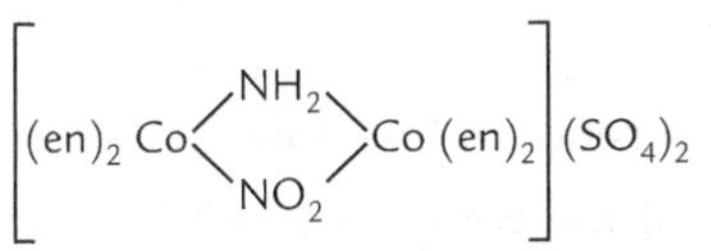 | Tetrakis-(ethylenediamine)-μ-amido-μ-hydroxodicobalt (III) sulphate |
| $[(NH_3)_5Co.NH_2.Co(NH_3)_5](NO_3)_5$ | Decaammine-μ-amidodicobalt (III) nitrate |

## 14.4 THEORIES OF COORDINATION COMPOUNDS

Several theories have been proposed for explaining bonding, hybridization and other properties of complex compounds. Some of them are discussed below:

### 14.4.1 Werner's Theory

*Alfred Werner* (1893), regarded as the father of coordination chemistry, put forward his well-known Werner's theory of coordination. According to this theory:

1. The metals possess two types of valencies, *primary* and *secondary*. The primary valency is ionizable and corresponds to the oxidation state of the metal forming the complex compound or ion. It is always satisfied by the negative ion. The attachment of the metal ion with the negative ligand is shown by dotted lines.

   The secondary valency is nonionizable and is satisfied either by negative ions or neutral molecules. The secondary valencies are shown by thick lines. It corresponds to the coordination number of the metal atom.
2. The central metal atom satisfies both the primary and secondary valencies. For this purpose, the negative ion may have dual behavior, *i.e.*, in satisfying the primary and the secondary valencies.
3. Every metal (atom or ion) has a fixed number of secondary valencies. This number is termed as the coordination number of the metal. The sphere enclosing only the metal and the coordinated group is known as the coordination sphere.
4. The ligands which satisfy the secondary valencies must point out in definite directions in space, whereas the primary valencies are nondirectional in nature.
5. The net charge on the complex is the algebraic sum of the charge borne by the central atom and the ligands. For example, $[Fe(CN)_6]^{3-}$ carries – 3 charge because $Fe^{3+}$ ion carries + 3 charge and 6 $CN^-$ carries – 6 charge, hence net charge is + 3 + (– 6) = – 3.
6. The secondary valencies attribute to the isomerism in coordination compounds.

The Werner's theory is mainly applied to give the structures of complexes of cobalt with ammonia. This may be illustrated as:

(*a*) **Luteo cobaltic chloride, $CoCl_3.6NH_3$**

(*i*) It is an orange yellow crystalline solid.

(*ii*) On treatment with $AgNO_3$, all three chlorine atoms are precipitated as AgCl. This indicates that all three chlorine atoms are attached to the metal with an ionic bond.

(*iii*) On treatment with concentrated $H_2SO_4$, it gives $Co_2(SO_4)_3$ and all three chlorine atoms are lost as HCl. This indicates that all six ammonia molecules are attached to the metal with strong bonds.

(*iv*) The conductivity measurement shows the presence of four ions.

On the basis of the above facts, the structure of the complex $[Co(NH_3)_6]Cl_3$ may be given as shown in Figure 14.1(*a*).

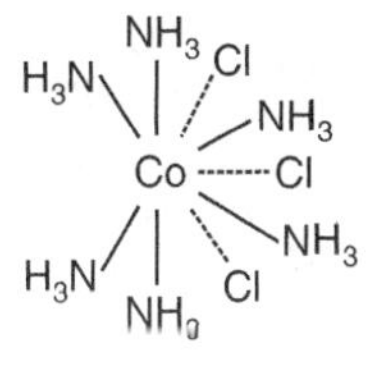


*FIGURE 14.1* **(*a*) Structrure of luteo cobaltic chloride.**

The primary valencies are represented by lines and the secondary valencies are represented by thick lines.

(*b*) **Roseo cobaltic chloride, $CoCl_3.5NH_3.H_2O$**

(*i*) It is a pink crystalline solid.

(*ii*) On treatment with $AgNO_3$, all three chlorine atoms are precipitated as AgCl. This indicates that all three chlorine atoms are attached to the metal with an ionic bond.

(*iii*) On treatment with concentrated $H_2SO_4$, it gives $Co_2(SO_4)_3$. $10NH_3.5H_2O$. This indicates that all five ammonia molecules and one water molecule are attached to the metal with a strong bond.

(*iv*) The conductivity measurement shows the presence of four ions.

On the basis of the above facts, the structure of the complex $[Co(NH_3)_5(H_2O)]Cl_3$ may be given as shown in Figure 14.1(*b*).

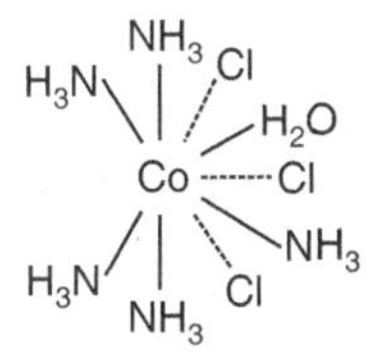


*FIGURE 14.1* **(*b*) Structure of roseo cobaltic chloride.**

(*c*) **Purpureo cobaltic chloride, $CoCl_3.5NH_3$**

(*i*) It is a violet crystalline solid.

(*ii*) On treatment with $AgNO_3$, only two chlorine atoms are precipitated as AgCl. This indicates that only two chlorine atoms are attached to the metal with an ionic bond.

(*iii*) On treatment with concentrated $H_2SO_4$, it gives $CoCl(SO_4).5NH_3$ and two chlorine atoms are lost as HCl. This compound does not give any precipitate with $AgNO_3$. This indicates that all five ammonia molecules and one chlorine atom are attached to the metal with a strong bond.

(*iv*) The conductivity measurement shows the presence of three ions. On the basis of the above facts, the structure of the complex $[Co(NH_3)_5Cl]Cl_2$, may be given as shown in Figure 14.1(*c*).

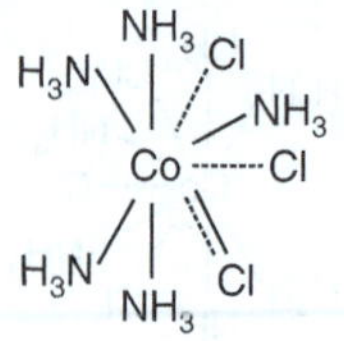

*FIGURE 14.1* **(*c*) Structure of purpureo cobaltic chloride.**

(*d*) **Praseo cobaltic chloride, $CoCl_3.4NH_3$**

(*i*) It is a green crystalline solid.

(*ii*) On treatment with $AgNO_3$, only one chlorine atom is precipitated as AgCl. This indicates that only one chlorine atom is attached to the metal with an ionic bond.

(*iii*) On treatment with concentrated $H_2SO_4$, only one chlorine atom is lost as HCl and no $(NH_4)_2SO_4$ is obtained. This indicates that all four ammonia molecules and two chlorine atoms are attached to the metal with a strong bond.

(*iv*) The conductivity measurement shows the presence of two ions. On the basis of the above facts, the structure of the complex $[Co(NH_3)_4Cl_2]Cl$ may be given as shown in Figure 14.1(*d*).

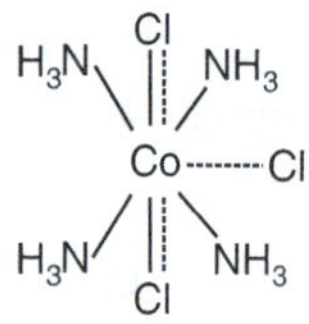

*FIGURE 14.1* **(*d*) Structure of praseo cobaltic chloride.**

### 14.4.2 Sidwick's Theory

According to *Sidwick,* all neutral molecules or anions which are capable of being coordinated to the central metal ion have atoms with at least one unshared (lone) pair of electrons in their valency shells. The ligands donate the electron pair to the central metal ion and thus form a coordinate bond, M ⟵ L which indicates that the ligand, L has donated an electron to the metal ion M. Thus, according to Sidwick's concept, the structure of the complex ion, $[Co(NH_3)_6]^{3+}$ can be shown as seen in Figure 14.2.

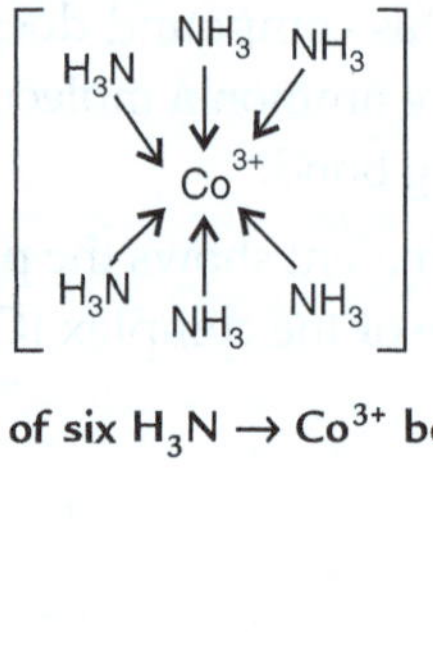

*FIGURE 14.2* **Formation of six $H_3N \rightarrow Co^{3+}$ bond in $[Co(NH_3)_6]^{3+}$ ion.**

### Sidwick's Effective Atomic Number (EAN) Rule

The effective atomic number of a metal in the complex compound is the sum of the electrons present in the metal ion and the total number of electrons gained from the ligands. According to Sidwick, the total number of electrons surrounding the coordinated metal ion is equal to the atomic number of the inert gas which follows the central metal atom in the periodic table. This is called *Effective Atomic Number Rule* or *Noble Gas Rule*. When EAN is 36(Kr), 54(Xe) or 86(Rn), the rule is said to be followed.

#### Calculation of EAN

EAN of the central metal atom or ion in a given complex ion is given by:

$$\text{EAN} = (Z - x) + n \times y$$

Here, Z = atomic number of the central metal atom, $x$ = oxidation state of the central metal ion, $n$ = number of ligands and $y$ = number of electrons donated by one ligand. With the help of this formula, EAN of the central metal atom or ion in some complexes has been calculated as shown in Table 14.6.

**TABLE 14.6 Calculation of EAN of the Central Metal Atom of Some Complex ions**

| *Complex Ion* | *Central Metal Atom* | *Atomic Number* | *Oxidation State of Metal in the Complex Ion* | *Number of Electrons Gained by Ligands* | *Effective Atomic Number* |
|---|---|---|---|---|---|
| $[Fe(CN_6]^{4-}$ | Fe | 26 | + 2 | 6 × 2 = 12 | 26 – 2 + 12 = 36 |
| $[Cu(CN)_4]^{3-}$ | Cu | 29 | + 1 | 4 × 2 = 8 | 29 – 1 + 8 = 36 |
| $[Co(NO_2)_6]^{3-}$ | Co | 27 | + 3 | 6 × 2 = 12 | 27 – 3 + 12 = 36 |
| $[Ni(CO)_4]$ | Ni | 28 | 0 | 4 × 2 = 8 | 28 – 0 + 8 = 36 |
| $[Pt(NH_3)_6]^{4+}$ | Pt | 78 | + 4 | 6 × 2 = 12 | 78 – 4 + 12 = 86 |
| *$[Fe(CN)_6]^{3-}$ | Fe | 26 | + 3 | 6 × 2 = 12 | 26 – 3 + 12 = 35 |
| *$[Cr(NH)_3]^{3+}$ | Cr | 24 | + 3 | 6 × 2 = 12 | 24 – 3 + 12 = 33 |
| $[Pt(NH_3)_4]^{2+}$ | Pt | 78 | + 2 | 4 × 2 = 8 | 72 – 2 + 8 = 78 |

From the above table, it is clear in the most of the cases the EAN of the central metal atom is equal to the atomic number of the next inert gas but has some exceptions (* complexes). Hence, the tendency to attain an inert gas configuration is not a necessary condition for the stability of complex compounds.

#### Limitations of Sidwick's Theory

Following are the limitations of Sidwick's Theory:

(*i*) There will be an accumulation of negative charge on the central metal ion by a donation of electron pairs from ligands. Thus, the central ion will become negatively charged with respect to ligands.

(*ii*) The electron pair available for donation of ammonia, water, and many other neutral molecules is the *s*-electron pair which has no bonding characteristics on the basis of available energy.

**Example 1.** *Using the rule, predict the molecular formulae of simple carbonyls of Fe (atomic number 26) and Cr (atomic number 24). The metals have zero oxidation state in the carbonyls.*

**Solution. Molecular formula of iron carbonyl**

Atomic number of iron = 26

Atomic Number of the next noble gas, Kr (*i.e.*, EAN) = 36

∴ Number of electrons to be provided by CO = 36 – 26 = 10

Since each CO provides 2 electrons, the total number of carbonyl (CO) groups $= \frac{10}{2} = 5$

∴ Molecular formula of simple iron carbonyl = $Fe(CO)_5$

**Molecular formula of chromium carbonyl**

Atomic number of chromium = 24

EAN of chromium = 36

∴ Number of electrons to be provided by CO = 36 – 24 = 12

Number of CO groups = 12 ÷ 2 = 6

Hence, the molecular formula of chromium carbonyl will be **$[Cr(CO)_6]$.**

**Example 2.** *Discuss the validity of the EAN rule for the complex ions $[Fe(CN)_6]^{4-}$ and $[Fe(CN)_6]^{3-}$ (Atomic number of Fe is 26)*

**Solution.** (*a*) **$[Fe(CN)_6]^{4-}$ ion**

Oxidation state of Fe in $[Fe(CN)_6]^{4-}$ = + 2

Number of electrons in Fe (II) of the complex = 26 – 2 = 24

Number of electrons donated by six $CN^-$ ligands = 6 × 2 = 12

∴ EAN of Fe(II) in $[Fe(CN_6)^{4-}$ = 24 + 12 = 36

Since 36 is equal to the nearest noble gas (Kr), the complex $[Fe(CN)_6]^{4-}$ follows the EAN rule.

(*b*) **$[Fe(CN_6)]^{3-}$ ion**

Oxidation state of Fe in $[Fe(CN_6)]^{3-}$ ion = + 3

Number of electrons in Fe = 26

∴ Number of electrons in Fe (III) of the complex = 26 – 3 = 23

Number of electrons donated by six $CN^-$ ligands = 6 × 2 = 12

∴ EAN of Fe (III) in $[Fe(CN_6)]^{3-}$ = 23 + 12 = 35

Since 35 does not coincide with the number of electrons in the atom of the nearest noble gas (Kr), so the complex $[Fe(CN_6)]^{3-}$ does not obey the EAN rule.

### 14.4.3 Valence Bond Theory (VBT)

This theory was proposed by *L. Pauling* and *J.L. Slator*. It deals with the electronic structure of central metal ion in its ground state, kind of bonding, shape, and magnetic properties of the complexes. The main assumptions of this theory are:

1. The central metal ion makes available a number of empty *s*, *p*, and *d* atomic orbitals. These vacant orbitals hybridize together to form hybrid orbitals which are equivalent in energy and have definite geometry. The most common hybridizations are given in Table 14.7.
2. The *d*-orbitals involved in the hybridization may be either inner *d*-orbitals, *i.e.*, $(n-1)d$ or outer *d*-orbitals, *i.e.*, *nd* orbitals. The complexes formed in these two ways are referred to as *low spin* and *high spin* complexes, respectively.
3. Each ligand has at least one sigma orbital containing a lone pair of electron.
4. The vacant hybrid orbital of the metal ion overlaps with the filled sigma orbital of the ligand forming *ligand → metal* coordinate bond.
5. If the complex contains unpaired electrons, it is paramagnetic in nature and if it contains all paired electrons, it is diamagnetic in nature.

#### Applications of Valence Bond Theory

#### I. Complexes with Coordination Number 4

**A. Square Planar Complexes.** The square planar coordination compounds result from $dsp^2$ hybridization involving mixing of $(n-1)d_{x^2-y^2}$, $ns$, $np_x$ and $np_y$ orbitals. For example, formation of $[Ni(CN)_4]^{2-}$, $[Ni(NH_3)]^{2+}$, $[Cu(NH_3)_4]^{2+}$, etc. occurs through $dsp^2$ hybridization.

TABLE 14.7 Some Common Hybridizations and their Geometries in Complexes

| *CN* | *Hybridization* | *Molecular Geometry* |
|---|---|---|
| 2 | $sp$ | 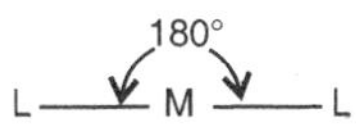 Linear $ML_2$ |
| 3 | $sp^2$ | 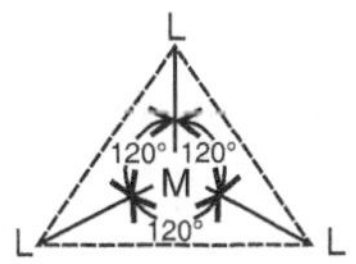 Equilateral triangular $ML_3$ |
| 4 | $sp^3$ | 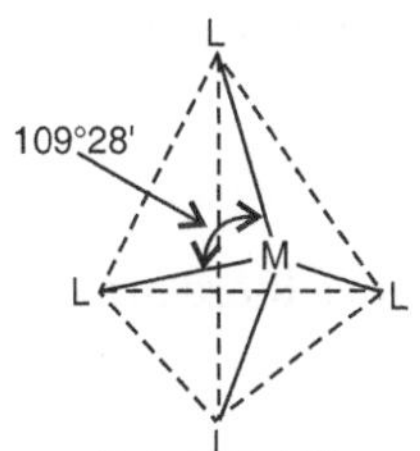 Tetrahedral $ML_4$ |

| | | |
|---|---|---|
| 4 | $dsp^2$ or $sp^2d$ | 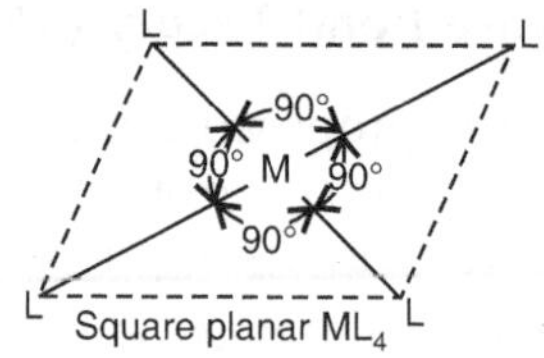 Square planar $ML_4$ |
| 4 | $dsp^3$ or $sp^3d$ | 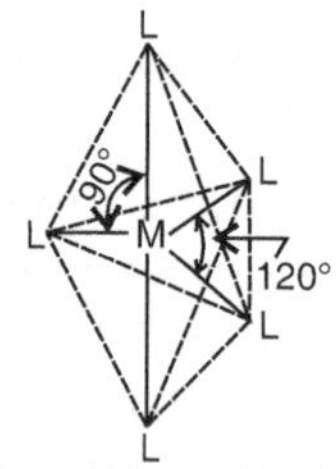 Trigonal bipyramidal $ML_5$ |
| 6 | $d^2sp^3$ or $sp^3d^2$ | 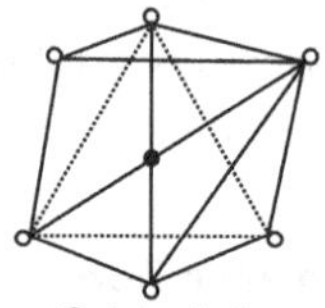 Octahedral |

**Formation of $[Ni(CN)_4]^{2-}$ Complex Ion**

The electronic configuration of nickel atom is $1s^2$, $2s^2$ $p^6$ $3s^2p^6d^84s^2$ and that of $Ni^{2+}$ is $1s^2$, $2s^2$ $p^6$ $3s^2$ $p^6d^8$ $4s^0$. When $4CN^-$ ions approach the $Ni^{2+}$ ion, pairing of electrons takes place such that the eight electrons in the $3d$-orbitals are paired and one empty $d$-orbital is available for $dsp^2$ hybridization and $[Ni(CN)_4]^{2-}$ complex ion is formed as shown:

(*i*) Ni atom

(*ii*) $Ni^{2+}$ ion

(*iii*) Pairing of $3d$ electrons

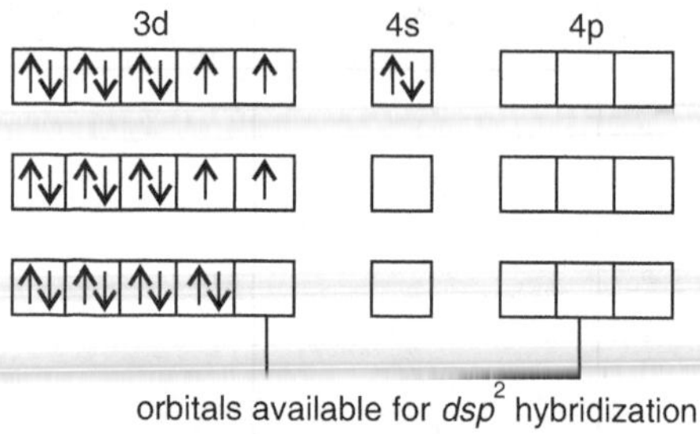


(*iv*) Formation of $[Ni(CN)_4]^{2-}$

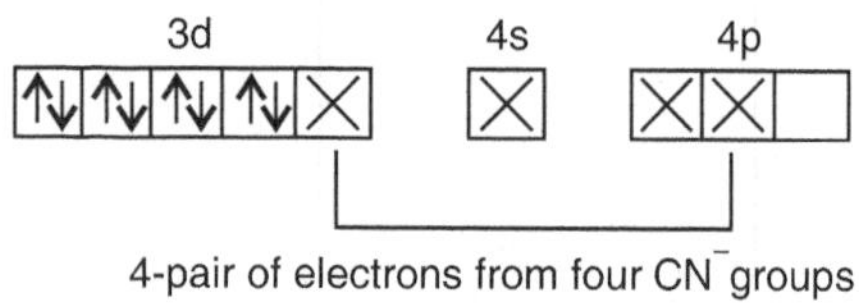


Since there are no unpaired electrons in the complex ion, it is diamagnetic in nature.

**B. Tetrahedral Complexes.** The tetrahedral complexes are formed as a result of $sp^3$ hybridization involving the mixing of $ns$, $np_x$, $np_y$, and $np_z$ orbitals. For example, $[Ni(CO)_4]$, $[Zn(NH_3)_4]^{2+}$, $[ZnCl_4]^{2-}$, $[MX_4]^{2-}$ where, M—Cu, Mn and Ni and X = Cl, Br, I, etc.

**Formation of $[MnCl_4]^{2-}$ Complex ion**

In $[MnCl_4]^{2-}$ complex, the ion Mn is in + 2 oxidation state, so its electronic configuration will be $3d^5$, $4s^0$, $4p^0$.

(*i*) Mn atom

(*ii*) $Mn^{2+}$ ion

(*iii*) Formation of $[MnCl_4]^{2-}$ ion

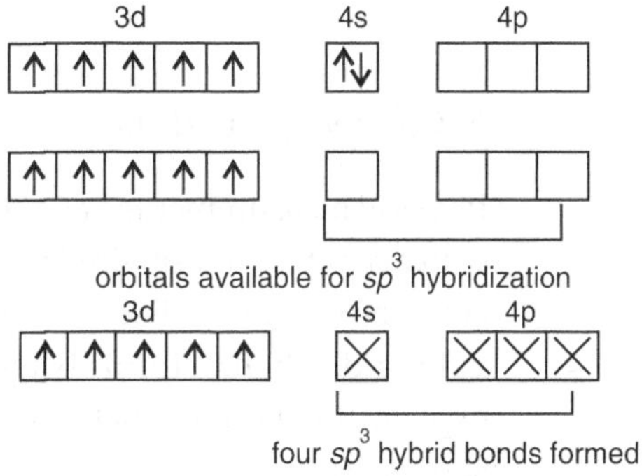


## II. Octahedral Complexes (CN = 6)

These are the most common type of complexes formed as a result of $(n-1)d^2.ns.np^3$ or $ns.np^3.nd^2$ type of hybridization. These are of two types:

**A. Inner-orbital Complexes.** These complexes are formed by the use of inner *d*-orbitals for hybridization. In these complexes, the electrons of the metal are forced to pair up and hence the complex will be diamagnetic and will have a lesser number of unpaired electrons. Such complexes are called *low spin complexes*. For example, consider the formation of $[Co(NH_3)_6]^{3+}$ complex ion.

(*i*) Co atom (Atomic number 27)

(*ii*) $Co^{3+}$ ion (ground state)

(*iii*) $Co^{3+}$ ion (after rearrangement in presence of $CN^-$ ions)

(*iv*) Formation of $[Co(NH_3)_6]^{3+}$ complex ion

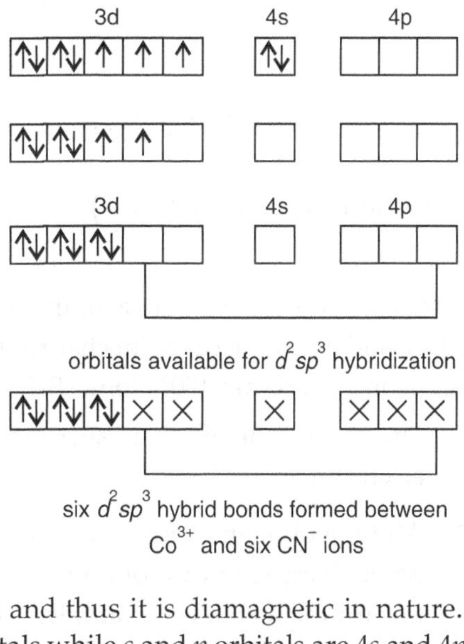


The complex ion has no unpaired electron and thus it is diamagnetic in nature. In this complex ion, the *d*-orbitals used are 3*d*-orbitals while *s* and *p* orbitals are 4*s* and 4*p*. Other examples of such complexes are $[Cr(NH_3)_6]^{3+}$, $[Mn(CN)_6]^{3}$, $[Fe(CN)_6]^{3-}$, etc.

**B. Outer-orbital Complexes.** These complexes are formed by the use of outer *d*-orbitals for hybridization. These complexes have a larger number of unpaired electrons. Such complexes are called *high spin complexes.* For example, consider the formation of $[CoF_6]^{3-}$ complex ion.

(*i*) Co atom

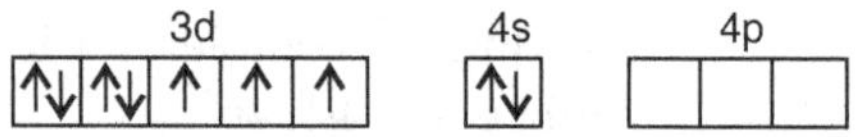


(*ii*) $Co^{3+}$ ion (ground state)

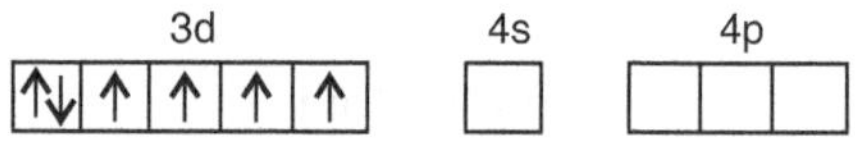


Physical measurements have shown that the $[CoF_6]^{3-}$ ion has a paramagnetic character equivalent to four unpaired electrons in the 3*d*-orbital. Thus none of the five 3*d*-orbitals could be disturbed. For such complexes, *J. Huggin* suggested that *d*-orbitals involved in hybridization to form six hybrid orbitals are 4*d*-orbitals and not 3*d*-orbitals, which explains the paramagnetic character of $[CoF_6]^{3-}$ ion. So, formation of $[CoF_6]^{3-}$ ion is shown as:

(*iii*)

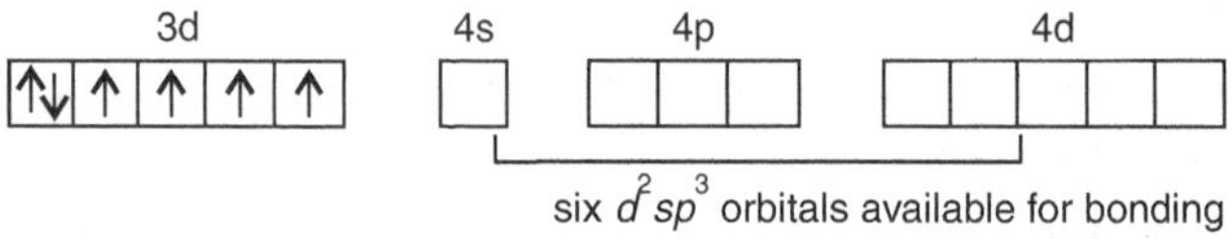


(*iv*) Formation of $[CoF_6]^{3-}$ ion

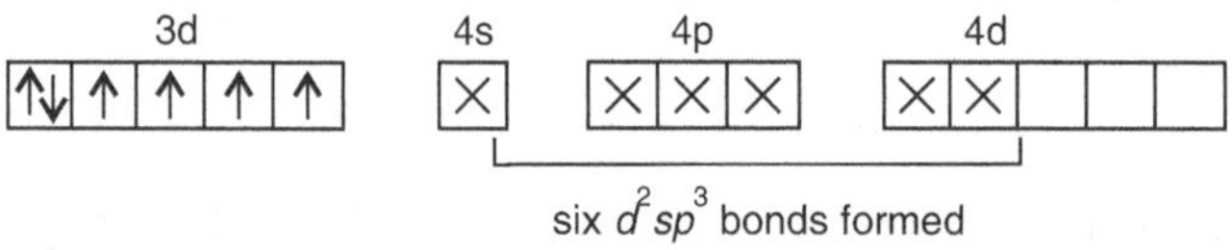


Other such complexes are $[FeF_6]^{3-}$, $[Cu(NH_3)_6]^{2+}$, $[Ni(NH_3)_6]^{2+}$, etc.

### 14.4.4 Crystal Field Theory (CFT)

This theory is based on the work of *Bethe* (1929) and *Van Vleck* (1931–35). This theory takes no consideration of the covalent nature of the metal-ligand bond and is based on the assumption that the metal ion and ligands act as point charges and the interaction between them is purely electrostatic, *i.e.*, metal-ligand bonds are 100% ionic. Salient features of this theory are:

1. The central metal ion is surrounded by ligands containing at least one lone pair of electrons.
2. All type of ligands, ionic ($F^-$, $Cl^-$, etc.] or molecules ($H_2O$, $NH_3$, etc.] are considered as point charges. In case of a neutral molecule, the negative end of the dipole is oriented toward the metal. For example,

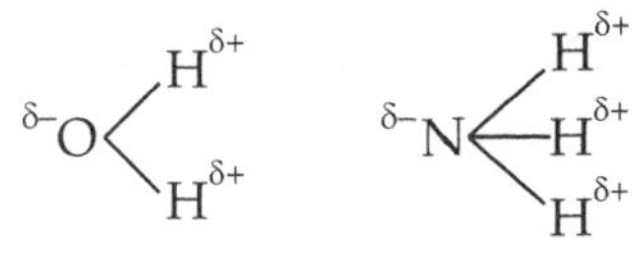


3. The attraction between the metal and the ligands is purely electrostatic.
4. The interaction between the valence electrons of the cation and those of the ligand is purely repulsive. These repulsive forces are responsible for the splitting of the *d*-orbitals of the metal cation.
5. The number of ligands and their arrangement around the central metal atom determines the crystal field.
6. CFT doesn't provide for electrons to enter the metal orbitals. Thus, the central metal ion and the ligands do not mix their orbitals or share electrons, *i.e.*, it does not consider any orbital overlap.
7. Complexes are thus presumed to form when centrally situated cations electrically attract ligands which may either be anions or dipole molecules.

**A. Crystal Field Splitting of *d*-orbitals in Octahedral Complexes**

On the basis of the orientation of the lobes, five *d*-orbitals (Figure 14.3) are divided into two groups:

(*a*) $e_g$ *group (doubly degenerate).* These orbitals have their lobes along the coordinate axis and are called *axial orbitals.* They include $d_{x^2-y^2}$ and $d_{z^2}$.

(*b*) $t_{2g}$ *group (triply degenerate).* The lobes of the *s*-orbitals lie between the axis and are called *non-axial orbitals.* They include $d_{xy}$, $d_{yz}$, and $d_{zx}$ orbitals.

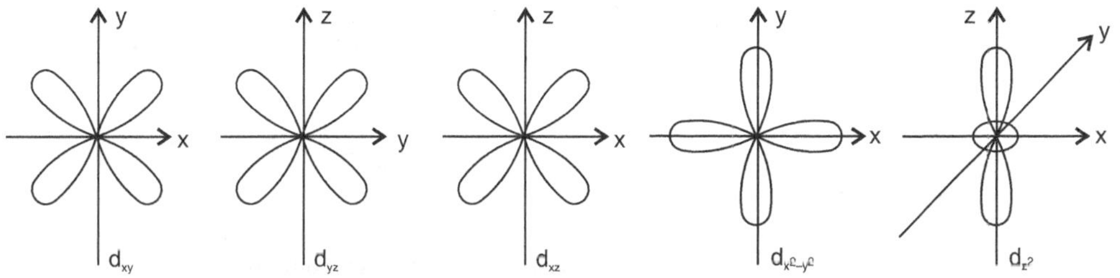


*FIGURE 14.3* **Shapes of *d*-orbitals.**

In a free metal ion, all five *d*-orbitals are degenerate, *i.e.*, are of same energy. Consider an octahedral complex, $[ML_6]^{n+}$ in which the central metal cation $M^{n+}$ is placed at the center of the octahedron and is surrounded by six ligands which reside at the six corners (Figure 14.4).

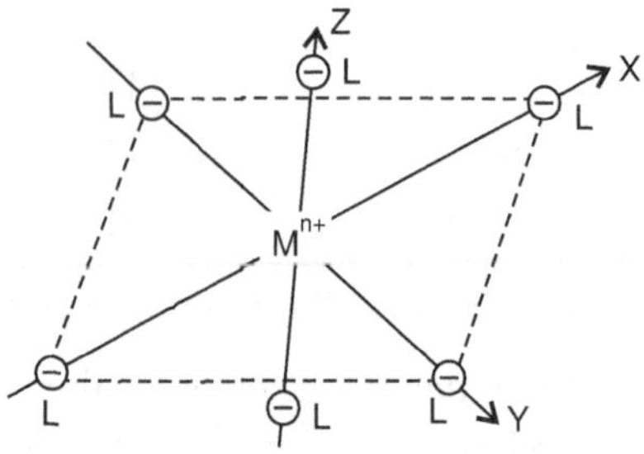


*FIGURE 14.4* **Position of the central metal cation $M^{n+}$ and six ligands, in an octahedral complex, $[ML_6]^{n+}$.**

Suppose that both ligands on each of the three axes are allowed to approach the metal cation, the electrons in *d*-orbitals of the metal cation are repelled by negative point charges by the negative end of the dipole of the ligands. This repulsion raises the energy of all five *d*-orbitals. If all the ligands approaching the central cation are at an equal distance from each of the *d*-orbitals, then the raised energy will be equal in all *d*-orbitals. But the lobes of two $e_g$ orbitals lie directly in the path of the approaching ligand, so the electrons in these orbitals experience greater force of repulsion than those in $t_{2g}$ orbitals, thus five originally degenerate *d*-orbitals split into two levels the $t_{2g}$ and $e_g$ levels. This separation of *d*-orbitals of the metal cation into two sets having different energies is called *crystal field splitting*. The energy gap between $t_{2g}$ and $e_g$ orbitals is denoted by $\Delta_o$ or 10 *Dq*. This energy difference, which arises due to the difference in electrostatic field exerted by ligands on $t_{2g}$ and $e_g$ orbitals of the central metal ion, is called *crystal field splitting energy*.

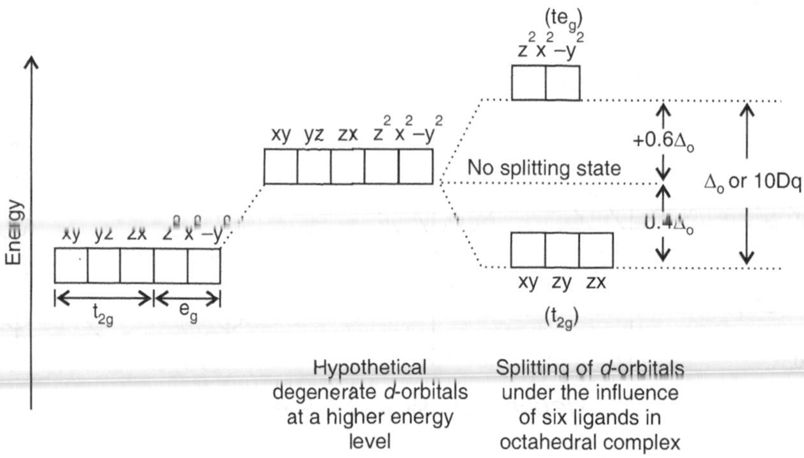


*FIGURE 14.5* **Crystal field splitting of d-orbitals in octahedral complexes.**

### Stronger and Weaker Ligands

Stronger ligands are those which exert a stronger field on the central metal ion and hence have higher splitting power. Weaker ligands exerts weak field on the central metal ion and consequently have lower splitting power. Thus, the stronger ligand ($CN^-$) gives a larger value of $\Delta_o$, while the weaker ligand ($F^-$) gives a smaller value of $\Delta_o$. The common ligands can be arranged in ascending order of crystal field splitting energy as:

$I^- < Br^- < S^{2-} < Cl^- < NO_3^- < F^- < OH^- < EtOH < C_2O_4^{2-} < H_2O < EDTA < (NH_3$ and Pyridine) $<$ ethylene diamine (en) $< NO_2^- < CN^- < CO$

The order remains practically constant for different metals and this series is called *spectrochemical series.*

Under the influence of a weak ligand, $\Delta_o$ is small and hence all five *d*-orbitals are almost degenerate and distribution of electrons takes place according to Hund's rule. Thus, in $[CoF_6]^{3-}$, $d^6$, electrons are filled as $t_{2g}^4, e_g^2$ while in $[Co(NH_3)_6]^{3+}$, $d^6$, electrons are filled as $t_{2g}^6, e_g^0$.

**Mean Pairing Energy (P)**

Mean pairing energy (P) is the energy required to pair two electrons against electron-electron repulsion in the same orbital. *P* is the pairing energy for one electron pair. *P* is generally expressed in $cm^{-1}$. Pairing energy depends on the principal energy level (*n*) of *d* electrons. For complexes having weak ligands, $\Delta_o < P$ while for complexes having stronger ligands, $\Delta_o > P$.

**Factors Affecting the Value of $\Delta_o$**

The value of $\Delta_o$ or 10Dq depends on the following factors.

(*i*) **Nature of the metal ion.** The effect of nature of the metal ion can be studied as follows:

(*a*) *Charge on the metal ion.* If the charge on the same metal ion increases, *e.g.*, $\Delta_o$ values for $[Cr(H_2O)_6]^{2+}$ and $[Cr(H_2O)_6]^{3+}$ are 13900 $cm^{-1}$ and 17400 $cm^{-1}$ while for $[Fe(H_2O)_6]^{2+}$ and $[Fe(H_2O)_6]^{3+}$ are 10400 $cm^{-1}$ and 13700 $cm^{-1}$ respectively.

Consider the case where the charge on a different metal ion is different but their electronic configuration is the same. In such cases, the values of $\Delta_o$ for $[V(H_2O_6]^{2+}$ and $[Cr(H_2O_6]^{2+}$ are 12400 $cm^{-1}$ and 17400 $cm^{-1}$, respectively.

(*b*) *Quantum number of the d-orbitals of the metal ion.* It is observed that $\Delta_o$ value increases about 30–50% from $3d^n$ to $4d^n$ whereas it remains almost same for $4d^n$ to $5d^n$ metal ion complexes *e.g.*, $\Delta_o$.

(*ii*) **Nature of the ligands.** If we compare the value of $\Delta_o$ for complexes of the same metal ion with different ligands, then we observe that the value of $\Delta_o$ vary regularly. It means the value of $\Delta_o$ depends upon the nature of the ligands. Thus, the ligands can be arranged in the order of increasing field strength and the series thus obtained is known as spectrochemical series.

(*iii*) **Geometry of the complex.** The 10Dq value for octahedral, tetrahedral, and square planar complexes are in the order of:

$$\Delta_{sp} > \Delta_o < \Delta_t$$

**III. Crystal Field Splitting in Tetrahedral Complexes**

A regular tetrahedron is related to a cube. One atom is at the center of the cube and four of the eight corners of the cube are occupied by ligands as shown in Figure 14.6.

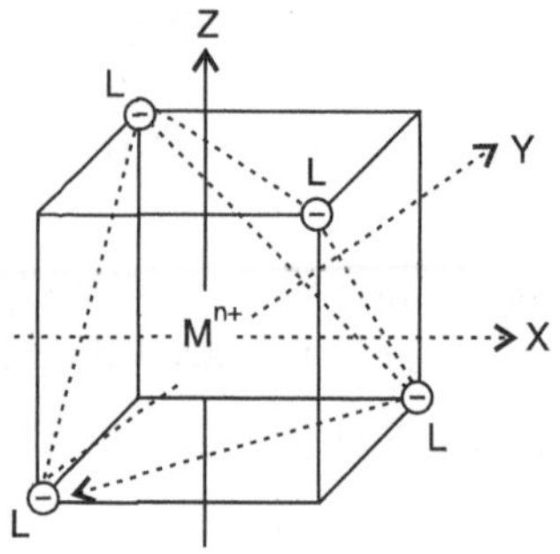


*FIGURE 14.6* **Tetrahedral arrangement of four ligands (L) around the metal ion ($M^{n+}$) in tetrahedral ion complex ion, $(ML_4)^{n+}$.**

The directions *x*, *y*, and *z* point to the centers of the faces of the cube. The $e_g$ orbitals point along *x*, *y*, and *z* axes and $t_{2g}$ orbitals point between *x*, *y*, and *z* axes. The direction of the approach of the ligands does not coincide exactly with either the $e_g$ or the $t_{2g}$ orbitals. The angle between an $e_g$ orbital, the central metal atom, and the ligand is 54.4°, while the angle between $t_{2g}$ orbital, the central atom, and the ligand is 35.16°. Thus, $t_{2g}$ orbitals are nearer to the direction of the ligands than the orbitals. The approach of the ligands raises the energies of both sets of orbitals. The energy of the $t_{2g}$ orbitals is raised most because they are closest to the ligands. The $t_{2g}$ orbitals are $0.4\Delta_t$ above the weighted average energy of the two groups and $e_g$ orbitals are $0.6\Delta_t$ below the average (Figure 14.7).

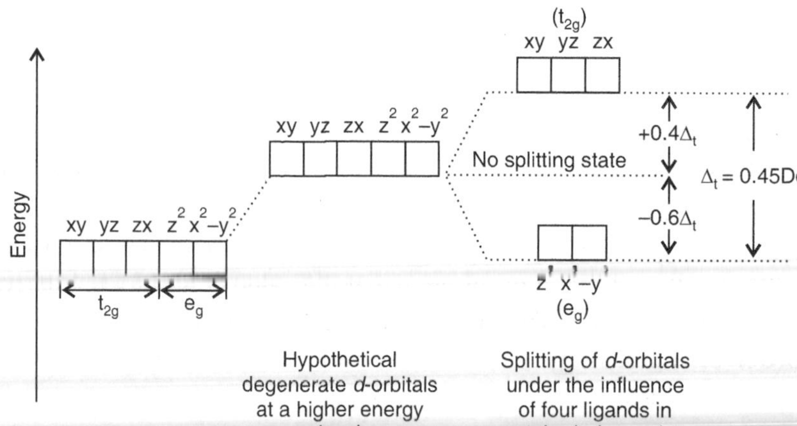


*FIGURE 14.7* **Crystal field splitting of *d*-orbitals in tetrahedral complexes.**

The magnitude of the crystal field splitting $\Delta_t$ in tetrahedral complexes is considerably less than in octahedral fields. The reasons for this are:

(*a*) There are only four ligands instead of six, so the ligand field is only two thirds the size.

(*b*) The direction of the orbitals doesn't coincide with the direction of the ligands. This also reduces the crystal field splitting by two thirds.

Thus, the tetrahedral crystal fields splitting energy,

$$\Delta_t = \frac{2}{3} \times \frac{2}{3} = \frac{4}{9} \Delta_o$$

Thus, it is never energetically favourable to pair electrons, and all tetrahedral complexes are thus high spin.

### IV. Crystal Field Splitting in Square Planar Complexes

Square planar and tetragonally distorted octahedral geometries are obtained from regular octahedral geometry. Consider a regular octahedral complex in which $L_a$ are two *trans*-ligands lying along the Z-axis and $L_b$ are basal ligands lying in the XY plane.

If the two ligands are moved slightly away from the Z-axis, then each of the two M-$L_a$ distances becomes slightly greater than each of the four coplanar M-$L_b$ distances and the symmetrical shape of the octahedral complex gets distorted to a tetragonally distorted octahedral shape. If the two $L_a$ are completely removed, then the tetragonal distorted becomes square planar as shown in Figure 14.8.

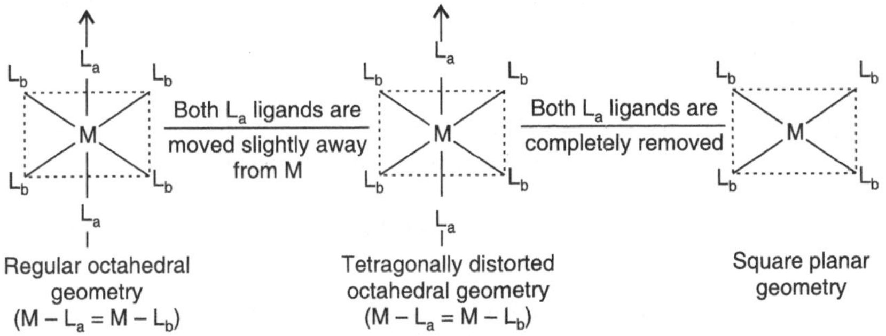


***FIGURE 14.8*** **Distortion in regular octahedral geometry leading to the formation of tetragonally distorted octahedral geometry and square planar geometry.**

Consider that the splitting in octahedral complexes, the energy of $t_{2g}$ orbitals $d_{xy}$, $d_{yz}$, and $d_{zx}$ is decreased, and the energy of $e_g$ orbitals ($d_{z^2}$ and $d_{x^2-y^2}$) is increased. However, in tetragonally distorted octahedron, $L_a$ ligands are moved away along the $z$-axis, so the $d$-orbitals along the $z$-axis, *i.e.*, $d_{z^2}$, $d_{x^2}$, and $d_{zx}$ planes experiences less repulsion than they do in an octahedral complex. The energies of $d_{z^2}$, $d_{yz}$, and $d_{zx}$ decrease while those of $d_{xy}$ and $d_{x^2-y^2}$ increase. The $d_{yz}$ and $d_{zx}$ orbitals still remain degenerate as they are in octrahedral complexes. In square planar complexes, the energies of $d_{z^2}$, $d_{yz}$, and $d_{zx}$ orbitals decreases, while those of $d_{x^2-y^2}$ and $d_{xy}$ orbitals increases. The splitting of $d$-orbitals in square planar complexes is shown in Figure 14.9.

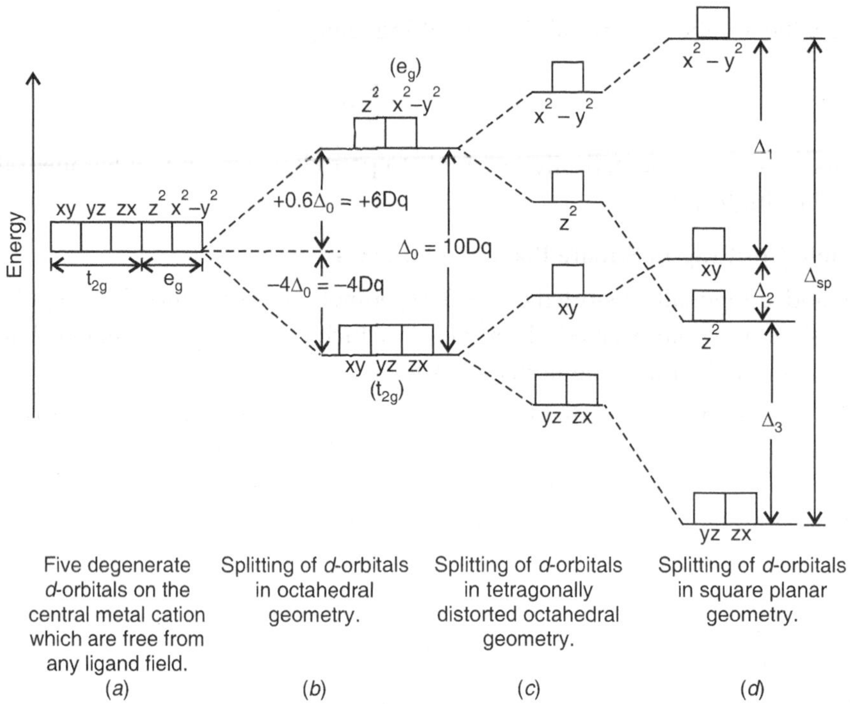


***FIGURE 14.9*** **Crystal field splitting of d-orbitals in octahedral, tetragonally distorted octahedral and square planar geometries.**

The crystal field splitting energy in square planar complexes is denoted by $\Delta_{sp}$ and is given by,

$$\Delta_{sp} = E(d_{x^2-y^2}) - E(d_{xz}\,d_{yz}) = \Delta_1 + \Delta_2 + \Delta_3$$

The value of $\Delta_{sp} > \Delta_o$ by 1.3 *i.e.*, $\Delta_{sp} = 1.3\Delta_o$. This is because $d_{xz}$ and $d_{yz}$ orbitals interact with only two ligands in square planar complexes, while in octahedral complexes, the interaction occurs with only four ligands.

### 14.4.5 Applications of CFT

Following are some important applications of CFT:

(*a*) **Color of the metal complexes.** According to CFT the color of the complex is due to the *d-d*-transition between the $t_{2g}$ and $e_g$ orbitals. If the $\Delta_o$ value for any complex is higher, it means that complex requires relatively large excitation energy and the complex is dark colored. Similarly, if the $\Delta_o$ values for any complex is lower, it means that the complex requires relatively low excitation energy and the complex is light in color. For example, $[Co(NH_3)_6]^{2+}$, $[Cu(H_2O)_4]^{2+}$, $[Cu(NH_3)_4]^{2+}$ are blue in color while $[Co(H_2O)_6]^{2+}$ is pink in color because former complexes require more excitation energy than the last one.

(*b*) **Magnetic properties of the metal complexes.** CFT can be successfully used in calculation of number of unpaired electrons in the high spin and low spin complexes with the help of the following formula:

$$\mu_s = \sqrt{n(n+2)}\ \text{BM}$$

where $n$ = number of unpaired electrons and $\mu_s$ is spin magnetic moment. Actually the value of $\mu_s$ are determined experimentally and then the value of '$n$' is calculated. The calculated and experimental values of $\mu_s$ for different metal ions are given in Table 14.8.

**TABLE 14.8 Value of $\mu_s$ and $\mu_{exp}$ in BM for 3*d*-Series Transition Metal Ions**

| *Metal Ion* | *$d^n$-Configuration* | *Number of unpaired Electron* | *$\mu_s$(cal.) BM* | *$\mu_s$(exp.) BM* |
|---|---|---|---|---|
| $Ti^{3+}$ | $3d^1$ | 1 | 1.73 | 1.73–1.85 |
| $V^{3+}$, $Ti^{2+}$ | $3d^2$ | 2 | 2.83 | 2.75–2.85 |
| $V^{2+}$ | $3d^3$ | 3 | 3.87 | 3.80–3.90 |
| $Cr^{3+}$ | – | 3 | 3.87 | 3.70–3.90 |
| $Cr^{2+}$ | $3d^4$ | 4 | 4.90 | 4.75–4.90 |
| $Mn^{3+}$ | – | 4 | 4.90 | 4.75–4.90 |
| $Mn^{2+}$ | $3d^5$ | 5 | 5.92 | 5.65–6.10 |
| $Fe^{3+}$ | – | 5 | 5.92 | 5.70–6.00 |
| $Fe^{2+}$ | $3d^6$ | 4 | 4.90 | 5.10–5.70 |
| $Co^{3+}$ | – | 4 | 4.90 | ≈ 5.4 |
| $Co^{2+}$ | $3d^7$ | 3 | 3.87 | 4.30–5.20 |
| $Ni^{2+}$ | $3d^8$ | 2 | 2.83 | 2.80–4.00 |
| $Cu^{2+}$ | $3d^9$ | 1 | 1.73 | 1.70–2.20 |

### 14.4.6 Limitations of CFT

Following are the limitations of CFT:

(*i*) The CFT ignores the attractive forces between the *d*-electrons of the metal ion and nuclear charge on the ligand atom. Therefore, all properties dependent upon the ligand orbitals and their interaction with metal orbitals are not explained.

(*ii*) In CFT model the partial covalency of the metal-ligand bond is not taken into consideration. According to CFT, metal-ligand bonding is purely electrostatic.

(*iii*) In CFT, only *d*-electrons of the metal ion are considered, the other metal orbitals such as $s$, $p_x$, $p_y$, and $p_z$ are not taken into consideration.

(*iv*) In CFT, π-orbitals of the ligand are not considered.

(*v*) This theory cannot explain the relative strength of the ligands, *i.e.*, it cannot explain why $H_2O$ is a stronger ligand than OH according to spectrochemical series.

(*vi*) It does not explain the charge transfer spectra on the intensities of the absorption bands.

### 14.4.7 Comparison Between Valence Bond Theory and Crystal Field Theory

(*i*) According to VBT, two 3*d*-orbitals, ($d_{z^2}$ and $d_{x^2-y^2}$) one 4*s*-orbital, and three 4*p*-orbital are involved in inner $d^2$-$sp^3$-hybridization, while the three 3*d*-orbitals ($d_{xy}$, $d_{yz}$, and $d_{xz}$) are not involved. In CFT, all five 3*d*-orbitals are taken into consideration.

(*ii*) In both VBT and CFT the nature of $d^2sp^3$ is the same. But in some inner-orbital octahedral complexes of VBT (*e.g.*, complex of $d^7$ ion), one electron of the 3*d*-orbital jumps to the 4*d*-orbital. Such promotions do not take place in CFT.

(*iii*) According to VBT, metal-ligand bonding in the complexes is only covalent, since in this theory the lone pair of electrons is denoted to metal ion. According to CFT, metal-ligand bonding in the complex is only electrovalent.

**Example 3.** *Give the number of unpaired electrons for the following complex ions:*

(*a*) $[Fe(CN)_6]^{4-}$ (*b*) $[Fe(CN)_6]^{3-}$ (*c*) $[VF_6]^{3-}$ (*d*) $[RhCl_6]^{3-}$

**Solution.** (*a*) In $[Fe(CN)_6]^{4-}$ ion, Fe is present as $Fe^{2+}$ with $3d^6$ configuration. Since $CN^-$ ion is a strong ligand, $[Fe(CN)_6]^{4-}$ is a low spin complex and hence the distribution of the six 3*d*-electrons in $t_{2g}$ and $e_g$ sets is:

$$Fe^{2+}(3d) \longrightarrow t_{2g}^6\, e_g^0, \text{ giving } \boldsymbol{n = 0.}$$

(*b*) In $[Fe(CN)_6]^{3-}$ where *n*, is the number of unpaired electrons in all cases.

$$Fe^{3+}(3d^5) \longrightarrow t_{2g}^5\, e_g^0, \text{ giving } \boldsymbol{n = 1}$$

(*c*) In $[VF_6]^{3-}$ vanadium is present as $V^{3+}$ ion which is a $3d^2$ system. This, in the question of the formation of high spin or low spin complex, does not arise and the distribution of the two 3*d* electrons in the $t_{2g}$ and $e_g$ sets is:

$$V^{3+}(3d^2) \longrightarrow t_{2g}^2\, e_g^0, \text{ giving } \boldsymbol{n = 2}$$

(*d*) $Cl^-$ is a weak ligand. Thus $[RhCl_6]^{3-}$ is a complex and thus

$$Rh^{3+}(4d^6) \longrightarrow t_{2g}^4\, e_g^2, \text{ giving } \boldsymbol{n = 4.}$$

## QUESTION BANK

1. Explain bonding in coordination compounds or explain crystal field theory of bonding of coordination complex.
2. Write the IUPAC names of the following:

(*i*) $K_4[Fe(CN)_6]$ (*ii*) $K_3[Al(C_2O_4)_3]$

(*iii*) $[Cr(en)_3]Cl_3$ (*iv*) $[(en)_2\,Co\overset{NO_2}{\underset{OH}{\lessgtr}}Co\,(en)_2]\,(SO_4)_2$

(*v*) $[CoCl(NH_3)_5]Cl_2$ (*vi*) $K[PtCl_3—NH_3]$

(*vii*) $[Au(CN)_4]$ (*viii*) $K_2[NiF_6]$

(*ix*) $(NH_4)_3[ZrF_7]$

(*x*) $[Pt(NH_3)_4Cl_2]^{2+}$

(*xi*) $Na_2[Co(H_2O)_2(OH)_4]$

(*xii*) $[Pt(NH_3)_2\ Cl(en)\ NO_2]\ Cl_2$

(*xiii*) $[V(H_2O)_6]Cl_3$

(*xiv*) $Na_2[ZnCl_4]$

(*xv*) $K_2[OsCl_5N_3]$

(*xvi*) $Na_3[Ag(S_2O_3)_2]$

(*xvii*) $[(NH_3)_4\ Co\langle{}^{NO_2}_{OH}\rangle Co\ (NH_3)_4]^{4+}$

(*xviii*) $Li[AlH_4]$

(*xix*) $[Co(en)\ (NH_3)_2\ ClBr]NO_3$

(*xx*) $(NH_4)_3\ [Co(C_2O_4)_2]$.

3. What is the difference between an inner orbital complex and an outer orbital complex? Discuss with one example each.
4. Fill in the blanks:
   (*i*) Crystal field theory was developed by ...... .
   (*ii*) Unit of magnetic moment is ...... .
   (*iii*) $t_{2g}$ denotes ...... .
   (*iv*) $\Delta_o$ denotes ...... and $\Delta_t$ denotes ...... .
5. Give reasons for the following:
   (*a*) $Fe(CN)_6{}^{4+}$ is diamagnetic and $Fe(H_2O)_6{}^{3+}$ is paramagnetic.
   (*b*) $Ni(CO)_4$ possesses tetrahedral geometry, while $Pt(NH_3)_2Cl$ square planar.
   (*c*) $Fe(CN)_6{}^{3-}$ is weekly paramagnetic while $[Fe(CN)_6]^{4-}$ is diamagnetic.
6. What do you understand by high spin complex and low spin complex?
7. Explain crystal field stabilization energy.
8. Write two differences between double salts and coordination compounds.
9. Define the following terms:
   (*a*) Chelates (*b*) Polynuclear complexes (*c*) Ligands
10. Write the rules of naming the coordination compounds.
11. What are the reasons for the formation of complexes by transition metal cations?
12. Explain the formation of $Fe(CN)_6{}^{4-}$ complex ions.
13. Explain the formation of $Fe(CN)_6{}^{3-}$ ions.
14. Write the molecular formula of the following compounds:
    (*i*) Potassium tetracyanonickelate (0)
    (*ii*) Potassium octacyano molybdate (IV)
    (*iii*) Penta carbonyliron (0)
    (*iv*) Benzenetricarbonylmanganese (I) ion.
15. Explain valence bond theory (VBT) using suitable examples.
16. Explain the following:
    (*i*) $[Fe(CN)_6]^{4-}$ ion is diamagnetic while $[FeF_6]^{-4}$ is paramagnetic.
    (*ii*) $[Ni(CO)_4]$ is tetrahedral while $[Ni(CN)_4]^{2-}$ is square planar.
    (*iii*) $[NiCl_4]^{2-}$ is tetrahedral while $[Ni(CN)_4]^{2-}$ is square planar.
    (*iv*) $[Cu(NH_3)_4]^{2+}$ is square planar in geometry.
    (*v*) $[CuCl_4]^{3-}$ is diamagnetic while $[Cu(NH_3)_4]^{2+}$ is paramagnetic.

17. What are the limitations of valence bond theory?
18. Explain crystal field theory (CFT) in detail.
19. Explain the following terms:
    (*i*) Effective atomic number
    (*ii*) Splitting of energy levels
    (*iii*) Crystal field splitting
    (*iv*) Crystal field stabilization energy.
20. What do you mean by monodentate, bidentate, tridentate, and hexadentate ligands? Give at least one example of each.
21. What are the important postulates of Werner's theory of complex compounds?
22. What is Sidwick's theory of coordination compounds?
23. How can you explain geometry and magnetic property of the following complex compounds/ions with the help of valence band theory?
    (*i*) $[NiCl_4]^{2-}$
    (*ii*) $[Cr(H_2O)_6]^{3+}$
    (*iii*) $Na_3[Co(NO_2)_6]$
    (*iv*) $[Fe(H_2O)_6]^{3+}$
    (*v*) $[PtCl_6]^{4-}$
    (*vi*) $[Pd(CN)_4]^{2-}$.
24. What is the order of energy levels of *d*-orbitals in square planar crystal field of ligands?
25. Explain the color of complex compounds on the basis of crystal field theory.
26. Calculate crystal field stabilization energy for the following systems:
    (*i*) $d^4$-tetrahedral
    (*ii*) $d^6$-tetrahedral
    (*iii*) $d^3$-low spin octahedral
    (*iv*) $d^4$-low spin octahedral.

Chapter 15

# SOLID STATE CHEMISTRY

## 15.1 INTRODUCTION

Solids are characterized by rigidity, incompressibility, slow diffusion, and mechanical strength. Also, unlike liquids and gases, solids have definite volume and shape.

The solid state has almost complete order. Solids are orderly because there is regular arrangement of the atoms, molecules or ions, and they are confined to small regions in the solids. They can not move very far away from their positions without disrupting the whole structure. The atoms, molecules or ions can have several different three dimensional arrangements and the result is a variety of geometrical structures. For example, diamond is composed of a single element, carbon, arranged in a three-dimensional network that has a melting point of 3550°C. In contrast, solid argon melts at – 189°C and has a cubic close-packed structure whose atoms are held by weak van der Waal forces. A solid crystal generally has sharp boundaries with clear-cut shapes and the beautiful shape and colour add to their value as gems.

The study of solid state begins with the observation of the shapes of crystals. This is known as *geometric crystallography*. In the mid-nineteenth century, crystal symmetry was described in terms of point lattice and this was followed by X-ray crystal structure determination by X-ray diffraction of crystals. Later on an important discovery in this field was the presence of lattice defects and their role in determining the properties of crystals.

## 15.2 TYPES OF SOLIDS

When a substance is at such a low temperature that the thermal agitation of its molecules is not strong enough to move them all around, the inter-molecular forces tend to hold the molecules together in more or less fixed positions. The material acquires a shape and is then said to be in solid state. The solid material may be divided into two distinct classes:

(*i*) *Crystalline solids.* These solids occur as small crystals, and are characterized by a definite crystal structure, a definite melting point, and a definite heat of fusion. Sugar and salt are two examples.

(*ii*) *Amorphous solids.* These solids do not have a definite geometrical structure because the constituent particles are not arranged in an orderly manner, but in a random manner. For example, rubber, plastic and glass. Strictly speaking these are regarded as supercooled liquids in which the forces of attraction holding the molecules together is so great that the material is rigid but there is no regularity of the structure. These substances loose their shape in time and flow under their own weight.

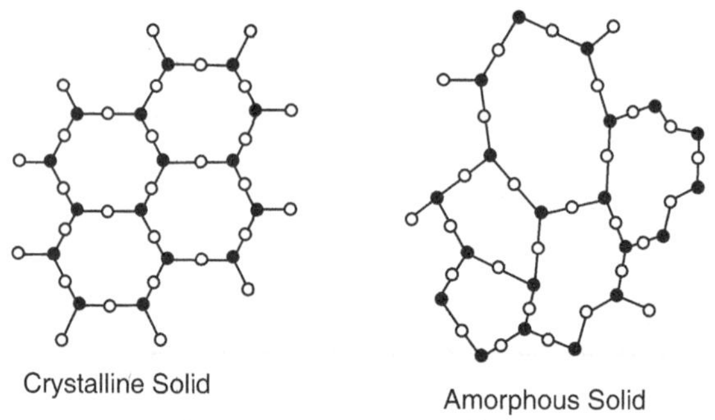


***FIGURE 15.1*** **Two-dimensional representation of a crystalline solid and an amorphous solid. Each substance has the formula $A_2O_3$. A is shown by • and O is shown by o.**

## Distinction between Crystalline and Amorphous Solids

1. *Isotropy and anisotropy.* Amorphous substances are said to be isotropic because they exhibit the same value of any property in all directions. Thus, refractive index, thermal and electrical conductivities, and coefficient of thermal expansion in amorphous solids are independent of the direction along which they are measured.

   Crystalline substances, on the other hand, are anisotropic and the magnitude of a physical property varies with directions. For example, in a crystal of silver iodide, the coefficient of thermal expansion is positive in one direction and negative in the other. Similarly, velocity of light in a crystal may vary with the direction in which it is measured. Thus, a ray of light passing through a Nicol prism splits up into two components, each having different velocities (double refraction).

   **Explanation of Isotropy and Anisotropy**

   In amorphous substances, as in liquids, the arrangement of particles is random and disordered. Therefore, all directions are equivalent and properties are independent of directions.

   On the other hand, the particles in a crystal are arranged and well ordered. Thus the arrangement of particles may be different in different directions. This is illustrated in

Figure 15.2 in which a simple two-dimensional arrangement of two different types of particles is shown.

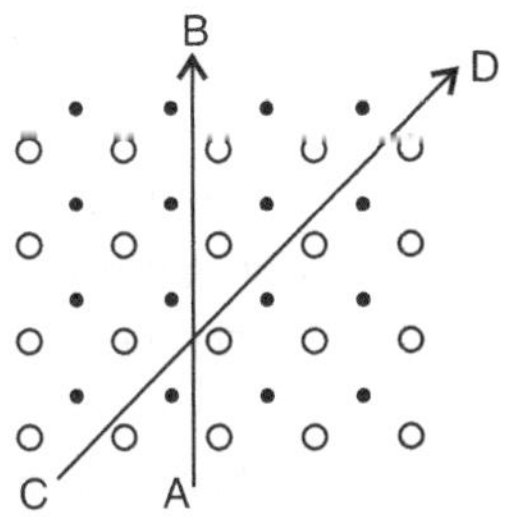


*FIGURE 15.2* **Anisotropy in crystals is due to different arrangements of particles in arrangements of particles in different directions.**

When a property is measured along AB and CD, the value along CD will be different from that along AB. This is so because there are two types of particles in alternate positions along CD. This explains why crystalline substances are anisotropic.

2. *Melting point.* A crystalline substance has a very sharp melting point while an amorphous substance does not. For example, as the temperature of glass is raised gradually, it softens and then starts flowing without undergoing any abrupt or sharp change from solid to liquid state.
3. *Characteristic geometry.* Crystalline solids have a regular arrangement of particles in typical geometrical forms, whereas amorphous solids have a complete random particle arrangement.
4. *Cooling curve.* The cooling curve for an amorphous substance is smooth, while the curve for a crystalline substance has two breaks (Figure 15.3), which corresponds to the beginning and the ends of the process of crystallization. The temperature remains constant during crystallization as the process is accompanied by some liberation of energy, which compensates for the loss of heat and causes the temperature to remain constant.

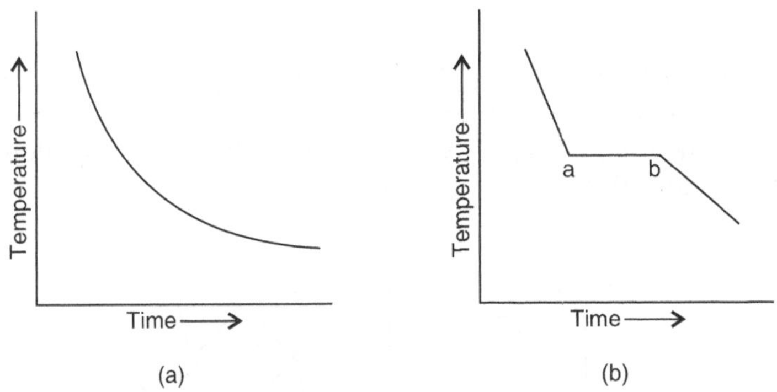


*FIGURE 15.3* **Cooling curve for (*a*) amorphous and (*b*) crystalline substances.**

The main points of distinction between amorphous and crystalline solids are summarized in Table 15.1.

TABLE 15.1 Distinction between Crystalline and Amorphous Solids

| Crystalline Solids | Amorphous Solids |
|---|---|
| (*i*) They have a definite and regular geometrical form. | They do not have definite geometrical shape. |
| (*ii*) They are rigid and incompressible. | They are also rigid and usually cannot be compressed to an appreciable extent. But graphite is soft. |
| (*iii*) They have a sharp melting point. | They melt over a wide range of temperature. |
| (*iv*) They have a definite heat of fusion. | They do not have a definite heat of fusion. |
| (*v*) They are anisotropic in nature. | They are isotropic in nature. |
| (*vi*) They have crystal symmetry. | They do not have symmetry. |
| (*vii*) They are true solids. | They are not true solids, but are supercooled liquids. |
| (*viii*) They give clean cleavage when cut into smaller crystals. | They give irregular cleavage. |
| (*ix*) Diamond, quartz, NaCl, calcite, etc. are common examples. | Rubber, glass, plastics, etc. are common examples. |

## 15.3 CRYSTAL STRUCTURE

A crystalline solid consists of a large number of small units, called *crystals*, each of which possesses a definite geometric shape bounded by plane faces. Thus, a crystal may be defined as *a solid substance having a definite geometrical shape with flat faces and sharp edges.* The branch of science which deals with geometry, properties and structures of crystals, and crystalline substances is called *crystallography*.

### 15.3.1 External Features of Crystals

The external geometry of a crystal is obtained as a result of the internal regular arrangements of atoms from which it is built. The different external features of crystals are briefly described below:

1. *Faces.* Crystals are bounded by a number of surfaces which are generally perfect flat. These surfaces are known as *faces.* Faces are of two types: like and unlike. There are some crystals bounded by faces which are all alike. For example, *fluorspar* is obtained in cubes in regular octahedron, whereas *galena* crystal exhibits a combination of a cube and an octahedron (Figure 15.4).

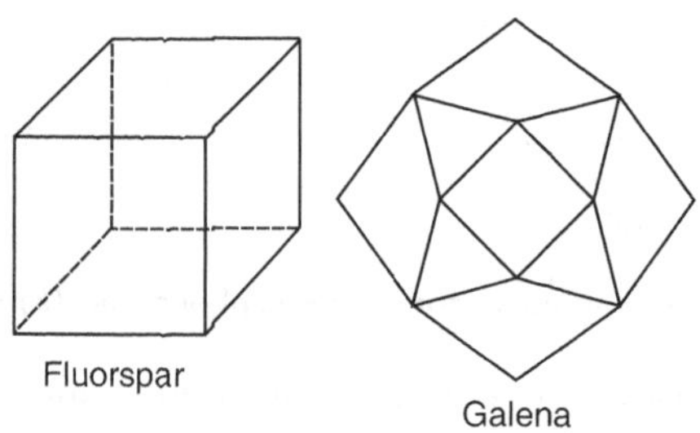


*FIGURE 15.4*

2. *Form.* All the faces corresponding to a crystal constitute a *form.* A simple form of a crystal is entirely made up of like faces, whereas a crystal, which consists of two or more simple forms, is known as a *combination.*
3. *Edges and interfacial angles.* An edge is formed by the intersection of two adjacent faces whereas the angle between any two faces of a crystal is known as the *interfacial angle* (Figure 15.5). The crystals are bounded by plane faces, straight edges, and interfacial angles. The relationship between these elements can be expressed by the formula:

$$f + c = e + 2$$

where $f$, $c$, and $e$ denote the number of faces, angles, and edges respectively.

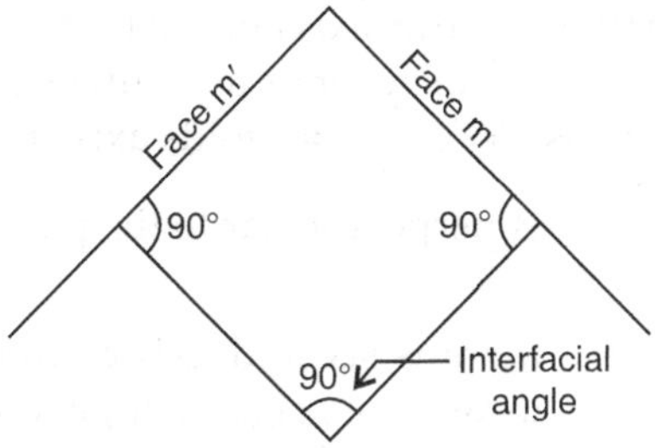


*FIGURE 15.5*

4. *Zone and zone axis.* Zone denotes the faces of a crystal occurring in sets meeting in parallel edges, or would do so if the planes of the faces are extended. Each zone forms a complete belt around the crystal. A line drawn through the center of a crystal in a direction parallel to the edge of a zone is called the *zone axis.*

### 15.3.2 Crystal Symmetry

Symmetry in crystals may be due to a plane, a line or a point. Accordingly, there are three types of symmetry elements associated with crystals. These are called the Elements of Symmetry.

1. *Plane of Symmetry.* A crystal is said to have a plane of symmetry if it can be divided by an imaginary plane into two equal parts, each of which is the exact mirror image of the other. For example, in case of water molecule, two planes of symmetry can be drawn along $z$ axis.

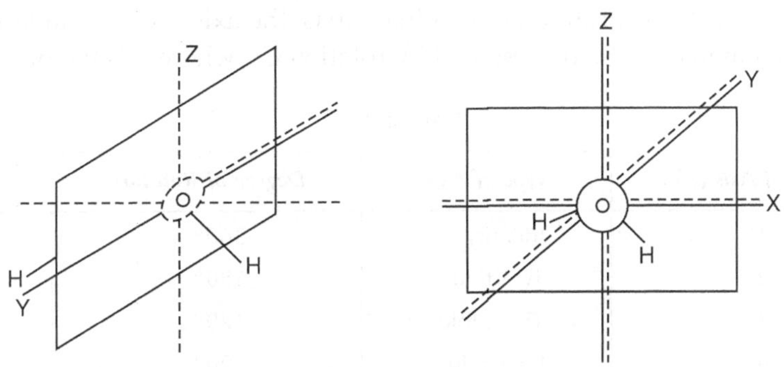


*FIGURE 15.6* **(*a*) Plane of symmetry.**

2. *Axis of Symmetry.* An axis of symmetry is an imaginary line drawn through the crystal such that during rotation of the crystal through 360°, the crystal presents exactly the same appearance more than once. If a similar view appears twice, it is called an axis of two fold symmetry or diad axis. If it appears thrice, it is an axis of three fold symmetry or triad axis, and so on. For example,

 (*i*) Consider the water molecule; an axis, passing through the center perpendicularly, is such that when a molecule is rotated through 180°, it represents exactly the same appearance two times in one revolution (rotation through an angle of 360°). Thus, the axis shown for the water molecule is 360°/180° = 2 fold rotation axis [Figure 15.6 (*b*) (*i*) and (*ii*)]

 (*ii*) Consider the molecule of $BCl_3$[Figure 15.6 (*b*) (*iii*)]. The geometry of this molecule is triangular planar. If we choose an axis normal to the plane of the molecule and passing through the boron atom, it will generate an equivalent appearance after every rotation of 120°. Thus, this axis is 360°/120° = 3 folds axis and there are three such 2 fold axis.

 (*iii*) $PtCl_4^{2-}$ has a four fold axis perpendicular to the plane of the ion and passing through the Pt atom.

 (*iv*) In the case of benzene, there is six fold axis of rotation passing through center of symmetry. Six fold axis means the original appearance is repeated after an angle of 60°. A hexagonal crystal has six fold axis of symmetry.

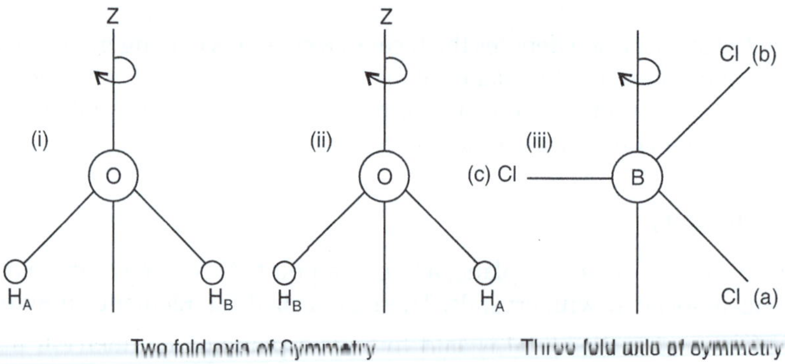


**FIGURE 15.6 (*b*) Axis of symmetry.**

In general, if the same appearance of a crystal is repeated on rotating through an angle of 360°/*n*, around an imaginary axis, the axis is called an *n*-fold axis. The axis of symmetry can be represented by rotation $C_n$, where *n* is the order of axis.

**TABLE 15.2**

| *Order of Axis (n)* | *Type of Axis* | *Degree of Rotation* | *Symbol* |
|---|---|---|---|
| 1 | Identity | 360° | 1 or E |
| 2 | Two fold | 180° | $C_2$ |
| 3 | Three fold | 120° | $C_3$ |
| 4 | Four fold | 90° | $C_4$ |
| 5 | Six fold | 60° | $C_6$ |

Where E is called identity axis, it means that rotation through a full 360° is necessary to obtain an indistinguishable structure. That is, there are no equivalent orientations around this axis. This structure is identical, rather than equivalent, to the original, and the axis is therefore 1 fold axis.

3. *Center of symmetry.* A point at the center of the crystal so that any line drawn through it will meet the surface of the crystal at equal distances on either side. It is noteworthy that a crystal may have a number of planes of symmetry or axis of symmetry, but it can have only one center of symmetry [Fig. 15.6 (*c*)]

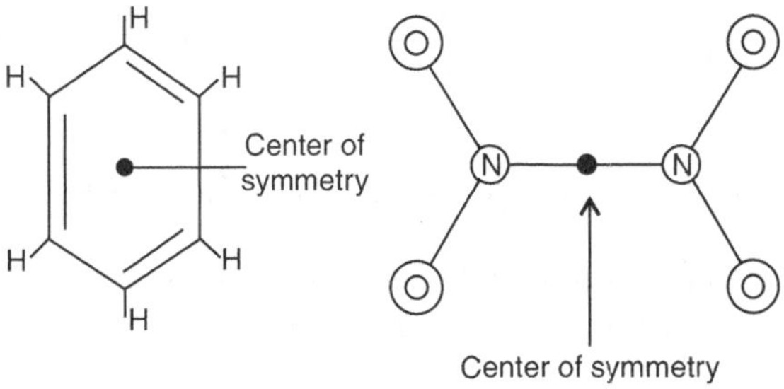


*FIGURE 15.6* **(*c*) Center of symmetry.**

### 15.3.3 Symmetry Elements of a Cubic Crystal

A crystal may have three different types of symmetry elements and there may be different numbers of each type of symmetry. The total number of planes, axis and center of symmetry possessed by a crystal is termed as elements of symmetry of the crystal. To explain this we may consider the elements of symmetry possessed by a cubic crystal such as NaCl crystal. A cubic crystal has 23 elements of symmetry. These elements of symmetry are:

1. **Plane of Symmetry**

   (*i*) *Rectangular plane of symmetry.* There are three rectangular planes of symmetry. On such rectangular plane of symmetry is shown in figure 15.7 (*a*).

   (*ii*) *Diagonal plane of symmetry.* This plane that passes diagonally through the cube is shown in Figure 15.7 (*b*). There can be a total of six such planes passing diagonally through the cube.

2. **Axis of Symmetry**

   (*i*) *Axis of four fold symmetry.* There are three such four fold axis of symmetry at right angles to one another [Fig. 15.7 (*c*)].

   (*ii*) *Axis of three fold symmetry.* There are a total of four such three fold axis. One such axis is shown in Fig. 15.7 (*d*).

   (*iii*) *Axis of two fold symmetry.* Axis of two fold symmetry emerging from opposite edges is shown in Fig. 15.7 (*e*). There are six such axes of two fold symmetry.

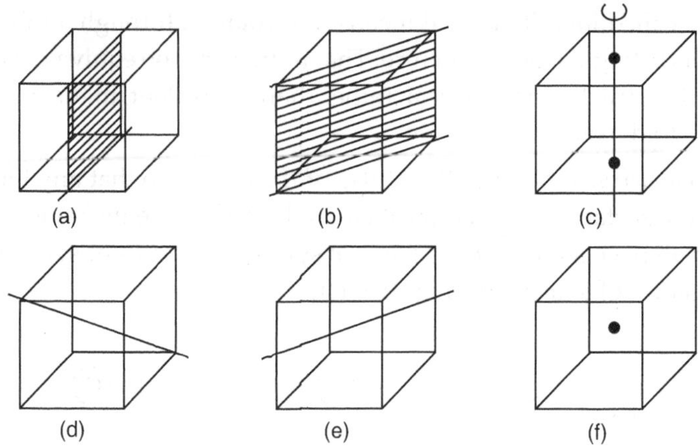


*FIGURE 15.7* **Elements of symmetry of a cube.**

3. **Center of Symmetry.** There is only one center of symmetry lying at the center of the cube. [Figure 15.7 (*f*)] Thus, the number of symmetry elements of various types in a cubic crystal are:

(*a*) Rectangular planes of symmetry = 3
(*b*) Diagonal planes of symmetry = 6
(*c*) Four-fold axes of symmetry = 3
(*d*) Three-fold axes of symmetry = 4
(*e*) Two-fold axes of symmetry = 6
(*f*) Center of symmetry
(located at the center of gravity of the cube) = 1

Total number of symmetry elements in a cubic crystal = 23

### 15.3.4 Crystal Lattice or Space Lattice

We can imagine a crystal as a periodic arrangement of points as shown in Figure 15.8. A point may represent an atom or a group of atoms arranged around it in a real crystal. Let us begin with a single point. Repeated translation of this point through a fixed distance will generate a linear array of points. This movement is denoted by vector $\hat{i}_1$. If we add a second translation $\hat{i}_2$ it will generate periodically repeating points on a plane and is known as two dimensional plane lattice [Figure 15.8 (*a*)]. If a third translation is added, we get a three dimensional arrangement of the points that is called a space lattice [Figure 15.8 (*b*)].

The lattice points are imaginary. In real crystal, they are occupied by atoms or groups of atoms that are arranged in a regular fashion about the lattice points. *This atom or group of atoms is the basis and the arrangement of the imaginary points is the lattice.* The real crystal is then

$$\text{Basis + Lattice = Crystal.}$$

*FIGURE 15.8* (*a*) **Two dimensional plane lattice.**

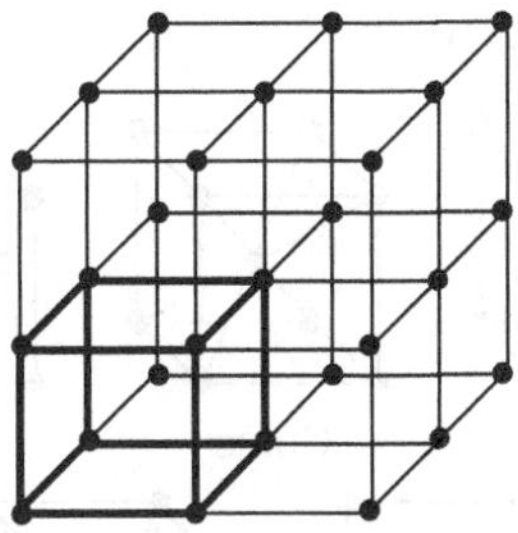

*FIGURE 15.8* (*b*) **Space lattice and unit cell.**

The space lattice may be defined as *the array of points showing how molecules, atoms, or ions are arranged at different points in three dimensional space.*

### 15.3.5 Unit Cell

The lattice points can be broken into a number of unit cells. This is done by lines joining the lattice points as shown in Figure 15.9. A repetition of three dimensional unit cell generates the space lattice.

A unit cell is the smallest repeating unit in space lattice which, when repeated over and over again, results in a crystal of the given substance. The three translations are selected along the three edges of the unit cell. The three selected transitions along the edges of the unit cell are called the crystallographic axis— *a*, *b*, and *c*— and angles between them are $\alpha$, $\beta$, and $\gamma$ as shown in Figure 15.9.

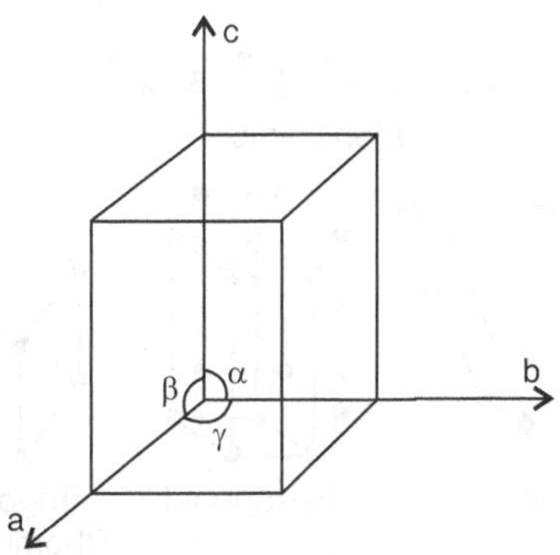


*FIGURE 15.9* **A unit cell showing the edges and the angles.**

A unit cell has one atom or ion at each corner of the lattice. Also, there may be atoms or ions in the faces and interior of the cell. A cell with an interior point is called the *body centered cell*. A cell which does not contain any interior points is known as the *primitive cell*. That is, a primitive cell is a regular three-dimensional unit cell with atoms or ions located at its corners only.

### 15.3.6 Parameters of the Unit Cells/Bravais Lattice

*Bravais*, in 1848, on the basis of the study of crystal symmetry, showed that all structures can be generated by using only 14 types of space lattice (14 types of unit cell). Thus there are only 14 types of space lattices, known as Bravais lattices. The 14 Bravais space lattices associated with various crystal systems are shown in Figure 15.10.

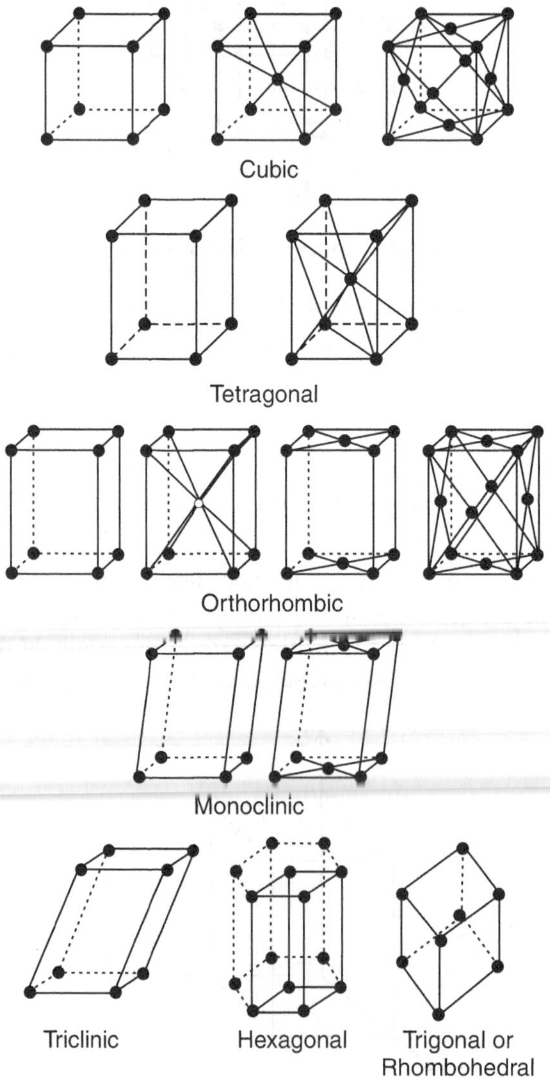


*FIGURE 15.10* **The 14 bravais lattices.**

These 14 types of unit cells will give rise to 230 types of lattice structure (space groups) by performing 32 symmetry operation as (*i*) translation, (*ii*) rotation, (*iii*) translation + rotation, and (*iv*) reflection. It was later on confirmed by *Federov* and *Barlow* independently that it is possible to have an arrangement of lattice points other than these 230 types that can repeat itself indefinitely in three dimensions. All real crystals belong to any one of the 230 space groups. The imaginary lattice points are occupied by atoms or groups of atoms. Table 15.3 shows the classification and the geometric properties of various Bravais lattices, belonging to the seven basic crystal systems.

**TABLE 15.3 The Seven Crystal Systems**

| *Crystal System* | *Bravais Lattices* | *Minimum Symmetry Elements* | *Parameters of Unit Cell* | | *Examples* |
|---|---|---|---|---|---|
| | | | *Intercepts* | *Crystal Angles* | |
| 1. Cubic | Primitive, Face centered Body centered = 3 | Four 3-fold (or $\bar{3}$) axes; Three 4-fold (or $\bar{4}$) axes | $a = b = c$ | $\alpha = \beta = \gamma = 90°$ | NaCl, KCl Diamond, Zinc blende |
| 2. Ortho-rhombic | Primitive, Face centered, Boby centered, End centered = 4 | Three mutually perpendicular 2-fold (or $\bar{2}$) axes | $a \neq b \neq c$ | $\alpha = \beta = \gamma = 90°$ | $KNO_3$, $K_2SO_4$ $BaSO_4$, $PbCO_3$ Rhombic sulphur |
| 3. Tetragonal | Primitive, Body centered = 2 | One 4-fold (or $\bar{4}$) axis | $a = b \neq c$ | $\alpha = \beta = \gamma = 90°$ | $ZnO_2$, $SnO_2$ $TiO_2$ |
| 4. Mono-clinic | Primitive, End centered = 2 | One 2-fold (or $\bar{2}$) axis | $a \neq b \neq c$ | $\alpha = \gamma = 90°$, $\beta \neq 90°$ | $Na_2SO_4.10H_2O$ Monoclinic sulphur |
| 5. Triclinic | Primitive = 1 | One 1-fold (or $\bar{1}$) axis | $a \neq b \neq c$ | $\alpha \neq \beta \neq \gamma \neq 90°$ | $CuSO_4.\ 5H_2O$ $K_2Cr_2O_7$ |
| 6. Hexagonal | Primitive = 1 | One 6-fold (or $\bar{6}$) axis | $a = b \neq c$ | $\alpha = \beta = 90°$ $\gamma = 120°$ | HgS, Ice, Graphite, Zn, Cd |
| 7. Rhombo-hedral | Primitive = 1 | One 3-fold (or $\bar{3}$) axis | $a = b = c$ | $\alpha = \beta = \beta \neq 90°$ | $NaNO_3$, As Sb, Bi |

The crystallographers have divided the 14 space lattices into seven crystal systems. The crystals belonging to the cubic system have three kinds of Bravais lattices, depending upon the shapes of the unit cell. These are:

1 The simple or primitive cubic lattice P, in which there are points only at the corners of each unit [Fig. 15.11 (*a*)].

2. The face centered cubic lattice F, in which there are points at the corners as well as the center of each of the six faces of the cubes [Fig. 15.11 (*b*)].

3. The body centered cubic lattice B, in which there are points at the corner as well as in the body center of each cube [Figure 15.11 (*c*)].

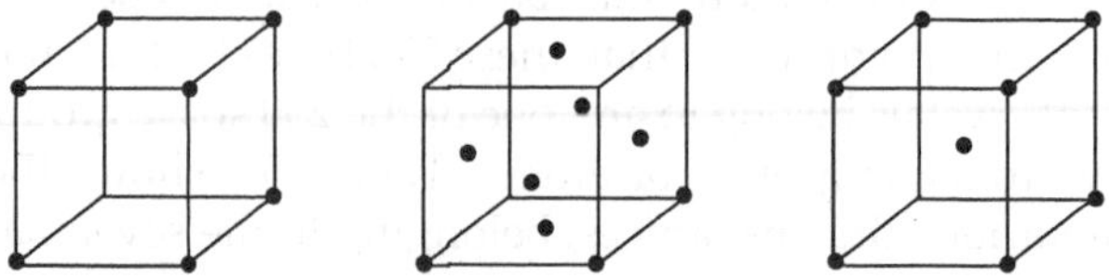

(a) Simple Cubic Lattice (SC) (b) Face Centered Cubic Lattice (FCC) (c) Body Centered Cubic Lattice (BCC)

***FIGURE 15.11*** **Cubic space lattices.**

### 15.3.7 Atoms or Points in a Unit Cell or Lattice Points

The points representing the atoms, molecules, or ions in a unit cell are known as lattice points. In the various unit cells there are three kinds of lattice points.

(*i*) *Points located at the corners of a unit cell:* Since such points lie at the corner of the unit cell, they are shared between eight such unit cells. Thus only 1/8th part of each such point contributes to any cell.

(*ii*) *Points located the center of the face of a unit cell:* Since such points are shared between two such cells, only 1/2 of each such point contributes to any cell.

(*iii*) *Points located at the center of the unit cell:* This point, being present entirely within the unit cell, wholly belongs to this unit cell.

Thus the total number of lattice points per units cell in three types of unit cells may be calculated as (Figure 15.12):

(*a*) Simple or primitive $8 \times \frac{1}{8} = 1$

(*b*) Face centered $8 \times \frac{1}{8} + 6 \times \frac{1}{2} = 4$

(*c*) Body centered $8 \times \frac{1}{8} + 1 = 2$

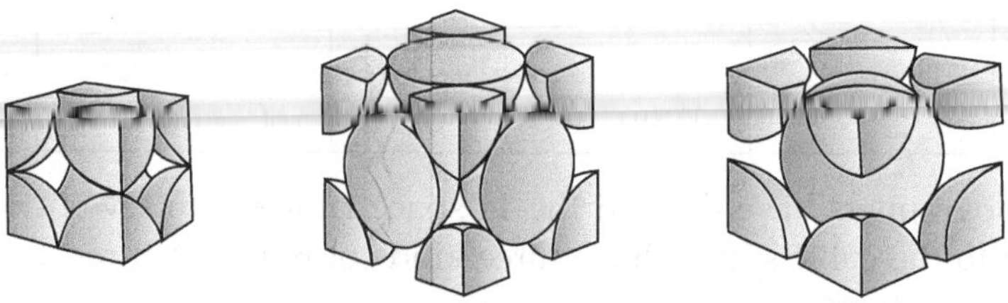

(a) (b) (c)

***FIGURE 15.12*** **(*a*) Simple cubic unit cell—1atom per unit cell, (*b*) Face centered cubic—4 atoms per unit cell, (*c*) Body centered cubic—2 atoms per unit cells.**

The number of atoms belonging to a unit cell is also termed as lattice sites and is denoted by Z.

### 15.3.8 Calculation of Mass of the Unit Cell

Mass of the unit cell can be calculated from the number of atoms in the unit cell. For this, multiply the mass of one atom by the number of atoms in the unit cell.

$$\text{Thus the mass of one atom} = \frac{\text{Molar mass of the substance}}{\text{Avogadro's number}}$$

**Example 1.** *The unit cell of metallic gold is face centered cubic*

(*a*) *How many atoms occupy the gold unit cell?*

(*b*) *What is the mass of a gold atom?*

*Molar mass of gold is 197 g.*

**Solution.** Eight corners at 1/8 atom each = 1 atom

six faces at 1/2 atom each = 3 atoms

∴ 4 atoms occupy the gold unit cell.

$$\text{Mass of gold atom} = \frac{\text{Molar mass}}{\text{Avogadro's number}}$$

$$\frac{197 \text{ g/mol}}{6.022 \times 10^{23} \text{ Au atoms/mol}} \times \frac{1}{4} = \mathbf{8.178 \times 10^{-23}} \text{ g}$$

**Example 2.** *By X-ray diffraction it is found that nickel crystals are face centered cubic. The edge of the unit cell is 3.5 Å. The atomic mass of nickel is 58.7 and its density is 8.94 g/cm$^3$. Calculate Avogadro's number from the data.*

**Solution.** Number of atoms in nickel unit cell = 4

Let Avogadro's number $= N$

Thus the mass of one Ni atom $= \dfrac{58.7}{N}$

the mass of one unit cell $= 4\left(\dfrac{58.7}{N}\right)$ g

the volume of one unit cell $= (3.52 \times 10^{-8})^3 \text{ cm}^3$

the density of nickel cell $= \dfrac{4(58.7/N)\text{g}}{(3.52 \times 10^{-8})^3 \text{ cm}^3} = 8.94 \text{ g/cm}^3$ (given)

∴ $\mathbf{N = 6.02 \times 10^{23}}$

### 15.3.9 Analysis of a Cubic System

1. **Atomic radius.** It is defined as *half the distance between the centers of two immediate neighbours in a unit cell.* It is represented by $r$ and is usually expressed in terms of length of edge in a unit cell. Radius of an atom is given as

(*i*) simple cubic structure (*sc*) = $a/2$

(*ii*) face centered cubic (*fcc*) $= \frac{\sqrt{2}}{4} a = 0.3535\, a$

(*iii*) body centered cubic (*bcc*) $= \frac{\sqrt{3}}{4} a = 0.433\, a$

2. **Number of atoms per unit cell.** Please refer to section 15.3.7.

3. **Coordination number (CN).** The number of spheres (atoms, molecules or ions) directly surrounding a single sphere in a crystal is called the *coordination number*. For example, in

   (*i*) *Simple cubic crystal structure (scc)*. There are six atoms which are the nearest neighbours for every corner atom. Thus, there are two nearest atoms, one along + $x$ axis and the other along – $x$ axis. Similarly there are two nearest neighbours along ± $y$ axis and two along ± $z$ axis thus, in all, there are 2 + 2 + 2 = 6 nearest neighbours. Hence, the CN is 6.

   (*ii*) *Body centered crystal structure (bcc)*. In this, an atom in the center (body of the cell) has all the eight corner atoms as its close neighbors. Hence, the CN is 8. CsCl, Cr, Fe, Mo, W crystallize in *bcc* lattice.

   (*iii*) *Face centered crystal structure (fcc)*. In this, each atom is in direct contact with the 12 nearest neighbors. Six of these lie in one layer in the same plane, 3 in a layer above and 3 in a layer below. Hence, the CN is 12. Cu, Au, Al crystallize in *fcc* lattice.

4. **Radius ratio.** It is defined as *the ratio of the cation radius to that of the anion in an ionic solid.* Thus radius ratio,

$$\frac{\text{Radius of cation}}{\text{Radius of anion}} = \frac{r_+}{r_-}$$

   **Significance.** The higher the radius ratio, the larger is the size of the cation and hence, the greater is its CN. Thus

   (*i*) when the radius ratio is greater than 0.732, a CN of eight is usually observed.

   (*ii*) when the radius ratio is greater than 0.414, but less than 0.72, a CN of six is usually observed.

   (*iii*) when the radius ratio is greater than 0.225, but less than 0.414, a CN of four is usually observed.

   (*iv*) When the radius ratio is greater than 0.155, but less than 0.225, a CN of three is observed. For example, BN, $B_2O_3$.

   (*v*) When the radius ratio is between 0 – 0.155, a CN of two is observed. For example, $HF_2^-$.

5. **Length of the unit cell and density of an element.** The X-Ray diffraction pattern of ionic crystals provides a convenient method of determining the length of unit cells, from which one can calculate the theoretical density of the unit cell, which will also be the density of the element (compound).

Let the length of the unit cell = $a$ pm = $a \times 10^{-10}$ cm  ($\because$ 1 pm = $10^{-10}$ cm)

Volume of the unit cell = $l^3$ cm$^3$ = $(a \times 10^{-10}\ \text{cm})^3 = a^3 \times 10^{-30}$ cm$^3$

$$\text{Density of the unit cell} = \frac{\text{Mass of the unit cell}}{\text{Volume of the unit cell}}$$

*Calculation of mass of the unit cell*

$$\text{Mass of the unit cell} = \begin{pmatrix}\text{Number of atoms in} \\ \text{the unit cell}\end{pmatrix} \times \begin{pmatrix}\text{Mass of each} \\ \text{atom}\end{pmatrix} = \text{Z} \times m$$

Mass of an atom ($m$), in turn, is given by the following relation

$$m = \frac{\text{Atomic mass}}{\text{Avogadro number}} = \frac{\text{M}}{\text{N}_\text{o}}\text{g}$$

$$\therefore \quad \text{Density of the unit cell} = \frac{\text{Mass of the unit cell}}{\text{Volume of the unit cell}}$$

$$= \text{Z} \times \frac{\text{M}}{\text{N}_\text{o}} g \times \frac{1}{a^3 \times 10^{-30}\ \text{cm}^3}$$

$$= \frac{\text{Z} \times \text{M}}{\text{N}_\text{o} \times a^3 \times 10^{-30}}\ \text{g cm}^{-3}$$

where Z is the number of formula units in the unit cell, M is the atomic mass of the element, and $a$ is the length (in pm) of the unit cell.

General formula for determining density of a substance will be

$$\text{Density} = \frac{\text{Z} \times \text{Formula mass of the substance}}{\text{N}_\text{o} \times a^3 \times 10^{-30}}\ \text{g/cm}^3$$

6. **Packing fraction, density of packing, or volume occupied in the lattice structure.** It is defined as the ratio of volumes occupied by atoms in unit cell ($v$) to the total volume of the unit cell (V), *i.e.*,

$$\text{Packing fraction} = \frac{v}{\text{V}}$$

(*i*) *For simple cubic structure (sc)*

Volume of atoms in the unit cell ($v$) = $1 \times \frac{4}{3}\pi r^3$

Since for *sc* structure, $\quad r = \frac{a}{2}$

$$\therefore \quad v = \frac{4}{3}\pi\left(\frac{a}{2}\right)^3 = \frac{\pi a^3}{6}$$

Total volume of unit cell (V) = $a^3$

$$\therefore \quad \text{Packing fraction} = \frac{\pi a^3}{6a^3} = \frac{\pi}{6} = 0.52$$

This shows that 52% of the unit cell is occupied by atoms and 48% is empty.

(ii) *For body centered cubic structure (bcc)*

Since the unit cell of *bcc* structure contains 2 atoms on the average,

$$\therefore \text{ Volume of atoms in the unit cell } (v) = 2 \times \frac{4}{3}\pi r^3$$

Since for *bcc* structure, $r = \dfrac{\sqrt{3}}{4}a$

$$\therefore \quad v = 2 \times \frac{4}{3}\pi\left(\frac{\sqrt{3}}{4}a\right)^3 = \frac{\sqrt{3}}{8} \times \pi a^3$$

Volume of unit cell (V) = $a^3$

$$\therefore \quad \text{Packing fraction} = \frac{\sqrt{3}\pi a^3}{8a^3} = \frac{\sqrt{3}\pi}{8} = 0.68$$

This shows that 68% of the space in unit cell is occupied by atoms and 32% is empty.

(iii) *For face centered cubic (fcc) structure*

The unit cell *fcc* structure contains 4 atoms on the average,

$$\therefore \quad \text{Volume of atoms in the unit cell } (v) = 4 \times \frac{4}{3}\pi r^3$$

Since for *fcc*, $r = \dfrac{\sqrt{2}}{4}a$

$$\therefore \quad v = \frac{4 \times 4\pi}{3} \times \frac{\sqrt{8}\,a^3}{64}$$

Volume of unit cell (V) $= a^3$

$$\therefore \quad \text{Packing fraction} = \frac{4 \times 4\pi}{3 \times a^3} \times \frac{\sqrt{8}\,a^3}{64} = 0.74$$

This shows that 74% of unit cell is occupied by atoms and 26% remains empty.

**Example 3.** *A compound, formed by elements X and Y, crystallizes in the cubic structure, where Y atoms are at the corner of a cube and X atoms are at alternate faces. What is the formula of the compound?*

**Solution.** Number of atoms 'Y' at eight corners of a cubic unit cell

$$= \left(8 \times \frac{1}{8}\right) = 1$$

Number of atoms 'X' present on alternate faces of unit cell = $\left(2 \times \frac{1}{2}\right) = 1$

$\therefore$ Formula of compound = XY

**Example 4.** *Sodium metal crystallizes in bcc lattice with the cell edge, a = 4.29 Å. What is the radius of sodium atom?*

**Solution.** For *bcc* lattice, $r = \frac{a\sqrt{3}}{4} = \frac{\sqrt{3}}{4} \times 4.29 = \mathbf{1.86\ Å}$

**Example 5.** *A crystal of lead (II) sulphide has NaCl structure. In this crystal the shortest distance between a $Pb^{2+}$ ion and a $S^{2-}$ ion is 297 pm. What is the length of the edge of the unit cell in lead sulphide? Also calculate the unit cell volume.*

**Solution.** Edge length = 2 × *d*

where *d* = shortest distance between $Pb^{+2}$ and $S^{2-}$ ions.

$\therefore$ Edge length, $a = 2 \times 297 \times 10^{-10}$ cm

$$a = 594 \times 10^{-10} \text{ cm}$$

Hence volume of unit cell *i.e.,*

$$a^3 = (594 \times 10^{-10} \text{ cm})^3 = \mathbf{209.6 \times 10^{-24}\ cm^3}$$

**Example 6.** *An element (Atomic mass = 60) having face centered cubic structure has a density of 6.23 g cm$^{-3}$. What is the edge length of the unit cell?*

**Solution.** Atomic mass of element = 60

Number of atoms per *fcc* unit cell = 4

Density of the element $= 6.23 \text{ g/cm}^3 = \frac{6.23}{10^{30}} \text{ g/(pm)}^3$

We know that, $\rho = \frac{Z \times M}{N_o \times a^3}$

or, $a^3 = \frac{Z \times M}{\rho \times N_o} = \frac{60 \times 4 \times 10^{30}}{6.23 \times 6.02 \times 10^{23}} = 64 \times 10^6 \text{ (pm)}^3$

$\therefore$ Edge of unit cell, $a = 4 \times 10^2 = \mathbf{400\ pm}$

**Example 7.** *A metal (Atomic mass = 50) has a bcc crystal structure. The density of the metal is 5.96 g cm$^{-3}$. Find the volume of its unit cell.*

**Solution.** Atomic mass of the metal = 50

Number of atoms per *bcc* unit cell = 2

Density of metal = 5.96 g/cm$^3$

$$\therefore \quad \text{Volume of the unit, } a^3 = \frac{M \times Z}{N_0 \times \rho} = \frac{50 \times 2}{6.02 \times 10^{23} \times 5.96}$$

$$= 2.787 \times 10^{-23} \text{ cm}^3.$$

## 15.4 LAWS OF CRYSTALLOGRAPHY

The branch of crystallography is based on the following three fundamental laws.

1. **The law of constancy of interfacial angles.** According to this law, *the size of the faces or even shapes of the crystals of a substance may vary widely with conditions of crystallization yet the interfacial angles between any two corresponding faces of the crystals are always constant.* For example, the interfacial angles in all sodium chloride crystals are found to be 90°, irrespective of the size and shapes of faces.

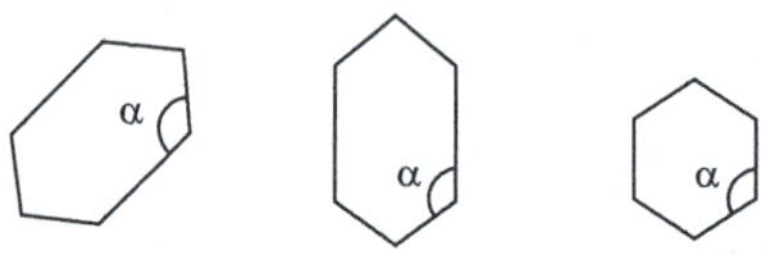


*FIGURE 15.13* **Shape and size may be different but interfacial angles in a crystal are same.**

Another example is the crystal of quartz (Figure 15.14). In all the possible distortions of the crystal of quartz the interfacial angles remain the same at 120°.

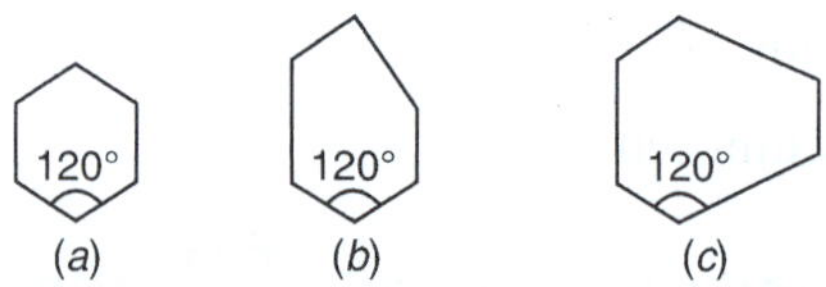


*FIGURE 15.14* **The constancy of interfacial angles. (*a*) Section of an ideal crystal (*b*) and (*c*) possible shapes of quartz crystal system.**

It follows from this law that a substance might be identified by the shape of the crystal, *i.e.*, the exact values for the angles that its faces and edges form.

2. **Law of constancy of symmetry.** According to this law, *all crystals of one and the same substance have the same symmetry.* The total number of planes, lines and centers of symmetry possessed by a crystal is termed as the *elements of symmetry* of the crystal. For details on elements of symmetry, please refer to section 15.3.2.

3. **Hauy's law of rational indices.** In a real crystal, the lattice points are occupied by atoms. Any plane passing through a large number of lattice points is called a crystal plane. The plane surfaces of a crystal are called the faces of a crystal. This law states that the intercepts on any face of a crystal along the crystallographic axis are either equal to unit intercepts ($a$, $b$, $c$) or some simple whole number multiples of them, $na$, $n'b$, $n''c$, where $n$, $n'$, and $n''$ etc. are whole numbers.

For example, suppose three axes are represented by OX, OY, and OZ and a unit plane cuts these axis at A, B, and C respectively, making the intercepts OA = $a$, OB = $b$, and OC = $c$. The ratio $a$: $b$: $c$ is known as *axial ratio.* This axial ratio represents ratio of the intercept made by the unit plane on the crystallographic axis. The intercepts made by the plane LMN are $2a$, $2b$, and $3c$, which are the simple whole number multiple of those of unit plane (Figure 15.15).

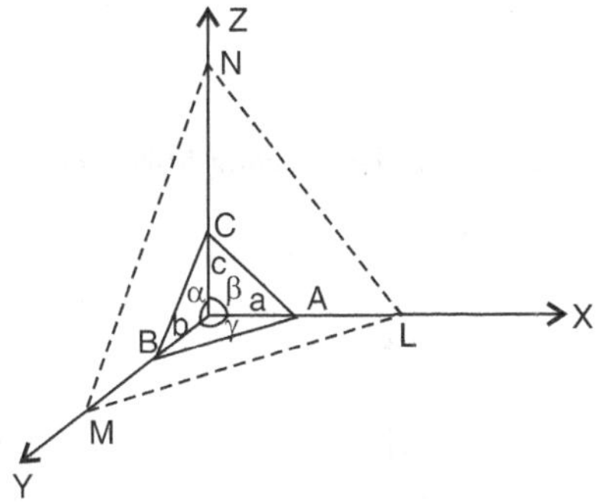


***FIGURE 15.15*** **Crystallographic axes and axial ratio.**

In general, if a face PQR makes intercepts $p$, $q$ and $r$ on the three axes and the intercepts made by the unit plane are $a$, $b$, and $c$, then we have $p = na$, $q = n'b$, and $r = n''c$, where $n$, $n'$ and $n''$ are simple whole numbers or some of these may be infinity (If some face does not cut some axes at all). The coefficients of $a$, $b$, $c$ ($n$, $n'$ and $n''$) are known as *Weiss indices* of plane. Weiss indices are not always simple intergral whole numbers but may have fractional values as well as infinity. Weiss indices have been replaced by Miller indices for simplicity and convenience.

In a real crystal, the lattice points are occupied by atoms. One can think of a very large number of lattice planes and one set of parallel planes can be distinguished from another by Miller indices. *Millier indices* of plane is *a set of three number h, k, l that defines a set of parallel planes in a crystal.* The Miller Indices of a face of crystal are inversely proportional to the intercepts of that face on the various axis.

The following procedure is generally followed to determine the Miller indices:

(*i*) Choose an origin.

(*ii*) Find the intercept that the first such plane makes with the three crystallographic axes.

(*iii*) Obtain the reciprocals.

(*iv*) Clear fractions.

The set of numbers thus generated in relation with the axes $a$, $b$, and $c$ are $h$, $k$, and $l$, respectively. Different sets of parallel planes have different $hkl$ or Miller indices. This is illustrated in Figure 15.16.

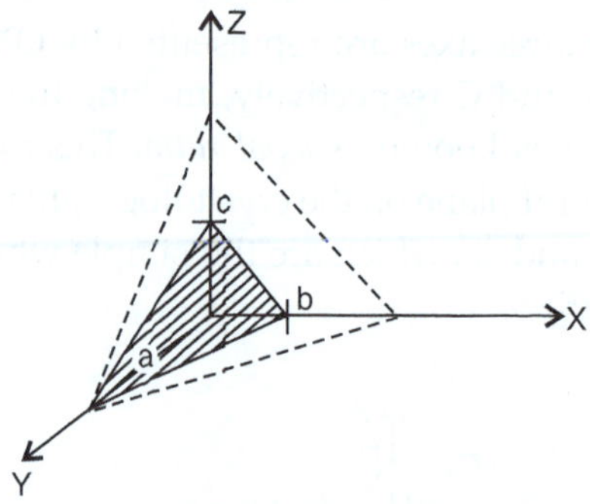


*FIGURE 15.16* **A plane having Miller indices 122.**

Various Miller indices of a cube are given in Figure 15.17.

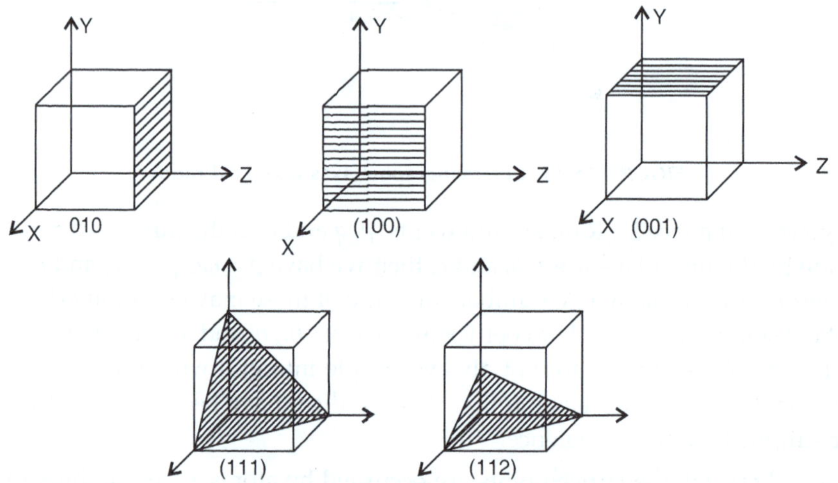


*FIGURE 15.17* **Miller planes of a cube.**

The major advantage of Miller indices is that they permit us to express interplanar distances of a set of *hkl* planes in terms of lattice parameters. The interplanar distances between the parallel planes are designated as $d_{hkl}$.

$$\frac{1}{(d_{hkl})^2} = \left(\frac{h}{a}\right)^2 + \left(\frac{k}{b}\right)^2 + \left(\frac{l}{c}\right)^2$$

For a cubic crystal $a = b = c$

$$d_{hkl} = \frac{a}{\sqrt{h^2 + k^2 + l^2}}$$

For a tetragonal system $a = b \neq c$

$$\frac{1}{(d_{hkl})^2} = \frac{h^2 + k^2}{a^2} + \frac{l^2}{c^2}$$

For orthorhombic system $a \neq b \neq c$

$$\frac{1}{(d_{hkl})^2} = \frac{h^2}{a^2} + \frac{k^2}{b^2} + \frac{l^2}{c^2}$$

where $h, k, l$ are the Miller indices of a plane.

## 15.5 PACKING IN METALLIC CRYSTALS

In the formation of crystals the constituent particles try to pack as closely as possible so as to attain a state of maximum possible density and stability. Since in different crystals, the units of pattern have different shapes and sizes, the actual mode of closest packing is different for different crystals. In this section, we shall focus on close packing modes of spherical particles of equal size. In metallic crystals the constituent particles are nearly of this type.

### 15.5.1 Close Packing in Two Dimensions

In order to understand the close packing of spheres, first of all, let us consider how the spheres can be arranged in a plane. The two possible arrangements are shown in Fig. 15.18. In the arrangement (*a*) the spheres in the adjacent rows lie just one over the other and show a horizontal as well as vertical alignment, and form squares. This type of packing is called *square close packing. Each sphere in this arrangement is in contact with four other spheres.*

In the arrangement (*b*), the spheres in every second row are seated in the depressions between the spheres of the first row. The spheres in the third row are vertically aligned with the spheres in the first row and those in fourth row vertically aligned with spheres in the second row.

The similar pattern is followed throughout. In this arrangement, each sphere is in direct contact with six other spheres. This arrangement gives rise to hexagonal patterns and is called *hexagonal close packing.*

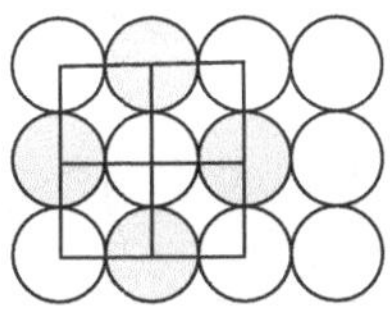

(a) Square close packing. (Coordination number 4)

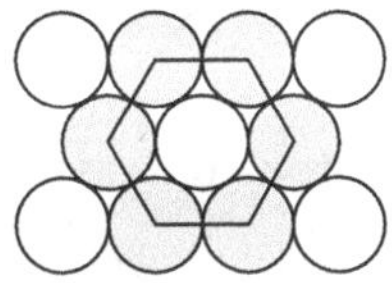

(b) Hexagonal close packing. (Coordination number 6)

*FIGURE 15.18* **Close packing in two dimensions.**

The arrangement (*b*) is obviously more closely packed than the arrangement (*a*). It can be shown by simple calculations that in the arrangement (*b*) 60.4 per cent of the available space is occupied by spheres whereas in arrangement (*a*) only 52.4 per cent of the space is occupied by spheres. If only one layer of spheres is to be packed, the spheres will arrange themselves according to hexagonal close packing because it is more efficient. In this arrangement there are a number of vacant spaces or voids. These vacant spaces are between three touching spheres whose centres lie at the corners of an equilateral triangle. These vacant spaces are, therefore, called *triangular voids.*

### 15.5.2 Close Packing in Three Dimensions

In order to develop three dimensional close packing, let us retain the hexagonal close packing in the first layer because it is the most efficient. If the spheres in the second layer are just placed over the spheres in the first layer so that the spheres in the two layers are vertically aligned then the voids in the two layers will also cover each other. This will amount to an inefficient filling of space. For close packing each sphere in the second layer rests in the hollow at the center of three touching spheres in the layer as shown in Fig. 15.19 (*b*). The spheres in the first layer are shown by solid lines while those in second layer are shown by broken lines. It may be noted that only half the triangular voids in the first layer are occupied by spheres in the second layer (*i.e.*, either *b* or *c*). The unoccupied hollows or voids in the first layer are indicated by (*c*) Fig. 15.19.

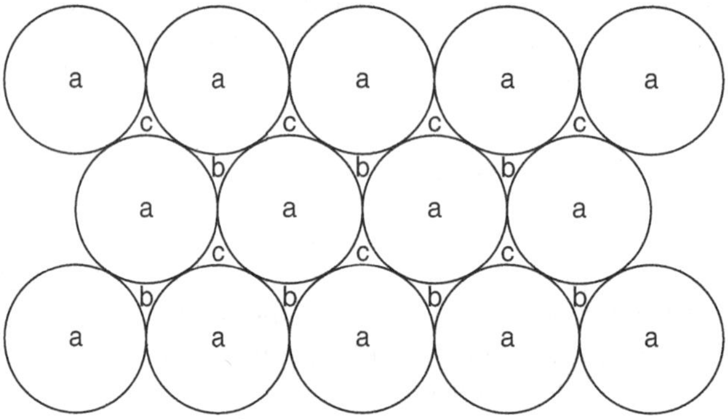


(a)

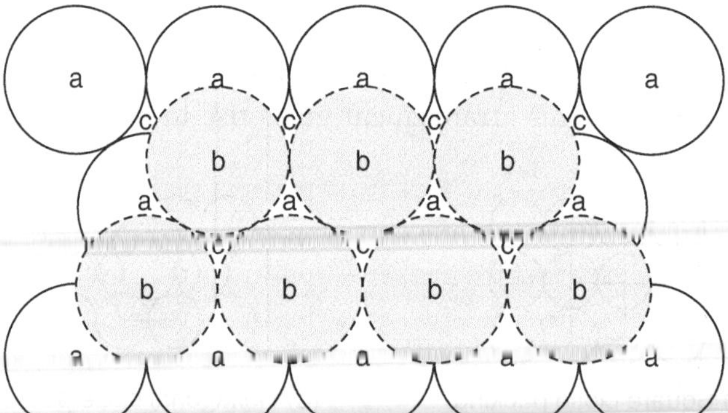


(b)

*FIGURE 15.19* **Close packing of spheres.**

There are two alternative ways in which spheres in the third layer can be arranged over the second layer. The third layer can lie vertically above the first if the spheres in the third layer rest in one set on hollows [corresponding to centers of the spheres (marked by *a*) in the first layer] on the top of the second layer, alternatively, the third layer can be different from the first if the spheres in the third layer lie on the other set of hollows which correspond to those marked '*c*' in the first layer. The first arrangement is called ABAB... type while the second alternative is called ABC ABC...

type. These two arrangements are equally efficient. In either type of close packing 74.0 percent of the available space is occupied by spheres. They differ from each other in overall symmetry. *The AB AB... arrangement has hexagonal symmetry and is known as hexagonal close packing (hcp)* (Fig. 15.20). *The ABC ABC... packing has cubic symmetry and is known as cubic close packing (ccp)* (Fig. 15.21). The cubic close packing has *face centered cubic* (*fcc*) unit cell.

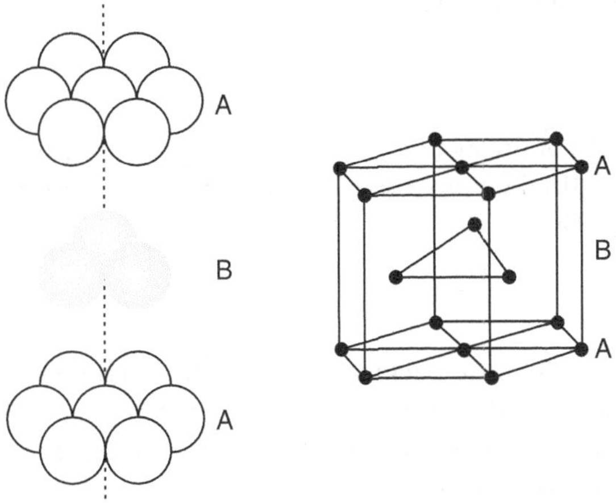


*FIGURE 15.20* **Hexagonal close packing (hcp).**

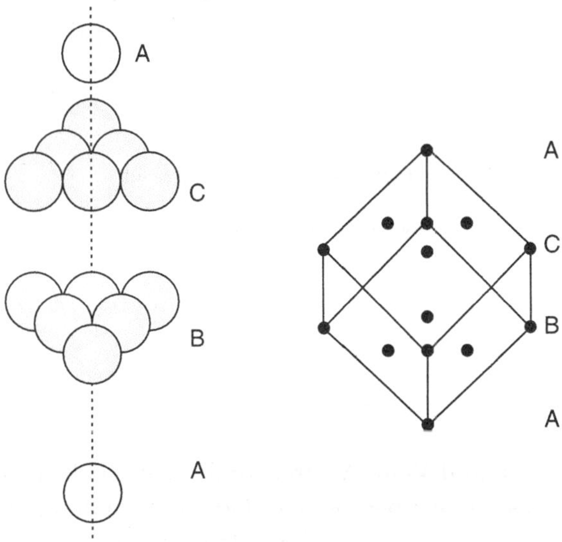


*FIGURE 15.21* **Cubic close packing (ccp).**

Some examples of metallic solids having *hcp*, *fcc* and *bcc* arrangements are given below :

**Hexagonal Close Packed:** Magnesium, Zinc, Cadmium and Titanium

**Face Centered Cubic:** Nickel, Copper, Silver, Gold and Aluminium

**Body Centered Cubic:** Sodium, Potassium, Rubidium, Tungsten, Chromium and Iron

## COORDINATION NUMBER

In the two types of three dimensional close packing (*hcp* and *ccp*) discussed above, each sphere is touching six other spheres in its own layer, three spheres in the layer above and three spheres in the layer below. Thus, in a close packing arrangement a sphere is touching 12 other spheres. *The number of nearest neighbours in a packing is called* **coordination number.** In close packing arrangement each sphere has a co-ordination number of 12.

### 15.5.3 Interstitial Voids

In *hcp* as well as *ccp* only 74% of the available space is occupied by spheres. The remaining space is vacant and constitutes interstitial voids or spaces. There are two types of interstitial spaces in three dimensional close packing. These are :

(A) *Tetrahedral voids*

(B) *Octahedral voids.*

(A) **Tetrahedral voids.** In close packing arrangement each sphere in the second layer rests on the hollow (triangular void) in the three touching spheres in the first layer. The centers of these four spheres are at the corners of a regular tetrahedron. *The vacant space between these four touching spheres is called tetrahedral void* (Fig. 15.22).

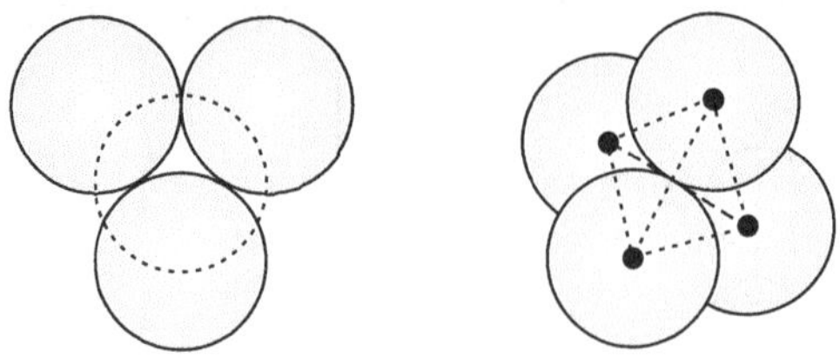

*FIGURE 15.22* **Tetrahedral void.**

In a close packing, the number of tetrahedral voids is double the number of spheres. So there are two tetradedral voids associated with each sphere, one above the sphere and one below the sphere. Radius of the tetrahedral void relative to the radius of the sphere is 0.225, *i.e.*, for tetrahedral void

$$\frac{r_{void}}{r_{sphere}} = 0.225.$$

**Radius Ratio of Tetrahedral Void.** A tetrahedral site in a cube is shown in Fig. 15.23. Here, the shaded sphere represents the tetrahedral void. Now if we assume that the face length of the cube = $a$ and $r_1$ and $r_2$ are the radii of the void and the sphere respectively then considering the right angled triangle ABC,

Face diagonal, $\quad AB = \sqrt{AC^2 + BC^2} = \sqrt{2^a}$

or $\quad 2r_2 = \sqrt{2^a} \quad$ [$\because$ Along length AB, the two spheres with radius $r_2$ touch each other.]

$\therefore \quad AB = r_2 + r_2 = 2r_2]$

or $$r_2 = \frac{\sqrt{2}}{2}\,a = \frac{1}{\sqrt{2}}\,a \qquad ...(i)$$

Similarly, considering the right angled triangle ABD, the body diagonal AD is given as

$$AD = \sqrt{AB^2 + BD^2}$$

$$= \sqrt{(\sqrt{2}a)^2 + a^2} = \sqrt{3}\,a$$

or $$2r_1 + 2r_2 = \sqrt{3}\,a \qquad [\because \text{ length AD} = 2r_1 + 2r_2]$$

or $$r_1 + r_2 = \frac{\sqrt{3}}{2}a \qquad ...(ii)$$

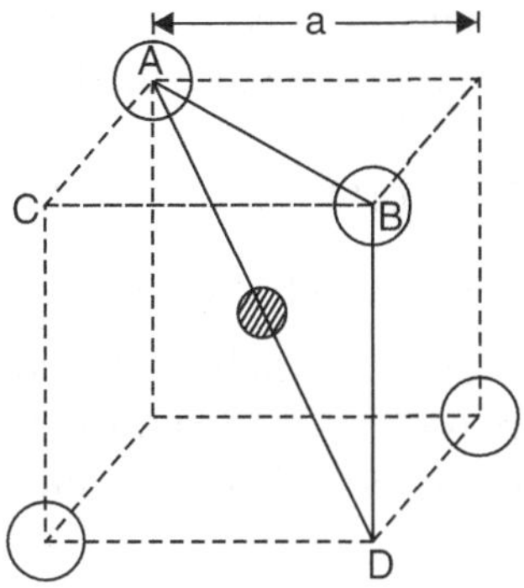


*FIGURE 15.23*

Dividing eqn. (*ii*) by eqn. (*i*), we get

$$\frac{r_1 + r_2}{r_2} = \frac{\sqrt{3}}{2}a \,.\, \frac{\sqrt{2}}{a} = \frac{\sqrt{3}}{\sqrt{2}}$$

or $$\frac{r_1}{r_2} + 1 = \frac{\sqrt{3}}{\sqrt{2}}$$

or $$\frac{r_1}{r_2} = \frac{\sqrt{3}}{\sqrt{2}} - 1 = \frac{\sqrt{3} - \sqrt{2}}{\sqrt{2}} = \frac{1.732 - 1.414}{1.414} = \mathbf{0.225.}$$

Thus, for a tetrahedral void $\frac{r_{void}}{r_{sphere}} = 0.225$.

(B) **Octahedral Voids.** As already discussed the spheres in the second layer rest on the triangular voids in the first layer. However, one half the triangular voids in the first layer are occupied by the spheres in the second layer while the other half marked '*c*' (Fig. 15.19) remain unoccupied. The triangular voids '*c*' in the first layer are overlapped by the triangular voids in the second layer.

*The interstitial void formed by combination of two triangular voids of the first and second layer is* called *octahedral void* because this is enclosed between six spheres, centres of which occupy corners of a regular octahedron. The octahedral void is shown in Fig. 15.24.

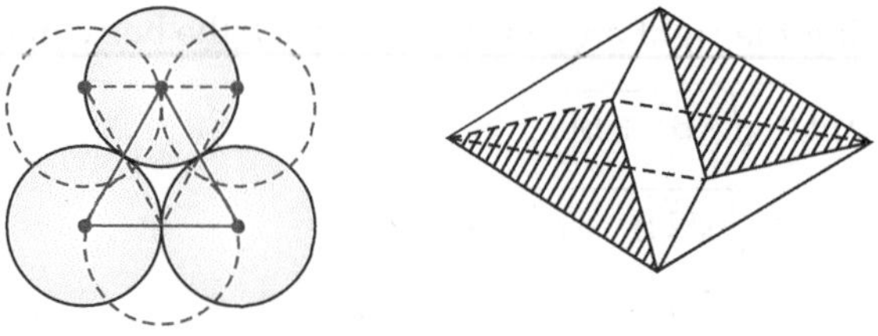


***FIGURE 15.24*** **Octahedral Void.**

In a close packing, the number of octahedral voids is equal to the number of spheres. Thus, there is only one octahedral void associated with each sphere. Radius of the octahedral void relative to the radius of the sphere is 0.414, *i.e.*, for octahedral void.

$$\frac{r_{void}}{r_{sphere}} = 0.414.$$

Octahedral void is larger than a tetrahedral void.

**Radius Ratio of Octahedral Void.** An octahderal void is shown in Fig. 15.25. The spheres present above and below the void are not shown in Fig. 15.25. Now, if we assume that each length of the square ABCD = $a$ and $r_1$ and $r_2$ are the radii of the void and the sphere respectively then for right angled triangle ABC

$$AC = \sqrt{AB^2 + BC^2} = \sqrt{2}\ a$$

But

$$AC = 2r_1 + 2r_2$$

$\therefore$

$$2r_1 + 2r_2 = \sqrt{2}\ a \qquad ...(i)$$

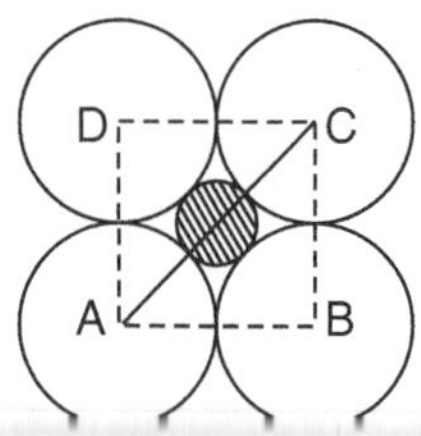


***FIGURE 15.25***

Now each side of the square is equal to $2r_2$, *i.e.*, $a = 2r_2$

Substituting in equation (*i*), we get

$\therefore$

$$2r_1 + 2r_2 = \sqrt{2}\ .\ 2r_2$$

$$\frac{r_1}{r_2} + 1 = \sqrt{2}$$

$$\frac{r_1}{r_2} = \sqrt{2} - 1 = 1.414 - 1 = \mathbf{0.414.}$$

Thus, for an octahedral void, $\dfrac{r_{void}}{r_{sphere}} = 0.414.$

### 15.5.4 Structures of Simple Ionic Compounds

In this section, we shall discuss the structures of a few simple ionic compounds of the type AB and $AB_2$. As discussed in the text, each solid tends to have its constituents closely packed to each other. Among the two ions constituting the binary compounds, the anions usually constitute the space lattice with *hcp* or *ccp* type of arrangement whereas the cations occupy the interstitial voids. From the knowledge of the packed structure and the voids occupied, we can have an idea about the structure of simple ionic compounds. For example :

(*i*) For a binary compound if the negative ions constitute the space lattice and all octahedral holes are occupied, then the formula of the ionic solid is AB.

(*ii*) Similarly, if all tetrahedral holes are occupied, the formula of the ionic solid is $A_2B$.

(*iii*) Now if half of the tetrahedral holes are occupied then the formula is AB.

Thus, from the above discussion, it is clear that the idea about arrangement of ions quite reasonably and satisfactorily explains the structures of many ionic solids.

In this section, we shall study structures of some binary compounds of AB type and $AB_2$ type. NaCl, CsCl, ZnS, etc. are examples of compounds of AB type while $CaF_2$, $SrF_2$ are examples of compounds of $AB_2$ type.

**1. Structure of Sodium Chloride (NaCl)—Rock-salt Structure**

(*i*) In sodium chloride, chloride ions (denoted by ●) can be considered as having cubic close packed structure (*i.e.*, face-centred cubic).

(*ii*) The sodium ions (represented by ●) occupy the octahedral holes in the cubic close packed type lattice of chloride ions (Fig. 15.26).

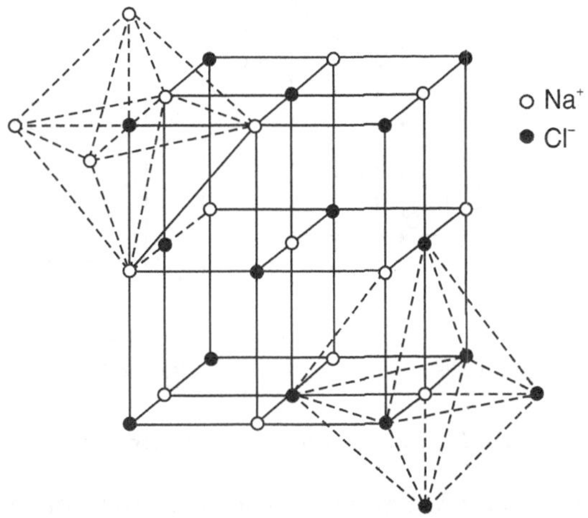


*FIGURE 15.26* **Sodium chloride structure.**

(*iii*) Since there are six octahedral holes around each chloride ion, therefore, each chloride ion is surrounded by six sodium ions. Similarly, each sodium ion is surrounded by six chloride ions (Fig. 15.26). Therefore, in NaCl co-ordination number of $Cl^-$ ion is 6 and that of $Na^+$ ions is also 6. This is called 6 : 6 co-ordination.

(*iv*) It may be pointed out here that for precise, fit the ratio of the radii of the two ions $\left(\frac{r_{Na^+}}{r_{Cl^-}}\right)$ should be equal to 0.414. Actually, the ratio $\left(\frac{r_{Na^+}}{r_{Cl^-}} = 0.525\right)$ exceeds this value and as a result the arrangement of $Cl^-$ ions slightly opens up to accommodate sodium ions. Thus, $Cl^-$ ions do not touch each other and form an expanded face centered lattice.

(*v*) The unit cell of sodium chloride has 4 sodium ions and 4 chloride ions as calculated below :

Number of sodium ions = 12 (At edge centers) × $\frac{1}{4}$ + 1 (At body center) × 1 = 4

Number of chloride ions = 8 (At corners) × $\frac{1}{8}$ + 6 (At face centers) × $\frac{1}{2}$ = 4

Thus, the number of NaCl units per unit cell is 4. The sodium chloride structure is also called *rock-salt structure.* Most of the halides lf alkali metals and oxides of alkaline-earth metals have this type of structure. Some of the common examples are NaI, KCl, RbI and RbF.

**2. Caesium Chloride (CsCl) Structure**

(*i*) In this structure, the $Cl^-$ ions are at the corners of a cube whereas $Cs^+$ ion is at the center of the cube or *vice versa* as shown in Fig. 15.27.

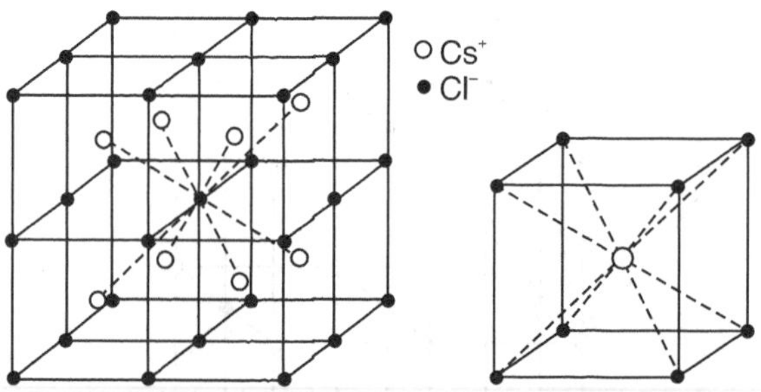


(a)
8 unit cells

(b)
One unit cell

***FIGURE 15.27*** **Caesium chloride structure.**

(*ii*) This structure has 8 : 8 co-ordination, *i.e.*, each $Cs^+$ ion is touching eight $Cl^-$ ions and each $Cl^-$ ion is touching eight $Cs^+$ ions.

(*iii*) For exact fitting of $Cs^+$ ions in the cubic voids the ratio $\frac{r_{Cs^+}}{r_{Cl^-}}$ should be equal to 0.732, however, actually the ratio is slightly larger (0.93). Therefore, packing of $Cl^-$ ions slightly opens up to accommodate $Cs^+$ ions.

(*iv*) The unit cell of caesium chloride has one $Cs^+$ ion and one $Cl^-$ ion as calculated below :

$$\text{No. of } Cl^- \text{ ions} = 8 \text{ (At corners)} \times \frac{1}{8} = 1$$

$$\text{No. of } Cs^+ \text{ ion} = 1 \text{ (At the body centre)} \times 1 = 1$$

Thus, no. of CsCl units per unit cell is 1.

Examples of the compounds having this type of structure are CsBr, CsI, TlCl and TlBr.

**3. Zinc Blende Structure or Sphalerite Structure (ZnS)**

(*i*) In zinc blende-structure, sulphide ions (represented by •) have cubic closed packed (or face-centered cubic) lattice and zinc ions (represented by O) occupy half of the total number of tetrahedral voids as shown in Fig. 15.28.

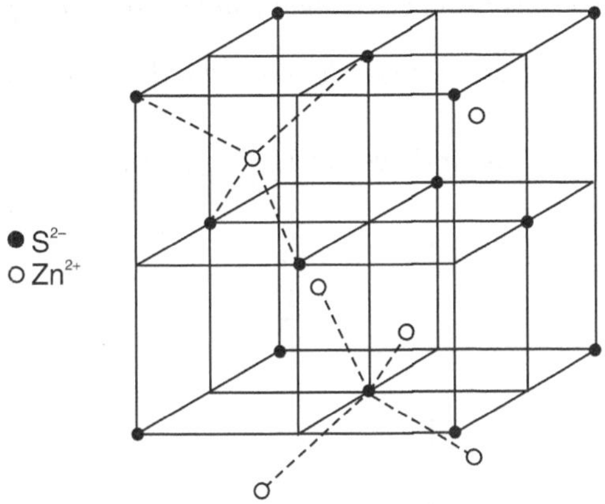


***FIGURE 15.28*** **Zinc blende structure of zinc sulphide.**

(*ii*) Each sulphide ion is tetrahedrally surrounded by four zinc ions and each zinc ion is surrounded by four sulphide ions. Thus, zinc sulphide has 4 : 4 co-ordination.

(*iii*) For exact fitting of $Zn^{2+}$ ions in the tetrahedral voids, in the close packing of $S^{2-}$ ions, the ratio $\frac{r_{Zn^{2+}}}{r_{S^{2-}}}$ should be 0.225. Actually this ratio is slightly large (0.40).

(*iv*) There are four $Zn^{2+}$ ions and four $S^{2-}$ ions per unit cell as calculated below :

$$\text{Number of } S^{2-} \text{ ions} = 8 \text{ (At corners)} \times \tfrac{1}{8} + 6 \text{ (At face centers)} \times \tfrac{1}{2} = 4$$

$$\text{Number of } Zn^{2+} \text{ ions} = 4 \text{ (Within the body)} \times 1 = 4$$

Thus, the number of ZnS units per unit cell is equal to 4.

It may be mentioned here that the bond between zinc and sulphur has appreciable covalent character due to small electronegativity difference. Electronegativity of zinc is 1.6 while that of sulphur is 2.5. The structure of sphalerite is similar to that of diamond. In diamond all the positions are occupied by carbon atoms.

Some more examples of ionic solids having zinc blende structure are CuCl, CuBr, CuI, AgI.

**Example 8.** *The planes in a crystalline solid intersect the crystal axes at 2a, 3b, c. Calculate the Weiss and the Miller indices for these planes.*

**Solution.** The Unit cell intercepts are $a, b, c$

| | | | |
|---|---|---|---|
| Intercepts | $2a,$ | $3b,$ | $c$ |
| Weiss Indices | 2 | 3 | 1 |
| Miller Indices: | | | |
| Reciprocal of Weiss Indices | $\frac{1}{2}$ | $\frac{1}{3}$ | $\frac{1}{1}$ |
| Miller Indices | 3 | 2 | 6 |

**Example 9.** *A crystal plane intercepts the three crystallographic axes at a, $\frac{1}{2}$b, and $\frac{3}{2}$c where a, b, and c are unit lengths along the X, Y, and Z-axes, respectively. What are the Miller indices of this plane?*

**Solution.** From the given data, Weiss indices are $1, \frac{1}{2}, \frac{3}{2}$, *i. e.,*

| | $a$ | $b$ | $c$ |
|---|---|---|---|
| | 1 | $\frac{1}{2}$ | $\frac{3}{2}$ |
| Taking reciprocals | 1 | 2 | $\frac{2}{3}$ |
| Multiplying by 3 | 3 | 6 | 2 |

Thus, Miller indices of the plane are 3, 6, 2 and hence the plane is designated as a (362) plane.

**Example 10.** *Calculate the Miller indices of the crystal plane which cut through the crystal axes at 2a, – 3b, – 3c.*

| **Solution.** Given | $a$ | $b$ | $c$ |
|---|---|---|---|
| | 2 | – 3 | – 3 |
| Taking reciprocals | $\frac{1}{2}$ | $-\frac{1}{3}$ | $-\frac{1}{3}$ |
| Multiplying by 6 | 3 | – 2 | – 2 |

Hence the Miller indices are 3, – 2, – 2 and thus the plane is designated as $(3, \bar{2}, \bar{2})$.

**Example 11.** *How do the spacings of the three planes 100, 110, and 111 of the cubic lattice vary?*

**Solution.** Applying the formula,

$$d_{hkl} = \frac{a}{\sqrt{h^2 + k^2 + l^2}}$$

$$d_{100} = \frac{a}{\sqrt{1^2 + 0^2 + 0^2}} = a$$

$$d_{110} = \frac{a}{\sqrt{1^2 + 1^2 + 0^2}} = \frac{a}{\sqrt{2}}$$

$$d_{111} = \frac{a}{\sqrt{1^2 + 1^2 + 1^2}} = \frac{a}{\sqrt{3}}$$

Thus $$d_{100}: d_{110}: d_{100} = 1: \frac{1}{\sqrt{2}} : \frac{1}{\sqrt{3}} = \mathbf{1: 0.707: 0.577}$$

## 15.6 DIFFRACTION OF X-RAYS BY CRYSTAL

*Van Laue* in 1912 suggested that the crystal lattice should act as a three dimensional diffraction grating for X-rays because the X-rays have wavelengths of the dimension of the interatomic distances in the crystal. Thus, when a beam of X-rays is allowed to fall on a crystal, a large number of images of different intensities are formed. If the diffracted waves are in the same phase, they reinforce each other and a series of bright spots is produced on a photographic plate placed in their path. On the other hand, if the diffracted waves are out of phase, dark spots are caused on the photographic plate. From the overall diffraction patterns produced by a crystal, we can arrive at the detailed information regarding the position of particles in the crystal. The study of crystal structure with the help of X-rays is called *X-ray crystallography.*

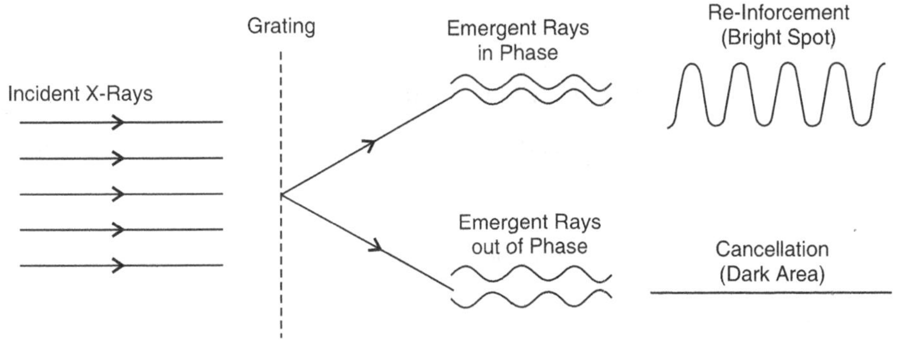


*FIGURE 15.29* **Diffraction patterns produced by crystals.**

### 15.6.1 Bragg's Law

X-rays are electromagnetic radiations with high energies and wavelengths of the order $10^{-10}$ m. These rays can be diffracted whenever they meet barrier with one or more openings that is about the same size as the wavelength. The layers of atoms or ions in crystals are separated by a distance of around 100 or 200 pm (pm-picometer), which is the right range to cause X-rays to be diffracted. Thus layers of atoms in a crystal act as a diffracting grating for X-rays.

### 15.6.2 Derivation of Bragg's Law

The waves that have matching crests and troughs reinforce constructively and this happens if the distance that the wave travels differ by a whole number of wavelength $\lambda$. Thus, constructive interference occurs when

$$\text{Path difference} = n\lambda,\ n \text{ is the whole number.}$$

Let us suppose that a narrow beam of monochromatic X-rays of wavelength $\lambda$ are incident at a glancing angle $\theta$ on a set of parallel atomic planes of crystal as shown in Figure 15.30. These rays are scattered by the atoms like B and E of two adjacent planes of separation '$d$', in all directions. Constructive interference will occur only between those scattered waves which are reflected specularly (*in specular reflection, the angle of incidence is equal to the scattering angle* $\theta = \theta'$ *and the incident ray, scattered ray, and normal to lattice planes all lie in the same plane*) from parallel planes of atoms in a similar way reflection of a light wave from a plane mirror, and have a path difference of $n\lambda$, where $\lambda$ is the wavelength of X-rays and $n$ is an integer. So two incident X-rays, AB and DE, are reflected specularly along BC and EF, respectively, from crystal planes XX and YY as shown in Figure 15.30. Draw normals BM on DE and BN on EF from B. The path difference between these two rays is

$$\Delta = ME + EN$$

From $\Delta$BME, $\quad \sin\theta = \dfrac{ME}{BE}$ or $ME = BE\sin\theta = d\sin\theta$

Similarly, from $\Delta$BNE, $\quad EN = BE\sin\theta = d\sin\theta$

$\therefore \quad \Delta = d\sin\theta + d\sin\theta = 2d\sin\theta$

Hence, $\quad 2d\sin\theta = n\lambda,\quad$ where $n = 0, 1, 2, 3, \ldots\ldots$

This is the well-known *Bragg's equation* and is called the mathematical expression of *Bragg's law*. $n = 0, 1, 2, 3, \ldots\ldots.$ etc. corresponds to first, second, third order spectrum, and so on.

From the Bragg's equation it is clear that if the wavelength of X-rays corresponds to maximum intensity at a glancing angle, $\theta$ is known and the lattice spacing or spacing between successive lattice plane $d$ may be found. Conversely, if the lattice spacing, $d$, is known, the wavelength $\lambda$ of X-rays may be determined.

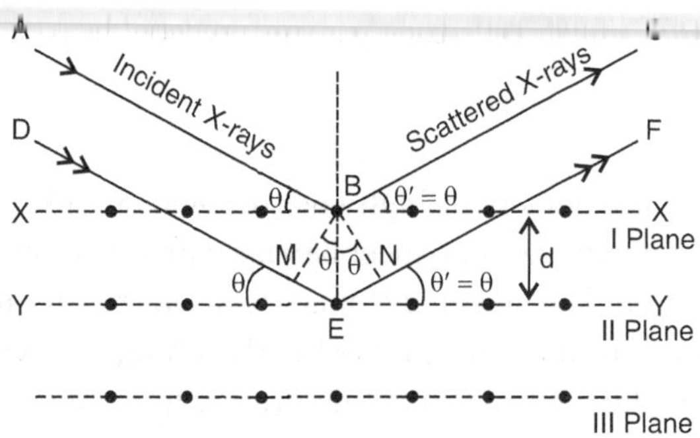


*FIGURE 15.30*

**Example 12.** *Find the interplanar distance in a crystal in which a series of planes produces a first order reflection from a copper X-ray tube (λ = 1.539 Å) at an angle of 22.5°.*

**Solution.** From Bragg's equation, $n\lambda = 2d \sin\theta$

We have $$d = \frac{\lambda}{2\sin\theta} \quad \text{when } n = 1$$

$$\therefore \quad d = \frac{1.5939}{2 \times \sin 22.5}$$

$$d = \frac{1.539}{2 \times 0.382}$$

$$= \mathbf{2.01\ Å}$$

### 15.6.3 Bragg's Spectrometer

For the study of X-ray diffraction by crystals, *W.H. Bragg* and his son *Lawrence Bragg* devised an X-ray spectrometer in which a crystal is used as a reflection grating instead of transmission grating. The experimental arrangement of Bragg's X-ray spectrometer is shown in Figure 15.31. A beam of monochromatic X-rays from a Coolidge tube is collimated into a sharp pencil by padding through two lead sheets, $S_1$ and $S_2$ that absorb all rays except those which pass through small slits provided in them. This collimated fine pencil of X-rays is made to fall upon a tiny crystal C mounted on the turn table T of the spectrometer (similar to that of an ordinary prism table) which is capable of rotation about a vertical axis passing through the center of the turn table. The rotation of the crystal can be read from the circular scale provided with a vernier $V_1$. The X-ray beam, after suffering a reflection from the crystal face, C, enters an ionization chamber I through a slit $S_3$. The ionization chamber I, which is filled with ethyl bromide, is mounted on an arm R which is capable of rotation about the same axis as the crystal or turn table. The position of the ioniz tion chamber can be recorded by a second vernier $V_2$. The chamber (or arm R) and circular table (or turn table) are so joined together that when this circular table (or crystal) rotates through an angle θ, the ionization chamber (arm R) automatically turns through an angle 2θ (that is, double the angle). By the coupling,

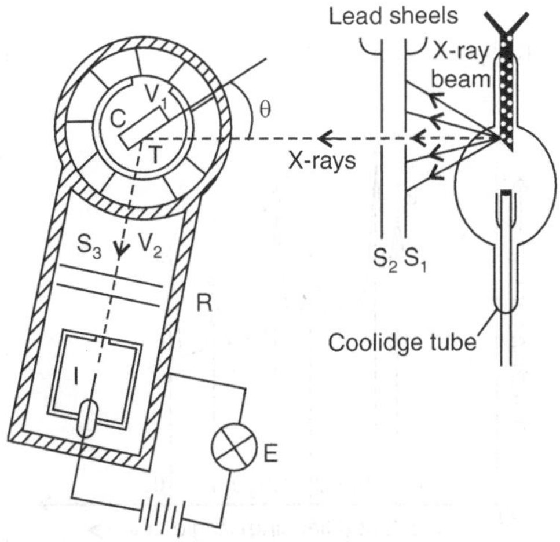


*FIGURE 15.31*

for every glancing angle ($\theta$) at the crystal surface, the X-ray beam will always be reflected into the ionization chamber. The X-rays entering the chamber ionize the gas causing a current in the associated circuit connected with the chamber. The resulting ionization current, which is a measure of the intensity of X-rays reflected by the crystal, is measured by an electrometer E.

For taking observations with the spectrometer, the glancing angle $\theta$ for the incident beam is increased in small steps (starting from zero). By rotating the crystal, the corresponding ionization current is recorded. Then a graph is plotted between the ionization current I and the glancing angle $\theta$. A curve obtained from sodium chloride crystal is shown in Figure 15.32. It is clear from the curve that for certain values of $\theta$, the ionization current I increases abruptly. The number of peaks ($A_1$, $A_2$, $A_3$, etc.) in the curve corresponds to those glancing angles which satisfy the Bragg's equation

$$2d \sin \theta = n\lambda$$

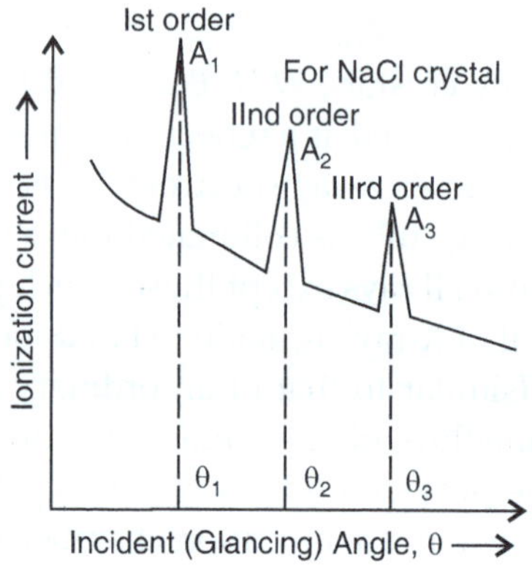


*FIGURE 15.32*

The different peaks correspond to different orders ($n$ = 1, 2, 3). Therefore, for first, second, and third order spectrums

$$2d \sin \theta_1 = \lambda,\ 2d \sin \theta_2 = 2\lambda \text{ and } 2d \sin \theta_3 = 3\lambda$$

or $\quad \sin \theta_1 : \sin \theta_2 : \sin \theta_3 = 1 : 2 : 3$

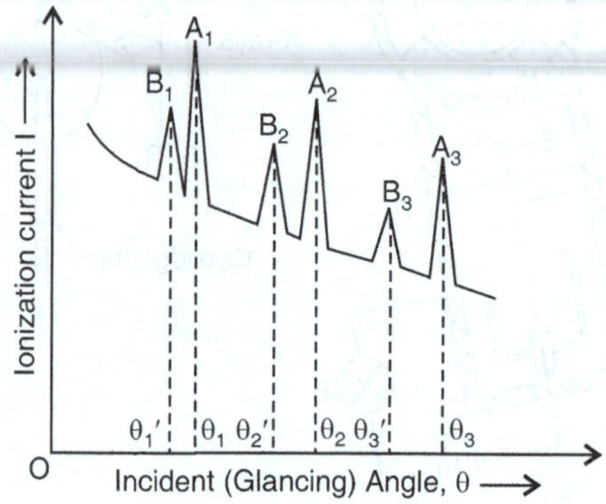


*FIGURE 15.33*

Hence, the glancing angles $\theta_1$, $\theta_2$, $\theta_3$ for the first, second, and third order reflection are measured directly from the curve. If the X-ray beam consists of two nearby wavelengths $\lambda_1$ and $\lambda_2$, in addition to the $A_1$, $A_2$, and $A_3$ corresponding to wavelength $\lambda_1$, the other peaks $B_1$, $B_2$, and $B_3$ corresponding to wavelength $\lambda_2$ also appear as shown in Figure 15.33. It is observed that for wavelength $\lambda_1$, Bragg's equation

$$\sin \theta_1 : \sin \theta_2 : \sin \theta_3 = 1: 2: 3$$

is satisfied. Similarly, if $\theta'_1$, $\theta'_2$, and $\theta'_3$ are the glancing angles for $\lambda_2$, then Bragg's equation

$$\sin \theta'_1 : \sin \theta'_2 : \sin \theta'_3 = 1: 2: 3$$

is again satisfied. From the measured values of the angle $\theta$ and known value of spacing $d$, the wavelength of X-rays can be calculated by using the Bragg's equation,

$$2d \sin \theta = \lambda.$$

### 15.6.4 Bragg's Law and Crystal Structure

Bragg's spectrometer can also be used to determine the structure of the crystals. For this purpose, the ratio of the lattice spacing for various groups of planes is obtained by using Bragg's equation, $2d \sin \theta = n\lambda$. These ratios would be different for different crystals. By comparing known standard ratios with experimentally determined ratios, crystal structure can be obtained.

Suppose, for a particular crystal, strong Bragg's reflections from the three sets of planes with interplanar spacings $d_1$, $d_2$, and $d_3$ are obtained for glancing angles $\theta_1$, $\theta_2$, and $\theta_3$, respectively, in the first order ($n = 1$). Then, from Bragg's equation, we have

$$2d_1 \sin \theta_1 = \lambda,\ 2d_2 \sin \theta_2 = \lambda \quad \text{and} \quad 2d_3 \sin \theta_3 = \lambda$$

$$\therefore \quad d_1 : d_2 : d_3 = 1/\sin \theta_1 : 1/\sin \theta_2 : 1/\sin \theta_3$$

For KCl crystal, Bragg obtained strong reflection of X-rays at glancing angles 5.23°, 7.37°, and 9.25° respectively, using three different reflection planes (100), (110), and (111).

Hence, for KCl

$$d_{100} : d_{110} : d_{111} = 1/\sin 5°23' : 1/\sin 7°37' : 1/\sin 9°25'$$

$$= 1/0.0938 : 1/0.1326 : 1/1.744$$

$$= 1 : 1/\sqrt{2} : 1/\sqrt{2}.$$

This corresponds to the theoretical result for a simple cubic (*sc*) lattice structure. Therefore, it is concluded that the **KCl crystal has a simple cubic structure.**

**Example 13.** *An X-ray tube with a copper target is operated at 25 kV. Calculate the glancing angle for a NaCl crystal for the CuK$_\alpha$ limit. (d for NaCl = 2.82 Å, h = 6.63 × $10^{34}$ J-sec, c = 3 × $10^8$ ms$^{-1}$ and e = 1.6 × $10^{-19}$ coulomb).*

**Solution.** If $d$ is the crystal spacing and $\theta$ the glancing angle for the NaCl crystal, then according to Bragg's equation for specular reflection, $2d \sin \theta = n\lambda$.

For the first order, $n = 1$

$$\lambda = 2d \sin\theta = 2 \times 2.82 \times \sin 15.8°$$

$$= 2 \times 2.82 \times 0.2723 = \mathbf{1.5358\ Å}$$

The wavelength of the X-ray photon at the shortest limit is given by

$$\lambda_{min} = \frac{hc}{eV} = \frac{6.63 \times 10^{-34} \times 3 \times 10^8}{1.6 \times 10^{-19} \times 25 \times 10^3} = 4.9725 \times 10^{-11} \text{ m} = 0.4972 \text{ Å}$$

Let $\theta_1$ be the glancing angle for photons at the shortest wavelength limit, $\lambda_{min}$. Then

$$\lambda_{min} = 2d \sin\theta_1 \quad \text{or} \quad \sin\theta_1 = \frac{\lambda_{min}}{2d}$$

or

$$\sin\theta_1 = \frac{0.4972}{2 \times 2.82} = 0.088 \quad \text{or} \quad \theta_1 = \mathbf{\sin^{-1}(0.088) = 5.05°}$$

**Example 14.** *X-rays having wavelengths of 2.6 Å are diffracted at an angle of 20° for the second order in Bragg's spectrometer. Find the spacing constant of the crystal.*

**Solution.** If $n$th order maximum in Bragg's spectrometer is given in a direction $\theta$, then according to Bragg's equation

$$2d \sin\theta = n\lambda \quad \text{or} \quad d = \frac{n\lambda}{2\sin\theta}$$

Here $\lambda = 2.6 \text{ Å} = 2.6 \times 10^{-10}$ m, and $\sin 20° = 0.3420$

$$d = \frac{2 \times 2.6 \times 10^{-10}}{2 \times 0.3420} = 7.602 \times 10^{-10} \text{ m} = \mathbf{7.60\ Å.}$$

**Example 15.** *Calculate the longest wavelength that can be analyzed by a rock salt crystal of spacing d = 2.82 Å in the first order.*

**Solution.** According to Bragg's equation for specular reflection

$$2d \sin\theta = n\lambda \quad \text{or} \quad \lambda = \frac{2d \sin\theta}{n}$$

For the longest wavelength, $(\sin\theta)_{max} = 1$

$$\lambda_{max} = \frac{2d}{n}$$

Here, $d = 2.82 \text{ Å} = 2.82 \times 10^{-10}$ m and $n = 1$

$$\therefore \quad \lambda_{max} = \frac{2.82 \times 10^{-10} \times 2}{1} = 5.64 \times 10^{-10} \text{ m} = \mathbf{5.64\ Å}$$

**Example 16.** *A beam of X-rays, λ = 0.842 Å is incident on a crystal at a glancing angle of 8°35′ when the first order Bragg's reflection occurs. Calculate the glancing angle for 3rd order reflection.*

**Solution.** According to Bragg's equation for specular reflection,

$$2d \sin \theta = n\lambda$$

For first order, $n = 1 \quad \therefore \quad 2d \sin \theta_1 = \lambda$ ... (*i*)

and for third order, $n = 3 \quad \therefore \quad 2d \sin \theta_3 = 3\lambda$ ... (*ii*)

Dividing equation (*ii*) by (*i*), we get

$$\frac{\sin \theta_3}{\sin \theta_1} = 3 \quad \text{or} \quad \sin \theta_3 = 3 \times \sin 8°35'$$

$$\therefore \quad 3 \times 0.15 = 0.45' \quad \text{or} \quad \theta_3 = \sin^{-1}(0.45) = \mathbf{26.7°}.$$

**Example 17.** *Diffraction angle 2θ equal to 16.8° for a crystal having interplanar distance in the crystal is 0.400 nm when second order diffraction is observed. Calculate the wavelength of X-rays used.*

**Solution.** Here $n = 2, d = 0.400 \text{ nm} = 0.4 \times 10^{-9} \text{ m}$

$$2\theta = 16.8° \quad \text{or} \quad \theta = 8.4°$$

Using Bragg's equation $n\lambda = 2d \sin \theta$

Substituting the values, we get

$$2 \times \lambda = 2 \times 0.4 \times 10^{-9} \times \sin 8.4$$

or $$\lambda = \frac{2 \times 0.4 \times 10^{-9} \times 0.146}{2}$$

or $$\lambda = 0.0584 \times 10^{-9} \text{ m}$$

or $$\lambda = 0.584 \times 10^{-10} \text{ m}$$

or $$\lambda = \mathbf{0.584 \text{ Å}}$$

## 15.7 ALLOTROPES OF CARBON

Carbon occurs in three allotropic forms: diamond, graphite, and fullerene.

### 15.7.1 Diamond

Diamond is the most hard, most dense, and the least reactive of all the allotropes. It has no color.

**Structure.** There is a tetrahedral ($sp^3$ hybridization) arrangement and every carbon atom is covalently bonded to four equidistant carbon atoms (Figure 15.34). The crystal is thus a three dimensional, giant compact, interlocking molecule without any limit on its extension. The structure may be looked as *fcc* lattice of carbon atoms, where alternate four out of eight tetrahedral holes are occupied by carbon atoms. The C-C bond distance is 1.54 Å and the bond angle is 109°28′. Since all the bounding orbitals are fully occupied, the crystal is an insulator. The extreme hardness, brittleness, and high melting point of diamond is due to the good strength of the bonds and their uniformity in all directions throughout the crystal lattice.

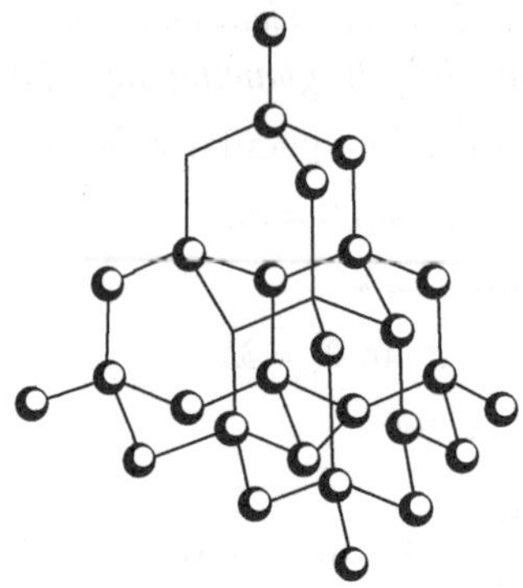

*FIGURE 15.34*

### 15.7.2 Graphite

Graphite derives its name from the Greek word "*graphein*", meaning *to write*. The material is generally greyish-black, opaque, and has a lustrous black shine. It is unique in that it has properties of both a metal and a nonmetal. It is flexible but not elastic, has a high thermal and electrical conductivity, and is highly refractory and chemically inert. Graphite has a low absorption of X-rays and neutrons, making it a particularly useful material in nuclear applications. World production of graphite was estimated to be about 602 000 tons in 2000, with China being the biggest producer followed by India, Brazil, Mexico, and then the Czech Republic.

**Structure.** The unusual combination of properties is due to its crystal structure. (Figure 15.24) There is a planar ($sp^2$ hybridization) arrangement and every carbon atom is bonded to three other carbon atoms, forming the hexagonal network sheets. The C-C bond distance is 1.42 Å. The hexagonal layers are held parallel by weak *van der Waals* forces. The distance between adjacent layers is 3.40 Å. The layers are stacked parallel to each other. The atoms within the rings are bonded covalently, while the layers are loosely bonded together by van der Waals forces. The high degree of anisotropy in graphite results from the two types of bonding acting in different crystallographic directions. For example, graphite's ability to form a solid film lubricant comes from these two contrasting chemical bonds. The fact that weak van der Waals forces govern the bonding between individual layers permits the layers to slide over one another, making it an ideal lubricant.

Moreover, each carbon atom is connected to only three other carbon atoms, and the fourth valency (or π-electrons) of every carbon atom is unsatisfied. These π-electrons are mobile throughout the structure and these contribute to electrical conductivity in the direction parallel to the hexagonal sheets. Hence, graphite is a good conductor of electricity.

**Graphite Classifications.** There are two main classifications of graphite: natural and synthetic.

1. *Natural Graphite.* Natural graphite is a mineral consisting of graphitic carbon. It varies considerably in crystallinity. Most commercial (natural) graphites are mined and often contain other minerals. Subsequent to mining, the graphite often requires a considerable amount of mineral processing such as froth flotation to concentrate the graphite. Natural graphite is an excellent conductor of heat and electricity. It is stable over a wide range of temperatures. Graphite is a highly refractory material with a high melting point (3650°C).

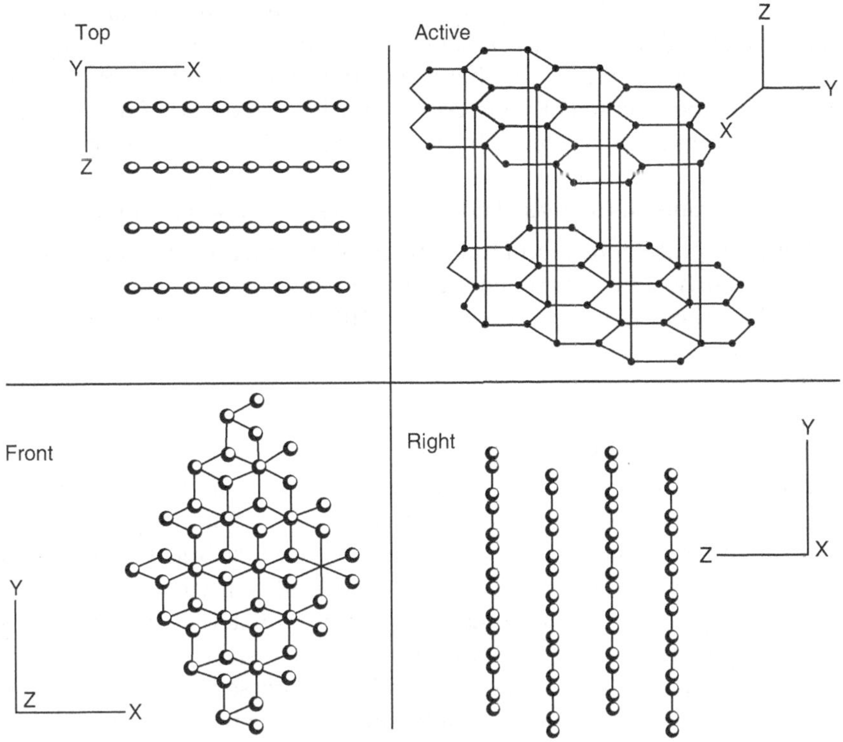


*FIGURE 15.35* **Crystal structure of graphite viewed from various angles.**

Natural graphite is subdivided into three types of material:

(*i*) *Amorphous Graphite*. Amorphous graphite is the least graphitic of the natural graphites. However, then term "amorphous" is a misnomer since the material is still crystalline. Amorphous graphite is found as minute particles in beds of mesomorphic rocks such as coal, slate, or shale deposits. The graphite content ranges from 25% to 85% dependent on the geological conditions.

Amorphous graphite is extracted using conventional mining techniques and occurs primarily in Mexico, North Korea, South Korea, and Austria.

(*ii*) *Flake Graphite.* Flake graphite is found in metamorphic rocks uniformly distributed through the body of the ore or in concentrated lens-shaped pockets. Carbon concentrations vary between 5% and 40%. Graphite flake occurs as a scaly or lemella form in certain metamorphic rocks such as limestone, gneisses, and schists.

Flake graphite is removed by froth flotation. "As floated" graphite contains between 80% and 90% graphite. Flake graphite is produced with > 98% through chemical beneficiation processes. Flake graphite occurs in most parts of the world.

(*iii*) *Crystalline Graphite.* Crystalline vein graphite is believed to originate from crude oil deposits that, through time, temperature and pressure have converted to graphite. Vein graphite fissures are typically between 1cm and 1m thick, and are typically > 90% pure. Although this form of graphite is found all over the world, it is only commercially mined in Sri Lanka by conventional shaft or surface mining techniques.

2. *Synthetic Graphite.* Synthetic graphite can be produced from coke and pitch. It tends to be of higher purity though not as crystalline as natural graphite. There are essentially two types of synthetic graphite. The first is electrographite, which is pure carbon produced from calcined petroleum coke and coal tar pitch in an electric furnace. The second type of synthetic graphite is produced by heating calcined petroleum pitch to 2800°C . On the whole, synthetic graphite tends to be of a lower density, higher porosity, and higher electrical resistance. Its increased porosity makes it unsuitable for refractory applications.

   Synthetic graphite consists mainly of graphitic carbon that has been obtained by graphitization, heat treatment of nongraphitic carbon, or by a chemical vapor deposition from hydrocarbons at temperatures above 2100 K.

### Applications

Following are some important applications of graphite:

1. *Refractory Materials.* Due to its high temperature, stability, and chemical inertness, graphite is a good candidate for a refractory material. It is used in the production of refractory bricks and in the production of "Mag-carbon" refractory bricks (Mg-C). Graphite is also used to manufacture crucibles, ladles, and molds for containing molten metals. Additionally, graphite is one of the most common materials used in the production of functional refractories for the continuous casting of steel. In this application, graphite flake is mixed with alumina and iron and then pressed to form components such as stopper rods, subentry nozzles, and ladle shrouds used in both regulating flow of molten steel and protecting against oxidation. This type of material may also be used as shielding for pyrometers.

   In the production of iron, graphite blocks are used to form part of the lining of the blast furnace. Its structural strength at temperature, thermal shock resistance, high thermal conductivity, low thermal expansion, and good chemical resistance are of paramount importance in this application.

   The electrodes used in many electrical metallurgical furnaces are manufactured from graphite such as the electric arc furnaces used for processing steel.

2. *Chemical Industry.* There are many high temperature uses for graphite in the chemical industry, such as in the production of phosphorus and calcium carbide in arc furnances. Graphite is used as anodes in some aqueous electrolytic processes, such as in the production of halogens (chlorine and fluorine).

3. *Electrical Applications.* The main application for graphite as an electrical material is in the manufacture of carbon brushes in electric motors. In this application, performance and lifetime of the component is very dependent on grade and structure.

4. *Nuclear Industry.* High purity electrographite is used in large amounts for the production of moderator rods and reflector components in nuclear reactors. Their suitability arises from their low absorption of neutrons, high thermal conductivity, and their strength at high temperature.

5. *Mechanical Application.* Graphite is used widely as an engineering material over a variety of applications. Applications include piston rings, thrust bearings, journal bearings, and vanes. Carbon based seals are used in the shafts and fuel pumps of many aircraft jet engines.

6. *Other Applications. Amorphous graphite* has applications in metallurgy, pencil production, refractories coatings, lubricants, and paint production. *Crystalline graphite* is used in batteries, lubricants, grinding wheels, and power metallurgy. *Synthetic graphites* are used in aerospace applications, batteries, carbon brushes, graphite electrodes for electric arc furnaces for metallurgical processing, and moderator rods in nuclear power plants.

## 15.8 FULLERENES

The fullerenes, hailed as one of the discoveries of the century, have created whole new fields of organic/organometallic chemistry and of physics. Together with the related nanotubes, they hold the promise of providing new materials with novel chemical and solid state properties.

Fullerenes were first identified in 1985 as products of experiments in which graphite was vaporized using a laser, work for which *R.F. Curl, Jr., R.E. Smally,* and *H.W. Kroto* shared the 1996 Nobel Prize in Chemistry. Fullerenes have since been discovered in nature as a result of lightning strikes, in the residue produced by carbon arc lamps, in interstellar dust, and in meteorites.

### Structure and Discovery

Any class of carbon molecules in which the carbon atoms are arranged into 12 pentagonal faces and 2 or more hexagonal faces to form a hollow sphere, cylinder, or similar figure are called fullerenes. The fullerene-like molecules (lacking a hexagonal face) with as few as 20 carbon atoms have been found.

The most common and most stable fullerene is *buckminsterfullerene,* a spheroidal molecule resembling a soccer ball, consisting of 60 carbon atoms (Figure 15.36). Buckminsterfullerene is the most abundant cluster of carbon atoms found in carbon soot. It is also the smallest carbon molecule whose pentagonal faces are isolated from each other. Other fullerenes that have been produced in macroscopic amounts have 70, 76, 84, 90, and 96 carbon atoms, and much larger fullerenes have been found, such as those that contain 180, 190, 240, and 540 carbon atoms.

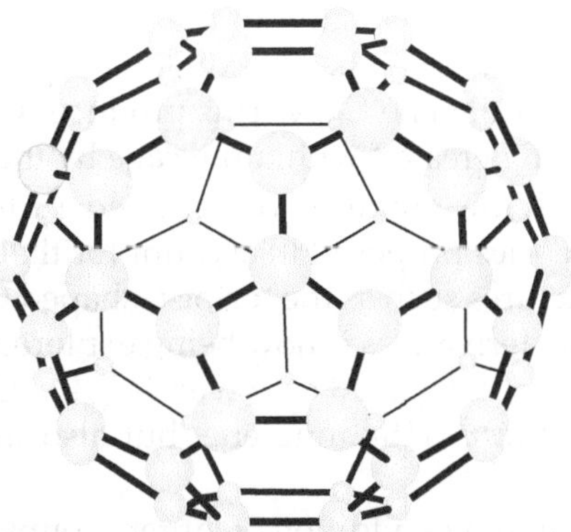

***FIGURE 15.36*** **Shape of a new Carbon molecule, buckminster-fullerene, discovered in 1985, belonging to the family of fullerenes.**

### Fullerene Chemistry

Fullerene chemistry involves substituting metal atoms for one or more carbon atoms in the molecule to produce compounds called fullerids, Buckyball, $C_{60}$ hollow cage carbon molecules named for *R. Buckminster Fuller* because of the resemblance of its molecular structure to his geodesic domes. Although buckminsterfullerene ($C_{60}$) was originally detected in soot in 1985, isolation was first reported in 1990. The soccerball-like molecules are prepared in helium by passing about 150 amps through a carbon rod and extracting the soot with benzene; the resulting magneta solution contains $C_{60}$ and $C_{70}$. Some of the reactions of fullerene are described below:

1. *Synthesis of phosphorylated methanofullerenes.* The reactions of fullerene $C_{60}$ with bromo (dimethoxyphosphoryl) acetates in the presence of NaH in toluene produced alkoxycarbonyl (dimethoxyphosphoryl) methanofullerenes whose structures were established by UV, IR, and $^1H$, $^{13}C$, and $^{31}P$ NMR spectroscopy.

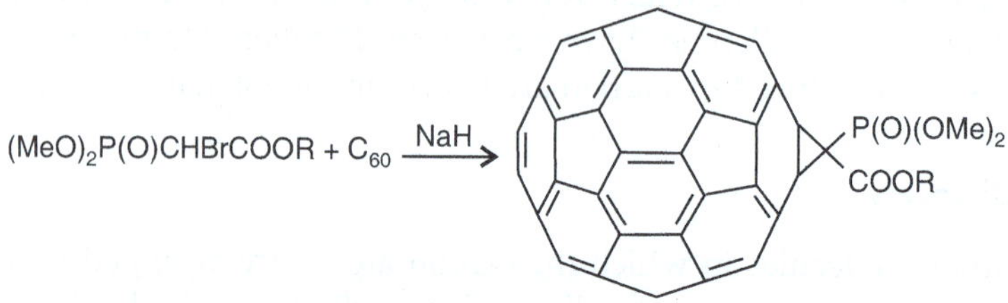


2. *Synthesis of sulphur containing organofullerenes.* Organosulphur fullerenes may be prepared by the following reaction.

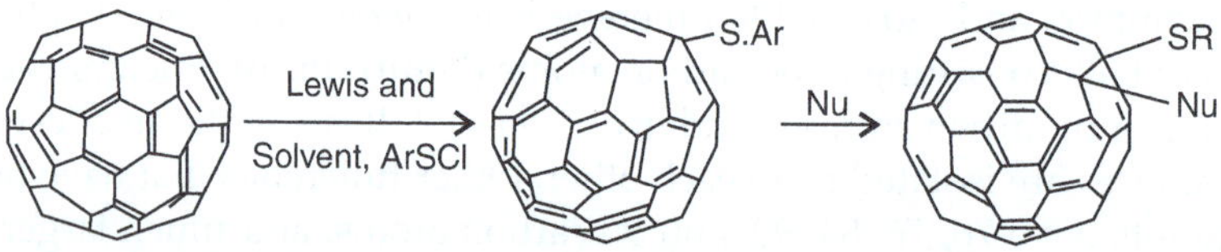


3. *Hydrogenation.* Hydrogenation of $C_{60}$ using a simple Zn/HCl technique yields $C_{60}H_{18}$ and $C_{60}H_{36}$ as the main products, due to the structures being more aromatic than the fullerene precursor. This work is being extended to $C_{70}$ and the higher fullerenes, and most recently to the azafullerene $C_{59}N$. This produces $C_{50}NH_5$ as the main product, due to an aromatic ring being created within the cage by the reduction.
4. *Fluorination.* A very wide range of fluorinated fullerenes are being isolated and characterized, and theories proposed to account for their formation. It has created some fascinating novel structures such as the tortoise-shaped $C_{60}F_{18}$ and the Saturn-like $C_{60}F_{20}$. The chemistry of these derivatives is now being explored and has led to unique fullerene structures, such as the triphenylated crown and $C_{60}F_{15}[C(CO_2Et)_2]_3$ which is not only the first example of an all *trans* [18] annulene, but also the first fullerene that is also an annulene.
5. *Alkylation.* Alkylation involves addition of larger groups than either of the two foregoing reactions, and whereas they are radical additions, alkylation involves reaction of fullerene anions with positive alkyl groups. This leads to different addition patterns, which vary also according to the alkylation technique, and lead to further information concerning the electronic and steric effects that participate in fullerene chemistry; alkylation of a range of fullerenes is being explored.

6. *Arylation.* This reaction differs from any of the above because it involves electrophilic substitution of the halogenofullerene into aromatics. Both fluoro-and chlorofullerenes, which are reasonably soluble have been used as electrophiles, but their different structures (arising from their different steric requirements), means that the arylated products are quite different.

### Applications

Following are some important applications of fullerenes:

1. Fullerenes find applications in preparing conducting films of alkali metal-doped fullerenes and superconductors (potassium-doped $T_c$ 18°K, rubidium-doped $T_c$ 30°K).
2. Fullerenes have also been used to produce tiny diamonds and thin diamond films.
3. Fullerenes research is expected to lead to new materials, lubricants, coatings, catalysts, electro-optical devices, and medical applications.
4. Organosulphur-fullerenes might offer new opportunities in the preparation of a new class of materials that might introduce interesting properties of potential importance of polymers, nanomaterials, and pharmaceuticals.
5. Fullerenes are very important in pharmaceutical applications. *Wudl* synthesized a derivative of a fullerene which is soluble in water. It seems to inhibit the activity of HIV and could be used in medicine for the treatment of AIDS.
6. Other possible applications of fullerenes include preparation of superconductors, electronic devices, micro electric devices, soft ferromagnets, etc.

## SOLVED PROBLEMS

**Problem 1.** *Calculate the value of Avogadro's constant from the following data. Density of NaCl = 2.165 g cm$^{-3}$. Distance between Na and Cl ions = 281 pm. Atomic weight of Na = 23 and Cl = 35.5.*

**Solution.** Density, $\rho = 2.165 \text{ g cm}^{-3}$

Edge length $= (2 \times 281) \text{ pm} = 2 \times 281 \times 10^{-10} \text{ cm}$

$= 562 \times 10^{-10} \text{ cm}$

NaCl crystal has *fcc* structure, therefore, Z = 4

Molecular mass of NaCl = 58.5 g

$$\therefore \quad N_0 = \frac{Z \times M}{(a)^3 \times \rho} = \frac{4 \times 58.5}{(562 \times 10^{-10})^3 \times 2.165}$$

$$= \frac{4 \times 58.5 \times 10^{30}}{(562)^3 \times 2.165} = \mathbf{6.09 \times 10^{23}}$$

**Problem 2.** *Calculate the density of silver which crystallizes in a face centered cubic lattice with unit cell length of 0.4086 nm (Atomic weight of Ag = 107.88).*

**Solution.** From the given data

Unit cell length, $a = 0.4086 \text{ nm} = 4.086 \times 10^{-10} \text{ m}$

In *fcc* lattice the number of atoms per unit cell, *i.e.*, $Z = 4$

$$M \text{ for Ag} = 107.88 \text{ g mol}^{-1} = 1.0788 \times 10^{-1} \text{ kg mol}^{-1}$$

$$N_o = 6.023 \times 10^{23}$$

$$\text{Now density of Ag, } \rho = \frac{ZM}{N_o a^3} = \frac{4 \times 1.0788 \times 10^{-1}}{6.023 \times 10^{23} (4.086 \times 10^{-10})^3}$$

$$= \mathbf{1.051 \times 10^4 \ kg \ m^{-3}}$$

**Problem 3.** *Calculate the number of atoms in cubic unit cell having one atom on each corner and two atoms on each diagonal.*

**Solution.**

(*a*) We know that the corner atom is shared by eight corners of the adjoining cubic unit cell. Therefore, the contribution of the atom at each corner is equal to 1/8 atom. Since a cubic unit cell has 8 corners, thus contribution of atoms at eight corners = $\frac{1}{8} \times 8 = 1$ atom.

(*b*) There will be 4 diagonals in a unit cube and these meet at the body center of the cube. If two atoms are present at that point, these will be shared by every diagonal. So the number of atoms within the cube = 2

Hence the total number of atoms per unit cube = 1 (corner) + 2 (diagonal) = 3

**Problem 4.** *Determine the density of cesium chloride which crystallizes out in bcc lattice. The length of a side of a unit cell is 412.1 pm; atomic masses of cesium and chlorine are 133 and 35.5, respectively.*

**Solution.** Length of the side of the unit cell = 412.1 pm = $412.1 \times 10^{-10}$ cm

$\therefore$ Volume of the unit cell = $(412.1 \times 10^{-10} \text{ cm})^3 = 7.00 \times 10^{-23}$ cm

Formula mass of CsCl = 133 + 35.5 = 168.5 amu

In a body centered cubic lattice of an ionic compound, the number of formula units per unit cell is one, while in a *bcc* lattice of a metal, the number of atoms per unit cell is two.

So here, $Z = 1$ formula unit

$$\therefore \quad \text{Density} = \frac{\text{Formula mass} \times Z}{\text{Unit cell volume} \times N_o}$$

$$= \frac{168.5 \text{ g mol}^{-1} \times 1 \text{ formula unit}}{7 \times 10^{-23} \text{ cm}^3 \times 6.023 \times 10^{23} \text{ formula units mol}^{-1}}$$

$$= \mathbf{4.0 \ g \ cm^{-3}}$$

**Problem 5.** *Rubidium (atomic mass 85.5) crystallizes in a body centered cubic lattice with a density of 1.51 g cm$^{-3}$. If the radius atom is 248 pm, calculate Avogadro's number.*

**Solution.** For a *bcc* unit cell, length of the unit cell and radius of the sphere are related as below.

$$l = \frac{4}{\sqrt{3}} \times r$$

$$l = \frac{4}{\sqrt{3}} \times 248 \text{ pm} = \frac{4}{1.732} \times 248 \times 10^{-10} \text{ cm}$$

$$\therefore \quad \text{Volume of the unit cell} = \left[\frac{4}{1.732} \times 248 \times 10^{-22} \text{ cm}\right]^3 = 1.88 \times 10^{-22} \text{ cm}^3$$

Again in a *bcc* lattice, the number of atoms per unit cell (Z) = 2

$$\text{Now since,} \qquad \text{Density} = \frac{\text{Mass of one unit cell}}{\text{Volume of one unit cell}}$$

$$= \frac{\text{Atomic mass} \times Z}{N_{Avog} \times \text{Volume of one unit cell}}$$

$$\therefore \qquad N_{Avog} = \frac{\text{Atomic mass} \times Z}{\text{Density} \times \text{Volume of unit cell}}$$

$$= \frac{85.5 \text{ g mol}^{-1} \times 2 \text{ atoms}}{1.5 \text{ g cm}^{-3} \times 1.88 \times 10^{-22} \text{ cm}^3} = \mathbf{6.023 \times 10^{23}\ atoms}$$

**Problem 6.** *KF has NaCl structure. What is the distance between $K^+$ and $F^-$ in KF, if the density of KF is 2.48 g $cm^{-3}$? [Atomic masses of K = 39.0, F = 19]*

**Solution.** Density of KF = 2.48 g $cm^{-3}$ = 2.48 × $10^{-30}$ g $pm^{-3}$

Molecular mass of KF = 39.0 + 19.0 = 58.0

KF or NaCl has *fcc* structure, so number of atoms in unit cell = 4

$$\therefore \qquad a^3 \frac{Z \times M}{\rho \times N_0} = \frac{4 \times 58}{2.48 \times 10^{-30} \times 6.023 \times 10^{23}}$$

$$a^3 = \frac{4 \times 58 \times 10^7}{2.48 \times 6.023}$$

$$\therefore \quad \text{Edge length,} \qquad a = \mathbf{538\ pm}$$

So, the distance between $K^+$ and $F^-$ ion = $\frac{a}{2}$ = **269.0 pm**

**Problem 7.** *Lead (II) sulphide crystal has NaCl structure. What is its density? The edge length of the unit cell of PbS crystal is 500 pm. [$N_o$ = 6.023 × $10^{23}$ $mole^{-1}$; Atomic masses of Pb = 207.2, S = 32].*

**Solution.** Given data: M = 239.2 amu, $a$ = 500 pm = 500 × $10^{-10}$ cm,
Z = 4, $N_o$ = 6.023 × $10^{23}$

$$\text{So,} \qquad \rho = \frac{239.2 \times 4}{6.023 \times (500 \times 10^{10})^3} = \mathbf{12.7\ g\ cm^{-3}}$$

**Problem 8.** *The unit cell of an element of atomic mass 96 and density 10.3 g cm$^{-3}$ is a cube with edge length of 314 pm. Find the structure of the crystal lattice, simple cubic, fcc, or bcc. [ $N_o$ = 6.023 × 10$^{23}$ atoms mole$^{-1}$ ].*

**Solution.** Given: $\rho = 10.3 \text{ g cm}^{-3}, M = 96$

$a = 314 \text{ pm} = 314 \times 10^{-10} \text{ cm}, N_o = 6.023 \times 10^{23}$

Formula used, $$Z = \frac{\rho \times N_o \times a^3}{M}$$

$$= \frac{10.3 \times 6.023 \times 10^{23} \times (314 \times 10^{-10})^3}{96} = \mathbf{2}$$

Therefore, the number of atoms per unit cell is two, which corresponds to *bcc* unit cell.

**Problem 9.** *A metal has a fcc crystal structure. The length of the unit cell is 404 pm. What is the molar mass of metal atoms if the density of the metal is 2.72 g cm$^{-3}$. [$N_o$ = 6.023 × 10$^{23}$ atoms mole$^{-1}$].*

**Solution.** Formula used, $M = \frac{a^3 N_o \rho}{Z}$

Given: $N_o = 6.023 \times 10^{23}$; $\rho = 2.72 \text{ g cm}^{-3}$; $Z = 4$

so, $$M = \frac{(404 \times 10^{-10})^3 \times 6.023 \times 10^{23} \times 2.72}{4} = \mathbf{27.00 \text{ g mole}^{-1}}$$

**Problem 10.** *Given for iron a = 288 pm, ρ = 7.86 cm$^{-3}$. Find the type of cubic lattice to which the crystal belongs. Also calculate the radius of iron. [Atomic mass of Fe = 55.85].*

**Solution.** Formula used, $Z = \frac{\rho \times a^3 . N_o}{M}$

Given: $\rho = 7.86 \text{ g cm}^{-3}$; $a = 288 \text{ pm} = 288 \times 10^{-10} \text{ cm}$; $M = 55.85$

$$Z = \frac{7.86 \times (288 \times 10^{-10})^3 \times 6.02 \times 10^{23}}{55.85} = \mathbf{2.023 \text{ or } 2}$$

Therefore, the crystal has *bcc* lattice.

**Problem 11.** *An atom has fcc crystal whose density is 10 g m$^{-3}$ and cell edge is 100 pm. How many atoms are present in its 100 g?*

**Solution.** Total number of atoms $= \frac{Z \times M}{a^3 d}$

where, $Z$ = Number of atoms in a unit cell

$M$ = Mass of metal

$a$ = Edge of unit cell

$$d = \text{Density of metal}$$

here, $$Z_{fcc} = 8 \times \frac{1}{8} + 6 \times \frac{1}{2} = 4$$

Substituting the values in the above relation,

$$\text{Number of atoms} = \frac{4 \times 100 \text{ g}}{(100 \times 10^{-12} \text{ m})^3 \times \text{g m}^{-3}} = \mathbf{4 \times 10^{31} \text{ atoms.}}$$

**Problem 12.** *In Bragg's reflection of X-ray, a reflection was found at 30° glancing angle with lattice planes of spacing 1.87 Å. If this is a second order reflection, calculate the wavelength of X-rays.*

**Solution.** According to Bragg's equation

$$2d \sin \theta = n\lambda$$

Here $\theta = 30°$, $n = 2$ and $d = 1.87$ Å

$$\therefore \quad \lambda = \frac{2d \sin \theta}{n} = \frac{2 \times 1.87 \times \sin 30°}{2} = 1.87 \times 0.5 = \mathbf{0.935 \text{ Å.}}$$

**Problem 13.** *X-rays having wavelength of 2.6 Å are diffracted at an angle of 20° for the second order in Bragg's spectrometer. Find the spacing constant of the crystal.*

**Solution.** If $n$th order maximum in Bragg's spectrometer is given in a direction $\theta$, then according to Bragg's equation:

$$2d \sin \theta = n\lambda$$

or, $$d = \frac{n\lambda}{2 \sin \theta}$$

Here $\lambda = 2.6$ Å $= 2.6 \times 10^{-10}$ m, $n = 2$, $\theta = 20°$ and $\sin 20° = 0.3420$

$$\therefore \quad d = \frac{2 \times 2.6 \times 10^{-10}}{2 \times 0.3420} = 7.602 \times 10^{-10} \text{ m} = \mathbf{7.60 \text{ Å.}}$$

**Problem 14.** *An X-ray beam of wavelength 0.97 Å is obtained in the third order after reflection at 60° from the crystal plane. Another beam is obtained in the first order after reflection at 30° from the same crystal plane. Find the wavelength of the second X-ray beam.*

**Solution.** According to Bragg's law of diffraction

$$2d \sin \theta = n\lambda \quad \textit{(i)}$$

For the first X-ray beam $\theta = 60°$, $n = 3$ and $\lambda = 0.97$ Å

$$\therefore \quad 2d \sin 60° = 3 \times 0.97 \quad \textit{(ii)}$$

Let $\lambda'$ be the wavelength of the second X-ray beam. For the same crystal plane, $d$ remains the same, therefore

$$2d \sin \theta' = n'\lambda'$$

Here $\theta' = 30°$ and $n' = 1$

$\therefore$ $2d \sin 30° = 1 \times \lambda'$ ...(iii)

Dividing equation (*iii*) by (*ii*), we get

$$\frac{\lambda'}{3 \times 0.97} = \frac{\sin 30°}{\sin 60°} \quad \text{or} \quad \lambda' = 3 \times 0.97 \times \frac{(1/2)}{(\sqrt{3}/2)}$$

or,
$$\lambda' = \frac{3 \times 0.97 \times 1}{\sqrt{3}} = \mathbf{1.68\ Å.}$$

**Problem 15.** *The first order reflection from the plane of NaCl is obtained at an angle $2\theta = 20°$ with incident beam. If $d_o = 2.82$ Å, calculate the wavelength of X-ray used.*

**Solution.** According to Bragg's equation for the diffraction of X-rays,

$$2d_o \sin \theta = n\lambda.$$

For first order $n = 1$,

$\therefore$ $2d_o \sin \theta = \lambda$

Here $2\theta = 20°$ or $\theta = 10°$ and $d_o = 2.82$ Å $= 2.82 \times 10^{-10}$ m

$\therefore$
$$\lambda = 2 \times 2.82 \times 10^{-10} \sin 10° = 2 \times 2.82 \times 10^{-10} \times 0.1736$$
$$= 9.791 \times 10^{-11} \text{ m} = \mathbf{0.9791\ Å.}$$

**Problem 16.** *A set of lattice planes reflects X-rays of wavelength 1.32 Å at a glancing angle of 9°30′. Deduce the possible spacing of this set of planes.*

**Solution.** Bragg's law for the diffraction of X-rays is

$$2d \sin \theta = n\lambda$$

For first order reflection, $n = 1$

$\therefore$ $2d \sin \theta_1 = \lambda$ ...(i)

For second order reflection, $n = 2$

$\therefore$ $2d \sin \theta_2 = 2\lambda$ ...(ii)

For third order reflection, $n = 3$

$\therefore$ $2d \sin \theta_3 = 3\lambda$ ...(iii)

Similarly, for fourth order reflection, $n = 4$

$\therefore$ $2d \sin \theta_4 = 4\lambda$ ...(iv)

Dividing equation (*ii*) by equation (*i*), we get

$$\frac{\sin \theta_2}{\sin \theta_1} = 2$$

or, $\sin \theta_2 = 2 \sin \theta_1$ ...(v)

Similarly, from equations (*iii*) and (*i*), and from equations (*iv*) and (*i*), we get

$$\sin \theta_3 = 3 \sin \theta_1$$

and

$$\sin \theta_4 = 4 \sin \theta_1$$

For the first order reflection the spacing *d* is

$$\frac{d}{n} = \frac{\lambda}{2 \sin \theta} = \frac{1.32 \times 10^{-10}}{2 \times \sin 9°30'}$$

or

$$d = \frac{1.32 \times 10^{-10}}{2 \times 0.1650} = \mathbf{4.0 \times 10^{-10}\ m}$$

If $n = 2$,

$$\frac{d}{n} = \frac{\lambda}{2 \sin \theta_2} = \frac{\lambda}{2 \times 2 \sin \theta_1} = \frac{1}{2}\left(\frac{\lambda}{2 \sin \theta_1}\right)$$

∴

$$\frac{d}{n} = \frac{4.0 \times 10^{-10}}{2}\ \text{m} = \mathbf{2.0 \times 10^{-10}\ m}$$

Similarly, if $n = 3$

$$\frac{d}{n} = \frac{\lambda}{2 \sin \theta_3} = \frac{\lambda}{2 \times 3 \sin \theta_1} = \frac{1}{3} \times 4.0 \times 10^{-10}\ \text{m}$$

or,

$$\frac{d}{n} = \mathbf{1.33 \times 10^{-10}\ m}$$

Similarly, if $n = 4$

$$\frac{d}{n} = \frac{\lambda}{2 \sin \theta_4} = \frac{\lambda}{2 \times 4 \sin \theta_1} = \frac{1}{4} \times 4.0 \times 10^{-10}$$

$$= \mathbf{1.0 \times 10^{-10}\ m} \text{ and so on.}$$

## QUESTION BANK

1. What is a unit cell? Give rules to determine the number of points/atoms/ions in one unit cell and find out how many cations and anions are present in sodium chloride unit cell.
2. (*a*) What is space lattice and unit cell? Draw unit cell of NaCl.

   (*b*) What is liquid crystal? Give two illustrations.
3. What is the meaning of internal structure of crystals? How the internal structure is known with the help of Bragg's equation?
4. (*a*) Define and explain the Law of interfacial angles.

   (*b*) Explain the following terms with suitable examples:

   (*i*) Axes of symmetry (*ii*) Plane of symmetry (*iii*) Center of symmetry.
5. (*a*) Distinguish between Weiss indices and Miller Indices.

   (*b*) What are the typical properties of ionic crystals?

6. What do you understand by the term lattice energy? Describe Born Haber cycle for finding the lattice energy of NaCl?
7. Aluminium forms face centered cubic crystals. The density of Al is 2.70 $g/cm^3$. Calculate the length of the side of the unit cell of Al. (Atomic weight of Al = 27)
8. What is meant by space lattice of a crystal? Derive Bragg's equation for the diffraction of X-rays by crystal lattice.
9. (*a*) Explain the terms:

   (*i*) Miller indices, (*ii*) Space lattice, (*iii*) Unit cell.

   (*b*) Draw a neat diagram of unit cell of sodium chloride crystal. Find the number of sodium and chloride ions in it.
10. (*a*) Explain the terms: lattice planes, space lattice and unit cell.

    (*b*) Show that lattice planes are described by Miller Indices.
11. (*a*) How many atoms are there per unit cell in: (*i*) a simple cubic arrangement of atoms; (*ii*) a body centered cubic arrangement of atoms?

    (*b*) State Bragg's law and explain the terms involved.
12. Explain Schottky and Frenkel defects. What are the consequences of these defects?
13. (*a*) What do you understand by the packing efficiency of a crystal?

    (*b*) Explain the terms point groups and space groups.

    (*c*) The ionic radii of Cs is 169 pm and Br is 195 pm. What kind of unit cell would be expected for CsBr crystal? Calculate the unit cell dimensions and the density of Cs Br crystal (Atomic weight of Cs = 133. Br = 80).
14. (*a*) Mention the factors on which the intensity of the diffracted beam from different sets of planes of a crystal depend.

    (*b*) An element crystallizes in the *fcc* structure. Calculate the percent void space.

    (*c*) Explain the symmetry elements of a crystal belonging to simple cubic systems.
15. Define and explain the following:

    (*a*) Plane of symmetry (*b*) Center of symmetry
16. (*a*) Describe the theory of Bragg's method of crystal analysis.

    (*b*) Differentiate between the cubic close packing and hexagonal close packing of spheres.
17. How is Avogadro's number determined from X-ray diffraction of crystals? Explain.
18. Gold has a face centered cubic structure with a unit length 4.07 Å and a density of 19.3 g $cm^3$. Calculate the Avogadro's number from the data (Atomic weight of Au = 197).
19. (*a*) Drive the relation $n\lambda = 2d \sin \theta$ in crystallography.

    (*b*) Draw diagrams to represent (*i*) FCC lattice. (*ii*) BCC lattice.
20. Polonium crystallizes in a simple cell cubic unit cell. It has an atomic mass = 209 and density = 91.5 kg $m^{-3}$. What is the edge length of its units cell?
21. (*a*) Write a short note on Bravais Lattices.

    (*b*) Calculate the angle at which first order diffraction will occur in an X-ray diffractometer when X-rays of wavelength 1.54 Å are diffracted by the atoms of a crystal, given that the interplanar distance is 4.04 Å.

22. (*a*) What do you understand by:

(*i*) Law of rational indices (*ii*) Axial ratios

(*iii*) Law of constancy of interfacial angles.

(*b*) How are crystals classified? Explain the various criteria.

23. Sketch the following types of lattices: (*a*) a simple cubic; (*b*) a face centered cubic; and (*c*) a body centered cubic. To which of these do sodium chloride and potassium chloride crystal belong?

24. (*a*) Explain the term: (*i*) Crystallographic axes, and (*ii*) Axial ratios.

(*b*) How are X-rays useful in crystal structure studies?

25. (*a*) What are the elements of symmetry expected in a cubic crystal?

(*b*) Give a brief account of 'point defects in crystals'.

26. Density of NaCl (at 20°C) = 2.17 gm $CC^{-1}$; molar mass = 58.5. Calculate the interplanar distance ($d_{200}$). The first order reflection maximum of the $K_a$ rays of Pd from the 200 planes of NaCl occurs at a glancing angle of 5.9°. Calculate the wavelength of the X-radiation.

27. What is the meaning of internal structure of crystals? How the internal structure is known with the help of Bragg's equation?

28. What do you understand by the term lattice energy? Describe Born-Haber cycle for finding the lattice energy of NaCl?

29. (*a*) Describe the theory of Bragg's method of crystal analysis.

(*b*) Differentiate between the cubic closepacking and hexagonal close packing of spheres.

30. (*a*) Derive a relation $n\lambda = 2d \sin \theta$ in crystallography.

(*b*) Draw diagram to represent (*i*) F.C.C. lattice (*ii*) B.C.C. lattice. Explain briefly Fenkel and Schottky defects in crystals.

31. Describe briefly how X-ray diffraction is used for structure determination?

32. Name the important defects in crystals. How are they caused? What is the effect of these defeciencies on the stability of the crystals?

33. What is lattice enthalpy? Discuss the use of Born-Haber cycle of the calculation of lattice enthaly.

34. (*a*) Derive Bragg's equation for the diffraction of X-rays by crystal lattice. How is this equation utilized to determine crystal structure?

(*b*) What is a liquid crystal? Name the different types of liquid crystals?

35. Polonium crystallizes in a simple cell cubic unit cell. It has an atomic mass = 209 and density = 91.5 kg $m^{-3}$. What is the edge length of its unit cell?

36. Calculate the angle at which first order diffraction will occur in an X-ray diffractometer when X-rays of wavelength 1.54 Å are diffracted by the atoms of a crystals, given that the interplanar distance is 4.04 Å.

37. Write short note on:

(*a*) Application of fullerenes (*b*) Applications of graphite

38. Discuss the structure of graphite. How it acts as a conductor of electricity?

39. Describe structure and properties of fullerenes.

Chapter 16

# GENERAL ORGANIC CHEMISTRY

## 16.1 INTRODUCTION

An organic reaction is believed to take place by the attack of a reagent on the compound containing carbon; designated as a *substrate* (meaning a substance lying below). An organic reaction may be represented as:

$$\text{Attacking reagent} + \text{Substrate} \longrightarrow \text{Products.}$$

A *substrate* may be defined as *the reactant that contains carbon atoms some of whose bonds with other atoms are broken and some new ones formed as a result of reaction with the attacking reagent.*

The carbon bonds in the substrate molecule are broken to give fragments which are very reactive and constitute *transitory intermediates.* They may react with other similar species or with the molecules present in their environments, thus establishing new bonds to give the products.

It is seen that most of the attacking reagents bear either a positive or a negative charge. These would not attack the substrate successfully unless the latter somehow possessed oppositely charged centers in the molecule. So the substrate molecule must develop polarity on some of its carbon atoms and substituents linked together. This is made possible by the displacement of the bonding electrons resulting in the development of polarity in the substrate molecule. Such effects involving the displacement of electrons in the substrate molecule are referred to as *electron displacement effects.*

## 16.2 ELECTRONIC DISPLACEMENTS IN ORGANIC MOLECULES

There are four types of electronic displacements in organic molecules:

### 16.2.1 Inductive Effect

*It is the permanent displacement of electrons from a covalent bond toward the more electronegative element or group.* When a carbon atom is joined, instead of an H-atom, to a substituent X having higher electronegativity then the shared pair of electrons will be displaced away from the carbon atom and toward the more electronegative substituent X. On the other hand, if a substituent Y having lower electronegativity is attached to it, the shared pair of the bond C-Y will be displaced toward the carbon atom and away from the substituent Y. As a result of these minor displacements of the bonding electron-pair, the joined atoms develop small fractional charges on them, as shown below:

| | | | |
|---|---|---|---|
| C : H | or | C—H | (standard) |
| C : X | or | $C^{\delta+}$—$X^{\delta-}$ | (– I) |
| C : Y | or | $C^{\delta-}$—$Y^{\delta+}$ | (+ I) |

The polarisation thus induced in the substrate molecule is permanent in nature.

*The permanent effect whereby polarity is induced on the carbon atom and the substituent attached to it due to minor displacement of bonding electron pairs caused by their different electronegativities, is known as inductive or I-effect.* When the substituent is electron attracting (X), it develops a negative charge and is said to exert a negative inductive effect or *–I effect*. If the substituent is electron donating, it acquires a positive charge and the inductive effect produced is called *+ I effect.*

Inductive effect is represented by the symbol →–, the arrow pointing toward the more electronegative substituent. It is always transmitted along the chain of the carbon atom but the extent of partial charge keeps on decreasing away from polar bond and at the third or fourth carbon atom it is almost zero for all practical purposes. The inductive effect relayed along the carbon chain may be represented as follows:

$$\overset{\delta\delta\delta+}{C}\rightarrow\!-\overset{\delta\delta+}{C}\rightarrow\!-\overset{\delta+}{C}\rightarrow\!-\overset{\delta-}{X}$$

With an electron donating substituent Y, the inductive effect will be reversed and may be represented as:

$$\overset{\delta\delta\delta-}{C}\!-\!\leftarrow\overset{\delta\delta-}{C}\!-\!\leftarrow\overset{\delta-}{C}\!-\!\leftarrow\overset{\delta+}{Y}$$

Some common groups which cause negative or positive I-effect are given below.

(*a*) – *I effect groups* (*electron-attracting*): For example

$$NO_2 > F > COOH > Cl > Br > I > OH > C_6H_5$$

(*b*) + *I effect groups* (*electron-donating*): For example

$$(CH_3)_3\,C— > (CH_3)_2\,CH— > CH_3CH_2— > CH_3—$$

**Applications of Inductive Effect**

(*i*) *Dipole moment.* The greater the inductive effect, the more will be the dipole moment in the molecule since inductive effect leads to a dipolar character in the molecule. For example, Dipole moments of

$$CH_3—F > CH_3—Cl > CH_3—Br > CH_3—I$$

since the –I effect of F > Cl > Br > I

(*ii*) *Bond length.* The greater the –I effect, the smaller will be the bond length in the molecule since inductive effect leads to ionic character in the bond. For example, bond length increases in the order

$$CH_3F < CH_3Cl < CH_3Br < CH_3I$$

since the –I effect decreases in the order F > Cl > Br > I

(*iii*) *Reactivity of alkyl halides.* Due to the presence of C–X polar bond in alkyl halides, alkyl halides are more reactive than the corresponding alkanes.

Reactivity increases with increase of (+ I) effect. We know (+ I) effect of

$$R_3C > R_2CH > RCH_2$$

Hence reactivity of $R_3C—X > R_2CH—X > RCH_2—X$. Reactivity is directly related to C—X bond polarity.

(*iv*) *Strength of carboxylic acids.* Acidity of carboxylic acid is due to the ease with which it ionises to give proton and this ease increases with the strength of the (– I) group attached at the α-carbon of carboxylic acid or increasing number of (– I) effect groups. For example, Acidic strength order is:

$$F—CH_2COOH > Cl—CH_2COOH > Br—CH_2COOH > I—CH_2COOH > H—CH_2COOH$$

and

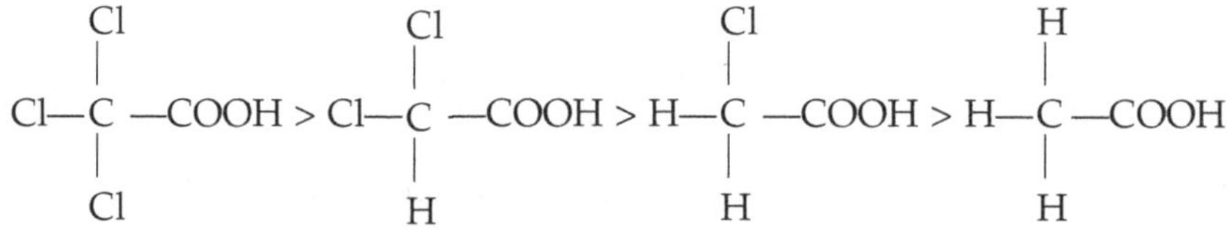

It is to be noted that carboxylic acids are acids because they release $H^+$ ions in solution. This proton release depends upon – I effect of oxygen

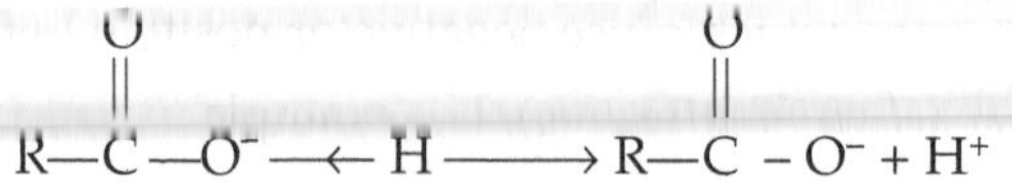

and resonance stabilization of carboxylate anion

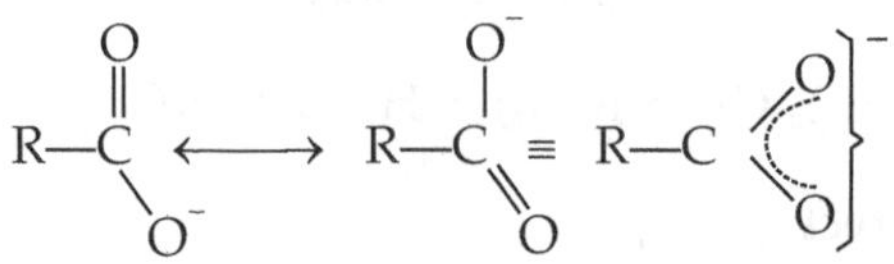

The (– I) effect of oxygen can be increased or decreased depending on the nature of the R-group. If the R-group also has (– I) effect, the net result is increase in acidity and vice-versa.

(*v*) *Basic Strength of Amines.* Amines are basic because they are electron pair donors or they can accept protons. The ease with which this happens determines the relative strength of amines. Relative basic strength of amines depends on inductive effect, steric effect and stabilization of cation by hydration. We will discuss here only the inductive effect.

Groups which have (+ I) effect, when attached to a nitrogen atom, make it simpler for the lone pair of electrons on nitrogen to be easily available, hence they increase the basic strength. But the groups with – I effect make it difficult for the easy availability of lone pairs of electrons on nitrogen, hence decreasing the basic strength.

Hence, increasing order of basic strength is as follows:

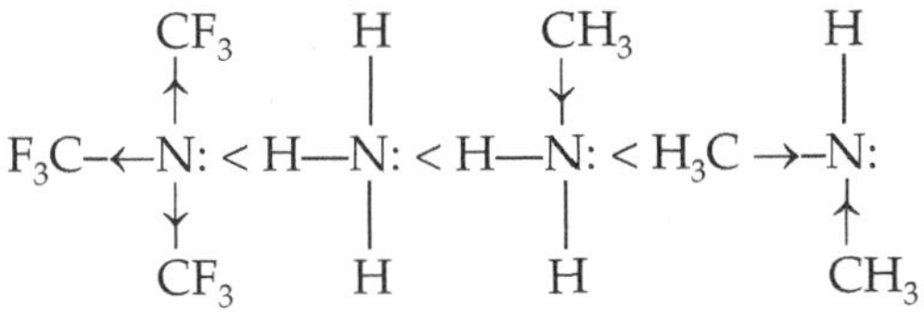

### 16.2.2 Electromeric Effect

It is a temporary effect which comes into play with the approach of an attacking reagent. When a double or a triple bond is attacked by an electrophilic reagent, a pair of bonding electrons is completely transferred from one atom to the other. The atom that takes charge of the electron-pair becomes negatively charged and the other positively charged. It can be shown as:

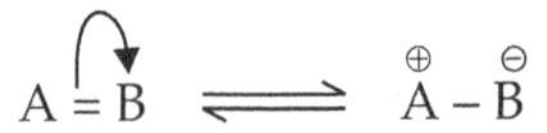

This is a purely temporary effect and vanishes as soon as the attacking reagent is removed. Thus, *the effect which causes a temporary polarization in the substrate molecule by shift of an electron-pair from one atom to the other under the influence of an attacking reagent, is called electromeric effect.* The electromeric effect is of two types:

(*a*) + **E effect** *involves the transfer of the electrons of the π-bond to an atom of the multiple bond to which the attacking reagent finally gets attached.* For example,

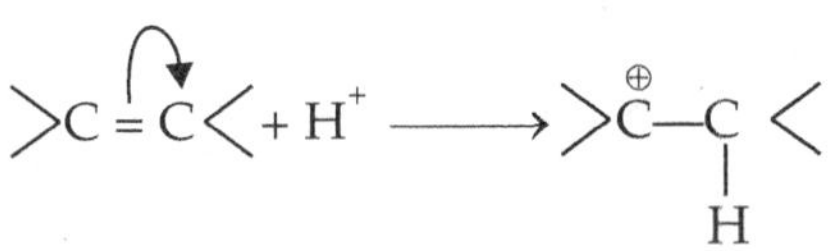

(*b*) – **E effect** *involves the transfer of the electrons of the π-bond to an atom of the multiple bond other than the one to which the attacking reagent finally gets attached.* For example,

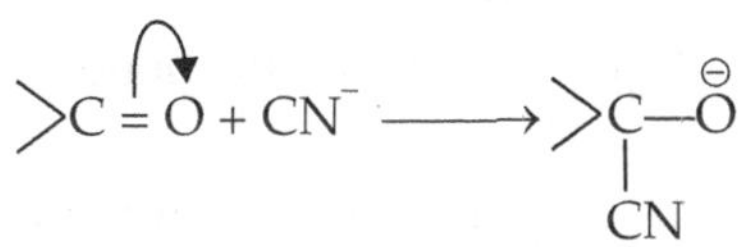

### 16.2.3 Mesomeric Effect

This effect also causes permanent displacement of electrons in the organic molecules. This effect is transmitted through the $\pi$-electrons of multiple bonds in the carbon chain of the molecule. It operates only in those systems which have an extended chain with conjugate $\pi$-bonds.

When an electron donating or withdrawing group is conjugated with a conjugated system, the electron displacement is transmitted through $\pi$-electrons in the chain. For example, let a carbonyl group ( $>C=O$ ) be linked with $CH_2=CH—CH=CH—CH=O$. The $\pi$-electrons of the carbon-oxygen bond get displaced towards the more electronegative oxygen atom, giving the following resonance structures as:

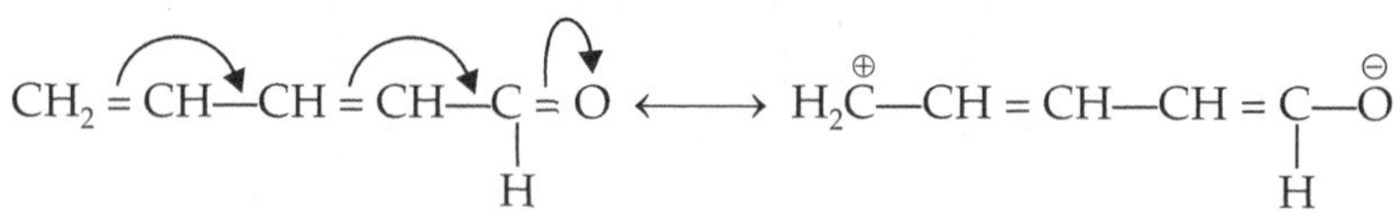

The mesomeric effect, like the inductive effect, is of two types:

(*a*) **+ M effect** *involves the transfer of electron pairs away from the atom or group attached.* Such groups have lone pairs of electrons and furnish the electron pair for conjugation with an attached unsaturated system. For example, $—\ddot{O}—H$, $—\ddot{O}—R$, $—\ddot{N}\begin{smallmatrix}H\\H\end{smallmatrix}$, $—\ddot{S}—R$, etc. groups have + M effect. The presence of a $—\ddot{N}H_2$ group at the end of a conjugate $\pi$-bond system brings about the following displacement of electrons.

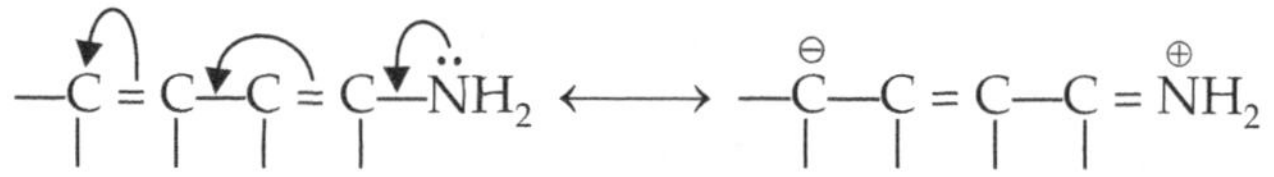

(*b*) **– M effect** *involves the transfer of electron pair towards the atom or group attached.* Groups such as $>C=O$, $—NO_2$, $—C\equiv N$, $—SO_3H$, etc. have *–M effect* due to the presence of highly electronegative atoms like O or N. It can be illustrated as:

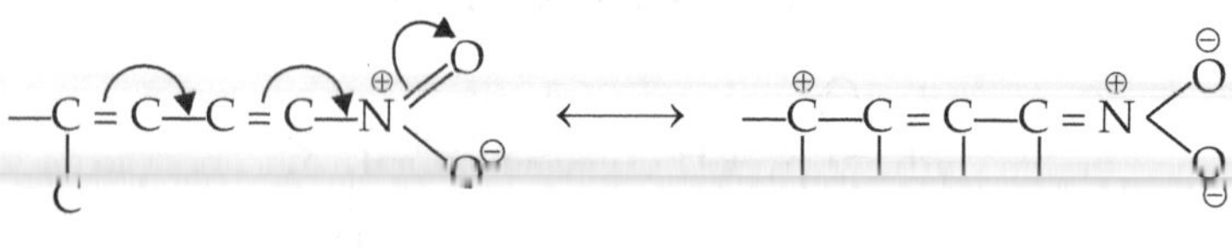

**Applications of Mesomeric Effect**

(*i*) *Dipole moment.* Dipole moment of $CH_3—CH_2—Cl > CH_2=CH—Cl$

Due to resonance in vinyl chloride.

(*ii*) *Bond Length.* Benzene ($C_6H_6$) has three double bonds and three single bonds within the ring. So if it is written as , it appears that there should be two different types of bond lengths corresponding to single (1.54 Å) and double bond (1.33 Å). But all the bond lengths in the benzene ring are equal. This can be explained by writing the following resonating structures of benzene:

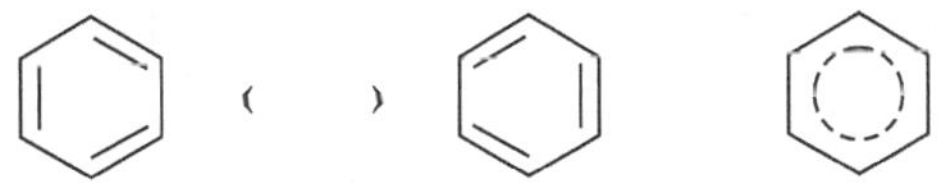

With the net result, the resonance hybrid has a structure where the position of the single and double bonds are delocalized, hence each bond has partial single and partial double bond character and equal bond length.

(*iii*) *Strength of acids.* Acidic strength of

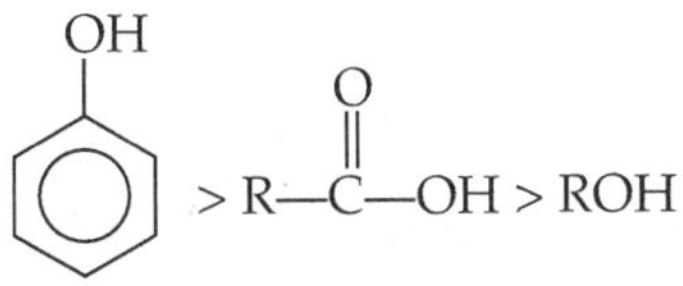


because the stability of the corresponding conjugate bases is in the order

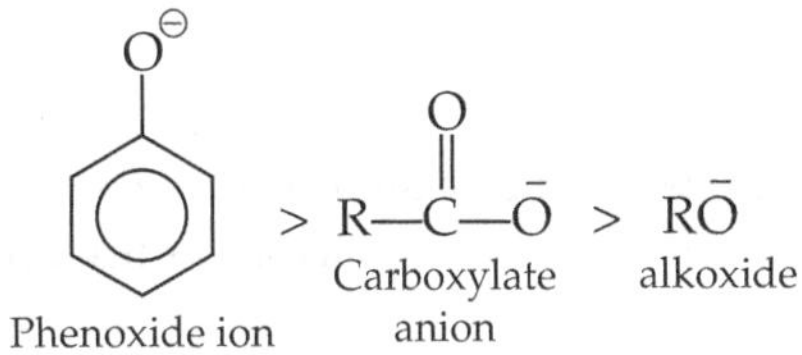


This is due to the more resonance stabilization of phenoxide ion.

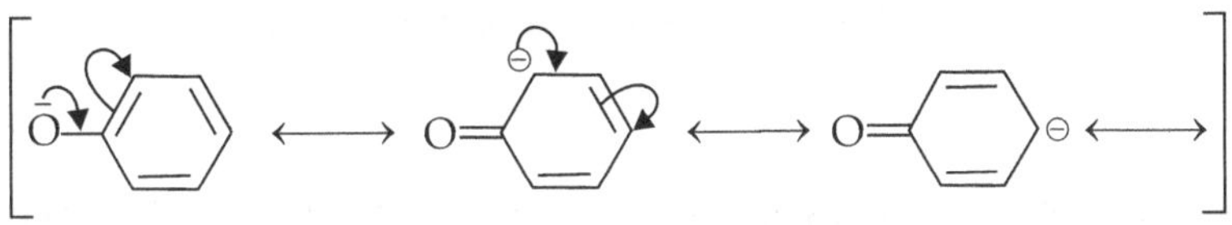

Compared to carboxylate anion

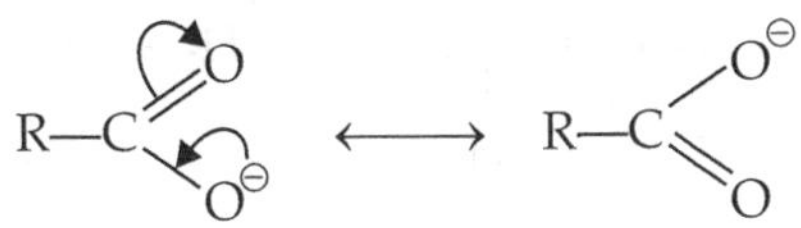

Alkoxide ion is not stabilized at all by resonance.

(*iv*) *Strength of Bases.* Base strength of $RNH_2 > NH_3 > C_6H_5—NH_2$. It is due to the (+ I) effect of R in $R \rightarrow \ddot{N}H_2$. Thus, a lone pair is more easily available. But in $C_6H_5—NH_2$, the phenyl group has (–M effect)

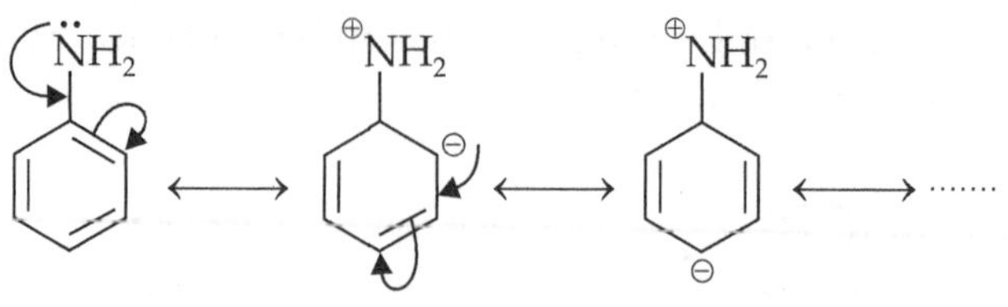

Hence, the lone pair present at the nitrogen in aniline is delocalized in the ring by resonance, so it is not free for protonation leading to a decrease of basic strength of aniline.

(*v*) *Stability of Intermediates.* The greater the number of resonating structures for a given intermediate, the more will be its stability.

For example, the stability of carbon free radicals is in the order:

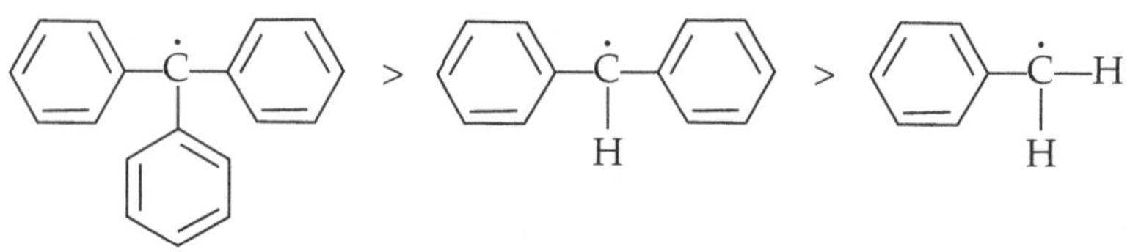

### 16.2.4 Hyperconjugative Effect

This effect is observed when an unsaturated system (>C = C<) is attached with the alkyl group. Hyperconjugative effect results in an electron displacement toward the double bond carbon in the following manner:

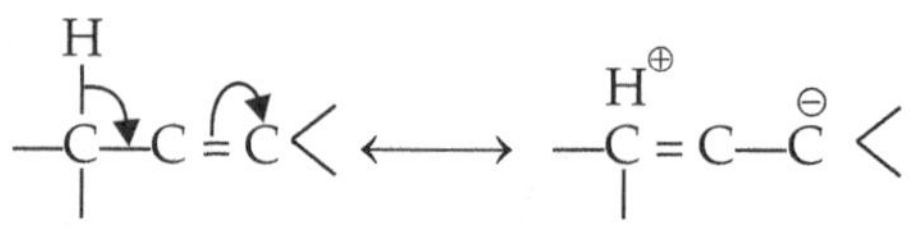

Since no bond exists between the free $H^+$ and C atom, so hyperconjugation is also called *no bond resonance. Baker and Nathan* suggested that the hyperconjugative effect takes place through the

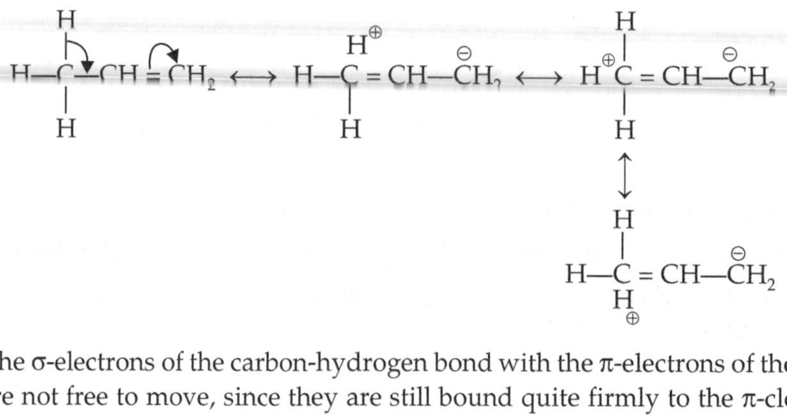

interaction of the σ-electrons of the carbon-hydrogen bond with the π-electrons of the double bond. The $H^+$ ions are not free to move, since they are still bound quite firmly to the π-cloud. The three C—H bonds of the methyl group in propylene contribute to this effect, as follows:

Thus, hyperconjugation occurs through the H-atom present on the α-carbon atom, *i.e.*, on the C-atom adjacent to the double bond. So, the greater the number of C—H bonds at the α-carbon, the greater would be the electron release. Consequently the release of electrons would be maximum with a methyl group and least for an isopropyl group, and for a *ter.*butyl group, this effect would be zero due to the absence of an α-H atom. The order of hyperconjugative electron release for the alkyl groups will be:

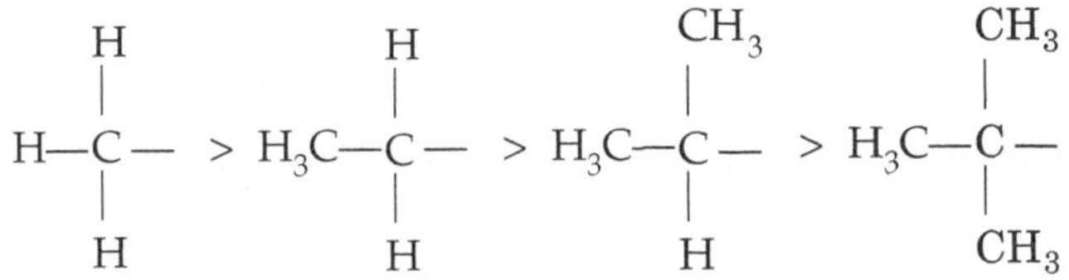

**Applications of Hyperconjugation**

(*i*) *Directive influence of alkyl groups in aromatic alkylbenzenes* can be explained by hyperconjugation, *e.g.*, toluene can be represented by

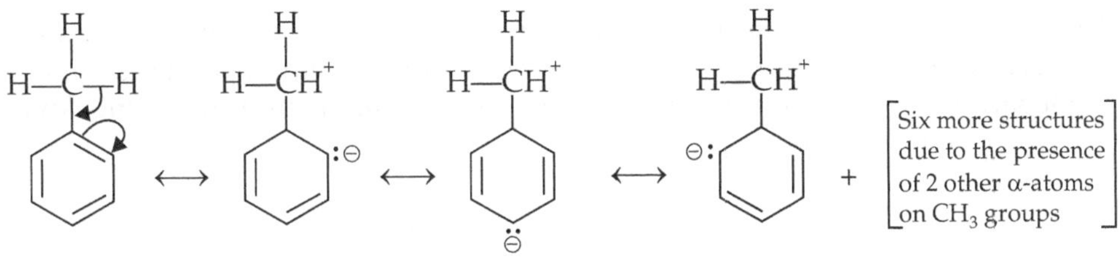


I　　II (*Ortho*)　　III (*Para*)　　IV (*Ortho*)

Thus, there can be 10 structures in which electron-density is maximum at either *ortho* or *para* positions. *Hence, methyl* (or *any alkyl*) *group is ortho and para-directing in electrophilic substitution reactions in aromatic alkylbenzenes.*

(*ii*) *Shortening the C—C single bond adjacent to multiple bonds.* For example, in propene due to hyperconjugation effect:

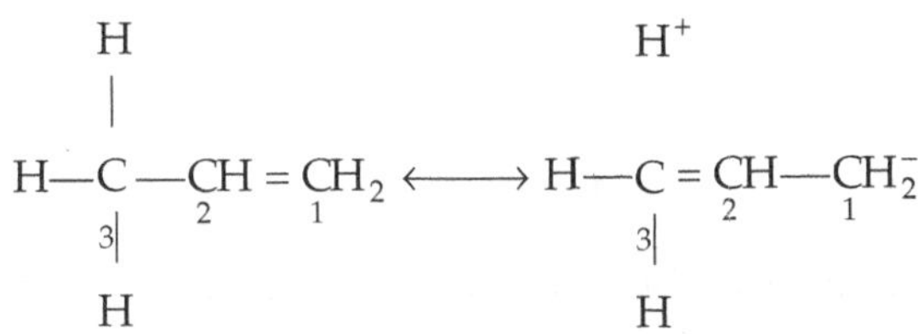

the $C_2$—$C_3$ single bond acquires a *partial double bond character* and consequently, it is *somewhat shorter* (149 pm) than the normal C—C single bond (154 pm).

(*iii*) *Explanation of relative stability of alkenes.* The greater the number of α-hydrogens in an alkene, the greater the number of hyperconjugation structures, and consequently, the greater is its stability. Let us illustrate this by considering the following alkenes:

$$CH_2=CH_2 \;;\; CH_3{-}CH=CH_2 \;;\; CH_3{-}CH=CH{-}CH_3\;;$$

(I) Ethene (No α-H) (II) Propene (3 α-H) (III) But-2-ene (6 α-H)

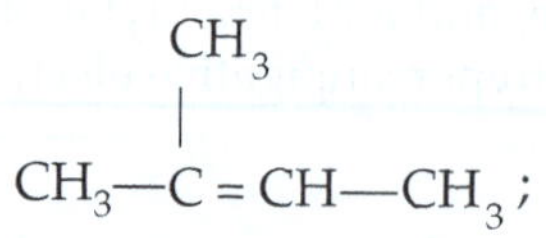

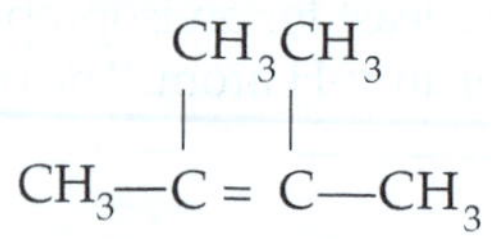

(IV) 2-methylbut-2-ene (9 α-H) (V) 2,3-methylbut-2-ene (12 α-H)

Hence, the relative stability of these alkenes is in the order:

(V) > (IV) > (III) > (II) > (I).

## 16.3 BOND FISSION IN ORGANIC MOLECULES

The bond fission in organic molecules takes place mainly in two alternative ways:

(*a*) *Homolytic fission or homolysis.* It takes place when the two atoms joined are of similar electronegativity. During fission one electron of the bonding pair goes with each of the departing atom or group resulting in two electrically neutral fragments known as *free radicals*, for example,

$$H_3C : X \longrightarrow H_3C^{\cdot} + X^{\cdot}$$

Due to the presence of an unpaired electron, the free radicals are highly reactive in nature. A free radical is a neutral entity.

(*b*) *Heterolytic fission or heterolysis.* When, during bond fission, both the shared electrons are taken over by one atom and the other atom goes without the electron, then such fission is called heterolytic fission. For example,

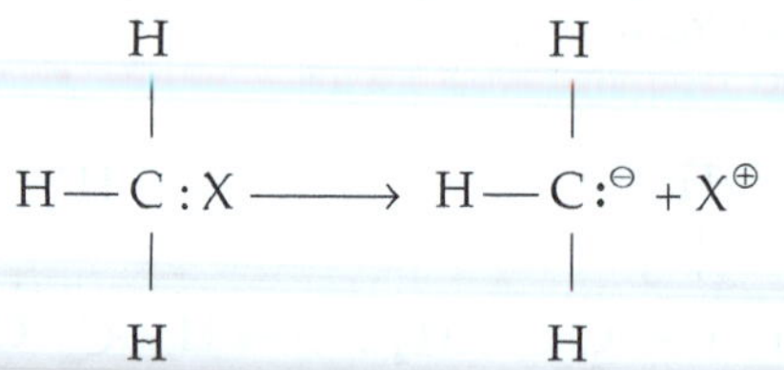

Such *organic molecules which have eight paired electron and a negative charge on one carbon are known as carbanion.*

## 16.4 REACTION INTERMEDIATES

Fission of a covalent bond results in the formation of three important reaction intermediates, *free radicals, carbonium ions* and *carbanions,* which then react with the reagent to form the products. Reaction intermediates are highly reactive species formed during the course of a chemical reaction.

### 16.4.1 Free Radicals

*A free radical is a species which has an odd or unpaired electron.* These are electrically neutral, but due to the presence of the odd electron, they are paramagnetic in nature. They are highly reactive due to the presence of the odd electron, which is in a constant search for another electron to pair up with.

**Structure.** A free radical may have a planar structure (Fig. 16.1(*a*)) with a carbon atom bearing the odd electron in $sp^2$ hybridized and the odd electron remains in the $p$ orbital or, the free radical may have an $sp^3$ hybridized carbon atom. (Fig. 16.1(*b*))

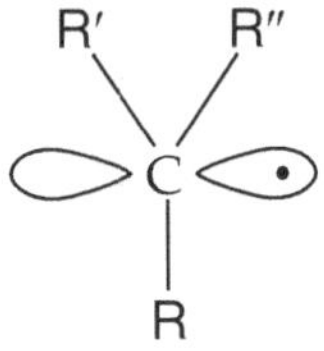


*FIGURE 16.1 (a)*

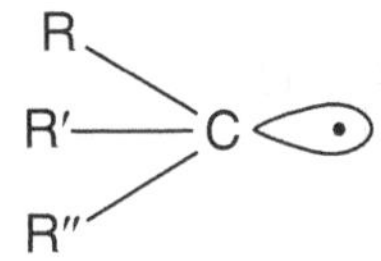


*FIGURE 16.1 (b)*

**Formation.** Free radicals are formed when a molecule is supplied with sufficient energy to effect homolysis of the covalent bond. Also, an oxidation-reduction reaction involving the gain or loss of a single electron can generate free radicals. For example,

$$(CH_3)_4\,Pb \xrightarrow{\Delta} 4CH_3^{\cdot} + Pb$$

$$CH_3—N = N—CH_3 \xrightarrow{\Delta} 2CH_3^{\cdot} + N_2$$

$$C_6H_5—\overset{\overset{\displaystyle O}{\|}}{C}—O—O—\overset{\overset{\displaystyle O}{\|}}{C}—C_6H_5 \xrightarrow{\Delta} 2C_6H_5COO\cdot \longrightarrow 2C_6H_5^{\cdot} + 2CO_2$$

$$CH_3COCH_3 \xrightarrow{h\nu} CH_3^{\cdot} + {}^{\cdot}COCH_3 \longrightarrow 2{}^{\cdot}CH_3 + CO$$

$$R—\overset{\overset{\displaystyle O}{\|}}{C}—O^{\ominus} \xrightarrow[-e^-]{} R—\overset{\overset{\displaystyle O}{\|}}{C}—O\cdot \longrightarrow R\cdot + CO_2$$

**Stability.** Free radicals are stabilized by hyperconjugation. Following is the relative order of stabilities of common free radicals.

*benzyl > allyl > tertiary > secondary > primary > methyl*

Allyl and benzyl radicals are more stable because of resonance involving $\pi$-electrons of the double bond or aromatic ring in addition to the hyperconjugative effect.

$$\underset{\text{Allyl free radical}}{CH_2 = CH—{}^{\cdot}CH_2} \longleftrightarrow \dot{C}H_2—CH = CH_2$$

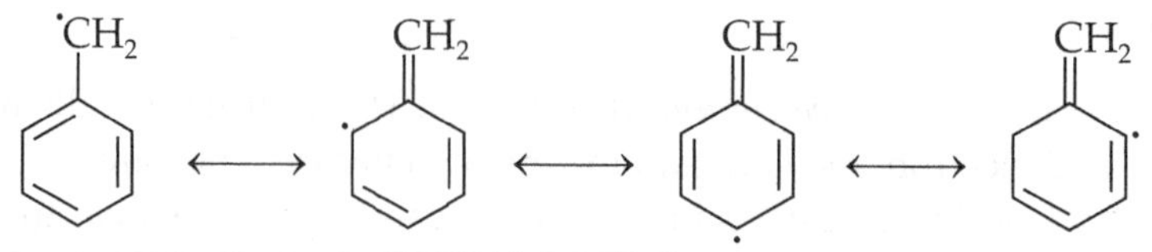

Benzyl free radical

The relative stability order of *ter*, *sec* and *pri*-free radicals is explained on the basis of hyperconjugation.

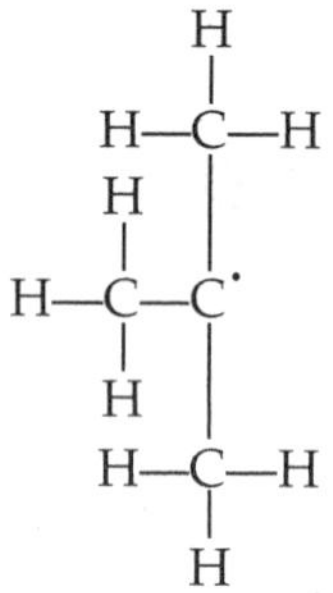

*ter*-Butyl free radical (having 9 H – C hyper-conjugative bonds)

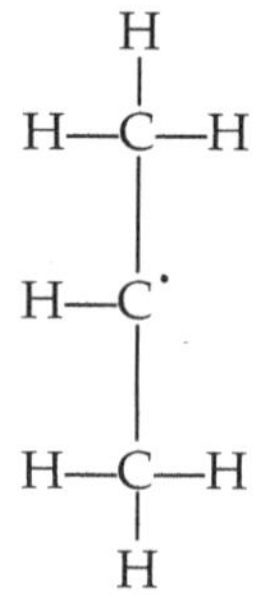

*sec*-Isopropyl free radical (having 6 H – C hyperconjugative bonds)

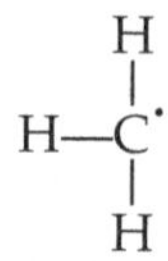

Methyl free radical (having no H – C hyper-conjugative bond)

### 16.4.2 Carbocations or Carbonium Ions

*An organic species having a carbon atom bearing six electrons only* is called a carbocation. Carbonium ions are classified as primary, secondary or tertiary depending upon the nature of the carbon atom bearing the positive charge.

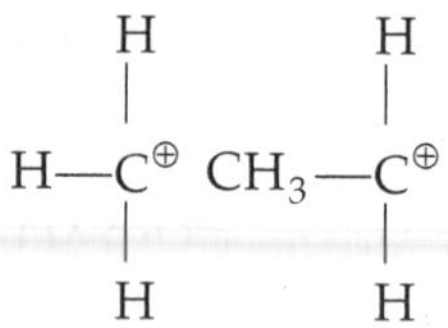

Primary carbocation

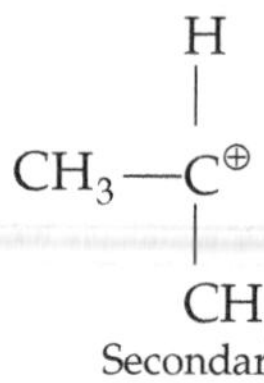

Secondary carbocation

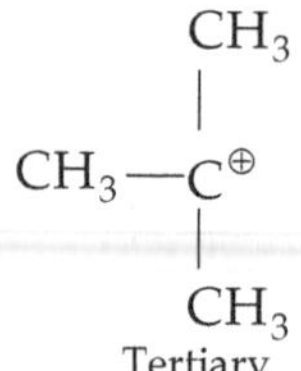

Tertiary carbocation

**Structure.** The carbon atom of the carbocation is $sp^2$ hybridized and it uses three hybrid orbitals forming three single bonds with substituents; the remaining *p*-orbital is empty. The carbocations thus have flat structure having all the three covalent bonds in one plane with the bond angles of 120° between them (Fig. 16.2). Carbocations have a strong tendency to complete the octet of carbon and thus are extensively reactive species.

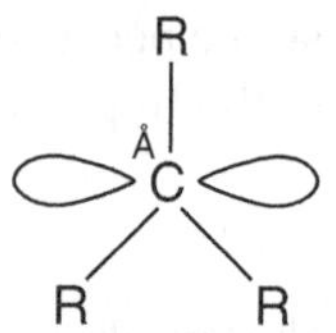

*FIGURE 16.2*

**Formation.** These can be generated in the following manner:

(*i*) Solvolysis of C—X bond (X = halogens)

$$R\text{—}X \longrightarrow R^{\oplus} + X^{\ominus}$$

(*ii*) Deamination of amines by nitrous acid.

$$R\text{—}NH_2 \xrightarrow{HNO_2} RN_2^{\oplus} \longrightarrow R^{\oplus} + N_2$$

(*iii*) Protonation of alcohols followed by dehydration.

$$R\text{—}OH \xrightarrow{H^{\oplus}} R^{\oplus}\,OH_2 \longrightarrow R^{\oplus} + H_2O$$

(*iv*) Protonation of alkenes.

$$R\text{—}CH = CH_2 \xrightarrow{H^{\oplus}} R\overset{\oplus}{C}H\text{—}CH_3$$

**Stability.** The stability of carbocations is explained both by resonance and inductive effect. Resonance explains the stability of allyl and benzyl carbocations over simple alkyl carbocations as in former carbocations positive charge can be dispersed on different carbon atoms.

$$CH_2 = CH\text{—}\overset{\oplus}{C}H_2 \longleftrightarrow \overset{\otimes}{C}H_2\text{—}CH = CH_2$$

Allyl carbonium ion

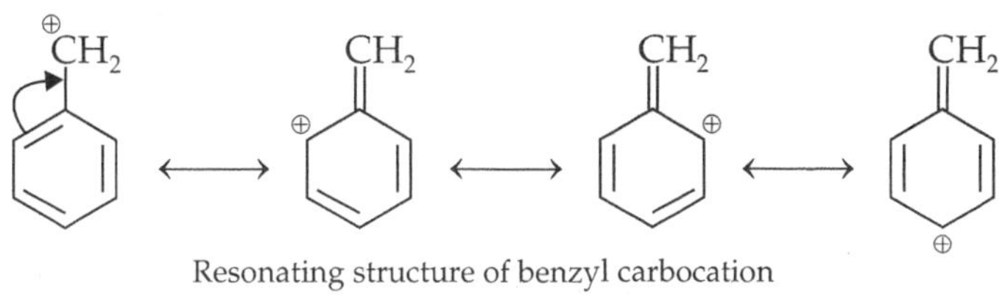

Resonating structure of benzyl carbocation

The stability of simple alkyl carbocations is explained on the basis of *inductive effect*. Due to the + I effect, the alkyl group releases electrons to the positive carbon and thus compensates for the electron-deficiency of the positive carbon. The positive charge thus gets dispersed all over the alkyl groups and this dispersal of charge increases the stability of whole system. Accordingly, tertiary carbocations are more stable than secondary ones which, in turn, are stabler than primary carbocations.

$$R\text{—}\underset{R}{\overset{R}{|}}C^{\oplus} > R\text{—}\underset{R}{\overset{H}{|}}C^{\oplus} > R\text{—}\underset{H}{\overset{H}{|}}C^{\oplus} > H\text{—}\underset{H}{\overset{H}{|}}C^{\oplus}$$

Conversely, electron withdrawing groups such as —CN, —$NO_2$, —Br, etc. will make a carbocation less stable.

The greater stability of alkyl substituted carbocations is ascribed to hyper-conjugation. For such carbocations, several hyperconjugative forms can also be written, thus spreading the charge all over the bonds. For example,

(*i*) 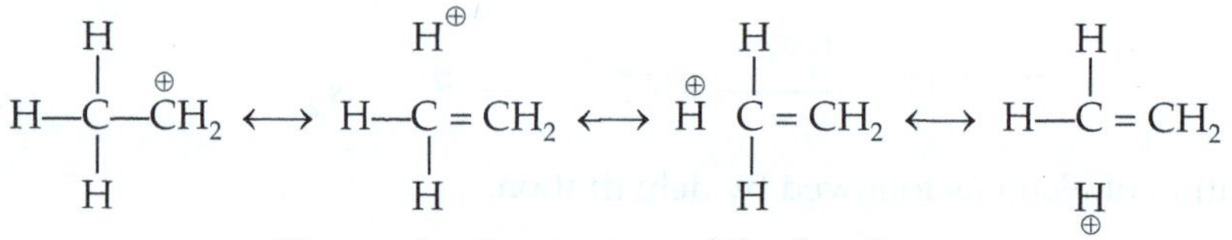

Hyperconjugative structures of 1° carbocation

(*ii*) In case of 2°-carbocations, more equivalent structures can be written.

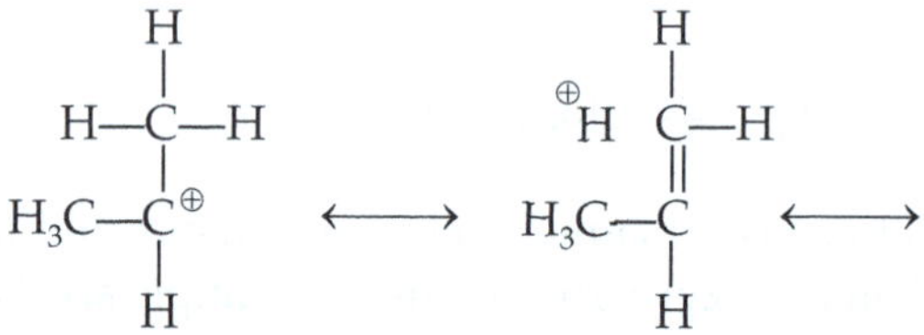


(*iii*) For 3°-carbocation, 9-such equivalent forms can be written.

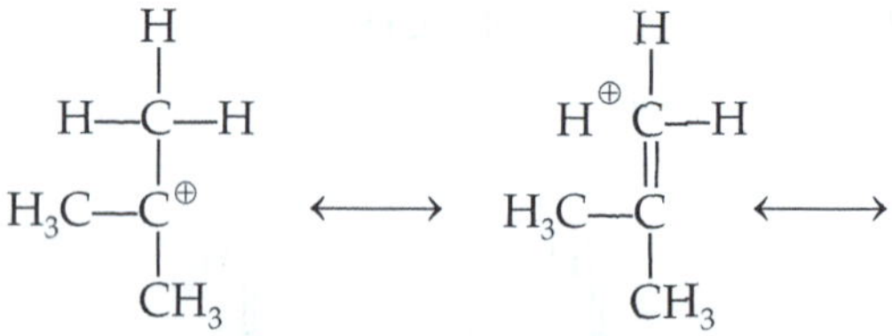


### 16.4.3 Carbanions

*A carbanion is a negatively charged organic species in which negative charge resides on the carbon atom.* Thus the carbon atom in carbanion contains four pairs of electrons, of which one pair is free. These can be represented as

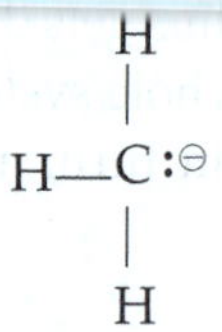

Methyl carbanion

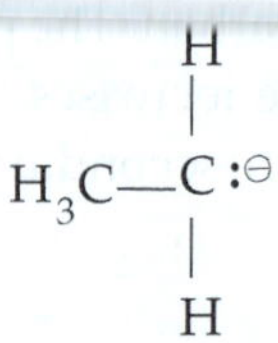

Ethyl carbanion

**Structure.** The carbon atom carrying the negative charge in carbanion is in $sp^3$ hybridized state. The three $sp^3$ hybrid orbitals form covalent bonds with the three atoms/groups while the fourth has a nonbonding pair of electrons. The configuration of simple carbanions such as methyl carbanion appears to be *pyramidal* just like that of ammonia, with which the methyl carbanion is *iso-electronic.*

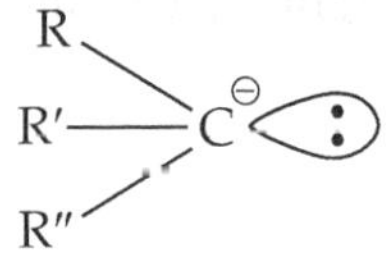

Simple carbanion

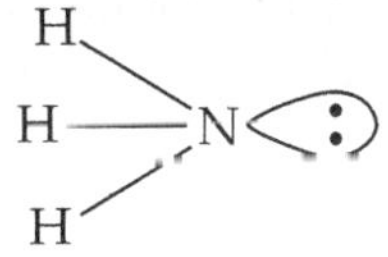

Ammonia

**Formation.** These can be formed as:

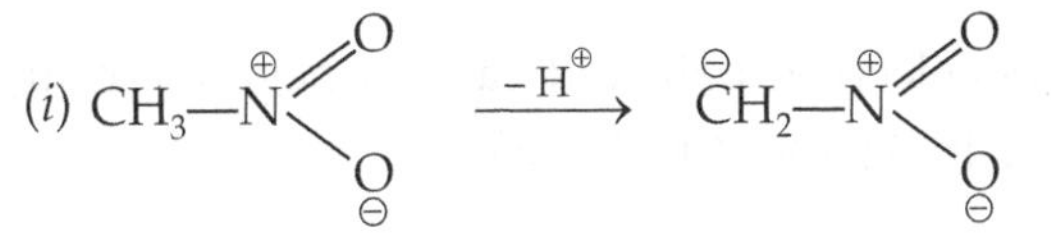

$$(ii)\ CH_3{-}C\equiv N \xrightarrow{-H^{\oplus}} \overset{\ominus}{C}H_2{-}C\equiv N$$

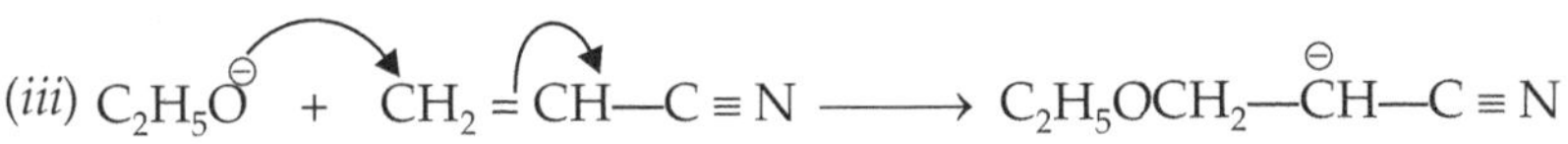

**Stability.** Carbanions are stabilized by electron withdrawing groups. Allylic and benzylic carbanions are stabilized by resonance.

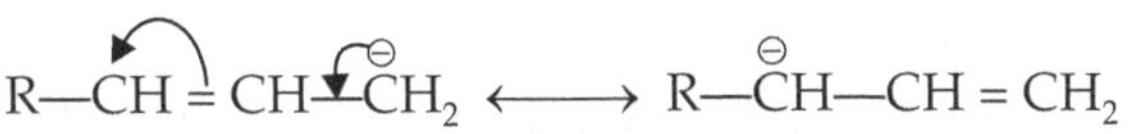

Allylic carbanion

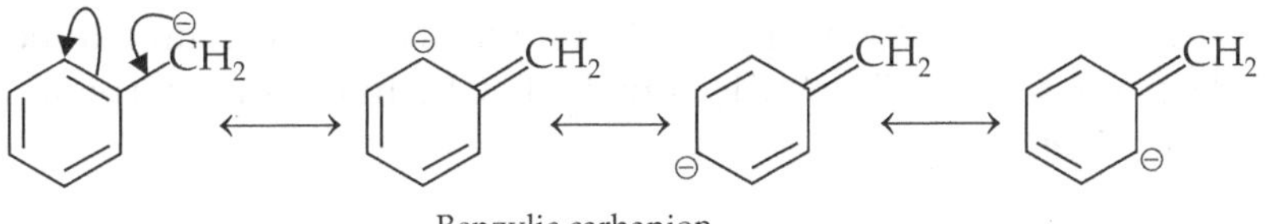

Benzylic carbanion

The order of stability of the simple carbanions is

*Methyl > primary > secondary > tertiary*

This order of stability is explained by inductive effect. The presence of + I groups releases the electrons making the carbon atom more negative, thus decreasing the stability. Hence, the greater the number of alkyl groups attached to the carbon atom bearing the negative charge, the lesser is its stability. Thus, methyl carbanion is most stable while the *ter*-alkyl carbanion is least stable.

Conversely, the presence of electron drawing groups ( – I groups) such as —C ≡ N, —$NO_2$, —Br, etc., will stabilize a carbanion because these groups will disperse the negative charge on the carbon by attracting electrons.

## 16.5 TYPES OF REAGENTS

Heterolytic fission in organic compounds is brought by certain reagents, broadly classified as *neucleophiles* and *electrophiles.*

### 16.5.1 Nucleophiles

*A nucleophile (nucleus loving) is an atom or a group of atoms that has an unshared pair of electrons.* Since these have high electron density, they attack the substrate at the point of minimum electron density, *i.e.*, nucleus. These are further classified as:

(*a*) *Negative nucleophiles* have an excess electron pair and carry a negative charge. For example, $RCH_2^{\ominus}$, $OH^{\ominus}$, $OR^{\ominus}$, $Br^{\ominus}$, $Cl^{\ominus}$, $HSO_3^{\ominus}$, $COO^{\ominus}$, etc.

(*b*) *Neutral nucleophiles* have at least one lone pair of electrons but no charge. For example,

$\ddot{N}H_3$, $H\ddot{\underset{..}{O}}H$, $R—\ddot{\underset{..}{O}}—H$, $R—\ddot{\underset{..}{O}}—R$, $R—\ddot{N}H_2$, R—MgX, R—Li, etc.

### 16.5.2 Electrophiles

These are electron seeking or loving species and thus attack the substrate molecule at the point of maximum electron density. These have incomplete valence shells and may carry a positive charge. These are further classified as:

(*a*) *Positive electrophiles* are deficient in two electrons and carry a positive charge. These include, $R—^{\oplus}CH_2$, $H^{\oplus}$, $Br^{\oplus}$, $H_3O^{\oplus}$, $NO_2^{\oplus}$, $NO^{\oplus}$, etc.

(*b*) *Neutral electrophiles* have a central atom with an incomplete octet but no charge. These include, $BF_3$, $AlCl_3$, $ZnCl_2$, $FeCl_3$, halogens, carbenes ($\ddot{C}Cl_2$), nitrenes ($:\ddot{N}—COOC_2H_5$), etc.

## 16.6 TYPES OF ORGANIC REACTIONS

Broadly, organic compounds may undergo five kinds of reactions, namely, substitution reactions, addition reactions, elimination reactions, molecular rearrangements, and molecular reactions.

### 16.6.1 Substitution Reactions

*When an atom or a group of atoms in a molecule is replaced by another atom or group of atoms,* such a reaction is called a substitution reaction. It is further classified as:

(*a*) **Nucleophilic substitution ($S_N$) reactions.** These reactions involve the attack of a nucleophile and are more common in aliphatic compounds. For example, hydrolysis of alkyl halides by aqueous KOH.

$$R—X + K—OH \longrightarrow R—OH + KX$$

In general it may be represented as:

$$\overset{\ominus}{Nu}: + R - X \longrightarrow R - Nu + \overset{\ominus}{X}:$$

These reactions may follow two mechanisms:

(*i*) *Unimolecular mechanism* ($S_N1$). The rate of such reactions depends only on the concentration of alkyl halide (substrate) and is independent of the nucleophile.

$$\text{Rate} \propto [RX] \tag{16.1}$$

Such reactions are believed to be completed in two steps. The first step is the rate determining step (slow) in which the alkyl halide ionizes to give carbonium ion. In the second step, the carbonium ion combines with the nucleophile to give the substituted product. This can be explained by the hydrolysis of *ter*-butyl bromide.

$$HO^{\ominus} + (CH_3)_3CBr \longrightarrow (CH_3)_3COH + Br^{\ominus}$$

The rate of the above reaction depends only on the concentration of *t*-butyl bromide.

$$\text{Rate} = k[(CH_3)_3C\text{—}Br] \tag{16.2}$$

Based on the kinetics, the reaction takes place in the following two steps:

**First step**

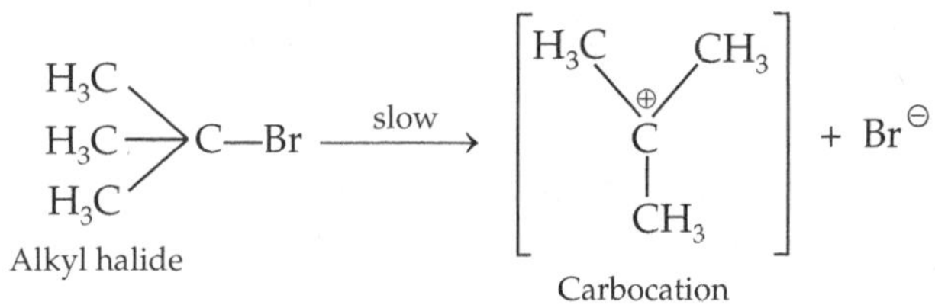


**Second step**

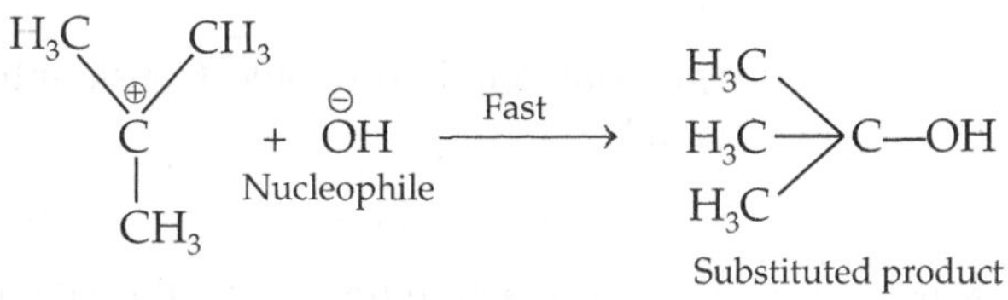


The energy profile diagram of a two step $S_N1$ reaction is shown in Figure 16.3.

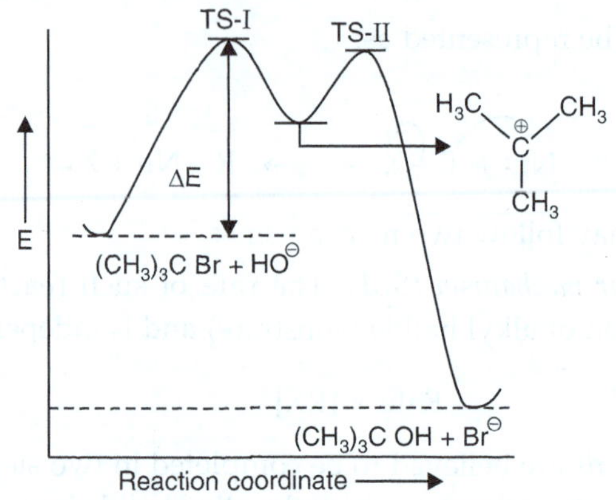


*FIGURE 16.3* **Energy profile for an $S_N1$ reaction.**

Solvents play a very important role in $S_N1$ reactions as the ionization is facilitated by solvation of carbocation.

(*ii*) *Bimolecular ($S_N2$) mechanism.* The rate of such reactions depends on the concentration of alkyl halide (substrate) as well as the base (nucleophile).

$$\text{Rate} \propto [\text{RX}]\,[\text{OH}^-] \tag{16.3}$$

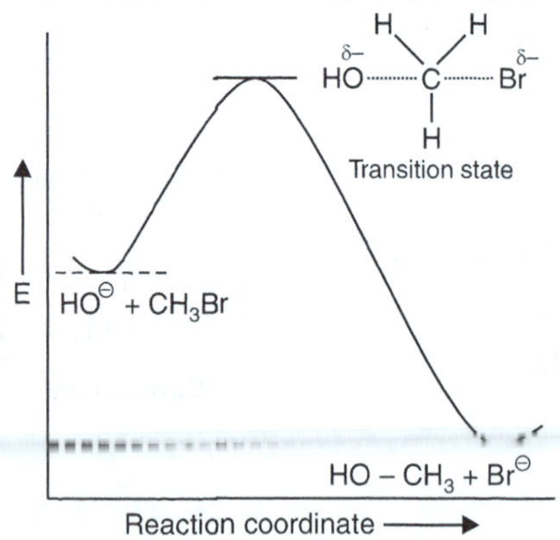


*FIGURE 16.4* **Energy profile for an $S_N2$ reaction.**

Such reactions are completed in one step. Consider the hydrolysis of methyl bromide by aqueous NaOH.

$$\text{HO}^{\ominus} + \text{CH}_3\text{—Br} \longrightarrow \text{HO—CH}_3 + \text{Br}^{\ominus}$$

Kinetic studies of this reaction reveal that the rate of such a reaction depends upon the concentration of alkyl halide as well as the hydroxide ion.

$$\text{Rate} \propto [\text{CH}_3\text{Br}]\,[\text{OH}^-] \tag{16.4}$$

If the concentration of any of them is doubled, then the rate of reaction is also doubled and when the rate of reaction of any one of them is halved, the rate of reaction is also halved.

In such a process the hydroxide ion (nucleophile) attacks the carbon atom from the side opposite that of the bromine (leaving group). This is because both the hydroxide ion and the bromine are electron-rich. The bond-breaking and bond-making are simultaneous processes and such a reaction, therefore, involves a *transition state* in which the carbon atom appears to be pentavalent. In transition state, the carbon atom is joined to three substituents by full covalent bonds and to the nucleophile and leaving group by partial bonds. The three substituents acquire a coplanar geometry.

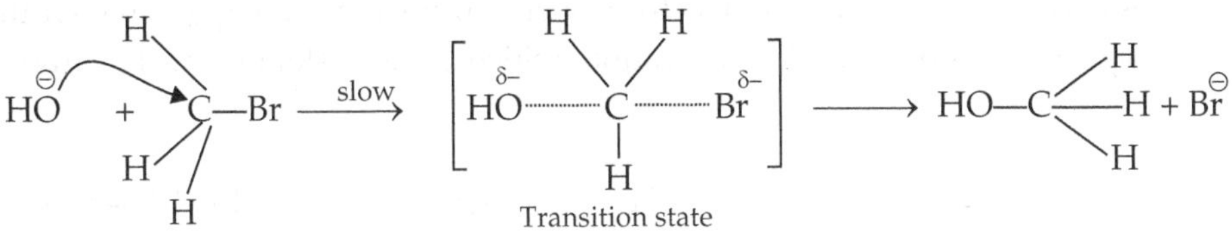


The energy profile of such a reaction is shown in Figure 16.4.

Comparison between $S_N1$ and $S_N2$ reaction is summarized in Table 16.1.

**TABLE 16.1**

| $S_N1$ | $S_N2$ |
|---|---|
| 1. These are first order reactions. | These are second order reactions. |
| 2. The nucleophile can attack from the back side as well as the front side. | The nucleophile can attack only from the back side. |
| 3. Reaction rate is influenced by electronic factors. | Reaction is influenced by steric factors. |
| 4. The order of reactivities of alkyl halides is: Benzyl > allyl > 3° > 2° > 1° halides. | The order of reactivities is: Methyl > 1° > 2° > 3° halides. |
| 5. Favored by low concentration of nucleophiles. | Favored by high concentration of nucleophiles. |
| 6. Favored by polar solvents. | Favored by nonpolar or low polar solvents. |

(*b*) **Electrophilic substitution reactions.** Electrophilic substitution reactions involve the attack by an electrophile. It is represented by $S_E$. A common example of electrophilic substitution is the bromination of benzene in the presence of $FeBr_3$.

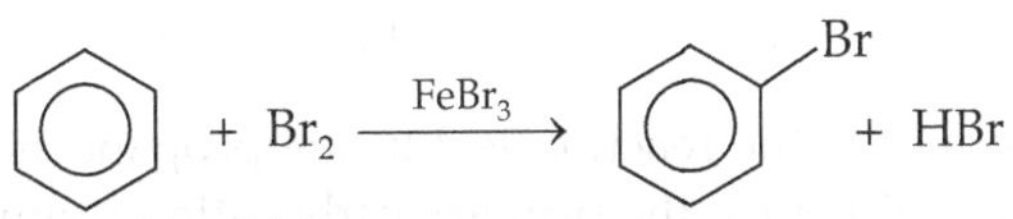


Other examples include Friedal-Craft reactions.

For mechanism, please refer to section 16.7.5.

### 16.6.2 Addition Reactions

An addition reaction is one *in which two substances combine to form a single compound.* Depending upon the type of attacking reagent an addition reaction may be of following types:

(*a*) **Electrophilic addition reaction.** These reactions are brought about by the attack of electrophiles and are generally given by compounds containing carbon-carbon double and triple bonds. For example *addition of HBr to propene.* This reaction takes place in the following steps:

**Step 1:** $H^+$ from HBr is added to propene to form a $\pi$-complex, which then forms isopropyl carbocation (2°-carbocation, which is more stable than 1°-carbocation)

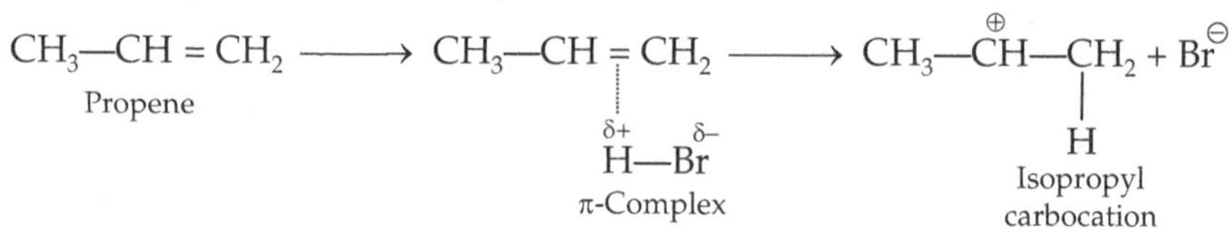


**Step 2:** The bromide ion (nucleophile) combines with the carbocation to give isopropyl bromide.

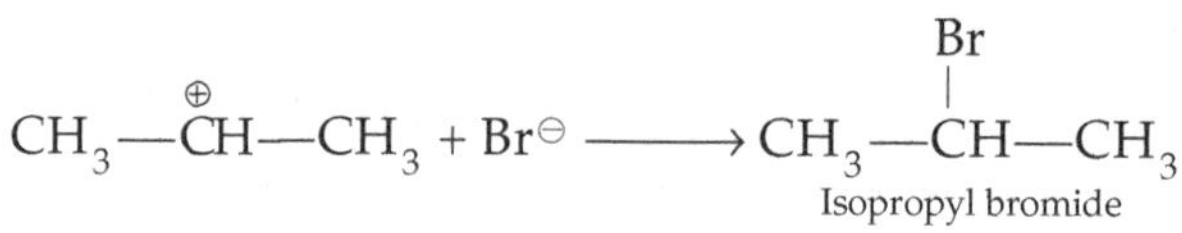


The product formed in the above reaction follows *Markownikoff's Rule,* according to which *the negative part of the addendum is added to that doubly bonded carbon atom which bears the minimum number of hydrogens.*

In the above case during the first step: two carbocations may be formed giving two different products in the following manner.

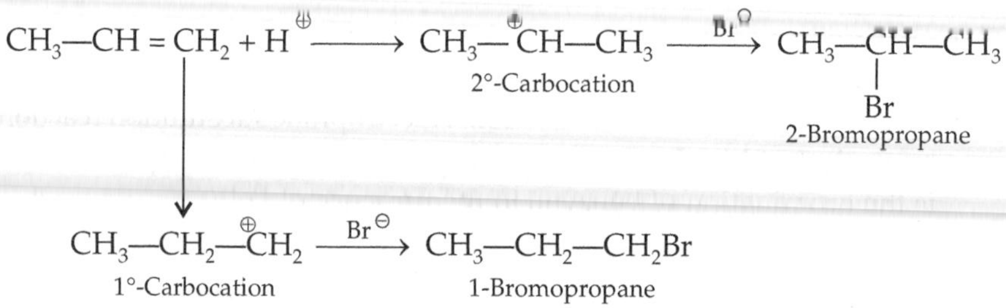


The major product of the reaction is 2-bromopropane as the secondary carbocation formed is more stable than the primary carbocation forming 1-bromopropane. Thus, Markownikoff's rule is obeyed.

*Anti-Markownikoff's Addition or Peroxide Effect.* According to this effect, the addition of HBr to propene in presence of peroxides gives 1-bromopropane as a major product rather than 2-bromopropane.

$$\underset{\substack{|\\ Br \\ \text{(Not formed)}}}{CH_3—CH—CH_3} \xleftarrow[\text{Peroxide}]{\text{HBr}} \underset{\text{Propene}}{CH_3—CH=CH_2} \xrightarrow[\text{Peroxide}]{\text{HBr}} \underset{\text{1-Bromopropane}}{CH_3—CH_2—CH_2—Br}$$

This is because in the presence of peroxides the addition involves a free radical mechanism. This effect was put forth by *Kharasch* and *Mayo*, so it is also called *Kharasch effect*.

(*b*) **Nucleophilic addition reactions.** These reactions involve the initial attack by a nucleophile. Additions on aldehydes and ketones, *i.e.*, a carbonyl group, are common examples. For example, base-catalyzed addition of HCN to a carbonyl group. It occurs in the following steps:

(*i*) HCN gives $H^{\oplus}$ and $CN^{\ominus}$ ions.

$$HCN \longrightarrow H^+ + CN^-$$

(*ii*) The carbonyl group undergoes polarization at the requirement of the attacking reagent.

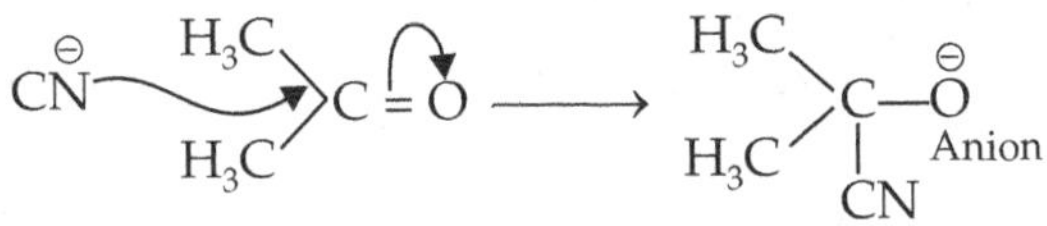


(*iii*) The proton, $H^{\oplus}$ attacks on the anion to form the addition product.

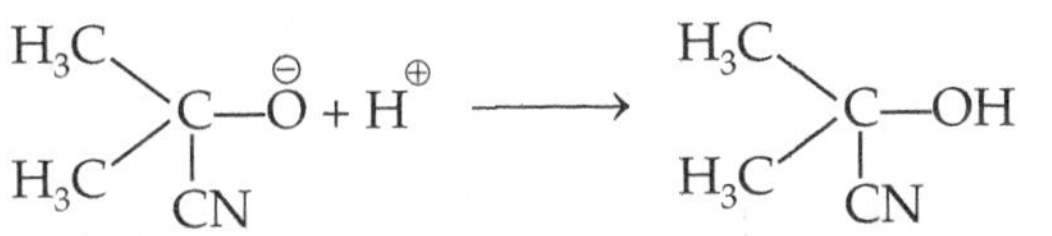


Acetone cyanohydrin

### 16.6.3 Elimination Reactions

These reactions are the reverse of the addition reaction and *involve the loss of atoms or groups from the same molecule to form multiple linkage.* On the basis of mechanism, elimination reactions may be of two types.

(*a*) α-**Elimination.** This type of elimination is very rare and involves elimination of two groups or atoms from the same carbon atom. For example,

$$\underset{\text{Chloroform}}{CHCl_3} \longrightarrow \underset{\text{Dichlorocarbene}}{:CCl_2} + HCl$$

(*b*) β-**Elimination.** This involves the removal of two groups or atoms from the two adjacent (α and β) carbon atoms of the molecule. For example,

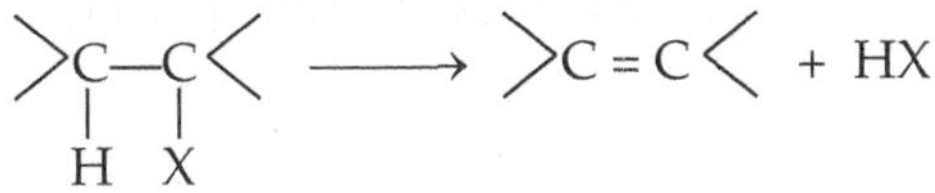

$$CH_3—CHBr—CH_3 \xrightarrow{alc.\ KOH} CH_3—CH = CH_2 + HBr$$

These reactions proceed *via* unimolecular ($E_1$) or bimolecular ($E_2$) mechanisms.

**Unimolecular ($E_1$) mechanism.** When the rate of an elimination reaction depends only on the concentration of the substrate. Like $S_N1$ reactions, these also occur in two steps. For example, dehydrohalogenation of *t*-alkyl halides. The first step involves the ionization of alkyl halide to give carbonium ion (slow) and the second step involves the fast abstraction of a proton by the base adjacent to β-carbon leading to the formation of alkene.

(*i*) 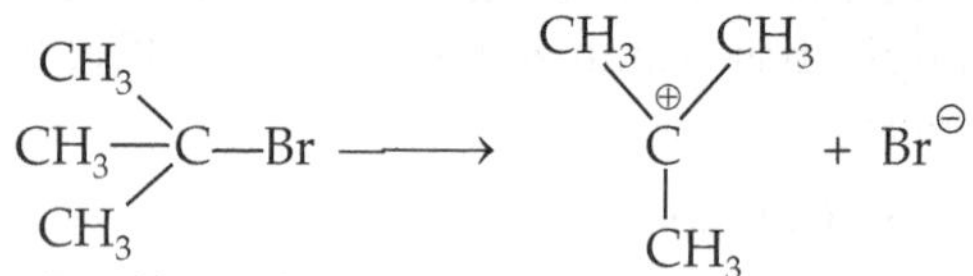

*t*-Butyl bromide — *t*-Butyl carbonium ion

(*ii*) 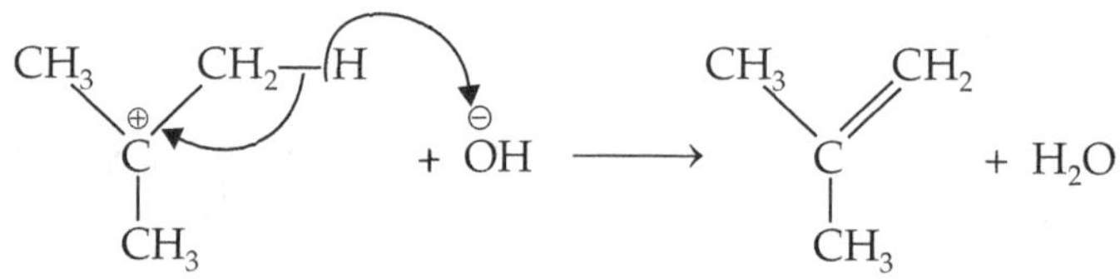

2-Methyl propene

When there is a possibility of formation of two types of alkenes, then the main product is most highly substituted alkene as per *Saytzeff Rule*. For example,

$$\underset{\underset{Br}{|}}{CH_3—CH_2—CH—CH_3} \xrightarrow{Alc.\ KOH} \underset{\substack{2\text{-Butene}\\(80\%)}}{CH_3CH = CHCH_3} + \underset{\substack{1\text{-Butene}\\(20\%)}}{CH_3CH_2CH = CH_2}$$

The other common example of $E_1$ reaction is dehydration of alcohols, which occurs as:

(*i*) $CH_3—CH_2—OH \xrightarrow{H^+} CH_3—\overset{\oplus}{C}H_2 + H_2O$

(*ii*) $CH_3—\overset{\oplus}{C}H_2 \longrightarrow CH_2 = CH_2 + H^{\oplus}$

**Bimolecular ($E_2$) mechanism.** When the rate of an elimination reaction is dependent both upon the concentration of substrate as well as the attacking reagent, the reaction is of the second order and is represented by $E_2$. For example, the reaction of alkyl halide with *alc.* KOH.

$$CH_3—CH_2—Br + \underset{(alc.)}{KOH} \longrightarrow CH_2 = CH_2 + H_2O + KBr$$

$$\text{Rate} = k[CH_3CH_2Br]\ [KOH] \qquad (16.5)$$

The mechanism is a one step process in which the abstraction of the proton from the β-carbon and the expulsion of the leaving group, *i.e.*, halide ion from the α-carbon atom, occur simultaneously.

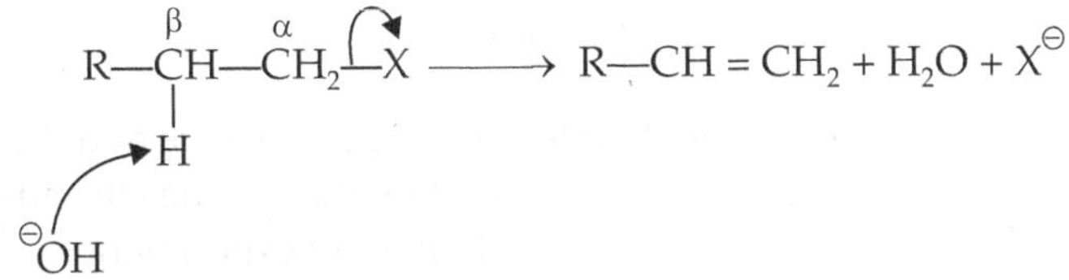

This is also known as *dehydrohalogenation* reaction.

## 16.7 ORGANIC NAME REACTIONS

### 16.7.1 Beckmann Rearrangement

Beckmann rearrangement and Hoffmann rearrangement involve migration of an alkyl or aryl group with its bonding pair of electrons from the carbon atom to the adjacent nitrogen atom.

When ketoximes and aldoximes are treated with acidic reagents such as conc. $H_2SO_4$, $PCl_5$, $P_2O_5$, $SOCl_2$, $SO_3$, $C_6H_5SO_2Cl$ and polyphosphoric acid (PPA), they rearrange to form a corresponding amide. This is known as *Beckmann's rearrangement.* For example,

(*i*)

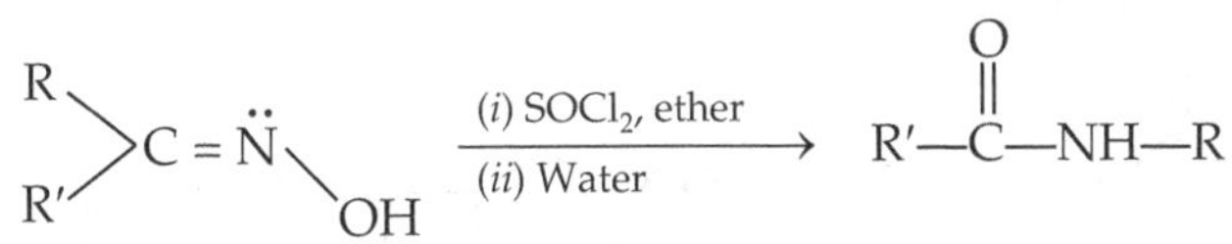

Ketoxime

(*ii*)

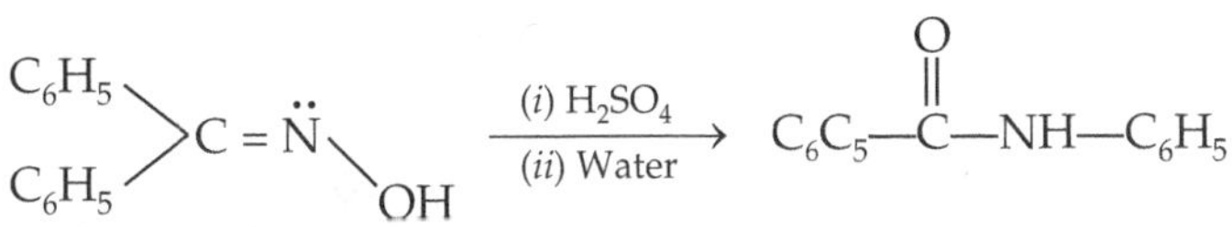

Benzophenone oxime — Benzanilide

(*iii*)

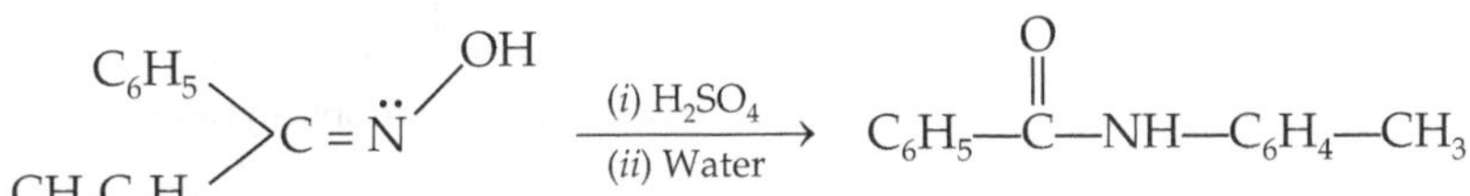

syn-phenyl-*p*-tolylketoxime — Benz-*p*-anilide

(iv) 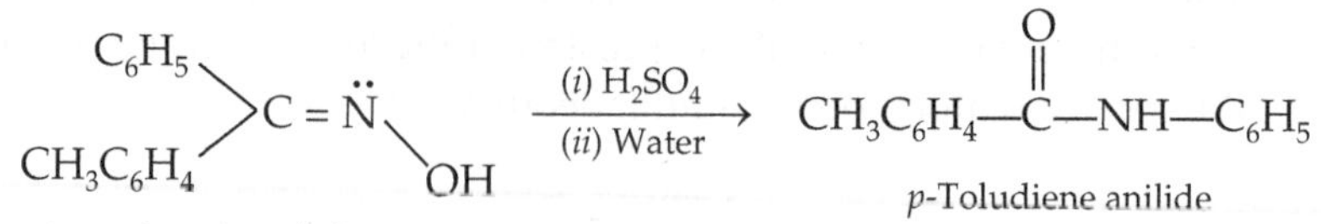


Anti-phenyl-*p*-tolylketoxime  *p*-Toludiene anilide

Beckmann's rearrangement is stereospecific in nature and the alkyl or aryl group *anti*- to the oxime hydroxyl group migrates preferentially. The *anti*-migration of the group was demonstrated by Meisenheimer by the following experiment:

Meisenheimer, in 1924 and 1926 showed that one of the isomer of 2-bromo-5-nitro acetophenone oxime was unaffected by sodium hydroxide whereas the other undergoes ring closure to form 3-methyl-5-nitro benzisoxazole, proving that, in the latter reacting groups, —OH and —Br are close together and hence it can be represented by the configuration (A), whereas the other is represented by (B).

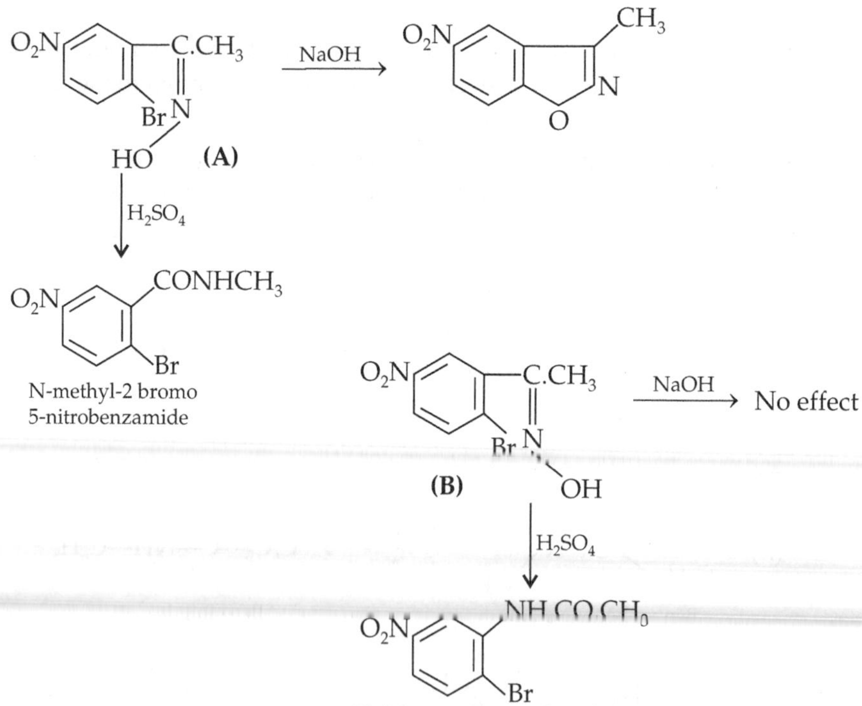

When the two above oximes (A) and (B) of known configuration are treated with acidic reagents, oxime (A) yields a substituted N-methyl benzamide whereas the B yields N-substituted benzamide, favoring migration of the *anti*-group.

**Mechanism**. The Beckmann rearrangement does not proceed by an intramolecular exchange of hydroxyl and phenyl groups. The mechanism involves the formation of a protonated oxime by the action of the proton ($H^+$) with

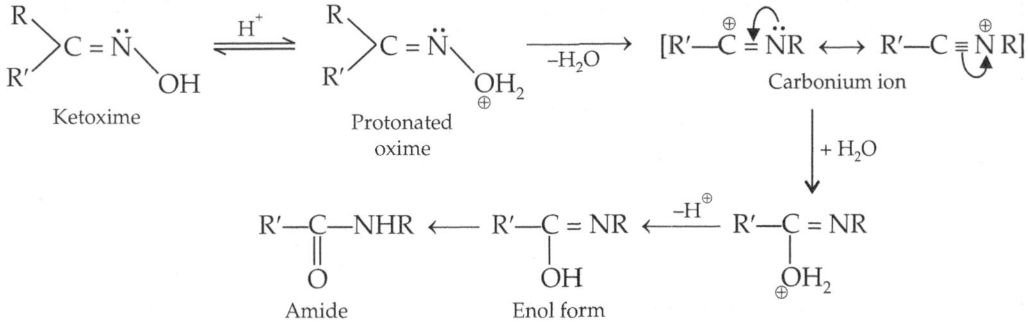


a simultaneous loss of a molecule of water and shift of an *anti* alkyl group leading to the formation of carbonium ion. The carbonium ion reacts with water to give an *enol* form of an amide which undergoes tautomerism to form an amide.

**Applications**

1. *Synthesis of ∈-caprolactum.* This rearrangement is used in the synthesis of ∈-caprolactum from cyclohexanone oxime and conc. sulphuric acid. ∈-Caprolactum is the raw material for nylon-type polymers, *nylon*-6.

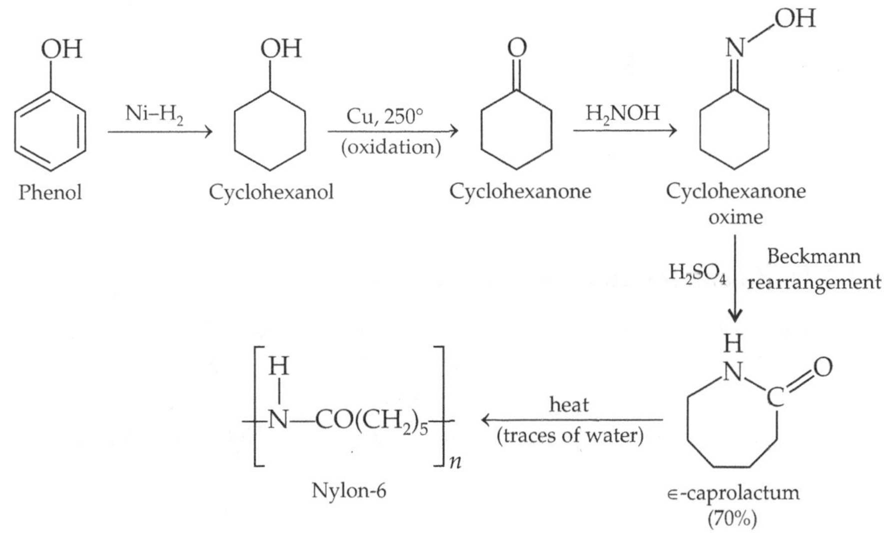

2. *Synthesis of isoquinolines*. The conversion of cinnamaldoxime into isoquinoline by heating with phosphorous pentaoxide involves Beckmann's rearrangement.

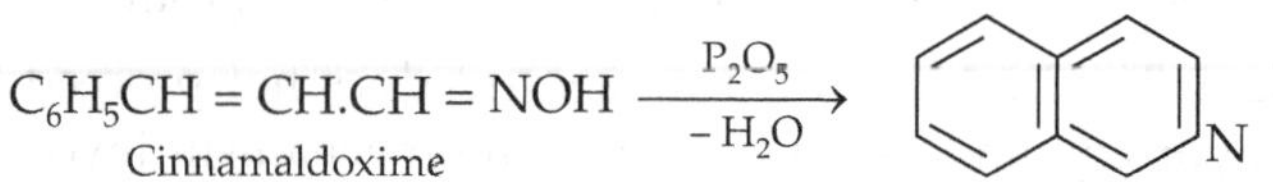


Cinnamaldoxime                Isoquinoline

3. *Synthesis of nitriles*. Aldoximes can be dehydrated readily by acid reagents to nitriles.

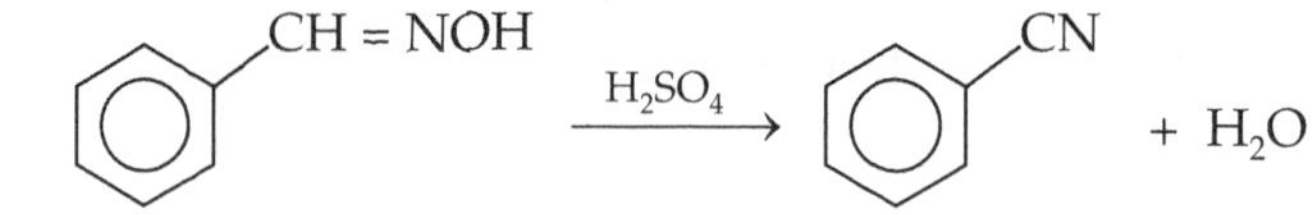


Benzaldoxime                Cyanobenzene

4. *Determination of configuration of ketoximes.* This determination is based on the fact that the two isomeric ketoximes, *syn* and *anti*, give different amides *via* Beckmann's rearrangement. The amides in turn can be identified by their hydrolysis products.

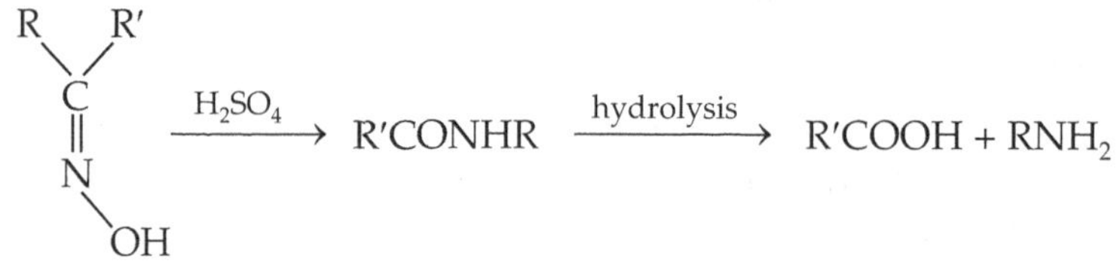


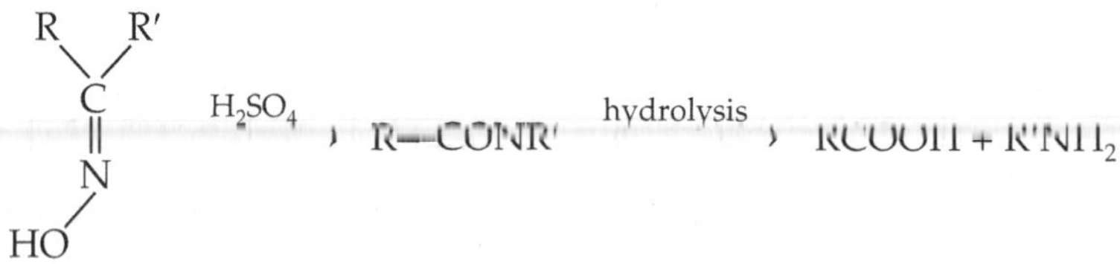


### 16.7.2 Hoffmann Reaction or Rearrangement

When amides of the fatty acids are treated with bromine and caustic soda (10%) in excess of about 70°C, primary amines free from secondary and tertiary amines are formed. This is known as *Hoffmann rearrangement.* The amine formed has one carbon atom less than the parent acid amide and so the reaction is also known as *Hoffmann degradation.*

$$\underset{\text{Acid amide}}{RCONH_2} + Br_2 + 4KOH \rightarrow \underset{1^\circ\text{-Amine}}{RNH_2} + 2KBr + K_2CO_3 + 2H_2O$$

**Mechanism.** The reaction is believed to proceed by the following steps:

(*i*) The hypobromite ion ($OBr^-$) formed by the action of KOH and $Br_2$ reacts with the amide forming N-bromamide.

(*ii*) In the second step, the base abstracts acidic hydrogen from the N-atom to form N-bromamide anion.

(*iii*) In third step, loss of $Br^-$ results in the formation of highly reactive acyl nitrene.

(*iv*) Acyl nitrene undergoes rearrangement of the alkyl group to the electron deficient nitrogen atom, forming more stable isocynate.

(*v*) Isocynate reacts with water forming carbamic acid which spontaneously eliminates $CO_2$ to yield the amine.

The mechanism may be represented as:

$$Br_2 + KOH \longrightarrow \underset{\text{Potassium hypobromite}}{KOBr} + HBr$$

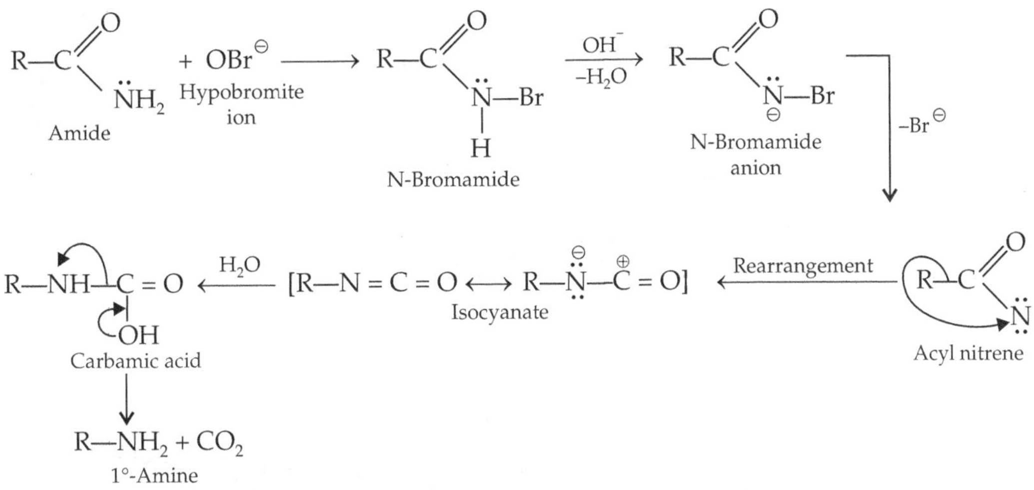


The formation of N-bromamide, its anion and isocyanate as the intermediate products take place and hence the above mechanism is proved by their isolation under suitable conditions. Elimination of bromide ion from N-bromamide anion forms a highly unstable neutral species $RCO\ddot{N}$ (acyl nitrene), which has only a sextet of electrons around the nitrogen atom. This species is therefore highly electron-deficient on nitrogen, and it gains some stability by the migration of the alkyl group with its pair of bonding electrons to form isocyanate. This rearrangement of the N-bromamide anion to isocyanate is known as *Hoffmann's rearrangement.*

## Applications

1. The reaction is widely used for the conversion of carboxylic acids and their derivatives into primary amines, which contain one carbon atom less than the starting material. For example,

$$(i)\quad \underset{\text{Isobutyric acid}}{(CH_3)_2CHCOOH} \xrightarrow[(b)\ NH_3]{(a)\ PCl_5} (CH_3)_2CHCONH_2 \xrightarrow[(b)\ KOH]{(a)\ Br_2} \underset{\text{Isopropylamine}}{(CH_3)_2CHNH_2}$$

(*ii*)

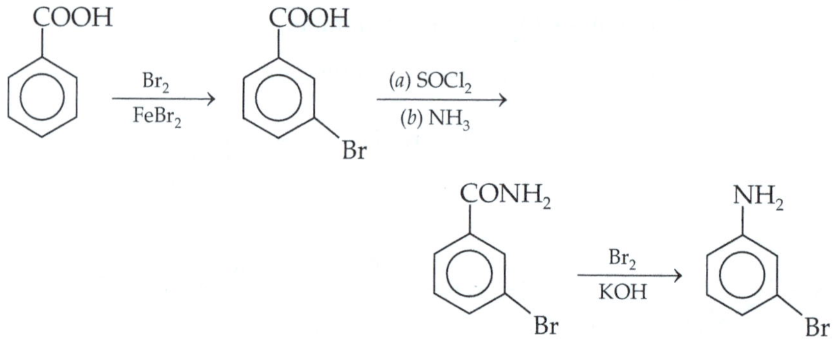


*m*-Bromoaniline

(*iii*)

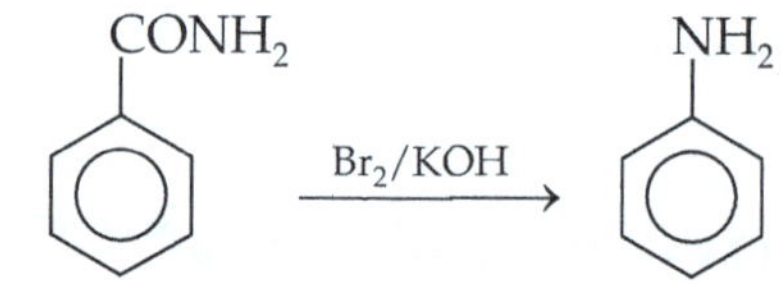


Benzamide Aniline

2. *Synthesis of anthralinic acid*

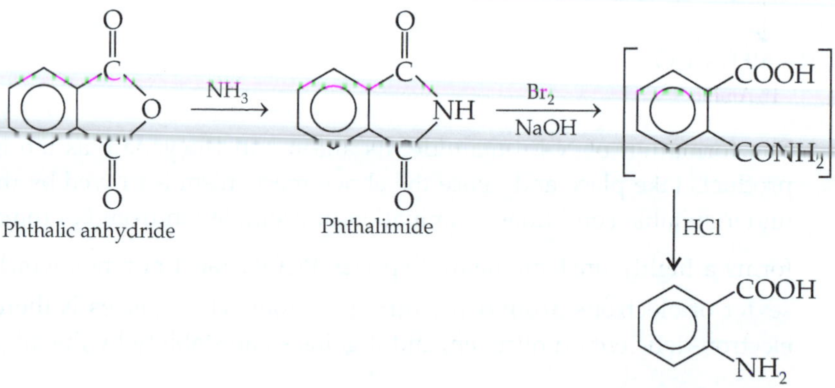


Phthalic anhydride Phthalimide Anthralinic acid

3. *Preparation of amino acids.* β-Alanine, can be obtained by treating succinimide with bromine and aqueous caustic potash.

Succinimide β-alanine

### 16.7.3 Cannizaro Reaction

Aldehydes containing no α-hydrogens in the presence of concentrated alkali undergo self-oxidation and reduction to yield a mixture of an alcohol and a salt of a carboxylic acid. This reaction is called a *Cannizaro reaction.* The important aldehydes that give a Cannizaro reaction are formaldehyde (HCHO), benzaldehyde ($C_6H_5CHO$), chloral ($Cl_3CHO$), trimethyl acectaldehyde $(CH_3)_3$ C.CHO and furaldehyde. For example,

(*i*) $\underset{\substack{\text{Formaldehyde} \\ \text{(no } \alpha\text{-hydrogen)}}}{2HCHO} + \underset{(50\%)}{NaOH} \longrightarrow \underset{\text{Methanol}}{CH_3OH} + \underset{\text{Sodium formate}}{HCOONa}$

(*ii*)

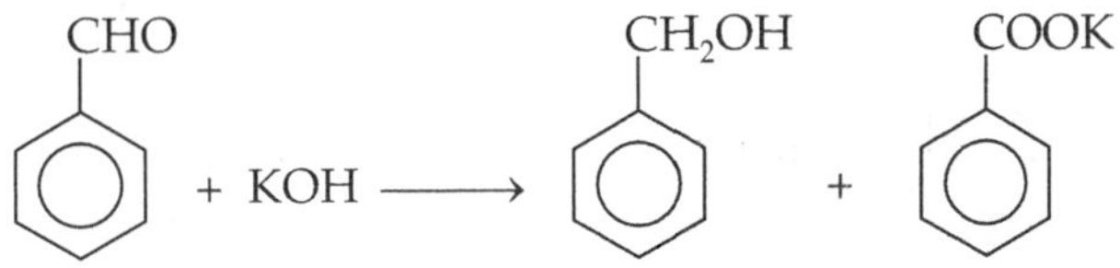


Benzyl alcohol Pot. benzoate

The reaction is generally brought about by allowing the aldehyde to stand at room temperature with concentrated aqueous or alcoholic alkali. Under these condition an aldehyde containing α-hydrogens undergoes aldol condensation.

The chief advantage of this reaction lies in its specific nature and only the carbonyl group is reduced.

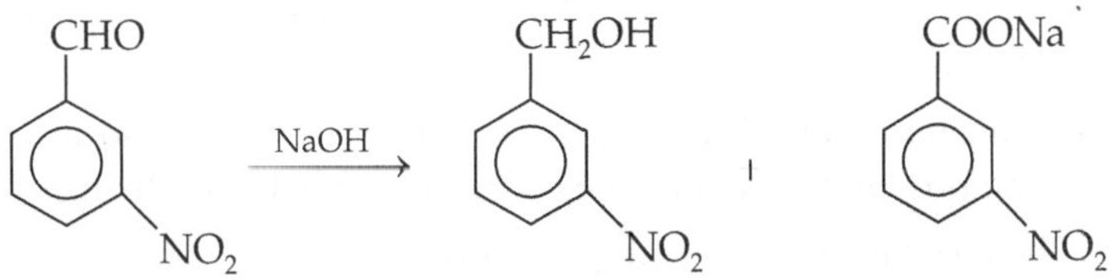


*m*-Nitro benzaldehyde *m*-Nitro benzylalcohol *m*-Nitro sodium benzoate

If one of the aldehydes in the reaction mixture is formaldehyde, the reaction yields almost exclusively sodium formate and the alcohol corresponding to the other aldehyde.

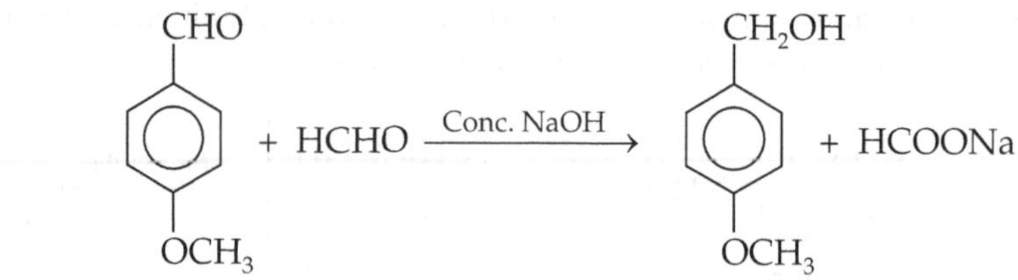

*p*-Methoxy benzylaldehyde → *p*-Methoxy benzyl alcohol

The Cannizzaro reaction can also take place between two different aldehydes and is then known as *crossed Cannizaro Reaction.*

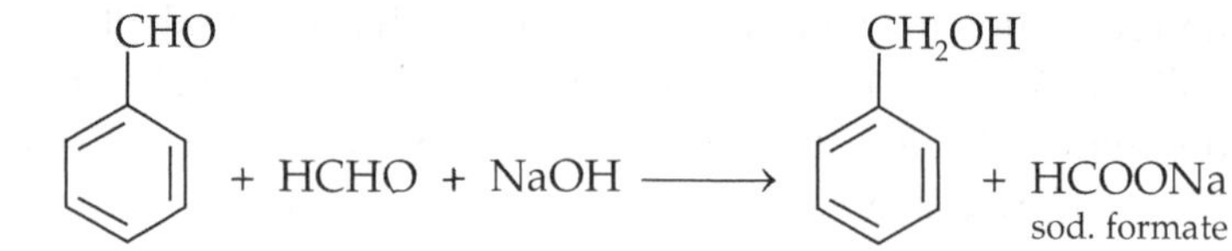

Benzaldehyde → Benzyl alcohol

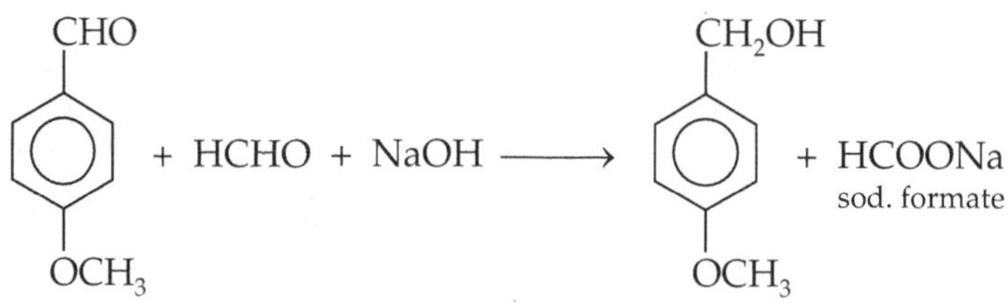

4-Methoxybenzaldehyde → 4-Methoxybenzyl alcohol

Cannizaro reaction sometimes fails due to steric hinderances, *e.g.*, *o*-disubstituted benzaldehyde fails to react normally.

**Mechanism.** The mechanism involves the following two steps:

(*i*) Nucleophilic addition of $OH^{\ominus}$ to one molecule of aldehyde forming hydroxy alkoxide anion (I).

(*ii*) Transfer of a hydride ion ($H^{\ominus}$) from (I) to the second molecule of aldehyde. The presence of a negative charge on (I) helps in the loss of the hydride ion.

The mechanism may be represented as:

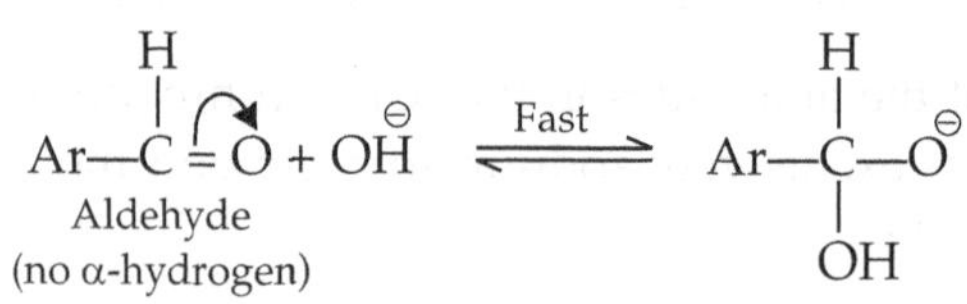

**(I)**

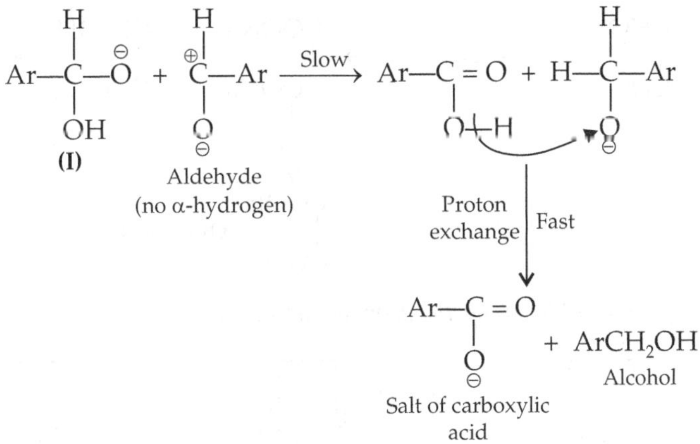


## Applications

1. *Preparation of pentaerythritol*

$$CH_3CHO \xrightarrow[\text{Aldol condensation}]{HCHO/OH^{\ominus}} C(CH_2OH)_3CHO \xrightarrow[\text{crossed cannizaro reaction}]{HCHO/OH^{\ominus}} \underset{\text{Pentaerythritol}}{C(CH_2OH)_4} + HCOO^{\ominus}$$

Pentaerythritol is used in the production of *pentaerythritol trinitrate,* an intermediate for explosive manufacture.

$$\underset{\text{Pentaerythritol}}{C(CH_2OH)_4} + 3HNO_2 \xrightarrow{H_2SO_4} \underset{\text{Pentaerythritol trinitrate}}{HOCH_2 . C(CH_2ONO_2)_3} \xrightarrow{H_2SO_4} \underset{\text{Pentaerythritol tetranitrate PETN (Explosive)}}{C(CH_2ONO_2)_4}$$

2. *Synthesis of glycollic and oxalic acid*

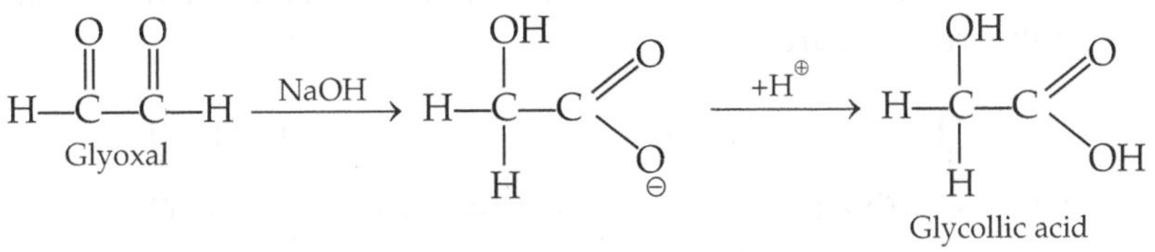

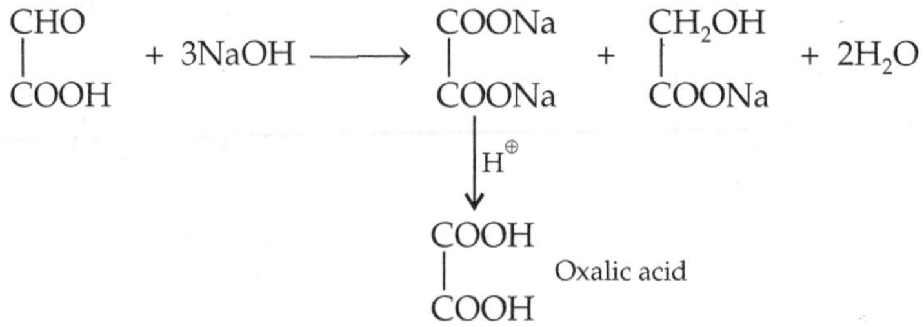


3. *Synthesis of heterocyclic alcohols and acids*

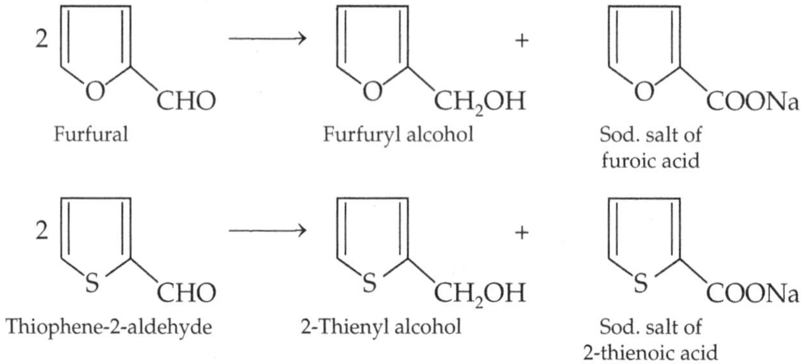


4. *Synthesis of β-hydroxy aldehyde*

$$RR'CH.CHO + HCHO \xrightarrow{NaOH} RR'C\begin{cases} CH_2OH \\ CHO \end{cases}$$

β-Hydroxy aldehyde

### 16.7.4 Aldol Condensation

When aldehydes or ketones containing at least one α-hydrogen undergo self- condensation in the presence of a dilute base to form a β-hydroxy aldehyde or a β-hydroxyketone, it is known as the *aldol condensation*. The condensation product is known as *aldol*. Some common examples of this reaction are:

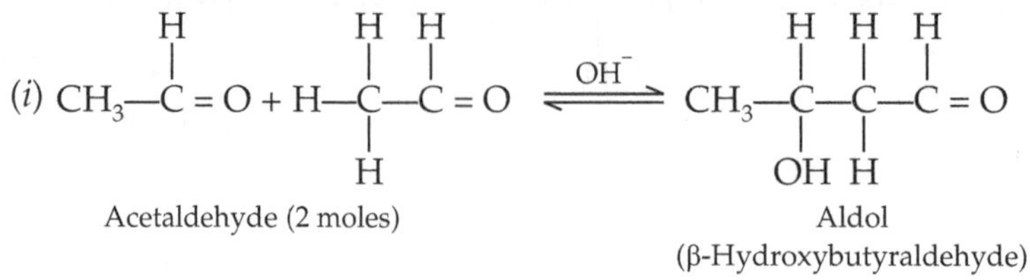

(ii) 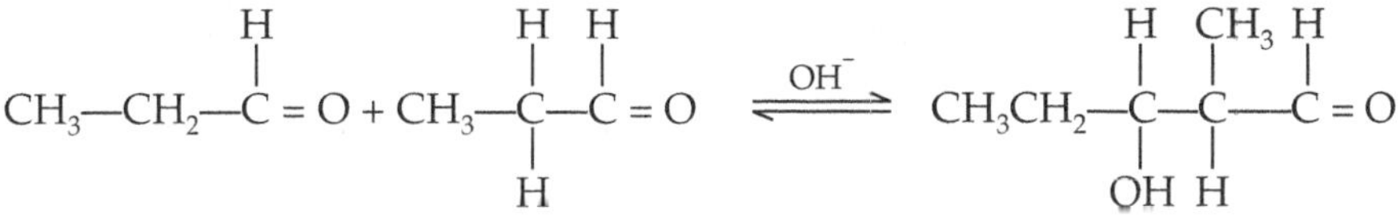

Propionaldehyde (2 moles) — 3-Hydroxy-2-methylpentanal

(iii) 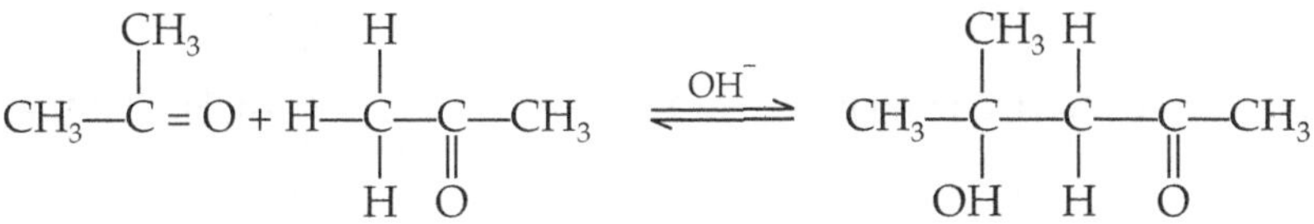

Acetone (2 moles) — Diacetone alcohol

Aldol condensation is possible only when the carbonyl compound contains at least one α-hydrogen atom. Thus the following aldehydes or ketones without α-hydrogens do not undergo aldol condensation:

$$\text{ArCHO, HCHO, R}_3\text{C.CHO, Ar—CO—Ar, Ar.CO.CR}_3 \xrightarrow[\text{dilute}]{\text{OH}^-} \text{No reaction}$$

However, these compounds undergo Cannizaro reaction when treated with a strong base.

Aldol condensation reactions are reversible. The reaction is not favorable for ketones for the following two reasons:

(i) the + I effect of an ethyl group renders the adduct less stable as compared to aldehydes.

(ii) the carbonyl group in ketones is more sterically hindered.

**Mechanism.** The mechanism of aldol condensation is quite clear and involves the following steps:

(i) The first step is the formation of a resonance stabilized carbanion (I) by the action of base on aldehyde.

(ii) The carbanion (nucleophile) now attacks on the carbonyl group of the second aldehyde to form an addition product (alkoxide), (II).

(iii) The alkoxide ion, II abstracts a hydrogen ion from water to form β-hydroxyaldehyde (aldol) III. The mechanism may be represented as:

Aldehyde — Resonance stabilized carbanions (I)

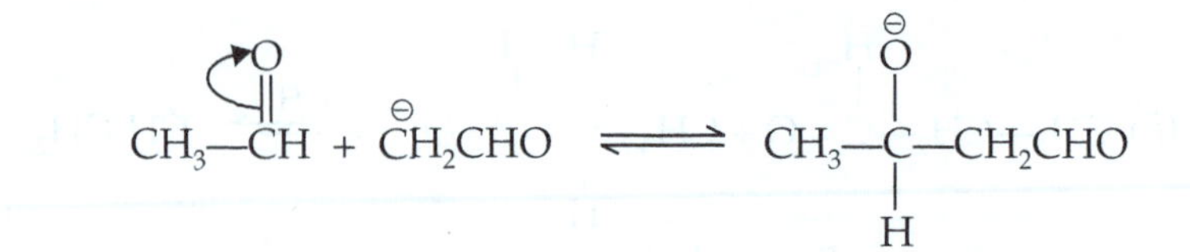

Alkoxide ion, (II)

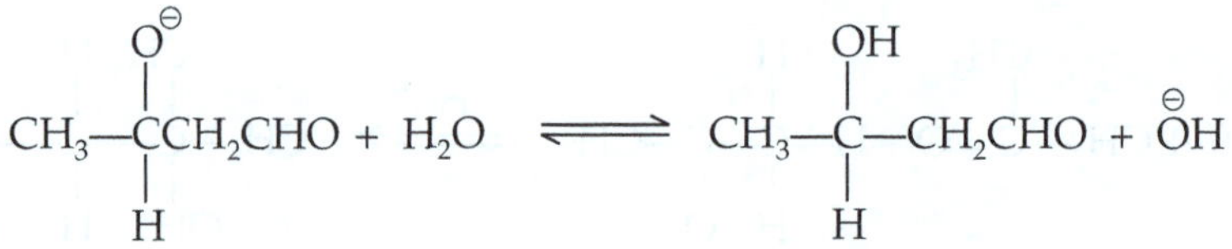

Aldol, (III)

*Crossed Aldol Condensation* involves the condensation of:

(*i*) *Two different aldehydes.* When a mixture of two different aldehydes, each having an α-hydrogen is treated with base, a mixture of all four possible products is formed, as each aldehyde can form a carbanion and can also act as the acceptor. But such reactions are of little synthetic value.

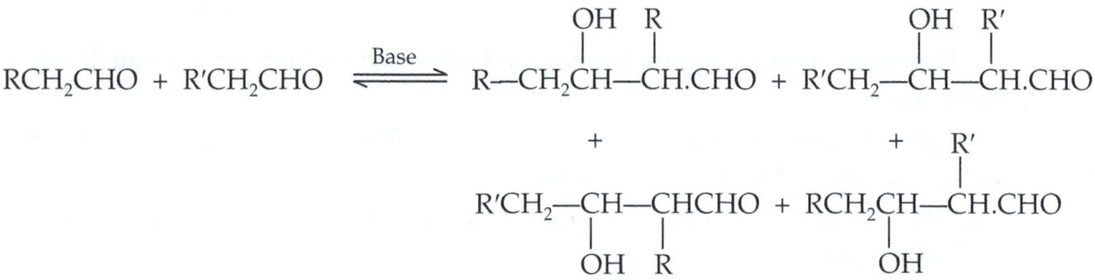

Where R and R′ are two different alkyl groups.

(*ii*) *An aldehyde and a ketone.* When an aldehyde or ketone, each having active hydrogens, is condensed, only two products are formed as the ketone itself does not undergo aldol condensation and also the carbonyl group of the aldehyde is more reactive. In such cases the ketones provide the carbanion and the aldehyde acts as the acceptor, *e.g.*,

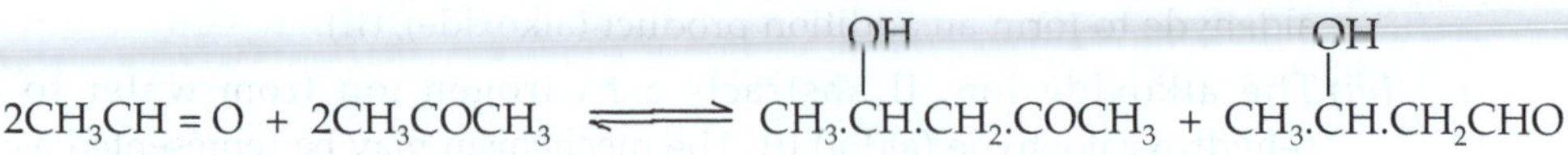

4-Hydroxypentan-2-one Aldol

*Acid Catalysed Aldol Condensation*

Aldol condensation is also catalyzed by acids, for example, condensation of benzaldehyde with acetophenone. Here the acid catalyzed the conversion of acetophenone into the enol form which then reacts with the protonated benzaldehyde.

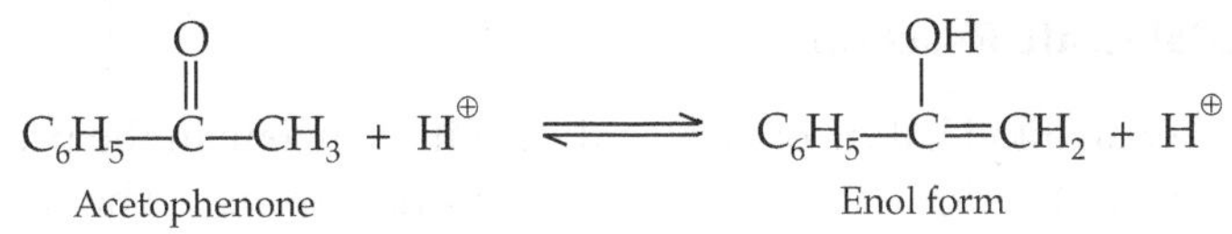

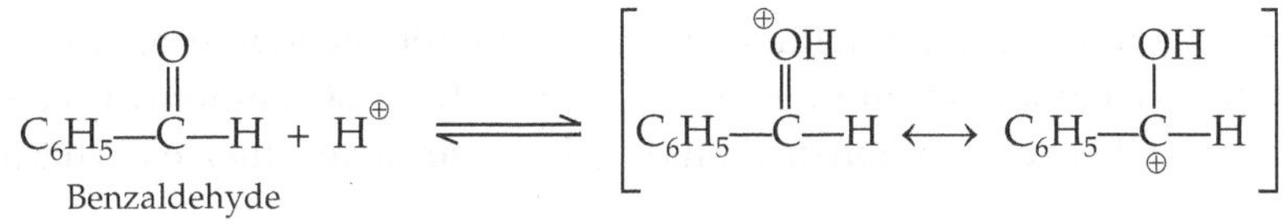

## Applications

1. *Synthesis of saturated alcohols on an industrial scale*

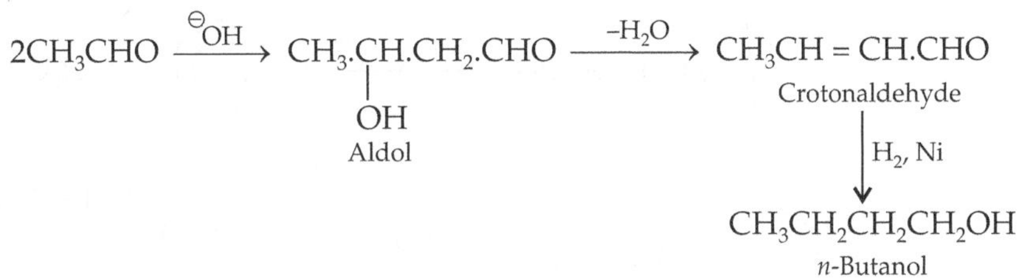

2. *Preparation of α, β-unsaturated aldehydes, ketones, and alcohols.* Unsaturated alcohols can be prepared by reducing the unsaturated aldehyde or ketone obtained by aldol condensation.

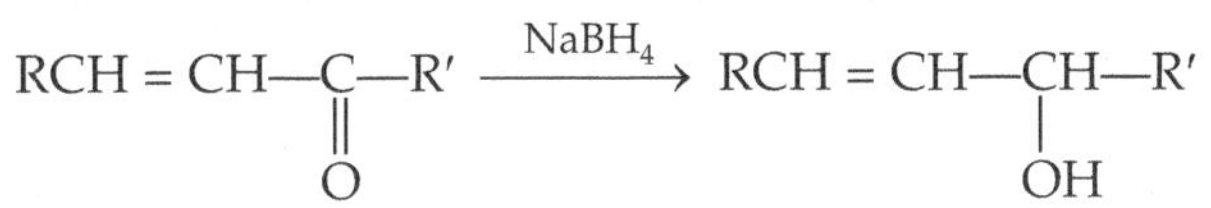

$NaBH_4$ does not attack the carbon-carbon double bond.

3. *Synthesis of sorbic acid, a food preservative*

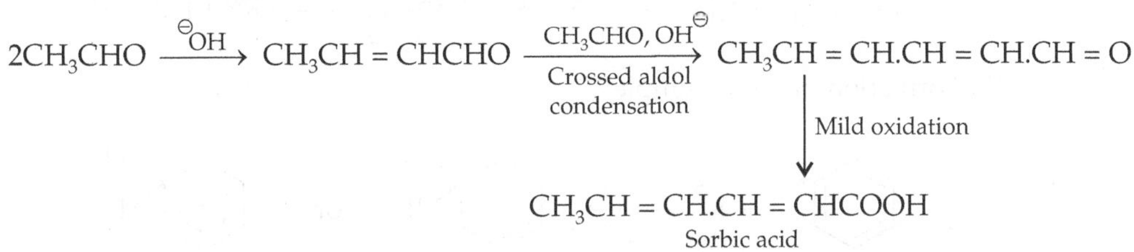

4. *Synthesis of pentaerythritol.* Please refer to application of Cannizaro reaction.

### 16.7.5 Friedal Craft Reactions

The reactions of an alkyl halide or acryl halide with benzene in the presence of anhydrous Lewis acid, generally $AlCl_3$, is known as *Friedal Craft reaction.* The other Lewis acids that may be used in the reaction include $AlBr_3$, $FeCl_3$, $SbCl_5$, $SnCl_4$, $BF_3$, etc. The alkylating agents may be alkyl halides, olefins, aliphatic alcohol, esters, ethers, etc. The acylating agents include acid chlorides and acid anhydrides. Commonly used solvents in the reaction are nitrobenzene, carbon disulphide, ether, petroleum ether, etc. This reaction may be studied as:

#### A. Friedal-Craft Alkylation

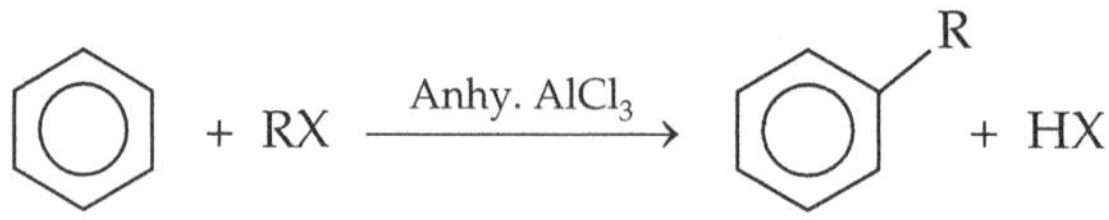


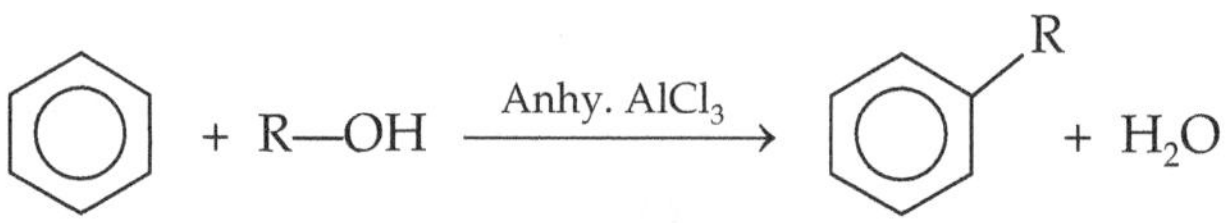

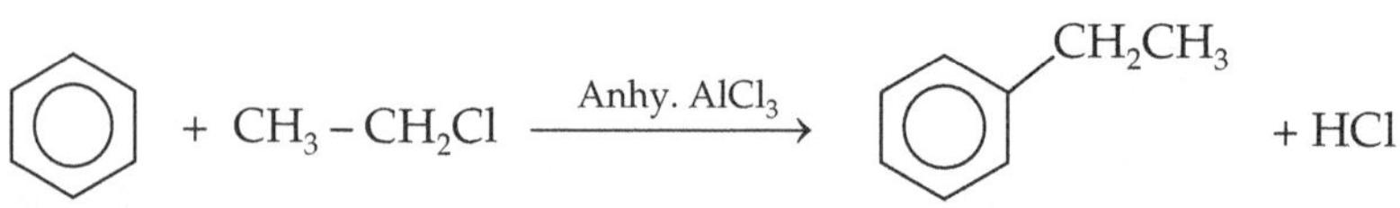


**Mechanism.** The mechanism of the reaction involves the following steps:

(*i*) Generation of electrophile

$$R—Cl + [AlCl_3] \longrightarrow \left[ \overset{\delta+}{R} \cdots Cl \cdots \overset{\delta-}{AlCl_3} \right] \longrightarrow R^{\oplus} + [AlCl_4^{\ominus}]$$

(*ii*) Formation of the σ-complex

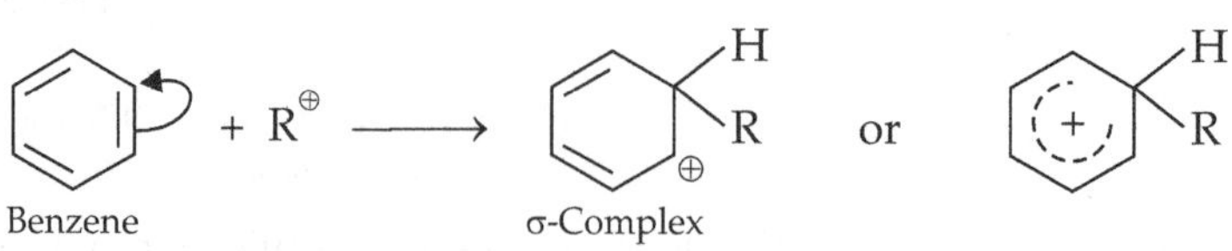

(*iii*) Elimination of proton from σ-complex

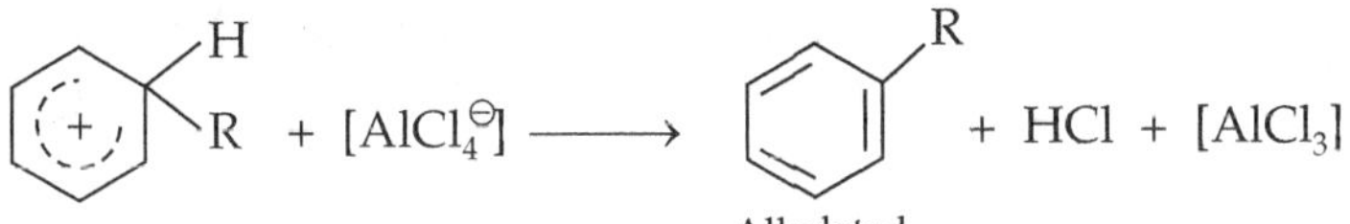

Alkylated benzene

**B. Friedal-Craft Acylation**

Acylation of benzene is brought about by acid chlorides or anhydrides in the presence of Lewis acids.

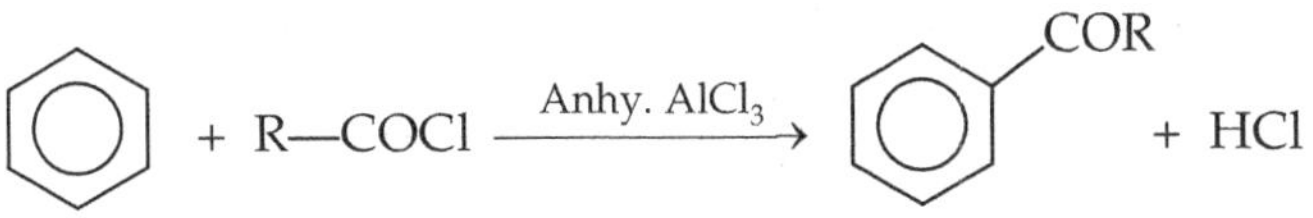

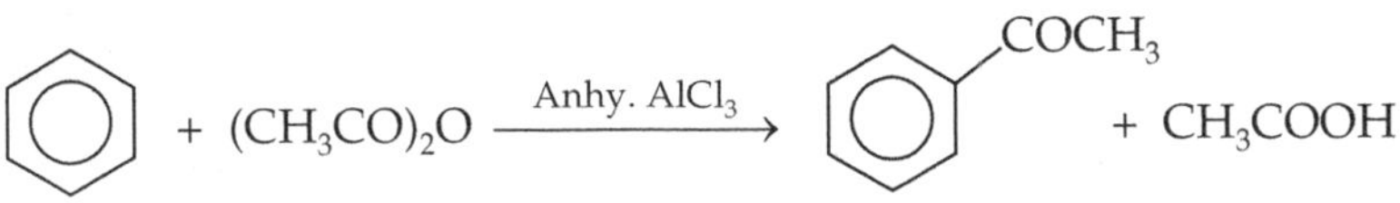

Acetophenone

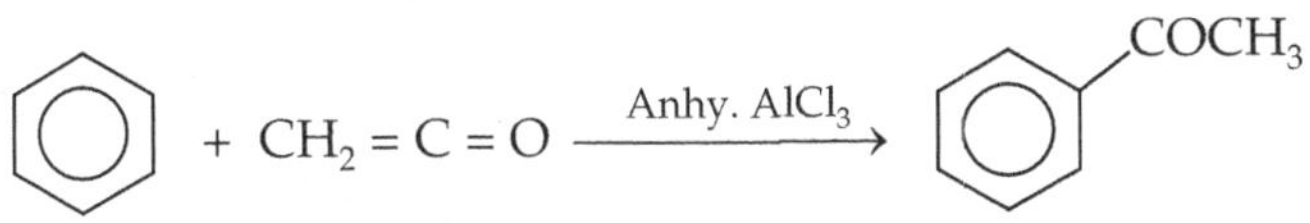

**Mechanism.** The mechanism involves the following steps:

(*i*) Generation of an electrophile

$$R—Cl + AlCl_3 \longrightarrow \underset{\text{Acylium ion}}{CH_3—\overset{\oplus}{C}=O} + [AlCl_4^{\ominus}]$$

(*ii*) Formation of σ-complex

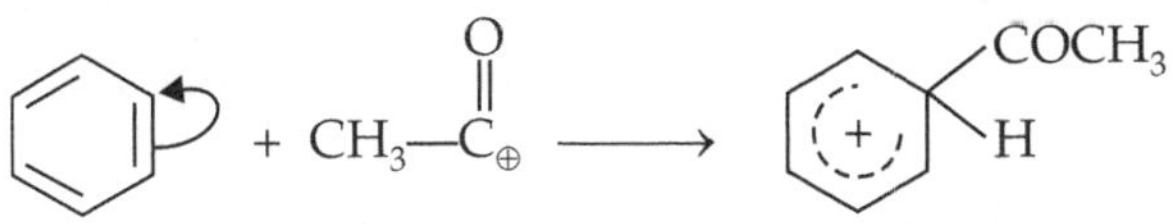

σ-complex

(*iii*) Elimination of proton

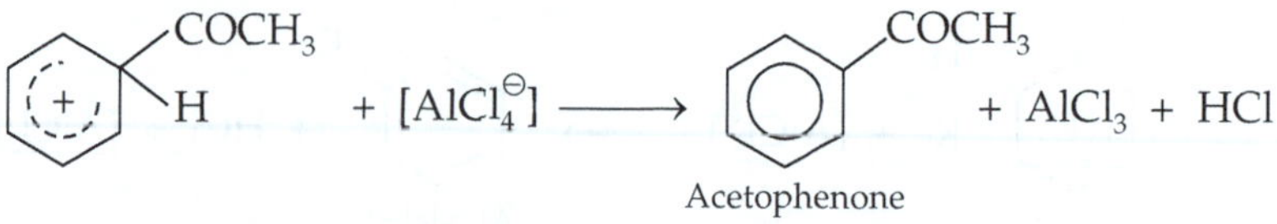


**Applications**

1. *Synthesis of hydrocarbons, ketones, acid chlorides*

(*a*)

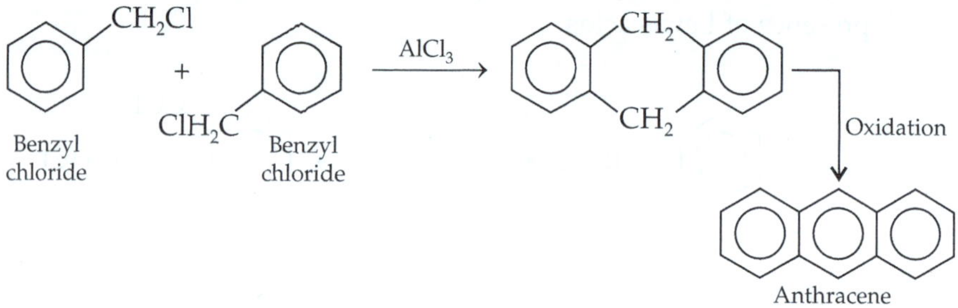


(*b*)

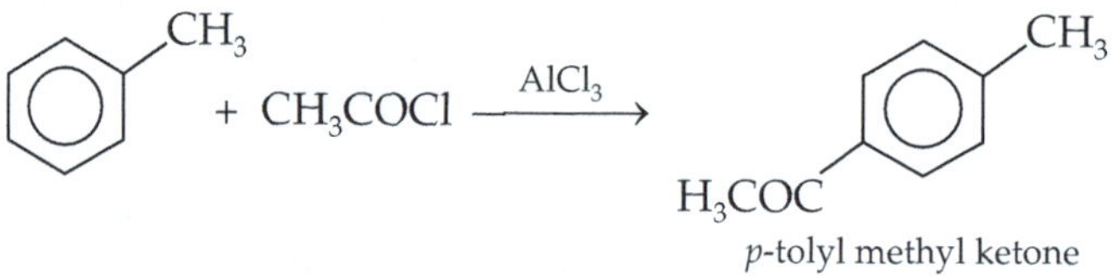


(*c*)

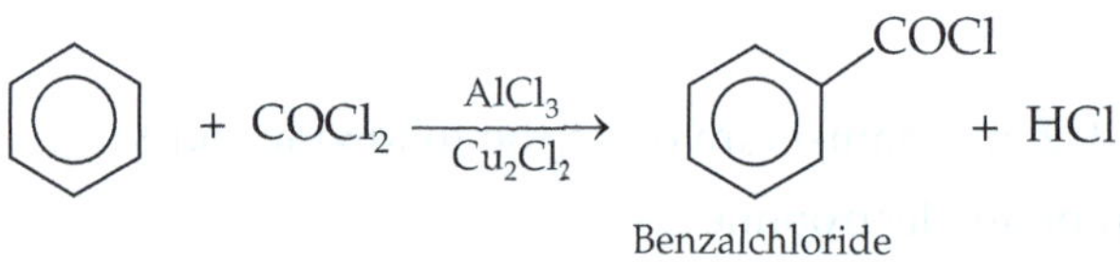


2. *Formylation*

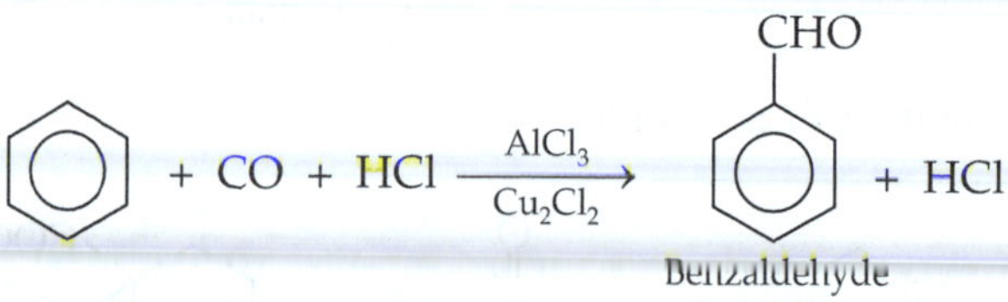


### 16.7.6 Diels-Alder Reaction

It involves the reaction between a *diene* and an unsaturated compound called *dienophile* to form a cyclic compound. It consists of 1, 4-addition conjugated diene to a second unsaturated molecule.

For example,

(*i*) 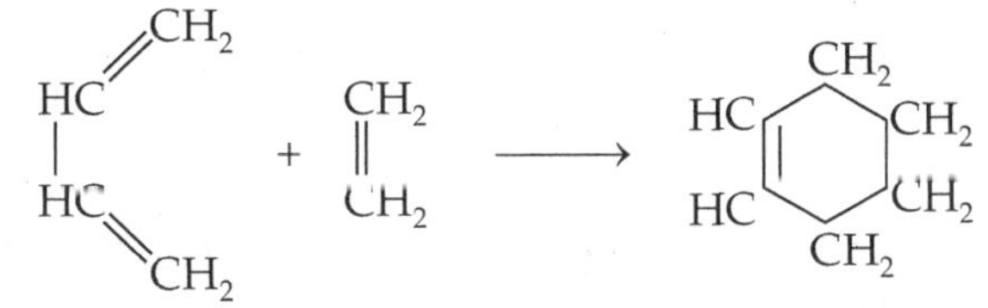

1, 3-Butadiene  Cyclohexene

(*ii*) 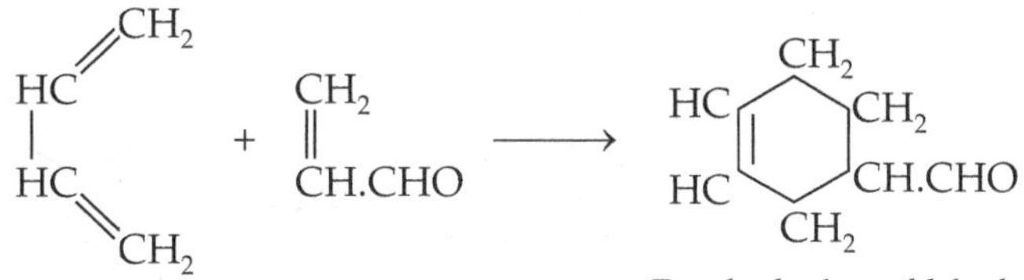

Tetrahydro benzaldehyde

**Components of Diels-Alder Reaction**

(*i*) *Dienes.* It may be conjugated diene or polyene, enyne or diyne. For example,

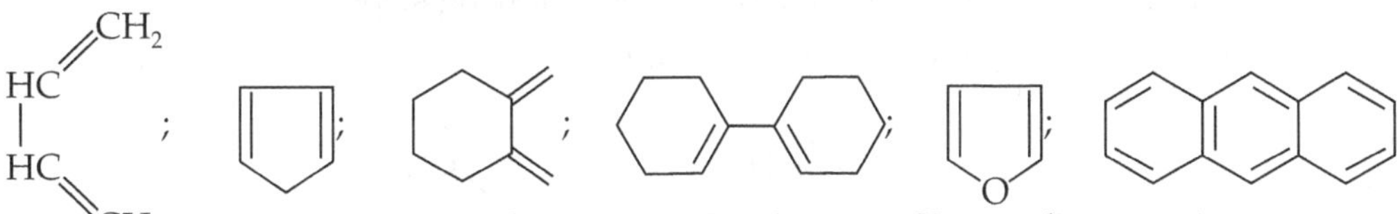

Acyclic; Alicyclic; Semicyclic; Bicyclic; Heterocyclic; Aromatic

(*ii*) *Dienophile.* These are usually α, β-unsaturated carbonyl compounds. For example,

$CH_2CH.CHO$; $C_6H_5CH = CH.CHO$; 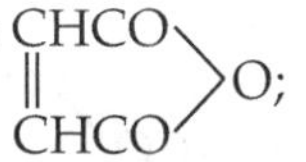; 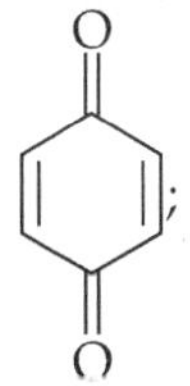; 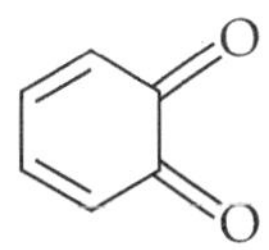

Acrylaldehyde; Cinnamaldehyde; Maleic anhydride; *p*-Benzoquinone; *o*-Benzoquinone

$CH_2 = CH.CO.CH_3$
Vinylmethylketone

### Some Examples of Diels-Alder Reaction

(*i*)

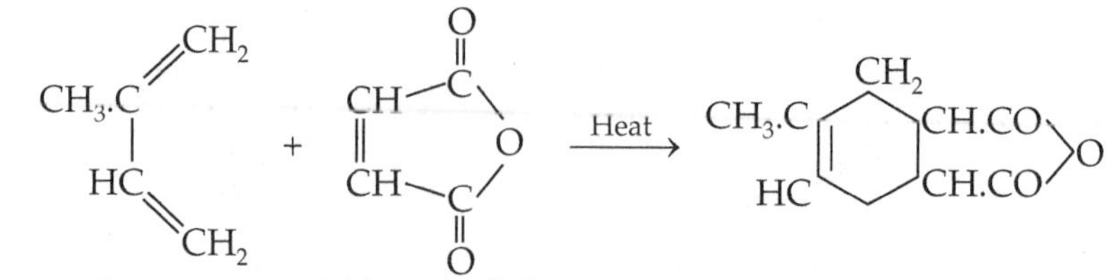

Isoprene (Diene) + Maleic anhydride (Dienophile) → Adduct

(*ii*)

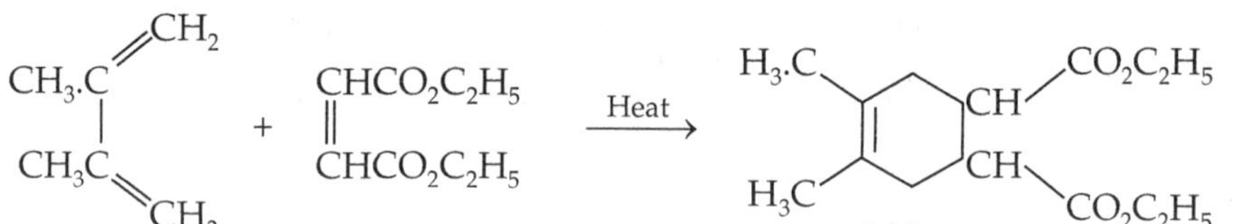

Adduct

**Mechanism.** Both diene and dienophile undergo electromeric shifts and such excited states of two molecules are created that they condense to form an adduct in 1, 4-position of the diene. It involves the following steps:

(*i*) Diene undergoing electromeric shifts

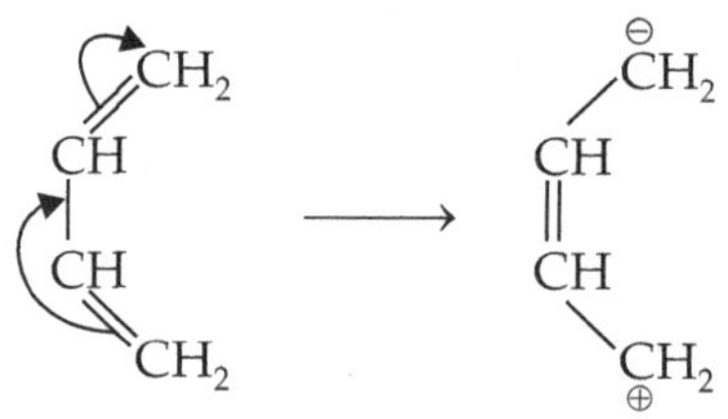

(*ii*) Dienophile undergoing electromeric shifts

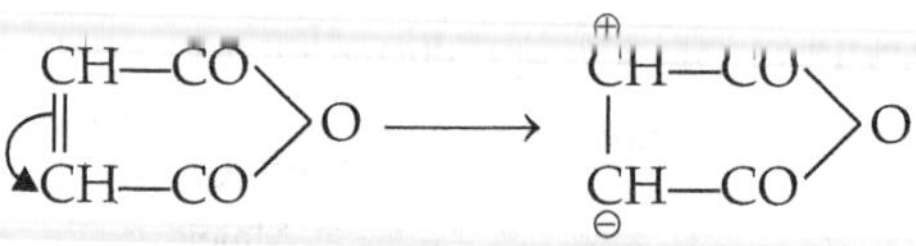

(*iii*) These excited forms of diene and dienophile combine in 1, 4-position of the diene to form an adduct, *i.e*, a six membered compound.

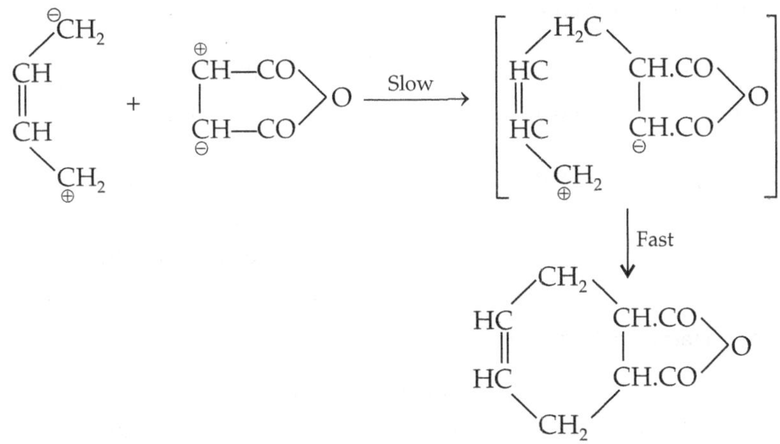

Tetrahydrophthalic anhydride (Adduct)

## Applications

1. *Synthesis of hydrocarbons*

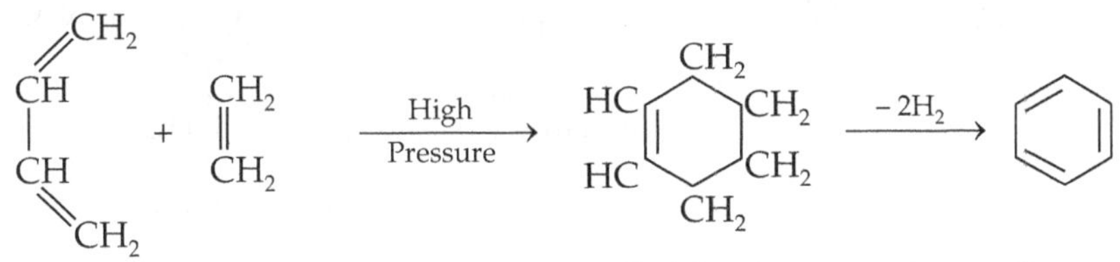

Pentadiene Cyclohexene Benzene

2. *Synthesis of aldehydes*

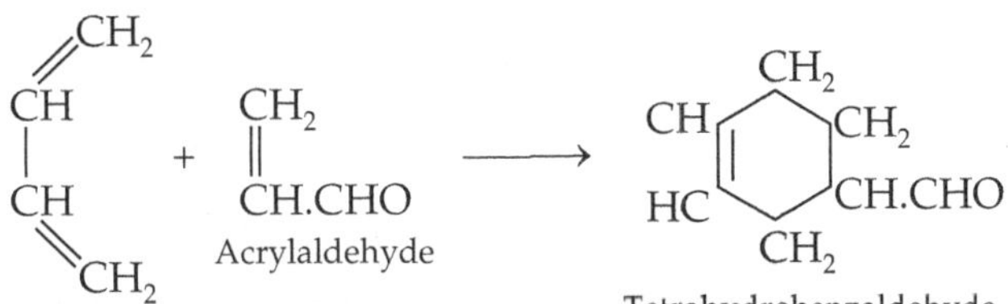

Acrylaldehyde Tetrahydrobenzaldehyde

3. *Synthesis of quinones*

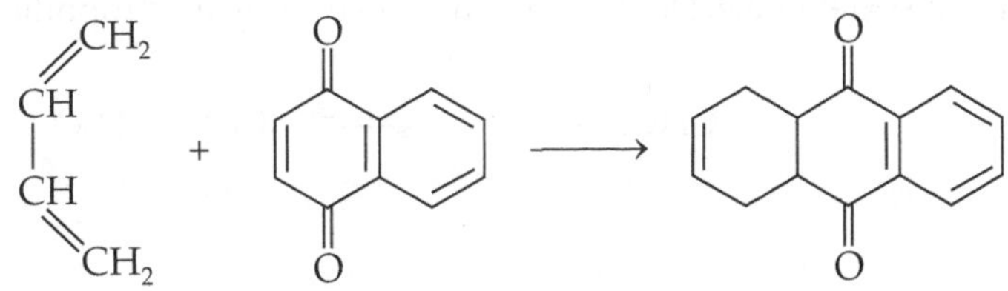

Naphthaquinone Tetrahydroanthraquinone

4. *Synthesis of acid anhydride*

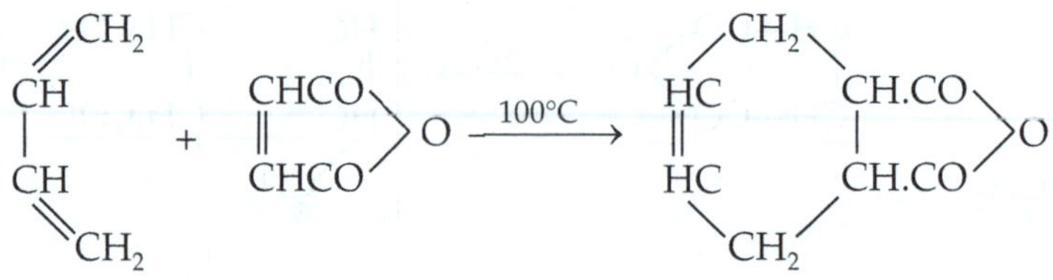

Tetrahydrophthalic anhydride

### 16.7.7 Reimer-Tiemann Reaction

When a phenol is treated with chloroform and aqueous alkali, an aldehyde group —CHO, is introduced into the aromatic ring, generally *ortho* to the —OH group. This reaction is known as *Reimer-Tiemann reaction.* For example,

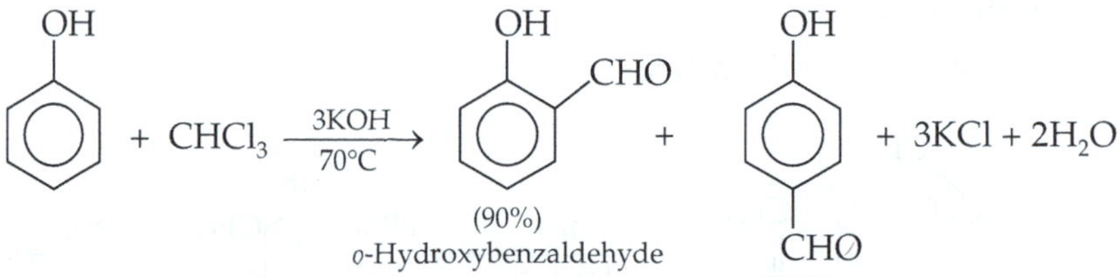

(90%) *o*-Hydroxybenzaldehyde

(10%) *p*-Hydroxybenzaldehyde

When carbon tetrachloride is used in the reaction, then a carboxylic group is introduced instead of an aldehyde group in the aromatic ring. For example,

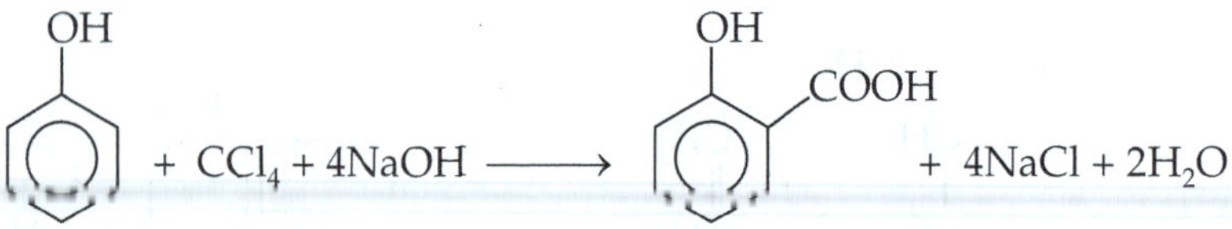

Salicylic acid

**Mechanism.** The Reimer-Tiemann reaction involves electrophilic substitution on the highly reactive phenoxide ring. The electrophilic reagent is dichlorocarbene, :$CCl_2$, generated by the action of base on chloroform. Dichlorocarbene is electrically neutral, contains a carbon atom with only a sextet of electrons, and hence is strongly electrophilic.

$$CHCl_3 + OH^{\ominus} \rightleftharpoons :CCl_3^{\ominus} + H_2O$$

$$:CCl_3^{\ominus} \longrightarrow \underset{\text{Dichloro carbene}}{:CCl_2} + Cl^{\ominus}$$

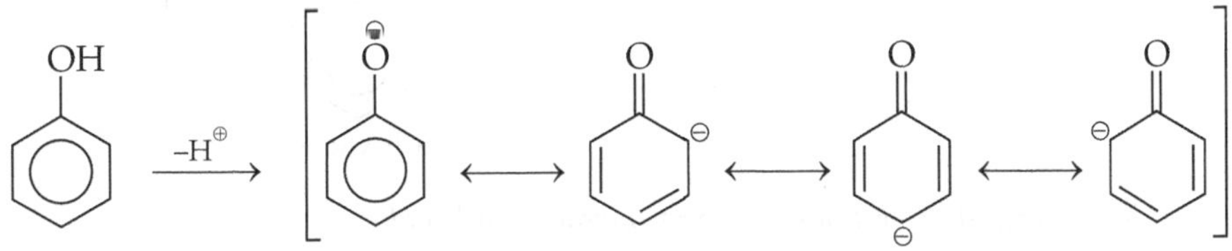

Resonating hybrids

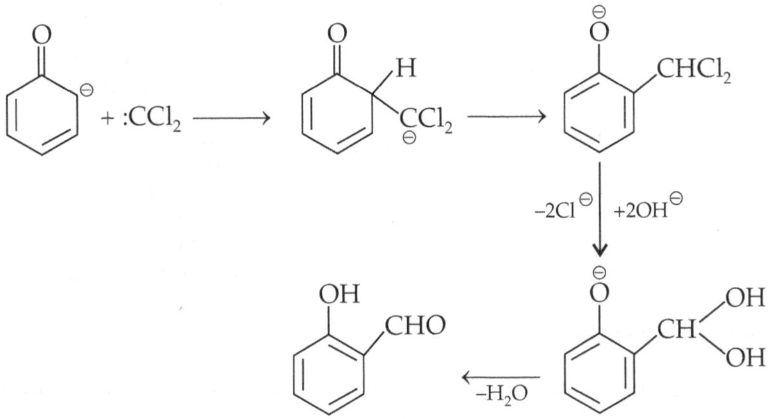

*o*-Hydroxybenzaldehyde (unstable)

The small amount of *p*-isomer formed can be separated by steam distillation.

## Applications

1. *Preparation of vanillin*

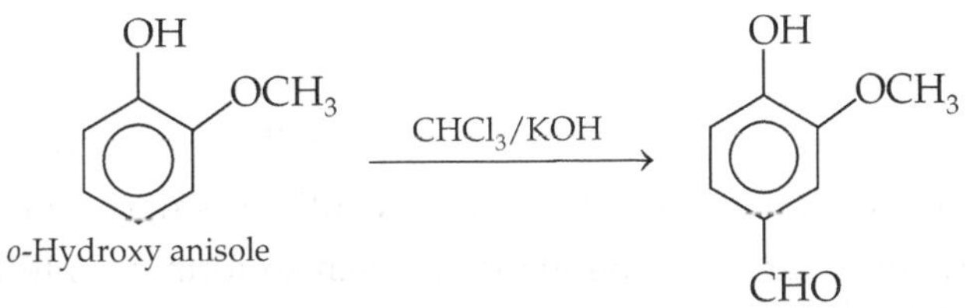

*o*-Hydroxy anisole Vanillin

2. *Preparation of salicylic acid*

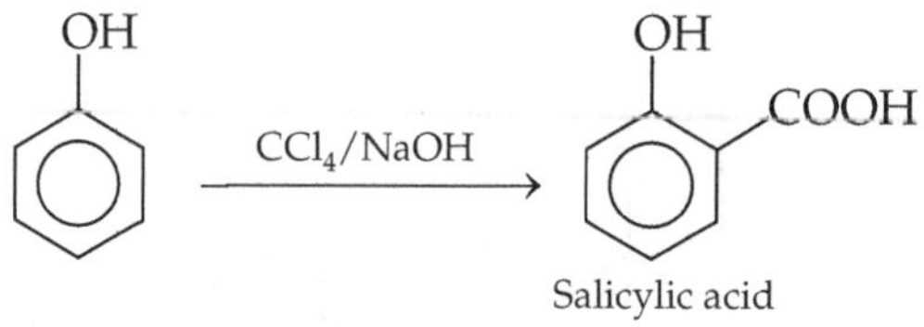


3. *Preparation of thiophene and pyrrole-2-aldehydes*

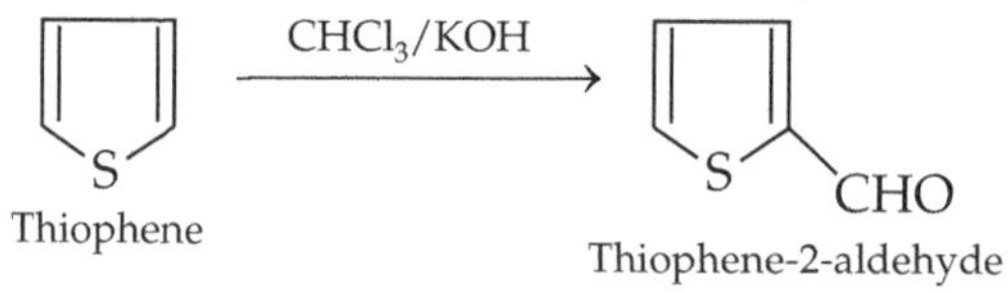


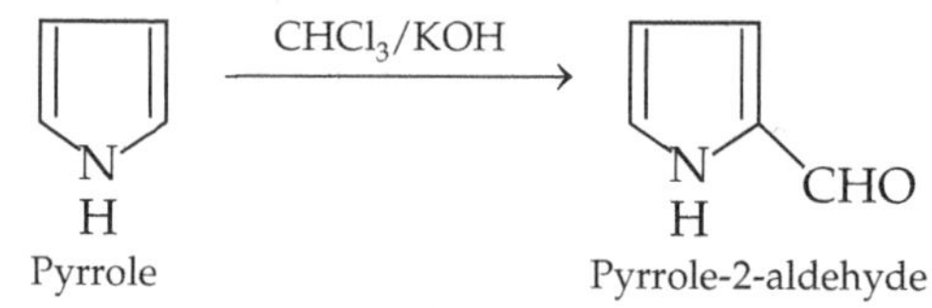


### 16.7.8 Skraup's Synthesis

This is the commercial method of preparation of quinoline and its derivatives. The reaction involves heating a primary aromatic amine with nitrobenzene or arsenic acid, glycerine, and concentrated sulphuric acid.

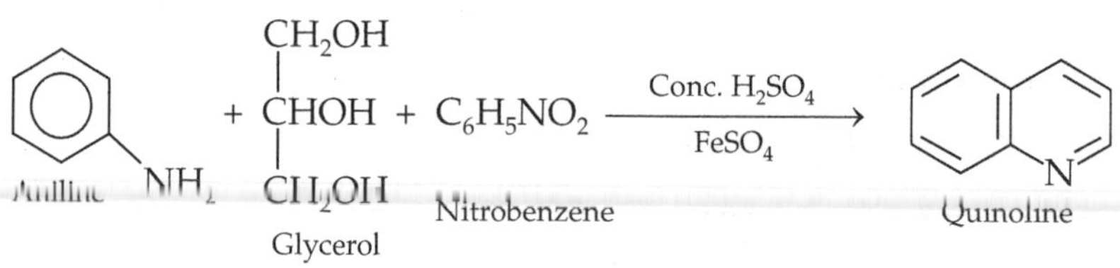


**Mechanism.** The reaction involves the following successive steps:

(*i*) Dehydration of glycerol (I) under the influence of conc. $H_2SO_4$ to form acrolein.

(*ii*) Addition of aromatic amine to acrolein to form an intermediate β-arylamino-aldeyde (II).

(*iii*) Ring closure by dehydration to form 1, 2-dihydroquinoline.

(*iv*) Oxidation of (III) to quinoline (IV).

(*i*) 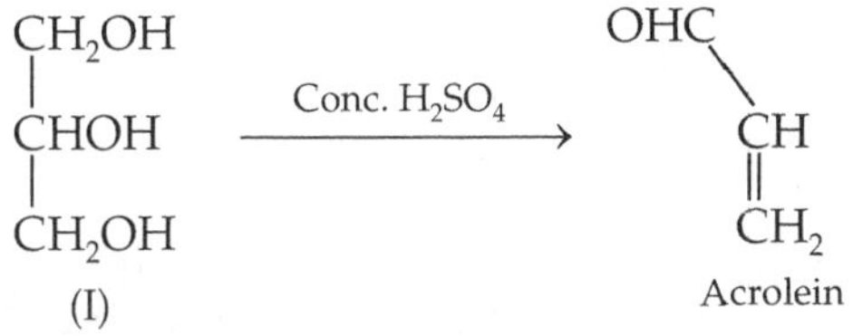

(*ii*) 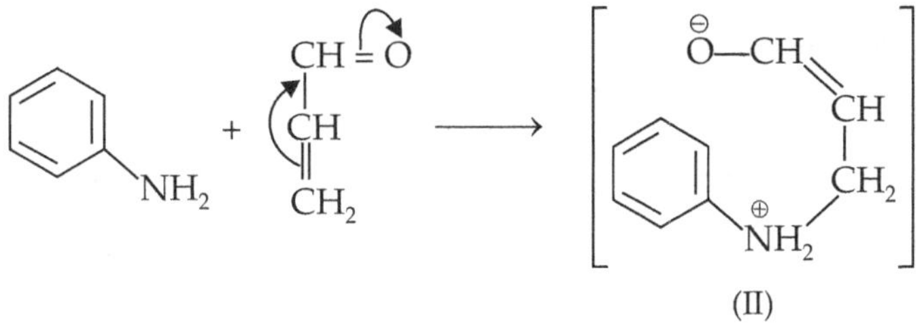

(*iii*) and (*iv*)

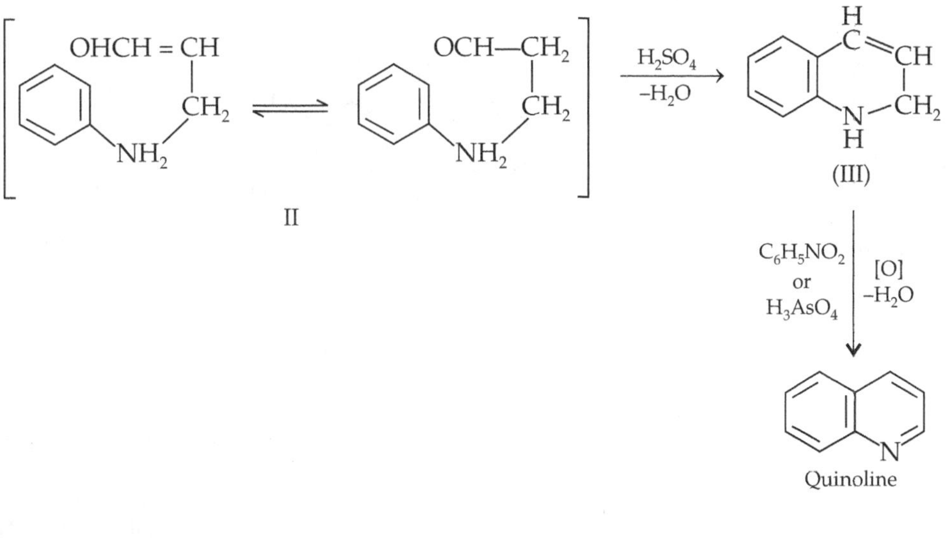

The function of nitrobenzene is to oxidize an intermediate dihydroquinoline formed during the synthesis. Glycerol acts as a condensing component. It condenses with aniline after dehydration, forming acrolein. $FeSO_4$ present in the reaction mixture reduces the violent nature of the reaction.

## Applications

1. *Synthesis of substituted quinolines*

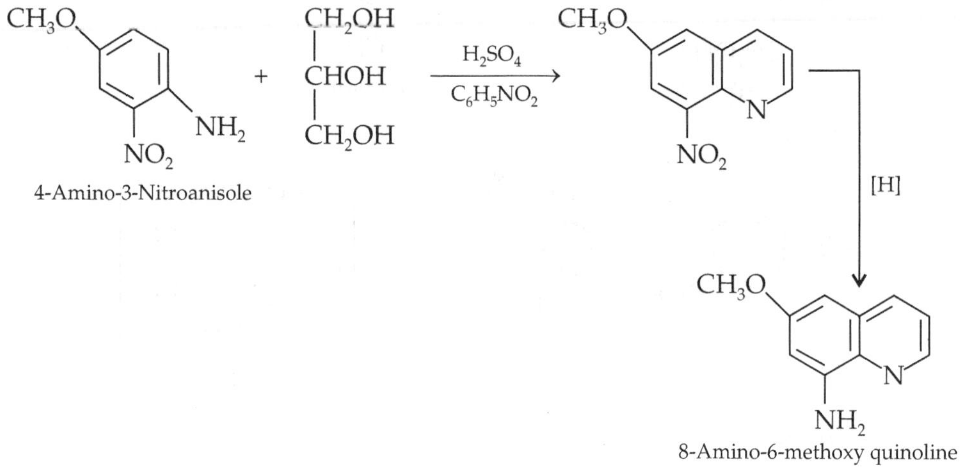


2. *Synthesis of alizarin blue*

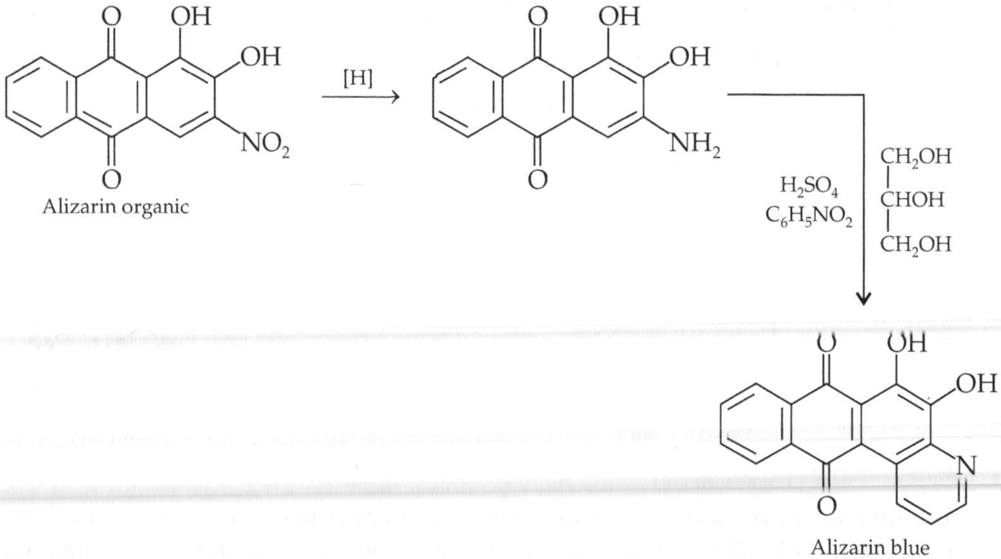

3. *Synthesis of anti-malarial pamaquin*

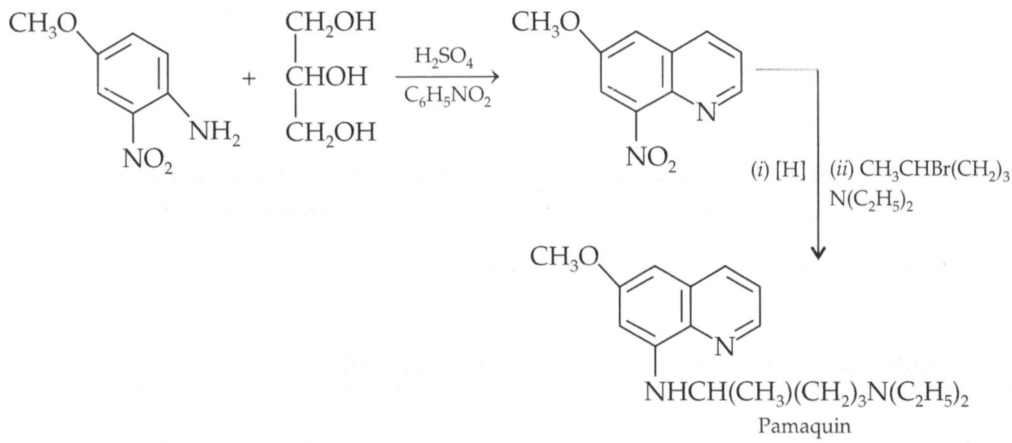


## 16.8 STEREOISOMERISM

Stereoisomerism is due to the difference in the relative positions of atoms or groups in space. *The isomers which differ in relative arrangement of atoms or groups in space around the carbon atom* are known as *stereoisomers* and the phenomenon is known as *stereoisomerism.*

*The branch of organic chemistry which deals with the three dimensional structure of the molecule and its effects on chemical behavior is known as stereochemistry.* Stereoisomers are roughly divided into two classes:

**Configurational stereoisomers,** are those *which interconvert with difficulty under normal conditions* and thus can be isolated. Interconversion in these stereoisomers usually involves a bond-breaking process. This type of isomerism exists either due to rigidity in a molecule (geometrical isomers) or due to right-or-left-handed arrangement of atoms around carbon atom (optical isomers).

**Conformational stereoisomers,** are those *which interconvert easily at room temperature* through rotations about single bonds.

Before studying the three types of stereoisomers (optical, geometric and conformers) we must consider the three dimensional structure of tetrahedral carbon.

*Kekule* (1858) proposed that four valencies of carbon are distributed in one plane and hence the carbon compound can be represented as a square with the carbon atom at the center of the square (Figure 16.5(*a*)). *Le Bel* and *Van't Hoff* (1874) proposed separately that the four valencies of carbon distributed in three dimensions and if we join the ends of these four valencies by straight lines, we get a regular *tetrahedron* (Figure 16.5*b*). The tetrahedral structure of carbon is supported by X-ray spectroscopy, electron diffraction methods, etc.

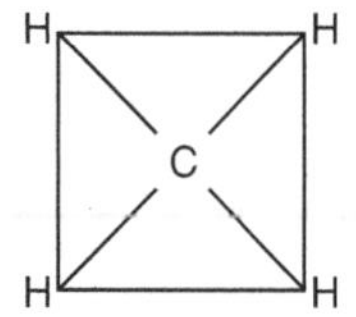


*FIGURE 16.5* **(*a*) Plane flat or two dimensional structure of methane**

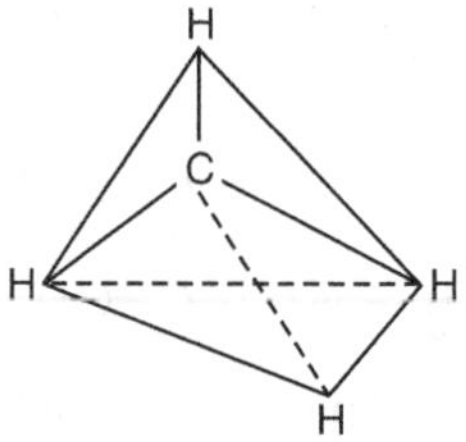


*FIGURE 16.5* **(*b*) Tetrahedral or three dimensional structure of methane**

Now let us discuss three types of isomerism.

## 16.9 ENANTIOMERISM OR OPTICAL ISOMERISM

When the two isomers are mirror images of each other and they differ in their behaviour towards the plane polarized light, such isomers are called *enantiomers* or *opitcal isomers* and the phenomenon is known as *enantiomerism* or *optical isomerism.*

### 16.9.1 Optical Activity

A beam of ordinary light consists of waves vibrating in all planes in space. When such light is passed through a Nicol prism, the vibrations in all other planes except one are cut off, such light is called *plane polarized light (PPL)* and the Nicol prism used to make ordinary light into *PPL* is called *polarizer.*

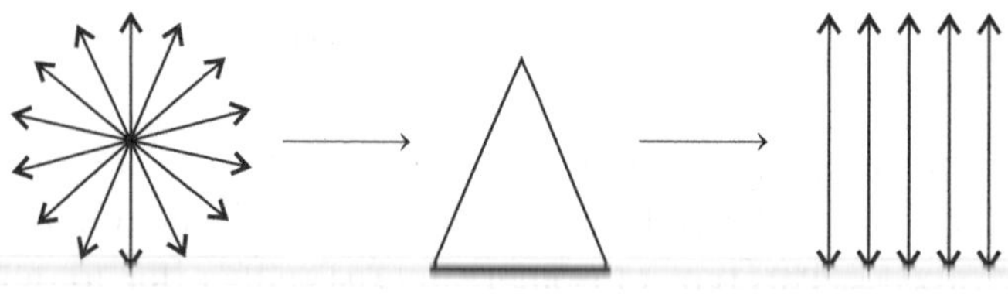

(*a*) Ordinary light waves vibrating in all directions (*b*) Nicol prism (*c*) Plane polarized light waves vibrating in one direction only

*FIGURE 16.6*

When the solutions of certain organic compounds are placed in the path of *PPL*, they rotate the path of *PPL* either to the left or to the right. Such substances which rotate the *PPL* are called *optically active* compounds and the property by virtue of which it rotates the *PPL* is called *optical activity*. For example, lactic acid, tartaric acid, glucose, fructose, etc. are optically active compounds.

An optically active compound may rotate the *PPL* in two directions, *i.e.*, toward the right or left side. If the rotation is toward the right, the compound is called *dextrorotatory* and is denoted by *d*- or (+) and if the rotation is toward the left, the compound is called *laevorotatory* and is denoted by *l*- or (–). The angle through which the plane of the polarized light is rotated is called the *angle of rotation.*

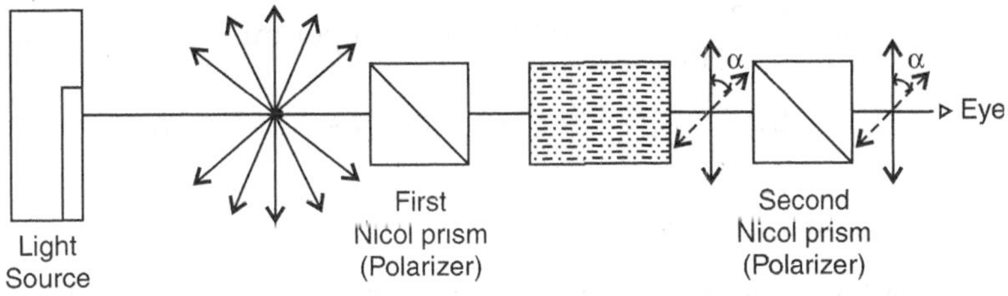


*FIGURE 16.7* **Diagrammatic representation of a polarimeter**

The specific rotation of a compound is measured by an apparatus called *polarimeter* (Figure 16.7) consisting of a source of light, Nicol prism (polarizer), a polarimeter tube containing solution of the compound, and another Nicol prism (analyzer). The optical activity is measured in terms of specific rotation (α) and is given by the equation:

$$[\alpha]_{\lambda}^{t} = \frac{\alpha}{l \times d}$$

where $t$ = temperature,

$\lambda$ = wavelength of light,

$\alpha$ = observed angle of rotation,

$l$ = length of the liquid or solution column,

$d$ = density of liquid in g/ml of the solution.

Sometimes, *molecular rotation (M)* is used in preference to specific rotation. It is related to specific rotation by the equation

$$[M]_{\lambda}^{t} = \frac{[\alpha]_{\lambda}^{t} \times M}{100}$$

where M is the molecular weight of the substance.

**Example 1.** *The concentration of cholesterol dissolved in chloroform is 6.15 gm per 100 ml of solution.*

(*i*) *A part of this solution in a 5 cm polarimeter tube causes an observed rotation of – 1.2°. Find the specific rotation of cholesterol.*

(*ii*) *What would be the observed rotation if the same solution were placed in a 10 cm tube?*

(*iii*) *What would be the observed rotation if 10 ml of the solution were diluted to 20 ml and placed in a 5 cm tube?*

**Solution.** (*i*) Specific rotation $[(\alpha)] = \frac{\alpha}{l \times d} = \frac{-1.2°}{0.5 \times 6.15 \times 10^{-2}} = -39°$

(*ii*) Observed rotation $\alpha = [\alpha] \times l \times d = -39° \times 1 \times 6.15 \times 10^{2} = -2.4°$

(*iii*) Given 6.15 g of cholesterol dissolved in 100 ml

so, 10 ml contains $61.5 \times 10^{-2}$ gm of cholesterol on dilution to 20 ml, $61.5 \times 10^{-2}$ gm of cholesterol present in 20 ml.

$$\therefore \qquad d = \frac{6.15 \times 10^{-2}}{20} = 3.075 \times 10^{-2} \text{ gm per ml}$$

$$\text{Now,} \qquad \alpha = [\alpha] \times l \times d = -39^\circ \times 0.5 \times 3.075 \times 10^{-2} = -6.0^\circ$$

### 16.9.2 Optical Activity and Chirality

An object which cannot be superimposed on its mirror image is known as chiral (*cheir* meaning handedness). This characteristic of nonsuperimposability of a structure on its mirror image is called *chirality*. The concept of chirality can be understood by the following examples:

(*i*) The most common example of a chiral object is the human hand (Figure 16.8). If we place our left hand in front of a mirror, the image looks like that of a right hand. When we try to superimpose left and right hands, it is observed that they cannot be superimposed and hence are considered chiral. Similarly, gloves and shoes are also chiral objects.

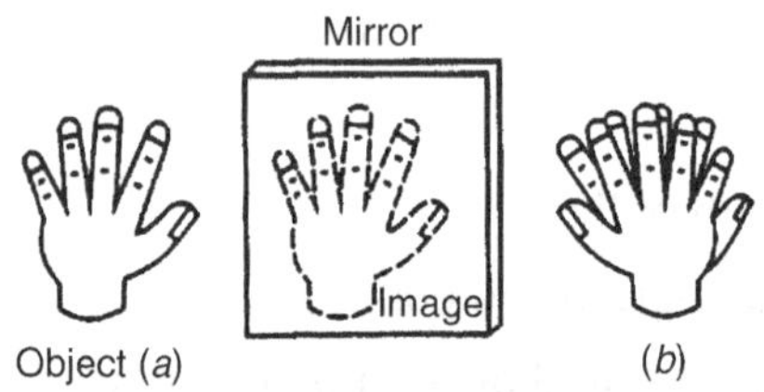


*FIGURE 16.8* **(*a*) Object and mirror image of a hand (*b*) one hand is placed over the other to illustrate nonsuperimposability of a chiral object.**

(*ii*) Consider the alphabet letters *'P'* and *'A'*. The letter *'P'* and its mirror image are not superimposable and it is chiral, while the letter *'A'* and its mirror image are superimposable on each other, hence are achiral.

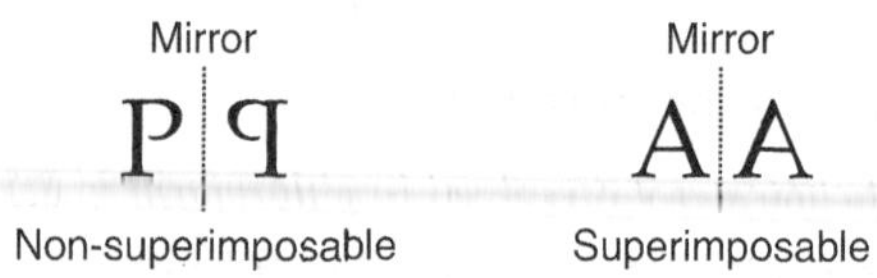


There is a close relationship between *chirality* and *optical activity* and quite often the descriptions optically active and chiral are used interchangeably. But it should be remembered that the optical activity refers to one property of chiral molecules, namely the rotation of plane polarized light.

Chirality in molecules is due to the presence of an $sp^3$ carbon atom with four different groups attached to it. Such a molecules has no plane of symmetry and exists as a pair of enantiomers. *A chiral carbon is thus one in which all the four substituents are different.*

### 16.9.3 Chirality and Elements of Symmetry

A chiral object does not have any elements of symmetry: (*i*) a plane of symmetry, (*ii*) a center of symmetry, and (*iii*) an alternating axis of symmetry. If there is at least one symmetry element present, the molecule will not show enantiomerism.

(*a*) **A plane of symmetry.** *A plane which divides an object or a molecule into two exactly identical halves is called plane of symmetry.* A molecule with at least a plane of symmetry can be superimposed on its mirror image and is a chiral. A molecule that does not have a plane of symmetry is usually chiral and it cannot be superimposed upon its mirror image. For example, *meso*-tartaric acid having plane of symmetry is optically inactive.

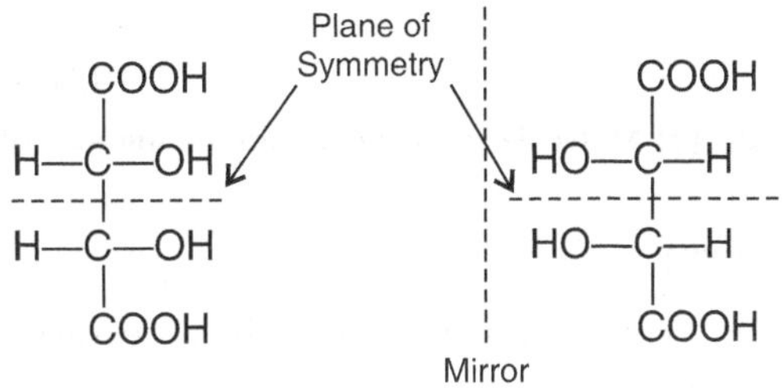


*FIGURE 16.9* **Molecules containing plane of symmetry are optically inactive.**

(*b*) **A center of symmetry.** *A point at which all the straight lines joining identical points in the molecule cross each other is called a center of symmetry.*

It is usually present only in an even-membered ring. For example,

(*i*) The molecule, 2, 4-dimethyl cyclobutane-1, 3-dicarboxylic acid has a center of symmetry (at center of cyclobutane ring) and is therefore optically inactive (Figure 16.6(*a*)).

(*ii*) In compound II of Fig. 16.10(*a*) the same group of atoms (like H) is present, showing the center of symmetry, hence the compound is optically inactive.

(*c*) **Alternating axis of symmetry.** A molecule is said to possess an alternating axis of symmetry if an arrangement identical to the original is obtained when the molecule is rotated through an angle $360°/n$ about an axis followed by reflection in a plane perpendicular to this axis (Fig. 16.10(*b*)). A molecule having *n*-fold axis of symmetry is optically inactive. For example, if the molecule *A* is rotated through 180° about the axis, passing through the center of the molecule, and then reflected in the mirror perpendicular to this axis, then an arrangement superimposable on '*A*' is obtained. Thus, the molecule (*A*) is said to possess $360°/180° = 2$ fold axis of symmetry and will be optically inactive.

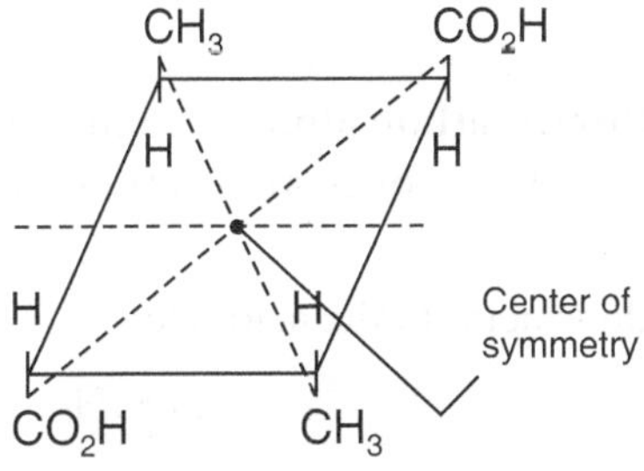


2,4-Dimethylcyclobutane-1,3-dicarboxylic acid

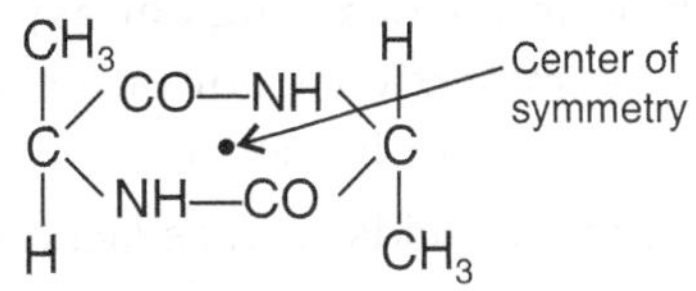


II

*FIGURE 16.10* (*a*) **Molecules containing a center of symmetry are optically inactive.**

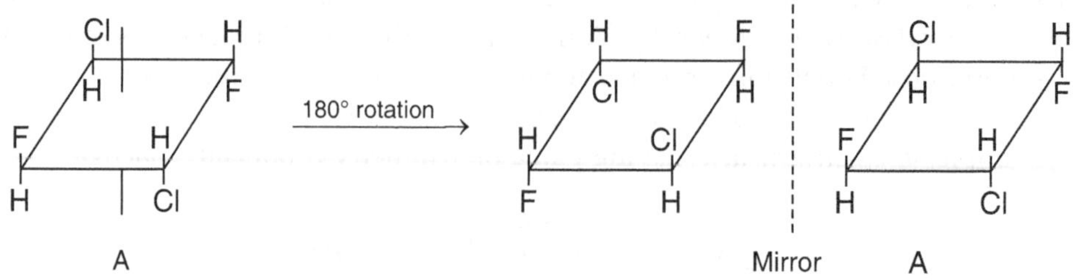


*FIGURE 16.10* **(*b*) Molecule '*A*' showing alternating axis of symmetry.**

### 16.9.4 Cause of Optical Activity

From the foregoing discussions it can be concluded that the necessary and sufficient conditions for a molecule to show optical activity are:

(*i*) **Molecular dissymmetry.** It is the fundamental condition for optical activity. A molecule that does not possess any elements of symmetry like plane, center, or alternating axis of symmetry is called *dissymmetric molecule* and is optically active. A molecule can exist in enantiomeric forms only when it has a molecular dissymmetry.

(*ii*) **Nonsuperimposability.** The molecule should be nonsuperimposable on its mirror image.

So, in describing optical activity, the concept of molecular dissymmetry and non-superimposability on mirror image are important as these are not related to chiral carbon atoms only. Even when a chiral carbon atom is present, a compound need not necessarily be optically active as in the case of *meso*-tartaric acid. On the other hand, there are optically active compounds having no chiral carbon atoms, such as allenes, biphenyls, spiranes, etc. (Please refer to section 16)

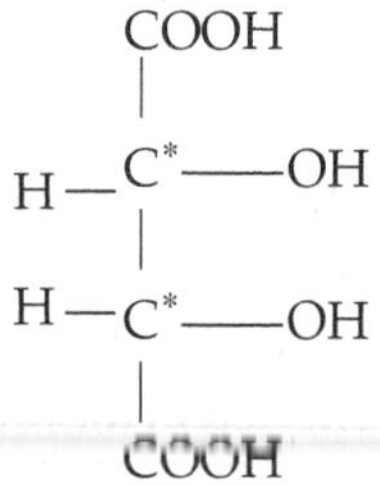


*Meso*-tartaric acid
(Optically inactive)

$$H_3C—HC=C=HC—CH_3$$

2, 3-Pentadiene (Optically active)

### 16.9.5 Examples of Optical Isomerism

1. **Compounds containing one asymmetric (chiral) carbon atom.** A compound containing one asymmetric carbon atom can exist in $2^1 = 2$ optically active isomers called enantiomorphs. For example,

   (*a*) Hydroxy acids such as lactic acid, mandelic acid, malic acid, etc.

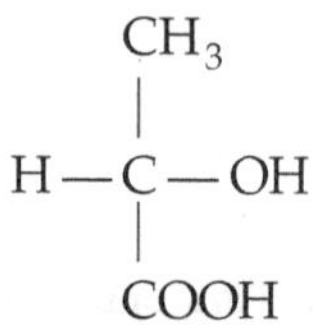


Lactic acid

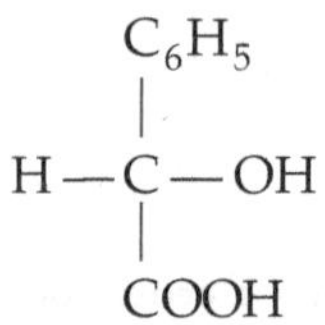


Mandelic acid

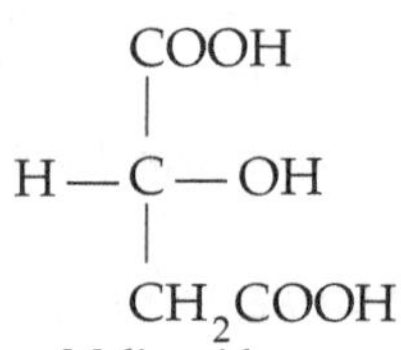


Malic acid

(*b*) Amino acids such as alanine 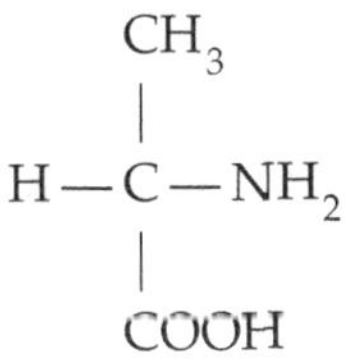

(*c*) $C_6H_5CHDCH_3$, α-Deutroethyl benzene

*Lactic acid.* It has one chiral carbon atom and according to Fischer projection formula, it can be represented as:

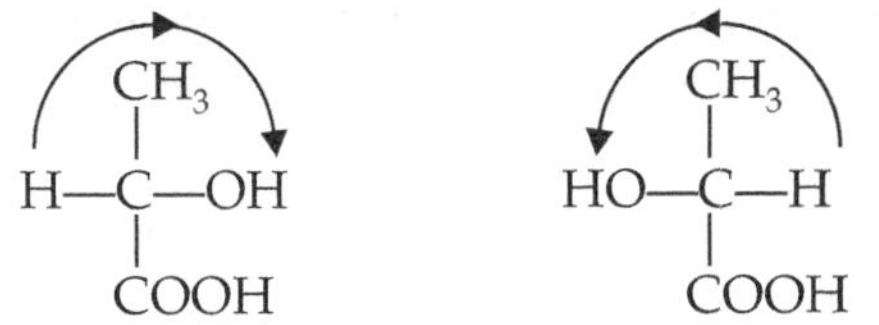

I — *d*- or (+) - Lactic Acid

II — *l*- or (–) - Lactic Acid

***FIGURE 16.11*** **Enantiomers of lactic acid.**

These two forms are nonsuperimposable mirror images of each other. Thus, lactic acid exists in the following three forms:

(*i*) *d- or (+)-lactic acid.* It rotates the *PPL* toward right as shown in model from H to OH *via* $CH_3$. It is obtained from the meat extract.

(*ii*) *l- or (–)-lactic acid.* It rotates the *PPL* toward the left side as shown in the model from the H to the OH group through the group of low molecular weight, *i.e.*, the $CH_3$ group. It can be obtained during fermentation of sucrose by *Bacillus aceti laevolactic.*

(*iii*) *dl- (±) racemic (r) lactic acid.* It is an equivalent mixture of *d*- and *l*-lactic acids. Since the rotation of *d*- and *l*-lactic acids are of the same degree but in opposite directions, thus rotation of one form is compensated for (neutralized) by the other form and the racemic mixture becomes optically inactive. This type of inactivity is said to be due to *external compensation*. It can be obtained from sour milk or fermentation of sucrose.

**TABLE 16.2 Physical Properties of Lactic Acid**

| *Form* | *mp(°C)* | *Density* | $[\alpha]_D^{25°C}$ |
|---|---|---|---|
| *d*- or (+) | 26 | 1.248 | +2.24 |
| *l*- or (–) | 26 | 1.246 | – 2.24 |
| *dl*– or (±) | 25 | 1.248 | 0 |

2. **Compounds containing two similar chiral carbon atoms.** The number of stereoisomers in a compound containing '*n*' number of similar carbon atoms is always less than $2^n$. For example, consider the case of tartaric acid. Theoretically, tartaric acid may be represented in the following four forms:

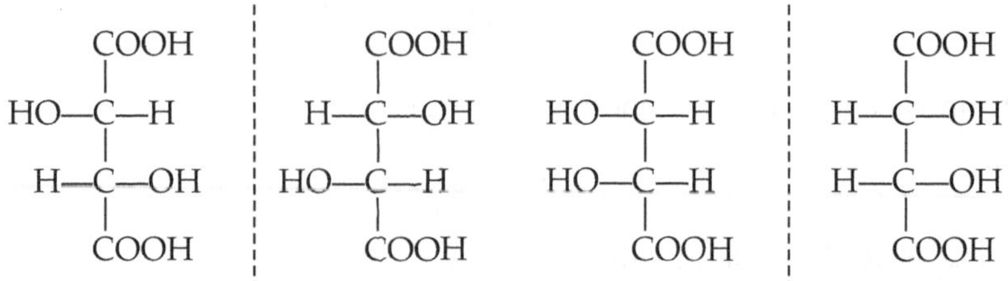

I II III IV

Nonsuperimposable enantiomers  Superimposable enantiomers

Forms I and II are the mirror images of each other and are nonsuperimposable and are not interconvertible by rotation about a C—C bond. Thus, these two form an enantiomorphic pair.

Structures III and IV do have an object mirror relationship, but if one of these is rotated through 180° within the plane of paper, the other structure is obtained, *i.e.*, III = IV. Thus, these two are identical forms. The above structures may be represented by the Fischer projection formula and are described below:

Tartaric acid occurs in three stereoisomeric forms, namely, dextro, laevo, and meso. The dextro and laevo (A and B) are optically active forms, as they rotate *PPL* in clockwise and anti-clockwise directions, respectively. In case of meso form (C) and (D), one asymmetric carbon is rotating *PPL* in a clockwise direction while the other asymmetric carbon rotates *PPL* in an anti-clockwise direction. Thus, dextro rotation of one half of the molecule is compensated, by the laevo rotation of the other half and thus the molecule is optically inactive. Since this compensation occurs within the molecule itself, so the molecule is said to be inactive due to *internal compensation.* Tartaric acid also occurs in a racemic form or *dl*- (±) form which is optically inactive due to *external compensation.*

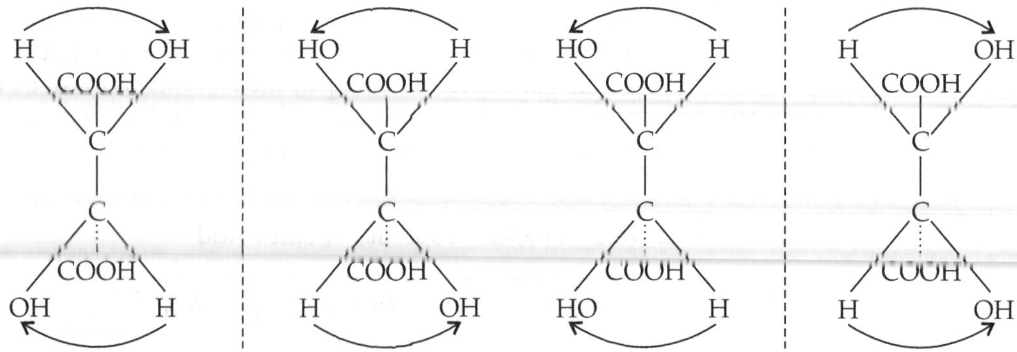

A *d*-(+) -(170°C)  B *l*-(–) -(170°C)  C D Meso-tartaric acid, *mp*-140°C

Thus, tartaric acid occurs in following stereoisomeric forms:

(*i*) *d-tartaric acid* (m.p. 170°C) occurs in free state in nature.

(*ii*) *l-tartaric acid* (m.p. 170°C) does not occur naturally, may be obtained by resolution of racemic form.

(*iii*) *meso-tartaric acid* (m.p. 140°C) obtained from mother liquor in preparation of *d*-tartaric acid.

(*iv*) *Racemic tartaric acid* (m.p. 206°C) obtained by heating *d*-tartaric acid with alkali.

3. **Compounds containing two dissimilar asymmetric carbon atoms (Distereomers).** Such isomers occur in four ($2^2 = 4$) stereoisomeric forms. For example, cinnamic acid dibromide, bromochloro succinic acid, etc. occur in four optically active isomers.

(*i*) *Cinnamic acid dibromide.* It occurs in the following enantiomers:

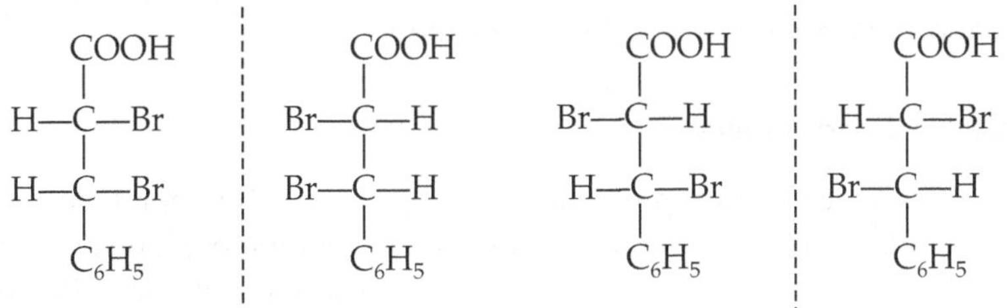

I II III IV

Enantiomers Enantiomers

Number of racemic modifications = $2^{2-1}$ = 2, racemic mixtures formed by mixing equivalent amounts of I and II, and III and IV.

(*ii*) *Bromochloro succinic acid*

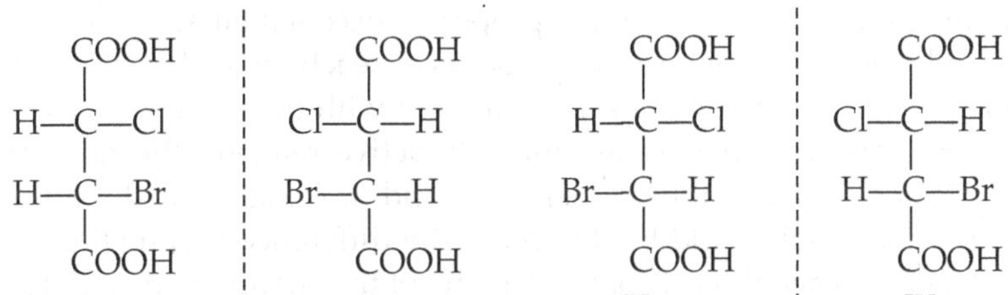

I II III IV

Enantiomers Enantiomers

Structures I and III, I and IV, II and III, II and IV are stereoisomers of the same compound but are not the mirror images of each other, such pairs are known as *distereomers.*

*Distereomers* are thus stereoisomers that are not enantiomers, and they have the same configuration at one chiral carbon atom but a different configuration on the other. Distereomers have different physical properties (it may be recalled that enantiomers have similar physical properties except the opposite sign of specific rotation) and similar but not identical chemical properties. They have different boiling points, melting points, refractive indices, solubilities, densities, molecular shape, and polarity, so they can be separated from each other by fractional distillation, chromatography, etc. and it may be recalled that enantiomers cannot be separated by these methods. Consider another example, 2-chloro-3-hydroxybutanedioic acid, exists in four stereoisomeric forms, *i.e.*, as two pairs of enantiomers.

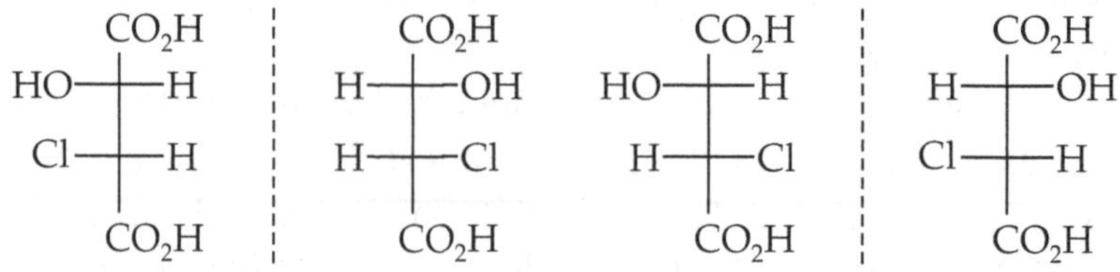

I II III IV

Of these structures, I, II and III, IV constitute enantiomorphic pairs, while the structures I, III and II, IV are related as distereoisomers.

### 16.9.6 Properties of Enantiomers

(*i*) Enantiomers have identical physical properties like melting point, boiling point, density, refractive index, etc. but differ in the direction of rotation of plane polarized light. One of the enantiomer rotates the plane polarized light to the right and the other to the same magnitude but to the left. The two enantiomers are designated as dextrorotatory, *d* or (+) and laevorotatory, *l* or (–) respectively. For example, *tartaric acids.*

| *Enantiomer* | *Specific rotation* | *MP* | *Density* |
|---|---|---|---|
| (+) | + 11.98° | 170° | 1.76 |
| (–) | – 11.98° | 170° | 1.76 |

(*ii*) Enantiomers have identical chemical properties except their action toward optically active compounds. For example, ordinary chemical reactions of lactic acid will be given by both of the enantiomers in the same way and with the same rate. However, when two enantiomers are treated with an optically active reagent, the rates are different. For example, rate of esterification of (+)-lactic acid or (–)-lactic acid with the same alcohol, (+)-sec-butyl alcohol would be different. The difference is sometimes so much that an optically active reagent reacts only with one of the enantiomers and does not react at all with the other. Such situations are very common in biological system; For example,

(*a*) (+)-glucose is easily metabolized in the animal system and is fermented by yeast while (–)-glucose is neither metabolized by animal system nor undergoes fermentation.

(*b*) (+)-tartaric acid is readily consumed by the mold *penicillium glaucum* while the (–)-tartaric acid is not.

(*iii*) When the two enantiomers are mixed in equimolecular quantities, it results in the formation of an optically inactive compound called *racemic, dl* or (±)-form.

**TABLE 16.3 Differences between Distereomers and Enantiomers**

| ***Distereomers*** | ***Enantiomers*** |
|---|---|
| (*i*) These have different physical properties. | (*i*) These have identical physical properties. |
| (*ii*) These show similar but not identical chemical properties. | (*ii*) Enantiomers have identical chemical properties. |
| (*iii*) These can be easily separated by fractional distillation, chromatography, etc. | (*iii*) These cannot be separated by physical methods. |

### 16.9.7 The Racemic Modification

A mixture of equimolecular amounts of a pair of enantiomers is known as racemic modification. A racemic modification will be optically inactive, because the rotation caused by one half of the mixture is exactly cancelled by the opposite rotation displayed by the other half. A racemic mixture is denoted by the prefix *dl*- or (±). It may be prepared in various ways as:

(*i*) *By mixing equal amounts of the two enantiomers.* It is the most trivial method of forming a racemic modification and is achieved by the intimate mixing of exactly equal amounts of the two enantiomers.

(*ii*) *By synthesis.* The synthesis of chiral compounds from achiral compounds in the absence of optically active reagent always produces a racemic mixture.

(*iii*) *By racemization.* It is the process of converting an optically active compound into the racemic modification by the application of heat, light or chemical reagents.

**Resolution or Separation of Racemic Modification.** The process of separation of enantiomers in a racemic modification is known as *resolution*. Some of methods of resolution are:

(*i*) *Mechanical separation.* It involves separation of two sets by hand picking with the help of a pair of tweezers and a lens. This is a painstaking method applicable to those cases which give well defined crystals on crystallization, that can be distinguished as enantiomers.

(*ii*) *Biochemical method.* This method involves the use of micro-organisms such as *Penicilium glaucum* which consumes only one type of enantiomer (*d*-tartaric acid) leaving behind tartaric acid. This method is not very satisfactory as one enantiomer is destroyed.

(*iii*) *Chemical method.* In this method a racemic modification is reacted with an optically active reagent to give a pair of non-identical derivatives known as distereomers which differ in physical properties and hence can be separated by fractional distillation, crystallization, etc. For example, racemic alcohol can be separated as:

$$\left.\begin{array}{l}(+)\text{ - alcohol}\\(-)\text{ - alcohol}\end{array}\right\} + (+)\text{ -acid} \rightarrow \begin{array}{l}(+)\text{ - }(+)\text{ ester} \xrightarrow{HO^{\ominus}} (+)\text{ - alcohol}\\(-)\text{ - }(+)\text{ ester} \xrightarrow{HO^{\ominus}} (-)\text{ - alcohol}\end{array}$$

Distereomers
(separable)

This method is widely used and has the advantage that both enantiomers are obtained.

### 16.9.8 R and S System of Nomenclature

This system was developed by *Cahn, Ingold,* and *Prelog* and is based on the actual three dimensional formula. This system involves the following rules:

(*i*) The four different groups/atoms attached to chiral carbon atoms are arranged in order of priority, which is determined by their atomic numbers. The atom of highest atomic number gets highest priority. For example, in IBrCHCl, I(Z = 53) gets highest priority and H(Z = 1) gets least priority. In case chiral carbon atom has two isotopic atoms, the isotope of higher mass number gets priority, *e.g.*, among deuterium (D) and hydrogen (H), the former gets priority.

(*ii*) If two same atoms are attached with the chiral carbon atom then priority is decided by the atomic number of the next atom, *e.g.*,

$$\underset{t.\text{butyl}}{(CH_3)_3C-} > \underset{\text{Isopropyl}}{(CH_3)_2CH-} > \underset{\text{ethyl}}{CH_3CH_2-} > \underset{\text{methyl}}{CH_3-}$$

(*iii*) A doubly or triply bonded atom 'X' is equivalent to two or three such atoms. For example,

$= C$ is equivalent to C—C

$\equiv C$ is equivalent to 

In the case of —CHO and —$CH_2O$ groups, the 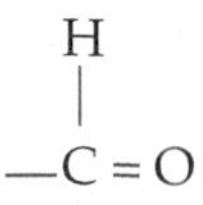 group has higher priority than the —$CH_2OH$ group. This is because in the —CHO group, the C-atom is attached with O, O, and H atoms while in the —$CH_2OH$ group, C-atom is attached with O, H, H atoms.

(*iv*) By application of these rules some common substituents are given the following priority sequence in decreasing order:

I, Br, Cl, $SO_3H$, SH, F, OR, OH, $NO_2$, $NR_2$, NHCOR, $NH_2$, $CCl_3$, COCl, COOR, COOH, $CONH_2$, CHO, $CH_2OH$, CN, $CR_3$, $C_6H_5$, $CH_2R$, $CH_3$, D and H.

(*v*) After assigning the priorities to the four groups/atoms attached to the chiral carbon, the molecule is imagined to be in position where the atom/group of the lowest is directed at the top or bottom. Now the arrangement of the remaining three atoms/groups is viewed in the decreasing order of their priorities. In looking so, if the eyes move in a clockwise direction, the configuration is assigned *R* (rectus meaning right), while if the eyes move in anti-clockwise direction, the configuration is assigned *S* (senister meaning left).

Consider the following examples.

(*i*) 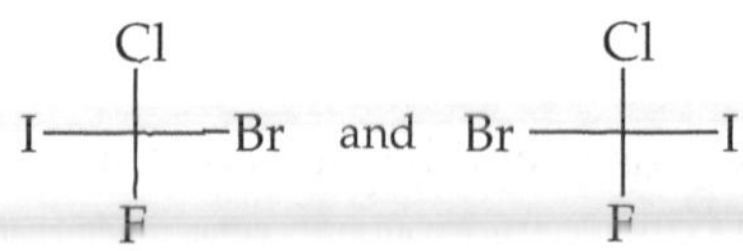

As per the previous rules, the priority sequence assigned to the different atoms attached to a chiral carbon atom is: I = 1, Br = 2, Cl = 3 and F = 4.

The molecule is then projected in such a way that the atom (a group) of lowest priority (H) is being pushed away from us. The outer three atoms (or groups) points towards us. We finally draw curved arrow that goes in the direction $1 \to 2 \to 3$. If this arrow is clockwise, the configuration is (R). If the arrow is anti-clockwise, the designation is (S).

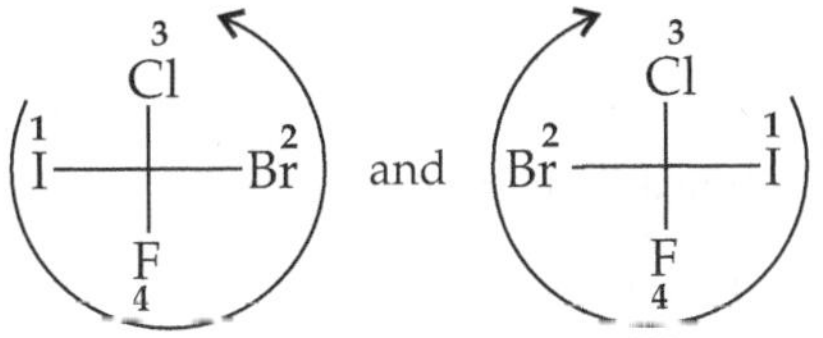

(S) and (R)

(*ii*)

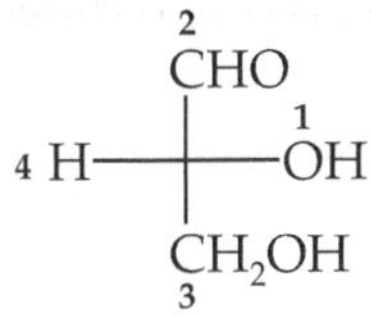

D-(+)-glyceraldehyde

In this compound, the OH group gets the highest priority followed by CHO(2), $CH_2OH$ (3) and H gets the least priority (4). However, in this Fischer projection the atom of lowest priority (H) is directed toward us as it is bonded horizontally and there may be difficulty in visualizing this molecule. In all such cases, the Fischer projection formula of a compound is converted into another equivalent projection formula by an even number of interchanges of the substituent atoms or groups such that the atom or group of lowest priority is placed vertically downwards. So (*ii*) becomes:

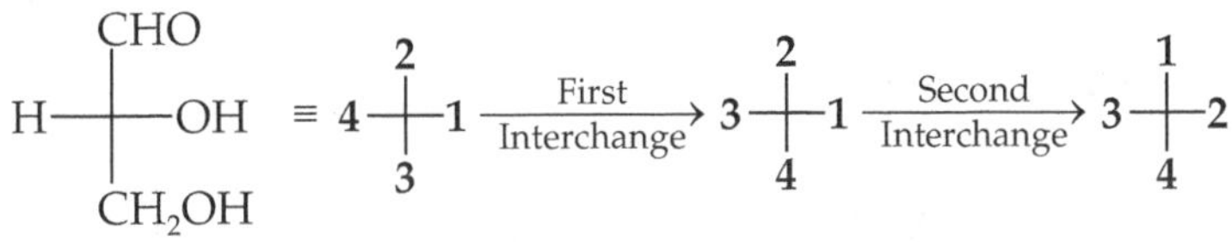

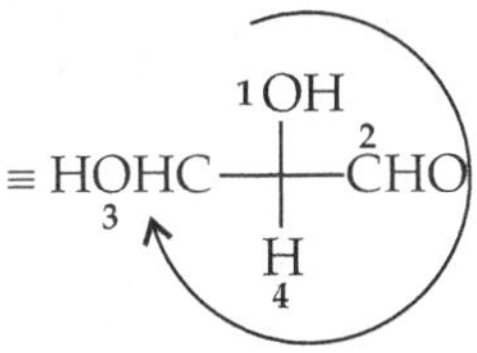

R

So, D-(+)-glyceraldehyde has (R) configuration.

It does not matter which groups are interchanged, in order to get the group of lowest priority (4) to the rear, but always an even number of exchanges are made.

(*iii*)

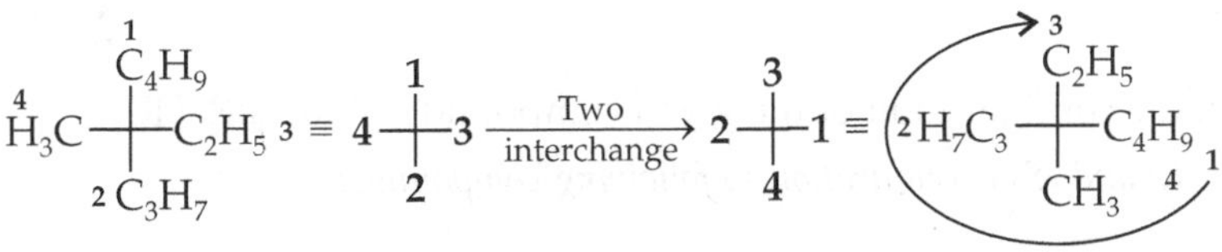

R

So, the given compound has (R) configuration.

(*iv*) 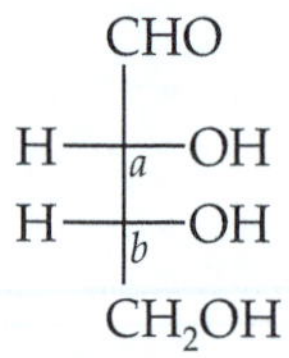 2,3,4-trihydroxybutanol or D-erythrose

When the compound possesses two different chiral centers, each of these may be either (R) or (S).

**For chiral center *a***

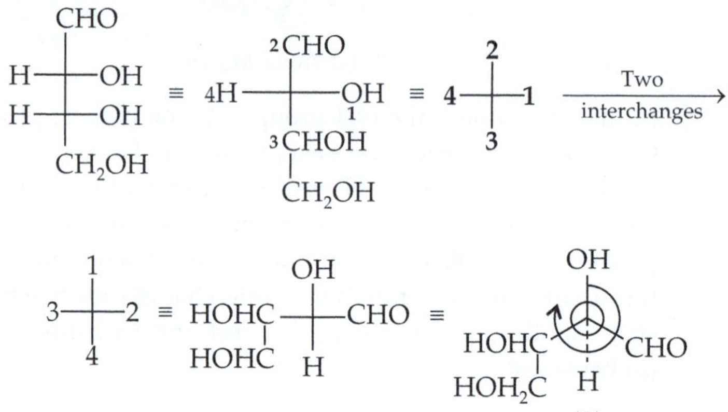

(R)

**For chiral center *b***

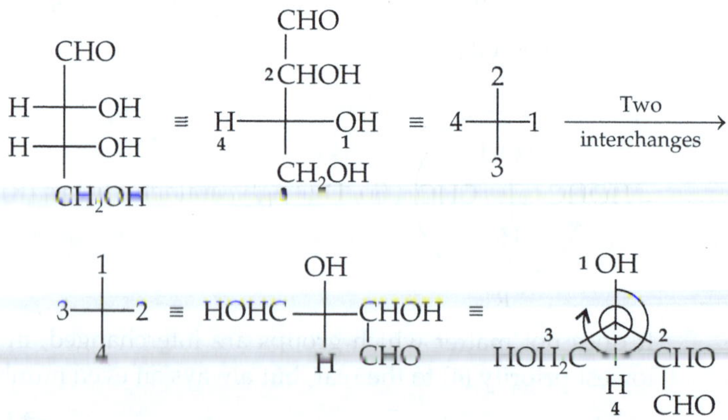

(R)

So, the configuration of the given compound will be (*a*R, *b*R).

**Example 2.** *Assign (R) and (S) configuration to following compounds:*

(*i*) 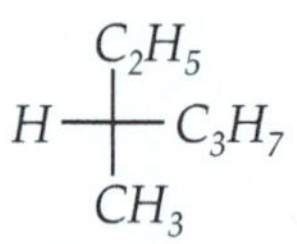 (*ii*) 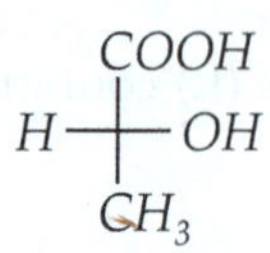 (*iii*) 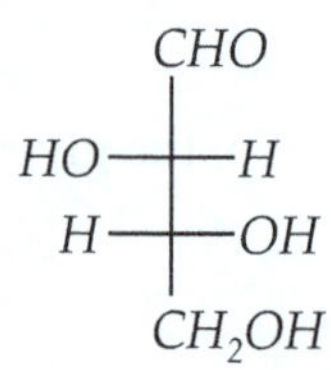

**Solution.**

(*i*) 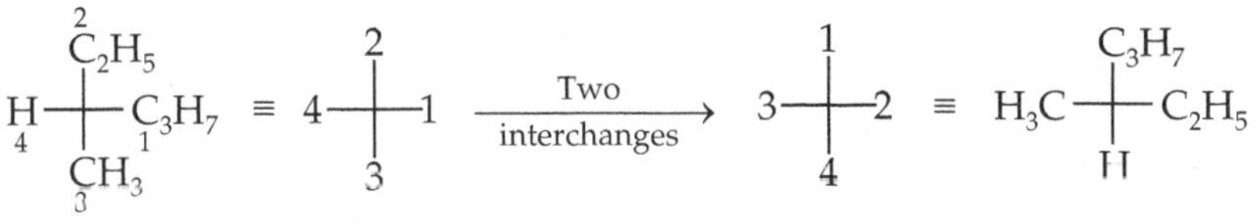

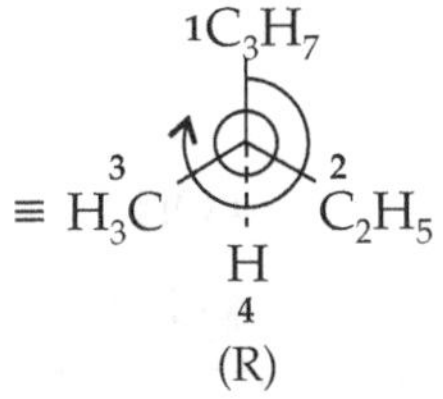 So, the compound has (R) configuration.

(*ii*) 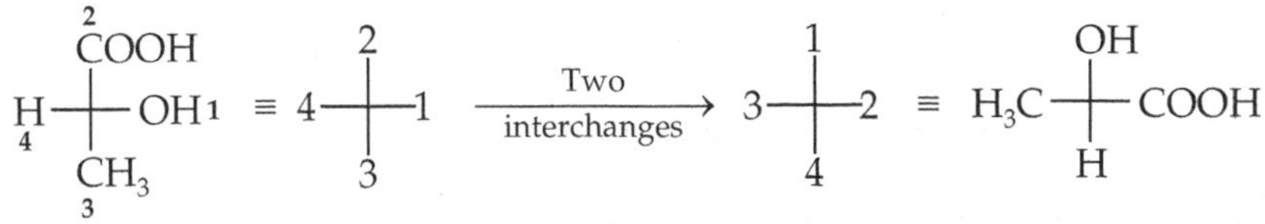

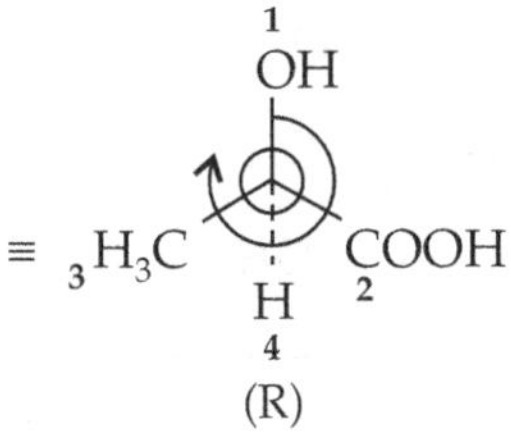 So, the compound has (R) configuration.

(*iii*) 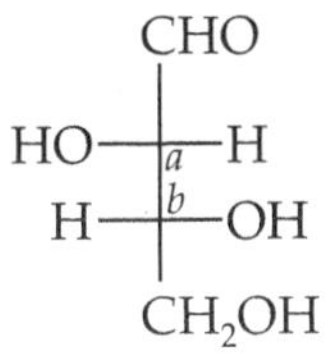

**For chiral center *a***

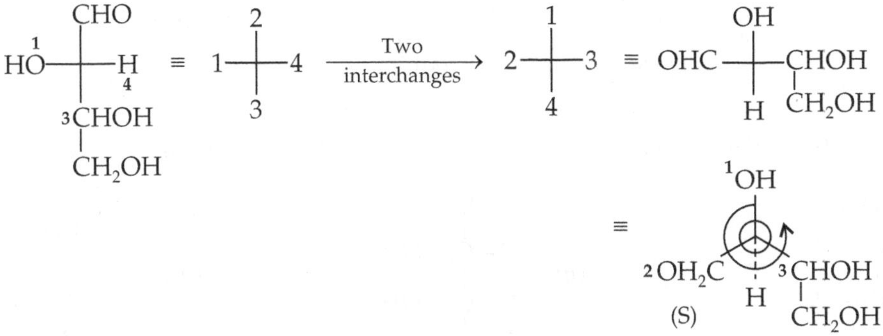

**For chiral center *b***

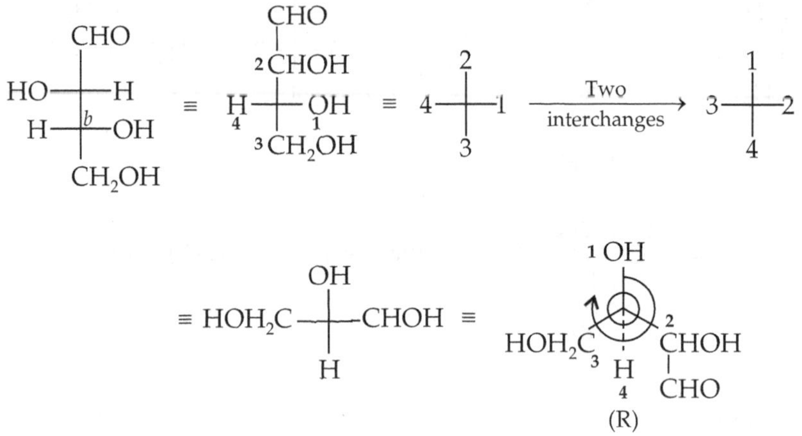

So, the given compound has (*a*S, *b*R) configuration.

**Example 3.** *Assign R or S configuration to the following compounds:*

**Solution.** (*i*) Order of Priority: Br(Z = 35) > Cl(Z = 17) > COOH(Z = 6) > H(Z = 1)

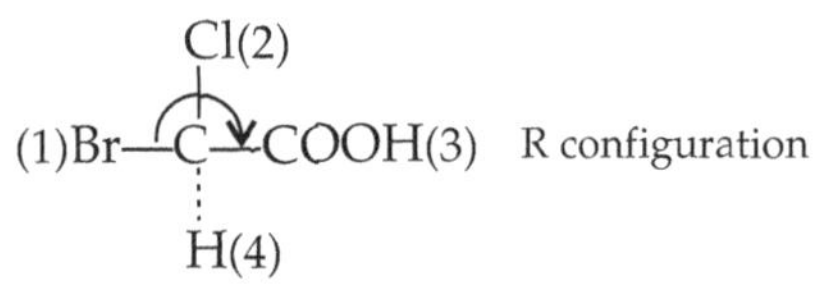

(*ii*) Order of Priority: OH > CHO > $CH_2OH$ > H

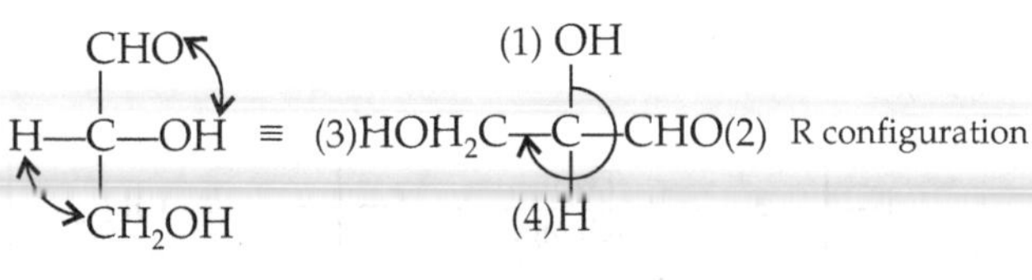

**Example 4.** *Assign R and S configuration to the chiral carbons in the following:*

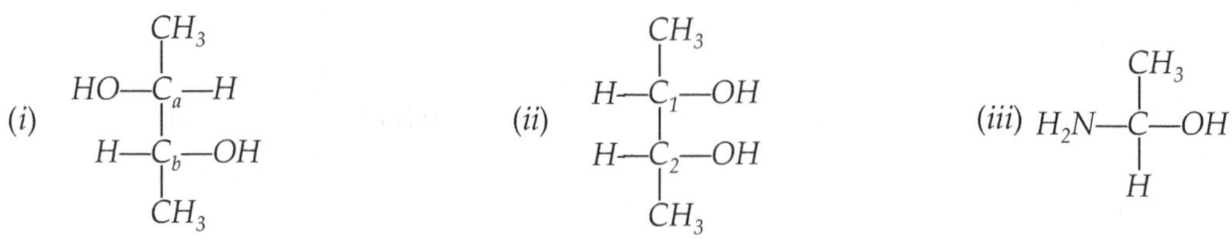

**Solution.** For the first carbon atom (*a*), the order of priority is:

$$OH > CH(OH)CH_3 > CH_3 > H$$

(*i*) 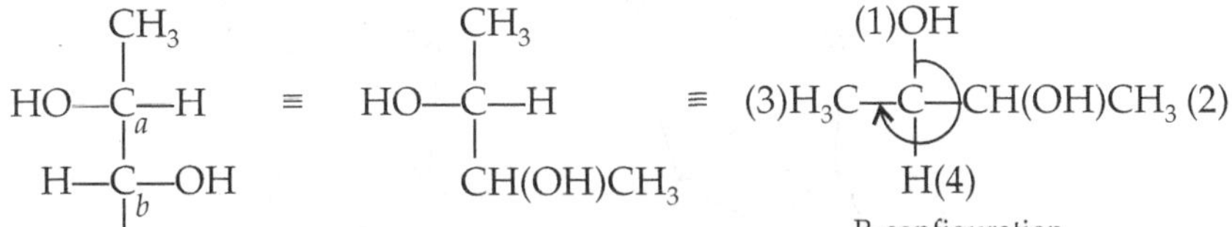

R-configuration

For the second C-atom (*b*), the order of priority is: $OH > CH(OH)\,CH_3 > CH_3 > H$

Thus, the compound has (*a*R, *b*R) configuration.

(*ii*) 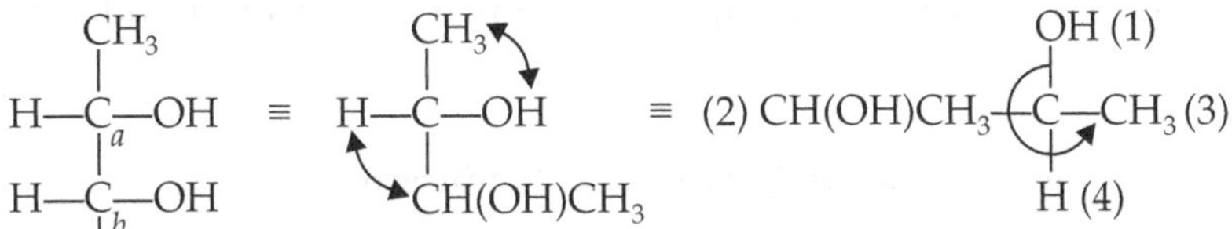

S-configuration

For the second C-atom (*b*)

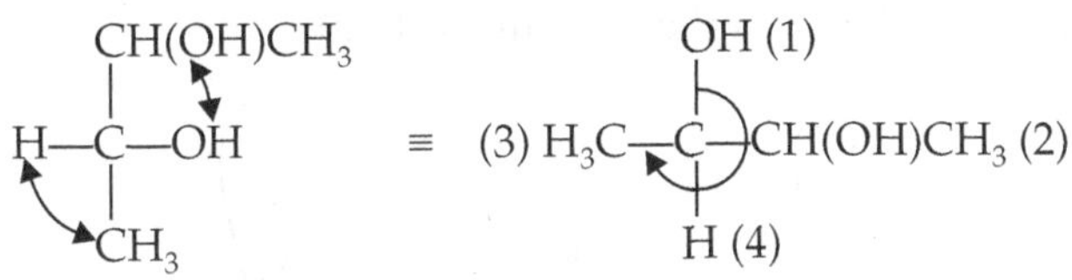

R-configuration

Thus, the compound has a (*a*S, *b*R) configuration.

(*iii*) 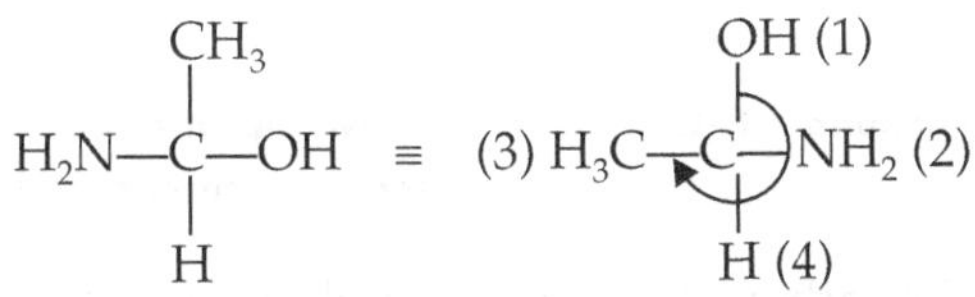

R-configuration

## 16.10 GEOMETRICAL ISOMERISM

This type of isomerism was first recognised in 1875, by *Lel Bel* and *Van't Hoff*, in organic molecules having two differently substituted atoms attached to one another by means of a double bond: *ba*C = C*ab*. According to them, the presence of a double bond makes the two carbon tetrahedra join to each other at two corners, which thus hinders the free rotation of the molecule. Now, if such a molecule has different groups on each carbon atom, there is a possibility of two types of spatial arrangement. So two geometrical isomers in compounds of the type *ba*C = C*ab* are possible (Figure 16.12).

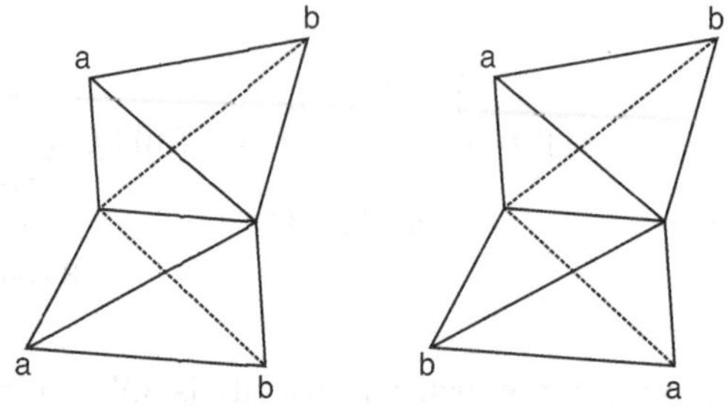


***FIGURE 16.12*** **Possibility of two geometrical isomers in compounds of type *ba*C = *Cab*.**

The molecule is flat and thus will superimpose on its mirror image and hence the compound is optically inactive.

### 16.10.1 Modern Interpretation of Geometrical Isomerism

The hindered rotation around a carbon-carbon double bond can be explained on the basis of molecular orbital theory. A double bond consists of a σ and a π-bond. The π-bond is formed by the sideways overlapping of two parallel π-orbitals of the two carbon atoms above and below the plane of carbon atoms. Maximum overlap will occur only if the molecule is planar, because only in this configuration are the π-orbitals parallel (Figure 16.13(*a*)).

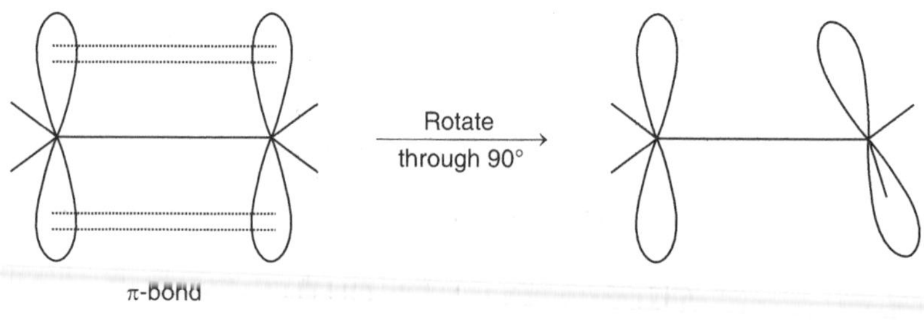


***FIGURE 16.13*** **(*a*) Rotation around a C-C double bond** **Overlap of *p*-orbitals not possible because they are perpendicular to each other.**

If the *p*-orbitals of one of the carbon atoms of the double bond are rotated with respect to the other, the *p*-orbitals will no longer be in a parallel position and thus will not overlap and the π-bond will break. Breaking of a π-bond requires 251 KJ/mol of energy which cannot be provided by collision of molecules at ordinary temperatures. Hence, the rotation about the carbon-carbon double bond is not free but strongly hindered or restricted. Due to this restricted rotation, the relative positions of the atoms or groups attached to the carbon atoms of the double bond get fixed and thus when each of the two double bonded carbon atom has two different atoms or groups, two different arrangements are possible (Figure 16.13(*b*)).

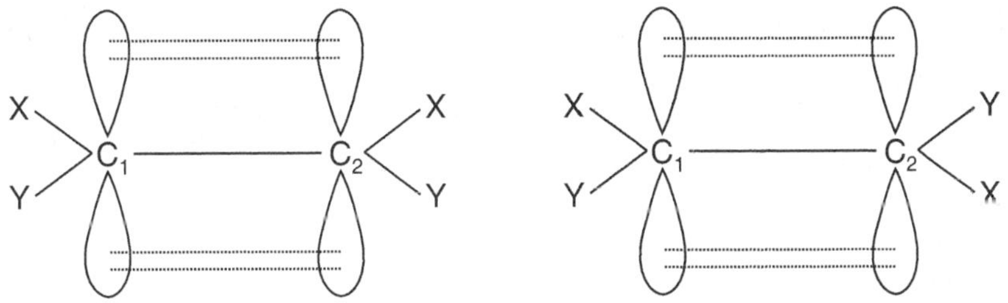


*FIGURE 16.13* **(*b*) Restricted rotation leading to two different arrangements in the compound, *yx*C = C*xy*.**

Thus, two necessary and sufficient conditions for a compound to show geometrical isomerism are:

(*i*) Presence of a carbon-carbon or carbon-nitrogen double bond.

(*ii*) each of the carbon atoms must have different substituents, as shown below.

$$\begin{array}{c} x-C-y \\ \| \\ x-C-y \end{array} \qquad \begin{array}{c} x-C-y \\ \| \\ x-C-b \end{array}$$

The following compounds will not show geometrical isomerism because they have similar substituents on one or both of the carbon atoms.

$$\begin{array}{c} a-C-a \\ \| \\ x-C-y \end{array} \qquad \begin{array}{c} a-C-b \\ \| \\ x-C-x \end{array} \qquad \begin{array}{c} a-C-a \\ \| \\ x-C-x \end{array}$$

Thus, *the isomers which possess the same structural formula but differ in the spatial arrangement of the groups around the double bond are* known as *geometrical isomers* and the phenomenon is known as *geometrical isomerism*. These are designated as *cis-* and *trans*-isomers in the following manner:

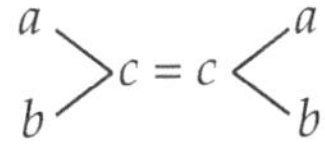


I
*cis*-isomer

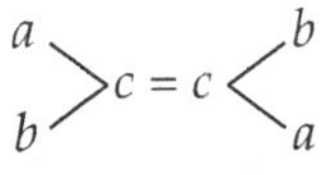


II
*trans*-isomer

The isomer having similar groups on the same side is called a *cis*-isomer (I) and the other having similar groups on the opposite side is called a *trans*-isomer (II). The phenomenon as a whole is called *cis-trans isomerism.*

## 16.10.2 Examples of Geometrical Isomerism

(*i*)

$$\begin{array}{c} H-C-COOH \\ \| \\ H-C-COOH \end{array} \qquad\qquad \begin{array}{c} H-C-COOH \\ \| \\ HOOC-C-H \end{array}$$

Maleic acid (*cis*-isomer)      Fumaric acid (*trans*-isomer) (*ii*) 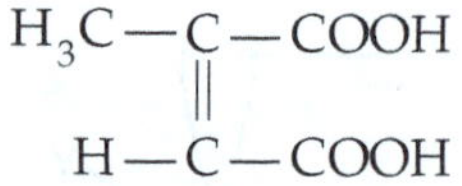


Citraconic acid
(*cis*-isomer)

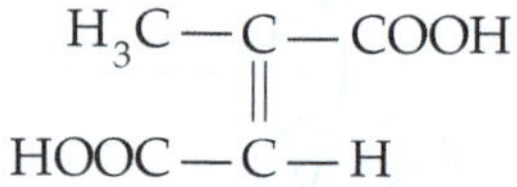


Mesaconic acid
(*trans*-isomer)

(*iii*) 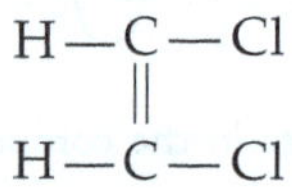


*cis*-1,2-Dichloroethylene

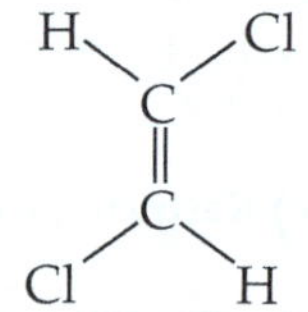


*trans*-1, 2-Dichloroethylene

(*iv*) 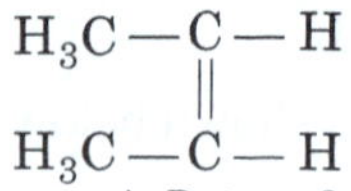


*cis*-Butene-2

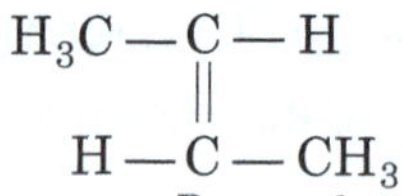


*trans*-Butene-2

(*v*) 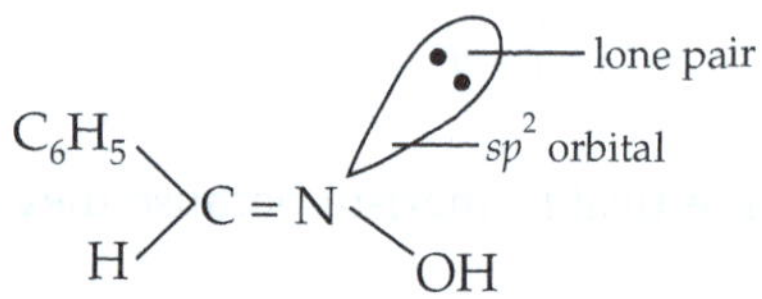


*syn*-Benzaldoxime

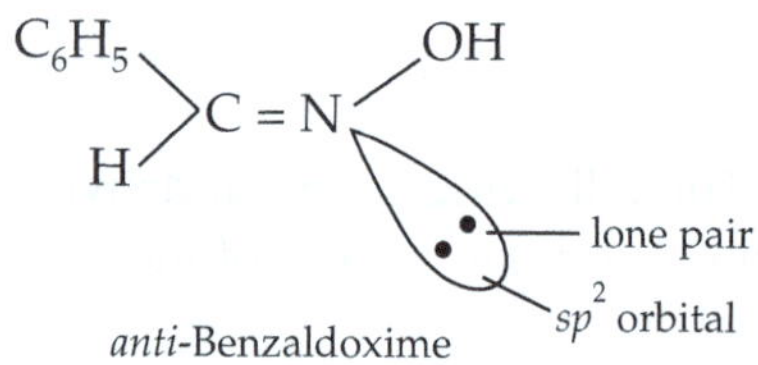


*anti*-Benzaldoxime

(*vi*) 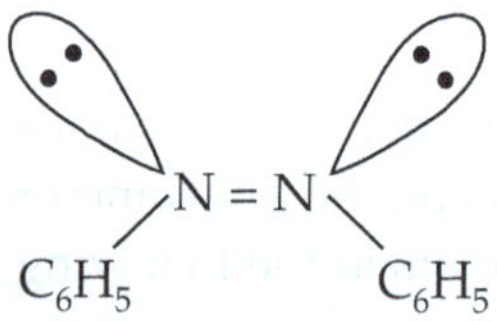


*syn*-Azobenzene

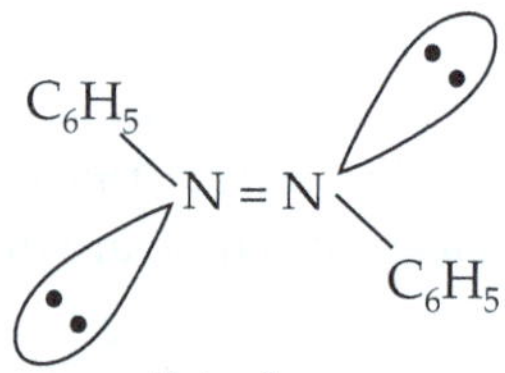


*anti*-Azobenzene

### 16.10.3 E and Z System of Nomenclature

In geometrical isomers, *cis*- and *trans*-designations have been used to indicate the arrangement of atoms or groups around a double bond. But if four different groups are attached with >C = C<, then *cis* and *trans* terms cannot be used. In such cases a new system of nomenclature of geometrical isomers called *E and Z system* is used. The following steps are used in this system:

(*i*) Assign a first (1) or second (2) priority to the atoms or groups attached to each carbon atom of the double bond. This is done in accordance with the rules described in R, S system of nomenclature.

(*ii*) If the higher priority atoms or groups are on opposite sides of the double bond, the configuration is designated as *E* (*entegen* meaning across). If the higher-priority atoms or groups are on the same side of the double bond, the configuration is designated as *Z* (*zusammen* meaning together).

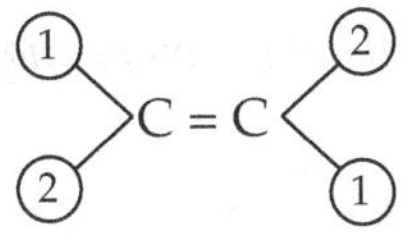

E

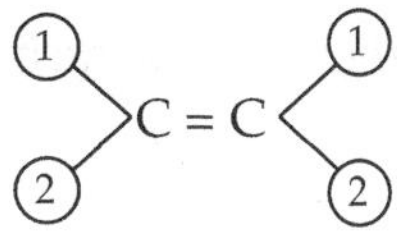

Z

Consider the following examples:

(*i*)

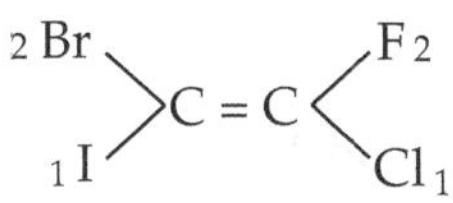

Z-isomer

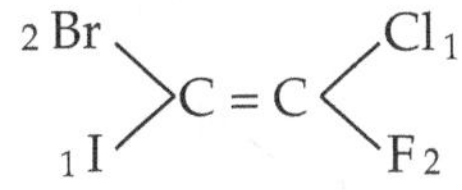

E-isomer

(*ii*)

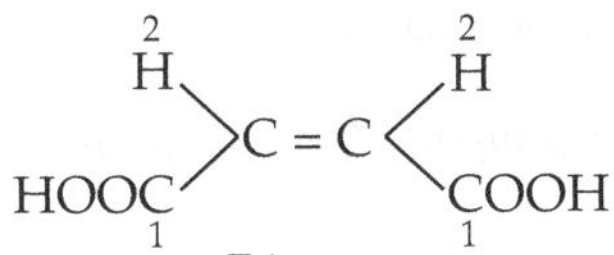

Z-isomer

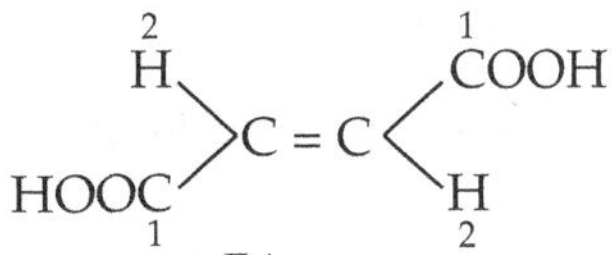

E-isomer

(*iii*)

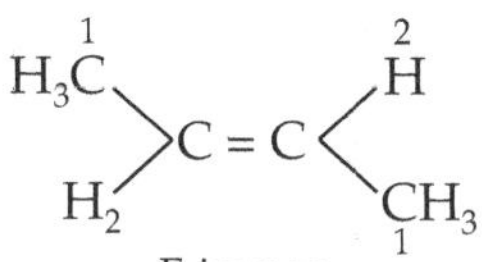

E-isomer

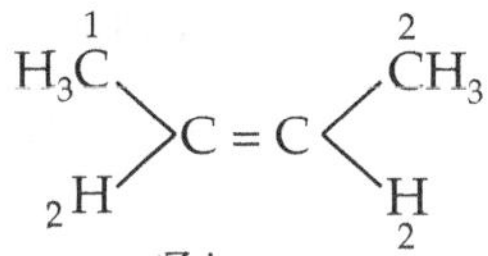

Z-isomer

(*iv*)

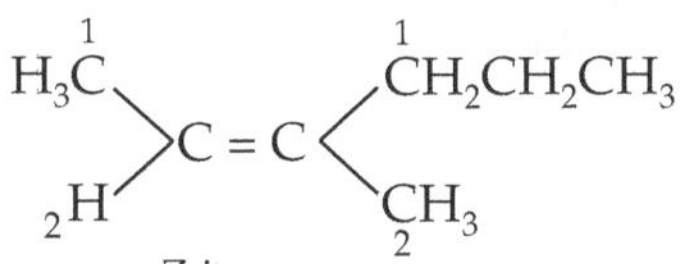

Z-isomer

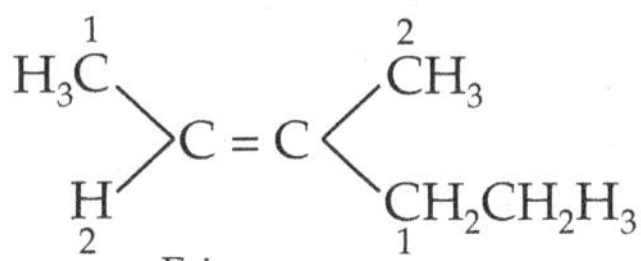

E-isomer

(*v*)

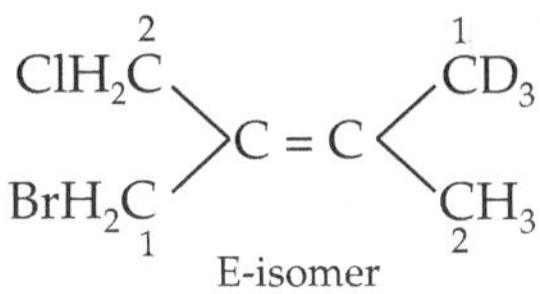

E-isomer

(*vi*)

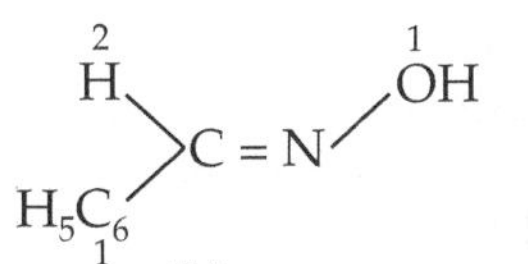

E-isomer

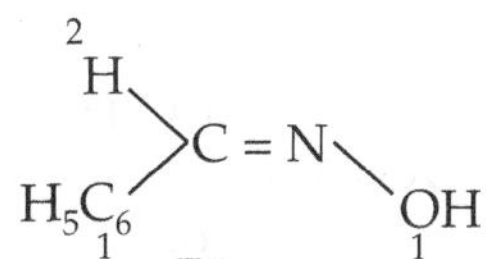

Z-isomer

**Example 5.** *Which of the following compounds will not exhibit with geometrical isomerism:*

(*i*) $CH_3—CH = CH_2$

(*ii*) $CH_3—C(Br) = C(Br)—CH_2CH_3$

(*iii*) $CH_3.CH_2.CH_2CH = CH.CH_3$

**Solution.** (*i*) $CH_3—CH = CH_2$

**Example 6.** *Assign E or Z configuration to each of the following compounds:*

(*i*) 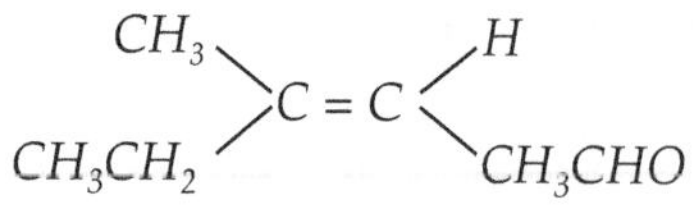


(*ii*) 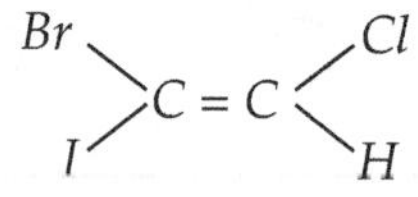


**Solution.**

(*i*) 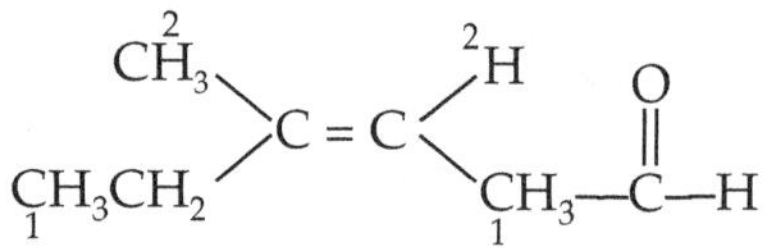


Z-configuration

(*ii*) 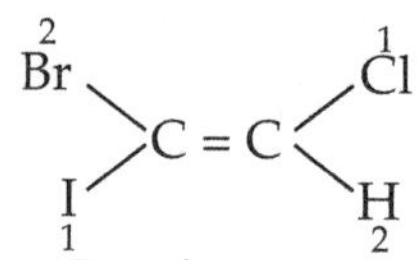


E-configuration

**Example 7.** *Allocate the symbol E or Z to each of the following compounds and explain:*

(*i*) 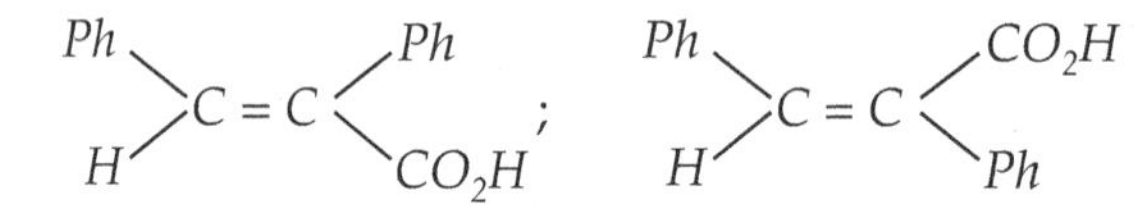


(*Ph = Phenyl = $C_6H_5$*)

(*ii*) 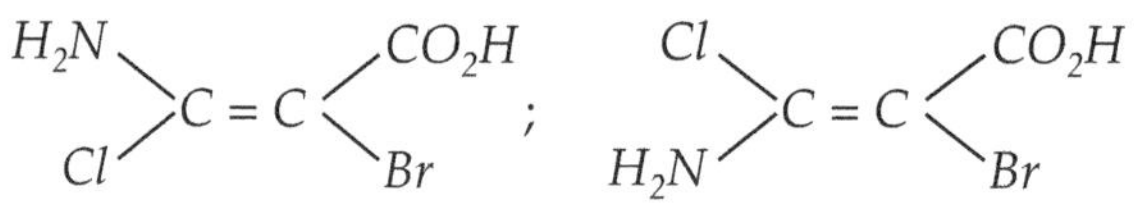


**Solution.**

(*i*) 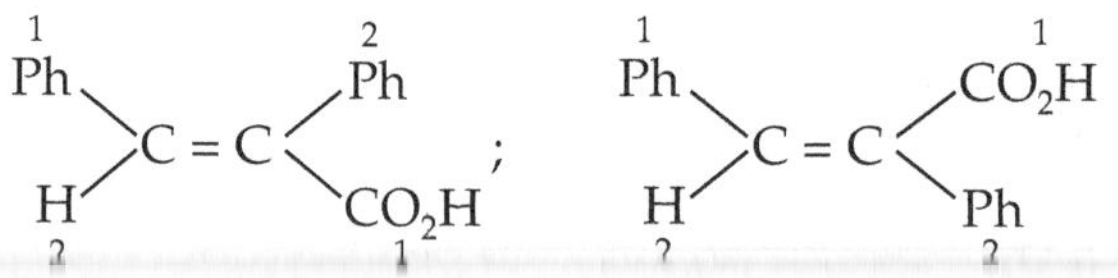


Z-configuration E-configuration

(*ii*) 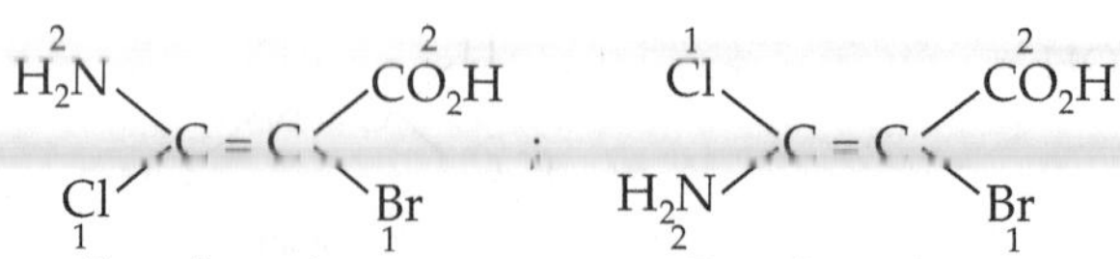


Z-configuration E-configuration

**Example 8.** *Assign E or Z configuration to the following alkenes:*

(*i*) *2-pentene* *or* 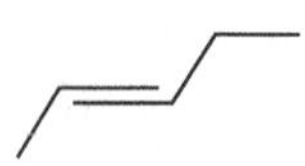

(*ii*) *3-methyl-2-pentene* *or* 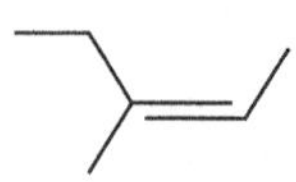

(*iii*) *4-methyl-2-pentene* *or* 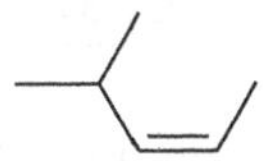

**Solution.**

(*i*) 2-pentene

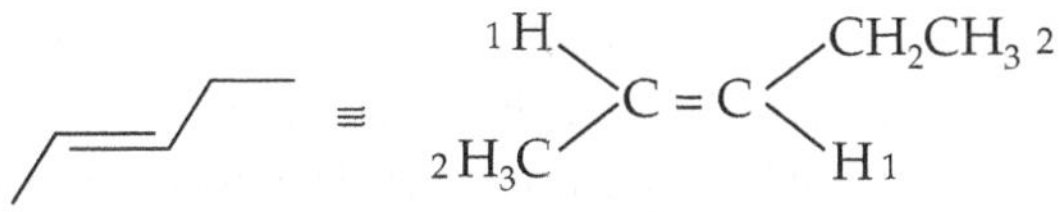


E-configuration

(*ii*) 3-methyl-2-pentene

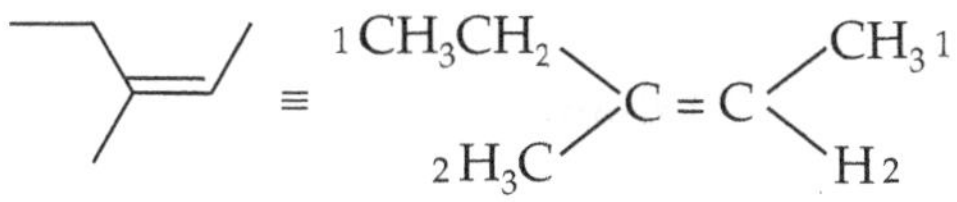


Z-configuration

(*iii*) 4-methyl-2-pentene

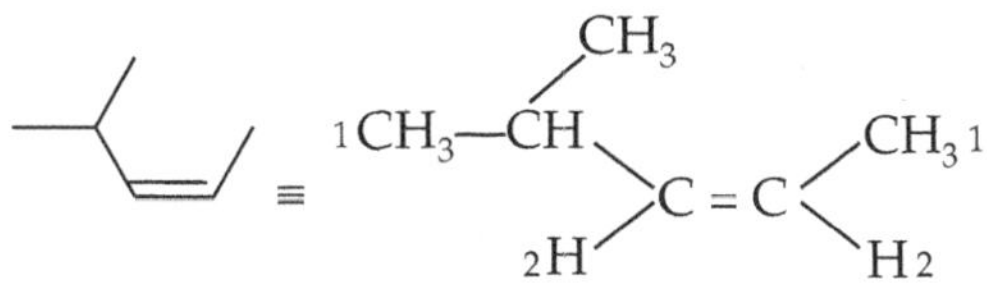


Z-configuration

**Example 9.** *Write E and Z configuration of the following compounds:*

(*i*) 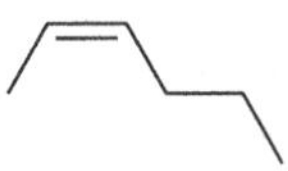

(*ii*) 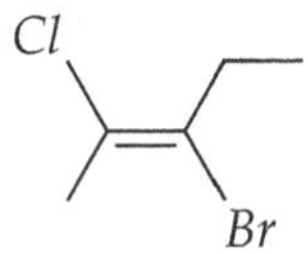


(*iii*) 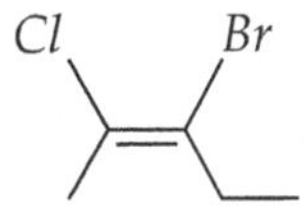


**Solution.**

(*i*) 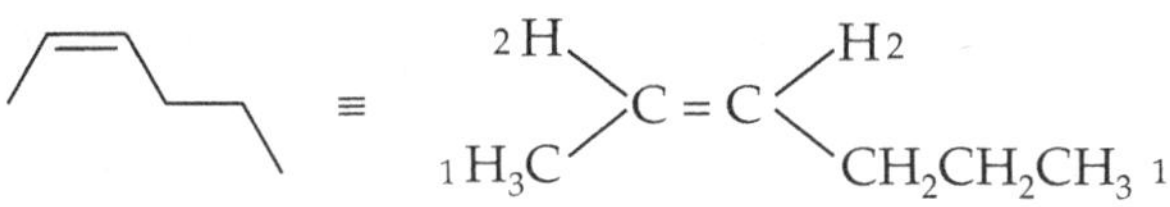


Z-configuration

(*ii*) 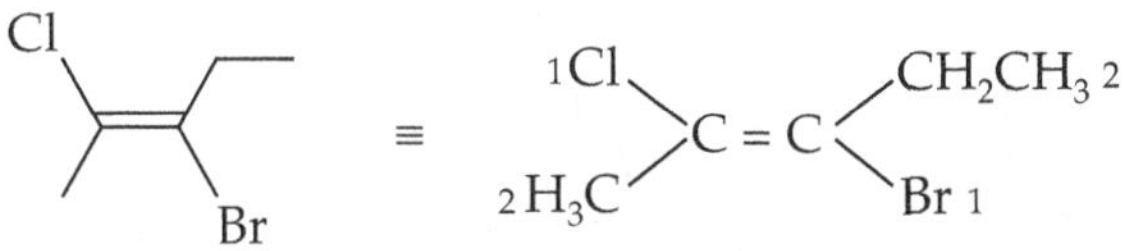


E-configuration

(*iii*) 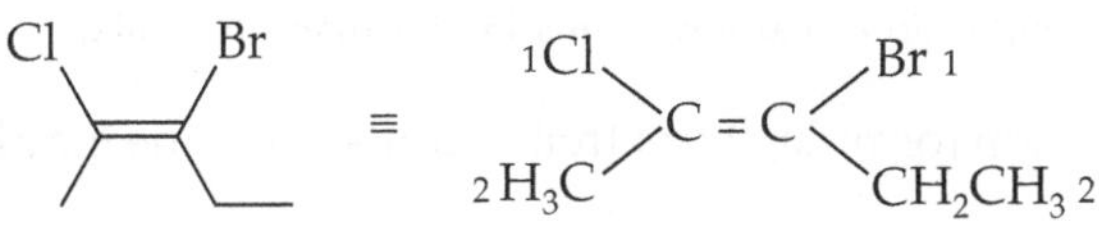


Z-configuration

## 16.11 CONFORMATIONAL ANALYSIS

Groups bonded to a carbon-carbon single bond can rotate freely about the C—C bond axis. *The different arrangement of atoms or groups in space that can be formed by the rotation of groups about the carbon-carbon bond axis are known as conformations.* This type of isomerism is found in alkanes, cycloalkanes and their substituted derivatives. Unlike optical and geometrical isomers, the conformers are interconvertible easily by rotation around single bond. But isolation of different conformers is not feasible because the potential energy barrier for their interconversions is too low. An energy barrier of about 16–20. kcal/mol is necessary for two conformers to exist independently at room temperature.

### 16.11.1 Conformations of Ethane

First consider various methods adopted for projecting a molecule (Figure 16.14) on a planar surface. These include planar projection, ball and stick model, Sawhorse formula, and Newmann projection formula.

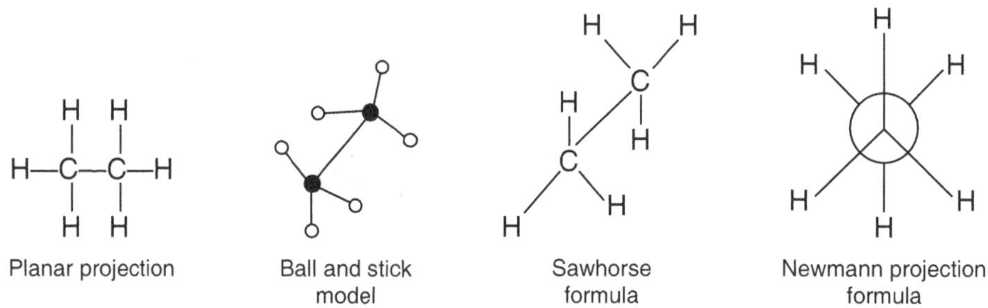


*FIGURE 16.14* **Different ways to represent conformations.**

Of these forms, Newmann projection formulae are generally used because eclipsing (crowding) of hydrogen atoms can best be represented by this formula (Figure 16.15). In this formula, the carbon atom nearer to the eye (the front carbon) and the groups attached to it are represented by equally spaced radii and the distant carbon atom (the rear carbon), and the group attached to it are represented by a circle with three equally spaced radical extensions.

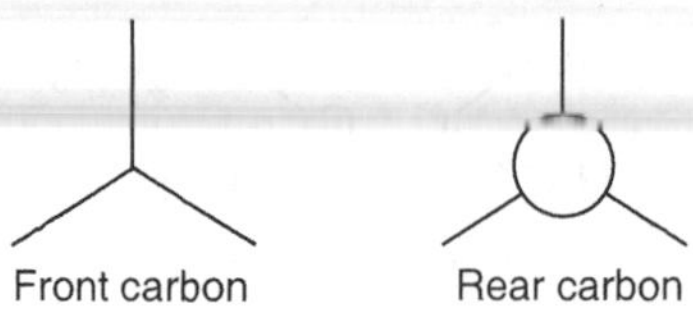


*FIGURE 16.15* **Representation of carbon atoms in Newmann formula.**

According to Newmann projection formula, the extreme forms of ethane may be represented as:

(*i*) *Eclipsed form.* It displays all the hydrogen atoms directly opposed to and eclipsed by the vicinal hydrogen atoms on the other carbon atom (Figure 16.16). Since the vicinal hydrogens are at a closer distance, they will experience repulsive interaction.

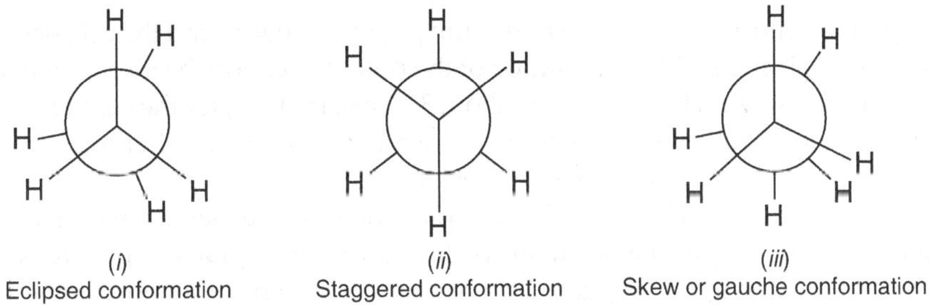


***FIGURE 16.16*** **Conformations of ethane.**

(*ii*) *Staggered form.* In this form all hydrogens are as far apart as possible and are obtained by rotating the eclipsed form by 60° in a clockwise or counter-clockwise direction.

(*iii*) *Skew form.* The infinite number of possible conformations between the eclipsed and staggered forms are called skew conformations. In these forms, hydrogens are closer than in staggered but farther away than in eclipsed form.

***Energy Considerations.*** All the above conformations of ethane are not equally stable. Among these forms, staggered conformation is the most stable while the eclipsed form is the least stable; and the stabilities of skew conformations lie between these two extreme limits. Hence the relative stabilities of various conformations of ethane are in the order:

Staggered > Skew > Eclipsed

This stability order is explained in terms of repulsive interactions (nonbounded) between bonding pairs of electrons, *i.e.*, electron pairs which form six C—H bonds in ethane. In the *staggered conformation*, the electron clouds of six carbon-hydrogen bonds are as far apart as possible, so there is minimum interaction between their electron clouds. Hence it is quite stable. In *eclipsed* or *skew* conformations, there is minimum separation of electron clouds of the six C—H bonds. Hence the potential energies of skew and eclipsed forms increase and thus their stability decreases with minimum in eclipsed forms. The repulsive interaction between the electron clouds which affect the stability of a conformation is known as *torsional strain*. Due to torsional strain, certain energy, called *torsional energy* is required to allow rotation around the C—C single bond. An ethane molecule having staggered conformations will have to cross an energy barrier equivalent to the torsional energy for being converted into eclipsed conformation. Figure 16.17 depicts an energy profile diagram for rotation about the C—C single bond in ethane.

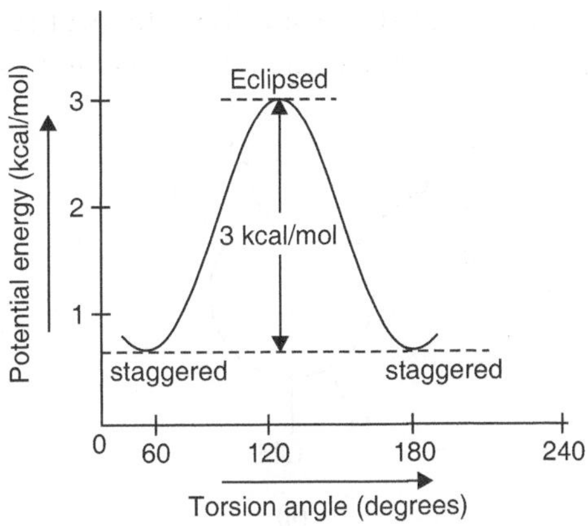


***FIGURE 16.17*** **Potential energy changes during rotation about C—C single bond of ethane.**

The staggered conformation has the minimum potential energy and the eclipsed conformation the maximum potential energy. The magnitude of the potential energy barrier to free rotation about C—C bond of ethane was found by Pitzer to be about 3.0 kcal/mol, *i.e.*, for the conversion of staggered to eclipsed conformation. However, this energy barrier is too small for either form to remain stable, *i.e.*, the two forms are interconvertible because even at ordinary temperature, ethane molecules have an energy of 16–20 kcal/mol which can easily overcome the small barrier of 3.0 kcal/mol. This means that rotation about a C—C single bond in ethane is almost free for all practical purposes. The molecule also spends most of its time in the staggered conformation, which is the most stable conformation. Under equilibrium conditions, any molecule spends most of its time in the most stable conformation.

### 16.11.2 Conformations of *n*-Butane

In butane, two hydrogens of ethane are replaced by methyl groups. The internal rotation of a substituted ethane is not usually symmetrical and therefore it may have maxima and minima of different heights. In *n*-butane, we have to consider several types of conformations obtained by rotating butane molecules six times, each time by 60° (Figure 16.18(*a*)).

(*i*) *Anti-staggered (I).* It is the most stable form in which the two methyl groups are as far apart as possible. It is a fully staggered conformation of lowest energy.

(*ii*) *Partially eclipsed (II).* In this form the two methyl groups are nearer and overlap the hydrogen atoms. It is less stable than anti-form and the energy difference is 3.4 kcal/mole energy.

(*iii*) *Gauche staggered (III).* In this form two methyl groups are nearer to each other but do not overlap. It is less stable than I but more stable than II. The energy difference with II is 0.8 kcal/mole energy.

(*iv*) *Fully eclipsed (IV).* In this form, the two methyl groups are eclipsed. It requires about 4.4 kcal/mole of energy.

(*v*) *Gauche staggered (V).* Here, two methyl groups are closer to each other. It requires 0.8 kcal/mole energy.

(*vi*) *Partially eclipsed (VI).* In this form the methyl groups overlap the hydrogens. It requires 3.4 kcal/mole energy.

Thus, the anti-form (I) is the most stable and the fully eclipsed form (IV) is the least stable conformation of *n*-butane. These can be represented by Newmann projection formula as:

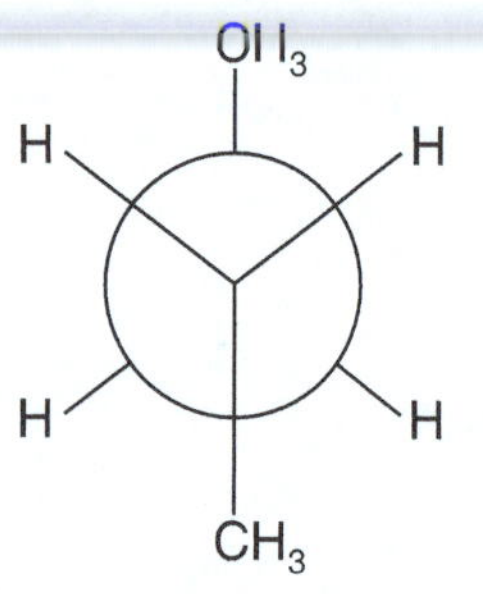


(*i*)
Anti form

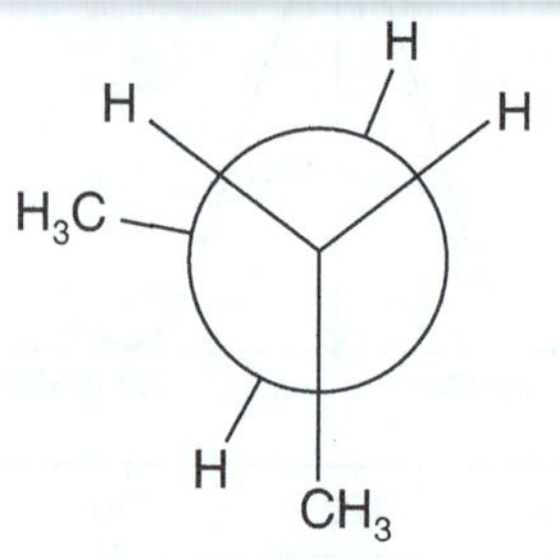


(*ii*)
Partially eclipsed form

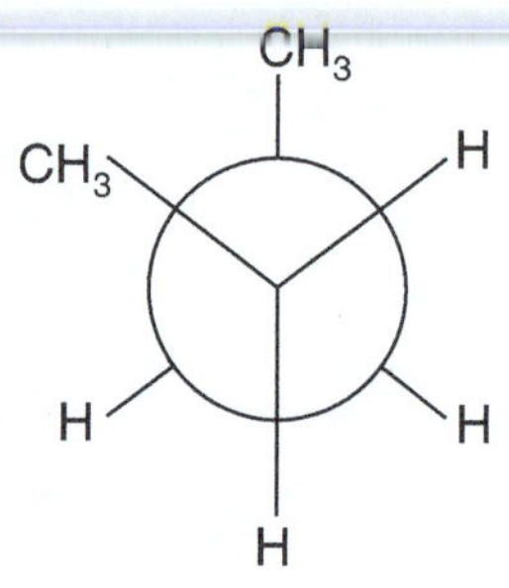


(*iii*)
Gauche form

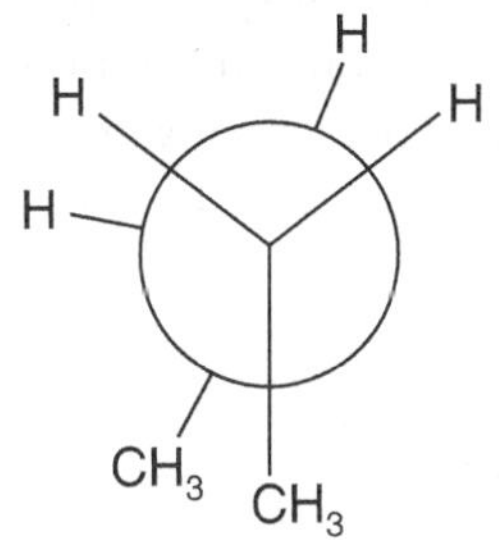


(iv)
Fully eclipsed form

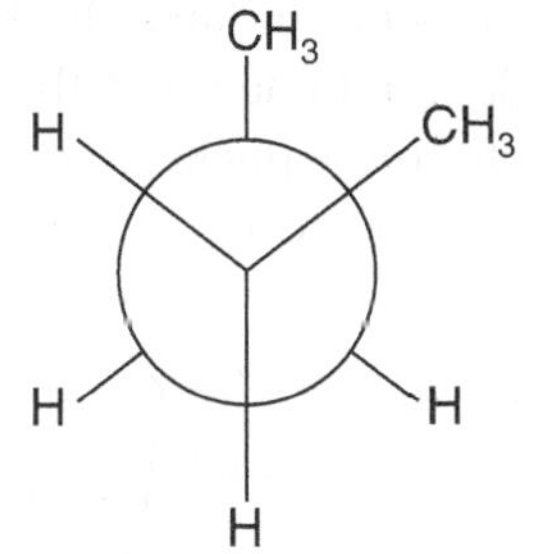


(v)
Gauche form

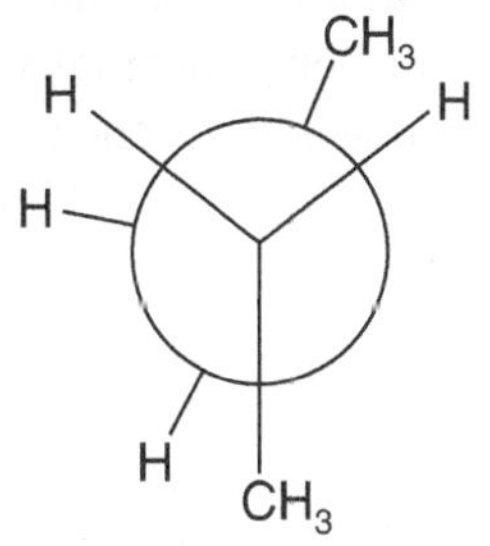


(vi)
Partially eclipsed form

*FIGURE 16.18* **(*a*) Potential energy changes during rotation around the carbon-carbon bond in *n*-butane.**

**Energy consideration.** Consider the energy profile diagram of *n*-butane conformations (Figure 16.18(*b*)) Gauche or Skew forms (III and V) are of somewhat higher energy than the anti-form (I) and this energy difference is computed at about 0.8 kcal/mole and is commonly known as *n-butane skew interaction.* In rotating the *n*-butane molecule, we come across two different eclipsed forms (IV and II or VI). The corresponding energy barriers are 4.4 kcal/mole and 3.4 kcal/mole, respectively.

Thus, at ordinary temperatures and in most reactions the molecule is expected to exist predominantly in anti form (I). Here also the conformers have never been isolated since the energy barriers are too small.

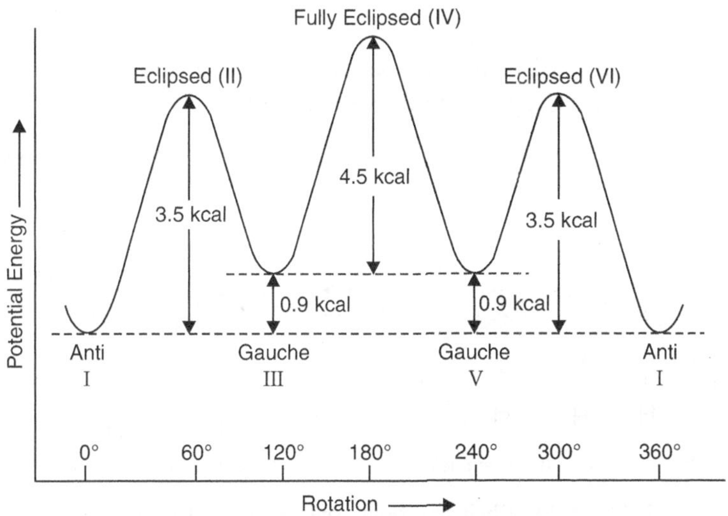


*FIGURE 16.18* **(*b*) Potential energy changes during rotation around the carbon-carbon bond in *n*-butane.**

### 16.11.3 Conformations of Cyclohexane

Cyclohexane is not a planar molecule. If it is assumed to be planar then the bond angle should be 120° which is quite different from normal tetrahedral bond angle of 109°28′. Hence, it should be strained and reactive but in practice it is found to be quite stable. This led *Sachse* and *Mohr* (1890) to propose that it is a nonplanar molecule or is multiplanar. In this form carbon atoms lie in different planes, and hence normal valency angle is retained and the ring is free from any strain and is called strainless ring. The two extreme strainless puckered conformations of cyclohexane are more stable *chair* and the less stable *boat* form. These forms may be represented as (Fig. 16.19):

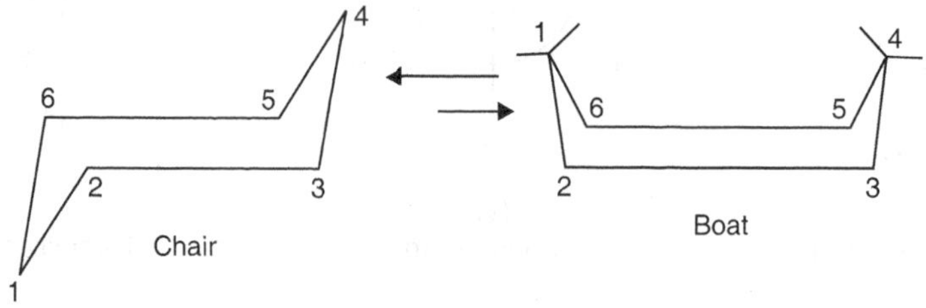


***FIGURE 16.19*** **Chair and boat conformations of cyclohexane.**

The chair form is more stable than the boat form and at room temperature the ratio of chair to boat form at equilibrium is about 1000 : 1.

*Relative stability of chair and boat forms.* As discussed earlier, the chair form of cyclohexane is more stable than the boat form. The relative instability of boat form is due to the following reasons:

(*i*) In *chair conformation,* adjacent hydrogens on $C_1$—$C_2$, $C_2$—$C_3$, $C_3$—$C_4$, and so on are staggered (Fig. 16.20) and hence no repulsion occurs among them, so it is of lower energy. While in boat conformation the hydrogens on $C_2$—$C_3$ and $C_5$—$C_6$ carbon atoms are eclipsed, so repulsion develops between them. Due to these steric repulsions between the hydrogen atoms in boat form, there is a torsional strain, hence it is less stable than chair form.

(*ii*) An additional strain arises in the boat conformation due to the crowding of flagpole hydrogens on $C_1$ and $C_4$, pointing toward each other.

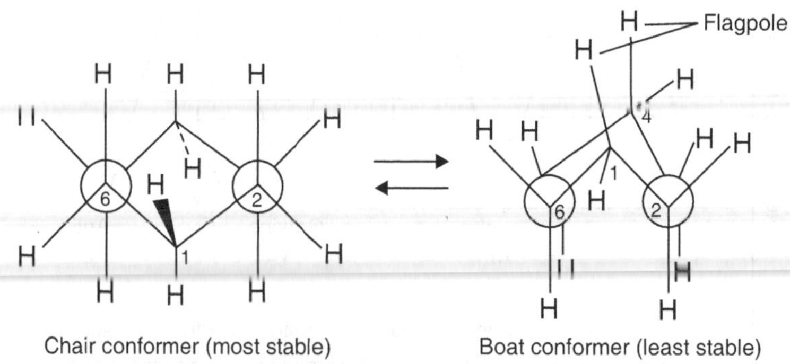


***FIGURE 16.20*** **Chair and boat conformation of cyclohexane.**

However, the energy difference between chair and boat forms has been found to be about 7.2 kcal/mole (Fig. 16.21) which is very small and thus the two conformations are readily interconvertible and cannot be isolated.

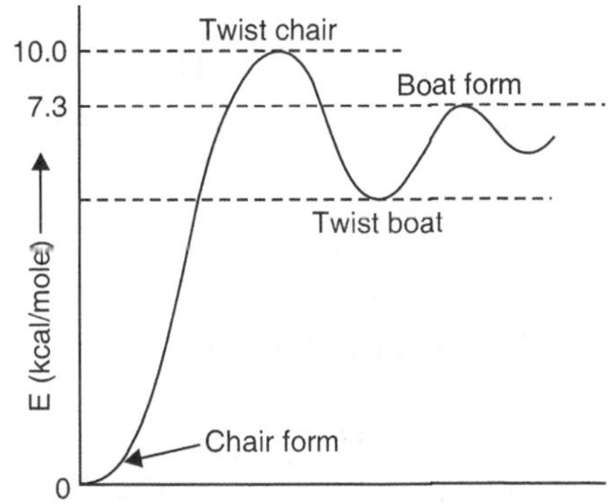


*FIGURE 16.21* **Potential energy of cyclohexane conformations.**

## 16.12 OPTICAL ACTIVITY WITHOUT CHIRALITY

Generally the compounds containing an asymmetric carbon show the optical activity. However, compounds which do not have asymmetric carbon can also show optical activity if the molecule is dissymmetric. For example:

(*a*) **Allenes.** Compounds in which a single carbon atom is connected to two other carbon atoms by a double bond are called allenes. Allene ($CH_2 = C = CH_2$) is the simplest member of the series. In allenes (Fig. 16.22) the two terminal carbon atoms are $sp^2$ hybridized and utilize these three hybrid orbitals for forming σ bonds with two hydrogens and a carbon, while the central carbon atom is *sp*-hybridized and is attracted to two carbons through these orbitals forming σ bonds. The central carbon atom still has two *p*-orbitals available for π-bonding. One of these can overlap to form a π-bond with the *p*-orbital of one of the methylene groups and the other will overlap in a perpendicular plane with the *p*-orbital of the methylene group.

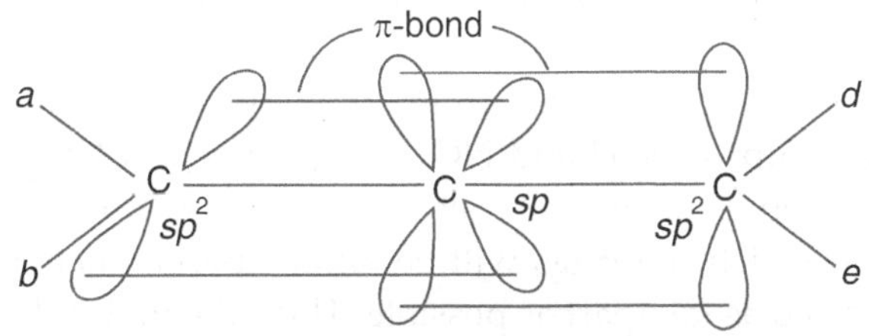


*FIGURE 16.22*

As a consequence of this special geometry of allenes, any allene with the substitution pattern *abc* = *c* = *cab* will exist in two enantiomeric forms. For example,

(*i*) 1, 3-diphenyl-propadiene exist in two enantiomeric forms which are not superimposable to its mirror images and are optically active.

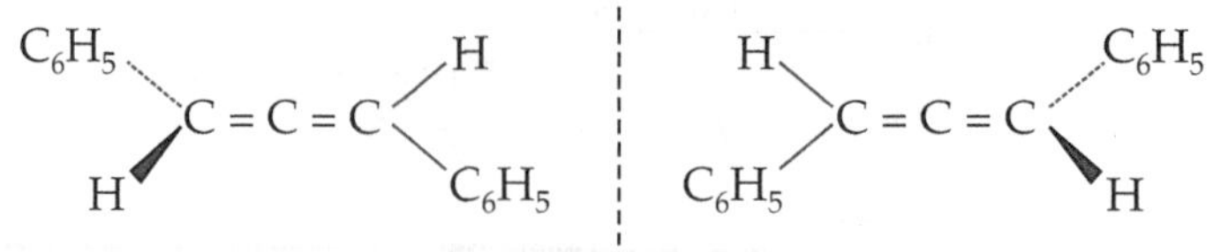

Mirror
1, 3-Diphenyl propadiene

(*ii*) 1-chloro-3, 4, 4-trimethyl Penta-1, 2-diene

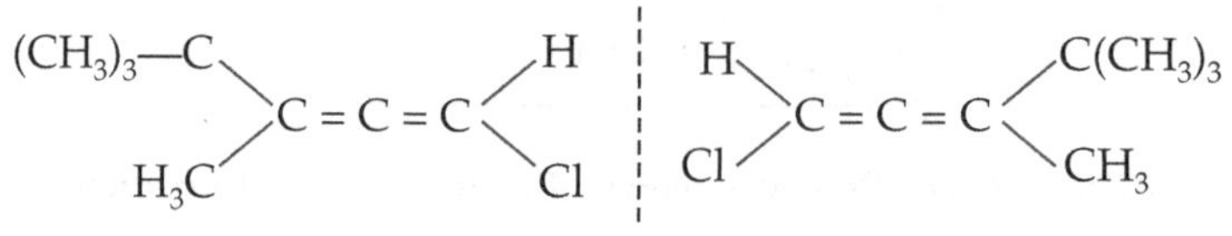

(*b*) **Biphenyl derivatives.** Biphenyl-2, 2′-disulphonic acid

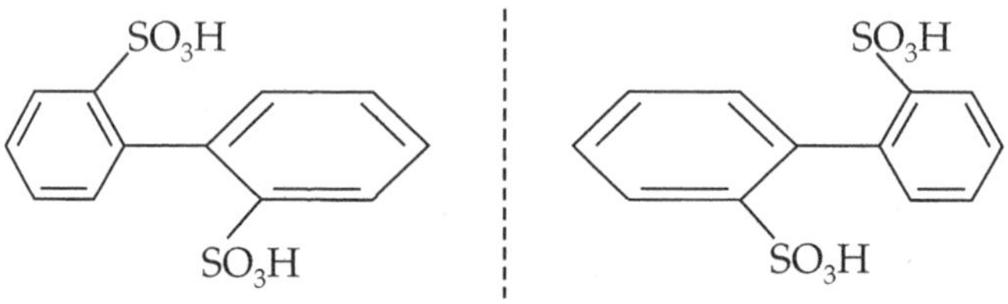

Mirror

These two forms are nonsuperimposable mirror images. They do not interconvert at room temperature because the energy required to twist one ring through 180° relative to the other is too high. This in turn is because during the twisting process, the two $SO_3H$ groups must come into very close proximity when the two benzene rings become coplanar and strong repulsive forces are introduced.

Thus, substituted biphenyls show optical isomerism when substituents in the 2-positions are large enough to prevent rotation about the bond joining the two benzene rings. For example, if we introduce large substituents such as a nitro group at position 6 and 6′ of diphenic acid, it will be prevented from attaining planar configuration because in that case the bulky groups COOH and $NO_2$ (or COOH) and $NO_2$ and $CO_2H$ (or $NO_2$) will experience steric repulsion against each other. In other words, the molecule will prefer a conformation in which the rings will be almost perpendicular to each other with the substituents located as far apart as possible. Thus, the molecule will have no elements of symmetry. In other words, it will not be superimposable on its mirror image and will exist as a pair of enantiomers I and II.

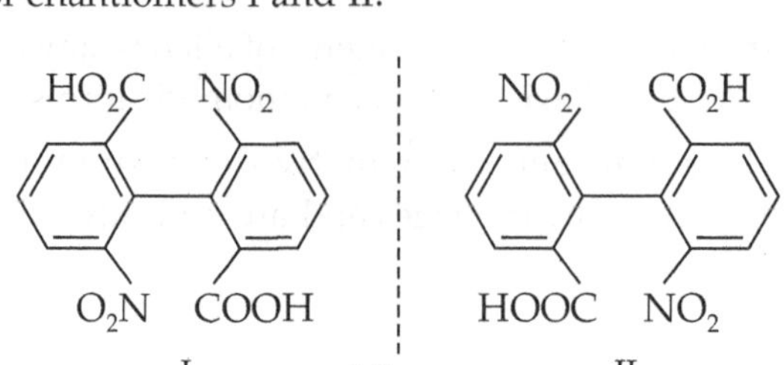

I Mirror II

(*c*) **Spiranes.** *Compounds having a carbon atom common to rings are* called spiranes. Such compounds may be considered similar to allenes in which both the double bonds are replaced by rings. These two rings are perpendicular to each other. Because of restricted rotation spiranes exhibit optical isomerism. For example,

(*i*)

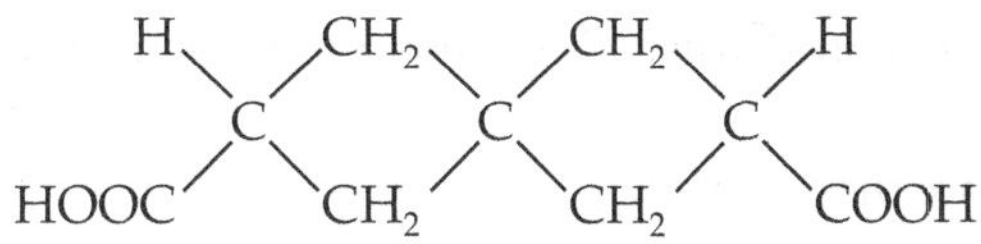

1, 7-Dicarboxylspirocycloheptane

(*ii*)

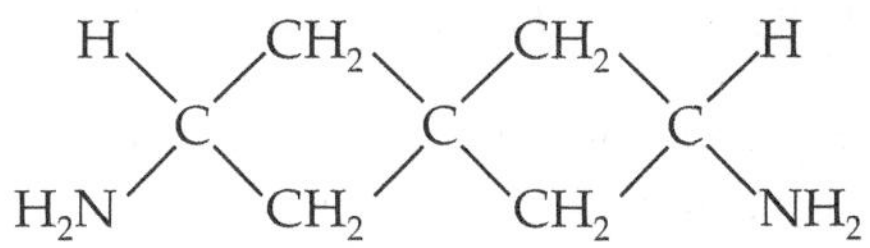

Diaminospiro cycloheptane

## QUESTION BANK

1. Write a note on carbonium ions.
2. Write a note on carbanions.
3. What are the different types of organic reactions? Explain them with one example each.
4. Write a note on inductive effect.
5. Write a note on electromeric effect.
6. What are electrophiles? Give two examples.
7. What are nucleophiles? Give two examples.
8. Which of the following is least stable carbanion?

   (*a*) $(CH_3)_3\overline{C}$ (*b*) $\overline{C}H_3$ (*c*) $(CH_3)_2\overline{C}H$ (*d*) $\overline{C}Cl_3$.
9. Classify the following into electrophilic and nucleophilic reagents.

   (*a*) $H^+$ (*b*) $Cl^-$ (*c*) $NO_2^+$ (*d*) $OH^-$.
10. (*a*) How many electrons are present in the outermost valency shell of carbon of the carbonium ion?

    (*b*) How many electrons are present in the outermost valency shell of carbon of the carbanion?
11. What do you mean by the reaction mechanism of a reaction?
12. Explain the reaction mechanism of the following reaction by equation only.

$$CH_2 = CH_2 + Br_2 \longrightarrow CH_2Br.CH_2Br$$

13. What is an addition reaction? Explain its reaction mechanism with an example.
14. Define the following with suitable examples:

    (*a*) Substitution reaction (*b*) Addition reaction.
15. What is an elimination reaction? Explain its reaction mechanism with an example.
16. Explain $S_N1$ reactions with examples.
17. Explain homolytic fission with an example.

18. Explain homolytic or heterolytic fission.
19. Is the following homolytic or heterolytic?

(*i*) $CH_3OH \longrightarrow CH_3^+ + OH^-$ (*ii*) $Cl_2 \longrightarrow Cl^\cdot + Cl^\cdot$

(*iii*) $CH_4 \longrightarrow CH_3^\cdot + H^\cdot$ (*iv*) $CH_3CH_2Br \longrightarrow CH_3CH_2^+ + \overline{Br}\cdot$

20. Write the names of the parts obtained when C—Cl bond of ethyl chloride undergoes heterolytic fission.
21. What are free radicals? How are they formed?
22. Consider the following reactions:

(*a*) $CH_3CH_2Cl \xrightarrow{\text{Aqueous NaOH}} CH_3CH_2OH$ (*b*) $(CH_3)_3C—Cl \xrightarrow{\text{Aqueous NaOH}} (CH_3)_3\ COH$

Which of these reactions follows $S_N1$ and which the $S_N2$ mechanism?

23. Write an example of electrophilic substitution.
24. Write an example of nucleophilic addition.
25. What is the difference between a free radical and an ion?
26. Define the following with suitable examples:

(*a*) Substitution reactions (*b*) Addition reactions (*c*) Elimination reactions.

27. What are:

(*a*) Electrophilic reagents

(*b*) Nucleophilic reagents? Give two examples of each.

28. Write notes on:

(*a*) Electronegativity (*b*) Polar and non-polar covalent bonds

(*c*) Inductive effect (*d*) Electromeric effect.

29. Explain $S_N1$ reactions.
30. Classify the following reactions as $S_N1$, $S_N2$, $E_1$ and $E_2$.

(*i*) A first order reaction between an alkyl halide and alkali to give an alcohol the order in alkali being zero.

(*ii*) The formation of an olefine from an alkyl halide and alkali, the reaction being of the first order in each of the reactants.

31. Complete the sentence:

Reagents that function as electron pair donors are called ...... .

32. Explain with examples what is meant by the homolytic fission of a molecule?
33. What is molecular rearrangement? Discuss the mechanism of conversion of caprolactum into nylon-6. Name the rearrangement involved.
34. What is cannizaro reaction? Discuss mechanism of formaldehyde with NaOH.
35. Complete the following:

(*i*) 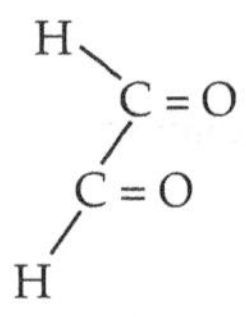 $\xrightarrow{\text{NaOH}}$ ? (*ii*) HCHO + 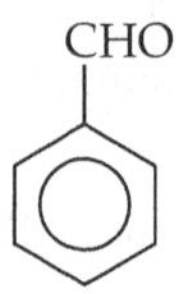 $\xrightarrow{\text{NaOH}}$ ?

(*iii*) 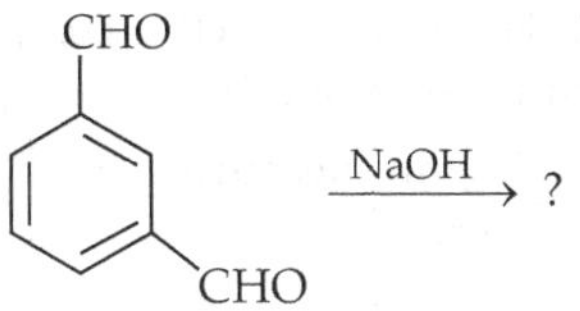


(*iv*) 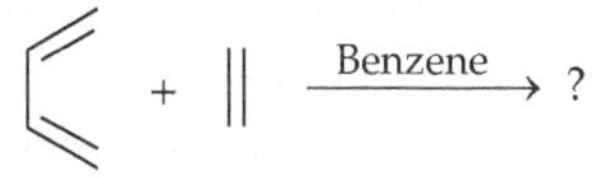


(*v*) 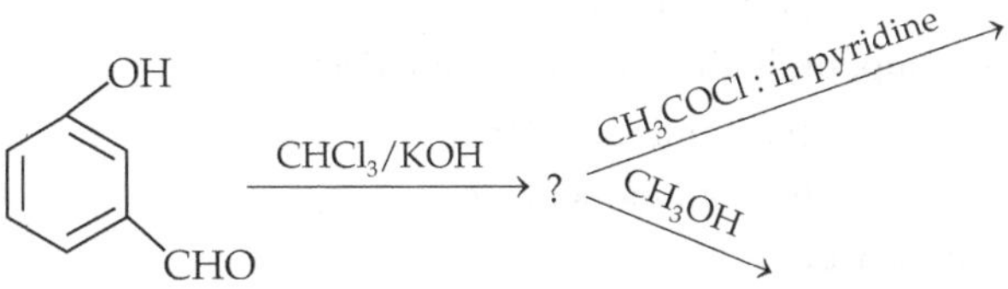


**36.** What is Diels-Alder reaction? Describe the mechanism of the reaction. What is the importance of the reaction in industry?

**37.** Write two applications of each of the following reactions:

Cannizaro reaction, Hoffmann rearrangement, Reimer-Tiemann reaction and Beckmann rearrangement.

**38.** What is Beckmann rearrangement? Discuss the mechanism and application of the reaction.

**39.** Convert indole into indole 3-aldehyde.

**40.** What is Hoffmann rearrangement? Discuss the mechanism and application of the reaction.

**41.** Complete the following reactions:

(*i*) 

(*ii*) 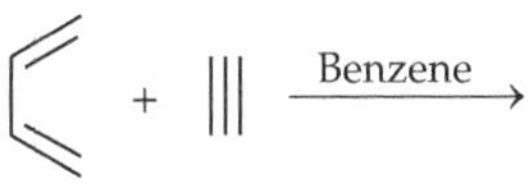


**42.** Write the mechanism of the following reactions:

(*i*) Cannizaro reaction (*ii*) Friedal-Craft alkylation

(*iii*) Skraup's synthesis.

**43.** What are the essential conditions for a compound to show geometrical isomerism?

**44.** Write three examples of geometrical isomerism.

**45.** Explain optical isomerism of chiral organic compounds.

**46.** Write five examples of organic compounds showing optical isomerism without chirality.

**47.** Define the terms: plane of symmetry, center of symmetry, alternating axis of symmetry.

**48.** State differences between enantiomers and diastereoisomers with suitable example.

**49.** An optical pure compound A gave an $[\alpha]^{25} = +20°$. A mixture of A and its enantioner B, gave $[\alpha]^{25} = +10°$. Calculate the ratio of A and B in the mixture.

**50.** (*a*) The specific rotation of a pure enantiomer is + 10°. If it is isolated from a reaction with 30° recemization and 70% reaction. What will be its observed rotation?

(*b*) What do you understand by altropisomerism?

51. An optically active compound A has a molecular formula $C_6H_{12}$ and an inactive compound B has a molecular formula $C_6H_{14}$ on hydrogenation. Write the structures of A and B.

52. The specific rotation of a solution of conine containing 0.35 *g*/ml measuring in 10 cm tube is + 16°. What will be the observed specific rotation of its enantiomer?

53. Explain clearly with examples:

    (*i*) optical activity (*ii*) isomerism.

54. Which of the following isomeric compounds show optical isomerism:

    (*a*) 2-aminopentane (*b*) 3-aminopentane
    (*c*) 2, 2-dimethyl propyl amine (*d*) 1-aminopentane

55. What is optical activity? How is it measured?

56. What is specific rotation?

57. Write names of two possible isomers of formula $C_4H_{10}O$.

58. Write a short note on optical isomerism of lactic acid.

59. Explain with examples:

    (*a*) R, S system of configuration of optical isomers.
    (*b*) E, Z system of configuration of geometrical isomers.

60. State necessary conditions for a compound to show optical isomerism. Illustrate your answer with examples.

61. What do you understand by E, Z notations?

62. Assign E or Z configuration to each of the following compounds:

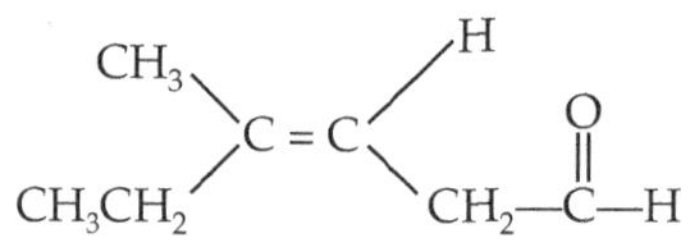


(*a*)

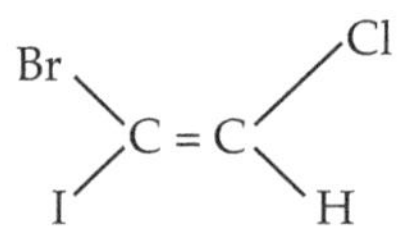


(*b*)

# Chapter 17 CHEMICAL THERMODYNAMICS

Thermodynamics means flow of heat and it deals with the quantitative relationship between heat and other forms of energy in physico-chemical transformations. So, thermodynamics may be defined as, *the branch of science which deals with energy changes in physical and chemical processes.*

Thermodynamics is based on three fundamental laws. They are applicable to all the phenomenon in nature. These laws are based on experimental facts and have been subjected to rigorous mathematical treatment and have yielded correlations between different observable properties of matter.

Thermodynamics has a great predicting power. It can predict whether a given process will occur simultaneously or not under a given set of conditions. The laws provide necessary criteria for predicting the feasibility of a process. However, it gives no information with regard to the rate at which a given change will proceed. Thermodynamics deals only with the states of the system and makes no mention of the mechanism of how the change is accomplished.

## 17.1 BASIC TERMS AND CONCEPTS IN THERMODYNAMICS

Some common terms frequently used in thermodynamics are discussed below.

1. **System.** *A specified part of the universe which is under observation* is called the *system.* A system is called *homogeneous* if physical properties and chemical composition are identical throughout the system. On the other hand, a system is said to be *heterogeneous* if it consists of parts (or phases) each of which has different physical and chemical properties. Thus, a homogeneous system consists of one phase while a *heterogeneous* system consists of two or more phases.

**Types of Systems.** Systems may be classified into three types :

(*i*) *Isolated system. A system which can neither exchange mass nor energy with the surroundings* is called an *isolated system.* For example, hot coffee in a corked thermos flask is an example of isolated system, because it can neither gain or lose energy nor gain or lose matter. Similarly, all chemical reactions carried out in close containers insulated from the surroundings are the examples of isolated systems.

(*ii*) *Closed system. A system which can exchange energy but not mass with its surroundings* is called a *closed system.* Hot coffee in a stainless flask is an example of closed system because energy can be gained or lost (through the steel walls) but not matter. Similarly, all chemical reactions carried out in closed containers are the examples of close systems. For example, heating of hydrogen iodide in a sealed glass bulb constiutes a closed system because there is exchange of heat energy from surroundings to the system but no exchange of mass takes place.

(*iii*) *Open system. A system which can exchange matter as well as energy with the surroudings* is called an *open system.* For example, hot coffee in an open flask is an example of open system because it can gain or lose matter as well as energy. Similarly, if we carry out a reaction between zinc granules and hydrochloric acid in an open beaker, it constitutes an open system because the $H_2$ gas evolved escapes and also the heat.

2. **Surroundings.** *The remaining portion of the universe which is not a part of the system* is called the surroundings.

The system and the surroundings are separated by real or imaginary boundaries. The boundary also defines the limits of the system. The system and the surroundings can interact across the boundary.

For example, if we wish to investigate the behaviour of alcohol in water, we take alcohol in a beaker and add water to it. Here, the contents of the beaker constitute the system, beaker serves as boundary and everything else is the surroundings.

Hydrogen evolved during the reaction (mass) goes out from the system to the surroundings and at the same time heat evolved during the reaction escapes from the system to the surroundings. These three type of systems for the vaporisation of a liquid are shown in Fig. 17.1.

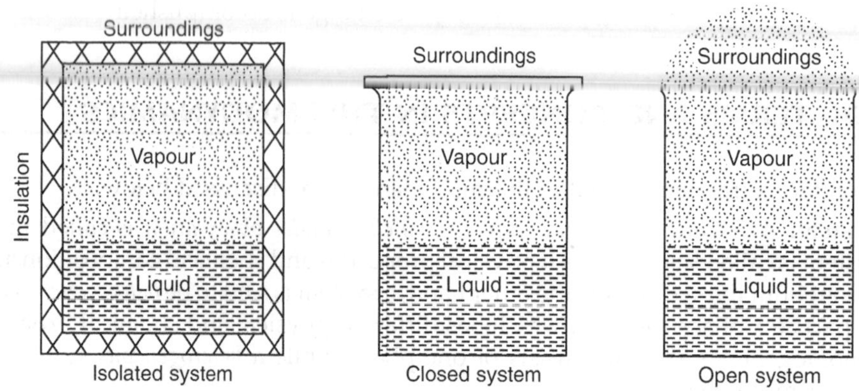


***FIGURE 17.1*** **Examples of isolated, closed and open systems.**

All physical and chemical processes taking place in open in our daily life are open systems because they are continuously exchanging matter and energy with the surroundings.

3. **Macroscopic Properties of the System.** Thermodynamics deals with matter in terms of bulk (large number of chemical species) behaviour. *The properties of the system which arise from the bulk behaviour of matter* are called *macroscopic properties.* The common examples of macroscopic properties are pressure, volume, temperature, surface tension, viscosity, density, refractive index etc. The macroscopic properties can be sub-divided into two types:

   (*i*) intensive properties and (*ii*) extensive properties.

   (*i*) **Intensive properties.** *The properties of the system whose value is independent of the amount of substance present in the system* are called *intensive properties.* For example, temperature, pressure, viscosity, surface tension, dielectric constant, specific heat capacity, vapour pressure, refractive index, etc.

   (*ii*) **Extensive properties.** *The properties of the system whose value depends upon the amount of substance present in the system* are called *extensive properties.* For example, mass, volume, surface area, energy, enthalpy, entropy, free energy, heat capacity, etc.

4. **State of a System.** The state of system means *the condition of existence of the system when its macroscopic properties have definite values.* If any of the macroscopic properties of the system changes, the state of the system is also said to change. Thus, the state of the system is fixed by its macroscopic properties.

5. **State Variables.** *The measurable properties required to describe the state of the system* are called *state variables.* For example, temperature, pressure, volume, composition, etc., are the state variables. It may be noted that for describing a system, it is not necessary to specify the values of all state variables. For any system, a certain minimum number of variables is sufficient to define its state because the other variables become automatically fixed. For example, consider a system consisting of an ideal gas. The state of this system may be defined by only three variables such as temperature (T), pressure (P) and volume (V). The values of other variables (such as amount of gas, density, etc.) will be definite and can be easily calculated.

6. **State Functions.** A *state function is a property of the system whose value depends only upon the state of the system and is independent of the path or manner by which the state is reached.* The change in the value of these properties (state functions) depends only upon the initial and final states of the system and not on the path by which the change from initial to final state is brought about. Some common state functions are pressure (P), volume (V), temperature (T), internal energy (E), enthalpy (H), entropy (S) etc.

7. **Types of Processes.** The state of a thermodynamic system can be changed by a process. In other words, a *process gives the path or operation by which a system changes from one state to another.* Some common types of processes are:

   (*i*) **Isothermal process.** *A process in which the temperature of the system remains constant.* In such a system, heat is either supplied to the system or removed from it.

   (*ii*) **Adiabatic process.** *A process in which the system does not exchange heat with the surroundings, i.e., no heat leaves or enters the system.* In such a process, temperature of the system always changes.

(*iii*) **Isobaric process.** *A process in which the pressure of the system remains constant.*

(*iv*) **Isochoric process.** *A process in which the volume of the system remains constant.*

(*v*) **Reversible process.** *A process in which the direction may be reversed at any stage by merely a small change in a variable like temperature, pressure, etc.* Therefore, in a reversible process, the driving force is only infinitesimally greater than the opposing force and is carried out infinitesimally slowly. In this process, at every instant, the system remains vitrually in a state of equilibrium with the surroundings.

(*vi*) **Irreversible process.** *A process which is not reversible* is called *an irreversible process.* The driving force is very different than the opposing force. All natural processes are irreversible.

(*vii*) **Cyclic process.** *A process in which the system undergoes a series of changes and ultimately returns to its original state* is called *cyclic process.*

## 17.2 INTERNAL ENERGY AND INTERNAL ENERGY CHANGE

Every substance possesses a fixed quantity of energy which depends upon its chemical nature and its state of existence. This is known as *intrinsic energy* or *internal energy* and is denoted by the symbol E (according to latest IUPAC recommendations internal energy is denoted by the symbol U). Every substance has a definite value of internal energy and is equal to the energies possessed by all its constituents namely atoms, ions or molecules. The various forms of energies which contribute towards the energy of the molecules are *translational energy, rotational energy, vibrational energy, electronic energy, nuclear energy of constituent atoms, potential energy of the molecules due to molecular interactions chemical bond energy due to existence of bonds between atoms within the molecules etc.* The sum of these different forms of energies associated with the molecules is called its *internal energy*. Different substances have different values of internal energy. For example, the internal energy of 1 mole of $CO_2$ at N.T.P. is different from the internal energy of 1 mole of $SO_2$ under similar conditions of temperature and pressure. It may be noted that the absolute value of internal energy of a substance cannot be calculated because it is not possible to determine the exact values for the constituent energies such as translational, vibrational, rotational energy, etc. However, we are interested mostly in change in internal energy which occurs during chemical reactions. The change in internal energy of a reaction may be considered as the difference between the internal energies of the two states.

Let $E_A$ and $E_B$ are the internal energies in states A and B respectively. Then the difference between the internal energies in the two states will be

$$\Delta E = E_B - E_A$$

The difference in internal energies ($\Delta E$) has a fixed value and will be independent of the path taken between two states A and B. For the chemical reactions, the change in internal energy may be considered as the difference between the internal energies of the products and that of the reactants, *i.e.*,

$$\Delta E = E_{products} - E_{reactants}$$

Thus, the internal energy, E is a *state function.* This means that $\Delta E$ depends only on the initial and final states and is independent of the path. In other words, $\Delta E$ will be same even if the change is brought about differently.

## 17.3 CONCEPT OF WORK AND HEAT

The energy of a system may increase or decrease in serveral ways but two common ways are :

(*i*) **Heat.** Heat is a form of energy which the system can exchange with the surroundings. If they are at different temperatures, the heat flows from higher temperature to lower temperature. Heat is expressed as *q*. As a convention, *the heat absorbed by the system is taken to be positive whereas the heat released by the system is taken to be nagative.*

(*ii*) **Work.** Work is said to be performed *if the point of application of force is displaced in the direction of the force.* It is equal to the force multiplied by the displacement (distance through which the force acts).

There are two main types of work which we generally come across. These are (*i*) electrical work and (*ii*) mechanical work.

Electrical work is important in systems where reaction takes place between ions whereas mechanical work is performed when a system changes its volume in the presence of external pressure. Mechanical work is important specially in systems that contain gases. This is also known as *pressure-volume work.*

**Pressure-volume Work.** It is the work done when the gas expands or contracts against the external pressure (usually atmospheric pressure). To understand this, consider a cylinder fitted with a frictionless and weightless piston having area of cross-section equal to A. Let it contains gas of volume V and let the pressure acting on the piston is P.

When the internal pressure of the gas is slightly more than the external pressure, P, the gas expands and the piston moves a very small distance *dl*. Then change in volume is given by

$$dV = A \, . \, dl$$

We know that $\text{Pressure} = \dfrac{\text{Force}}{\text{Area}}$

or $$\text{Force} = \text{Pressure} \times \text{Area}$$

Therefore, force on the piston,

$$F = P \, . \, A$$

If the small work done by the movement of the piston is *dw*, then

$$\text{Work} = \text{Force} \times \text{Distance}$$

$$dw = -P \, . \, A \, . \, dl$$

(negative sign is used for work done by system as convention; discussed later).

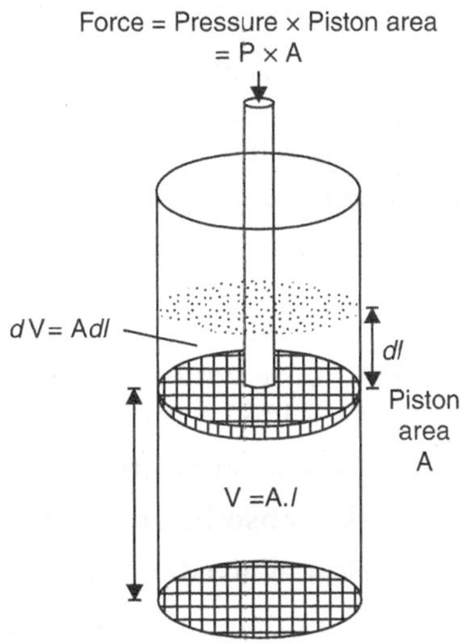

*FIGURE 17.2* **Pressure-volume work.**

But A × *dl* = *d*V is a small change in volume of the gas so that

$$dw = -P \, . \, dV$$

If the volume of the system changes by a finite quantity from volume $V_1$ to $V_2$, then total work done, $w$ can be obtained by integrating $-P \cdot dV$ (when P is constant) over the volume change from $V_1$ and $V_2$. Thus,

$$w = -\int_{V_1}^{V_2} P dV$$

$$= -P(V_2 - V_1)$$

$$= -P\Delta V \qquad (\because \Delta V = V_2 - V_1)$$

If the external pressure (P) is slightly more than the pressure of the gas, the gas will contract *i.e.*, the work will be done by the surroundings on the system. In that case, $V_2$ will be less than $V_1$ and $\Delta V$ is negative and therefore,

$$w = P\Delta V$$

Thus,

(*i*) *If the gas expands,* $V_2 > V_1$ and work is done by the system and $w$ is negative.

(*ii*) *If the gas contracts,* $V_2 < V_1$ and work is done on the system and $w$ is positive.

**Sign conventions for heat and work.** It is very important to understand the sign conventions for $q$ and $w$.

The signs of $w$ or $q$ are related to the internal energy change. When $w$ or $q$ is positive, it means that energy has been supplied to the system as work or as heat. The internal energy of the system in such a case increases.

When $w$ or $q$ is negative, it means that energy has left the system as work or as heat. The internal energy of the system in such a case decreases. Let us illustrate these conventions :

If heat is absorbed by a system, $q$ is *positive* (since there is increase in energy of the system) but if heat is evolved by a system $q$ is given *negative* sign (since there is loss in energy of the system). For example, if we supply 25 kJ of heat energy to the system, we write $q = +25$ kJ but if same amount of heat is given out to the surroundings, we write $q = -25$ kJ.

When work is done on the system, $w$ is *positive* (energy of the system increases). But if system does work on the surroundings, $w$ is *negative* (energy of the system decreases because work is done at the cost of energy). For example, if 15 kJ of energy is transferred to the system by doing work on the system, we write $w = +15$ kJ. But if the system does 15 kJ of work on the surroundings, we write, $w = -15$ kJ because the system has lost 15 kJ of energy.

Thus, to summarise :

| | |
|---|---|
| Heat **absorbed** by the system | = $q$ **positive** |
| Heat **evolved** by the system | = $q$ **negative.** |
| Work done **on** the system | = $w$ **positive** |
| Work done **by** the system | = $w$ **negative.** |

**Units of Energy.** Work, heat and energy have same units. The units of energy are erg, calorie of joule. The S.I. units of energy is the Joule (J) which is equal to Newton-metre (1 J = 1 N m). The various common units of energy are related as :

$$1 \text{ cal} = 4.184 \times 10^7 \text{ erg}$$

$$1 \times 10^7 \text{ erg} = 1 \text{ J (or 1 N m)} \qquad (\because 1 \text{ cal} = 4.184 \text{ J})$$

## 17.4 FIRST LAW OF THERMODYNAMICS OR CONSERVATION OF ENERGY

The first law of thermodynamics is the law of conservation of energy. The law has been stated in various forms but the fundamental implication is that, *although energy may be transformed from one form into another, it can neither be created nor destroyed.* In other words, whenever energy of a particular form disappears an exactly equivalent amount of another form must be produced.

The first law of thermodynamics has no theoretical proof. It is a law based on human experience and has not yet been violated. The following observations justify the validity of this law :

(*i*) It is not possible to construct a perpetual machine which can do work without the expenditure of energy. If the law were not true, it would have been possible to construct such a machine.

(*ii*) James Joule (1850) conducted a large number of experiments regarding the conversion of work into heat energy. He concluded that for every 4.183 joule of work done on the system, one calorie of heat is produced. He also pointed out that the same amount of work done always produces same amount of heat irrespective of how the work is done.

(*iii*) Energy is conserved in chemical reactions also. For example, the electrical energy equivalent to 286.4 kJ $mol^{-1}$ of energy is consumed when one mole of water decomposes into gaseous hydrogen and oxygen. On the other hand, the same amount of energy in the form of heat is liberated when one mole of liquid water is produced from gases hydrogen and oxygen.

$$H_2O(l) + 286.4 \text{ kJ} \longrightarrow H_2(g) + \tfrac{1}{2}O_2(g)$$

$$H_2(g) + \tfrac{1}{2}O_2(g) \longrightarrow H_2O(l) + 286.4 \text{ kJ}$$

These examples justify that energy is always conserved though it may change its form.

**Mathematical Formulation of the First Law of Thermodynamics**

According to first law of thermodynamics, the energy can neither be created nor destroyed. The internal energy of the system can be changed in two ways :

(*i*) *by allowing heat to flow into the system or out of the system.*

(*ii*) *by work done on the system or work done by the system.*

Let us consider a system whose internal energy is $E_1$. Now, if the system is supplied $q$ amount of heat, then the internal energy of the system increases and becomes $E_1 + q$. Now, if work ($w$) is done on the system, then its intenal energy further increases and becomes $E_2$. The energy $E_2$ is the energy in the final state. It will be given as :

$$E_2 = E_1 + q + w$$

or

$$E_2 - E_1 = q + w$$

or

$$\Delta E = q + w \qquad ...(17.1)$$

*i.e.*, Change in internal energy = Heat added to the system + Work done on the system

The relationship between internal energy, work and heat (Eq. 17.1) is a *mathematical statement of first law of thermodynamics.*

It should be noted that energy is a state function and $\Delta E$ depends only on the initial and final states whereas $q$ and $w$ are not state functions. The values of $q$ and $w$ depend upon the way in which the change is carried out. However, whatever may be the process, $\Delta E$ is always equal to $q + w$ and, therefore, $q + w$ is also a state function.

If the internal energy of the system does not change during a change, *i.e.*, $\Delta E = 0$, then

$$0 = q + w \quad \text{or} \quad q = -w \qquad ...(17.2)$$

*This means that heat absorbed by the system is equal to the negative of the work done on the system or equal to work done by the system.*

## 17.5 CHANGE IN INTERNAL ENERGY

The significance of change in internal energy can be understood from first law of thermodynamics. According to first law of themodynamics.

$$\Delta E = q + w$$

For a system involving only pressure-volume work, $w = -P\Delta V$ so that

$$\Delta E = q - P\Delta V$$

If the process is carried out at constant volume, *i.e.*, $\Delta V = 0$, then the product $P\Delta V$ is also zero. Thus,

$$\Delta E = q_v$$

The subscript $v$ is $q_v$ denotes that volume is kept constant. Thus, *the change in internal energy is equal to heat absorbed or evolved at constant temperature and constant volume.*

The change in internal energy in a chemical reaction is determined with the help of *bomb calorimeter* as described in Section 11.4.1.

## 17.6 ENTHALPY AND ENTHALPY CHANGE

Chemical reactions are generally carried out at constant pressure *i.e.*, at atmospheric pressure. $\Delta E$ gives the change in internal energy at constant volume. To express the energy changes at constant pressure, a new term called enthalpy change was used. *Enthalpy* may be defined as *the sum of the internal energy and the pressure-volume energy of the system.*

It is denoted by H and is given by

$$H = E + PV \qquad ...(17.3)$$

Every substance has a definite value of enthalpy in a particular state. Like internal energy, the absolute value of enthalpy cannot be measured. However, the change in enthalpy accompanying a process can be determined. Thus, for a change represented by the process :

$$A \longrightarrow B$$

the change in enthalpy may be expressed as

$$\Delta H = H_B - H_A$$

where $H_A$ is the enthalpy in state A and $H_B$ is the enthalpy in state B and ΔH is the enthalpy change. It may be noted that like internal energy, enthalpy is also a function of the state of the system and therefore, ΔH is independent of the manner in which the final state is achieved.

Change in enthalpy, ΔH may be expressed as

$$\Delta H = \Delta E + \Delta(PV) \qquad ...(17.4)$$

$$\Delta H = \Delta E + P\Delta V + V\Delta P$$

If the process is carried out at constant pressure, *i.e.*, ΔP = 0, then

$$\Delta H = \Delta E + P\Delta V \qquad ...(17.5)$$

Now, $\Delta E = q + w$

If work done during the change is only expansion work,

$w = -P\Delta V$ so that

$$\Delta E = q - P\Delta V$$

Substituting in Eq. (17.5), the value of ΔE, we get

$$\Delta H = q - P\Delta V + P\Delta V = q$$

∴ $\Delta H = q_p$ (at constant pressure) ...(17.6)

where $q_p$ indicates that heat change has taken place at constant pressure. Thus, *the enthalpy change is a measure of heat change (evolved or absorbed) taking place during a process at constant pressure.*

Thus, we can sum up :

If $q$ is the amount of heat absorbed by the system.

$q$ (at constant volume) = ΔE

$q$ (at constant pressure) = ΔH

**Relationship between ΔH and ΔE**

As already studied (Eq. 17.5), ΔH and ΔE are related as :

$$\Delta H = \Delta E + P\Delta V$$

The difference between ΔH and ΔE is not usually significant for solids and liquids. However, when the systems involve gases, the difference becomes significant. Consider a chemical reaction taking place at a constant temperature (T) and pressure (P). Let $V_r$ is the total volume of the gaseous reactants and $V_p$ is the total volume of the gaseous products, $n_r$ is the number of moles of gaseous reactants and $n_p$ is the number of moles of gaseous products. According to ideal gas equation,

$$PV = nRT$$

For reactants, $PV_r = n_r RT$ (at constant T and P) ...(17.7)

For products, $PV_p = n_p RT$ (at constant T and P) ...(17.8)

Subtracting Eq. (17.7) from Eq. (17.8), we get

$$P(V_p - V_r) = n_p RT - n_r RT$$

or $(n_p - n_r) RT$

or $P\Delta V = \Delta n_g RT$ ...(17.9)

where $\Delta n_g$ is the change in the number of gaseous moles of products and gaseous moles of reactants.

Thus, Eq. (17.5) becomes

$$\Delta H = \Delta E + \Delta n_g\, RT$$

$$\begin{bmatrix}\text{Energy} \\ \text{change at} \\ \text{constant P}\end{bmatrix} = \begin{bmatrix}\text{Energy change} \\ \text{at constant V}\end{bmatrix} + \begin{bmatrix}\text{Change in the} \\ \text{number of} \\ \text{gaseous moles}\end{bmatrix} \times RT$$

This relation is very useful for converting ΔH into ΔE or ΔE into ΔH.

Thus,
$$\Delta H = \Delta E + P\Delta V$$
$$\Delta H = \Delta E + \Delta n_g\, RT$$

It is important to note that ΔH will be equal to ΔE under the following conditions :

(i) *Reactions which do not involve any gaseous components.* For example, reaction between aqueous solution of hydrochloric acid and sodium hydroxide

$$HCl(aq) + NaOH(aq) \longrightarrow NaCl(aq) + H_2O(l)$$

(ii) *Reactions carried out in closed vessels of fixed volume i.e.,* $\Delta V = 0$.

(iii) *Reactions which involve gaseous reactants and products but there is no change in the number of moles of the gaseous components.*

For example,

$$H_2(g) + Cl_2(g) \longrightarrow 2HCl(g)$$
$$C(s) + O_2(g) \longrightarrow CO_2(g)$$

In all such reactions the volume change is zero *i.e.,* $\Delta n_g = 0$

$$\therefore \quad \Delta H = \Delta E + \Delta n_g\, RT = \Delta E + 0 \times RT = \Delta E$$

However, in gaseous reactions having $\Delta n_g$ not equal to zero, ΔH and ΔE will be different. For example :

(i) *Reactions in which there is an increase in the number of moles of the gaseous components.* In such cases, $\Delta n_g$ is positive and consequently, the enthalpy change (ΔH) is greater than the internal energy change (ΔE), *i.e.,* ΔH > ΔE. Some examples of such reactions are given below :

$$C(s) + \tfrac{1}{2}O_2(g) \longrightarrow CO(g)\ ; \qquad [\Delta n_g = (1 - ½) = + ½]$$
$$PCl_5(g) \longrightarrow PCl_3(g) + Cl_2(g)\ ; \qquad [\Delta n_g = (2 - 1) = + 1]$$

(ii) *Reactions in which there is a decrease in the number of moles of the gaseous components.* In such cases, $\Delta n_g$ is negative and consequently, the enthalpy change (ΔH) is less than the internal energy change (ΔE), *i.e.,* ΔH < ΔE. Some examples of such reactions are given below :

$$CO(g) + ½O_2(g) \longrightarrow CO_2(g)\ ; \qquad [\Delta n_g = (1 - 1½) = - ½]$$
$$N_2(g) + 3H_2(g) \longrightarrow 2NH_3(g)\ ; \qquad [\Delta n_g = (2 - 4) = - 2]$$

Thus, for gaseous reactions, the difference between ΔH and ΔE depends upon the value of $\Delta n_g$.

For example,

$$\Delta n_g = 0 \quad ; \Delta H = \Delta E$$
$$\Delta n_g = -ve \; ; \Delta H < \Delta E$$
$$\Delta n_g = +ve \; ; \Delta H > \Delta E$$

## 17.7 ENTHALPIES OF REACTIONS

*The enthalpy change (amount of heat evolved or absorbed) during a chemical reaction when, the moles of reactants and products being same as indicated by the balanced chemical equation,* is called the enthalpy **of the reaction.** For example, the enthalpy change for the reaction of one mole of methane with two moles of oxygen to form one mole of carbon dioxide and two moles of water is – 890.3 kJ *i.e.,* 890.3 kJ of heat is evolved during this reaction. Thus,

$$CH_4(g) + 2O_2(g) \longrightarrow CO_2(g) + 2H_2O(g); \qquad \Delta H = -890.3 \text{ kJ}$$

**Standard state.** It may be noted that the enthalpy of a reaction depends upon the conditions (such as temperature, pressure, etc.) under which the reaction is carried out. Therefore, it is convenient to fix up a *standard* or *a reference state* for reporting enthalpy data. According to thermodynamic convention, *a substance is in its standard state when it is at a pressure of 1 atm and temperature of 298 K.*

Therefore, the *standard enthalpy* of a reaction is the value of enthalpy change when the reaction is carried out at 1 atm pressure and 298 K. It is represented as $\Delta H°$ where the superscript indicates the standard state.

**Enthalpy of formation.** The *enthalpy of formation is defined as the enthalpy change when one mole of a compound is formed from its elements.*

It is denoted by $\Delta H_f$. For example, enthalpies of formation of carbon dioxide and water ($H_2O$) may be expressed as :

$$C(s) + O_2(g) \longrightarrow CO_2(g) \; ; \qquad \Delta H = \Delta H_f = -393.5 \text{ kJ}$$
$$H_2(g) + \tfrac{1}{2} O_2(g) \longrightarrow H_2O(l) \; ; \qquad \Delta H = \Delta H_f = -286.0 \text{ kJ}$$

If all the species of the chemical reaction are in their standard state, (*i.e.,* at 1 atm pressure and 298 K) the enthalpy of formation is called *standard enthalpy of formation,* expressed as $\Delta H_f°$. It may be noted that *the standard state of an element is the pure element in its most stable form at 298 K and one atmosphere pressure.* For example, standard state of carbon is C (graphite), sulphur is S (rhombic), oxygen is $O_2(g)$, bromine is $Br_2$ ($l$), iodine is $I_2(s)$, etc.

*The enthalpy of formation of every element in its standard state is arbitrarily assumed to be zero.*

Thus, *the enthalpy change accompanying the formation of one mole of a compound from its elements, all the substances being in their standard states (1 atm pressure and 298 K) is called standard enthalpy of formation.*

The standard enthalpies of formation (or also called standard enthalpies) of various compounds are given in Table 17.1.

**TABLE 17.1 Standard enthalpies of formation of various compounds**

| *Substance* | $\Delta H_f^\circ$ *(298 K) (kJ mol$^{-1}$)* | *Substance* | $\Delta H_f^\circ$ *(298 K) (kJ mol$^{-1}$)* |
|---|---|---|---|
| $H_2(g)$ | 0.0 | $I_2(g)$ | 62.26 |
| $H(g)$ | 217.7 | $Cl(g)$ | 121.4 |
| $H_2(l)$ | - 285.8 | $Br(g)$ | 111.8 |
| $H_2O(g)$ | - 241.8 | $I(g)$ | 106.6 |
| $H_2O_2(l)$ | - 188.0 | $HF(g)$ | - 268.6 |
| $CO(g)$ | - 110.5 | $HCl(g)$ | - 92.30 |
| $CO_2(g)$ | - 393.5 | $HBr(g)$ | - 36.23 |
| $CH_4(g)$ | - 74.85 | $HI(g)$ | 25.44 |
| $C_2H_6(g)$ | - 84.68 | $H_2S(g)$ | - 20.17 |
| $C_2H_4(g)$ | 52.30 | $HNO_3(l)$ | - 174.1 |
| $C_2H_2(g)$ | 226.7 | $H_2SO_4(l)$ | - 811.3 |
| $C_6H_6(l)$ | 49.04 | S(rhombic) | 0.0 |
| $C_2H_6(g)$ | 82.93 | S(monoclinic) | - 297.1 |
| $N(g)$ | 472.7 | $SO_2(g)$ | - 296.9 |
| $NO(g)$ | 90.37 | $Hg_2Cl_2(s)$ | - 264.9 |
| $NO_2(g)$ | 33.85 | $NaCl(s)$ | - 411.0 |
| $N_2O(g)$ | 81.85 | $Na_2CO_3(s)$ | - 113.1 |
| $NH_3(g)$ | - 46.19 | $KCl(s)$ | - 435.9 |
| $O_3(g)$ | 142.3 | $KNO_3(s)$ | - 492.7 |
| $O(g)$ | 247.5 | $AgCl(s)$ | - 127.0 |
| $OF_2(g)$ | 23.0 | $CH_3OH(l)$ | - 239.0 |
| $Br_2(g)$ | 30.71 | $C_2H_5OH(l)$ | - 277.0 |
| $I_2(s)$ | 0.0 | $CH_3COOH(l)$ | - 484.2 |

*Standard enthalpy of formation of elements = 0.0.

**Calculation of standard enthalpy of reaction ($\Delta H^\circ$).** The knowledge of standard enthalpies of formation of various substances can be used to calculate the enthalpies of reactions. In general, the standard enthalpy of any reaction ($\Delta H^\circ$) is equal to the difference between the $\Delta H_f^\circ$ values of all products and $\Delta H_f^\circ$ values of all the reactants, *i.e.*,

$$\underset{\text{for a reaction}}{\Delta H^\circ} = \begin{bmatrix}\text{Standard enthalpies}\\ \text{of formation of all}\\ \text{products}\end{bmatrix} - \begin{bmatrix}\text{Standard enthalpies}\\ \text{of formation of all}\\ \text{reactants}\end{bmatrix}$$

*i.e.*,

$$\Delta H^\circ = \Sigma \Delta H_f^\circ(\text{products}) - \Sigma \Delta H_f^\circ(\text{reactants})$$

where $\Sigma$ is known as summation and means sum of. For the reaction,

$$aA + bB \longrightarrow cC + dD$$

$$\Delta H^\circ = \Sigma\, \Delta H_f^\circ(\text{products}) - \Sigma\, \Delta H_f^\circ\ (\text{reactants})$$

$$= [c\, \Delta H_f^\circ\ (C) + d\, \Delta H_f^\circ(D)] - [a\, \Delta H_f^\circ(A) + b\, \Delta H_f^\circ(B)]$$

**Enthalpy of combustion.** *The enthalpy of combustion of a substance is the amount of heat evolved when 1 mole of a substance is completely burnt or oxidised.* For example,

$$C(s) + O_2(g) \longrightarrow CO_2(g); \qquad \Delta H = \Delta H_c = -393.5 \text{ kJ}$$

$$CH_4(g) + 2O_2(g) \longrightarrow CO_2(g) + 2H_2O(l); \qquad \Delta H = -890.3 \text{ kJ}$$

**Example 1.** *A system absorbs 500 J of heat and does work equivalent to 200 J on its surroundings. Calculate change in internal energy.*

**Solution.** Heat absorbed by the system = 500 J or $q$ = 500 J.

Work done by the system = 200 J or $w = -200$ J

From first law of thermodynamics,

$$\Delta E = q + w$$

$$\Delta E = 500 + (-200) = \mathbf{300\ J.}$$

**Example 2.** *A sample of gas present in a cylinder fitted with a frictionless piston expands against a constant pressure of 1 atm from a volume of 2 L to 12 L. During the process, it absorbs 600 J of heat from the surroundings. Calculate the change in internal energy of the system.*

**Solution.** During the process,

$$q = 600 \text{ J}, \quad \Delta V = 12 - 2 = 10 \text{ L}, \quad P = 1 \text{ atm}$$

$$w = -P\Delta V$$

$$= -1 \times 10 = -10 \text{ L atm}$$

Now, $\quad 1 \text{ L atm} = 101.3 \text{ J}$

$\therefore \quad w = -10 \times 101.3 = -1013 \text{ J}$

According to first law of thermodynamics,

$$\Delta E = q + w$$

$$= 600 - 1013 = \mathbf{-413\ J.}$$

**Example 3.** *The heat of combustion of gaseous methane ($CH_4$) at constant volume is measured in a bomb calorimeter at 298 K and is found to be – 885.4 kJ mol$^{-1}$. Find the value of enthalpy change.*

**Solution.** Combustion of methane gives $CO_2(g)$ and $H_2O(l)$ as :

$$CH_4(g) + 2O_2(g) \longrightarrow CO_2(g) + 2H_2O(l)$$

$$\Delta E = -885.4 \text{ kJ mol}^{-1} = 885400 \text{ J mol}^{-1}$$

$$\Delta n_g = 1 - (1 + 2) = -2 \text{ mol}$$

$$T = 298 \text{ K},\ R = 8.314 \text{ J mol}^{-1} \text{ K}^{-1}$$

Now, $\quad \Delta H = \Delta E + \Delta n_g RT$

$$= -885400 + (-2 \text{ mol}) \times (8.314 \text{ J mol}^{-1} \text{ K}^{-1}) \times (298 \text{ K})$$

$$= -885400 - 4955$$

$$= -890355 = \mathbf{-890.355\ kJ.}$$

**Example 4.** *The enthalpy change ($\Delta H$) for the reaction :*

$$N_2(g) + 3H_2(g) \longrightarrow 2NH_3(g)$$

*is – 92.38 kJ at 298 K. What is $\Delta E$ at 298 K ?*

**Solution.** $\Delta H$ and $\Delta E$ are related as

$$\Delta H = \Delta E + \Delta n_g RT$$

For the reaction,

$$N_2(g) + 3H_2(g) \longrightarrow 2NH_3(g)$$

$$\Delta n_g = 2 - (1 + 3) = -2 \text{ mol, } T = 298 \text{ K}$$

$$\Delta H = -92.38 \text{ kJ} = -92380 \text{ J, } R = 8.314 \text{ JK}^{-1} \text{ mol}^{-1}$$

$$-92380 = \Delta E + (-2 \text{ mol}) \times (8.314 \text{ J mol}^{-1} \text{ K}^{-1}) \times (298 \text{ K})$$

$$-92380 = \Delta E - 4955$$

$$\Delta E = -92380 + 4955$$

$$= -87425 \text{ J} = \mathbf{-87.425 \text{ kJ}}.$$

**Example 5.** *Calculate the standard enthalpy change for the reaction :*

$$CH_4(g) + 2O_2(g) \longrightarrow CO_2(g) + 2H_2O(l)$$

*The standard enthalpies of formation of $CH_4(g)$, $CO_2(g)$ and $H_2O(l)$ are – 74.8 kJ mol$^{-1}$, – 393.5 kJ mol$^{-1}$ and – 285.8 kJ mol$^{-1}$ respectively.*

**Solution.** The standard enthalpy change for the reaction, $\Delta H°$ is given as :

$$\Delta H° = \Sigma\Delta H_f° \text{ (products)} - \Sigma\Delta H_f° \text{ (reactants)}$$

$$= \{\Delta H_f°[CO_2(g)] + 2\Delta H_f°[H_2O(l)]\} - \{\Delta H_f°[CH_4(g)] + 2\Delta H_f°[O_2(g)]\}$$

$$\Delta H_f°[CO_2(g)] = -393.5 \text{ kJ mol}^{-1}, \Delta H_f°[H_2O(l)] = -285.8 \text{ kJ mol}^{-1}$$

$$\Delta H_f°[CH_4(g)] = -74.8 \text{ kJ mol}^{-1}, \Delta H_f°[O_2(g)] = 0 \quad \text{(by convention)}$$

$$\therefore \quad \Delta H° = [(-393.5) + 2(-285.8)] - [(-74.8) + 2 \times 0]$$

$$= -965.1 + 74.8 = \mathbf{-890.3 \text{ kJ mol}^{-1}}.$$

**Example 6.** *Calculate standard enthalpy change ($\Delta H°$) and standard internal energy change ($\Delta E°$) for the reaction :*

$$OF_2(g) + H_2O(g) \longrightarrow O_2(g) + 2HF(g)$$

*Given that standard enthalpies of formation, $\Delta H_f°$ of $OF_2(g)$, $H_2O(g)$ and $HF(g)$ are + 23.0, – 241.8 and – 268.6 kJ mol$^{-1}$ respectively.*

**Solution.** The standard enthalpy change for the reaction is :

$$\Delta H° = \Sigma\Delta H_f° \text{ (products)} - \Sigma\Delta H_f° \text{ (reactants)}$$

$$= [\Delta H_f°\{O_2(g)\} + 2\Delta H_f°\{HF(g)\}] - [\Delta H_f°\{OF_2(g)\} + \Delta H_f°\{H_2O(g)\}]$$

$$= [0 + 2(-268.6)] - [(+23.0) + (-241.8)]$$

$$= -573.2 + 218.8 = \mathbf{-318.4 \text{ kJ}}$$

Now, $$\Delta H° = \Delta E° + \Delta n_g RT$$

or $$\Delta E° = \Delta H° - \Delta n_g RT$$

$$\Delta H° = -318.4 \text{ kJ}, \Delta n_g = (1 + 2) - (1 + 1) = 1 \text{ mol}$$
$$R = 8.314 \text{ JK}^{-1} \text{ mol}^{-1}, T = 298 \text{ K},$$
$$\therefore \quad \Delta E° = -318.4 \text{ kJ} - (1 \text{ mol}) \times (8.314 \times 10^{-3} \text{ kJ K}^{-1} \text{ mol}^{-1}) \times (298 \text{ K})$$
$$= -318.4 \text{ kJ} - 2.5 \text{ kJ} = \mathbf{-320.9 \text{ kJ}}.$$

## 17.8 ENTHALPY CHANGE FOR REACTIONS OCCURRING IN SEVERAL STAGES : HESS'S LAW

We know that reaction can occur in several stages. To calculate the overall enthalpy change, G.H. Hess (1840) proposed an important generalisation regarding the enthalpies of reactions known as *Hess's law.* It states that *the enthalpy change in a particular reaction is always constant and is independent of the path in which the reaction takes place.*

This means that the amount of heat evolved or absorbed in a chemical reaction is always same whether the reaction takes place in one step or in several steps. Suppose in a reaction, the reactant A changes to product B in one step and the enthalpy change during this process is ΔH. Now, suppose the same process is carried out in two steps (Fig. 17.3).

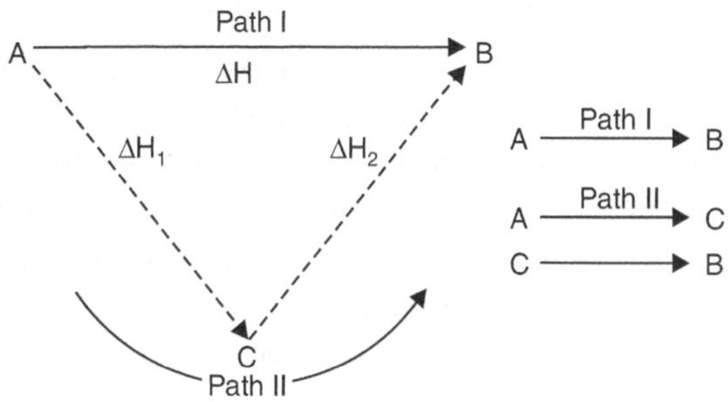


*FIGURE 17.3* **Illustration of Hess's law.**

(*i*) The reactant A changes to C (intermediate stage) and the enthalpy change during this step is $\Delta H_1$.

(*ii*) C changes to the product B and the enthalpy change during this step is $\Delta H_2$.

Then, according to Hess's law, the enthalpy change for the reaction, whether the reaction follows path I or path II, is same *i.e.,*

$$\Delta H = \Delta H_1 + \Delta H_2$$

**Theoretical Justification of the Hess's Law.** The theoretical justification for the Hess's law can be obtained from the following considerations :

Ths substance A can be converted into B by two paths. In path I, A is converted directly into B and let heat evolved in this step be $q$. On the other hand, when path II is followed, let the heats evolved during the steps A to C and C to B be $q_1$ and $q_2$ respectively. The total heat evolved in path II. $q' = q_1 + q_2$. According to Hess's law $q'$ must be equal to $q$.

Let us suppose that $q'$ is greater than $q$. This means that by converting A to B by path II and then converting B back to A by path I, the heat energy equal to $q' - q$ can be created. This directly comes in conflict with the law of conservation of energy (first law of thermodynamics) according to which the energy can neither be created nor destroyed. Hence, $q$ must be equal to $q'$. In other words, heat changes during the process must be independent of the path. This is Hess's law.

### 17.8.1 Illustration of Hess's Law

Let us illustrate the law by considering the formation of carbon dioxide from carbon and oxygen. Carbon can be converted into $CO_2$ by two ways :

**Path 1.** $C(s) + O_2(g) \longrightarrow CO_2(g) \quad \Delta H = -393.5 \text{ kJ}$

**Path 2.** $C(s) + ½O_2(g) \longrightarrow CO(g) \quad \Delta H_1 = -110.5 \text{ kJ}$

$CO(g) + ½O_2(g) \longrightarrow CO_2(g) \quad \Delta H_2 = -283.0 \text{ kJ}$

---

$C(s) + O_2(g) \longrightarrow CO_2(g) \quad \Delta H = -393.5 \text{ kJ}$

$\therefore \quad \Delta H = \Delta H_1 + \Delta H_2$

*Hence, it may be concluded that according to Hess's law, the thermochemical equations can be added, subtracted or multiplied like algebraic equations to obtain the desired equation.*

### 17.8.2 Applications of Hess's Law

Hess's law has been found to be very useful in calculating the enthalpy changes for chemical reactions which are not suitable for direct measurements with calorimeter. Therefore, enthalpy changes for these reactions can be calculated from the experimentally determined values for some other appropriate reactions. Some of the important applications of Hess's law are discussed below :

1. **Determination of enthalpy of formation of substances which cannot be measured experimentally.** There are a large number of compounds such as methane ($CH_4$), carbon monoxide (CO), benzene ($C_6H_6$) etc., which cannot be synthesised directly from their elements. Therefore, their enthalpies of formation cannot be determined experimentally by calorimetric method. The enthalpies of formation of such substances can be determined by an indirect method based on Hess's law. This is illustrated by the following examples :

**Example 7.** *Calculate the enthalpy of formation of benzene represented by the following reaction :*

$$6C(s) + 3H_2(g) \longrightarrow C_6H_6(l)$$

*The standard enthalpy of combustion of benzene is – 3268.0 kJ and the standard enthalpies of formation of $CO_2(g)$ and $H_2O(l)$ are – 393.5 and – 285.8 kJ respectively.*

**Solution.** The desired equation is

(*i*) $\quad 6C(s) + 3H_2(g) \longrightarrow C_6H_6(l) \qquad \Delta H = ?$

Other equations representing various reactions can be written as :

(*ii*) $\quad C_6H_6(l) + 7\frac{1}{2}O_2(g) \longrightarrow 6CO_2(g) + 3H_2O(l); \qquad \Delta H = -3268.0 \text{ kJ}$

(*iii*) $\quad C(s) + O_2(g) \longrightarrow CO_2(g); \qquad \Delta H = -393.5 \text{ kJ}$

(*iv*) $\quad H_2(g) + \frac{1}{2}O_2(g) \longrightarrow H_2O(l); \qquad \Delta H = -285.8 \text{ kJ}$

Multiply Eq. (*iii*) by 6 and Eq. (*iv*) by 3 and add

(*iii*) × 6, $\quad 6C(s) + 6\,O_2(g) \longrightarrow 6CO_2(g); \qquad \Delta H = -2361.0 \text{ kJ}$

(*iv*) × 3, $\quad 3H_2(g) + \frac{3}{2}\,O_2(g) \longrightarrow 3H_2O(l); \qquad \Delta H = -857.4 \text{ kJ}$

---

(*v*) $\quad 6C(s) + 3H_2(g) + \frac{15}{2}\,O_2(g) \longrightarrow 6CO_2(g) + 3H_2O(l); \quad \Delta H = -3218.4 \text{ kJ}$

Subtract eq. (*ii*) from eq. (*v*)

$$C_6H_6(l) + \frac{15}{2} O_2(g) \longrightarrow 6CO_2(g) + 3H_2O(l); \qquad \Delta H = -3268.0 \text{ kJ}$$

$$6C(s) + 3H_2(g) \longrightarrow C_6H_6(l); \qquad \Delta H = +49.6 \text{ kJ}$$

∵ Enthalpy of formation of benzene = + **49.6 kJ.**

2. **Determination of enthalpy of transition.** Hess's law is useful in determining the enthalpies of transition of the allotropic modifications such as graphite do diamond, rhombic sulphur to monoclinic sulphur, yellow phosphorus to red phosphorus etc. This is illustrated by the following example :

**Example 8.** *The enthalpies of combution of diamond and graphite are – 395.4 kj and – 393.5 kJ respectively. Calculate the enthalpy of transformation from diamond to graphite.*

**Solution.** The enthalpies of combustion may be represented as :

(*i*) $C(\text{diamond}) + O_2(g) \longrightarrow CO_2(g); \qquad \Delta H = -395.4 \text{ kJ}$

(*ii*) $C(\text{graphite}) + O_2(g) \longrightarrow CO_2(g); \qquad \Delta H = -393.5 \text{ kJ}$

The conversion can be written as :

$$C\ (\text{diamond}) \longrightarrow C(\text{graphite}) \qquad \Delta H = ?$$

This can be obtained by subtracting Eq. (*ii*) from Eq. (*i*)

$$C\ (\text{diamond}) + O_2(g) \longrightarrow CO_2(g); \qquad \Delta H = -395.4 \text{ kJ}$$
$$C\ (\text{graphite}) + O_2(g) \longrightarrow CO_2(g); \qquad \Delta H = -393.5 \text{ kJ}$$

$$C\ (\text{diamond}) \longrightarrow C\ (\text{graphite}) \qquad \mathbf{\Delta H = -1.9\ kJ}$$

3. **Predicting enthalpies of various reactions.** The practical utility of Hess's law lies in the fact that *thermochemical equations may be added, subtracted, multiplied or divided by numerical factors like algebraic equations.* As a result, it is possible to calculate enthalpies of many reactions whose direct measurements would be difficult or even quite impossible.

**Example 9.** *Calculate the enthalpy of formation of anhydrous aluminium chloride, $Al_2Cl_6$ from the following data :*

(*i*) $2Al(s) + 6HCl(aq) \longrightarrow Al_2Cl_6(aq) + 3H_2(g); \quad \Delta H = -1004.0 \text{ kJ}$

(*ii*) $H_2(g) + Cl_2(g) \longrightarrow 2HCl(g); \qquad \Delta H = -183.9 \text{ kJ}$

(*iii*) $HCl(g) + (aq) \longrightarrow HCl(aq); \qquad \Delta H = -73.2 \text{ kJ}$

(*iv*) $Al_2Cl_6(s) + (aq) \longrightarrow Al_2Cl_6(aq); \qquad \Delta H = -643 \text{ kJ}$

**Solution.** The formation of $Al_2Cl_6$ from its elements may be represented as :

$$2Al(s) + 3Cl_2(g) \longrightarrow Al_2Cl_6(s) \qquad \Delta H = ?$$

Multiply Eq. (*ii*) by 3 and Eq. (*iii*) by 6 and add

(*ii*) × 3, $3H_2(g) + 3\ Cl_2(g) \longrightarrow 6HCl(g); \qquad \Delta H = -551.7 \text{ kJ}$

(*iii*) × 6, $6HCl(g) + (aq) \longrightarrow 6HCl(aq); \qquad \Delta H = -439.2 \text{ kJ}$

(*v*) $3H_2(g) + 3Cl_2(g) + (aq) \longrightarrow 6HCl\ (aq); \qquad \mathbf{\Delta H = -990.0\ kJ}$

Add Eq. (*i*) to Eq. (*v*)

$$(i) \quad 2Al(s) + 6HCl(aq) \longrightarrow Al_2Cl_6(aq) + 3H_2(g); \quad \Delta H = -1004.0 \text{ kJ}$$

Add (*vi*)

$$2Al(s) + 3Cl_2(g) + (aq) \longrightarrow Al_2Cl_6(aq); \quad \Delta H = -1994.9 \text{ kJ}$$

Subtract Eq. (*iv*) from Eq. (*vi*)

$$(vi) \quad Al_2Cl_6(s) + aq \longrightarrow Al_2Cl_6(aq); \quad \Delta H = -643 \text{ kJ}$$

Subtract $\quad 2Al(s) + 3Cl_2(g) \longrightarrow Al_2Cl_6(g); \quad \Delta H = -1351.9 \text{ kJ}$

Thus, enthalpy of formation of anhydrous $Al_2Cl_6$ = **– 1351.9 kJ**

## 17.9 BOND ENERGIES

*The amount of energy required to break a bond between the atoms in a molecule in the gaseous state* is called the *bond dissociation energy*. For example, 435.4 kJ of energy is required to break or dissociate the H—H bond in one mole of hydrogen molecules. This may be expressed as :

$$H_2(g) \longrightarrow 2H(g); \quad \Delta H = 435.4 \text{ kJ mol}^{-1}$$

(Positive value of ΔH means that bond breaking is an endothermic process). Similalry , the bond energy of Cl—Cl bond in $Cl_2$ is 243 kJ mol$^{-1}$ and I—I in $I_2$ is 15.1 kJ mol$^{-1}$.

The bond energies or bond dissociation energies for diatomic molecules containing two different atoms are calculated by using Hess's law.

For example, for HCl molecule,

$$HCl(g) \longrightarrow H(g) + Cl(g)\ ; \quad \Delta H = ?$$

Here, enthalpy of formation of HCl(*g*), H(*g*) and Cl(*g*) are :

$$HCl(g) = -92.2 \text{ kJ mol}^{-1} \qquad H(g) = 217.7 \text{ kJ mol}^{-1}$$

$$Cl(g) = 121.4 \text{ kJ mol}^{-1}$$

$$\Delta H^\circ = \Sigma\Delta H_f^\circ \text{(products)} - \Sigma\Delta H_f^\circ \text{(reactants)}$$

$$= \Sigma\{\Delta H_f^\circ [H(g)] + \Delta H_f^\circ [Cl(g)]\} - \Delta H_f^\circ [HCl(g)]$$

$$= 217.7 + 121.4 - (-92.2) = 431.3 \text{ kJ mol}^{-1}$$

Thus, bond dissociation energy of gaseous HCl is 431.3 kJ mol$^{-1}$. In case of *polyatomic molecules* (which contain more than one bond of the same type) the average of bond dissociation energies of the bonds are taken. *The average value of bond dissociation energies is* called bond energy. For example, the successive bond dissociation energies of the C—H bonds in $CH_4$ are :

$$H_3C\text{—}H(g) \longrightarrow CH_3(g) + H(g); \quad \Delta H = 425 \text{ kJ mol}^{-1}$$

$$H_2C\text{—}H(g) \longrightarrow CH_2(g) + H(g); \quad \Delta H = 470 \text{ kJ mol}^{-1}$$

$$HC\text{—}H(g) \longrightarrow CH(g) + H(g); \quad \Delta H = 416 \text{ kJ mol}^{-1}$$

$$CH \longrightarrow C(g) + H(g); \quad \Delta H = 335 \text{ kJ mol}^{-1}$$

The bond energy is the average of these four dissociation energies *i.e.*,

$$\text{Bond energy of C—H bond} = \frac{425 + 470 + 416 + 335}{4} = \frac{1646}{4} = 411.5 \text{ kJ}$$

Thus, **bond energy** may be defined as *the average energy required to break bonds of a given type in one mole of the gaseous molecules.*

Obviously, *for diatomic molecules (such as $H_2$, $Cl_2$, $Br_2$ etc.) the bond dissociation energy is same as bond energy.*

**Use of bond energies to calculate enthalpy change of reactions.** The bond energy data can also be used to calculate the enthalpy change of many reactions. The chemical reaction involves the cleavage of old bonds in reactants and formation of new bonds in products. The enthalpy change of the reaction may be given as :

$$\textbf{Enthalpy of reaction} = \begin{bmatrix}\textbf{Sum of bond energies}\\ \textbf{of reactants}\end{bmatrix} - \begin{bmatrix}\textbf{Sum of bond}\\ \textbf{energies of products}\end{bmatrix}$$

For example, consider the reaction,

$$H_2(g) + Cl_2(g) \longrightarrow 2HCl(g)$$

In this reaction, it is clear that one H—H bond and one Cl—Cl bond break while two H—Cl bonds are formed.

$$\therefore \quad \Delta H = \text{B.E. (H—H)} + \text{B.E. (Cl—Cl)} - 2\text{B.E. (H—Cl)}$$

The bond energies of different bonds have been found to be

H—H bond = 435 kJ, Cl—Cl bond = 243 kJ, H—Cl bond = 431 kJ

$$\therefore \quad \Delta H = 435 + 243 - 2 \times 431 = \mathbf{-\ 184\ kJ.}$$

The average bond energies of some common bonds are given in Table 17.2.

**Table 17.2. Average bond energies of some common bonds**

| *Bond* | *Average bond energy ($kJ\ mol^{-1}$)* | *Bond* | *Average bond energy ($kJ\ mol^{-1}$)* |
|---|---|---|---|
| H—H | 435 | N ≡ N | 940 |
| H—F | 565 | C—C | 347 |
| H—Cl | 431 | C = C | 619 |
| H—Br | 364 | C ≡ C | 812 |
| H—I | 297 | O—H | 463 |
| F—F | 155 | N—H | 389 |
| Cl—Cl | 243 | C—H | 413 |
| Br—Br | 190 | C—Cl | 326 |
| I—I | 149 | C—O | 335 |
| O—O | 138 | C = O | 707 |
| O = O | 494 | C—N | 293 |
| N—N | 159 | C = N | 616 |
| N = N | 418 | C ≡ N | 879 |

**Example 10.** *Calculate ΔH for the reaction :*

$$\begin{array}{c} H \\ | \\ H\text{—}C\text{—}Cl(g) \\ | \\ Cl \end{array} \longrightarrow C(g) + 2H(g) + 2Cl(g)$$

*The average bond energies of C—H bond and C—Cl bond are 413.0 kJ mol$^{-1}$ and 326.0 kJ mol$^{-1}$ respectively.*

**Solution.** In the reaction,

$$\begin{array}{c} \mathrm{H} \\ | \\ \mathrm{H\text{—}C\text{—}Cl}(g) \\ | \\ \mathrm{Cl} \end{array} \longrightarrow \mathrm{C}(g) + 2\mathrm{H}(g) + 2\mathrm{Cl}(g)$$

two C—H and two C—Cl bonds are to be broken so that

$$\Delta H = 2\text{ B.E. (C—H)} + 2\text{ B.E. (C—Cl)}$$

$$\Delta H = 2(413.0) + 2(326.0)$$

$$= 826.0 + 652.0 = \mathbf{1482\ kJ.}$$

**Example 11.** *Calculate the C—C bond energy from the following data :*

(*i*) $2C(graphite) + 3H_2(g) \longrightarrow C_2H_6(g)$; $\Delta H = -84.67$ kJ mol$^{-1}$

(*ii*) $C(graphite) \longrightarrow C(g)$; $\Delta H = 716.7$ kJ mol$^{-1}$

(*iii*) $H_2(g) \longrightarrow 2H(g)$; $\Delta H = 435.9$ kJ mol$^{-1}$

(*iv*) Bond energy of C—H bond is 416 kJ mol$^{-1}$.

**Solution.** Let us first of all, calculate ΔH for the reaction containing C—C bond.

$$2C(g) + 6H(g) \longrightarrow C_2H_6(g)$$

Multiply Eq. (*ii*) by 2 and subtract from Eq. (*i*)

(*i*) $2C(graphite) + 3H_2(g) \longrightarrow C_2H_6(g)$; $\Delta H = -84.67$ kJ mol$^{-1}$

(*ii*) × 2, $2C(graphite) \longrightarrow 2C(g)$; $\Delta H = 1433.4$ kJ mol$^{-1}$

---

(*iv*) $2C(g) + 3H_2(g) \longrightarrow C_2H_6(g)$; $\Delta H = -1518.07$ kJ mol$^{-1}$

Multiply Eq. (*iii*) by 3 and subtract from Eq. (*iv*)

$3H_2(g) \longrightarrow 6H(g)$; $\Delta H = -1307.7$ kJ mol$^{-1}$

---

$2C(g) + 6H(g) \longrightarrow C_2H_6(g)$ $\Delta H = -2825.77$ kJ mol$^{-1}$

In this case, one C—C bond and six C—H bonds are formed so that

$$\Delta H = \Sigma\text{B.E. of reactants} - \Sigma\text{B.E. of products}$$

or $$\Delta H = 0 - \text{B.E. (C—C)} - 6\text{ B.E. (C—H)}$$

$$-2825.77 = -\text{B.E. (C—C)} - 6 \times 416$$

or $$\text{B.E. (C—C)} = 2825.77 - 2496 = \mathbf{329.77\ kJ\ mol^{-1}.}$$

**Example 12.** *Calculate $\Delta H°$ for the reaction,*

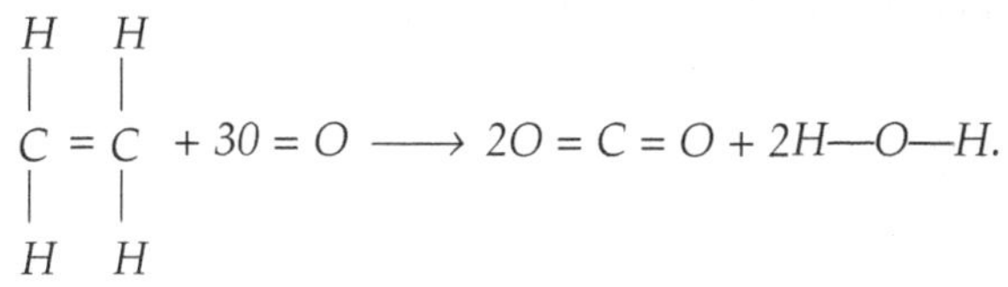


*The average bond energies of various bonds are :*

| *Bond* | *C—H* | *O = O* | *C = O* | *O—H* | *C = C* |
|---|---|---|---|---|---|
| *Bond energy (kJ mol$^{-1}$)* | *414* | *499* | *724* | *460* | *619* |

**Solution.** In this reaction, four C—H bonds, one C = C bond and three O=O bonds are broken while four C=O bonds and four O—H bonds are formed. Thus,

$$\Delta H° = 4 \text{ B.E. (C—H)} + \text{B.E. (C=C)} + 3 \text{ B.E. (O=O)} - 4 \text{ B.E. (C=O)} - 4\text{B.E. (O—H)}$$

$$= 4 \times 414 + 619 + 3 \times 499 - 4 \times 724 - 4 \times 460$$

$$= \mathbf{964 \ kJ \ mol^{-1}}.$$

## 17.10 HEAT CAPACITY

*Heat capacity* (*c*) of a substance is *the amount of heat needed to raise the temperature of 1 g of it through 1°C ; whereas molar heat capacity* (C) *is the amount of heat required to raise the temperature of 1 mole of it through* 1°C. Molar heat capacity is of *two* types :

(*i*) *Molar heat capacity at constant volume* ($C_v$) is measured by keeping *volume constant.* It is given by :

$$C_v = \left(\frac{\partial E}{\partial T}\right)_v$$

(*ii*) *Molar heat capacity at constant pressure* ($C_p$) is measured by keeping *pressure constant.* It is given by :

$$C_p = \left(\frac{\partial H}{\partial T}\right)_p = \frac{dq}{dT}$$

**Relation between $C_p$ and $C_v$** : $H = E + PV$

When the temperature is changed ; then $\Delta H = \Delta E + \Delta(PV)$

or $$\frac{\Delta H}{T} = \frac{\Delta E}{\Delta T} + \frac{\Delta(PV)}{\Delta T}$$

or $$C_p = C_v + \frac{\Delta(PV)}{\Delta T} \quad ...(17.10)$$

For 1 mole of any ideal gas $P_1V_2 = RT_2$ and $P_1V_1 = RT_1$, so that $\Delta(PV) = R(T_2 - T_2) = R\,\Delta\,T$

∴ For 1 mole of gas $C_p = C_v + R$ ...(17.11)

For *solids and liquids*, $\Delta H = \Delta E$ and $\Delta (PV) \approx 0$, so that $C_p = C_v$.

## 17.11 KIRCHOFF'S EQUATION

The *variation of heat of reaction with temperature* can be derived with the help of first law of thermodynamics. Consider a simple hypothetical reaction in which the reactant in state A at the temperature $T_1$ yields the product in state B at the temperature $T_2$. We may proceed from state A at $T_1$ to B at $T_2$ by *two paths* as shown in Fig. 17.4.

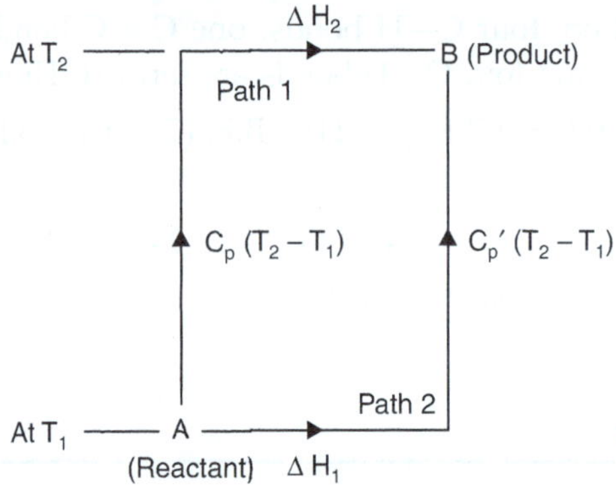


***FIGURE 17.4*** **Alternative paths for changing reactants into products.**

Let $\Delta H_1$ be the heat of reaction at temperature $T_1$ and $\Delta H_2$ be the heat of reaction at temperature $T_2$. Again let $C_p$ and $C'_p$ represent respectively the mean molar heat capacity of the reactants and product at *constant pressure*.

(*i*) In the ***first path***, the reactants are first heated from $T_1$ to $T_2$ and then allowed to react at the higher temperature, $T_2$. Therefore, the total heat change involved is given by :

$$\Delta H(\text{path 1}) = C_p (T_2 - T_1) + \Delta H_2 \qquad ...(17.12)$$

(*ii*) In the ***second path***, the reactants are allowed to react all the lower temperature, $T_1$, and then the products are heated to $T_2$. The value of total heat change ($\Delta H$) for this path is as follows :

$$\Delta H\ (\text{path 2}) = \Delta H_1 + C'_p (T_2 - T_1) \qquad ...(17.13)$$

According to the first law of thermodynamics,

$$\Delta H(\text{path 1}) = \Delta H(\text{path 2})$$

or

$$C_p(T_2 - T_1) + \Delta H_2 = \Delta H_1 + C'_p(T_2 - T_1)$$

∴

$$\boxed{\frac{\Delta H_2 - \Delta H_1}{(T_2 - T_1)} = C'_p - C_p = \Delta C_p} \qquad ...(17.14)$$

where $\Delta C_p$ = difference in the heat capacities of the products and the reactants at constant pressure.

This equation (*iii*), is called **Kirchoff's equation**

Above equation may also be written as :

$$\frac{\Delta H_2 - \Delta H_1}{\Delta T} = \Delta C_p \qquad [\text{where } \Delta T = T_2 - T_1]$$

If $\Delta T$ is *small*, then it takes the form :

$$\left(\frac{\partial(\partial H)}{\partial T}\right)_P = \Delta C_p \qquad ...(17.15)$$

Similarly, we can obtain the equation for the variation of $\Delta E$ with temperature, namely :

$$\frac{\Delta E_2 - \Delta E_1}{\Delta T} = \Delta C_v \qquad ...(17.16)$$

or if $\Delta T$ is *small*, then

$$\left(\frac{\partial(\partial E)}{\partial T}\right)_V = \Delta C_v \qquad ...(17.17)$$

## 17.12 LIMITATIONS OF FIRST LAW OF THERMODYNAMICS

There are three main limitations of first law of thermodynamics. These are as follows :

1. The first law of thermodynamics says that *all* physical and chemical processes take place in such a manner that the total energy of the universe is constant and are form of energy can be transformed into other form. But this law does not tell us under what conditions and to what extent it is possible to bring about conversion of one form of energy into the other. This can be seen in following examples :

   (*i*) Consider that heat is flowing from hot to cold body. First law simply tells us that the quantity of heat lost by the hot body is equal to the heat gained by the cold body. It never tells us that heat can only flow from the hot to the cold body and not in the reverse direction.

   (*ii*) Zinc dissolves spontaneously in sulphuric acid, a definite amount of energy is liberated. The reverse process is not spontaneous. Without external influence the reaction proceeds in only one direction.

2. The first law of thermodynamics is a qualitative statement which does not preclude the existance of either a heat engine or a refrigerator. The first law does not contradict the existance of a 100% efficient heat engine or a self acting refrigerator. In practice these two are not attainable.

3. The first law fails to explain why chemical reactions do not proceed to completion.

## 17.13 RANDOMNESS AND SPONTANEITY

In order to find out some other factor which may be responsible for the feasibility of a process, let us examine endothermic reactions. The simplest process occurring without any energy change (*i.e.*, $\Delta H$ is almost zero) may be illustrated with the help of following experiment.

Let us consider two containers connected by means of a valve as shown in Fig. 17.5. The containers are perfectly insulated so that no heat enters or leaves the system. While one container is filled with a litre of helium gas, the other is filled with a litre of neon gas. The two gases are non-reacting (being inert gases). When the valve is opened, we observe that the two gases get mixed together spontaneously. During the mixing, there is a negligible energy change and, therefore, the criterion of decrease of energy for feasibility of process does not help us. If we look closely to the final and initial states of system, we observe that in the initial state the two gases are orderly filled separately in the two containers. On the other hand, after mixing, helium and neon are distributed between both the containers. In other words, each gas occupies a large volume in the final state and, thereby, number of possible locations for the molecules of each gas is increased. This means that there is more disorder or randomness on mixing. Thus, the gases mix to achieve more randomness. Once mixed the gases cannot separate because to do so they will involve decrease in randomness due to lesser volume. Thus, it may be concluded that *the process proceeds spontaneously in a direction in which the randomness or disorder of the system increases.*

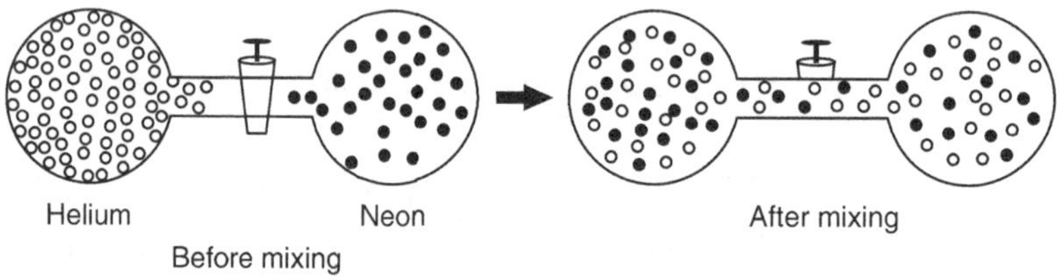


***FIGURE 17.5*** **Process of mixing of two gases. Gases mix spontaneously because of increase in randomness.**

Some common examples of processes which occur in the direction of increased randomness are given below :

(*i*) *Conversion of solids into liquids.* The melting of solids into liquids (ΔH, +ve) results in the increase of randomness.

$$H_2O(s) \longrightarrow H_2O(l)$$

(*ii*) *Evaporation of water.* The evaporation of water also results into increase of randomness because the molecules in the vapour state have more randomness than in the liquid state.

$$H_2O(l) \longrightarrow H_2O(g)$$

(*iii*) *Dissolution of ammonium chloride in water.* Solid ammonium chloride has less randomness while in solution ammonium chloride particles move freely as $NH_4^+$ and $Cl^-$ ions and hence, randomness increases.

$$NH_4Cl(s) + \text{Water} \longrightarrow NH_4^+(aq) + Cl^-(aq)$$

Thus, we find that in all the above endothermic processes, there is always increase in randomness. Since the above reactions are spontaneous, *the tendency to achieve maximum randomness is another factor which determines the spontaneity of a process.*

### Driving Force as the Overall Tendency for a Process

We have learnt that the spontaneous processes occur because of the two tendencies :

(*i*) *Tendency of a system to acquire a state of minimum energy.*

(*ii*) *Tendency of a system to acquire a state of maximum randomness.*

The *overall tendency for a process to take place by itself* is called *the driving force.*

Regarding these two tendencies it is very important to keep in mind the following points :

(*i*) *The two tendencies act independent of each other.*

(*ii*) *The two tendencies may work in the same or in opposite directions in a process.*

Thus, when there is no change in energy the second tendency is the governing factor *i.e.*, reaction will be spontaneous or non-spontaneous depending upon whether randomness factor favours or opposes. On the other hand, if there is no change in randomness, the energy factor is the controlling factor. However, this is not so simple in actual practice as both the tendencies operate simultaneously. In these cases, the significance of the magnitude of these tendencies becomes important. Obviously, *the tendency with the greater magnitude determines whether the process is feasible or not.*

## 17.14 ENTROPY

The extent of disorder or randomness in a system may be expressed by a property known as entropy. ***Entropy*** may be defined as *the property of a system which measures the degree of disorder or randomness in the system.*

It is generally expressed by the symbol, S. Entropy like internal energy and enthalpy, is a *state function* and change in entropy therefore, depends only on the initial and final states of the system. The change in entropy during the process when a system undergoes a change from one state to another is represented by ΔS. Thus,

$$\Delta S = S \text{ (final state)} - S \text{ (initial state)}$$

and for chemical reactions,

$$\Delta S = S \text{ (products)} - S \text{ (reactants)}$$

For a reversible process at equilibrium, the change in entropy may be expressed as :

$$\Delta S = \frac{q_{rev}}{T} \qquad ...(17.18)$$

where $q_{rev}$ represents the heat supplied at temperature T in a reversible process.

**Units of entropy.** Since entropy change is expressed as heat divided by temperature, the units of entropy are calories per degree (expressed as entropy units eu) or joules per kelvin, $JK^{-1}$ (expressed as EU).

### Physical Meaning of Entropy

The physical meaning of entropy is that entropy is a measure of degree of disorder (or randomness) of a system. The relation between entropy and disorder provides a suitable explanation for entropy change in various processes. The greater the disorder in a system, the higher is the entropy. Obviously, for a given substance the solid state is the state of lowest entropy (most ordered state), the gaseous state is the state of highest entropy and the liquid state is intermediate between the two. In the case of mixing of two gases when the stop cock is opened, the gases mix to achieve more randomness or disorder. In this case, there is no exchange of matter or energy between the system and the surroundings. The change occurs from ordered state (less entropy) to disordered state (higher entropy). Thus the change in entropy is positive.

## 17.15 SECOND LAW OF THERMODYNAMICS

The second law of thermodynamics introduces the concept of entropy and its relation with spontaneous processes. In an isolated system such as mixing of gases, there is no exchange of energy or matter between the system and the surroundings. But due to increase in randomness, there is increase in entropy. Thus, we can say that for a spontaneous process in an isolated system, the change in entropy is positive, *i.e.*, $\Delta S > 0$. However, if the system is not isolated, we have to take into account the entropy changes of the system and the surroundings. Then, the total entropy change ($\Delta S_{total}$) will be equal to the sum of the change in entropy of the system ($\Delta S_{system}$) and the change in entropy of the surroundings ($\Delta S_{surroundings}$) *i.e.*,

$$\Delta S_{total} = \Delta S_{system} + \Delta S_{surroundings}$$

For a spontaneous process, $\Delta S_{total}$ must be positive, *i.e.*,

$$\Delta S_{total} = \Delta S_{system} + \Delta S_{surroundings} > 0 \quad ...(17.19)$$

But system and surroundings constitute *universe* for thermodynamic point of view so that for spontaneous change

$$\Delta S_{universe} > 0$$

**Statement of Second Law of Thermodynamics**

The above discussion forms the basis of the second law of thermodynamics. This states that *the entropy of the universe always increases in the course of every spontaneous (natural) change.*

Thus, combining the first and second laws of thermodynamics, the combined statement becomes :

*The energy of the universe is conserved whereas the entropy of the universe always increase in any natural process*

### Entropy and Equilibrium State

During a spontaneous process, the entropy of the system goes on increasing. When the system reaches the equilibrium state, the entropy of the system becomes maximum and, therefore, no more increase in the entropy of the system is possible. Thus, the condition for any system to achieve equilibrium state is that its entropy becomes maximum. The mathematical condition for entropy (S) to be maximum is that the change in entropy ($\Delta S$) is zero, *i.e.*,

$$\Delta S = 0 \text{ (at equilibrium)}$$

Let us apply the entropy criterion to decide the feasibility of the process of conversion of water to ice, at 1 atm pressure,

$$H_2O(l) \longrightarrow H_2O(s)$$

The entropy changes for three different temperatures 272, 273 and 274 K are given in Table 17.3.

*TABLE 17.3* **Entropy changes for the transformation**

$$H_2O(l) \longrightarrow H_2O(s) \text{ at 1 atm. pressure}$$

| Temperature °C | Temperature K | ΔS system ($JK^{-1}$ $mol^{-1}$) | ΔS surroundings ($JK^{-1}$ $mol^{-1}$) | ΔS total ($JK^{-1}$ $mol^{-1}$) |
|---|---|---|---|---|
| - 1 | 272 | - 21.85 | + 21.93 | + 0.08 |
| 0 | 273 | - 21.99 | + 21.99 | 0.00 |
| + 1 | 274 | - 22.13 | + 22.05 | - 0.08 |

(*i*) At 272 K,

$$\Delta S_{total} = \Delta S_{system} + \Delta S_{surroundings}$$
$$= -21.85 + 21.93 = +0.08 \text{ JK}^{-1} \text{ mol}^{-1}$$

Since $\Delta S_{total}$ is positive at 272 K, the freezing of liquid water to ice is spontaneous.

$$H_2O(l) \longrightarrow H_2O(s) \qquad \textbf{Spontaneous at 272 K}$$

(*ii*) At 274 K

$$\Delta S_{total} = \Delta S_{system} + \Delta S_{surroundings}$$
$$= -22.13 + 22.05 = -0.08 \text{ JK}^{-1} \text{ mol}^{-1}$$

Since $\Delta S_{total}$ is negative, the freezing of water is not spontaneous at 274 K.

$$H_2O(l) \longrightarrow H_2O(s) \qquad \textbf{Non-spontaneous at 274 K}$$

But for the reverse process, $\Delta S_{total}$ will be positive, *i.e.*, $\Delta S_{total} = +0.08$ $JK^{-1}$ $mol^{-1}$, *i.e.*, melting is spontaneous at 274 K.

$$H_2O(s) \longrightarrow H_2O(l) \qquad \textbf{Spontaneous at 274 K}$$

(*iii*) At 273 K

$$\Delta S_{total} = \Delta S_{system} + \Delta S_{surroundings}$$
$$= -21.99 + 21.99 = 0$$

Since $\Delta S_{total}$ is zero, the process is at equilibrium, *i.e.*, neither freezing nor melting is spontaneous. At this temperature, water and ice are in equilibrium and no net change is observed.

$$H_2O(l) \longrightarrow H_2O(s) \qquad \textbf{Equilibrium at 273 K}$$

Thus, we observe $\Delta S_{total}$ is a criterion for spontaneity of a change.

### Characteristics of Entropy

The important characteristics of entropy are summed up below :

(*i*) Entropy is an extensive property. Its value depends upon the amount of the substance present in the system.

(*ii*) Entropy of a system is a state function. It depends upon the state variables (T, P, V, *n*).

Thus, the change in entopy is given as :

$$\Delta S = S_{\text{final state}} - S_{\text{initial state}}$$

(*iii*) The change in entropy in going from one state to another is independent of the path.

(*iv*) The change in entropy for a cyclic process is always zero.

(*v*) The total entropy change of an isolated system is equal to the entropy change of system and entropy change of the surroundings. The sum is called **entropy change of universe.**

$$\Delta S = \Delta S_{\text{system}} + \Delta S_{\text{surroundings}}$$

(*vi*) In a reversible process, $\Delta S_{\text{universe}} = 0$ and therefore,

$$\Delta\, S_{\text{system}} = -\,\Delta\, S_{\text{surroundings}}$$

(*vii*) In a irreversible process, $\Delta S_{\text{universe}} > 0$. This means that there is increase in entropy of universe in spontaneous changes.

## 17.16 ENTROPY CHANGE DURING PHASE TRANSITION

The change of matter from one state (solid, liquid or gas) to another is called **phase transition.** Such changes occur at definite temperatures such as melting point (solid to liquid), boiling point (liquid to vapours) etc., and are accompanied by absorption or evolution of heat.

The entropy change for these may be calculated as :

$$\Delta S_{\text{trans}} = \frac{q_{\text{rev}}}{T}$$

where $q$ is the heat evolved or absorbed during transition and T is the temperature. For the transformation of one mole of a substance at constant pressure, $q_{\text{rev}}$ is equal to molar enthalpy change for the transformation *i.e.*, $q_{\text{rev}} = \Delta H_{\text{trans}}$. Let us consider some examples :

1. **Entropy of fusion.** It may be defined as *the entropy change when 1 mole of the solid substance changes into liquid form at its melting point.* For example, when ice melts as :

$$\text{Water}(s) \rightleftharpoons \text{Water}(l)$$

The change in entropy is given by

$$S_{\text{water}} - S_{\text{ice}} = \Delta S_{\text{fusion}} = \frac{\Delta H_{\text{fusion}}}{T_f} \qquad \text{...(17.20)}$$

where $\Delta H_{\text{fusion}}$ is the enthalpy of fusion and $T_f$ is the fusion temperature.

2. **Entropy of vaporisation.** Entropy of vaporisation may be defined as *the entropy change when 1 mole of a liquid changes into vapours at its boiling point.* The entropy of vaporisation for a liquid at its boiling point is :

$$\Delta S_{vaporisation} = \frac{\Delta H_{vap}}{T_b} \quad ...(17.21)$$

where $\Delta H_{vap}$ is the enthalpy of vaporisation and $T_b$ is the boiling point.

**Example 13.** *Calcualte the entropy of fusion of ice, its latent heat of fusion is 6.025 kJ mol$^{-1}$.*

**Solution.** The entropy of fusion

$$\Delta S_{fusion} = \frac{\Delta H_{fusion}}{T_f}$$

Molar latent heat of fusion, $\Delta H_{fusion} = 6.025 \text{ kJ mol}^{-1}$

$$= 6.025 \times 1000 \text{ J mol}^{-1}$$

$$T_f = 0 + 273 = 273\text{K}$$

$$\Delta S_{fusion} = \frac{6.025 \times 1000 \text{ J mol}^{-1}}{273\text{K}} = \mathbf{22.07 \text{ JK}^{-1} \text{ mol}^{-1}}.$$

**Example 14.** *The enthalpy change for the transition of liquid water to steam is 40.8 kJ mol$^{-1}$ at 373 K. Calculate the entropy change for the process.*

**Solution.** The entropy change when liquid water changes to steam is :

$$\Delta S = \frac{\Delta H}{T}$$

$$\Delta S = 40.8 \text{ kJ mol}^{-1}, T = 373 \text{ K}$$

$$\therefore \quad \Delta S = \frac{40.8 \times 10^3 \text{ J mol}^{-1}}{373\text{K}} = \mathbf{109.4 \text{ JK}^{-1} \text{ mol}^{-1}}.$$

## 17.17 FREE ENERGY

The free energy of a system is defined as *the maximum amount of energy available to a system during a process that can be converted into useful work.*

In other words, it is a measure of capacity of a system to do useful work. It is denoted by symbol G and is given by

$$G = H - TS \quad ...(17.22)$$

where H is the enthalpy of the system, S is its entropy and T is the absolute temperature.

Now $\quad H = E + PV$

$$G = E + PV - TS$$

**Free energy change.** The change in free energy may be expressed as

$$\Delta G = \Delta E + \Delta(PV) - \Delta(TS)$$

If the process is carried out at *constant temperature* and *pressure*, the terms Δ(PV) and Δ(TS) become

$$\Delta(PV) = P\Delta V \text{ and}$$
$$\Delta(TS) = T\Delta S$$

∴ $$\Delta G = \Delta E + P\Delta V - T\Delta S$$

But $$\Delta E + P\Delta V = \Delta H$$

or $$\Delta G = \Delta H - T\Delta S$$

**Physical significance of free energy :** From the law of conservation of energy, the total heat energy supplied to its system,

$$q = \Delta E + W_{expansion} + W_{non\text{-}expansion}$$
$$= \Delta E + P\Delta V + W_{non\text{-}expansion}$$
$$= \Delta H + W_{non\text{-}expansion} \quad ...(i)$$

Now in a reversible change at constant temperature T, we have

$$\Delta S = q/T$$
$$q = T\Delta S \quad ...(ii)$$

∴ From (*i*) and (*ii*), we get :

$$T\Delta S = \Delta H + W_{non\text{-}expansion}$$

or $$\Delta H - T\Delta S = -W_{non\text{-}expansion}$$
$$= (\Delta G)_{P.T}$$

or $$-(\Delta G)_{P.T} = W_{non\text{-}expansion} \quad ...(17.23)$$

*i.e., – ΔG is a measure of non-expansion or useful work obtainable in a reversible reaction at constant temperature and pressure.* In other words, ***decrease in free energy***, *during a process, is equal to the "**useful work**" obtainable from the process.*

### 17.17.1 Standard Free Energy Change

*Standard state* for a substance is defined as *the pure material in its most stable form (gas, liquid or solid) at the external pressure of 1 atm and at the specified temperature* (*most usually* = 298 *K*). The *standard state conventions* are :

**For a solid** : *The pure substance at 1 atm external pressure.*

**For a liquid** : *The pure substance at 1 atm external pressure.*

**For a gas** : *An ideal gas at 1 atm partial pressure.*

**For a solution** : *An ideal solution at 1 M concentration.*

**Standard free energy change** (ΔG°) is defined as *the free energy change for a process at 298 K in which the reactants in their standard states are converted into the products in their standard states.* Thus :

$$\Delta G^\circ = \Sigma\, G^\circ_{(products)} - \Sigma\, G^\circ_{(reactants)}$$

The value of ΔG° for a reaction can be derived from the standard free energies of formation ($\Delta G^\circ_f$). The *standard free energy of formation* ($\Delta G^\circ_f$) of a compound is defined as *the change in free energy, when 1 mol of the compound is formed from its constituent elements in their standard states.* Thus :

$$\Delta G^\circ = \Sigma\, G^\circ_{f(products)} - \Sigma\, G^\circ_{f(reactants)}$$

$$= \begin{bmatrix} \text{Sum of the standard} \\ \text{free energies of} \\ \text{formation of } products \end{bmatrix} - \begin{bmatrix} \text{Sum of the standard} \\ \text{free energies of} \\ \text{formation of } reactants \end{bmatrix}$$

It may be pointed here that *the standard free energy of formation of an element in its standard state is* ***zero***.

**Relation between ($\Delta G^\circ$) and equilibrium constant ($K_{eq}$)** ; The standard free energy change ($\Delta G^\circ$) of a reaction is related to the equilibrium constant ($K_{eq}$) by the relation :

$$\Delta G^\circ = -\,2.303\, RT \log K_{eq}$$

where $R$ = gas constant (= 8.314 J $K^{-1}$ $mol^{-1}$); T = temperature in Kelvin.

**Relation between standard free energy change and cell potential :** We know that decrease in free energy is equal to maximum useful work that can be obtained from a process, *i.e.*,

$$-\,\Delta G = w_{max}$$

If the useful work involved is the *electrical work* (*e.g.*, in Daniel cell), then :

$$-\,\Delta G = w_{max} = \text{electrical work} = nFE$$

where $n$ = number of electrons involved in the cell reaction ; $E$ = emf (or potential) of the cell, and $F$ = Faraday (= 96,500 C).

If reactants as well products of the cell reaction are in their standard states (*i.e.*, 298 K and 1 atm), then :

$$-\,\Delta G^\circ = nFE$$

where $\Delta G^\circ$ = standard free energy change, and $E^\circ$ = *standard EMF* (or potential) of cell.

**Feasibility of cell reaction :** The cell reaction is *feasible* if $-\,\Delta G^\circ$ is *positive* (or *standard free energy change is negative*). A *positive* value of $\Delta G^\circ$ show that the reaction may be feasible in *reverse direction*.

### 17.17.2 Gibbs-Helmholtz Equation

We know : $G = H - TS$ and $H = E + PV$

$$G = E + PV - TS$$

Upon differentiation we get :

$$dG = dE + PdV + VdP - TdS - SdT$$

But $dq = dE + dw = dE + PdV$ ...(1st law)

$\therefore$ $dG = dq + VdP - TdS - SdT$

For reversible process :

$$dS = dq/T \quad \text{or} \quad T.dS = dq$$

$\therefore$ $dG = VdP - S.dT$

If pressure remains constant, *i.e.*, $dP = 0$ then :

$$dG = -\,SdT \qquad ...(17.24)$$

Let $G_1$ = initial free energy of a system at $T$; and $G_1 + dG_1$ = initial free energy of the system at $T + dT$, where $dT$ is infinitesimally small and pressure is constant,

$$\therefore \qquad dG_1 = -S_1 dT \qquad ...(17.25)$$

where $S_1$ is the entropy of the system in the *initial state.*

Now suppose that the free energy of the system is final state is $G_2$ at $T$. Let $G_2 + dG_2$ is the free energy of the system at $T + dT$ in the final state, then :

$$dG_2 = -S_2\, dT \qquad ...(17.26)$$

where $S_2$ is the entropy of the system in the *final state.*

Subtracting (*ii*) from (*iii*), we get :

$$dG_2 - dG_1 = -(S_2 - S_1)\, dT$$

or

$$d(\Delta G) = -\Delta S\, dT \qquad ...(17.27)$$

$\therefore$ At constant pressure,

$$\left[\frac{\partial(\Delta G)}{\partial T}\right] = -\Delta S \qquad ...(17.28)$$

We know :

$$\Delta G = \Delta H - T\Delta S$$

or

$$-\Delta S = \frac{\Delta G - \Delta H}{T} \qquad ...(17.29)$$

From (17.28) and (17.29), we get :

$$\frac{\Delta G - \Delta H}{T} = \left(\frac{\partial(\Delta G)}{\partial T}\right)_P$$

or

$$\Delta G = \Delta H + T\left(\frac{\partial(\Delta G)}{\partial T}\right)_P$$

This equation is called **Gibbs-Helmholtz equation** in terms of *free energy and enthalpy changes at constant pressure.*

***Application of Gibbs Helmholtz Equations* : Calculation of EMF of a reversible cell :** Suppose the e.m.f. of a reversible cell, when operated at constant pressure and constant temperature is $E$. Let $n$ faraday, *i.e.*, $nF$ coulombs of electricity is yielded by the cell, then :

Electrical work obtained = $nFE$

We know : Net electrical work = decrease in free energy

$$\therefore \qquad nFE = -\Delta G$$

Substituting in Gibbs-Helmholtz equation, we have :

$$-nFE = \Delta H - T\left(\frac{\partial(nFE)}{\partial T}\right)_P$$

$$= \Delta H - nFT\left(\frac{\partial E}{\partial T}\right)_P \qquad (\because \; n \text{ and } F \text{ are constant})$$

or

$$E = \frac{-\Delta H}{nF} + T\left(\frac{\partial E}{\partial T}\right)_P$$

Thus, be knowing $E$(e.m.f. of the cell) and $\frac{\partial E}{\partial T}$ (*i.e.*, variation of e.m.f. with temperature at constant pressure), the value of heat of reaction ($\Delta H$) occurring in the cell can be calculated.

***Significance of Gibbs-Helmholtz Equation :*** Gibbs-Helmholtz equation helps us in better understanding of the force that causes the chemical reaction to proceed forward. We *know that if* $\Delta G$ *is negative, the reaction occurs spontaneously, and if free energy of the reactants and the products are equal, i.e.,* $\Delta G = 0$*, the reaction is in equilibrium.* From the relation,

$$\Delta G = \Delta H - T.\Delta S$$

it is clear that $\Delta G$ is made up of *two opposing parts, viz.,*

(1) $\Delta H$ : When the reaction is exothermic, $\Delta H$ is negative, so $\Delta G$ tends to be negative and consequently, the reaction tends to the *spontaneous*. This is supported by the fact that generally exothermic reactions at ordinary temperature and pressure are spontaneous.

(2) $T.\Delta S$ : If $\Delta S$ positive, $\Delta G$ tends to be negative. Thus, physical and chemical *changes proceed spontaneously in a direction in which the entropy increases.* For example, a solid substance dissolves in the solvent, because entropy (or randomness) of the system increases.

***Criteria of spontaneity of process :*** *For a reaction to be spontaneous,* $\Delta G$ *must be negative.* The $\Delta G$ will be *negative* under the following condition :

(*i*) If $\Delta H$ is *negative,* and $\Delta S$ is *positive,* since both factors will make $\Delta G$ negative.

(*ii*) If both $\Delta H$ and $\Delta S$ are *positive,* and the *magnitude of* $T\Delta S > \Delta H$.

(*iii*) If both $\Delta H$ and $\Delta S$ are *negative,* and the *magnitude* of $\Delta H > T\Delta G$.

**TABLE 17.4 Spontaneous and non-spontaneous reactions**

| $\Delta H$ | $\Delta S$ | $-T\Delta S$ | $\Delta G$ | *Comment* |
|---|---|---|---|---|
| –ve | +ve | –ve | –ve at all $T$ | Spontaneous at all $T$ |
| –ve | –ve | +ve | –ve at low $T$ | Spontaneous at low $T$ |
| | | | +ve at high $T$ | Non-spontaneous at high $T$ |
| +ve | +ve | –ve | –ve at high $T$ | Spontaneous at high $T$ |
| | | | +ve at low $T$ | Non-spontaneous at low $T$ |
| +ve | –ve | +ve | +ve at all $T$ | Non-spontaneous at all $T$ |

### 17.17.3 Free Energy Change of An Ideal Gas in Isothermal Reversible Change

We know : $G = H - TS$ and $H = E + PV$

$\therefore$ $G = E + PV - TS$

Upon differentiation, we get :

$$dG = dE + P\,.\,dV + V\,.\,dP - T\,.\,dS - S\,.\,dT \quad ...(17.30)$$

From 1st law, we have $dq = dE + dw = dE + PdV$ ...(17.31)

$\therefore$ From (17.30) and (17.31), we get :

$$dG = dq + V\,.\,dP - T\,.\,dS - S\,.\,dT \quad ...(17.32)$$

We know that for a *reversible* process,

$$dS = \frac{dq}{T} \quad \text{or} \quad T\,.\,dS = dq$$

or $$dq - T\,.\,dS = 0 \qquad ...(17.33)$$

From (17.32) and (17.33), we get :

$$dG = V\,.\,dP - S\,.\,dT \qquad ...(17.34)$$

## 17.18 CLAPEYRON-CLAUSIUS EQUATION

Consider a system consisting of only 1 mole of a substance, which exists in two phases $A$ and $B$. Let the temperature and pressure of the system be $T$ and $P$ respectively. Suppose free energy/mole of the substance in two phases $A$ and $B$ be $G_A$ and $G_B$. The system being *in equilibrium,* so there is *no free energy change, i.e.,*

$$G_A = G_B \qquad ...(17.35)$$

Let the temperature of the system be raised to $T + dT$, so that pressure becomes $P + dP$. Suppose the new free energies per mole of A and B phases be $G_A + dG_A$ and $G_B + dG_B$ respectively. As the two phases are *still in equilibrium,* so :

$$G_A + dG_A = G_B + dG_B \qquad ...(17.36)$$

We know : $$G = H - TS = E + PV - TS \qquad ...(17.37)$$

$\therefore$ $$dG = dE + PdV + VdP - TdS - SdT \qquad ...(17.38)$$

But $$dq = dE + dw = dE + PdV \qquad ...(17.39)$$

and $$dq/T = dS \text{ or } TdS = dq \qquad ...(17.40)$$

From (17.38), (17.39) and (17.40), we get :

$$dG = V.dP - S.dT \qquad ...(17.41)$$

As the various equilibria considered here are such in which *work done is due to volume change only,* so equation (17.41) may be applied to phase A as well as B.

$\therefore$ $$dG_A = V_A dP - S_A dT \qquad ...(17.42)$$

and $$dG_B = V_B dP - S_B dT \qquad ...(17.43)$$

where $V_A$ and $V_B$ are the molar volumes of phase $A$ and $B$ respectively, and $S_A$ and $S_B$ are their molar entropies.

From (17.41), (17.42) and (17.43), we get :

$$V_A.dP - S_A.dT = V_B.dP - S_B.dT$$

or $$\frac{dP}{dT} = \frac{S_B - S_A}{V_B - V_A} = \frac{\Delta S}{V_B - V_A} \qquad ...(17.44)$$

where $\Delta S$ = molar entropy change, and $V_B - V_A$ is the change in volume when 1 mole of substance changes from $A$ to $B$.

If molar heat of transition from phase A to B is $q$, then

$$\Delta S = q/T \qquad ...(17.45)$$

From (17.43) and (17.44), we get :

$$\frac{dP}{dT} = \frac{q}{T(V_B - V_A)} \qquad ...(17.46)$$

This is the **Clapeyron-Clausius equation.**

### 17.18.1 *Applications*

1. **Determination of latent heat from vapour pressure :** If the vapour pressure at two different temperature of the two phases *in equilibrium* are known, then the latent heat can be calculated by using the Clapeyron-Clausius equation.

   In case of vaporization, $V_v >> V_l$

   or $$V_v - V_l \approx V_v$$

   (The molar volume $V_l$ is neglected in comparison with $V_v$)

   ∴ The Clapeyron-Clausius equation takes the form :

   $$\frac{dP}{dT} = \frac{L_v}{T\,V_v}$$

   Assuming that the gas law ($PV = RT$) is applicable to vapour, we may substitute, $V = RT/P$, in above equation.

   ∴ $$\frac{dP}{dT} = \frac{L_v P}{R\,T^2}$$

   or $$\frac{dP}{P} = \frac{L_v\,.\,dT}{R\,T^2}$$

   Integrating, on the assumption than $L_v$, is constant, over a small temperature change, we get :

   $$\ln\frac{P_2}{P_1} = \frac{L_v}{R}\left[\frac{1}{T_1} - \frac{1}{T_2}\right]$$

   or $$\ln\frac{P_2}{P_1} = \frac{L_v}{R}\left[\frac{T_2 - T_1}{T_1 T_2}\right] \qquad ...(17.47)$$

   where $P_1$ and $P_2$ are the vapour pressure at $T_1$ and $T_2$ respectively. By substituting these values, the molar heat of vaporization, $L_v$, can be computed with the help of *Integrated Clapeyron-Clausius equation*. (17.47).

2. **To study the effect of pressure on boiling point :** If boiling points of a liquid at one pressure and its latent heat of vaporization are known, then boiling point of liquid at another pressure can be computed with the help of *Integrated Clapeyron-Clausius equation* (17.47).
3. **To study the effect of temperature on vapour pressure of liquid :** If latent heat ($L_v$) and vapour pressure of a liquid at one temperature are known, then vapour pressure of liquid at another temperature can be calculated with the help of *Integrated Clapeyron-Clausius equation* (17.47).

## 17.19 ABSOLUTE ENTROPIES AND THIRD LAW OF THERMODYNAMICS

The third law of thermodynamics deals with the entropies of the perfect crystalline substances at absolute zero of temperature. According to third law of thermodynamics :

*At absolute zero the entropy of a perfectly crystalline substance is taken as zero.*

This law was first formulated by Nernst in 1906. Since entropy is related to disorder, therefore, according to third law, at absolute zero, there is least disorder *i.e.*, there is perfect order and, therefore, the entropy at perfect order is taken as zero. It may be noted that this law is true only for those substances which exist in the perfectly crystalline form at 0 K. However, if there are imperfections of any type in the perfect cyrstalline arrangement at 0 K, then the entropy will be larger than zero.

### Utility of Third Law of Thermodynamics

The third law helps to calculated the absolute entropies of pure substances at different temperatures. The entropy (S) of the substance at different temperature T may be calculated by the measurement of heat capacity changes. If $S_0$ be the entropy of the substance at 0 K and S be its entropy at T K, then

$$\Delta S = S - S_0 = \int_0^T \frac{C_p}{T} dT$$

Since $S_0 = 0$, according to third law

$$S = \int_0^T \frac{C_p}{T} dT$$

The value of integral can be calculated from the graph of $C_p/T$ against temperature ($T$). The area under the curve between 0 K and $T$ K gives the value of integral and, thus, the value of $S$ at temperature $T$.

*When the entropy of one mole of the susbtance is expressed at 298K and 1 atm pressure, it is* called *standard entropy* of the substance and is denoted as S°. The standard entropies of various substances are given in Table 17.5. The standard entropy change (ΔS°) for a chemical reaction can be calculated from the standard entropies of various substances in products and reactants as :

$$\Delta S^\circ = \begin{bmatrix}\text{Sum of the standard} \\ \text{entropies of products}\end{bmatrix} - \begin{bmatrix}\text{Sum of the standard} \\ \text{entropies of reactants}\end{bmatrix}$$

or $$\Delta S^\circ = \Sigma S^\circ \text{ (products} - \Sigma S^\circ \text{ (reactants)}$$

**Table 17.5. Absolute entropies of substances at 298 K and 1 atm pressure**

| *Substance* | *S°* ($JK^{-1}$ $mol^{-1}$) | *Substance* | *S°* ($JK^{-1}$ $mol^{-1}$) |
|---|---|---|---|
| $H_2(g)$ | 130.6 | $HBr(g)$ | 198.5 |
| $Cl_2(g)$ | 203.3 | $HI(g)$ | 206.3 |
| $Br_2(l)$ | 152.3 | $NH_3(g)$ | 192.5 |
| $I_2(s)$ | 116.7 | $CH_4(g)$ | 186.2 |
| $O_2(g)$ | 205.03 | $C_2H_4(g)$ | 219.5 |
| $N_2(g)$ | 191.5 | $C_6H_6(g)$ | 173.3 |
| $Al(s)$ | 28.3 | $SO_2(g)$ | 248.5 |
| C(graphite) | 5.69 | $CO_2(g)$ | 213.6 |
| C(diamond) | 2.4 | $NO_2(g)$ | 240.5 |
| $H_2O(g)$ | 188.7 | $NaCl\ (s)$ | 72.38 |
| $H_2O(l)$ | 69.96 | $CO(g)$ | 197.6 |
| $HCl(g)$ | 186.7 | | |

## 17.20 COUPLING–DRIVING NON-SPONTANEOUS REACTIONS

It is well known that *only those chemical/biological reactions which involve loss of free energy (i.e.,* $\Delta G$ = *negative) are thermodynamically favoured*. Many reactions such as reduction of copper (II) sulphide to copper and anabolism (in cells) are non-spontaneous (*i.e.,* $\Delta G$ = positive) and they need input of energy to proceed. The required energy for such a reaction is provided by coupling it with another reaction, which has $\Delta G$ *of much larger negative value than the positive* $\Delta G$ *value of the original non-spontaneous reaction*. Thus, a non-spontaneous reaction :

$$A \longrightarrow B$$

with $\Delta G_1 > 0$ (*i.e.,* $\Delta G$ = positive) can proceed if it is coupled with another spontaneous reaction : $S \rightarrow P$ in which $\Delta G_2$ is of much larger negative value than positive value of $\Delta G_1$ ;

| | | | |
|---|---|---|---|
| (*i*) | $A \longrightarrow B$ ; | $\Delta G_1 = +\text{ ve}$ | (*original reaction*) |
| (*ii*) | $S \longrightarrow P$ ; | $\Delta G_2 = -\text{ ve}$ | (*coupled reaction*) |
| Net : | $A + S \longrightarrow B + P$ ; | $\Delta G = \Delta G_1 + \Delta G_2$ = Negative | (*spontaneous*) |

In such coupled reactions, the energy lost in reaction (*ii*) is *transferred* to energy needed in the reaction (*i*), thereby making the net reaction spontaneous. This can be explained by considering following reactions :

1. *Extraction of copper from its ore* : If we look the reaction :

$$Cu_2S(s) \longrightarrow 2\,Cu\,(s) + S(s)\,; \qquad \Delta\,G° = +\,88.2\text{ KJ}$$

we find that it is *highly non-spontaneous*. Evidently, we cannot hope to obtain copper from $Cu_2S$ merely by trying to catalyse this reaction. Actually, we need to *"do work"* on the reaction in some way, in order to force the above reaction to occur. This can be done by coupling this reaction so that we arrive at an overall reaction is spontaneous. Now consider the reaction :

$$S\,(s) + O_2\,(g) \longrightarrow SO_2\,(g)\,; \qquad \Delta G = -\,300.1\text{ kJ}$$

This is a *highly spontaneous* reaction. On coupling reaction (*i*) with reaction (*ii*), we get:

$$Cu_2\,(s) \longrightarrow 2\,Cu(s) + S\,(s)\,; \qquad \Delta G° = +\,88.2\text{ kJ}$$

$$S\,(s) + O_2\,(g) \longrightarrow SO_2\,(g)\quad ; \qquad \Delta G° = -\,300.1\text{ kJ}$$

$$Cu_2S\,(s) + O_2 \longrightarrow 3\,Cu\,(s) + SO_2\,(g)\,; \quad \Delta G° = -\,211.9\text{ kJ}$$

Thus, free energy change for the overall coupled reaction is *negative*, so the coupled reaction proceeds *spontaneously*.

2. *Biochemical reaction* : Many biochemical reactions that are essential to the maintenance of life do not occur spontaneously within the human body. However, these reactions are made to occur by coupling them with reactions that are spontaneous and release energy. The energy release that accompany the metabolism of food-stuffs provide the primary source of necessary free energy. For example, glucose ($C_6H_{12}O_6$) is oxidized in the body releasing a substantial amount of energy.

$$(C_6H_{12}O_6\,(s) + 6\,O_2\,(g) \longrightarrow 6\,CO_2\,(g) + 6\,H_2O\,(l)\,; \quad \Delta G° = -\,2{,}880\text{ kJ}$$

This released energy "does work" in the body. However, some means is necessary to couple (or connect) this released energy to the reaction that needs energy. One means of accomplishing this aim is shown in Fig. 17.6. We know that ATP (adenosine triphosphate) is a high-energy molecule, so when ATP is converted into a lower-energy molecule, ADP (adenosine diphosphate), the energy so-released drive other chemical reactions such as conversion of simpler molecules into cell constituents.

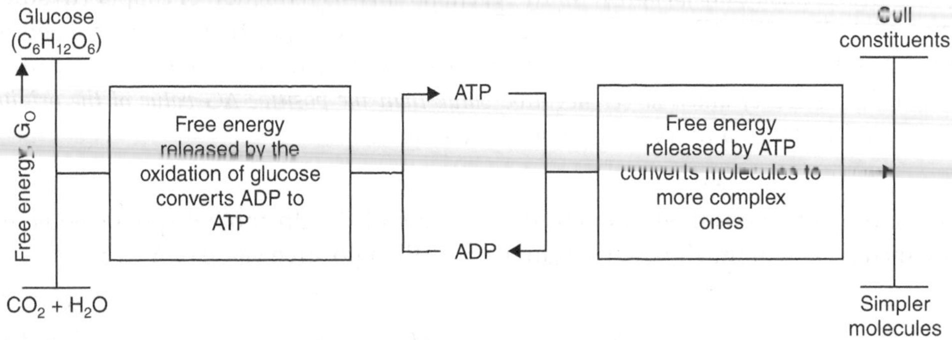


*FIGURE 17.6* **Schematic representation of the free energy changes occurring in cell metabolism. The oxidation of glucose to $CO_2$ and $H_2O$ releases free energy, which is used to convert ADP into the more energetic ATP. ATP is then used as an energy source to convert simpler molecules into more complex cell constituents. During this process ATP is converted to ADP. In this way reactions occur is cell metabolism.**

The energy released is glucose oxidation is utilized in part to convert ADP back to ATP. Consequently, the ATP $\Leftrightarrow$ ADP interconversions act as a means of storing energy and releasing it to drive the required biological reactions. It may be pointed here that coupling of bioreactions, so that the free energy released in one reaction may be used in another reaction requires particular *enzymes* as catalysts.

## 17.21 VAN'T HOFF ISOTHERM

Consider a general reaction :

$$aA + bB \longrightarrow cC + dD \qquad ...(17.48)$$

We know : $G = H - TS = E + PV - TS$

or $dG = dE + PdV + VdP - TdS - SdT$

But $dq = dE + PdV$ and $dS = dq/T$

$\therefore$ $dG = V\,.\,dP - SdT$

At constant temperature, $(dG)_T = V\,.\,dP$

$\therefore$ Free energy change for 1 mole of any gas at a constant temperature is given by :

$$dG = V.dP = RT\,.\,\frac{dP}{P} \qquad [\because \quad PV = RT]$$

Integrating : $\int dG = RT \int \frac{dP}{P}$

$$G = G^\circ + RT \ln P \qquad ...(17.49)$$

where $G^\circ$ integration constant and is known as *standard free energy, i.e.,* $G^\circ = G$ when $P = 1$ atm.

Let the energy/mole of $A$, $B$, $C$ and $D$ at their respective pressures $P_A$, $P_B$, $P_C$ and $P_D$ are $G_A$, $G_B$, $G_C$ and $G_D$ respectively. Then, free energy change for the reaction (*i*) is given by :

$$\Delta G = G_{products} - G_{reactants} = [cG_C + dG_D] - [aG_A + bG_B] \qquad ...(17.50)$$

Substituting the value of $G_A$, $G_B$, $G_C$ and $G_D$ in (17.50) from (17.49), we get :

$$\Delta\, G = [cG^\circ_c + cRT \ln P_C + dG^\circ_D + dRT \ln P_D] - [aG^\circ_{A+}\, aRT \ln P_A + bG^\circ_B + bRT \ln P_B]$$

$$= [cG^\circ_C + dG^\circ_D - aG_A{}^\circ - bG^\circ_B] + RT \ln \frac{(P_C)^c\,(P_D)^d}{(P_A)^a\,(P_B)^b}$$

or

$$\Delta\, G = \Delta G^\circ + RT \ln \frac{(P_C)^a\,(P_D)^d}{(P_A)^a\,(P_B)^b} \qquad ...(17.51)$$

where $\Delta\, G^\circ$ = standard free energy of reaction (*i*).

We know at *equilibrium*, $\Delta G = 0$

$\therefore$

$$0 = \Delta G^\circ + RT \ln \left[\frac{[P_C]^a\,[P_D]^d}{[P_A]^a\,[P_B]^b}\right]_{eq}$$

$$= \Delta G^\circ + RT \ln \dot{K}_{eq} \qquad \left[ \because \left[ \frac{[P_C]^c\,[P_D]^d}{[P_A]^a\,[P_B]^b} \right]_{eq} = K_{eq} \right]$$

or
$$\Delta G^\circ = -\,RT \ln K_{eq} \qquad \text{...(17.52)}$$

From (17.51) and (17.52), we get :

$$\Delta G = -\,RT \ln K_{eq} + RT \ln \frac{[P_C]^c\,[P_D]^d}{[P_A]^a\,[P_B]^b}$$

or
$$-\,\Delta G = RT \ln K_{eq} - RT \ln \frac{[P_C]^c\,[P_D]^d}{[P_A]^a\,[P_B]^b} \qquad \text{...(17.53)}$$

This expression, called **Van't Hoff isotherm,** is very important and *gives a quantative relation for free energy change accompanying a chemical reaction.*

**Van't Hoff isochore :** $\Delta G^\circ = -\,\text{RT} \ln K_{eq}$

or
$$\ln K_{eq} = \frac{-\,\Delta G^\circ}{RT} = \frac{-\,(\Delta H^\circ - T\,\Delta S^\circ)}{RT} = \frac{-\,\Delta H^\circ}{RT} + \frac{\Delta S^\circ}{R}$$

$$\ln K_1 = \frac{-\,\Delta H^\circ}{RT_1} + \frac{\Delta S^\circ}{R} \text{ and } \ln K_2 = \frac{-\,\Delta H^\circ}{RT_2} + \frac{\Delta S^\circ}{R}$$

or
$$\ln K_2 - \ln K_1 = \frac{\Delta H^\circ}{R}\left[\frac{1}{T_1} - \frac{1}{T_2}\right] = \frac{\Delta H^\circ}{R}\left[\frac{T_2 - T_1}{T_2 T_2}\right]$$

or
$$\ln \frac{K_2}{K_1} = \frac{\Delta H^\circ}{R}\left[\frac{T_2 - T_1}{T_1 T_2}\right]$$

or
$$\log \frac{K_2}{K_1} = \frac{\Delta H^\circ}{2.303\,R}\left(\frac{T_2 - T_1}{T_1 T_2}\right) \qquad \text{...(17.54)}$$

This expression, called the *Van't Hoff isochore,* can be used to calculate equilibrium constant at $T_2$, if its value at $T_1$ is known.

## SOLVED PROBLEMS

### TYPE-I : Problems on Enthalpy and Hess Law

**Problem 1.** *Use enthalpies of formation to calculate the value of* $\Delta H^\circ$ *for the reaction :*

$$2H_2S(g) + 3O_2(g) \longrightarrow 2H_2O(l) + 2SO_2(g)$$

$$\Delta H_f^\circ \text{ of } H_2S(g) = -\,20.17 \text{ kJ mol}^{-1}$$

$$\Delta H_f^\circ \text{ of } H_2O(l) = -\,286.0 \text{ kJ mol}^{-1}$$

$$\Delta H_f^\circ \text{ of } SO_2(g) = -\,296.9 \text{ kJ mol}^{-1}$$

**Solution.** $2H_2S(g) + 3O_2(g) \longrightarrow 2H_2O(l) + 2SO_2(g)$

$$\Delta H^\circ_{f,reaction} = \Sigma\Delta H^\circ_{f,products} - \Sigma\Delta H^\circ_{f,reactants}$$

$$\Delta H^\circ = 2(\Delta H^\circ_{f,H_2O} + 2(\Delta H^\circ_{f,SO_2}) - [2(\Delta H^\circ_{f,H_2S}) + 3(\Delta H^\circ_{f,O_2})]$$

$$= 2 \text{ mol} \times (-286 \text{ kJ mol}^{-1}) + 2 \text{ mol} \times (-296.9 \text{ kJ mol}^{-1}) - [2 \text{ mol} \times 20.17 \text{ kJ mol}^{-1} + 3 \text{ mol} \times 0]$$

$$= -572 \text{ kJ} - 593.8 \text{ kJ} + 40.30 \text{ kJ}$$

$$= \mathbf{-1125.46 \text{ kJ}.}$$

**Problem 2.** *Calculate the standard internal energy change for the reaction*

$$OF_2(g) + H_2O(g) \longrightarrow O_2(g) + 2HF(g) \text{ at } 298 \text{ K}$$

*The standard enthalpies of formation in kJ mol$^{-1}$ are :*

*$OF_2(g)$, + 20 ; $H_2O$, –250 and HF, – 270.*

$$R = 8.134 \text{ JK}^{-1} \text{ mol}^{-1}.$$

**Solution.** $OF_2(g) + H_2O(g) \longrightarrow O_2(g) + 2HF(g)$ at 298 K

$$\Delta H^\circ_{f,reaction} = \Sigma\Delta H^\circ_{f,products} - \Sigma\Delta H^\circ_{f,reactants}$$

$$\Delta H^\circ = \Delta H^\circ_{f,O_2} + 2\Delta H^\circ_{f,HF} - \left[\Delta H^\circ_{f,OF_2} + \Delta H^\circ_{f,H_2O}\right]$$

$$= 1 \text{ mol} \times 0 + 2 \text{ mol} \times (-270 \text{ kJ mol}^{-1}) - [1 \text{ mol} \times (+20 \text{ kJ mol}^{-1}) + 1 \text{ mol} \times (-250 \text{ kJ mol}^{-1})]$$

$$= -540 \text{ kJ} - [-230 \text{ kJ}] \approx \mathbf{-310 \text{ kJ}}$$

$$\Delta n = \text{Moles of gaseous products} - \text{Moles of gaseous reactants}$$

$$= (1 \text{ mol} + 2 \text{ mol}) - (1 \text{ mol} + 1 \text{ mol}) = 1 \text{ mol}$$

$$\Delta nRT = 1 \text{ mol} \times 8.134 \text{ JK}^{-1} \text{ mol}^{-1} \times 298 \text{ K}$$

$$= 2423.9 \text{ J} \approx 2.42 \text{ kJ}$$

$$\Delta H = \Delta E + \Delta nRT$$

$$\Delta E = \Delta H - \Delta nRT$$

$$= -310 \text{ kJ} - 2.42 \text{ kJ} = \mathbf{-312.42 \text{ kJ}.}$$

**Problem 3.** *The standard enthalpy change for the combustion of fructose*

$$C_6H_{12}O_6(s) + 6O_2(g) \longrightarrow 6CO_2(g) + 6H_2O(l)$$

*is –2826.7 kJ. Calculated the enthalpy of formation of fructose. Given*

*$\Delta H^\circ_f$ for $CO_2(g)$ = –393.5 kJ mol$^{-1}$*

*$\Delta H^\circ_f$ for $H_2O(l)$ = –285.8 kJ mol$^{-1}$.*

**Solution.**

$$\Delta H^\circ = 6\Delta H^\circ_{f,CO_2(g)} + 6\Delta H^\circ_{f,H_2O(l)} - [\Delta H^\circ_{f,C_6H_{12}O_6} + 6\Delta H^\circ_{f,O_2}]$$

$$-2826.7 \text{ kJ} = 6 \text{ mol} \times (-393.5 \text{ kJ mol}^{-1}) + 6 \text{ mol} \times (-285.8 \text{ kJ mol}^{-1}) - [1 \text{ mol} \times \Delta H^\circ_{f,C_6H_{12}O_6} + \text{mol} \times 0]$$

$$= -2361 \text{ kJ} - 1714.8 \text{ kJ} - \Delta H^\circ_{f,C_6H_{12}O_6}$$

$$\Delta H^\circ_{f,C_6H_{12}O_6} = -2361 \text{ kJ} - 1714.8 \text{ kJ} + 2826.7 \text{ kJ}$$

$$= \mathbf{1249.1 \text{ kJ mol}^{-1}.}$$

**Problem 4.** *The standard enthalpy change for the reaction*

$$3Fe_2O_3(S) \longrightarrow 2Fe_3O_4(s) + \frac{1}{2}O_2(g)$$

*is +229.3 kJ. Calculate the standard molar enthalpy of formation of $Fe_3O_4(s)$. The standard enthalpy of formation of $Fe_2O_3(s)$ is – 823.4 kJ mol$^{-1}$.*

**Solution.**

$$\Delta H^\circ = 2\text{ mol} \times \Delta H^\circ_{f,\mathrm{Fe_3O_4}} + \frac{1}{2}\text{mol} \times \Delta H^\circ_{f,\mathrm{O_2}} - [3\text{ mol} \times \Delta H^\circ_{f,\mathrm{Fe_3O_4}}]$$

$$+ 229.3 \text{ kJ} = 2 \text{ mol} \times \Delta H^\circ_{f,\mathrm{Fe_3O_4}} + \frac{1}{2}\text{mol} \times 0 - 3 \text{ mol} \times (-823.4 \text{ kJ mol}^{-1})$$

$$2 \text{ mol} \times \Delta H^\circ_{f,\mathrm{Fe_3O_4}} = 229.3 \text{ kJ} + 0 \text{ kJ} - 2470.2 \text{ kJ}$$

$$= -2240.9 \text{ kJ}$$

$$\Delta H^\circ_{f,\mathrm{Fe_3O_4}} = \frac{-2240.9 \text{ kJ}}{2 \text{ mol}} = \mathbf{-1120.5 \text{ kJ mol}^{-1}}.$$

**Problem 5.** *Given the enthalpy change for the following reactions :*

(*a*) $SO_2(g) \longrightarrow S(s) + O_2(g)$ $\quad \Delta H^\circ_a = +296.8$ kJ

(*b*) $2S(s) + 3O_2(g) \longrightarrow 2SO_3(g)$ $\quad \Delta H^\circ_b = -791.4$ kJ

*Calculate ΔH° for the reaction*

$$2SO_2(g) + O_2(g) \longrightarrow 2SO_3(g)$$

**Solution.** Multiply eq. (*a*) by 2

(*c*) $2SO_2(g) \longrightarrow 2S(s) + 2O_2(g)$ $\quad \Delta H^\circ_c = +593.6$ kJ

Add eq. (*c*) and eq. (*b*)

(*c*) $2SO_2(g) \longrightarrow 2S(s) + 2O_2(g)$ $\quad \Delta H^\circ_c = +593.6$ kJ

(*b*) $2S(s) + 3O_2(g) \longrightarrow 2SO_3(g)$ $\quad \Delta H^\circ_b = -791.4$ kJ

$2SO_2(g) + O_2(g) \longrightarrow 2SO_3(g)$ $\quad \Delta H^\circ = \mathbf{-197.8}$ **kJ**

**Problem 6.** *Given that*

(*a*) $4NH_3(g) + 5O_2(g) \longrightarrow 4NO(g) + 6H_2O(l)$ $\quad \Delta H^\circ_a = -1170$ kJ

(*b*) $4NH_3(g) + 3O_2(g) \longrightarrow 2N_2(g) + 6H_2O(l)$ $\quad \Delta H^\circ_b = -1530$ kJ

*Calculate ΔH° for the equation*

$$N_2(g) + O_2(g) \longrightarrow 2NO(g)$$

**Solution.** Subtract eq. (*b*) from eq. (*a*)

(*a*) $4NH_3(g) + 5O_2(g) \longrightarrow 4NO(g) + 6H_2O(l)$ $\quad \Delta H^\circ_a = -1170$ kJ

(*b*) $4NH_3(g) + 3O_2(g) \longrightarrow 2N_2(g) + 6H_2O(l)$ $\quad \Delta H^\circ_b = -1530$ kJ

– – – – +

(*c*) $2O_2(g) \longrightarrow -2N_2(g) + 4NO(g)$ $\quad \Delta H^\circ = +360$ kJ

or $2N_2(g) + 2O_2(g) \longrightarrow 4NO(g)$ $\quad \Delta H^\circ = +360$ kJ

Divide eq. (*c*) by 2

$$\frac{2}{2}N_2(g) + \frac{2}{2}O_2(g) \longrightarrow \frac{4}{2}NO(g) \qquad \Delta H^\circ = \frac{+360}{2}\text{ kJ}$$

$$N_2(g) + O_2(g) \longrightarrow 2NO(g) \qquad \mathbf{\Delta H^\circ = +180\ kJ}$$

**Problem 7.** *Given the following standard heats of reactions :*

(*i*) *heat of formation of water = – 285.8 kJ*

(*ii*) *heat of combustion of acetylene = – 1299.6 kJ*

(*iii*) *heat of combustion of ethylene = – 1410.8 kJ*

*Calculate the heat of reaction for the hydrogenation of acetylene at constant volume (25°C).*

**Solution.**

(*a*) $$H_2(g) + \tfrac{1}{2}O_2(g) \longrightarrow H_2O(l), \qquad \Delta H_a = -285.8\text{ kJ}$$

(*b*) $$C_2H_2(g) + \tfrac{5}{2}O_2(g) \longrightarrow H_2O(l) + 2CO_2(g), \qquad \Delta H_b = -1299.6\text{ kJ}$$

(*c*) $$C_2H_4(g) + 3O_2(g) \longrightarrow 2H_2O(l) + 2CO_2(g), \qquad \Delta H_c = -1410.8\text{ kJ}$$

Add equation (*a*) and equation (*b*)

$$H_2(g) + \tfrac{1}{2}O_2(g) \longrightarrow H_2O(l), \qquad \Delta H_a = -285.8\text{ kJ}$$

$$C_2H_2(g) + \tfrac{5}{2}O_2(g) \longrightarrow H_2O(l) + 2CO_2(g) \qquad \Delta H_b = -1299.6\text{ kJ}$$

(*d*) $$H_2(g) + C_2H_2(g) + 3O_2(g) \longrightarrow 2H_2O(l) + 2CO_2(g), \qquad \Delta H_d = -1585.4\text{ kJ}$$

Subract equation (*c*) from equation (*d*)

$$H_2(g) + C_2H_2(g) + 3O_2(g) \longrightarrow 2H_2O(l) + 2CO_2(g), \qquad \Delta H_d = -1585.4\text{ kJ}$$

$$C_2H_4(g) + 3O_2(g) \longrightarrow 2H_2O(l) + 2CO_2(g), \qquad \Delta H_c = -1410.8\text{ kJ}$$

$$- \quad - \quad - \qquad\qquad - \qquad\qquad +$$

$$H_2(g) + C_2H_2(g) \longrightarrow C_2H_4(g) \qquad \mathbf{\Delta H = -174.6\ kJ}$$

$\Delta n$ = Moles of the gaseous products – Moles of the gaseous reactants

$$= 1\text{ mol} - (1\text{ mol} + 1\text{ mol}) = -1\text{ mol}$$

$$\Delta H = \Delta E + \Delta nRT$$

$$\Delta E = \Delta H - \Delta nRT$$

$$= -174.6\text{ kJ} \times \frac{1000\text{ J}}{1\text{ kJ}} - (-1\text{ mol} \times 8.314\text{ J K}^{-1}\text{ mol}^{-1} \times 298\text{ K})$$

$$= -174.6\text{ kJ} + 2477.6\text{ kJ}$$

$$= -174.6\text{ kJ} + 2.4776\text{ kJ}$$

$$\approx \mathbf{-172.1\ kJ.}$$

**Problem 8.** *The enthalpy change for the equation*

$$OF_2(g) \longrightarrow O\,(g) + 2F(g)$$

*is 368 kJ. Calculate the average oxygen fluorine bond energy of* $OF_2$.

**Solution.** Dissociation of $OF_2$ into $O(g)$ and $2F(g)$ involves the breaking of two O—F bonds.

$$\therefore \quad \text{Average O—F bond energy} = \frac{\text{Enthalpy change for the equation}}{2}$$

$$= \frac{368\ \text{kJ}}{2} = \mathbf{184\ kJ\ mol^{-1}}.$$

**Problem 9.** *Calculate* $\Delta H°$ *for the reaction*

$$CH_3OH(g) + F_2(g) \longrightarrow FCH_2OH(g) + HF(g)$$

*The bonding in* $CH_3OH$ *and* $FCH_2OH$ *is*

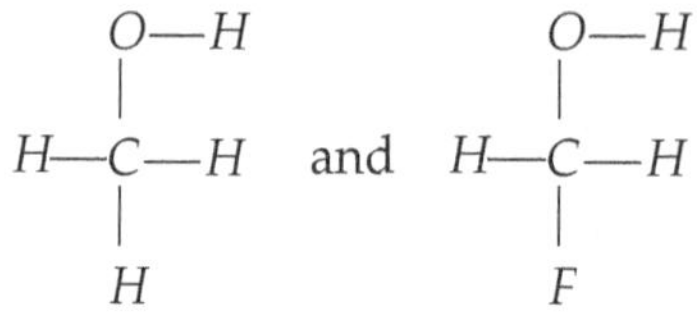

*The molar bond enthalpies (bond energy) are :*

$H_{C—H} = 414$ kJ mol$^{-1}$ $\qquad H_{C—F} = 439$ kJ mol$^{-1}$

$H_{C—O} = 351$ kJ mol$^{-1}$ $\qquad H_{F—F} = 155$ kJ mol$^{-1}$

$H_{O—H} = 464$ kJ mol$^{-1}$ $\qquad H_{H—F} = 565$ kJ mol$^{-1}$.

**Solution.** This reaction involves the rupture of three C—H bonds, one C—O bond, one O—H bond and one F—F bond.

$\therefore$ Sum of molar bond enthalpies of reactants

$= 3 \times H_{C—H} + (1 \times H_{C—O}) + (1 \times H_{O—H}) + (1 \times H_{F—F})$

$= 3\ \text{mol} \times 414\ \text{kJ mol}^{-1} + 1\ \text{mol} \times 351\ \text{kJ mol}^{-1} + 1\ \text{mol} \times 464\ \text{kJ mol}^{-1}$
$+ 1\ \text{mol} \times 155\ \text{kJ mol}^{-1}$

$= 1242\ \text{kJ} + 351\ \text{kJ} + 464\ \text{kJ} + 155\ \text{kJ}$

$= 2212$ kJ.

This reaction also involves the formation of two C—H bonds, one C—F bond, one C—O bond and one O—H bond.

$\therefore$ Sum of molar bond enthalpies of products

$= 2\ \text{mol} \times H_{C—H} + (1\ \text{mol} \times H_{C—F}) + (1\ \text{mol} \times H_{C—O}) + (1\ \text{mol} \times H_{O—H})$

$= 2\ \text{mol} \times 414\ \text{kJ mol}^{-1} + (1\ \text{mol} \times 439\ \text{kJ mol}^{-1}) + (1\ \text{mol} \times 351\ \text{kJ mol}^{-1})$
$+ (1\ \text{mol} \times 464\ \text{kJ mol}^{-1})$

$= 828\ \text{kJ} + 439\ \text{kJ} + 351\ \text{kJ} + 464\ \text{kJ}$

$= \mathbf{2082\ kJ}$.

$\Delta H° =$ **Sum of molar bond enthalpies of reactant** $-$ **Sum of molar bond enthalpies of products**

$= 2212\ \text{kJ} - 2082\ \text{kJ}$

$= \mathbf{+\ 130\ kJ}$.

**Problem 10.** *Calculate the enthalpy of combustion for the following reaction*

$$2HC \equiv CH(g) + SO_2(g) \longrightarrow 4CO_2(g) + 2H_2O(g)$$

*The bond energies of the various bonds involved are:*

$$H_{C—H} = 414 \text{ kJ mol}^{-1} \qquad H_{O—H} = 464 \text{ kJ mol}^{-1}$$
$$H_{C\equiv C} = 811 \text{ kJ mol}^{-1} \qquad H_{O=O} = 502 \text{ kJ mol}^{-1}$$
$$H_{C=O} = 730 \text{ kJ mol}^{-1}.$$

**Solution.** This reaction involves the rupture of four C—H bonds, two C≡C bonds and five O=O bonds.

∴ Sum of molar bond enthalpies of reactants

$$= (4 \times H_{C—H}) + (2 \times H_{C\equiv C}) + (5 \times H_{O=O})$$
$$= (4 \text{ mol} \times 414 \text{ kJ mol}^{-1}) + (2 \text{ mol} \times 811 \text{ kJ mol}^{-1}) + (5 \text{ mol} \times 502 \text{ kJ mol}^{-1})$$
$$= 1656 \text{ kJ} + 1622\text{kJ} + 2510 \text{ kJ} = \mathbf{5788 \text{ kJ}}$$

This reaction also involves the formation of eight C=O bonds, and four O—H bonds.

∴ Sum of molar bond enthalpies of products

$$= (8 \times H_{C=O}) + (4 \times H_{O—H})$$
$$= (8 \text{ mol} \times 730 \text{ kJ mol}^{-1}) + (4 \text{ mol} \times 464 \text{ mol}^{-1})$$
$$= 5840 \text{ kJ} + 1856 \text{ kJ} = \mathbf{7696 \text{ kJ}.}$$

$$\Delta H^\circ = \begin{matrix}\textbf{Sum of molar bond} \\ \textbf{enthalpies of reactants}\end{matrix} - \begin{matrix}\textbf{Sum of molar bond} \\ \textbf{enthalpies of reactants}\end{matrix}$$

$$= 5788 \text{ kJ} - 7696 \text{ kJ} = \mathbf{-1908 \text{ kJ}.}$$

**TYPE–II. Problems based on Bond Energy**

**Problem 11.** *Calculate the bond energy of the C—H bond from the following data :*

| | | |
|---|---|---|
| (*a*) | $C(s) + 2H_2(g) \longrightarrow CH_4(g)$ | $\Delta H_a = -94.0$ kJ |
| (*b*) | $H_2(g) \longrightarrow 2H(g)$ | $\Delta H_b = +433.0$ kJ |
| (*c*) | $C(s) \longrightarrow C(g)$ | $\Delta H_c = +703$ kJ |

**Solution.** Multiply Eq. (*b*) by 2 and subtract the resultant equation from Eq. (*a*).

| | | |
|---|---|---|
| (*a*) | $C(s) + 2H_2(g) \longrightarrow CH_4(g)$ | $\Delta H_a = -94.0$ kJ |
| (*b*) | $2H_2(g) \longrightarrow 4H(g)$ | $\Delta H_b = +866$ kJ |
| (*c*) | –    – | – |
| (*d*) | $C(s) + 4(g) \longrightarrow CH_4(g)$ | $\Delta H_d = -960$ kJ |

Now subtract Eq. (*c*) from Eq. (*d*).

| | |
|---|---|
| $C(s) + 4H(g) \longrightarrow CH_4(g)$ | $\Delta H_d = -960$ kJ |
| $C(s) \longrightarrow C(g)$ | $\Delta H_c = +703$ kJ |
| –    – | – |
| $C(g) + 4H(g) \longrightarrow CH_4(g)$ | $\Delta H = -1663$ kJ |

$$\text{Average bond energy (C—H)} = \frac{\text{Heat of formation}}{\text{Number of bonds}}$$

$$= \frac{1663 \text{ kJ}}{4} = \mathbf{+ 415.75 \text{ kJ mol}^{-1}}.$$

**Problem 12.** *Given :*

| | | |
|---|---|---|
| (*a*) | $2C(s) + 3H_2(g) \longrightarrow C_2H_6(g)$ | $\Delta H = -84.7$ kJ |
| (*b*) | $2C(s) \longrightarrow 2C(g)$ | $\Delta H = +1430$ kJ |
| (*c*) | $3H_2(g) \longrightarrow 6H(g)$ | $\Delta H = +1308$ kJ |

*The average bond-energy of C—H bond is 413 kJ mol$^{-1}$. Calculate the single C—C bond energy.*

**Solution.** Subtract eq. (*a*) from the sum of eq. (*b*) and eq. (*c*).

$$2C(s) + 3H_2(g) - 2C(s) - 3H_2(g) \longrightarrow 2C(g) + 6H(g) - C_2H_6(g)$$

$$\Delta H = 1430 \text{ kJ} + 1308 \text{ kJ} - (-84.7 \text{ kJ})$$

$$C_2H_6(g) \longrightarrow 2C(g) + 6H(g) \qquad \Delta H = +2822.7 \text{ kJ}$$

Energy required in breaking 6 moles of C—H bonds in 1 mole of ethane = 6 × 413 = 2278 kJ

Bond energy of single C—C bond = 2822.7 kJ – 2478 kJ = **+ 344.7 kJ mol$^{-1}$.**

**Problem 13.** *Calculate the resonance energy of $NO_2$ (O—N = O). The measure enthalpy of formation of $NO_2$, $\Delta H°_f$ is 34 kJ mol$^{-1}$. The bond energies given are :*

$$H_{N—O} = 222 \text{ kJ mol}^{-1} \qquad H_{O—O} = 498 \text{ kJ mol}^{-1}$$
$$H_{N=N} = 946 \text{ kJ mol}^{-1} \qquad H_{N=O} = 607 \text{ kJ mol}^{-1}.$$

**Solution.** The enthalpy of formation of $NO_2$ is given by the following equation.

$$\tfrac{1}{2}N_2(g) + O_2(g) \longrightarrow NO_2(g)$$

Sum of bond energies of reactants

$$= \frac{1}{2} \times H_{N\equiv N} + H_{O=O}$$

$$= \frac{1}{2} \text{ mol} \times 946 \text{ kJ mol}^{-1} + 1 \text{ mol} \times 498 \text{ kJ mol}^{-1}$$

$$= 473 \text{ kJ} + 498 \text{ kJ} = \mathbf{971 \text{ kJ}}.$$

Sum of bond energies of products

$$= H_{N—O} + H_{N=O}$$

$$= 1 \text{ mol} \times 222 \text{ kJ mol}^{-1} + 1 \text{ mol} \times 607 \text{ kJ mol}^{-1}$$

$$= 222 \text{ kJ} + 607 \text{ kJ} = \mathbf{829 \text{ kJ}}$$

$$\Delta H°_{f,\text{calculated}} = \textbf{Sum of bond energies of reactants} - \textbf{Sum of bond energies of products}$$

$$= 971 \text{ kJ} - 829 \text{ kJ} = \mathbf{142 \text{ kJ}}$$

$$\text{Resonance-energy} = \Delta H^\circ_{f,\text{calculated}} - \Delta H^\circ_f$$
$$= 142 \text{ kJ mol}^{-1} - 34 \text{ kJ mol}^{-1}$$
$$= \mathbf{108 \text{ kJ mol}^{-1}}.$$

**Problem 14.** *The standard molar enthalpy of formation of ammonia is – 46.2 kJ mol*$^{-1}$*. Given* $H_{H—H}$ = *435 kJ mol*$^{-1}$ *and* $H_{N-H}$ = *390 kJ mol*$^{-1}$*. Calculate the bond energy in* $N_2(g)$*.*

**Solution.** $\frac{1}{2}N_2(g) + \frac{3}{2}H_2(g) \longrightarrow NH_3(g)$ $\qquad \Delta H^\circ_f = -46.2 \text{ kJ mol}^{-1}$

Multiply this equation by 2

$N_2(g) + 3H_2(g) \longrightarrow 2NH_3(g)$ $\qquad \Delta H = -92.4 \text{ kJ}$

Sum of bond energies of reactants = $H_{N\equiv N} + (3 \times H_{H—H})$

$= H_{N\equiv N} + 3 \text{ mol} \times 435 \text{ kJ mol}^{-1}$

$= H_{N\equiv N} + 1305 \text{ kJ}$

Sum of bond energies of products = $(2 \times 3H_{N—H}) = 6H_{N—H}$

$= 6 \text{ mol} \times 390 \text{ kJ mol}^{-1} = 2340 \text{ kJ}.$

$\Delta H$ = **Sum of bond energies of reactants** – **Sum of bond energies of products**

$$-92.4 \text{ kJ} = H_{N\equiv N} + 1305 \text{ kJ} - 2340 \text{ kJ}$$
$$H_{N\equiv N} = -92.4 \text{ kJ} - 1305 \text{ kJ} + 2340 \text{ kJ}$$
$$= \mathbf{+\,942.6 \text{ kJ mol}^{-1}}.$$

**Problem 15.** *Calculate the bond energy of HCl, given that H—H bond energy is 433 kJ mol*$^{-1}$*. Cl—Cl bond energy is 242 kJ mol*$^{-1}$ *and* $\Delta H_f$ *for HCl is – 91 kJ mol*$^{-1}$*.*

**Solution.** $\frac{1}{2}H_2(g) + \frac{1}{2}Cl_2(g) \longrightarrow HCl(g)$ $\qquad \Delta H^\circ_f = -91 \text{ kJ mol}^{-1}$

Multiply this equation by 2

$H_2(g) + Cl_2(g) \longrightarrow 2HCl(g)$ $\qquad \Delta H = -182 \text{ kJ mol}^{-1}$

This equation inovlves rupture of one H—H bond, one Cl—Cl bond and formation of two H—Cl bonds.

∴ $\Delta H$ = **Sum of bond energies of reactants** – **Sum of bond energies of products**

$$= H_{H—H} + H_{Cl—Cl} - 2H_{H—Cl}$$
$$-182 \text{ kJ} = (1 \text{ mol} \times 433 \text{ kJ}) + (1 \text{ mol} \times 242 \text{ kJ mol}^{-1}) - (2 \text{ mol} \times H_{H—Cl})$$
$$+2 \text{ mol} \times H_{H—Cl} = +182 \text{ kJ} + 433 \text{ kJ} + 242 \text{ kJ} = 857 \text{ kJ}$$
$$H_{H—Cl} = \frac{857 \text{ kJ}}{2 \text{ mol}} = \mathbf{428.5 \text{ kJ mol}^{-1}}.$$

**TYPE–III. Problems based on Entropy**

**Problems 16.** (*i*) *Calculate the entropy change in 600 g of water at 30°C when 60 J heat flow out of it into the surroundings at 25°C.*

*(ii) What is the entropy change in the surroundings ?*

*(iii) What is the total entropy change in the universe due to this process ?*

**Solution.**

$$T_{system} = 30°C + 273 = 310 \text{ K}$$

$$T_{surroundings} = 25°C + 273 = 298 \text{ K}$$

$$\Delta S_{total} = \Delta S_{system} + \Delta S_{surroundings}$$

$$= \frac{-q}{T_{system}} + \frac{+q}{T_{surroundings}}$$

∵ Heat flows from the system into the surroundings.

$$= -\frac{60 \text{ J}}{310 \text{ K}} + \frac{60 \text{ J}}{298 \text{ K}}$$

$$= -0.19 \text{ JK}^{-1} + 0.20 \text{ JK}^{-1}$$

$$\mathbf{= +0.1 \text{ JK}^{-1}.}$$

**Problem 17.** *200 Joules heat flow from the surroundings at 25°C into a large copper block at –10°C. What is the total entropy change ? Assume that the flow of heat does not change the temperature of the copper block and the surroundings.*

**Solution.**

$$T_{system} = -10°C + 273 = 263 \text{ K}$$

$$T_{surroundings} = 25°C + 273 = 298 \text{ K}$$

$$\Delta S_{total} = \Delta S_{system} + \Delta S_{surroundings}$$

$$= \frac{+q}{T_{system}} + \frac{-q}{T_{surroundings}}$$

∵ Heat flows from the surroundings into the system.

$$\Delta S_{total} = \frac{+200 \text{ J}}{263 \text{ K}} + \frac{-200 \text{ J}}{298 \text{ K}}$$

$$= +0.76 \text{ JK}^{-1} - 0.67 \text{ JK}^{-1}$$

$$\mathbf{= +0.09 \text{ JK}^{-1}.}$$

**Problem 18.** *The enthalpy change for the transition of liquid water to steam is 40.8 kJ mol$^{-1}$ at 373 K. Calculate ΔS° for the process.*

**Solution.**

$$\Delta S_{vap} = \frac{\Delta H_{vap}}{T}$$

$$= \frac{40.8 \times 1000 \text{ J mol}^{-1}}{373 \text{ K}} \qquad (1 \text{ kJ} = 1000 \text{ J})$$

$$\mathbf{= 109 \text{ J K}^{-1} \text{ mol}^{-1}.}$$

**Problem 19.** *(i) The melting point of ice is 273 K. At this temperature enthalpy of fusion of ice is 6.00 kJ mol$^{-1}$. Calculate the entropy change for melting one mole of ice. (ii) State giving a reason whether the entropy change for vaporization of one mole of water will be more or less than the entropy change per mole for fusion of ice.*

**Solution.** (*i*) $\Delta H_{fusion} = 6.00 \text{ kJ mol}^{-1}$

Melting point $= 273 \text{ K}$

$$\text{Molar entropy of fusion} = \frac{q_{rev}}{T} = \frac{\Delta H_{fusion}}{T}$$

$$= \frac{(6.00 \times 1000) \text{ J mol}^{-1}}{273 \text{ K}} = \mathbf{22 \text{ JK}^{-1} \text{ mol}^{-1}}$$

(*ii*) The entropy change for vaporization of one mole of water will be more than entropy change per mole of fusion of ice because gaseous state is more disordered than the liquid state.

**Problem 20.** *Calculate the temperature at which liquid water will be in equilibrium with water vapour.* $\Delta H_{vap} = 40.73 \text{ kJ mol}^{-1}$ *and* $\Delta S_{vap} = 109 \text{ J mol}^{-1} \text{ K}^{-1}$.

**Solution.** For a change at equilibrium :

$$\Delta S = \frac{\Delta H}{T}$$

$$T = \frac{\Delta H}{\Delta S} = \frac{40.73 \times 1000 \text{ J mol}^{-1}}{109 \text{ J mol}^{-1} \text{ K}^{-1}}$$

$$= 373.7 \text{ K}$$

$$t°\text{C} = 373.7 \text{ K} - 273 = \mathbf{100.7°C.}$$

**Problem 21.** *The molar entropy change for conversion of liquid ethane into ethane vapour at its boiling point of – 88.6°C is 84.9 J* $K^{-1}$ $mol^{-1}$. *Calculate the molar enthalpy of vaporization.*

**Solution.** Boiling point of ethane,

$$T = -88.6°\text{C} = 273 + (-88.6°\text{C}) = 184.4 \text{ K}$$

$$\Delta S_{vap} = \frac{\Delta H_{vap}}{T}$$

$$\Delta H_{vap} = T\,\Delta S_{vap}$$

$$= 184.4 \text{ K} \times 84.9 \text{ J K}^{-1} \text{ mol}^{-1}$$

$$= 15655 \text{ J mol}^{-1} = \mathbf{15.655 \text{ kJ mol}^{-1}.}$$

**Problem 22.** *Calculate* $\Delta S°$ *reaction for the following reaction at 25°C.*

$$2NaBr(s) + Cl_2(g) \longrightarrow 2NaCl(s) + Br_2(l)$$

$$S°_{NaBr(s)} = 86.8 \text{ J K}^{-1} \text{ mol}^{-1} \qquad S°_{NaCl(s)} = 72.1 \text{ J K}^{-1} \text{ mol}^{-1}$$

$$S°_{Cl_2(g)} = 222.9 \text{ J K}^{-1} \text{ mol}^{-1} \qquad S°_{Br_2(l)} = 152.2 \text{ J K}^{-1} \text{ mol}^{-1}.$$

**Solution.** $2NaBr(s) + Cl_2(g) \longrightarrow 2NaCl(s) + Br_2(l)$

$$\Delta S° = \Sigma S°_{products} - \Sigma S°_{reactants}$$

$$= 2S°_{NaCl} + S°_{Br_2} - [2S°_{NaBr} + S°_{Cl_2}]$$

$$= 2 \text{ mol} \times 72.1 \text{ JK}^{-1} \text{ mol}^{-1} + (1 \text{ mol} \times 152.2 \text{ JK}^{-1} \text{ mol}^{-1}) - [2 \text{ mol} \times 86.65 \text{ JK}^{-1} \text{ mol}^{-1} + (1 \text{ mol} \times 222.9 \text{ JK}^{-1} \text{ mol}^{-1})]$$

$$= 144.2 \text{ JK}^{-1} + 152.2 \text{ JK}^{-1} - [173.6 \text{ JK}^{-1} + 222.9 \text{ JK}^{-1}]$$

$$= 296.4 \text{ JK}^{-1} - 396.5 \text{ JK}^{-1}$$

$$= -\mathbf{100.1\ JK^{-1}}.$$

**TYPE IV. Problems based on Free Energy**

**Problem 23.** *Calculate standard free energy of the following reaction*

$$2CH_4(g) + O_2(g) \longrightarrow 2CH_3OH(l)$$

*The standard molar free energy of formation of $CH_4(g)$ and $CH_3OH(l)$ are –50.8 kJ mol$^{-1}$ and –166.3 kJ mol$^{-1}$ respectively.*

**Solution.** $2CH_4(g) + O_2(g) \longrightarrow 2CH_3OH(l)$

$$\Delta G^\circ = \Sigma\Delta G^\circ_{f,\text{products}} - \Sigma\Delta G^\circ_{f,\text{reactants}}$$

$$= 2(\Delta G^\circ_{f,CH_3OH}) - [2(\Delta G^\circ_{f,CH_4}) + \Delta G^\circ_{f,O_2}]$$

$$= 2 \text{ mol} \times (-166.3 \text{ kJ mol}^{-1}) - [2 \text{ mol} \times (-50.8 \text{ kJ mol}^{-1}) + 1 \text{ mol} \times (0)]$$

$$= -332.6 \text{ kJ mol}^{-1} + 101.6 \text{ kJ mol}^{-1}$$

$$= -\mathbf{231.0\ kJ\ mol^{-1}}.$$

**Note.** The standard free energies of formation of elements are zero.

**Problem 24.** *Calculate molar free energy change for the reaction.*

$$Fe_3O_4(s) + 2C(s, \text{graphite}) \longrightarrow 3Fe(s) + 2CO_2(g)$$

$$\Delta G^\circ_{f,Fe_3O_4(s)} = -1015 \text{ kJ mol}^{-1} \quad \Delta G^\circ_{f,CO_2(g)} = -394.4 \text{ kJ mol}^{-1}.$$

**Solution.** $Fe_3O_4(s) + 2C(s, \text{graphite}) \longrightarrow 3Fe(s) + 2CO_2(g)$

$$\Delta G^\circ = \Sigma\Delta G^\circ_{f,\text{products}} - \Sigma\Delta G^\circ_{f,\text{reactants}}$$

$$= 3\Delta G^\circ_{f,Fe} + 2\Delta G^\circ_{f,CO_2} - [\Delta G^\circ_{f,Fe_3O_4} + 2(\Delta G^\circ_{f,C})]$$

$$= 3 \text{ mol} \times (0 \text{ kJ mol}^{-1}) + 2 \text{ mol} \times (-394.4 \text{ kJ mol}^{-1}) - [1 \text{ mol} \times (-1015 \text{ kJ mol}^{-1}) + 2 \text{ mol} \times (0 \text{ kJ mol}^{-1})]$$

$$= -798.8 \text{ kJ mol}^{-1} + 1015 \text{ kJ mol}^{-1}$$

$$= +\mathbf{216.2\ kJ\ mol^{-1}}.$$

**Problem 25.** *Calculate $\Delta G^\circ$ of the reaction*

$$CH_4(g) + 4Cl_2(g) \longrightarrow CCl_4(l) + 4HCl(g)$$

$$\Delta G^\circ_{f,CH_4(g)} = \mathbf{-50.8\ kJ\ mol^{-1}}; \qquad \Delta G^\circ_{f,CCl_4(l)} = \mathbf{-65.3\ kJ\ mol^{-1}};$$

$$\Delta G^\circ_{f,HCl(g)} = \mathbf{-95.3\ kJ\ mol^{-1}}.$$

**Solution.** $CH_4(g) + 4Cl_2(g) \longrightarrow CCl_4(l) + 4HCl(g)$

$$\Delta G^\circ = \Sigma\Delta G^\circ_{f,\text{products}} - \Sigma\Delta G^\circ_{f,\text{reactants}}$$

$$= \Delta G^\circ_{f,CCl_4} + 4\Delta G^\circ_{f,HCl} - [\Delta G^\circ_{f,CH_4} + 4\Delta G^\circ_{f,Cl_2}]$$

$$= 1\text{mol} \times (-65.3 \text{ kJ mol}^{-1}) + 4 \text{ mol} \times (-95.3 \text{ kJ mol}^{-1})$$

$$- [1 \text{ mol} \times (-50.8 \text{ kJ mol}^{-1}) + 4 \text{ mol} \times (0 \text{ kJ mol}^{-1})]$$

$$= -65.3 \text{ kJ mol}^{-1} - 381.2 \text{ kJ mol}^{-1} + 50.8 \text{ kJ mol}^{-1}$$

$$= \mathbf{-395.7 \text{ kJ mol}^{-1}}.$$

**Problem 26.** *For the reaction*

$$2NO(g) + O_2(g) \longrightarrow 2NO_2(g)$$

*Calculate $\Delta G$ at 700 K, when enthalpy and entropy changes ($\Delta H$ and $\Delta S$) are $-113.0$ kJ mol$^{-1}$ and $-145$ J K$^{-1}$, respectively.*

**Solution.**

$$\Delta G = \Delta H - T\Delta S$$

$$= -113.0 \text{ kJ mol}^{-1} - 700 \text{ K} \times (-145 \text{ J K}^{-1} \text{ mol}^{-1})$$

$$= -113.0 \text{ kJ mol}^{-1} + 101500 \text{ J mol}^{-1}$$

$$= -113.0 \text{ kJ mol}^{-1} + 101.5 \text{ kJ mol}^{-1}$$

$$= \mathbf{-11.5 \text{ kJ mol}^{-1}}$$

**Problem 27.** *From the following values of $\Delta H$ and $\Delta S$ decide whether or not these reactions will be spontaneous at 298 K.*

(*i*) *Reaction A :* $\Delta H = -10.5 \times 10^3$ J mol$^{-1}$ ;

$\Delta S = +31$ J K$^{-1}$ mol$^{-1}$.

(*ii*) *Reaction B :* $\Delta H = -11.7 \times 10^3$ J mol$^{-1}$

$\Delta S = -105$ J K$^{-1}$ mol$^{-1}$.

**Solution.** $\Delta G = \Delta H - T\Delta S$

(*i*)

$$\Delta G = -10.5 \times 10^3 \text{ J mol}^{-1} - 298 \text{ K} \times (+31 \text{ J K}^{-1} \text{ mol}^{-1})$$

$$= -10.5 \times 10^3 \text{ J mol}^{-1} - 9.238 \times 10^3 \text{ J mol}^{-1}$$

$$= \mathbf{-19.738 \times 10^3 \text{ J mol}^{-1}}.$$

As $\Delta G$ is negative, this reaction is spontaneous at 298 K.

(*ii*)

$$\Delta G = -11.7 \times 10^3 \text{ J mol}^{-1} - 298 \text{ K} \times (-105 \text{ J K}^{-1} \text{ mol}^{-1})$$

$$= -11.7 \times 10^3 \text{ J mol}^{-1} + 31.29 \times 10^3 \text{ J mol}^{-1}$$

$$= \mathbf{+19.59 \times 10^3 \text{ J mol}^{-1}}.$$

As $\Delta G$ is positive, the reaction is non-spontaneous at 298 K.

**Problem 28.** *For the reaction, $Ag_2O(s) \rightleftharpoons 2Ag(s) + \frac{1}{2}O_2(g)$, calculate the temperature at which free energy change ($\Delta H$) is equal to zero. $\Delta H = +30.56$ kJ and $\Delta S = +0.066$ kJ at 1 atm. pressure.*

(*i*) *Predict also the nature of the reaction at this temperature and below it.*

(*ii*) *Calculate the temperature at which rates of forward and backward reactions will be the same.*

**Solution.** (*i*) $$\Delta S = \frac{\Delta H}{T}$$

or $$T = \frac{\Delta H}{\Delta S} = \frac{30.56 \text{ kJ mol}^{-1}}{0.066 \text{ kJ K}^{-1} \text{ mol}^{-1}} = 463.1 \text{ K}$$

(*a*) At $T < 463.1$ K. $\Delta G$ is positive. Therefore, the forward reaction is non-spontaneous or backward reaction is spontaneous.

(*b*) At $T > 463.1$ K, $\Delta G$ is negative. Therefore, the forward reaction is spontaneous.

(*ii*) When rates of forward and backward reactions are the same, the reaction is at equilibrium and therefore $\Delta G = 0$

$$\Delta G = \Delta H - T\Delta S = 0$$

$$T = \frac{\Delta H}{\Delta S} = \frac{30.56 \text{ kJ mol}^{-1}}{0.066 \text{ JK}^{-1} \text{ mol}^{-1}} = \mathbf{463.1 \text{ K}.}$$

**Problem 29.** *You are given normal boiling points and standard enthalpies of vaporization. Calculate the entropy of vaporisation of liquids listed below :*

| *Liquid* | *b.p.* (°C) | $\Delta H°_{vap}$ (*kJ mol*$^{-1}$) |
|---|---|---|
| (*a*) *Ethanol* ($C_2H_5OH$) | 78.4 | 42.4 |
| (*b*) *Toluene* ($CH_3C_6H_5$) | 110.6 | 35.2 |

**Solution.** (*a*) $\Delta S = \frac{\Delta H}{T}$; $T = 78.4°\text{C} + 273 = 351.4 \text{ K}$

$$= \frac{42.4 \text{ kJ mol}^{-1}}{351.4 \text{ K}} = 0.1206 \text{ kJ K}^{-1} \text{ mol}^{-1}$$

$$= \mathbf{120.6 \text{ J K}^{-1} \text{ mol}^{-1}}$$

(*b*) $$\Delta S = \frac{\Delta H}{T};\ T = 110.6°\text{C} + 273 = 383.6 \text{ K}$$

$$= \frac{35.2 \text{ kJ mol}^{-1}}{383.6 \text{ K}} = 0.09175 \text{ kJ K}^{-1} \text{ mol}^{-1}$$

$$= \mathbf{91.75 \text{ J K}^{-1} \text{ mol}^{-1}.}$$

**Problem 30.** *Calculate* $\Delta G°$ *of the following reaction.*

$$CO(g) + \tfrac{1}{2}O_2(g) \longrightarrow CO_2(g)\ ;\quad \Delta H° = -282.84 \text{ kJ}$$

***Given :***

$$S°_{CO_2(g)} = 213.8 JK^{-1} mol^{-1}$$

$$S°_{CO(g)} = 197.9 JK^{-1} mol^{-1}$$

$$S°_{O_2(g)} = 205.0\ JK^{-1} mol^{-1}$$

**Solution.**

$$\Delta S^\circ = \Sigma S^\circ_{\text{products}} - \Sigma S^\circ_{\text{reactants}}$$

$$= 1 \text{ mol} \times S^\circ_{CO_2} - [1 \text{ mol} \times S^\circ_{CO} + \tfrac{1}{2} \text{ mol} \times S^\circ_{O_2}]$$

$$= 1 \text{ mol} \times 213.8 \text{ J K}^{-1} \text{ mol}^{-1}$$

$$- [1 \text{ mol} \times 197.9 \text{ J K}^{-1} \text{ mol}^{-1} + \tfrac{1}{2} \text{ mol} \times 205.0 \text{ JK}^{-1} \text{ mol}^{-1}]$$

$$= 213.8 \text{ JK}^{-1} - [197.9 \text{ JK}^{-1} + 102.5 \text{ JK}^{-1}]$$

$$= 213.8 \text{ JK}^{-1} - 300.4 \text{ JK}^{-1}$$

$$= -86.6 \text{ JK}^{-1} = \mathbf{-0.0866 \text{ kJ K}^{-1}}$$

$$\Delta G^\circ = \Delta H^\circ - T\Delta S^\circ \qquad T = 25°C = 298 \text{ K}$$

($\because$ Standard enthalpy and entropy values are given).

$$= -282.84 \text{ kJ} - 298 \text{ K} \times (-0.0866 \text{ kJ K}^{-1})$$

$$= -282.84 \text{ kJ} - 25.8 \text{ kJ}$$

$$= \mathbf{-308.64 \text{ kJ}.}$$

**Problem 31.** *Define Gibbs free energy. How is the change in free energy related to spontaneity ?For a reaction*

$$M_2O(s) \longrightarrow 2M(s) + \tfrac{1}{2}O_2(g)$$

$$\Delta H = 30 \text{ kJ mol}^{-1} \text{ and } \Delta S = 0.07 \text{ kJ K}^{-1} \text{ mol}^{-1} \text{ (at 1 atm.)}$$

*Calculate upto which temperature the reaction would not be spontaneous.*

**Solution.** For the reaction at equilibrium

$$\Delta G = \Delta H - T\Delta S = 0$$

$$T = \frac{\Delta H}{\Delta S}$$

$$= \frac{30 \text{ kJ mol}^{-1}}{0.07 \text{ kJ K}^{-1} \text{ mol}^{-1}}$$

$$= \mathbf{428.6 \text{ K}}$$

At $T < 428.6$ K. Say at 420 K

$$\Delta G = 30 \text{ kJ mol}^{-1} - 420 \text{ K} \times 0.07 \text{ kJ K}^{-1} \text{ mol}^{-1}$$

$$= 30 \text{ kJ mol}^{-1} - 29.4 \text{ kJ K}^{-1} \text{ mol}^{-1}$$

$$= \mathbf{+0.6 \text{ kJ mol}^{-1}}$$

This indicates that the reaction would not be spontaneous upto 428.6 K.

**Problem 32.** *At what temperature, $\Delta G$ of a reaction is – 500 kJ if $\Delta S$ and $\Delta H$ for the reaction are – 430 J $K^{-1}$ and – 630 kJ respectively.*

**Solution.**

$$\Delta G = \Delta H - T\Delta S$$

$$T = \frac{\Delta H - \Delta G}{\Delta S}$$

$$= \frac{-630 \times 1000 \text{ J} - (-500 \times 1000 \text{ J})}{-430 \text{ J K}^{-1}}$$

$$= \frac{-130 \times 1000 \text{ J}}{-430 \text{ J K}^{-1}}$$

$$= \mathbf{302.3 \text{ K}}.$$

**Problem 33.** *Calculate the entropy change $\Delta S$ per mole for the following reactions :*

*(i) Combustion of hydrogen in a fuel cell at 298 K*

$$H_2(g) + \tfrac{1}{2} O_2(g) \longrightarrow H_2O(g)$$

$$\Delta H = -241.60 \text{ kJ} \qquad \Delta G = -228.40 \text{ kJ}$$

*(ii) Vaporization of methanol at its normal boiling point (338 K)*

$$\text{Methanol } (l) \longrightarrow \text{Methanol } (g) \qquad \Delta H = 23.9 \text{ kJ}.$$

**Solution.** (*i*)

$$\Delta G = \Delta H - T\Delta S$$

$$\Delta S = \frac{\Delta H - \Delta G}{T}$$

$$= \frac{-241.60 \text{ kJ} - (-228.40 \text{ kJ})}{298 \text{ K}}$$

$$= \mathbf{-0.044 \text{ kJ K}^{-1}}$$

(*ii*)

$$\Delta G = \Delta H - T\Delta S = 0$$

$$\Delta S = \frac{\Delta H}{T} = \frac{23.9 \text{ kJ}}{338 \text{ K}} = \mathbf{0.071 \text{ kJ K}^{-1}}$$

**TYPE V. Problems based on Free Energy, Equilibrium Constant and Cell Potential.**

**Problem 34.** *For the equilibrium*

$$PCl_5(g) \rightleftharpoons PCl_3(g) + Cl_2(g)$$

*at 298 K, $K = 1.8 \times 10^{-7}$. What is $\Delta G°$ for the reaction ?*

**Solution.**

$$\Delta G° = -2.303\, RT \log K$$

$$= -2.303 \times 8.31 \text{ J K}^{-1} \text{ mol}^{-1} \times 298 \text{ K} \times \log (1.8 \times 10^{-7})$$

$$= -2.303 \times 8.31 \text{ J K}^{-1} \text{ mol}^{-1} \times 298 \text{ K} \times (0.0334 - 7)$$

$$= -2.303 \times 8.31 \times 298 \times (-6.9664) \text{ J mol}^{-1}$$

$$= +39730 \text{ J mol}^{-1}$$

$$= \mathbf{+39.73 \text{ kJ mol}^{-1}}.$$

**Problem 35.** *$\Delta G°$ at 25°C for the chemical equation :*

$$CH_3COOH(aq) + H_2O(l) \rightleftharpoons H_3O^+(aq) + CH_3COO^-(aq) \text{ is } + 2.71 \times 10^4 \text{ J}.$$

*Calculate the value of equilibrium constant at 25°C.*

**Solution.**

$$\Delta G° = -2.303\, RT \log K$$

$$\log K = \frac{-\Delta G^\circ}{2.303\ RT}$$

$$= \frac{-(+2.71 \times 10^4\ J)}{2.303 \times 8.314\ J\ K^{-1} \times 298\ K}$$

$$= -4.7495 = 0.2505 - 5$$

$$K = \text{Antilog}\ (0.2505 - 5)$$

$$= \mathbf{1.78 \times 10^{-5}}.$$

**Problem 36.** *Given that $\Delta G^\circ_f$, for $Br_2(g)$ at 25°C is 3.14 k J mol$^{-1}$. Calculate the vapour pressure of bromine at 25°C.*

**Solution.** $\frac{1}{2} H_2(g) + \frac{1}{2} Br_2(q) \longrightarrow HBr(g)$

$$\Delta G^\circ = -2.303\ RT \log K$$

$$\log K = \frac{-\Delta G^\circ}{2.303\ RT}$$

$$= \frac{-(+3.14 \times 10^3\ J)}{2.303 \times 8.314\ J\ K^{-1} \times 298\ K}$$

$$= 0.55$$

$$K = \text{Antilog}\ 0.55 = 3.548$$

$$K = \frac{HBr\ (g)}{H_2(g)^{1/2}(Br_2(g))^{1/2}}$$

$$= \frac{p_{HBr}}{(1\ atm)^{1/2}\ (1\ atm)^{1/2}} = p_{HBr}$$

$$p_{HBr} = \mathbf{K = 3.548\ atm.}$$

**Problem 37.** *What is the equilibrium constant, K, for the following reaction at 400 K ?*

$$2NOCl(g) \rightleftharpoons 2NO(g) + Cl_2(g)$$

*$\Delta H^\circ = 77.2$ kJ mol$^{-1}$ and $\Delta S^\circ = 122$ J K$^{-1}$ mol$^{-1}$ at 400 K.*

**Solution.**

$$\Delta G = \Delta H - T\Delta S$$

$$= 77.2\ kJ\ mol^{-1} - 400\ K \times (122\ J\ K^{-1}\ mol^{-1})$$

$$= 77.2\ kJ\ mol^{-1} - 48800\ J\ mol^{-1}$$

$$= 77200\ J\ mol^{-1} - 48800\ J\ mol^{-1}$$

$$= 28400\ J\ mol^{-1}$$

$$\Delta G = 2.303\ RT \log K$$

$$\log K = \frac{-\Delta G}{2.303\ RT}$$

$$= \frac{-28400 \text{ J mol}^{-1}}{2.303 \times 8.31 \text{ J K}^{-1} \text{ mol}^{-1} \times 400 \text{ K}}$$

$$= \frac{-28400 \text{ J mol}^{-1}}{7655.2 \text{ J mol}^{-1}} = -3.70$$

$$K = \text{Antilog } (-3.70)$$
$$= \text{Antilog } (0.30 - 4)$$
$$= \text{Antilog } 0.30 \times 10^{-4}$$
$$= \mathbf{1.995 \times 10^{-4}}.$$

**Problem 38.** *The measured voltage of the electrochemical cell*

$$Cu(s) + 2Ag^{+} (1\ M) \rightleftharpoons 2Ag(s) + Cu^{2+} (1\ M)$$

*is 0.54 V at 25°C. Calculate the free energy change of the reaction.*

**Solution.** $n$ is 2 because 1 mol of copper loses 2 mol of electrons

$$\Delta G^\circ = -nFE^\circ$$
$$= -2 \text{ mol} \times 96500 \text{ C/mol} \times 0.54 \text{ V}$$
$$= \mathbf{-104220 \text{ J}}$$

**Problem 39.** *The Standard voltage of the cell*

$$2H_2(g) + O_2(g) \rightleftharpoons 2H_2O(l)$$

*is 1.23 V at 25°C. The standard enthalpy of the reaction is – 571.6 kJ mol*$^{-1}$

(*i*) *Calculate the standard free energy change for the reaction.*

(*ii*) *Calculate the standard entropy change for the reaction.*

(*iii*) *Calculate the equilibrium constant of the reaction.*

**Solution.** (*i*) Anode reaction : $[2H^+(aq) + 2e^- \longrightarrow H_2(g)] \times 2$

Cathode reaction $2H_2O(l) \rightarrow O_2(g) + 4H^+ (aq) + 4e^-$

$$\therefore \quad n = 4$$

$$\Delta G^\circ = -nFE^\circ$$
$$= -4 \text{ mol} \times 96500 \text{ C/mol} \times 1.23 \text{ V}$$
$$= \mathbf{-474780 \text{ J}}.$$

(*ii*)
$$\Delta G^\circ = \Delta H^\circ - T\Delta S^\circ$$

$$\Delta S^\circ = \frac{\Delta H^\circ - \Delta G^\circ}{T}$$

$$= \frac{-571.6 \times 10^3 \text{ J} - (-474780 \text{ J})}{298 \text{ K}}$$

$$= \mathbf{-324.9 \text{ JK}^{-1}}.$$

(*iii*)

$$\Delta G^\circ = -2.303\ RT \log K$$

$$\log K = \frac{-\Delta G^\circ}{2.303\ RT}$$

$$= \frac{-(-474780\ J)}{2.303 \times 8.314\ J\ K^{-1} \times 298\ K}$$

$$= 83.2$$

$$K = \text{Antilog } 83.2$$

$$= \mathbf{1.28 \times 10^{83}}.$$

## QUESTION BANK

1. Define the terms :
   System; State of a system; State variable; Macroscopic properties; Isothermal process; Heat of reaction; Enthalpy.
2. Distinguish between :
   (*i*) Open, closed and isolated systems
   (*ii*) Extensive and intensive properties
   (*iii*) Isothermal and adiabatic changes
   (*iv*) Reversible and irreversible processes.
3. Define $C_v$ and $C_p$ in terms of properties of the system $E$ and $H$ respectively.
4. Define the term bond energy. How would you estimate the bond energy of C—C bond in ethane ?
5. Derive an expression for the variation of heat of reaction with temperature with the help of the first law of thermodynamics.
6. Derive Gibbs-Helmholtz equation.
7. Derive Clapeyron-Clausius equation. Mention its *two* applications.
8. State and explain second law of thermodynamics. Explain the term entropy.
9. Explain the terms "energy content" and "heat energy".
10. Show that the rate of variation of the heat content with temperature at constant pressure is equal to the increase in the heat capacity of the reaction.
11. Write short notes on : (*i*) heat of neutralization, and (*ii*) heat of combustion.
12. What is meant by the term entropy ?
13. Write short note on : (*i*) heat of solution, (*ii*) Van't Hoff isochore.
14. Derive mathematically expression for the adiabatic expansion of an ideal gas.
15. Explain : (*i*) work function, (*ii*) Gibb's energy function.
16. Derive Clausius-Clapeyron equation and discuss its applications.
17. Comment on the statement "the entropy of the universe is always increasing" ?
18. What is meant by Van't Hoff reaction isotherm ? Derive an expression for the reaction isotherm of the general reaction :

$$aA + bB + .... = lL + mM + ....$$

19. Discuss the criteria for a chemical reaction to be spontaneous.
20. (*a*) Show that $C_p - C_v = R$.
    (*b*) Explain the term entropy.
    (*c*) State and derive Kirchoff's equation.
21. Define molar heat capacity of a thermodynamic system. Also define the molar heat capacity of a gas $C_v$ and $C_p$. What is relation between the two ?
22. Explain the concepts of Helmholtz free energy and Gibb's free energy. Describe the relationship between the two.
23. Define lattice energy. How will you determine the lattice energy of an ionic crystal using Born-Haber cycle ?
24. Explain with suitable illustrations : (*i*) Joule-Thomson effect, (*ii*) heat capacities.
25. Explain thermodynamic equilibrium.
26. Define the term heat of formation.
27. What are exothermic and endothermic reactions ? How do you account for the exothermicity or endothermicity of the reaction ?
28. What is the relation between heat of reaction at constant volume and at constant pressure ?
29. Derive the Kirchoff's equation. How does the heat of the reaction vary with temperature ?
30. Derive the Clapeyron-Classius equation. What are the applications of this equation ?
31. Derive an expression for the variation of heat of a reaction with temperature. What is the significance of entropy ?
32. What is the criteria for spontaneity ? Derive the Gibbs-Helmholtz equation and mention its applications ?
33. State Hess's law and mention its important applications with suitable examples.
34. Distinguish between reversible and irreversible processes. State the criteria for reversibility and irreversibility in terms of *H*, *E* and *S*.
35. Write short note on internal energy.
36. State and explain first law of thermodynamics.
37. Define : (*i*) internal energy, (*ii*) heat of combustion, (*iii*) heat of reaction.
38. State and explain Hess's law of constant heat summation. Define heat of combustion with example. Why is the heat of neutralization between a strong acid and a strong base a constant ?

## NUMERICAL PROBLEMS

1. In a certain process, 430 J of work is done on a system which gives off 120 J of heat. What is ΔE for the process ? **[Ans.** 310 J]
2. During a process, a system absorbs 710 J of heat and does work. The change in ΔE for the process is 460 J. What is the work done by the system ? **[Ans.** $w = -250$ J]
3. During a process, the internal energy of the system increases by 245 kJ while the system performed 96 kJ of work on its surroundings. How much heat was transferred between the system and the surroundings during this process. In which direction did the heat flow ?

   **[Ans.** 341 kJ; heat flows from surroundings to the system]

4. One mole of a gas occupying 3L is expanded against a constant pressure of 1 atm to a volume of 12 L. Calculate *w*. **[Ans.** – 911.7 J]

5. What will be the volume change if 607.8 J of work is done by a system containing an ideal gas ? The surrounding exert constant pressure of 10 atm. (1 atm = 101. 3 J). **[Ans.** 0.6 L]

6. 500 cm$^3$ of a sample of an ideal gas is compressed by an average pressure of 0.5 atm to 250 cm$^3$. During this process 10 J of heat flows out to the surroundings. Calculate the change in internal energy of the system. **[Ans.** 2.67 J]

7. The enthalpy changes for the following reactions at 298 K and 1 atm are given below :

(*i*) $CH_3COOH(l) + 2O_2(g) \longrightarrow 2CO_2(g) + 2H_2O(l)$; $\Delta H = -874$ kJ

(*ii*) $CH_3CH_2OH(l) + 3O_2(g) \longrightarrow 2CO_2(g) + 3H_2O(l)$; $\Delta H = -1363$ kJ

Calculate the internal energy changes for these reactions :

**[Ans.** (*i*) – 874 kJ (*ii*) – 1360.53 kJ]

8. The enthalpy of combustion of benzoic acid ($C_6H_5COOH$) at 298 K and 1 atm pressure is – 2546.0 kJ mol$^{-1}$. What is ΔE for the reaction ? **[Ans.** – 2544.8 kJ mol$^{-1}$]

9. The heat produced by combustion of 1.0 g of benzene in a bomb calorimeter was found to be 41.8 kJ at 25°C. Calculate the enthalpy of combustion of benzene at 25°C. **[Ans.** – 3264.1 kJ mol$^{-1}$]

10. The enthalpy of formation of $Fe_2O_3(s)$ is – 824.2 kJ mol$^{-1}$. Calculate the enthalpy change for the reaction :

$$4Fe(s) + 3O_2(g) \longrightarrow 2Fe_2O_3(s)$$ **[Ans.** – 1648.4 kJ]

11. Calculate ΔH° for the reaction.

$$CO_2(g) + H_2(g) \longrightarrow CO(g) + H_2O(g).$$

12. Given that $\Delta H_f^\circ$ for $CO_2(g)$, $CO(g)$ and $H_2O(g)$ is – 393.5, – 111.3 and – 241.8 kJ mol$^{-1}$ respectively. **[Ans.** 40.4 kJ]

13. The molar heat of formation of $NH_4NO_3(s)$ is – 367.54 kJ and those of $N_2O(g)$ and $H_2O(l)$ are 81.46 kJ and 285.8 kJ respectively at 25°C and 1 atm pressure. Calculate ΔE for the reaction :

$$NH_4NO_3(s) \longrightarrow N_2O(g) + 2H_2O(l)$$

14. The enthalpies of formation of $C_2H_5OH(l)$, $CO_2(g)$ and $H_2O(l)$ are – 277, – 393.5 and – 285.8 kJ/mol respectively. Calculate the enthalpy change for the reaction.

$$C_2H_5OH(l) + 3O_2(g) \longrightarrow 2CO_2(g) + 3H_2O(l)$$ **[Ans.** – 1367.4 kJ]

15. Calculate the standard enthalpy change for the oxidation of $NH_3$ according to the equation.

$$4NH_3(g) + 5O_2(g) \longrightarrow 4NO(g) + 6H_2O(g)$$

Given that $\Delta H_f^\circ$ for $NH_3(g)$, $NO(g)$ and $H_2O(g)$ are – 46.2, 90.3 and – 241.8 kJ mol$^{-1}$ respectively.

**[Ans.** – 904.8 kJ]

16. When 2 moles of $C_2H_6(g)$ are completely burnt, 3129 kJ of heat is liberated. Calculate the heat of formation of $C_2H_6(g)$. $\Delta H_f$ for $CO_2(g)$ and $H_2O(l)$ are – 395 and – 286 kJ mol$^{-1}$ respectively.

**[Ans.** – 83.5 kJ]

17. Calculate $\Delta H_f^\circ$ for chloride ion from the following data :

$$\tfrac{1}{2}H_2(g) + \tfrac{1}{2}Cl_2(g) \longrightarrow HCl(g); \quad \Delta H_f^\circ = -92.4 \text{ kJ}$$

$$HCl(g) + nH_2O \longrightarrow H^+(aq) + Cl^-(aq); \quad \Delta H^\circ = -74.8 \text{ kJ}$$

$$\Delta H_f^\circ \; H^+(aq) = 0.0 \text{ kJ}$$ **[Ans.** – 167.2 kJ]

**18.** Calculate the enthalpy of formation of acetic acid ($CH_3COOH$) if its enthalpy of combustion is – 874 kJ $mol^{-1}$. The enthalpy of formation of $CO_2(g)$ and $H_2O(l)$ are – 393.5 and – 285.8 kJ $mol^{-1}$ respectively. **[Ans.** – 484.6 kJ]

**19.** Calculate the standard enthalpy of formation of propane ($C_3H_8$) if its enthalpy of combustion is –2220.2 kJ $mol^{-1}$. The enthalpy of formation of $CO_2(g)$ and $H_2O(l)$ are – 393.5 and – 285.8 kJ $mol^{-1}$ respectively. [ **Ans.** – 103.5 kJ]

**20.** Calculate the enthalpy of allotropic transformation from monoclinic to rhombic sulphur from the following data :

$$\text{S(rhombic)} + O_2(g) \longrightarrow SO_2(g)\ ; \qquad \Delta H = -295.1 \text{ kJ}$$

$$\text{S(monoclinic)} + O_2(g) \longrightarrow SO_2(g)\ ; \qquad \Delta H = -296.4 \text{ kJ}$$

**[Ans.** – 1.3 kJ]

**21.** Calculate the enthalpy of formation of ethane from the following data :

$$C(s) + O_2(g) \longrightarrow CO_2(g); \qquad \Delta H = -393.5 \text{ kJ}$$

$$H_2(g) + \tfrac{1}{2}O_2(g) \longrightarrow H_2O(l); \qquad \Delta H = -285.8 \text{ kJ}$$

$$C_2H_6(g) + \tfrac{7}{2}O_2(g) \longrightarrow 2CO_2(g) + 3H_2O(l); \quad \Delta H = -1560.0 \text{ kJ}$$

**[Ans.** – 484.6 kJ]

**22.** The standard enthalpy of formation of $SO_2(g)$ and $SO_3(g)$ are – 296.6 kJ and – 395.6 kJ respectively. Calculate ΔH for the reaction :

$$SO_2(g) + \tfrac{1}{2}\ O_2(g) \longrightarrow SO_3(g); \qquad \Delta H = -295.1 \text{ kJ}$$

**[Ans.** – 99.0 kJ]

**23.** Calculate the entropy change involved in the conversion of one mole of liquid water at 373 K to vapours at the same temperature (Latent heat of vaporisation = 2.275 kJ/g).

[**Hint.** Molar enthalpy of vaporisation = 2.275 × 18 = 40.95 kJ $mol^{-1}$]

$$\therefore \qquad \Delta S_{vap} = 40950 \text{ J mol}^{-1}/373 \text{ K} = 109.78 \text{ JK}^{-1} \text{ mol}^{-1}.$$

30.4 kJ of heat is required to melt 1 mole of sodium chloride. The entropy change during melting is 28.4 J $mol^{-1}$ $K^{-1}$. Calculate the melting point of sodium chloride : [**Ans.** 1070.4 K]

**24.** Calculate entropy changes of fusion and vaporisation for chlorine from the following data :

$$\Delta H_{fusion} = 6.40 \text{ kJ mol}^{-1} \qquad \text{m.p.} = -102°\text{C}$$

$$\Delta H_{vap} = 20.4 \text{ kJ mol}^{-1} \qquad \text{b.p.} = -34°\text{C}$$

[**Ans.** $\Delta S_{fusion}$ = 37.4 $JK^{-1}$ $mol^{-1}$, $\Delta S_{vap}$ = 85.3 $JK^{-1}$ $mol^{-1}$]

**25.** $\Delta S_{vap}$ of acetone is 93.0 $JK^{-1}$ $mol^{-1}$. If boiling point of acetone is 56°C. calculate the heat required for the vaporisation of 1 g of acetone. [**Ans.** 527.5 J]

**26.** Enthalpy and entropy changes of a reaction are 40.63 kJ $mol^{-1}$ and 108.8 $JK^{-1}$ $mol^{-1}$ respectively. Predict the feasibility of the reaction at 27°C.

[**Ans.** ΔG = 7.99 kJ $mol^{-1}$, Not feasible in the forward direction.]

**27.** Calculate the free energy change on dissolving one mole of sodium chloride at 25°C.

Lattice energy = + 777.8 kJ $mol^{-1}$

Hydrogen energy of NaCl = – 774.1 kJ $mol^{-1}$

ΔS at 25°C = 43 $JK^{-1}$ $mol^{-1}$

[**Ans.** ΔH = 777.8 – 774.1 = 3.7 kJ $mol^{-1}$ ∴ ΔG = – 9.11 kJ $mol^{-1}$]

**28.** For the reaction :

$$2NO(g) + O_2(g) \longrightarrow 2NO_2(g)$$

calculate $\Delta G$ at 700 K when enthalpy and entropy changes are – 113.0 kJ mol$^{-1}$ and – 145 JK$^{-1}$ mol$^{-1}$ respectively. **[Ans.** 11.5 kJ mol$^{-1}$]

**29.** The values of $\Delta H$ and $\Delta S$ for two reactions are given below :

*Reaction A :* $\Delta H = -10.5 \times 10^3$ J mol$^{-1}$

$\Delta S = +31$ J K$^{-1}$ mol$^{-1}$

*Reaction B :* $\Delta H = -11.7 \times 10^3$ J mol$^{-1}$

$\Delta S = -105$ J K$^{-1}$ mol$^{-1}$.

**30.** Decide whether these reactions are spontaneous or not at 298 K.

**[Ans.** Reaction A : Spontaneous and Reaction B : Non-spontaneous]

At what temperature does the reduction of lead oxide to lead by carbon becomes spontaneous ?

$$PbO + C \longrightarrow Pb(s) + CO(g)$$

For the reaction, $\Delta H$ and $\Delta S$ at 25°C are 108.4 kJ mol$^{-1}$ and 190 Jk-mol$^{-1}$ respectively.

**[Ans.** 570.5 K]

**31.** Calculate $\Delta G°$ for the reaction

$$NO(g) \rightleftharpoons \frac{1}{2} N_2(g) + \frac{1}{2} O_2(g)$$

if K = $1.55 \times 10^{15}$ at 298 K. **[Ans.** – 86.7 kJ mol$^{-1}$]

**32.** The equilibrium constant for the reaction :

$$CO_2(g) + H_2(g) \rightleftharpoons CO(g) + H_2O(g)$$

is 73 at 25°C. Calculate standard free energy change. **[Ans.** 10.63 kJ mol$^{-1}$]

**33.** The standard free energy change for the reaction :

$$2H_2 + O_2 \rightleftharpoons 2H_2O(l)$$

at 25°C is – 474.3 kJ. Calculate the equilibrium constant. **[Ans.** $1.32 \times 10^{33}$]

**34.** The equilibrium constant for the reaction :

$$CH_3COOH(l) + C_2H_5OH(l) \rightleftharpoons CH_3COOC_2H_5(l) + H_2O(l)$$

has been found to be equal to 4. Calculate free energy change for the reaction.

**[Ans.** – 3.435 kJ mol$^{-1}$]

**35.** Calculate the entropy change for a reaction :

$$X \longrightarrow Y$$

Given that $\Delta H° = 28.40$ kJ mol$^{-1}$ and equilibrium constant is $1.8 \times 10^{-7}$ at 298 K.

**[Ans.** – 33.8 J mol$^{-1}$ K$^{-1}$]

**36.** Both ethane and methane are used as fuel. Predict which of these is a better fuel ? Given that heat of formation of $CH_4(g)$, $C_2H_6(g)$, $CO_2(g)$ and $H_2O(l)$ are – 74.85, – 84.68, – 393.5 and – 286 kJ mol$^{-1}$ respectively.

**37.** At 0°C, ice and water are in equilibrium and $\Delta H = 6.0$ kJ mol$^{-1}$ for the process,

$$H_2O(s) \longrightarrow H_2O(l)$$

What will be $\Delta S$ and $\Delta G$ for the conversion of ice into liquid water ?

**38.** Calculate the resonance energy of $N_2O$ from the following data :

$\Delta H_f^\circ$ of $N_2O$ = 82 kJ mol$^{-1}$. Bond energies of $N \equiv N$, $N = N$, $O = O$ and $N = O$ bonds are 946, 418, 498 and 607 kJ mol$^{-1}$ respectively.

**39.** A gas expands from a volume of 3.0 dm$^3$ to 5.0 dm$^3$ against a constant external pressure of 3.0 atm. The work done during the expansion is used to heat 10.0 ml of water of temperature 290.0 K. Calculate the final temperature of water. (Specific heat of water = 4.184 J g$^{-1}$ K$^{-1}$).

**40.** 0.16 g of methane was subjected to combustion at 27°C in a bomb calorimeter. The temperature of calorimeter system (including water) was found to rise by 0.5°C. Calculate the heat of combustion of methane at (*i*) constant volume and (*ii*) at constant pressure. The thermal capacity of calorimeter system is 17.7 kJ K$^{-1}$ (R = 8.31 JK$^{-1}$ mol$^{-1}$)

**41.** Calculate the maximum work obtained when 0.75 mole of an ideal gas expands isothermally and reversibly at 27°C from a volume of 15L and 25L.

**42.** When 1.0 g of propane was burnt in excess of oxygen in a bomb calorimeter 52.5 kJ of heat was evolved at 25°C. Find ΔH for the reaction.

**43.** Using the data (all values in k cal per mole at 25°C) given below, calculate the bond energy of C—C and C—H bonds :

$$\Delta H^\circ_{\text{combustion}} \text{(ethane)} = -372.0; \quad \Delta H^\circ_{\text{combustion}} \text{(propane)} = -530.0$$

$$\Delta H^\circ \text{ for C(graphite)} \longrightarrow C(g) = 172.0$$

Bond energy of H—H = 104.0

$$\Delta H_f^\circ \text{ of } H_2O(l) = -68.0$$

$$\Delta H_f^\circ \text{ of } CO_2 = -94.0.$$

**44.** The standard Gibb's free energy change for the reaction

$$N_2(g) + 3H_2(g) \rightleftharpoons 2NH_3(g)$$

is – 33.2 kJ mol$^{-1}$ at 298 K.

(*a*) Calculate the equilibrium constant for the above reaction.

(*b*) What would be the equilibrium constant if the reaction is written as

$$\tfrac{1}{2} N_2(g) + \tfrac{3}{2} H_2(g) \rightleftharpoons NH_3(g)$$

(*c*) What will be the equilibrium constant if the reaction is :

$$NH_3(g) \rightleftharpoons \tfrac{1}{2} N_2(g) + \tfrac{3}{2} H_2(g)$$

**45.** Calculate ΔG and ΔG° for the reaction :

$$A + B \rightleftharpoons C + D$$

at 27°C for which K = $10^2$.

# APPENDICES

## APPENDIX-I

INTERNATIONAL ATOMIC WEIGHTS OF
ELEMENTS BASED ON C-12

| Element | Symbol | Atomic Number | Atomic Weight |
|---|---|---|---|
| Aluminium | Al | 13 | 26.9815 |
| Antimony | Sb | 51 | 121.75 |
| Argon | Ar | 184 | 39.948 |
| Arsenic | As | 33 | 74.9216 |
| Barium | Ba | 56 | 137.34 |
| Beryllium | Be | 4 | 9.0122 |
| Bismuth | Bi | 83 | 208.980 |
| Boron | B | 5 | 10.811 |
| Bromine | Br | 35 | 79.909 |
| Cadmium | Cd | 48 | 112.40 |
| Calcium | Ca | 20 | 40.08 |
| Carbon | C | 6 | 12.01115 |
| Cerium | Ce | 58 | 140.12 |
| Chlorine | Cl | 17 | 35.453 |
| Chromium | Cr | 24 | 51.996 |
| Cobalt | Co | 27 | 58.9332 |
| Copper | Cu | 29 | 63.54 |
| Fluorine | F | 9 | 18.9984 |
| Germanium | Ge | 32 | 72.59 |
| Gold | Au | 79 | 196.967 |
| Hafnium | Hf | 72 | 178.49 |

| Element | Symbol | Atomic Number | Atomic Weight |
|---|---|---|---|
| Helium | He | 2 | 4.0026 |
| Hydrogen | H | 1 | 1.00797 |
| Iodine | I | 53 | 126.9044 |
| Iridium | Ir | 77 | 192.2 |
| Iron | Fe | 26 | 55.847 |
| Lanthanum | La | 57 | 138.91 |
| Lead | Pb | 82 | 207.19 |
| Lithium | Li | 3 | 6.939 |
| Magnesium | Mg | 12 | 24.312 |
| Manganese | Mn | 25 | 54.9380 |
| Mercury | Hg | 80 | 200.59 |
| Molybdenum | Mo | 42 | 95.94 |
| Nickel | Ni | 28 | 58.71 |
| Niobium | Nb | 41 | 92.906 |
| Nitrogen | N | 7 | 14.0067 |
| Osmium | Os | 6 | 190.2 |
| Oxygen | O | 8 | 15.9994 |
| Palladium | Pd | 46 | 106.09 |
| Phosphorus | P | 15 | 30.9738 |
| Platinum | Pt | 78 | 195.09 |
| Potassium | K | 19 | 39.102 |
| Radium | Ra | 88 | 226.05 |
| Rhenium | Re | 75 | 186.21 |
| Rhodium | Rh | 45 | 102.905 |
| Ruthenium | Ru | 44 | 101.07 |
| Scandium | Sc | 21 | 44.956 |
| Selenium | Se | 34 | 78.96 |
| Silicon | Si | 14 | 28.086 |
| Silver | Ag | 47 | 107.870 |
| Sodium | Na | 11 | 22.9898 |
| Strontium | Sr | 38 | 87.62 |
| Sulphur | S | 16 | 32.064 |
| Tantalum | Ta | 73 | 180.948 |
| Tellurium | Te | 52 | 127.60 |
| Thallium | Tl | 81 | 204.37 |
| Thorium | Th | 90 | 232.038 |
| Tin | Sn | 50 | 118.69 |
| Titanium | Ti | 22 | 47.90 |
| Tungsten | W | 74 | 183.85 |
| Uranium | U | 92 | 238.03 |
| Vanadium | V | 23 | 50.942 |
| Zinc | Zn | 30 | 65.37 |
| Zirconium | Zr | 40 | 91.22 |

# APPENDIX-II

## LABORATORY REAGENTS

### 1. Concentrated Acids

| Name and Formula of Acid | Formula Weight | Equivalent Weight | Weight % | Specific Gravity | Aprox. Strength |
|---|---|---|---|---|---|
| Acetic acid ($CH_3COOH$) glacial | 60.05 | 60.05 | 99–100 or 99.5 | 1.05 | 17.4 N or 17.4 M |
| Hydrochloric acid (HCl) | 36.46 | 36.46 | 36 | 1.18 | 11.6 ≈ 12 N or M |
| Hydrobromic acid (HBr) | 80.92 | 80.92 | 47.8 | 1.49 | 9 N or M |
| Hydrofluoric acid (HF) | 20.01 | 20.01 | 46 | 1.15 | 26.5 N or M |
| Hydriodic acid (HI) | 127.91 | 127.91 | 47.475 | 1.50 | 5.5 N or M |
| Nitric acid ($HNO_3$) | 63.01 | 63.01 | 69.5 or 70 | 1.42 | 15.6 N ≈ 16 N or M |
| Phosphoric acid ($H_3PO_4$) | 98.00 | | 85 | 1.69 | 40.5 N or 13.5 M |
| Perchloric acid ($HClO_4$) | 100.46 | | 70 | 1.66 | 9 N or 18 M |
| Sulphuric acid ($H_2SO_4$) | 98.08 | 49.0 | 96 | 1.84 | 35.9 ≈ 36 N or 18.0 M |

### 2. Dilute Acids

| Acid | Equivalent Weight | Approx. Strength | Preparation of 1 liter of Solution |
|---|---|---|---|
| Acetic acid ($CH_3COOH$) | 60.05 | 5 N (143 ml acid per 500 ml because 58 ml/l will give N solution). | Add 286 ml of glacial acid to distilled water to make up of volume of 1 liter. |
| Hydrochloric acid (HCl) | 36.46 | 5 N (85.5 ml/l will be N solution and 214 ml/500 ml will make 5 N solution). | Add 428 ml of concentrated HCl to water to make up the volume of 1 liter. |
| Nitric acid ($HNO_3$) | 63.01 | 5 N (On diluting 65 ml to 1 liter we get N while 160 ml to 500 ml we get 5 N solution). | Add 320 ml of concentrated $HNO_3$ to distilled water to make up the volume of 1 liter. |
| Sulphuric acid ($H_2SO_4$) | 49.08 | 5 N (27.9 ml acid per liter is N and 70 ml/500 ml is 5 N solution). | Add 140 ml of concentrated $H_2SO_4$ to distilled water to make up the volume of 1 liter. |

### 3. Bases

| Acid | Equivalent Weight | Approx. Strength | Preparation of 1 liter of Solution |
|---|---|---|---|
| Concentrated ammonium hydroxide ($NH_4OH$) | 35.05 | 1.51 N or 58.6% by weight of $NH_4OH$ or 28.30% $NH_3$. | Highest concentration with specific gravity 0.90. |
| Dilute ammonium hydroxide ($NH_4OH$) | 35.05 | 5 N (66.5 ml concentrated alkali per liter will give N solution and 166 ml/500 will give 5 N solution). | Dilute 332 ml of concentrated ammonia (specific gravity 0.90 with distilled water by making it to 1 liter. |
| Sodium hydroxide (NaOH) | 40.00 | 5 N | Dissolve 220 g of caustic soda pellets in distilled water and dilute to 1 liter. |
| Lime water, $Ca(OH)_2$ | $\frac{74.09}{2}$ = 37.05 | 0.04 N | Shake 10 g of lime (sufficient to saturate 1 liter of water) with 1 liter of water and allow it to stand for sometime. Filter and keep it in a stoppered bottle. |

# APPENDIX-III

## EQUIVALENT WEIGHT OF SUBSTANCES REQUIRED IN VOLUMETRIC ANALYSIS

| Titration and Substances used | Chemical Formula | Equivalent Weight |
|---|---|---|
| **Acid-Alkali Titration** | | |
| Sodium carbonate (Washing Soda) | $Na_2CO_3$ | M.W./2 = 53.06 |
| Sodium bicarbonate (Baking soda) | $NaHCO_3$ | M.W./1 = 84.00 |
| Potassium carbonate (Pearl ash) | $K_2CO_3$ | M.W./2 = 69.00 |
| Potassium bicarbonate | $KHCO_3$ | M.W./1 = 100.00 |
| Sodium hydroxide (Caustic soda) | NaOH | M.W./1 = 40.00 |
| Barium hydroxide | $Ba(OH)_2.8H_2O$ | M.W./2 = 157.75 |
| Oxalic acid (hydrated) | $H_2C_2O_4.2H_2O$ | M.W./2 = 63.04 |
| Hydrochloric acid | HCl | M.W./1 = 35.46 |
| Sulphuric acid | $H_2SO_4$ | M.W./2 = 49.04 |
| **Redox Titration** | | |
| Potassium permanganate | $KMnO_4$ | M.W./5 = 31.606 |
| Potassium dichromate | $K_2Cr_2O_7$ | M.W./6 = 49.03 |
| Ferrous sulphate (hydrated) {Green vitriol} | $FeSO_4.\ 7H_2O$ | M.W./1 = 278.00 |
| Ferrous sulphate (anhydrous) | $FeSO_4$ | M.W./1 = 152.00 |
| Ferrous ammonium sulphate (Mohr's salt) | $FeSO_4.\ (NH_4)_2SO_4$ $6H_2O$ | M.W./1 = 392.14 |
| Oxalic acid (hydrated) | $H_2C_2O_4.2H_2O$ | M.W./2 = 63.04 |
| Iron (ferrous) | Fe | At. W./1 = 55.84 |
| **Iodine Titration** | | |
| Iodine | $I_2$ | M.W./2 = 126.92 |
| Hypo (Sodium thiosulphate) Hydrated | $Na_2S_2O_3.5H_2O$ | M.W./1 = 248.19 |
| Anhydrous Sodium thiosulphate | $Na_2S_2O_3$ | M.W./1 = 158.09 |
| Arsenious oxide | $As_2O_3$ | M.W./4 = 49.45 |
| Copper sulphate (hydrated) {Blue vitriol} | $CuSO_4.5H_2O$ | M.W./1 = 249.68 |
| **Argentometric Titration** | | |
| Silver nitrate (caustic lunar) | $AgNO_3$ | M.W./1 = 169.89 |
| Silver | Ag | At.W./1 = 107.88 |
| Sodium chloride (common salt) | NaCl | M.W./1 = 58.45 |
| Chloride ion | $Cl^-$ | At.W./1 = 35.46 |
| Potassium sulphocyanide | KCNS | M.W./1 = 97.17 |
| Ammonium sulphocyanide | $NH_4CNS$ | M.W./1 = 76.12 |

# APPENDIX-IV

## MOLECULAR WEIGHTS, EQUIVALENT WEIGHTS AND WEIGHTS OF VARIOUS SUBSTANCES REQUIRED TO PREPARE ONE LITER SOLUTION OF A PARTICULAR STRENGTH, FOR USE IN VOLUMETRIC ANALYSIS

| Substance | Molecular Weight | Equivalent Weight | Amount in gms per liter for Solutions of Strength | | |
|---|---|---|---|---|---|
| | | | N/10 | N/20 | N/40 |
| Ammonium thiocyanate [$NH_4CNS$] | 76.10 | 76.10 | 7.61 | 3.805 | 1.9025 |
| Copper sulphate [$CuSO_4.5H_2O$] | 249.71 | 249.71 | 24.971 | 12.4855 | 6.4225 |
| Ferrous ammonium sulphate [$FeSO_4(NH_4)_2SO_4.6H_2O$] | 392.10 | 392.10 | 39.2100 | 19.6050 | 9.8025 |
| Ferrous sulphate [$FeSO_4.7H_2O$] | 278.00 | 278.00 | 27.8000 | 13.9000 | 6.9500 |
| Iodine [$I_2$] | 253.80 | 126.9 | 12.6900 | 6.3450 | 3.1725 |
| Oxalic acid [$H_2C_2O_4.2H_2O$] | 126.08 | 63.04 | 6.3040 | 3.1520 | 2.5760 |
| Potassium carbonate [$K_2CO_3$] | 138.0 | 69.00 | 6.9000 | 3.4500 | 1.7250 |
| Potassium dichromate [$K_2Cr_2O_7$] | 294.19 | 49.03 | 4.9030 | 2.4515 | 1.2257 |
| Potassium permanganate [$KMnO_4$] | 158.00 | 31.60* | 3.1600 | 1.5800 | .7900 |
| Potassium thiocyanate [KCNS] | 97.16 | 97.16 | 9.7160 | 4.8580 | 2.4290 |
| Sodium chloride [NaCl] | 58.46 | 58.46 | 5.8460 | 2.9230 | 1.4615 |
| Sodium carbonate [$Na_2CO_3$] | 106.30 | 53.00 | 5.3000 | 2.6500 | 1.3250 |
| Sodium hydroxide [NaOH] | 40.00 | 40.00 | 4.0000 | 2.0000 | 1.0000 |
| Sodium thiosulphate [$Na_2S_2O_3.5H_2O$] | 248.20 | 248.20 | 24.8200 | 12.4100 | 6.2050 |
| Silver nitrate [$AgNO_3$] | 169.90 | 169.90 | 16.9900 | 8.4950 | 4.2475 |
| Nitric acid, (Sp. Gr. 1.16) | 63.00 | 63.00 | 8.3 ml | 4.65 ml | 2.33 ml |
| Hydrochloric acid, (Sp. Gr. 1.18) | 36.50 | 36.50 | 1.0 ml | 5 ml | 2.5 ml |
| Sulphuric acid, (Sp. Gr. 1.184) | 98.00 | 49.00 | 2.75 ml | 1.83 ml | 0.68 ml |

# *INDEX*

## C

## R



## S

**X**



**Y**



**Z**